Cytochrome P450 2D6

Structure, Function, Regulation, and Polymorphism

Cytochrome P450 2D6

Structure, Function, Regulation, and Polymorphism

Shufeng Zhou

CRC Press
Taylor & Francis Group
Boca Raton London New York

CRC Press is an imprint of the
Taylor & Francis Group, an **informa** business

CRC Press
Taylor & Francis Group
6000 Broken Sound Parkway NW, Suite 300
Boca Raton, FL 33487-2742

First issued in paperback 2020

© 2016 by Taylor & Francis Group, LLC
CRC Press is an imprint of Taylor & Francis Group, an Informa business

No claim to original U.S. Government works

ISBN-13: 978-1-4665-9787-7 (hbk)
ISBN-13: 978-0-367-65859-5 (pbk)

Visit the Taylor & Francis Web site at
http://www.taylorandfrancis.com

and the CRC Press Web site at
http://www.crcpress.com

Contents

Preface

This book focuses on how genetic, epigenetic, physiological, pathological, and structural factors govern the highly variable metabolism of a number of drugs in clinical use by human cytochrome P450 2D6 (CYP2D6), an important enzyme in the liver. The book covers in depth our current knowledge about the structure, function, polymorphism, and regulation of CYP2D6. It highlights the functional consequences and relevance of the structural and polymorphic changes of CYP2D6, the impact on clinical practice and drug development, and implications in precise medicine.

The human CYP superfamily contains 57 functional genes and 58 pseudogenes within 14 subfamilies. Four families of CYPs, CYP1, CYP2, CYP3, and CYP4, are the main ones contributing to the oxidative metabolism of more than 95% of clinical drugs, with CYPs in other families mainly being involved in the metabolism of important endogenous compounds such as steroids, fatty acids, retinoids, and arachidonic acids. CYP2D6, the focus of this book, accounts for only a small percentage of total hepatic CYP content (~2%–4%). However, it metabolizes ~25% of all medications and more than 160 drugs in the human liver. CYP2D6 is highly polymorphic and subject to inhibition by a number of drugs, therefore resulting in a large interindividual variability in drug clearance and response, possibly the occurrence of adverse drug reactions, and harmful drug–drug interactions. Notably, this enzyme is not induced by all known conventional CYP inducers, although a genome-wide association study indicated that more than 1600 genes can regulate the expression of CYP2D6 in humans.

This book consists of eight chapters that cover an overview of the human CYP superfamily, mammalian CYP2D members, substrate specificity of CYP2D6, inhibitor selectivity of CYP2D6, structural features of CYP2D6, regulation of CYP2D6, and pharmacogenomics of CYP2D6. It is unique because it is the first book that addresses a single but important CYP enzyme in detail, including its structure, function, regulation, and polymorphism that plays a very important role in individualized pharmacotherapy, drug development, and toxicology. This book will be useful for clinicians, pharmacists, pharmaceutical scientists, pharmacologists, geneticists, and toxicologists. It tells us how important this enzyme is and how we can use drugs more reasonably and safely. Since the book has updated knowledge about this single enzyme, it can be used as an important reference book by medical, pharmacy, biomedical, pharmaceutical, and nursing students. The book is also a good source of knowledge about CYP2D6 for public, private, and university libraries.

I acknowledge my appreciation for the great support from my wife Judy Liang, my daughter Helen Chew, and my son Kevin Chow. I also appreciate the support from Dr. Judy L. Genshaft, Professor Charles (Charly) J. Lockwood, Professor Kevin B. Sneed, Professor Howard L. McLeod, Professor Guangji Wang, Professor Lin He, Professor Kaitai Yao, Professor Kaixian Chen, Professor Yixin Zeng, Professor Mark S. Kindy, Professor Zhixu He, Professor Tianxin Yang, Professor Wei Duan, Professor William T. Beck, Professor Yinxue Yang, Professor Zhi-Min Yuan, Professor Xiaowu Chen, Professor Xueji Zhang, Professor Shuming He, Professor Jiaxuan Qiu, Professor Eric B. Haura, Professor Min Huang, and Professor Penghui Zhou. In addition, I appreciate the technical support from Dr. Zhi-Wei Zhou, Cameron T. Durlacher, Christopher J. Menzie, and Tommy Rogers.

I hope that this book will become a good resource to advocate further studies on human CYP2D6 with regard to its regulation and the functional impacts of its polymorphisms.

Author

Professor Shufeng Zhou is presently the associate vice president of Global Medical Development and associate dean of International Research, College of Pharmacy, University of South Florida (USF), Tampa, Florida. He conducted his clinical medical training in China at Nanchang University in 1986 and Sun Yat-sen Medical University in 1989. He earned his PhD in pharmacology in 2001 from the School of Medicine, the University of Auckland, New Zealand. Since 2002, Professor Zhou has served as an assistant professor, senior lecturer, associate professor, and professor for the National University of Singapore, Queensland University of Technology, RMIT University in Australia, and now at USF. His major research interests include systems pharmacology, drug metabolism and drug transport, pharmacokinetics/pharmacometrics, pharmacogenomics, nanomedicine, and Chinese medicine.

Professor Zhou is one of the Highly Cited Researchers 2014, according to Thomson Reuter. He has published more than 450 peer-reviewed papers in various pharmacology and medical journals including *New England Journal of Medicine*, *Cancer Research*, *Journal of Biological Chemistry*, *Clinical Pharmacology and Therapeutics*, and *Clinical Cancer Research*. He has also published 20 books and book chapters and more than 440 conference abstracts. His work has been cited more than 13,500 times by other colleagues with an H-index of 58. Professor Zhou has given more than 150 invited seminars/keynote presentations to a variety of academic institutes, government agencies, and high-profile international conferences. He has received more than US$40 million in grants from various funding bodies and industries.

To date, Professor Zhou has trained 28 PhD students, 18 MSc/Honors students, 22 postdoctoral fellows, and more than 150 visiting doctors. He serves as an editor in chief or editor for at least 21 biomedical journals, including *Drug Design, Development and Therapy*, *Drug, Healthcare and Patient Safety*, *Clinical & Experimental Physiology & Pharmacology*, *Clinical Pharmacology & Biopharmaceutics*, *Journal of Pharmacogenomics and Pharmacoproteomics*, *Drug Metabolism Letters*, *Journal of Pharmacy and Nutrition Sciences*, *Current Drug Metabolism*, etc., and he is an editorial board member of more than 75 medical and pharmacological journals.

Professor Zhou is a recipient of several reputable national and international awards, a voting member of US Pharmacopeia and consultant for the World Health Organization and the Food and Drug Administration, a council member or chair of several national and international professional societies, and a visiting/adjunct professor for more than 40 reputable research universities around the world.

1 Introduction to Human Cytochrome P450 Superfamily

1.1 THE CYTOCHROME P450s IN NATURE

The cytochrome P450s (CYPs), a large diverse heme-containing enzyme superfamily with a large number of members, are found across all organisms in prokaryotic and eukaryotic worlds from animals, plants, fungi, protists, bacteria, and archaea to viruses (Bolwell et al. 1994; Danielson 2002; Feyereisen 1999; Gillam and Hayes 2013; Kelly and Kelly 2013; Munro and Lindsay 1996; Podust and Sherman 2012; Roberts 1999; Werck-Reichhart and Feyereisen 2000). These proteins were first discovered in 1958 by their unusual reduced carbon monoxide difference spectrum that exhibits a Soret peak at 450 nm, thus called "Pigment at 450 nm" or "P450" (Omura and Sato 1964). The unique spectral peak is produced by a thiolate anion acting as the fifth ligand to the heme. This peak is a unique feature only observed in four classes of hemoproteins, namely, P450s, nitric oxide synthases, chloroperoxidases, and protein H450 (Omura 2005; Poulos 2014; van Rantwijk and Sheldon 2000). The nomenclature of CYPs is based on the protein sequences, with similar sequences being clustered into families and subfamilies. *CYP* genes form a multigene family and encode proteins with amino acid sequence identities >40%. Each family comprises subfamilies with amino acid sequence identities >55%. In the classification of CYPs, a clan is defined as a higher-order category of CYP families. To date, the *CYP* genes from all organisms consist of at least 700 families and 800 subfamilies.

Bacterial CYPs are soluble with approximately 400 amino acids (Munro and Lindsay 1996), while eukaryotic CYPs are larger, consisting of ~500 amino acids. Eukaryotic CYPs are usually membrane bound through an N-terminal hydrophobic peptide and other less well-understood mechanisms. The two locations of these eukaryotic proteins are the endoplasmic reticulum (ER) membrane and the mitochondrial inner membrane. As of August 13, 2013, 21,039 CYPs had been reported, with 6313 from animals (1056 from mammals, 3452 from insects, 922 from other vertebrates, and 883 from non-insect vertebrates), 7446 from plants, 5729 from fungi, 247 from protozoa, 1254 from bacteria, 48 from archaea, and 2 from viruses.

A multigene family may result from gene duplication/gene amplification, exon shuffling, expression of overlapping genes, programmed frameshifting, alternative splicing RNA editing, and gene sharing (Boutanaev et al. 2015; Good et al. 2014; Hancks et al. 2015; Kawashima and Satta 2014; Yasukochi and Satta 2015). Gene duplication can arise from ectopic homologous recombination, retrotransposition event, aneuploidy, whole genome duplication/polyploidy, and replication slippage (Copley 2012; Taylor and Raes 2004). It is believed that modern P450s originate from an ancestral gene that existed approximately three and a half billion years ago before the advent of eukaryotes and before the existence of an oxygen-rich atmosphere (Danielson 2002; Degtyarenko and Kulikova 2001; Gillam and Hayes 2013; Omura 2010; Podust and Sherman 2012; Roberts 1999; Tralau and Luch 2013; Werck-Reichhart and Feyereisen 2000). Under anaerobic conditions, the first CYP may have served as nitroreductases or endoperoxide isomerases. Once the earth's atmosphere had accumulated a substantial amount of molecular oxygen approximately 2.8 billion years ago, CYPs might have been employed to protect early life forms from oxygen toxicity. Over its evolutionary history, the CYP superfamily is considered to have undergone repeated rounds of expansion by gene and genome duplication (Danielson 2002; Degtyarenko and Kulikova 2001; Gillam and Hayes 2013; Omura 2010; Podust and Sherman 2012; Roberts 1999; Tralau and Luch 2013; Werck-Reichhart and Feyereisen 2000). It is postulated that approximately one and a half billion years ago, the first of these expansions gave rise to the families of CYPs that are primarily involved in the metabolism of endogenous fatty acids and cholesterol (e.g., CYP4 and 11 families). Around 900 million years ago, another expansion of the gene family is speculated to have resulted in several of the endogenous steroid-synthesizing CYP families (e.g., CYP19, 21 and 27 families). A dramatic expansion of several CYP families, including those known or suspected of being involved in xenobiotic metabolism (e.g., CYP2, 3, 4, and 6), commenced approximately 400 million years ago. Phylogenetic analyses of CYPs suggest that they are also among the most rapidly evolving of genes, which is a characteristic that is needed to protect the cells from the injuries when exposed to increasing toxic xenobiotic compounds (Gillam and Hayes 2013; Omura 2010; Roberts 1999; Tralau and Luch 2013).

CYPs catalyze various types of oxidation including hydroxylation, *N*-, *O*-, and *S*-dealkylation, sulfoxidation, epoxidation, deamination, desulfuration, dehalogenation, peroxidation, and *N*-oxide reduction of a large number of endogenous (e.g., fatty acids and retinoids) and exogenous (e.g., drugs and procarcinogens) compounds (Newsome et al. 2013; Werck-Reichhart and Feyereisen 2000). In addition to these classical reactions, CYPs also catalyze many uncommon reactions such as one- and two-electron reductions, one-electron oxidation, oxidative cleavage of carboxylic acid esters, desaturation, deformylation of aldehydes, ring formation, ipso mechanisms for aryl dehalogenation and *O-* and *N*-dearylation, rearrangements of oxidized eicosanoids, aldoxime dehydration, and hydrolysis of phosphatidylcholine (Guengerich 2001).

Most CYPs are monooxygenases or mixed function oxidases, and electrons for reduction of the heme and later the oxygen substrate are provided by protein partners that bind to the face of the protein proximal to the heme (Hollenberg 1992; Iyanagi et al. 2012; Meunier et al. 2004). The heme prosthetic group is the catalytic center of the enzyme, where a reactive hypervalent oxo-iron protoporphyrin IX radical cation intermediate is formed for subsequent insertion of the iron-bound oxygen atom into a substrate bond. Substrates bind in a cavity or cleft above the surface of the heme in proximity to the reactive intermediate. The thiolate side chain of a conserved Cys residue binds to the axial coordination site of the iron opposite to the bound oxygen, giving rise to the unique spectral and functional properties of CYP enzymes. Microsomal CYPs transfer electrons from nicotinamide adenine dinucleotide phosphate (NADPH) via cytochrome P450 reductase while cytochrome b_5 can also contribute reducing power to this system after being reduced by cytochrome b_5 reductase (Henderson et al. 2015; Iyanagi et al. 2012; Locuson et al. 2007; Meunier et al. 2004; Pandey and Fluck 2013). Mitochondrial CYPs use adrenodoxin reductase and adrenodoxin (i.e., ferredoxin) to transfer electrons from NADPH to CYP (Neve and Ingelman-Sundberg 2010). However, CYP5A1 (thromboxane X_2 synthase, TBXAS1), CYP8A1 (prostacyclin H_2 synthase), and CYP74A (allene oxide synthase) do not need protein partners for their catalysis (Brash 2009). The catalytic mechanism of CYPs is similar owing to a conserved heme-thiolate functionality, but amino acid variations in the substrate binding sites confer compound selectivity, regioselectivity, and stereoselectivity of metabolism (Werck-Reichhart and Feyereisen 2000).

Eukaryotic CYPs generally have sizes ranging from approximately 480 to 560 amino acids and can be grouped into three broad categories on the basis of their subcellular localization. CYPs found in bacteria and eukaryotic mitochondria are called "type I," whereas those found in eukaryotic ER are termed "type II" (Neve and Ingelman-Sundberg 2008). A third type of CYPs is the cytosolic form. While all P450s in prokaryotes are cytosolic, soluble CYPs are extremely rare in eukaryotic cells. Animal CYPs are localized primarily in the membrane of the ER, but some of them are also present in other cellular compartments, such as the cell surface and mitochondria, where the enzymes display catalytic activity toward CYP-specific substrates (Ahn and Yun 2010; Anandatheerthavarada et al. 2009; Avadhani et al. 2011; Neve and Ingelman-Sundberg 2010). Microsomal CYPs are targeted to the ER by an N-terminal leader that includes a transmembrane helix linked by a polar connector to the catalytic domain, which is sequestered to the cytoplasmic side of the membrane. In the ER, NADPH donates electrons to the diflavin protein P450 oxidoreductase, which then passes them on to the type II CYPs.

Identification of CYPs in organisms is solely based on primary structure analysis of protein sequences, especially for the presence of two CYP signature motifs, FXXGXRXCXG (also called CXG) in the heme-binding domain and the EXXR motif in the K helix (Oezguen and Kumar 2011; Oezguen et al. 2008; Seifert and Pleiss 2009; Sirim et al. 2010; Syed and Mashele 2014). The motifs EXXR and PER form the E–R–R triad, which is important for locking the heme pocket into position and to assure stabilization of the core structure of CYPs. The Cys residue of the CYP signature CXG motif is conserved in all CYPs, whereas the two Gly residues and one Phe residue are generally conserved. The Cys residue in the CXG motif located in the β-bulge region (also named as Cyspocket) acts as a fifth ligand to the heme iron. The first of the two Gly residues, which occurs four amino acids before the Cys residue, allows for the formation of the β-hairpin turn; the second Gly residue, which occurs two amino acids after the Cys residue, allows for a sharp turn in the backbone into the L helix and for its positioning in proximity to the heme (Hasemann et al. 1995; Mestres 2005). The Glu and Arg residues of EXXR motif are also conserved in CYPs with few exceptions (Oezguen et al. 2008; Seifert and Pleiss 2009; Sirim et al. 2010). The EXXR motif is important for the stabilization of the meander loop and probably for the maintenance of the CYP tertiary structure. Furthermore, the conserved motif AGXDTT contributes to oxygen binding and activation. Although CYPs all preserve the basic structural fold, in response to the enormously wide range of substrate specificities, their substrate-binding regions are much more variable, yet may possess a signature motif. In addition, most CYPs show significant substrate promiscuity, and therefore, their substrate-binding pockets are well known for the high structural plasticity and the ability to alter shape and volume depending on the chemical structure they accommodate. Six putative substrate recognition sites (SRSs) for eukaryotic CYPs have been proposed on the basis of the analysis of the CYP2 family and CYPs structure (Gotoh 1992). The position of the SRSs in eukaryotic CYPs correlates well with substrate binding sites that have been identified in the crystal structures of prokaryotic CYPs in complex with the substrates (Kirischian and Wilson 2012; Lindberg and Negishi 1989). For CYP2A4 and 2A5, amino acid differences in SRS1, SRS2, and SRS5 account for the differences in substrate specificity between these two enzymes. Differences in the enzyme kinetics of CYPs that metabolize the same substrate have also been associated with residues located in SRSs.

Plant cytochromes P450 are involved in a wide range of biosynthetic reactions, forming various fatty acid conjugates, plant hormones, defensive compounds, or medically important drugs (Bolwell et al. 1994; O'Keefe 1995). Terpenoids, which represent the largest class of characterized natural plant compounds, are often good substrates for plant CYPs. Prokaryotic CYPs are involved in the biosynthesis of antibiotics such as erythromycin and mycinamicin and catalyze key reactions required for environmental bioremediation through the degradation of a variety of hydrocarbons. Despite their lower P450 gene counts, fungal genomes show the highest CYP diversity, with at least 400 CYP families found across the 4034 annotated fungal P450s (as of 2014) (W. Chen et al. 2014) while only 129 P450 families found across 4267 annotated plant P450s (Nelson 2011). Currently, fungal CYP families are grouped into CYP51-69, CYP501-699, and CYP5001-6999. Fungal CYPs extensively participate in a wide variety of physiologically important reactions in fungi that contribute to the fitness and fecundity of fungi in various ecological niches. Fungal P450s play an essential role in many pathways in the primary and secondary metabolism of fungal species, including membrane ergosterol biosynthesis, outer spore wall components biosynthesis, alkane and fatty acids degradation, fatty acids hydroxylation, mycotoxins (i.e., aflatoxins, trichothecenes, and fumonisins), and plant hormones biosynthesis (gibberellin biosynthesis) (W. Chen et al. 2014). Microbial cytochrome P450s are often soluble enzymes and are involved in diverse metabolic processes. In bacteria, the distribution of P450s is very variable with many bacteria having no identified P450s (e.g., *Escherichia coli*). Cytochrome P450cam (CYP101) originally from *Pseudomonas putida* has been used as a model for many cytochromes P450 and is the first cytochrome P450 three-dimensional protein structure solved by x-ray crystallography (Poulos et al. 1987). This enzyme is part of a camphor-hydroxylating catalytic cycle consisting of two electron transfer steps from putidaredoxin, a 2Fe–2S cluster-containing protein cofactor.

Since the early 2000s, there have been several studies focused on the CYP genome complements (CYPomes) in metazoans, with studies completed on vertebrates, hemichordates, insects, crustaceans, *Capitella teleta*, and Cnidaria (Dejong and Wilson 2014; Feyereisen 2011; Kelly et al. 2009; Khatri et al. 2010; Lah et al. 2013; Lamb et al. 2002, 2003; Newsome et al. 2013; Sezutsu et al. 2013). The smallest number of genes in a metazoan CYPome is found in the sponge *Amphimedon queenslandica* (35 *CYP* genes) and the largest metazoan CYPome identified so far contains ~235 genes in the lancelet, *Branchiostoma floridae* (D.R. Nelson et al. 2013). Vertebrate genomes typically contain 57–102 *CYP* genes. Many animals have as many or more *CYP* genes than humans do. For example, mice have 102 *Cyp* genes and 88 pseudogenes, rats contain 89 *CYP* genes and 79 pseudogenes, dogs have 48 *CYP* genes, rhesus monkey have 114 *CYP* genes, pigs contain 58 *CYP* genes, zebrafish has 81 *Cyp* genes, sea squirt has 97 *Cyps*, rice (*Oryza sativa*) has 457 *Cyp* genes, *Arabidopsis thaliana* has 272 *Cyp* genes, *Aspergillus oryzae* has 159 *Cyp* genes, *Caenorhabditis elegans* contains 83 *Cyp* genes, *Drosophila* has 84 P450 genes and 6 pseudogenes, and sea urchins have 120 *Cyp* genes. In contrast, there are relatively a small number of *CYP* genes in eubacteria or archaea, ranging from none in *E. coli* to 33 in *Streptomyces avermitilis* (D.R. Nelson et al. 2013). In mammals, CYPs are critical for drug metabolism, blood homeostasis, cholesterol biosynthesis, steroidogenesis, and immune response. The classes of CYPs most often investigated in animals are those either involved in development (e.g., retinoic acid or hormone metabolism) or involved in the metabolism of toxic compounds such as heterocyclic amines or polyaromatic hydrocarbons. Most CYP enzymes in animals are presumed to have monooxygenase activity, as is the case for most mammals.

Mammalian CYPs are mainly expressed in the liver, but these enzymes are also found in extrahepatic tissues such as intestine, kidney, lung, heart, brain, and adrenal gland where CYPs can play a substantial role in the metabolism of endogenous compounds and xenobiotics (Seliskar and Rozman 2007; Woodland et al. 2008). Extrahepatic tissues with high P450 expression levels are the respiratory and gastrointestinal tracts that are exposed to foreign compounds entering the body where xenobiotics can be activated or inactivated (Thelen and Dressman 2009). The extrahepatic P450 enzymes can be important for tissue-specific metabolic activation or inactivation of xenobiotic compounds including drugs and procarcinogens and biosynthesis of endogenous compounds that are important for essential cellular functions. Several CYPs are often overexpressed in tumor cells and may play a role in the regulation of cancer growth, development, and metastasis (Baer-Dubowska and Szaefer 2013; Harvey and Morgan 2014; Oyama et al. 2012; Xu et al. 2011).

CYPs have been extensively studied in mice, rats, dogs, and less so in zebrafish and other species, in order to facilitate use of these model organisms in biomedical research, drug discovery, and toxicological evaluation. This chapter will highlight the tissue distribution, function, regulation, and polymorphism of human CYP superfamily.

1.2 HUMAN CYP SUPERFAMILY

1.2.1 MEMBERS OF THE CYP SUPERFAMILY: FAMILY, SUBFAMILY, CLAN, AND MOTIF

The human CYP superfamily represents the most important Phase I drug-metabolizing enzymes that oxidize a number of endogenous substances and xenobiotics, including more than 90% of all medications into more hydrophilic metabolites (Lewis 2004; Nebert et al. 2013). There are as many as 57 functional *CYP* genes and 58 pseudogenes within 18 families (i.e., *CYP1, 2, 3, 4, 5, 7, 8, 11, 17, 19-21, 24, 26, 27, 39, 46,* and *51*; see Figure 1.1 and Tables 1.1 and 1.2) and 43 subfamilies in humans (http://drnelson.uthsc.edu/cytochromeP450 .html). Human CYPs are classified into families, subfamilies, and individual enzymes by amino acid sequence homology. Family and subfamily designations reflect >35% and >70% amino acid sequence identity, respectively. Orthologs typically exhibit 70% or greater sequence identity. The CYP2, 3, and 4 families contain far more members than the other 15 families, while genes in the remaining 14 families often have only a single member. In particular, the CYP2 family is the largest one of human CYPs comprising 13 subfamilies and 16 functional genes. The CYP4 family has five subfamilies within which there are 12 functional genes.

Human *CYP* genes often occur in clusters, with several related genes, pseudogenes, and detritus exons aligned in tandem (D.R. Nelson et al. 2004, 2013). Human *CYP* genes include seven clusters: *CYP2ABFGST, CYP2C, CYP2C, CYP2J, CYP3A, CYP4ABXZ,* and *CYP4F*. Genes in subfamilies are sometimes clustered with genes of other subfamilies. Mouse and human each have 30 *CYP* genes that lie outside the

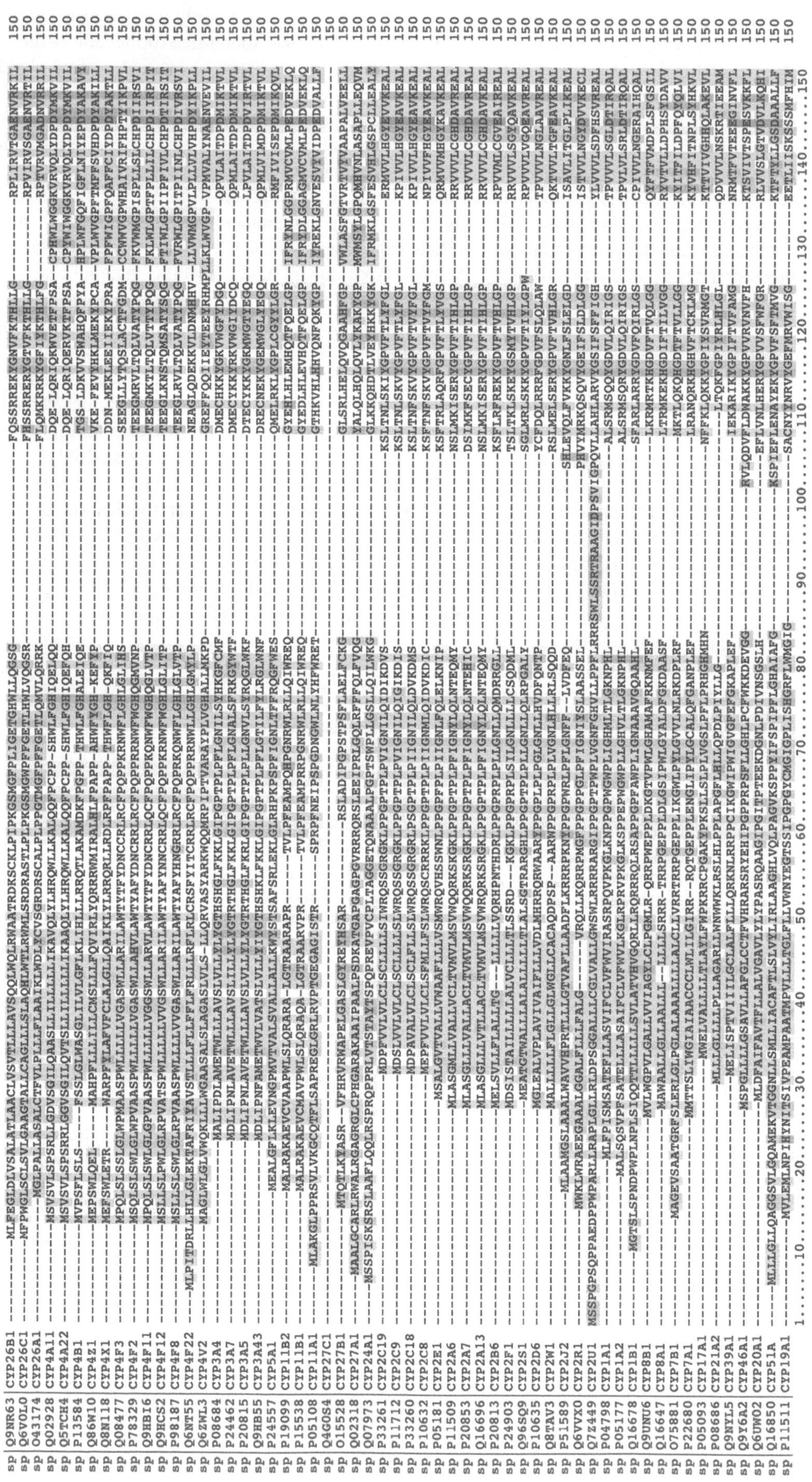

(a)

FIGURE 1.1 **(See color insert.)** (a) Complete sequence alignment of human CYPs. Protein sequences of 57 human CYP enzymes are obtained from the Swiss-Prot database (http://www .uniprot.org/). Multiple sequence alignment of the CYPs is performed using the program Clustal W v2.0 (http://www.clustal.org/clustal2/) with default parameters. *(Continued)*

The figure shows a multiple sequence alignment (positions 1–150) for the following Swiss-Prot accessions and CYP enzymes:

sp			sp	
Q9NR63	CYP26B1		P33261	CYP2C19
Q6V0L0	CYP26C1		P11712	CYP2C9
O43174	CYP26A1		P33260	CYP2C18
Q02928	CYP4A11		P10632	CYP2C8
Q5TCH4	CYP4A22		P05181	CYP2E1
P13584	CYP4B1		P11509	CYP2A6
Q86W10	CYP4Z1		P20853	CYP2A7
Q8N118	CYP4X1		Q16696	CYP2A13
Q08477	CYP4F3		P20813	CYP2B6
P78329	CYP4F2		P24903	CYP2F1
Q9HB16	CYP4F11		Q96SQ9	CYP2S1
Q9HCS2	CYP4F12		P10635	CYP2D6
P98187	CYP4F8		Q8TAV3	CYP2W1
Q6NT55	CYP4F22		P51589	CYP2J2
Q6ZWL3	CYP4V2		Q6VVX0	CYP2R1
P08684	CYP3A4		Q7Z449	CYP2U1
P24462	CYP3A7		P04798	CYP1A1
P20815	CYP3A5		P05177	CYP1A2
Q9HB55	CYP3A43		Q16678	CYP1B1
P24557	CYP5A1		Q9UNU6	CYP8B1
P19099	CYP11B2		Q16647	CYP8A1
P15538	CYP11B1		O75881	CYP7B1
P05108	CYP11A1		P22680	CYP7A1
Q4G0S4	CYP27C1		P05093	CYP17A1
O15528	CYP27B1		P08686	CYP21A2
Q02318	CYP27A1		Q9NYL5	CYP39A1
Q07973	CYP24A1		Q9Y6A2	CYP46A1
			Q60M02	CYP20A1
			Q16850	CYP51A1
			P11511	CYP19A1

(a)

FIGURE 1.1 (CONTINUED) **(See color insert.)** (a) Complete sequence alignment of human CYPs. Protein sequences of 57 human CYP enzymes are obtained from the Swiss-Prot database (http://www.uniprot.org/). Multiple sequence alignment of the CYPs is performed using the program Clustal W v2.0 (http://www.clustal.org/clustal2/) with default parameters. *(Continued)*

FIGURE 1.1 (CONTINUED) (See color insert.) (a) Complete sequence alignment of human CYPs. Protein sequences of 57 human CYP enzymes are obtained from the Swiss-Prot database (http://www.uniprot.org/). Multiple sequence alignment of the CYPs is performed using the program Clustal W v2.0 (http://www.clustal.org/clustal2/) with default parameters.

(Continued)

(a)

(a)

FIGURE 1.1 (CONTINUED) **(See color insert.)** (a) Complete sequence alignment of human CYPs. Protein sequences of 57 human CYP enzymes are obtained from the Swiss-Prot database (http://www.uniprot.org/). Multiple sequence alignment of the CYPs is performed using the program Clustal W v2.0 (http://www.clustal.org/clustal2/) with default parameters.

(Continued)

(a)

FIGURE 1.1 (CONTINUED) **(See color insert.)** (a) Complete sequence alignment of human CYPs. Protein sequences of 57 human CYP enzymes are obtained from the Swiss-Prot database (http://www.uniprot.org/). Multiple sequence alignment of the CYPs is performed using the program Clustal W v2.0 (http://www.clustal.org/clustal2/) with default parameters.

(Continued)

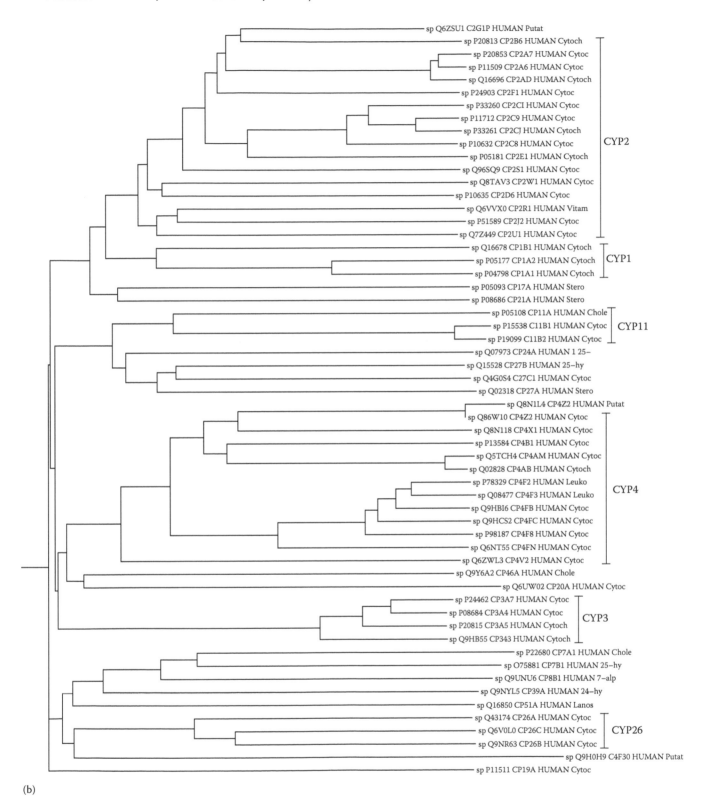

(b)

FIGURE 1.1 (CONTINUED) (b) The phylogenetic tree of human CYPs. From the output of the alignment, the phylogenetic tree is generated to infer the evolutionary relationships between the P450 enzyme sequences. CYP2 family on chromosome 19q; the CYP2C subfamily on chromosome 10q; the CYP3A subfamily on chromosome 7q; and the CYP4 family on chromosome 1p. *(Continued)*

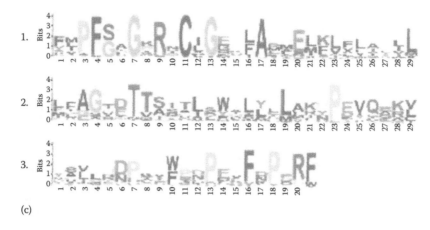

(c)

FIGURE 1.1 (CONTINUED) **(See color insert.)** (c) The MEME program version 4.10.1 is used to discover and analyze sequence motifs that occur repeatedly in the human P450s (http://meme-suite.org).

seven gene clusters. These CYP genes are distributed on all chromosomes except chromosomes 5, 16, and 17. Five clusters of closely related genes are located on chromosomes 1, 7, and 10 (one cluster each) and chromosome 19 (two clusters). Studies have shown that an initial tandem duplication occurs in an early mammalian ancestor and that gene duplications and rearrangements frequently occur in a lineage-specific manner. Clusters of related *CYP* families are called clans. There are 10 *CYP* clans in humans: clans 2, 3, 4, 7, 19, 20, 26, 46, and 51 and the mitochondrial clan (Nelson et al. 2004). CYP families within a single clan have likely been diverged from a common ancestor gene.

Human CYPs can be classified into two types, the detoxification type (D-type) and the biosynthesis type (B-type), on the basis of their substrates and function (Nebert and Dalton 2006). In humans, the D-type detoxifies xenobiotics such as plant alkaloids, aromatic compounds, fatty acids, and clinical drugs. On the other hand, the B-type is involved in the biosynthesis of physiologically important endogenous compounds such as steroids, cholesterols, vitamin D, and bile acids. Among the 57 functional *CYP* genes in the human genome, 35 are D-type genes and 22 are B-type genes. This classification is based on the description of the enzyme substrate specificity and subfamily or family classification. D-type genes constitute four *CYP* families: *CYP1* (3 genes), *CYP2* (16 genes), *CYP3* (4 genes), and *CYP4* (12 genes). B-type genes contain 14 families: *CYP5* (1 gene), *CYP7* (2 genes), *CYP8* (2 genes), *CYP11* (3 genes), *CYP17* (1 gene), *CYP19* (1 gene), *CYP20* (1 gene), *CYP21* (1 gene), *CYP24* (1 gene), *CYP26* (3 genes), *CYP27* (3 genes), *CYP39* (1 gene), *CYP46* (1 gene), and *CYP51* (1 gene) (Table 1.1). The birth and death (pseudogenization) rates of B- and D-type genes differ in magnitude: the rates in B-type genes are 0.7 and 0.2 per 100 million years (myr), respectively, whereas those in D-type genes are 12.7 and 6.9 per 100 myr, respectively (Kawashima and Satta 2014). Compared with D-type genes, the evolution of B-type genes is highly conserved with regard to their mode of birth and death processes as well as amino acid substitutions. The substrates of B-type enzymes are often endogenous compounds such as vitamin D, steroids, bile acids,

prostaglandins (PGs), and cholesterol with important physiological functions. In contrast, the substrates of D-type enzymes are xenobiotics such as plant alkaloids and synthetic drugs. The differences in the evolutionary mode between B- and D-type genes may reflect differences in their respective substrates.

The amino acid alignment of the 57 functional *CYP* genes by us and Kawashima and Satta (2014) shows that four amino acids are conserved: Glu242, Arg245, Phe310, and Cys316 (Figure 1.1a). Two of these, namely, Phe310 and Cys316, are located near the heme-binding region. Cys316C is known to be structurally close to the iron ion in the heme-binding region and to serve as an active center of the enzyme (Meunier et al. 2004). The other two residues including Glu242 and Arg245 are located approximately 80 amino acids upstream from the proximal Cys. Although it is unclear whether these two amino acids are involved in any specific function, their conservation suggests some evolutionary or functional importance. In the phylogenetic tree (Figure 1.1b), members of each family form monophyletic groups with respect to other families, and each monophyletic group is supported by a relatively high bootstrap value. An ancestral gene of *CYP19A1* appears to be duplicated, generating the ancestor of the CYP1 and 2 families. The ancestor of CYP11A1 and 11B1 is duplicated, generating the ancestors of the *CYP1*, *CYP2*, *CYP3*, *CYP4*, *CYP11*, and *CYP26* families. The ancestors of *CYP46A* and *22A* are duplicated to form *CYP4V2*, which is then duplicated to result in the whole *CYP4* family (Figure 1.1b). The ancestors of *CYP17A* and *CYP21A* are duplicated and generate *CYP1* and *CYP2* families. Furthermore, *CYP3* family members are the ancestors of the *CYP4* family.

Sequence alignments feature a number of conserved sequence motifs located primarily in the C-terminal half of the CYP proteins. At least three well-conserved motifs in human CYPs have been identified (Figure 1.1c). The first motif is "FXXGXRXCXG" (also called "CXG"), located in the heme-binding domain. The Cys residue of this signature "CXG" motif is conserved in all human CYPs, while the two Gly residues and one Phe residue are generally well conserved. The Cys residue in the "CXG" motif is located in the β-bulge region serving as a fifth ligand to the heme iron.

TABLE 1.1
A List of 57 Human Functional *CYP* Genes

Gene	Alias	Names	Chromosomal Location	Substrates/ Function	Number of Amino Acids	Number of Exons
CYP1A1	*AHH*; *AHRR*; *CP11*; *CYP1*; *P1-450*; *P450-C*; *P450DX*	Cytochrome P450 1A1; CYPIA1; cytochrome P450-C; cytochrome P450-P1; cytochrome P450 form 6; xenobiotic monooxygenase; aryl hydrocarbon hydroxylase; flavoprotein-linked monooxygenase; dioxin-inducible cytochrome P1-450	15q24.1	Drugs, procarcinogens, steroids, and fatty acids	512	7
CYP1A2	*CP12*; *P3-450*; *P450(PA)*	Cytochrome P450 1A2; CYPIA2; P450 form 4; cytochrome P450 4; cytochrome P450; cytochrome P450-P3; dioxin-inducible P3-450; microsomal monooxygenase; xenobiotic monooxygenase; aryl hydrocarbon hydroxylase; flavoprotein-linked monooxygenase	15q24.1	Drugs, fatty acids, and steroids	516	7
CYP1B1	*CP1B*; *GLC3A*; *CYPIB1*; *P4501B1*	Cytochrome P450 1B1; microsomal monooxygenase; xenobiotic monooxygenase; aryl hydrocarbon hydroxylase; flavoprotein-linked monooxygenase	2p22.2	Drugs, procarcinogens, steroids, and fatty acids	543	3
CYP2A6	*CPA6*; *CYP2A*; *CYP2A3*; *P450PB*; *CYPIIA6*; *P450C2A*	Cytochrome P450 2A6; cytochrome P450(I); cytochrome P450 IIA3;coumarin 7-hydroxylase; xenobiotic monooxygenase; 1,4-cineole 2-exo-monooxygenase; flavoprotein-linked monooxygenase	19q13.2	Drugs and steroids	494	9
CYP2A7	*CPA7*; *CPAD*; *CYP2A*; *CYPIIA7*; *P450-IIA4*	Cytochrome P450 2A7; cytochrome P450IIA4; cytochrome P450 IIA4	19q13.2	Xenobiotics, steroids, and fatty acids	494	9
CYP2A13	*CPAD*; *CYP2A*; *CYPIIA13*	Cytochrome P450 2A13	19q13.2	Drugs and other xenobiotics	494	9
CYP2B6	*CYP 2B*; *CYP 2B7*; *CYP 2B7P*	Cytochrome P450 2B6; cytochrome P450 IIB; 11,4-cineole 2-exo-monooxygenase	19q13.2	Drugs, steroids, and fatty acids	491	9
CYP2C8	CYP 2C8	Cytochrome P450 2C8; P450 form 1; cytochrome P450 IIC2; cytochrome P450 MP-12; cytochrome P450 MP-20; cytochrome P450 form 1; microsomal monooxygenase; xenobiotic monooxygenase; *S*-mephenytoin 4-hydroxylase; flavoprotein-linked monooxygenase	10q23.33	Drugs, steroids, and fatty acids	490	9
CYP2C9	*CPC9*; *CYP2C*; *CYP2C10*; *CYPIIC9*; *P450IIC9*	Cytochrome P450 2C9; cytochrome P-450MP; cytochrome P450 PB-1; microsomal monooxygenase; xenobiotic monooxygenase; flavoprotein-linked monooxygenase	10q24	Drugs, steroids, and fatty acids	490	9
CYP2C18	*CYP 2C*; *CYP 2C17*; *CPCI*; *P450IIC17*; *P450-6B/29C*	Cytochrome P450 2C18; microsomal monooxygenase; unspecific monooxygenase; flavoprotien-linked monooxygenase; *S*-mephenytoin hydroxylase associated cytochrome P450	10q24	Drugs, steroids, and fatty acids	490	9

(Continued)

TABLE 1.1 (CONTINUED)
A List of 57 Human Functional *CYP* Genes

Gene	Alias	Names	Chromosomal Location	Substrates/ Function	Number of Amino Acids	Number of Exons
CYP2C19	*CYP2C*; *CYPJ*; *P450C2C*; *CYPIIC17*; *CYPIIC19*; *P450IIC19*	Cytochrome P450 2C19; cytochrome P450-11A; cytochrome P450-254C; cytochrome P450 II C; microsomal monooxygenase; xenobiotic monooxygenase; mephenytoin 4′-hydroxylase; *S*-mephenytoin 4-hydroxylase	10q24.1-q24.3	Drugs	490	9
CYP2D6	*CYD6*; *CYP2D*; *CYP2D7AP*; *CYP2D7BP*; *CYP2D7P2*; *CYP2DL1*; *CYPIID6*; *P450C2D*; *P450DB1*; *CYP2D8P2*; *P450-DB1*	Cytochrome P450 2D6; cytochrome P45-DB1; microsomal moooxygenase; xenobiotic monooxygenase; debrisoquine 4-hydroxylase	22q13.1	Drugs	497	10
CYP2E1	*CYP2E*; *CYE1*; *P450-J*; *P450C2E*	Cytochrome P450 2E1; cytochrome P450-J; microsomal monooxygenase; xenobiotic monooxygenase; 4-nitrophenol 2-hydroxylase; flavoprotein-linked monooxygenase	10q26.3-	Drugs, ethanol, and procarcinogens	493	9
CYP2F1	*CYP2F*; *C2F1*	Cytochrome P450 2F1; CYPIIF1; microsomal monooxygenase; xenobiotic monooxygenase; flavoprotein-linked monooxygenase	19q13.2	Drugs and coumarins	491	11
CYP2J2	CP J2	Cytochrome P450 2J2; CYPIIJ2; microsomal monooxygenase, arachidonic acid epoxygenase; flavoprotein-linked monooxygenase	1p31.3-p31.2	Fatty acid (e.g., AA)	502	9
CYP2R1	CYP 2R1	Vitamin D 25-hydroxylase; cytochrome P450 2R1	11p15.2	Vitamin D (D-25-hydroxylase activity)	501	8
CYP2S1	CYP 2S1	Cytochrome P450 2S1; CYPIIS1	19q13.1	Xenobiotics	504	9
CYP2U1	*SPG49*; *SPG56*; *P450TEC*	Cytochrome P450 2U1; spastic paraplegia 49	4q25	AA, DHEA, and long-chain fatty acids	544	5
CYP2W1	CYP 2W1	Cytochrome P450 2W1; CYPIIW1	7p22.3	Unknown	490	10
CYP3A4	*HLP*; *CP33*; *CP34*; *CYP3A*; *NF-25*; *CYP3A3*; *P450C3*; *CYPIIIA3*; *CYPIIIA4*; *P450PCN1*	Cytochrome P450 3A4; 1,8-cineole 2-exo-monooxygenase; P450-III, steroid inducible; albendazole monooxygenase; albendazole sulfoxidase; cytochrome P450 3A3; cytochrome P450 HLp; cytochrome P450 NF-25; cytochrome P450, subfamily IIIA (nifedipine oxidase), polypeptide 3; cytochrome P450, subfamily IIIA (nifedipine oxidase), polypeptide 4; cytochrome P450-PCN1; glucocorticoid-inducible P450; nifedipine oxidase; quinine 3-monooxygenase; taurochenodeoxycholate 6α-hydroxylase	7q21.1	Drugs, steroids, and fatty acids	503	14

(Continued)

TABLE 1.1 (CONTINUED)
A List of 57 Human Functional *CYP* Genes

Gene	Alias	Names	Chromosomal Location	Substrates/ Function	Number of Amino Acids	Number of Exons
CYP3A5	*CP 35; CYPIIIA5; P450PCN3; PCN3*	Cytochrome P450 3A5; cytochrome P450 HLp2; cytochrome P450-PCN3; microsomal monooxygenase; xenobiotic monooxygenase; aryl hydrocarbon hydroxylase; flavoprotein-linked monooxygenase	7q21.1	Drugs, steroids, and fatty acids	502	18
CYP3A7	*CP 37; CYPIIIA7; P450(HFL33); P450-HFLA; P-450111A7*	Cytochrome P450 3A7; cytochrome P450-HFLA; microsomal monooxygenase; xenobiotic monooxygenase; aryl hydrocarbon hydroxylase; flavoprotein-linked monooxygenase	7q21-q22.1	Drugs, steroids, and fatty acids	503	13
CYP3A43	CYP 3A43	Cytochrome P450 3A43	7q21.1	Low level of testosterone 6β-hydroxylase activity	503	15
CYP4A11	*CP4Y; CYP4A2; CYP4AII*	Cytochrome P450 4A11; CYPIVA11P450HL-ω; 20-HETE synthase; alkane-1 monooxygenase; cytochrome P450HL-ω; cytochrome P-450; HK-ω fatty acid ω-hydroxylase; lauric acid ω-hydroxylase; 20-hydroxyeicosatetraenoic acid synthase	1p33	Medium-chain fatty acids such as laurate and myristate	519	12
CYP4A22		Cytochrome P450 4A22; CYPIVA22; cytochrome P450 4A22K; fatty acid ω-hydroxylase; lauric acid ω-hydroxylase	1p33	Unknown	519	12
CYP4B1	*CYPIVB1; P-450HP*	Cytochrome P450 4B1; cytochrome P450-HP; microsomal monooxygenase	1p33	Xenobiotics, steroids, and fatty acids	511	12
CYP4F2	CP F2	Leukotriene-B$_4$ ω-hydroxylase 1; CYPIVF2; cytochrome P450 4F2; cytochrome P450-LTB-ω; leukotriene-B$_4$ 20-monooxygenase 1	p13.12	Eicosanoids	520	13
CYP4F3	*CP F3; CYP4F; LTB4H*	Leukotriene-B$_4$ ω-hydroxylase 2; CYPIVF3; cytochrome P450 3F3; cytochrome P450-LTB-ω; leukotriene B$_4$ 20-monooxygenase; leukotriene B$_4$ ω hydroxylase; leukotriene-B$_4$ 20-monooxygenase 2	19p13.2	Eicosanoids (ω-hydroxylation of LTB$_4$)	520	15
CYP4F8	*CYPIVF8; CPF8*	Cytochrome P450 4F8; microsomal monooxygenase; flavoprotein-linked monooxygenase	19p13.1	Eicosanoids (form 19*R*-hydroxy-prostaglandins)	520	13
CYP4F11	CYPIVF11	Cytochrome P450 4F11	19p13.1	Flavoprotein hydroxylation	524	13
CYP4F12	*CYPIVF12; F22329_1*	Cytochrome P450 4F12	19p13.1	Fatty acids	524	14
CYP4F22	*LI3; ARC15; INLNE*	Cytochrome P450 4F22	19p13.12	Fatty acids	531	14
CYP4V2	*CYP 4AH1; BCD*	Cytochrome P450 4V2	4q35.2	Fatty acids	525	11
CYP4X1	CYPIVX1	Cytochrome P450 4X1	1p33	Flavoprotein hydroxylation	509	12
CYP4Z1	CYP 4A20	Cytochrome P450 4Z1	1p33	Flavoprotein hydroxylation	505	12

(Continued)

TABLE 1.1 (CONTINUED)
A List of 57 Human Functional *CYP* Genes

Gene	Alias	Names	Chromosomal Location	Substrates/ Function	Number of Amino Acids	Number of Exons
CYP5A1	*TBXAS1; TS; TXS; CYP5; TBXAS1; TXAS; THAS; GHOSAL; BDPLT14*	Thromboxane A-synthase; TXA synthase; cytochrome P450 5A1	7q34-q35	Thromboxane synthesis	534	18
CYP7A1	*CP7A; CYP7; CYPVII*	Cholesterol 7α-monooxygenase; cytochrome P450 7A1; cholesterol 7α-hydroxylase	8q11-q12	Conversion of cholesterol to bile acids	504	6
CYP7B1	*CP7B; CBAS3; SPG5A*	25-Hydroxycholesterol 7α-hyroxylase; cytochrome P450 7B1; oxysterol 7α-hydroxylase	8q21.3	Conversion of cholesterol to bile acids	506	6
CYP8A1	*CYP 8; PGIS; PTGI; PTGIS*	Prostacyclin synthase; prostaglandin I_2 synthase	20q13.13	Isomerisation of PGH_2 to prostacyclin	500	10
CYP8B1	*CYP 12; CP8B*	7α-Hydroxycholest-4-en-3-one 12α-hydroxylase; CYPVIIIB1; cytochrome P450 8B1; sterol 12α-hydroxylase	3p22-p21.3	Steroids	501	1
CYP11A1	*CYP 11A; CYPXIA1; P450SCC*	Cholesterol side-chain cleavage enzyme, mitochondrial; steroid 20-22-lyase; cytochrome P450 11A1	15q23-q24	Side-chain cleavage of cholesterol pregnenolone	521	12
CYP11B1	*CYP 11B; FHI; CPN1; P450C11*	Cytochrome P450 11B1, mitochondrial; CYPXIB1; cytochrome P450C11; steroid 11β-hydroxylase	8q21	Steroids	503	11
CYP11B2	*CYP 11BL; CYP 11B; CPN2; ALDOS; CYOXIB2; P450C18; P450aldo*	Cytochrome P450 11B2, mitochondrial; cytochrome P450C18; aldosterone synthase; steroid 11β-monooxygenase; steroid 11β/18-hydroxylase; aldosterone-synthesizing enzyme	8q21-q22	Steroids, especially production of aldosterone	503	9
CYP17A1	*CYP 17; CPT7; S17AH; P450C17*	Steroid 17α-hydroxylase/17,20 lyase; cytochrome P450c17; steroid 17α-monooxygenase; 17α-hydroprogesterone aldolase	10q24.3	Steroid metabolism, especially the conversion of pregnenolone and progesterone	508	8
CYP19A1	*CYP 19; ARO; ARO1; CPV1; CYAR; CYPXIX; P-450AROM*	Aromatase; estrogen synthase; cytochrome P-450AROM; cytochrome P450 19A1; microsomal monooxygenase; flavoprotein-linked monooxygenase	15q21.1	Steroid metabolism, formation of aromatic C18 estrogens and C19 androgens	503	14
CYP20A1	CYP-M	Cytochrome P450 20A1; cytochrome P450 monooxygenase	2q33.2	Unknown	462	14
CYP21A2	*CYP21; CYP21B; CAH1; CPS1; CA21H; P450c21B*	Steroid 21-hydroxylase; 21-OHase; cytochrome P450 XXI; cytochrome P450-C21B	6p21.3	21-Hydroxylation of steroids; required for adrenal synthesis of mineralocorticoids and glucocorticoids	495	11

(Continued)

TABLE 1.1 (CONTINUED)
A List of 57 Human Functional *CYP* Genes

Gene	Alias	Names	Chromosomal Location	Substrates/ Function	Number of Amino Acids	Number of Exons
CYP24A1	*CYP 24; CP24; HCAI; P450-CC24*	1,25-Dihydroxyvitamin D 24-hydroxylase, mitochondrial; 24-OHase; cytochrome P450-CC24; cytochrome P450 24A1; vitamin D 24-hydroxylase; exo-mitochondrial protein	20q13	Vitamin D hydroxylation	514	12
CYP26A1	*CYP 26; CP26; P450RAI; P450RAI1*	Cytochrome P450 26A1; hP450RAO; retinoic acid 4-hydroxylase; retinoic acid–metabolizing cytochrome; cytochrome P450 retinoic acid–inactivating 1	10q23-q24	Retinoic acid metabolism	497	8
CYP26B1	*CYP 26A2; RHFCA; P450RAI2*	Cytochrome P450 26B1; retinoic acid–metabolizing cytochrome; cytochrome P450 retinoic acid–inactivating 2	2p13.2	Retinoic acid metabolism	512	7
CYP26C1	FFDD4	Cytochrome P450 26C1	10q23.33	Retinoic acid metabolism	522	4
CYP27A1	*CYP 27; CTX; CP27*	Sterol 26-hydroxylase, mitochondrial; cytochrome P-450C27/25; sterol 27-hydroxylase; vitamin D 25-hydroxylase; cholestanetriol 26-monooxygenase; 5β-chlestane-3-alpha, 7α, 12αtriol 27-hydroxylase	2q35	Steroid metabolism, catalyzing first step in oxidation of side chain of sterol intermediates	531	9
CYP27B1	*CYP 1; CYP 1α; CYP 27B; VDR; CP2B; PDDR; VDD1; VDDRI; P450c1*	25-Hydroxyvitamin D-1α hydroxylase, mitochondrial; 1α(OH)ase; VD3 1A hydroxylase; cytochrome P450C1α; cytochrome P450VD1-α; calcidiol 1-monooxygenase; 25-OHD_1α-hydroxylase	12q14.1	Vitamin D metabolism	508	9
CYP27C1	CYP 27C1	Cytochrome P450 27C1	2q14.3	Unknown	372	9
CYP39A1	CYP 39A1	24-Hydroxycholesterol 7α-hydroxylase	6p21.1-p11.2	Cholesterol	469	12
CYP46A1	*CY46; CYP46*	Cholesterol 24-hydroxylase; CH24H; cytochrome P450 46A1	14q32.1	Cholesterol	500	16
CYP51A1	*LDM; CP51; CYP51; CYPL1; P450L1; P450-14DM*	Lanosterol 14α-demethylase; CYPL1; cytochrome P450LI; cytochrome P45014DM; sterol 14α-demethylase	7q21.2-	Sterols	509	11

The second conserved motif in the human CYP superfamily is "AGXDTT," which is involved in oxygen binding and activation. The terminal Thr in this motif participates in the formation of the enzyme's critical oxygen-binding pocket. In addition, the third motif is "EXXR," located in helix K in the C-terminal, which interacts with residues in the "Meander," a loop of approximately 14 amino acids C-terminal to the cysteinyl heme-binding loop. The conserved residues in helix K appear to be essential for enzyme function since mutation of either residue causes a complete loss of enzyme activity.

1.2.2 SUBCELLULAR LOCATION OF CYPs

CYP11A1, 11B1, 11B2, 24A1, 27A1, 27B1, and 27C1 are the seven CYP enzymes exclusively located in mitochondria

that are therefore type I. In the mitochondrion, NADH or NADPH can donate electrons to the membrane-bound flavoprotein ferredoxin reductase, which then passes them to a soluble iron–sulfur protein ferredoxin in order to donate the electrons to type I CYPs. The cryptic mitochondrial targeting signals of CYP2B1, 2D6, and 2E1 require activation by protein kinase A or protein kinase C–mediated phosphorylation at sites immediately flanking the targeting signal or membrane anchoring regions (Avadhani et al. 2011). The cryptic mitochondria targeting signal of CYP1A1 requires activation by endoproteolytic cleavage by a cytosolic endoprotease, which exposes the mitochondrial signal. Heme iron in both types of CYPs ultimately receives the electrons and donates them to molecular oxygen as the terminal electron acceptor, mediating catalysis.

TABLE 1.2

A List of 58 Human *CYP* Pseudogenes

Pseudogene	Alias	Genomic Location
CYP1D1P	*CYP1A8P, CYP1D1*	9q21.13
CYP2A7P1	*CYP2A7P2, CYP2A18PC, CYP2A18PN, CYP2A7PT*	19q13.2
CYP2AB1P	CYP2D31P	3q27.1
CYP2AC1P	CYP2C57P	6p12.3
CYP2B7P1	*CYP2B7, CYP2B, CYP2BP*	19q13.2
CYP2C115P	CYP2C9-de1b	10q23.33
CYP2C23P	CYP2C62P	10q24.2
CYP2C56P		2q24.3
CYP2C58P		10q23.33
CYP2C59P		10q23.33
CYP2C60P		10q23.33
CYP2C61P		10q21.3
CYP2C63P		21q21.3
CYP2C64P		Xq28
CYP2D7P1	*CYP2D, CYP2D7, CYP2D@, CYP2D7AP, P450C2D, P450DB1*	22q13.2
CYP2D8P1		22q13.2
CYP2F1P		19q13.2
CYP2G1P	*CYP2G1, CYP2GP1, CYP2G2P*	19q13.2
CYP2G2P	*CYP2G2, CYP2GP2*	19q13.2
CYP2T2P		19q13.2
CYP2T3P		19q13.2
CYP21A1P	*CYP21A, CYP21P*	6p21.33
CYP3A137P		7q22.1
CYP3A51P	CYP3A5-del3c	7q22.1
CYP3A52P		7q22.1
CYP3A54P		7p22.1
CYP3AP1	CYP3A5P1	7q22.1
CYP3AP2	CYP3A5P2	7q21.3-q22.1
CYP4A26P		1p33
CYP4A27P		1p33
CYP4A43P		1p33
CYP4A44P		1p33
CYP4F9P		19p13.12
CYP4F10P		19p13.12
CYP4F23P		19p13.12
CYP4F24P		19p13.12
CYP4F25P		9q21.11
CYP4F26P		9p13.3
CYP4F27P		2q21.1
CYP4F29P	*C21orf15, CYP4F3LP*	21q11
CYP4F30P	C2orf14	2q21.2
CYP4F31P		2q21.1
CYP4F32P		2q11.1
CYP4F33P		9p13.1
CYP4F34P		13q11
CYP4F35P		18p11.21
CYP4F36P	CYP2C9-de1b	10q23.33
CYP4F43P		2q21.1
CYP4F44P		8p11.1
CYP4F45P		9p12

(Continued)

TABLE 1.2 (CONTINUED)

A List of 58 Human *CYP* Pseudogenes

Pseudogene	Alias	Genomic Location
CYP4F59P		9p12
CYP4F60P		9p11.2
CYP4F61P		9q13
CYP4Z2P		1p33
CYP46A4P		1p33
CYP51P1		3p12.2
CYP51P2		13q12.3
CYP51P3	–	6q24.3

1.2.3 CYP-MEDIATED METABOLISM OF XENOBIOTICS AND ENDOGENOUS COMPOUNDS

Three families of CYPs, namely, CYP1s, CYP2s, and CYP3s, are the main ones contributing to the oxidative metabolism of more than 90% of clinical drugs with CYPs in other families being involved in the metabolism of endogenous compounds such as steroids, fatty acids, bilirubin, and arachidonic acids (AAs) (Lewis and Ito 2010; Nebert and Dalton 2006; Nebert and Russell 2002; Nebert et al. 2013; Niwa et al. 2009). The relative abundance of the hepatic CYPs has been determined as CYP1A2 (13%), 2A6 (4%), 2C (20%), 2D6 (2%), 2E1 (7%), and 3A4 (30%) (Shimada et al. 1994). A study has demonstrated that of the 110 drugs tested, 66% are metabolized by one or more CYP enzymes; of these, 44% are metabolized by CYP3A4, 41% by 2D6, 26% by 2C19, 9% by 1A2, and 4% by 2C9 (Stringer et al. 2009).

CYP4 members are major fatty acid ϖ-hydroxylases (Hsu et al. 2007). These enzymes remove excess free fatty acids to prevent lipotoxicity and catabolize leukotrienes (LTs) and prostanoids including PGs, thromboxanes, and prostacyclins, and result in bioactive metabolites from AA ϖ-hydroxylation. In addition to endogenous substrates, CYP4s can also metabolize xenobiotics, including therapeutic drugs and toxicants. CYP5A1 is a TBXAS1. CYP7A1 and 7B1 are 7α-hydroxylases of steroids. CYP11A1, 11B1, and 11B2 are involved in steroid biosynthesis. CYP17A1 is present in adrenal cortex and has steroid 17α-hydroxylase and 17,20-lyase activities for steroids. CYP19A1 is an aromatase present in gonads, brain, and adipose tissue that catalyzes aromatization of androgens to estrogens. CYP21A2 is detected in adrenal cortex and has 21-hydroxylase activity toward steroids (Speiser 2001; Wedell 2011). CYP26A1, 26B1, and 26C1 are retinoid acid (RA) hydroxylases (Topletz et al. 2012). CYP39A1 catalyzes 7α-hydroxylation of 24-hydroxycholesterol (Wikvall 2001). Moreover, CYP46A1 is a cholesterol 24-hydroxylase (Lund et al. 1999) while CYP51A1 is a lanosterol 14α-demethylase (Lepesheva and Waterman 2011).

The majority of CYPs play a role in the metabolism of endogenous compounds and xenobiotics, but some of them do not show clear function with endogenous or xenobiotic substrates (Guengerich et al. 2005; Nebert et al. 2013; Stark and Guengerich 2007). These so-called orphan CYPs could include CYP2A7, 2S1, 2U1, 2W1, 3A43, 4A22, 4F11, 4F22, 4V2, 4X1, 4Z1, 20A1, and 27C1 (Guengerich et al. 2005; Stark and Guengerich 2007). However, there are increasing studies that report the identification of new substrates for some of these CYPs. For example, CYP2S1 has been shown to participate in the synthesis and metabolism of bioactive lipids such as PGs and retinoids (Bui et al. 2011; Fromel et al. 2013; Madanayake et al. 2012). CYP2U1 is found to metabolize AA, docosahexaenoic acid (DHA), and other long-chain fatty acids (Chuang et al. 2004). CYP2W1 has been found to metabolize AA and may activate procarcinogens (Eun et al. 2010; Tan et al. 2011; Travica et al. 2013). As such, there are still about eight so-called orphan CYPs that currently have not been unequivocally assigned a biological function: for example, CYP2A7, 4A22, 4F11, 4F22, 4V2, 4X1, 20A1, and 27C1. CYP4X1 has been found to metabolize anandamide (an endogenous cannabinoid neurotransmitter) in brain (Stark et al. 2008a).

CYPs have many physiologically relevant functions including regulation of vascular tone in the cardiovascular system, ion transport in the kidney and the inflammation and immune system, the secretion of pancreatic peptide hormones, cell proliferation, and programmed cell death (Fleming 2008; Go et al. 2015; Nebert and Russell 2002; Nebert et al. 2013; Sato et al. 2011). CYPs participate in cellular functions such as the metabolism of eicosanoids, the biosynthesis of cholesterol and bile acids, synthesis and metabolism of steroids and vitamin D, synthesis and degradation of biogenic amines, and the hydroxylation of RA and presumably other morphogens (Bishop-Bailey et al. 2014; Ferguson and Tyndale 2011; Hsu et al. 2007). The metabolites of these endogenous compounds often have important physiological activities that regulate cellular metabolism, death, and survival.

For example, CYP2R1, 3A4, 2J3, 24A1, 27A1, and 27B1 all metabolize vitamin D_3 (Jones et al. 2014; Omdahl et al. 2003; Prosser and Jones 2004; Sakaki et al. 2005). During activation, vitamin D first undergoes hydroxylation by hepatic 25-hydroxylase (CYP2R1), resulting in $25(OH)D_3$. A second hydroxylation by renal 1α-hydroxylase (CYP27B1) forms the active form $1,25(OH)_2D_3$, which exerts its biological effects by binding to the vitamin D receptor (VDR). This active form of vitamin D is inactivated by 24-hydroxylase (CYP24A1), an enzyme that is responsible for the five-step 24-oxidation pathway from $1,25(OH)_2D_3$ to calcitroic acid (Sakaki et al. 2005). Vitamin D plays a central role in calcium homeostasis and bone metabolism (St-Arnaud 2008). Multiple CYPs play an important role in maintenance of cholesterol homeostasis. CYP51 is involved in cholesterol biosynthesis, whereas CYP 7A1, 27A1, 46A1, 7B1, 39A1, and 8B1 are the key enzymes in cholesterol catabolism to bile acids, the major route of cholesterol elimination (Pikuleva 2006). Conversion of cholesterol to steroids is initiated by CYP11A1, and CYP3A4 contributes to bile acid biosynthesis as well (Pikuleva 2006). Six CYPs including the CYP11 family and three type II CYPs including CYP17A1, 19A1, and 21A2 play indispensable roles in the biosynthesis of steroids. The key CYP enzymes in the bile acid biosynthetic pathways are CYP7A1, 8B1, 27A1, and 7B1. Biosynthesis and metabolism of cholesterol, bile acids, and oxysterols involve CYP3A4, 7A1, 7B1, 8B1, 27A1, 39A1, 46A1, and 51A1.

Multiple CYPs convert AA to bioactive eicosanoids such as epoxyeicosatrienoic acids (EETs), midchain hydroxyeicosatetraenoic acids (HETEs), and ω-terminal HETES (Alsaad et al. 2013; Bellien and Joannides 2013; Chen and Wang 2013; Elbekai and El-Kadi 2006; Fleming 2008, 2011a,b; Imig 2013; Imig et al. 2011; Nebert and Dalton 2006; O'Donnell et al. 2009; Panigrahy et al. 2010; Sato et al. 2011; Wu et al. 2014; Xu et al. 2011; Zordoky and El-Kadi 2010). EETs and HETEs show vasodilatory, anti-inflammatory, fibrinolytic, cardioprotective, and angiogenic activities and provide protection from ischemia in brain, heart, and lung (Elbekai and El-Kadi 2006; Fleming 2008; Imig et al. 2011; Sato et al. 2011; Wu et al. 2014; Xu et al. 2011). Epoxide hydrolases hydrolyze EETs to their dihydroxyeicosatrienoic acids, which are physiologically less active.

1.2.4 STRUCTURAL FEATURES OF CYPS

Rabbit CYP2C5 was the first mammalian structure to be determined (P.A. Williams et al. 2000) and showed significantly higher homology to the human CYPs than to their bacterial counterparts. Thereafter, the structures of CYP2C5 in complex with diclofenac (Protein Data Bank [PDB] ID: 1NR6) or a dimethyl derivative of sulfaphenazole (1N6B) have been reported (Wester et al. 2004). The structures of the rabbit CYP2B4 in a ligand-free form (1PO5) or in complex with 1-(4-cholorophenyl)imidazole (2Q6N), bifonazole (2BDM), or 4-(4-chlorophenyl)imidazole (1SUO) have also been determined (Scott et al. 2004; Zhao et al. 2006). The structure of rat CYP2B4 has also been reported (Shah et al. 2013; Wilderman et al. 2010). All structural information for the mammalian CYPs before the publication of the rabbit CYP2C5 structure is gained mainly from homology modeling studies based on published microbial P450 structures and site-directed mutagenesis studies.

To date, the crystal structures of many human CYPs have been resolved by x-ray crystallography (Table 1.3). These include CYP1A1 (Walsh et al. 2013), CYP1B1 (A. Wang et al. 2011), CYP1A2 (Sansen et al. 2007b), CYP2A6 (DeVore and Scott 2012a; DeVore et al. 2008, 2012; Sansen et al. 2007a; Yano et al. 2005, 2006), CYP2B6 (Gay et al. 2010b; Shah et al. 2011, 2012; Wilderman et al. 2013), CYP2A13 (DeVore et al. 2012; DeVore and Scott 2012a; Smith et al. 2007), CYP2B6 (Gay et al. 2010b; Shah et al. 2011), CYP2C8 (Schoch et al. 2004, 2008), CYP2C9 (Williams et al. 2003), CYP2C19 (Thelen and Dressman 2009), CYP2D6 (Nguyen and Conley 2008; Rowland et al. 2006; A. Wang et al. 2012), CYP2E1 (Porubsky et al. 2008), CYP2R1 (Strushkevich et al. 2008), CYP3A4 (Williams et al. 2004; Yano et al. 2004), CYP7A1 (Tempel et al. 2014), CYP8A1 (also called prostacyclin synthase) (Li et al. 2008), CYP11A1 (Oyama et al. 2012), CYP11B2 (Strushkevich et al. 2013), CYP17A1 (also called steroid 17-α-hydroxylase) (DeVore and Scott 2012b; Petrunak

TABLE 1.3
Reported 3D Structures of Human CYPs

CYP Enzyme	PDB IDs	Resolution (Å)	Ligands in Complex	References
CYP1A1	4I8V	2.60	ANF	Walsh et al. 2013
CYP1A2	2HI4	1.95	ANF	Sansen et al. 2007b
CYP1B1	3PM0	2.70	ANF	A. Wang et al. 2011
CYP2A6	4RUI, 4EJJ, 3T3R, 3T3Q, 3EBS, 2PG7, 2PG6, 2PG5, 2FDY, 2FDW, 2FDV, 2FDU, 1Z11, 1Z10	1.65–2.80	Sabinene, phenacetin, pilocarpine, nicotine, coumarin, methoxsalen, (5-(pyridin-3-yl)furan-2-yl) methanamine, aldrithiol	DeVore and Scott 2012a; DeVore et al. 2008, 2012; Sansen et al. 2007a; Shah et al. 2015; Yano et al. 2005, 2006
CYP2A13	4EJI, 4EJH, 4EJG, 3T3S, 2P85	1.90–3.00	Nicotine, indole, pilocarpine, 4-(methylnitrosamino)-1-(3-puridyl)-1-butanone (2 molecules)	DeVore and Scott 2012a; DeVore et al. 2012; Smith et al. 2007
CYP2B6	4RRT, 4RQL, 4I91, 3UA5, 3QU8, 3QOA, 3IBD	2.00–2.80	α-Pinene, amlodipine (2 molecules), sabinene, 4-(4-chlorophenyl)imidazole, 4-benzylpyridine, (+)-3-carene, 4-(4-nitrobenzyl)pyridine	Gay et al. 2010b; Shah et al. 2011, 2012, 2015; Wilderman et al. 2013
CYP2C8	2VN0, 2NNJ, 2NNI, 2NNH, 1PQ2	2.28–2.80	Felodipine, montelukast, 9-cis-RA, troglitazone	Schoch et al. 2004, 2008
CYP2C9	1R9O, 1OG5, 1OG2	2.00–2.60	Flurbiprofen, S-warfarin	Branden et al. 2014; Wester et al. 2004; Williams et al. 2003
CYP2C19	4GQS	2.87	Ligand-free	Thelen and Dressman 2009
CYP2D6	4WNV, 4WNU, 4WNT, 4WNW, 3TDA, 3TBG, 3QM4, 2F9Q	2.10–3.00	Prinomastat, thioridazine, quinine, quinidine, ajmalicine	Nguyen and Conley 2008; Rowland et al. 2006; A. Wang et al. 2012
CYP2E1	3T3Z, 3GPH, 3KOH, 3LC4, 3E6I, 3E4E	2.20–3.10	Pilocarpine, indazole, 4-methylpyrazole, ω-imidazolyl-dodecanoic acid, ω-imidazolyl octanoic acid	DeVore et al. 2012; Porubsky et al. 2008
CYP2R1	3C6G, 3CZH, 3DL9, 3RUK	2.30–2.80	Vitamin D_3, vitamin D_2, 1α-hydroxy-vitamin D_2	DeVore and Scott 2012b; Strushkevich et al. 2008, 2013
CYP3A4	4NY4, 4K9T, 4K9V, 4K9W, 4K9U, 4K9X, 4I3Q, 4I4H, 4I4G, 1TQN, 3UA1, 2J0D, 3NXU, 3TJS, 2V0M	2.00–2.95	Ketoconazole, desoxyritonavir analog, erythromycin, desthiazolylmethyloxycarbonyl ritonavir, pyridine-substituted desoxyritonavir, oxazole-substituted desoxyritonavir, bromoergocryptine	Branden et al. 2014; Ekroos and Sjogren 2006; Sevrioukova and Poulos 2010, 2012a,b, 2013c; Williams et al. 2004; Yano et al. 2004
CYP7A1	3V8D, 3SN5, 3DAX	1.90–2.75	7-Ketocholesterol, cholest-4-en-3-one	Tempel et al. 2014
CYP8A1	3B6H, 3B98, 3B99, 2IAG	1.62–2.50	Minoxidil, U51605 (substrate analog)	Chiang et al. 2006; Li et al. 2008
CYP11A1	3NA1, 3N9Z, 3N9Y, 3NA0	2.25	20-Hydroxycholesterol, cholesterol 20,22-dihydroxycholesterol	Oyama et al. 2012
CYP11B2	4DVQ, 4FDH	2.49	Deoxycorticosterone, fadrozole	Strushkevich et al. 2013
CYP17A1	4NKZ, 4NKY, 4NKX, 4NKW, 4NKV, 3RUK, 3SWZ	2.40–2.60	Abiraterone, 17α-hydroxypregnenolone, pregnenolone, progesterone, pregnenolone, TOK-001	DeVore and Scott 2012b
CYP19A1	3S79, 4GL7, 4GL5, 3S7S, 4KQ8	2.75–3.90	Androstenedione, exemestane, Hddg046 (Compound 5, a designed inhibitor), Hddg029 (Compound 4, a designed inhibitor)	DeVore and Scott 2012b; Ghosh et al. 2012; J. Lo et al. 2013
CYP21A2	4Y8W, 2RFB, 2RFC, 3QZ1	2.50–3.10	Progesterone	W.W. Ho et al. 2008; Pallan et al. 2015; Zhao et al. 2012
CYP46A1	4J14, 4ENH, 4FIA, 3MDM, 3MDR, 3MDT, 3MDV, 2Q9G, 2Q9F	1.60–2.50	Posaconazole, fluvoxamine, bicalutamide, clotrimazole, voriconazole, tranylcypromine, thioperamide, cholesterol-3-sulfate	Mast et al. 2008, 2010, 2013
CYP51A1	3JUS, 3JUV, 3LD6	2.80–3.12	Ketoconazole, econazole	Strushkevich et al. 2013

Note: ANF, α-naphthoflavone (7,8-benzoflavone).

et al. 2014), CYP19A1 (also called aromatase), CYP21A2 (also called steroid 21-hydroxylase) (Pallan et al. 2015; Zhao et al. 2012), CYP46A1 (also called cholesterol 24-hydroxylase) (Mast et al. 2008, 2010, 2013), and CYP51A1 (Hargrove et al. 2012).

In addition to the ligand-free structures, a number of complex structures of human CYPs have been reported (see Table 1.3): CYP1A1 complexed with β-naphthoflavone (4I8V) (Sansen et al. 2007b); CYP1A2 in complex with α-naphthoflavone (2HI4) (Sansen et al. 2007b); CYP1B1 complexed with β-naphthoflavone (3PM0) (A. Wang et al. 2011); CYP2A6 complexed with coumarin (1Z10), methoxsalen (1Z11) (Yano et al. 2005), the inhibitor aldrithiol (2FDY), N-methyl(5-(pyridin-3-yl)furan-2-yl)methanamine (2FDV), or 5-(pyridin-3-yl)furan-2-yl)methanamine (2FDW) (Schoch et al. 2004, 2008); several CYP2A6 mutants including N297Q (2PG5), L240C/N297Q (2PG6), N297Q/I300V (2PG7), and I208S/I300F/G301A/S369G (3EBS); CYP2A13 in complex with indole (2P85) (Smith et al. 2007); CYP2B6 in complex with 4-benzylpyridine (3QOA) (Shah et al. 2011), α-pinene (4I91) (Wilderman et al. 2013), 4-(4-nitrobenzyl)pyridine (3QU8) (Shah et al. 2011), amlodipine (3UA5) (Shah et al. 2012), or 4-(4-chlorophenyl)imidazole (3IBD) (Gay et al. 2010b); CYP2C8 in a free form (1PG2) and in complex with montelukast (2NNI), felodipine (2NNJ), two molecules of 9-cis RA (2NNH), and troglitazone (2VN0) (Schoch et al. 2004, 2008); CYP2C9 in free form (1OG2 and 1OG5) (Williams et al. 2003) and in complex with flurbiprofen (1R9O) (Wester et al. 2004); CYP2D6 (2F9Q, 4WNU, 4WNT, 4WNW, and 4WNV) (Nguyen and Conley 2008; Rowland et al. 2006); CYP2E1 in complex with 4-methylpyrazole (3E4E) or indazole (3E6I) (Porubsky et al. 2008); CYP2R1 in complex with vitamin D₃ (3C6G), 1-α-hydroxy-vitamin D₂ (3DL9), or vitamin D₂ (3CZH) (Strushkevich et al. 2008); CYP3A4 in a ligand-free form (1TQN [Yano et al. 2004] and 1W0E, 1W0F, and 1W0G [Williams et al. 2004]) and in complex with erythromycin (2J0D) (Ekroos and Sjogren 2006) or ketoconazole (2V0M) (Ekroos and Sjogren 2006), 7A1 (3DAX); CYP8A1 in complex with the inhibitor minoxidil (3B6H) (Li et al. 2008); CYP17A1 in complex with steroids (4NKZ, 4NKY, 4NKX, 4NKW, and 4NKV) (Petrunak et al. 2014); and CYP46A1 in a free form (2Q9G) or in complex with cholesterol-3-sulfate (2Q9F) (Mast et al. 2008).

The known structures of human CYPs are very similar, albeit with some subtle differences that affect their substrate binding specificity, and all include the common CYP fold known from structures of microbial CYPs (Gay et al. 2010a; Johnson and Stout 2005, 2013; Wang et al. 2009). Helices C, D, and I–L, together with β-sheets 1 and 2, comprise the structural core that forms portions of the heme-binding site and the proximal surface where protein partners bind. Helix F–G, helix B–C, and the N- and C-terminal regions, which form the outer boundaries of the substrate-binding cavity, are more dynamic and exhibit more varied secondary and tertiary structures. The flexibility of this architecture was first demonstrated for P450 102A1, which exhibited an open channel to the active site when crystallized without a substrate and

a closed form when a substrate was bound. Several solvent access channels that can expand, contract, and merge for substrate access and product exit have been defined from structures and molecular dynamics studies.

1.2.5 Inhibition of CYPs

The expression levels of human CYPs are subject to inhibition by a number of endogenous compounds and xenobiotics including drugs (Lin and Lu 1998; Mackenzie et al. 2003). This contributes to the large interindividual variability in drug clearance, drug response, and the occurrence of adverse drug reactions (ADRs) (Schuetz et al. 1993; Wilkinson 2005). Indeed, many clinically significant drug–drug interactions are caused by inhibition of CYP-mediated drug metabolism (Wilkinson 2005).

The inhibition of human CYP activities can occur by three main mechanisms: reversible inhibition, mechanism-based inactivation (including quasi-irreversible and irreversible inhibition), and induction. Reversible inhibition refers to competition of two drugs for a CYP (Feng and He 2013; Fontana et al. 2005; Grime et al. 2009; Korzekwa 2014; Murray 1997; Riley et al. 2007; Zhou 2008b; Zhou et al. 2004, 2005, 2007; Zhou and Zhou 2009). Mechanism-based inhibition of CYPs involves the inactivation of the enzyme via the formation of metabolic intermediates that bind tightly and irreversibly to the enzyme. The metabolic intermediate can then exert its inhibitory effect by forming a direct covalent interaction with an amino acid in the active site, a noncovalent tight binding complex, or an inactive enzyme product that is released from the inhibitor. Mechanism-based inhibition of CYPs is characterized by NADPH-, time-, and concentration-dependent enzyme inactivation. A number of therapeutic drugs have been identified as mechanism-based CYP inhibitors. Mechanism-based inhibition of CYPs can decrease a drug's first-pass clearance in the liver and gut, thus greatly altering the kinetic behavior. Moreover, the decrease in functional catalytic activity of the damaged CYP can lead to enhanced exposure of other drugs that are normally cleared by the CYPs, thus setting the stage for potential drug–drug interactions (Zhou et al. 2007; Zhou and Zhou 2009).

1.2.6 Regulation of CYPs via Nuclear Receptors, miRNAs, Inflammation, and Cytokines

Most CYP1, CYP2, CYP3, and CYP4 members are regulated at transcriptional, posttranscriptional, and epigenetic levels (Zanger et al. 2014). Besides nuclear receptors, phenotypic variations of human CYP genes can be regulated by other factors, such as cis- and trans-acting transcriptional factors, alternative splicing, RNA and protein stability, expression of regulatory RNAs (e.g., miRNAs), and epigenetics such as DNA methylation and histone modifications (Wray et al. 2003). With the discovery of the regulating role of miRNAs in the expression of a number of important target genes, there is increased interest in delineating the role of miRNAs in the regulation of human CYPs.

Many human CYPs are inducible by a variety of compounds (Dickins 2004; Hewitt et al. 2007; Lin and Lu 1998, 2001; Liu et al. 2007; Matoulkova et al. 2014; Pelkonen et al. 1998; Sinz et al. 2008; Smutny et al. 2013; Tompkins and Wallace 2007; Waxman 1999; Zhu 2010). Induction of hepatic CYPs by drugs and other compounds can occur by increasing the intrinsic clearance, thereby increasing the extraction ratio, or by increasing the hepatic blood flow. There are three main mechanisms for hepatic enzyme induction: (a) increased anabolism of the enzyme(s) by upregulation of gene expression, (b) increased anabolism through stabilization of mRNA molecules, and (c) decreased rate of catabolism (degradation of the protein) (e.g., posttranslational stabilization) (Dickins 2004; Lin and Lu 1998, 2001; Tompkins and Wallace 2007; Waxman 1999). Unlike hepatic enzyme inhibition, induction takes time (hours or days) and is a function of chronic exposure. Inducers of CYPs can be separated into five classes: (a) archetypical—phenobarbital-like inducers (e.g., phenobarbital, phenytoin), (b) polycyclic aromatic hydrocarbon (PAH)–like inducers (e.g., cigarette smoke and omeprazole), (c) pregnenolone 16-α carbonite and glucocorticoid-type inducers (e.g., dexamethasone, rifampin, and erythromycin), (d) ethanol-like (e.g., ethanol and isoniazid); and (e) peroxisome proliferators type (e.g., clofibrate and phthalates used in plasticizers) (Lin and Lu 1998, 2001; Zhu 2010).

The nuclear receptors, aryl hydrocarbon receptor (AhR), pregnane X receptor (PXR), constitutive androstane receptor (CAR), glucocorticoid receptor (GR), VDR, retinoid X receptor (RXR), RA receptor (RAR), hepatocyte nuclear factor-4α (HNF-4α), farnesoid X receptor (FXR), liver X receptor (LXR), peroxisome proliferator-activated receptors (PPARs), and dosage-sensitive sex reversal, adrenal hypoplasia critical region, on chromosome X, gene 1 (DAX-1), represent an important group of major regulators of CYPs (Amacher 2010; Handschin and Meyer 2003; Matoulkova et al. 2014; Monostory and Dvorak 2011; Smutny et al. 2013). Ligands that bind to and activate these nuclear receptors include lipophilic substances such as endogenous hormones, vitamins A and D, and xenobiotic endocrine disruptors. PXR is activated by a large number of endogenous and exogenous chemicals including steroids, antibiotics, antimycotics, bile acids, and hyperforin, a constituent of the herbal antidepressant St. John's wort (Handschin and Meyer 2003). PXR regulates CYP3A4, 3A5, 2B6, and 2C8; carboxylesterases; and dehydrogenases (Handschin and Meyer 2003). In addition, PXR also upregulates the expression of Phase II conjugating enzymes such as uridine 5′-diphospho-glucuronosyltransferases (UGTs), glutathione S-transferases, and transporting proteins such as P-glycoprotein (P-gp) (Handschin and Meyer 2003). Treatment with PXR activators can lead to the repression or attenuation of other biochemical pathways in liver and intestine including both energy metabolism and the inflammatory response in humans.

miRNAs are a family of small noncoding RNAs of ~22 nucleotides that regulate gene expression by targeting mRNAs in a sequence-specific manner, leading to induction of translational repression or mRNA degradation (Ambros 2004; Bartel 2009; Carthew and Sontheimer 2009; Mendell and Olson 2012; Winter et al. 2009; Yates et al. 2013). Majority of miRNAs are originated after a series of processes from their precursor molecules termed as primary miRNAs (approximately 70 nt in length), which are generated by transcription of the corresponding gene by RNA polymerase II (Ambros 2004; Bartel 2009; Carthew and Sontheimer 2009; Winter et al. 2009). Primary miRNAs have stem loop structures, and they are capped at the 5′-end, polyadenylated, and are usually spliced. Mature miRNAs act on its target mRNAs by complementary base pairings with the miRNA-recognition elements in the 3′-UTR of target mRNAs. On the basis of miRBase version 21 released in June 2014 (http://www.mirbase.org/), there are 1881 miRNA precursors and 2588 mature miRNAs in humans. These RNAs are expressed in a cell- and tissue-specific manner. It is estimated that miRNAs account for ~1% of predicted genes in higher eukaryotic genomes and that up to 30% of genes within the human genome might be regulated by miRNAs. A line of evidence indicates that miRNAs are involved in the regulation of a number of genes that are involved in development, cell proliferation and apoptosis, and carcinogenesis (He and Hannon 2004; Lee and Dutta 2009; Lieberman et al. 2013; Mendell and Olson 2012; Xiao and Rajewsky 2009; Yates et al. 2013). Many interesting findings related to differential expression of miRNA in various human diseases including various cancers, neurodegenerative diseases, and metabolic diseases have been reported. There are increasing studies that indicate the involvement of miRNAs in the regulation of human CYP enzymes (Gomez and Ingelman-Sundberg 2009; Ikemura et al. 2014; Ingelman-Sundberg et al. 2013; Klaassen et al. 2011; Nakajima and Yokoi 2011; Smutny et al. 2013). For example, Tsuchiya et al. (2006) have reported that CYP1B1 is a target of miR-27b and CYP1B1 and is posttranscriptionally regulated by miR-27b. CYP2E1 is regulated by miR-378 (Mohri et al. 2009). miR-892a also regulates CYP1A1 (Choi et al. 2012). miR-130b downregulates CYP2C9, CAR, and FXRα (Rieger et al. 2015). Takagi et al. (2008) have further reported that miR-148a regulated PXR at a posttranscriptional level and this can alter the inducible or constitutive levels of CYP3A4 in human liver. This group of researchers also reported that HNF-4α was regulated by miR-24 and miR-34a (Takagi et al. 2009). The downregulation of HNF-4α by these two miRNAs suppressed the expression of CYP7A1 and 8B1 as well as morphological changes and the decrease of S-phase population in HepG2 cells. miRNAs regulate CYP3A4 at both posttranscriptional and transcriptional levels via direct targeting of the 3′-UTR of CYP3A4 and indirect targeting of the 3′-UTR of VDR, respectively (Pan et al. 2009). Luciferase reporter assays showed that CYP3A4 3′-UTR-luciferase activity was significantly decreased in HEK293 cells transfected with plasmid expressing miR-27b or mmu-miR-298, whereas the activity remained unchanged in cells transfected with plasmid expressing miR-122a or miR-328 (Pan et al. 2009).

Inflammation and various cytokines are able to alter the expression and activities of CYPs (Harvey and Morgan 2014; J.I. Lee et al. 2010; Morgan et al. 2008; Renton 2001,

2005; Shah and Smith 2015). Cytokines, acting through their respective receptors, are a broad group of low-molecular-weight proteins that are generated by macrophages, lymphocytes, neutrophils, mast cells, endothelial cells, and fibroblasts (Headland and Norling 2015; McNab et al. 2015; Sokol and Luster 2015). A variety of cytokines such as chemokines, interleukins (ILs), tumor necrosis factor-α (TNF-α), and interferons (IFNs) are released in response to infection and other inflammatory diseases (Headland and Norling 2015; McNab et al. 2015; Sokol and Luster 2015). Typical proinflammatory cytokines include IL-1β, IL-2, IL-6, IL-12, IL-18, TNF-α, IFN-α, and IFN-γ, while anti-inflammatory cytokines include IL-4, IL-5, IL-10, IL-11, and IL-13. It appears that certain specific cytokines are primarily responsible for the downregulation of some CYPs as part of the inflammatory process (Shah and Smith 2015). In cultured human hepatocytes, IL-1 and TNF-α downregulate CYP2C8 and 3A4 mRNA expression by 75% and 95%, respectively, but have no effect on CYP2B6, 2C9, and 2C19 (Aitken and Morgan 2007). In contrast, IL-6 causes a decrease in CYP2B6, 2C8, 2C9, 2C19, and 3A4 mRNA expression levels. CYP2C18, typically expressed at very low levels in liver, is unaffected by cytokine treatment. Another in vitro study demonstrates that IL-1 and IL-6 downregulate CYP3A4 while IL-12 and IL-23 have no effect on the expression or activity of CYPs (Dallas et al. 2013). IL-1β decreases CAR expression and decreases phenobarbital- or bilirubin-mediated induction of CYP2B6, 2C9, and 3A4 in human hepatocytes (Assenat et al. 2004). There is a transcriptional suppression of CYP2A13 expression by lipopolysaccharide in cultured human lung cells (H. Wu et al. 2013). IL-6 treatment significantly induces CYP1B1 and 2E1, but not 1A1, in colorectal cancer HCT116 and SW480 cells (Patel et al. 2014). The regulation of CYP2E1 expression by IL-6 is via a transcriptional mechanism involving signal transducer and activator of transcription 3 (STAT3)–mediated pathway while IL-6 downregulates the CYP1B1-targeting miR27b through modulation of DNA methylation (Patel et al. 2014). A study in healthy volunteers has shown that inflammatory response to even a very low dose of lipopolysaccharide significantly decreases CYP1A2 and 2C19 activities (Shedlofsky et al. 1994). However, administration of IL-10 to healthy volunteers does not alter CYP1A2, 2C9, or 2D6 activities, but it reduces CYP3A activity (Gorski et al. 2000). A study in patients with rheumatoid arthritis has shown that the plasma concentrations of EXP 3174, the active CYP2C9-generated metabolite of losartan, are significantly decreased (Daneshtalab et al. 2006). In a study of 128 patients, there is a significantly increased voriconazole plasma trough concentrations owing to inhibition of CYP2C19 (van Wanrooy et al. 2014). Chronic inflammation downregulates CYP3A4 activity in patients with renal failure and undergoing hemodialysis (Molanaei et al. 2012). The systemic exposure to simvastatin, a CYP3A4 substrate, is 4- to 10-fold higher in patients with rheumatoid arthritis (Schmitt et al. 2011). Notably, there is an upregulation of CYP2C19 and 3A4 after administration with tocilizumab, a humanized monoclonal antibody against IL-6 receptor (Schmitt et al. 2011). There is preliminary

evidence that CYP downregulation during inflammation may be partly mediated by modulation of nuclear receptors including PXR, CAR, RXRα, PPARα, and FXR via downregulation of receptor expression and/or inhibition of receptor activation (Gerbal-Chaloin et al. 2013). In response to inflammation, the effects of cytokines on CYP enzymes are regulated independently and may have a critical effect on clinical responses to certain drugs in disease states although clinical studies investigating the correlation between downregulation of CYPs and plasma cytokine levels have not been conclusive.

1.2.7 PHENOTYPES AND POLYMORPHISMS OF CYPs

Phenotypically, a population with various CYP activities contain extensive metabolizers (EMs), intermediate metabolizers (IMs), and poor metabolizers (PMs). For CYP2D6 and 2C19, there is a fourth phenotype, the ultrarapid metabolizers (UMs). The distribution of these phenotypes is different in distinct ethnic groups. In many cases, the phenotype–genotype relationships for *CYP* genes can be established but there is a discordance when accurate genotyping is difficult or phenol conversion occurs. Therefore, the results from many genotype-focused association studies are inconsistent and often conflicting. Although a PM genotype correctly predicts the PM phenotype, quantitative prediction of drug-metabolizing capacity among EM patients is complex and difficult. For example, 16 cancer patients have the CYP2C19 EM genotype, but 4 of them (25%) show PM phenotype, and in the remaining 12, there is a general shift toward a slower metabolic activity (M.L. Williams et al. 2000). Similar results are observed in another study in patients with advanced cancer (Helsby et al. 2008). A study in 87 Caucasian patients treated with antidepressant monotherapy with other coadministered drugs reveals genotype–phenotype mismatch in 10 (11.4%) of these patients and thus the CYP2D6 phenotype, rather than the genotype, can better predict clinical response to the antidepressant (Gressier et al. 2015).

Genetic mutations play an important role in the enzyme activity variation of many CYPs, in particular CYP2C9, 2C19, and 2D6 (Chen and Wei 2015; Daly 2012; Ingelman-Sundberg et al. 2007; Polimanti et al. 2012; Stingl and Viviani 2015; Zhou et al. 2008, 2009b). Genetic polymorphisms within CYPs mainly affect the metabolism of drugs that are substrates for those particular enzymes, probably leading to differences in drug response in addition to an altered risk for ADRs (Ingelman-Sundberg et al. 2007; Kirchheiner and Seeringer 2007; Tomalik-Scharte et al. 2008). Many members of the *CYP* families are polymorphic and allelic variants resulting in altered protein expression or activity have significant effects on the disposition of drugs. Because of the important role of CYPs in the disposition and biosynthesis of endogenous compounds such as lanosterol, bile acids, vitamin D, eicosanoids, fatty acids, AA, RA, and steroids, *CYP* mutations may affect organ development and function and therefore cause disease phenotypes or increased risk for certain diseases.

1.2.8 CYP-Related Diseases

Given the important role of CYPs in the metabolism of a series of endogenous compounds, CYP-mediated diseases comprise those caused by aberrant steroidogenesis, defects in fatty acid, cholesterol and bile acid pathways, vitamin D dysregulation, and retinoid (as well as putative eicosanoid) dysregulation during fertilization, implantation, embryogenesis, fetogenesis, and neonatal development (Nebert et al. 2013; Rowland and Mangoni 2014). These diseases include the following:

- 21-Hydroxylase deficiency
- Alzheimer's disease
- Aromatase deficiency
- Aromatase excess syndrome
- Autoimmune Addison disease
- Bietti's crystalline dystrophy
- Cerebrotendinous xanthomatosis
- Congenital adrenal hyperplasia attributed to 11β-hydroxylase deficiency
- Corticosterone methyloxidase deficiency
- Early-onset glaucoma
- Essential hypertension
- Familial hyperaldosteronism
- Ghosal hematodiaphyseal dysplasia
- Hereditary spastic paraplegia type 5
- Multiple sclerosis
- Peters anomaly
- Vitamin D–dependent rickets

Probably because of redundancy, allelic variants leading to absence of activity in any specific CYP2 or CYP3 members do not appear to be specifically associated with serious human diseases. Since CYPs can bioactivate many procarcinogens and some of them are highly expressed in tumor tissues, CYPs are associated with carcinogenesis (Go et al. 2015). A number of published genome-wide association studies have shown that one or several genes in the *CYP1*, *CYP2*, *CYP3*, or *CYP4* families are significantly associated with increased risk of a particular complex disease. Variants in the CYP1, 2A6, 2B6, 2C, 2D6, 2E1, and 3A members have been associated with increased risk of cancer, birth defects, or ADRs. Mutations in the CYP1 and 4 families can cause autosomal recessive disorders.

1.3 HUMAN CYP1 FAMILY

The human CYP1 family contains CYP1A1, 1A2, and 1B1. The *CYP1A* gene cluster has been mapped to chromosome 15q24.1, with close link between *CYP1A1* and *1A2* sharing a common 5′-flanking region (Corchero et al. 2001). The *CYP1A1* and *1A2* genes are separated by a 23-kb segment that contains no open-reading frames (Corchero et al. 2001). The *CYP1A1* and *1A2* genes are oriented head to head on chromosome 15. The 23.3-kb spacer region might contain distinct regulatory regions for one or the other of these genes, or the regulatory regions for the two genes may overlap

one another. The *CYP1A2* gene may arise via duplication of *CYP1A1* approximately 350 myr ago during the evolution of mammals and birds based on phylogenetic analysis of *CYP1A* genes (Morrison et al. 1995). CYP1A2 is one of the major CYPs in the liver (~13%–15%), while CYP1A1, a well-known aryl hydrocarbon hydroxylase, is expressed in the liver at very low levels. CYP1A1 is mainly expressed in human extrahepatic tissues such as intestine, lung, placenta, and lymphocytes (Androutsopoulos et al. 2009; Zhou et al. 2009a,c). In humans, CYP1A2 shares 80% amino acid sequence and 74% DNA sequence identity with CYP1A1 and approximately 40% with 1B1, and the substrate specificities of these enzymes often overlap. Like CYP1A1, 1B1 is an extrahepatic enzyme but has low amino acid sequence identity with both CYP1A1 (38%) and 1A2 (37%). This is less than the 40% required to be classified to the same CYP family, but the grouping was nonetheless assigned on the basis of their common induction by AhR and similar substrate specificities toward such compounds as PAHs, HCAs, and estradiol. CYP1A1 as an extrahepatic enzyme is considered to play a minor role in the elimination of these drugs in vivo, and thus the polymorphism of the *CYP1A1* gene may have a minor effect on their metabolic clearance (Ingelman-Sundberg et al. 2007; Zhou et al. 2008, 2009b). CYP1A1 and 1B1 catalyze the oxidation of procarcinogens to carcinogenic reactive intermediates. As a result, the expression of CYP1A1 and 1B1 is an important contributor to carcinogenesis. The CYP1 family is also involved in the metabolism of endogenous hormones.

The crystal structures of CYP1A1, 1A2, and 1B1 have been resolved (Sansen et al. 2007b; Walsh et al. 2013; A. Wang et al. 2011), and the overall structures of the three human CYP1 family enzymes are very similar but have several differences in specific secondary-structure elements. All three enzymes have narrow active-site cavities, consistent with the planar ligand topology favored by this family of P450 enzymes. The active-site volume is 524 $Å^3$ for CYP1A1, 469 $Å^3$ for CYP1A2, and 441 $Å^3$ for CYP1B1 (Sansen et al. 2007b; Walsh et al. 2013; A. Wang et al. 2011). Comparison of the active-site residues of CYP1A1 to 1A2 and 1B1 reveals a number of similarities and differences. The overall positioning of side chains lining the active site is more similar for CYP1A1 and 1B1 than for CYP1A1 and 1A2, despite the higher sequence identity between the latter two enzymes. The residues that normally compose the two strands of the β4 system in other human CYPs do not adopt this secondary structure in either CYP1A1 or 1A2 but do adopt it in 1B1. Two short two-residue beta strands in the HI loop of CYP1A2 are not present in the single molecule of CYP1A1 that could be modeled in this region. The three-residue disruption of helix F observed only in CYP1A2 and 1B1 is clearly extended to five residues in CYP1A1 (Sansen et al. 2007b; Walsh et al. 2013; A. Wang et al. 2011).

The expression of *CYP1* is under control of the cytosolic AhR, which is activated upon binding to typical ligands such as dioxin and 2,3,7,8-tetrachlorodibenzo-*p*-dioxin (TCDD) (Abel and Haarmann-Stemmann 2010; Barouki et al. 2012; Beischlag et al. 2008; Bock 2014; Denison et al. 2002;

Haarmann-Stemmann and Abel 2006; Hahn 2002; Kohle and Bock 2007; Kung et al. 2009; Nguyen and Bradfield 2008; Pascussi et al. 2008; Petrulis and Perdew 2002; Van Voorhis et al. 2013). Endogenous agonist AhR ligands (such as bilirubin, tryptophan-N-formylated derivatives, and lipoxin A_4) have already been identified, together with an endogenous antagonist, 7-ketocholesterol (Bock 2014; Denison et al. 2002; Nguyen and Bradfield 2008). The activated AhR translocates into the nucleus, dissociates from the chaperone complex, and binds with the AhR nuclear translocator. This heterodimer forms a high-affinity DNA binding complex that binds to specific target DNA sequences known as AhR-response elements in the regulatory areas of the responsive *CYP1* genes (Abel and Haarmann-Stemmann 2010; Barouki et al. 2012; Beischlag et al. 2008; Bock 2014; Denison et al. 2002; Haarmann-Stemmann and Abel 2006; Hahn 2002; Kohle and Bock 2007; Kung et al. 2009; Nguyen and Bradfield 2008; Pascussi et al. 2008; Petrulis and Perdew 2002; Van Voorhis et al. 2013). AhR is considered to have important physiological roles. In $AhR^{-/-}$ mice, vascular differentiation is severely disrupted, reproductive ability is impaired, and liver and immune abnormalities are observed (Vasquez et al. 2003).

1.3.1 CYP1A1 (Aryl Hydrocarbon Hydroxylase)

The *CYP1A1* gene is located on chromosome 15 near the mannose phosphate isomerase (*MPI*) locus at 15q22-24. *MPI* is located in the region 15q22-qter. This gene is conserved in chimpanzee, rhesus monkey, dog, cow, mouse, rat, chicken, zebrafish, *A. thaliana*, rice, and frog. CYP1A1 is expressed in fetal liver but at a low level in adult liver, placenta, peripheral blood cells, skin, intestine, and lung (Drahushuk et al. 1998; Lampen et al. 1998; Mace et al. 1998; Shimada et al. 1996b, 1992; Yang et al. 1995). At the mRNA level, *CYP1A1* is expressed in pancreas, thymus, prostate, small intestine, colon, uterus, and mammary gland (Androutsopoulos et al. 2009; Zhou et al. 2009a,c). CYP1A1 mRNA and protein are also detected in neurons of cerebral cortex, Purkinje and granule cell layers of cerebellum, and pyramidal neurons of the hippocampus (Chinta et al. 2005). There is a large interindividual variation in the expression level of CYP1A1. CYP1A1 possesses functional similarities and differences with CYP1A2 and 1B1. Many of the substrate and inhibitor probes do not clearly differentiate between CYP1A1 and 1A2 (Tassaneeyakul et al. 1993).

As an extrahepatic enzyme, CYP1A1 is involved in the metabolism of many drugs; activates certain procarcinogens, toxicants, and environmental pollutants; and metabolizes several endogenous substrates (Hildebrandt et al. 2007). Like CYP1A2, 1A1 can metabolize a number of therapeutic drugs, including acetaminophen (Laine et al. 2009; Roe et al. 1993), phenacetin (Huang et al. 2012; Tassaneeyakul et al. 1993), caffeine (Tassaneeyakul et al. 1992), propranolol (Ching et al. 1996), theophylline (Sarkar and Jackson 1994), warfarin (Zhang et al. 1995), chlorzoxazone (Carriere et al. 1993), aminoflavone (Kuffel et al. 2002), verlukast (Grossman et al. 1993), toremifene (Berthou et al. 1994), and riluzole

(Sanderink et al. 1997). Phenacetin is primarily metabolized to acetaminophen via O-deethylation, with the formation of a minor metabolite acetol (~5%) via acetyl hydroxylation (Yun et al. 2000). Acetaminophen can be further metabolized by CYP1A1 and 1A2 to a toxic metabolite, N-acetyl-p-benzoquinone imine, along with 3-hydroxy acetaminophen (Laine et al. 2009; Roe et al. 1993). The N-demethylation of theophylline to 3-methylxanthine and 1-methylxanthine is mediated by CYP1A1 and 1A2 (Sarkar and Jackson 1994). The 8-hydroxylation of R-warfarin is a pathway that is selective for CYP1A1 activity, while CYP2A6, 2B6, 2C9, 2D6, 2E1, and 3A4 do not yield either 6- or 8-OH-warfarin from R-warfarin (Zhang et al. 1995). Verlukast, a potent LTD_4 antagonist, is epoxidized by CYP1A1 but not CYP1A2 (Grossman et al. 1993). In addition, the resorufin substrates 7-ethoxyresorufin and 7-methoxyresorufin undergo O-dealkylation by CYP1A1 and 1A2 (Nerurkar et al. 1993; Tassaneeyakul et al. 1993). 7-Ethoxyresorufin O-deethylation is used as a marker reaction for CYP1A1, while phenacetin O-deethylation is for CYP1A2 (Bourrie et al. 1996). Importantly, CYP1A1 can bioactivate certain procarcinogens such as PAHs (e.g., benzo[a]pyrene [BaP] in tobacco smoke), heterocyclic aromatic amines (HCAs), industrial arylamines, and aflatoxin B_1 (AFB_1) to generate reactive metabolites that cause tumor development. CYP1A1 and 1A2 are also involved in the bioactivation and detoxification of the herbal nephrotoxic and carcinogenic compound aristolochic acid (Levova et al. 2011; Sistkova et al. 2008; Stiborova et al. 2012). CYP1A1 can metabolize endogenous substrates including the endogenous fatty acid precursors of PGs such as AA and eicosapentaenoic acid (EPA) (Arnold et al. 2010; Schwarz et al. 2004, 2005), 17β-estradiol (A.J. Lee et al. 2003), and melatonin (X. Ma et al. 2005; Yeleswaram et al. 1999). CYP1A1 together with 1A2, 1B1, and 3A4 catalyze the 2-, 4-, and 16α-hydroxylation of 17β-estradiol (Badawi et al. 2001; A.J. Lee et al. 2003). In addition, CYP1A1, 1A2, and 1B1 all 6-hydroxylate melatonin, with CYP2C19 playing a minor role (X. Ma et al. 2005). CYP1A1 exhibits hydroxylase activity toward AA, whereas toward EPA, it is an epoxygenase (Arnold et al. 2010; Schwarz et al. 2004, 2005). CYP1A1 metabolizes EPA and DHA primarily to 17,18-epoxyeicosatetraenoic acid and 19,20-epoxydocosapentaenoic acid, respectively. These bioactive metabolites of ω-3 polyunsaturated fatty acids (e.g., EPA and DHA) play important roles in the regulation of vascular tone and of renal, pulmonary, and cardiac functions. Elevated blood pressure in *Cyp1a1* knockout mice is associated with reduced vasodilation to ω-3 polyunsaturated fatty acids (Agbor et al. 2012). Therefore, CYP1A1 may play a role in the pathogenesis of cardiovascular diseases. Clinical studies have indicated that polymorphisms of *CYP1A1* may modulate the risk of hypertension, coronary artery disease, and hyperlipidemia (Bailon-Soto et al. 2014; Kopf et al. 2010; X.L. Wang et al. 2002; Zou et al. 2014).

Many CYP1A1 substrates are inhibitors of this enzyme. AA significantly inhibits CYP1A1- and 1A2-dependent 7-ethoxycoumarin O-deethylation (Yamazaki and Shimada 1999). Retinol, RA, and cholecalciferol are all inhibitors of CYP1A1 (Yamazaki and Shimada 1999). Caffeine, paraxanthine, propranolol, and theophylline similarly inhibit CYP1A1- and 1A2-catalyzed phenacetin *O*-deethylation (Tassaneeyakul et al. 1993). α-Naphthoflavone (ANF), a common inhibitor used for CYP1A1 with a K_i of 0.01 μM, is more effective against CYP1A2 (Bourrie et al. 1996; Tassaneeyakul et al. 1993). ANF and 7-ethoxycoumarin are approximately 10-fold more potent as inhibitors of CYP1A2 than CYP1A1 (Tassaneeyakul et al. 1993). Mollugin originally isolated from *Rubia cordifolia* inhibits CYP1A2 with an IC_{50} of 3.55 μM (Kim et al. 2013). Moreover, the dietary flavonoids including acacetin, diosmetin, eupatorin, and chrysin are inhibitors of CYP1A1 and 1A2 (Androutsopoulos et al. 2011). Quercetin and kaempferol are also inhibitors of CYP1A2 (Savai et al. 2015).

In the determined structure of CYP1A1 refined to a resolution of 2.6 Å, ANF binds within an enclosed active site, with the planar benzochromen-4-one core packed flat against the I helix that composes one wall of the active site, and the 2-phenyl substituent oriented toward the catalytic heme iron (Walsh et al. 2013). Canonical helices A through L are present, including short F′ and G′ helices typically thought to be buried in the membrane and often involved in active-site access for hydrophobic ligands. CYP1A1 binds ANF adjacent to the heme within an enclosed active site. ANF forms a very complementary π–π stacking interaction with the side chain of Phe224, adjacent to the break in the F helix. This interaction with Phe224 and the packing of the hydrophobic residues Phe123, Ile115, L496, Val382, and Ile386 constrain the front and sides of the ANF binding pocket. Additional hydrophobic residues, Phe319, Phe258, and Leu312, form the rear of the active site (Walsh et al. 2013).

Constitutive CYP1A1 expression is generally low, but this enzyme can be greatly induced by many of its own substrates, particularly the PAHs and HCAs, via AhR. CYP1A1 mRNA levels are very high in the lung cells of smokers but typically undetectable in nonsmokers. CYP1A1 protein is also higher in smokers compared to nonsmokers (lung median levels of 16 vs. 6.5 pmol/mg microsomal protein, respectively). The induction and procarcinogen activation make CYP1A1 a target for development of potential chemopreventative agents that inhibit CYP1A1. In contrast, in certain malignancies that overexpress CYP1A1, another therapeutic approach utilizes this enzyme to activate anticancer prodrugs to electrophilic species that kill cancer cells.

1.3.2 CYP1A2 (Aryl Hydrocarbon Hydroxylase)

This gene maps to chromosome 15q24.1 that spans almost 7.8 kb and contains seven exons (Ikeya et al. 1989). Between *CYP1A2* and *1A1*, exons 2, 4, 6, and especially 5 are strikingly conserved in both nucleotides and total number of bases. It encodes a 515-residue protein with a molecular mass of 583 kDa. It is conserved in chimpanzee, dog, cow, mouse, rat, *A. thaliana*, and rice. CYP1A2 accounts for approximately 15% of the total CYP content in human liver (Shimada et al. 1994) and metabolizes a variety of

clinically important drugs, such as clozapine (Bertilsson et al. 1994), ropivacaine (Oda et al. 1995), olanzapine (Ring et al. 1996), theophylline (Ha et al. 1995), tacrine (Spaldin et al. 1994), aminopyrine (Niwa et al. 1999), zolmitriptan (a selective serotonin receptor 1B/1D agonist used for treatment of migraine) (Wild et al. 1999), triamterene (a potassium-sparing diuretic) (Fuhr et al. 2005), mexiletine (a class IB antiarrhythmic drug) (Nakajima et al. 1998), phenacetin (Tassaneeyakul et al. 1993), paracetamol (Tassaneeyakul et al. 1993), flutamide (Shet et al. 1997), lidocaine (Orlando et al. 2004), imipramine (Koyama et al. 1997), paraxanthine (Tassaneeyakul et al. 1992), propranolol (Masubuchi et al. 1994), verapamil (Kroemer et al. 1993), propafenone (a class Ic antiarrhythmic agent) (Botsch et al. 1993), lorcaserin (a novel selective 5-HT_{2C} agonist) (Usmani et al. 2012), tizanidine (Granfors et al. 2004a), 5,6-dimethylxanthenone-4 acetic acid (an anticancer drug in Phase III trials) (Zhou et al. 2000, 2002), and leflunomide (a novel disease-modifying antirheumatic drug that inhibits de novo synthesis of pyrimidine ribonucleotides and leads to a decrease in lymphocyte proliferation) (Kalgutkar et al. 2003). CYP1A2 metabolizes nicotine via C-oxidation but only at high substrate concentration (Yamazaki et al. 1999). Caffeine is predominantly metabolized by CYP1A2 and thus is usually used as a "gold standard" probe for determining CYP1A2 activity (Carrillo et al. 2000; Kalow and Tang 1991; Ryu et al. 2007; Tassaneeyakul et al. 1992). The 1-, 3-, and 7-demethylation of caffeine forms theobromine, paraxanthine, and theophylline, respectively. CYP1A2 also metabolizes several important endogenous substrates, such as melatonin (Skene et al. 2001), bilirubin (Zaccaro et al. 2001), uroporphyrinogen (Lambrecht et al. 1992), estrone, and estradiol (Yamazaki et al. 1998b). Like CYP1A1, 1A2 can convert a number of procarcinogens including PAHs, HCAs, and AFB_1 to reactive electrophiles that can attack DNA and induce tumor formation (Eaton et al. 1995). Many drugs, including oral contraceptives (Abernethy and Todd 1985) and fluoroquinolones, such as levofloxacin and ciprofloxacin (Granfors et al. 2004b; Parker et al. 1994), inhibit CYP1A2 activity. Fluvoxamine is a very potent CYP1A2 inhibitor (Becquemont et al. 1997; Brosen et al. 1993). Furafylline (a mechanism-based inhibitor) and ANF are commonly used as selective inhibitors for CYP1A2 in reaction phenotyping studies (Eagling et al. 1998; Newton et al. 1995; Tassaneeyakul et al. 1993). Isoniazid is also a mechanism-based inhibitor of CYP1A2 (Wen et al. 2002). Cudratricusxanthone A isolated from the roots of *Cudrania tricuspidata* Bureau is a potent inhibitor of CYP1A2, 2C8, and 2C9 with K_i values of 1.3, 2.2, and 1.5 μM (Sim et al. 2015). Generally, CYP1A2 ligands contain a planar ring that can fit the narrow and planar active site of the enzyme (Sansen et al. 2007b).

Tobacco smoking and dietary constituents such as cruciferous vegetables and charcoal-broiled meat can induce CYP1A2 activity (Kalow and Tang 1991). Induction of CYP1A2 activity may also be influenced by treatment with high-dose (120 mg/day) omeprazole (Han et al. 2002). Rifampicin is only a weak inducer of CYP1A2 (Backman et al. 2006). Like CYP1A1 and 1B1, 1A2 is induced through AhR-mediated transactivation after ligand binding and nuclear translocation, but the extent of induction is generally lower than that of CYP1A1. Many substrates for CYP1A1 and 1A2 such as PAHs are also AhR ligands.

To date, at least 20 alleles (*1B through to *21) of the CYP1A2 gene have been identified (http://www.cypalleles.ki.se/cyp1a2.htm). *1A is referred as the wild type. The most extensively studied polymorphisms are −3860G>A (*1C), −2467delT (*1D), −739T>G (*1E), and −163C>A (*1F), which were first reported in a Japanese population (Chida et al. 1999). *1C is reported to cause decreased inducibility of the enzyme in Japanese smokers, probably owing to decreased enzyme expression (Nakajima et al. 1999). The −163C>A in intron 1 causes increased enzyme inducibility in the presence of an inducer (e.g., smoking) (Sachse et al. 1999), although this association is controversial (Aklillu et al. 2003; Chida et al. 1999; Larsen and Brosen 2005; Nordmark et al. 2002; Shimoda et al. 2002). Smokers with the −163C/C genotype have 40% lower plasma 17X/137X ratios compared with those with the −163A/A genotype in smokers (Sachse et al. 1999). *1J (−163C>A; −7391T>G) and *1K (−163C>A; −739T>G; −729C>T) have been detected in Ethiopians (Aklillu et al. 2003). The *1K haplotype is associated with 40% lower inducibility in vitro, and nonsmokers heterozygous for *1K have significantly lower CYP1A2 activity compared with the wild type (Aklillu et al. 2003). *1G, *1H, *1L, *1M, *1N, *1P, *1Q, *1R, *1S, *1T, *1U, *1V, and *1W are relatively rare and do not alter enzyme activity (Chevalier et al. 2001a; Ghotbi et al. 2007; Soyama et al. 2005). The −3113A>G polymorphism, with a frequency of 10% in a Chinese population, has been reported to be associated with decreased CYP1A2 activity (Chen et al. 2005). There are significant ethnic differences in the distribution of common and rare CYP1A2 mutations. The −3860G>A and the −2467delT mutations are lower in Caucasians compared with Asians, while −739G is frequent in Ethiopians and Saudi Arabians (Chida et al. 1999; Larsen and Brosen 2005; Nordmark et al. 2002; Sachse et al. 1999). The −163C>A SNP has similar frequencies in all populations studied, with the highest frequency in Africans. The CYP1A2*1F allele is more frequent in Caucasians and Africans, while *1D, *1L, *1M, and *1N are more common in Asians (Chida et al. 1999; Larsen and Brosen 2005; Nordmark et al. 2002; Sachse et al. 1999). CYP1A2*1J, *1K, and *1W are rare in all populations studied.

1.3.3 CYP1B1 (ARYL HYDROCARBON HYDROXYLASE)

The CYP1B1 gene is located on chromosome 2p22.2; the gene contains three exons and two introns, but only exons 2 and 3 encode the 543–amino acid protein (Tang et al. 1996). The CYP1B1 expression level is high in lung but low in kidney, spleen, thymus, prostate, ovary, lung, small intestine, colon, uterus, mammary bland, and particularly in liver (Shimada et al. 1996a). CYP1B1 is also expressed in the iris, trabecular

meshwork, and ciliary body of the eye. CYP1B1 can activate a broad range of chemical procarcinogens such as PAHs and their oxygenated derivatives, HCAs, aromatic amines, and nitropolycyclic hydrocarbons. The enzyme can also catalyze the conversion of BaP to the 7,8-dihydrodiol, the first step in the formation of diol epoxide. Cyp1b1 knockout mice do not form tumors treated with the procarcinogen 7,12-dimethylbenz[α]anthracene (Buters et al. 1999). The enzyme is the most efficient catalyst of 17β-estradiol 4-hydroxylation. 4-Hydroxyestradiol is more reactive with oxygen and more likely to oxidize to O-quinone to bind DNA (Bolton et al. 1998). Thus, 4-hydroxyestrogens are considered to be candidates for causing estrogen-dependent tumor (Liehr et al. 1995). ANF and 2-ethynylpyrene are potent and selective inhibitors of CYP1B1 (Shimada et al. 1998). Resveratrol, a polyphenol found in red grapes, is a noncompetitive inhibitor of CYP1B1 with a K_i of 23 µM (Chun et al. 1999). 2,4,3′,5′-Tetramethoxystilbene is also a potent and selective competitive inhibitor of CYP1B1 with a K_i of 3 nM (Chun et al. 2001). Both dietary flavonoids quercetin and myricetin are potent inhibitors of CYP1B1 (Androutsopoulos et al. 2011). Moreover, berberine is a very potent and selective inhibitor of CYP1B1 with a K_i of 44 nM, and berberine-mediated inhibition is abolished by a mutation of Asn228 to Thr in CYP1B1 (S.N. Lo et al. 2013).

Mutations of CYP1B1 are strongly associated with primary congenital glaucoma (Badeeb et al. 2014; X. Chen et al. 2014; Gong et al. 2015; Huang et al. 2014; Li et al. 2011; Lopez-Garrido et al. 2013; Micheal et al. 2015; Sheikh et al. 2014; Vasiliou and Gonzalez 2008). It is thought that 1B1 can metabolize a key signaling molecule involved in eye development, possibly a steroid. Glaucoma is the second leading cause of blindness worldwide and is characterized by visual field defects, retinal ganglion cell death, and progressive degeneration of the optic nerve. To date, at least 25 alleles have been identified (*2 through to *26) (http://www.cypalleles.ki.se/cyp1b1.htm), which accounts for 50% of primary congenital glaucoma cases (Li et al. 2011; Vasiliou and Gonzalez 2008). Asn453Ser (*4) and Leu432Val (present in *3, *5–*8) have been reported to be associated with primary open-angle glaucoma in French and Indian population, respectively (Acharya et al. 2006; Melki et al. 2004). A large study conducted on French patients with primary open-angle glaucoma reported the CYP1B1 mutations in 4.6% (11 out of 236) of the cohort. It has been reported that CYP1B1 is expressed in the trabecular meshwork and in the posterior segment of the eye and CYP1B1 mutations could cause reduction of protein activity or abundance (Li et al. 2011; Vasiliou and Gonzalez 2008). Cyp1b1-deficient mice have ocular drainage structure abnormalities resembling those reported in human primary congenital glaucoma patients (Libby et al. 2003). The severe dysgenesis in eyes lacking both Cyp1b1 and the tyrosinase gene is alleviated by administration of the tyrosinase product dihydroxyphenylalanine. Further studies are needed to further explore the role of CYP1B1 mutations in the pathogenesis of primary glaucoma.

1.4 HUMAN CYP2ABFGST CLUSTER: CYP2A6, 2A7, 2A13, 2B6, 2F1, AND 2S1

The human *CYP2ABFGST* cluster is located on chromosome 19q13.2 (Hoffman et al. 1995). This cluster includes loci from the *CYP2A, 2B, 2F, 2G, 2S,* and *2T* subfamilies with six functional genes (*CYP2A6, 2A7, 2B6, 2A13, 2F1,* and *2S1*) and seven pseudogenes including *CYP2A18P-T, 2A18P-N, 2B7P, 2G1P, 2G2P, 2T2P,* and *2T3P. CYP2G1P* contains a single nucleotide deletion in exon 2 and a 2.4-kbp deletion between exons 3 and 7, whereas *CYP2G2P* contains two nonsense mutations in exons 1 and 3, respectively (Sheng et al. 2000). This cluster is considered to be created by duplication events in evolution.

1.4.1 CYP2A6 (COUMARIN 7-HYDROXYLASE)

The *CYP2A6* gene spans ~6 kb in size consisting of nine exons and has been mapped to chromosome 19q13.2 (Fernandez-Salguero et al. 1995; Miles et al. 1989). It is located within a 350-kb gene cluster together with the *CYP2A7* and *2A13* genes, two *CYP2A7P* pseudogenes, and genes in the *CYP2B* and *CYP2F* subfamilies on chromosome 19q (Fernandez-Salguero et al. 1995; Hoffman et al. 1995). CYP2A6 is mainly expressed in the liver accounting for approximately 1%–10% of total CYPs and is an important enzyme since it can metabolize approximately 3% of clinical drugs and is subject to polymorphism with significant clinical impacts (Di et al. 2009; Mwenifumbo and Tyndale 2007; Raunio and Rahnasto-Rilla 2012; Satarug et al. 2006). Only trace amounts of 2A6 are found in extrahepatic tissues such as nasal mucosa and lung (Chiang et al. 2012; Koskela et al. 1999). CYP2A6 has also been found to be expressed in steroid-related tissues such as adrenal gland, testis, ovary, and breast (Nakajima et al. 2006b).

Typical substrates for CYP2A6 include coumarin and valproic acid, as well as environmental and toxic agents including nicotine and AFB_1 (Mahavorasirikul et al. 2009; Raunio and Rahnasto-Rilla 2012). In humans, coumarin is primarily (70%–80%) metabolized to 7-hydroxycoumarin by CYP2A6 (Li et al. 1997; Miles et al. 1990; Soucek 1999; Yun et al. 1991); therefore, coumarin is the marker substance for determining CYP2A6 activity both in vitro and in vivo. Approximately 80% of a nicotine dose is eliminated by CYP2A6, and there is a clear link between *CYP2A6* genotypes, smoking behavior, and lung cancer risk. CYP2A6 metabolizes more than 30 drugs (Lewis 2003; Rendic 2002; Zanger et al. 2007). Valproic acid, an anticonvulsant agent, is metabolized by CYP2A6 with substantial contributions from CYP2B6 and 2C9 (Sadeque et al. 1997). Another anticonvulsant drug, losigamone, is also metabolized by CYP2A6 (Torchin et al. 1996). Halothane is metabolized by CYP2A6 with a major contribution from 3A4 (Spracklin et al. 1996). In addition, CYP2A6 is responsible for the sulfoxidation and thiono-oxidation of diethyldithiocarbamate methyl ester to form *S*-methyl-*N,N*-diethylthiolcarbamate sulfoxide, the putative active metabolite responsible for the

alcohol deterrent effects of disulfiram (Madan et al. 1998). CYP2A6, 2B6, and 3A4 are the high K_m components for cyclophosphamide and its isomeric analog ifosfamide 4-hydroxylation, while CYP2C8 and 2C9 are the low K_m components (Chang et al. 1993). Tegafur, a prodrug, is converted to its active metabolite 5-fluorouracil largely by hepatic CYP2A6, with considerable contributions from CYP1A2 and 2C8 (Ikeda et al. 2000; Komatsu et al. 2000). Formation of 3-hydroxypilocarpine from pilocarpine, a cholinergic agonist, is mainly metabolized by CYP2A6 (Endo et al. 2007). Although CYP3A4 and 2C8 are the major CYPs involved in the *N*-dealkylation of cisapride, a gastrointestinal prokinetic agent (withdrawn from the market) (Desta et al. 2001), CYP2A6 and 2B6 contribute to its metabolism to some extent (Bohets et al. 2000). Hydroxylation of 2n-propylquinoline, a newly developed drug for the treatment of visceral leishmaniasis, is mainly by CYP2A6 with contributions from CYP2E1 and 2C19 (Belliard et al. 2003). Moreover, CYP2A6 metabolizes the neuroprotective drug chlormethiazole, the oral nonsteroidal aromatase inhibitors letrozole and fadrozole, and the inhalational anesthetic halothane, sevoflurane, and methoxyflurane (Oscarson 2001). Notably, CYP2A6 plays an essential role in the metabolic activation of a number of procarcinogens and environmental compounds (Rendic 2002). These include tobacco-specific procarcinogenic nitrosamines such as 4-(methylnitrosamino)-1-(3-pyridyl)-1-butanone (NNK) and *N'*-nitrosonornicotine (NNN) (Brown et al. 2007), 1,3-butadiene (Duescher and Elfarra 1994), 2,6-dichlorobenzonitrile (a herbicide), and AFB_1 (Le Gal et al. 2003; Oscarson 2001). CYP2A6 can bioactivate hexamethylphosphoramide, a nasal procarcinogen (Liu et al. 1996). CYP2A6 has also been found to metabolize gasoline additives such as methyl *tert*-butyl ether, ethyl *tert*-butyl ether, and *tert*-amyl methyl ether (Hong et al. 1997, 1999; Le Gal et al. 2001). Naphthalene, an environmental pollutant and a component of jet fuel, is mainly metabolized to *trans*-1,2-dihydro-1,2-naphthalenediol (dihydrodiol), 1-naphthol, and 2-naphthol by CYP3A4 and 2A6 (Cho et al. 2006). Moreover, CYP2A6 is involved in the metabolism of several endogenous compounds (Rendic 2002). CYP2A6 together with 1A2, 2C8, 3A4/5, and 2B6 catalyze the formation of 4-OH and 4-oxo-RA from atRA (Marill et al. 2000; McSorley and Daly 2000). CYP2A6 metabolizes steroids such as testosterone (Yamazaki et al. 1994) and estradiol (A.J. Lee et al. 2001, 2003), but the contribution is minor. CYP2A6 catalyzes testosterone 7α-hydroxylation (Yamazaki et al. 1994), progesterone 6β-hydroxylation (Niwa et al. 1998), and 17β-estradiol 16β-hydroxylation (A.J. Lee et al. 2001, 2003). Bilirubin, the breakdown product of heme, has been suggested to be the endogenous substrate of CYP2A6.

Pilocarpine, a substrate of CYP2A6, is commonly used as a selective competitive inhibitor of CYP2A6 (Bourrie et al. 1996; Kimonen et al. 1995; Kinonen et al. 1995; Xia et al. 2002). Tranylcypromine, a nonhydrazine monoamine oxidase inhibitor used in psychiatry, is a potent and relatively selective inhibitor of CYP2A6 (Draper et al. 1997; Taavitsainen et al. 2001). Many natural compounds inhibit CYP2A6 activity.

Several drugs and compounds are mechanism-based inhibitors of CYP2A6, including selegiline (Siu and Tyndale 2008), valproic acid (Wen et al. 2001), isoniazid (Wen et al. 2002), and several furanocoumarins such as 5-methoxypsoralen, 8-methoxypsoralen, and psoralen (Draper et al. 1997; Koenigs and Trager 1998).

The preference of CYP2A6 for small-molecule substrates such as coumarin and nicotine suggests that its active site may be small in comparison with those belonging to members of the CYP2C and 3A subfamilies. Of all human CYPs with known structures, CYP2A6 has the second smallest active-site cavity with a volume of 260 Å3. This is approximately four-fold smaller than those of CYP2C8, 2C9, or 3A4 (Yano et al. 2005). To date, there are 14 structures of CYP2A6 available in the PDB. These include the structures of CYP2A6 in complex with coumarin (1Z10) (Yano et al. 2005), pilocarpine (3T3R and 3T3Q) (DeVore et al. 2012), methoxsalen (1Z11) (Yano et al. 2005), sabinene (4RUI) (Shah et al. 2015), nicotine (4EJJ) (DeVore and Scott 2012a), and synthetic 3-heteroaromatic analogs of nicotine as inhibitors (2FDY, 2FDW, 2FDV, and 2FDU) (Yano et al. 2006). Several structures of the CYP2A6 Asn297Gln (2PG5), Leu240Cys/Asn297Gln (2PG6), and Asn297Gln/Ile300Val (2PG7) mutants have also been solved to resolutions of 1.95, 2.50, and 2.80 Å, respectively (Sansen et al. 2007a). DeVore et al. (2009) have reported the structure of CYP2A6 Ile208Ser/Ile300Phe/Gly301Ala/Ser369Gly mutant in complex with phenacetin (3EBS). The CYP2A6 structure shows a clearly well-adapted enzyme for the oxidation of small, planar substrates that can be accommodated within the compact, small, and hydrophobic active site. Inside the active site, Asn297 serves as one hydrogen bond donor and thus orients ligands such as coumarin for regioselective oxidation. The active-site residues Gln104, Val113, Phe18, Phe205, Phe209, Asn290, Ile293, Asn297, Thr298, Thr305, and His477 play a critical role in the orientation and anchoring of coumarin, while neighboring residues such as Thr212 are involved in directing the access of coumarin to the binding site (Sansen et al. 2007a; Yano et al. 2005). The small active-site volume of CYP2A6 may be associated with rather tight packing of the secondary structural units. Other active-site residues such as Phe118, Asn297, and Thr305 also play an important role in ligand orientation (Yano et al. 2005).

Both in vitro and in vivo studies have demonstrated a wide (20- to >100-fold) interindividual variation in CYP2A6 expression and activity (Hadidi et al. 1998; Iscan et al. 1994; Rautio et al. 1992; Yamano et al. 1990), which is due primarily to genetic polymorphisms in the CYP2A6 gene. CYP2A6 activity is also modified by certain drugs as well as pathological and environmental factors. To date, at least 45 alleles of the CYP2A6 gene have been identified (*1B1 through to *46) (http://www.cypalleles.ki.se/cyp2a6.htm). These mutations may increase the interindividual variability in CYP2A6 activity and thus affect smoking behavior or cancer susceptibility (Rossini et al. 2008). *1A is the wild type, and initially two defective alleles, CYP2A6*2 (51G>A; 479T>A) and *3, have been identified (Fernandez-Salguero et al. 1995; Yamano et al. 1990). *2 is a null allele having no enzyme activity toward probe substrates because it does not incorporate heme (Yamano et al. 1990). Individuals carrying *2 had significantly increased 3-hydroxylated coumarin formation, probably attributed to the variant enzyme with Leu160His substitution having regiospecificity that catalyzes coumarin 3-hydroxylation instead of the usual 7-hydroxylation (Hadidi et al. 1997). There are multiple mutations in *3 that resemble the alterations found in the neighboring CYP2A7P pseudogene. The genotyping method, functional impact, and frequency of *3 are controversial. *4 presents a whole gene deletion that responds for the majority of PMs in all of CYP2A6 variants (Liu et al. 2013). *4 may be associated with susceptibility of lung cancer for Asian smokers (Liu et al. 2013). In a large-scale epidemiological study, Kamataki et al. (2005) have discovered that the smokers carrying the *4C allele have less risk of tobacco-related cancers. The subjects may lack the CYP2A6 gene to activate N-nitrosamines that are nicotine-derived carcinogens found in tobacco smokes. *5 is a null allele where a highly conserved Gly479 is replaced with Val owing to a point mutation in exon 9 (1436G>T) (Oscarson et al. 1999). *6 contains a 383G>A mutation leading to an Arg128Gln substitution (Kitagawa et al. 2001). *7 contains a 1412T>C (Ile471Thr) mutation and a conversion with the CYP2A7 sequence in the 3'-UTR of *7 (Ariyoshi et al. 2001). Coumarin 7-hydroxylation in the CYP2A6.7 is decreased by 63%, whereas nicotine C-oxidation is not detected at a certain substrate concentration (Ariyoshi et al. 2001). *8 has a point mutation of 1454G>T → Arg485Leu in exon 9 (Ariyoshi et al. 2001). The frequency of *8 in Japanese and Koreans is 2.2% and 1.4%, respectively (Yoshida et al. 2002). *9 contains a −48T-to-G nucleotide substitution in the TATA box of the 5'-flanking region of CYP2A6 (Pitarque et al. 2001). This mutation reduces the expression levels and the catalytic activities toward coumarin (Kiyotani et al. 2003). *15 carries a Lys194Glu substitution, while *16 harbors a Arg203Ser mutation, with a 1.2% frequency in the Korean population and a 1.5%–2.2% frequency in Japanese people (Nakajima et al. 2006a; Tiong et al. 2010). *17 contains multiple SNPs: 209C>T, 1779G>A, 4489C>T, 5065G>A (Val365Met), 5163G>A, 5717C>T, and 5825A>G (Fukami et al. 2004). The allele frequency in black subjects is 9.4%, but not found in Caucasian, Japanese, and Korean subjects. *20 possesses the deletion of two nucleotides in exon 4 (2141_2142delAA) resulting in a frameshift from codon 196 and an early stop codon at 220 in exon 5 as well as three synonymous SNPs of G51A, T1191C, and C1546G in the 3'-UTR, which produces a truncated protein with abolished functional activity (Fukami et al. 2005). The frequency of this allele in African Americans is 1.6%. *21 contains 1427A>G (Lys476Arg) and one synonymous SNP 51G>A (Haberl et al. 2005; Tiong et al. 2010). *15, *16, and *21 do not have marked functional impact on the enzyme activities for coumarin 7-hydroxylase. The lys194Glu substitution in *15 is not located within any SRSs but adjacent to helix F while Arg203Ser in *16 is located within helix F and is directly positioned within the highly conserved SRS2 of the CYP2A family. *22 carries two SNPs: Asp158Glu and Leu160Ile, resulting in significantly reduced enzyme activity (Tiong et al. 2010). Both Asp158Glu and Leu160Ile substitutions are located in the D helix, which appears to be exterior to the putative active site of CYP2A6 affecting the folding and conformational changes in the protein distant regions involved in ligand egress, binding, and orientation as well as heme binding. Eighteen 2A6 alleles including *6, *7, *10, *11, *12, *17, *18, *19, *23 (Arg203Cys), *24 (Val110Leu and Asn438Tyr), *26 (Phe118Leu, Arg128Leu, and Ser131Ala), *27 (Phe118Leu and Arg203Frameshift), *35 (Asn438Tyr), *41 (51G>A, 507C>T, and 3515G>A → Arg265Gln), *42 (51G>A, 3524T>C → Ile268Thr, and 5684T>C), *43 (4406C>T → Thr303Ile), *44 (51G>A, 5661G>A → Glu390Lys, 5738C>T, 5745A>G → Asn418Asp, and 5750G>C → Glu419Asp), and *45 (51G>A, 4464G>A, and 6531T>C → Lyeu462Pro) have been reported to produce enzymes with reduced activities, whereas polymorphisms in the promoter region of CYP2A6 have been implicated for the decreased transcriptional activity observed in *1D (−1013A>G), *1H (−745A>G), and *9 (−48T>G) (Al Koudsi et al. 2009; Fernandez-Salguero et al. 1995; Fukami et al. 2004; Haberl et al. 2005; M.K. Ho et al. 2008; Mwenifumbo et al. 2008; Mwenifumbo and Tyndale 2007; Oscarson et al. 2002; Piliguian et al. 2014; Satarug et al. 2006; Xu et al. 2002; Yoshida et al. 2002). There is marked ethnic difference in the frequency of CYP2A6 alleles. CYP2A6*4, *7, *9, and *10 alleles are popular in the Japanese population with a frequency of 19.8%, 12.6%, 20.7%, and 4.3%, respectively (Fujieda et al. 2004). CYP2A6*4, *7, and *9 are prevalent (10%–20%) in Asian populations (Yoshida et al. 2003) but have a lower frequency (1%–7%) in the white people (Schoedel et al. 2004). CYP2A6*16 is prominent among whites (0.3%–3.6%) and African Americans (1.7%), whereas this allele is undetected in the Asian population (Nakajima et al. 2006a). CYP2A6*21 also has a higher occurrence in white subjects at 0.5% to 7.0% compared with Chinese (3.4%) and black subjects (0.6%), whereas Japanese people are unaffected (Haberl et al. 2005; Nakajima et al. 2006a). A systemic study of CYP2A6 polymorphisms in four ethnic populations does not identify any individuals carrying CYP2A6*22 within the populations studied (Nakajima et al. 2006a), and the variant has a low frequency (0.3%) among whites (Haberl et al. 2005).

CYP2A6 is induced by phenobarbital and is regulated at transcriptional, translational, and epigenetic levels (Abu-Bakar et al. 2013; Raunio and Rahnasto-Rilla 2012). CYP2A6 activity is increased by dietary cadmium or cruciferous vegetables. CYP2A6 are induced during pathological conditions associated with liver injury when the function of most other CYP enzymes is compromised. Regulation of CYP2A6 is complex where the promoters interact with multiple stress-activated transcription factors. CYP2A6 is regulated via nuclear factor-erythroid 2 related factor 2 (Yokota et al. 2011), suggesting that CYP2A6 is induced under oxidative stress. miR-126 downregulates CYP2A6 in human hepatocytes (Nakano et al. 2015). Interestingly, CYP2A7P is another target of miR-126 and restores the miR-126–mediated downregulation of CYP2A6 by acting as a decoy for miR-126 in the liver (Nakano et al. 2015).

1.4.2 CYP2A7

The CYP2A7 gene is located approximately 25 kb upstream of the CYP2A6 gene with 96.5% identity in the nucleotide sequence (Yamano et al. 1990). It has been mapped to chromosome 19q13.2 and spans 350 kb containing nine exons (Fernandez-Salguero et al. 1995). This gene, which produces two transcript variants (Ding et al. 1995), is part of a large cluster of P450 genes from the CYP2A, CYP2B, and CYP2F subfamilies on chromosome 19q. The CYP2A7 gene is conserved in rhesus monkey, mouse, rat, C. elegans, and frog. CYP2A7 is expressed in the liver and localizes to the ER, but its substrate has yet to be determined. The polymorphic alleles of CYP2A7 including CYP2A7*1B, *1C, and *1D have a gene conversion with the CYP2A6 sequence and confound the genotyping of the CYP2A6*4 allele (Fukami et al. 2006). The CYP2A7 alleles are found in European Americans with a moderate frequency but very rare in African Americans, Japanese, and Koreans. Because of the similarity, CYP2A6*4 is often misgenotyped by CYP2A7 alleles.

1.4.3 CYP2A13

The CYP2A13 gene is located in the CYP2 cluster on chromosome 19q13.2 (Fernandez-Salguero and Gonzalez 1995; Hoffman et al. 2001). The gene contains nine exons and encodes a protein with 494 amino acids. The nucleotide and protein sequences of CYP2A13 are highly similar to 2A6 with 95.3% and 93.5% identity, respectively (Fernandez-Salguero and Gonzalez 1995). In the olfactory mucosa and respiratory tract and lung, CYP2A13 is highly expressed as a functional protein (Chiang et al. 2012, 2013; Gu et al. 2000; Koskela et al. 1999; Su et al. 1996; Xiang et al. 2015; Zhu et al. 2006). CYP2A13 mRNA is also detected in a number of tissues, including brain, mammary gland, prostate, testis, and uterus, but not in heart, kidney, colon, small intestine, bone marrow, spleen, stomach, thymus, and skeletal muscle (Koskela et al. 1999). The CYP2A13-transgenic mice are normal in gross morphology, development, and fertility, and CYP2A13 expresison is limited to the respiratory tract only where it can bioactivate NNK (Jia et al. 2014).

CYP2A13 has similar substrate specificity to 2A6 with some differences. Both CYP2A13 and 2A6 show high activity toward coumarin (von Weymarn and Murphy 2003), nicotine (Murphy et al. 2005; von Weymarn et al. 2012), and cotinine (Bao et al. 2005). CYP2A13 shows metabolic activities for theophylline 8-hydroxylation and 3-demethylation (Fukami et al. 2007). CYP2A6 also has activity for theophylline 8-hydroxylation. CYP2A13 also metabolizes phenacetin and theophylline (Fukami et al. 2007), two typical substrates of CYP1A2. Compared with CYP2A6, 2A13 shows much lower activity for coumarin and much higher activity for NNK (Su et al. 2000). CYP2A13 metabolizes various procarcinogens such as tobacco-specific nitrosamines NNK and NNN, AFB$_1$, and 4-aminobiphenyl (Brown et al. 2007; Jalas et al. 2003; Megaraj et al. 2014; Smith et al. 2003). In human lung, CYP2A13 has a significant activity in metabolizing AFB$_1$

to AFB$_1$-8,9-epoxide and AFM$_1$-8,9-epoxide (Dohnal et al. 2014; Yang et al. 2012, 2013; Zhang et al. 2013). CYP2A13 expressed in the respiratory tract (Zhu et al. 2006) can convert NNK into carcinogenic species that cross-link DNA and consequently induce carcinogenesis (Su et al. 2000). CYP2A13 is 30–215 times more efficient at activating NNK into its carcinogenic metabolites than 2A6 (He et al. 2004; Su et al. 2000). CYP2A13 also efficiently metabolizes the air pollutants such as naphthalene, styrene, and toluene, and the pneumotoxin 3-methylindole (D'Agostino et al. 2009; Fukami et al. 2008). The expression of CYP2A13 has been observed in most lung carcinomas, squamous lung carcinomas, and large cell lung carcinomas (Chiang et al. 2013; Fukami et al. 2010). The cancer risk is threefold higher in CYP2A13-positive subjects. CYP2A13 expression is linked to the early occurrence of lung cancer but not with cancer progression (Chiang et al. 2013).

To date, nine variants and several subvariants of *CYP2A13* (*1B* through to *10*) have been described (http://www.cypal leles.ki.se/cyp2a13.htm). *2* contains 3375C>T (Arg257Cys) and 74G>A (Arg25Gln) (Fujieda et al. 2004; Zhang et al. 2002). The frequency of the 3375C>T SNP in white, black, Hispanic, and Asian individuals is 1.9%, 14.4%, 5.8%, and 7.7%, respectively (Zhang et al. 2002). Fujieda et al. (2003) have identified *3* (133_134insT; 1706C>G, Asp158Glu), *4* (579G>A, Arg101Gln), *5* (7343T>A, Phe453Tyr), and *6* (7465C>T, Arg494Cys) in a Japanese population. The frequency of 579G>A, 1706C>G, 7343T>A, and 7465C>T is 0.3%, 4.9%, 0.3%, and 1.0%, respectively. *7* contains 578C>T (Arg101stop) (Cauffiez et al. 2004c; Zhang et al. 2003). It has a frequency of 2.0%–3.4% in a Finland population (Cauffiez et al. 2004c). *8* and *9* contain 1706C>G (Asp158Glu) and 5294C>T (Val323Leu), respectively (Cauffiez et al. 2004c). Several alleles of *CYP2A13* found in Caucasian, African, and Asian populations have functional impact on substrate metabolism and cancer risk (Cauffiez et al. 2004c, 2005; D'Agostino et al. 2008; Jiang et al. 2004; Schlicht et al. 2007; S.L. Wang et al. 2006; Zhang et al. 2002, 2003). *2* is associated with a substantially reduced risk for lung cancer with an odds ratio of 0.41 (Wang et al. 2003c).

1.4.4 CYP2B6

CYP2B6 is an important hepatic protein that has important functional relevance (Hodgson and Rose 2007; Mo et al. 2009a; Turpeinen and Zanger 2012; Wang and Tompkins 2008). The *CYP2B6* gene is located in tandem head-to-tail arrangement within a large *CYP2* gene cluster on chromosome 19q (Zanger and Klein 2013). The gene maps to chromosome 19q13.2, which is located together with the pseudogene *CYP2B7P* and consists of nine exons encoding 491 amino acids (Yamano et al. 1989). Transcript variants for this gene have been described; however, it has not been resolved whether these transcripts are produced by this gene or by the closely related pseudogene *CYP2B7P*. Both the gene and the pseudogene are located in the middle of a *CYP2A* pseudogene found in a large cluster of *CYP* genes from the *CYP2A, 2B*, and *2F* subfamilies on chromosome 19q. CYP2B6 is mainly expressed in the liver, accounting for 6% of total microsomal CYPs based on a study with a small size of liver samples ($n = 13$) (Stresser and Kupfer 1999), in contrast to earlier estimates that it represented less than 1% (Shimada et al. 1994) and various extrahepatic tissues including the kidney, skin, brain, intestine, and lung (Code et al. 1997; Gervot et al. 1999).

The substrates of CYP2B6 are usually nonplanar molecules, neutral or weakly basic, highly lipophilic with one or two hydrogen bond acceptors (Lewis 2000; Zhao and Halpert 2007). CYP2B6 is known to metabolize 8%–10% of clinically used drugs ($n > 60$) to some extent. The drugs metabolized by CYP2B6 include aminopyrine, amitriptyline, antipyrine, arteether, artelinic acid, artemisinin, bupropion, carbamazepine, cinnarizine, cisapride, clomethiazole, clopidogrel, cyclophosphamide, dextromethorphan, diazepam, diclofenac, dimemorfan, disulfiram, diphenhydramine, efavirenz, erythromycin, ethylmorphine, etodolac, flunarizine, flunitrazepam, fluoxetine, haloperidol, halothane, ifosfamide, imipramine, ketamine, lidocaine, meperidine (pethidine), *S*-mephenytoin, *S*-mephobarbital, methadone, methoxyflurane, mexiletine, mianserin, midazolam, nevirapine, nicotine, perhexiline, prasugrel, promazine, promethazine, propofol, ropivacaine, seratrodast, sertraline, sibutramine, selegiline, sevoflurane, tamoxifen, tazofelone, *N*-methyl,*N*-propargyl-2-phenylethylamine, temazepam, thiotepa, tramadol, tretinoin, trofosfamide, valproic acid, verapamil, and zotepine (Kharasch et al. 2015; Liu et al. 2015; Mo et al. 2009a; Turpeinen and Zanger 2012). CYP2B6 can also metabolize a number of procarcinogens and environmental compounds (Hodgson and Rose 2007; Rendic 2002). Known procarcinogens as CYP2B6 substrates include AFB$_1$, 4-(methylnitrosamino)-1-(3-pyridyl)-1-butanone, 6-aminochrysene, BaP, dibenzo[*a,h*]anthracene, dibenzo[*a,l*]pyrene, dibenzo[*a*]pyrene, and 7,12-dimethylbenz[*a*]anthracene. CYP2B6 also metabolizes several herbicides including acetochlor, ametryne, atrazine, butachlor, metolachlor, and terbutryne. This enzyme also metabolizes several pesticides including methoxychlor, chlorpyrifos, endosulfan, parathion, and *o*-phenylphenol (Hodgson and Rose 2007; Rendic 2002). Several central nervous system (CNS) stimulants are metabolized by CYP2B6, such as 3,4-methylenedioxymethamphetamine, 3,4-methylenedioxyamphetamine, 3,4-methylenedioxyethylamphetamine, and benzphetamine. Additional substrates known to be partially metabolized by CYP2B6 include ethanol, chloroform, acetonitrile, *n*-hexane, isoprene, naphthalene, phenanthrene, toluene, SKF-525A, and nicotine (Rendic 2002). CYP2B6 catalyzes the metabolism of several endogenous compounds (Mo et al. 2009a; Rendic 2002; Turpeinen and Zanger 2012). These include AA, lauric acid, 17β-estradiol, estrone, ethinylestradiol, and testosterone. The role of CYP2B6 in the metabolism of these endogenous compounds is small, while other CYPs such as CYP4 members play a more important role in the oxidative metabolism of these endogenous substances. Compared with other members of the CYP2B subfamily, CYP2B6 hydroxylates testosterone and androstenedione at a much lower activity (Domanski et al. 1999; Mo et al. 2009a; Turpeinen and Zanger 2012).

The determined crystal structures of CYP2B6 K262R variant in complex with 4-(4-chlorophenyl)imidazole (4CPI), 4-benzylpyridine, and 4-(4-nitrobenzyl)pyridine, the monoterpene sabinene, and the calcium channel blocker amlodipine show important information about the rearrangement of active-site residues to accommodate small inhibitors (Gay et al. 2010b; Shah et al. 2011, 2012, 2015). There are at least 17 residues located at the entrance channel in CYP2B6, including Ser207, Glu301, Arg308, Val477, Pro38, Leu43, Met46, Arg48, Val212, Leu216, Leu219, Gly222, Leu70, Arg73, Ala102, Ser221, and Gly222. Binding of amlodipine to CYP2B6 is accompanied by conformational shifts of the G–H loop and helices A′, A, B′, F, F′, G′, G, and H. The active site of the CYP2B6–amlodipine complex extends to Phe115, Ser210, Gly299, Thr300, Glu301, Leu362, Gly366, Pro368, and Gly478, which are not found to be a part of the CYP2B6-4CPI binding pocket. The estimated volume of the active-site cavity is 755 Å^3 (Shah et al. 2015). The structure of CYP2B6 in complex with amlodipine provides valuable insight into how active-site topology is altered to accommodate large or multiple drug molecules. In addition to Glu218 and Glu387 or Gln215, the residues located at the respective substrate access channel entrance in CYP2B6 may play a crucial role in substrate recognition.

A number of compounds have been found to inhibit CYP2B6 activity (Rendic 2002). These include orphenadrine (Ekins et al. 1997; Guo et al. 1997), 2-phenyl-2-(1-piperidinyl) propane (Chun et al. 2000), ritonavir (Hesse et al. 2001), efavirenz (Hesse et al. 2001), and nelfinavir (Hesse et al. 2001). Thiotepa, a substrate of CYP2B6, is a selective inhibitor of CYP2B6-catalyzed S-mephenytoin N-demethylation (Rae et al. 2002). Paroxetine, fluvoxamine, sertraline, desmethylsertraline, fluoxetine, norfluoxetine, and nefazodone all inhibit CYP2B6-mediated bupropion hydroxylation (Hesse et al. 2000). Several drugs and other compounds have been identified as the mechanism-based inhibitors of CYP2B6. Both clopidogrel and ticlopidine inhibit CYP2B6-mediated bupropion hydroxylation in a mechanism-based manner (Richter et al. 2004). Mifepristone (Lin et al. 2009), duloxetine (Paris et al. 2009), phencyclidine (Shebley et al. 2009), 17α-ethynylestradiol (Kent et al. 2002b, 2008), desethylamiodarone (Ohyama et al. 2000), n-propylxanthate and several xanthates (Kent et al. 1999; Yanev et al. 1999), imperatorin (Zheng et al. 2015), ticlopidine (Nishiya et al. 2009), clopidogrel (Nishiya et al. 2009), the thiolactone metabolite of prasugrel (Nishiya et al. 2009), and glabridin (an active ingredient from licorice) (Kent et al. 2002a) are all mechanism-based inhibitors of CYP2B6. Inhibition of CYP2B6 may result in drug–drug interactions. For example, clopidogrel and ticlopidine significantly increase bupropion exposure by 60%–85% and decrease the ratio of hydroxybupropion/bupropion by 71%–93% in humans (Turpeinen et al. 2004, 2005).

CYP2B6 is highly inducible and is closely regulated by CAR and PXR. Transactivation of PXR occurs only through direct ligand binding, while CAR can be stimulated by both direct interaction with a ligand and indirect activation. It has been found that activation or translocation of CAR

mediates induction of rodent Cyp2b genes by phenobarbital-type inducers (Kawamoto et al. 1999). The most effective inducer of human CYP2B6 is phenobarbital (Chang et al. 1997; Moore et al. 2000; Qatanani and Moore 2005; Xiong et al. 2002). Phenobarbital induction of CYP2B6 is mediated by CAR, although phenobarbital is not a ligand of CAR (Goodwin et al. 2001; Pascussi et al. 2000). Induction of CYP2B6 is often a CAR-mediated event in which the inducer triggers nuclear translocation of CAR, possibly via receptor dephosphorylation. Subsequently, CAR forms a heterodimer with RXRα. Metformin suppresses CAR-mediated CYP2B6 induction via indirect CAR inhibition at the level of phosphorylation and nuclear translocation, but it does not decrease constitutive CYP2B6 expression. Instead, LY2090314 (3-[9-fluoro-2-(piperidin-1-ylcarbonyl)-1,2,3,4-tetrahydro[1,4]diazepino[6,7,1-hi]indol-7-yl]-4-imidazo[1,2-a] pyridin-3-yl-1H-pyrrole-2,5-dione) specifically suppresses constitutive expression of the CYP2B6 gene, which is at least partially attributed to direct CAR inhibition. The binding of the CAR/RXRα heterodimer to the nuclear receptor binding sites (direct repeat-4 motifs) in the 5′-flanking sequence of the CYP2B genes results in the activation of a 51-bp phenobarbital-responsive enhancer module (PBREM). CAR is mostly expressed in the liver, which transactivates both the PBREM of the CYP2B6 gene and the CYP3A4 xenobiotic response element. The −1.7-kb promoter region of the CYP2B6 gene contains PBREM consisting of two imperfect DR4 motifs (NR1 and NR2) (Swales et al. 2005). In addition, PXR has been shown to mediate transactivation of the CYP2B6 PBREM in response to rifampin (Goodwin et al. 2001). Both CAR and PXR can bind and activate response elements in the promoter region of CYP2B6 and 3A4 genes (Makinen et al. 2002; Wang et al. 2003a). In addition, dexamethasone-activated GR can enhance CYP2B6 induction in the presence of PXR and CAR activators (Pascussi et al. 2000; Wang et al. 2003b). Induction of CYP2B6 can result in increased clearance and diminished pharmacological activity of substrate drugs. For example, chronic rifampin treatment in humans decreased bupropion exposure threefold.

There are large interindividual differences in hepatic CYP2B6 protein and mRNA levels ranging from 20- to 278-fold (Chang et al. 2003; Code et al. 1997; Ekins et al. 1998; Gervot et al. 1999; Hanna et al. 2000; Stresser and Kupfer 1999). Such a large variability is probably attributed to effects of genetic polymorphisms or exposure to drugs that are inducers or inhibitors of CYP2B6 (Zanger et al. 2007). CYP2B6 is one of the most polymorphic CYP genes in humans. To date, at least 62 variants and subvariants (*1B through to *38) of the CYP2B6 gene have been identified (http://www.cypalleles .ki.se/cyp2b6.htm). Many of these variants affect transcriptional regulation, splicing, mRNA and protein expression, and catalytic activity. *1A is referred to as the wild type. The most commonly observed SNPs of CYP2B6 yield Gln172His in *6A, *6B, *6C, *7A, *7B, *9, *13A, *13B, *19, *20, *26, *34, *36, *37, and *38; Lys262Arg in *4A, *4B, *4D, *4C, *6A, *6B, *6C, *7A, *7B, *16, *19, *20, *26, *34, *36, *37, and *38; and Arg487Cys in *5A, *5B, and *5C. *2 (64C>T →

Arg22Cys) is extremely rare and carriers of this allele have increased risk of ADRs in the CNS when receiving efavirenz therapy (Usami et al. 2007). *4 (785A>G → Lys262Arg) carriers have increased total clearance of bupropion, whereas other alleles are not different from the wild type (Kirchheiner et al. 2003). Consistently, the CYP2B6*1/*4 genotype is associated with a 20% higher area under the plasma concentration curve (AUC) of hydroxybupropion than the *1/*1 (wild type) genotype. *4 carriers have increased total clearance of bupropion (Kirchheiner et al. 2003). There is a clinical report where *5 (1459C>T → Arg487Cys) is associated with altered response or survival in patients with proliferative lupus nephritis when treated with pulsed cyclophosphamide (Takada et al. 2004). The most common variant is *6 (516G>T → Gln172His and 785A>G → Lys262Arg), which occurs from 15% to more than 60% in different populations. Women carrying 516G>T SNP have higher mean nevirapine plasma levels compared to those with heterozygous 516GT and wild-type 516GG (Oluka et al. 2015). Among smokers with the *6 genotype, bupropion produces significantly higher abstinence rates than placebo at the end of treatment and at the 6-month follow-up (A.M. Lee et al. 2007). In contrast, bupropion is no more effective than placebo for smokers in the *1 group at the end of treatment or at the 6-month follow-up. *18 (983T>C → Ile328Thr) is another important variant that occurs mostly in Africans (4%–12%) and does not express functional protein (Zanger and Klein 2013). The 983T>C SNP, part of *18, is also associated with nevirapine or efavirenz plasma concentration in 225 Caucasians and 146 blacks (Wyen et al. 2008). CYP2B6*4 and *6 mutants show reduced catalytic ability for ifosfamide whereas CYP2B6*7 and *9 have enhanced catalytic ability toward ifosfamide (Calinski et al. 2015). In vitro analysis reveals no enzymatic activity in CYP2B6.8 (Lys139Glu), 2B6.12 (Gly99Glu), 2B6.18 (Ile328Thr), 2B6.21 (Pro428Thr), and 2B6.24 (Gly476Asp); lower activity in CYP2B6.10 (Gln21Leu and Arg22Cys), 2B6.11 (Met46Val), 2B6.14 (Arg140Gln), 2B6.15 (Ile391Asn), 2B6.16 (Lys262Arg and Ile328Thr), 2B6.20 (Thr168Ile, Gln172His, and Lys262Arg), and 2B6.27 (Met198Thr); and higher activity in CYP2B6.2 (Arg22Cys), 2B6.4 (Lys262Arg), 2B6.6 (Gln172His and Lys262Arg), and 2B6.19 (Gln172His, Lys262Arg, and Arg336Cys), compared with that of wild-type CYP2B6.1 with artemether demethylation as the marker reaction (M. Honda et al. 2011). The 1282C>A (Pro428Thr) substitution in *21 also decreases protein stability. Another in vitro study shows that CYP2B6.1, 2B6.10 (Gln21Leu and Arg22Cys), and 2B6.14 (Arg140Gln) have significantly lower V_{max}/K_m values for selegiline N-demethylation, while CYP2B6.8 (Lys139Glu), 2B6.11 (Met46Val), 2B6.12 (Gly99Glu), 2B6.13 (Lys139Glu, Gln172His, and Lys262Arg), 2B6.15 (Ile391Asn), 2B6.18 (Ile328Thr), 2B6.21 (Pro428Thr), 2B6.24 (Gly476Asp), and 2B6.28 (Thr306Ser and Arg378X) are totally inactive in the deethylation of 7-ethoxy-4-trifluoromethylcoumarin and the N-demethylation and N-depropargylation of selegiline (Watanabe et al. 2010). Rotger et al. (2005) have reported that the CYP2B6 516T/T genotype is associated with greater (1.7-fold) plasma levels of nevirapine in HIV-infected patients.

Penzak et al. (2007) have found that the median nevirapine trough concentration in individuals homozygous for the variant allele (516T/T) is 1.8-fold higher than that in 516G/G carriers in HIV-infected African patients in Uganda. A study has found a significant effect of the CYP2B6 genotype on thiotepa's clearance (Ekhart et al. 2009). The increased thiotepa and tepa clearance caused by CYP2B6*7 (1459C>T → Arg487Cys) results in a reduction of the combined exposure to thiotepa and tepa of 20% in homozygous patients.

1.4.5 CYP2F1

This gene is part of a large cluster of cytochrome P450 genes from the CYP2A, CYP2B, and CYP2F subfamilies on chromosome 19q. It maps to chromosome 19q13.2 and contains nine exons. CYP2F1 is primarily expressed in lung with little or no hepatic expression (Carr et al. 2003; Wei et al. 2012). The −152 to −182 5′ region of CYP2F1 is identified as a specific promoter element that binds to a protein (Carr et al. 2003). This protein localizes to the ER and is known to dehydrogenate 3-methylindole, an endogenous toxin derived from the fermentation of tryptophan. CYP2F1 catalyzes the 7-ethoxy-and-propoxycoumarin O-dealkylation and 7-pentoxyresorufin O-depentylation. CYP2F1 can activate the lung toxicants 4-ipomeanol (Czerwinski et al. 1991), 3-methylindole, naphthalene (Lanza et al. 1999), and styrene (Nakajima et al. 1994).

Presently, seven variants of CYP2F1 have been documented (*2A through to *6) (http://www.cypalleles.ki.se/cyp2f1.htm). A study has identified 24 mutations distributed in the promoter region of the gene, as well as in the coding regions and their flanking intronic sequences (Tournel et al. 2007). In addition to the wild-type CYP2F1*1 allele, seven allelic variants, CYP2F1*2A, *2B, *3, *4, *5A, *5B, and *6, are identified. The most frequent allelic variant, CYP2F1*2A (25.6%), harbors a combination of nine mutations, including two missense mutations (Asp218Asn and Gln266His) and a 1-bp insertion (c.14_15insC) that creates a premature stop codon in exon 2, probably leading to the synthesis of a severely truncated protein with no catalytic activity (Tournel et al. 2007).

1.4.6 CYP2S1

The CYP2S1 gene is located in chromosome 19q13.2 within a cluster including CYP2 family members CYP2A6, 2A13, 2B6, and 2F1. The gene consists of nine exons (Rylander et al. 2001). The mRNA expression level of CYP2S1 is highest in trachea, lung, stomach, small intestine, and spleen (Deb and Bandiera 2009; Rylander et al. 2001; Saarikoski et al. 2005). There are also moderate expression of mRNA CYP2S1 in colon, appendix, liver, kidney, thymus, substantia nigra, peripheral leukocytes, and placenta. CYP2S1 metabolizes naphthalene, suggesting the involvement of CYP2S1 in the metabolism of toxic and carcinogenic compounds, similar to other dioxin-inducible CYPs. CYP2S1 is involved in the synthesis and metabolism of bioactive lipids including PGs and retinoids (Bui et al. 2011; Fromel et al. 2013; Madanayake et al. 2012).

CYP2S1 metabolizes PGG_2 into several products including 12S-hydroxy-5Z,8E,10E-heptadecatrienoic acid (12-HHT). It also metabolizes PGH_2 into malondialdehyde, 12-HHT, and thromboxane A_2. CYP2S1 efficiently metabolizes the hydroperoxyeicosatetraenoic acids (5S-, 12S-, and 15S-HPETE) and 13S-hydroperoxyoctadecadienoic acid (HPODE) into 5-oxo-eicosatetraenoic acid (ETA), 12-oxo-ETA 1, 15-oxo-ETA, and 13-octadecadienoic acid, respectively (Bui et al. 2011). CYP2S1 can metabolize eicosanoids in the absence of both NADPH and NADPH cytochrome P450 reductase (Bui et al. 2011) and can also activate the anticancer agent AQ4N (banoxantrone, a highly selective bioreductive drug that is activated in, and is preferentially toxic to, hypoxic cells in tumors) (Nishida et al. 2010; Tan et al. 2011).

TCDD induces *CYP2S1* mRNA transcription (Rivera et al. 2002, 2007), suggesting that the induction of CYP2S1 is mediated by AhR in a manner typical for CYP1 family members. The synthetic GR ligand dexamethasone, DEX, represses CYP2S1 expression in human cell lines (Bebenek et al. 2012). CYP2S1 expression is elevated in multiple epithelial-derived cancers as well as in the chronic hyperproliferative disease psoriasis.

To date, 12 variants of *CYP2S1* have been documented (*1B* through to *5B*) (http://www.cypalleles.ki.se/cyp2s1 .htm). In a Finnish Caucasian population, eight SNPs of CYP2S1 have been identified, with seven of the SNPs in the protein-coding region and one in the proximal 3′-UTR (Saarikoski et al. 2004). Two of these SNPs (10347C>T and 13106C>T) result in nonconservative amino acid substitutions, that is, Arg380Cys and Pro466Leu, respectively. The respective allelic variants, *CYP2S1*2* (10347C>T) and *3* (13106C>T; 13255A>G), occur in the study population at frequencies of 0.50% and 3.75%, respectively. The most common of the variant alleles is *CYP2S1*1H* (23.8%), harboring a 13255A>G substitution located in the 3′-UTR (Saarikoski et al. 2004). Jang et al. (2007) have reported two new variants: *CYP2S1*4* carrying 1284G>A (Ser61Asn, 0.3%) and *5* with 5479T>G (Leu230Arg, 0.8%) in 50 Korean subjects.

1.5 HUMAN CYP2C CLUSTER: CYP2C8, 2C9, 2C18, AND 2C19

The *CYP2C* cluster in humans is small, with only 4 genes, whereas in mice, it has expanded to 15 genes. The human *CYP2C* genes map to chromosome 10q24 in the following order: Cen-*2C18-2C19-2C9-2C8*-Tel (Gray et al. 1995). The human *CYP2C* genes have a strong potential to recombine, because of many L1 LINE repetitive DNA sequences, which are located principally in intron 5. *CYP2C9* and *2C19* share L1PA7, L1M4, L1MB5, and L1PA16 repeats in this intron. *CYP2C18* and *2C19* share L1PA5 repeats. *CYP2C8* and *2C19* share an L1P repeat, although the two genes are on opposite strands.

CYP2C8, 2C9, and 2C19 are primarily located in the liver, accounting for ~20% of total CYP contents (Shimada et al. 1994), whereas the CYP2C18 protein seems to be primarily expressed in the skin (Zaphiropoulos 1997). CYP2C9 is the

second most prevalent CYP enzyme in the small intestine of humans (Paine et al. 2006). In the human liver, CYP2C9 is present at a much higher level than CYP2C19 (Shimada et al. 1994), with molar ratios of CYP2C9 to 2C19 reported as 13:1 to 17:1 (Inoue et al. 1997; Shon et al. 2002). Low to moderate levels of other CYP2C mRNAs and proteins have also been detected in small intestine and other extrahepatic tissues (Klose et al. 1999). CYP2C8, 2C9, 2C18, and 2C19 share >82% amino acid identity, but they have distinct substrate specificity (Miners and Birkett 1998).

1.5.1 CYP2C8

The *CYP2C8* gene maps to chromosome 10q24.1, spanning 31 kb with nine exons (Inoue et al. 1994; Klose et al. 1999). CYP2C8 accounts for ~7% of total hepatic CYP contents (Shimada et al. 1994) and metabolizes ~5%–8% of drugs (Chen and Goldstein 2009; Holstein et al. 2012; Lai et al. 2009; Niwa and Yamazaki 2012; Totah and Rettie 2005; Xiaoping et al. 2013). CYP2C8 is mainly expressed in the liver and kidney, although its expression level is relatively low in liver compared to kidney. CYP2C8 is also expressed in the adrenal gland, brain, uterus, mammary gland, ovary, and duodenum.

CYP2C8 is responsible for the oxidation metabolism of many clinical drugs including thiazolidinediones, meglitinides, 3-hydroxy-3-methylglutaryl-coenzyme A (HMG-CoA) reductase inhibitors, chemotherapeutic agents (e.g., paclitaxel), antimalarials, antiarrhythmics, and retinoid derivatives (Lai et al. 2009; Totah and Rettie 2005; Xiaoping et al. 2013). The enzyme also plays an intermediate or minor role in metabolism of other agents such as nonsteroidal anti-inflammatory drugs (NSAIDs) (ibuprofen, diclofenac, and tenoxicam), fluvastatin, simvastatin acid, carbamazepine, cyclophosphamide, dapsone, diltiazem, ifosfamide, loperamide, methadone, morphine, torsemide, loratadine, anastrozole, fenretinide, verapamil, and zopiclone (Illingworth et al. 2011; Kamdem et al. 2010; Totah and Rettie 2005). The antihistamine drug loratadine is converted to desloratadine through dealkylation, 3-hydroxydesloratadine, 5-hydroxydesloratadine, and 6-hydroxydesloratadine by CYP2C8 with major contribution from 3A4 (Kazmi et al. 2015; Yumibe et al. 1996).

Notably, several acyl glucuronides are good substrates of CYP2C8 (Delaforge et al. 2005). The 4′-hydroxylation of diclofenac acyl glucuronide is catalyzed by CYP2C8 (Kumar et al. 2002). The acyl glucuronide of naproxen (Delaforge et al. 2005) and gemfibrozil 1-*O*-β-glucuronide (Baer et al. 2009) are also metabolized by CYP2C8. Desloratadine, a second-generation, nonsedating selective H_1-receptor antagonist (McClellan and Jarvis 2001), is oxidized to form 3-hydroxydesloratadine, which is then conjugated to its corresponding *O*-glucuronide by UGT2B10 followed by CYP2C8 oxidation (Kazmi et al. 2015). Estradiol-17β-glucuronide is oxidized on the aromatic C2-position by CYP2C8 (Delaforge et al. 2005).

The prototypical substrate for CYP2C8 in vitro is paclitaxel, and its 6α-hydroxylation has been widely used as a standard marker reaction with K_m of 5.4–19 μM (Cresteil et al.

2002; Harris et al. 1994; Rahman et al. 1994). Amodiaquine N-deethylation may serve as an alternative marker reaction for CYP2C8 in vitro because of its high affinity and high turnover rate (Li et al. 2002). Rosiglitazone p-hydroxylation has also been recommended by the Food and Drug Administration (FDA) as the marker reaction for in vitro CYP2C8 activity determination (Huang et al. 2007, 2008). Both repaglinide and rosiglitazone have been recommended by the FDA as suitable probe substrates of CYP2C8 for in vivo studies (Huang et al. 2007, 2008).

The enzyme catalyzes the conversion of endogenous compounds such as AA to biologically active epoxide metabolites (e.g., 11-, 13-, or 15-HETE) that are critical for the regulation of hepatic glycogenolysis, peptide hormone secretion in the pancreas and pituitary, and platelet aggregation, as well as sodium transport, water reabsorption, and vascular tone in the kidney (Lai et al. 2009; Theken and Lee 2007; Totah and Rettie 2005; Xiaoping et al. 2013). Several steroids are metabolized by CYP2C8. Both 17β-estradiol and estrogen are substrates of CYP2C8 in vitro. Estradiol 17-O-β-glucuronide is also a substrate for CYP2C8. CYP2C8 together with 2C9 and 3A4 have been identified as major contributors for the metabolism of 9-cis-RA, and CYP2C8 and 3A4 play a major role in 13-cis-RA metabolism (Marill et al. 2002).

Montelukast and quercetin have been receeommended by the FDA as the selective inhibitors for CYP2C8 used in in vitro studies (Huang et al. 2007, 2008). Trimethoprim, gemfibrozil, rosiglitazone, and pioglitazone are acceptable inhibitors only used for in vitro studies. For in vivo inhibition studies, gemfibrozil is preferred. Mechanism-based CYP2C8 inhibitors include nortriptyline (Polasek et al. 2004), phenelzine (Polasek et al. 2004), isoniazid (Polasek et al. 2004), amiodarone (Polasek et al. 2004), desethylamiodarone (Ohyama et al. 2000), 17α-ethynylestradiol (Polasek et al. 2004), and verapamil (Polasek et al. 2004). The β-O-glucuronide of gemfibrozil is also a mechanism-based inhibitor of CYP2C8 (Ogilvie et al. 2006).

CYP2C8 is one of the most inducible members of the CYP2C family in primary cultures of human hepatocytes and is significantly induced by the prototypical inducers including rifampin, phenobarbital, and dexamethasone (Chen et al. 1994; Ferguson et al. 2005; Gerbal-Chaloin et al. 2001; Madan et al. 2003; Raucy et al. 2002). The transcription of the CYP2C8 gene is thus regulated by CAR, PXR, GR, and HNF-4α (Ferguson et al. 2005; Johnson and Stout 2005; Kojima et al. 2007). PXR is involved at least partly in the control of the inducible expression of CYP2C8 and 2C9 in response to rifampicin. CAR appears to be involved in the phenobarbital-mediated induction of CYP2C8 and 2C9.

To date, at least 13 variants of CYP2C8 have been identified and named CYP2C8*1B through to *14 (http://www.cypalleles .ki.se/cyp2c8.htm). CYP2C8*1A is a wild type. CYP2C8*1B and *1C contain −271C>A and −370T>G in the 5′-flanking region, respectively (Bahadur et al. 2002). CYP2C8*2, *3, *4, *8, and *14 have decreased enzymatic activity. *2 carries an 805A>T (Ile296Phe) SNP (Dai et al. 2001). *3 contains 416G>A and 1196A>G mutations causing Arg139Lys and Lys399Arg,

respectively (Dai et al. 2001). CYP2C8.3 maintains catalytic efficiency toward amiodarone but exhibits only 50% activity toward paclitaxel 6α-hydroxylation compared with the wild-type enzyme (Bahadur et al. 2002; Soyama et al. 2001, 2002). There is an incomplete linkage disequilibrium between CYP2C8*3 and 2C9*2, with 96% of Swedish subjects carrying the CYP2C8*3 together with the 2C9*2 allele and 85% of subjects with the 2C9*2 allele also carrying the 2C8*3 allele (Yasar et al. 2002b). CYP2C8*4 has a 792C>G substitution leading to Ile264Met (Bahadur et al. 2002). The *5 allele, containing an A base deletion at position 475, generating frameshift at codon 159 and an early stop codon at 177, is expected to result in an inactive enzyme owing to loss of two-thirds of its encoded enzyme (Nakajima et al. 2003; Soyama et al. 2001). This allelic variant has been identified in a patient suffering from rhabdomyolysis after administration of cerivastatin at a normal dose (Ishiguro et al. 2004). The activity of *8 carrying Arg186Gly is only 10%–20% of that of the wild-type enzyme for paclitaxel 6α-hydroxylation, while the Gly171Ser, Lys247Arg, and Lys383Asn variants exhibit expression and activity similar to those of the wild-type enzyme (Hichiya et al. 2005).

There is a marked ethnic difference in the frequencies of alleles of CYP2C8. *2 is found primarily in Africans with a frequency of 18% (Dai et al. 2001) but very rare in Caucasians (0.4%) (Bahadur et al. 2002). *3 has a high frequency in Caucasians (13%) (Solus et al. 2004) but very low in Africans (~2%) and appears to be absent from Japanese (Bahadur et al. 2002; Dai et al. 2001; Nakajima et al. 2003). *4 has a frequency of 7.5% in Caucasians (Bahadur et al. 2002), while *5 has a very low frequency in Japanese (~0.25%) (Nakajima et al. 2003; Soyama et al. 2001).

1.5.2 CYP2C9

This gene is located within a cluster of CYP genes on chromosome 10q24 (Gray et al. 1995). CYP2C9 encodes a protein of 490 amino acids, with a molecular weight of 55.6 kDa. The CYP2C9 is the most abundant enzyme among all the CYPs in the liver except for CYP3A4. It is responsible for Phase I metabolism of 15% of clinically used drugs (He et al. 2011; Mo et al. 2009b,c; Rettie and Jones 2005; Van Booven et al. 2010; Zhou et al. 2010). The enzyme is absent in fetal liver but is quickly expressed after the first month of birth (Treluyer et al. 1997). CYP2C9 is also expressed in the small intestine (Obach et al. 2001).

S-Flurbiprofen (4′-hydroxylation) (Yamazaki et al. 1998a), S-warfarin (7-hydroxylation) (Yamazaki et al. 1998a), tolbutamide (methylhydroxylation), phenytoin (4′-hydroxylation) (Giancarlo et al. 2001), losartan (oxidation) (C.R. Lee et al. 2003), and diclofenac (4′-hydroxylation) (Yamazaki et al. 1998a) have been commonly used as probe substrates for CYP2C9 (Kumar et al. 2006). The substrates of CYP2C9 include oral sulfonylurea hypoglycemics (e.g., tolbutamide, glyburide, glimepiride, gliclazide, and glipizide), NSAIDs (e.g., diclofenac, ibuprofen, ketoprofen, suprofen, naproxen, flurbiprofen, indomethacin, meloxicam, piroxicam, tenoxicam, and lornoxicam), selective cyclooxygenase 2 inhibitors (e.g., celecoxib, lumiracoxib, etoricoxib, and valdecoxib), diuretics (e.g., torasemide and sulfinpyrazone), antiepileptics (e.g., phenytoin and phenobarbital), angiotensin II receptor inhibitors (e.g., losartan, irbesartan, and candesartan), anticancer drugs (e.g., cyclophosphamide and tamoxifen), and anticoagulants (e.g., S-acenocoumarol, phenprocoumon, and S-warfarin) (He et al. 2011; Miners and Birkett 1998; Rettie and Jones 2005). Vicriviroc (SCH 417690), a CCR5 receptor antagonist, is metabolized by CYP3A4/5 and 2C9 (Ghosal et al. 2007). CYP2C9 is also involved in the metabolism of the procarcinogen BaP, although CYP1A1/1A2 are the major enzymes responsible for its bioactivation (Ma and Lu 2007). Typical substrates of CYP2C9 such as celecoxib, ibuprofen, flurbiprofen, and diclofenac are relatively small, are lipophilic, and contain acidic groupings with pK_a values in the range 3.8–8.1, which will be ionized at physiological pH, forming anions in vivo (He et al. 2011). The carboxylate groups of tienilic acid and diclofenac have been shown to be responsible for substrate preference and orientation in the active site of CYP2C9. Therefore, a typical CYP2C9 substrate should contain an anionic site and a hydrophobic site. However, neutral or positively charged compounds may also be substrates of CYP2C9.

Inhibitors of CYP2C9 can cause serious drug–drug interactions when added to a therapeutic regimen with low therapeutic index drugs like S-warfarin, tolbutamide, or phenytoin. Life-threatening bleeding episodes, hypoglycemia, and neurotoxicity are the results of eliminating the enzymatic activity of CYP2C9 in these patients. One of the highly selective CYP2C9 competitive inhibitor is sulfaphenazole, which has poor affinity to other CYP2C subfamily (Ha-Duong et al. 2001a). Tienilic acid is a mechanism-based inactivator of CYP2C9 (Beaune et al. 1987; Dansette et al. 1992).

The first crystal structure of human CYP2C9 refined to a 2.55-Å resolution has been resolved by Williams et al. (2003). Currently, there are three crystal structures of the human CYP2C9 in the PDB: one ligand-free protein (1OG2) and two in complex with S-warfarin (1OG5) or flurbiprofen (1R9O) (Wester et al. 2004; Williams et al. 2003). The molecule used to obtain the 1OG2 and 1OG5 structures was a chimeric enzyme containing seven mutations in the F–G loop including Lys206Glu, Ile215Val, Cys216Tyr, Ser220Pro, Pro221Ala, Ile222Leu, and Ile223Leu (Williams et al. 2003). CYP2C9 is a two-domain protein with typical fold characteristics of the CYPs (Williams et al. 2003) whose structures have been reported previously. The B–C loop forms part of the active site and contributes to substrate specificity, with the Phe114Ile mutant being unable to catalyze diclofenac hydroxylation and the high affinity of sulfaphenazole to CYP2C9 being completely lost when the phenyl moiety of Phe114 was replaced with the alkyl group of Ile or when the phenyl substituent of sulfaphenazole was replaced with a cyclohexyl group (Melet et al. 2003). In the structures of CYP2C9 without ligand bound or with bound S-warfarin, residues 101–106 in the B–C loop form helix B′ (Williams et al. 2003). In addition, residues 212–222 in the F–G loop form helices F′ and G′, which was not observed in rabbit CYP2C5 (P.A. Williams et

al. 2000) and bacterial CYPs (Ravichandran et al. 1993). The heme group is positioned between helices I and L and the iron is pentacoordinated with Cys435. As in other CYP structures, a water molecule located 7 Å above the heme iron is hydrogen bonded to Thr301, which is highly conserved and considered to play a role in the proton transfer process during catalytic cycle (Haines et al. 2001). In the 1OG2 and 1OG5 structures, the heme is stabilized by hydrogen bonds between the propionates and the side chains of Trp120, Arg124, His368, and Arg433 (Williams et al. 2003). In addition, Arg97 forms hydrogen bonds to the propionates as well as the carbonyl oxygen atoms of Val113 and Pro367. The active-site cavity of CYP2C9 extends up and away from helix I, with Phe114 and Phe476 located on opposite sides of the channel, and the very top of the channel being formed by helix B′ and the B–C and F–G loops. Phe69, Phe100, Leu102, Leu208, Leu362, Leu366, and Phe476 generate a hydrophobic patch in the active site while Arg105 and Arg108 point away from the cavity (Williams et al. 2003). Asp293 is close to Phe110 and Phe114 and forms a hydrogen bond to the backbone nitrogen of Ile112 and consequently is well ordered, whereas Glu300 points into the active site but shows some flexibility in the ligand-free structure 1OG1. In addition, both Gln214 and Asn217 are near Phe476 and may provide potential hydrogen-bonding interactions with ligands.

Drugs and xenobiotics including rifampicin, hyperforin, phenytoin, phenobarbital, and dexamethasone can induce the CYP2C9 expression leading to enhanced metabolism of many therapeutic drugs and other CYP2C substrates (Chen et al. 2004; Gerbal-Chaloin et al. 2001, 2002; Rettie and Jones 2005). The induction of CYP2C9 by rifampicin, hyperforin, and phenobarbital is mediated by transactivation of PXR (Chen et al. 2004), but CAR and GR are also involved.

The *CYP2C9* gene is highly polymorphic. To date, more than 59 variants and a series of subvariants of *CYP2C9* (*1B through to *60) have been identified (http://www.cypalleles .ki.se/cyp2c9.htm). *CYP2C9*2* contains a missense mutation in exon 3 resulting in an Arg144Cys substitution (Rettie and Jones 2005). A clinical study has shown a modest decrease (up to 35% in V_{max} but no change in K_m) in drug clearance because of this mutation (Kirchheiner et al. 2002). Arg144 is responsible for mapping the exterior helix C of the protein, which forms part of the putative CYP2C9 reductase binding site (Scott et al. 2004). Loss of activity may alter the affinity for the coenzyme P450 reductase (Higashi et al. 2002). *CYP2C9*3* has a missense mutation in exon 7 that results in an Ile359Leu substitution. There is substantial loss of enzymatic activity, which is the combination of decreased V_{max} and increased K_m for CYP2C9 substrates (Rettie et al. 1999). Low warfarin dosage and increased risk of bleeding during the warfarin stabilization is observed in a Caucasian population carrying *3/*3 homozygotes (Higashi et al. 2002). *4 is an extremely rare missense mutation of 1076T>C originally identified in a Japanese epileptic patient with an ADR to phenytoin (Imai et al. 2000). *5 contains the 1080C>G transversion in exon 7 causing an Asp360Glu change, which has been found almost exclusively in African Americans (Allabi et al. 2004, 2005;

Dickmann et al. 2001). *13 has been identified in a Chinese PM of lornoxicam and the allele has a 269T>C transversion in exon 2 that leads to a Leu90Pro substitution (Si et al. 2004). *16 (Thr299Ala) and *19 (Gln454His) are identified in 125 Asians from Singapore (Zhao et al. 2004). An in vitro study shows that CYP2C9.16 exhibits 80%–90% lower catalytic activity for tolbutamide, whereas CYP2C9.19 shows 30%–40% lower catalytic activity toward tolbutamide (DeLozier et al. 2005). The CYP2C9.26 variant exhibits a 90% decrease in the V_{max} value with largely unchanged K_m value toward diclofenac. CYP2C9.27 and CYP2C9.29 discovered in 263 Japanese individuals exhibits a slight increase in catalytic activities toward tolbutamide. Both CYP2C9.28 and CYP2C9.30 show twofold higher K_m values and threefold lower V_{max} values than the wild-type enzyme, suggesting an important role of Gln214 and Ala477 for substrate recognition. *34 (Arg335Gln) discovered in 724 Japanese do not have a functional impact on enzyme activity in one study (Maekawa et al. 2006), but other studies have observed significantly decreased enzymatic activities toward tolbutamide, glimepiride, and losartan (Dai et al. 2014; Y.H. Wang et al. 2014). Moreover, *42 (Arg124Gln), *45 (Arg132Trp), *46 (Ala149Thr), and *48 (Ile207Thr) produce enzymes with remarkably reduced enzymatic activities toward tolbutamide (Hu et al. 2015).

There is a large variation in CYP2C9 polymorphisms between different ethnic groups. African Americans carry the *2 and *3 variants at much lower frequency (1%–2%) than Caucasians (40%) (Sullivan-Klose et al. 1996). There is no *2 detected in Korean and East Asia populations, although *3 is present at low frequencies (1%–2%) (Jin et al. 2015; Yoon et al. 2001). In the Chinese Han population, *1, *2, and *3 have a frequency of 0.963, 0.001, and 0.036, respectively (Yang et al. 2003). *5 is only expressed in African Americans with 1% allele frequency (Dickmann et al. 2001; Yasar et al. 2002a). The rare null allele *6 is detected in African Americans (Kidd et al. 2001). In addition, *16 is another important defective allele for Chinese Han individuals, except for the commonly studied alleles *2, *3, and *13 (Dai et al. 2014).

1.5.3 CYP2C18 (S-MEPHENYTOIN HYDROXYLASE)

This gene is located within a cluster of CYP2C genes on chromosome 10q24 (Gray et al. 1995). An additional gene, CYP2C17, was once thought to exist; however, CYP2C17 is now considered an artifact based on a chimera of CYP2C18 and 2C19. Alternatively spliced transcript variants encoding different enzymes have been found for CYP2C18. The CYP2C18 is predominantly expressed in lung and skin at a significant level (Ding and Kaminsky 2003; Mace et al. 1998; Zaphiropoulos 1999). However, the protein and mRNA expression levels of CYP2C18 are the lowest in the liver among the four CYP2C subfamily members (Minoletti et al. 1999; Romkes et al. 1991). CYP2C18 is expressed in the heart and may play a role in verapamil metabolism (Thum and Borlak 2000). CYP2C18 has a low catalytic activity toward tolbutamide (Zhu-Ge et al. 2002) as well as other drugs. However, CYP2C18 is active in phenytoin metabolism and

bioactivation of a reactive intermediate (Kinobe et al. 2005). Sulfaphenazole is not an effective inhibitor for CYP2C18 while many sulfaphenazole derivatives are moderate inhibitors of this enzyme in vitro (Ha-Duong et al. 2001a,b).

1.5.4 CYP2C19

CYP2C9 is encoded by the CYP2C19 gene consisting of nine exons, which maps to chromosome 10q24 (Gray et al. 1995; Romkes et al. 1991). The predominant expression organ of CYP2C19 is in the liver and only accounts for less than 5% of total hepatic CYPs. CYP2C19 is responsible for the metabolism of ~10% of commonly used drugs, including proton pump inhibitors (e.g., omeprazole, lansoprazole, pantoprazole, and rabeprazole), tricyclic antidepressants (e.g., imipramine, amitriptyline, and nortriptyline), selective serotonin reuptake inhibitors (e.g., citalopram, fluoxetine, and sertraline), moclobemide (an antidepressant), benzodiazepines (e.g., diazepam, flunitrazepam, quazepam, and clobazam), barbiturates (e.g., hexobarbital, mephobarbital, and phenobarbital), phenytoin, S-mephenytoin, voriconazole, vorapaxar (a potent oral thrombin protease-activated receptor 1 antagonist), bupropion, and proguanil (an antimalarial drug) (Desta et al. 2002; Gardiner and Begg 2006; Ghosal et al. 2011). CYP2C19 favors 5-hydroxylation of the pyridine group of R-omeprazole with 10-fold greater intrinsic activity than toward S-omeprazole, whereas it catalyzes 5-O-demethylation of S-omeprazole in the benzimidazole group (Abelo et al. 2000). The AUC of both omeprazole and lansoprazole in PMs is 4- to 15-fold higher compared to homozygous EMs, whereas the values in heterozygous EMs are only 2- to 3-fold higher than homozygous EMs (Cho et al. 2002; Furuta et al. 1999a,b, 2001; Ieiri et al. 2001; Kim et al. 2002; Shirai et al. 2001, 2002). With multiple dosing, the increase in the AUC of omeprazole, but not of lansoprazole or pantoprazole, decreases to approximately twofold in EMs, owing to inhibition of its own metabolism (Andersson et al. 1998; Shirai et al. 2001). Because of significantly impaired metabolism of proton pump inhibitors, PMs have superior acid inhibition with conventional doses of omeprazole and lansoprazole and possibly pantoprazole (Furuta et al. 1999b; Hunfeld et al. 2008; Sagar et al. 2000; Shirai et al. 2001). CYP2C19 is responsible for the 5 and 5'-hydroxylation of thalidomide, an immunomodulatory drug. Besides, steroids, including progesterone 21-hydroxylation and testosterone 17-oxidation, are oxidized by CYP2C19. Diazinon, an organophosphate insecticide, is also activated in human liver by the enzyme. Rifampicin, dexamethasone, and phenobarbital all induce the expression of CYP2C19 mRNA in human hepatocytes (Raucy et al. 2002). The enzyme can also be induced by rifampicin in vivo (Zhou et al. 1990).

To date, at least 33 variants of CYP2C19 have been identified (http://www.cypalleles.ki.se/cyp2c9.hpm). *1A represents the wild type with an EM phenotype. The first CYP2C19 variant discovered is *2A containing 681G>A on exon 5 that causes a splicing defect (De Morais et al. 1994b). *2B and *2C also carry this mutation and additional SNPs (99C>T; 990C>T; and 991A>G) (Ibeanu et al. 1998b). *3A and *3B share the 636G>A SNP, resulting in a premature stop codon in exon 4 together with 991A>G and 1251A>C (*3B also contains 1078G>A) (Fukushima-Uesaka et al. 2005). *2A, *2B, *2C, *3A, and *3B are null alleles, resulting in complete loss of enzyme activity (De Morais et al. 1994a). The majority of PMs of CYP2C19 are attributed to these variant alleles (Desta et al. 2002). *4 is an initiation codon variant of 1A>G, resulting in GTG initiation codon and also carries 99C>T and 991A>G (Ferguson et al. 1998). *5A, *5B, *6, and *8 harbor 1297C>T, 1297C>T, 395G>A, and 358T>C, respectively, that cause single amino acid changes and affect the structure and stability of the protein (Ibeanu et al. 1998a, 1999). *7 causes a splicing defect. The *6 allele also contains 99C>T and 991A>G, which are shared by *1B, *2A, *2B, *2C, *10, *11, and *12. *4, *6, and *7 result in no enzyme activity, and *5A and *5B cause greatly reduced enzyme activity. *8 encodes a protein with decreased enzyme activity (Ibeanu et al. 1998b). Because these variant alleles (*4 to *8) only account for a minor percentage of 2C19 defective alleles, it is unlikely that these alleles will result in clinically significant consequences. Other potentially defective variant alleles of CYP2C19 have been discovered and these alleles are designated as *9 through to *19. *17 contains two SNPs (99C>T; 991A>G) causing the Ile331Val substitution and a UM phenotype, with frequencies of 21% in Caucasians, 16% in African Americans, and 3% in Chinese (Scott et al. 2011; Sim et al. 2006).

There is racial difference in the distribution of the polymorphic CYP2C19 gene. The polymorphism of the PM phenotype in Caucasians and black Africans is 1%–8% but higher in Asians (13%–23%) with at least eight alleles and highest in some Pacific Islanders up to 75% (Desta et al. 2002). The frequency of CYP2C19*2 has been shown to be ~17% in African Americans, 30% in Chinese, and ~15% in Caucasians (Desta et al. 2002). *2 is the dominant defective allele and accounts for around 75%–85% of PM phenotype in Chinese and Caucasian populations (Desta et al. 2002). *3 is more frequent in Chinese (5%) and less frequent in African Americans (0.4%) and Caucasians (0.04%). Almost all PMs in the Asians and Africans can be attributed to *2 and *3. *9, *10, *12, *13, *14, and *15 have been detected at low frequencies in African Americans (Blaisdell et al. 2002). *16 has a frequency of 0.6% in Japanese (Morita et al. 2004). *17 has been reported in the Chinese, Swedish, and Ethiopian populations with frequencies of 4%–19% (Desta et al. 2002). *6 through to *15 have not yet been detected in any Asian populations, but *18 and *19 have been found in the Japanese population with very low frequencies of 0.2%–0.3% (Hanioka et al. 2007).

1.6 OTHER HUMAN CYP2 FAMILY MEMBERS

1.6.1 CYP2D6

The CYP2D6 gene maps to chromosome 22q13.1 and consists of nine exons with an open-reading frame of 1491 bp coding for a 497–amino acid protein (Eichelbaum et al.

1987; Gough et al. 1993; Heim and Meyer 1990; Kimura et al. 1989). CYP2D6 belongs to a gene cluster of highly homologous inactive pseudogenes CYP2D7P and 2D8P (Heim and Meyer 1992; Kimura et al. 1989; Steen et al. 1995). The CYP2D7P contains a T-insertion in exon 1, disrupting the reading frame, while CYP2D8P encompasses multiple deletions and insertions in its exons (Kimura et al. 1989). Interestingly, a common variant of CYP2D7P with the frameshift mutation 138delT in exon 1 has been found to express an active form in the human brain, which could convert codeine to morphine exclusively (Pai et al. 2004). This appears to be a brain-specific CYP expressed in the brain only. However, another study could not identify functional CYP2D7P transcripts in Asian, Caucasian, or African American individuals (Gaedigk et al. 2005). The subcellular location of CYP2D7P is also unusual. When it was expressed in insect cells, the protein was detected mainly in mitochondrial fractions but not the microsomes (W.Y. Zhang et al. 2009).

There is a significant difference between rodents and humans in the number of active CYP2D genes. Whereas the mouse has nine different active Cyp2d genes (Nelson et al. 2004), the rat harbors six functional Cyp2D genes, while the human carries only one, which indeed is absent from 7% of the Caucasian population. It is reasonable to assume that the mouse and rat have retained their Cyp2d genes active because of a need for a dietary detoxification potential, whereas the more restricted food taken by humans in the past including the intellectual capability to transfer information regarding suitable food between generations has resulted in the loss of a selection pressure as to keep the genes active. The evolution of the human CYP2D locus has involved removal of three genes, inactivation of two (CYP2D7P and 2D8P), and partial inactivation of one (CYP2D6) (Heim and Meyer 1992; Ingelman-Sundberg 2005). On the basis of the identification and characterization of a nonfunctional CYP2D7P gene and a 2D8P pseudogene, Kimura et al. (1989) have suggested that gene duplication events give rise to CYP2D6 and CYP2D7P and that gene conversion events occur later to generate CYP2D8P.

CYP2D6 accounts for only a small percentage of all hepatic CYPs (<2%); however, it metabolizes ~25% of all medications in the human liver (Cascorbi 2003; Gardiner and Begg 2006; Ingelman-Sundberg 2005; Ingelman-Sundberg et al. 2007; Zhou et al. 2009b). CYP2D6 has been identified in the human kidney (Nishimura et al. 2003), intestine (Madani et al. 1999; Nishimura et al. 2003; Prueksaritanont et al. 1995), breast (Huang et al. 1997), lung (Bernauer et al. 2006; Guidice et al. 1997), placenta (Hakkola et al. 1996), and brain (Chinta et al. 2002; Miksys et al. 2002; Siegle et al. 2001) at low to moderate levels. CYP2D6 protein and enzyme activity toward bufuralol have been detected at low levels in the human intestine and are differentially expressed along the length of the gastrointestinal tract (de Waziers et al. 1990; Madani et al. 1999; Prueksaritanont et al. 1995). There is a large interindividual variation in the enzyme activity of CYP2D6 (Sachse et al. 1997).

The details of human CYP2D6 with regard to structure, function, regulation, and polymorphisms will be addressed in individual chapters of this book.

1.6.2 CYP2E1

The *CYP2E1* gene maps to chromosome 10q26.3 and contains nine exons. The *CYP2E1* gene is conserved in chimpanzee, rhesus monkey, dog, cow, mouse, and rat. The predominant expression site of CYP2E1 is in the liver with a moderate abundance. Other expression sites include lung (Hukkanen et al. 2002), esophagus, small intestine (Warner and Gustafsson 1994), brain (Upadhya et al. 2000), nasal mucosa (Kazakoff et al. 1994), and pancreas (Norton et al. 1998). There is an interindividual variation of an order of magnitude as well as a racial difference.

CYP2E1 metabolizes both endogenous substrates, such as ethanol, acetone, and acetal, and exogenous substrates including benzene, carbon tetrachloride, ethylene glycol, and nitrosamines, which are premutagens found in cigarette smoke. It is an ethanol-oxidizing enzyme that is involved in the metabolism of many low-molecular-weight xenobiotics including acetaminophen, benzene, and chlorzoxazone, as well as halothane (Koop 1992; Lieber 1997). Chlorzoxazone 6-hydroxylation has been shown to be catalyzed specifically by CYP2E1. Chlorzoxazone assay is used to estimate CYP2E1 activity in vivo noninvasively (O'Shea et al. 1994). CYP2E1 also catalyzes the 11-hydroxylation of lauric acid as well as oxidizes indole via 3-hydroxylation. Many studies have shown that 2E1 is involved in oxidizing of low-molecular-weight potential carcinogens such as nitrosamines, benzene, styrene, CCl_4, $CHCl_3$, CH_2Cl_2, CH_3CCl_3, 1,2-dichloropropane, ethylene dichloride, ethylene dibromide, vinyl bromide, acrylonitrile, vinyl carbamate, ethyl carbamate, and trichloroethylene (Hoffler et al. 2003; H. Wang et al. 2002).

The regulation of CYP2E1 involves transcriptional activation, mRNA stabilization, increased mRNA translation efficiency, and decreased protein degradation. The *CYP2E1* gene is regulated transcriptionally by HNF-1. Studies have shown that insulin attenuates mRNA levels while glucagon or dibutyryl cAMP upregulates mRNA of CYP2E1 in vitro (Woodcroft and Novak 1999). Cytokines such as IL-6 (Siewert et al. 2000), IL-4 (Lagadic-Gossmann et al. 2000), IL-1β, and TNF-α (Hakkola et al. 2003) also attenuate mRNA levels in vitro and in vivo. miR-378 regulates CYP2E1 (Mohri et al. 2009).

Alcohol and aldehyde dehydrogenase inhibitors also inhibit CYP2E1 activity. 4-Methylpyrazole is a highly selective inhibitor for CYP2E1 for in vitro experiments (Pernecky et al. 1990). 3-Amino-1,2,4-trizole and diethyldithiocarbamate are mechanism-based inhibitors with the oxidized form (disulfiram) used in patients with alcohol aversion therapy (Koop 1990).

To date, at least 13 variants of *CYP2E1* have been reported (*1B* through to *7C*) (http://www.cypalleles.ki.se/cyp2e1.htm). *2, *3, and *4 carry Arg76His, Val389Ile, and Val179Ile,

respectively (Fairbrother et al. 1998; Hu et al. 1997). Some studies have linked *CYP2E1* polymorphisms to increased risk of lung, oral cavity, and stomach cancers but the results are still inconclusive (Ghoshal et al. 2014; Shen et al. 2015; X.H. Ye et al. 2015). Some connections have been made between *CYP2E1* polymorphism and p53 mutations in the presence of vinyl chloride, which is a substrate for CYP2E1 (Wong et al. 2002). Alcohol intake could be correlated to the polymorphism in the 5′-flanking region of *CYP2E1*, which is related to the binding of a transcription factor (Watanabe et al. 1994). Studies in *Cyp2e1* knockout mice indicate that Cyp2e1 is the major factor for acetaminophen-induced hepatotoxicity (Lee et al. 1996). Alcoholics produce autoantibodies against CYP2E1 owing to hydroxyethyl radicals (Clot et al. 1996). CYP2E1 is also an autoantigen associated with halothane-induced hepatitis and alcoholics (Bourdi et al. 1996; Clot et al. 1996; Eliasson and Kenna 1996; Miyakawa et al. 2000).

1.6.3 CYP2J2

CYP2J subfamily members have been identified in many species, including rabbits, rats, mice, and humans. The mouse *Cyp2j* cluster has seven functional genes including *Cyp2j5*, *Cyp2j6*, *Cyp2j8*, *Cyp2j9*, *Cyp2j11*, *Cyp2j12*, and *Cyp2j13* (Graves et al. 2013; Ma et al. 1998, 1999, 2002; Qu et al. 2001; Zhou et al. 2013), compared with a single *CYP2J2* gene in humans (Ma et al. 1998). The mouse also contains three pseudogenes (*Cyp2j7-ps*, *Cyp2j14-ps*, and *Cyp2j15-ps*) (Wu et al. 1996). *Cyp2j7-ps* lacks an open-reading frame and is expected to encode a nonfunctional protein. This cluster has the unusual property that all the genes and pseudogene fragments are oriented in the same direction, which is not the case for the other six *CYP* gene clusters. CYP2J members have been shown to be involved in the metabolism of AA and linoleic acid as well as various xenobiotics (Ma et al. 1999).

The *CYP2J2* gene maps to chromosome 1p31.3-p31.2 (Ma et al. 1998). The human CYP2J2 was first isolated in the liver but found most abundant in heart (DeLozier et al. 2007; Evangelista et al. 2013; Michaud et al. 2010; Wu et al. 1996; Yamazaki et al. 2006). mRNA expression of *CYP2J2* is also found in the placenta, brain, kidney, pancreas, skeletal muscle, bladder, lung, and gastrointestinal tract (Bieche et al. 2007; Zeldin et al. 1996, 1997). CYP2J2 is highly expressed in various human tumor cells (Chen et al. 2011; Jiang et al. 2005), and the EET metabolites resulting from AA are demonstrated to be implicated in cancer development. CYP2J2 is found not readily induced by known CYP inducers. CYP2J2 is differentially regulated in various cell types and different organs through the activator protein-1 (AP-1), the AP-1-like element, and miRNA let-7b (Chen et al. 2012; Xu et al. 2013).

CYP2J2 is thought to be the predominant enzyme responsible for epoxidation of AA in cardiac tissue (Chen and Wang 2013; Evangelista et al. 2013; Xu et al. 2013). CYP2J2 converts AA into regioisomeric 5,6-, 8,9-, 11,12-, or 14,15-EETs (Scarborough et al. 1999). EETs are found in heart tissue and important for a number of biological functions including angiogenesis, protection of endothelial cells against hypoxia-reoxygenation injury, regulation of vasodilation, inhibition of cytokine-induced endothelial cell adhesion-molecule expression, inhibition of vascular smooth muscle cell migration, upregulation of endothelial nitric oxide biosynthesis, protection from ischemic–reperfusion injury, activation of the ion channels sensitive to ATP, and protection of doxorubicin-induced cardiotoxicity (Chen and Wang 2013; Shahabi et al. 2014; X. Wang et al. 2014; Xu et al. 2011, 2013).

Like CYP3A4, CYP2J2 metabolizes antihistamine drugs such as terfenadine, astemizole, and ebastine; tyrosine kinase inhibitors such as dasatinib, imatinib, nilotinib, sorafenib, and sunitinib; tamoxifen; amiodarone; vorapaxar; thioridazine; mesoridazine; danazol; albendazole; fenbendazole; ritonavir; and cyclosporine (Hashizume et al. 2002; Kaspera et al. 2014; C.A. Lee et al. 2010; Matsumoto et al. 2002; Narjoz et al. 2014; Z. Wu et al. 2013). CYP2J2 and 2C19 are the major enzymes responsible for metabolism of albendazole and fenbendazole (Z. Wu et al. 2013). CYP2J2 plays a dominant role in the first-pass intestinal metabolism of ebastine to its pharmacologically active metabolite carebastine (Hashizume et al. 2002; J.I. Lee et al. 2010). A number of drugs are potent and selective CYP2J2 inhibitors (Ren et al. 2013). In particular, telmisartan and flunarizine inhibit CYP2J2 with K_i values of 0.19 μM and 0.13 μM, respectively. Norfloxacin and metoprolol are selective CYP2J2 inhibitors (Ren et al. 2013). Danazol is a potent CYP2J2 inhibitor, with a K_i value of 20 nM, although it also inhibits CYP2C9 and 2D6 (Lee et al. 2012).

To date, at least nine variants of *CYP2J2* have been reported (http://www.cypalleles.ki.se/cyp2j2.htm). King et al. (2002) have cloned and sequenced the entire *CYP2J2* gene containing nine exons and eight introns. A variety of polymorphisms have been found: one is in the promoter region, eight were exonic regions, five are in introns, and four are in the 3′-UTR with only four SNPs resulting in amino acid changes including Arg158Cys, Ile192Asn, Asp342Asn, and Asn404Tyr. A fifth variant (Thr143Ala) is identified by screening a human heart cDNA library (King et al. 2002). Three variants (Asn404Tyr, Arg158Cys, and Thr143Ala) show significantly reduced metabolism of both AA and linoleic acid. The Ile192Asn variant shows a significantly reduced activity toward AA while the Asp342Asn variant shows similar metabolism to wild-type CYP2J2 for both endogenous substrates (King et al. 2002).

1.6.4 CYP2R1

The *CYP2R1* gene maps to chromosome 11p15.2 (Cheng et al. 2003). It contains five exons and spans approximately 15.5 kb. The gene encodes a 501–amino acid protein with a molecular weight of 57 kDa. CYP2R1 (also known as vitamin D 25-hydroxylase) is mainly found in the liver, converting vitamin D into 25-hydroxyvitamin D (calcidiol), which is the major circulatory form of the vitamin (Cheng et al. 2004; Jones et al. 2014; Zhu et al. 2013). Calcidiol is subsequently converted by CYP27B1 (i.e., 25-hydroxyvitamin D_3 1-α-hydroxylase) to calcitriol, the active form of vitamin D_3

that binds to VDR, which mediates most of the physiological actions of the vitamin (Bikle 2014). Mitochondrial CYP27A1 is regarded as a physiologically important vitamin D_3 25-hydroxylase, but CYP2R1 is thought to play a major role because of the absence of sex and species differences and catalytic activity toward both vitamin D_2 and D_3 (Zhu et al. 2013). CYP2R1 is regiospecific to the C25-position of a secosteroid in contrast to other CYP enzymes with vitamin D 25-hydroxylase activity (e.g., CYP27A1, 2C11, and 3A4). Indeed, there is a clear difference between CYP2R1 and 27A1 in the metabolism of vitamin D (Shinkyo et al. 2004). CYP2R1 shows hydroxylase activity at the C25-position, whereas CYP27A1 has hydroxylase activity at the C24- and C27-positions. CYP2R1 also shows significantly higher affinity and C25 hydroxylation activity toward vitamin D_3 than CYP27A1 (Shinkyo et al. 2004).

Presently, there are three crystal structures of CYP2R1 (3C6G, 3CZH, and 3DL9). The CYP2R1 has a typical CYP fold, consisting of α-helices (A–L) with β-sheets mostly on one side of the molecule and with the heme buried deep inside the protein (Strushkevich et al. 2008). The heme is tightly bound to CYP2R1 via hydrogen-bonding interactions of the D-ring propionate with Trp133, Arg137, and Arg446 and the A-ring propionate with Arg109, His381, and Ser442. The CYP2R1 structure adopts a closed conformation with the substrate access channel being covered by the ordered B′ helix and slightly opened to the surface, which defines the substrate entrance point (Strushkevich et al. 2008). The active site is lined by conserved, mostly hydrophobic residues. The active site is formed by amino acid residues of the B′ helix: Leu114, Phe115, and Met118; B–C loop: Leu125 and Asn126; F helix: Phe214, Asn217, and Ala221; G helix: Ala250, Val253, and Tyr254; I helix: Phe302, Glu306, Ala310, and Thr314; the loop between helix K and β-sheet 1: Val375 and Ile379; and the loop between C-terminal β-sheet 4: Met487 and Thr488 (Strushkevich et al. 2008). The size of the CYP2R1 active site is approximately 60% that of CYP2C9 (1OG5) (e.g., 979 Å³ of CYP2R1 vs. 1667 Å³ of CYP2C9). Vitamin D_3 is bound in an elongated conformation with the aliphatic side chain pointing toward the heme. The conjugated triene of vitamin D_3 forms van der Waals interactions with Phe214 of the F helix. Phe214 is highly conserved in human vitamin D–related CYPs (e.g., CYP24A1, 27A1, 27B1, and 11A1) and has been identified as an active-site residue (Strushkevich et al. 2008). The bound vitamin D_3 is nearly buried at the active site with the exception that its 3-OH group is visible through a small opening in the channel. The structure of CYP2R1 shows the exact binding site of vitamin D_3 and provides a new insight into the gating function of the B′ helix for substrate entrance to the active site.

Diseases associated with *CYP2R1* mutations include rickets attributed to a defect in vitamin D 25-hydroxylation and vitamin D–dependent rickets type I (Afzal et al. 2014; Al Mutair et al. 2012; Barry et al. 2014; Cheng et al. 2004; Elkum et al. 2014; Li et al. 2014; Nissen et al. 2014; Z. Ye et al. 2015). *CYP2R1*2* carries a substitution of Leu99Pro that eliminates the enzyme activity and is associated with low circulating levels of 25-hydroxyvitamin D and classic symptoms of vitamin D deficiency (Cheng et al. 2004).

1.6.5 CYP2U1

The *CYP2R1* gene maps to chromosome 4q25. There is a high mRNA expression of *CYP2U1* in human thymus, with lesser expression in the heart, brain, and platelets, whereas similar mRNA levels were detected in the thymus and brain in rats (Jarrar et al. 2013; Karlgren et al. 2004; Toselli et al. 2015). CYP2U1 is mainly detected in the amygdala and prefrontal cortex (Toselli et al. 2015). There is a much higher CYP2U1 protein expression in rat brain than in the thymus, particularly in limbic structures and in the cortex. CYP2U1 metabolizes AA, DHA, and other long-chain fatty acids (Chuang et al. 2004), suggesting that CYP2U1 may play a role in brain and immune functions. CYP2U1 also metabolizes endogenous *N*-arachidonoylserotonin (Siller et al. 2014). CYP2U1 metabolizes propanone, acetone, and 2-oxypropane. Moreover, debrisoquine and terfenadine derivatives, substrates of CYP2D6 and 2J2, are hydroxylated by recombinant CYP2U1 with regioselectivity different from those for CYP2D6 and 2J2 (Ducassou et al. 2015).

Autosomal recessive spastic paraplegia-56 (SPG56) is caused by homozygous or compound heterozygous mutations in *CYP2U1* (Citterio et al. 2014; Fink 2013; Tesson et al. 2012). SPG56 is an autosomal recessive neurodegenerative disorder characterized by early-onset progressive lower-limb spasticity resulting in walking difficulties (Fink 2013). Upper limbs are often also affected, and some patients may have a subclinical axonal neuropathy. In affected members of two consanguineous Saudi Arabian families with autosomal recessive spastic paraplegia, a homozygous mutation, Asp316Val, in the *CYP2U1* gene has been identified (Tesson et al. 2012).

1.6.6 CYP2W1

CYP2W1 is an ancient member of the CYP superfamily of enzymes, which are involved in myriad metabolic and biosynthetic reactions (Nelson et al. 2004). The *CYP2W1* gene maps to choromosome 7p22.3 and spans over 5.5 kb (Karlgren et al. 2006; Nelson et al. 2004). The 9-exon gene encodes a 490–amino acid protein. The 5′-flanking region, first exon, and first intron of the *CYP2W1* gene contain abundant CpG dinucleotides, including two CpG islands (Karlgren et al. 2006). CYP2W1 is mainly expressed in colorectal, hepatic, and adrenal gland tumors, whereas the levels are barely detectable in normal adult tissues (Karlgren et al. 2006; Ronchi et al. 2014). High CYP2W1 levels are observed in ~30% of human colorectal specimens (Stenstedt et al. 2012), and this enzyme is significantly expressed in liver cancer with higher expression in liver metastases than in the parent tumor (Stenstedt et al. 2014). In both colon and liver cancers, the survival is poor in patients with higher levels of CYP2W1 expression (Edler et al. 2009; Stenstedt et al. 2012). Low expression levels of CYP2W1 are also detected in adrenal gland, gastric, and lung tumors (Gomez et al. 2007; Karlgren et al. 2006). A recent study has detected a high level of CYP2W1 expresison in a normal adrenal gland (Ronchi et al. 2014). In adrenocortical cancer patients treated with mitotane only, high CYP2W1 expression is associated with longer overall survival and time to progression and a better response to therapy (Ronchi et al. 2014).

Unlike other CYPs, CYP2W1 has a unique luminal orientation in the ER but still retains catalytic activity (Gomez et al. 2010). CYP2W1 catalyzes the oxidation of indole and shows monooxygenase activity toward 3-methylindole and chlorzoxazone, but not AA (Yoshioka et al. 2006). CYP2W1 metabolizes certain lipids including lysolecithin and their stereoisomers (Karlgren et al. 2006; Xiao and Guengerich 2012). CYP2W1 expressed in HEK 293 cells is active in the metabolism of indoline substrates and is able to activate AFB$_1$ into a cytotoxic product (Gomez et al. 2010). Lysophospholipids including oleyl (18:1) lysophosphatidylcholine (lysolecithin), lysophosphatidylinositol, lysophosphatidylserine, lysophosphatidylglycerol, lysophosphatidylethanolamine, and lysophosphatidic acid, but not diacylphospholipids, are substrates for CYP2W1 (Xiao and Guengerich 2012). CYP2W1-mediated epoxidation and hydroxylation of 18:1 lysolecithin are considerably more efficient than for the C18:1 free fatty acid. Tumor-sepcific CYP2W1 converts duocarmycin analogs to cytotoxic metabolites and kill the cancer cells via induction of DNA damage (Travica et al. 2013). This might allow the development of a novel combined therapy of colorectal cancer that would include a tumor-specific induction of CYP2W1 followed by the treatment with CYP2W1-activated prodrug. Both CYP2W1 and 2S1 catalyze the reductive activation of the anticancer prodrug AQ4N (banoxantrone) to the topoisomerase II inhibitor AQ4 (Nishida et al. 2010). In addition, CYP2W1 can bioactivate heterocyclic amines such as 2-amino-3,4-dimethylimidazo[4,5-*f*]quinoline and 2-amino-3methylimidazo[4,5-*f*]quinoline (Eun et al. 2010), suggesting a role for CYP2W1 in carcinogenesis.

Using a human fetal mRNA multiple-tissue cDNA panel lacking gastrointestinal tissues, a transient *CYP2W1* mRNA expression in lungs, liver, skeletal muscle, and kidney at gestational week 30 is observed (Choudhary et al. 2005). Another study has observed the expression of *CYP2W1* in the colon and small intestine at early stages of embryonic life but is completely silenced shortly after the birth (Choong et al. 2015). Immunohistochemical analysis of human fetal colon reveals that CYP2W1 expression is restricted to the crypt cells. The silencing of *CYP2W1* after birth correlates with the increased methylation of CpG-rich regions in both murine and human *CYP2W1* genes. In colorectal tumors, higher expression of CYP2W1 is associated with increased demethylation of the CpG island in the exon 1/intron 1 junction (Gomez et al. 2007). This implies interesting parallels between the developmental and cancer regulatory mechanisms of CYP2W1 expression. CYP2W1 and 2S1 are selectively induced in breast cancer cells only after treatment with 5F-203 (an AhR agonist, 2-(4-amino-3-methylphenyl)-5-fluorobenzothiazole) or GW-610 (2-(3,4-dimethoxyphenyl)-5-fluorobenzothiazole), two new anticancer agents (Tan et al. 2011). CYP2W1 is important for the bioactivation of GW-610 in colorectal cancer cells, while CYP2S1 appears to be involved in the deactivation of benzothiazoles. CYP2W1 is also induced by imatinib (a BCR-Abl inhibitor used in the treatment of Philadelphia chromosome–positive [Ph+] chronic myelogenous leukemia), linoleic acid, and its derivatives in colon adenocarcinoma HCC2998 cells (Choong et al. 2015). The imatinib-mediated induction of CYP2W1 suggests an adjuvant therapy to treatment with duocarmycins that thus would involve induction of tumor CYP2W1 levels followed by the 2W1-activated duocarmycin prodrugs. However, CYP2W1 is not induced by the CAR agonist CITCO (6-(4-chlorophenyl)imidazo[2,1-b][1,3]thiazole-5-carbaldehyde O-(3,4-dichlorobenzyl)oxime) and the PPARγ agonist ciglitazone and TCDD, a potent AhR ligand (Choong et al. 2015).

To date, six variants of *CYP2W1* have been reported (*1B* through to *6) (http://www.cypalleles.ki.se/cyp2w1.htm). A systemic study in 200 Japanese subjects has identified six SNPs including 166C>T (Leu56Leu) in exon 1, 173A>C (Glu58Ala) in exon 1, 2008G>A (Ala181Thr) in exon 4, 5432G>A (Val432Ile) in exon 9, 5584G>C (Gln482His) in exon 9, and 5601C>T (Pro488Leu) in exon 9 (Hanzawa et al. 2008). The allele carrying the silent SNP (166C>T, Leu56Leu) is termed *1B. The most frequent variant allele is *6 (5601C>T, Pro488Leu), followed by *1B and *1A (the wild type), and their frequencies in Japanese are 0.368, 0.318, and 0.295, respectively. *2 and *3 carry Ala181Thr and Glu58Ala, respectively. *5 harbors Val432Ile and Gln482His. The Ala181Thr polymorphism is associated with increased colorectal cancer risk (Gervasini et al. 2010).

1.7 HUMAN CYP3A CLUSTER: CYP3A4, 3A5, 3A7, AND 3A43

Gellner et al. (2001) have found that, on chromosome 7q21.1, where the *CYP3A4* gene had previously been mapped, there is a 231-kb region containing three *CYP3A* genes: *CYP3A4*, *3A5*, and *3A7*, as well as two pseudogenes (*CYP3A5P1* and *3A5P2*) and another *CYP3A* gene, which they termed *CYP3A43*. The CYP3A subfamily, CYP3A4, 3A5, 3A7 and 3A43, is one of the most versatile of the enzyme systems that are responsible for the elimination of 37% of drugs out of the 200 most frequently prescribed drugs in the United States. CYP3A4 and 3A5 together account for approximately 30% of total hepatic CYPs, and approximately half of the drugs that are oxidatively metabolized by CYPs are CYP3A substrates. Both CYP3A4 and 3A5 are expressed in the liver and intestine, with CYP3A5 being the predominant form expressed in extrahepatic tissues. CYP3A4 is the most abundantly expressed form in the liver while CYP3A5 expression at the protein level is only approximately 10.6% of that of CYP3A4 (Wang and Tompkins 2008). CYP3A7 is considered a fetal-specific CYP enzyme, and CYP3A43 has extremely low expression in the liver and only plays a minor role in drug metabolism.

1.7.1 CYP3A4

The *CYP3A4* gene is located on chromosome 7q22.1 and is approximately 27 kb long consisting of 13 exons and 12 introns (Brooks et al. 1988). CYP3A4 is the most abundant CYP in human liver, accounting for 25%–30% total CYPs, and in the small intestine. The enzyme is also present in the lung, stomach, colon, and adrenal gland but not in the kidney, prostate, testis, or thymus. The enzyme has a dominant role in drug metabolism. There is a large intraindividual variation in *CYP3A4* gene expression as much as 10-folds (Haddad et al. 2007) and up to 90-fold in CYP3A4 protein expression in liver samples (Schuetz 2004). CYP3A4 varies with gender, with 25% higher enzymatic activity in females (Hunt et al. 1992; Wolbold et al. 2003). There is no significant difference with age (Haddad et al. 2007). The level of CYP3A4 is low in fetal liver, although the expression rapidly increases after birth and reaches 50% of adult levels between 6 and 12 months of age.

The substrate specificity of the CYP3A4 enzymes is very broad, with an extremely large number of structurally divergent chemicals being metabolized often in a regio- and stereoselective fashion (Sevrioukova and Poulos 2013d; Thummel and Wilkinson 1998; Zhou 2008a). CYP3A4 is known to metabolize a large variety of compounds varying in molecular weight from lidocaine (M_r = 234 Da) to cyclosporine (M_r = 1203 Da). CYP3A4 exhibits a relatively large substrate-binding cavity that is consistent with its capacity to oxidize bulky substrates such as cyclosporine, statins, taxanes, and macrolide antibiotics. Although the active-site volume is similar to that of CYP2C8, the shape of the active-site cavity differs considerably because of differences in the folding and packing of portions of the protein that form the cavity. Compared with CYP2C8, the active-site cavity of 3A4 is much larger near the heme iron (Yano et al. 2004). A small number of drugs are used as model substrates of CYP3A4 when assessing in vitro and in vivo phenotyping activity and drug interactions. Testosterone is the most commonly used in vitro CYP3A4 probe (50% of reported studies) in contrast to midazolam (15%–20% of in vitro estimates of CYP3A4 activity), whereas nifedipine, felodipine, and erythromycin are used in <10% of studies (Yuan et al. 2002). In vivo CYP3A4 function can be determined noninvasively by the erythromycin breath test (Chiou et al. 2001). Midazolam is considered as one of the best in vivo probe drugs for the study of CYP3A4 activity. A considerable number, but not all (e.g., benzodiazepines), of CYP3A substrates interact with P-gp either as substrates or as inhibitors (calcium channel blockers, azole antifungal agents, immunosuppressants, natural product anticancer agents, and macrolide antibiotics). CYP3A4 catalyzes 6α-hydroxylation of the hepatotoxic lithocholic acid and 25-hydroxylation of the cholic acid precursor 5β-cholestane-3α,7α,12α-triol (Araya and Wikvall 1999; Bodin et al. 2005). These substrates of CYP3A4 are ligands for PXR. Oxysterol 4β-hydroxycholesterol is formed by CYP3A4, and this oxysterol has been used as a marker reaction for determining CYP3A4/5 activity (Diczfalusy et al. 2009; A. Honda et al. 2011).

Chemicals used as selective inhibitors of CYP3A4 include a small number of compounds inhibiting CYP3A4 in an irreversible (e.g., triacetyloleandomycin, gestodene) or reversible (e.g., ketoconazole) manner (Zhou 2008a). Ketoconazole is the most widely used, but its selectivity for CYP3A4 is often a concern. A number of drugs with widely differing structures and therapeutic targets have been reported to be mechanism-based inhibitors of CYP3A4 (Zhou et al. 2004). To date, the identified clinically important mechanism-based CYP3A4 inhibitors mainly include macrolide antibiotics (e.g., clarithromycin and erythromycin), anti-HIV agents (e.g., ritonavir and delavirdine), antidepressants (e.g., fluoxetine and fluvoxamine), calcium channel blockers (e.g., verapamil and diltiazem), steroids and their modulators (e.g., gestodene and mifepristone), and several herbal and dietary components (Zhou et al. 2004). Large numbers of acetylenes, particularly those synthetic steroids such as gestodene, norethisterone, ethinyl estradiol, and norgestrel, are also mechanism-based inactivators of CYPs (Guengerich 1990). Marked inhibition of CYP3A4 may cause severe ADRs owing to enhanced exposure of coadministered drugs (Huang et al. 2007; Riley et al. 2007; Zhou 2008a; Zhou et al. 2004, 2005). For example, when CYP3A4 inhibitors such as erythromycin or clarithromycin are coadministered with terfenadine, astemizole, or pimozide, patients may experience Torsades de pointes (a life-threatening ventricular arrhythmia associated with QT prolongation) (Dresser et al. 2000; Michalets and Williams 2000; Spinler et al. 1995). Terfenadine is a CYP3A4 substrate that undergoes extensive first-pass metabolism after oral administration (Honig et al. 1992; Jurima-Romet et al. 1994). Normally, the carboxylate metabolite of terfenadine is the principal circulating entity in plasma, whereas unchanged terfenadine is not present at measurable concentrations (Honig et al. 1992, 1993). Furthermore, rhabdomyolysis has been observed when simvastatin was combined with erythromycin or ritonavir (Williams and Feely 2002). Symptomatic hypotension may occur when CYP3A4 inhibitors are combined with some dihydropyridine calcium antagonists (Anderson and Nawarskas 2001) as well as with the phosphodiesterase inhibitor sildenafil (Simonsen 2002). In addition, ataxia can occur when carbamazepine is coadministered with mechanism-based CYP3A4 inhibitors such as macrolide antibiotics, isoniazid, verapamil, or diltiazem (Patsalos et al. 2002; Spina et al. 1996). Several drugs (e.g., mibefradil) that have been identified as CYP3A4 inactivators have been withdrawn from the market because of reported toxic drug–drug toxicities. Thus, caution should be taken when these drugs are used, in particular when used in combination with other drugs that are CYP3A4 substrates and have narrow therapeutic indices. The identification of drugs causing marked CYP3A4 inhibition and the mechanisms involved are important in terms of rational use of therapeutic drugs to avoid harmful drug–drug interactions.

CYP3A4 possesses an unexpected peripheral binding site located above a Phe-rich cluster forming a highly ordered hydrophobic core, which may be an effector site involved in the initial recognition of substrates or allosteric effectors (Williams et al. 2004; Yano et al. 2004). CYP3A4 has been shown to display cooperative behavior with the binding of substrates that does not follow typical Michaelis–Menten kinetics (Domanski et al. 2001b; He et al. 2003; Shou et al. 1999), and this is often explained by a flexible and large CYP3A4 active site, which can accommodate two or more ligand molecules simultaneously. To date, there have been 17 crystal structures of human CYP3A4 (4K9X, 4K9W, 4K9V, 4K9U, 4K9T, 4I4H, 4I4G, 4I3Q, 3TJS, 3UA1, 3NXU, 2V0M, 2J0D, 1W0G, 1TQN, 1W0E, and 1W0F) available (Ekroos and Sjogren 2006; Sevrioukova and Poulos 2010, 2012a, 2013a,b; Williams et al. 2004; Yano et al. 2004). These structures have similar secondary structures with some differences in the conformation of the flexible regions at residues 282–285 (between helices H and I), residues 261–270 (between G and H helices), and Arg212. The CYP3A4 x-ray structures retain the tertiary structure with the common CYP fold as observed for all other CYPs. The CYP3A4 structures contain a closed, buried active site connected to bulk solvent via several tunnels. Yano et al. (2004) have reported an active-site volume of 1386 Å^3 while Williams et al. (2004) found a small volume of 520 Å^3. Although the active-site volume of CYP3A4 is similar to that of CYP2C8, the shape of the active-site cavity differs considerably because of differences in the folding and packing of portions of the protein that form the cavity (Yano et al. 2004). Compared with CYP2C8, the active-site cavity of CYP3A4 is much larger near the heme iron (Yano et al. 2004). CYP3A4 contains an unexpected peripheral binding site located above a Phe7 residue cluster, which may be involved in the initial recognition of substrates or allosteric effectors (Williams et al. 2004). The progesterone molecule resides in the peripheral "nest" formed by loops between the F and F' helices and the G' and G helices, that is, in the F–G loop region. This resembles that of palmitate binding in the CYP2C8 (1PQ2) structure (Schoch et al. 2004). In CYP3A4, helices F and G are short, resulting in extended unstructured linkers to the intervening short F0 and G0 helices. Three Phe residues from these linker regions interact with the Phe residue from helix F0, the B–C region, and helix I, resulting in a unique Phe cluster on the ceiling of the active-site cavity. Phe213 and Phe215 point toward the active site and form the Phe cluster together with other five Phe residues including Phe108, Phe219, Phe220, Phe241, and Phe304, which are involved in the initial recognition of the substrates. Binding of the small substrate metyrapone (M_r = 226) does not alter the structure of CYP3A4, in contrast to changes caused by substrate binding active sites of the other CYPs such as rabbit CYP2B4 (Scott et al. 2001, 2003, 2004) and rabbit CYP2C5 (Wester et al. 2003a,b). Large ligands, when used in crystallization, either stay outside of the active-site pocket (e.g., for progesterone) or have yet to be cocrystallized with the protein (e.g., erythromycin). In contrary to above reports, a study indicates that CYP3A4 undergoes dramatic conformational changes upon binding to ketoconazole or erythromycin with a differential but substantial (>80%) increase in the active-site volume (Ekroos and Sjogren 2006), providing a structural basis for ligand promiscuity of CYP3A4. The binding of two molecules of ketoconazole to the active site of CYP3A4 and multiple binding modes for erythromycin may provide an explanation for the atypical kinetics of CYP3A4.

The *CYP3A4* gene has two 5′-promoter sites for increasing enzyme expression through induction (Finta and Zaphiropoulos 2000). PXR interacts with steroids (such as dexamethasone), statin drugs, and macrolide antibiotics and can regulate the expression of CYP3A4. Compared to substrates, a much smaller group of drugs behave as CYP3A4 inducers (Zhou 2008a). The most potent inducers of CYP3A4 are rifampin and rifabutin, whose coadministration can reduce a drug's plasma concentration 20- to 40-fold, effectively negating drug efficacy (Thummel and Wilkinson 1998; Zhou 2008a). Commonly used anticonvulsants such as carbamazepine and, to a lesser extent, phenytoin, primidone, and phenobarbital, can also significantly increase CYP3A4 activity. Drugs that induce CYP3A4 and other CYPs are structurally diverse; however, all have one major similarity, lipophilicity.

Induction of CYP3A4 is another important reason for clinically observed drug–drug interactions. Rifabutin and rifampin are potent CYP3A4 inducers and cause clinically significant interactions with warfarin, oral contraceptives, cyclosporine, glucocorticoids, ketoconazole or itraconazole, theophylline, quinidine sulfate, digitoxin or digoxin, verapamil hydrochloride, HIV-related protease inhibitors, zidovudine, delavirdine mesylate, nifedipine, and midazolam (Finch et al. 2002). The induction of hepatic and intestinal CYP3A4 by St John's wort may partly explain that St. John's wort increases the clearance of many coadministered drugs such as indinavir (Durr et al. 2000; Piscitelli et al. 2000; Wang et al. 2001), cyclosporine (Breidenbach et al. 2000; Moschella and Jaber 2001; Wentworth et al. 2000), and oral contraceptives (Fugh-Berman 2000), as all these drugs are substrates of CYP3A4.

Clinicians should adopt proper strategies when using drugs that are potent CYP3A4 inhibitors or inducers. Early identification of drugs behaving as potent CYP3A4 inhibitors, inactivators, or inducers and the mechanisms involved are important. If these drugs have to be used in some instances, rational use of such drugs and close monitoring become necessary, including use of a safe drug combination regimen, dose adjustment, or discontinuation of therapy when toxic drug interactions occur. In particular, when combined with drugs with narrow therapeutic indices, close monitoring of plasma drug concentrations and observing of potential toxicities should be conducted and dosage is accordingly adjusted.

Significant interindividual variability in the expression and activity of CYP3A4 has been observed and such a substantial variability is considered a result of interplays of environmental, physiological, and genetic factors (Werk and Cascorbi 2014). To date, 25 variants have been identified in the *CYP3A4* gene (http://www.cypalleles.ki.se/cyp3a4.htm). *1B (i.e., *CYP3A4-V*) contains a −392A>G mutation in the nifedipine-specific response element of the 5′-regulatory region of the gene (Hashimoto et al. 1993). *2 containing a 664T>C SNP (leading to Ser222Pro change) is found at a frequency of 2.7% in Caucasians but is absent in blacks and Chinese (Sata et al. 2000). *2 encodes a CYP with substrate-dependent altered kinetics compared with the wild-type enzyme. The frequency of *3 with a 1334T>C mutation and subsequently a Met445Thr change in Caucasians is 1.1% (van Schaik et al. 2001). *4 and *5 contain 352A>G (Ile118Val) and 653C>G (Pro218Arg) in exons 4 and 5, respectively (Lamba et al. 2002b). *6 represents a frameshift mutation (831insA) arising from an A17776 insertion in exon 9, resulting in an early stop codon. Functionally, *4, *5, and *6 are associated with decreased enzyme activity (Garcia-Martin et al. 2002; Hsieh et al. 2001). In addition, *7 with the 167G>A SNP (Eiselt et al. 2001) through to *20 with an insertion of A between 1461 and 1462 nucleotides resulting in completely defective enzyme (Westlind-Johnsson et al. 2006) have been identified. Genetic polymorphisms in *CYP3A4* seem to be more prevalent in the Caucasian population than in Asians (Werk and Cascorbi 2014). In genotyping studies comparing the frequencies of *1B, *4, *5, and *6 in healthy subjects, *4 is detected in 2 of 110 Chinese subjects and *6 is found in 1 of 104 Malays and 1 of 101 Indians, whereas *1B and *5 are absent in all three ethnic groups (Chowbay et al. 2003). The frequency of *1B carrying −392A>G is higher in white and Hispanic subjects (3.6%–11.0%) and much higher in blacks (53.0%–69.0%). Like other Asian ethnic groups, *1B is absent in Japanese (Naoe et al. 2000). Notably, polymorphisms of *CYP3A4* do not translate into significant interindividual variability in vivo, probably because of the induction of CYP3A4 on exposure to substrates or compensation from CYP3A5 (Werk and Cascorbi 2014). Most *CYP3A4* alleles have been reported to have minimal to moderate function compared to wild type. The majority of functional SNPs reported for *CYP3A4* are infrequent and thus can only account for a small proportion of the observed interindividual differences in CYP3A4 activity. The more common SNP reported for *CYP3A4* being associated with drug response or disease is *1B. However, results reported to date have been inconsistent and its functionality remains controversial. In addition, *1B is in strong linkage disequilibrium with the *CYP3A5*1* functional allele, raising the possibility that the observed association could have been confounded by a high CYP3A5 expression in *CYP3A4*1B* carriers (Kuehl et al. 2001). *22 carrying an intron 6 SNP (rs35599367, C>T) is reported to be associated with CYP3A4 expression and activity, as well as response to statin therapy, as reflected by the lower dose requirement to achieve optimal lipid control among variant allele carriers (D. Wang et al. 2011). *1B is associated with early puberty (Kadlubar et al. 2003). High activity of CYP3A4, but not CYP3A5, which primarily metabolizes testosterone, shows a striking association with the onset of puberty (Kadlubar et al. 2003). *1B has been associated with high-grade and the advanced stage of prostate cancers in Caucasians, Hispanics, and African Americans (Zeigler-Johnson et al. 2002). In addition, there are reports on the association of *1B with therapy-related leukemia (Blanco et al. 2002; Felix et al. 1998). Individuals with the *1B genotype may be at increased risk for epipodophyllotoxin-related leukemia and that epipodophyllotoxin metabolism by CYP3A4 may contribute to the secondary cancer risk (Felix et al. 1998).

1.7.2 CYP3A5

The *CYP3A5* gene is localized in a cluster on chromosome 7q21-q22.1 and consists of 13 exons (Finta and Zaphiropoulos 2000; Schuetz and Guzelian 1995; Spurr et al. 1989). Sequence analysis of the 5′-flanking region has identified putative estrogen-response element, progesterone-response element, AP3, GF1 (GATA1), and basic transcription element binding sites as well as a CAAT box. CYP3A5 is a 503–amino acid protein that shares 87% sequence identity with CYP3A4. The enzyme is expressed in multiple organs including the liver, small intestine, kidney, lung, prostate, adrenal gland, and pituitary (Hukkanen et al. 2003; Murray et al. 1995; Yamakoshi et al. 1999). Some studies have observed the expression in peripheral blood cells (Janardan et al. 1996) but others have not (Koch et al. 2002). The total mRNA expression level of *CYP3A5* is approximately 2% of the CYP3A subfamily (Koch et al. 2002). Fetal liver has CYP3A5 expression but in a polymorphic manner (Hakkola et al. 2001). The expression of CYP3A5 can be found in only 10%–20% of Caucasians, 33% in Japanese, and 55% in African Americans (Kuehl et al. 2001), and the level is usually less than CYP3A4. Polymorphism has been linked to the variation in expression levels. The activity of CYP3A5 is similar or reduced compared to CYP3A4 with different oxidation sites such as AFB_1 3α-hydroxylation versus 8,9-epoxidation (Xue et al. 2001). CYP3A4 and 3A5 are considered to have similar substrate specificity, but the contribution of CYP3A5 to the total metabolic clearance of CYP3A substrates in the liver in vivo has yet to be determined. Because of their similarities, CYP3A4 inhibitors including ketoconazole and fluconazole as well as the mechanism-based inactivator gestodene all inhibit CYP3A5. The regulation of *CYP3A5* gene is similar to the *CYP3A4* gene. CYP3A4 and CYP3A5 are coregulated in the liver and intestine in terms of transcription control although other factors can alter the expression (Thummel et al. 1996).

At least 10 variants of *CYP3A5* have been identified (http://www.cypalleles.ki.se/cyp3a5.htm). Approximately 50% of the liver sample shows alternate splicing of *CYP3A5*. The Caucasian population has the *3 allele with an inserted intron (Hustert et al. 2001; Lamba et al. 2002a). Africans have a higher incidence of *1 but the expressed CYP3A5 is functional. Since the functional *CYP3A5*1* allele is in linkage disequilibrium with *CYP3A4*1B* and CYP3A5 does not have a major function in statin metabolism, the clinical data should be interpreted properly. *2 contains a 27289C>A SNP, resulting in a Thr398Asn substitution in 5%–10% of Caucasians (Jounaidi et al. 1996). *3 has 10 subvariants, designated as *3A to *3J, sharing a 6986A>G SNP in intron 3 (Kuehl et al. 2001) with additional SNPs (e.g., 31611C>T; 3705C>T; 31551T>C;

and/or 31611C>T) (Saeki et al. 2003). Individuals homozygous for *3 appear to be defective in CYP3A5 because of the creation of a cryptic splice site, which results in the incorporation of the intronic sequence in the mature mRNA and the production of a truncated protein as a result of premature termination of translation (Kuehl et al. 2001). *3 is correlated with the simvastatin systemic exposure and response in humans (Kim et al. 2007; Kivisto et al. 2004). However, other studies do not find any association between *3 and lipid-lowering response to statins (Fiegenbaum et al. 2005; Hu et al. 2013). *4 harbors a 14763A>G mutation, leading to Gln200Arg change, while *5 carries a 12952T>C change at intron 5 splicing donor site (Chou et al. 2001). *6 and *7 are rare alleles resulting from abnormal intronic or exonic splicing and frameshift mutation (14690G>A, 27131-32insT), respectively (Kuehl et al. 2001). Analysis of the 5′-flanking region of the *3A5* gene has identified two linked polymorphisms, −369T>G and −45A>G, which are associated with increased expression and activity of the enzyme (Paulussen et al. 2000). They are from the pseudogene *CYP3AP1* (Finta and Zaphiropoulos 2000; Gellner et al. 2001); the −45A>G polymorphism in *CYP3AP1* is linked to the *3 defect allele in Caucasians (Kuehl et al. 2001). There are marked interethnic differences in the frequencies of *CYP3A5* mutations. The frequency of *CYP3A5*3* varies from ~50% in African Americans, 70% in Chinese, to 90% in Caucasians. It is functionally important as the expressed protein is a truncated species and is associated with extremely low enzymatic activity (Kuehl et al. 2001). *6 is absent in the Asian (Fukuen et al. 2002) and Caucasian (Kuehl et al. 2001) populations. However, this allele is predominant in Africans with a frequency of ~7.5% (Kuehl et al. 2001).

1.7.3 CYP3A7

The *CYP3A7* gene has been mapped to chromosome 7q22.1 and contains 13 exons and spans at least 17.1 kb (Itoh et al. 1992). The *CYP3A7* gene has ~90% sequence identity with *3A4* in the coding region (Komori et al. 1989). CYP3A7 is the major human fetal CYP and is expressed strongly during the embryonic and fetal stage but decreased quickly during the first week of birth (Fanni et al. 2014; Hakkola et al. 2001; Leeder et al. 2005). CYP3A7 is the most dominant CYP expressed in fetal liver and can catalyze several reactions. It only accounts for less than 2% of all CYPs in adult liver. The enzyme is also expressed in human placenta and elevated in endometrium during pregnancy or during the secretory phase of the menstrual cycle (Schuetz et al. 1993). The expression of CYP3A7 mRNA and protein has been observed in placentas and thus may play a role in placenta-mediated drug disposition (Hakkola et al. 2001; Maezawa et al. 2010; O'Shaughnessy et al. 2013; Schuetz et al. 1993). CYP3A7 expression is varied approximately 5-fold in fetal tissue and up to 77-fold in mRNA (Hakkola et al. 2001). CYP3A7 catalytic activities are similar to 3A4 and 3A5 but mostly occurs in fetal liver. In fetal liver, CYP3A4 shows the highest metabolic capacity toward glyburide, followed by CYP3A7 and 3A5 (Shuster et al. 2014). Testosterone 6β-hydroxylation and 16α-hydroxylation of DHEA 3-sulfate are catalyzed by CYP3A7. The enzyme is also responsible for several other reactions including activation of AFB$_1$ and heterocyclic amines and RA 4-hydroxylation (Chen et al. 2000; Hashimoto et al. 1995). The conversion rate of CYP3A7 is generally lower than that of 3A4 and 3A5 (Williams et al. 2002). Regulation of the CYP3A7 gene is complex. The main regulators are Sp1, Sp3, HNF-3β, nuclear factor-1, and upstream stimulator factor 1 (Riffel et al. 2009). A distal xenobiotic response enhance module has been reported in the CYP3A7 gene (Bertilsson et al. 2001). CYP3A7 has a functional PXR element similar to CYP3A4. Therefore, CYP3A4 inducers should be able to induce CYP3A7 including rifampicin. Similarly, CYP3A4 inhibitors such as ketoconazole and troleandomycin are also inhibitors for 3A7.

To date, at least six variants and subvariants of the CYP3A7 gene (*1B through to *3) have been reported (http://www.cypalleles.ki.se/cyp3a7.htm). *1C is unusual because a part of the CYP3A4 promoter replaces the corresponding region of 3A7 that leads to high levels of expression (Smit et al. 2005). This SNP results in lower circulating DHEA and estrone levels. *2 carries 1226C>G in exon 11 causing a Thr409Arg substitution (Rodriguez-Antona et al. 2005). CYP3A7.2 is a functional enzyme with a significantly higher catalytic constant compared with the wild-type enzyme. There is a linkage disequilibrium between *2 and CYP3A5*1 that is subject to interethnic differences.

1.7.4   CYP3A43

The CYP3A43 gene has been mapped to chromosome 7q21.1 with 16 exons. CYP3A43 has 75% sequence identity as CYP3A4 and 3A5 and 71% identical to CYP3A7 (Domanski et al. 2001a; Gellner et al. 2001). The enzyme can be found in the liver, kidney, pancreas, brain, and prostate. The mRNA expression level of CYP3A43 in the liver is very low (approximately 0.1% of CYP3A4 and 2% of CYP3A5 in the liver) (Westlind et al. 2001), suggesting its minor role in drug metabolism. It has a low testosterone 6β-hydroxylation activity (Domanski et al. 2001a). In the brain, CYP3A43 can metabolize alprazolam to both α-OH-alprazolam and 4-hydroxy alprazolam while CYP3A4 metabolizes alprazolam predominantly to its inactive metabolite, 4-hydroxy alprazolam (Agarwal et al. 2008). Rifampicin is a CYP3A43 inducer.

To date, four variants of CYP3A43 have been reported (http://www.cypalleles.ki.se/cyp3a43.htm). Three polymorphisms in the coding region of CYP3A43 have been identified, comprising two nucleotide substitutions, one silent (1047C>T) and one missense mutation (1018C>G leading to Pro340Ala, present in CYP3A43*2B and *3), and a frameshift mutation (74delA, present in CYP3A43*2A and *2B), leading to a premature stop codon and, presumably, to a severely truncated protein (Cauffiez et al. 2004b). Genetic variation in CYP3A43 is associated with response to olanzapine (Brandl et al. 2015; Soderberg and Dahl 2013).

1.8   HUMAN CYP4ABXZ CLUSTER: CYP4A11, 4A22, 4B1, 4X1, AND 4Z1

The CYP4 family consists of 11 subfamilies, which encode constitutive and inducible enzymes expressed in both mammals and insects (Simpson 1997). They are the major fatty acid ω-hydroxylases. There is a cluster of human CYP4 genes on chromosome 1p33, including CYP4A11, 4A22, 4B1, 4X1, and 4Z1 (Hsu et al. 2007). These CYP4 members are capable of hydroxylating the terminal ω-carbon and, to a lesser extent, the (ω-1) position of saturated and unsaturated fatty acids, as well as enzymes active in the ω-hydroxylation of various PGs. The CYP4 subfamily is also involved in the metabolism of AA and LTs, leading to the formation of physiologically important metabolites (Hsu et al. 2007). In addition, CYP4 members can metabolize xenobiotics including therapeutic drugs and other compounds.

1.8.1   CYP4A11 (20-Hydroxyeicosatetraenoic Acid Synthase, Fatty Acid ϖ-Hydroxylase, and Lauric Acid ϖ-Hydroxylase)

The human CYP4A11 gene is located on chromosome 1p33, spans approximately 12.6 kb, and contains 12 exons (Imaoka et al. 1993). The CYP4A11 gene is conserved in chimpanzee, mouse, and rat. CYP4A11 is responsible for the major lauric acid ω-hydroxylases in human liver and kidney (Imaoka et al. 1993). It also catalyzes ω- and ω-1 hydroxylation of myristic and palmitic acids (Dhar et al. 2008; Hoch et al. 2000; Kawashima et al. 2000). CYP4A11 oxidizes endogenous AA to 20-HETE, a renal vasoconstrictor and natriuretic (Lasker et al. 2000). The expression has also been observed in human keratinocytes (Gonzalez et al. 2001). CYP4A11 expression varies approximately 10-fold in humans (Powell et al. 1996; Savas et al. 2003). The regulation of CYP4A11 expression is still not well understood but may play a role in peroxisome proliferation system as well as carcinogenesis. Peroxisome proliferators and dexamethasone could induce CYP4A11 expression in cell cultures. Substituted imidazoles and acetylenic fatty acids are inhibitors for CYP4A11.

More than 850 SNPs of the CYP4A11 gene are listed in the National Center for Biotechnology Information SNP database (http://www.ncbi.nlm.nih.gov/snp/). These SNPs have not been assigned to specific haplotypes by the Human Cytochrome P450 Allele Nomenclature. The Leu131Phe mutant catalyzes only ω-1 hydroxylation and not ω-hydroxylation of lauric acid (Dierks et al. 1998). Leu131 controls access to substituted imidazole inhibitors. Another coding SNP, rs1126742 (T8590C) in exon 11, which results in a Phe434Ser substitution, has been found to be associated with essential hypertension in several ethnic groups (Dhar et al. 2008; Gainer et al. 2008; Liang et al. 2014; Sugimoto et al. 2008; Williams et al. 2011; Yan et al. 2013; Yang et al. 2014). Gainer et al. (2005) have reported that the T8590C (Phe434Ser) SNP of CYP4A11 causes a lower catalytic activity for AA ω-hydroxylation to form 20-HETE. Other SNPs of CYP4A11 have shown no or markedly lower activity for AA ω-hydroxylation compared to wild-type CYP4A11 (Saito et al. 2015).

1.8.2 CYP4A22 (Fatty Acid ϖ-Hydroxylase and Lauric Acid ϖ-Hydroxylase)

This gene encodes a 519–amino acid protein. The *CYP4A22* gene has 95% sequence identity as the *CYP4A11* gene. CYP4A22 is expressed in the liver and kidney with a lower level than 4A11 (Savas et al. 2003). However, the expression of CYP4A22 could not be observed in HepG2 cells or PPARα-overexpressing cells (Savas et al. 2003). Like CYP4A11, 4A22 is a fatty acid ω-hydroxylase. CYP4A22 shows ω-hydroxylation activity against laurate and palmitate, but not against AA or PGA$_1$.

To date, 20 variants of *CYP4A22* have been reported (*2 through to *15) (http://www.cypalleles.ki.se/cyp4a22.htm). One study in a Japanese population has identified 13 polymorphisms in the *CYP4A22* coding region: two are silent mutations located in exons 8 (His323His) and 9 (Gly390Gly); nine are missense mutations located in exons 1 (Arg11Cys), 3 (Arg126Trp), 4 (Gly130Ser and Asn152Tyr), 5 (Val185Phe), 6 (Cys231Arg), 7 (Lys276Thr), 10 (Leu428Pro), and 12 (Leu509Phe); one is a nonsense mutation located in exon 9 (Gln368stop); and one is a nucleotide deletion (G7067del) that causes a frameshift and consequently results in a stop codon 80 nucleotides downstream (Hiratsuka et al. 2006). The functional impact of these SNPs is unclear.

1.8.3 CYP4B1

This gene maps to chromosome 1p33 (Nhamburo et al. 1989). CYP4B1 is mainly expressed in the lung, kidney, breast, prostate, and bladder. The expression is also detected in bladder and breast tumors (Imaoka et al. 2000). The enzyme catalyzes the *N*-hydroxylation of 2-aminofluorene and ω-hydroxylation of lauric acid. Activation of carcinogens such as 2-aminofluorene could be a risk factor in bladder cancer. There are approximately two orders of magnitudes in its expression variation in bladder (Imaoka et al. 2000). CYP4B1 can be used as a drug delivery vehicle as well as activation system including the activation of 4-ipomeanol and 2-aminoanthracene in vitro (Frank et al. 2002). CYP4B1 could not be induced and no inhibitor has been found (Guengerich 2005). Functional CYP4B1 could be expressed as a fusion protein with NADPH-P450 reductase since the presence of auxiliary proteins may stabilize the enzyme (Imaoka et al. 2001).

Presently, six variants of *CYP4B1* have been reported (*2A through to *7) (http://www.cypalleles.ki.se/cyp4b1.htm). *CYP4B1*2A* harbors three missense mutations including Met331Ile, Arg340Cys, and Arg375Cys and a double nucleotide deletion (AT881-882del) that causes a frameshift and premature stop codon in the second third of the coding sequence of the gene, leading to the synthesis of a severely truncated protein (Lo-Guidice et al. 2002). *3, *4, and *5 contain missense mutations Arg173Trp, Ser322Gly, and Met331Ile, respectively. *6 carries 517C>T (R173W) and 1033G>A (V345I) and *7 harbors AT881-882-del, 993G>A, and 1018C>T (Hiratsuka et al. 2004).

1.9 HUMAN CYP4F CLUSTER: CYP4F2, 4F3, 4F8, 4F11, 4F12, AND 4F22

Mice have nine *Cyp4f* genes and humans have six *CYP4F* genes encoding CYP4F2, 4F3, 4F8, 4F11, 4F12, and 4F22. The *CYP4F* gene family is clustered in a 0.5-Mb stretch of genomic DNA on the p13 region of chromosome 19 (Kirischian and Wilson 2012). CYP4F2 and 4F3B have 93% amino acid sequence identity and both share a high degree of identity with other CYP4F enzymes. CYP4F enzymes account for 15% of the total hepatic CYPs (Michaels and Wang 2014). Members of the CYP4F subfamily are important enzymes involved in the biotransformation of endogenous eicosanoids (e.g., AA, PGs, and LTB$_4$) and are involved in the regulation of many physiological functions, such as inflammation and vasoconstriction (Fer et al. 2008; Hardwick 2008; Kikuta et al. 1998, 2004, 2007). CYP4F2 is the principal hepatic ω-hydroxylase for LTB$_4$ and AA, resulting in deactivation of LTB$_4$ and formation of the potent vasoconstrictor 20-HETE, respectively (Jin et al. 1998; Powell et al. 1998). Both CYP4F3A and 4F3B catalyze the ω-hydroxylation of LTB$_4$ and AA; however, CYP4F3A has 30-fold greater affinity (K_m) for LTB$_4$ and conversely 8.4-fold lower affinity for AA than 4F3B (Christmas et al. 2001). As ω-hydroxylases, CYP4Fs have the potential to convert AA to 20-HETE that has potent actions on renal tubular and vascular functions (Hardwick 2008; Kikuta et al. 2007). The ω-hydroxylated LTB$_4$ undergoes further metabolism to 20-carboxy-LTB$_4$, which can undergo B-oxidation from its ω-side along with traditional β-oxidation from the C1 carbon, leading to the inactivation of this potent proinflammatory agent.

Besides endogenous substrates, CYP4F enzymes are involved in the metabolism of xenobiotics including some drugs, but it is considered that this family seems to play a minor role in drug metabolism. CYP4F11 metabolizes erythromycin and ethylmorphine (Kalsotra et al. 2004), and 4F12 metabolizes ebastine with contribution from 2J2 (Hashizume et al. 2002). The new immunomodulatory agent used for the treatment of relapsing multiple sclerosis, fingolimod (FTY720; 2-amino-2-[2-(4-octylphenyl)ethyl]-1,3-propanediol), is eliminated predominantly by CYP4F2 or 4F3B-mediated ω-hydroxylation of the aliphatic chain (Jin et al. 2011). CYP4F2 and 4F3B also catalyze the *O*-demethylation of the antitrypanosomal methamidoxime prodrugs pafuramidine (DB289; 2,5-bis(4-amidinophenyl)furan-bis-*O*-methylamidoxime), DB844, and DB868 (Ju et al. 2014; M.Z. Wang et al. 2006).

1.9.1 CYP4F2 (20-Hydroxyeicosatetraenoic Acid Synthase, Arachidonic Acid ϖ-Hydroxylase, and Leukotriene-B4 ϖ-Hydroxylase 1)

CYP4F2 can be found in the liver, the S2 and S3 segments of proximal tubules in the cortex, and the outer medulla of the kidney (Kikuta et al. 2000, 2002). The expression of CYP4F2 in human can range within fivefold (Jin et al. 1998). The enzyme is responsible for the ω-hydroxylation of several lipids (Jin et al. 1998), AA (Lasker et al. 2000), LTB$_4$ (Kikuta et al. 2000), lipoxin A$_4$, 8-HETE, 12-HETE, and 12-hydroxystearic acid (Kikuta et al. 2000). In addition, CYP4F2 catalyzes the ω-hydroxylation of tocopherol (vitamin E) and phylloquinone (vitamin K_1) phytyl side chains, suggesting a role in the regulation of vitamin E status and synthesis of clotting factors in the liver (McDonald et al. 2009; Sontag and Parker 2002).

To date, only two variants of *CYP4F2* have been documented (http://www.cypalleles.ki.se/cyp4f2.htm). *2 carries a Trp12Gly replacement and *3 harbors a Val433Met substitution (Stec et al. 2007). Both of these variants are frequent in both African Americans and European Americans. These two SNPs caused a reduced production of 20-HETE from AA but had no effect on the ω-hydroxylation of LTB_4 (Stec et al. 2007). Several clinical studies have shown that *CYP4F2* mutations affect the required maintenance dosage and clinical outcome of the anticoagulant warfarin in various ethnic groups (Danese et al. 2012; Liang et al. 2012a,b; McDonald et al. 2009; Scott et al. 2012; Tatarunas et al. 2014; J.E. Zhang et al. 2009).

1.9.2 CYP4F3

The *CYP4F3* gene contains 13 exons and spans approximately 22.2 kb (Kikuta et al. 1998). The gene produces two tissue-specific splice variants, CYP4F3A (myeloid form) and CYP4F3B (liver form), but both transcripts are present in other tissues (Christmas et al. 2001). The *CYP4F3* gene is shown to be tissue-specific splicing and alternate promoters (Corcos et al. 2012). Exon 4 containing CYP4F3A is expressed in neutrophils, whereas the CYP4F3B form containing exon 3 is expressed in the liver, kidney, trachea, and gastrointestinal tract (Christmas et al. 1999, 2001). CYP4F3 expression results from the synthesis of two distinct enzymes, CYP4F3A and 4F3B, which originate from the alternative splicing of a single pre-mRNA precursor molecule. Although the catalytic activity of CYP4F3 is lower than CYP4F2 ($K_m = 0.7$ μM), the enzyme still plays a part in LTB_4 ω-hydroxylation. CYP4F3B catalyzes ω-hydroxylation of AA (Christmas et al. 2001). Apart from the ω-hydroxylation of LTB_4 and PGs, CYP4F3 is the main enzyme for the oxidation of fatty acid epoxides (Fer et al. 2008; Le Quere et al. 2004). There are potential clinical implication for this enzyme because of the physiological and pathological importance of LTB_4 and 20-HETE in inflammation and regulation of blood pressure (Wu et al. 2014).

1.9.3 CYP4F8

This gene has been mapped to chromosome 19p13.1, encoding a protein with 520 amino acids (Bylund et al. 1999). CYP4F8 (also called PGH_{19} hydroxylase) has 81.2% and 76.7% amino acid identity with CYP4F2 and 4F3, respectively. It is mainly expressed in the epidermis, hair follicles, sweat glands, corneal epithelium, proximal renal tubules, and epithelial lining of the gut and urinary tract (Bylund et al. 1999, 2000). Overexpression has been shown in patients with psoriasis (Stark et al. 2003). The CYP4F8 catalyzes the ω-2 hydroxylation of AA and three stable PGH_2 analogs but not PGD_2, E_1, E_2, $F_{2\alpha}$, and LTB_4 (Bylund et al. 1999, 2000; Stark

et al. 2005). PGH_1 and H_2 ω-2 hydroxylation and isomerization to 19R-hydroxy PGE are catalyzed in seminal vesicles by CYP4F8 (Bylund et al. 2000). The exact physiological role of CYP4F8 is unclear but may be related to the regulation of inflammatory responses. Mutations of *CYP4F8* is associated to gemcitabine response (Harris et al. 2014).

1.9.4 CYP4F11 (PHYLLOQUINONE ω-HYDROXYLASE AND 3-HYDROXY FATTY ACID ω-HYDROXYLASE)

This gene belongs to the *CYP4* gene cluster found on chromosome 19, located 16 kb upstream of the *CYP4F2* gene (Cui et al. 2000). It contains 12 exons and encodes an enzyme with 524 amino acids. The CYP4F11 amino acid sequence has 80.0%, 82.3%, and 79.2% identity to CYP4F2, CYP4F3, and 4F8 amino acid sequences, respectively (Cui et al. 2000). CYP4F11 is expressed primarily in the liver, kidney, heart, intestine, and skeletal muscle (Uehara et al. 2015). A stimulation of the c-Jun NH_2-terminal kinase pathway causes an increase in *CYP4F11* mRNA, whereas retinoids downregulate the expression of CYP4F11 mRNA (Wang et al. 2010). The NF-κB signaling pathway negatively regulates the *CYP4F11* gene (Bell and Strobel 2012). CYP4F11 can metabolize LTB_4 and AA but at a much lower activity than CYP4F2 or 4F3A/B (Kalsotra et al. 2004). 3-Hydroxystearate and 3-hydroxypalmitate are converted to ω-hydroxylated 3-OH dicarboxylic acid precursors in human liver and CYP4F11 is the predominant catalyst of this reaction (Dhar et al. 2008). Both CYP4F2 and 4F11 catalyze the ω-hydroxylation of the menaquinone form of vitamin K (Edson et al. 2013). CYP4F2, but not CYP4F11, catalyzes sequential metabolism of the menaquinone form of vitamin K to the ω-acid without apparent release of the intermediate aldehyde (Edson et al. 2013). Polymorphism of *CYP4F2* and *4F11* does not affect warfarin response and side effects in American Indian and Alaska Native people (Fohner et al. 2015).

1.9.5 CYP4F12

The *CYP4F12* gene contains 13 exons and maps to chromosome 19p13.1. CYP4F12 is mainly detected in the liver, kidney, colon, small intestine, urogenital tract, and heart (Bylund et al. 2001; Hashizume et al. 2001; Stark et al. 2004). It has a 78%–83% amino acid identity to CYP4F2, 4F3, 4F8 and 4F11. CYP4F12 catalyzes the hydroxylation of AA at C18 and C20 (Bylund et al. 2001; Hashizume et al. 2001), ω1 through ω3-hydroxylations of 9,11-diazo-PGH_2 and 9,11-methanoepoxy-PGH_2 (Stark et al. 2005), and ω-oxidation of LTB_4 (Bylund et al. 2001). In addition, CYP4F12 contributes to astemizole *O*-demethylation (Eksterowicz et al. 2014) and hydroxylation of ebastine and terfenadine, with contributions from CYP3A4 and 2J2 (Hashizume et al. 2002). Amiodarone is a very potent inhibitor for CYP4F12 with an IC_{50} of 78 nM. CYP4F12 is inducible through the activation of the p53 pathway, a sequence-specific transcription factor known to regulate the expression of multiple CYPs, and also exhibits an increased binding to PXR in the presence of rifampicin (Goldstein et al. 2013; Hariparsad et al. 2009).

Ten SNPs of *CYP4F12* have been reported, comprising seven missense mutations, 31C>T (Leu11Phe), 38C>T (Pro13Leu), 47C>T (Met16Thr), 4759G>A (Asp76Asn), 4801G>A (Val90Leu), 8896C>T (Arg188Cys), and 23545G>A (Gly522Ser) (Cauffiez et al. 2004a). Five out of seven variants do not exhibit any significant difference in the catalytic activity, whereas two variants, Val90Ile and Arg188Cys, display significant changes in the Michaelis–Menten parameters toward ebastine hydroxylation (Cauffiez et al. 2004a).

1.9.6 CYP4F22

This gene includes 12 coding exons and the cDNA spans 2.6 kb in length. All *CYP4F22* mutations reported to date are predicted to abolish the function of the encoded CYP protein and to compromise the 12*R*-lipoxygenase (hepoxilin) pathway. 12*R*-lipoxygenase is expressed only in the epidermis and the tonsils and is upregulated in psoriatic lesions. It transforms 20:4n-6 to 12*R*-HPETE, which is important for the development of the water permeability barrier function in the epidermis. CYP4F22 catalyzes the ω-3 hydroxylation of 20:4n-6; however, oxygenation of 8*R*,11*R*,12*R*-HETE is not observed (Nilsson et al. 2010). An additional function of CYP4F22 is to synthesize the ω-hydroxy fatty acids in the ceramide. Mutations in the *CYP4F22* gene have been linked to lamellar ichthyosis (Kelly et al. 2011; Rodriguez-Pazos et al. 2011; Sugiura et al. 2013). CYP4F22 is involved in the metabolism of lipid substrates that are important to differentiation/keratinization of epidermal keratinocytes, at least during the fetal period (Sasaki et al. 2012).

1.10 OTHER HUMAN CYP4 FAMILY MEMBERS

1.10.1 CYP4V2 (DOCOSAHEXAENOIC ACID ω-HYDROXYLASE)

CYP4V2 is an unusual *CYP4* member in that it resides on chromosome 4q35.2, separate from the *CYP4ABXZ* and *CYP4F* gene clusters on chromosomes 1 and 19. The *CYP4V2* gene is conserved in chimpanzee, rhesus monkey, dog, cow, mouse, rat, chicken, fruit fly, mosquito, *C. elegans*, and frog. The gene contains 11 exons spanning 19 kb (Li et al. 2004). The protein has very low sequence identity (31%–37%) to other CYP4 enzymes. The mRNA expression of CYP4V2 is found in the heart, brain, placenta, lung, liver, skeletal muscle, kidney, pancreas, retina, retinal pigment epithelium, and lymphocytes (Stark and Guengerich 2007). In particular, CYP4V2 is widely distributed and detected in the eye with highest expression in retina. The expression of CYP4V2 in a wide variety of tissues is consistent with inclusions and their clinical sequelae found in the cornea, retina, and lymphocytes. CYP4V2 is a selective ω-hydroxylase of saturated, medium-chain fatty acids with relatively high catalytic efficiency toward myristic acid (Hsu et al. 2007). CYP4V2 also hydroxylates the ω-3 polyunsaturated fatty acids, including DHA and EPA. HET0016

(*N*-hydroxy-*N'*-(4-n-butyl-2-methylphenyl formamidine) is a nanomolar inhibitor of this enzyme (IC$_{50}$ = 38 nM).

The mutations in the *CYP4V2* gene such as Trp44Arg, Gly61Ser, Phe73Leu, Ile111Thr, Arg390Cys, Val458Met, and Arg508His are associated with Bietti's crystalline corneoretinal dystrophy (Gekka et al. 2005; Kelly et al. 2011; Lai et al. 2007, 2010; Lee et al. 2005; Li et al. 2004; Lin et al. 2005; Mackay and Halford 2012; Mamatha et al. 2011; Nakamura et al. 2006; Nakano et al. 2012; Shan et al. 2005; Wada et al. 2005; Xiao et al. 2011; Yokoi et al. 2010, 2011; Zenteno et al. 2008). This severe ocular disease is a progressive disorder that leads to atrophy of the retinal epithelium, constriction of the visual field, and night blindness. The small glistening crystals can also occur in the corneal limbus and circulating lymphocytes. Several mutations in *CYP4V2* are also linked to retinitis pigmentosa, an ocular disease characterized typically by pigmentation and atrophy in the midperiphery of the retina (Y. Wang et al. 2012). The Gln259Lys mutation in CYP4V2 is associated with deep vein thrombosis and factor XI level (Bezemer et al. 2008). Defective ω-oxidation of ocular fatty acids/lipids secondary to mutations in the *CYP4V2* gene could contribute to the occurrence of these ocular diseases.

1.10.2 CYP4X1

CYP4X1 maps to chromosome 1p33 and contains 12 exons (Bylund et al. 2002). CYP4X1 is predominantly expressed in the trachea, aorta, heart, liver, prostate, breast, and brain such as the cerebellum, amygdala, and basal ganglia (Bylund et al. 2002; Savas et al. 2005; Stark et al. 2008a). PPARα activation can transactivate CYP4X1. CYP4X1 converts the natural endocannabinoid anandamide to a single monooxygenated product, 14,15-EET ethanolamide, and AA is oxidized by this enzyme more slowly to 14,15- and 8,9-EETs but only in the presence of cytochrome b_5 (Stark et al. 2008a). Homology and docking of CYP4X1 with AA has revealed that Tyr112, Ala126, Ile222, Ile223, Thr312, Leu315, Ala316, Asp319, Thr320, Phe491, and Ile492 are actively participating in the interaction, while docking of CYP4X1 with anandamide shows that Tyr112, Gln114, Pro118, Ala126, Ile222, Ile223, Ser251, Leu315, Ala316, and Phe491 are key residues for ligand binding (Kumar 2015).

1.10.3 CYP4Z1

The *CYP4Z1* gene is part of a cluster of cytochrome P450 genes on chromosome 1p33 (Rieger et al. 2004). CYP4Z1 mRNA is preferentially expressed in mammary tissue (Savas et al. 2005). CYP4Z1 is overexpressed in breast cancer cells (Rieger et al. 2004) and may promote tumor angiogenesis and growth in human breast cancer (W. Yu et al. 2012). In T47-D breast cancer cells, CYP4Z1 mRNA levels are induced by dexamethasone (14-fold) or by progesterone (10-fold) (Savas et al. 2005), but its levels are not affected by 17β-estradiol treatment. This enzyme catalyzes the in-chain hydroxylation of lauric acid and myristic acid, resulting in four different monohydroxylated products at positions ω-2, ω-3, ω-4, and ω-5, respectively (Zollner et al. 2009).

1.11 HUMAN CYP5 FAMILY

1.11.1 CYP5A1 (Thromboxane A Synthase 1, TBXAS1)

CYP5A1 (also called thromboxane A synthase 1, TBXAS1) is a 60-kDa protein with 533 amino acids and a heme prosthetic group. The *CYP5A1* gene has been mapped to chromosome 7q34 and spans more than 75 kb consisting of 13 exons and 12 introns (Chase et al. 1993; Miyata et al. 1994b). The enzyme is expressed in platelets, erythroleukemia cells (Yokoyama et al. 1991), human monocytes (Nusing and Ullrich 1992), leukocytes (Young et al. 2002), and kidney interstitial dendritic reticulum cells (Nusing et al. 1994). Some expression is also seen in the lung and liver (Nusing and Ullrich 1992). CYP5A1 catalyzes the conversion of PGH_2 to thromboxane, which causes vasoconstriction and platelet aggregation. PGH_2 is converted to thromboxane A_2 and 12-HTT. In concert with prostacyclin, thromboxane A_2 plays a pivotal role in the maintenance of homeostasis. CYP5A1 has a heme group coordinated to the thiolate group of a cysteine residue, specifically Cys480 (Ohashi et al. 1992).

Mutagenesis studies that made substitutions at that position result in loss of catalytic activity and loss of heme binding. Other residues that show similar results were Trp133, Arg478, Asn110, and Arg413 (Wang and Kulmacz 2002). Located near the heme propionate groups or the distal face of the heme, these residues are also important for proper integration of heme into the apoprotein. The enzyme plays a role in the pathogenesis of a number of diseases including myocardial infraction, unstable angina, pregnancy-induced hypertension and preeclampsia, thrombosis and thrombotic disorders, pulmonary hypertension, asthma, septic shock, atherosclerosis, lupus nephritis, and Raynaud's phenomenon. Ghosal hematodiaphyseal dysplasia is a rare autosomal recessive disorder characterized by increased bone density with predominant diaphyseal involvement and aregenerative corticosteroid-sensitive anemia. Genevieve et al. (2008) have identified mutations in the *CYP5A1* gene in the two families originally reported by Isidor et al. (2007) and in two more families of Tunisian and Pakistani origin. The mutations involved residues conserved across species, and three of the four were located near the heme-binding domain. In patients with Ghosal hematodiaphyseal syndrome, Genevieve et al. (2008) have detected T248C (Leu83Pro), G1238A (Arg413Glu), G1444T (Gly482Trp), and T1463C (Leu488Pro) transitions in the *CYP5A1* gene. An overexpression of CYP5A1 has been observed in a variety of cancers, such as papillary thyroid carcinoma, prostate cancer, and renal cancer, and CYP5A1 can regulate the growth, apoptosis, angiogenesis, and metastasis of cancers (Cathcart et al. 2010).

Because of their role in thromboxane synthesis, TBXAS1 inhibitors and thromboxane A_2 receptor antagonists have been a great interest for many years for their potential use in preventing platelet aggregation. Drugs with dual action, such as dual TBXAS1 inhibitors/thromboxane A_2 receptor antagonist and dual COX-1/thromboxane A_2 receptor antagonists, are currently in clinical development with the hope of providing a better, more complete inhibition of the thromboxane A_2 pathway. For example, ifetroban is a potent and selective thromboxane A_2 receptor antagonist; dipyridamole antagonizes this receptor too; picotamide has activity both as a TBXAS1 inhibitor and as a thromboxane A_2 receptor antagonist. These inhibitors often contain a basic nitrogen atom that binds to the CYP5A1 heme. Terutroban is an orally active drug in clinical development for use in secondary prevention of thrombotic events in cardiovascular disease. Despite great expectations on this drug supported by a large body of preclinical and clinical evidence and pathophysiological rationale, the PERFORM trial fails to demonstrate the superiority of terutroban over aspirin in secondary prevention of cerebrovascular and cardiovascular events among ~20,000 patients with stroke (Bousser et al. 2011).

1.12 HUMAN CYP7 FAMILY

The *CYP7A1* gene encodes cholesterol 7α-hydroxylase, catalyzing the first and major rate-limiting step in the classical, neutral pathway for bile acid biosynthesis. The *CYP7B1* gene codes for a 7α-hydroxylase involved in metabolism of oxysterols, sex hormones, and neurosteroids. CYP7B1 shares 40% amino acid sequence identity with CYP7A1.

1.12.1 CYP7A1 (Cholesterol 7α-Hydroxylase)

The *CYP7A1* gene maps to chromosome 8q11-q12 and contains recognition sequences for a number of liver-specific transcription factors (Cohen et al. 1992; Nishimoto et al. 1993; Wang and Chiang 1994). The gene spans 10 kb and contains six exons. The *CYP7A1* gene is conserved in chimpanzee, rhesus monkey, dog, cow, mouse, rat, chicken, zebrafish, and frog. CYP7A1 (also called cholesterol 7α-hydroxylase) is only expressed in the liver with a low to moderate expression level compared to other xenobiotic-metabolizing enzymes. The enzyme is responsible for cholesterol 7α-hydroxylation, which is a rate-limiting step in bile acid synthesis (Chiang 2009; Gilardi et al. 2007; Lorbek et al. 2012). The reaction is essential for proper absorption of dietary lipids and fat-soluble vitamins but not for maintaining cholesterol and lipid levels. CYP7A1 also catalyzes 7α-hydroxylation of 27-hydroxycholesterol and other oxysterols (Norlin et al. 2000). CYP7A1 has a very short half-life ($t_{1/2}$: 1–2 h).

To date, there are three crystal structures of CYP7A1 in the PDB (3V8D, 3SN5, and 3DAX). The structure of CYP7A1 reveals a motif of residues that promote cholest-4-en-3-one binding parallel to the heme, thus positioning the C7 atom for hydroxylation (Tempel et al. 2014). Additional regions of the binding cavity are involved to accommodate the elongated conformation of the aliphatic side chain. Structural complex with 7-ketocholesterol shows an active-site rigidity and provides an explanation for its inhibitory effect on CYP7A1 (Tempel et al. 2014).

The regulation of CYP7A1 expression is complex since the catabolism of cholesterol into bile acids is regulated by oxysterols and bile acids, which induce or repress transcription of CYP7A1. The enzyme is mainly regulated at the transcriptional level, and bile acids are the most important physiological inhibitors of CYP7A1 transcription (Gilardi et al. 2007). Multiple mechanisms are involved in the control of CYP7A1 transcription, and a variety of transcription factors and nuclear receptors participate in sophisticated regulatory networks. Dietary cholesterol elevates the expression level of CYP7A1 in animal models (Li et al. 2008). Competitive inhibitor 7-oxocholesterol could reduce bile acid synthesis and increase in CYP7A1 synthesis (Breuer et al. 1993). There is a bile acid–responsive element in the *CYP7A1* promoter. The orphan steroid receptors are also involved in *CYP7A1* gene regulation. The proximal promoter region interacts with LXRα. The oxysterols 24*S*-hydroxycholesterol and 24*S*-epoxycholesterol activate LXRα and LXRβ. *CYP7A1* transcription was stopped in mice lacking LXRα (Peet et al. 1998). Chenodeoxycholate, a bile acid derived from cholesterol, interacts with FXR to suppress *CYP7A1* transcription (Chiang et al. 2000). PXR binds lithocholic acid and downregulates *CYP7A1*. Therefore, *CYP7A1* negative feedback controls the synthesis of cholesterol metabolites (Russell 1999). PPARα also plays a role in CYP7A1 downregulation as a result of competition with HNF-4 for the DR-1 sequence (Marrapodi and Chiang 2000). TNF-α also downregulates CYP7A1 transcription through MEKK1, affecting HNF-4 (De Fabiani et al. 2001). Insulin and phorbol esters also repress *CYP7A1* expression (Wang et al. 1996). Coffee terpenes (cafestol) can inhibit CYP7A1 expression and raise cholesterol levels (Post et al. 1997). Overexpression of CYP7A1 in HepG2 cells increased bile synthesis but decreased HMG-CoA reductase activity (Pandak et al. 2001). Moreover, retinoic acid represses CYP7A1 expression in human hepatocytes and HepG2 cells via FXR/RXR-dependent and -independent mechanisms (Cai et al. 2010). miR-122a and miR-422a may destabilize CYP7A1 mRNA to inhibit CYP7A1 expression (Song et al. 2010). Polymorphisms of *CYP7A1* have been reported, with some associated with clinical outcomes. Deficiency of this enzyme will increase the risk of hypercholesterolemia and cholesterol gallstones (Pullinger et al. 2002).

1.12.2 CYP7B1 (Oxysterol 7α-Hydroxylase 1)

This gene has been mapped to chromosome 8q12.3. CYP7B1 (also called oxysterol 7α-hydroxylase 1) is widely expressed in the brain, particularly in the hippocampus, and also in the liver and kidney, albeit at much lower levels. The enzyme catalyzes the 7α-hydroxylation of neurosteroids DHEA, pregnenolone, cholesterol derivatives 25-hydroxycholesterol and 27-hydroxycholesterol, and other steroids including estrogen receptor ligands such as 5α-androstane-3β,17β-diol (anediol) and 5α-androstene-3β,17β-diol (enediol) (Cui et al. 2013; Rose

et al. 1997; Yantsevich et al. 2014). DHEA and pregnenolone have similar structures and metabolic roles, and DHEA is one of the most abundant neurosteroids and performs important functions. Hepatic CYP7B1 activity is crucial for the inactivation of steroids, and loss of the enzyme activity is associated with liver failure in children. Like CYP7A1, 7B1 also participates in 7α-hydroxylation of 27-hydroxycholesterol and other oxysterols (Pandak et al. 2002).

A homology modeling study has highlighted 15 residues that are consistently involved in binding with 25-hydroxycholesterol within the active site: Phe114, Ser115, Leu118, Leu119, Arg267, Gly288, Phe289, Trp291, Ala292, Ser366, Thr367, Ile369, Leu488, Phe489, and Gly490 (Cui et al. 2013). The following key residues contribute to the binding of the three substrates enediol, DHEA, and anediol: Phe114, Ser115, Leu118, Trp291, Ala292, Ala295, Asn296, Thr367, Phe489, and Gly490. Knockout of the *Cyp7a1* gene in mice results in disruption of 7α-hydroxylation of DHEA and 25-hydroxycholesterol in the brain, spleen, thymus, heart, lung, prostate, uterus, and mammary gland (Rose et al. 2001), although the animals appear to be viable and normal.

Spastic paraplegia type 5 is an autosomal, recessive, hereditary spastic paraparesis, often showing periventricular and subcortical white matter degeneration. The patients have increased levels of 27-hydroxycholesterol in plasma and cerebrospinal fluid. A number of mutations in *CYP7B1* have been documented in patients with hereditary spastic paraplegia type 5 in different ethnic groups (Arnoldi et al. 2012; Biancheri et al. 2009; Cao et al. 2011; Criscuolo et al. 2009; Goizet et al. 2009; Ishiura et al. 2014; Lan et al. 2015; Mignarri et al. 2014; Noreau et al. 2012; Roos et al. 2014; Schule et al. 2009, 2010; Tsaousidou et al. 2008). Biancheri et al. (2009) have identified three mutations in the *CYP7B1* gene in patients with autosomal recessive hereditary spastic paraplegia. Goizet et al. (2009) have identified three nonsense (Arg63X, Arg112X, and Tyr275X) and five missense mutations (Thr297Ala, Arg417His, Arg417Cys, Phe470Ile, and Arg486Cys), with the last four clustering in exon 6 at the C-terminal end of the protein in patients with hereditary spastic paraplegia type 5.

1.13 HUMAN CYP8 FAMILY

1.13.1 CYP8A1 (Prostaglandin I₂ Synthase, PGIS/PTGIS)

The *CYP8A1* gene contains 10 exons and consensus sequence for Sp1, AP-2, GATA, NF-κB, a CACCC box, GR, and a shear stress responsive element (GAGACC) (Nakayama et al. 1996; Wang et al. 1996; Yokoyama et al. 1996). The mRNA of CYP8A1 is expressed in the ovary, heart, skeletal muscle, lung, prostate (Miyata et al. 1994a), umbilical vein (Spisni et al. 1995), and neurons (Mehl et al. 1999). Another site of expression is fallopian tubes in luminal epithelial, tubal smooth muscle, vascular endothelial cells, and vascular muscle cells (Huang et al. 2002; Lee et al. 2002). CYP8A1 converts PGH₂ into prostacyclin (i.e., PGI₂), which is a potent vasodilator and platelet adhesion inhibitor. Imbalance between prostacyclin and thromboxane A_2 may lead to several cardiovascular diseases including myocardial infarction, stoke, atherosclerosis, and pulmonary hypertension. PGG_2, H_2, 13S-hydroxyl H_2, and 15-keto H_2 and H_3 are isomerized to the corresponding prostacyclins (Lin et al. 1998). Unlike most CYPs, CYP8A1 does not require molecular oxygen. Instead, it uses its heme cofactor to catalyze the isomerization of PGH_2 to prostacyclin. CYP8A1 is slowly deactivated during the normal process itself by one of the intermediates in the catalytic reaction. There are only a few inhibitors including phenylbutazone and peroxynitrite (K_i = 50 nM) for this enzyme. The determined structures of CYP8A1 provide useful information on ligand-specific heme conformational change (Chiang et al. 2006; Li et al. 2008). The 2IAG structure exhibits a typical triangular prism-shaped CYP fold with the conserved acid–alcohol pair in the I helix of CYPs replaced by Gly286 and Asn287, while the distinctive disruption of the I helix and the presence of a nearby water channel remain conserved (Chiang et al. 2006). The side chain of Asn287 appears to be positioned to facilitate the endoperoxide bond cleavage, suggesting a functional conservation of this residue in O–O bond cleavage. A combination of bent I helix and tilted B' helix creates a channel extending from the heme distal pocket, accommodating various ligands. The inhibitor (minoxidil)-bound and substrate analog U51605-bound structures of CYP8A1 show a stereo-specific substrate binding and suggest features of the enzyme that facilitate isomerization (Li et al. 2008). Unlike most microsomal CYPs, where large substrate-induced conformational changes occur at the distal side of the heme, conformational changes in CYP8A1 are observed at the proximal side and in the heme itself. The conserved and extensive heme propionate–protein interactions seen in all other P450s, which are largely absent in the ligand-free CYP8A1, are recovered upon U51605 binding accompanied by water exclusion from the active site. In contrast, when minoxidil binds, the propionate–protein interactions are not recovered and water molecules are largely retained (Li et al. 2008). These findings suggest that CYP8A1 represents a divergent evolution of the P450 family, in which a heme barrier has evolved to ensure strict binding specificity for PGH_2, leading to a radical-mediated isomerization.

Polymorphisms of *CYP8A1* have been reported with a variety of tandem repeats (VNTR) with four to six tandem repeats of a 9-bp unit containing a putative Spl transcriptional factor binding site (Chevalier et al. 2001b). The gene has at least 13 alleles identified (http://www.cypalleles.ki.se/cyp8a1.htm). Three rare missense mutations (Pro38Leu, Ser118Arg, and Arg379Ser) have been reported in the coding sequence of the gene. An SNP in exon 8 (1135C>A → Arg379Ser) is linked to increased risk for myocardial infarction (Bousoula et al. 2012; Nakayama et al. 2002a). Hypertension and splicing variation leading to skipping of exon 9 have been linked together (Nakayama et al. 2002b). The VNTR polymorphisms in the promoter region of *CYP8A1* may result in decreased CYP8A1 expression, which is associated with severe pulmonary hypertension (Stearman et al. 2014; Tuder et al. 1999).

1.13.2 CYP8B1 (STEROL 12α-HYDROXYLASE)

The *CYP8B1* gene maps to chromosome 3p22.1 and it lacks introns (Gafvels et al. 1999). CYP8B1 (also called sterol 12α-hydroxylase) is responsible for catalyzing sterol 12α-hydroxylation in the liver (Ishida et al. 1999). CYP8B1 catalyzes the synthesis of cholic acid and controls the ratio of cholic acid to chenodeoxycholic acid in the bile (Zhang and Chiang 2001). Regulation of CYP8B1 is involved in bile acid feedback inhibition. Transcription of CYP8B1 is inhibited by bile acids, cholesterol, and insulin. The gene is activated by HNF-4α in HepG2 cells, which can be downregulated by bile acids and FXR. C-36, a derived peptide, can interact with the α_1-fetoprotein transcription factor site in the *CYP8B1* promoter, resulting in a conformation change and suppressing the transcription of the gene. HNF-4α can block the inhibitory effect of FTF and bile acids (Yang et al. 2002).

1.14 HUMAN CYP11 FAMILY

In mammals, cholesterol is the common precursor of all steroid hormones, and its conversion to pregnenolone is the initial and rate-limiting step in hormone biosynthesis in steroidogenic tissues such as gonads and adrenal glands (Norlin and Wikvall 2007; Pikuleva 2006). The production of glucocorticoids and mineralocorticoids occurs in the adrenal gland and the final steps are catalyzed by three mitochondrial CYPs, namely, CYP11A1 (cholesterol side-chain cleavage enzyme), 11B1 (steroid 11β-hydroxylase), and 11B2 (aldosterone synthase, steroid 11β/18-hydroxylase) (Maezawa et al. 2010; Miyakawa et al. 2000; Petrunak et al. 2014). The CYP11 members are important enzymes that participate in steroid biosynthesis and metabolism (Maezawa et al. 2010). *CYP11B1* shows close homology to the *CYP11B2* gene, which encodes aldosterone synthase and is normally expressed only in the zona glomerulosa. Both *CYP11B* genes map to chromosome 8q21, while CYP11A1 is located on chromosome 15q24.1 (Morohashi et al. 1987).

1.14.1 CYP11A1 (CHOLESTEROL SIDE-CHAIN CLEAVAGE ENZYME)

The *CYP11A1* gene contains nine exons and eight introns encoding a 521–amino acid protein with a molecular weight of 60 kDa (Morohashi et al. 1987). CYP11A1 (also called cholesterol side-chain cleavage enzyme) is responsible for the conversion of cholesterol to pregnenolone by side-chain cleavage, which initiates steroid synthesis. The reaction involves a three-step sequence, with generation of (22R)-20α, 22-dihydroxycholesterol as an intermediate (Tuckey and Cameron 1993). Oxidative cleavage of the diol to pregnenolone and 4-methylpentanal ends the overall reaction. The enzyme is expressed primarily in steroidogenic tissues including the adrenal cortex, gonads, ovary, brain, and pancreas (Morales et al. 1999). The protein is imported into the mitochondria without processing of the amino-terminal extension peptide. The mature form is produced in the mitochondria inner membrane after cleavage. The activity of CYP11A1 is limited by the supply of cholesterol in the inner membrane. Many inhibitors have been studied for the bovine CYP11A1 including acetylenic mechanism–based inactivators (Olakanmi and Seybert 1990). For human CYP11A1, the use of inhibitors can potentially become treatment for prostatic cancer. Anticonvulsants have been reported as weak CYP11A1 inhibitors (Ohnishi and Ichikawa 1997).

Corticotropin (ACTH) and cAMP can regulate CYP11A1. ACTH is a hormone that is released from the anterior pituitary in response to stress. In bovine, the two Sp1-binding sites in the *CYP11A1* gene control cAMP transcription though the protein kinase A signaling pathway. The steroidogenic factor 1 (SF1) activates *CYP11A1* transcription through interaction with protein factors upstream (Chung et al. 1997). An upstream CREB region and an AP-1 site are involved in cAMP response. The TATA box drives cell type–specific cAMP transcription. SF1 also interacts with Sp1. The expression of CYP11A1 is inhibited by the nuclear receptor DAX-1. All these factors are involved in the regulation of the *CYP11A1* gene.

Mutations of *CYP11A1* can lead to the rare disorder lipoid congenital adrenal hyperplasia (al Kandari et al. 2006; Hiort et al. 2005; Katsumata et al. 2002; Kim 2014; Kim et al. 2008; Rubtsov et al. 2009; Sahakitrungruang et al. 2011; Tajima et al. 2001). Affected patients typically present with signs of severe adrenal failure in early infancy and 46,XY genetic males are phenotypic females as a result of disrupted testicular androgen secretion. ACTH and plasma renin activity are elevated and adrenal steroids are inappropriately low or absent; the 46,XY patients have female external genitalia, sometimes with clitoromegaly. Notably, CYP11A1 autoantibodies are often detected in patients with autoimmune polyglandular syndromes type I and II (Seissler et al. 1999) and Addison's disease (Chen et al. 1996).

1.14.2 CYP11B1 (Steroid 11β-Hydroxylase)

The *CYP11B1* gene contains nine exons and spans 6.5 kb encoding a 503–amino acid protein with a molecular weight of 57.6 kDa. CYP11B1 (also called steroid 11β-hydroxylase) only catalyzes the 11β-hydroxylation of deoxycortisol to form cortisol, which is the main glucocorticoid in the body. In addition to 11β-hydroxylase activity, CYP11B1 also catalyzes 18- or 19-hydroxylation of steroids and the aromatization of androstenedione to estrone. Overproduction of glucocorticoids can lead to overactive CYP11B1 and is associated with Cushing's disease (Bureik et al. 2002). The protein is expressed in the adrenal cortex (Bureik et al. 2002). CYP11B1 is synthesized in the cytosol and transported into the mitochondria with a 24-residue-*N*-terminal targeting sequence, which is lost after entry. CYP11B1 receives electron from adrenodoxin instead of NADPH-P450 reductase.

The regulation of CYP11B1 involves ACTH and six *cis*-acting regulatory elements. ACTH could activate *CYP11B1* transcription by changing composition in AP-1 factors in rat

(Hashimoto et al. 1992; Mukai et al. 1998). The *CYP11B1* gene also contains a cAMP response element (CRE). CRE-binding protein interacts with an Ad1 element to activate activating transcription factor 1 (ATF-1) and ATF-2. SF1 binds to the Ad4 site, which is required for transcription (Bassett et al. 2002; Wang et al. 2000).

Mutations of *CYP11B1* include a 5'-base duplication (Skinner and Rumsby 1994), cluster of mutations in exons 6–8 (Curnow et al. 1993), and splice donor site mutations (Chabre et al. 2000). Defects in CYP11B1 are the cause of adrenal hyperplasia type 4, which is a form of congenital adrenal hyperplasia, a common recessive disease attributed to defective synthesis of cortisol. Congenital adrenal hyperplasia is characterized by androgen excess leading to ambiguous genitalia in affected females, rapid somatic growth during childhood in both sexes with premature closure of the epiphyses, and short adult stature. Patients with adrenal hyperplasia type 4 usually have hypertension. Defects in CYP11B1 are a cause of familial hyperaldosteronism type 1. It is a disorder characterized by hypertension, variable hyperaldosteronism, and abnormal adrenal steroid production, including 18-oxocortisol and 18-hydroxycortisol.

1.14.3 CYP11B2 (Aldosterone Synthase/ Steroid 11β/18-Hydroxylase)

CYP11B2 (also known as aldosterone synthase, previously known as corticosterone methyloxidase) has 503 amino acids with a molecular weight of 57.6 kDa. It is responsible for the three-step conversion from 11-deoxycorticosterone to aldosterone. The enzyme catalyzes the conversion of the 11β-hydroxylation of 11-deoxycorticosterone followed by 18-hydroxylation and two-electron oxidation of the 18-alcohol to an aldehyde. CYP11B2 is expressed in the adrenal cortex (zona glomerulosa). It is also a mitochondrial CYP as other CYP11 family. Regulation of *CYP11B2* can be overlapped with *CYP11B1*, which involves CRE/Ad1 element and ARF-1. However, SF1 does not involve the regulating process.

Mutations of *CYP11B2* can result in corticosterone methyloxidase deficiency and hyperaldosteronism (Mitsuuchi et al. 1993; Pascoe et al. 1992). Defect of CYP11B2 can cause corticosterone methyloxidase (CMO) deficiency types 1 and 2. The mutations lead to insufficient production of aldosterone, which impairs the kidneys' ability to reabsorb salt into the blood and release potassium in the urine. The resulting imbalance of ions in the body underlies the signs and symptoms of this disorder, which include nausea, vomiting, dehydration, low blood pressure, extreme tiredness, and muscle weakness. CMO-1 deficiency is also known as aldosterone deficiency owing to a defect in 18-hydroxylase or aldosterone deficiency I. CMO-1 deficiency is an autosomal recessive disorder of aldosterone biosynthesis. CMO-2 is an autosomal recessive disorder of aldosterone biosynthesis. In CMO-2 deficiency, aldosterone can be low or normal, but the secretion of 18-hydroxycorticosterone is increased. Consequently, patients have a greatly increased ratio of 18-hydroxycorticosterone to aldosterone and a low ratio of corticosterone to 18-hydroxycorticosterone in serum. Furthermore, polymorphisms of this gene have been linked to idiopathic hyperaldosteronism (Mulatero et al. 2000). A polymorphism in the promoter region of *CYP11B2* is linked to essential hypertension (Tsukada et al. 2002). A genetic change in the *CYP11B2* gene causes familial hyperaldosteronism type I owing to an anti-Lepore-type fusion of the *CYP11B1* and *11B2* genes, a disorder that leads to hypertension, variable hyperaldosteronism, and abnormal adrenal steroid production, including an increase in 18-oxocortisol and 18-hydroxycortisol. The hybrid gene has the promoting part of *CYP11B1*, ACTH-sensitive, and the coding part of *CYP11B2*.

1.15 HUMAN CYP17 FAMILY

1.15.1 CYP17A1 (Steroid 17α-Monooxygenase, 17α-Hydroxylase, 17,20 Lyase, and 17,20 Desmolase)

This gene is mapped to chromosome 10q24.3 and contains eight exons (Fan et al. 1992). It encodes a 508–amino acid protein with a molecular weight of 57.4 kDa. CYP17A1 (also called steroid 17α-monooxygenase, 17α-hydroxylase, 17,20 lyase, or 17,20 desmolase) is mainly expressed in steroidogenic tissues such as adrenals and gonads (Yoshimoto and Auchus 2014). The enzyme is also expressed in other organs including the heart and adipose tissue as well as in fetal kidney, thymus, and spleen. CYP17A1 is a microsomal enzyme that is responsible for the 17α-hydroxylation and the 17,20-lyase reactions. It is a key enzyme in the steroidogenic pathway that produces progestins, mineralocorticoids, glucocorticoids, androgens, and estrogens (Yoshimoto and Auchus 2014). CYP17A1 catalyzes 17α-hydroxylation of pregnenolone to form 17-hydroxypregnenolone. The enzyme also catalyzes the conversion of progesterone to 17-hydroxyprogesterone. CYP17A1 acts upon 17-hydroxyprogesterone and 17-hydroxypregnenolone to split the side chain off the steroid nucleus (Yoshimoto and Auchus 2014). CYP17A1 also plays a role in the 16-en-synthase reaction, which converts pregnenolone into 5,16-androstadien-3β-ol. Abiraterone is used to treat castration-resistant advanced prostate cancer that blocks the biosynthesis of androgens by inhibiting CYP17A1 (Yin and Hu 2014). The enantiomer of progesterone is a competitive inhibitor of CYP17A1 with a K_i of 0.2 μM. Another important inhibitor is ketoconazole, which inhibits the lyase activity of CYP17A1 but not the 17-hydroxylation activity. Two acidic residues, Glu48 and Glu49 in cytochrome b_5, are essential for stimulating the 17,20-lyase activity of CYP17A1 (Miyakawa et al. 2000). Substitution of Ala, Gly, Cys, or Gln for these two Glu residues abrogates the capacity to stimulate 17,20-lyase activity. Mutations Glu49Asp and Glu48Asp/Glu49/Asp retain 23% and 38% of wild-type activity, respectively. Residues including Lys88, Arg347, and Arg358/Arg449 of CYP17A1 interact with Glu61, Glu42, and Glu48/Glu49 of cytochrome b_5, respectively (Miyakawa et al. 2000).

The regulation of the *CYP17A1* gene is complex. Induction of CYP17A1 has been found to be cAMP mediated and testosterone can suppress the induction of the enzyme. It has been shown that the regulation of homeodomain protein Pbx1 and protein kinase A interaction at −250/−241 is cAMP dependent. The 5′-flanking region of the *CYP17A1* gene contains three functional SF1 elements that collectively mediate 25-fold or greater induction of promoter activity by SF1 (Hanley et al. 2001). In constructs containing all three functional SF1 elements, DAX1 inhibited this activation by at least 55%. In the presence of only one or two SF1 elements, DAX1 inhibition was lost although SF1 transactivation persisted.

To date, there are seven crystal structures of CYP17A1 in the PDB in complex with an inhibitor or substrate (4NKZ, 4NKY, 4NKX, 4NKW, 4NKV, 3SWZ, and 3RUK). The structures in complex with either abiraterone or TOK-001 (another CYP17A1 inhibitor) were first reported in 2012 (DeVore and Scott 2012b). Both of these inhibitors bind the heme iron, forming a 60° angle above the heme plane and packing against the central I helix with the 3β-OH interacting with Asn202 in the F helix. Similar binding modes are observed with CYP17A1 substrates such as pregnenolone, progesterone, and 17α-hydroxypregnenolone (Petrunak et al. 2014). This binding mode differs substantially from those that are predicted by homology models and from steroids in other CYP enzymes with known structures, and some features of this binding mode are more similar to steroid receptors (DeVore and Scott 2012b). The overall structures of CYP17A1 provide a rationale for understanding many mutations that are associated with steroidogenic diseases, and the active site reveals multiple steric and hydrogen-bonding features that will facilitate a better understanding of the enzyme's dual hydroxylase and lyase catalytic activities.

A number of mutations in *CYP17A1* are linked to 17α-hydroxylase deficiency, 17α-hydroxylase/17,20-lyase deficiency, pseudohermaphroditism, and adrenal hyperplasia type 5 (Dhir et al. 2009; Keskin et al. 2015; Kim et al. 2014; Krone and Arlt 2009; Rubtsov et al. 2015; Sahakitrungruang et al. 2009; Zhang et al. 2015). Lack of lyase activity can cause pseudohermaphroditism. Congenital adrenal hyperplasia is characterized by androgen excess leading to ambiguous genitalia in affected females, rapid somatic growth during childhood in both sexes with premature closure of the epiphyses, and short adult stature. SNPs in the coding region are the most common mutations observed in *CYP17A1*. Other mutations include a frameshift and premature stop codon attributed to a 2-bp deletion, changing the C-terminal 28 amino acids owing to a 4-bp duplication, and a 5′-splice site mutation. Some studies have reported that CYP17A1 loses its 17,20-lyase activity but not the 17-hydroxylation in patients presenting with symptoms owing to the change in Arg346Ala and Phe417Cys. The mutation at Thr306, which may involve protonation of Fe–OO− or O–O cleavage, affects the 17α-hydroxylation but not the lyase activity.

1.16 HUMAN CYP19 FAMILY

1.16.1 CYP19A1 (AROMATASE)

The *CYP19A1* gene maps to chromosome 15q21.1 and contains 13 exons that encode a 503–amino acid protein (Toda and Shizuta 1993) and a number of alternative noncoding first exons that regulate tissue-specific expression. CYP19A1 (also called aromatase) is localized to the ER and catalyzes the last steps of estrogen biosynthesis. It is responsible for oxidizing the androgens androstenedione and testosterone to the estrogens estrone and 17β-estradiol (Simpson et al. 2002). The enzyme also converts 16-hydroxyl DHEA to estriol. CYP19A1 is expressed in multiple human tissues including the ovaries, testes, placenta, fetal liver, adipose tissue, chondrocytes and osteoblasts of bone, vasculature smooth muscle, and brain (Simpson et al. 2002).

Inhibition of CYP19A1 has been of interest because of the enzyme role in estrogen regulation for treatment of estrogen-dependent cancers (Gobbi et al. 2014). The third-generation inhibitors that are in current clinical use include anastrozole, exemestane, and letrozole. Aromatase inhibitors are also beginning to be prescribed to men on testosterone replacement therapy as a way to keep estrogen levels from spiking once doses of testosterone are introduced to their systems. Exemestane is the only third-generation steroidal aromatase inhibitor and its efficacy as a first-line treatment in metastatic breast cancer has been demonstrated (Van Asten et al. 2014). Exemestane is mostly used as part of sequential adjuvant treatment after tamoxifen, but in this setting, it is also active in monotherapy.

The crystal structure of CYP19A1 has been determined and there are six structures of this enzyme in the PDB (4KQ8, 4GL7, 4GL5, 3S7S, 3S79, and 3EQM). The 2.90-Å-resolution crystal structure of aromatase purified from term human placenta in complex with androstenedione (androst-4-ene-3,17-dione) shows the characteristic CYP fold (Ghosh et al. 2009). Unlike the active sites of many microsomal CYPs that metabolize drugs and other xenobiotics, aromatase has an androgen-specific cleft that binds the androstenedione molecule. The volume of the binding pocket is <400 Å$_3$. The residues comprising the catalytic cleft include Ile305, Ala306, Asp309, and Thr310 from the I helix; Phe221 and Trp224 from the F helix; Ile133 and Phe134 from the B–C loop; Val370, Leu372, and Val373 from the K helix–β3 loop; Met374 from β3; and Leu477 and Ser478 from the β8–β9 loop. Hydrophobic and polar residues exquisitely complement the steroid backbone. The side chains of Arg115, Ile133, Phe134, Phe221, Trp224, Ala306, Thr310, Val370, Val373, Met374, and Leu477 have direct van der Waals contacts with androstenedione. Ile133, Phe134, Phe221, Trp224, and Leu477 approach the substrate from the α-face and follow the contour and puckering of the steroid backbone, whereas the side chains of Arg115, Ala306, and Met374 make contacts at its edge, and Thr310, Val370, and Val373, on the β-face (Ghosh et al. 2009). Replacing the active-site residue Asp309 with an Asn yields an inactive enzyme, consistent with its key role in aromatization (J. Lo et al. 2013). Spectroscopic studies have demonstrated a pH-dependent binding of androstenedione and exemestane to the enzyme, but aromatase binding to anastrozole is pH independent (Lytton et al. 2002). Mutation of Arg192 at the lipid interface, pivotal to the proton relay network in the access channel, results in the loss of enzyme activity. In addition to the distal catalytic residues, mutation of Lys440 and Tyr361 of the heme-proximal region critically interferes with substrate binding, enzyme activity, and heme stability (J. Lo et al. 2013). The D–E loop deletion mutant Del7 that disrupts the intermolecular interaction significantly reduces enzyme activity. The relative juxtaposition of the hydrophobic amino-terminal region and the opening to the catalytic cleft displays why membrane anchoring is necessary for the lipophilic substrates to gain access to the active site (Ghosh et al. 2009). The structural basis for the CYP19A1's androgenic specificity and unique catalytic mechanism can be used for developing next-generation aromatase inhibitors.

CYP19A1 is localized in the ER where it is regulated by tissue-specific promoters that are in turn controlled by hormones, cytokines, and other factors. The regulation of the *CYP19A1* gene is primarily dependent on the four-tissue promoters I.1, I.4, I.f, and I.6 (Simpson et al. 2002). The sequence is read as exon I and spliced into mRNA but no coding is involved. Therefore, the final product, CYP19A1, is always the same. The exon I.4 promoter is employed in adipose tissue, bone, skin, fetal liver, and leiomyoma tissue derived from the myometrium. The glucose regulatory factor Sp1, STAT3, and PPARγ are all involved in regulating this system. The strong promoter exon I.1, an 89-kb upstream element, is employed in the placenta. Exon I.6 is utilized to regulate the gene in bone with 1,25-dihydroxycholecalciferol, ILs, TNF-α, and TGF-β$_1$ as stimulators. Exon I.f is involved in the CYP19A1 in the brain but has not been studied extensively. Androgens do not affect the expression level of CYP19A1 in the brain. Exon IIa encodes the transcription of the 5′-UTR of the *CYP19A1* gene in preovulatory follicles and corpora lutea of human ovary (Simpson et al. 2002).

To date, four variants of *CYP19A1* have been reported (http://www.cypalleles.ki.se/cyp19a1.htm). The *2, *3, *4, and *5 carry Trp39Arg, Thr201Met, Arg264Cys, and Met364Thr, respectively. The Arg264Cys, Met364Thr, and double Trp39Arg/Arg264Cys mutants exhibit significant decreases in levels of activity and immunoreactive protein when compared with the wild-type enzyme in COS-1 cells (C.X. Ma et al. 2005). A slight decrease in protein level was observed for the Trp39Arg mutant, while Thr201Met has no significant changes in either activity or protein level. There was also a fourfold increase in apparent K_m value for Met364Thr toward androstenedione (C.X. Ma et al. 2005). Some studies have suggested the link between *CYP19A1* polymorphisms with breast cancer but the evidence is still conflicted and inconclusive (Kristensen et al. 2000; Surekha et al. 2014; Zhang et al. 2011).

1.17 HUMAN CYP20 FAMILY

1.17.1 CYP20A1

The *CYP20A1* gene maps to chromosome 2q33.2 and contains 13 exons. The *CYP20A1* gene is conserved in chimpanzee, rhesus monkey, dog, cow, mouse, rat, chicken, zebrafish, and frog. CYP20A1 is detected in human liver, and extrahepatic expression is observed in several human brain regions, including substantia nigra, hippocampus, and amygdala (Stark et al. 2008b). *CYP20A1* mRNA expression has also been found in rat brain, heart, and liver. The enzyme does not show any catalytic activity toward a number of potential steroids and biogenic amines. This protein lacks one amino acid of the conserved heme-binding site and the conserved I helix motif AGX(D,E)T, suggesting that its substrate may carry its own oxygen. As an orphan CYP, the substrates and functions of CYP20A1 are still unknown.

1.18 HUMAN CYP21 FAMILY

1.18.1 CYP21A2 (STEROID 21-HYDROXYLASE)

This gene maps to chromosome 6p21.3 and contains 10 exons. The *CYP21A1* gene is conserved in human, chimpanzee, rhesus monkey, dog, cow, rat, chicken, zebrafish, and frog. It shows relatively low homology with other members of the CYP superfamily. *CYP21A2* is not located in the large *CYP* gene clusters but resides in a multiallelic, complex, and tandem copy number variation of the major histocompatibility complex region (Horton et al. 2004). CYP21A2 is expressed primarily in the adrenal cortex, and low amounts have been observed in human lymphocytes and brain. This protein localizes to the ER and hydroxylates steroids at the 21-position. Its activity is required for the synthesis of steroid hormones including cortisol and aldosterone. CYP21A2 is an important enzyme that is required for glucocorticoid and mineralocorticoid synthesis (Speiser 2001; Wedell 2011). It is involved in converting 21-hydroxylation of progesterone and 17-hydroxyprogesterone into deoxycorticosterone and 11-deoxycortisol, respectively. The most recent study showed that the enantiomeric form of progesterone is a competitive inhibitor of CYP21A2 but not as effective as CYP17A1.

The protein expression level is regulated by ACTH in a similar way as CYP17A1 (Coulter and Jaffe 1998). Adrenal-specific protein factor and an Ad4-like sequence are used by the cAMP responsive sequence in the 5′-flanking region. The nonfunctional *CYP21A1P* pseudogene can compete for the transcription factors and other regulatory proteins with the *CYP21A2* gene (Szabo et al. 2013), which leads to some regulatory issues for the *CYP21A2* gene. The *C4* gene intron, spanning 35 kb, contains some of the transcriptional regulatory factors for the *CYP21A2* gene since it resides closely to the major histocompatibility locus, 2.3 kb downstream from the *C4* gene.

To date, more than 180 variants of *CYP21A2* have been reported (http://www.cypalleles.ki.se/cyp21a1.htm). Deficiencies in the 21-hydroxylation reaction often result in "salt-wasting syndrome." Mutations of *CYP21A2* cause congenital adrenal

hyperplasia because of 21-hydroxylase deficiency (Al-Agha et al. 2012; Baradaran-Heravi et al. 2007; Bleicken et al. 2009; Fluck and Pandey 2011; Goncalves et al. 2007; Grischuk et al. 2006; Haider et al. 2013; Krone and Arlt 2009; Massimi et al. 2014; Minutolo et al. 2011; Riepe et al. 2005; Speiser 2001; Witchel and Azziz 2011). A related pseudogene, *CYP21A1P*, is located near this gene (Araujo et al. 2007; Koppens et al. 2002; Tsai and Lee 2012; Wedell 2011); gene conversion events involving the functional gene and the pseudogene are thought to account for many cases of steroid 21-hydroxylase deficiency. Mutations of *CYP21A2* including IVS2-13A/C>G, Arg356Trp, and Arg149Pro are associated with congenital adrenal hyperplasia owing to 21-hydroxylase deficiency. Treatment of this disease involves administration of mineralocorticoids and glucocorticoids (Krone and Arlt 2009). Studies in mice have shown feasibility of transfecting active *CYP21A2* genes as a gene therapy treatment for the disease.

1.19 HUMAN CYP24 FAMILY

1.19.1 CYP24A1 (1,25-DIHYDROXYVITAMIN D₃ 24-HYDROXYLASE, VITAMIN D 24-HYDROXYLASE/ VITAMIN D₃ 24-HYDROXYLASE)

This gene maps to chromosome 20q13.2 and contains 10 exons encoding a protein with a relative molecular mass of 55 kDa (Chen et al. 1993; Hahn et al. 1993; Labuda et al. 1993). The *CYP24A1* gene is conserved in chimpanzee, rhesus monkey, dog, cow, mouse, rat, chicken, zebrafish, fruit fly, mosquito, *C. elegans*, and frog. CYP24A1 (also called 1,25-dihydroxyvitamin D_3 24-hydroxylase, or vitamin D 24-hydroxylase) involves deactivation of the active form of vitamin D through the C24-oxidation pathway (Prosser and Jones 2004; Sakaki et al. 2005; Tieu et al. 2014). The C24-oxidation pathway is initiated by hydroxylation at C24, resulting in $1,24,25(OH)_3D_3$. This is followed by oxidation of the 24-hydroxyl group, forming 24-oxo-$1,25(OH)_2D_3$ and hydroxylation at C23 giving 24-oxo-$1,23,25(OH)_3D_3$. This intermediate undergoes cleavage of the carbon–carbon bond between C23 and C24, with the cleavage product reported to be tetranor-$1,23(OH)_2D_3$. The precursor to $1,25(OH)_2D_3$, 25-$25(OH)$ D_3, is also a substrate of CYP24A1 and is metabolized through similar pathways. The enzyme catalyzes the 24-hydroxylation of $25(OH)D_3$ in the kidney mitochondrial membrane (Chen et al. 1993; Prosser and Jones 2004; Sakaki et al. 2005). In regulating the level of vitamin D_3, this enzyme plays a role in calcium homeostasis and the vitamin D endocrine system (Jones et al. 2014). The flavoprotein adrenodoxin reductase reaction transfers electrons to the proteins from the iron sulfur protein adrenodoxin. Electrons are derived from NADPH and transferred in sequence through NAPDH-adrenodoxin reductase. The specificity of this catalytic reaction is determined by substrate orientation within the enzyme's active site, where the target carbon atom must be in proximity to the heme-oxyferryl center (St-Arnaud 2011; Tieu et al. 2014). CYP24A1 can also catalyze other side-chain reactions such as the C23-hydroxylation pathway. The enzyme is present in both proximal and distal kidney tubules as well as human non-small cell lung carcinomas, keratinocytes, colon carcinomas, and prostatic cancer cells with low amounts (Omdahl et al. 2003). Unregulated expression level of CYP24A1 could lead to vitamin D deficiency. A recent study has found that overexpressing CYP24A1 had a low plasma level of $24,25(OH)_2D_3$ in transgenic rats (Kasuga et al. 2002). The rats also developed atherosclerotic lesions in the aorta as well as albuminuria and hyperlipidemia (Kasuga et al. 2002). These results indicate that CYP24A1 also involves other pathways besides vitamin D metabolism.

The crystal structure of CYP24A1 at 2.5 Å resolution has been resolved (3K9V) (Annalora et al. 2010). The hydrophobic amino acid residues surrounding the substrate access channel are associated with the hydrophobic domain of the inner mitochondrial membrane. This indicates that substrates enter the active site of CYP24A1 from the hydrophobic membrane domain. Phospholipid vesicles provide a membrane environment similar to the inner mitochondrial membrane and have been used extensively to characterize the properties of other mitochondrial CYPs, such as CYP11A1 (Headlam et al. 2003), 27A1 (Tieu et al. 2012), and 27B1 (Tang et al. 2012). Analysis of CYP24A1's proximal surface identifies the determinants of adrenodoxin recognition as a constellation of conserved residues from helices K, K″, and L that converge with an adjacent lysine-rich loop for binding the redox protein (Annalora et al. 2010).

The activity of CYP24A1 is predominantly controlled by levels of $1,25(OH)_2D_3$, serum calcium, and parathyroid hormone. The promoter region of *CYP24A1* contains a TATA box, a CAAT box, GC boxes, two vitamin D–responsive elements, and AP1- and AP2-binding sites (Chen and DeLuca 1995). The expression of CYP24A1 is induced by the level of vitamin D probably attributed to a VDR element in the 5′-region of the *CYP24A1* gene. CAR/PXR and VDR bind to and transactivate the same response elements in the promoter of *CYP24A1* (Moreau et al. 2007). The induction of the enzyme is also enhanced by the parathyroid hormone and cAMP via VDR-mediated pathways. The mRNA level of *CYP24A1* is induced by the presence of $1,25(OH)_2D_3$ and vitamin D–responsive elements. The antitumor effects of the vitamin D hormone may be reduced by the activity of CYP24A1 owing to the catabolic pathway regulation of $1,25(OH)_2D_3$ (St-Arnaud 2011). CYP21A4 can be induced by $1,25(OH)_2D_3$ via the Ets-1 phosphorylation at Thr38. However, the $1,25(OH)_2D_3$ stimulation of Erk1/2 requires RXRα phosphorylation on Ser260.

Several studies have shown the inversely proportional correlation between the growth inhibition induced by $1,25(OH)_2D_3$ and the level of *CYP24A1* or CYP24A1 activity (Anderson et al. 2006; Lou et al. 2004; Mimori et al. 2004; Parise et al. 2006). The expression level of *CYP24A1* has been reported to be elevated in human tumors including breast cancer (Albertson et al. 2000), esophageal cancer (Mimori et al. 2004), skin cancer (Mitschele et al. 2004; Reichrath et al. 2004), ovarian cancer (Anderson et al. 2006), colon adenocarcinomas (Anderson et al. 2006; Cross et al. 2005), and primary lung cancer (Anderson et al. 2006; Parise et al. 2006).

CYP24A1 mutations are associated with congenital hypercalcemia and nephrocalcinosis (Schlingmann et al. 2011). Patients with idiopathic infantile hypercalcemia have an exaggerated and prolonged increase in plasma levels of active $1,25(OH)_2D_3$ after receiving prophylactic vitamin D (Nguyen et al. 2010; Schlingmann et al. 2011). The physiologic importance of CYP24A1 in the catabolism of $1,25(OH)_2D_3$ and $25(OH)D_3$ has already been demonstrated in *Cyp24a1* knockout mice, which show severe hypercalcemia leading to perinatal death in approximately 50% of the mice (Masuda et al. 2005). Long-term vitamin D treatment in *Cyp24a1* knockout mice results in renal calcium deposition compatible with nephrocalcinosis. The administration of exogenous $1,25(OH)_2D_3$ and $25(OH)D_3$ to the knockout mice leads to a significant increase in $1,25(OH)_2D_3$ levels, indicating an inability to clear the active vitamin D hormone from the circulation (Masuda et al. 2005).

1.20 HUMAN CYP26 FAMILY

In the human genome, there are three members of the CYP26 family: 26A1, 26B1, and 26C1. These three members are all RA hydroxylases with similar substrate specificity but different tissue-specific expression patterns (Ross and Zolfaghari 2011; Thatcher et al. 2011). The CYP26 family members are primarily responsible for atRA clearance, resulting in the hydroxylated metabolites 4-OH-atRA, 18-OH-atRA, and perhaps 16-OH-atRA. CYP26A1 and 26B1 are qualitatively similar RA hydroxylases with overlapping expression profiles; CYP26A1 has higher catalytic activity for RA than CYP26B1 (Topletz et al. 2012). CYP26A1 and 26B1 metabolize atRA with less activity for 9-*cis*-RA (White et al. 2000). atRA and its major metabolites 4-OH-RA, 18-OH-RA, and 4-oxo-RA are all substrates of CYP26A1 and CYP26B1, and CYP26A1 has a 2- to 10-fold higher catalytic activity toward these substrates (Topletz et al. 2012). CYP26C1 metabolizes both atRA and 9-*cis*-RA (Taimi et al. 2004). Among these three genes, *CYP26A1* has been the most extensively studied in humans and mice (Pennimpede et al. 2010). CYP26A1 is the predominant atRA hydroxylase in the adult human liver. *Cyp26a1*-deficient mice show anomalies that include caudal agenesis, similar to those induced by administering excess RA, and in the tailbud, the *T* (brachyury) and *Wnt3a* genes are downregulated (Abu-Abed et al. 2001).

The intracellular processing of retinol involves lecithin retinol acyl transferase, which is responsible for the esterification of retinol (Jahng et al. 2003), while hydroxylation of retinol is catalyzed by the retinoic acid hydroxylases including CYP26A1, 26B1, and 26C1 (Topletz et al. 2012). Vitamin A (retinol) is metabolized in a two-step reaction into atRA, which is the most important bioactive metabolite. In the first step, vitamin A is reversibly oxidized to retinal by retinol dehydrogenases, while step 2 irreversibly transforms retinal into atRA via retinal dehydrogenases (Jones et al. 2014). atRA is a critical regulator of gene expression during embryonic development and in the maintenance of adult epithelial tissues. It also plays a role in regulation of apoptosis and immune system surveillance. atRA is a ligand for nuclear retinoic acid receptors (RARα, RARβ, or RARγ) and binds specifically to certain RA response elements (RAREs) throughout dimerization with retinoic X receptors (RXRα, RXRβ, or RXRγ), thus regulating gene expression during cell proliferation, embryonic development, and differentiation of adult epithelial tissues (Rhinn and Dolle 2012). CYP26A1 is essential for embryonic development, while CYP26B1 is essential for postnatal survival as well as germ cell development (Ross and Zolfaghari 2011). The CYP26 subfamily is important for the elimination of atRA and thereby crucial for the control of the atRA homeostasis in the cell.

1.20.1   CYP26A1 (Retinoic Acid 4-Hydroxylase)

CYP26A1 maps to chromosome 10q23-24 and has eight exons (White et al. 1998). The *CYP26A1* gene is conserved in chimpanzee, rhesus monkey, dog, cow, mouse, rat, chicken, zebrafish, *A. thaliana*, rice, and frog. Several disease-related loci including hand-split foot and infantile onset spinocerebellar atrophy have also been mapped to this region. In humans, a syndrome known as caudal regression syndrome is characterized by congenital sacral agenesis and involvement of the nervous, urinary, and gastrointestinal systems (De Marco et al. 2006). CYP26A1 (also called retinoic acid 4-hydroxylase) has been detected in different cell lines with different tissue origins including kidney, liver, breast, intestine, and lung. It is also expressed in fetal liver and brain tissue. Cyp26a1 is expressed early in the mouse embryo in a stage- and region-specific manner: in the neural plate, neural crest cells for the cranial ganglia, hindgut, and tailbud mesoderm (Fujii et al. 1997). atRA is converted to less active and more polar 4-OH, 4-oxo, 16-OH, and 18-OH metabolites by CYP26A1 (Thatcher et al. 2011), but the biological role of atRA metabolites is largely unknown. 9-*cis*-RA and 13-*cis*-RA are also substrates of CYP26A1. atRA, 4-OH-atRA, 18-OH-atRA, and 4-oxo-atRA are substrates of CYP26A1, forming a variety of biologically inactive diols and oxo-alcohols from these metabolites (Topletz et al. 2015). *Cyp26a1*-deficient mice die during mid-late gestation, with spina bifida, severe caudal agenesis, abnormalities of the kidneys and hindgut, and hindbrain (Abu-Abed et al. 2001; Fujii et al. 1997). These findings indicate that CYP26A1 protects fetal brain and other tissues from excess RA, which is teratogenic (Pennimpede et al. 2010).

Several imidazole- and triazole-containing compounds such as R116010 and R115866 are potent inhibitors of CYP26A1 with IC$_{50}$ values of 4.3 and 5.1 nM, respectively (Thatcher et al. 2011). Liarozole and ketoconazole are less potent inhibitors with IC$_{50}$ values of 2.1 and 0.550 μM, respectively. The RARγ agonist CD1530 is a potent inhibitor for CYP26A1 as ketoconazole with an IC$_{50}$ of 530 nM, whereas the RARα and RARβ agonists tested did not significantly inhibit CYP26A1 (Thatcher et al. 2011). The pan-RAR agonist 4-[(*E*)-2-(5,6,7,8-tetrahydro-5,5,8,8-tetramethyl-2-naphthalenyl)-1-propenyl]benzoic acid and the PPAR ligands rosiglitazone and pioglitazone inhibit CYP26A1 with IC$_{50}$

values of 3.7, 4.2, and 8.6 μM, respectively (Thatcher et al. 2011).

The *CYP26A1* gene is highly inducible by atRA via RAR- and PPAR-mediated pathways (Topletz et al. 2015; S. Yu et al. 2012; Zhang et al. 2010). This autoinduction not only allows for efficient feedback for atRA regulating its own clearance but also suggests that atRA would induce the formation of potentially active metabolites by CYP26A1. The biological activity of atRA is largely mediated by its binding to the nuclear ligand–activated RARs and PPARβ/δ (S. Yu et al. 2012). Analysis of a 2.2-kb 5′-flanking region upstream of the *CYP26A1* transcription start site has identified three conserved hexameric direct repeat-5 elements, RARE1, -2, and -3, and a half site, RARE4. atRA rapidly induces upregulation of *CYP26A1* mRNA expression in a dose-dependent manner in Caco-2, HepG2, HUVEC, and NB4 cells (Ozpolat et al. 2005). Other retinoids including retinol, 9-*cis*-RA, and 13-*cis*-RA also induce significant CYP26A1 expression in HepG2 and NB4 cells. RARα plays a major role in CYP26A1 expression in HepG2 cells. Treatment of HepG2 cells with atRA, 4*S*-OH-RA, 4*R*-OH-RA, 4-oxo-RA, and 18-OH-RA resulted in mRNAs of CYP26A1 and RARβ increasing 300- to 3000-fold, with 4-oxo-RA and atRA being the most potent inducers (Ozpolat et al. 2005). The presence of multiple RAREs may account for the remarkable inducibility of CYP26A1 in liver by atRA and its metabolites, which represents an important mechanism for restoring retinoid homeostasis when the concentration of RA rises.

A total of 13 SNPs have been identified in *CYP26A1*, with 3 of them being nonsynonymous: 517C>A (Arg173Ser), 558C>A (Phe186Leu), and 1072T>C (Cys358Arg) (S.J. Lee et al. 2007). These alleles have been named as *CYP26A1*2*, *3*, and *4*, respectively (http://www.cypalleles.ki.se/cyp26a1 .htm). Wild-type CYP26A1 protein metabolized atRA to less active 4-oxo-RA, 4-OH-RA, and water-soluble metabolites. CYP26A1.3 (F186L) and CYP26A1.4 (Cys358Arg) alloenzymes showed significantly lower (40%–80%) activity toward atRA (S.J. Lee et al. 2007).

CYP26A1 is associated with cancer development and survival, presenting oncogenic and pro-survival properties (Osanai et al. 2010). This enzyme is often overexpressed in various cancers (Brown et al. 2014; Osanai and Lee 2014). Nicotine significantly suppresses the constitutive expression of CYP26A1 in breast cancer cells, and cells treated with nicotine exhibit enhanced sensitivity to apoptosis (Osanai and Lee 2011). CYP26A1 may represent a therapeutic target in cancer treatment (C.H. Nelson et al. 2013).

1.20.2   CYP26B1

This gene maps to chromosome 2p13.2 and contains eight exons. The *CYP26B1* gene has 44% sequence identity with *CYP26A1*. The gene expression pattern is different from *CYP26A1* in mice and human fetal brain. In adult human brain, a higher mRNA level of *CYP26B1* is found in the cerebellum compared to *CYP26A1* (Trofimova-Griffin and Juchau

2002). The enzymatic characteristics of CYP26B1 are similar to CYP26A1 in which it catalyzes the formation of 4-OH-RA, 4-oxo-RA, and 18-OH-RA (Topletz et al. 2012). Knockout of the mouse Cyp26b1 gene causes embryonic lethality.

Mutations in this gene are associated with radiohumeral fusions and other skeletal and craniofacial anomalies, and increased levels of the encoded protein are associated with atherosclerotic lesions. Vascular cells express the spliced variant of CYP26B1 lacking exon 2 and it is also increased in atherosclerotic lesions (Elmabsout et al. 2012). The rs2241057 polymorphism encoding a Leu264Ser substitution of CYP26B1 results in enhanced metabolizing activity for atRA (Krivospitskaya et al. 2012). This SNP is associated with increased risk of atherosclerosis in mice and Crohn's disease in humans (Fransen et al. 2013). In one family, CYP26B1 Arg363Leu homozygosity causes embryonic lethality and is devoid of virtually all activity in vitro. In a second family, CYP26B1 Ser146Pro homozygosity results in an Antley–Bixler skeletal phenotype. atRA induces chemokine receptor 9 in mouse T cells in a Cyp26b1-dependent manner (Takeuchi et al. 2011), suggesting a link of CYP26B1 polymorphisms and inflammation.

1.20.3 CYP26C1

Like CYP26A1, this gene maps to chromosome 10q23.33. The CYP26C1 gene is conserved in chimpanzee, cow, mouse, rat, chicken, zebrafish, A. thaliana, and frog. Both CYP26C1 and 26A1 are present in the ER and share 43% amino acid identity, while CYP26C1 and 26B1 share 51% amino acid identity. Cyp26c1 differs from Cyp26a1 and Cyp26b1 in that it is less sensitive to the inductive effects of RA and relatively resistant to the inhibitory effects of ketoconazole CYP26C1 converting atRA to polar water-soluble metabolites similar to those generated by CYP26A1 and 26B1 (Taimi et al. 2004). atRA is the preferred substrate for CYP26C1. Although CYP26C1 shares sequence similarity with CYP26A1 and 26B1, its catalytic activity appears distinct from two other CYP26 family members. Specifically, CYP26C1 can metabolize 9-cis-RA and is much less sensitive than other CYP26 family members to the inhibitory effects of ketoconazole. CYP26C1 is not widely expressed in the adult but is inducible by atRA in human keratinocytes (Taimi et al. 2004). CYP26C1 differs from CYP26A1 and 26B1 in that it is less sensitive to the inductive effects of RA and relatively resistant to the inhibitory effects of ketoconazole.

Loss-of-function mutations in CYP26C1 cause focal facial dermal dysplasia type IV characterized by facial lesions resembling aplasia cutis in a preauricular distribution along the line of fusion of the maxillary and mandibular prominences (Slavotinek et al. 2013). The maternally inherited duplication 844_851dupCCATGCA results in a frameshift mutation that predicts the incorporation of 128 different amino acids from wild-type residue 284 and then premature truncation before residue 412. The duplication mutation occurs in a position 5′ to the heme thiolate domain and eliminates this domain and its catalytic activity. No patient with focal facial dermal dysplasia type II or III was found to have a CYP26C1 mutation (Slavotinek et al. 2013).

Cyp26c1 knockout mice do not manifest overt anatomic abnormalities (Uehara et al. 2007). However, the Cyp26a1/26c1 double knockout mice mimic the teratogenic defects seen with RA (Uehara et al. 2007), suggesting functional redundancy between these two genes. These mice have microcephaly, hypoplasia of the first and second pharyngeal arches, and deficient migration of the cranial neural crest cells with lethality.

1.21 HUMAN CYP27 FAMILY

This family contains three members: CYP27A1, 27B1, and 27C1. Both sterol 27-hydroxylase and 25-hydroxy-D_3 1α-hydroxylase are assigned to the CYP27 family since they share greater than 40% sequence identity, while sterol 27-hydroxylase is assigned to the A subfamily and 25-hydroxy-D_3 1α-hydroxylase to the B subfamily of CYP27 since their protein sequences are <55% identical. CYP27A1 and 27B1 catalyze the 25-hydroxylation of vitamin D_3 and the 1α-hydroxylation of 25(OH)D_3, respectively (Wikvall 2001). However, CYP27C1 does not metabolize vitamin D_3, 1α- or 25(OH)D_3, and cholesterol (Wu et al. 2006).

1.21.1 CYP27A1 (STEROL 27-HYDROXYLASE)

CYP27A1 (also called sterol 27-hydroxylase) is mainly expressed in the liver and located in mitochondria. CYP27A1 mRNA is also observed in macrophages (Gottfried et al. 2006), leukocytes (Shiga et al. 1999), skin fibroblasts (Garuti et al. 1997), kidney (Gascon-Barre et al. 2001), and the arterial wall (Shanahan et al. 2001). The enzyme catalyzes the 27-hydroxylation of cholesterol and 25-hydroxylation of vitamin D_3 (Jones et al. 2014; Norlin and Wikvall 2007; Pikuleva 2006; Pikuleva et al. 1997; Tieu et al. 2012; Wikvall 2001). Bioactivation of cholesterol into bile acids is crucial for regulation of cholesterol homeostasis. The "classic" pathway of bile acid formation starts with a 7α-hydroxylation of cholesterol by CYP7A1 in the liver, while the "acidic" pathway starts with a hepatic or extrahepatic 27-hydroxylation by CYP27A1 (Norlin and Wikvall 2007). Formation of cholic acid requires insertion of a 12α-hydroxyl group catalyzed by CYP8B1. Oxysterols are precursors to bile acids, participate in cholesterol transport, and are known to affect the expression of several genes in cholesterol homeostasis. CYP27A1 is attached to the inner mitochondrial membrane and substrates appear to reach the active site through the membrane phase. The distance between the hydroxylation site and the end of the site chain is proportional to the regioselectivity of the enzyme (Dilworth et al. 1995).

Macrophages can convert vitamin D_3 into the active metabolite 1,25(OH)$_2$$D_3$ by CYP27A1 (Gottfried et al. 2006). In macrophages, 27-hydroxycholesterol may inhibit the production of proinflammatory factors associated with cardiovascular diseases (Taylor et al. 2010). The expression of

CYP27A1 is upregulated in atherosclerosis, which suggests that CYP27A1 has a protective effect owing to the ability of 27-hydroxycholesterol to act as an endogenous ligand for LXR (Fu et al. 2001; Shanahan et al. 2001). To maintain cholesterol homeostasis, CYP27A1 expression is downregulated by immune complexes and IFN-γ in human aortic endothelial cells, peripheral blood mononuclear cells, and monocyte-derived macrophages (Reiss et al. 2001). TGF-β1 upregulates CYP27A1 in macrophages (Hansson et al. 2005). Induction of CYP27A1 by TGF-β1 may be responsible for some of the antiatherogenic activity of this cytokine. *CYP27A1* is under coupled regulation by retinoids and ligands of PPARs via a PPAR–retinoic acid receptor response element in its promoter, resulting in increased 27-hydroxycholesterol level and upregulation of LXR-mediated processes in macrophages (Szanto et al. 2004).

Mutations in the *CYP27A1* gene are known to be associated with cerebrotendinous xanthomatosis (CTX) (Bjorkhem 2013; Moghadasian 2004; Sundaram et al. 2008). CTX is a rare autosomal recessive lipid storage disorder in which cholestanol and cholesterol are accumulated in many tissues. CTX usually leads to tendon xanthoma, premature cataracts, juvenile atherosclerosis, and a progressive neurological syndrome involving mental retardation, cerebellar ataxia, pyramidal tract signs, myopathy, and peripheral neuropathy (Bjorkhem 2013; Moghadasian 2004; Sundaram et al. 2008). *CYP27A1* is also a susceptibility gene for sporadic amyotrophic lateral sclerosis, which is a neurodegenerative disease characterized by progressive muscle weakness caused by loss of central and peripheral motor neurons (Diekstra et al. 2012).

1.21.2 CYP27B1 (MITOCHONDRIAL 25-HYDROXYVITAMIN D 1α-HYDROXYLASE)

The *CYP27B1* gene maps to chromosome 12q14.1 and has nine exons spanning ~6.5 kb in length (Fu et al. 1997). The *CYP27B1* gene is conserved in chimpanzee, rhesus monkey, dog, cow, mouse, rat, zebrafish, and frog. The gene encodes a 508–amino acid protein. CYP27B1 (also called mitochondrial 25-hydroxyvitamin D 1α hydroxylase) localizes to the inner mitochondrial membrane where it hydroxylates 25(OH)D$_3$ at the 1α-position to form active 1α,25(OH)$_2$D$_3$ (Zehnder et al. 1999). This is the rate-limiting step in the bioactivation of vitamin D. CYP27B1 is present in the distal convoluted tubule, the cortical and medullary part of the collecting duct, and the papillary epithelia of kidney, and lower expression is found in the loop of Henle and Bowman's capsule, glomeruli, and vasculature structures (Zehnder et al. 1999). CYP27B1 is also expressed in skin, lymph nodes, colon (epithelial cells and parasympathetic ganglia), pancreas (islets), adrenal medulla, brain (cerebellum and cerebral cortex) (Zehnder et al. 2001), placenta (decidual and trophoblastic cells) (Diaz et al. 2002; Zehnder et al. 2001), cervix (Friedrich et al. 2002), and parathyroid glands (Segersten et al. 2002). 1α-Hydroxylase activity in extrarenal tissues is an important source of 1,25(OH)$_2$D$_3$ for

its autocrine and paracrine functions in bone, skin, macrophages, prostate, parathyroid, and several other tissues. The expression of CYP27B1 was increased in parathyroid adenomas but lost in carcinomas compared to normal tissues (Segersten et al. 2002). Studies have shown a decrease in the expression of CYP27B1 in nonsmall cell lung carcinomas (Jones et al. 1999), colon tumors (Cross et al. 2005; Ogunkolade et al. 2002; Tangpricha et al. 2001), and prostate cancer (Hsu et al. 2001).

CYP27B1 is negatively regulated by the concerted action of fibroblast growth factor-23 (FGF-23) and klotho (Chanakul et al. 2013; Kocelak et al. 2012; Krajisnik et al. 2007; Martin et al. 2012), a process that closely links vitamin D metabolism to phosphate homeostasis. MAPK signaling via MEK/ERK1/2 plays a critical role in the transcriptional regulation of *CYP27B1* by FGF-23 in the kidney. The regulatory region for FGF-23 lies within the first 200 bp of 5′-flanking DNA of *CYP27B1*. The parathyroid gland expresses klotho and is a target for FGF-23 action, which is to inhibit parathyroid hormone secretion and stimulate CYP27B1 expression (Ritter et al. 2012). The stimulation of CYP27B1 expression by FGF-23 in the parathyroid gland is in sharp contrast with its inhibitory effect on CYP27B1 expression in the kidney. Furthermore, calcium, parathyroid hormone, and 1α,25(OH)$_2$D$_3$ tightly regulate the expression of CYP27B1 in kidney and extrarenal tissues (Bland et al. 1999). The parathyroid hormone regulates CYP27B1 activity via a cAMP-mediated mechanism in the proximal convoluted tubule, whereas calcitonin regulates enzyme activity via a non–cAMP-mediated mechanism in the proximal straight tubule.

Mutations of *CYP27B1* result in autosomal recessive 1α-hydroxylase deficiency, also known as vitamin D–dependent rickets type I (Kitanaka et al. 2001). Wang et al. (1998) have documented 13 missense mutations such as Arg389His and Thr409Ile from patients with 1α-hydroxylase deficiency from 17 families representing various ethnic groups and found that none of them encode an active protein. The Leu343Phe and Glu189Gly mutants only retain 2% and 22% of wild-type activity, respectively. Some of the mutations are splicing defects (Porcu et al. 2002). *CYP27B1* −1260C>A (rs10877012) and +2838T>C (rs4646536) polymorphisms are associated with type 1 diabetes (Bailey et al. 2007). *Cyp27b1* knockout mice have showed skeletal, reproductive, and immune dysfunction besides rickets (Panda et al. 2001).

1.21.3 CYP27C1

This maps to chromosome 2q14.3 and contains 14 exons. The *CYP27C1* gene is conserved in chimpanzee, rhesus monkey, dog, cow, chicken, zebrafish, *Magnaporthe oryzae*, and frog. CYP27C1 mRNA is expressed in the liver, kidney, pancreas, and several other human tissues (Wu et al. 2006). In the presence of recombinant human adrenodoxin and adrenodoxin reductase, recombinant CYP27C1 does not catalyze the oxidation of vitamin D3, 1α- or 25-hydroxyvitamin D$_3$, and cholesterol. Further studies are needed to identify its substrate specificity, regulation, and polymorphisms.

1.22 HUMAN CYP39A1, 46A1, AND 51A1

1.22.1 CYP39A1 (Oxysterol 7α-Hydroxylase 2)

In liver, 7α-hydroxylation of cholesterols, mediated by CYP7A1 and 39A1, is the rate-limiting step of bile acid synthesis and metabolic elimination (Lathe 2002). In brain and other tissues, both sterols and some steroids including DHEA are prominently 7α-hydroxylated by CYP7B1. CYP39A1 is also known as oxysterol 7α-hydroxylase 2 (CYP7B1 is called oxysterol 7α-hydroxylase 1). The *CYP39A1* gene contain 14 exons and are located on chromosome 6p12.3 (Li-Hawkins et al. 2000). The enzyme is an ER protein involved in the conversion of cholesterol to bile acids. It catalyzes the 7α-hydroxylation of 24-hydroxycholesterol, 25-hydroxycholesterol, and 27-hydroxycholesterol (Li-Hawkins et al. 2000). CYP39A1 is a microsomal CYP enzyme that has preference for 24-hydroxycholesterol and is expressed in the liver (Li-Hawkins et al. 2000). The levels of hepatic CYP39A1 mRNA do not change in response to dietary cholesterol, bile acids, or a bile acid–binding resin, unlike those encoding other sterol 7α-hydroxylases (CYP7A1 and 7B1). In mice, hepatic Cyp39a1 expression is sexually dimorphic (female > male), which is opposite that of Cyp7b1 (male > female) (Li-Hawkins et al. 2000). A recent genome-wide association study have identified 16 loci that are associated with levels of 19 sterols and 25-hydroxylated derivatives of vitamin D, including *CYP39A1* (Stiles et al. 2014). Several pharmacogenetic studies have indicated that *CYP39A1* mutations change the clinical response to anticancer drugs (Aslibekyan et al. 2014; ten Brink et al. 2013; Uchiyama et al. 2012). CYP39A1 genotypes are found to be associated with busulfan clearance in pediatric patients undergoing hematopoietic stem cell transplantation (ten Brink et al. 2013). Mutation of this gene is associated with methotrexate-induced ADRs in patients with rheumatoid arthritis (Aslibekyan et al. 2014). Interestingly, one *CYP39A1* SNP (rs7761731, 972T>A → Asn324Lys) is found to be the only SNP significantly associated ($P = 0.049$; odds ratio = 9.0) with the incidence of grade 4 neutropenia induced by docetaxel in 42 Japanese patients with gynecological cancers such as ovarian cancer and endometrial cancer of the uterus (Uchiyama et al. 2012).

1.23 HUMAN CYP46 FAMILY

1.23.1 CYP46A1 (Cholesterol 24-Hydroxylase)

This gene contains 15 exons and is located on chromosome 14q32.1 (Lund et al. 1999). The *CYP46A1* gene is conserved in chimpanzee, rhesus monkey, dog, cow, mouse, rat, chicken, *M. oryzae*, *A. thaliana*, rice, and frog. The encoded protein contains 500 amino acids with a molecular weight of 56.8 kDa. CYP46A1 (also called cholesterol 24-hydroxylase) is predominantly expressed in the brain and converts cholesterol into 24S-hydroxycholesterol and, to a lesser extent, 25-hydroxycholesterol (Lund et al. 1999). The unique and conserved features of the enzyme implied that it plays important roles in the regulation of brain cholesterol metabolism. CYP46A1 resides in the ER. The LXRβ and CYP46A1 expression overlaps in the brain. 24-Hydroxycholesterol is a better substrate for CYP46A1 than cholesterol. The expression of CYP46A1 is inducible by epigenetic modifiers (Nunes et al. 2010; Shafaati et al. 2009). CYP46A is downregulated by the histone deacetylase inhibitor trichostatin A (Shafaati et al. 2009). In the 1.9-Å structure of CYP46A1 complexed with cholesterol 3-sulfate, the substrate is bound in the productive orientation and occupies the entire length of the banana-shaped hydrophobic active-site cavity (Mast et al. 2008). Structures of the co-complexes demonstrate that each drug binds in a single orientation to the active site of CYP46A1 with tranylcypromine, thioperamide, and voriconazole coordinating the heme iron via their nitrogen atoms and clotrimazole being at a 4-Å distance from the heme iron (Mast et al. 2010). A unique helix B′–C loop insertion containing residues 116–120 contributes to positioning cholesterol for oxygenation catalyzed by CYP46A1. In the structure of CYP46A1 in complex with posaconazole, the long antifungal molecule coordinates the P450 heme iron with the nitrogen atom of its terminal azole ring and adopts a linear configuration occupying the whole length of the substrate access channel and extending beyond the protein surface (Mast et al. 2013). Numerous drug–protein interactions determine the submicromolar K_d of posaconazole for CYP46A1.

Mutations of *CYP46A1* are associated with Alzheimer's disease. A T-C polymorphism located 151 bases 5′ to exon 3, *CYP46A1*TT*, is associated with increased risk of Alzheimer's disease (Papassotiropoulos et al. 2003). Neuropathologic examination of the patients and controls have shown that brain β-amyloid load and the levels of soluble β-amyloid-42 and phosphorylated tau in cerebrospinal fluid are significantly higher in subjects with the *CYP46A1*TT* genotype. It is unknown why *CYP46A1* polymorphisms can affect the development of Alzheimer's disease. It is speculated that functional change of cholesterol 24-hydroxylase may alter cholesterol concentrations in vulnerable neurons, thereby affecting amyloid precursor protein processing and β-amyloid production. In *Cyp46a1* knockout mice, hepatic cholesterol and bile acid metabolism remain unchanged compared to wild-type controls, but synthesis of new cholesterol in the brain is reduced by approximately 40%, despite steady-state levels of cholesterol being similar in the knockout mice (Lund et al. 2003).

1.24 HUMAN CYP51 FAMILY

1.24.1 CYP51A1 (Lanosterol 14α-Demethylase/ Sterol 14α-Demethylase)

Distribution of CYP51A1 is very widespread among living organisms. Genes encoding CYP51A1 (also called lanosterol 14α-demethylase/sterol 14α-demethylase) are found in yeast, plants, fungi, animals, and even prokaryotes, suggesting that this is among the oldest of the *CYP* genes. CYP51A1 is a common target of antifungal drugs (e.g., miconazole and ketoconazole), which inhibit CYP51A1 activity and formation of ergosterol. This gene has seven exons and maps to chromosome 7q21.2. It encodes a 503–amino acid protein that has a molecular weight of 56.8 kDa. CYP51A1 catalyzes the 14α-demethylation of lanosterol, an important intermediate in cholesterol synthesis. This demethylation step is the initial checkpoint in the transformation of lanosterol to other sterols. CYP51A1 is the most evolutionarily conserved member of the CYP superfamily (Lepesheva and Waterman 2011); it is conserved in chimpanzee, rhesus monkey, dog, cow, mouse, rat, chicken, zebrafish, *Saccharomyces cerevisiae*, *Kluyveromyces lactis*, *Eremothecium gossypii*, *Schizosaccharomyces pombe*, *M. oryzae*, *A. thaliana*, rice, and frog. CYP51A1 catalyzes a complex 14α-demethylation reaction with the aid of cytochrome P450 reductase. In humans, CYP51A1 converts lanosterol to 4,4-dimethyl-5α-cholesta-8,14,24-triene-3β-ol and 24,25-dihydrolanosterol to dihydro-4,4-dimethyl-5α-cholesta-8,14,24-triene-3β-ol. The enzyme mRNA expression levels are highest in testis (both round and elongated spermatids), ovary, adrenal, prostate, live, kidney, and lung (Rozman et al. 1996). CYP51A1 in mammals is also responsible for production of the follicular fluid meiosis-activating sterol. Together with the testis meiosis-activating sterol, the product of the downstream sterol Δ14-reductase reaction, these sterols are present at elevated levels in gonads and are involved in oocyte maturation and spermatogenesis (Keber et al. 2013). The *Cyp51* knockout mice show embryonic lethality at day 15 postcoitum with features similar to Antley–Bixler syndrome (Keber et al. 2011). Among mammalian P450s, human CYP51 is structurally similar to cholesterol 24-hydroxylase CYP46A1 with 20% sequence identity and CYP3A4 with 21% sequence identity.

To date, three structures of CYP51A1 (3JUV, 3JUS, and 3LD6) have been reported with or without complex with ketoconazole or econazole (Strushkevich et al. 2010). CYP51A1 has a typical P450 fold with a well-conserved structural core formed by helices E, I, J, K, and L around the heme. Ketoconazole binds to the protein mostly through hydrophobic interactions with Tyr131 and Leu134 (B′ helix), Tyr145 (B–C loop), Ala311 and Thr315 (I helix), and Ile377 (K helix–β1–4 loop). In both inhibitor-bound structures, the active-site residues Tyr131 and Tyr145 form hydrogen bonds to the heme propionates. The substantial conformational changes in the B′ helix and F–G loop regions are induced upon ligand binding, consistent with the membrane nature of the protein and its substrate (Strushkevich et al. 2010). In the B′ helix conformation of the inhibitor-bound structures, Leu134 and Thr135 point toward the active site wherein Leu134 makes a hydrophobic contact with ketoconazole and econazole. Trp239 of the F″ helix that interacts with Phe77 (A′ helix) in the ligand-free structure is repositioned to interact with Phe105 (β1–2 strand) and the distal portion of the ketoconazole in the azole-bound structure. The access channel is typical for mammalian sterol-metabolizing P450 enzymes but is different from that observed in *Mycobacterium tuberculosis* CYP51. The observed access channel formed by the F–G loop, A′ helix, and β4 loop is consistent with the idea that a nonpolar lanosterol enters the active site from the membrane without being exposed to cytosol.

The predominant transcription initiation sites of the *CYP51A1* gene are in the liver, lung, kidney, and placenta 250 and 249 bp upstream from the translation site. A second site is at −100 bp with the absence of TATA and CAAT regions and is a GC-rich region in the promoter sequence (Guengerich 2005; Rozman et al. 1996). A sterol/sterol-regulatory element binding protein-dependent pathway regulates the expression of *CYP51A1* in liver and other tissues. cAMP/cAMP-responsive element modulator regulates the expression of *CYP51A1* in testis (Halder et al. 2002). LXRα regulates cholesterol biosynthesis by directly silencing the expression of CYP51A1 and squalene synthase (Wang et al. 2008).

Mutations of *CYP51A1* are associated with pregnancy pathologies (Lewinska et al. 2013). There is an association of *CYP51A1* SNP rs6465348 CC genotype with lower total cholesterol and LDL-c during the second trimester of pregnancy. In a case study by Kelley et al. (2002), a patient with Antley–Bixler syndrome and ambiguous genitalia appears to have a defect in *CYP51A1*.

1.25 HIGHLIGHTS OF THIS BOOK

The human CYP superfamily contains 57 functional genes and 58 pseudogenes within 14 subfamilies. Three families of CYPs, namely, CYP1s, CYP2s, and CYP3s, are the main ones contributing to the oxidative metabolism of more than 90% of clinical drugs with CYPs in other families being involved in the metabolism of endogenous molecules such as steroids, fatty acids, bilirubin, and AAs. CYP2D6, one of the most investigated CYPs, accounts for only a small percentage of total hepatic CYP content (~2%–4%). However, it metabolizes ~25% of all medications in the human liver.

This book will discuss the structure, function, regulation, and polymorphism of CYP2D6 in depth to update our knowledge on this important enzyme. The book consists of eight chapters. This book is unique since it is the first book to discuss in detail an important CYP enzyme (including its structure, function, regulation, and polymorphism) that plays a very important role in individualized pharmacotherapy, personalized medicine, drug development, and toxicology. This book covers a hot topic about CYP2D6 that has drawn the attention of clinicians, pharmacists, pharmaceutical scientists, and pharmacologists. Since the book will update our knowledge about this single enzyme, it can be used as an important reference book by MD, PharmD, biomedical, and nursing students.

REFERENCES

Abel, J., Haarmann-Stemmann, T. 2010. An introduction to the molecular basics of aryl hydrocarbon receptor biology. *Biol Chem* 391:1235–1248.

Abelo, A., Andersson, T. B., Antonsson, M., Naudot, A. K., Skanberg, I., Weidolf, L. 2000. Stereoselective metabolism of omeprazole by human cytochrome P450 enzymes. *Drug Metab Dispos* 28:966–972.

Abernethy, D. R., Todd, E. L. 1985. Impairment of caffeine clearance by chronic use of low-dose oestrogen-containing oral contraceptives. *Eur J Clin Pharmacol* 28:425–428.

Abu-Abed, S., Dolle, P., Metzger, D., Beckett, B., Chambon, P., Petkovich, M. 2001. The retinoic acid-metabolizing enzyme, CYP26A1, is essential for normal hindbrain patterning, vertebral identity, and development of posterior structures. *Genes Dev* 15:226–240.

Abu-Bakar, A., Hakkola, J., Juvonen, R., Rahnasto-Rilla, M., Raunio, H., Lang, M. A. 2013. Function and regulation of the *Cyp2a5/CYP2A6* genes in response to toxic insults in the liver. *Curr Drug Metab* 14:137–150.

Acharya, M., Mookherjee, S., Bhattacharjee, A., Bandyopadhyay, A. K., Daulat Thakur, S. K., Bhaduri, G., Sen, A., Ray, K. 2006. Primary role of CYP1B1 in Indian juvenile-onset POAG patients. *Mol Vis* 12:399–404.

Afzal, S., Brondum-Jacobsen, P., Bojesen, S. E., Nordestgaard, B. G. 2014. Genetically low vitamin D concentrations and increased mortality: Mendelian randomisation analysis in three large cohorts. *BMJ* 349:g6330.

Agarwal, V., Kommaddi, R. P., Valli, K., Ryder, D., Hyde, T. M., Kleinman, J. E., Strobel, H. W., Ravindranath, V. 2008. Drug metabolism in human brain: High levels of cytochrome P4503A43 in brain and metabolism of anti-anxiety drug alprazolam to its active metabolite. *PLoS One* 3:e2337.

Agbor, L. N., Walsh, M. T., Boberg, J. R., Walker, M. K. 2012. Elevated blood pressure in cytochrome P4501A1 knockout mice is associated with reduced vasodilation to omega-3 polyunsaturated fatty acids. *Toxicol Appl Pharmacol* 264:351–360.

Ahn, T., Yun, C. H. 2010. Molecular mechanisms regulating the mitochondrial targeting of microsomal cytochrome P450 enzymes. *Curr Drug Metab* 11:830–838.

Aitken, A. E., Morgan, E. T. 2007. Gene-specific effects of inflammatory cytokines on cytochrome P450 2C, 2B6 and 3A4 mRNA levels in human hepatocytes. *Drug Metab Dispos* 35:1687–1693.

Aklillu, E., Carrillo, J. A., Makonnen, E., Hellman, K., Pitarque, M., Bertilsson, L., Ingelman-Sundberg, M. 2003. Genetic polymorphism of CYP1A2 in Ethiopians affecting induction and expression: Characterization of novel haplotypes with single-nucleotide polymorphisms in intron 1. *Mol Pharmacol* 64:659–669.

Al-Agha, A. E., Ocheltree, A. H., Al-Tamimi, M. D. 2012. Association between genotype, clinical presentation, and severity of congenital adrenal hyperplasia: A review. *Turk J Pediatr* 54:323–332.

al Kandari, H., Katsumata, N., Alexander, S., Rasoul, M. A. 2006. Homozygous mutation of P450 side-chain cleavage enzyme gene (CYP11A1) in 46, XY patient with adrenal insufficiency, complete sex reversal, and agenesis of corpus callosum. *J Clin Endocrinol Metab* 91:2821–2826.

Al Koudsi, N., Ahluwalia, J. S., Lin, S. K., Sellers, E. M., Tyndale, R. F. 2009. A novel *CYP2A6* allele (*CYP2A6*35*) resulting in an amino-acid substitution (Asn438Tyr) is associated with lower CYP2A6 activity *in vivo*. *Pharmacogenomics J* 9:274–282.

Al Mutair, A. N., Nasrat, G. H., Russell, D. W. 2012. Mutation of the CYP2R1 vitamin D 25-hydroxylase in a Saudi Arabian family with severe vitamin D deficiency. *J Clin Endocrinol Metab* 97:E2022–E2025.

Albertson, D. G., Ylstra, B., Segraves, R., Collins, C., Dairkee, S. H., Kowbel, D., Kuo, W. L., Gray, J. W., Pinkel, D. 2000. Quantitative mapping of amplicon structure by array CGH identifies CYP24 as a candidate oncogene. *Nat Genet* 25:144–146.

Allabi, A. C., Gala, J. L., Horsmans, Y., Babaoglu, M. O., Bozkurt, A., Heusterspreute, M., Yasar, U. 2004. Functional impact of *CYP2C9*5*, *CYP2C9*6*, *CYP2C9*8*, and *CYP2C9*11 in vivo* among black Africans. *Clin Pharmacol Ther* 76:113–118.

Allabi, A. C., Gala, J. L., Horsmans, Y. 2005. *CYP2C9, CYP2C19, ABCB1 (MDR1)* genetic polymorphisms and phenytoin metabolism in a Black Beninese population. *Pharmacogenet Genomics* 15:779–786.

Alsaad, A. M., Zordoky, B. N., Tse, M. M., El-Kadi, A. O. 2013. Role of cytochrome P450-mediated arachidonic acid metabolites in the pathogenesis of cardiac hypertrophy. *Drug Metab Rev* 45:173–195.

Amacher, D. E. 2010. The effects of cytochrome P450 induction by xenobiotics on endobiotic metabolism in pre-clinical safety studies. *Toxicol Mech Methods* 20:159–166.

Ambros, V. 2004. The functions of animal microRNAs. *Nature* 431:350–355.

Anandatheerthavarada, H. K., Sepuri, N. B., Avadhani, N. G. 2009. Mitochondrial targeting of cytochrome P450 proteins containing NH$_2$-terminal chimeric signals involves an unusual TOM20/TOM22 bypass mechanism. *J Biol Chem* 284:17352–17363.

Anderson, J. R., Nawarskas, J. J. 2001. Cardiovascular drug–drug interactions. *Cardiol Clin* 19:215–234.

Anderson, M. G., Nakane, M., Ruan, X., Kroeger, P. E., Wu-Wong, J. R. 2006. Expression of VDR and CYP24A1 mRNA in human tumors. *Cancer Chemother Pharmacol* 57:243–240.

Andersson, T., Holmberg, J., Rohss, K., Walan, A. 1998. Pharmacokinetics and effect on caffeine metabolism of the proton pump inhibitors, omeprazole, lansoprazole, and pantoprazole. *Br J Clin Pharmacol* 45:369–375.

Androutsopoulos, V. P., Tsatsakis, A. M., Spandidos, D. A. 2009. Cytochrome P450 CYP1A1: Wider roles in cancer progression and prevention. *BMC Cancer* 9:187.

Androutsopoulos, V. P., Papakyriakou, A., Vourloumis, D., Spandidos, D. A. 2011. Comparative CYP1A1 and CYP1B1 substrate and inhibitor profile of dietary flavonoids. *Bioorg Med Chem* 19:2842–2849.

Annalora, A. J., Goodin, D. B., Hong, W. X., Zhang, Q., Johnson, E. F., Stout, C. D. 2010. Crystal structure of CYP24A1, a mitochondrial cytochrome P450 involved in vitamin D metabolism. *J Mol Biol* 396:441–451.

Araujo, R. S., Mendonca, B. B., Barbosa, A. S., Lin, C. J., Marcondes, J. A., Billerbeck, A. E., Bachega, T. A. 2007. Microconversion between *CYP21A2* and *CYP21A1P* promoter regions causes the nonclassical form of 21-hydroxylase deficiency. *J Clin Endocrinol Metab* 92:4028–4034.

Araya, Z., Wikvall, K. 1999. 6α-Hydroxylation of taurochenodeoxycholic acid and lithocholic acid by CYP3A4 in human liver microsomes. *Biochim Biophys Acta* 1438:47–54.

Ariyoshi, N., Sawamura, Y., Kamataki, T. 2001. A novel single nucleotide polymorphism altering stability and activity of CYP2A6. *Biochem Biophys Res Commun* 281:810–814.

Arnold, C., Konkel, A., Fischer, R., Schunck, W. H. 2010. Cytochrome P450-dependent metabolism of omega-6 and omega-3 long-chain polyunsaturated fatty acids. *Pharmacol Rep* 62:536–547.

Arnoldi, A., Crimella, C., Tenderini, E., Martinuzzi, A., D'Angelo, M. G., Musumeci, O., Toscano, A. et al. 2012. Clinical phenotype variability in patients with hereditary spastic paraplegia type 5 associated with *CYP7B1* mutations. *Clin Genet* 81:150–157.

Aslibekyan, S., Brown, E. E., Reynolds, R. J., Redden, D. T., Morgan, S., Baggott, J. E., Sha, J. et al. 2014. Genetic variants associated with methotrexate efficacy and toxicity in early rheumatoid arthritis: Results from the treatment of early aggressive rheumatoid arthritis trial. *Pharmacogenomics J* 14:48–53.

Assenat, E., Gerbal-Chaloin, S., Larrey, D., Saric, J., Fabre, J. M., Maurel, P., Vilarem, M. J., Pascussi, J. M. 2004. Interleukin 1β inhibits CAR-induced expression of hepatic genes involved in drug and bilirubin clearance. *Hepatology* 40:951–960.

Avadhani, N. G., Sangar, M. C., Bansal, S., Bajpai, P. 2011. Bimodal targeting of cytochrome P450s to endoplasmic reticulum and mitochondria: The concept of chimeric signals. *FEBS J* 278:4218–4229.

Backman, J. T., Granfors, M. T., Neuvonen, P. J. 2006. Rifampicin is only a weak inducer of CYP1A2-mediated presystemic and systemic metabolism: Studies with tizanidine and caffeine. *Eur J Clin Pharmacol* 62:451–461.

Badawi, A. F., Cavalieri, E. L., Rogan, E. G. 2001. Role of human cytochrome P450 1A1, 1A2, 1B1, and 3A4 in the 2-, 4-, and 16alpha-hydroxylation of 17beta-estradiol. *Metabolism* 50:1001–1003.

Badeeb, O. M., Micheal, S., Koenekoop, R. K., den Hollander, A. I., Hedrawi, M. T. 2014. *CYP1B1* mutations in patients with primary congenital glaucoma from Saudi Arabia. *BMC Med Genet* 15:109.

Baer, B. R., DeLisle, R. K., Allen, A. 2009. Benzylic oxidation of gemfibrozil-1-*O*-β-glucuronide by P450 2C8 leads to heme alkylation and irreversible inhibition. *Chem Res Toxicol* 22:1298–1309.

Baer-Dubowska, W., Szaefer, H. 2013. Modulation of carcinogen-metabolizing cytochromes P450 by phytochemicals in humans. *Expert Opin Drug Metab Toxicol* 9:927–941.

Bahadur, N., Leathart, J. B., Mutch, E., Steimel-Crespi, D., Dunn, S. A., Gilissen, R., Houdt, J. V. et al. 2002. *CYP2C8* polymorphisms in Caucasians and their relationship with paclitaxel 6α-hydroxylase activity in human liver microsomes. *Biochem Pharmacol* 64:1579–1589.

Bailey, R., Cooper, J. D., Zeitels, L., Smyth, D. J., Yang, J. H., Walker, N. M., Hypponen, E. et al. 2007. Association of the vitamin D metabolism gene CYP27B1 with type 1 diabetes. *Diabetes* 56:2616–2621.

Bailon-Soto, C. E., Galaviz-Hernandez, C., Lazalde-Ramos, B. P., Hernandez-Velazquez, D., Salas-Pacheco, J., Lares-Assef, I., Sosa-Macias, M. 2014. Influence of *CYP1A1*2C* on high triglyceride levels in female Mexican indigenous Tarahumaras. *Arch Med Res* 45:409–416.

Bao, Z., He, X. Y., Ding, X., Prabhu, S., Hong, J. Y. 2005. Metabolism of nicotine and cotinine by human cytochrome P450 2A13. *Drug Metab Dispos* 33:258–261.

Baradaran-Heravi, A., Vakili, R., Robins, T., Carlsson, J., Ghaemi, N., A'Rabi, A., Abbaszadegan, M. R. 2007. Three novel CYP21A2 mutations and their protein modelling in patients with classical 21-hydroxylase deficiency from northeastern Iran. *Clin Endocrinol (Oxf)* 67:335–341.

Barouki, R., Aggerbeck, M., Aggerbeck, L., Coumoul, X. 2012. The aryl hydrocarbon receptor system. *Drug Metabol Drug Interact* 27:3–8.

Barry, E. L., Rees, J. R., Peacock, J. L., Mott, L. A., Amos, C. I., Bostick, R. M., Figueiredo, J. C. et al. 2014. Genetic variants in *CYP2R1*, *CYP24A1*, and *VDR* modify the efficacy of vitamin D₃ supplementation for increasing serum 25-hydroxyvitamin D levels in a randomized controlled trial. *J Clin Endocrinol Metab* 99:E2133–E2137.

Bartel, D. P. 2009. MicroRNAs: Target recognition and regulatory functions. *Cell* 136:215–233.

Bassett, M. H., Zhang, Y., Clyne, C., White, P. C., Rainey, W. E. 2002. Differential regulation of aldosterone synthase and 11beta-hydroxylase transcription by steroidogenic factor-1. *J Mol Endocrinol* 28:125–135.

Beaune, P., Dansette, P. M., Mansuy, D., Kiffel, L., Finck, M., Amar, C., Leroux, J. P., Homberg, J. C. 1987. Human anti-endoplasmic reticulum autoantibodies appearing in a drug-induced hepatitis are directed against a human liver cytochrome P-450 that hydroxylates the drug. *Proc Natl Acad Sci U S A* 84:551–555.

Bebenek, I. G., Solaimani, P., Bui, P., Hankinson, O. 2012. CYP2S1 is negatively regulated by corticosteroids in human cell lines. *Toxicol Lett* 209:30–34.

Becquemont, L., Ragueneau, I., Le Bot, M. A., Riche, C., Funck-Brentano, C., Jaillon, P. 1997. Influence of the CYP1A2 inhibitor fluvoxamine on tacrine pharmacokinetics in humans. *Clin Pharmacol Ther* 61:619–627.

Beischlag, T. V., Luis Morales, J., Hollingshead, B. D., Perdew, G. H. 2008. The aryl hydrocarbon receptor complex and the control of gene expression. *Crit Rev Eukaryot Gene Expr* 18:207–250.

Bell, J. C., Strobel, H. W. 2012. Regulation of cytochrome P450 4F11 by nuclear transcription factor-kappaB. *Drug Metab Dispos* 40:205–211.

Belliard, A. M., Baune, B., Fakhfakh, M., Hocquemiller, R., Farinotti, R. 2003. Determination of the human cytochrome P450s involved in the metabolism of 2n-propylquinoline. *Xenobiotica* 33:341–355.

Bellien, J., Joannides, R. 2013. Epoxyeicosatrienoic acid pathway in human health and diseases. *J Cardiovasc Pharmacol* 61:188–196.

Bernauer, U., Heinrich-Hirsch, B., Tonnies, M., Peter-Matthias, W., Gundert-Remy, U. 2006. Characterisation of the xenobiotic-metabolizing Cytochrome P450 expression pattern in human lung tissue by immunochemical and activity determination. *Toxicol Lett* 164:278–288.

Berthou, F., Dreano, Y., Belloc, C., Kangas, L., Gautier, J. C., Beaune, P. 1994. Involvement of cytochrome P450 3A enzyme family in the major metabolic pathways of toremifene in human liver microsomes. *Biochem Pharmacol* 47:1883–1895.

Bertilsson, L., Carrillo, J. A., Dahl, M. L., Llerena, A., Alm, C., Bondesson, U., Lindstrom, L., Rodriguez de la Rubia, I., Ramos, S., Benitez, J. 1994. Clozapine disposition covaries with CYP1A2 activity determined by a caffeine test. *Br J Clin Pharmacol* 38:471–473.

Bertilsson, G., Berkenstam, A., Blomquist, P. 2001. Functionally conserved xenobiotic responsive enhancer in cytochrome P450 3A7. *Biochem Biophys Res Commun* 280:139–144.

Bezemer, I. D., Bare, L. A., Doggen, C. J., Arellano, A. R., Tong, C., Rowland, C. M., Catanese, J. et al. 2008. Gene variants associated with deep vein thrombosis. *JAMA* 299:1306–1314.

Biancheri, R., Ciccolella, M., Rossi, A., Tessa, A., Cassandrini, D., Minetti, C., Santorelli, F. M. 2009. White matter lesions in spastic paraplegia with mutations in *SPG5/CYP7B1*. *Neuromuscul Disord* 19:62–65.

Bieche, I., Narjoz, C., Asselah, T., Vacher, S., Marcellin, P., Lidereau, R., Beaune, P., de Waziers, I. 2007. Reverse transcriptase-PCR quantification of mRNA levels from cytochrome (CYP)1, CYP2 and CYP3 families in 22 different human tissues. *Pharmacogenet Genomics* 17:731–742.

Bikle, D. D. 2014. Vitamin D metabolism, mechanism of action, and clinical applications. *Chem Biol* 21:319–329.

Bishop-Bailey, D., Thomson, S., Askari, A., Faulkner, A., Wheeler-Jones, C. 2014. Lipid-metabolizing CYPs in the regulation and dysregulation of metabolism. *Annu Rev Nutr* 34:261–279.

Bjorkhem, I. 2013. Cerebrotendinous xanthomatosis. *Curr Opin Lipidol* 24:283–287.

Blaisdell, J., Mohrenweiser, H., Jackson, J., Ferguson, S., Coulter, S., Chanas, B., Xi, T., Ghanayem, B., Goldstein, J. A. 2002. Identification and functional characterization of new potentially defective alleles of human *CYP2C19*. *Pharmacogenetics* 12:703–711.

Blanco, J. G., Edick, M. J., Hancock, M. L., Winick, N. J., Dervieux, T., Amylon, M. D., Bash, R. O. et al. 2002. Genetic polymorphisms in *CYP3A5*, *CYP3A4* and *NQO1* in children who developed therapy-related myeloid malignancies. *Pharmacogenetics* 12:605–611.

Bland, R., Walker, E. A., Hughes, S. V., Stewart, P. M., Hewison, M. 1999. Constitutive expression of 25-hydroxyvitamin D3-D1alpha-hydroxylase in a transformed human proximal tubule cell line: Evidence for direct regulation of vitamin D metabolism by calcium. *Endocrinology* 140:2027–2034.

Bleicken, C., Loidi, L., Dhir, V., Parajes, S., Quinteiro, C., Dominguez, F., Grotzinger, J. et al. 2009. Functional characterization of three CYP21A2 sequence variants (p.A265V, p.W302S, p.D322G) employing a yeast co-expression system. *Hum Mutat* 30:E443–E450.

Bock, K. W. 2014. Homeostatic control of xeno- and endobiotics in the drug-metabolizing enzyme system. *Biochem Pharmacol* 90:1–6.

Bodin, K., Lindbom, U., Diczfalusy, U. 2005. Novel pathways of bile acid metabolism involving CYP3A4. *Biochim Biophys Acta* 1687:84–93.

Bohets, H., Lavrijsen, K., Hendrickx, J., van Houdt, J., van Genechten, V., Verboven, P., Meuldermans, W., Heykants, J. 2000. Identification of the cytochrome P450 enzymes involved in the metabolism of cisapride: *In vitro* studies of potential co-medication interactions. *Br J Pharmacol* 129:1655–1667.

Bolton, J. L., Pisha, E., Zhang, F., Qiu, S. 1998. Role of quinoids in estrogen carcinogenesis. *Chem Res Toxicol* 11:1113–1127.

Bolwell, G. P., Bozak, K., Zimmerlin, A. 1994. Plant cytochrome P450. *Phytochemistry* 37:1491–1506.

Botsch, S., Gautier, J. C., Beaune, P., Eichelbaum, M., Kroemer, H. K. 1993. Identification and characterization of the cytochrome P450 enzymes involved in N-dealkylation of propafenone: Molecular base for interaction potential and variable disposition of active metabolites. *Mol Pharmacol* 43:120–126.

Bourdi, M., Chen, W., Peter, R. M., Martin, J. L., Buters, J. T., Nelson, S. D., Pohl, L. R. 1996. Human cytochrome P450 2E1 is a major autoantigen associated with halothane hepatitis. *Chem Res Toxicol* 9:1159–1166.

Bourrie, M., Meunier, V., Berger, Y., Fabre, G. 1996. Cytochrome P450 isoform inhibitors as a tool for the investigation of metabolic reactions catalyzed by human liver microsomes. *J Pharmacol Exp Ther* 277:321–332.

Bousoula, E., Kolovou, V., Vasiliadis, I., Karakosta, A., Xanthos, T., Johnson, E. O., Skandalakis, P., Kolovou, G. D. 2012. *CYP8A1* gene polymorphisms and left main coronary artery disease. *Angiology* 63:461–465.

Bousser, M. G., Amarenco, P., Chamorro, A., Fisher, M., Ford, I., Fox, K. M., Hennerici, M. G. et al., PERFORM Study Investigators. 2011. Terutroban versus aspirin in patients with cerebral ischaemic events (PERFORM): A randomised, double-blind, parallel-group trial. *Lancet* 377:2013–2022.

Boutanaev, A. M., Moses, T., Zi, J., Nelson, D. R., Mugford, S. T., Peters, R. J., Osbourn, A. 2015. Investigation of terpene diversification across multiple sequenced plant genomes. *Proc Natl Acad Sci U S A* 112:E81–E88.

Branden, G., Sjogren, T., Schnecke, V., Xue, Y. 2014. Structure-based ligand design to overcome CYP inhibition in drug discovery projects. *Drug Discov Today* 19:905–911.

Brandl, E. J., Chowdhury, N. I., Tiwari, A. K., Lett, T. A., Meltzer, H. Y., Kennedy, J. L., Muller, D. J. 2015. Genetic variation in *CYP3A43* is associated with response to antipsychotic medication. *J Neural Transm* 122:29–34.

Brash, A. R. 2009. Mechanistic aspects of CYP74 allene oxide synthases and related cytochrome P450 enzymes. *Phytochemistry* 70:1522–1531.

Breidenbach, T., Kliem, V., Burg, M., Radermacher, J., Hoffmann, M. W., Klempnauer, J. 2000. Profound drop of cyclosporin A whole blood trough levels caused by St. John's wort (*Hypericum perforatum*). *Transplantation* 69:2229–2230.

Breuer, O., Sudjana-Sugiaman, E., Eggertsen, G., Chiang, J. Y., Bjorkhem, I. 1993. Cholesterol 7 alpha-hydroxylase is upregulated by the competitive inhibitor 7-oxocholesterol in rat liver. *Eur J Biochem* 215:705–710.

Brooks, B. A., McBride, O. W., Dolphin, C. T., Farrall, M., Scambler, P. J., Gonzalez, F. J., Idle, J. R. 1988. The gene *CYP3* encoding P-450pcn1 (nifedipine oxidase) is tightly linked to the gene *COL1A2* encoding collagen type 1 alpha on 7q21-q22.1. *Am J Hum Genet* 43:280–284.

Brosen, K., Skjelbo, E., Rasmussen, B. B., Poulsen, H. E., Loft, S. 1993. Fluvoxamine is a potent inhibitor of cytochrome P4501A2. *Biochem Pharmacol* 45:1211–1214.

Brown, P. J., Bedard, L. L., Reid, K. R., Petsikas, D., Massey, T. E. 2007. Analysis of CYP2A contributions to metabolism of 4-(methylnitrosamino)-1-(3-pyridyl)-1-butanone in human peripheral lung microsomes. *Drug Metab Dispos* 35:2086–2094.

Brown, G. T., Cash, B. G., Blihoghe, D., Johansson, P., Alnabulsi, A., Murray, G. I. 2014. The expression and prognostic significance of retinoic acid metabolising enzymes in colorectal cancer. *PLoS One* 9:e90776.

Bui, P., Imaizumi, S., Beedanagari, S. R., Reddy, S. T., Hankinson, O. 2011. Human CYP2S1 metabolizes cyclooxygenase- and lipoxygenase-derived eicosanoids. *Drug Metab Dispos* 39:180–190.

Bureik, M., Lisurek, M., Bernhardt, R. 2002. The human steroid hydroxylases CYP1B1 and CYP11B2. *Biol Chem* 383:1537–1551.

Buters, J. T., Sakai, S., Richter, T., Pineau, T., Alexander, D. L., Savas, U., Doehmer, J., Ward, J. M., Jefcoate, C. R., Gonzalez, F. J. 1999. Cytochrome P450 CYP1B1 determines susceptibility to 7, 12-dimethylbenz[a]anthracene-induced lymphomas. *Proc Natl Acad Sci U S A* 96:1977–1982.

Bylund, J., Finnstrom, N., Oliw, E. H. 1999. Gene expression of a novel cytochrome P450 of the CYP4F subfamily in human seminal vesicles. *Biochem Biophys Res Commun* 261:169–174.

Bylund, J., Hidestrand, M., Ingelman-Sundberg, M., Oliw, E. H. 2000. Identification of CYP4F8 in human seminal vesicles as a prominent 19-hydroxylase of prostaglandin endoperoxides. *J Biol Chem* 275:21844–21849.

Bylund, J., Bylund, M., Oliw, E. H. 2001. cDna cloning and expression of CYP4F12, a novel human cytochrome P450. *Biochem Biophys Res Commun* 280:892–897.

Bylund, J., Zhang, C., Harder, D. R. 2002. Identification of a novel cytochrome P450, CYP4X1, with unique localization specific to the brain. *Biochem Biophys Res Commun* 296:677–684.

Cai, S. Y., He, H., Nguyen, T., Mennone, A., Boyer, J. L. 2010. Retinoic acid represses CYP7A1 expression in human hepatocytes and HepG2 cells by FXR/RXR-dependent and independent mechanisms. *J Lipid Res* 51:2265–2274.

Calinski, D., Zhang, H., Ludeman, S. M., Dolan, M. E., Hollenberg, P. F. 2015. Hydroxylation and *N*-dechloroethylation of ifosfamide and deuterated ifosfamide by the human cytochrome P450s and their commonly occurring polymorphisms. *Drug Metab Dispos* 43:1084–1090.

Cao, L., Fei, Q. Z., Tang, W. G., Liu, J. R., Zheng, L., Xiao, Q., He, S. B., Fu, Y., Chen, S. D. 2011. Novel mutations in the *CYP7B1* gene cause hereditary spastic paraplegia. *Mov Disord* 26:1354–1356.

Carr, B. A., Wan, J., Hines, R. N., Yost, G. S. 2003. Characterization of the human lung CYP2F1 gene and identification of a novel lung-specific binding motif. *J Biol Chem* 278:15473–15483.

Carriere, V., Goasduff, T., Ratanasavanh, D., Morel, F., Gautier, J. C., Guillouzo, A., Beaune, P., Berthou, F. 1993. Both cytochromes P450 2E1 and 1A1 are involved in the metabolism of chlorzoxazone. *Chem Res Toxicol* 6:852–857.

Carrillo, J. A., Christensen, M., Ramos, S. I., Alm, C., Dahl, M. L., Benitez, J., Bertilsson, L. 2000. Evaluation of caffeine as an in vivo probe for CYP1A2 using measurements in plasma, saliva, and urine. *Ther Drug Monit* 22:409–417.

Carthew, R. W., Sontheimer, E. J. 2009. Origins and mechanisms of miRNAs and siRNAs. *Cell* 136:642–655.

Cascorbi, I. 2003. Pharmacogenetics of cytochrome p4502D6: Genetic background and clinical implication. *Eur J Clin Invest* 33:17–22.

Cathcart, M. C., Reynolds, J. V., O'Byrne, K. J., Pidgeon, G. P. 2010. The role of prostacyclin synthase and thromboxane synthase signaling in the development and progression of cancer. *Biochim Biophys Acta* 1805:153–166.

Cauffiez, C., Klinzig, F., Rat, E., Tournel, G., Allorge, D., Chevalier, D., Pottier, N. et al. 2004a. Human *CYP4F12* genetic polymorphism: Identification and functional characterization of seven variant allozymes. *Biochem Pharmacol* 68:2417–2425.

Cauffiez, C., Lo-Guidice, J. M., Chevalier, D., Allorge, D., Hamdan, R., Lhermitte, M., Lafitte, J. J., Colombel, J. F., Libersa, C., Broly, F. 2004b. First report of a genetic polymorphism of the cytochrome P450 3A43 (*CYP3A43*) gene: Identification of a loss-of-function variant. *Hum Mutat* 23:101.

Cauffiez, C., Lo-Guidice, J. M., Quaranta, S., Allorge, D., Chevalier, D., Cenee, S., Hamdan, R. et al. 2004c. Genetic polymorphism of the human cytochrome CYP2A13 in a French population: Implication in lung cancer susceptibility. *Biochem Biophys Res Commun* 317:662–669.

Cauffiez, C., Pottier, N., Tournel, G., Lo-Guidice, J. M., Allorge, D., Chevalier, D., Migot-Nabias, F., Kenani, A., Broly, F. 2005. CYP2A13 genetic polymorphism in French Caucasian, Gabonese and Tunisian populations. *Xenobiotica* 35:661–669.

Chabre, O., Portrat-Doyen, S., Vivier, J., Morel, Y., Defaye, G. 2000. Two novel mutations in splice donor sites of CYP11B1 in congenital adrenal hyperplasia due to 11beta-hydroxylase deficiency. *Endocr Res* 26:797–801.

Chanakul, A., Zhang, M. Y., Louw, A., Armbrecht, H. J., Miller, W. L., Portale, A. A., Perwad, F. 2013. FGF-23 regulates CYP27B1 transcription in the kidney and in extra-renal tissues. *PLoS One* 8:e72816.

Chang, T. K., Weber, G. F., Crespi, C. L., Waxman, D. J. 1993. Differential activation of cyclophosphamide and ifosphamide by cytochromes P-450 2B and 3A in human liver microsomes. *Cancer Res* 53:5629–5637.

Chang, T. K., Yu, L., Maurel, P., Waxman, D. J. 1997. Enhanced cyclophosphamide and ifosfamide activation in primary human hepatocyte cultures: Response to cytochrome P-450 inducers and autoinduction by oxazaphosphorines. *Cancer Res* 57:1946–1954.

Chang, T. K., Bandiera, S. M., Chen, J. 2003. Constitutive androstane receptor and pregnane X receptor gene expression in human liver: Interindividual variability and correlation with CYP2B6 mRNA levels. *Drug Metab Dispos* 31:7–10.

Chase, M. B., Baek, S. J., Purtell, D. C., Schwartz, S., Shen, R. F. 1993. Mapping of the human thromboxane synthase gene (*TBXAS1*) to chromosome 7q34-q35 by two-color fluorescence in situ hybridization. *Genomics* 16:771–773.

Chen, K. S., DeLuca, H. F. 1995. Cloning of the human 1α,25-dihydroxyvitamin D-3 24-hydroxylase gene promoter and identification of two vitamin D-responsive elements. *Biochim Biophys Acta* 1263:1–9.

Chen, Y., Goldstein, J. A. 2009. The transcriptional regulation of the human CYP2C genes. *Curr Drug Metab* 10:567–578.

Chen, C., Wang, D. W. 2013. CYP epoxygenase derived EETs: From cardiovascular protection to human cancer therapy. *Curr Top Med Chem* 13:1454–1469.

Chen, Q., Wei, D. 2015. Human cytochrome P450 and personalized medicine. *Adv Exp Med Biol* 827:341–351.

Chen, K. S., Prahl, J. M., DeLuca, H. F. 1993. Isolation and expression of human 1,25-dihydroxyvitamin D_3 24-hydroxylase cDNA. *Proc Natl Acad Sci U S A* 90:4543–4547.

Chen, D., Lepar, G., Kemper, B. 1994. A transcriptional regulatory element common to a large family of hepatic cytochrome P450 genes is a functional binding site of the orphan receptor HNF-4. *J Biol Chem* 269:5420–5427.

Chen, S., Sawicka, J., Betterle, C., Powell, M., Prentice, L., Volpato, M., Rees Smith, B., Furmaniak, J. 1996. Autoantibodies to steroidogenic enzymes in autoimmune polyglandular syndrome, Addison's disease, and premature ovarian failure. *J Clin Endocrinol Metab* 81:1871–1876.

Chen, H., Fantel, A. G., Juchau, M. R. 2000. Catalysis of the 4-hydroxylation of retinoic acids by cyp3a7 in human fetal hepatic tissues. *Drug Metab Dispos* 28:1051–1057.

Chen, Y., Ferguson, S. S., Negishi, M., Goldstein, J. A. 2004. Induction of human CYP2C9 by rifampicin, hyperforin, and phenobarbital is mediated by the pregnane X receptor. *J Pharmacol Exp Ther* 308:495–501.

Chen, X., Wang, L., Zhi, L., Zhou, G., Wang, H., Zhang, X., Hao, B., Zhu, Y., Cheng, Z., He, F. 2005. The G-113A polymorphism in CYP1A2 affects the caffeine metabolic ratio in a Chinese population. *Clin Pharmacol Ther* 78:249–259.

Chen, C., Wei, X., Rao, X., Wu, J., Yang, S., Chen, F., Ma, D. et al. 2011. Cytochrome P450 2J2 is highly expressed in hematologic malignant diseases and promotes tumor cell growth. *J Pharmacol Exp Ther* 336:344–355.

Chen, F., Chen, C., Yang, S., Gong, W., Wang, Y., Cianflone, K., Tang, J., Wang, D. W. 2012. Let-7b inhibits human cancer phenotype by targeting cytochrome P450 epoxygenase 2J2. *PLoS One* 7:e39197.

Chen, W., Lee, M. K., Jefcoate, C., Kim, S. C., Chen, F., Yu, J. H. 2014. Fungal cytochrome p450 monooxygenases: Their distribution, structure, functions, family expansion, and evolutionary origin. *Genome Biol Evol* 6:1620–1634.

Chen, X., Chen, Y., Wang, L., Jiang, D., Wang, W., Xia, M., Yu, L., Sun, X. 2014. *CYP1B1* genotype influences the phenotype in primary congenital glaucoma and surgical treatment. *Br J Ophthalmol* 98:246–251.

Cheng, J. B., Motola, D. L., Mangelsdorf, D. J., Russell, D. W. 2003. De-orphanization of cytochrome P450 2R1: A microsomal vitamin D 25-hydroxylase. *J Biol Chem* 278:38084–38093.

Cheng, J. B., Levine, M. A., Bell, N. H., Mangelsdorf, D. J., Russell, D. W. 2004. Genetic evidence that the human CYP2R1 enzyme is a key vitamin D 25-hydroxylase. *Proc Natl Acad Sci U S A* 101:7711–7715.

Chevalier, D., Cauffiez, C., Allorge, D., Lo-Guidice, J. M., Lhermitte, M., Lafitte, J. J., Broly, F. 2001a. Five novel natural allelic variants-951A>C, 1042G>A (D348N), 1156A>T (I386F),

1217G>A (C406Y) and 1291C>T (C431Y)-of the human CYP1A2 gene in a French Caucasian population. *Hum Mutat* 17:355–356.

Chevalier, D., Cauffiez, C., Bernard, C., Lo-Guidice, J. M., Allorge, D., Fazio, F., Ferrari, N. et al. 2001b. Characterization of new mutations in the coding sequence and 5′-untranslated region of the human prostacylcin synthase gene (CYP8A1). *Hum Genet* 108:148–155.

Chiang, J. Y. 2009. Bile acids: Regulation of synthesis. *J Lipid Res* 50:1955–1966.

Chiang, J. Y., Kimmel, R., Weinberger, C., Stroup, D. 2000. Farnesoid X receptor responds to bile acids and represses cholesterol 7alpha-hydroxylase gene (CYP7A1) transcription. *J Biol Chem* 275:10918–10924.

Chiang, C. W., Yeh, H. C., Wang, L. H., Chan, N. L. 2006. Crystal structure of the human prostacyclin synthase. *J Mol Biol* 364:266–274.

Chiang, H. C., Wang, C. K., Tsou, T. C. 2012. Differential distribution of CYP2A6 and CYP2A13 in the human respiratory tract. *Respiration* 84:319–326.

Chiang, H. C., Lee, H., Chao, H. R., Chiou, Y. H., Tsou, T. C. 2013. Pulmonary CYP2A13 levels are associated with early occurrence of lung cancer—Its implication in mutagenesis of non-small cell lung carcinoma. *Cancer Epidemiol* 37:653–659.

Chida, M., Yokoi, T., Fukui, T., Kinoshita, M., Yokota, J., Kamataki, T. 1999. Detection of three genetic polymorphisms in the 5′-flanking region and intron 1 of human CYP1A2 in the Japanese population. *Jpn J Cancer Res* 90:899–902.

Ching, M. S., Bichara, N., Blake, C. L., Ghabrial, H., Tukey, R. H., Smallwood, R. A. 1996. Propranolol 4- and 5-hydroxylation and N-desisopropylation by cloned human cytochrome P4501A1 and P4501A2. *Drug Metab Dispos* 24:692–694.

Chinta, S. J., Pai, H. V., Upadhya, S. C., Boyd, M. R., Ravindranath, V. 2002. Constitutive expression and localization of the major drug metabolizing enzyme, cytochrome P450 2D in human brain. *Brain Res Mol Brain Res* 103:49–61.

Chinta, S. J., Kommaddi, R. P., Turman, C. M., Strobel, H. W., Ravindranath, V. 2005. Constitutive expression and localization of cytochrome P-450 1A1 in rat and human brain: Presence of a splice variant form in human brain. *J Neurochem* 93:724–736.

Chiou, W. L., Jeong, H. Y., Wu, T. C., Ma, C. 2001. Use of the erythromycin breath test for *in vivo* assessments of cytochrome P4503A activity and dosage individualization. *Clin Pharmacol Ther* 70:305–310.

Cho, J. Y., Yu, K. S., Jang, I. J., Yang, B. H., Shin, S. G., Yim, D. S. 2002. Omeprazole hydroxylation is inhibited by a single dose of moclobemide in homozygotic EM genotype for CYP2C19. *Br J Clin Pharmacol* 53:393–397.

Cho, T. M., Rose, R. L., Hodgson, E. 2006. *In vitro* metabolism of naphthalene by human liver microsomal cytochrome P450 enzymes. *Drug Metab Dispos* 34:176–183.

Choi, Y. M., An, S., Lee, E. M., Kim, K., Choi, S. J., Kim, J. S., Jang, H. H., An, I. S., Bae, S. 2012. CYP1A1 is a target of miR-892a-mediated post-transcriptional repression. *Int J Oncol* 41:331–336.

Choong, E., Guo, J., Persson, A., Virding, S., Johansson, I., Mkrtchian, S., Ingelman-Sundberg, M. 2015. Developmental regulation and induction of cytochrome P450 2W1, an enzyme expressed in colon tumors. *PLoS One* 10:e0122820.

Chou, F. C., Tzeng, S. J., Huang, J. D. 2001. Genetic polymorphism of cytochrome P450 3A5 in Chinese. *Drug Metab Dispos* 29:1205–1209.

Choudhary, D., Jansson, I., Stoilov, I., Sarfarazi, M., Schenkman, J. B. 2005. Expression patterns of mouse and human CYP orthologs (families 1–4) during development and in different adult tissues. *Arch Biochem Biophys* 436:50–61.

Chowbay, B., Cumaraswamy, S., Cheung, Y. B., Zhou, Q., Lee, E. J. 2003. Genetic polymorphisms in *MDR1* and *CYP3A4* genes in Asians and the influence of *MDR1* haplotypes on cyclosporin disposition in heart transplant recipients. *Pharmacogenetics* 13:89–95.

Christmas, P., Ursino, S. R., Fox, J. W., Soberman, R. J. 1999. Expression of the CYP4F3 gene. Tissue-specific splicing and alternative promoters generate high and low K(m) forms of leukotriene B(4) omega-hydroxylase. *J Biol Chem* 274:21191–21199.

Christmas, P., Jones, J. P., Patten, C. J., Rock, D. A., Zheng, Y., Cheng, S. M., Weber, B. M. et al. 2001. Alternative splicing determines the function of CYP4F3 by switching substrate specificity. *J Biol Chem* 276:38166–38172.

Chuang, S. S., Helvig, C., Taimi, M., Ramshaw, H. A., Collop, A. H., Amad, M., White, J. A., Petkovich, M., Jones, G., Korczak, B. 2004. CYP2U1, a novel human thymus- and brain-specific cytochrome P450, catalyzes ω- and ω-1-hydroxylation of fatty acids. *J Biol Chem* 279:6305–6314.

Chun, Y. J., Kim, M. Y., Guengerich, F. P. 1999. Resveratrol is a selective human cytochrome P450 1A1 inhibitor. *Biochem Biophys Res Commun* 262:20–24.

Chun, J., Kent, U. M., Moss, R. M., Sayre, L. M., Hollenberg, P. F. 2000. Mechanism-based inactivation of cytochromes P450 2B1 and P450 2B6 by 2-phenyl-2-(1-piperidinyl)propane. *Drug Metab Dispos* 28:905–911.

Chun, Y. J., Kim, S., Kim, D., Lee, S. K., Guengerich, F. P. 2001. A new selective and potent inhibitor of human cytochrome P450 1B1 and its application to antimutagenesis. *Cancer Res* 61:8164–8170.

Chung, B. C., Guo, I. C., Chou, S. J. 1997. Transcriptional regulation of the CYP11A1 and ferrodoxin genes. *Steriods* 62:37–42.

Citterio, A., Arnoldi, A., Panzeri, E., D'Angelo, M. G., Filosto, M., Dilena, R., Arrigoni, F. et al. 2014. Mutations in *CYP2U1*, *DDHD2* and *GBA2* genes are rare causes of complicated forms of hereditary spastic paraparesis. *J Neurol* 261:373–381.

Clot, P., Albano, E., Eliasson, E., Tabone, M., Arico, S., Israel, Y., Moncada, C., Ingelman-Sundberg, M. 1996. Cytochrome P4502E1 hydroxyethyl radical adducts as the major antigen in autoantibody formation among alcoholics. *Gastroenterology* 111:206–216.

Code, E. L., Crespi, C. L., Penman, B. W., Gonzalez, F. J., Chang, T. K., Waxman, D. J. 1997. Human cytochrome P4502B6: Interindividual hepatic expression, substrate specificity, and role in procarcinogen activation. *Drug Metab Dispos* 25:985–993.

Cohen, J. C., Cali, J. J., Jelinek, D. F., Mehrabian, M., Sparkes, R. S., Lusis, A. J., Russell, D. W., Hobbs, H. H. 1992. Cloning of the human cholesterol 7 α-hydroxylase gene (CYP7) and localization to chromosome 8q11-q12. *Genomics* 14:153–161.

Copley, S. D. 2012. Toward a systems biology perspective on enzyme evolution. *J Biol Chem* 287:3–10.

Corchero, J., Pimprale, S., Kimura, S., Gonzalez, F. J. 2001. Organization of the *CYP1A* cluster on human chromosome 15: Implications for gene regulation. *Pharmacogenetics* 11:1–6.

Corcos, L., Lucas, D., Le Jossic-Corcos, C., Dreano, Y., Simon, B., Plee-Gautier, E., Amet, Y., Salaun, J. P. 2012. Human cytochrome P450 4F3: Structure, functions, and prospects. *Drug Metabol Drug Interact* 27:63–71.

Coulter, C. L., Jaffe, R. B. 1998. Functional maturation of the primate fetal adrenal *in vivo*: 3. Specific zonal localization and developmental regulation of CYP21A2 (P450c21) and CYP11B1/CYP11B2 (P450c11/aldosterone synthase) lead to integrated concept of zonal and temporal steroid biosynthesis. *Endocrinology* 139:5144–5150.

Cresteil, T., Monsarrat, B., Dubois, J., Sonnier, M., Alvinerie, P., Gueritte, F. 2002. Regioselective metabolism of taxoids by human CYP3A4 and 2C8: Structure-activity relationship. *Drug Metab Dispos* 30:438–445.

Criscuolo, C., Filla, A., Coppola, G., Rinaldi, C., Carbone, R., Pinto, S., Wang, Q. et al. 2009. Two novel *CYP7B1* mutations in Italian families with SPG5: A clinical and genetic study. *J Neurol* 256:1252–1257.

Cross, H. S., Bises, G., Lechner, D., Manhardt, T., Kallay, E. 2005. The Vitamin D endocrine system of the gut—Its possible role in colorectal cancer prevention. *J Steroid Biochem Mol Biol* 97:121–128.

Cui, X., Nelson, D. R., Strobel, H. W. 2000. A novel human cytochrome P450 4F isoform (CYP4F11): cDNA cloning, expression, and genomic structural characterization. *Genomics* 68:161–166.

Cui, Y. L., Zhang, J. L., Zheng, Q. C., Niu, R. J., Xu, Y., Zhang, H. X., Sun, C. C. 2013. Structural and dynamic basis of human cytochrome P450 7B1: A survey of substrate selectivity and major active site access channels. *Chemistry* 19:549–557.

Curnow, K. M., Slutsker, L., Vitek, J., Cole, T., Speiser, P. W., New, M. I., White, P. C., Pascoe, L. 1993. Mutations in the CYP11B1 gene causing congenital adrenal hyperplasia and hypertension cluster in exons 6, 7, and 8. *Proc Natl Acad Sci U S A* 90:4552–4556.

Czerwinski, M., McLemore, T. L., Philpot, R. M., Nhamburo, P. T., Korzekwa, K., Gelboin, H. V., Gonzalez, F. J. 1991. Metabolic activation of 4-ipomeanol by complementary DNA-expressed human cytochromes P-450: Evidence for species-specific metabolism. *Cancer Res* 51:4636–4638.

D'Agostino, J., Zhang, X., Wu, H., Ling, G., Wang, S., Zhang, Q. Y., Liu, F., Ding, X. 2008. Characterization of *CYP2A13*2*, a variant cytochrome P450 allele previously found to be associated with decreased incidences of lung adenocarcinoma in smokers. *Drug Metab Dispos* 36:2316–2323.

D'Agostino, J., Zhuo, X., Shadid, M., Morgan, D. G., Zhang, X., Humphreys, W. G., Shu, Y. Z., Yost, G. S., Ding, X. 2009. The pneumotoxin 3-methylindole is a substrate and a mechanism-based inactivator of CYP2A13, a human cytochrome P450 enzyme preferentially expressed in the respiratory tract. *Drug Metab Dispos* 37:2018–2027.

Dai, D., Zeldin, D. C., Blaisdell, J. A., Chanas, B., Coulter, S. J., Ghanayem, B. I., Goldstein, J. A. 2001. Polymorphisms in human *CYP2C8* decrease metabolism of the anticancer drug paclitaxel and arachidonic acid. *Pharmacogenetics* 11:597–607.

Dai, D. P., Xu, R. A., Hu, L. M., Wang, S. H., Geng, P. W., Yang, J. F., Yang, L. P. et al. 2014. *CYP2C9* polymorphism analysis in Han Chinese populations: Building the largest allele frequency database. *Pharmacogenomics J* 14:85–92.

Dallas, S., Chattopadhyay, S., Sensenhauser, C., Batheja, A., Singer, M., Silva, J. 2013. Interleukins-12 and -23 do not alter expression or activity of multiple cytochrome P450 enzymes in cryopreserved human hepatocytes. *Drug Metab Dispos* 41:689–693.

Daly, A. K. 2012. Genetic polymorphisms affecting drug metabolism: Recent advances and clinical aspects. *Adv Pharmacol* 63:137–167.

Danese, E., Montagnana, M., Johnson, J. A., Rettie, A. E., Zambon, C. F., Lubitz, S. A., Suarez-Kurtz, G. et al. 2012. Impact of the *CYP4F2* p.V433M polymorphism on coumarin dose requirement: Systematic review and meta-analysis. *Clin Pharmacol Ther* 92:746–756.

Daneshtalab, N., Lewanczuk, R. Z., Russell, A. S., Jamali, F. 2006. Drug-disease interactions: Losartan effect is not downregulated by rheumatoid arthritis. *J Clin Pharmacol* 46:1344–1355.

Danielson, P. B. 2002. The cytochrome P450 superfamily: Biochemistry, evolution and drug metabolism in humans. *Curr Drug Metab* 3:561–597.

Dansette, P. M., Thang, D. C., el Amri, H., Mansuy, D. 1992. Evidence for thiophene-S-oxide as a primary reactive metabolite of thiophene in vivo: Formation of a dihydrothiophene sulfoxide mercapturic acid. *Biochem Biophys Res Commun* 186:1624–1630.

De Fabiani, E., Mitro, N., Anzulovich, A. C., Pinelli, A., Galli, G., Crestani, M. 2001. The negative effects of bile acids and tumor necrosis factor-alpha on the transcription of cholesterol 7alpha-hydroxylase gene (CYP7A1) converge to hepatic nuclear factor-4: A novel mechanism of feedback regulation of bile acid synthesis mediated by nuclear receptors. *J Biol Chem* 276:30708–30716.

De Marco, P., Merello, E., Mascelli, S., Raso, A., Santamaria, A., Ottaviano, C., Calevo, M. G., Cama, A., Capra, V. 2006. Mutational screening of the *CYP26A1* gene in patients with caudal regression syndrome. *Birth Defects Res A Clin Mol Teratol* 76:86–95.

De Morais, S. M., Wilkinson, G. R., Blaisdell, J., Meyer, U. A., Nakamura, K., Goldstein, J. A. 1994a. Identification of a new genetic defect responsible for the polymorphism of S-mephenytoin metabolism in Japanese. *Mol Pharmacol* 46:594–598.

De Morais, S. M., Wilkinson, G. R., Blaisdell, J., Nakamura, K., Meyer, U. A., Goldstein, J. A. 1994b. The major genetic defect responsible for the polymorphism of S-mephenytoin metabolism in humans. *J Biol Chem* 269:15419–15422.

de Waziers, I., Cugnenc, P. H., Yang, C. S., Leroux, J. P., Beaune, P. H. 1990. Cytochrome P450 isoenzymes, epoxide hydrolase and glutathione transferases in rat and human hepatic and extrahepatic tissues. *J Pharmacol Exp Ther* 253:387–394.

Deb, S., Bandiera, S. M. 2009. Characterization and expression of extrahepatic CYP2S1. *Expert Opin Drug Metab Toxicol* 5:367–380.

Degtyarenko, K. N., Kulikova, T. A. 2001. Evolution of bioinorganic motifs in P450-containing systems. *Biochem Soc Trans* 29:139–147.

Dejong, C. A., Wilson, J. Y. 2014. The cytochrome P450 superfamily complement (CYPome) in the annelid *Capitella teleta*. *PLoS One* 9:e107728.

Delaforge, M., Pruvost, A., Perrin, L., Andre, F. 2005. Cytochrome P450-mediated oxidation of glucuronide derivatives: Example of estradiol-17β-glucuronide oxidation to 2-hydroxy-estradiol-17β-glucuronide by CYP 2C8. *Drug Metab Dispos* 33:466–473.

DeLozier, T. C., Lee, S. C., Coulter, S. J., Goh, B. C., Goldstein, J. A. 2005. Functional characterization of novel allelic variants of *CYP2C9* recently discovered in southeast Asians. *J Pharmacol Exp Ther* 315:1085–1090.

DeLozier, T. C., Kissling, G. E., Coulter, S. J., Dai, D., Foley, J. F., Bradbury, J. A., Murphy, E., Steenbergen, C., Zeldin, D. C., Goldstein, J. A. 2007. Detection of human CYP2C8, CYP2C9, and CYP2J2 in cardiovascular tissues. *Drug Metab Dispos* 35:682–688.

Denison, M. S., Pandini, A., Nagy, S. R., Baldwin, E. P., Bonati, L. 2002. Ligand binding and activation of the Ah receptor. *Chem Biol Interact* 141:3–24.

Desta, Z., Soukhova, N., Morocho, A. M., Flockhart, D. A. 2001. Stereoselective metabolism of cisapride and enantiomer-enantiomer interaction in human cytochrome P450 enzymes: Major role of CYP3A. *J Pharmacol Exp Ther* 298:508–520.

Desta, Z., Zhao, X., Shin, J. G., Flockhart, D. A. 2002. Clinical significance of the cytochrome P450 2C19 genetic polymorphism. *Clin Pharmacokinet* 41:913–958.

DeVore, N. M., Scott, E. E. 2012a. Nicotine and 4-(methylnitrosamino)-1-(3-pyridyl)-1-butanone binding and access channel in human cytochrome P450 2A6 and 2A13 enzymes. *J Biol Chem* 287:26576–26585.

DeVore, N. M., Scott, E. E. 2012b. Structures of cytochrome P450 17A1 with prostate cancer drugs abiraterone and TOK-001. *Nature* 482:116–119.

DeVore, N. M., Smith, B. D., Urban, M. J., Scott, E. E. 2008. Key residues controlling phenacetin metabolism by human cytochrome P450 2A enzymes. *Drug Metab Dispos* 36:2582–2590.

DeVore, N. M., Smith, B. D., Wang, J. L., Lushington, G. H., Scott, E. E. 2009. Key residues controlling binding of diverse ligands to human cytochrome P450 2A enzymes. *Drug Metab Dispos* 37:1319–1327.

DeVore, N. M., Meneely, K. M., Bart, A. G., Stephens, E. S., Battaile, K. P., Scott, E. E. 2012. Structural comparison of cytochromes P450 2A6, 2A13, and 2E1 with pilocarpine. *FEBS J* 279:1621–1631.

Dhar, M., Sepkovic, D. W., Hirani, V., Magnusson, R. P., Lasker, J. M. 2008. Omega oxidation of 3-hydroxy fatty acids by the human *CYP4F* gene subfamily enzyme CYP4F11. *J Lipid Res* 49:612–624.

Dhir, V., Reisch, N., Bleicken, C. M., Lebl, J., Kamrath, C., Schwarz, H. P., Grotzinger, J. et al. 2009. Steroid 17α-hydroxylase deficiency: Functional characterization of four mutations (A174E, V178D, R440C, L465P) in the *CYP17A1* gene. *J Clin Endocrinol Metab* 94:3058–3064.

Di, Y. M., Chow, V. D., Yang, L. P., Zhou, S. F. 2009. Structure, function, regulation and polymorphism of human cytochrome P450 2A6. *Curr Drug Metab* 10:754–780.

Diaz, L., Arranz, C., Avila, E., Halhali, A., Vilchis, F., Larrea, F. 2002. Expression and activity of 25-hydroxyvitamin D-1α-hydroxylase are restricted in cultures if human syncytiotrophoblast cells from preeclamptic pregancies. *J Clin Endocrinol Metab* 87:3876–3882.

Dickins, M. 2004. Induction of cytochromes P450. *Curr Top Med Chem* 4:1745–1766.

Dickmann, L. J., Rettie, A. E., Kneller, M. B., Kim, R. B., Wood, A. J., Stein, C. M., Wilkinson, G. R., Schwarz, U. I. 2001. Identification and functional characterization of a new *CYP2C9* variant (*CYP2C9*5*) expressed among African Americans. *Mol Pharmacol* 60:382–387.

Diczfalusy, U., Kanebratt, K. P., Bredberg, E., Andersson, T. B., Bottiger, Y., Bertilsson, L. 2009. 4β-Hydroxycholesterol as an endogenous marker for CYP3A4/5 activity. Stability and half-life of elimination after induction with rifampicin. *Br J Clin Pharmacol* 67:38–43.

Diekstra, F. P., Saris, C. G., van Rheenen, W., Franke, L., Jansen, R. C., van Es, M. A., van Vught, P. W. et al. 2012. Mapping of gene expression reveals *CYP27A1* as a susceptibility gene for sporadic ALS. *PLoS One* 7:e35333.

Dierks, E. A., Zhang, Z., Johnson, E. F. de Montellano, P. R. 1998. The catalytic site of cytochrome P4504A11 (CYP4A11) and its L131F mutant. *J Biol Chem* 273:23055–23061.

Dilworth, F. J., Scott, I., Green, A., Strugnell, S., Guo, Y. D., Roberts, E. A., Kremer, R., Calverley, M. J., Makin, H. L., Jones, G. 1995. Different mechanisms of hydroxylation site selection by liver and kidney cytochrome P450 species (CYP27 and CYP24) involved in vitamin D metabolism. *J Biol Chem* 270:16766–16774.

Ding, X., Kaminsky, L. S. 2003. Human extrahepatic cytochromes P450: Function in xenobiotic metabolism and tissue-selective chemical toxicity in the respiratory and gastrointestinal tracts. *Annu Rev Pharmacol Toxicol* 43:149–173.

Ding, S., Lake, B. G., Friedberg, T., Wolf, C. R. 1995. Expression and alternative splicing of the cytochrome P-450 CYP2A7. *Biochem J* 306 (Pt 1):161–166.

Dohnal, V., Wu, Q., Kuca, K. 2014. Metabolism of aflatoxins: Key enzymes and interindividual as well as interspecies differences. *Arch Toxicol* 88:1635–1644.

Domanski, T. L., Schultz, K. M., Roussel, F., Stevens, J. C., Halpert, J. R. 1999. Structure-function analysis of human cytochrome P450 2B6 using a novel substrate, site-directed mutagenesis, and molecular modeling. *J Pharmacol Exp Ther* 290:1141–1147.

Domanski, T. L., Finta, C., Halpert, J. R., Zaphiropoulos, P. G. 2001a. cDNA cloning and initial characterization of CYP3A43, a novel human cytochrome P450. *Mol Pharmacol* 59:386–392.

Domanski, T. L., He, Y. A., Khan, K. K., Roussel, F., Wang, Q., Halpert, J. R. 2001b. Phenylalanine and tryptophan scanning mutagenesis of CYP3A4 substrate recognition site residues and effect on substrate oxidation and cooperativity. *Biochemistry* 40:10150–10160.

Drahushuk, A. T., McGarrigle, B. P., Larsen, K. E., Stegeman, J. J., Olson, J. R. 1998. Detection of CYP1A1 protein in human liver and induction by TCDD in precision-cut liver slices incubated in dynamic organ culture. *Carcinogenesis* 19:1361–1368.

Draper, A. J., Madan, A., Parkinson, A. 1997. Inhibition of coumarin 7-hydroxylase activity in human liver microsomes. *Arch Biochem Biophys* 341:47–61.

Dresser, G. K., Spence, J. D., Bailey, D. G. 2000. Pharmacokinetic-pharmacodynamic consequences and clinical relevance of cytochrome P450 3A4 inhibition. *Clin Pharmacokinet* 38:41–57.

Ducassou, L., Jonasson, G., Dhers, L., Pietrancosta, N., Ramassamy, B., Xu-Li, Y., Loriot, M. A. et al. 2015. Expression in yeast, new substrates, and construction of a first 3D model of human orphan cytochrome P450 2U1: Interpretation of substrate hydroxylation regioselectivity from docking studies. *Biochim Biophys Acta* 1850:1426–1437.

Duescher, R. J., Elfarra, A. A. 1994. Human liver microsomes are efficient catalysts of 1,3-butadiene oxidation: Evidence for major roles by cytochromes P450 2A6 and 2E1. *Arch Biochem Biophys* 311:342–349.

Durr, D., Stieger, B., Kullak-Ublick, G. A., Rentsch, K. M., Steinert, H. C., Meier, P. J., Fattinger, K. 2000. St John's wort induces intestinal P-glycoprotein/MDR1 and intestinal and hepatic CYP3A4. *Clin Pharmacol Ther* 68:598–604.

Eagling, V. A., Tjia, J. F., Back, D. J. 1998. Differential selectivity of cytochrome P450 inhibitors against probe substrates in human and rat liver microsomes. *Br J Clin Pharmacol* 45:107–114.

Eaton, D. L., Gallagher, E. P., Bammler, T. K., Kunze, K. L. 1995. Role of cytochrome P4501A2 in chemical carcinogenesis: Implications for human variability in expression and enzyme activity. *Pharmacogenetics* 5:259–274.

Edler, D., Stenstedt, K., Ohrling, K., Hallstrom, M., Karlgren, M., Ingelman-Sundberg, M., Ragnhammar, P. 2009. The expression of the novel CYP2W1 enzyme is an independent prognostic factor in colorectal cancer—A pilot study. *Eur J Cancer* 45:705–712.

Edson, K. Z., Prasad, B., Unadkat, J. D., Suhara, Y., Okano, T., Guengerich, F. P., Rettie, A. E. 2013. Cytochrome P450-dependent catabolism of vitamin K: ω-hydroxylation catalyzed by human CYP4F2 and CYP4F11. *Biochemistry* 52:8276–8285.

Eichelbaum, M., Baur, M. P., Dengler, H. J., Osikowska-Evers, B. O., Tieves, G., Zekorn, C., Rittner, C. 1987. Chromosomal assignment of human cytochrome P-450 (debrisoquine/sparteine type) to chromosome 22. *Br J Clin Pharmacol* 23:455–458.

Eiselt, R., Domanski, T. L., Zibat, A., Mueller, R., Presecan-Siedel, E., Hustert, E., Zanger, U. M. et al. 2001. Identification and functional characterization of eight CYP3A4 protein variants. *Pharmacogenetics* 11:447–458.

Ekhart, C., Doodeman, V. D., Rodenhuis, S., Smits, P. H., Beijnen, J. H., Huitema, A. D. 2009. Polymorphisms of drug-metabolizing enzymes (GST, CYP2B6 and CYP3A) affect the pharmacokinetics of thiotepa and tepa. *Br J Clin Pharmacol* 67:50–60.

Ekins, S., VandenBranden, M., Ring, B. J., Wrighton, S. A. 1997. Examination of purported probes of human CYP2B6. *Pharmacogenetics* 7:165–179.

Ekins, S., Vandenbranden, M., Ring, B. J., Gillespie, J. S., Yang, T. J., Gelboin, H. V., Wrighton, S. A. 1998. Further characterization of the expression in liver and catalytic activity of CYP2B6. *J Pharmacol Exp Ther* 286:1253–1259.

Ekroos, M., Sjogren, T. 2006. Structural basis for ligand promiscuity in cytochrome P450 3A4. *Proc Natl Acad Sci U S A* 103:13682–13687.

Eksterowicz, J., Rock, D. A., Rock, B. M., Wienkers, L. C., Foti, R. S. 2014. Characterization of the active site properties of CYP4F12. *Drug Metab Dispos* 42:1698–1707.

Elbekai, R. H., El-Kadi, A. O. 2006. Cytochrome P450 enzymes: Central players in cardiovascular health and disease. *Pharmacol Ther* 112:564–587.

Eliasson, E., Kenna, J. G. 1996. Cytochrome P450 2E1 is a cell surface autoantigen in halothane hepatitis. *Mol Pharmacol* 50:573–582.

Elkum, N., Alkayal, F., Noronha, F., Ali, M. M., Melhem, M., Al-Arouj, M., Bennakhi, A., Behbehani, K., Alsmadi, O., Abubaker, J. 2014. Vitamin D insufficiency in Arabs and South Asians positively associates with polymorphisms in *GC* and *CYP2R1* genes. *PLoS One* 9:e113102.

Elmabsout, A. A., Kumawat, A., Saenz-Mendez, P., Krivospitskaya, O., Savenstrand, H., Olofsson, P. S., Eriksson, L. A. et al. 2012. Cloning and functional studies of a splice variant of CYP26B1 expressed in vascular cells. *PLoS One* 7:e36839.

Endo, T., Ban, M., Hirata, K., Yamamoto, A., Hara, Y., Momose, Y. 2007. Involvement of CYP2A6 in the formation of a novel metabolite, 3-hydroxypilocarpine, from pilocarpine in human liver microsomes. *Drug Metab Dispos* 35:476–483.

Eun, C. Y., Han, S., Lim, Y. R., Park, H. G., Han, J. S., Cho, K. S., Chun, Y. J., Kim, D. 2010. Bioactivation of aromatic amines by human CYP2W1, an orphan cytochrome P450 enzyme. *Toxicol Res* 26:171–175.

Evangelista, E. A., Kaspera, R., Mokadam, N. A., Jones, J. P., 3rd Totah, R. A. 2013. Activity, inhibition, and induction of cytochrome P450 2J2 in adult human primary cardiomyocytes. *Drug Metab Dispos* 41:2087–2094.

Fairbrother, K. S., Grove, J., de Waziers, I., Steimel, D. T., Day, C. P., Crespi, C. L., Daly, A. K. 1998. Detection and characterization of novel polymorphisms in the *CYP2E1* gene. *Pharmacogenetics* 8:543–552.

Fan, Y. S., Sasi, R., Lee, C., Winter, J. S., Waterman, M. R., Lin, C. C. 1992. Localization of the human *CYP17* gene (cytochrome P45017α) to 10q24.3 by fluorescence *in situ* hybridization and simultaneous chromosome banding. *Genomics* 14:1110–1111.

Fanni, D., Fanos, V., Ambu, R., Lai, F., Gerosa, C., Pampaloni, P., Van Eyken, P., Senes, G., Castagnola, M., Faa, G. 2014. Overlapping between CYP3A4 and CYP3A7 expression in the fetal human liver during development. *J Matern Fetal Neonatal Med* 1–5.

Felix, C. A., Walker, A. H., Lange, B. J., Williams, T. M., Winick, N. J., Cheung, N. K., Lovett, B. D., Nowell, P. C., Blair, I. A., Rebbeck, T. R. 1998. Association of *CYP3A4* genotype with treatment-related leukemia. *Proc Natl Acad Sci U S A* 95:13176–13181.

Feng, S., He, X. 2013. Mechanism-based inhibition of CYP450: An indicator of drug-induced hepatotoxicity. *Curr Drug Metab* 14:921–945.

Fer, M., Corcos, L., Dreano, Y., Plee-Gautier, E., Salaun, J. P., Berthou, F., Amet, Y. 2008. Cytochromes P450 from family 4 are the main omega hydroxylating enzymes in humans: CYP4F3B is the prominent player in PUFA metabolism. *J Lipid Res* 49:2379–2389.

Ferguson, C. S., Tyndale, R. F. 2011. Cytochrome P450 enzymes in the brain: Emerging evidence of biological significance. *Trends Pharmacol Sci* 32:708–714.

Ferguson, R. J., De Morais, S. M., Benhamou, S., Bouchardy, C., Blaisdell, J., Ibeanu, G., Wilkinson, G. R. et al. 1998. A new genetic defect in human *CYP2C19*: Mutation of the initiation codon is responsible for poor metabolism of *S*-mephenytoin. *J Pharmacol Exp Ther* 284:356–361.

Ferguson, S. S., Chen, Y., LeCluyse, E. L., Negishi, M., Goldstein, J. A. 2005. Human CYP2C8 is transcriptionally regulated by the nuclear receptors constitutive androstane receptor, pregnane X receptor, glucocorticoid receptor, and hepatic nuclear factor 4α. *Mol Pharmacol* 68:747–757.

Fernandez-Salguero, P., Gonzalez, F. J. 1995. The *CYP2A* gene subfamily: Species differences, regulation, catalytic activities and role in chemical carcinogenesis. *Pharmacogenetics* 5 Spec No:S123–S128.

Fernandez-Salguero, P., Hoffman, S. M., Cholerton, S., Mohrenweiser, H., Raunio, H., Rautio, A., Pelkonen, O. et al. 1995. A genetic polymorphism in coumarin 7-hydroxylation: Sequence of the human *CYP2A* genes and identification of variant *CYP2A6* alleles. *Am J Hum Genet* 57:651–660.

Feyereisen, R. 1999. Insect P450 enzymes. *Annu Rev Entomol* 44:507–533.

Feyereisen, R. 2011. Arthropod CYPomes illustrate the tempo and mode in P450 evolution. *Biochim Biophys Acta* 1814:19–28.

Fiegenbaum, M., da Silveira, F. R., Van der Sand, C. R., Van der Sand, L. C., Ferreira, M. E., Pires, R. C., Hutz, M. H. 2005. The role of common variants of *ABCB1*, *CYP3A4*, and *CYP3A5* genes in lipid-lowering efficacy and safety of simvastatin treatment. *Clin Pharmacol Ther* 78:551–558.

Finch, C. K., Chrisman, C. R., Baciewicz, A. M., Self, T. H. 2002. Rifampin and rifabutin drug interactions: An update. *Arch Intern Med* 162:985–992.

Fink, J. K. 2013. Hereditary spastic paraplegia: Clinico-pathologic features and emerging molecular mechanisms. *Acta Neuropathol* 126:307–328.

Finta, C., Zaphiropoulos, P. G. 2000. The human cytochrome P450 3A locus. Gene evolution by capture of downstream exons. *Gene* 260:13–23.

Fleming, I. 2008. Vascular cytochrome p450 enzymes: Physiology and pathophysiology. *Trends Cardiovasc Med* 18:20–25.

Fleming, I. 2011a. Cytochrome P450-dependent eicosanoid production and crosstalk. *Curr Opin Lipidol* 22:403–409.

Fleming, I. 2011b. The cytochrome P450 pathway in angiogenesis and endothelial cell biology. *Cancer Metastasis Rev* 30:541–555.

Fluck, C. E., Pandey, A. V. 2011. Clinical and biochemical consequences of p450 oxidoreductase deficiency. *Endocr Dev* 20:63–79.

Fohner, A. E., Robinson, R., Yracheta, J., Dillard, D. A., Schilling, B., Khan, B., Hopkins, S. et al. 2015. Variation in genes controlling warfarin disposition and response in American Indian and Alaska Native people: CYP2C9, VKORC1, CYP4F2, CYP4F11, GGCX. *Pharmacogenet Genomics* 25:343–353.

Fontana, E., Dansette, P. M., Poli, S. M. 2005. Cytochrome p450 enzymes mechanism based inhibitors: Common sub-structures and reactivity. *Curr Drug Metab* 6:413–454.

Frank, S., Steffens, S., Fischer, U., Tlolko, A., Rainov, N. G., Kramm, C. M. 2002. Differential cytotoxicity and bystander effect of the rabbit cytochrome P450 4B1 enzyme gene by two different prodrugs: Implications for pharmacogene therapy. *Cancer Gene Ther* 9:178–188.

Fransen, K., Franzen, P., Magnuson, A., Elmabsout, A. A., Nyhlin, N., Wickbom, A., Curman, B. et al. 2013. Polymorphism in the retinoic acid metabolizing enzyme CYP26B1 and the development of Crohn's Disease. *PLoS One* 8:e72739.

Friedrich, M., Villena-Heinsen, C., Axt-Fliedner, R., Meyberg, R., Tilgen, W., Schmidt, W., Reichrath, J. 2002. Analysis of 25-hydroxyvitamine D_3-1alpha-hydroxylase in cervical tissue. *Anticancer Res* 22:183–186.

Fromel, T., Kohlstedt, K., Popp, R., Yin, X., Awwad, K., Barbosa-Sicard, E., Thomas, A. C., Lieberz, R., Mayr, M., Fleming, I. 2013. Cytochrome P4502S1: A novel monocyte/macrophage fatty acid epoxygenase in human atherosclerotic plaques. *Basic Res Cardiol* 108:319.

Fu, G. K., Portale, A. A., Miller, W. L. 1997. Complete structure of the human gene for the vitamin D 1α-hydroxylase, P450c1α. *DNA Cell Biol* 16:1499–1507.

Fu, X., Menke, J. G., Chen, Y., Zhou, G., MacNaul, K. L., Wright, S. D., Sparrow, C. P., Lund, E. G. 2001. 27-hydroxycholesterol is an endogenous ligand for liver X receptor in cholesterol-loaded cells. *J Biol Chem* 276:38378–38387.

Fugh-Berman, A. 2000. Herb-drug interactions. *Lancet* 355:134–138.

Fuhr, U., Kober, S., Zaigler, M., Mutschler, E., Spahn-Langguth, H. 2005. Rate-limiting biotransformation of triamterene is mediated by CYP1A2. *Int J Clin Pharmacol Ther* 43:327–334.

Fujieda, M., Yamazaki, H., Kiyotani, K., Muroi, A., Kunitoh, H., Dosaka-Akita, H., Sawamura, Y., Kamataki, T. 2003. Eighteen novel polymorphisms of the CYP2A13 gene in Japanese. *Drug Metab Pharmacokinet* 18:86–90.

Fujieda, M., Yamazaki, H., Saito, T., Kiyotani, K., Gyamfi, M. A., Sakurai, M., Dosaka-Akita, H. et al. 2004. Evaluation of *CYP2A6* genetic polymorphisms as determinants of smoking behavior and tobacco-related lung cancer risk in male Japanese smokers. *Carcinogenesis* 25:2451–2458.

Fujii, H., Sato, T., Kaneko, S., Gotoh, O., Fujii-Kuriyama, Y., Osawa, K., Kato, S., Hamada, H. 1997. Metabolic inactivation of retinoic acid by a novel P450 differentially expressed in developing mouse embryos. *EMBO J* 16:4163–4173.

Fukami, T., Nakajima, M., Yoshida, R., Tsuchiya, Y., Fujiki, Y., Katoh, M., McLeod, H. L., Yokoi, T. 2004. A novel polymorphism of human *CYP2A6* gene *CYP2A6*17* has an amino acid substitution (V365M) that decreases enzymatic activity *in vitro* and *in vivo*. *Clin Pharmacol Ther* 76:519–527.

Fukami, T., Nakajima, M., Higashi, E., Yamanaka, H., McLeod, H. L., Yokoi, T. 2005. A novel *CYP2A6*20* allele found in African-American population produces a truncated protein lacking enzymatic activity. *Biochem Pharmacol* 70:801–808.

Fukami, T., Nakajima, M., Sakai, H., McLeod, H. L., Yokoi, T. 2006. *CYP2A7* polymorphic alleles confound the genotyping of CYP2A6*4A allele. *Pharmacogenomics J* 6:401–412.

Fukami, T., Nakajima, M., Sakai, H., Katoh, M., Yokoi, T. 2007. CYP2A13 metabolizes the substrates of human CYP1A2, phenacetin, and theophylline. *Drug Metab Dispos* 35:335–339.

Fukami, T., Katoh, M., Yamazaki, H., Yokoi, T., Nakajima, M. 2008. Human cytochrome P450 2A13 efficiently metabolizes chemicals in air pollutants: Naphthalene, styrene, and toluene. *Chem Res Toxicol* 21:720–725.

Fukami, T., Nakajima, M., Matsumoto, I., Zen, Y., Oda, M., Yokoi, T. 2010. Immunohistochemical analysis of CYP2A13 in various types of human lung cancers. *Cancer Sci* 101:1024–1028.

Fukuen, S., Fukuda, T., Maune, H., Ikenaga, Y., Yamamoto, I., Inaba, T., Azuma, J. 2002. Novel detection assay by PCR-RFLP and frequency of the *CYP3A5* SNPs, *CYP3A5*3* and *6, in a Japanese population. *Pharmacogenetics* 12:331–334.

Fukushima-Uesaka, H., Saito, Y., Maekawa, K., Ozawa, S., Hasegawa, R., Kajio, H., Kuzuya, N. et al. 2005. Genetic variations and haplotypes of *CYP2C19* in a Japanese population. *Drug Metab Pharmacokinet* 20:300–307.

Furuta, T., Ohashi, K., Kobayashi, K., Iida, I., Yoshida, H., Shirai, N., Takashima, M. et al. 1999a. Effects of clarithromycin on the metabolism of omeprazole in relation to *CYP2C19* genotype status in humans. *Clin Pharmacol Ther* 66:265–274.

Furuta, T., Ohashi, K., Kosuge, K., Zhao, X. J., Takashima, M., Kimura, M., Nishimoto, M., Hanai, H., Kaneko, E., Ishizaki, T. 1999b. *CYP2C19* genotype status and effect of omeprazole on intragastric pH in humans. *Clin Pharmacol Ther* 65:552–561.

Furuta, T., Shirai, N., Xiao, F., Ohashi, K., Ishizaki, T. 2001. Effect of high-dose lansoprazole on intragastric pH in subjects who are homozygous extensive metabolizers of cytochrome P4502C19. *Clin Pharmacol Ther* 70:484–492.

Gaedigk, A., Gaedigk, R., Leeder, J. S. 2005. *CYP2D7* splice variants in human liver and brain: Does *CYP2D7* encode functional protein? *Biochem Biophys Res Commun* 336:1241–1250.

Gafvels, M., Olin, M., Chowdhary, B. P., Raudsepp, T., Andersson, U., Persson, B., Jansson, M., Bjorkhem, I., Eggertsen, G. 1999. Structure and chromosomal assignment of the sterol 12α-hydroxylase gene (CYP8B1) in human and mouse: Eukaryotic cytochrome P-450 gene devoid of introns. *Genomics* 56:184–196.

Gainer, J. V., Bellamine, A., Dawson, E. P., Womble, K. E., Grant, S. W., Wang, Y., Cupples, L. A. et al. 2005. Functional variant of CYP4A11 20-hydroxyeicosatetraenoic acid synthase is associated with essential hypertension. *Circulation* 111:63–69.

Gainer, J. V., Lipkowitz, M. S., Yu, C., Waterman, M. R., Dawson, E. P., Capdevila, J. H., Brown, N. J., Group, A. S. 2008. Association of a CYP4A11 variant and blood pressure in black men. *J Am Soc Nephrol* 19:1606–1612.

Garcia-Martin, E., Martinez, C., Pizarro, R. M., Garcia-Gamito, F. J., Gullsten, H., Raunio, H., Agundez, J. A. G. 2002. *CYP3A4* variant alleles in white individuals with low CYP3A4 enzyme activity. *Clin Pharmacol Ther* 71:196–204.

Gardiner, S. J., Begg, E. J. 2006. Pharmacogenetics, drug-metabolizing enzymes, and clinical practice. *Pharmacol Rev* 58:521–590.

Garuti, R., Croce, M. A., Tiozzo, R., Dotti, M. T., Federico, A., Bertolini, S., Calandra, S. 1997. Four novel mutations of dterol 27-hydroxylase gene in Italian patients with cerebrotendinous xanthomatosis. *J Lipid Res* 38:2322–2334.

Gascon-Barre, M., Demers, C., Ghrab, O., Theodoropoulos, C., Lapointe, R., Jones, G., Valiquette, L., Menard, D. 2001. Expression of CYP27A, a gene encoding a vitamin D-25 hydroxylase in human liver and kidney. *Clin Endocrinol (Oxf)* 54:107–115.

Gay, S. C., Roberts, A. G., Halpert, J. R. 2010a. Structural features of cytochromes P450 and ligands that affect drug metabolism as revealed by X-ray crystallography and NMR. *Future Med Chem* 2:1451–1468.

Gay, S. C., Shah, M. B., Talakad, J. C., Maekawa, K., Roberts, A. G., Wilderman, P. R., Sun, L. et al. 2010b. Crystal structure of a cytochrome P450 2B6 genetic variant in complex with the inhibitor 4-(4-chlorophenyl)imidazole at 2.0-A resolution. *Mol Pharmacol* 77:529–538.

Gekka, T., Hayashi, T., Takeuchi, T., Goto-Omoto, S., Kitahara, K. 2005. CYP4V2 mutations in two Japanese patients with Bietti's crystalline dystrophy. *Ophthalmic Res* 37:262–269.

Gellner, K., Eiselt, R., Hustert, E., Arnold, H., Koch, I., Haberl, M., Deglmann, C. J. et al. 2001. Genomic organization of the human *CYP3A* locus: Identification of a new, inducible *CYP3A* gene. *Pharmacogenetics* 11:111–121.

Genevieve, D., Proulle, V., Isidor, B., Bellais, S., Serre, V., Djouadi, F., Picard, C. et al. 2008. Thromboxane synthase mutations in an increased bone density disorder (Ghosal syndrome). *Nat Genet* 40:284–286.

Gerbal-Chaloin, S., Pascussi, J. M., Pichard-Garcia, L., Daujat, M., Waechter, F., Fabre, J. M., Carrere, N., Maurel, P. 2001. Induction of *CYP2C* genes in human hepatocytes in primary culture. *Drug Metab Dispos* 29:242–251.

Gerbal-Chaloin, S., Pascussi, J. M., Pichard-Garcia, L., Vilarem, M. J., Maurel, P. 2002. Transcriptional regulation of CYP2C9 gene: Role of glucocorticoid receptor and constitutive androstane receptor. *J Biol Chem* 277:209–217.

Gerbal-Chaloin, S., Iankova, I., Maurel, P., Daujat-Chavanieu, M. 2013. Nuclear receptors in the cross-talk of drug metabolism and inflammation. *Drug Metab Rev* 45:122–144.

Gervasini, G., de Murillo, S. G., Ladero, J. M., Agundez, J. A. 2010. CYP2W1 variant alleles in Caucasians and association of the *CYP2W1* G541A (Ala181Thr) polymorphism with increased colorectal cancer risk. *Pharmacogenomics* 11:919–925.

Gervot, L., Rochat, B., Gautier, J. C., Bohnenstengel, F., Kroemer, H., de Berardinis, V., Martin, H., Beaune, P. de Waziers, I. 1999. Human CYP2B6: Expression, inducibility and catalytic activities. *Pharmacogenetics* 9:295–306.

Ghosal, A., Ramanathan, R., Yuan, Y., Hapangama, N., Chowdhury, S. K., Kishnani, N. S., Alton, K. B. 2007. Identification of human liver cytochrome P450 enzymes involved in biotransformation of vicriviroc, a CCR5 receptor antagonist. *Drug Metab Dispos* 35:2186–2195.

Ghosal, A., Lu, X., Penner, N., Gao, L., Ramanathan, R., Chowdhury, S. K., Kishnani, N. S., Alton, K. B. 2011. Identification of human liver cytochrome P450 enzymes involved in the metabolism of SCH 530348 (Vorapaxar), a potent oral thrombin protease-activated receptor 1 antagonist. *Drug Metab Dispos* 39:30–38.

Ghosh, D., Griswold, J., Erman, M., Pangborn, W. 2009. Structural basis for androgen specificity and oestrogen synthesis in human aromatase. *Nature* 457:219–223.

Ghosh, D., Lo, J., Morton, D., Valette, D., Xi, J., Griswold, J., Hubbell, S. et al. 2012. Novel aromatase inhibitors by structure-guided design. *J Med Chem* 55:8464–8476.

Ghoshal, U., Tripathi, S., Kumar, S., Mittal, B., Chourasia, D., Kumari, N., Krishnani, N., Ghoshal, U. C. 2014. Genetic polymorphism of cytochrome P450 (CYP) 1A1, CYP1A2, and CYP2E1 genes modulate susceptibility to gastric cancer in patients with Helicobacter pylori infection. *Gastric Cancer* 17:226–234.

Ghotbi, R., Christensen, M., Roh, H. K., Ingelman-Sundberg, M., Aklillu, E., Bertilsson, L. 2007. Comparisons of CYP1A2 genetic polymorphisms, enzyme activity and the genotype–phenotype relationship in Swedes and Koreans. *Eur J Clin Pharmacol* 63:537–546.

Giancarlo, G. M., Venkatakrishnan, K., Granda, B. W., von Moltke, L. L., Greenblatt, D. J. 2001. Relative contributions of CYP2C9 and 2C19 to phenytoin 4-hydroxylation *in vitro*: Inhibition by sulfaphenazole, omeprazole, and ticlopidine. *Eur J Clin Pharmacol* 57:31–36.

Gilardi, F., Mitro, N., Godio, C., Scotti, E., Caruso, D., Crestani, M., De Fabiani, E. 2007. The pharmacological exploitation of cholesterol 7α-hydroxylase, the key enzyme in bile acid synthesis: From binding resins to chromatin remodelling to reduce plasma cholesterol. *Pharmacol Ther* 116:449–472.

Gillam, E. M., Hayes, M. A. 2013. The evolution of cytochrome P450 enzymes as biocatalysts in drug discovery and development. *Curr Top Med Chem* 13:2254–2280.

Go, R. E., Hwang, K. A., Choi, K. C. 2015. Cytochrome P450 1 family and cancers. *J Steroid Biochem Mol Biol* 147:24–30.

Gobbi, S., Rampa, A., Belluti, F., Bisi, A. 2014. Nonsteroidal aromatase inhibitors for the treatment of breast cancer: An update. *Anticancer Agents Med Chem* 14:54–65.

Goizet, C., Boukhris, A., Durr, A., Beetz, C., Truchetto, J., Tesson, C., Tsaousidou, M. et al. 2009. *CYP7B1* mutations in pure and complex forms of hereditary spastic paraplegia type 5. *Brain* 132:1589–1600.

Goldstein, I., Rivlin, N., Shoshana, O. Y., Ezra, O., Madar, S., Goldfinger, N., Rotter, V. 2013. Chemotherapeutic agents induce the expression and activity of their clearing enzyme CYP3A4 by activating p53. *Carcinogenesis* 34:190–198.

Gomez, A., Ingelman-Sundberg, M. 2009. Epigenetic and microRNA-dependent control of cytochrome P450 expression: A gap between DNA and protein. *Pharmacogenomics* 10:1067–1076.

Gomez, A., Karlgren, M., Edler, D., Bernal, M. L., Mkrtchian, S., Ingelman-Sundberg, M. 2007. Expression of CYP2W1 in colon tumors: Regulation by gene methylation. *Pharmacogenomics* 8:1315–1325.

Gomez, A., Nekvindova, J., Travica, S., Lee, M. Y., Johansson, I., Edler, D., Mkrtchian, S., Ingelman-Sundberg, M. 2010. Colorectal cancer-specific cytochrome P450 2W1: Intracellular localization, glycosylation, and catalytic activity. *Mol Pharmacol* 78:1004–1011.

Goncalves, J., Friaes, A., Moura, L. 2007. Congenital adrenal hyperplasia: Focus on the molecular basis of 21-hydroxylase deficiency. *Expert Rev Mol Med* 9:1–23.

Gong, B., Qu, C., Li, X., Shi, Y., Lin, Y., Zhou, Y., Shuai, P. et al. 2015. Mutation spectrum of *CYP1B1* in Chinese patients with primary open-angle glaucoma. *Br J Ophthalmol* 99:425–430.

Gonzalez, M. C., Marteau, C., Franchi, J., Migliore-Samour, D. 2001. Cytochrome P450 4A11 expression in human keratinocytes: Effects of ultraviolet irradiation. *Br J Dermatol* 145:749–757.

Good, R. T., Gramzow, L., Battlay, P., Sztal, T., Batterham, P., Robin, C. 2014. The molecular evolution of cytochrome P450 genes within and between drosophila species. *Genome Biol Evol* 6:1118–1134.

Goodwin, B., Moore, L. B., Stoltz, C. M., McKee, D. D., Kliewer, S. A. 2001. Regulation of the human *CYP2B6* gene by the nuclear pregnane X receptor. *Mol Pharmacol* 60:427–431.

Gorski, J. C., Hall, S. D., Becker, P., Affrime, M. B., Cutler, D. L., Haehner-Daniels, B. 2000. *In vivo* effects of interleukin-10 on human cytochrome P450 activity. *Clin Pharmacol Ther* 67:32–43.

Gotoh, O. 1992. Substrate recognition sites in cytochrome P450 family 2 (CYP2) proteins inferred from comparative analyses of amino acid and coding nucleotide sequences. *J Biol Chem* 267:83–90.

Gottfried, E., Rehli, M., Hahn, J., Holler, E., Andreesen, R., Kreutz, M. 2006. Monocyte-derived cells express CYP27A1 and convert vitamin D₃ into its active metabolite. *Biochem Biophys Res Commun* 349:209–213.

Gough, A. C., Smith, C. A., Howell, S. M., Wolf, C. R., Bryant, S. P., Spurr, N. K. 1993. Localization of the *CYP2D* gene locus to human chromosome 22q13.1 by polymerase chain reaction, in situ hybridization, and linkage analysis. *Genomics* 15:430–432.

Granfors, M. T., Backman, J. T., Laitila, J., Neuvonen, P. J. 2004a. Tizanidine is mainly metabolized by cytochrome p450 1A2 in vitro. *Br J Clin Pharmacol* 57:349–353.

Granfors, M. T., Backman, J. T., Neuvonen, M., Neuvonen, P. J. 2004b. Ciprofloxacin greatly increases concentrations and hypotensive effect of tizanidine by inhibiting its cytochrome P450 1A2-mediated presystemic metabolism. *Clin Pharmacol Ther* 76:598–606.

Graves, J. P., Edin, M. L., Bradbury, J. A., Gruzdev, A., Cheng, J., Lih, F. B., Masinde, T. A. et al. 2013. Characterization of four new mouse cytochrome P450 enzymes of the CYP2J subfamily. *Drug Metab Dispos* 41:763–773.

Gray, I. C., Nobile, C., Muresu, R., Ford, S., Spurr, N. K. 1995. A 2.4-megabase physical map spanning the *CYP2C* gene cluster on chromosome 10q24. *Genomics* 28:328–332.

Gressier, F., Verstuyft, C., Hardy, P., Becquemont, L., Corruble, E. 2015. Response to CYP2D6 substrate antidepressants is predicted by a CYP2D6 composite phenotype based on genotype and comedications with CYP2D6 inhibitors. *J Neural Transm* 122:35–42.

Grime, K. H., Bird, J., Ferguson, D., Riley, R. J. 2009. Mechanism-based inhibition of cytochrome P450 enzymes: An evaluation of early decision making in vitro approaches and drug–drug interaction prediction methods. *Eur J Pharm Sci* 36:175–191.

Grischuk, Y., Rubtsov, P., Riepe, F. G., Grotzinger, J., Beljelarskaia, S., Prassolov, V., Kalintchenko, N. et al. 2006. Four novel missense mutations in the CYP21A2 gene detected in Russian patients suffering from the classical form of congenital adrenal hyperplasia: Identification, functional characterization, and structural analysis. *J Clin Endocrinol Metab* 91:4976–4980.

Grossman, S. J., Herold, E. G., Drey, J. M., Alberts, D. W., Umbenhauer, D. R., Patrick, D. H., Nicoll-Griffith, D., Chauret, N., Yergey, J. A. 1993. CYP1A1 specificity of verlukast epoxidation in mice, rats, rhesus monkeys, and humans. *Drug Metab Dispos* 21:1029–1036.

Gu, J., Su, T., Chen, Y., Zhang, Q. Y., Ding, X. 2000. Expression of biotransformation enzymes in human fetal olfactory mucosa: Potential roles in developmental toxicity. *Toxicol Appl Pharmacol* 165:158–162.

Guengerich, F. P. 1990. Mechanism-based inactivation of human liver microsomal cytochrome P-450 IIIA4 by gestodene. *Chem Res Toxicol* 3:363–371.

Guengerich, F. P. 2001. Uncommon P450-catalyzed reactions. *Curr Drug Metab* 2:93–115.

Guengerich, P. F. 2005. *Human Cytochrome P450 Enzymes*, Kluwer Academic/Plenum Publishers, New York.

Guengerich, F. P., Wu, Z. L., Bartleson, C. J. 2005. Function of human cytochrome P450s: Characterization of the orphans. *Biochem Biophys Res Commun* 338:465–469.

Guidice, J. M., Marez, D., Sabbagh, N., Legrand-Andreoletti, M., Spire, C., Alcaide, E., Lafitte, J. J., Broly, F. 1997. Evidence for CYP2D6 expression in human lung. *Biochem Biophys Res Commun* 241:79–85.

Guo, Z., Raeissi, S., White, R. B., Stevens, J. C. 1997. Orphenadrine and methimazole inhibit multiple cytochrome P450 enzymes in human liver microsomes. *Drug Metab Dispos* 25:390–393.

Ha-Duong, N. T., Dijols, S., Marques-Soares, C., Minoletti, C., Dansette, P. M., Mansuy, D. 2001a. Synthesis of sulfaphenazole derivatives and their use as inhibitors and tools for comparing the active sites of human liver cytochromes P450 of the 2C subfamily. *J Med Chem* 44:3622–3631.

Ha-Duong, N. T., Marques-Soares, C., Dijols, S., Sari, M. A., Dansette, P. M., Mansuy, D. 2001b. Interaction of new sulfaphenazole derivatives with human liver cytochrome p450 2Cs: Structural determinants required for selective recognition by CYP 2C9 and for inhibition of human CYP 2Cs. *Arch Biochem Biophys* 394:189–200.

Ha, H. R., Chen, J., Freiburghaus, A. U., Follath, F. 1995. Metabolism of theophylline by cDNA-expressed human cytochromes P-450. *Br J Clin Pharmacol* 39:321–326.

Haarmann-Stemmann, T., Abel, J. 2006. The arylhydrocarbon receptor repressor (AhRR): Structure, expression, and function. *Biol Chem* 387:1195–1199.

Haberl, M., Anwald, B., Klein, K., Weil, R., Fuss, C., Gepdiremen, A., Zanger, U. M., Meyer, U. A., Wojnowski, L. 2005. Three haplotypes associated with CYP2A6 phenotypes in Caucasians. *Pharmacogenet Genomics* 15:609–624.

Haddad, A., Davis, M., Lagman, R. 2007. The pharmacological importance of cytochrome CYP3A4 in the palliation of symptoms: Review and recommendations for avoiding adverse drug interactions. *Suport Care Cancer* 15:251–257.

Hadidi, H., Zahlsen, K., Idle, J. R., Cholerton, S. 1997. A single amino acid substitution (Leu160His) in cytochrome P450 CYP2A6 causes switching from 7-hydroxylation to 3-hydroxylation of coumarin. *Food Chem Toxicol* 35:903–907.

Hadidi, H., Irshaid, Y., Vagbo, C. B., Brunsvik, A., Cholerton, S., Zahlsen, K., Idle, J. R. 1998. Variability of coumarin 7- and 3-hydroxylation in a Jordanian population is suggestive of a functional polymorphism in cytochrome P450 CYP2A6. *Eur J Clin Pharmacol* 54:437–441.

Hahn, M. E. 2002. Aryl hydrocarbon receptors: Diversity and evolution. *Chem Biol Interact* 141:131–160.

Hahn, C. N., Baker, E., Laslo, P., May, B. K., Omdahl, J. L., Sutherland, G. R. 1993. Localization of the human vitamin D 24-hydroxylase gene (CYP24) to chromosome 20q13.2→q13.3. *Cytogenet Cell Genet* 62:192–193.

Haider, S., Islam, B., D'Atri, V., Sgobba, M., Poojari, C., Sun, L., Yuen, T., Zaidi, M., New, M. I. 2013. Structure-phenotype correlations of human *CYP21A2* mutations in congenital adrenal hyperplasia. *Proc Natl Acad Sci U S A* 110:2605–2610.

Haines, D. C., Tomchick, D. R., Machius, M., Peterson, J. A. 2001. Pivotal role of water in the mechanism of P450BM-3. *Biochemistry* 40:13456–13465.

Hakkola, J., Raunio, H., Purkunen, R., Pelkonen, O., Saarikoski, S., Cresteil, T., Pasanen, M. 1996. Detection of cytochrome P450 gene expression in human placenta in first trimester of pregnancy. *Biochem Pharmacol* 52:379–383.

Hakkola, J., Raunio, H., Purkunen, R., Saarikoski, S., Vahakangas, K., Pelkonen, O., Edwards, R. J., Boobis, A. R., Pasanen, M. 2001. Cytochrome P450 3A expression in the human fetal liver: Evidence that CYP3A5 is expressed in only a limited number of fetal livers. *Biol Neonate* 80:193–201.

Hakkola, J., Hu, Y., Ingelman-Sundberg, M. 2003. Mechanisms of down-regulation of CYP2E1 expression by inflammatory cytokines in rat hepatoma cells. *J Pharmacol Exp Ther* 304:1048–1054.

Halder, S. K., Fink, M., Waterman, M. R., Rozman, D. 2002. A cAMP-responsive element binding site is essential for sterol regulation of the human lanosterol 14alpha-demethylase gene (CYP51). *Mol Endocrinol* 16:1853–1863.

Han, X. M., Ouyang, D. S., Chen, X. P., Shu, Y., Jiang, C. H., Tan, Z. R., Zhou, H. H. 2002. Inducibility of CYP1A2 by omeprazole in vivo related to the genetic polymorphism of CYP1A2. *Br J Clin Pharmacol* 54:540–543.

Hancks, D. C., Hartley, M. K., Hagan, C., Clark, N. L., Elde, N. C. 2015. Overlapping patterns of rapid evolution in the nucleic acid sensors cGAS and OAS1 suggest a common mechanism of pathogen antagonism and escape. *PLoS Genet* 11:e1005203.

Handschin, C., Meyer, U. A. 2003. Induction of drug metabolism: The role of nuclear receptors. *Pharmacol Rev* 55:649–673.

Hanioka, N., Tsuneto, Y., Saito, Y., Sumada, T., Maekawa, K., Saito, K., Sawada, J., Narimatsu, S. 2007. Functional characterization of two novel *CYP2C19* variants (*CYP2C19*18* and *CYP2C19*19*) found in a Japanese population. *Xenobiotica* 37:342–355.

Hanley, N. A., Rainey, W. E., Wilson, D. I., Ball, S. G., Parker, K. L. 2001. Expression profiles of SF-1, DAX1, and CYP17 in the human fetal adrenal gland: Potential interactions in gene regulation. *Mol Endocrinol* 15:57–68.

Hanna, I. H., Reed, J. R., Guengerich, F. P., Hollenberg, P. F. 2000. Expression of human cytochrome P450 2B6 in *Escherichia coli*: Characterization of catalytic activity and expression levels in human liver. *Arch Biochem Biophys* 376:206–216.

Hansson, M., Wikvall, K., Babiker, A. 2005. Regulation of sterol 27-hydroxylase in human monocyte-derived macrophages: Up-regulation by transforming growth factor β1. *Biochim Biophys Acta* 1687:44–51.

Hanzawa, Y., Sasaki, T., Mizugaki, M., Ishikawa, M., Hiratsuka, M. 2008. Genetic polymorphisms and haplotype structures of the human *CYP2W1* gene in a Japanese population. *Drug Metab Dispos* 36:349–352.

Hardwick, J. P. 2008. Cytochrome P450 omega hydroxylase (CYP4) function in fatty acid metabolism and metabolic diseases. *Biochem Pharmacol* 75:2263–2275.

Hargrove, T. Y., Kim, K., de Nazare Correia Soeiro, M., da Silva, C. F., Batista, D. D., Batista, M. M., Yazlovitskaya, E. M., Waterman, M. R., Sulikowski, G. A., Lepesheva, G. I. 2012. CYP51 structures and structure-based development of novel, pathogen-specific inhibitory scaffolds. *Int J Parasitol Drugs Drug Resist* 2:178–186.

Hariparsad, N., Chu, X., Yabut, J., Labhart, P., Hartley, D. P., Dai, X., Evers, R. 2009. Identification of pregnane-X receptor target genes and coactivator and corepressor binding to promoter elements in human hepatocytes. *Nucleic Acids Res* 37:1160–1173.

Harris, J. W., Rahman, A., Kim, B. R., Guengerich, F. P., Collins, J. M. 1994. Metabolism of taxol by human hepatic microsomes and liver slices: Participation of cytochrome P450 3A4 and an unknown P450 enzyme. *Cancer Res* 54:4026–4035.

Harris, M., Bhuvaneshwar, K., Natarajan, T., Sheahan, L., Wang, D., Tadesse, M. G., Shoulson, I. et al. 2014. Pharmacogenomic characterization of gemcitabine response—A framework for data integration to enable personalized medicine. *Pharmacogenet Genomics* 24:81–93.

Harvey, R. D., Morgan, E. T. 2014. Cancer, inflammation, and therapy: Effects on cytochrome p450-mediated drug metabolism and implications for novel immunotherapeutic agents. *Clin Pharmacol Ther* 96:449–457.

Hasemann, C. A., Kurumbail, R. G., Boddupalli, S. S., Peterson, J. A., Deisenhofer, J. 1995. Structure and function of cytochromes P450: A comparative analysis of three crystal structures. *Structure* 3:41–62.

Hashimoto, T., Morohashi, K., Takayama, K., Honda, S., Wada, T., Handa, H., Omura, T. 1992. Cooperative transcription activation between Ad1, a CRE-like element, and other elements in the CYP11B gene promoter. *J Biochem* 112:573–575.

Hashimoto, H., Toide, K., Kitamura, R., Fujita, M., Tagawa, S., Itoh, S., Kamataki, T. 1993. Gene structure of *CYP3A4*, an adult specific form of cytochrome P450 in human livers, and its transcriptional control. *Eur J Biochem* 218:585–595.

Hashimoto, H., Yanagawa, Y., Sawada, M., Itoh, S., Deguchi, T., Kamataki, T. 1995. Simultaneous expression of human CYP3A7 and N-acetyltransferase in Chinese hamster CHL cells results in high cytotoxicity for carcinogenic heterocyclic amines. *Arch Biochem Biophys* 320:323–329.

Hashizume, T., Imaoka, S., Hiroi, T., Terauchi, Y., Fujii, T., Miyazaki, H., Kamataki, T., Funae, Y. 2001. cDNA cloning and expression of a novel cytochrome p450 (cyp4f12) from human small intestine. *Biochem Biophys Res Commun* 280:1135–1141.

Hashizume, T., Imaoka, S., Mise, M., Terauchi, Y., Fujii, T., Miyazaki, H., Kamataki, T., Funae, Y. 2002. Involvement of CYP2J2 and CYP4F12 in the metabolism of ebastine in human intestinal microsomes. *J Pharmacol Exp Ther* 300:298–304.

He, L., Hannon, G. J. 2004. MicroRNAs: Small RNAs with a big role in gene regulation. *Nat Rev Genet* 5:522–531.

He, Y. A., Roussel, F., Halpert, J. R. 2003. Analysis of homotropic and heterotropic cooperativity of diazepam oxidation by CYP3A4 using site-directed mutagenesis and kinetic modeling. *Arch Biochem Biophys* 409:92–101.

He, X. Y., Shen, J., Ding, X., Lu, A. Y., Hong, J. Y. 2004. Identification of critical amino acid residues of human CYP2A13 for the metabolic activation of 4-(methylnitrosamino)-1-(3-pyridyl)-1-butanone, a tobacco-specific carcinogen. *Drug Metab Dispos* 32:1516–1521.

He, S. M., Zhou, Z. W., Li, X. T., Zhou, S. F. 2011. Clinical drugs undergoing polymorphic metabolism by human cytochrome P450 2C9 and the implication in drug development. *Curr Med Chem* 18:667–713.

Headlam, M. J., Wilce, M. C., Tuckey, R. C. 2003. The F–G loop region of cytochrome P450scc (CYP11A1) interacts with the phospholipid membrane. *Biochim Biophys Acta* 1617:96–108.

Headland, S. E., Norling, L. V. 2015. The resolution of inflammation: Principles and challenges. *Semin Immunol*.

Heim, M., Meyer, U. A. 1990. Genotyping of poor metabolisers of debrisoquine by allele-specific PCR amplification. *Lancet* 336:529–532.

Heim, M. H., Meyer, U. A. 1992. Evolution of a highly polymorphic human cytochrome P450 gene cluster: *CYP2D6*. *Genomics* 14:49–58.

Helsby, N. A., Lo, W. Y., Sharples, K., Riley, G., Murray, M., Spells, K., Dzhelai, M., Simpson, A., Findlay, M. 2008. *CYP2C19* pharmacogenetics in advanced cancer: Compromised function independent of genotype. *Br J Cancer* 99:1251–1255.

Henderson, C. J., McLaughlin, L. A., Scheer, N., Stanley, L. A., Wolf, C. R. 2015. Cytochrome b_5 is a major determinant of human cytochrome P450 CYP2D6 and CYP3A4 activity *in vivo*. *Mol Pharmacol* 87:733–739.

Hesse, L. M., Venkatakrishnan, K., Court, M. H., von Moltke, L. L., Duan, S. X., Shader, R. I., Greenblatt, D. J. 2000. CYP2B6 mediates the *in vitro* hydroxylation of bupropion: Potential drug interactions with other antidepressants. *Drug Metab Dispos* 28:1176–1183.

Hesse, L. M., von Moltke, L. L., Shader, R. I., Greenblatt, D. J. 2001. Ritonavir, efavirenz, and nelfinavir inhibit CYP2B6 activity *in vitro*: Potential drug interactions with bupropion. *Drug Metab Dispos* 29:100–102.

Hewitt, N. J., LeCluyse, E. L., Ferguson, S. S. 2007. Induction of hepatic cytochrome P450 enzymes: Methods, mechanisms, recommendations, and in vitro-in vivo correlations. *Xenobiotica* 37:1196–1224.

Hichiya, H., Tanaka-Kagawa, T., Soyama, A., Jinno, H., Koyano, S., Katori, N., Matsushima, E. et al. 2005. Functional characterization of five novel CYP2C8 variants, G171S, R186X, R186G, K247R, and K383N, found in a Japanese population. *Drug Metab Dispos* 33:630–636.

Higashi, M. K., Veenstra, D. L., Kondo, L. M., Wittkowsky, A. K., Srinouanprachanh, S. L., Farin, F. M., Rettie, A. E. 2002. Association between CYP2C9 genetic variants and anticoagulation-related outcomes during warfarin therapy. *JAMA* 287:1690–1698.

Hildebrandt, A. G., Schwarz, D., Krusekopf, S., Kleeberg, U., Roots, I. 2007. Recalling P446. P4501A1 (CYP1A1) opting for clinical application. *Drug Metab Rev* 39:323–341.

Hiort, O., Holterhus, P. M., Werner, R., Marschke, C., Hoppe, U., Partsch, C. J., Riepe, F. G., Achermann, J. C., Struve, D. 2005. Homozygous disruption of P450 side-chain cleavage (CYP11A1) is associated with prematurity, complete 46,XY sex reversal, and severe adrenal failure. *J Clin Endocrinol Metab* 90:538–541.

Hiratsuka, M., Nozawa, H., Konno, Y., Saito, T., Konno, S., Mizugaki, M. 2004. Human *CYP4B1* gene in the japanese population analyzed by denaturing HPLC. *Drug Metab Pharmacokinet* 19:114–119.

Hiratsuka, M., Nozawa, H., Katsumoto, Y., Moteki, T., Sasaki, T., Konno, Y., Mizugaki, M. 2006. Genetic polymorphisms and haplotype structures of the *CYP4A22* gene in a Japanese population. *Mutat Res* 599:98–104.

Ho, M. K., Mwenifumbo, J. C., Zhao, B., Gillam, E. M., Tyndale, R. F. 2008. A novel *CYP2A6* allele, *CYP2A6*23*, impairs enzyme function *in vitro* and *in vivo* and decreases smoking in a population of Black-African descent. *Pharmacogenet Genomics* 18:67–75.

Ho, W. W., Li, H., Nishida, C. R., Ortiz de Montellano, P. R., Poulos, T. L. 2008. Crystal structure and properties of CYP231A2 from the thermoacidophilic archaeon Picrophilus torridus. *Biochemistry* 47:2071–2079.

Hoch, U., Zhang, Z., Kroetz, D. L., Ortiz de Montellano, P. R. 2000. Structural determination of the substrate specificities and regioselectivities of the rat and human fatty acid omega-hydroxylases. *Arch Biochem Biophys* 373:63–71.

Hodgson, E., Rose, R. L. 2007. The importance of cytochrome P450 2B6 in the human metabolism of environmental chemicals. *Pharmacol Ther* 113:420–428.

Hoffler, U., El-Masri, H. A., Ghanayem, B. I. 2003. Cytochrome P450 2E1 (CYP2E1) is the principal enzyme responsible for urethane metabolism: Comparative studies using CYP2E1-null and wild-type mice. *J Pharmacol Exp Ther* 305:557–564.

Hoffman, S. M., Fernandez-Salguero, P., Gonzalez, F. J., Mohrenweiser, H. W. 1995. Organization and evolution of the cytochrome P450 *CYP2A-2B-2F* subfamily gene cluster on human chromosome 19. *J Mol Evol* 41:894–900.

Hoffman, S. M., Nelson, D. R., Keeney, D. S. 2001. Organization, structure and evolution of the CYP2 gene cluster on human chromosome 19. *Pharmacogenetics* 11:687–698.

Hollenberg, P. F. 1992. Mechanisms of cytochrome P450 and peroxidase-catalyzed xenobiotic metabolism. *FASEB J* 6:686–694.

Holstein, A., Beil, W., Kovacs, P. 2012. CYP2C metabolism of oral antidiabetic drugs—Impact on pharmacokinetics, drug interactions and pharmacogenetic aspects. *Expert Opin Drug Metab Toxicol* 8:1549–1563.

Honda, A., Miyazaki, T., Ikegami, T., Iwamoto, J., Maeda, T., Hirayama, T., Saito, Y., Teramoto, T., Matsuzaki, Y. 2011. Cholesterol 25-hydroxylation activity of CYP3A. *J Lipid Res* 52:1509–1516.

Honda, M., Muroi, Y., Tamaki, Y., Saigusa, D., Suzuki, N., Tomioka, Y., Matsubara, Y., Oda, A., Hirasawa, N., Hiratsuka, M. 2011. Functional characterization of *CYP2B6* allelic variants in demethylation of antimalarial artemether. *Drug Metab Dispos* 39:1860–1865.

Hong, J. Y., Yang, C. S., Lee, M., Wang, Y. Y., Huang, W. Q., Tan, Y., Patten, C. J., Bondoc, F. Y. 1997. Role of cytochromes P450 in the metabolism of methyl *tert*-butyl ether in human livers. *Arch Toxicol* 71:266–269.

Hong, J. Y., Wang, Y. Y., Bondoc, F. Y., Lee, M., Yang, C. S., Hu, W. Y., Pan, J. 1999. Metabolism of methyl *tert*-butyl ether and other gasoline ethers by human liver microsomes and heterologously expressed human cytochromes P450: Identification of CYP2A6 as a major catalyst. *Toxicol Appl Pharmacol* 160:43–48.

Honig, P. K., Woosley, R. L., Zamani, K., Conner, D. P., Cantilena, L. R. 1992. Changes in the pharmacokinetics and electrocardiographic pharmacodynamics of terfenadine with concomitant administration of erythromycin. *Clin Pharmacol Ther* 52:231–238.

Honig, P. K., Wortham, D. C., Zamani, K., Conner, D. P., Mullin, J. C., Cantilena, L. R. 1993. Terfenadine-ketoconazole interaction: Pharmacokinetic and electrocardiographic consequences. *JAMA* 269:1513–1518.

Horton, R., Wilming, L., Rand, V., Lovering, R. C., Bruford, E. A., Khodiyar, V. K., Lush, M. J. et al. 2004. Gene map of the extended human MHC. *Nat Rev Genet* 5:889–899.

Hsieh, K. P., Lin, Y. Y., Cheng, C. L., Lai, M. L., Lin, M. S., Siest, J. P., Huang, J. D. 2001. Novel mutations of *CYP3A4* in Chinese. *Drug Metab Dispos* 29:268–273.

Hsu, J. Y., Feldman, D., McNeal, J. E., Peehl, D. M. 2001. Reduced 1alpha-hydroxylase activity in human prostate cancer cells correlates with decreased susceptibility to 25-hydroxyvitamin D3-induced growth inhibition. *Cancer Res* 61:2852–2856.

Hsu, M. H., Savas, U., Griffin, K. J., Johnson, E. F. 2007. Human cytochrome p450 family 4 enzymes: Function, genetic variation and regulation. *Drug Metab Rev* 39:515–538.

Hu, Y., Oscarson, M., Johansson, I., Yue, Q. Y., Dahl, M. L., Tabone, M., Arinco, S., Albano, E., Ingelman-Sundberg, M. 1997. Genetic polymorphism of human *CYP2E1*: Characterization of two variant alleles. *Mol Pharmacol* 51:370–376.

Hu, M., Mak, V. W., Xiao, Y., Tomlinson, B. 2013. Associations between the genotypes and phenotype of CYP3A and the lipid response to simvastatin in Chinese patients with hypercholesterolemia. *Pharmacogenomics* 14:25–34.

Hu, G. X., Pan, P. P., Wang, Z. S., Yang, L. P., Dai, D. P., Wang, S. H., Zhu, G. H. et al. 2015. *In vitro* and *in vivo* characterization of 13 *CYP2C9* allelic variants found in Chinese Han population. *Drug Metab Dispos* 43:561–569.

Huang, Z., Fasco, M. J., Kaminsky, L. S. 1997. Alternative splicing of CYP2D mRNA in human breast tissue. *Arch Biochem Biophys* 343:101–108.

Huang, J. C., Arbab, F., Tumbusch, K. J., Goldsby, J. S., Matijevic-Aleksic, N., Wu, K. K. 2002. Human fallopian tubes express prostacyclin (PGI) synthase and cyclooxygenases and synthesize abundant PGI. *J Clin Endocrinol Metab* 87:4361–4368.

Huang, S. M., Temple, R., Throckmorton, D. C., Lesko, L. J. 2007. Drug interaction studies: Study design, data analysis, and implications for dosing and labeling. *Clin Pharmacol Ther* 81:298–304.

Huang, S. M., Strong, J. M., Zhang, L., Reynolds, K. S., Nallani, S., Temple, R., Abraham, S. et al. 2008. New era in drug interaction evaluation: US Food and Drug Administration update on CYP enzymes, transporters, and the guidance process. *J Clin Pharmacol* 48:662–670.

Huang, Q., Deshmukh, R. S., Ericksen, S. S., Tu, Y., Szklarz, G. D. 2012. Preferred binding orientations of phenacetin in CYP1A1 and CYP1A2 are associated with isoform-selective metabolism. *Drug Metab Dispos* 40:2324–2331.

Huang, X., Li, M., Guo, X., Li, S., Xiao, X., Jia, X., Liu, X., Zhang, Q. 2014. Mutation analysis of seven known glaucoma-associated genes in Chinese patients with glaucoma. *Invest Ophthalmol Vis Sci* 55:3594–3602.

Hukkanen, J., Pelkonen, O., Hakkola, J., Raunio, H. 2002. Expression and regulation of xenobiotic-metabolizing cytochrome P450 (CYP) enzymes in human lung. *Crit Rev Toxicol* 32:391–411.

Hukkanen, J., Vaisanen, T., Lassila, A., Piipari, R., Anttila, S., Pelkonen, O., Raunio, H., Hakkola, J. 2003. Regulation of CYP3A5 by glucocorticoids and cigarette smoke in human lung-derived cells. *J Pharmacol Exp Ther* 304:745–752.

Hunfeld, N. G., Mathot, R. A., Touw, D. J., van Schaik, R. H., Mulder, P. G., Franck, P. F., Kuipers, E. J., Geus, W. P. 2008. Effect of *CYP2C19*2* and *17* mutations on pharmacodynamics and kinetics of proton pump inhibitors in Caucasians. *Br J Clin Pharmacol* 65:752–760.

Hunt, C. M., Westerkam, W. R., Stave, G. M. 1992. Effect of age and gender on the activity of human hepatic CYP3A4. *Biochem Pharmacol* 44:275–283.

Hustert, E., Haberl, M., Burk, O., Wolbold, R., He, Y. Q., Klein, K., Nuessler, A. C. et al. 2001. The genetic determinants of the CYP3A5 polymorphism. *Pharmacogenetics* 11:773–779.

Ibeanu, G. C., Blaisdell, J., Ghanayem, B. I., Beyeler, C., Benhamou, S., Bouchardy, C., Wilkinson, G. R., Dayer, P., Daly, A. K., Goldstein, J. A. 1998a. An additional defective allele, *CYP2C19*5*, contributes to the *S*-mephenytoin poor metabolizer phenotype in Caucasians. *Pharmacogenetics* 8:129–135.

Ibeanu, G. C., Goldstein, J. A., Meyer, U., Benhamou, S., Bouchardy, C., Dayer, P., Ghanayem, B. I., Blaisdell, J. 1998b. Identification of new human *CYP2C19* alleles (*CYP2C19*6* and *CYP2C19*2B*) in a Caucasian poor metabolizer of mephenytoin. *J Pharmacol Exp Ther* 286:1490–1495.

Ibeanu, G. C., Blaisdell, J., Ferguson, R. J., Ghanayem, B. I., Brosen, K., Benhamou, S., Bouchardy, C., Wilkinson, G. R., Dayer, P., Goldstein, J. A. 1999. A novel transversion in the intron 5 donor splice junction of *CYP2C19* and a sequence polymorphism in exon 3 contribute to the poor metabolizer phenotype for the anticonvulsant drug *S*-mephenytoin. *J Pharmacol Exp Ther* 290:635–640.

Ieiri, I., Kishimoto, Y., Okochi, H., Momiyama, K., Morita, T., Kitano, M., Morisawa, T. et al. 2001. Comparison of the kinetic disposition of and serum gastrin change by lansoprazole versus rabeprazole during an 8-day dosing scheme in relation to *CYP2C19* polymorphism. *Eur J Clin Pharmacol* 57:485–492.

Ikeda, K., Yoshisue, K., Matsushima, E., Nagayama, S., Kobayashi, K., Tyson, C. A., Chiba, K., Kawaguchi, Y. 2000. Bioactivation of tegafur to 5-fluorouracil is catalyzed by cytochrome P450 2A6 in human liver microsomes *in vitro*. *Clin Cancer Res* 6:4409–4415.

Ikemura, K., Iwamoto, T., Okuda, M. 2014. MicroRNAs as regulators of drug transporters, drug-metabolizing enzymes, and tight junctions: Implication for intestinal barrier function. *Pharmacol Ther* 143:217–224.

Ikeya, K., Jaiswal, A. K., Owens, R. A., Jones, J. E., Nebert, D. W., Kimura, S. 1989. Human *CYP1A2*: Sequence, gene structure, comparison with the mouse and rat orthologous gene, and differences in liver 1A2 mRNA expression. *Mol Endocrinol* 3:1399–1408.

Illingworth, N. A., Boddy, A. V., Daly, A. K., Veal, G. J. 2011. Characterization of the metabolism of fenretinide by human liver microsomes, cytochrome P450 enzymes and UDP-glucuronosyltransferases. *Br J Pharmacol* 162:989–999.

Imai, J., Ieiri, I., Mamiya, K., Miyahara, S., Furuumi, H., Nanba, E., Yamane, M. et al. 2000. Polymorphism of the cytochrome P450 (CYP) 2C9 gene in Japanese epileptic patients: Genetic analysis of the *CYP2C9* locus. *Pharmacogenetics* 10:85–89.

Imaoka, S., Ogawa, H., Kimura, S., Gonzalez, F. J. 1993. Complete cDNA sequence and cDNA-directed expression of CYP4A11, a fatty acid omega-hydroxylase expressed in human kidney. *DNA Cell Biol* 12:893–899.

Imaoka, S., Yoneda, Y., Sugimoto, T., Hiroi, T., Yamamoto, K., Nakatani, T. 2000. CYP4B1 is a possible risk factor for bladder cancer in humans. *Biochem Biophys Res Commun* 277:776–780.

Imaoka, S., Hayashi, K., Hiroi, T., Yabusaki, Y., Kamataki, T., Funae, Y. 2001. A transgenic mouse expressing human CYP4B1 in the liver. *Biochem Biophys Res Commun* 284:757–762.

Imig, J. D. 2013. Epoxyeicosatrienoic acids, 20-hydroxyeicosatetraenoic acid, and renal microvascular function. *Prostaglandins Other Lipid Mediat* 104–105:2–7.

Imig, J. D., Simpkins, A. N., Renic, M., Harder, D. R. 2011. Cytochrome P450 eicosanoids and cerebral vascular function. *Expert Rev Mol Med* 13:e7.

Ingelman-Sundberg, M. 2005. Genetic polymorphisms of cytochrome P450 2D6 (*CYP2D6*): Clinical consequences, evolutionary aspects and functional diversity. *Pharmacogenomics J* 5:6–13.

Ingelman-Sundberg, M., Sim, S. C., Gomez, A., Rodriguez-Antona, C. 2007. Influence of cytochrome P450 polymorphisms on drug therapies: Pharmacogenetic, pharmacoepigenetic and clinical aspects. *Pharmacol Ther* 116:496–526.

Ingelman-Sundberg, M., Zhong, X. B., Hankinson, O., Beedanagari, S., Yu, A. M., Peng, L., Osawa, Y. 2013. Potential role of epigenetic mechanisms in the regulation of drug metabolism and transport. *Drug Metab Dispos* 41:1725–1731.

Inoue, K., Inazawa, J., Suzuki, Y., Shimada, T., Yamazaki, H., Guengerich, F. P., Abe, T. 1994. Fluorescence *in situ* hybridization analysis of chromosomal localization of three human cytochrome P450 2C genes (*CYP2C8*, *2C9*, and *2C10*) at 10q24.1. *Jpn J Hum Genet* 39:337–343.

Inoue, K., Yamazaki, H., Imiya, K., Akasaka, S., Guengerich, F. P., Shimada, T. 1997. Relationship between *CYP2C9* and *2C19* genotypes and tolbutamide methyl hydroxylation and *S*-mephenytoin 4′-hydroxylation activities in livers of Japanese and Caucasian populations. *Pharmacogenetics* 7:103–113.

Iscan, M., Rostami, H., Iscan, M., Guray, T., Pelkonen, O., Rautio, A. 1994. Interindividual variability of coumarin 7-hydroxylation in a Turkish population. *Eur J Clin Pharmacol* 47:315–318.

Ishida, H., Kuruta, Y., Gotoh, O., Yamashita, C., Yoshida, Y., Noshiro, M. 1999. Structure, evolution, and liver-specific expression of sterol 12α-hydroxylase P450 (CYP8B). *J Biochem* 126:19–25.

Ishiguro, A., Kubota, T., Ishikawa, H., Iga, T. 2004. Metabolic activity of dextromethorphan O-demethylation in healthy Japanese volunteers carrying duplicated CYP2D6 genes: Duplicated allele of CYP2D6*10 does not increase CYP2D6 metabolic activity. Clin Chim Acta 344:201–204.

Ishiura, H., Takahashi, Y., Hayashi, T., Saito, K., Furuya, H., Watanabe, M., Murata, M. et al. 2014. Molecular epidemiology and clinical spectrum of hereditary spastic paraplegia in the Japanese population based on comprehensive mutational analyses. J Hum Genet 59:163–172.

Isidor, B., Dagoneau, N., Huber, C., Genevieve, D., Bader-Meunier, B., Blanche, S., Picard, C. et al. 2007. A gene responsible for Ghosal hemato-diaphyseal dysplasia maps to chromosome 7q33-34. Hum Genet 121:269–273.

Itoh, S., Yanagimoto, T., Tagawa, S., Hashimoto, H., Kitamura, R., Nakajima, Y., Okochi, T., Fujimoto, S., Uchino, J., Kamataki, T. 1992. Genomic organization of human fetal specific P-450IIIA7 (cytochrome P-450HFLa)-related gene(s) and interaction of transcriptional regulatory factor with its DNA element in the 5′ flanking region. Biochim Biophys Acta 1130:133–138.

Iyanagi, T., Xia, C., Kim, J. J. 2012. NADPH-cytochrome P450 oxidoreductase: Prototypic member of the diflavin reductase family. Arch Biochem Biophys 528:72–89.

Jahng, W. J., Xue, L., Rando, R. R. 2003. Lecithin retinol acyltransferase is a founder member of a novel family of enzymes. Biochemistry 42:12805–12812.

Jalas, J. R., Ding, X., Murphy, S. E. 2003. Comparative metabolism of the tobacco-specific nitrosamines 4-(methylnitrosamino)-1-(3-pyridyl)-1-butanone and 4-(methylnitrosamino)-1-(3-pyridyl)-1-butanol by rat cytochrome P450 2A3 and human cytochrome P450 2A13. Drug Metab Dispos 31:1199–1202.

Janardan, S. K., Lown, K. S., Schmiedlin-Ren, P., Thummel, K. E., Watkins, P. B. 1996. Selective expression of CYP3A5 and not CYP3A4 in human blood. Pharmacogenetics 6:379–385.

Jang, Y. J., Cha, E. Y., Kim, W. Y., Park, S. W., Shon, J. H., Lee, S. S., Shin, J. G. 2007. CYP2S1 gene polymorphisms in a Korean population. Ther Drug Monit 29:292–298.

Jarrar, Y. B., Cho, S. A., Oh, K. S., Kim, D. H., Shin, J. G., Lee, S. J. 2013. Identification of cytochrome P450s involved in the metabolism of arachidonic acid in human platelets. Prostaglandins Leukot Essent Fatty Acids 89:227–234.

Jia, K., Li, L., Liu, Z., Hartog, M., Kluetzman, K., Zhang, Q. Y., Ding, X. 2014. Generation and characterization of a novel CYP2A13-transgenic mouse model. Drug Metab Dispos 42:1341–1348.

Jiang, J. H., Jia, W. H., Chen, H. K., Feng, B. J., Qin, H. D., Pan, Z. G., Shen, G. P. et al. 2004. Genetic polymorphisms of CYP2A13 and its relationship to nasopharyngeal carcinoma in the Cantonese population. J Transl Med 2:24.

Jiang, J. G., Chen, C. L., Card, J. W., Yang, S., Chen, J. X., Fu, X. N., Ning, Y. G., Xiao, X., Zeldin, D. C., Wang, D. W. 2005. Cytochrome P450 2J2 promotes the neoplastic phenotype of carcinoma cells and is up-regulated in human tumors. Cancer Res 65:4707–4715.

Jin, R., Koop, D. R., Raucy, J. L., Lasker, J. M. 1998. Role of human CYP4F2 in hepatic catabolism of the proinflammatory agent leukotriene B₄. Arch Biochem Biophys 359:89–98.

Jin, Y., Zollinger, M., Borell, H., Zimmerlin, A., Patten, C. J. 2011. CYP4F enzymes are responsible for the elimination of fingolimod (FTY720), a novel treatment of relapsing multiple sclerosis. Drug Metab Dispos 39:191–198.

Jin, T., Geng, T., He, N., Shi, X., Wang, L., Yuan, D., Kang, L. 2015. Genetic polymorphism analysis of the drug-metabolizing enzyme CYP2C9 in a Chinese Tibetan population. Gene 567:196–200.

Johnson, E. F., Stout, C. D. 2005. Structural diversity of human xenobiotic-metabolizing cytochrome P450 monooxygenases. Biochem Biophys Res Commun 338:331–336.

Johnson, E. F., Stout, C. D. 2013. Structural diversity of eukaryotic membrane cytochrome p450s. J Biol Chem 288:17082–17090.

Jones, G., Ramshaw, H., Zhang, A., Cook, R., Byford, V., White, J., Petkovich, M. 1999. Expression and activity of vitamin D-metabolizing cytochrome P450s (CYP1alpha and CYP24) in human nonsmall cell lung carcinomas. Endocrinology 140:3303–3310.

Jones, G., Prosser, D. E., Kaufmann, M. 2014. Cytochrome P450-mediated metabolism of vitamin D. J Lipid Res 55:13–31.

Jounaidi, Y., Hyrailles, V., Gervot, L., Maurel, P. 1996. Detection of CYP3A5 allelic variant: A candidate for the polymorphic expression of the protein? Biochem Biophys Res Commun 221:466–470.

Ju, W., Yang, S., Ansede, J. H., Stephens, C. E., Bridges, A. S., Voyksner, R. D., Ismail, M. A. et al. 2014. CYP1A1 and CYP1B1-mediated biotransformation of the antitrypanosomal methamidoxime prodrug DB844 forms novel metabolites through intramolecular rearrangement. J Pharm Sci 103:337–349.

Jurima-Romet, M., Crawford, K., Cyr, T., Inaba, T. 1994. Terfenadine metabolism in human liver. In vitro inhibition by macrolide antibiotics and azole antifungals. Drug Metab Dispos 22:849–857.

Kadlubar, F. F., Berkowitz, G. S., Delongchamp, R. R., Wang, C., Green, B. L., Tang, G., Lamba, J., Schuetz, E., Wolff, M. S. 2003. The CYP3A4*1B variant is related to the onset of puberty, a known risk factor for the development of breast cancer. Cancer Epidemiol Biomarkers Prev 12:327–331.

Kalgutkar, A. S., Nguyen, H. T., Vaz, A. D., Doan, A., Dalvie, D. K., McLeod, D. G., Murray, J. C. 2003. In vitro metabolism studies on the isoxazole ring scission in the anti-inflammatory agent lefluonomide to its active alpha-cyanoenol metabolite A771726: Mechanistic similarities with the cytochrome P450-catalyzed dehydration of aldoximes. Drug Metab Dispos 31:1240–1250.

Kalow, W., Tang, B. K. 1991. Caffeine as a metabolic probe: Exploration of the enzyme-inducing effect of cigarette smoking. Clin Pharmacol Ther 49:44–48.

Kalsotra, A., Turman, C. M., Kikuta, Y., Strobel, H. W. 2004. Expression and characterization of human cytochrome P450 4F11: Putative role in the metabolism of therapeutic drugs and eicosanoids. Toxicol Appl Pharmacol 199:295–304.

Kamataki, T., Fujieda, M., Kiyotani, K., Iwano, S., Kunitoh, H. 2005. Genetic polymorphism of CYP2A6 as one of the potential determinants of tobacco-related cancer risk. Biochem Biophy Res Com 338:306–310.

Kamdem, L. K., Liu, Y., Stearns, V., Kadlubar, S. A., Ramirez, J., Jeter, S., Shahverdi, K. et al. 2010. In vitro and in vivo oxidative metabolism and glucuronidation of anastrozole. Br J Clin Pharmacol 70:854–869.

Karlgren, M., Backlund, M., Johansson, I., Oscarson, M., Ingelman-Sundberg, M. 2004. Characterization and tissue distribution of a novel human cytochrome P450-CYP2U1. Biochem Biophys Res Commun 315:679–685.

Karlgren, M., Gomez, A., Stark, K., Svard, J., Rodriguez-Antona, C., Oliw, E., Bernal, M. L., Ramon y Cajal, S., Johansson, I., Ingelman-Sundberg, M. 2006. Tumor-specific expression of the novel cytochrome P450 enzyme, CYP2W1. Biochem Biophys Res Commun 341:451–458.

Kaspera, R., Kirby, B. J., Sahele, T., Collier, A. C., Kharasch, E. D., Unadkat, J. D., Totah, R. A. 2014. Investigating the contribution of CYP2J2 to ritonavir metabolism in vitro and in vivo. Biochem Pharmacol 91:109–118.

Kasuga, H., Hosogane, N., Matsuoka, K., Mori, I., Sakura, Y., Shimakawa, K., Shinki, T., Suda, T., Taketomi, S. 2002. Characterization of transgenic rats constitutively expressing vitamin D-24-hydroxylase gene. *Biochem Biophys Res Commun* 297:1332–1338.

Katsumata, N., Ohtake, M., Hojo, T., Ogawa, E., Hara, T., Sato, N., Tanaka, T. 2002. Compound heterozygous mutations in the cholesterol side-chain cleavage enzyme gene (CYP11A) cause congenital adrenal insufficiency in humans. *J Clin Endocrinol Metab* 87:3808–3813.

Kawamoto, T., Sueyoshi, T., Zelko, I., Moore, R., Washburn, K., Negishi, M. 1999. Phenobarbital-responsive nuclear translocation of the receptor CAR in induction of the *CYP2B* gene. *Mol Cell Biol* 19:6318–6322.

Kawashima, A., Satta, Y. 2014. Substrate-dependent evolution of cytochrome P450: Rapid turnover of the detoxification-type and conservation of the biosynthesis-type. *PLoS One* 9:e100059.

Kawashima, H., Naganuma, T., Kusunose, E., Kono, T., Yasumoto, R., Sugimura, K., Kishimoto, T. 2000. Human fatty acid omega-hydroxylase, CYP4A11: Determination of complete genomic sequence and characterization of purified recombinant protein. *Arch Biochem Biophys* 378:333–339.

Kazakoff, K., Iversen, P., Lawson, T., Baron, J., Guengerich, F. P., Pour, P. M. 1994. Involvement of cytochrome P450 2E1-like isoform in the activation of N-nitrosobis(2-oxopropyl)amine in the rat nasal mucosa. *Eur J Cancer B Oral Oncol* 30b:179–185.

Kazmi, F., Barbara, J. E., Yerino, P., Parkinson, A. 2015. A long-standing mystery solved: The formation of 3-hydroxydesloratadine is catalyzed by CYP2C8 but prior glucuronidation of desloratadine by UDP-glucuronosyltransferase 2B10 is an obligatory requirement. *Drug Metab Dispos* 43:523–533.

Keber, R., Motaln, H., Wagner, K. D., Debeljak, N., Rassoulzadegan, M., Acimovic, J., Rozman, D., Horvat, S. 2011. Mouse knockout of the cholesterogenic cytochrome P450 lanosterol 14α-demethylase (Cyp51) resembles Antley–Bixler syndrome. *J Biol Chem* 286:29086–29097.

Keber, R., Rozman, D., Horvat, S. 2013. Sterols in spermatogenesis and sperm maturation. *J Lipid Res* 54:20–33.

Kelley, R. I., Kratz, L. E., Glaser, R. L., Netzloff, M. L., Wolf, L. M., Jabs, E. W. 2002. Abnormal sterol metabolism in a patient with Antley–Bixler syndrome and ambiguous genitalia. *Am J Med Genet* 110:95–102.

Kelly, S. L., Kelly, D. E. 2013. Microbial cytochromes P450: Biodiversity and biotechnology. Where do cytochromes P450 come from, what do they do and what can they do for us? *Philos Trans R Soc Lond B Biol Sci* 368:20120476.

Kelly, D. E., Krasevec, N., Mullins, J., Nelson, D. R. 2009. The CYPome (Cytochrome P450 complement) of *Aspergillus nidulans*. *Fungal Genet Biol* 46 Suppl 1:S53–S61.

Kelly, E. J., Nakano, M., Rohatgi, P., Yarov-Yarovoy, V., Rettie, A. E. 2011. Finding homes for orphan cytochrome P450s: CYP4V2 and CYP4F22 in disease states. *Mol Interv* 11:124–132.

Kent, U. M., Yanev, S., Hollenberg, P. F. 1999. Mechanism-based inactivation of cytochromes P450 2B1 and P450 2B6 by n-propylxanthate. *Chem Res Toxicol* 12:317–322.

Kent, U. M., Aviram, M., Rosenblat, M., Hollenberg, P. F. 2002a. The licorice root derived isoflavan glabridin inhibits the activities of human cytochrome P450S 3A4, 2B6, and 2C9. *Drug Metab Dispos* 30:709–715.

Kent, U. M., Mills, D. E., Rajnarayanan, R. V., Alworth, W. L., Hollenberg, P. F. 2002b. Effect of 17α-ethynylestradiol on activities of cytochrome P450 2B (P450 2B) enzymes: Characterization of inactivation of P450s 2B1 and 2B6 and identification of metabolites. *J Pharmacol Exp Ther* 300:549–558.

Kent, U. M., Sridar, C., Spahlinger, G., Hollenberg, P. F. 2008. Modification of serine 360 by a reactive intermediate of 17α-ethynylestradiol results in mechanism-based inactivation of cytochrome P450s 2B1 and 2B6. *Chem Res Toxicol* 21:1956–1963.

Keskin, M., Ugurlu, A. K., Savas-Erdeve, S., Sagsak, E., Akyuz, S. G., Cetinkaya, S., Aycan, Z. 2015. 17α-Hydroxylase/17,20-lyase deficiency related to *P.Y27**(c.81C>A) mutation in *CYP17A1* gene. *J Pediatr Endocrinol Metab* 28:919–921.

Kharasch, E. D., Friedel, C., Gadel, S. 2015. Differences in methadone metabolism by CYP2B6 variants. *Drug Metab Dispos* 43:994–1001.

Khatri, Y., Hannemann, F., Ewen, K. M., Pistorius, D., Perlova, O., Kagawa, N., Brachmann, A. O., Muller, R., Bernhardt, R. 2010. The CYPome of *Sorangium cellulosum* So ce56 and identification of CYP109D1 as a new fatty acid hydroxylase. *Chem Biol* 17:1295–1305.

Kidd, R. S., Curry, T. B., Gallagher, S., Edeki, T., Blaisdell, J., Goldstein, J. A. 2001. Identification of a null allele of CYP2C9 in an African-American exhibiting toxicity to phenytoin. *Pharmacogenomics* 11:803–808.

Kikuta, Y., Kato, M., Yamashita, Y., Miyauchi, Y., Tanaka, K., Kamada, N., Kusunose, M. 1998. Human leukotriene B4 omega-hydroxylase (CYP4F3) gene: Molecular cloning and chromosomal localization. *DNA Cell Biol* 17:221–230.

Kikuta, Y., Kusunose, E., Kusunose, M. 2000. Characterization of human liver leukotriene B$_4$ ω-hydroxylase P450 (CYP4F2). *J Biochem* 127:1047–1052.

Kikuta, Y., Kusunose, E., Kusunose, M. 2002. Prostaglandin and leukotriene omega-hydroxylases. *Prostaglandins Other Lipid Mediat* 68–69:345–362.

Kikuta, Y., Yamashita, Y., Kashiwagi, S., Tani, K., Okada, K., Nakata, K. 2004. Expression and induction of CYP4F subfamily in human leukocytes and HL60 cells. *Biochim Biophys Acta* 1683:7–15.

Kikuta, Y., Mizomoto, J., Strobel, H. W., Ohkawa, H. 2007. Expression and physiological function of CYP4F subfamily in human eosinophils. *Biochim Biophys Acta* 1771:1439–1445.

Kim, C. J. 2014. Congenital lipoid adrenal hyperplasia. *Ann Pediatr Endocrinol Metab* 19:179–183.

Kim, K. A., Shon, J. H., Park, J. Y., Yoon, Y. R., Kim, M. J., Yun, D. H., Kim, M. K., Cha, I. J., Hyun, M. H., Shin, J. G. 2002. Enantioselective disposition of lansoprazole in extensive and poor metabolizers of CYP2C19. *Clin Pharmacol Ther* 72:90–99.

Kim, K. A., Park, P. W., Lee, O. J., Kang, D. K., Park, J. Y. 2007. Effect of polymorphic *CYP3A5* genotype on the single-dose simvastatin pharmacokinetics in healthy subjects. *J Clin Pharmacol* 47:87–93.

Kim, C. J., Lin, L., Huang, N., Quigley, C. A., AvRuskin, T. W., Achermann, J. C., Miller, W. L. 2008. Severe combined adrenal and gonadal deficiency caused by novel mutations in the cholesterol side chain cleavage enzyme, P450scc. *J Clin Endocrinol Metab* 93:696–702.

Kim, H., Choi, H. K., Jeong, T. C., Jahng, Y., Kim, D. H., Lee, S. H., Lee, S. 2013. Selective inhibitory effects of mollugin on CYP1A2 in human liver microsomes. *Food Chem Toxicol* 51:33–37.

Kim, Y. M., Kang, M., Choi, J. H., Lee, B. H., Kim, G. H., Ohn, J. H., Kim, S. Y., Park, M. S., Yoo, H. W. 2014. A review of the literature on common *CYP17A1* mutations in adults with 17-hydroxylase/17,20-lyase deficiency, a case series of such mutations among Koreans and functional characteristics of a novel mutation. *Metabolism* 63:42–49.

Kimonen, T., Juvonen, R. O., Alhava, E., Pasanen, M. 1995. The inhibition of CYP enzymes in mouse and human liver by pilocarpine. *Br J Pharmacol* 114:832–836.

Kimura, S., Umeno, M., Skoda, R. C., Meyer, U. A., Gonzalez, F. J. 1989. The human debrisoquine 4-hydroxylase (*CYP2D*) locus: Sequence and identification of the polymorphic *CYP2D6* gene, a related gene, and a pseudogene. *Am J Hum Genet* 45:889–904.

King, L. M., Ma, J., Srettabunjong, S., Graves, J., Bradbury, J. A., Li, L., Spiecker, M., Liao, J. K., Mohrenweiser, H., Zeldin, D. C. 2002. Cloning of CYP2J2 gene and identification of functional polymorphisms. *Mol Pharmacol* 61:840–852.

Kinobe, R. T., Parkinson, O. T., Mitchell, D. J., Gillam, E. M. 2005. P450 2C18 catalyzes the metabolic bioactivation of phenytoin. *Chem Res Toxicol* 18:1868–1875.

Kinonen, T., Pasanen, M., Gynther, J., Poso, A., Jarvinen, T., Alhava, E., Juvonen, R. O. 1995. Competitive inhibition of coumarin 7-hydroxylation by pilocarpine and its interaction with mouse CYP 2A5 and human CYP 2A6. *Br J Pharmacol* 116:2625–2630.

Kirchheiner, J., Seeringer, A. 2007. Clinical implications of pharmacogenetics of cytochrome P450 drug metabolizing enzymes. *Biochim Biophys Acta* 1770:489–494.

Kirchheiner, J., Bauer, S., Meineke, I., Rohde, W., Prang, V., Meisel, C., Roots, I., Brockmoller, J. 2002. Impact of CYP2C9 and CYP2C19 polymorphisms on tolbutamide kinetics and the insulin and glucose response in healthy volunteers. *Pharmacogenomics* 12:101–109.

Kirchheiner, J., Klein, C., Meineke, I., Sasse, J., Zanger, U. M., Murdter, T. E., Roots, I., Brockmoller, J. 2003. Bupropion and 4-OH-bupropion pharmacokinetics in relation to genetic polymorphisms in CYP2B6. *Pharmacogenetics* 13:619–626.

Kirischian, N. L., Wilson, J. Y. 2012. Phylogenetic and functional analyses of the cytochrome P450 family 4. *Mol Phylogenet Evol* 62:458–471.

Kitagawa, K., Kunugita, N., Kitagawa, M., Kawamoto, T. 2001. *CYP2A6*6*, a novel polymorphism in cytochrome P450 2A6, has a single amino acid substitution (R128Q) that inactivates enzymatic activity. *J Biol Chem* 276:17830–17835.

Kitanaka, S., Takeyama, K., Murayama, A., Kato, S. 2001. The molecular basis of vitamin D-dependent rickets type I. *Endocr J* 48:427–432.

Kivisto, K. T., Niemi, M., Schaeffeler, E., Pitkala, K., Tilvis, R., Fromm, M. F., Schwab, M., Eichelbaum, M., Strandberg, T. 2004. Lipid-lowering response to statins is affected by CYP3A5 polymorphism. *Pharmacogenetics* 14:523–525.

Kiyotani, K., Yamazaki, H., Fujieda, M., Iwano, S., Matsumura, K., Satarug, S., Ujjin, P. et al. 2003. Decreased coumarin 7-hydroxylase activities and CYP2A6 expression levels in humans caused by genetic polymorphism in *CYP2A6* promoter region (*CYP2A6*9*). *Pharmacogenetics* 13:689–695.

Klaassen, C. D., Lu, H., Cui, J. Y. 2011. Epigenetic regulation of drug processing genes. *Toxicol Mech Methods* 21:312–324.

Klose, T. S., Blaisdell, J. A., Goldstein, J. A. 1999. Gene structure of *CYP2C8* and extrahepatic distribution of the human CYP2Cs. *J Biochem Mol Toxicol* 13:289–295.

Kocelak, P., Olszanecka-Glinianowicz, M., Chudek, J. 2012. Fibroblast growth factor 23—Structure, function and role in kidney diseases. *Adv Clin Exp Med* 21:391–401.

Koch, I., Weil, R., Wolbold, R., Brockmoller, J., Hustert, E., Burk, O., Nuessler, A. et al. 2002. Interindividual variability and tissue-specificity in the expression of cytochrome P450 3A mRNA. *Drug Metab Dispos* 30:1108–1114.

Koenigs, L. L., Trager, W. F. 1998. Mechanism-based inactivation of P450 2A6 by furanocoumarins. *Biochemistry* 37:10047–10061.

Kohle, C., Bock, K. W. 2007. Coordinate regulation of Phase I and II xenobiotic metabolisms by the Ah receptor and Nrf2. *Biochem Pharmacol* 73:1853–1862.

Kojima, K., Nagata, K., Matsubara, T., Yamazoe, Y. 2007. Broad but distinct role of pregnane x receptor on the expression of individual cytochrome p450s in human hepatocytes. *Drug Metab Pharmacokinet* 22:276–286.

Komatsu, T., Yamazaki, H., Shimada, N., Nakajima, M., Yokoi, T. 2000. Roles of cytochromes P450 1A2, 2A6, and 2C8 in 5-fluorouracil formation from tegafur, an anticancer prodrug, in human liver microsomes. *Drug Metab Dispos* 28:1457–1463.

Komori, M., Nishio, K., Ohi, H., Kitada, M., Kamataki, T. 1989. Molecular cloning and sequence analysis of cDNA containing the entire coding region for human fetal liver cytochrome P-450. *J Biochem* 105:161–163.

Koop, D. R. 1990. Inhibition of ethanol-inducible cytochrome P450IIE1 by 3-amino-1,2,4-triazole. *Chem Res Toxicol* 3:377–383.

Koop, D. R. 1992. Oxidative and reductive metabolism by cytochrome P450 2E1. *FASEB J* 6:724–730.

Kopf, P. G., Scott, J. A., Agbor, L. N., Boberg, J. R., Elased, K. M., Huwe, J. K., Walker, M. K. 2010. Cytochrome P4501A1 is required for vascular dysfunction and hypertension induced by 2,3,7,8-tetrachlorodibenzo-p-dioxin. *Toxicol Sci* 117:537–546.

Koppens, P. F., Hoogenboezem, T., Degenhart, H. J. 2002. Duplication of the *CYP21A2* gene complicates mutation analysis of steroid 21-hydroxylase deficiency: Characteristics of three unusual haplotypes. *Hum Genet* 111:405–410.

Korzekwa, K. 2014. Enzyme kinetics of oxidative metabolism: Cytochromes P450. *Methods Mol Biol* 1113:149–166.

Koskela, S., Hakkola, J., Hukkanen, J., Pelkonen, O., Sorri, M., Saranen, A., Anttila, S., Fernandez-Salguero, P., Gonzalez, F., Raunio, H. 1999. Expression of *CYP2A* genes in human liver and extrahepatic tissues. *Biochem Pharmacol* 57:1407–1413.

Koyama, E., Chiba, K., Tani, M., Ishizaki, T. 1997. Reappraisal of human CYP isoforms involved in imipramine N-demethylation and 2-hydroxylation: A study using microsomes obtained from putative extensive and poor metabolizers of S-mephenytoin and eleven recombinant human CYPs. *J Pharmacol Exp Ther* 281:1199–1210.

Krajisnik, T., Bjorklund, P., Marsell, R., Ljunggren, O., Akerstrom, G., Jonsson, K. B., Westin, G., Larsson, T. E. 2007. Fibroblast growth factor-23 regulates parathyroid hormone and 1α-hydroxylase expression in cultured bovine parathyroid cells. *J Endocrinol* 195:125–131.

Kristensen, V. N., Harada, N., Yoshimura, N., Haraldsen, E., Lonning, P. E., Erikstein, B., Karesen, R., Kristensen, T., Borresen-Dale, A. L. 2000. Genetic variants of *CYP19* (aromatase) and breast cancer risk. *Oncogene* 19:1329–1333.

Krivospitskaya, O., Elmabsout, A. A., Sundman, E., Soderstrom, L. A., Ovchinnikova, O., Gidlof, A. C., Scherbak, N. et al. 2012. A *CYP26B1* polymorphism enhances retinoic acid catabolism and may aggravate atherosclerosis. *Mol Med* 18:712–718.

Kroemer, H. K., Gautier, J. C., Beaune, P., Henderson, C., Wolf, C. R., Eichelbaum, M. 1993. Identification of P450 enzymes involved in metabolism of verapamil in humans. *Naunyn Schmiedebergs Arch Pharmacol* 348:332–337.

Krone, N., Arlt, W. 2009. Genetics of congenital adrenal hyperplasia. *Best Pract Res Clin Endocrinol Metab* 23:181–192.

Kuehl, P., Zhang, J., Lin, Y., Lamba, J., Assem, M., Schuetz, J., Watkins, P. B. et al. 2001. Sequence diversity in *CYP3A* promoters and characterization of the genetic basis of polymorphic CYP3A5 expression. *Nat Genet* 27:383–391.

Kuffel, M. J., Schroeder, J. C., Pobst, L. J., Naylor, S., Reid, J. M., Kaufmann, S. H., Ames, M. M. 2002. Activation of the antitumor agent aminoflavone (NSC 686288) is mediated by induction of tumor cell cytochrome P450 1A1/1A2. *Mol Pharmacol* 62:143–153.

Kumar, S. 2015. Computational identification and binding analysis of orphan human cytochrome P450 4X1 enzyme with substrates. *BMC Res Notes* 8:9.

Kumar, S., Samuel, K., Subramanian, R., Braun, M. P., Stearns, R. A., Chiu, S. H., Evans, D. C., Baillie, T. A. 2002. Extrapolation of diclofenac clearance from in vitro microsomal metabolism data: Role of acyl glucuronidation and sequential oxidative metabolism of the acyl glucuronide. *J Pharmacol Exp Ther* 303:969–978.

Kumar, V., Wahlstrom, J. L., Rock, D. A., Warren, C. J., Gorman, L. A., Tracy, T. S. 2006. CYP2C9 inhibition: Impact of probe selection and pharmacogenetics on *in vitro* inhibition profiles. *Drug Metab Dispos* 34:1966–1975.

Kung, T., Murphy, K. A., White, L. A. 2009. The aryl hydrocarbon receptor (AhR) pathway as a regulatory pathway for cell adhesion and matrix metabolism. *Biochem Pharmacol* 77:536–546.

Labuda, M., Lemieux, N., Tihy, F., Prinster, C., Glorieux, F. H. 1993. Human 25-hydroxyvitamin D 24-hydroxylase cytochrome P450 subunit maps to a different chromosomal location than that of pseudovitamin D-deficient rickets. *J Bone Miner Res* 8:1397–1406.

Lagadic-Gossmann, D., Lerche, C., Rissel, M., Joannard, F., Galisteo, M., Guillouzo, A., Corcos, L. 2000. The induction of the human hepatic CYP2E1 gene by interleukin 4 is transcriptional and regulated by protein kinase C. *Cell Biol Toxicol* 16:221–233.

Lah, L., Haridas, S., Bohlmann, J., Breuil, C. 2013. The cytochromes P450 of *Grosmannia clavigera*: Genome organization, phylogeny, and expression in response to pine host chemicals. *Fungal Genet Biol* 50:72–81.

Lai, T. Y., Ng, T. K., Tam, P. O., Yam, G. H., Ngai, J. W., Chan, W. M., Liu, D. T., Lam, D. S., Pang, C. P. 2007. Genotype phenotype analysis of Bietti's crystalline dystrophy in patients with CYP4V2 mutations. *Invest Ophthalmol Vis Sci* 48:5212–5220.

Lai, X. S., Yang, L. P., Li, X. T., Liu, J. P., Zhou, Z. W., Zhou, S. F. 2009. Human CYP2C8: Structure, substrate specificity, inhibitor selectivity, inducers and polymorphisms. *Curr Drug Metab* 10:1009–1047.

Lai, T. Y., Chu, K. O., Chan, K. P., Ng, T. K., Yam, G. H., Lam, D. S., Pang, C. P. 2010. Alterations in serum fatty acid concentrations and desaturase activities in Bietti crystalline dystrophy unaffected by CYP4V2 genotypes. *Invest Ophthalmol Vis Sci* 51:1092–1097.

Laine, J. E., Auriola, S., Pasanen, M., Juvonen, R. O. 2009. Acetaminophen bioactivation by human cytochrome P450 enzymes and animal microsomes. *Xenobiotica* 39:11–21.

Lamb, D. C., Skaug, T., Song, H. L., Jackson, C. J., Podust, L. M., Waterman, M. R., Kell, D. B., Kelly, D. E., Kelly, S. L. 2002. The cytochrome P450 complement (CYPome) of *Streptomyces coelicolor* A3(2). *J Biol Chem* 277:24000–24005.

Lamb, D. C., Ikeda, H., Nelson, D. R., Ishikawa, J., Skaug, T., Jackson, C., Omura, S., Waterman, M. R., Kelly, S. L. 2003. Cytochrome p450 complement (CYPome) of the avermectin-producer *Streptomyces avermitilis* and comparison to that of *Streptomyces coelicolor* A3(2). *Biochem Biophys Res Commun* 307:610–619.

Lamba, J. K., Lin, Y. S., Schuetz, E. G., Thummel, K. E. 2002a. Genetic contribution to variable human CYP3A-mediated metabolism. *Adv Drug Deliv Rev* 54:1271–1294.

Lamba, J. K., Lin, Y. S., Thummel, K., Daly, A., Watkins, P. B., Strom, S., Zhang, J., Schuetz, E. G. 2002b. Common allelic variants of cytochrome P4503A4 and their prevalence in different populations. *Pharmacogenetics* 12:121–132.

Lambrecht, R. W., Sinclair, P. R., Gorman, N., Sinclair, J. F. 1992. Uroporphyrinogen oxidation catalyzed by reconstituted cytochrome P450IA2. *Arch Biochem Biophys* 294:504–510.

Lampen, A., Bader, A., Bestmann, T., Winkler, M., Witte, L., Borlak, J. T. 1998. Catalytic activities, protein- and mRNA-expression of cytochrome P450 isoenzymes in intestinal cell lines. *Xenobiotica* 28:429–441.

Lan, M. Y., Yeh, T. H., Chang, Y. Y., Kuo, H. C., Sun, H. S., Lai, S. C., Lu, C. S. 2015. Clinical and genetic analysis of Taiwanese patients with hereditary spastic paraplegia type 5. *Eur J Neurol* 22:211–214.

Lanza, D. L., Code, E., Crespi, C. L., Gonzalez, F. J., Yost, G. S. 1999. Specific dehydrogenation of 3-methylindole and epoxidation of naphthalene by recombinant human CYP2F1 expressed in lymphoblastoid cells. *Drug Metab Dispos* 27:798–803.

Larsen, J. T., Brosen, K. 2005. Consumption of charcoal-broiled meat as an experimental tool for discerning CYP1A2-mediated drug metabolism in vivo. *Basic Clin Pharmacol Toxicol* 97:141–148.

Lasker, J. M., Chen, W. B., Wolf, I., Bloswick, B. P., Wilson, P. D., Powell, P. K. 2000. Formation of 20-hydroxyeicosatetraenoic acid, a vasoactive and natriuretic eicosanoid, in human kidney. Role of Cyp4F2 and Cyp4A11. *J Biol Chem* 275:4118–4126.

Lathe, R. 2002. Steroid and sterol 7-hydroxylation: Ancient pathways. *Steroids* 67:967–977.

Le Gal, A., Dreano, Y., Gervasi, P. G., Berthou, F. 2001. Human cytochrome P450 2A6 is the major enzyme involved in the metabolism of three alkoxyethers used as oxyfuels. *Toxicol Lett* 124:47–58.

Le Gal, A., Dreano, Y., Lucas, D., Berthou, F. 2003. Diversity of selective environmental substrates for human cytochrome P450 2A6: Alkoxyethers, nicotine, coumarin, *N*-nitrosodiethylamine, and *N*-nitrosobenzylmethylamine. *Toxicol Lett* 144:77–91.

Le Quere, V., Plee-Gautier, E., Potin, P., Madec, S., Salaun, J. P. 2004. Human CYP4F3s are the main catalysts in the oxidation of fatty acid epoxides. *J Lipid Res* 45:1446–1458.

Lee, Y. S., Dutta, A. 2009. MicroRNAs in cancer. *Annu Rev Pathol* 4:199–227.

Lee, S. S. T., Buters, J. T., Pineau, T., Fernandez-Salguero, P., Gonzalez, F. J. 1996. Role of CYP2E1 in the hepatotoxicity of acetaminophen. *J Biol Chem* 271:12063–12067.

Lee, A. J., Kosh, J. W., Conney, A. H., Zhu, B. T. 2001. Characterization of the NADPH-dependent metabolism of 17β-estradiol to multiple metabolites by human liver microsomes and selectively expressed human cytochrome P450 3A4 and 3A5. *J Pharmacol Exp Ther* 298:420–432.

Lee, M. J., Huang, Y. C., Sweeney, M. G., Wood, N. W., Reilly, M. M., Yip, P. K. 2002. Mutation of the sterol 27-hydroxylase gene (CYP27A1) in a Taiwanese family wirh cerebrotendinous xanthomatosis. *J Neurol* 246.

Lee, A. J., Cai, M. X., Thomas, P. E., Conney, A. H., Zhu, B. T. 2003. Characterization of the oxidative metabolites of 17β-estradiol and estrone formed by 15 selectively expressed human cytochrome P450 isoforms. *Endocrinology* 144:3382–3398.

Lee, C. R., Pieper, J. A., Frye, R. F., Hinderliter, A. L., Blaisdell, J. A., Goldstein, J. A. 2003. Tolbutamide, flurbiprofen, and losartan as probes of CYP2C9 activity in humans. *J Clin Pharmacol* 43:84–91.

Lee, K. Y., Koh, A. H., Aung, T., Yong, V. H., Yeung, K., Ang, C. L., Vithana, E. N. 2005. Characterization of Bietti crystalline dystrophy patients with CYP4V2 mutations. *Invest Ophthalmol Vis Sci* 46:3812–3816.

Lee, A. M., Jepson, C., Hoffmann, E., Epstein, L., Hawk, L. W., Lerman, C., Tyndale, R. F. 2007. CYP2B6 genotype alters abstinence rates in a bupropion smoking cessation trial. *Biol Psychiatry* 62:635–641.

Lee, S. J., Perera, L., Coulter, S. J., Mohrenweiser, H. W., Jetten, A., Goldstein, J. A. 2007. The discovery of new coding alleles of human *CYP26A1* that are potentially defective in the metabolism of all-*trans* retinoic acid and their assessment in a recombinant cDNA expression system. *Pharmacogenet Genomics* 17:169–180.

Lee, C. A., Neul, D., Clouser-Roche, A., Dalvie, D., Wester, M. R., Jiang, Y., Jones, J. P., 3rd, Freiwald, S., Zientek, M., Totah, R. A. 2010. Identification of novel substrates for human cytochrome P450 2J2. *Drug Metab Dispos* 38:347–356.

Lee, J. I., Zhang, L., Men, A. Y., Kenna, L. A., Huang, S. M. 2010. CYP-mediated therapeutic protein-drug interactions: Clinical findings, proposed mechanisms and regulatory implications. *Clin Pharmacokinet* 49:295–310.

Lee, C. A., Jones, J. P., 3rd, Katayama, J., Kaspera, R., Jiang, Y., Freiwald, S., Smith, E., Walker, G. S., Totah, R. A. 2012. Identifying a selective substrate and inhibitor pair for the evaluation of CYP2J2 activity. *Drug Metab Dispos* 40:943–951.

Leeder, J. S., Gaedigk, R., Marcucci, K. A., Gaedigk, A., Vyhlidal, C. A., Schindel, B. P., Pearce, R. E. 2005. Variability of CYP3A7 expression in human fetal liver. *J Pharmacol Exp Ther* 314:626–635.

Lepesheva, G. I., Waterman, M. R. 2011. Structural basis for conservation in the CYP51 family. *Biochim Biophys Acta* 1814:88–93.

Levova, K., Moserova, M., Kotrbova, V., Sulc, M., Henderson, C. J., Wolf, C. R., Phillips, D. H. et al. 2011. Role of cytochromes P450 1A1/2 in detoxication and activation of carcinogenic aristolochic acid I: Studies with the hepatic NADPH: cytochrome P450 reductase null (HRN) mouse model. *Toxicol Sci* 121:43–56.

Lewinska, M., Zelenko, U., Merzel, F., Golic Grdadolnik, S., Murray, J. C., Rozman, D. 2013. Polymorphisms of CYP51A1 from cholesterol synthesis: Associations with birth weight and maternal lipid levels and impact on CYP51 protein structure. *PLoS One* 8:e82554.

Lewis, D. F. 2000. On the recognition of mammalian microsomal cytochrome P450 substrates and their characteristics: Towards the prediction of human p450 substrate specificity and metabolism. *Biochem Pharmacol* 60:293–306.

Lewis, D. F. 2003. Human cytochromes P450 associated with the Phase 1 metabolism of drugs and other xenobiotics: A compilation of substrates and inhibitors of the CYP1, CYP2 and CYP3 families. *Curr Med Chem* 10:1955–1972.

Lewis, D. F. 2004. 57 varieties: The human cytochromes P450. *Pharmacogenomics* 5:305–318.

Lewis, D. F., Ito, Y. 2010. Human CYPs involved in drug metabolism: Structures, substrates and binding affinities. *Expert Opin Drug Metab Toxicol* 6:661–674.

Li-Hawkins, J., Lund, E. G., Bronson, A. D., Russell, D. W. 2000. Expression cloning of an oxysterol 7α-hydroxylase selective for 24-hydroxycholesterol. *J Biol Chem* 275:16543–16549.

Li, Y., Li, N. Y., Sellers, E. M. 1997. Comparison of CYP2A6 catalytic activity on coumarin 7-hydroxylation in human and monkey liver microsomes. *Eur J Drug Metab Pharmacokinet* 22:295–304.

Li, X. Q., Bjorkman, A., Andersson, T. B., Ridderstrom, M., Masimirembwa, C. M. 2002. Amodiaquine clearance and its metabolism to *N*-desethylamodiaquine is mediated by CYP2C8: A new high affinity and turnover enzyme-specific probe substrate. *J Pharmacol Exp Ther* 300:399–407.

Li, A., Jiao, X., Munier, F. L., Schorderet, D. F., Yao, W., Iwata, F., Hayakawa, M. et al. 2004. Bietti crystalline corneoretinal dystrophy is caused by mutations in the novel gene *CYP4V2*. *Am J Hum Genet* 74:817–826.

Li, Y. C., Chiang, C. W., Yeh, H. C., Hsu, P. Y., Whitby, F. G., Wang, L. H., Chan, N. L. 2008. Structures of prostacyclin synthase and its complexes with substrate analog and inhibitor reveal a ligand-specific heme conformation change. *J Biol Chem* 283:2917–2926.

Li, N., Zhou, Y., Du, L., Wei, M., Chen, X. 2011. Overview of cytochrome P450 1B1 gene mutations in patients with primary congenital glaucoma. *Exp Eye Res* 93:572–579.

Li, L. H., Yin, X. Y., Wu, X. H., Zhang, L., Pan, S. Y., Zheng, Z. J., Wang, J. G. 2014. Serum 25(OH)D and vitamin D status in relation to VDR, GC and CYP2R1 variants in Chinese. *Endocr J* 61:133–141.

Liang, R., Li, L., Li, C., Gao, Y., Liu, W., Hu, D., Sun, Y. 2012a. Impact of *CYP2C9*3*, VKORC1-1639, CYP4F2rs2108622 genetic polymorphism and clinical factors on warfarin maintenance dose in Han-Chinese patients. *J Thromb Thrombolysis* 34:120–125.

Liang, R., Wang, C., Zhao, H., Huang, J., Hu, D., Sun, Y. 2012b. Influence of *CYP4F2* genotype on warfarin dose requirement—A systematic review and meta-analysis. *Thromb Res* 130:38–44.

Liang, J. Q., Yan, M. R., Yang, L., Suyila, Q., Cui, H. W., Su, X. L. 2014. Association of a CYP4A11 polymorphism and hypertension in the Mongolian and Han populations of China. *Genet Mol Res* 13:508–517.

Libby, R. T., Smith, R. S., Savinova, O. V., Zabaleta, A., Martin, J. E., Gonzalez, F. J., John, S. W. 2003. Modification of ocular defects in mouse developmental glaucoma models by tyrosinase. *Science* 299:1578–1581.

Lieber, C. S. 1997. Cytochrome P-4502E1: Its physiological and pathological role. *Physiol Rev* 77:517–544.

Lieberman, J., Slack, F., Pandolfi, P. P., Chinnaiyan, A., Agami, R., Mendell, J. T. 2013. Noncoding RNAs and cancer. *Cell* 153:9–10.

Liehr, J. G., Ricci, M. J., Jefcoate, C. R., Hannigan, E. V., Hokanson, J. A. B. T., Z. 1995. 4-Hydroxylation of estradiol by human uterine myometrium and myoma microsomes: Implications for the mechanism of uterine tumorigenesi. *Proc Natl Acad Sci U S A* 92:9220–9224.

Lin, J. H., Lu, A. Y. 1998. Inhibition and induction of cytochrome P450 and the clinical implications. *Clin Pharmacokinet* 35:361–390.

Lin, J. H., Lu, A. Y. 2001. Interindividual variability in inhibition and induction of cytochrome P450 enzymes. *Annu Rev Pharmacol Toxicol* 41:535–567.

Lin, Y., Wu, K. K., Ruan, K. H. 1998. Characterization of the secondary structure and membrane interaction of the putative membrane anchor domains of prostaglandin I2 synthase and cytochrome P450 2C1. *Arch Biochem Biophys* 352:78–84.

Lin, J., Nishiguchi, K. M., Nakamura, M., Dryja, T. P., Berson, E. L., Miyake, Y. 2005. Recessive mutations in the CYP4V2 gene in East Asian and Middle Eastern patients with Bietti crystalline corneoretinal dystrophy. *J Med Genet* 42:e38.

Lin, H. L., Zhang, H., Hollenberg, P. F. 2009. Metabolic activation of mifepristone [RU486; 17β-hydroxy-11β-(4-dimethylaminophenyl)-17α-(1-propynyl)-estra-4,9-dien-3-one] by mammalian cytochromes P450 and the mechanism-based inactivation of human CYP2B6. *J Pharmacol Exp Ther* 329:26–37.

Lindberg, R. L., Negishi, M. 1989. Alteration of mouse cytochrome P450coh substrate specificity by mutation of a single amino-acid residue. *Nature* 339:632–634.

Liu, C., Zhuo, X., Gonzalez, F. J., Ding, X. 1996. Baculovirus-mediated expression and characterization of rat CYP2A3 and human CYP2A6: Role in metabolic activation of nasal toxicants. *Mol Pharmacol* 50:781–788.

Liu, Y. T., Hao, H. P., Liu, C. X., Wang, G. J., Xie, H. G. 2007. Drugs as CYP3A probes, inducers, and inhibitors. *Drug Metab Rev* 39:699–721.

Liu, Y. L., Xu, Y., Li, F., Chen, H., Guo, S. L. 2013. CYP2A6 deletion polymorphism is associated with decreased susceptibility of lung cancer in Asian smokers: A meta-analysis. *Tumour Biol* 34:2651–2657.

Liu, Z., Li, L., Wu, H., Hu, J., Ma, J., Zhang, Q. Y., Ding, X. 2015. Characterization of CYP2B6 in a CYP2B6-humanized mouse model: Inducibility in the liver by phenobarbital and dexamethasone and role in nicotine metabolism *in vivo*. *Drug Metab Dispos* 43:208–216.

Lo-Guidice, J. M., Allorge, D., Cauffiez, C., Chevalier, D., Lafitte, J. J., Lhermitte, M., Broly, F. 2002. Genetic polymorphism of the human cytochrome P450 CYP4B1: Evidence for a nonfunctional allelic variant. *Pharmacogenetics* 12:367–374.

Lo, J., Di Nardo, G., Griswold, J., Egbuta, C., Jiang, W., Gilardi, G., Ghosh, D. 2013. Structural basis for the functional roles of critical residues in human cytochrome p450 aromatase. *Biochemistry* 52:5821–5829.

Lo, S. N., Chang, Y. P., Tsai, K. C., Chang, C. Y., Wu, T. S., Ueng, Y. F. 2013. Inhibition of CYP1 by berberine, palmatine, and jatrorrhizine: Selectivity, kinetic characterization, and molecular modeling. *Toxicol Appl Pharmacol* 272:671–680.

Locuson, C. W., Wienkers, L. C., Jones, J. P., Tracy, T. S. 2007. CYP2C9 protein interactions with cytochrome b_5: Effects on the coupling of catalysis. *Drug Metab Dispos* 35:1174–1181.

Lopez-Garrido, M. P., Medina-Trillo, C., Morales-Fernandez, L., Garcia-Feijoo, J., Martinez-de-la-Casa, J. M., Garcia-Anton, M., Escribano, J. 2013. Null *CYP1B1* genotypes in primary congenital and nondominant juvenile glaucoma. *Ophthalmology* 120:716–723.

Lorbek, G., Lewinska, M., Rozman, D. 2012. Cytochrome P450s in the synthesis of cholesterol and bile acids—From mouse models to human diseases. *FEBS J* 279:1516–1533.

Lou, Y. R., Qiao, S., Talonpoika, R., Syvala, H., Tuohimaa, P. 2004. The role of Vitamin D3 metabolism in prostate cancer. *J Steroid Biochem Mol Biol* 92:317–325.

Lund, E. G., Guileyardo, J. M., Russell, D. W. 1999. cDNA cloning of cholesterol 24-hydroxylase, a mediator of cholesterol homeostasis in the brain. *Proc Natl Acad Sci U S A* 96:7238–7243.

Lund, E. G., Xie, C., Kotti, T., Turley, S. D., Dietschy, J. M., Russell, D. W. 2003. Knockout of the cholesterol 24-hydroxylase gene in mice reveals a brain-specific mechanism of cholesterol turnover. *J Biol Chem* 278:22980–22988.

Lytton, S. D., Berg, U., Nemeth, A., Ingelman-Sundberg, M. 2002. Autoantibodies against cytochrome P450s in sera of children treated with immunosuppressive drugs. *Clin Exp Immunol* 127:293–302.

Ma, Q., Lu, A. Y. 2007. CYP1A induction and human risk assessment: An evolving tale of *in vitro* and *in vivo* studies. *Drug Metab Dispos* 35:1009–1016.

Ma, J., Ramachandran, S., Fiedorek, F. T., Jr. Zeldin, D. C. 1998. Mapping of the CYP2J cytochrome P450 genes to human chromosome 1 and mouse chromosome 4. *Genomics* 49:152–155.

Ma, J., Qu, W., Scarborough, P. E., Tomer, K. B., Moomaw, C. R., Maronpot, R., Davis, L. S., Breyer, M. D., Zeldin, D. C. 1999. Molecular cloning, enzymatic characterization, developmental expression, and cellular localization of a mouse cytochrome P450 highly expressed in kidney. *J Biol Chem* 274:17777–17788.

Ma, J., Bradbury, J. A., King, L., Maronpot, R., Davis, L. S., Breyer, M. D., Zeldin, D. C. 2002. Molecular cloning and characterization of mouse CYP2J6, an unstable cytochrome P450 isoform. *Biochem Pharmacol* 64:1447–1460.

Ma, C. X., Adjei, A. A., Salavaggione, O. E., Coronel, J., Pelleymounter, L., Wang, L., Eckloff, B. W. et al. 2005. Human aromatase: Gene resequencing and functional genomics. *Cancer Res* 65:11071–11082.

Ma, X., Idle, J. R., Krausz, K. W., Gonzalez, F. J. 2005. Metabolism of melatonin by human cytochromes p450. *Drug Metab Dispos* 33:489–494.

Mace, K., Bowman, E. D., Vautravers, P., Shields, P. G., Harris, C. C., Pfeifer, A. M. 1998. Characterisation of xenobiotic-metabolising enzyme expression in human bronchial mucosa and peripheral lung tissues. *Eur J Cancer* 34:914–920.

Mackay, D. S., Halford, S. 2012. Focus on molecules: Cytochrome P450 family 4, subfamily V, polypeptide 2 (CYP4V2). *Exp Eye Res* 102:111–112.

Mackenzie, P. I., Gregory, P. A., Gardner-Stephen, D. A., Lewinsky, R. H., Jorgensen, B. R., Nishiyama, T., Xie, W., Radominska-Pandya, A. 2003. Regulation of UDP glucuronosyltransferase genes. *Curr Drug Metab* 4:249–257.

Madan, A., Parkinson, A., Faiman, M. D. 1998. Identification of the human P450 enzymes responsible for the sulfoxidation and thiono-oxidation of diethyldithiocarbamate methyl ester: Role of P450 enzymes in disulfiram bioactivation. *Alcohol Clin Exp Res* 22:1212–1219.

Madan, A., Graham, R. A., Carroll, K. M., Mudra, D. R., Burton, L. A., Krueger, L. A., Downey, A. D. et al. 2003. Effects of prototypical microsomal enzyme inducers on cytochrome P450 expression in cultured human hepatocytes. *Drug Metab Dispos* 31:421–431.

Madanayake, T. W., Fidler, T. P., Fresquez, T. M., Bajaj, N., Rowland, A. M. 2012. Cytochrome P450 2S1 depletion enhances cell proliferation and migration in bronchial epithelial cells, in part, through modulation of prostaglandin E_2 synthesis. *Drug Metab Dispos* 40:2119–2125.

Madani, S., Paine, M. F., Lewis, L., Thummel, K. E., Shen, D. D. 1999. Comparison of CYP2D6 content and metoprolol oxidation between microsomes isolated from human livers and small intestines. *Pharm Res* 16:1199–1205.

Maekawa, K., Fukushima-Uesaka, H., Tohkin, M., Hasegawa, R., Kajio, H., Kuzuya, N., Yasuda, K. et al. 2006. Four novel defective alleles and comprehensive haplotype analysis of CYP2C9 in Japanese. *Pharmacogenet Genomics* 16:497–514.

Maezawa, K., Matsunaga, T., Takezawa, T., Kanai, M., Ohira, S., Ohmori, S. 2010. Cytochrome P450 3As gene expression and testosterone 6β-hydroxylase activity in human fetal membranes and placenta at full term. *Biol Pharm Bull* 33:249–254.

Mahavorasirikul, W., Tassaneeyakul, W., Satarug, S., Reungweerayut, R., Na-Bangchang, C., Na-Bangchang, K. 2009. *CYP2A6* genotypes and coumarin-oxidation phenotypes in a Thai population and their relationship to tobacco smoking. *Eur J Clin Pharmacol* 65:377–384.

Makinen, J., Frank, C., Jyrkkarinne, J., Gynther, J., Carlberg, C., Honkakoski, P. 2002. Modulation of mouse and human phenobarbital-responsive enhancer module by nuclear receptors. *Mol Pharmacol* 62:366–378.

Mamatha, G., Umashankar, V., Kasinathan, N., Krishnan, T., Sathyabaarathi, R., Karthiyayini, T., Amali, J., Rao, C., Madhavan, J. 2011. Molecular screening of the CYP4V2 gene in Bietti crystalline dystrophy that is associated with choroidal neovascularization. *Mol Vis* 17:1970–1977.

Marill, J., Cresteil, T., Lanotte, M., Chabot, G. G. 2000. Identification of human cytochrome P450s involved in the formation of all-*trans*-retinoic acid principal metabolites. *Mol Pharmacol* 58:1341–1348.

Marill, J., Capron, C. C., Idres, N., Chabot, G. G. 2002. Human cytochrome P450s involved in the metabolism of 9-*cis*- and 13-*cis*-retinoic acids. *Biochem Pharmacol* 63:933–943.

Marrapodi, M., Chiang, J. Y. 2000. Peroxisome proliferator-activated receptor alpha (PPARalpha) and agonist inhibit cholesterol 7alpha-hydroxylase gene (CYP7A1) transcription. *J Lipid Res* 41:514–520.

Martin, A., David, V., Quarles, L. D. 2012. Regulation and function of the FGF23/klotho endocrine pathways. *Physiol Rev* 92:131–155.

Massimi, A., Malaponti, M., Federici, L., Vinciguerra, D., Manca Bitti, M. L., Vottero, A., Ghizzoni, L. et al. 2014. Functional and structural analysis of four novel mutations of *CYP21A2* gene in Italian patients with 21-hydroxylase deficiency. *Horm Metab Res* 46:515–520.

Mast, N., White, M. A., Bjorkhem, I., Johnson, E. F., Stout, C. D., Pikuleva, I. A. 2008. Crystal structures of substrate-bound and substrate-free cytochrome P450 46A1, the principal cholesterol hydroxylase in the brain. *Proc Natl Acad Sci U S A* 105:9546–9551.

Mast, N., Charvet, C., Pikuleva, I. A., Stout, C. D. 2010. Structural basis of drug binding to CYP46A1, an enzyme that controls cholesterol turnover in the brain. *J Biol Chem* 285:31783–31795.

Mast, N., Zheng, W., Stout, C. D., Pikuleva, I. A. 2013. Antifungal azoles: Structural insights into undesired tight binding to cholesterol-metabolizing CYP46A1. *Mol Pharmacol* 84:86–94.

Masubuchi, Y., Hosokawa, S., Horie, T., Suzuki, T., Ohmori, S., Kitada, M., Narimatsu, S. 1994. Cytochrome P450 isozymes involved in propranolol metabolism in human liver microsomes. The role of CYP2D6 as ring-hydroxylase and CYP1A2 as N-desisopropylase. *Drug Metab Dispos* 22:909–915.

Masuda, S., Byford, V., Arabian, A., Sakai, Y., Demay, M. B., St-Arnaud, R., Jones, G. 2005. Altered pharmacokinetics of 1α,25-dihydroxyvitamin D_3 and 25-hydroxyvitamin D_3 in the blood and tissues of the 25-hydroxyvitamin D-24-hydroxylase (*Cyp24a1*) null mouse. *Endocrinology* 146:825–834.

Matoulkova, P., Pavek, P., Maly, J., Vlcek, J. 2014. Cytochrome P450 enzyme regulation by glucocorticoids and consequences in terms of drug interaction. *Expert Opin Drug Metab Toxicol* 10:425–435.

Matsumoto, S., Hirama, T., Matsubara, T., Nagata, K., Yamazoe, Y. 2002. Involvement of CYP2J2 on the intestinal first-pass metabolism of antihistamine drug, astemizole. *Drug Metab Dispos* 30:1240–1245.

McClellan, K., Jarvis, B. 2001. Desloratadine. *Drugs* 61:789–796; discussion 797.

McDonald, M. G., Rieder, M. J., Nakano, M., Hsia, C. K., Rettie, A. E. 2009. CYP4F2 is a vitamin K_1 oxidase: An explanation for altered warfarin dose in carriers of the V433M variant. *Mol Pharmacol* 75:1337–1346.

McNab, F., Mayer-Barber, K., Sher, A., Wack, A. O'Garra, A. 2015. Type I interferons in infectious disease. *Nat Rev Immunol* 15:87–103.

McSorley, L. C., Daly, A. K. 2000. Identification of human cytochrome P450 isoforms that contribute to all-*trans*-retinoic acid 4-hydroxylation. *Biochem Pharmacol* 60:517–526.

Megaraj, V., Zhou, X., Xie, F., Liu, Z., Yang, W., Ding, X. 2014. Role of CYP2A13 in the bioactivation and lung tumorigenicity of the tobacco-specific lung procarcinogen 4-(methylnitrosamino)-1-(3-pyridyl)-1-butanone: *In vivo* studies using a CYP2A13-humanized mouse model. *Carcinogenesis* 35:131–137.

Mehl, M., Bidmon, H. J., Hilbig, H., Zilles, K., Dringen, R., Ullrich, V. 1999. Prostacyclin synthase is localized in rat, bovine and human neuronal brain cells. *Neurosci Lett* 271:187–190.

Melet, A., Assrir, N., Jean, P., Pilar Lopez-Garcia, M., Marques-Soares, C., Jaouen, M., Dansette, P. M., Sari, M. A., Mansuy, D. 2003. Substrate selectivity of human cytochrome P450 2C9: Importance of residues 476, 365, and 114 in recognition of diclofenac and sulfaphenazole and in mechanism-based inactivation by tienilic acid. *Arch Biochem Biophys* 409:80–91.

Melki, R., Colomb, E., Lefort, N., Brezin, A. P., Garchon, H. J. 2004. *CYP1B1* mutations in French patients with early-onset primary open-angle glaucoma. *J Med Genet* 41:647–651.

Mendell, J. T., Olson, E. N. 2012. MicroRNAs in stress signaling and human disease. *Cell* 148:1172–1187.

Mestres, J. 2005. Structure conservation in cytochromes P450. *Proteins* 58:596–609.

Meunier, B., de Visser, S. P., Shaik, S. 2004. Mechanism of oxidation reactions catalyzed by cytochrome p450 enzymes. *Chem Rev* 104:3947–3980.

Michaels, S., Wang, M. Z. 2014. The revised human liver cytochrome P450 "Pie": Absolute protein quantification of CYP4F and CYP3A enzymes using targeted quantitative proteomics. *Drug Metab Dispos* 42:1241–1251.

Michalets, E. L., Williams, C. R. 2000. Drug interactions with cisapride: Clinical implications. *Clin Pharmacokinet* 39:49–75.

Michaud, V., Frappier, M., Dumas, M. C., Turgeon, J. 2010. Metabolic activity and mRNA levels of human cardiac CYP450s involved in drug metabolism. *PLoS One* 5:e15666.

Micheal, S., Ayub, H., Zafar, S. N., Bakker, B., Ali, M., Akhtar, F., Islam, F., Khan, M. I., Qamar, R. den Hollander, A. I. 2015. Identification of novel *CYP1B1* gene mutations in patients with primary congenital and primary open-angle glaucoma. *Clin Exp Ophthalmol* 43:31–39.

Mignarri, A., Malandrini, A., Del Puppo, M., Magni, A., Monti, L., Ginanneschi, F., Tessa, A., Santorelli, F. M., Federico, A., Dotti, M. T. 2014. Hereditary spastic paraplegia type 5: A potentially treatable disorder of cholesterol metabolism. *J Neurol* 261:617–619.

Miksys, S., Rao, Y., Hoffmann, E., Mash, D. C., Tyndale, R. F. 2002. Regional and cellular expression of CYP2D6 in human brain: Higher levels in alcoholics. *J Neurochem* 82:1376–1387.

Miles, J. S., Bickmore, W., Brook, J. D., McLaren, A. W., Meehan, R., Wolf, C. R. 1989. Close linkage of the human cytochrome P450IIA and P450IIB gene subfamilies: Implications for the assignment of substrate specificity. *Nucleic Acids Res* 17:2907–2917.

Miles, J. S., McLaren, A. W., Forrester, L. M., Glancey, M. J., Lang, M. A., Wolf, C. R. 1990. Identification of the human liver cytochrome P450 responsible for coumarin 7-hydroxylase activity. *Biochem J* 267:365–371.

Mimori, K., Tanaka, Y., Yoshinaga, K., Masuda, T., Yamashita, K., Okamoto, M., Inoue, H., Mori, M. 2004. Clinical significance of overexpression of the candidate oncogene CYO24 in esophageal cancer. *Ann Oncol* 15:236–241.

Miners, J. O., Birkett, D. J. 1998. Cytochrome P450 2C9: An enzyme of major importance in human drug metabolism. *Br J Clin Pharmacol* 45:525–538.

Minoletti, C., Dijols, S., Dansette, P. M., Mansury, D. 1999. Comparison of the substrate specificities of human liver cytochrome P450s 2C9 and 2C18: Application to the design of a specific substrate of CYP 2C18. *Biochemistry* 38:7828–7836.

Minutolo, C., Nadra, A. D., Fernandez, C., Taboas, M., Buzzalino, N., Casali, B., Belli, S., Charreau, E. H., Alba, L., Dain, L. 2011. Structure-based analysis of five novel disease-causing mutations in 21-hydroxylase-deficient patients. *PLoS One* 6:e15899.

Mitschele, T., Diesel, B., Friedrich, M., Meineke, V., Maas, R. M., Gartner, B. C., Kamradt, J., Meese, E., Tilgen, W., Reichrath, J. 2004. Analysis of the vitamin D system in basal cell carcinomas (BCCs). *Lab Invest* 84:693–702.

Mitsuuchi, Y., Kawamoto, T., Miyahara, K., Ulick, S., Morton, D. H., Naiki, Y., Kuribayashi, I. et al. 1993. Congenitally defective aldosterone biosynthesis in humans: Inactivation of the P-450C18 gene (CYP11B2) due to nucleotide deletion in CMO I deficient patients. *Biochem Biophys Res Commun* 190:864–869.

Miyakawa, H., Kitazawa, E., Kikuchi, K., Fujikawa, H., Kawaguchi, N., Abe, K., Matsushita, M. et al. 2000. Immunoreactivity to various human cytochrome P450 proteins of sera from patients with autoimmune hepatitis, chronic hepatitis B, and chronic hepatitis C. *Autoimmunity* 33:23–32.

Miyata, A., Hara, S., Yokoyama, C., Inoue, H., Ullrich, V., Tanabe, T. 1994a. Molecular cloning and expression of human prostacyclin synthase. *Biochem Biophys Res Commun* 200:1728–1734.

Miyata, A., Yokoyama, C., Ihara, H., Bandoh, S., Takeda, O., Takahashi, E., Tanabe, T. 1994b. Characterization of the human gene (*TBXAS1*) encoding thromboxane synthase. *Eur J Biochem* 224:273–279.

Mo, S. L., Liu, Y. H., Duan, W., Wei, M. Q., Kanwar, J. R., Zhou, S. F. 2009a. Substrate specificity, regulation, and polymorphism of human cytochrome P450 2B6. *Curr Drug Metab* 10:730–753.

Mo, S. L., Zhou, Z. W., Yang, L. P., Wei, M. Q., Zhou, S. F. 2009b. New insights into the structural features and functional relevance of human cytochrome P450 2C9. Part I. *Curr Drug Metab* 10:1075–1126.

Mo, S. L., Zhou, Z. W., Yang, L. P., Wei, M. Q., Zhou, S. F. 2009c. New insights into the structural features and functional relevance of human cytochrome P450 2C9. Part II. *Curr Drug Metab* 10:1127–1150.

Moghadasian, M. H. 2004. Cerebrotendinous xanthomatosis: Clinical course, genotypes and metabolic backgrounds. *Clin Invest Med* 27:42–50.

Mohri, T., Nakajima, M., Fukami, T., Takamiya, M., Aoki, Y., Yokoi, T. 2009. Human CYP2E1 is regulated by miR-378. *Biochem Pharmacol*.

Molanaei, H., Stenvinkel, P., Qureshi, A. R., Carrero, J. J., Heimburger, O., Lindholm, B., Diczfalusy, U., Odar-Cederlof, I., Bertilsson, L. 2012. Metabolism of alprazolam (a marker of CYP3A4) in hemodialysis patients with persistent inflammation. *Eur J Clin Pharmacol* 68:571–577.

Monostory, K., Dvorak, Z. 2011. Steroid regulation of drug-metabolizing cytochromes P450. *Curr Drug Metab* 12:154–172.

Moore, L. B., Parks, D. J., Jones, S. A., Bledsoe, R. K., Consler, T. G., Stimmel, J. B., Goodwin, B. et al. 2000. Orphan nuclear receptors constitutive androstane receptor and pregnane X receptor share xenobiotic and steroid ligands. *J Biol Chem* 275:15122–15127.

Morales, A., Cuellar, A., Ramirez, J., Vilchis, F., Diaz-Sanchez, V. 1999. Synthesis of steroids in pancreas: Evidence of cytochrome P-450scc activity. *Pancreas* 19:39–44.

Moreau, A., Maurel, P., Vilarem, M. J., Pascussi, J. M. 2007. Constitutive androstane receptor-vitamin D receptor crosstalk: Consequence on *CYP24* gene expression. *Biochem Biophys Res Commun* 360:76–82.

Morgan, E. T., Goralski, K. B., Piquette-Miller, M., Renton, K. W., Robertson, G. R., Chaluvadi, M. R., Charles, K. A. et al. 2008. Regulation of drug-metabolizing enzymes and transporters in infection, inflammation, and cancer. *Drug Metab Dispos* 36:205–216.

Morita, J., Kobayashi, K., Wanibuchi, A., Kimura, M., Irie, S., Ishizaki, T., Chiba, K. 2004. A novel single nucleotide polymorphism (SNP) of the *CYP2C19* gene in a Japanese subject with lowered capacity of mephobarbital 4'-hydroxylation. *Drug Metab Pharmacokinet* 19:236–238.

Morohashi, K., Sogawa, K., Omura, T., Fujii-Kuriyama, Y. 1987. Gene structure of human cytochrome P-450(SCC), cholesterol desmolase. *J Biochem* 101:879–887.

Morrison, H. G., Oleksiak, M. F., Cornell, N. W., Sogin, M. L., Stegeman, J. J. 1995. Identification of cytochrome P450 1A (*CYP1A*) genes from two teleost fish, toadfish (*Opsanus tau*) and scup (*Stenotomus chrysops*), and phylogenetic analysis of *CYP1A* genes. *Biochem J* 308 (Pt 1):97–104.

Moschella, C., Jaber, B. L. 2001. Interaction between cyclosporine and *Hypericum perforatum* (St. John's wort) after organ transplantation. *Am J Kidney Dis* 38:1105–1107.

Mukai, K., Mitani, F., Agake, R., Ishimura, Y. 1998. Adrenocorticotropic hormone stimulates CYP11B1 gene transcription through a mechanism involving AP-1 factors. *Eur J Biochem* 256:190–200.

Mulatero, P., Schiavone, D., Fallo, F., Rabbia, F., Pilon, C., Chiandussi, L., Pascoe, L., Veglio, F. 2000. CYP11B2 Gene Polymorphisms in Idiopathic Hyperaldosteronism. *Hypertenison* 35:694–698.

Munro, A. W., Lindsay, J. G. 1996. Bacterial cytochromes P-450. *Mol Microbiol* 20:1115–1125.

Murphy, S. E., Raulinaitis, V., Brown, K. M. 2005. Nicotine 5'-oxidation and methyl oxidation by P450 2A enzymes. *Drug Metab Dispos* 33:1166–1173.

Murray, M. 1997. Drug-mediated inactivation of cytochrome P450. *Clin Exp Pharmacol Physiol* 24:465–470.

Murray, G. I., Pritchard, S., Melvin, W. T., Burke, M. D. 1995. Cytochrome P450 CYP3A5 in the human anterior pituitary gland. *FEBS Lett* 364:79–82.

Mwenifumbo, J. C., Tyndale, R. F. 2007. Genetic variability in CYP2A6 and the pharmacokinetics of nicotine. *Pharmacogenomics* 8:1385–1402.

Mwenifumbo, J. C., Al Koudsi, N., Ho, M. K., Zhou, Q., Hoffmann, E. B., Sellers, E. M., Tyndale, R. F. 2008. Novel and established *CYP2A6* alleles impair *in vivo* nicotine metabolism in a population of Black African descent. *Hum Mutat* 29:679–688.

Nakajima, M., Yokoi, T. 2011. MicroRNAs from biology to future pharmacotherapy: Regulation of cytochrome P450s and nuclear receptors. *Pharmacol Ther* 131:330–337.

Nakajima, T., Elovaara, E., Gonzalez, F. J., Gelboin, H. V., Raunio, H., Pelkonen, O., Vainio, H., Aoyama, T. 1994. Styrene metabolism by cDNA-expressed human hepatic and pulmonary cytochromes P450. *Chem Res Toxicol* 7:891–896.

Nakajima, M., Kobayashi, K., Shimada, N., Tokudome, S., Yamamoto, T., Kuroiwa, Y. 1998. Involvement of CYP1A2 in mexiletine metabolism. *Br J Clin Pharmacol* 46:55–62.

Nakajima, M., Yokoi, T., Mizutani, M., Kinoshita, M., Funayama, M., Kamataki, T. 1999. Genetic polymorphism in the 5'-flanking region of human CYP1A2 gene: Effect on the CYP1A2 inducibility in humans. *J Biochem* 125:803–808.

Nakajima, M., Fujiki, Y., Noda, K., Ohtsuka, H., Ohkuni, H., Kyo, S., Inoue, M., Kuroiwa, Y., Yokoi, T. 2003. Genetic polymorphisms of *CYP2C8* in Japanese population. *Drug Metab Dispos* 31:687–690.

Nakajima, M., Fukami, T., Yamanaka, H., Higashi, E., Sakai, H., Yoshida, R., Kwon, J. T., McLeod, H. L., Yokoi, T. 2006a. Comprehensive evaluation of variability in nicotine metabolism and *CYP2A6* polymorphic alleles in four ethnic populations. *Clin Pharmacol Ther* 80:282–297.

Nakajima, M., Itoh, M., Sakai, H., Fukami, T., Katoh, M., Yamazaki, H., Kadlubar, F. F., Imaoka, S., Funae, Y., Yokoi, T. 2006b. CYP2A13 expressed in human bladder metabolically activates 4-aminobiphenyl. *Int J Cancer* 119:2520–2526.

Nakamura, M., Lin, J., Nishiguchi, K., Kondo, M., Sugita, J., Miyake, Y. 2006. Bietti crystalline corneoretinal dystrophy associated with *CYP4V2* gene mutations. *Adv Exp Med Biol* 572:49–53.

Nakano, M., Kelly, E. J., Wiek, C., Hanenberg, H., Rettie, A. E. 2012. CYP4V2 in Bietti's crystalline dystrophy: Ocular localization, metabolism of ω-3-polyunsaturated fatty acids, and functional deficit of the p.H331P variant. *Mol Pharmacol* 82:679–686.

Nakano, M., Fukushima, Y., Yokota, S., Fukami, T., Takamiya, M., Aoki, Y., Yokoi, T., Nakajima, M. 2015. CYP2A7 pseudogene transcript affects CYP2A6 expression in human liver by acting as a decoy for miR-126. *Drug Metab Dispos* 43:703–712.

Nakayama, T., Soma, M., Izumi, Y., Kanmatsuse, K. 1996. Organization of the human prostacyclin synthase gene. *Biochem Biophys Res Commun* 221:803–806.

Nakayama, T., Soma, M., Saito, S., Honye, J., Yajima, J., Rahmutula, D., Kaneko, Y. et al. 2002a. Association of a novel single nucleotide polymorphism of the prostacyclin synthase gene with myocardial infarction. *Am Heart J* 143:797–801.

Nakayama, T., Soma, M., Watanabe, Y., Hasimu, B., Sato, M., Aoi, N., Kosuge, K. et al. 2002b. Splicing mutation of the prostacyclin synthase gene in a family associated with hypertension. *Biochem Biophys Res Commun* 297:1135–1139.

Naoe, T., Takeyama, K., Yokozawa, T., Kiyoi, H., Seto, M., Uike, N., Ino, T. et al. 2000. Analysis of genetic polymorphism in *NQO1*, *GST-M1*, *GST-T1*, and *CYP3A4* in 469 Japanese patients with therapy-related leukemia/myelodysplastic syndrome and de novo acute myeloid leukemia. *Clin Cancer Res* 6:4091–4095.

Narjoz, C., Favre, A., McMullen, J., Kiehl, P., Montemurro, M., Figg, W. D., Beaune, P., de Waziers, I., Rochat, B. 2014. Important role of CYP2J2 in protein kinase inhibitor degradation: A possible role in intratumor drug disposition and resistance. *PLoS One* 9:e95532.

Nebert, D. W., Russell, D. W. 2002. Clinical importance of the cytochromes P450. *Lancet* 360:1155–1162.

Nebert, D. W., Dalton, T. P. 2006. The role of cytochrome P450 enzymes in endogenous signalling pathways and environmental carcinogenesis. *Nat Rev Cancer* 6:947–960.

Nebert, D. W., Wikvall, K., Miller, W. L. 2013. Human cytochromes P450 in health and disease. *Philos Trans R Soc Lond B Biol Sci* 368:20120431.

Nelson, D. R. 2011. Progress in tracing the evolutionary paths of cytochrome P450. *Biochim Biophys Acta* 1814:14–18.

Nelson, D. R., Zeldin, D. C., Hoffman, S. M., Maltais, L. J., Wain, H. M., Nebert, D. W. 2004. Comparison of cytochrome P450 (CYP) genes from the mouse and human genomes, including nomenclature recommendations for genes, pseudogenes and alternative-splice variants. *Pharmacogenetics* 14:1–18.

Nelson, C. H., Buttrick, B. R., Isoherranen, N. 2013. Therapeutic potential of the inhibition of the retinoic acid hydroxylases CYP26A1 and CYP26B1 by xenobiotics. *Curr Top Med Chem* 13:1402–1428.

Nelson, D. R., Goldstone, J. V., Stegeman, J. J. 2013. The cytochrome P450 genesis locus: The origin and evolution of animal cytochrome P450s. *Philos Trans R Soc Lond B Biol Sci* 368:20120474.

Nerurkar, P. V., Park, S. S., Thomas, P. E., Nims, R. W., Lubet, R. A. 1993. Methoxyresorufin and benzyloxyresorufin: Substrates preferentially metabolized by cytochromes P4501A2 and 2B, respectively, in the rat and mouse. *Biochem Pharmacol* 46:933–943.

Neve, E. P., Ingelman-Sundberg, M. 2008. Intracellular transport and localization of microsomal cytochrome P450. *Anal Bioanal Chem* 392:1075–1084.

Neve, E. P., Ingelman-Sundberg, M. 2010. Cytochrome P450 proteins: Retention and distribution from the endoplasmic reticulum. *Curr Opin Drug Discov Devel* 13:78–85.

Newsome, A. W., Nelson, D., Corran, A., Kelly, S. L., Kelly, D. E. 2013. The cytochrome P450 complement (CYPome) of *Mycosphaerella graminicola*. *Biotechnol Appl Biochem* 60:52–64.

Newton, D. J., Wang, R. W., Lu, A. Y. 1995. Cytochrome P450 inhibitors. Evaluation of specificities in the in vitro metabolism of therapeutic agents by human liver microsomes. *Drug Metab Dispos* 23:154–158.

Nguyen, L. P., Bradfield, C. A. 2008. The search for endogenous activators of the aryl hydrocarbon receptor. *Chem Res Toxicol* 21:102–116.

Nguyen, A. D., Conley, A. J. 2008. Adrenal androgens in humans and nonhuman primates: Production, zonation and regulation. *Endocr Dev* 13:33–54.

Nguyen, M., Boutignon, H., Mallet, E., Linglart, A., Guillozo, H., Jehan, F., Garabedian, M. 2010. Infantile hypercalcemia and hypercalciuria: New insights into a vitamin D-dependent mechanism and response to ketoconazole treatment. *J Pediatr* 157:296–302.

Nhamburo, P. T., Gonzalez, F. J., McBride, O. W., Gelboin, H. V., Kimura, S. 1989. Identification of a new P450 expressed in human lung: Complete cDNA sequence, cDNA-directed expression, and chromosome mapping. *Biochemistry* 28:8060–8066.

Nilsson, T., Ivanov, I. V., Oliw, E. H. 2010. LC-MS/MS analysis of epoxyalcohols and epoxides of arachidonic acid and their oxygenation by recombinant CYP4F8 and CYP4F22. *Arch Biochem Biophys* 494:64–71.

Nishida, C. R., Lee, M. de Montellano, P. R. 2010. Efficient hypoxic activation of the anticancer agent AQ4N by CYP2S1 and CYP2W1. *Mol Pharmacol* 78:497–502.

Nishimoto, M., Noshiro, M., Okuda, K. 1993. Structure of the gene encoding human liver cholesterol 7 alpha-hydroxylase. *Biochim Biophys Acta* 1172:147–150.

Nishimura, M., Yaguti, H., Yoshitsugu, H., Naito, S., Satoh, T. 2003. Tissue distribution of mRNA expression of human cytochrome P450 isoforms assessed by high-sensitivity real-time reverse transcription PCR. *Yakugaku Zasshi* 123:369–375.

Nishiya, Y., Hagihara, K., Ito, T., Tajima, M., Miura, S., Kurihara, A., Farid, N. A., Ikeda, T. 2009. Mechanism-based inhibition of human cytochrome P450 2B6 by ticlopidine, clopidogrel, and the thiolactone metabolite of prasugrel. *Drug Metab Dispos* 37:589–593.

Nissen, J., Rasmussen, L. B., Ravn-Haren, G., Andersen, E. W., Hansen, B., Andersen, R., Mejborn, H., Madsen, K. H., Vogel, U. 2014. Common variants in *CYP2R1* and *GC* genes predict vitamin D concentrations in healthy Danish children and adults. *PLoS One* 9:e89907.

Niwa, T., Yamazaki, H. 2012. Comparison of cytochrome P450 2C subfamily members in terms of drug oxidation rates and substrate inhibition. *Curr Drug Metab* 13:1145–1159.

Niwa, T., Yabusaki, Y., Honma, K., Matsuo, N., Tatsuta, K., Ishibashi, F., Katagiri, M. 1998. Contribution of human hepatic cytochrome P450 isoforms to regioselective hydroxylation of steroid hormones. *Xenobiotica* 28:539–547.

Niwa, T., Sato, R., Yabusaki, Y., Ishibashi, F., Katagiri, M. 1999. Contribution of human hepatic cytochrome P450s and steroidogenic CYP17 to the N-demethylation of aminopyrine. *Xenobiotica* 29:187–193.

Niwa, T., Murayama, N., Yamazaki, H. 2009. Oxidation of endobiotics mediated by xenobiotic-metabolizing forms of human cytochrome. *Curr Drug Metab* 10:700–712.

Nordmark, A., Lundgren, S., Ask, B., Granath, F., Rane, A. 2002. The effect of the CYP1A2 *1F mutation on CYP1A2 inducibility in pregnant women. *Br J Clin Pharmacol* 54:504–510.

Noreau, A., Dion, P. A., Szuto, A., Levert, A., Thibodeau, P., Brais, B., Dupre, N., Rioux, M. F., Rouleau, G. A. 2012. *CYP7B1* mutations in French-Canadian hereditary spastic paraplegia subjects. *Can J Neurol Sci* 39:91–94.

Norlin, M., Wikvall, K. 2007. Enzymes in the conversion of cholesterol into bile acids. *Curr Mol Med* 7:199–218.

Norlin, M., Toll, A., Bjorkhem, I., Wikvall, K. 2000. 24-hydroxycholesterol is a substrate for hepatic cholesterol 7α-hydroxylase (CYP7A). *J Lipid Res* 41:1629–1639.

Norton, I. D., Apte, M. V., Haber, P. S., McCaughan, G. W., Pirola, R. C., Wilson, J. S. 1998. Cytochrome P4502E1 is present in rat pancreas and is induced by chronic ethanol administration. *Gut* 42:426–430.

Nunes, M. J., Milagre, I., Schnekenburger, M., Gama, M. J., Diederich, M., Rodrigues, E. 2010. Sp proteins play a critical role in histone deacetylase inhibitor-mediated derepression of CYP46A1 gene transcription. *J Neurochem* 113:418–431.

Nusing, R., Ullrich, V. 1992. Regulation of cyclooxygenase and thromboxane synthase in human monocytes. *Eur J Biochem* 206:131–136.

Nusing, R., Fehr, P. M., Gudat, F., Kemeny, E., Mihatsch, M. J., Ullrich, V. 1994. The localization of thromboxane synthase in normal and pathological human kidney tissue using a monoclonal antibody Tu 300. *Virchows Arch* 424:69–74.

O'Donnell, V. B., Maskrey, B., Taylor, G. W. 2009. Eicosanoids: Generation and detection in mammalian cells. *Methods Mol Biol* 462:5–23.

O'Keefe, D. P. 1995. Plants as hosts for heterologous cytochrome P450 expression. *Drug Metabol Drug Interact* 12:383–389.

O'Shaughnessy, P. J., Monteiro, A., Bhattacharya, S., Fraser, M. J., Fowler, P. A. 2013. Steroidogenic enzyme expression in the human fetal liver and potential role in the endocrinology of pregnancy. *Mol Hum Reprod* 19:177–187.

O'Shea, D., Davis, S. N., Kim, R. B., Wilkinson, G. R. 1994. Effect of fasting and obesity in humans on the 6-hydroxylation of chlorzoxazone: A putative probe of CYP2E1 activity. *Clin Pharmacol Ther* 56:359–367.

Obach, R. S., Zhang, Q. Y., Dunbar, D., Kaminsky, L. S. 2001. Metabolic characterization of the major human small intestinal cytochrome p450s. *Drug Metab Dispos* 29:347–352.

Oda, Y., Furuichi, K., Tanaka, K., Hiroi, T., Imaoka, S., Asada, A., Fujimori, M., Funae, Y. 1995. Metabolism of a new local anesthetic, ropivacaine, by human hepatic cytochrome P450. *Anesthesiology* 82:214–220.

Oezguen, N., Kumar, S. 2011. Analysis of cytochrome P450 conserved sequence motifs between Helices E and H: Prediction of critical motifs and residues in enzyme functions. *J Drug Metab Toxicol* 2:1000110.

Oezguen, N., Kumar, S., Hindupur, A., Braun, W., Muralidhara, B. K., Halpert, J. R. 2008. Identification and analysis of conserved sequence motifs in cytochrome P450 family 2. Functional and structural role of a motif 187RFDYKD192 in CYP2B enzymes. *J Biol Chem* 283:21808–21816.

Ogilvie, B. W., Zhang, D., Li, W., Rodrigues, A. D., Gipson, A. E., Holsapple, J., Toren, P., Parkinson, A. 2006. Glucuronidation converts gemfibrozil to a potent, metabolism-dependent inhibitor of CYP2C8: Implications for drug–drug interactions. *Drug Metab Dispos* 34:191–197.

Ogunkolade, B. W., Boucher, B. J., Fairclough, P. D., Hitman, G. A., Dorudi, S., Jenkins, P. J., Bustin, S. A. 2002. Expression of 25-hydroxyvitamin D-1-alpha-hydroxylase mRNA in individuals with colorectal cancer. *Lancet* 359:1831–1832.

Ohashi, K., Ruan, K. H., Kulmacz, R. J., Wu, K. K., Wang, L. H. 1992. Primary structure of human thromboxane synthase determined from the cDNA sequence. *J Biol Chem* 267:789–793.

Ohnishi, T., Ichikawa, Y. 1997. Direct inhibitions of the activities of steroidgenic cytochrome P-450 monooxygenase systems by anticonvulsants. *J Steriod Biochem Mil Biol* 60:77–85.

Ohyama, K., Nakajima, M., Suzuki, M., Shimada, N., Yamazaki, H., Yokoi, T. 2000. Inhibitory effects of amiodarone and its N-deethylated metabolite on human cytochrome P450 activities: Prediction of *in vivo* drug interactions. *Br J Clin Pharmacol* 49:244–253.

Olakanmi, O., Seybert, D. W. 1990. Modified acetylenic steroids as potent mechanism-based inhibitors of cytochrome P-450SCC. *J Steroid Biochem* 36:273–280.

Oluka, M. N., Okalebo, F. A., Guantai, A. N., McClelland, R. S., Graham, S. M. 2015. Cytochrome P450 2B6 genetic variants are associated with plasma nevirapine levels and clinical response in HIV-1 infected Kenyan women: A prospective cohort study. *AIDS Res Ther* 12:10.

Omdahl, J. L., Bobrovnikova, E. V., Annalora, A., Chen, P., Serda, R. 2003. Expression, structure-function, and molecular modeling of vitamin D P450s. *J Cell Biochem* 88:356–362.

Omura, T. 2005. Heme-thiolate proteins. *Biochem Biophys Res Commun* 338:404–409.

Omura, T. 2010. Structural diversity of cytochrome P450 enzyme system. *J Biochem* 147:297–306.

Omura, T., Sato, R. 1964. The carbon monoxide-binding pigment of liver microsomes. I. Evidence for its hemoprotein nature. *J Biol Chem* 239:2370–2378.

Orlando, R., Piccoli, P., De Martin, S., Padrini, R., Floreani, M., Palatini, P. 2004. Cytochrome P450 1A2 is a major determinant of lidocaine metabolism in vivo: Effects of liver function. *Clin Pharmacol Ther* 75:80–88.

Osanai, M., Lee, G. H. 2011. Nicotine-mediated suppression of the retinoic acid metabolizing enzyme CYP26A1 limits the oncogenic potential of breast cancer. *Cancer Sci* 102:1158–1163.

Osanai, M., Lee, G. H. 2014. Increased expression of the retinoic acid-metabolizing enzyme CYP26A1 during the progression of cervical squamous neoplasia and head and neck cancer. *BMC Res Notes* 7:697.

Osanai, M., Sawada, N., Lee, G. H. 2010. Oncogenic and cell survival properties of the retinoic acid metabolizing enzyme, CYP26A1. *Oncogene* 29:1135–1144.

Oscarson, M. 2001. Genetic polymorphisms in the cytochrome P450 2A6 (*CYP2A6*) gene: Implications for interindividual differences in nicotine metabolism. *Drug Metab Dispos* 29:91–95.

Oscarson, M., McLellan, R. A., Gullsten, H., Agundez, J. A., Benitez, J., Rautio, A., Raunio, H., Pelkonen, O., Ingelman-Sundberg, M. 1999. Identification and characterisation of novel polymorphisms in the *CYP2A* locus: Implications for nicotine metabolism. *FEBS Lett* 460:321–327.

Oscarson, M., McLellan, R. A., Asp, V., Ledesma, M., Bernal Ruiz, M. L., Sinues, B., Rautio, A., Ingelman-Sundberg, M. 2002. Characterization of a novel *CYP2A7/CYP2A6* hybrid allele (*CYP2A6*12*) that causes reduced CYP2A6 activity. *Hum Mutat* 20:275–283.

Oyama, T., Uramoto, H., Kagawa, N., Yoshimatsu, T., Osaki, T., Nakanishi, R., Nagaya, H. et al. 2012. Cytochrome P450 in non-small cell lung cancer related to exogenous chemical metabolism. *Front Biosci (Schol Ed)* 4:1539–1546.

Ozpolat, B., Mehta, K., Lopez-Berestein, G. 2005. Regulation of a highly specific retinoic acid-4-hydroxylase (CYP26A1) enzyme and all-*trans*-retinoic acid metabolism in human intestinal, liver, endothelial, and acute promyelocytic leukemia cells. *Leuk Lymphoma* 46:1497–1506.

Pai, H. V., Kommaddi, R. P., Chinta, S. J., Mori, T., Boyd, M. R., Ravindranath, V. 2004. A frameshift mutation and alternate splicing in human brain generate a functional form of the pseudogene cytochrome P4502D7 that demethylates codeine to morphine. *J Biol Chem* 279:27383–27389.

Paine, M. F., Hart, H. L., Ludington, S. S., Haining, R. L., Rettie, A. E., Zeldin, D. C. 2006. The human intestinal cytochrome P450 "pie." *Drug Metab Dispos* 34:880–886.

Pallan, P. S., Wang, C., Lei, L., Yoshimoto, F. K., Auchus, R. J., Waterman, M. R., Guengerich, F. P., Egli, M. 2015. Human cytochrome P450 21A2, the major steroid 21-hydroxylase: Structure of the enzyme-progesterone substrate complex and rate-limiting C-H bond cleavage. *J Biol Chem* 290: 13128–13143.

Pan, Y. Z., Gao, W., Yu, A. M. 2009. MicroRNAs regulate CYP3A4 expression via direct and indirect targeting. *Drug Metab Dispos* 37:2112–2117.

Panda, D. K., Miao, D., Tremblay, M. L., Sirois, J., Farookhi, R., Hendy, G. N., Goltzman, D. 2001. Targeted ablation of the 25-hydroxyvitamin D 1alpha -hydroxylase enzyme: Evidence for skeletal, reproductive, and immune dysfunction. *Proc Natl Acad Sci U S A* 98:7498–7503.

Pandak, W. M., Schwarz, C., Hylemon, P. B., Mallonee, D., Valerie, K., Heuman, D. M., Fisher, R. A., Redford, K., Vlahcevic, Z. R. 2001. Effects of CYP7A1 overexpression on cholesterol and bile acid homeostasis. *Am J Physiol Gastrointest Liver Physiol* 281:G878–G889.

Pandak, W. M., Hylemon, P. B., Ren, S., Marques, D., Gil, G., Redford, K., Mallonee, D., Vlahcevic, Z. R. 2002. Regulation of oxysterol 7α-hydroxylase (CYP7B1) in primary cultures of rat hepatocytes. *Hepatology* 35:1400–1408.

Pandey, A. V., Fluck, C. E. 2013. NADPH P450 oxidoreductase: Structure, function, and pathology of diseases. *Pharmacol Ther* 138:229–254.

Panigrahy, D., Kaipainen, A., Greene, E. R., Huang, S. 2010. Cytochrome P450-derived eicosanoids: The neglected pathway in cancer. *Cancer Metastasis Rev* 29:723–735.

Papassotiropoulos, A., Streffer, J. R., Tsolaki, M., Schmid, S., Thal, D., Nicosia, F., Iakovidou, V. et al. 2003. Increased brain β-amyloid load, phosphorylated tau, and risk of Alzheimer disease associated with an intronic CYP46 polymorphism. *Arch Neurol* 60:29–35.

Paris, B. L., Ogilvie, B. W., Scheinkoenig, J. A., Ndikum-Moffor, F., Gibson, R., Parkinson, A. 2009. *In vitro* inhibition and induction of human liver cytochrome P450 (CYP) enzymes by milnacipran. *Drug Metab Dispos* 37:2045–2054.

Parise, R. A., Egorin, M. J., Kanterewicz, B., Taimi, M., Petkovich, M., Lew, A. M., Chuang, S. S., Nichols, M., El-Hefnawy, T., Hershberger, P. A. 2006. CYP24, the enzyme that catabolizes the antiproliferative agent vitamin D, is increased in lung cancer. *Int J Cancer* 119:1819–1828.

Parker, A. C., Preston, T., Heaf, D., Kitteringham, N. R., Choonara, I. 1994. Inhibition of caffeine metabolism by ciprofloxacin in children with cystic fibrosis as measured by the caffeine breath test. *Br J Clin Pharmacol* 38:573–576.

Pascoe, L., Curnow, K. M., Slutsker, L., Rosler, A., White, P. C. 1992. Mutations in the human CYP11B2 (aldosterone synthase) gene causing corticosterone methyloxidase II deficiency. *Proc Natl Acad Sci U S A* 89:4996–5000.

Pascussi, J. M., Gerbal-Chaloin, S., Fabre, J. M., Maurel, P., Vilarem, M. J. 2000. Dexamethasone enhances constitutive androstane receptor expression in human hepatocytes: Consequences on cytochrome P450 gene regulation. *Mol Pharmacol* 58:1441–1450.

Pascussi, J. M., Gerbal-Chaloin, S., Duret, C., Daujat-Chavanieu, M., Vilarem, M. J., Maurel, P. 2008. The tangle of nuclear receptors that controls xenobiotic metabolism and transport: Crosstalk and consequences. *Annu Rev Pharmacol Toxicol* 48:1–32.

Patel, S. A., Bhambra, U., Charalambous, M. P., David, R. M., Edwards, R. J., Lightfoot, T., Boobis, A. R., Gooderham, N. J. 2014. Interleukin-6 mediated upregulation of CYP1B1 and CYP2E1 in colorectal cancer involves DNA methylation, miR27b and STAT3. *Br J Cancer* 111:2287–2296.

Patsalos, P. N., Froscher, W., Pisani, F., van Rijn, C. M. 2002. The importance of drug interactions in epilepsy therapy. *Epilepsia* 43:365–385.

Paulussen, A., Lavrijsen, K., Bohets, H., Hendrickx, J., Verhasselt, P., Luyten, W., Konings, F., Armstrong, M. 2000. Two linked mutations in transcriptional regulatory elements of the CYP3A5 gene constitute the major genetic determinant of polymorphic activity in humans. *Pharmacogenetics* 10:415–424.

Peet, D. J., Turley, S. D., Ma, W., Janowski, B. A., Lobaccaro, J. M., Hammer, R. E., Mangelsdorf, D. J. 1998. Cholesterol and bile acid metabolism are impaired in mice lacking the nuclear oxysterol receptor LXR alpha. *Cell* 93:693–704.

Pelkonen, O., Maenpaa, J., Taavitsainen, P., Rautio, A., Raunio, H. 1998. Inhibition and induction of human cytochrome P450 (CYP) enzymes. *Xenobiotica* 28:1203–1253.

Pennimpede, T., Cameron, D. A., MacLean, G. A., Li, H., Abu-Abed, S., Petkovich, M. 2010. The role of CYP26 enzymes in defining appropriate retinoic acid exposure during embryogenesis. *Birth Defects Res A Clin Mol Teratol* 88:883–894.

Penzak, S. R., Kabuye, G., Mugyenyi, P., Mbamanya, F., Natarajan, V., Alfaro, R. M., Kityo, C., Formentini, E., Masur, H. 2007. Cytochrome P450 2B6 (*CYP2B6*) G516T influences nevirapine plasma concentrations in HIV-infected patients in Uganda. *HIV Med* 8:86–91.

Pernecky, S. J., Porter, T. D., Coon, M. J. 1990. Expression of rabbit cytochrome P-450IIE2 in yeast and stabilization of the enzyme by 4-methylpyrazole. *Biochem Biophys Res Commun* 172:1331–1337.

Petrulis, J. R., Perdew, G. H. 2002. The role of chaperone proteins in the aryl hydrocarbon receptor core complex. *Chem Biol Interact* 141:25–40.

Petrunak, E. M., DeVore, N. M., Porubsky, P. R., Scott, E. E. 2014. Structures of human steroidogenic cytochrome P450 17A1 with substrates. *J Biol Chem* 289:32952–32964.

Pikuleva, I. A. 2006. Cytochrome P450s and cholesterol homeostasis. *Pharmacol Ther* 112:761–773.

Pikuleva, I. A., Bjorkhem, I., Waterman, M. R. 1997. Expression, purification, and enzymatic properties of recombinant human cytochrome P450c27 (CYP27). *Arch Biochem Biophys* 343:123–130.

Piliguian, M., Zhu, A. Z., Zhou, Q., Benowitz, N. L., Ahluwalia, J. S., Sanderson Cox, L., Tyndale, R. F. 2014. Novel *CYP2A6* variants identified in African Americans are associated with slow nicotine metabolism *in vitro* and *in vivo*. *Pharmacogenet Genomics* 24:118–128.

Piscitelli, S. C., Burstein, A. H., Chaitt, D., Alfaro, R. M., Falloon, J. 2000. Indinavir concentrations and St John's wort. *Lancet* 355:547–548.

Pitarque, M., von Richter, O., Oke, B., Berkkan, H., Oscarson, M., Ingelman-Sundberg, M. 2001. Identification of a single nucleotide polymorphism in the TATA box of the *CYP2A6* gene: Impairment of its promoter activity. *Biochem Biophys Res Commun* 284:455–460.

Podust, L. M., Sherman, D. H. 2012. Diversity of P450 enzymes in the biosynthesis of natural products. *Nat Prod Rep* 29:1251–1266.

Polasek, T. M., Elliot, D. J., Lewis, B. C., Miners, J. O. 2004. Mechanism-based inactivation of human cytochrome P450 2C8 by drugs *in vitro*. *J Pharmacol Exp Ther* 311:996–1007.

Polimanti, R., Piacentini, S., Manfellotto, D., Fuciarelli, M. 2012. Human genetic variation of CYP450 superfamily: Analysis of functional diversity in worldwide populations. *Pharmacogenomics* 13:1951–1960.

Porcu, L., Meloni, A., Casula, L., Asunis, I., Marini, M. G., Cao, A., Moi, P. 2002. A novel splicing defect (IVS6+1G>T) in a patient with pseudovitamin D deficiency rickets. *J Endocrinol Invest* 25:557–560.

Porubsky, P. R., Meneely, K. M., Scott, E. E. 2008. Structures of human cytochrome P-450 2E1. Insights into the binding of inhibitors and both small molecular weight and fatty acid substrates. *J Biol Chem* 283:33698–33707.

Post, S. M., de Wit, E. C., Princen, H. M. 1997. Cafestol, the cholesterol-raising factor in boiled coffee, suppresses bile acid synthesis by downregulation of cholesterol 7 alpha-hydroxylase and sterol 27-hydroxylase in rat hepatocytes. *Arterioscler Thromb Vasc Biol* 17:3064–3070.

Poulos, T. L. 2014. Heme enzyme structure and function. *Chem Rev* 114:3919–3962.

Poulos, T. L., Finzel, B. C., Howard, A. J. 1987. High-resolution crystal structure of cytochrome P450cam. *J Mol Biol* 195:687–700.

Powell, P. K., Wolf, I., Lasker, J. M. 1996. Identification of CYP4A11 as the major lauric acid omega-hydroxylase in human liver microsomes. *Arch Biochem Biophys* 335:219–226.

Powell, P. K., Wolf, I., Jin, R., Lasker, J. M. 1998. Metabolism of arachidonic acid to 20-hydroxy-5,8,11, 14-eicosatetraenoic acid by P450 enzymes in human liver: Involvement of CYP4F2 and CYP4A11. *J Pharmacol Exp Ther* 285:1327–1336.

Prosser, D. E., Jones, G. 2004. Enzymes involved in the activation and inactivation of vitamin D. *Trends Biochem Sci* 29:664–673.

Prueksaritanont, T., Dwyer, L. M., Cribb, A. E. 1995. (+)-bufuralol 1′-hydroxylation activity in human and rhesus monkey intestine and liver. *Biochem Pharmacol* 50:1521–1525.

Pullinger, C. R., Eng, C., Salen, G., Shefer, S., Batta, A. K., Erickson, S. K., Verhagen, A. et al. 2002. Human cholesterol 7α-hydroxylase (CYP7A1) deficiency has a hypercholesterolemic phenotype. *J Clin Invest* 110:109–117.

Qatanani, M., Moore, D. D. 2005. CAR, the continuously advancing receptor, in drug metabolism and disease. *Curr Drug Metab* 6:329–339.

Qu, W., Bradbury, J. A., Tsao, C. C., Maronpot, R., Harry, G. J., Parker, C. E., Davis, L. S. et al. 2001. Cytochrome P450 CYP2J9, a new mouse arachidonic acid omega-1 hydroxylase predominantly expressed in brain. *J Biol Chem* 276:25467–25479.

Rae, J. M., Soukhova, N. V., Flockhart, D. A., Desta, Z. 2002. Triethylenethiophosphoramide is a specific inhibitor of cytochrome P450 2B6: Implications for cyclophosphamide metabolism. *Drug Metab Dispos* 30:525–530.

Rahman, A., Korzekwa, K. R., Grogan, J., Gonzalez, F. J., Harris, J. W. 1994. Selective biotransformation of taxol to 6α-hydroxytaxol by human cytochrome P450 2C8. *Cancer Res* 54:5543–5546.

Raucy, J. L., Mueller, L., Duan, K., Allen, S. W., Strom, S., Lasker, J. M. 2002. Expression and induction of CYP2C P450 enzymes in primary cultures of human hepatocytes. *J Pharmacol Exp Ther* 302:475–482.

Raunio, H., Rahnasto-Rilla, M. 2012. CYP2A6: Genetics, structure, regulation, and function. *Drug Metabol Drug Interact* 27:73–88.

Rautio, A., Kraul, H., Kojo, A., Salmela, E., Pelkonen, O. 1992. Interindividual variability of coumarin 7-hydroxylation in healthy volunteers. *Pharmacogenetics* 2:227–233.

Ravichandran, K. G., Boddupalli, S. S., Hasermann, C. A., Peterson, J. A., Deisenhofer, J. 1993. Crystal structure of hemoprotein domain of P450BM-3, a prototype for microsomal P450's. *Science* 261:731–736.

Reichrath, J., Rafi, L., Rech, M., Mitschele, T., Meineke, V., Gartner, B. C., Tilgen, W., Holick, M. F. 2004. Analysis of the vitamin D system in cutaneous squamous cell carcinomas. *J Cutan Pathol* 31:224–231.

Reiss, A. B., Awadallah, N. W., Malhotra, S., Montesinos, M. C., Chan, E. S., Javitt, N. B., Cronstein, B. N. 2001. Immune complexes and IFN-gamma decrease cholesterol 27-hydroxylase in human arterial endothelium and macrophages. *J Lipid Res* 42:1913–1922.

Ren, S., Zeng, J., Mei, Y., Zhang, J. Z., Yan, S. F., Fei, J., Chen, L. 2013. Discovery and characterization of novel, potent, and selective cytochrome P450 2J2 inhibitors. *Drug Metab Dispos* 41:60–71.

Rendic, S. 2002. Summary of information on human CYP enzymes: Human P450 metabolism data. *Drug Metab Rev* 34:83–448.

Renton, K. W. 2001. Alteration of drug biotransformation and elimination during infection and inflammation. *Pharmacol Ther* 92:147–163.

Renton, K. W. 2005. Regulation of drug metabolism and disposition during inflammation and infection. *Expert Opin Drug Metab Toxicol* 1:629–640.

Rettie, A. E., Jones, J. P. 2005. Clinical and toxicological relevance of CYP2C9: Drug–drug interactions and pharmacogenetics. *Annu Rev Pharmacol Toxicol* 45:477–494.

Rettie, A. E., Haining, R. L., Bajpai, M., Levy, R. H. 1999. A common genetic basis for idiosyncratic toxicity of warfarin and phenytoin. *Epilepsy Res* 35:253–255.

Rhinn, M., Dolle, P. 2012. Retinoic acid signalling during development. *Development* 139:843–858.

Richter, T., Murdter, T. E., Heinkele, G., Pleiss, J., Tatzel, S., Schwab, M., Eichelbaum, M., Zanger, U. M. 2004. Potent mechanism-based inhibition of human CYP2B6 by clopidogrel and ticlopidine. *J Pharmacol Exp Ther* 308:189–197.

Rieger, M. A., Ebner, R., Bell, D. R., Kiessling, A., Rohayem, J., Schmitz, M., Temme, A., Rieber, E. P., Weigle, B. 2004. Identification of a novel mammary-restricted cytochrome P450, CYP4Z1, with overexpression in breast carcinoma. *Cancer Res* 64:2357–2364.

Rieger, J. K., Reutter, S., Hofmann, U., Schwab, M., Zanger, U. M. 2015. Inflammation-associated microRNA-130b down-regulates cytochrome P450 activities and directly targets CYP2C9. *Drug Metab Dispos* 43:884–888.

Riepe, F. G., Tatzel, S., Sippell, W. G., Pleiss, J., Krone, N. 2005. Congenital adrenal hyperplasia: The molecular basis of 21-hydroxylase deficiency in H-2(aw18) mice. *Endocrinology* 146:2563–2574.

Riffel, A. K., Schuenemann, E., Vyhlidal, C. A. 2009. Regulation of the CYP3A4 and CYP3A7 promoters by members of the nuclear factor I transcription factor family. *Mol Pharmacol* 76:1104–1114.

Riley, R. J., Grime, K., Weaver, R. 2007. Time-dependent CYP inhibition. *Expert Opin Drug Metab Toxicol* 3:51–66.

Ring, B. J., Catlow, J., Lindsay, T. J., Gillespie, T., Roskos, L. K., Cerimele, B. J., Swanson, S. P., Hamman, M. A., Wrighton, S. A. 1996. Identification of the human cytochromes P450 responsible for the in vitro formation of the major oxidative metabolites of the antipsychotic agent olanzapine. *J Pharmacol Exp Ther* 276:658–666.

Ritter, C. S., Haughey, B. H., Armbrecht, H. J., Brown, A. J. 2012. Distribution and regulation of the 25-hydroxyvitamin D3 1α-hydroxylase in human parathyroid glands. *J Steroid Biochem Mol Biol* 130:73–80.

Rivera, S. P., Saarikoski, S. T., Hankinson, O. 2002. Identification of a novel dioxin-inducible cytochrome P450. *Mol Pharmacol* 61:255–259.

Rivera, S. P., Wang, F., Saarikoski, S. T., Taylor, R. T., Chapman, B., Zhang, R., Hankinson, O. 2007. A novel promoter element containing multiple overlapping xenobiotic and hypoxia response elements mediates induction of cytochrome P4502S1 by both dioxin and hypoxia. *J Biol Chem* 282:10881–10893.

Roberts, G. C. 1999. The power of evolution: Accessing the synthetic potential of P450s. *Chem Biol* 6:R269–R272.

Rodriguez-Antona, C., Jande, M., Rane, A., Ingelman-Sundberg, M. 2005. Identification and phenotype characterization of two CYP3A haplotypes causing different enzymatic capacity in fetal livers. *Clin Pharmacol Ther* 77:259–270.

Rodriguez-Pazos, L., Ginarte, M., Fachal, L., Toribio, J., Carracedo, A., Vega, A. 2011. Analysis of TGM1, ALOX12B, ALOXE3, NIPAL4 and CYP4F22 in autosomal recessive congenital ichthyosis from Galicia (NW Spain): Evidence of founder effects. *Br J Dermatol* 165:906–911.

Roe, A. L., Snawder, J. E., Benson, R. W., Roberts, D. W., Casciano, D. A. 1993. HepG2 cells: An in vitro model for P450-dependent metabolism of acetaminophen. *Biochem Biophys Res Commun* 190:15–19.

Romkes, M., Faletto, M. B., Blaisdell, J. A., Raucy, J. L., Goldstein, J. A. 1991. Cloning and expression of complementary DNAs for multiple members of the human cytochrome P450IIC subfamily. *Biochemistry* 30:3247–3255.

Ronchi, C. L., Sbiera, S., Volante, M., Steinhauer, S., Scott-Wild, V., Altieri, B., Kroiss, M., Bala, M., Papotti, M., Deutschbein, T., Terzolo, M., Fassnacht, M., Allolio, B. 2014. CYP2W1 is highly expressed in adrenal glands and is positively associated with the response to mitotane in adrenocortical carcinoma. *PLoS One* 9:e105855.

Roos, P., Svenstrup, K., Danielsen, E. R., Thomsen, C., Nielsen, J. E. 2014. CYP7B1: Novel mutations and magnetic resonance spectroscopy abnormalities in hereditary spastic paraplegia type 5A. *Acta Neurol Scand* 129:330–334.

Rose, K. A., Stapleton, G., Dott, K., Kieny, M. P., Best, R., Schwarz, M., Russell, D. W., Bjorkhem, I., Seckl, J., Lathe, R. 1997. Cyp7b, a novel brain cytochrome P450, catalyzes the synthesis of neurosteroids 7alpha-hydroxy dehydroepiandrosterone and 7alpha-hydroxy pregnenolone. *Proc Natl Acad Sci U S A* 94:4925–4930.

Rose, K., Allan, A., Gauldie, S., Stapleton, G., Dobbie, L., Dott, K., Martin, C. et al. Neurosteroid hydroxylase CYP7B: Vivid reporter activity in dentate gyrus of gene-targeted mice and abolition of a widespread pathway of steroid and oxysterol hydroxylation. *J Biol Chem* 276:23937–23944.

Ross, A. C., Zolfaghari, R. 2011. Cytochrome P450s in the regulation of cellular retinoic acid metabolism. *Annu Rev Nutr* 31:65–87.

Rossini, A., de Almeida Simao, T., Albano, R. M., Pinto, L. F. 2008. CYP2A6 polymorphisms and risk for tobacco-related cancers. *Pharmacogenomics* 9:1737–1752.

Rotger, M., Colombo, S., Furrer, H., Bleiber, G., Buclin, T., Lee, B. L., Keiser, O., Biollaz, J., Decosterd, L., Telenti, A. 2005. Influence of CYP2B6 polymorphism on plasma and intracellular concentrations and toxicity of efavirenz and nevirapine in HIV-infected patients. *Pharmacogenet Genomics* 15:1–5.

Rowland, A., Mangoni, A. A. 2014. Cytochrome P450 and ischemic heart disease: Current concepts and future directions. *Expert Opin Drug Metab Toxicol* 10:191–213.

Rowland, P., Blaney, F. E., Smyth, M. G., Jones, J. J., Leydon, V. R., Oxbrow, A. K., Lewis, C. J. et al. 2006. Crystal structure of human cytochrome P450 2D6. *J Biol Chem* 281:7614–7622.

Rozman, D., Stromstedt, M., Waterman, M. R. 1996. The three human cytochrome P450 lanosterol 14 alpha-demethylase (CYP51) genes reside on chromosomes 3, 7, and 13: Structure of the two retrotransposed pseudogenes, association with a line-1 element, and evolution of the human CYP51 family. *Arch Biochem Biophys* 333:466–474.

Rubtsov, P., Karmanov, M., Sverdlova, P., Spirin, P., Tiulpakov, A. 2009. A novel homozygous mutation in *CYP11A1* gene is associated with late-onset adrenal insufficiency and hypospadias in a 46,XY patient. *J Clin Endocrinol Metab* 94:936–939.

Rubtsov, P., Nizhnik, A., Dedov, I., Kalinchenko, N., Petrov, V., Orekhova, A., Spirin, P., Prassolov, V., Tiulpakov, A. 2015. Partial deficiency of 17α-hydroxylase/17,20-lyase caused by a novel missense mutation in the canonical cytochrome heme-interacting motif. *Eur J Endocrinol* 172:K19–K25.

Russell, D. W. 1999. Nuclear orphan receptors control cholesterol catabolism. *Cell* 97:539–542.

Rylander, T., Neve, E. P., Ingelman-Sundberg, M., Oscarson, M. 2001. Identification and tissue distribution of the novel human cytochrome P450 2S1 (CYP2S1). *Biochem Biophys Res Commun* 281:529–535.

Ryu, J. Y., Song, I. S., Sunwoo, Y. E., Shon, J. H., Liu, K. H., Cha, I. J., Shin, J. G. 2007. Development of the "Inje cocktail" for high-throughput evaluation of five human cytochrome P450 isoforms in vivo. *Clin Pharmacol Ther* 82:531–540.

Saarikoski, S. T., Suitiala, T., Holmila, R., Impivaara, O., Jarvisalo, J., Hirvonen, A., Husgafvel-Pursiainen, K. 2004. Identification of genetic polymorphisms of *CYP2S1* in a Finnish Caucasian population. *Mutat Res* 554:267–277.

Saarikoski, S. T., Rivera, S. P., Hankinson, O., Husgafvel-Pursiainen, K. 2005. CYP2S1: A short review. *Toxicol Appl Pharmacol* 207:62–69.

Sachse, C., Brockmoller, J., Bauer, S., Roots, I. 1997. Cytochrome P450 2D6 variants in a Caucasian population: Allele frequencies and phenotypic consequences. *Am J Hum Genet* 60:284–295.

Sachse, C., Brockmoller, J., Bauer, S., Roots, I. 1999. Functional significance of a C→A polymorphism in intron 1 of the cytochrome P450 CYP1A2 gene tested with caffeine. *Br J Clin Pharmacol* 47:445–449.

Sadeque, A. J., Fisher, M. B., Korzekwa, K. R., Gonzalez, F. J., Rettie, A. E. 1997. Human CYP2C9 and CYP2A6 mediate formation of the hepatotoxin 4-ene-valproic acid. *J Pharmacol Exp Ther* 283:698–703.

Saeki, M., Saito, Y., Nakamura, T., Murayama, N., Kim, S. R., Ozawa, S., Komamura, K. et al. 2003. Single nucleotide polymorphisms and haplotype frequencies of *CYP3A5* in a Japanese population. *Hum Mutat* 21:653–659.

Sagar, M., Tybring, G., Dahl, M. L., Bertilsson, L., Seensalu, R. 2000. Effects of omeprazole on intragastric pH and plasma gastrin are dependent on the *CYP2C19* polymorphism. *Gastroenterology* 119:670–676.

Sahakitrungruang, T., Tee, M. K., Speiser, P. W., Miller, W. L. 2009. Novel P450c17 mutation H373D causing combined 17α-hydroxylase/17,20-lyase deficiency. *J Clin Endocrinol Metab* 94:3089–3092.

Sahakitrungruang, T., Tee, M. K., Blackett, P. R., Miller, W. L. 2011. Partial defect in the cholesterol side-chain cleavage enzyme P450scc (CYP11A1) resembling nonclassic congenital lipoid adrenal hyperplasia. *J Clin Endocrinol Metab* 96:792–798.

Saito, T., Honda, M., Takahashi, M., Tsukada, C., Ito, M., Katono, Y., Hosono, H. et al. 2015. Functional characterization of 10 *CYP4A11* allelic variants to evaluate the effect of genotype on arachidonic acid omega-hydroxylation. *Drug Metab Pharmacokinet* 30:119–122.

Sakaki, T., Kagawa, N., Yamamoto, K., Inouye, K. 2005. Metabolism of vitamin D$_3$ by cytochromes P450. *Front Biosci* 10: 119–134.

Sanderink, G. J., Bournique, B., Stevens, J., Petry, M., Martinet, M. 1997. Involvement of human CYP1A isoenzymes in the metabolism and drug interactions of riluzole in vitro. *J Pharmacol Exp Ther* 282:1465–1472.

Sansen, S., Hsu, M. H., Stout, C. D., Johnson, E. F. 2007a. Structural insight into the altered substrate specificity of human cytochrome P450 2A6 mutants. *Arch Biochem Biophys* 464:197–206.

Sansen, S., Yano, J. K., Reynald, R. L., Schoch, G. A., Griffin, K. J., Stout, C. D., Johnson, E. F. 2007b. Adaptations for the oxidation of polycyclic aromatic hydrocarbons exhibited by the structure of human P450 1A2. *J Biol Chem* 282:14348–14355.

Sarkar, M. A., Jackson, B. J. 1994. Theophylline N-demethylations as probes for P4501A1 and P4501A2. *Drug Metab Dispos* 22:827–834.

Sasaki, K., Akiyama, M., Yanagi, T., Sakai, K., Miyamura, Y., Sato, M., Shimizu, H. 2012. CYP4F22 is highly expressed at the site and timing of onset of keratinization during skin development. *J Dermatol Sci* 65:156–158.

Sata, F., Sapone, A., Elizondo, G., Stocker, P., Miller, V. P., Zheng, W., Raunio, H., Crespi, C. L., Gonzalez, F. J. 2000. *CYP3A4* allelic variants with amino acid substitutions in exons 7 and 12: Evidence for an allelic variant with altered catalytic activity. *Clin Pharmacol Ther* 67:48–56.

Satarug, S., Tassaneeyakul, W., Na-Bangchang, K., Cashman, J. R., Moore, M. R. 2006. Genetic and environmental influences on therapeutic and toxicity outcomes: Studies with CYP2A6. *Curr Clin Pharmacol* 1:291–309.

Sato, M., Yokoyama, U., Fujita, T., Okumura, S., Ishikawa, Y. 2011. The roles of cytochrome p450 in ischemic heart disease. *Curr Drug Metab* 12:526–532.

Savai, J., Varghese, A., Pandita, N., Chintamaneni, M. 2015. *In vitro* assessment of CYP1A2 and 2C9 inhibition potential of *Withania somnifera* and *Centella asiatica* in human liver microsomes. *Drug Metabol Personal Ther* 30:137–141.

Savas, U., Hsu, M. H., Johnson, E. F. 2003. Differential regulation of human CYP4A genes by peroxisome proliferators and dexamethasone. *Arch Biochem Biophys* 409:212–220.

Savas, U., Hsu, M. H., Griffin, K. J., Bell, D. R., Johnson, E. F. 2005. Conditional regulation of the human *CYP4X1* and *CYP4Z1* genes. *Arch Biochem Biophys* 436:377–385.

Scarborough, P. E., Ma, J., Qu, W., Zeldin, D. C. 1999. P450 subfamily CYP2J and their role in the bioactivation of arachidonic acid in extrahepatic tissues. *Drug Metab Rev* 31:205–234.

Schlicht, K. E., Michno, N., Smith, B. D., Scott, E. E., Murphy, S. E. 2007. Functional characterization of CYP2A13 polymorphisms. *Xenobiotica* 37:1439–1449.

Schlingmann, K. P., Kaufmann, M., Weber, S., Irwin, A., Goos, C., John, U., Misselwitz, J. et al. 2011. Mutations in *CYP24A1* and idiopathic infantile hypercalcemia. *N Engl J Med* 365: 410–421.

Schmitt, C., Kuhn, B., Zhang, X., Kivitz, A. J., Grange, S. 2011. Disease-drug–drug interaction involving tocilizumab and simvastatin in patients with rheumatoid arthritis. *Clin Pharmacol Ther* 89:735–740.

Schoch, G. A., Yano, J. K., Wester, M. R., Griffin, K. J., Stout, C. D., Johnson, E. F. 2004. Structure of human microsomal cytochrome P450 2C8. Evidence for a peripheral fatty acid binding site. *J Biol Chem* 279:9497–9503.

Schoch, G. A., Yano, J. K., Sansen, S., Dansette, P. M., Stout, C. D., Johnson, E. F. 2008. Determinants of cytochrome P450 2C8 substrate binding: Structures of complexes with montelukast, troglitazone, felodipine, and 9-*cis*-retinoic acid. *J Biol Chem* 283:17227–17237.

Schoedel, K. A., Hoffmann, E. B., Rao, Y., Sellers, E. M., Tyndale, R. F. 2004. Ethnic variation in *CYP2A6* and association of genetically slow nicotine metabolism and smoking in adult Caucasians. *Pharmacogenetics* 14:615–626.

Schuetz, E. G. 2004. Lessons from the *CYP3A4* promoter. *Mol Pharmacol* 65:279–281.

Schuetz, J. D., Guzelian, P. S. 1995. Isolation of CYP3A5P cDNA from human liver: A reflection of a novel cytochrome P-450 pseudogene. *Biochim Biophys Acta* 1261:161–165.

Schuetz, J. D., Kauma, S., Guzelian, P. S. 1993. Identification of the fetal liver cytochrome CYP3A7 in human endometrium and placenta. *J Clin Invest* 92:1018–1024.

Schule, R., Brandt, E., Karle, K. N., Tsaousidou, M., Klebe, S., Klimpe, S., Auer-Grumbach, M. et al. 2009. Analysis of *CYP7B1* in non-consanguineous cases of hereditary spastic paraplegia. *Neurogenetics* 10:97–104.

Schule, R., Siddique, T., Deng, H. X., Yang, Y., Donkervoort, S., Hansson, M., Madrid, R. E., Siddique, N., Schols, L., Bjorkhem, I. 2010. Marked accumulation of 27-hydroxycholesterol in SPG5 patients with hereditary spastic paresis. *J Lipid Res* 51:819–823.

Schwarz, D., Kisselev, P., Ericksen, S. S., Szklarz, G. D., Chernogolov, A., Honeck, H., Schunck, W. H., Roots, I. 2004. Arachidonic and eicosapentaenoic acid metabolism by human CYP1A1: Highly stereoselective formation of 17(R),18(S)-epoxyeicosatetraenoic acid. *Biochem Pharmacol* 67:1445–1457.

Schwarz, D., Kisselev, P., Chernogolov, A., Schunck, W. H., Roots, I. 2005. Human CYP1A1 variants lead to differential eicosapentaenoic acid metabolite patterns. *Biochem Biophys Res Commun* 336:779–783.

Scott, E. E., Spatzenegger, M., Halpert, J. R. 2001. A truncation of 2B subfamily cytochromes P450 yields increased expression levels, increased solubility, and decreased aggregation while retaining function. *Arch Biochem Biophys* 395:57–68.

Scott, E. E., He, Y. A., Wester, M. R., White, M. A., Chin, C. C., Halpert, J. R., Johnson, E. F., Stout, C. D. 2003. An open conformation of mammalian cytochrome P450 2B4 at 1.6-Å resolution. *Proc Natl Acad Sci U S A* 100:13196–13201.

Scott, E. E., White, M. A., He, Y. A., Johnson, E. F., Stout, C. D., Halpert, J. R. 2004. Structure of mammalian cytochrome P450 2B4 complexed with 4-(4-chlorophenyl)imidazole at 1.9-Å resolution: Insight into the range of P450 conformations and the coordination of redox partner binding. *J Biol Chem* 279:27294–27301.

Scott, S. A., Sangkuhl, K., Gardner, E. E., Stein, C. M., Hulot, J. S., Johnson, J. A., Roden, D. M., Klein, T. E., Shuldiner, A. R., Clinical Pharmacogenetics Implementation Consortium. 2011. Clinical Pharmacogenetics Implementation Consortium guidelines for cytochrome P450–P452C19 (CYP2C19) genotype and clopidogrel therapy. *Clin Pharmacol Ther* 90:328–332.

Scott, S. A., Patel, M., Martis, S., Lubitz, S. A., van der Zee, S., Yoo, C., Edelmann, L., Halperin, J. L., Desnick, R. J. 2012. Copy number variation and warfarin dosing: Evaluation of CYP2C9, VKORC1, CYP4F2, GGCX and CALU. *Pharmacogenomics* 13:297–307.

Segersten, U., Correa, P., Hewison, M., Hellman, P., Dralle, H., Carling, T., Akerstrom, G., Westin, G. 2002. 25-hydroxyvitamin D(3)-1alpha-hydroxylase expression in normal and pathological parathyroid glands. *J Clin Endocrinol Metab* 87:2967–2972.

Seifert, A., Pleiss, J. 2009. Identification of selectivity-determining residues in cytochrome P450 monooxygenases: A systematic analysis of the substrate recognition site 5. *Proteins* 74:1028–1035.

Seissler, J., Schott, M., Steinbrenner, H., Peterson, P., Scherbaum, W. A. 1999. Autoantibodies to adrenal cytochrome P450 antigens in isolated Addison's disease and autoimmune polyendocrine syndrome type II. *Exp Clin Endocrinol Diabetes* 107:208–213.

Seliskar, M., Rozman, D. 2007. Mammalian cytochromes P450—importance of tissue specificity. *Biochim Biophys Acta* 1770: 458–466.

Sevrioukova, I. F., Poulos, T. L. 2010. Structure and mechanism of the complex between cytochrome P4503A4 and ritonavir. *Proc Natl Acad Sci U S A* 107:18422–18427.

Sevrioukova, I. F., Poulos, T. L. 2012a. Interaction of human cytochrome P4503A4 with ritonavir analogs. *Arch Biochem Biophys* 520:108–116.

Sevrioukova, I. F., Poulos, T. L. 2012b. Structural and mechanistic insights into the interaction of cytochrome P4503A4 with bromoergocryptine, a type I ligand. *J Biol Chem* 287:3510–3517.

Sevrioukova, I. F., Poulos, T. L. 2013a. Dissecting cytochrome P450 3A4-ligand interactions using ritonavir analogues. *Biochemistry* 52:4474–4481.

Sevrioukova, I. F., Poulos, T. L. 2013b. Pyridine-substituted desoxyritonavir is a more potent inhibitor of cytochrome P450 3A4 than ritonavir. *J Med Chem* 56:3733–3741.

Sevrioukova, I. F., Poulos, T. L. 2013c. Pyridine-substituted desoxyritonavir is a more potent inhibitor of cytochrome P450 3A4 than ritonavir. *Biochemistry* 52:4474–4481.

Sevrioukova, I. F., Poulos, T. L. 2013d. Understanding the mechanism of cytochrome P450 3A4: Recent advances and remaining problems. *Dalton Trans* 42:3116–3126.

Sezutsu, H., Le Goff, G., Feyereisen, R. 2013. Origins of P450 diversity. *Philos Trans R Soc Lond B Biol Sci* 368:20120428.

Shafaati, M., O'Driscoll, R., Bjorkhem, I., Meaney, S. 2009. Transcriptional regulation of cholesterol 24-hydroxylase by histone deacetylase inhibitors. *Biochem Biophys Res Commun* 378:689–694.

Shah, R. R., Smith, R. L. 2015. Inflammation-induced phenoconversion of polymorphic drug metabolizing enzymes: Hypothesis with implications for personalized medicine. *Drug Metab Dispos* 43:400–410.

Shah, M. B., Pascual, J., Zhang, Q., Stout, C. D., Halpert, J. R. 2011. Structures of cytochrome P450 2B6 bound to 4-benzylpyridine and 4-(4-nitrobenzyl)pyridine: Insight into inhibitor binding and rearrangement of active site side chains. *Mol Pharmacol* 80:1047–1055.

Shah, M. B., Wilderman, P. R., Pascual, J., Zhang, Q., Stout, C. D., Halpert, J. R. 2012. Conformational adaptation of human cytochrome P450 2B6 and rabbit cytochrome P450 2B4 revealed upon binding multiple amlodipine molecules. *Biochemistry* 51:7225–7238.

Shah, M. B., Jang, H. H., Zhang, Q., David Stout, C., Halpert, J. R. 2013. X-ray crystal structure of the cytochrome P450 2B4 active site mutant F297A in complex with clopidogrel: Insights into compensatory rearrangements of the binding pocket. *Arch Biochem Biophys* 530:64–72.

Shah, M. B., Wilderman, P. R., Liu, J., Jang, H. H., Zhang, Q., Stout, C. D., Halpert, J. R. 2015. Structural and biophysical characterization of human cytochromes P450 2B6 and 2A6 bound to volatile hydrocarbons: Analysis and comparison. *Mol Pharmacol* 87:649–659.

Shahabi, P., Siest, G., Visvikis-siest, S. 2014. Influence of inflammation on cardiovascular protective effects of cytochrome P450 epoxygenase-derived epoxyeicosatrienoic acids. *Drug Metab Rev* 46:33–56.

Shan, M., Dong, B., Zhao, X., Wang, J., Li, G., Yang, Y., Li, Y. 2005. Novel mutations in the CYP4V2 gene associated with Bietti crystalline corneoretinal dystrophy. *Mol Vis* 11:738–743.

Shanahan, C. M., Carpenter, K. L., Cary, N. R. 2001. A potential role for sterol 27-hydroxylase in atherogenesis *Atherosclerosis* 154:269–276.

Shebley, M., Kent, U. M., Ballou, D. P., Hollenberg, P. F. 2009. Mechanistic analysis of the inactivation of cytochrome P450 2B6 by phencyclidine: Effects on substrate binding, electron transfer, and uncoupling. *Drug Metab Dispos* 37:745–752.

Shedlofsky, S. I., Israel, B. C., McClain, C. J., Hill, D. B., Blouin, R. A. 1994. Endotoxin administration to humans inhibits hepatic cytochrome P450-mediated drug metabolism. *J Clin Invest* 94:2209–2214.

Sheikh, S. A., Waryah, A. M., Narsani, A. K., Shaikh, H., Gilal, I. A., Shah, K., Qasim, M., Memon, A. I., Kewalramani, P., Shaikh, N. 2014. Mutational spectrum of the *CYP1B1* gene in Pakistani patients with primary congenital glaucoma: Novel variants and genotype–phenotype correlations. *Mol Vis* 20:991–1001.

Shen, Z. T., Wu, X. H., Li, B., Shen, J. S., Wang, Z., Li, J., Zhu, X. X. 2015. CYP2E1 Rsa Iota/Pst Iota polymorphism and lung cancer susceptibility: A meta-analysis involving 10,947 subjects. *J Cell Mol Med*.

Sheng, J., Guo, J., Hua, Z., Caggana, M., Ding, X. 2000. Characterization of human *CYP2G* genes: Widespread loss-of-function mutations and genetic polymorphism. *Pharmacogenetics* 10:667–678.

Shet, M. S., McPhaul, M., Fisher, C. W., Stallings, N. R., Estabrook, R. W. 1997. Metabolism of the antiandrogenic drug (Flutamide) by human CYP1A2. *Drug Metab Dispos* 25:1298–1303.

Shiga, K., Fukuyama, R., Kimura, S., Nakajima, K., Fushiki, S. 1999. Mutation of the sterol 27-hydroxylase gene (CYP27) results in truncation of mRNA expressed in leucocytes in a Japanese family with cerebrotendinous xanthomatosis. *J Neurol Neurosurg Psychiatry* 67:675–677.

Shimada, T., Yun, C. H., Yamazaki, H., Gautier, J. C., Beaune, P. H., Guengerich, F. P. 1992. Characterization of human lung microsomal cytochrome P-450 1A1 and its role in the oxidation of chemical carcinogens. *Mol Pharmacol* 41:856–864.

Shimada, T., Yamazaki, H., Mimura, M., Inui, Y., Guengerich, F. P. 1994. Interindividual variations in human liver cytochrome P450 enzymes involved in the oxidation of drugs, carcinogens and toxic chemicals. *J Pharmacol Exp Ther* 270:414–423.

Shimada, T., Hayes, C. L., Yamazaki, H., Amin, S., Hecht, S. S., Guengerich, F. P., Sutter, T. R. 1996a. Activation of chemically diverse procarcinogens by human cytochrome P-450 1B1. *Cancer Res* 56:2979–2984.

Shimada, T., Yamazaki, H., Mimura, M., Wakamiya, N., Ueng, Y. F., Guengerich, F. P., Inui, Y. 1996b. Characterization of microsomal cytochrome P450 enzymes involved in the oxidation of xenobiotic chemicals in human fetal liver and adult lungs. *Drug Metab Dispos* 24:515–522.

Shimada, T., Yamazaki, H., Foroozesh, M., Hopkins, N. E., Alworth, W. L., Guengerich, F. P. 1998. Selectivity of polycyclic inhibitors for human cytochrome P450s 1A1, 1A2, and 1B1. *Chem Res Toxicol* 11:1048–1056.

Shimoda, K., Someya, T., Morita, S., Hirokane, G., Yokono, A., Takahashi, S., Okawa, M. 2002. Lack of impact of CYP1A2 genetic polymorphism (C/A polymorphism at position 734 in intron 1 and G/A polymorphism at position -2964 in the 5′-flanking region of CYP1A2) on the plasma concentration of haloperidol in smoking male Japanese with schizophrenia. *Prog Neuropsychopharmacol Biol Psychiatry* 26:261–265.

Shinkyo, R., Sakaki, T., Kamakura, M., Ohta, M., Inouye, K. 2004. Metabolism of vitamin D by human microsomal CYP2R1. *Biochem Biophys Res Commun* 324:451–457.

Shirai, N., Furuta, T., Moriyama, Y., Okochi, H., Kobayashi, K., Takashima, M., Xiao, F. et al. 2001. Effects of *CYP2C19* genotypic differences in the metabolism of omeprazole and rabeprazole on intragastric pH. *Aliment Pharmacol Ther* 15:1929–1937.

Shirai, N., Furuta, T., Xiao, F., Kajimura, M., Hanai, H., Ohashi, K., Ishizaki, T. 2002. Comparison of lansoprazole and famotidine for gastric acid inhibition during the daytime and night-time in different *CYP2C19* genotype groups. *Aliment Pharmacol Ther* 16:837–846.

Shon, J. H., Yoon, Y. R., Kim, K. A., Lim, Y. C., Lee, K. J., Park, J. Y., Cha, I. J., Flockhart, D. A., Shin, J. G. 2002. Effects of *CYP2C19* and *CYP2C9* genetic polymorphisms on the disposition of and blood glucose lowering response to tolbutamide in humans. *Pharmacogenetics* 12:111–119.

Shou, M., Mei, Q., Ettore, M. W., Jr., Dai, R., Baillie, T. A., Rushmore, T. H. 1999. Sigmoidal kinetic model for two co-operative substrate-binding sites in a cytochrome P450 3A4 active site: An example of the metabolism of diazepam and its derivatives. *Biochem J* 340 (Pt 3):845–853.

Shuster, D. L., Risler, L. J., Prasad, B., Calamia, J. C., Voellinger, J. L., Kelly, E. J., Unadkat, J. D. et al. 2014. Identification of CYP3A7 for glyburide metabolism in human fetal livers. *Biochem Pharmacol* 92:690–700.

Si, D., Guo, Y., Zhang, Y., Yang, L., Zhou, H., Zhong, D. 2004. Identification of a novel variant *CYP2C9* allele in Chinese. *Pharmacogenetics* 14:465–469.

Siegle, I., Fritz, P., Eckhardt, K., Zanger, U. M., Eichelbaum, M. 2001. Cellular localization and regional distribution of CYP2D6 mRNA and protein expression in human brain. *Pharmacogenetics* 11:237–245.

Siewert, E., Bort, R., Kluge, R., Heinrich, P. C., Castell, J., Jover, R. 2000. Hepatic cytochrome P450 down-regulation during aseptic inflammation in the mouse is interleukin 6 dependent. *Hepatology* 32:49–55.

Siller, M., Goyal, S., Yoshimoto, F. K., Xiao, Y., Wei, S., Guengerich, F. P. 2014. Oxidation of endogenous N-arachidonoylserotonin by human cytochrome P450 2U1. *J Biol Chem* 289:10476–10487.

Sim, S. C., Risinger, C., Dahl, M. L., Aklillu, E., Christensen, M., Bertilsson, L., Ingelman-Sundberg, M. 2006. A common novel *CYP2C19* gene variant causes ultrarapid drug metabolism relevant for the drug response to proton pump inhibitors and antidepressants. *Clin Pharmacol Ther* 79:103–113.

Sim, J., Choi, E., Lee, Y. M., Jeong, G. S., Lee, S. 2015. *In vitro* inhibition of human cytochrome P450 by cudratricusxanthone A. *Food Chem Toxicol* 81:171–175.

Simonsen, U. 2002. Interactions between drugs for erectile dysfunction and drugs for cardiovascular disease. *Int J Impot Res* 14:178–188.

Simpson, A. E. 1997. The cytochrome P450 4 (CYP4) family. *Gen Pharmacol* 28:351–359.

Simpson, E. R., Clyne, C., Rubin, G., Boon, W. C., Robertson, K., Britt, K., Speed, C., Jones, M. 2002. Aromatase—A brief overview. *Annu Rev Physiol* 64:93–127.

Sinz, M., Wallace, G., Sahi, J. 2008. Current industrial practices in assessing CYP450 enzyme induction: Preclinical and clinical. *AAPS J* 10:391–400.

Sirim, D., Widmann, M., Wagner, F., Pleiss, J. 2010. Prediction and analysis of the modular structure of cytochrome P450 monooxygenases. *BMC Struct Biol* 10:34.

Sistkova, J., Hudecek, J., Hodek, P., Frei, E., Schmeiser, H. H., Stiborova, M. 2008. Human cytochromes P450 1A1 and 1A2 participate in detoxication of carcinogenic aristolochic acid. *Neuro Endocrinol Lett* 29:733–737.

Siu, E. C., Tyndale, R. F. 2008. Selegiline is a mechanism-based inactivator of CYP2A6 inhibiting nicotine metabolism in humans and mice. *J Pharmacol Exp Ther* 324:992–999.

Skene, D. J., Papagiannidou, E., Hashemi, E., Snelling, J., Lewis, D. F., Fernandez, M., Ioannides, C. 2001. Contribution of CYP1A2 in the hepatic metabolism of melatonin: Studies with isolated microsomal preparations and liver slices. *J Pineal Res* 31:333–342.

Skinner, C. A., Rumsby, G. 1994. Steroid 11 beta-hydroxylase deficiency caused by a five base pair duplication in the CYP11B1 gene. *Hum Mol Genet* 3:377–378.

Slavotinek, A. M., Mehrotra, P., Nazarenko, I., Tang, P. L., Lao, R., Cameron, D., Li, B. et al. 2013. Focal facial dermal dysplasia, type IV, is caused by mutations in *CYP26C1*. *Hum Mol Genet* 22:696–703.

Smit, P., van Schaik, R. H., van der Werf, M., van den Beld, A. W., Koper, J. W., Lindemans, J., Pols, H. A., Brinkmann, A. O., de Jong, F. H., Lamberts, S. W. 2005. A common polymorphism in the *CYP3A7* gene is associated with a nearly 50% reduction in serum dehydroepiandrosterone sulfate levels. *J Clin Endocrinol Metab* 90:5313–5316.

Smith, G. B., Bend, J. R., Bedard, L. L., Reid, K. R., Petsikas, D., Massey, T. E. 2003. Biotransformation of 4-(methylnitrosamino)-1-(3-pyridyl)-1-butanone (NNK) in peripheral human lung microsomes. *Drug Metab Dispos* 31:1134–1141.

Smith, B. D., Sanders, J. L., Porubsky, P. R., Lushington, G. H., Stout, C. D., Scott, E. E. 2007. Structure of the human lung cytochrome P450 2A13. *J Biol Chem* 282:17306–17313.

Smutny, T., Mani, S., Pavek, P. 2013. Post-translational and post-transcriptional modifications of pregnane X receptor (PXR) in regulation of the cytochrome P450 superfamily. *Curr Drug Metab* 14:1059–1069.

Soderberg, M. M., Dahl, M. L. 2013. Pharmacogenetics of olanzapine metabolism. *Pharmacogenomics* 14:1319–1336.

Sokol, C. L., Luster, A. D. 2015. The chemokine system in innate immunity. *Cold Spring Harb Perspect Biol* 7.

Solus, J. F., Arietta, B. J., Harris, J. R., Sexton, D. P., Steward, J. Q., McMunn, C., Ihrie, P., Mehall, J. M., Edwards, T. L., Dawson, E. P. 2004. Genetic variation in eleven Phase I drug metabolism genes in an ethnically diverse population. *Pharmacogenomics* 5:895–931.

Song, K. H., Li, T., Owsley, E., Chiang, J. Y. 2010. A putative role of micro RNA in regulation of cholesterol 7α-hydroxylase expression in human hepatocytes. *J Lipid Res* 51:2223–2233.

Sontag, T. J., Parker, R. S. 2002. Cytochrome P450 ω-hydroxylase pathway of tocopherol catabolism. Novel mechanism of regulation of vitamin E status. *J Biol Chem* 277:25290–25296.

Soucek, P. 1999. Expression of cytochrome P450 2A6 in *Escherichia coli*: Purification, spectral and catalytic characterization, and preparation of polyclonal antibodies. *Arch Biochem Biophys* 370:190–200.

Soyama, A., Saito, Y., Hanioka, N., Murayama, N., Nakajima, O., Katori, N., Ishida, S., Sai, K., Ozawa, S., Sawada, J. I. 2001. Non-synonymous single nucleotide alterations found in the

CYP2C8 gene result in reduced *in vitro* paclitaxel metabolism. *Biol Pharm Bull* 24:1427–1430.

Soyama, A., Hanioka, N., Saito, Y., Murayama, N., Ando, M., Ozawa, S., Sawada, J. 2002. Amiodarone *N*-deethylation by CYP2C8 and its variants, *CYP2C8*3* and CYP2C8 P404A. *Pharmacol Toxicol* 91:174–178.

Soyama, A., Saito, Y., Hanioka, N., Maekawa, K., Komamura, K., Kamakura, S., Kitakaze, M. et al. 2005. Single nucleotide polymorphisms and haplotypes of CYP1A2 in a Japanese population. *Drug Metab Pharmacokinet* 20:24–33.

Spaldin, V., Madden, S., Pool, W. F., Woolf, T. F., Park, B. K. 1994. The effect of enzyme inhibition on the metabolism and activation of tacrine by human liver microsomes. *Br J Clin Pharmacol* 38:15–22.

Speiser, P. W. 2001. Molecular diagnosis of *CYP21* mutations in congenital adrenal hyperplasia: Implications for genetic counseling. *Am J Pharmacogenomics* 1:101–110.

Spina, E., Pisani, F., Perucca, E. 1996. Clinically significant pharmacokinetic drug interactions with carbamazepine. An update. *Clin Pharmacokinet* 31:198–214.

Spinler, S. A., Cheng, J. W., Kindwall, K. E., Charland, S. L. 1995. Possible inhibition of hepatic metabolism of quinidine by erythromycin. *Clin Pharmacol Ther* 57:89–94.

Spisni, E., Bartolini, G., Orlandi, M., Belletti, B., Santi, S., Tomasi, V. 1995. Prostacyclin (PGI2) synthase is a constitutively expressed enzyme in human endothelial cells. *Exp Cell Res* 219:507–513.

Spracklin, D. K., Thummel, K. E., Kharasch, E. D. 1996. Human reductive halothane metabolism *in vitro* is catalyzed by cytochrome P450 2A6 and 3A4. *Drug Metab Dispos* 24:976–983.

Spurr, N. K., Gough, A. C., Stevenson, K., Wolf, C. R. 1989. The human cytochrome P450 *CYP3A* locus: Assignment to chromosome 7q22-qter. *Hum Genet* 81:171–174.

St-Arnaud, R. 2008. The direct role of vitamin D on bone homeostasis. *Arch Biochem Biophys* 473:225–230.

St-Arnaud, R. 2011. *CYP24A1: Structure, Function, and Physiological Role*, Elsevier, Waltham, MA.

Stark, K., Guengerich, F. P. 2007. Characterization of orphan human cytochromes P450. *Drug Metab Rev* 39:627–637.

Stark, K., Torma, H., Cristea, M., Oliw, E. H. 2003. Expression of CYP4F8 (prostaglandin H 19-hydroxylase) in human epithelia and prominent induction in epidermis of psoriatic lesions. *Arch Biochem Biophys* 409:188–196.

Stark, K., Schauer, L., Sahlen, G. E., Ronquist, G., Oliw, E. H. 2004. Expression of CYP4F12 in gastrointestinal and urogenital epithelia. *Basic Clin Pharmacol Toxicol* 94:177–183.

Stark, K., Wongsud, B., Burman, R., Oliw, E. H. 2005. Oxygenation of polyunsaturated long chain fatty acids by recombinant CYP4F8 and CYP4F12 and catalytic importance of Tyr-125 and Gly-328 of CYP4F8. *Arch Biochem Biophys* 441:174–181.

Stark, K., Dostalek, M., Guengerich, F. P. 2008a. Expression and purification of orphan cytochrome P450 4X1 and oxidation of anandamide. *FEBS J* 275:3706–3717.

Stark, K., Wu, Z. L., Bartleson, C. J., Guengerich, F. P. 2008b. mRNA distribution and heterologous expression of orphan cytochrome P450 20A1. *Drug Metab Dispos* 36:1930–1937.

Stearman, R. S., Cornelius, A. R., Lu, X., Conklin, D. S., Del Rosario, M. J., Lowe, A. M., Elos, M. T. et al. 2014. Functional prostacyclin synthase promoter polymorphisms. Impact in pulmonary arterial hypertension. *Am J Respir Crit Care Med* 189:1110–1120.

Stec, D. E., Roman, R. J., Flasch, A., Rieder, M. J. 2007. Functional polymorphism in human CYP4F2 decreases 20-HETE production. *Physiol Genomics* 30:74–81.

Steen, V. M., Andreassen, O. A., Daly, A. K., Tefre, T., Borresen, A. L., Idle, J. R., Gulbrandsen, A. K. 1995. Detection of the poor metabolizer-associated *CYP2D6(D)* gene deletion allele by long-PCR technology. *Pharmacogenetics* 5:215–223.

Stenstedt, K., Hallstrom, M., Johansson, I., Ingelman-Sundberg, M., Ragnhammar, P., Edler, D. 2012. The expression of CYP2W1: A prognostic marker in colon cancer. *Anticancer Res* 32:3869–3874.

Stenstedt, K., Hallstrom, M., Ledel, F., Ragnhammar, P., Ingelman-Sundberg, M., Johansson, I., Edler, D. 2014. The expression of CYP2W1 in colorectal primary tumors, corresponding lymph node metastases and liver metastases. *Acta Oncol* 53:885–891.

Stiborova, M., Levova, K., Barta, F., Shi, Z., Frei, E., Schmeiser, H. H., Nebert, D. W., Phillips, D. H., Arlt, V. M. 2012. Bioactivation versus detoxication of the urothelial carcinogen aristolochic acid I by human cytochrome P450 1A1 and 1A2. *Toxicol Sci* 125:345–358.

Stiles, A. R., Kozlitina, J., Thompson, B. M., McDonald, J. G., King, K. S., Russell, D. W. 2014. Genetic, anatomic, and clinical determinants of human serum sterol and vitamin D levels. *Proc Natl Acad Sci U S A* 111:E4006–E4014.

Stingl, J., Viviani, R. 2015. Polymorphism in CYP2D6 and CYP2C19, members of the cytochrome P450 mixed-function oxidase system, in the metabolism of psychotropic drugs. *J Intern Med* 277:167–177.

Stresser, D. M., Kupfer, D. 1999. Monospecific antipeptide antibody to cytochrome P-450 2B6. *Drug Metab Dispos* 27:517–525.

Stringer, R. A., Strain-Damerell, C., Nicklin, P., Houston, J. B. 2009. Evaluation of recombinant cytochrome P450 enzymes as an in vitro system for metabolic clearance predictions. *Drug Metab Dispos* 37:1025–1034.

Strushkevich, N., Usanov, S. A., Plotnikov, A. N., Jones, G., Park, H. W. 2008. Structural analysis of CYP2R1 in complex with vitamin D_3. *J Mol Biol* 380:95–106.

Strushkevich, N., Usanov, S. A., Park, H. W. 2010. Structural basis of human CYP51 inhibition by antifungal azoles. *J Mol Biol* 397:1067–1078.

Strushkevich, N., Gilep, A. A., Shen, L., Arrowsmith, C. H., Edwards, A. M., Usanov, S. A., Park, H. W. 2013. Structural insights into aldosterone synthase substrate specificity and targeted inhibition. *Mol Endocrinol* 27:315–324.

Su, T., Sheng, J. J., Lipinskas, T. W., Ding, X. 1996. Expression of CYP2A genes in rodent and human nasal mucosa. *Drug Metab Dispos* 24:884–890.

Su, T., Bao, Z., Zhang, Q. Y., Smith, T. J., Hong, J. Y., Ding, X. 2000. Human cytochrome P450 CYP2A13: Predominant expression in the respiratory tract and its high efficiency metabolic activation of a tobacco-specific carcinogen, 4-(methylnitrosamino)-1-(3-pyridyl)-1-butanone. *Cancer Res* 60:5074–5079.

Sugimoto, K., Akasaka, H., Katsuya, T., Node, K., Fujisawa, T., Shimaoka, I., Yasuda, O., Ohishi, M., Ogihara, T., Shimamoto, K., Rakugi, H. 2008. A polymorphism regulates CYP4A11 transcriptional activity and is associated with hypertension in a Japanese population. *Hypertension* 52:1142–1148.

Sugiura, K., Takeichi, T., Tanahashi, K., Ito, Y., Kosho, T., Saida, K., Uhara, H., Okuyama, R., Akiyama, M. 2013. Lamellar ichthyosis in a collodion baby caused by *CYP4F22* mutations in a non-consanguineous family outside the Mediterranean. *J Dermatol Sci* 72:193–195.

Sullivan-Klose, T. H., Ghanayem, B. I., Bell, D. A., Zhang, Z. Y., Kaminsky, L. S., Shenfield, G. M., Miners, J. O., Birkett, D. J., Goldstein, J. A. 1996. The role of the CYP2C9-Leu359 allelic variant in the tolbutamide polymorphism. *Pharmacogenomics* 6:341–349.

Sundaram, S. S., Bove, K. E., Lovell, M. A., Sokol, R. J. 2008. Mechanisms of disease: Inborn errors of bile acid synthesis. *Nat Clin Pract Gastroenterol Hepatol* 5:456–468.

Surekha, D., Sailaja, K., Rao, D. N., Padma, T., Raghunadharao, D., Vishnupriya, S. 2014. Association of CYP19 polymorphisms with breast cancer risk: A case-control study. *J Nat Sci Biol Med* 5:250–254.

Swales, K., Kakizaki, S., Yamamoto, Y., Inoue, K., Kobayashi, K., Negishi, M. 2005. Novel CAR-mediated mechanism for synergistic activation of two distinct elements within the human cytochrome P450 2B6 gene in HepG2 cells. *J Biol Chem* 280:3458–3466.

Syed, K., Mashele, S. S. 2014. Comparative analysis of P450 signature motifs EXXR and CXG in the large and diverse kingdom of fungi: Identification of evolutionarily conserved amino acid patterns characteristic of P450 family. *PLoS One* 9:e95616.

Szabo, J. A., Szilagyi, A., Doleschall, Z., Patocs, A., Farkas, H., Prohaszka, Z., Racz, K., Fust, G., Doleschall, M. 2013. Both positive and negative selection pressures contribute to the polymorphism pattern of the duplicated human *CYP21A2* gene. *PLoS One* 8:e81977.

Szanto, A., Benko, S., Szatmari, I., Balint, B. L., Furtos, I., Ruhl, R., Molnar, S. et al. 2004. Transcriptional regulation of human CYP27 integrates retinoid, peroxisome proliferator-activated receptor, and liver X receptor signaling in macrophages. *Mol Cell Biol* 24:8154–8166.

Taavitsainen, P., Juvonen, R., Pelkonen, O. 2001. *In vitro* inhibition of cytochrome P450 enzymes in human liver microsomes by a potent CYP2A6 inhibitor, *trans*-2-phenylcyclopropylamine (tranylcypromine), and its nonamine analog, cyclopropylbenzene. *Drug Metab Dispos* 29:217–222.

Taimi, M., Helvig, C., Wisniewski, J., Ramshaw, H., White, J., Amad, M., Korczak, B., Petkovich, M. 2004. A novel human cytochrome P450, CYP26C1, involved in metabolism of 9-*cis* and all-*trans* isomers of retinoic acid. *J Biol Chem* 279:77–85.

Tajima, T., Fujieda, K., Kouda, N., Nakae, J., Miller, W. L. 2001. Heterozygous mutation in the cholesterol side chain cleavage enzyme (p450scc) gene in a patient with 46,XY sex reversal and adrenal insufficiency. *J Clin Endocrinol Metab* 86:3820–3825.

Takada, K., Arefayene, M., Desta, Z., Yarboro, C. H., Boumpas, D. T., Balow, J. E., Flockhart, D. A., Illei, G. G. 2004. Cytochrome P450 pharmacogenetics as a predictor of toxicity and clinical response to pulse cyclophosphamide in lupus nephritis. *Arthritis Rheum* 50:2202–2210.

Takagi, S., Nakajima, M., Mohri, T., Yokoi, T. 2008. Posttranscriptional regulation of human pregnane X receptor by micro-RNA affects the expression of cytochrome P450 3A4. *J Biol Chem* 283:9674–9680.

Takagi, S., Nakajima, M., Kida, K., Yamaura, Y., Fukami, T., Yokoi, T. 2009. MicroRNAs regulate human hepatocyte nuclear factor 4α modulating the expression of metabolic enzymes and cell cycle. *J Biol Chem*.

Takeuchi, H., Yokota, A., Ohoka, Y., Iwata, M. 2011. Cyp26b1 regulates retinoic acid-dependent signals in T cells and its expression is inhibited by transforming growth factor-β. *PLoS One* 6:e16089.

Tan, B. S., Tiong, K. H., Muruhadas, A., Randhawa, N., Choo, H. L., Bradshaw, T. D., Stevens, M. F., Leong, C. O. 2011. CYP2S1 and CYP2W1 mediate 2-(3,4-dimethoxyphenyl)-5-fluorobenzothiazole (GW-610, NSC 721648) sensitivity in breast and colorectal cancer cells. *Mol Cancer Ther* 10:1982–1992.

Tang, Y. M., Wo, Y. Y., Stewart, J., Hawkins, A. L., Griffin, C. A., Sutter, T. R., Greenlee, W. F. 1996. Isolation and characterization of the human cytochrome P450 *CYP1B1* gene. *J Biol Chem* 271:28324–28330.

Tang, E. K., Tieu, E. W., Tuckey, R. C. 2012. Expression of human CYP27B1 in *Escherichia coli* and characterization in phospholipid vesicles. *FEBS J* 279:3749–3761.

Tangpricha, V., Flanagan, J. N., Whitlatch, L. W., Tseng, C. C., Chen, T. C., Holt, P. R., Lipkin, M. S., Holick, M. F. 2001. 25-hydroxyvitamin D-1alpha-hydroxylase in normal and malignant colon tissue. *Lancet* 357:1673–1674.

Tassaneeyakul, W., Mohamed, Z., Birkett, D. J., McManus, M. E., Veronese, M. E., Tukey, R. H., Quattrochi, L. C., Gonzalez, F. J., Miners, J. O. 1992. Caffeine as a probe for human cytochromes P450: Validation using cDNA-expression, immunoinhibition and microsomal kinetic and inhibitor techniques. *Pharmacogenetics* 2:173–183.

Tassaneeyakul, W., Birkett, D. J., Veronese, M. E., McManus, M. E., Tukey, R. H., Quattrochi, L. C., Gelboin, H. V., Miners, J. O. 1993. Specificity of substrate and inhibitor probes for human cytochromes P450 1A1 and 1A2. *J Pharmacol Exp Ther* 265:401–407.

Tatarunas, V., Lesauskaite, V., Veikutiene, A., Grybauskas, P., Jakuska, P., Jankauskiene, L., Bartuseviciute, R., Benetis, R. 2014. The effect of *CYP2C9*, *VKORC1* and *CYP4F2* polymorphism and of clinical factors on warfarin dosage during initiation and long-term treatment after heart valve surgery. *J Thromb Thrombolysis* 37:177–185.

Taylor, J. S., Raes, J. 2004. Duplication and divergence: The evolution of new genes and old ideas. *Annu Rev Genet* 38:615–643.

Taylor, J. M., Borthwick, F., Bartholomew, C., Graham, A. 2010. Overexpression of steroidogenic acute regulatory protein increases macrophage cholesterol efflux to apolipoprotein AI. *Cardiovasc Res* 86:526–534.

Tempel, W., Grabovec, I., MacKenzie, F., Dichenko, Y. V., Usanov, S. A., Gilep, A. A., Park, H. W., Strushkevich, N. 2014. Structural characterization of human cholesterol 7α-hydroxylase. *J Lipid Res* 55:1925–1932.

ten Brink, M. H., van Bavel, T., Swen, J. J., van der Straaten, T., Bredius, R. G., Lankester, A. C., Zwaveling, J., Guchelaar, H. J. 2013. Effect of genetic variants *GSTA1* and *CYP39A1* and age on busulfan clearance in pediatric patients undergoing hematopoietic stem cell transplantation. *Pharmacogenomics* 14:1683–1690.

Tesson, C., Nawara, M., Salih, M. A., Rossignol, R., Zaki, M. S., Al Balwi, M., Schule, R. et al. 2012. Alteration of fatty-acid-metabolizing enzymes affects mitochondrial form and function in hereditary spastic paraplegia. *Am J Hum Genet* 91:1051–1064.

Thatcher, J. E., Buttrick, B., Shaffer, S. A., Shimshoni, J. A., Goodlett, D. R., Nelson, W. L., Isoherranen, N. 2011. Substrate specificity and ligand interactions of CYP26A1, the human liver retinoic acid hydroxylase. *Mol Pharmacol* 80:228–239.

Theken, K. N., Lee, C. R. 2007. Genetic variation in the cytochrome P450 epoxygenase pathway and cardiovascular disease risk. *Pharmacogenomics* 8:1369–1383.

Thelen, K., Dressman, J. B. 2009. Cytochrome P450-mediated metabolism in the human gut wall. *J Pharm Pharmacol* 61:541–558.

Thum, T., Borlak, J. 2000. Gene expression in distinct regions of the heart. *Lancet* 355:979–983.

Thummel, K. E., Wilkinson, G. R. 1998. *In vitro* and *in vivo* drug interactions involving human CYP3A. *Annu Rev Pharmacol Toxicol* 38:389–430.

Thummel, K. E., O'Shea, D., Paine, M. F., Shen, D. D., Kunze, K. L., Perkins, J. D., Wilkinson, G. R. 1996. Oral first-pass elimination of midazolam involves both gastrointestinal and hepatic CYP3A-mediated metabolism. *Clin Pharmacol Ther* 59:491–502.

Tieu, E. W., Li, W., Chen, J., Baldisseri, D. M., Slominski, A. T., Tuckey, R. C. 2012. Metabolism of cholesterol, vitamin D$_3$ and 20-hydroxyvitamin D$_3$ incorporated into phospholipid vesicles by human CYP27A1. *J Steroid Biochem Mol Biol* 129:163–171.

Tieu, E. W., Tang, E. K., Tuckey, R. C. 2014. Kinetic analysis of human CYP24A1 metabolism of vitamin D via the C24-oxidation pathway. *FEBS J* 281:3280–3296.

Tiong, K. H., Yiap, B. C., Tan, E. L., Ismail, R., Ong, C. E. 2010. Functional characterization of cytochrome P450 2A6 allelic variants *CYP2A6*15*, *CYP2A6*16*, *CYP2A6*21*, and *CYP2A6*22*. *Drug Metab Dispos* 38:745–751.

Toda, K., Shizuta, Y. 1993. Molecular cloning of a cDNA showing alternative splicing of the 5′-untranslated sequence of mRNA for human aromatase P-450. *Eur J Biochem* 213:383–389.

Tomalik-Scharte, D., Lazar, A., Fuhr, U., Kirchheiner, J. 2008. The clinical role of genetic polymorphisms in drug-metabolizing enzymes. *Pharmacogenomics J* 8:4–15.

Tompkins, L. M., Wallace, A. D. 2007. Mechanisms of cytochrome P450 induction. *J Biochem Mol Toxicol* 21:176–181.

Topletz, A. R., Thatcher, J. E., Zelter, A., Lutz, J. D., Tay, S., Nelson, W. L., Isoherranen, N. 2012. Comparison of the function and expression of CYP26A1 and CYP26B1, the two retinoic acid hydroxylases. *Biochem Pharmacol* 83:149–163.

Topletz, A. R., Tripathy, S., Foti, R. S., Shimshoni, J. A., Nelson, W. L., Isoherranen, N. 2015. Induction of CYP26A1 by metabolites of retinoic acid: Evidence that CYP26A1 is an important enzyme in the elimination of active retinoids. *Mol Pharmacol* 87:430–441.

Torchin, C. D., McNeilly, P. J., Kapetanovic, I. M., Strong, J. M., Kupferberg, H. J. 1996. Stereoselective metabolism of a new anticonvulsant drug candidate, losigamone, by human liver microsomes. *Drug Metab Dispos* 24:1002–1008.

Toselli, F., Booth Depaz, I. M., Worrall, S., Etheridge, N., Dodd, P. R., Wilce, P. A., Gillam, E. M. 2015. Expression of CYP2E1 and CYP2U1 proteins in amygdala and prefrontal cortex: Influence of alcoholism and smoking. *Alcohol Clin Exp Res* 39:790–797.

Totah, R. A., Rettie, A. E. 2005. Cytochrome P450 2C8: Substrates, inhibitors, pharmacogenetics, and clinical relevance. *Clin Pharmacol Ther* 77:341–352.

Tournel, G., Cauffiez, C., Billaut-Laden, I., Allorge, D., Chevalier, D., Bonnifet, F., Mensier, E., Lafitte, J. J., Lhermitte, M., Broly, F., Lo-Guidice, J. M. 2007. Molecular analysis of the *CYP2F1* gene: Identification of a frequent non-functional allelic variant. *Mutat Res* 617:79–89.

Tralau, T., Luch, A. 2013. The evolution of our understanding of endo-xenobiotic crosstalk and cytochrome P450 regulation and the therapeutic implications. *Expert Opin Drug Metab Toxicol* 9:1541–1554.

Travica, S., Pors, K., Loadman, P. M., Shnyder, S. D., Johansson, I., Alandas, M. N., Sheldrake, H. M., Mkrtchian, S., Patterson, L. H., Ingelman-Sundberg, M. 2013. Colon cancer-specific cytochrome P450 2W1 converts duocarmycin analogues into potent tumor cytotoxins. *Clin Cancer Res* 19:2952–2961.

Treluyer, J. M., Gueret, G., Cheron, G., Sonnier, M., Cresteil, T. 1997. Developmental expression of CYP2C and CYP2C-dependent activities in the human liver: In-vivo/in-vitro correlation and inducibility. *Pharmacogenetics* 7:441–452.

Trofimova-Griffin, M. E., Juchau, M. R. 2002. Developmental expression of cytochrome CYP26B1 (P450RAI-2) in human cephalic tissues. *Brain Res Dev Brain Res* 136:175–178.

Tsai, L. P., Lee, H. H. 2012. Analysis of *CYP21A1P* and the duplicated *CYP21A2* genes. *Gene* 506:261–262.

Tsaousidou, M. K., Ouahchi, K., Warner, T. T., Yang, Y., Simpson, M. A., Laing, N. G., Wilkinson, P. A. et al. 2008. Sequence alterations within *CYP7B1* implicate defective cholesterol homeostasis in motor-neuron degeneration. *Am J Hum Genet* 82:510–515.

Tsuchiya, Y., Nakajima, M., Takagi, S., Taniya, T., Yokoi, T. 2006. MicroRNA regulates the expression of human cytochrome P450 1B1. *Cancer Res* 66:9090–9098.

Tsukada, K., Ishimitsu, T., Teranishi, M., Saitoh, M., Yoshii, M., Inada, H., Ohta, S. et al. 2002. Positive association of CYP11B2 gene polymorphism with genetic predisposition to essential hypertension. *J Hum Hypertens* 16:789–793.

Tuckey, R. C., Cameron, K. J. 1993. Human placental cholesterol side-chain cleavage: Enzymatic synthesis of (22R)-20 alpha,22-dihydroxycholesterol. *Steroids* 58:230–233.

Tuder, R. M., Cool, C. D., Geraci, M. W., Wang, J., Abman, S. H., Wright, L., Badesch, D., Voelkel, N. F. 1999. Prostacyclin synthase expression is decreased in lungs from patients with severe pulmonary hypertension. *Am J Respir Crit Care Med* 159:1925–1932.

Turpeinen, M., Zanger, U. M. 2012. Cytochrome P450 2B6: Function, genetics, and clinical relevance. *Drug Metabol Drug Interact* 27:185–197.

Turpeinen, M., Nieminen, R., Juntunen, T., Taavitsainen, P., Raunio, H., Pelkonen, O. 2004. Selective inhibition of CYP2B6-catalyzed bupropion hydroxylation in human liver microsomes in vitro. *Drug Metab Dispos* 32:626–631.

Turpeinen, M., Tolonen, A., Uusitalo, J., Jalonen, J., Pelkonen, O., Laine, K. 2005. Effect of clopidogrel and ticlopidine on cytochrome P450 2B6 activity as measured by bupropion hydroxylation. *Clin Pharmacol Ther* 77:553–559.

Uchiyama, T., Kanno, H., Ishitani, K., Fujii, H., Ohta, H., Matsui, H., Kamatani, N., Saito, K. 2012. An SNP in *CYP39A1* is associated with severe neutropenia induced by docetaxel. *Cancer Chemother Pharmacol* 69:1617–1624.

Uehara, M., Yashiro, K., Mamiya, S., Nishino, J., Chambon, P., Dolle, P., Sakai, Y. 2007. CYP26A1 and CYP26C1 cooperatively regulate anterior-posterior patterning of the developing brain and the production of migratory cranial neural crest cells in the mouse. *Dev Biol* 302:399–411.

Uehara, S., Murayama, N., Nakanishi, Y., Nakamura, C., Hashizume, T., Zeldin, D. C., Yamazaki, H., Uno, Y. 2015. Immunochemical quantification of cynomolgus CYP2J2, CYP4A and CYP4F enzymes in liver and small intestine. *Xenobiotica* 45:124–130.

Upadhya, S. C., Tirumalai, P. S., Boyd, M. R., Mori, T., Ravindranath, V. 2000. Cytochrome P4502E (CYP2E) in brain: Constitutive expression, induction by ethanol and localization by fluorescence in situ hybridization. *Arch Biochem Biophys* 373:23–34.

Usami, O., Ashino, Y., Komaki, Y., Tomaki, M., Irokawa, T., Tamada, T., Hayashida, T., Teruya, K., Hattori, T. 2007. Efavirenz-induced neurological symptoms in rare homozygote CYP2B6 *2/*2 (C64T). *Int J STD AIDS* 18:575–576.

Usmani, K. A., Chen, W. G., Sadeque, A. J. 2012. Identification of human cytochrome P450 and flavin-containing monooxygenase enzymes involved in the metabolism of lorcaserin, a novel selective human 5-hydroxytryptamine 2C agonist. *Drug Metab Dispos* 40:761–771.

Van Asten, K., Neven, P., Lintermans, A., Wildiers, H., Paridaens, R. 2014. Aromatase inhibitors in the breast cancer clinic: Focus on exemestane. *Endocr Relat Cancer* 21:R31–R49.

Van Booven, D., Marsh, S., McLeod, H., Carrillo, M. W., Sangkuhl, K., Klein, T. E., Altman, R. B. 2010. Cytochrome P450 2C9-CYP2C9. *Pharmacogenet Genomics* 20:277–281.

van Rantwijk, F., Sheldon, R. A. 2000. Selective oxygen transfer catalysed by heme peroxidases: Synthetic and mechanistic aspects. *Curr Opin Biotechnol* 11:554–564.

van Schaik, R. H., de Wildt, S. N., Brosens, R., van Fessem, M., van den Anker, J. N., Lindemans, J. 2001. The *CYP3A4*3* allele: Is it really rare? *Clin Chem* 47:1104–1106.

Van Voorhis, M., Fechner, J. H., Zhang, X., Mezrich, J. D. 2013. The aryl hydrocarbon receptor: A novel target for immunomodulation in organ transplantation. *Transplantation* 95:983–990.

van Wanrooy, M. J., Span, L. F., Rodgers, M. G., van den Heuvel, E. R., Uges, D. R., van der Werf, T. S., Kosterink, J. G., Alffenaar, J. W. 2014. Inflammation is associated with voriconazole trough concentrations. *Antimicrob Agents Chemother* 58:7098–7101.

Vasiliou, V., Gonzalez, F. J. 2008. Role of CYP1B1 in glaucoma. *Annu Rev Pharmacol Toxicol* 48:333–358.

Vasquez, A., Atallah-Yunes, N., Smith, F. C., You, X., Chase, S. E., Silverstone, A. E., Vikstrom, K. L. 2003. A role for the aryl hydrocarbon receptor in cardiac physiology and function as demonstrated by AhR knockout mice. *Cardiovasc Toxicol* 3:153–163.

von Weymarn, L. B., Murphy, S. E. 2003. CYP2A13-catalysed coumarin metabolism: Comparison with CYP2A5 and CYP2A6. *Xenobiotica* 33:73–81.

von Weymarn, L. B., Retzlaff, C., Murphy, S. E. 2012. CYP2A6- and CYP2A13-catalyzed metabolism of the nicotine $\Delta5'(1')$ iminium ion. *J Pharmacol Exp Ther* 343:307–315.

Wada, Y., Itabashi, T., Sato, H., Kawamura, M., Tada, A., Tamai, M. 2005. Screening for mutations in CYP4V2 gene in Japanese patients with Bietti's crystalline corneoretinal dystrophy. *Am J Ophthalmol* 139:894–899.

Walsh, A. A., Szklarz, G. D., Scott, E. E. 2013. Human cytochrome P450 1A1 structure and utility in understanding drug and xenobiotic metabolism. *J Biol Chem* 288:12932–12943.

Wang, D. P., Chiang, J. Y. 1994. Structure and nucleotide sequences of the human cholesterol 7 alpha-hydroxylase gene (CYP7). *Genomics* 20:320–323.

Wang, L. H., Kulmacz, R. J. 2002. Thromboxane synthase: Structure and function of protein and gene. *Prostaglandins Other Lipid Mediat* 68–69:409–422.

Wang, H., Tompkins, L. M. 2008. CYP2B6: New insights into a historically overlooked cytochrome P450 isozyme. *Curr Drug Metab* 9:598–610.

Wang, D. P., Stroup, D., Marrapodi, M., Crestani, M., Galli, G., Chiang, J. Y. 1996. Transcriptional regulation of the human cholesterol 7 alpha-hydroxylase gene (CYP7A) in HepG2 cells. *J Lipid Res* 37:1831–1841.

Wang, J. T., Lin, C. J., Burridge, S. M., Fu, G. K., Labuda, M., Portale, A. A., Miller, W. L. 1998. Genetics of vitamin D 1α-hydroxylase deficiency in 17 families. *Am J Hum Genet* 63:1694–1702.

Wang, X. L., Bassett, M., Zhang, Y., Yin, S., Clyne, C., White, P. C., Rainey, W. E. 2000. Transcriptional regulation of human 11beta-hydroxylase (hCYP11B1). *Endocrinology* 141:3587–3594.

Wang, Z. Q., Gorski, C., Hamman, M. A., Huang, S. M., Lesko, L. J., Hall, S. D. 2001. The effects of St John's wort (*Hypericum perforatum*) on human cytochrome P450 activity. *Clin Pharmacol Ther* 70:317–326.

Wang, H., Chanas, B., Ghanayem, B. I. 2002. Cytochrome P450 2E1 (CYP2E1) is essential for acrylonitrile metabolism to cyanide: Comparative studies using CYP2E1-null and wild-type mice. *Drug Metab Dispos* 30:911–917.

Wang, X. L., Greco, M., Sim, A. S., Duarte, N., Wang, J., Wilcken, D. E. 2002. Effect of CYP1A1 MspI polymorphism on cigarette smoking related coronary artery disease and diabetes. *Atherosclerosis* 162:391–397.

Wang, H., Faucette, S., Sueyoshi, T., Moore, R., Ferguson, S., Negishi, M., LeCluyse, E. L. 2003a. A novel distal enhancer module regulated by pregnane X receptor/constitutive androstane receptor is essential for the maximal induction of *CYP2B6* gene expression. *J Biol Chem* 278:14146–14152.

Wang, H., Faucette, S. R., Gilbert, D., Jolley, S. L., Sueyoshi, T., Negishi, M., LeCluyse, E. L. 2003b. Glucocorticoid receptor enhancement of pregnane X receptor-mediated CYP2B6 regulation in primary human hepatocytes. *Drug Metab Dispos* 31:620–630.

Wang, H., Tan, W., Hao, B., Miao, X., Zhou, G., He, F., Lin, D. 2003c. Substantial reduction in risk of lung adenocarcinoma associated with genetic polymorphism in CYP2A13, the most active cytochrome P450 for the metabolic activation of tobacco-specific carcinogen NNK. *Cancer Res* 63:8057–8061.

Wang, M. Z., Saulter, J. Y., Usuki, E., Cheung, Y. L., Hall, M., Bridges, A. S., Loewen, G. et al. 2006. CYP4F enzymes are the major enzymes in human liver microsomes that catalyze the *O*-demethylation of the antiparasitic prodrug DB289 [2,5-bis(4-amidinophenyl)furan-bis-*O*-methylamidoxime]. *Drug Metab Dispos* 34:1985–1994.

Wang, S. L., He, X. Y., Shen, J., Wang, J. S., Hong, J. Y. 2006. The missense genetic polymorphisms of human CYP2A13: Functional significance in carcinogen activation and identification of a null allelic variant. *Toxicol Sci* 94:38–45.

Wang, Y., Rogers, P. M., Su, C., Varga, G., Stayrook, K. R., Burris, T. P. 2008. Regulation of cholesterologenesis by the oxysterol receptor, LXRα. *J Biol Chem* 283:26332–26339.

Wang, J. F., Zhang, C. C., Chou, K. C., Wei, D. Q. 2009. Structure of cytochrome p450s and personalized drug. *Curr Med Chem* 16:232–244.

Wang, Y., Bell, J. C., Keeney, D. S., Strobel, H. W. 2010. Gene regulation of *CYP4F11* in human keratinocyte HaCaT cells. *Drug Metab Dispos* 38:100–107.

Wang, A., Savas, U., Stout, C. D., Johnson, E. F. 2011. Structural characterization of the complex between α-naphthoflavone and human cytochrome P450 1B1. *J Biol Chem* 286:5736–5743.

Wang, D., Guo, Y., Wrighton, S. A., Cooke, G. E., Sadee, W. 2011. Intronic polymorphism in CYP3A4 affects hepatic expression and response to statin drugs. *Pharmacogenomics J* 11:274–286.

Wang, A., Savas, U., Hsu, M. H., Stout, C. D., Johnson, E. F. 2012. Crystal structure of human cytochrome P450 2D6 with prinomastat bound. *J Biol Chem* 287:10834–10843.

Wang, Y., Guo, L., Cai, S. P., Dai, M., Yang, Q., Yu, W., Yan, N. et al. 2012. Exome sequencing identifies compound heterozygous mutations in *CYP4V2* in a pedigree with retinitis pigmentosa. *PLoS One* 7:e33673.

Wang, X., Ni, L., Yang, L., Duan, Q., Chen, C., Edin, M. L., Zeldin, D. C., Wang, D. W. 2014. CYP2J2-derived epoxyeicosatrienoic acids suppress endoplasmic reticulum stress in heart failure. *Mol Pharmacol* 85:105–115.

Wang, Y. H., Pan, P. P., Dai, D. P., Wang, S. H., Geng, P. W., Cai, J. P., Hu, G. X. 2014. Effect of 36 *CYP2C9* variants found in the Chinese population on losartan metabolism *in vitro*. *Xenobiotica* 44:270–275.

Warner, M., Gustafsson, J. A. 1994. Effect of ethanol on cytochrome P450 in the rat brain. *Proc Natl Acad Sci U S A* 91:1019–1023.

Watanabe, J., Hayashi, S., Kawajiri, K. 1994. Different regulation and expression of the human CYP2E1 gene due to the RsaI polymorphism in the 5′-flanking region. *J Biochem* 116:321–326.

Watanabe, T., Sakuyama, K., Sasaki, T., Ishii, Y., Ishikawa, M., Hirasawa, N., Hiratsuka, M. 2010. Functional characterization of 26 *CYP2B6* allelic variants (CYP2B6.2-CYP2B6.28, except CYP2B6.22). *Pharmacogenet Genomics* 20:459–462.

Waxman, D. J. 1999. P450 gene induction by structurally diverse xenochemicals: Central role of nuclear receptors CAR, PXR, and PPAR. *Arch Biochem Biophys* 369:11–23.

Wedell, A. 2011. Molecular genetics of 21-hydroxylase deficiency. *Endocr Dev* 20:80–87.

Wei, Y., Wu, H., Li, L., Liu, Z., Zhou, X., Zhang, Q. Y., Weng, Y. et al. 2012. Generation and characterization of a CYP2A13/2B6/2F1-transgenic mouse model. *Drug Metab Dispos* 40:1144–1150.

Wen, X., Wang, J. S., Kivisto, K. T., Neuvonen, P. J., Backman, J. T. 2001. *In vitro* evaluation of valproic acid as an inhibitor of human cytochrome P450 isoforms: Preferential inhibition of cytochrome P450 2C9 (CYP2C9). *Br J Clin Pharmacol* 52:547–553.

Wen, X., Wang, J. S., Neuvonen, P. J., Backman, J. T. 2002. Isoniazid is a mechanism-based inhibitor of cytochrome P450 1A2, 2A6, 2C19 and 3A4 isoforms in human liver microsomes. *Eur J Clin Pharmacol* 57:799–804.

Wentworth, J. M., Agostini, M., Love, J., Schwabe, J. W., Chatterjee, V. K. 2000. St John's wort, a herbal antidepressant, activates the steroid X receptor. *J Endocrinol* 166:R11–R16.

Werck-Reichhart, D., Feyereisen, R. 2000. Cytochromes P450: A success story. *Genome Biol* 1:REVIEWS3003.

Werk, A. N., Cascorbi, I. 2014. Functional gene variants of CYP3A4. *Clin Pharmacol Ther* 96:340–348.

Wester, M. R., Johnson, E. F., Marques-Soares, C., Dansette, P. M., Mansuy, D., Stout, C. D. 2003a. Structure of a substrate complex of mammalian cytochrome P450 2C5 at 2.3 Å resolution: Evidence for multiple substrate binding modes. *Biochemistry* 42:6370–6379.

Wester, M. R., Johnson, E. F., Marques-Soares, C., Dijols, S., Dansette, P. M., Mansuy, D., Stout, C. D. 2003b. Structure of mammalian cytochrome P450 2C5 complexed with diclofenac at 2.1 Å resolution: Evidence for an induced fit model of substrate binding. *Biochemistry* 42:9335–9345.

Wester, M. R., Yano, J. K., Schoch, G. A., Yang, C., Griffin, K. J., Stout, C. D., Johnson, E. F. 2004. The structure of human cytochrome P450 2C9 complexed with flurbiprofen at 2.0-Å resolution. *J Biol Chem* 279:35630–35637.

Westlind, A., Malmebo, S., Johansson, I., Otter, C., Andersson, T. B., Ingelman-Sundberg, M., Oscarson, M. 2001. Cloning and tissue distribution of a novel human cytochrome p450 of the CYP3A subfamily, CYP3A43. *Biochem Biophys Res Commun* 281:1349–1355.

Westlind-Johnsson, A., Hermann, R., Huennemeyer, A., Hauns, B., Lahu, G., Nassr, N., Zech, K., Ingelman-Sundberg, M., von Richter, O. 2006. Identification and characterization of *CYP3A4*20*, a novel rare *CYP3A4* allele without functional activity. *Clin Pharmacol Ther* 79:339–349.

White, J. A., Beckett, B., Scherer, S. W., Herbrick, J. A., Petkovich, M. 1998. P450RAI (CYP26A1) maps to human chromosome 10q23-q24 and mouse chromosome 19C2-3. *Genomics* 48:270–272.

White, J. A., Ramshaw, H., Taimi, M., Stangle, W., Zhang, A., Everingham, S., Creighton, S., Tam, S. P., Jones, G., Petkovich, M. 2000. Identification of the human cytochrome P450, P450RAI-2, which is predominantly expressed in the adult cerebellum and is responsible for all-*trans*-retinoic acid metabolism. *Proc Natl Acad Sci U S A* 97:6403–6408.

Wikvall, K. 2001. Cytochrome P450 enzymes in the bioactivation of vitamin D to its hormonal form (review). *Int J Mol Med* 7:201–209.

Wild, M. J., McKillop, D., Butters, C. J. 1999. Determination of the human cytochrome P450 isoforms involved in the metabolism of zolmitriptan. *Xenobiotica* 29:847–857.

Wilderman, P. R., Shah, M. B., Liu, T., Li, S., Hsu, S., Roberts, A. G., Goodlett, D. R. et al. 2010. Plasticity of cytochrome P450 2B4 as investigated by hydrogen-deuterium exchange mass spectrometry and X-ray crystallography. *J Biol Chem* 285:38602–38611.

Wilderman, P. R., Shah, M. B., Jang, H. H., Stout, C. D., Halpert, J. R. 2013. Structural and thermodynamic basis of (+)-α-pinene binding to human cytochrome P450 2B6. *J Am Chem Soc* 135:10433–10440.

Wilkinson, G. R. 2005. Drug metabolism and variability among patients in drug response. *N Engl J Med* 352:2211–2221.

Williams, D., Feely, J. 2002. Pharmacokinetic-pharmacodynamic drug interactions with HMG-CoA reductase inhibitors. *Clin Pharmacokinet* 41:343–370.

Williams, M. L., Bhargava, P., Cherrouk, I., Marshall, J. L., Flockhart, D. A., Wainer, I. W. 2000. A discordance of the cytochrome P450 2C19 genotype and phenotype in patients with advanced cancer. *Br J Clin Pharmacol* 49:485–488.

Williams, P. A., Cosme, J., Sridhar, V., Johnson, E. F., McRee, D. E. 2000. Mammalian microsomal cytochrome P450 monooxygenase: Structural adaptations for membrane binding and functional diversity. *Mol Cell* 5:121–131.

Williams, J. A., Ring, B. J., Cantrell, V. E., Jones, D. R., Eckstein, J., Ruterbories, K., Hamman, M. A., Hall, S. D., Wrighton, S. A. 2002. Comparative metabolic capabilities of CYP3A4, CYP3A5, and CYP3A7. *Drug Metab Dispos* 30:883–891.

Williams, P. A., Cosme, J., Ward, A., Angove, H. C., Matak Vinkovic, D., Jhoti, H. 2003. Crystal structure of human cytochrome P450 2C9 with bound warfarin. *Nature* 424:464–468.

Williams, P. A., Cosme, J., Vinkovic, D. M., Ward, A., Angove, H. C., Day, P. J., Vonrhein, C., Tickle, I. J., Jhoti, H. 2004. Crystal structures of human cytochrome P450 3A4 bound to metyrapone and progesterone. *Science* 305:683–686.

Williams, J. S., Hopkins, P. N., Jeunemaitre, X., Brown, N. J. 2011. CYP4A11 T8590C polymorphism, salt-sensitive hypertension, and renal blood flow. *J Hypertens* 29:1913–1918.

Winter, J., Jung, S., Keller, S., Gregory, R. I., Diederichs, S. 2009. Many roads to maturity: MicroRNA biogenesis pathways and their regulation. *Nat Cell Biol* 11:228–234.

Witchel, S. F., Azziz, R. 2011. Congenital adrenal hyperplasia. *J Pediatr Adolesc Gynecol* 24:116–126.

Wolbold, R., Klein, K., Burk, O., Nussler, A. K., Neuhaus, P., Eichelbaum, M., Schwab, M., Zanger, U. M. 2003. Sex is a major determinant of CYP3A4 in human liver. *Hepatology* 37:978–988.

Wong, R. H., Du, C. L., Wang, J. D., Chan, C. C., Luo, J. C., Cheng, T. J. 2002. XRCC1 and CYP2E1 polymorphisms as susceptibility factors of plasma mutant p53 protein and anti-p53 antibody expression in vinyl chloride monomer-exposed polyvinyl chloride workers. *Cancer Epidemiol Biomarkers Prev* 11:475–482.

Woodcroft, K. J., Novak, R. F. 1999. The role of phosphatidylinositol 3-kinase, Src kinase, and protein kinase A signaling pathways in insulin and glucagon regulation of CYP2E1 expression. *Biochem Biophys Res Commun* 266:304–307.

Woodland, C., Huang, T. T., Gryz, E., Bendayan, R., Fawcett, J. P. 2008. Expression, activity and regulation of CYP3A in human and rodent brain. *Drug Metab Rev* 40:149–168.

Wray, G. A., Hahn, M. W., Abouheif, E., Balhoff, J. P., Pizer, M., Rockman, M. V., Romano, L. A. 2003. The evolution of transcriptional regulation in eukaryotes. *Mol Biol Evol* 20:1377–1419.

Wu, S., Moomaw, C. R., B., T. K., Falck, J. R., Zeldin, D. C. 1996. Molecular cloning and expression of CYP2J2, a human cytochrome P450 arachidonic acid epoxygenase higly expressed in heart. *J Biol Chem* 271:3460–3468.

Wu, Z. L., Bartleson, C. J., Ham, A. J., Guengerich, F. P. 2006. Heterologous expression, purification, and properties of human cytochrome P450 27C1. *Arch Biochem Biophys* 445:138–146.

Wu, H., Liu, Z., Ling, G., Lawrence, D., Ding, X. 2013. Transcriptional suppression of CYP2A13 expression by lipopolysaccharide in cultured human lung cells and the lungs of a CYP2A13-humanized mouse model. *Toxicol Sci* 135:476–485.

Wu, Z., Lee, D., Joo, J., Shin, J. H., Kang, W., Oh, S., Lee do, Y. et al. 2013. CYP2J2 and CYP2C19 are the major enzymes responsible for metabolism of albendazole and fenbendazole in human liver microsomes and recombinant P450 assay systems. *Antimicrob Agents Chemother* 57:5448–5456.

Wu, C. C., Gupta, T., Garcia, V., Ding, Y., Schwartzman, M. L. 2014. 20-HETE and blood pressure regulation: Clinical implications. *Cardiol Rev* 22:1–12.

Wyen, C., Hendra, H., Vogel, M., Hoffmann, C., Knechten, H., Brockmeyer, N. H., Bogner, J. R. et al. 2008. Impact of CYP2B6 983T>C polymorphism on non-nucleoside reverse transcriptase inhibitor plasma concentrations in HIV-infected patients. *J Antimicrob Chemother* 61:914–918.

Xia, X. Y., Peng, R. X., Yu, J. P., Wang, H., Wang, J. 2002. *In vitro* metabolic characteristics of cytochrome P450 2A6 in Chinese liver microsomes. *Acta Pharmacol Sin* 23:471–476.

Xiang, C., Wang, J., Kou, X., Chen, X., Qin, Z., Jiang, Y., Sun, C. et al. 2015. Pulmonary expression of CYP2A13 and ABCB1 is regulated by FOXA2, and their genetic interaction is associated with lung cancer. *FASEB J* 29:1986–1998.

Xiao, C., Rajewsky, K. 2009. MicroRNA control in the immune system: Basic principles. *Cell* 136:26–36.

Xiao, Y., Guengerich, F. P. 2012. Metabolomic analysis and identification of a role for the orphan human cytochrome P450 2W1 in selective oxidation of lysophospholipids. *J Lipid Res* 53:1610–1617.

Xiao, X., Mai, G., Li, S., Guo, X., Zhang, Q. 2011. Identification of CYP4V2 mutation in 21 families and overview of mutation spectrum in Bietti crystalline corneoretinal dystrophy. *Biochem Biophys Res Commun* 409:181–186.

Xiaoping, L., Zhong, F., Tan, X. 2013. Cytochrome P450 2C8 and drug metabolism. *Curr Top Med Chem* 13:2241–2253.

Xiong, H., Yoshinari, K., Brouwer, K. L., Negishi, M. 2002. Role of constitutive androstane receptor in the *in vivo* induction of Mrp3 and CYP2B1/2 by phenobarbital. *Drug Metab Dispos* 30:918–923.

Xu, C., Rao, Y. S., Xu, B., Hoffmann, E., Jones, J., Sellers, E. M., Tyndale, R. F. 2002. An *in vivo* pilot study characterizing the new *CYP2A6*7, *8, and *10 alleles. *Biochem Biophys Res Commun* 290:318–324.

Xu, X., Zhang, X. A., Wang, D. W. 2011. The roles of CYP450 epoxygenases and metabolites, epoxyeicosatrienoic acids, in cardiovascular and malignant diseases. *Adv Drug Deliv Rev* 63:597–609.

Xu, M., Ju, W., Hao, H., Wang, G., Li, P. 2013. Cytochrome P450 2J2: Distribution, function, regulation, genetic polymorphisms and clinical significance. *Drug Metab Rev* 45:311–352.

Xue, L., Wang, H. F., Wang, Q., Szklarz, G. D., Domanski, T. L., Halpert, J. R., Correia, M. A. 2001. Influence of P450 3A4 SRS-2 residues on cooperativity and/or regioselectivity of aflatoxin B(1) oxidation. *Chem Res Toxicol* 14:483–491.

Yamakoshi, Y., Kishimoto, T., Sugimura, K., Kawashima, H. 1999. Human prostate CYP3A5: Identification of a unique 5′-untranslated sequence and characterization of purified recombinant protein. *Biochem Biophys Res Commun* 260:676–681.

Yamano, S., Nhamburo, P. T., Aoyama, T., Meyer, U. A., Inaba, T., Kalow, W., Gelboin, H. V., McBride, O. W., Gonzalez, F. J. 1989. cDNA cloning and sequence and cDNA-directed expression of human P450 2B1: Identification of a normal and two variant cDNAs derived from the *CYP2B* locus on chromosome 19 and differential expression of the 2B mRNAs in human liver. *Biochemistry* 28:7340–7348.

Yamano, S., Tatsuno, J., Gonzalez, F. J. 1990. The *CYP2A3* gene product catalyzes coumarin 7-hydroxylation in human liver microsomes. *Biochemistry* 29:1322–1329.

Yamazaki, H., Shimada, T. 1999. Effects of arachidonic acid, prostaglandins, retinol, retinoic acid and cholecalciferol on xenobiotic oxidations catalysed by human cytochrome P450 enzymes. *Xenobiotica* 29:231–241.

Yamazaki, H., Mimura, M., Sugahara, C., Shimada, T. 1994. Catalytic roles of rat and human cytochrome P450 2A enzymes in testosterone 7α- and coumarin 7-hydroxylations. *Biochem Pharmacol* 48:1524–1527.

Yamazaki, H., Inoue, K., Chiba, K., Ozawa, N., Kawai, T., Suzuki, Y., Goldstein, J. A., Guengerich, F. P., Shimada, T. 1998a. Comparative studies on the catalytic roles of cytochrome P450 2C9 and its Cys- and Leu-variants in the oxidation of warfarin, flurbiprofen, and diclofenac by human liver microsomes. *Biochem Pharmacol* 56:243–251.

Yamazaki, H., Shaw, P. M., Guengerich, F. P., Shimada, T. 1998b. Roles of cytochromes P450 1A2 and 3A4 in the oxidation of estradiol and estrone in human liver microsomes. *Chem Res Toxicol* 11:659–665.

Yamazaki, H., Inoue, K., Hashimoto, M., Shimada, T. 1999. Roles of CYP2A6 and CYP2B6 in nicotine C-oxidation by human liver microsomes. *Arch Toxicol* 73:65–70.

Yamazaki, H., Okayama, A., Imai, N., Guengerich, F. P., Shimizu, M. 2006. Inter-individual variation of cytochrome P4502J2 expression and catalytic activities in liver microsomes from Japanese and Caucasian populations. *Xenobiotica* 36:1201–1209.

Yan, H. C., Liu, J. H., Li, J., He, B. X., Yang, L., Qiu, J., Li, L., Ding, D. P., Shi, L., Zhao, S. J. 2013. Association between the CYP4A11 T8590C variant and essential hypertension: New data from Han Chinese and a meta-analysis. *PLoS One* 8:e80072.

Yanev, S., Kent, U. M., Pandova, B., Hollenberg, P. F. 1999. Selective mechanism-based inactivation of cytochromes P-450 2B1 and P-450 2B6 by a series of xanthates. *Drug Metab Dispos* 27:600–604.

Yang, H. Y., Namkung, M. J., Juchau, M. R. 1995. Expression of functional cytochrome P4501A1 in human embryonic hepatic tissues during organogenesis. *Biochem Pharmacol* 49:717–726.

Yang, Y., Zhang, M., Eggertsen, G., Chiang, J. Y. 2002. On the mechanism of bile acid inhibition of rat sterol 12alpha-hydroxylase gene (CYP8B1) transcription: Roles of alpha-fetoprotein transcription factor and hepatocyte nuclear factor 4alpha. *Biochim Biophys Acta* 1583:63–73.

Yang, J. Q., Morin, S., Verstuyft, C., Fan, L. A., Zhang, Y., Xu, C. D., Barbu, V., Funck-Brentano, C., Jaillon, P., Becquemont, L. 2003. Frequency of cytochrome P450 2C9 allelic variants in the Chinese and French populations. *Fundam Clin Pharmacol* 17:373–376.

Yang, X. J., Lu, H. Y., Li, Z. Y., Bian, Q., Qiu, L. L., Li, Z., Liu, Q., Li, J., Wang, X., Wang, S. L. 2012. Cytochrome P450 2A13 mediates aflatoxin B₁-induced cytotoxicity and apoptosis in human bronchial epithelial cells. *Toxicology* 300:138–148.

Yang, X., Zhang, Z., Wang, X., Wang, Y., Zhang, X., Lu, H., Wang, S. L. 2013. Cytochrome P450 2A13 enhances the sensitivity of human bronchial epithelial cells to aflatoxin B_1-induced DNA damage. *Toxicol Appl Pharmacol* 270:114–121.

Yang, H., Fu, Z., Ma, Y., Huang, D., Zhu, Q., Erdenbat, C., Xie, X., Liu, F., Zheng, Y. 2014. CYP4A11 gene T8590C polymorphism is associated with essential hypertension in the male western Chinese Han population. *Clin Exp Hypertens* 36:398–403.

Yano, J. K., Wester, M. R., Schoch, G. A., Griffin, K. J., Stout, C. D., Johnson, E. F. 2004. The structure of human microsomal cytochrome P450 3A4 determined by X-ray crystallography to 2.05-Å resolution. *J Biol Chem* 279:38091–38094.

Yano, J. K., Hsu, M. H., Griffin, K. J., Stout, C. D., Johnson, E. F. 2005. Structures of human microsomal cytochrome P450 2A6 complexed with coumarin and methoxsalen. *Nat Struct Mol Biol* 12:822–823.

Yano, J. K., Denton, T. T., Cerny, M. A., Zhang, X., Johnson, E. F., Cashman, J. R. 2006. Synthetic inhibitors of cytochrome P-450 2A6: Inhibitory activity, difference spectra, mechanism of inhibition, and protein cocrystallization. *J Med Chem* 49:6987–7001.

Yantsevich, A. V., Dichenko, Y. V., Mackenzie, F., Mukha, D. V., Baranovsky, A. V., Gilep, A. A., Usanov, S. A., Strushkevich, N. V. 2014. Human steroid and oxysterol 7α-hydroxylase CYP7B1: Substrate specificity, azole binding and misfolding of clinically relevant mutants. *FEBS J* 281:1700–1713.

Yasar, U., Aklillu, E., Canaparo, R., Sandberg, M., Sayi, J., Roh, H. K., Wennerholm, A. 2002a. Analysis of CYP2C9*5 in Caucasian, Oriental and black-African populations. *Eur J Clin Pharmacol* 58:555–558.

Yasar, U., Lundgren, S., Eliasson, E., Bennet, A., Wiman, B., de Faire, U., Rane, A. 2002b. Linkage between the *CYP2C8* and *CYP2C9* genetic polymorphisms. *Biochem Biophys Res Commun* 299:25–28.

Yasukochi, Y., Satta, Y. 2015. Molecular evolution of the CYP2D subfamily in primates: Purifying selection on substrate recognition sites without the frequent or long-tract gene conversion. *Genome Biol Evol* 7:1053–1067.

Yates, L. A., Norbury, C. J., Gilbert, R. J. 2013. The long and short of microRNA. *Cell* 153:516–519.

Ye, X. H., Song, L., Peng, L., Bu, Z., Yan, S. X., Feng, J., Zhu, X. L., Liao, X. B., Yu, X. L., Yan, D. 2015. Association between the CYP2E1 polymorphisms and lung cancer risk: A meta-analysis. *Mol Genet Genomics* 290:545–558.

Ye, Z., Sharp, S. J., Burgess, S., Scott, R. A., Imamura, F., InterAct, C., Langenberg, C., Wareham, N. J., Forouhi, N. G. 2015. Association between circulating 25-hydroxyvitamin D and incident type 2 diabetes: A mendelian randomisation study. *Lancet Diabetes Endocrinol* 3:35–42.

Yeleswaram, K., Vachharajani, N., Santone, K. 1999. Involvement of cytochrome P-450 isozymes in melatonin metabolism and clinical implications. *J Pineal Res* 26:190–191.

Yin, L., Hu, Q. 2014. CYP17 inhibitors—A biraterone, C17,20-lyase inhibitors and multi-targeting agents. *Nat Rev Urol* 11:32–42.

Yokoi, Y., Nakazawa, M., Mizukoshi, S., Sato, K., Usui, T., Takeuchi, K. 2010. Crystal deposits on the lens capsules in Bietti crystalline corneoretinal dystrophy associated with a mutation in the CYP4V2 gene. *Acta Ophthalmol* 88:607–609.

Yokoi, Y., Sato, K., Aoyagi, H., Takahashi, Y., Yamagami, M., Nakazawa, M. 2011. A novel compound heterozygous mutation in the *CYP4V2* gene in a Japanese patient with Bietti's crystalline corneoretinal dystrophy. *Case Rep Ophthalmol* 2:296–301.

Yokota, S., Higashi, E., Fukami, T., Yokoi, T., Nakajima, M. 2011. Human CYP2A6 is regulated by nuclear factor-erythroid 2 related factor 2. *Biochem Pharmacol* 81:289–294.

Yokoyama, C., Miyata, A., Ihara, H., Ullrich, V., Tanabe, T. 1991. Molecular cloning of human platelet thromboxane A synthase. *Biochem Biophys Res Commun* 178:1479–1484.

Yokoyama, C., Yabuki, T., Inoue, H., Tone, Y., Hara, S., Hatae, T., Nagata, M., Takahashi, E. I., Tanabe, T. 1996. Human gene encoding prostacyclin synthase (PTGIS): Genomic organization, chromosomal localization, and promoter activity. *Genomics* 36:296–304.

Yoon, Y. R., Shon, J. H., Kim, M. K., Lee, H. R., Park, J. Y., Cha, I. J., Shin, J. G. 2001. Frequency of cytochrome P4502C9 mutant alleles in a Korean population. *Br J Clin Pharmacol* 51:277–280.

Yoshida, R., Nakajima, M., Watanabe, Y., Kwon, J. T., Yokoi, T. 2002. Genetic polymorphisms in human *CYP2A6* gene causing impaired nicotine metabolism. *Br J Clin Pharmacol* 54:511–517.

Yoshida, R., Nakajima, M., Nishimura, K., Tokudome, S., Kwon, J. T., Yokoi, T. 2003. Effects of polymorphism in promoter region of human *CYP2A6* gene (*CYP2A6*9*) on expression level of messenger ribonucleic acid and enzymatic activity *in vivo* and *in vitro*. *Clin Pharmacol Ther* 74:69–76.

Yoshimoto, F. K., Auchus, R. J. 2014. The diverse chemistry of cytochrome P450 17A1 (P450c17, CYP17A1). *J Steroid Biochem Mol Biol*.

Yoshioka, H., Kasai, N., Ikushiro, S., Shinkyo, R., Kamakura, M., Ohta, M., Inouye, K., Sakaki, T. 2006. Enzymatic properties of human CYP2W1 expressed in *Escherichia coli*. *Biochem Biophys Res Commun* 345:169–174.

Young, V., Ho, M., Vosper, H., Belch, J. J., Palmer, C. N. 2002. Elevated expression of the genes encoding TNF-alpha and thromboxane synthase in leucocytes from patients with systemic sclerosis. *Rheumatology (Oxford)* 41:869–875.

Yu, S., Levi, L., Siegel, R., Noy, N. 2012. Retinoic acid induces neurogenesis by activating both retinoic acid receptors (RARs) and peroxisome proliferator-activated receptor β/δ (PPARβ/δ). *J Biol Chem* 287:42195–42205.

Yu, W., Chai, H., Li, Y., Zhao, H., Xie, X., Zheng, H., Wang, C. et al. 2012. Increased expression of CYP4Z1 promotes tumor angiogenesis and growth in human breast cancer. *Toxicol Appl Pharmacol* 264:73–83.

Yuan, R., Madani, S., Wei, X. X., Reynolds, K., Huang, S. M. 2002. Evaluation of cytochrome p450 probe substrates commonly used by the pharmaceutical industry to study *in vitro* drug interactions. *Drug Metab Dispos* 30:1311–1319.

Yumibe, N., Huie, K., Chen, K. J., Snow, M., Clement, R. P., Cayen, M. N. 1996. Identification of human liver cytochrome P450 enzymes that metabolize the nonsedating antihistamine loratadine. Formation of descarboethoxyloratadine by CYP3A4 and CYP2D6. *Biochem Pharmacol* 51:165–172.

Yun, C. H., Shimada, T., Guengerich, F. P. 1991. Purification and characterization of human liver microsomal cytochrome P450 2A6. *Mol Pharmacol* 40:679–685.

Yun, C. H., Miller, G. P., Guengerich, F. P. 2000. Rate-determining steps in phenacetin oxidations by human cytochrome P450 1A2 and selected mutants. *Biochemistry* 39:11319–11329.

Zaccaro, C., Sweitzer, S., Pipino, S., Gorman, N., Sinclair, P. R., Sinclair, J. F., Nebert, D. W., De Matteis, F. 2001. Role of cytochrome P450 1A2 in bilirubin degradation Studies in Cyp1a2 (-/-) mutant mice. *Biochem Pharmacol* 61:843–849.

Zanger, U. M., Klein, K. 2013. Pharmacogenetics of cytochrome P450 2B6 (CYP2B6): Advances on polymorphisms, mechanisms, and clinical relevance. *Front Genet* 4:24.

Zanger, U. M., Klein, K., Saussele, T., Blievernicht, J., Hofmann, M. H., Schwab, M. 2007. Polymorphic *CYP2B6*: Molecular mechanisms and emerging clinical significance. *Pharmacogenomics* 8:743–759.

Zanger, U. M., Klein, K., Thomas, M., Rieger, J. K., Tremmel, R., Kandel, B. A., Klein, M., Magdy, T. 2014. Genetics, epigenetics, and regulation of drug-metabolizing cytochrome p450 enzymes. *Clin Pharmacol Ther* 95:258–261.

Zaphiropoulos, P. G. 1997. Exon skipping and circular RNA formation in transcripts of the human cytochrome P450 2C18 gene in epidermis and of the rat androgen binding protein gene in testis. *Mol Cell Biol* 17:2985–2993.

Zaphiropoulos, P. G. 1999. RNA molecules containing exons originating from different members of the cytochrome P450 2C gene subfamily (CYP2C) in human epidermis and liver. *Nucleic Acids Res* 27:2585–2590.

Zehnder, D., Bland, R., Walker, E. A., Bradwell, A. R., Howie, A. J., Hewison, M., Stewart, P. M. 1999. Expression of 25-hydroxyvitamin D_3-1alpha-hydroxylase in the human kidney. *J Am Soc Nephrol* 10:2465–2473.

Zehnder, D., Bland, R., Williams, M. C., McNinch, R. W., Howie, A. J., Stewart, P. M., Hewison, M. 2001. External expression of 25-hydroxyvitamin D_3-1α hydroxylase. *J Clin Endocrinol Metab* 86:888–894.

Zeigler-Johnson, C. M., Walker, A. H., Mancke, B., Spangler, E., Jalloh, M., McBride, S., Deitz, A., Malkowicz, S. B., Ofori-Adjei, D., Gueye, S. M., Rebbeck, T. R. 2002. Ethnic differences in the frequency of prostate cancer susceptibility alleles at SRD5A2 and CYP3A4. *Hum Hered* 54:13–21.

Zeldin, D. C., Foley, J., Ma, J., Boyle, J. E., Pascual, J. M., Moomaw, C. R., Tomer, K. B., Steenbergen, C., Wu, S. 1996. CYP2J subfamily P450s in the lung: Expression, localization, and potential functional significance. *Mol Pharmacol* 50:1111–1117.

Zeldin, D. C., Foley, J., Goldsworthy, S. M., Cook, M. E., Boyle, J. E., Ma, J., Moomaw, C. R., Tomer, K. B., Steenbergen, C., Wu, S. 1997. CYP2J subfamily cytochrome P450s in the gastrointestinal tract: Expression, localization, and potential functional significance. *Mol Pharmacol* 51:931–943.

Zenteno, J. C., Ayala-Ramirez, R., Graue-Wiechers, F. 2008. Novel CYP4V2 gene mutation in a Mexican patient with Bietti's crystalline corneoretinal dystrophy. *Curr Eye Res* 33:313–318.

Zhang, M., Chiang, J. Y. 2001. Transcriptional regulation of the human sterol 12alpha-hydroxylase gene (CYP8B1): Roles of heaptocyte nuclear factor 4alpha in mediating bile acid repression. *J Biol Chem* 276:41690–41699.

Zhang, Z., Fasco, M. J., Huang, Z., Guengerich, F. P., Kaminsky, L. S. 1995. Human cytochromes P4501A1 and P4501A2: R-warfarin metabolism as a probe. *Drug Metab Dispos* 23:1339–1346.

Zhang, X., Su, T., Zhang, Q. Y., Gu, J., Caggana, M., Li, H., Ding, X. 2002. Genetic polymorphisms of the human *CYP2A13* gene: Identification of single-nucleotide polymorphisms and functional characterization of an Arg257Cys variant. *J Pharmacol Exp Ther* 302:416–423.

Zhang, X., Chen, Y., Liu, Y., Ren, X., Zhang, Q. Y., Caggana, M., Ding, X. 2003. Single nucleotide polymorphisms of the human cyp2a13 gene: Evidence for a null allele. *Drug Metab Dispos* 31:1081–1085.

Zhang, J. E., Jorgensen, A. L., Alfirevic, A., Williamson, P. R., Toh, C. H., Park, B. K., Pirmohamed, M. 2009. Effects of *CYP4F2* genetic polymorphisms and haplotypes on clinical outcomes in patients initiated on warfarin therapy. *Pharmacogenet Genomics* 19:781–789.

Zhang, W. Y., Tu, Y. B., Haining, R. L., Yu, A. M. 2009. Expression and functional analysis of CYP2D6.24, CYP2D6.26, CYP2D6.27, and CYP2D7 isozymes. *Drug Metab Dispos* 37:1–4.

Zhang, Y., Zolfaghari, R., Ross, A. C. 2010. Multiple retinoic acid response elements cooperate to enhance the inducibility of *CYP26A1* gene expression in liver. *Gene* 464:32–43.

Zhang, B., Beeghly-Fadiel, A., Long, J., Zheng, W. 2011. Genetic variants associated with breast-cancer risk: Comprehensive research synopsis, meta-analysis, and epidemiological evidence. *Lancet Oncol* 12:477–488.

Zhang, Z., Yang, X., Wang, Y., Wang, X., Lu, H., Zhang, X., Xiao, X., Li, S., Wang, X., Wang, S. L. 2013. Cytochrome P450 2A13 is an efficient enzyme in metabolic activation of aflatoxin G_1 in human bronchial epithelial cells. *Arch Toxicol* 87:1697–1707.

Zhang, M., Sun, S., Liu, Y., Zhang, H., Jiao, Y., Wang, W., Li, X. 2015. New, recurrent, and prevalent mutations: Clinical and molecular characterization of 26 Chinese patients with 17α-hydroxylase/17,20-lyase deficiency. *J Steroid Biochem Mol Biol* 150:11–16.

Zhao, Y., Halpert, J. R. 2007. Structure-function analysis of cytochromes P450 2B. *Biochim Biophys Acta* 1770:402–412.

Zhao, F., Loke, C., Rankin, S. C., Guo, J. Y., Lee, H. S., Wu, T. S., Tan, T. et al. 2004. Novel *CYP2C9* genetic variants in Asian subjects and their influence on maintenance warfarin dose. *Clin Pharmacol Ther* 76:210–219.

Zhao, Y., White, M. A., Muralidhara, B. K., Sun, L., Halpert, J. R., Stout, C. D. 2006. Structure of microsomal cytochrome P450 2B4 complexed with the antifungal drug bifonazole: Insight into P450 conformational plasticity and membrane interaction. *J Biol Chem* 281:5973–5981.

Zhao, B., Lei, L., Kagawa, N., Sundaramoorthy, M., Banerjee, S., Nagy, L. D., Guengerich, F. P., Waterman, M. R. 2012. Three-dimensional structure of steroid 21-hydroxylase (cytochrome P450 21A2) with two substrates reveals locations of disease-associated variants. *J Biol Chem* 287:10613–10622.

Zheng, L., Cao, J., Lu, D., Ji, L., Peng, Y., Zheng, J. 2015. Imperatorin is a mechanism-based inactivator of CYP2B6. *Drug Metab Dispos* 43:82–88.

Zhou, S. F. 2008a. Drugs behave as substrates, inhibitors and inducers of human cytochrome P450 3A4. *Curr Drug Metab* 9:310–322.

Zhou, S. F. 2008b. Potential strategies for minimizing mechanism-based inhibition of cytochrome P450 3A4. *Curr Pharm Des* 14:990–1000.

Zhou, Z. W., Zhou, S. F. 2009. Application of mechanism-based CYP inhibition for predicting drug–drug interactions. *Expert Opin Drug Metab Toxicol* 5:579–605.

Zhou, H. H., Anthony, L. B., Wood, A. J., Wikinson, G. R. 1990. Induction of polumorphic 4′-hydroxylation of S-mephenytoin by rifampicin. *Br J Clin Pharmacol* 30:471–475.

Zhou, S., Paxton, J. W., Tingle, M. D., Kestell, P. 2000. Identification of the human liver cytochrome P450 isoenzyme responsible for the 6-methylhydroxylation of the novel anticancer drug 5,6-dimethylxanthenone-4-acetic acid. *Drug Metab Dispos* 28:1449–1456.

Zhou, S., Kestell, P., Paxton, J. W. 2002. Predicting pharmacokinetics and drug interactions in patients from in vitro and in vivo models: The experience with 5,6-dimethylxanthenone-4-acetic acid (DMXAA), an anti-cancer drug eliminated mainly by conjugation. *Drug Metab Rev* 34:751–790.

Zhou, S., Chan, E., Lim, L. Y., Boelsterli, U. A., Li, S. C., Wang, J., Zhang, Q., Huang, M., Xu, A. 2004. Therapeutic drugs that behave as mechanism-based inhibitors of cytochrome P450 3A4. *Curr Drug Metab* 5:415–442.

Zhou, S., Yung Chan, S., Cher Goh, B., Chan, E., Duan, W., Huang, M., McLeod, H. L. 2005. Mechanism-based inhibition of cytochrome P450 3A4 by therapeutic drugs. *Clin Pharmacokinet* 44:279–304.

Zhou, S. F., Xue, C. C., Yu, X. Q., Li, C., Wang, G. 2007. Clinically important drug interactions potentially involving mechanism-based inhibition of cytochrome P450 3A4 and the role of therapeutic drug monitoring. *Ther Drug Monit* 29:687–710.

Zhou, S. F., Di, Y. M., Chan, E., Du, Y. M., Chow, V. D., Xue, C. C., Lai, X. et al. 2008. Clinical pharmacogenetics and potential application in personalized medicine. *Curr Drug Metab* 9:738–784.

Zhou, S. F., Chan, E., Zhou, Z. W., Xue, C. C., Lai, X., Duan, W. 2009a. Insights into the structure, function, and regulation of human cytochrome P450 1A2. *Curr Drug Metab* (published online).

Zhou, S. F., Liu, J. P., Chowbay, B. 2009b. Polymorphism of human cytochrome P450 enzymes and its clinical impact. *Drug Metab Rev* 41:89–295.

Zhou, S. F., Liu, Y. H., Mo, S. L., Chan, E. 2009c. Insights into the substrate specificity, inhibitors, and polymorphism and the clinical impact of human cytochrome P450 1A2. *AAPS J* (ePub ahead of printing).

Zhou, S. F., Zhou, Z. W., Huang, M. 2010. Polymorphisms of human cytochrome P450 2C9 and the functional relevance. *Toxicology* 278:165–188.

Zhou, G. L., Beloiartsev, A., Yu, B., Baron, D. M., Zhou, W., Niedra, R., Lu, N. et al. 2013. Deletion of the murine cytochrome P450 Cyp2j locus by fused BAC-mediated recombination identifies a role for Cyp2j in the pulmonary vascular response to hypoxia. *PLoS Genet* 9:e1003950.

Zhu, B. T. 2010. On the general mechanism of selective induction of cytochrome P450 enzymes by chemicals: Some theoretical considerations. *Expert Opin Drug Metab Toxicol* 6:483–494.

Zhu, L. R., Thomas, P. E., Lu, G., Reuhl, K. R., Yang, G. Y., Wang, L. D., Wang, S. L., Yang, C. S., He, X. Y., Hong, J. Y. 2006. CYP2A13 in human respiratory tissues and lung cancers: An immunohistochemical study with a new peptide-specific antibody. *Drug Metab Dispos* 34:1672–1676.

Zhu, J. G., Ochalek, J. T., Kaufmann, M., Jones, G., Deluca, H. F. 2013. CYP2R1 is a major, but not exclusive, contributor to 25-hydroxyvitamin D production in vivo. *Proc Natl Acad Sci U S A* 110:15650–15655.

Zhu-Ge, J., Yu, Y. N., Qian, Y. L., Li, X. 2002. Establishment of a transgenic cell line stably expressing human cytochrome P450 2C18 and identification of a CYP2C18 clone with exon 5 missing. *World J Gastroenterol* 8:888–892.

Zollner, A., Dragan, C. A., Pistorius, D., Muller, R., Bode, H. B., Peters, F. T., Maurer, H. H., Bureik, M. 2009. Human CYP4Z1 catalyzes the in-chain hydroxylation of lauric acid and myristic acid. *Biol Chem* 390:313–317.

Zordoky, B. N., El-Kadi, A. O. 2010. Effect of cytochrome P450 polymorphism on arachidonic acid metabolism and their impact on cardiovascular diseases. *Pharmacol Ther* 125:446–463.

Zou, J. G., Ma, Y. T., Xie, X., Yang, Y. N., Pan, S., Adi, D., Liu, F., Chen, B. D. 2014. The association between CYP1A1 genetic polymorphisms and coronary artery disease in the Uygur and Han of China. *Lipids Health Dis* 13:145.

2 Mammalian CYP2D Members
A Comparison of Structure, Function, and Regulation

2.1 INTRODUCTION

CYPs are heme-containing enzymes that are responsible for metabolizing numerous endogenous and exogenous compounds, including hydroxytryptamines, steroids, drugs, procarcinogens, neurotoxins, and environmental compounds (Nebert and Dalton 2006; Nebert and Russell 2002; Nebert et al. 2013; Nelson et al. 2013). There are clear differences between humans and other animal species with regard to the tissue distribution, regulation, substrate specificity, and inhibitor selectivity of CYPs (Emoto et al. 2013; Graham and Lake 2008; Marathe and Rodrigues 2006; Martignoni et al. 2006). The human *CYP2D6* gene consists of nine exons coding for a coding for a 497–amino acid protein with a molecular weight of 55.8 kDa (Eichelbaum et al. 1987; Heim and Meyer 1990; Kimura et al. 1989). *CYP2D6* belongs to a gene cluster containing five highly homologous inactive pseudogenes (Table 2.1) (Heim and Meyer 1992; Kimura et al. 1989; Steen et al. 1995). The *CYP2D6* gene is conserved in chimpanzee, rhesus monkey, rat, chicken, and frog. In humans, CYP2D6 is involved in the metabolism of more than 150 drugs and approximately 25% of commonly prescribed drugs, although it only constitutes 2%–4% of the total hepatic CYPs (Zhou et al. 2009). Its substrates include antidepressants, antipsychotics, analgesics and antitussives, β-blocking agents, antiarrhythmics, and antiemetics. Human CYP2D6 is largely not inducible but is subject to inhibition by a large number of drugs and other compounds (Zhou et al. 2009). Classic xenobiotic inducers that transactivate pregnane X receptors, aryl hydrocarbon receptors, and constitutive androstane receptors do not induce CYP2D6. However, the orphan nuclear receptor HNF-4α regulates CYP2D6 transcription via binding to a direct repeat site (DR1) on the *CYP2D6* promoter (Cairns et al. 1996). In CYP2D6-humanized mice lacking Hnf-4α in the liver, a 50% decrease in CYP2D6 mRNA and activity is observed (Corchero et al. 2001). However, HNF-4α is a common regulator of many hepatic P450 enzymes, which all change in a different manner during pregnancy, and thus it is unlikely that changes in CYP2D6 activity during human pregnancy can be completely explained by changes in HNF-4α-mediated transcriptional activity. Furthermore, human CYP2D6 is highly polymorphic, which often causes altered or abolished enzyme activities, leading not only to severe adverse effects in pharmacotherapy but also to therapeutic failure (Gaedigk 2013; Haertter 2013; Teh and Bertilsson 2012; Zhou 2009a,b). The genetic variations in *CYP2D6* results in four different drug metabolism phenotypes: poor metabolizers (PMs), intermediate metabolizers (IMs), extensive metabolizers (EMs), and ultrarapid metabolizers (UMs). The latter is the result of gene duplication/multiplication and occurs with inheritance of more than two copies of the fully functional *CYP2D6* alleles. The crystal structure of human CYP2D6 has been determined and shows the characteristic CYP fold observed for other members of the CYP superfamily (Nguyen and Conley 2008; Rowland et al. 2006; Wang et al. 2012). The lengths and orientations of the individual secondary structural elements in the CYP2D6 structure are similar to those seen in other human CYP2 members, such as CYP2C8 and 2C9. Many important amino acid residues in the active site of human CYP2D6 are implicated in substrate recognition and binding, including Phe120, Glu216, Asp301, Phe481, and Phe483.

Orthologs are genes in different species that originated by vertical descent from a single gene of the last common ancestor. Human *CYP2D6* has at least 68 orthologs present in 68 different species. These include polar bear, carmine bee-eater, golden-collared manakin, American crow, Sunda flying lemur, Anna's hummingbird, green monkey, Cape elephant shrew, Cape golden mole, white-throated sparrow, long-tailed chinchilla, naked mole-rat, nine-banded armadillo, southern white rhinoceros, Florida manatee, western gorilla, Bolivian squirrel monkey, pygmy chimpanzee, and so on. Yasukochi and Satta (2011) have examined the evolution of the *CYP2D* gene cluster and found that the number of members within the CYP2D subfamily varies between species. For example, primates have two to three *CYP2D* genes, whereas rodents, rabbits, and horses have seven, five, and six *CYP2D* genes, respectively. It has been suggested that the expansion of members within the CYP2D subfamily in herbivores might be related to the fact that several plant toxins are substrates for CYP2Ds. A further window analysis and statistical tests have revealed that entire genomic sequences of paralogous genes are extensively homogenized by gene conversion during molecular evolution of *CYP2D* genes in primates (Yasukochi and Satta 2015). A neighbor-joining tree based on genomic sequences at the nonsubstrate recognition sites shows that *CYP2D6* and *2D8* genes are clustered together because of gene conversion. In contrast, a phylogenetic tree using amino acid sequences at substrate recognition sites does not cluster the *CYP2D6* and *2D8* genes, suggesting that the functional constraint on substrate specificity is one of the causes for purifying selection at the substrate recognition sites. It appears that the *CYP2D* gene subfamily in primates has evolved to

TABLE 2.1

Human CYP2D Pseudogenes

Pseudogene	Chromosomal Location	Genomic Location	Alias	Name	RefSeq mRNA
CYP2D7P	22q13	NC_000022.11	CYP2D; CYP2D7; CYP2D@; P450C2D; P450DB1; CYP2D7AP; CYP2D7P1	Cytochrome P450, family 2, subfamily D, polypeptide 7, pseudogene	–
LOC101929829	22q13	NW_004504305.1	–	Cytochrome P450, family 2, subfamily D, polypeptide 6 pseudogene	XM_006726320.1; XM_005278357.2
CYP2D8P	22q13.2	NC_000022.11	CYP2DP1; CYP2D8P1	Cytochrome P450, family 2, subfamily D, polypeptide 8, pseudogene	–
CYP2D7BP	22q13.2	AC_000065.1	–	Cytochrome P450, subfamily IID (debrisoquine, sparteine, etc., -metabolizing), polypeptide 7b (pseudogene)	–
LOC100631371	–	–	–	Cytochrome P450, family 2, subfamily D, polypeptide 6 pseudogene	–

maintain the regioselectivity for a substrate hydroxylation activity between individual enzymes, although extensive gene conversion has occurred across *CYP2D* coding sequences.

In the past years, many *CYP2D/Cyp2d* genes from nonhuman animals have been cloned and characterized (Table 2.2; for the sequence alignment and phylogenetic tree, please see Figure 2.1). Because of the differences in the multiplicity and substrate specificity of CYP2D family members among species, it is difficult to predict pathways of human CYP2D6-dependent drug metabolism on the basis of animal studies. To better understand the similarities and differences between human *CYP2D6* with these counterparts, we will discuss them with regard to tissue distribution, regulation, and function in this chapter.

2.2 RAT Cyp2d SUBFAMILY: Cyp2d1, 2d2, 2d3, 2d4, AND 2d5

The rat is one of the most common animal models used in the study of drug metabolism and toxicity. The genome in rats have 89 *Cyp* gene and 79 pseudogenes. Rats contain five Cyp2d enzymes, namely, Cyp2d1–2d5 (Gerhard et al. 2004; Nelson et al. 2004), which have been identified by genomic analysis (Gonzalez et al. 1987; Kawashima and Strobel 1995; Matsunaga et al. 1989). Rat Cyp2d5 has >95% similarities in amino acid sequence to Cyp2d1 and Cyp2d4 (Gonzalez et al. 1987; Kawashima and Strobel 1995; Matsunaga et al. 1989). Rat Cyp2d3, but not Cyp2d1, 2, or 4, is considered the homolog of human CYP2D6. Cyp2d18 is previously thought a functional member of rat Cyp2d family, but it is found a variant of Cyp2d4. The rat Cyp2ds and human CYP2D6 have a sequence identity of ~56%, but this is significantly lower in the active-site region (~34%) (Nelson et al. 2004). Like human

CYP2D6, the six Cyp2ds in rats are expressed in various tissues such as the liver, kidney, and brain. Rat Cyp2d2 and 2d3 mRNAs are mainly expressed in the liver, kidney, and small intestinal mucosa. In contrast, rat Cyp2d1/5 mRNAs are expressed in various tissues. Cyp2d4/18 mRNAs are expressed in the brain, adrenal glands, ovary, and testis, in addition to the liver, kidney, and small intestinal mucosa.

2.2.1 TISSUE DISTRIBUTION OF RAT CYP2DS

Rat Cyp2ds are mainly detected in the liver, and other tissues including kidney and brain at low to moderate levels (Hiroi et al. 1998; Lindell et al. 2003; Masubuchi et al. 1996). Cyp2d1 is detected in the duodenum, jejunum, and ileum at least at the mRNA level (Lindell et al. 2003). Cyp2d1 is only expressed at lower levels in rat intestine (Mitschke et al. 2008). The pattern of Cyp2d expression in most brain regions in rats is similar to that in humans (Miksys et al. 2000). For example, there is a high expression of Cyp2ds/CYP2D6 in pyramidal neurons of the frontal cortex layers II through VI and a lower expression in globus pallidus than in the caudate and putamen in rats and humans (Miksys et al. 2000, 2005). In contrast, in cerebellum of rats, Cyp2ds are expressed at a moderate to high level in Purkinje cells (Miksys et al. 2000), whereas the Purkinje cells express very low levels of CYP2D6 in cerebellum of humans (Miksys et al. 2002).

Cyp2d4 is highly expressed in rat brain (Miksys et al. 2000), suggesting a possible role in brain biosynthesis of important endogenous compounds and disposition of drugs. Cyp2d4 has also been identified in rat breast (Hellmold et al. 1995). Cyp2d5 is abundant in the basal ganglia of male and female rat brains, whereas Cyp2d1 is not detected and Cyp2d2 is only weakly expressed (Riedl et al. 1999). Cyp2d2 and Cyp2d3 are mainly expressed in

TABLE 2.2
Reported Mammalian CYP2D/Cyp2d Genes in Various Species

CYP2D/Cyp2d Gene	Chromosomal Location	Genomic Location	Alias	Names	RefSeq mRNA	Homologs
Rat (*Rattus norvegicus*)						
Cyp2d1	7q34	NC_005106.4	Cyp2d9	Cytochrome P450 2D1; CYPIID1; P450-DB1; P450-UT-7; P450-CMF1A; cytochrome P450-DB1; cytochrome P450-UT-7; cytochrome P450-CMF1A; debrisoquine 4-hydroxylase	NM_153313.1	
Cyp2d2	7q34	NC_005106.4	Cyp2d26	Cytochrome P450 2D26; CYPIID26P450-DB2; P450-CMF2; cytochrome P450-DB2; cytochrome P450-CMF2; debrisoquine 4-hydroxylase; cytochrome P450, subfamily IID2	NM_012730.1	
Cyp2d3	7q34	NC_005106.4	Cyp2d6; Cyp2d13	Cytochrome P450 2D3; CYPIID3; P450-DB3; cytochrome P450-DB3; debrisoquine 4-hydroxylase; cytochrome P450, subfamily IID3; cytochrome P450, family 2, subfamily d, polypeptide 6; cytochrome P450, family 2, subfamily d, polypeptide 13	NM_173093.1	Human, chimpanzee, rhesus monkey, chicken, and frog
Cyp2d4	7q34	NC_005106.4	Cyp2d6; Cyp2d18; Cyp2d22; Cyp2d4v1; Cyp2d4v2	Cytochrome P450, family 2, subfamily d, polypeptide 4; CYPIID4; CYPIID18P450-DB4P450-CMF3; P450 2D-29/2D-35; cytochrome P450 2D4; cytochrome P450-DB4; cytochrome P450 2d18; cytochrome P450-CMF3; cytochrome P450 2D-29; cytochrome P450 2D-35; debrisoquine 4-hydroxylase; cytochrome P450, subfamily IID4; cytochrome P450, subfamily 2D, polypeptide 6; cytochrome P450, subfamily IID, polypeptide 6; cytochrome P450, family 2, subfamily d, polypeptide 22	NM_138515.2	Chimpanzee, dog, cow, and mouse
Cyp2d5	7q34	NC_005106.4	Cyp2d10	Cytochrome P450 2D10; CYPIID10; P450-DB5; cytochrome P450-DB5; cytochrome P450CMF1b; cytochrome P450-CMF1B; debrisoquine 4-hydroxylase; cytochrome P450, family 2, subfamily d, polypeptide 10	NM_173304.2	
Cat (*Felis catus*)						
CYP2D6	B4	NC_018729.1		Cytochrome P450 2D6	NM_001195406.1	
Rhesus Monkey (*Macaca mulatta*)						
CYP2D6	10	NC_007867.1	CYP2D42	Cytochrome P450 2D6; cytochrome P450 CYP2D42	NM_001040218.1	Human, chimpanzee, rat, chicken, and frog
Chimpanzee (*Pan troglodytes*)						
CYP2D6	22	NC_006489.3		Cytochrome P450 2D6; CYPIID6	NM_001040622.1	Human, rhesus monkey, rat, chicken, and frog

(Continued)

TABLE 2.2 (CONTINUED)
Reported Mammalian *CYP2D/Cyp2d* Genes in Various Species

CYP2D/Cyp2d Gene	Chromosomal Location	Genomic Location	Alias	Names	RefSeq mRNA	Homologs
CYP2D7	22	NC_006489.3		Cytochrome P450 2D7	XM_001170370.2	Dog, cow, mouse, and rat
Amur Tiger (*Panthera tigris altaica*)						
CYP2D6	Unknown	NW_006711815.1		Cytochrome P450, family 2, subfamily D, polypeptide 6; cytochrome P450 2D6	NM_001290578.1	
Sheep (*Ovis aries*)						
CYP2D6				Cytochrome P450 2D6	NM_001245974.1	
Chicken (*Gallus gallus*)						
CYP2D6	1	NC_006088.3	CYP2D49; CYP2D6c	Cytochrome P450 2D3-like; cytochrome P450 CYP2D49	NM_001195557.1	Human, chimpanzee, rhesus monkey, rat, and frog
Human (*Homo sapiens*)						
CYP2D6	22q13.1	NC_000022.11	CPD6; CYP2D; CYP2DL1; CYPIID6; P450C2D; P450DB1; CYP2D7AP; CYP2D7BP; CYP2D7P2; CYP2D8P2; P450-DB1	Cytochrome P450 2D6; cytochrome P450-DB1; microsomal monooxygenase; xenobiotic monooxygenase; debrisoquine 4-hydroxylase; flavoprotein-linked monooxygenase; cytochrome P450, family 2, subfamily D, polypeptide 7 pseudogene 2; cytochrome P450, family 2, subfamily D, polypeptide 8 pseudogene 2; cytochrome P450, subfamily IID (debrisoquine, sparteine, etc., -metabolizing)-like 1; cytochrome P450, subfamily IID (debrisoquine, sparteine, etc., -metabolizing), polypeptide 6; cytochrome P450, subfamily II (debrisoquine, sparteine, etc., -metabolising), polypeptide 7 pseudogene 2	NM_000106.5	Chimpanzee, rhesus monkey, rat, chicken, and frog
Mouse (*Mus musculus*)						
Cyp2d9	15 E1; 15 38.6 cM	NC_000081.6	Cyp2d; P450-2D	Cytochrome P450 2D9;CYPIID9; cytochrome P450CA; cytochrome P450, 2d9; cytochrome P450-16-α; testosterone 16α-hydroxylase; testosterone 16-α hydroxylase	NM_010006.2	
Cyp2d10	15 E1; 15 38.58 cM	NC_000081.6	Cyp2d; P450-2D; AI303445	Cytochrome P450 2D10; CYPIID10; cytochrome P450, 2d10; cytochrome P450-16-α; testosterone 16-α hydroxylase	NM_010005.3	
Cyp2d11	15 E1; 15 38.58 cM	NC_000081.6	Cyp2d; P450-2D	Cytochrome P450 2D11; CYPIID11; cytochrome P450CC; cytochrome P450, 2d11; cytochrome P450-16-α; testosterone 16-α hydroxylase	NM_001104531.1	
Cyp2d12	15 E1; 15 38.64 cM	NC_000081.6	9030605E09Rik	Cytochrome P450, family 2, subfamily d, polypeptide 12; cytochrome P450, 2d12	NM_201360.1	

(Continued)

TABLE 2.2 (CONTINUED)
Reported Mammalian CYP2D/Cyp2d Genes in Various Species

CYP2D/Cyp2d Gene	Chromosomal Location	Genomic Location	Alias	Names	RefSeq mRNA	Homologs
Cyp2d13	15 E1; 15 38.67 cM	NC_000081.6	1300007K12Rik	May be a pseudogene	NR_003552.1 (noncoding RNA)	
Cyp2d22	15 E2; 15	NC_000081.6	2D22	Cytochrome P450, 2d22	NM_001163472.1	Chimpanzee, dog, cow, and rat
Cyp2d26	15 E1; 15	NC_000081.6	1300006E06Rik	Cytochrome P450 2D26; CYPIID26; cytochrome P450, 2d26	NM_029562.2	
Cyp2d34	15 E1; 15	NC_000081.6		Cytochrome P450, family 2, subfamily d, polypeptide 34	NM_145474.2	
Cyp2d40	15 E1; 15	NC_000081.6	CYPIID3; P450-DB3; 1300013D18Rik	Cytochrome P450, family 2, subfamily d, polypeptide 40	NM_023623.2	
Cattle (Bos taurus)						
CYP2D14	5	AC_000162.1		Cytochrome P450 2D14; CYPIID14	NM_174529.2	Chimpanzee, dog, mouse, and rat
Dog (Canis lupus familiaris)						
CYP2D15	10	NC_006592.3	CYP2D	Cytochrome P450 2D15; CYPIID15; cytochrome-P450-2D15; cytochrome P450 DUT2	NM_001003333.1	Chimpanzee, cow, mouse, and rat
Guinea Pig (Cavia porcellus)						
Cyp2d16	Unknown	NT_176391.1		Cytochrome P450 2D16; CYPIID16	NM_001173036.1	
Cynomolgus Monkey (Crab-Eating Macaque, Macaca fascicularis)						
CYP2D17	10	NC_022281.1	CYPIID17	Cytochrome P450 2D17; cytochrome P450 2D17	NM_001283966.1	
CYP2D44	10	NC_022281.1		Cytochrome P450 2D17-like; cytochrome P450 2D44	NM_001284773.1	
Marmoset (Callithrix jacchus)						
CYP2D19	Unknown	NW_003187933.1	CYPIID19	Cytochrome P450 2D19; cytochrome P450 CM2D-1	NM_001204438.1	
Rabbit (Oryctolagus cuniculus)						
Cyp2d23	Unknown	NW_003159360.1		Cytochrome P450 2D/I	NM_001168395.1	
Cyp2d24	Unknown	NW_003159360.1		Cytochrome P450 2D/II	NM_001168397.1	
Pig (Sus scrofa)						
CYP2D25	5; 5	NC_010447.4	CYPIID25	Vitamin D_3 25-hydroxylase; cytochrome P-450; cytochrome P450 2D25	NM_214394.1	
Horse (Equus caballus)						
CYP2D50	28	NC_009171.2		Cytochrome P450 2D50	NM_001111306.1	

```
sp_Q7Z449_CP2U1_HUMAN_Cytoc        MSSPGPSQPPAEDPPWPARLLRAPLGLLRLDPSGGALLLCGLVALLGWSW
sp_Q0IIF9_CP2U1_BOVIN_Cytoc        MASPGLPQPPTEDAAWPLRLLHAPPGLLRLDPTGGALLLLVLAALLGWSW
sp_Q9CX98_CP2U1_MOUSE_Cytoc        MSSLGDQRPAAGE---------QPGARLHVRATGGALLLCLLAVLLGWVW
sp_O46658_CP2DP_PIG_Vitamin        MGLLTGDLLGI----------------------LALAMVIFLLLVDLMH
sp_Q01361_CP2DE_BOVIN_Cytoc        MGLLSGDTLGP----------------------LAVALLIFLLLLDLMH
sp_Q29473_CP2DF_CANFA_Cytoc        MGLLTGDTLGP----------------------LAVAVAIFLLLVDLMH
sp_Q2XNC9_CP2D6_PANPA_Cytoc        MGL---EALVP----------------------LAVIVTIFLLLVDLMH
sp_Q2XNC8_CP2D6_PANTR_Cytoc        MGL---EALVP----------------------LAVIVTIFLLLVDLMH
sp_P10635_CP2D6_HUMAN_Cytoc        MGL---EALVP----------------------LAVIVAIFLLLVDLMH
sp_A0A087X1C5_CP2D7_HUMAN_P        MGL---EALVP----------------------LAMIVAIFLLLVDLMH
sp_Q29488_CP2DH_MACFA_Cytoc        MEL---DALVP----------------------LAVTVAIFLLLVDLMH
sp_O18992_CP2DJ_CALJA_Cytoc        MGL---DALVP----------------------LAVTVAIFVLLVDLMH
sp_Q9QYG6_CP2DR_MESAU_Cytoc        MALLIGDGLWS----------------------GVIFTALFLLLVDLMH
sp_Q9QYG5_CP2DK_MESAU_Cytoc        MVLLIGDGLWS----------------------GVIFTALFLLLVDLMH
sp_Q8CIM7_CP2DQ_MOUSE_Cytoch       MGLLVGDDLWA----------------------VVIFTAIFLLLVDLVH
sp_P10634_CP2DQ_RAT_Cytochro       MGLLIGDDLWA----------------------VVIFTAIFLLLVDLVH
sp_P24457_CP2DB_MOUSE_Cytoch       MELLTGAGLWS----------------------VAIFTVIFILLVDLMH
sp_P24456_CP2DA_MOUSE_Cytoch       MELLTGAGLWS----------------------VAIFTVIFILLVDLMH
sp_P11714_CP2D9_MOUSE_Cytoch       MELLTGTDLWP----------------------VAIFTVIFILLVDLTH
sp_P12939_CP2DA_RAT_Cytochro       MELLNGTGLWP----------------------MAIFTVIFILLVDLMH
sp_P10633_CP2D1_RAT_Cytochro       MELLNGTGLWS----------------------MAIFTVIFILLVDLMH
sp_P12938_CP2D3_RAT_Cytochro       MELLAGTGLWP----------------------MAIFTVIFILLVDLMH
sp_Q9QUJ1_CP2DS_MESAU_Cytoc        MELLTGHGLWP----------------------VAMFTVILILLVDLLH
sp_Q64680_CP2DI_RAT_Cytochro       MRMPTGSELWP----------------------IAIFTIIFLLLVDLMH
sp_P13108_CP2D4_RAT_Cytochro       MRMPTGSELWP----------------------IAIFTIIFLLLVDLMH
sp_Q64403_CP2DG_CAVPO_Cytoc        MGLLTGDALFS----------------------VAVAVAIFLLLVDLMH
sp_P19225_CP270_RAT_Cytochr        MALFIFLGIWLS--------------------------CLVFLFLWNQHH
sp_P50170_RDH2_RAT_Retinol_        ------------------------------------MWLYLLALVGLWN
sp_Q811S7_UBIP1_MOUSE_Upstr        MAWVLSMDEVIESGLVHDFDSSLSGIGQELGAGAYSMSDVLALPIFKQED
                                                                                    :
```

```
sp_Q7Z449_CP2U1_HUMAN_Cytoc        LRRRRARGIPPGPTPWPLVGNFGHVLLPPFLRRRSWLSSRTRAAGIDPSV
sp_Q0IIF9_CP2U1_BOVIN_Cytoc        LWRLPERGIPPGPAPWPVVGNFGFVLLPRFLRRKSWPYRRARNGGMNASG
sp_Q9CX98_CP2U1_MOUSE_Cytoc        LRRQRACGIPPGPKPRPLVGNFGHLLVPRFLRPQFWLGS-----GSQTDT
sp_O46658_CP2DP_PIG_Vitamin        RRSRWAPRYPPGPMPLPGLGNL-----------------------LQVNF
sp_Q01361_CP2DE_BOVIN_Cytoc        RRSRWAPRYPPGPTPLPVLGNL-----------------------LQVDF
sp_Q29473_CP2DF_CANFA_Cytoc        RRRRWATRYPPGPTPVPMVGNL-----------------------LQMDF
sp_Q2XNC9_CP2D6_PANPA_Cytoc        RRQRWAARYPPGPLPLPGLGNL-----------------------LHVDF
sp_Q2XNC8_CP2D6_PANTR_Cytoc        RRQRWAARYPPGPLPLPGLGNL-----------------------LHVDF
sp_P10635_CP2D6_HUMAN_Cytoc        RRQRWAARYPPGPLPLPGLGNL-----------------------LHVDF
sp_A0A087X1C5_CP2D7_HUMAN_P        RHQRWAARYPPGPLPLPGLGNL-----------------------LHVDF
sp_Q29488_CP2DH_MACFA_Cytoc        RRQRWAARYPPGPLPLPGLGNL-----------------------LHVDF
sp_O18992_CP2DJ_CALJA_Cytoc        RRQRWAARYPPGPMPLPGLGNL-----------------------LHVDF
sp_Q9QYG6_CP2DR_MESAU_Cytoc        RRKFWRARYPPGPMPLPGLGNL-----------------------LQVDF
sp_Q9QYG5_CP2DK_MESAU_Cytoc        RRKFWRARYPPGPMPLPGLGNL-----------------------LQVDF
sp_Q8CIM7_CP2DQ_MOUSE_Cytoch       RRQRWTACYPPGPVPFPGLGNL-----------------------LQVDF
sp_P10634_CP2DQ_RAT_Cytochro       RHKFWTAHYPPGPVPLPGLGNL-----------------------LQVDF
sp_P24457_CP2DB_MOUSE_Cytoch       RHQHWTSRCPPGPVPWPVLGNL-----------------------LQVDL
sp_P24456_CP2DA_MOUSE_Cytoch       RHQRWTSRYPPGPVPWPVLGNL-----------------------LQVDL
sp_P11714_CP2D9_MOUSE_Cytoch       QRQRWTSRYPPGPVPWPVLGNL-----------------------LQVDL
sp_P12939_CP2DA_RAT_Cytochro       RHQRWTSRYPPGPVPWPVLGNL-----------------------LQVDP
sp_P10633_CP2D1_RAT_Cytochro       RRHRWTSRYPPGPVPWPVLGNL-----------------------LQVDL
sp_P12938_CP2D3_RAT_Cytochro       RRQRWTSRYPPGPVPWPVLGNL-----------------------LQVDL
sp_Q9QUJ1_CP2DS_MESAU_Cytoc        RRQRWASRYPPGPVPLPLLGNL-----------------------LQVDL
sp_Q64680_CP2DI_RAT_Cytochro       RRQRWTSRYPPGPVPWPVLGNL-----------------------LQIDF
sp_P13108_CP2D4_RAT_Cytochro       RRQRWYPPGPVPWPVLGNL-----------------------LQIDF
sp_Q64403_CP2DG_CAVPO_Cytoc        RRQRWAARYPPGPVPVPGLGNL-----------------------LQVDF
sp_P19225_CP270_RAT_Cytochr        VRR----KLPPGPTPLPIFGNI-----------------------LQVGV
sp_P50170_RDH2_RAT_Retinol_        LLRLFRERK------------------------------------VVSHL
sp_Q811S7_UBIP1_MOUSE_Upstr        SSLSLEDEAKHPPFQYVMCAATS----------------------PAVKL
```

(a)

FIGURE 2.1 (a) Complete sequence alignment of CYP2D members. Protein sequences of the CYP2D enzymes from different species are obtained from the Swiss-Prot database (http://www.uniprot.org/). Multiple sequence alignment of the CYP2Ds is performed using the program Clustal W v2.0 (http://www.clustal.org/clustal2/) with default parameters.   *(Continued)*

```
sp_Q7Z449_CP2U1_HUMAN_Cytoc       IGPQVLLAHLARVYGSIFSFFIGHYLVVVLSD---FHSVREALVQQAEVF
sp_Q0IIF9_CP2U1_BOVIN_Cytoc       QGVQLLLADLGRVYGNIFSFLIGHYLVVVLND---FHSVREALVQQAEVF
sp_Q9CX98_CP2U1_MOUSE_Cytoc       VGQHVYLARMARVYGNIFSFFIGHRLVVVLSD---FHSVREALVQQAEVF
sp_O46658_CP2DP_PIG_Vitamin       QDPRLSFIQLRRRFGDVFSLQQIWRPVVVLNG---LAAVREALVSHSHET
sp_Q01361_CP2DE_BOVIN_Cytoc       EDPRPSFNQLRRRFGNVFSLQQVWTPVVVLNG---LAAVREALVYRSQDT
sp_Q29473_CP2DF_CANFA_Cytoc       QEPICYFSQLQGRFGNVFSLELAWTPVVVLNG---LEAVREALVHRSEDT
sp_Q2XNC9_CP2D6_PANPA_Cytoc       QNTPYCFDQLRRRFGDVFSLQLAWTPVVVLNG---LAAVREALVTHGEDT
sp_Q2XNC8_CP2D6_PANTR_Cytoc       QNTPYCFDQLRRRFGDVFSLQLAWTPVVVLNG---LAAVREALVTHGEDT
sp_P10635_CP2D6_HUMAN_Cytoc       QNTPYCFDQLRRRFGDVFSLQLAWTPVVVLNG---LAAVREALVTHGEDT
sp_A0A087X1C5_CP2D7_HUMAN_P       QNTPYCFDQLRRRFGDVFSLQLAWTPVVVLNG---LAAVREAMVTRGEDT
sp_Q29488_CP2DH_MACFA_Cytoc       KNTPYCFDQLRRRFGNVFSLQLAWTPVVVLNG---LAAVREALVTCGEDT
sp_O18992_CP2DJ_CALJA_Cytoc       QNTPNSFNQLRRRFGDVFSLQLAWTPVVVLNG---LEAVREALVTRGEDT
sp_Q9QYG6_CP2DR_MESAU_Cytoc       EHMPYSLYKFRQRYGDVFSLQMAWKPVVVING---LKAVREVLVNCGEDT
sp_Q9QYG5_CP2DK_MESAU_Cytoc       ENMPYSLYKFQQRYGDVFSLQMAWKPVVVING---LKAVREVLVNCGEDT
sp_Q8CIM7_CP2DQ_MOUSE_Cytoch      ENIPYSFYKLQNRYGNVFSLQMAWKPVVVVNG---LKAVRELLVTYGEDT
sp_P10634_CP2DQ_RAT_Cytochro      ENMPYSLYKLRSRYGDVFSLQIAWKPVVVING---LKAVRELLVTYGEDT
sp_P24457_CP2DB_MOUSE_Cytoch      GNMPYSLYKLQNRYGDVFSLQMGWKPMVVING---LKAMKEVLLTCGEDT
sp_P24456_CP2DA_MOUSE_Cytoch      DNMPYSLYKLQNRYGDVFSLQMGWKPMVVING---LKAMKEVLLTCGEDT
sp_P11714_CP2D9_MOUSE_Cytoch      GNMPYSLYKLQNRYGDVFSLQMAWKPMVVING---LKAMKEMLLTCGEDT
sp_P12939_CP2DA_RAT_Cytochro      SNMPYSMYKLQHRYGDVFSLQMGWKPMVIVNR---LKAVQEVLVTHGEDT
sp_P10633_CP2D1_RAT_Cytochro      SNMPYSLYKLQHRYGDVFSLQKGWKPMVIVNR---LKAVQEVLVTHGEDT
sp_P12938_CP2D3_RAT_Cytochro      CNMPYSMYKLQNRYGDVFSLQMGWKPVVVING---LKAVQELLVTCGEDT
sp_Q9QUJ1_CP2DS_MESAU_Cytoc       KNMQYSVHKLQQHYGDVFSLQMAWKHMVMING---LKAVREVLVNYGEYT
sp_Q64680_CP2DI_RAT_Cytochro      QNMPAGFQKLRCRFGDLFSLQLAFESVVVLNG---LPALREALVKYSEDT
sp_P13108_CP2D4_RAT_Cytochro      QNMPAGFQKLRCRFGDLFSLQLAFESVVVLNG---LPALREALVKYSEDT
sp_Q64403_CP2DG_CAVPO_Cytoc       ENMAYSCDKLRHQFGDVFSLQFVWTPVVVVNG---LLAVREALVNNSTDT
sp_P19225_CP270_RAT_Cytochr       KNISKSMCMLAKEYGPVFTMYLGMKPTVVLYG---YEVLKEALIDRGEEF
sp_P50170_RDH2_RAT_Retinol_       QDKYVFITGCDSGFGNLLARQLDRRGMRVLAA---CLTEKGAEQLRS-KT
sp_Q811S7_UBIP1_MOUSE_Upstr       HDETLTYLNQGQSYEIRMLDNRKMGDMPELSGKLVKSIIRVVFHDRRLQY
                                          :  :              :  :

sp_Q7Z449_CP2U1_HUMAN_Cytoc       SDRPRVPLIS---IVTKEKGVVFAHYGPVWRQQRKFSHSTLRHFGLGKLS
sp_Q0IIF9_CP2U1_BOVIN_Cytoc       SDRPRVPLTS---IMTKGKGIVFAHYGPVWRQQRKFSHSTLRHFGLGKLS
sp_Q9CX98_CP2U1_MOUSE_Cytoc       SDRPRMPLIS---IMTKEKGIVFAHYGPIWKQQRRFSHSTLRHFGLGKLS
sp_O46658_CP2DP_PIG_Vitamin       SDRPPVFILEHLGYGPRSEGVILARYGKAWREQRRFSVSTLRNFGLGKKS
sp_Q01361_CP2DE_BOVIN_Cytoc       ADRPPPAVYEHLGYGPRAEGVILARYGDAWREQRRFSLTTLRNFGLGKKS
sp_Q29473_CP2DF_CANFA_Cytoc       ADRPPMPIYDHLGLGPESQGLFLARYGRAWREQRRFSLSTLRNFGLGRKS
sp_Q2XNC9_CP2D6_PANPA_Cytoc       ADRPPVPITQILGFGPRSQGVFLARYGPAWREQRRFSVSTLRNLGLGKKS
sp_Q2XNC8_CP2D6_PANTR_Cytoc       ADRPPVPITQILGFGPRSQGVFLARYGPAWREQRRFSVSTLRNLGLGKKS
sp_P10635_CP2D6_HUMAN_Cytoc       ADRPPVPITQILGFGPRSQGVFLARYGPAWREQRRFSVSTLRNLGLGKKS
sp_A0A087X1C5_CP2D7_HUMAN_P       ADRPPAPIYQVLGFGPRSQGVILSRYGPAWREQRRFSVSTLRNLGLGKKS
sp_Q29488_CP2DH_MACFA_Cytoc       ADRPPVPINQVLGFGPRSQGVFLARYGPAWREQRRFSVSTLRNLGLGKKS
sp_O18992_CP2DJ_CALJA_Cytoc       ADRPPVPITEMLGFGPHSQGLFLARYGPAWREQRRFSVSTLRNLGLGKKS
sp_Q9QYG6_CP2DR_MESAU_Cytoc       ADRPPVPIFNHVGFGHNSQVAFARYGPQWREQRRFCVSTMRDFGVGKKS
sp_Q9QYG5_CP2DK_MESAU_Cytoc       ADRPPVPIFNHLGYRPKSQGVVFARYGPQWREQRRFSVSTMRDFGVGKKS
sp_Q8CIM7_CP2DQ_MOUSE_Cytoch      SDRPLMPIYNHIGYGHKSKGVILAPYGPEWREQRRFSVSTLRDFGLGKKS
sp_P10634_CP2DQ_RAT_Cytochro      ADRPLLPIYNHLGYGNKSKGVVLAPYGPEWREQRRFSVSTLRDFGVGKKS
sp_P24457_CP2DB_MOUSE_Cytoch      ADRPQVPIFEYLGVKPGSQGVVLAPYGPEWQEQRRFSVSTLRNFGLGKKS
sp_P24456_CP2DA_MOUSE_Cytoch      ADRPQVPIFEYLGVKPGSQGVVLAPYGPEWREQRRFSVSTLRNFGLGKKS
sp_P11714_CP2D9_MOUSE_Cytoch      ADRPPVPIFEYLGVKPGSQGVVLAPYGPEWREQRRFSVSTLRNFGLGKKS
sp_P12939_CP2DA_RAT_Cytochro      ADRPPVPIFKCLGVKPRSQGVVFASYGPEWREQRRFSVSTLRTFGMGKKS
sp_P10633_CP2D1_RAT_Cytochro      ADRPPVPIFKCLGVKPRSQGVILASYGPEWREQRRFSVSTLRTFGMGKKS
sp_P12938_CP2D3_RAT_Cytochro      ADRPEMPIFQHIGYGHKAKGVVLAPYGPEWREQRRFSVSTLRNFGLGKKS
sp_Q9QUJ1_CP2DS_MESAU_Cytoc       ADRPKIPIYEHGSLGPKARGVILAPYGPEWREQRRFSVSTLRNLGLGKKS
sp_Q64680_CP2DI_RAT_Cytochro      ADRPPLHFNDQSGFGPRSQGVVLARYGPAWRQQRRFSVSTFRHFGLGKKS
sp_P13108_CP2D4_RAT_Cytochro      ADRPPLHFNDQSGFGPRSQGVVLARYGPAWRQQRRFSVSTFRHFGLGKKS
sp_Q64403_CP2DG_CAVPO_Cytoc       SDRPTLPTNALLGFGPKAQGVIGAYYGPAWREQRRFSVSSLRNFGLGKKS
sp_P19225_CP270_RAT_Cytochr       SDKMHSSMLSKVSQG---LGIVFS-NGEIWKQTRRFSLMVLRSMGMGKRT
sp_P50170_RDH2_RAT_Retinol_       SDR-----------------------------------------LETVI
sp_Q811S7_UBIP1_MOUSE_Upstr       TEHQQLEGWKWNRPGDRLLDLDIPMSVGIIDTRTNPSQLNAVEFLWDPAK
                                          : ::
```

(a)

FIGURE 2.1 (CONTINUED) (a) Complete sequence alignment of CYP2D members. Protein sequences of the CYP2D enzymes from different species are obtained from the Swiss-Prot database (http://www.uniprot.org/). Multiple sequence alignment of the CYP2Ds is performed using the program Clustal W v2.0 (http://www.clustal.org/clustal2/) with default parameters. *(Continued)*

```
sp_Q7Z449_CP2U1_HUMAN_Cytoc     LEPKIIEEFKYVKAEMQKHGEDPFCPFSIISNAVSNIICSLCFGQRFDYT
sp_Q0IIF9_CP2U1_BOVIN_Cytoc     LEPKIIEEFRYVKEEMQKHGDAPFNPFPIVNNAVSNIICSLCFGRRFDYT
sp_Q9CX98_CP2U1_MOUSE_Cytoc     LEPRIIEEFAYVKEAMQKHGEAPFSPFPIISNAVSNIICSLCFGQRFDYT
sp_O46658_CP2DP_PIG_Vitamin     LEEWVTQEASCLCAAFADQAGRPFSPNNLLNKAVSNVIASLTFARRFEYN
sp_Q01361_CP2DE_BOVIN_Cytoc     LEQWVEEASCLCAAFADQAGRPFSPMDLLNKAVSNVIASLTFGCRFEYN
sp_Q29473_CP2DF_CANFA_Cytoc     LEQWVTEEASCLCAAFAEQAGRPFGPGALLNKAVSNVISSLTYGRRFEYD
sp_Q2XNC9_CP2D6_PANPA_Cytoc     LEQWVTEEAACLCAAFANHSGRPFRPNGLLDKAVSNVIASLTCGRRFEYD
sp_Q2XNC8_CP2D6_PANTR_Cytoc     LEQWVTEEAACLCAAFANHSGRPFRPNGLLDKAVSNVIASLTCGRRFEYD
sp_P10635_CP2D6_HUMAN_Cytoc     LEQWVTEEAACLCAAFANHSGRPFRPNGLLDKAVSNVIASLTCGRRFEYD
sp_A0A087X1C5_CP2D7_HUMAN_P     LEQWVTEEAACLCAAFADQAGRPFRPNGLLDKAVSNVIASLTCGRRFEYD
sp_Q29488_CP2DH_MACFA_Cytoc     LEQWVTEEAACLCAAFTDQAGRPFRPNSLLDKAVSNVIASLTYGRRFEYD
sp_O18992_CP2DJ_CALJA_Cytoc     LEQWVTEEATYLCAAFADHAGRPFRPNGLLDKAVSNVIASLTCRRRFEYN
sp_Q9QYG6_CP2DR_MESAU_Cytoc     LEQWVTEEAGHLCDAFTQEAGHPFNPTTLLNKSVCNVISSLIYAHRFDYE
sp_Q9QYG5_CP2DK_MESAU_Cytoc     LEQWVTEEAGHLCDAFTQEAGHPFNPITLLNKSVCNVISSLIYAHRFDYE
sp_Q8CIM7_CP2DQ_MOUSE_Cytoch    LEQWVTEEAGHLCDAFTKEAEHPFNPSPLLSKAVSNVIASLIYARRFEYE
sp_P10634_CP2DQ_RAT_Cytochro    LEQWVTEEAGHLCDTFAKEAEHPFNPSILLSKAVSNVIASLVYARRFEYE
sp_P24457_CP2DB_MOUSE_Cytoch    LEDWVTKEARHLCDAFTAQAGQSINPNTMLNNAVCNVIASLIFARRFEYE
sp_P24456_CP2DA_MOUSE_Cytoch    LEDWVTKEARHLCDAFTAQAGQPINPNTMLNNAVCNVIASLIFARRFEYE
sp_P11714_CP2D9_MOUSE_Cytoch    LEDWVTKEANHLCDAFTAQAGQPINPNMLNKSTCNVIASLIFARRFEYE
sp_P12939_CP2DA_RAT_Cytochro    LEEWVTKEAGHLCDAFTAQNGRSINPKAMLNKALCNVIASLIFARRFEYE
sp_P10633_CP2D1_RAT_Cytochro    LEEWVTKEAGHLCDAFTAQAGQSINPKAMLNKALCNVIASLIFARRFEYE
sp_P12938_CP2D3_RAT_Cytochro    LEQWVTDEASHLCDALTAEAGRPLDPYTLLNKAVCNVIASLIYARRFDYG
sp_Q9QUJ1_CP2DS_MESAU_Cytoc     LEQWVTDEAGHLCDAFKDQAGRPFNPSTLLNKAVCNVITSLIFARRFEYE
sp_Q64680_CP2DI_RAT_Cytochro    LEQWVTEEARCLCAAFADHSGFPFSPNTLLDKAVCNVIASLLFACRFEYN
sp_P13108_CP2D4_RAT_Cytochro    LEQWVTEEARCLCAAFADHSGFPFSPNTLLDKAVCNVIASLLFACRFEYN
sp_Q64403_CP2DG_CAVPO_Cytoc     LEQWVTEEAACLCAAFTNHAGQPFCPKALLNKAVCNVISSLIYARRFDYD
sp_P19225_CP270_RAT_Cytochr     IENRIQEEVVYLLEALRKTNGSPCDPSFLLACVPCNVISSVIFQHRFDYS
sp_P50170_RDH2_RAT_Retinol_     LDVTKTESIVAATQWVKERVGNTGLWGLVNNAGISGHLGPNEWMNKQNIA
sp_Q811S7_UBIP1_MOUSE_Upstr     RTSAFIQVHCISTEFTPRKHGGEKGVPFRIQVDTFKQNENGEYTDHLHSA
                                                .                                 :  .


sp_Q7Z449_CP2U1_HUMAN_Cytoc     NSEFKKMLGFMSRGLEICLNSQVLLVNICPWLYYLPFGPFKELRQIEKDI
sp_Q0IIF9_CP2U1_BOVIN_Cytoc     NSEFKQMLNFMSRALEVCLNTQLLLVNICSWLYYLPFGPFKELRQIEKDL
sp_Q9CX98_CP2U1_MOUSE_Cytoc     NKEFKKVLDFMSRGLEICLHSQLFLINICPWFYYLPFGPFKELRQIERDI
sp_O46658_CP2DP_PIG_Vitamin     DPRMLKLLDLVLEGLKEEVGLMRQVLEAMPVLRHIP-GLCAKLFPRQKAF
sp_Q01361_CP2DE_BOVIN_Cytoc     DPRIIKLLDLTEDGLKEEFNLVRKVVEAVPVLLSIP-GLAARVFPAQKAF
sp_Q29473_CP2DF_CANFA_Cytoc     DPRLLQLLELTQQALKQDSGFLREALNSIPVLLHIP-GLASKVFSAQKAI
sp_Q2XNC9_CP2D6_PANPA_Cytoc     DPRFLRLLDLAQEGLKEESGFLREVLNAVPVLLHIP-ALAGKVLRFQKAF
sp_Q2XNC8_CP2D6_PANTR_Cytoc     DPRFLRLLDLAQEGLKEESGFLREVLNAIPVLLHIP-ALAGKVLRFQKAF
sp_P10635_CP2D6_HUMAN_Cytoc     DPRFLRLLDLAQEGLKEESGFLREVLNAVPVLLHIP-ALAGKVLRFQKAF
sp_A0A087X1C5_CP2D7_HUMAN_P     DPRFLRLLDLAQEGLKEESGFLREVLNAVPVLPHIP-ALAGKVLRFQKAF
sp_Q29488_CP2DH_MACFA_Cytoc     DPRFLRLFDLTHEALKEESGFLREVLNAIPLLLRIP-GLAGKVLRSQKAF
sp_O18992_CP2DJ_CALJA_Cytoc     DPCLLRLLDLTMEGLKEESGLLREVLNAIPVLLRIP-GLAGKVLRSQKAF
sp_Q9QYG6_CP2DR_MESAU_Cytoc     DPFFNSLLKMLQESFGEDTGFIAEVLNAVPVLLRIP-GLPGKAFPKLTAF
sp_Q9QYG5_CP2DK_MESAU_Cytoc     DPFFNKLLKTLQESFGEDSGFIAEVLNAVPVLLRIP-GLPGKAFPKLTAF
sp_Q8CIM7_CP2DQ_MOUSE_Cytoch    DPFFNRMLKTLKESLGEDTGFVGEVLNAIPMLLHIP-GLPDKAFPKLNSF
sp_P10634_CP2DQ_RAT_Cytochro    DPFFNRMLKTLKESFGEDTGFMAEVLNAIPILLQIP-GLPGKVFPKLNSF
sp_P24457_CP2DB_MOUSE_Cytoch    DPYLIRMLKMLKECFTEISGFIPGVLNEFPIFLRIP-GLADMVFQGGQKSF
sp_P24456_CP2DA_MOUSE_Cytoch    DPYLIRMQKVLEDSLTEISGLIPEVLNMFPILLRIP-GLPGKVFQGGQKSL
sp_P11714_CP2D9_MOUSE_Cytoch    DPFLIRMLKVLEQSLTEVSGLIPEVLNAFPILLRIP-RLADKALQGGQKSF
sp_P12939_CP2DA_RAT_Cytochro    DPYLIRMLTLVEESLIEVSGFIPEVLNTFPALLRIP-GLADKVFQGGQKTF
sp_P10633_CP2D1_RAT_Cytochro    DPYLIRMVKLVEESLTEVSGFIPEVLNTFPALLRIP-GLADKVFQGGQKTF
sp_P12938_CP2D3_RAT_Cytochro    DPDFIKVLKILKESMGEQTGLFPEVLNMFPVLLRIP-GLADKVFPGGQKTF
sp_Q9QUJ1_CP2DS_MESAU_Cytoc     DANLIRMLRLLEEALTNISGFIPEILNTFPVLLHIP-GLFDKVFSGQKTF
sp_Q64680_CP2DI_RAT_Cytochro    DPRFIRLLDLLKDTLEEESGFLPMLLNVFPMLLHIP-GLLGKVFSGKKAF
sp_P13108_CP2D4_RAT_Cytochro    DPRFIRLLDLLKDTLEEESGFLPMLLNVFPMLLHIP-GLLGKVFSGKKAF
sp_Q64403_CP2DG_CAVPO_Cytoc     DPMVLRLLEFLEETLRENSSLKIQVLNSIPLLLRIP-CVAAKVLSAQRSF
sp_P19225_CP270_RAT_Cytochr     DEKFQKFIENFHTKIEILASPWAQLCSAYPVLYYLP-GIHNKFLKDVTEQ
sp_P50170_RDH2_RAT_Retinol_     SVLDVNLLGMIEVTLSTVP--------------------LVRKARGR
sp_Q811S7_UBIP1_MOUSE_Upstr     SCQIKVFKPKGADRKQKNDREKMEKRTAHEKEKYQP--SYDTTILTEMRL
                                       .                    .
```

(a)

FIGURE 2.1 (CONTINUED)   (a) Complete sequence alignment of CYP2D members. Protein sequences of the CYP2D enzymes from different species are obtained from the Swiss-Prot database (http://www.uniprot.org/). Multiple sequence alignment of the CYP2Ds is performed using the program Clustal W v2.0 (http://www.clustal.org/clustal2/) with default parameters.   (*Continued*)

```
sp_Q7Z449_CP2U1_HUMAN_Cytoc     TSFLKKIIKDHQESLDREN-PQDFIDMYLLHMEEERKNNSNSSFDEEYLF
sp_Q0IIF9_CP2U1_BOVIN_Cytoc     TLFLKKIIKDHRESLDVEN-PQDFIDMYLLHVEEEKKNNSNSGFDEDYLF
sp_Q9CX98_CP2U1_MOUSE_Cytoc     SCFLKNIIREHQESLDASN-PQDFIDMYLLHMEEEQGASRRSSFDEDYLF
sp_O46658_CP2DP_PIG_Vitamin     LVMIDELITEHKMTRDLAQPPRDLTDAFLDEMKEAKG-NPESSFNDENLR
sp_Q01361_CP2DE_BOVIN_Cytoc     MALIDELIAEQKMTRDPTQPPRHLTDAFLDEVKEAKG-NPESSFNDENLR
sp_Q29473_CP2DF_CANFA_Cytoc     ITLTNEMIQEHRKTRDPTQPPRHLIDAFVDEIEKAKG-NPKTSFNEENLC
sp_Q2XNC9_CP2D6_PANPA_Cytoc     LTQLDELLTEHRMTWDPAQPPRDLTEAFLAEMEKAKG-NPESSFNDENLR
sp_Q2XNC8_CP2D6_PANTR_Cytoc     LTQLDELLTEHRMTWDPAQPPRDLTEAFLAEMEKAKG-NPESSFNDENLR
sp_P10635_CP2D6_HUMAN_Cytoc     LTQLDELLTEHRMTWDPAQPPRDLTEAFLAEMEKAKG-NPESSFNDENLR
sp_A0A087X1C5_CP2D7_HUMAN_P     LTQLDELLTEHRMTWDPAQPPRDLTEAFLAKKEKAKG-SPESSFNDENLR
sp_Q29488_CP2DH_MACFA_Cytoc     LTQLDELLTEHRMTWDPAQPPRDLTEAFLAEMEKAKG-NPESSFNEENLR
sp_O18992_CP2DJ_CALJA_Cytoc     LAQLDELLTEHRMTWDPAQPPRDLTEAFLAEMEKTKG-NPESSFNDENLH
sp_Q9QYG6_CP2DR_MESAU_Cytoc     MDSLYKMLIEHKTTWDPAQPPRGLTDAFLAEVEKAKG-RPESSFNDENLR
sp_Q9QYG5_CP2DK_MESAU_Cytoc     MDSLYKMLIEHKTTWDPAQPPRGLTDAFLAEVEKAKG-RPESSFNDENLH
sp_Q8CIM7_CP2DQ_MOUSE_Cytoch    IALVNKMLIEHDLTWDPAQPPRDLTDAFLAEVEKAKG-NPESSFNDKNLR
sp_P10634_CP2DQ_RAT_Cytochro    IALVDKMLIEHKKSWDPAQPPRDMTDAFLAEMQKAKG-NPESSFNDENLR
sp_P24457_CP2DB_MOUSE_Cytoch    MAILDNLLTENRTTWDPDQPPRNLTDAFLAEIEKAKG-NPESSFNDENLR
sp_P24456_CP2DA_MOUSE_Cytoch    LAIVENLLTENRNTWDPDQPPRNLTDAFLAEIEKVKG-NAESSFNDENLR
sp_P11714_CP2D9_MOUSE_Cytoch    IAILDNLLTENRTTWDPVQAPRNLTDAFLAEIEKAKG-NPESSFNDENLL
sp_P12939_CP2DA_RAT_Cytochro    MAFLDNLLAENRTTWDPAQPPRNLTDAFLAEVEKAKG-NPESSFNDENLR
sp_P10633_CP2D1_RAT_Cytochro    MALLDNLLAENRTTWDPAQPPRNLTDAFLAEVEKAKG-NPESSFNDENLR
sp_P12938_CP2D3_RAT_Cytochro    LTMVDNLVTEHKKTWDPDQPPRDLTDAFLAEIEKAKG-NPESSFNDANLR
sp_Q9QUJ1_CP2DS_MESAU_Cytoc     AAIVDNLLTENRRTWDPEQPPRGLTDAFLAEMEKAKG-NPESSFNDQNLR
sp_Q64680_CP2DI_RAT_Cytochro    VAMLDELLTEHKVTWDPAQPPRDLTDAFLAEVEKAKG-NPESSFNDENLR
sp_P13108_CP2D4_RAT_Cytochro    VAMLDELLTEHKVTWDPAQPPRDLTDAFLAEVEKAKG-NPESSFNDENLR
sp_Q64403_CP2DG_CAVPO_Cytoc     IALNDKLLAEHNTGWAPDQPPRDLTDAFLTEMHKAQG-NSESSFNDENLR
sp_P19225_CP270_RAT_Cytochr     KKFILMEINRHRASLNLSN-PQDFIDYFLIKMEKEKH-NEKSEFTMDNLI
sp_P50170_RDH2_RAT_Retinol_     VVNVASIAGRLSFCGGGYCISKYGVEAFSDSLRRELS------YFGVKVA
sp_Q811S7_UBIP1_MOUSE_Upstr     EPIIEDAVEHEQKKSSKRTLPADYGDSLAKRGSCSPWPDTPTAYVN-NSP
                                                 .           :                :
```

```
sp_Q7Z449_CP2U1_HUMAN_Cytoc     YIIGDLFIAGTDTTTNSLLWCLLYMSLNPDVQE-----------------
sp_Q0IIF9_CP2U1_BOVIN_Cytoc     YIIGDLFIAGTDTTTNSLLWCLLYMSLHPNIQE-----------------
sp_Q9CX98_CP2U1_MOUSE_Cytoc     YIIGDLFIAGTDTTTNSLLWCLLYMSLNPDVQK-----------------
sp_O46658_CP2DP_PIG_Vitamin     LVVAHLFSAGMITTSTTLAWALLLMILHPDVQR-----------------
sp_Q01361_CP2DE_BOVIN_Cytoc     LVVADLFSAGMVTTSTTLAWALLLMILHPDVQR-----------------
sp_Q29473_CP2DF_CANFA_Cytoc     MVTSDLFIAGMVTSITLTWALLLMILHPDVQR-----------------
sp_Q2XNC9_CP2D6_PANPA_Cytoc     IVVADLFSAGMVTTSTTLAWGLLLMILHPDVQR-----------------
sp_Q2XNC8_CP2D6_PANTR_Cytoc     IVVADLFSAGIVTTSTTLAWGLLLMILHPDVQR-----------------
sp_P10635_CP2D6_HUMAN_Cytoc     IVVADLFSAGMVTTSTTLAWGLLLMILHPDVQR-----------------
sp_A0A087X1C5_CP2D7_HUMAN_P     IVVGNLFLAGMVTTSTTLAWGLLLMILHLDVQRGRRVSPGCPIVGTHVCP
sp_Q29488_CP2DH_MACFA_Cytoc     MVVADLFSAGMVTTSTTLAWGLLLMILHPDVQR-----------------
sp_O18992_CP2DJ_CALJA_Cytoc     LVVADLFSAGMVTTSITLAWGLLLMILHPDVQR-----------------
sp_Q9QYG6_CP2DR_MESAU_Cytoc     MVVADMFIAGMVTTSTTLSWALLLMILHPDVQS-----------------
sp_Q9QYG5_CP2DK_MESAU_Cytoc     VVVADLFIAGMVTTSTTLSWALLLMILHPDVQS-----------------
sp_Q8CIM7_CP2DQ_MOUSE_Cytoch    IVVIDLFMAGMVTTSTTLSWALLLMILHPDVQR-----------------
sp_P10634_CP2DQ_RAT_Cytochro    LVVIDLFMAGMVTTSTTLSWALLLMILHPDVQR-----------------
sp_P24457_CP2DB_MOUSE_Cytoch    MVVGDLFTAGMVTTSTTLSWALLLMILHPDVQR-----------------
sp_P24456_CP2DA_MOUSE_Cytoch    MVVLDLFTAGMVTTSTTLSWALLLMILHPDVQR-----------------
sp_P11714_CP2D9_MOUSE_Cytoch    MVVRDLFGAGMLTTSTTLSWALMLMILHPDVQR-----------------
sp_P12939_CP2DA_RAT_Cytochro    MVVVDLFTAGMVTTATTLTWALLLMILYPDVQR-----------------
sp_P10633_CP2D1_RAT_Cytochro    MVVVDLFTAGMVTTATTLTWALLLMILYPDVQR-----------------
sp_P12938_CP2D3_RAT_Cytochro    LVVNDLFGAGMVTTSITLTWALLLMILHPDVQC-----------------
sp_Q9QUJ1_CP2DS_MESAU_Cytoc     MVVNDLFIAGTVSTSTTLSWALLFMIQYPDVQR-----------------
sp_Q64680_CP2DI_RAT_Cytochro    VVVADLFMAGMVTTSTTLTWALLFMILRPDVQC-----------------
sp_P13108_CP2D4_RAT_Cytochro    VVVADLFMAGMVTTSTTLTWALLFMILHPDVQC-----------------
sp_Q64403_CP2DG_CAVPO_Cytoc     LLVSDLFGAGMVTTSVTLSWALLLMILHPDVQR-----------------
sp_P19225_CP270_RAT_Cytochr     VTIGDLFGAGTETTSSTIKYGLLLLLKYPEVTA-----------------
sp_P50170_RDH2_RAT_Retinol_     IVEPGFFRT---DVTNGVTLSSNFQMLWDQTSS-----------------
sp_Q811S7_UBIP1_MOUSE_Upstr     SPAPTFTSSQPSTCSVPDSNSSSPNHQGDGAAQ-----------------
                                       :   :       :
```

(a)

FIGURE 2.1 (CONTINUED) (a) Complete sequence alignment of CYP2D members. Protein sequences of the CYP2D enzymes from different species are obtained from the Swiss-Prot database (http://www.uniprot.org/). Multiple sequence alignment of the CYP2Ds is performed using the program Clustal W v2.0 (http://www.clustal.org/clustal2/) with default parameters.   *(Continued)*

```
sp_Q7Z449_CP2U1_HUMAN_Cytoc    -KVHEEIERVIGANRAPSLTDKAQMPYTEATIMEVQ--RLTVVVPLAIPH
sp_Q0IIF9_CP2U1_BOVIN_Cytoc    -KIHEEIARVIGADRAPSLTDKAQMPYTEATIMEVQ--RLSTVVPLSIPH
sp_Q9CX98_CP2U1_MOUSE_Cytoc    -KVHEEIERVIGCDRAPSLTDKAQMPYTEATIMEVQ--RLSMVVPLAIPH
sp_O46658_CP2DP_PIG_Vitamin    -RVQQEIDEVIGHVRQPEIKDQALMPFTLAVLHEVQ--RFGDIVPLGVAH
sp_Q01361_CP2DE_BOVIN_Cytoc    -RVQQEIDEVIGQVRRPEMGDQALMPFTVAVVHEVQ--RFADIVPLGLPH
sp_Q29473_CP2DF_CANFA_Cytoc    -RVQQEIDEVIGREQLPEMGDQTRMPFTVAVIHEVQ--RFGDIVPLGVPH
sp_Q2XNC9_CP2D6_PANPA_Cytoc    -RVQQEIDDVIGQVRRPEMGDQARMPYTTAVIHEVQ--RFGDIVPLGVTH
sp_Q2XNC8_CP2D6_PANTR_Cytoc    -RVQQEIDDVIGQVRRPEMGDQARMPYTTAVIHEVQ--RFGDIVPLGVTH
sp_P10635_CP2D6_HUMAN_Cytoc    -RVQQEIDDVIGQVRRPEMGDQAHMPYTTAVIHEVQ--RFGDIVPLGVTH
sp_A0A087X1C5_CP2D7_HUMAN_P    VRVQQEIDDVIGQVRRPEMGDQAHMPCTTAVIHEVQ--HFGDIVPLGVTH
sp_Q29488_CP2DH_MACFA_Cytoc    -RVQQEIDDVIGQVRRPEMGDQARMPYTTAVIHEVQ--RFGDIVPLGVTH
sp_O18992_CP2DJ_CALJA_Cytoc    -RVQQEIDDVIGRVRRPEMGDQTYMPYTTAVIHEVQ--RFADIVPLGVTH
sp_Q9QYG6_CP2DR_MESAU_Cytoc    -RVQQEIDDVIGQVRRPEMADQARMPYTNAVIHEVQ--RFGDIAPVNIPH
sp_Q9QYG5_CP2DK_MESAU_Cytoc    -RVQQEIDDVIGQVRRPEMADQARMPYTNAVIHEVQ--RFGDIAPVNVPH
sp_Q8CIM7_CP2DQ_MOUSE_Cytoch   -RVHQEIDEVIGHVRHPEMADQARMPYTNAVIHEVQ--RFADIVPTNLPH
sp_P10634_CP2DQ_RAT_Cytochro   -RVHEEIDEVIGQVRRPEMADQARMPFTNAVIHEVQ--RFADIVPTNIPH
sp_P24457_CP2DB_MOUSE_Cytoch   -RVQQEIDAVIGQVQHPEMADQARMPYTNAVIHEVQ--RFGDIAPLPLPR
sp_P24456_CP2DA_MOUSE_Cytoch   -RVQQEIDAVIGQVRHPEMADQARMPYTNAVIHEVQ--RFGDIAPLNLPR
sp_P11714_CP2D9_MOUSE_Cytoch   -RVQQEIDEVIGQVRHPEMADQAHMPYTNAVIHEVQ--RFGDIVPVNLPR
sp_P12939_CP2DA_RAT_Cytochro   -RVQQEIDEVIGQVRCPEMTDQAHMPYTNAVIHEVQ--RFGDIAPLNLPR
sp_P10633_CP2D1_RAT_Cytochro   -RVQQEIDEVIGQVRCPEMTDQAHMPYTNAVIHEVQ--RFGDIAPLNLPR
sp_P12938_CP2D3_RAT_Cytochro   -RVQQEIDEVIGQVRHPEMADQAHMPFTNAVIHEVQ--RFADIVPMNLPH
sp_Q9QUJ1_CP2DS_MESAU_Cytoc    -RVQQEIDDILGPGRSPEMADQARMPYTNAVIHEVQ--RFADIAPLNLPC
sp_Q64680_CP2DI_RAT_Cytochro   -RVQQEIDEVIGQVRRPEMADQARMPFTNAVIHEVQ--RFADILPLGVPH
sp_P13108_CP2D4_RAT_Cytochro   -RVQQEIDEVIGQVRRPEMADQARMPFTNAVIHEVQ--RFADILPLGVPH
sp_Q64403_CP2DG_CAVPO_Cytoc    -HVQQEIDEVIGQVRCPEMADQAHMPFTNAVIHEVQ--RFADIVPMGVPH
sp_P19225_CP270_RAT_Cytochr    -KIQEEITRVIGRHRRPCMQDRNHMPYTDAVLHEIQ--RYIDFVPIPLPR
sp_P50170_RDH2_RAT_Retinol_    -----EVREVYGEN------------YLASYLKMLN--GLDQRCNKDLSL
sp_Q811S7_UBIP1_MOUSE_Upstr    -ASGEQIQPSATTQETQQWLLKNRFSSYTRLFSNFSGADLLKLTKEDLVQ
                                             ::                    .  ..            :


sp_Q7Z449_CP2U1_HUMAN_Cytoc    MTSENTVLQGYTIPKGTLILPNLWSVHRDPAIWEKPEDFYPNRFLDDQGQ
sp_Q0IIF9_CP2U1_BOVIN_Cytoc    MTSEKTVLQGFTIPKGTIILPNLWSVHRDPAIWEKPNDFYPDRFLDDQGQ
sp_Q9CX98_CP2U1_MOUSE_Cytoc    MTSEKTVLQGFTIPKGTVVLINLWSVHRDPAIWEKPDDFCPHRFLDDQGQ
sp_O46658_CP2DP_PIG_Vitamin    MTSCDIEVQGFLIPKGTTLITNLTSVLKDETVWKKPFRFYPEHFLDAQGR
sp_Q01361_CP2DE_BOVIN_Cytoc    MTSRDIEVQGFHIPKGTTLITNLSSVLKDETVWEKPFRFHPEHFLDAQGR
sp_Q29473_CP2DF_CANFA_Cytoc    MTSRDTEVQGFLIPKGTTLITNLSSVLKDEKVWKKPFRFYPEHFLDAQGH
sp_Q2XNC9_CP2D6_PANPA_Cytoc    MTSRDIEVQGFRIPKGTTLFTNLSSVLKDEAVWEKPFRFHPEHFLDAQGH
sp_Q2XNC8_CP2D6_PANTR_Cytoc    MTSRDIEVQGFRIPKGTTLFTNLSSVLKDKAVWEKPFRFHPEHFLDAQGH
sp_P10635_CP2D6_HUMAN_Cytoc    MTSRDIEVQGFRIPKGTTLITNLSSVLKDEAVWEKPFRFHPEHFLDAQGH
sp_A0A087X1C5_CP2D7_HUMAN_P    MTSRDIEVQGFRIPKGTTLITNLSSVLKDEAVWKKPFRFHPEHFLDAQGH
sp_Q29488_CP2DH_MACFA_Cytoc    MTSRDIELQGFLIPKGTTLFTNLSSVLKDEAVWEKPFRFHPEHFLDAQGH
sp_O18992_CP2DJ_CALJA_Cytoc    MTSRDIEVQGFLIPKGTTLFTNLSSVLKDEANWEKPFRFHPEHFLDAQGR
sp_Q9QYG6_CP2DR_MESAU_Cytoc    MTSHDVEVQGFLIPKGTTLIPNLSSVLKDETVWEKPLHFHPEHFLDAQGR
sp_Q9QYG5_CP2DK_MESAU_Cytoc    MTSRDVEVQGFLIPKGTTLIPNLSSVLKDETVWEKPLHFHPEHFLDAQGR
sp_Q8CIM7_CP2DQ_MOUSE_Cytoch   MTSRDIKFQDFFIPKGTTLIPNLSSVLKDETVWEKPLRFYPEHFLDAQGH
sp_P10634_CP2DQ_RAT_Cytochro   MTSRDIKFQGFLIPKGTTLIPNLSSVLKDETVWEKPLRFHPEHFLDAQGH
sp_P24457_CP2DB_MOUSE_Cytoch   ITSRDIEVQDFLVTKGSTLIPNMSSVLKDETVWEKPLRFHPEHFLDAQGH
sp_P24456_CP2DA_MOUSE_Cytoch   ITSRDIEVQDFLIPKGSILIPNMSSVLKDETVWEKPLRFHPEHFLDAQGH
sp_P11714_CP2D9_MOUSE_Cytoch   ITSHDIEVQDFLIPKGTILLPNMSSMLKDESVWEKPLRFHPEHFLDAQGH
sp_P12939_CP2DA_RAT_Cytochro   ITSCDIEVQDFVIPKGTTLIINLSSVLKDETVWEKPLRFHPEHFLDAQGN
sp_P10633_CP2D1_RAT_Cytochro   FTSCDIEVQDFVIPKGTTLIINLSSVLKDETVWEKPHRFHPEHFLDAQGN
sp_P12938_CP2D3_RAT_Cytochro   KTSRDIEVQGFLIPKGTTLIPNLSSVLKDETVWEKPLRFHPEHFLDAQGN
sp_Q9QUJ1_CP2DS_MESAU_Cytoc    ITSRDIEVQGFLIPKGTTLITNLSSVLKDETVWEKPLHFHPEHFLDAQGC
sp_Q64680_CP2DI_RAT_Cytochro   KTSRDIEVQGFLIPKGTTLIINLSSVLKDETVWEKPLRFHPEHFLDAQGN
sp_P13108_CP2D4_RAT_Cytochro   KTSRDIEVQGFLIPKGTTLITNLSSVLKDETVWEKPLRFHPEHFLDAQGN
sp_Q64403_CP2DG_CAVPO_Cytoc    MTSRDVEVQGFLIPKGTMLFTNLSSVLKDETVWEKPLHFHPGHFLDAEGR
sp_P19225_CP270_RAT_Cytochr    KTTQDVEFRGYHIPKGTSVMACLTSALHDDKEFPNPEKFDPGHFLDEKGN
sp_P50170_RDH2_RAT_Retinol_    VTDCMEHALTSCHPR------TRYSAGWDAKFFYLPMSYLPTFLVDALFY
sp_Q811S7_UBIP1_MOUSE_Upstr    ICGAADGIRLYNSLKSRSVRPRLTIYVCQEQPSSTALQGQPQAAGSGGES
                                        :            :        .     *
```

(a)

FIGURE 2.1 (CONTINUED)   (a) Complete sequence alignment of CYP2D members. Protein sequences of the CYP2D enzymes from different species are obtained from the Swiss-Prot database (http://www.uniprot.org/). Multiple sequence alignment of the CYP2Ds is performed using the program Clustal W v2.0 (http://www.clustal.org/clustal2/) with default parameters.   *(Continued)*

```
sp_Q7Z449_CP2U1_HUMAN_Cytoc      LIKKETFIPFGIGKRVCMGEQLAKMELFLMFVSLMQSFAFA-LPEDSKKP
sp_Q0IIF9_CP2U1_BOVIN_Cytoc      LIKKETFIPFGIGKRVCMGEQLAKMELFLMFVSLMQSFTFV-LPKDSK-P
sp_Q9CX98_CP2U1_MOUSE_Cytoc      LLKRETFIPFGIGKRVCMGEQLAKMELFLMFVSLMQTFTFA-LPEGSEKP
sp_O46658_CP2DP_PIG_Vitamin      FTKQEAFMPFSAGRRSCLGEPLARMELFLFTTLLQAFSFS-VPTGQPCP
sp_Q01361_CP2DE_BOVIN_Cytoc      FVKQEAFIPFSAGRRACLGEPLARMELFLFTTSLLQHFSFS-VPAGQPRP
sp_Q29473_CP2DF_CANFA_Cytoc      FVKHEAFMPFSAGRRVCLGEPLARMELFLFFTCLLQRFSFS-VPAGQPRP
sp_Q2XNC9_CP2D6_PANPA_Cytoc      FVKPEAFLPFSAGRRACLGEPLARMELFLFFTSLLQHFSFS-VPTGQPRP
sp_Q2XNC8_CP2D6_PANTR_Cytoc      FVKPEAFLPFSAGRRACLGEPLARMELFLFFTSLLQHFSFS-VPTGQPRP
sp_P10635_CP2D6_HUMAN_Cytoc      FVKPEAFLPFSAGRRACLGEPLARMELFLFFTSLLQHFSFS-VPTGQPRP
sp_A0A087X1C5_CP2D7_HUMAN_P      FVKPEAFLPFSAGRRACLGEPLARMELFLFFTSLLQHFSFS-VAAGQPRP
sp_Q29488_CP2DH_MACFA_Cytoc      FVKPEAFLPFSAGRRACLGEPLARMELFLFFTCLLQRFSFS-VPAGQPRP
sp_O18992_CP2DJ_CALJA_Cytoc      FVKPEAFLPFSAGRRACLGEPLARMELFLFFTCLLQRFSFS-VPAGQPRP
sp_Q9QYG6_CP2DR_MESAU_Cytoc      FVKHEAFMPFSAGRRACLGEPLARMELFLFFTCLLQRFSFS-VPAGQPRP
sp_Q9QYG5_CP2DK_MESAU_Cytoc      FVKQEAFMPFSAGRRACLGEPLARMELFLFFTCLLQRFSFS-VPAGQPRP
sp_Q8CIM7_CP2DQ_MOUSE_Cytoch     FVKHEAFMPFSAGRRSCLGEPLARMELFLFFTCLLQRFSFS-VPDGQPRP
sp_P10634_CP2DQ_RAT_Cytochro     FVKHEAFMPFSAGRRACLGEPLARMELFLFFTCLLQRFSFS-VLAGRPRP
sp_P24457_CP2DB_MOUSE_Cytoch     FVKPEAFMPFSAGHRSCLGEALARMELFLFFTCLLQRFSIS-VPDGQPQP
sp_P24456_CP2DA_MOUSE_Cytoch     FVKPEAFMPFSAGRRSCLGEPLARMELFLFFTCLLQHFSFS-VPNGQPRP
sp_P11714_CP2D9_MOUSE_Cytoch     FVKPEAFMPFSAGRRSCLGEALARMELFLFFTCLLQRFSFS-VPDGQPQP
sp_P12939_CP2DA_RAT_Cytochro     FVKHEAFMPFSAGRRACLGEPLARMELFLFFTCLLQHFSFS-VPAGQPRP
sp_P10633_CP2D1_RAT_Cytochro     FVKHEAFMPFSAGRRACLGEPLARMELFLFFTCLLQRFSFS-VPVGQPRP
sp_P12938_CP2D3_RAT_Cytochro     FVKHEAFMPFSAGRRACLGEPLARMELFLFFTCLLQRFSFS-VPTGQPRP
sp_Q9QUJ1_CP2DS_MESAU_Cytoc      FVKQEAFMPFSAGRRACLGEPLARMELFLFFTCLLQRFSFS-VPAGQPRP
sp_Q64680_CP2DI_RAT_Cytochro     FVKHEAFMPFSAGRRACLGEPLARMELFLFFTCLLQRFSFS-VPAGQPRP
sp_P13108_CP2D4_RAT_Cytochro     FVKHEAFMPFSAGRRACLGEPLARMELFLFFTCLLQRFSFS-VPTGQPRP
sp_Q64403_CP2DG_CAVPO_Cytoc      FVKREAFMPFSAGPRICLGEPLARMELFLFFTSLLQRFSFS-VPEGQPRP
sp_P19225_CP270_RAT_Cytochr      FKKSDYFMAFSAGRRACIGEGLARMEMFLILTSILQHFTLKPLVNPEDID
sp_P50170_RDH2_RAT_Retinol_      WTSPKPEKAL----------------------------------------
sp_Q811S7_UBIP1_MOUSE_Upstr      GGGTPSVYHAIYLEEMVASEVARKLASVFNIPFHQINQVYRQGPTGIHIL


sp_Q7Z449_CP2U1_HUMAN_Cytoc      LLTGRFGLTLAPHPFNITISRR-----------
sp_Q0IIF9_CP2U1_BOVIN_Cytoc      ILTGKYGLTLAPHPFNIIISKR-----------
sp_Q9CX98_CP2U1_MOUSE_Cytoc      VMTGRFGLTLAPHPFNVTISKR-----------
sp_O46658_CP2DP_PIG_Vitamin      SDHGVFAFLLFPSPYQLCAVPR-----------
sp_Q01361_CP2DE_BOVIN_Cytoc      SEHGVFAFLVTPAPYQLCAVPR-----------
sp_Q29473_CP2DF_CANFA_Cytoc      SDHGVFTFLKVPAPFQLCVEPR-----------
sp_Q2XNC9_CP2D6_PANPA_Cytoc      SHHGVFAFLVTPSPYELCAVPR-----------
sp_Q2XNC8_CP2D6_PANTR_Cytoc      SHHGVFAFLVTPSPYELCAVPR-----------
sp_P10635_CP2D6_HUMAN_Cytoc      SHHGVFAFLVSPSPYELCAVPR-----------
sp_A0A087X1C5_CP2D7_HUMAN_P      SHSRVVSFLVTPSPYELCAVPR-----------
sp_Q29488_CP2DH_MACFA_Cytoc      SHHGVFAFLVTPSPYELCAVPR-----------
sp_O18992_CP2DJ_CALJA_Cytoc      SPHGVFAFLVTPSPYELCAVPR-----------
sp_Q9QYG6_CP2DR_MESAU_Cytoc      SDQGIFALPVTPTPYELCAVVR-----------
sp_Q9QYG5_CP2DK_MESAU_Cytoc      SDQGVFALPVTPTPYELCAVVR-----------
sp_Q8CIM7_CP2DQ_MOUSE_Cytoch     SDYGIYTMPVTPEPYQLCAVAR-----------
sp_P10634_CP2DQ_RAT_Cytochro     STHGVYALPVTPQPYQLCAVAR-----------
sp_P24457_CP2DB_MOUSE_Cytoch     SNYRVHAIPVAPFPYQLCAVMREQGH-------
sp_P24456_CP2DA_MOUSE_Cytoch     RNLGVFPFPVAPYPYQLCAVMREQGH-------
sp_P11714_CP2D9_MOUSE_Cytoch     SNSGVYGILVAPSPYQLCAVVRDQGH-------
sp_P12939_CP2DA_RAT_Cytochro     STLGNFAISVAPLPYQLCAAVREQGH-------
sp_P10633_CP2D1_RAT_Cytochro     STHGFFAFPVAPLPYQLCAVVREQGL-------
sp_P12938_CP2D3_RAT_Cytochro     SDYGVFAFLLSPSPYQLCAFKR-----------
sp_Q9QUJ1_CP2DS_MESAU_Cytoc      SDHVVLGVLKSPAPYQLCAVPR-----------
sp_Q64680_CP2DI_RAT_Cytochro     SNYGVFGALTTPRPYQLCASPR-----------
sp_P13108_CP2D4_RAT_Cytochro     SDYGIFGALTTPRPYQLCASPR-----------
sp_Q64403_CP2DG_CAVPO_Cytoc      SDRGAPYLVVLPSPYQLCAVLR-----------
sp_P19225_CP270_RAT_Cytochr      TTPVQPGLLSVPPPFELCFIPV-----------
sp_P50170_RDH2_RAT_Retinol_      ---------------------------------
sp_Q811S7_UBIP1_MOUSE_Upstr      VSDQMVQNFQDETCFLFSTVKAENNDGIHIILK
```

(a)

FIGURE 2.1 (CONTINUED) (a) Complete sequence alignment of CYP2D members. Protein sequences of the CYP2D enzymes from different species are obtained from the Swiss-Prot database (http://www.uniprot.org/). Multiple sequence alignment of the CYP2Ds is performed using the program Clustal W v2.0 (http://www.clustal.org/clustal2/) with default parameters. *(Continued)*

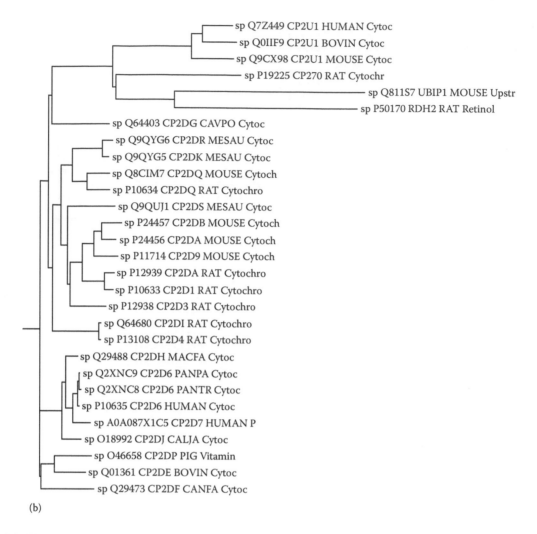

sp Q7Z449 CP2U1 HUMAN Cytoc
sp Q0IIF9 CP2U1 BOVIN Cytoc
sp Q9CX98 CP2U1 MOUSE Cytoc
sp P19225 CP270 RAT Cytochr
sp Q811S7 UBIP1 MOUSE Upstr
sp P50170 RDH2 RAT Retinol
sp Q64403 CP2DG CAVPO Cytoc
sp Q9QYG6 CP2DR MESAU Cytoc
sp Q9QYG5 CP2DK MESAU Cytoc
sp Q8CIM7 CP2DQ MOUSE Cytoch
sp P10634 CP2DQ RAT Cytochro
sp Q9QUJ1 CP2DS MESAU Cytoc
sp P24457 CP2DB MOUSE Cytoch
sp P24456 CP2DA MOUSE Cytoch
sp P11714 CP2D9 MOUSE Cytoch
sp P12939 CP2DA RAT Cytochro
sp P10633 CP2D1 RAT Cytochro
sp P12938 CP2D3 RAT Cytochro
sp Q64680 CP2DI RAT Cytochro
sp P13108 CP2D4 RAT Cytochro
sp Q29488 CP2DH MACFA Cytoc
sp Q2XNC9 CP2D6 PANPA Cytoc
sp Q2XNC8 CP2D6 PANTR Cytoc
sp P10635 CP2D6 HUMAN Cytoc
sp A0A087X1C5 CP2D7 HUMAN P
sp O18992 CP2DJ CALJA Cytoc
sp O46658 CP2DP PIG Vitamin
sp Q01361 CP2DE BOVIN Cytoc
sp Q29473 CP2DF CANFA Cytoc

(b)

FIGURE 2.1 (CONTINUED) (b) The phylogenetic tree of CYP2Ds. The *CYP2D6* gene is conserved in chimpanzee, rhesus monkey, rat, chicken, and frog.

the liver, kidney, and small intestinal mucosa, while Cyp2d1 and Cyp2d5 are expressed systemically in various tissues (Hiroi et al. 1998; Kawashima et al. 1996). Cyp2d4 is found in the brain, adrenal gland, ovary, testis, and gonecystis, in addition to the liver, kidney, and small intestinal mucosa. Cyp2d4 has also been detected in rat breast (Hellmold et al. 1995). The tissue-specific distribution of rat Cyp2ds suggests that each enzyme may have tissue-specific catalytic properties. Each rat Cyp2d shows a specific tissue distribution pattern, suggesting that each Cyp2d has specific catalytic properties and plays specific roles in various tissues.

2.2.2 SUBSTRATES OF RAT CYP2DS

Dextromethorphan is a commonly used probe substrate of human CYP2D6, which is a synthetic analog of narcotic analgesics used as an over-the-counter antitussive agent (Bolser 2006). The formation of the active metabolite dextrorphan from dextromethorphan is primarily mediated by CYP2D6 via *O*-demethylation with minor contribution from CYP3A4 in humans (Dayer et al. 1989; Jacqz-Aigrain et al. 1993). Rat liver microsomes can metabolize dextromethorphan with

similar K_m values to humans (2.5 vs. 5 μM) (Dayer et al. 1989; Kerry et al. 1993). Rat liver also has a lower-affinity enzyme with a K_{m2} of 158 μM; 98% of total intrinsic clearance is contributed by the high-affinity K_{m1} site (Kerry et al. 1993). Rat brain microsomes are able to *O*-demethylate codeine (Chen et al. 1990) and hydroxylate amphetamines (Lin et al. 1992) and *O*-demethylate dextromethorphan (Jolivalt et al. 1995; Tyndale et al. 1999). However, the K_m of dextromethorphan conversion to dextrorphan in rat brain microsomes is 400 μM, ~100-fold higher than that of rat liver microsomes (Kerry et al. 1993).

In rats, dextromethorphan is metabolized to dextrorphan, 3-methoxymorphinan, and 3-hydroxymorphinan, with dextrorphan being the primary one (Kerry et al. 1993). Additionally, the *O*-demethylated metabolites dextrorphan and 3-hydroxy-morphinan are subject to subsequent Phase II conjugating reaction (Kerry et al. 1993). Inhibition studies using known CYP2D6 inhibitors such as quinine, dextropropoxyphene, methadone, and propafenone have confirmed that the metabolism of dextromethorphan to dextrorphan is mostly via Cyp2ds in rats (Kerry et al. 1993). Conversion of dextromethorphan to dextrorphan via *O*-demethylation appears to be a suitable marker for Cyp2ds in rats.

Heterologously expressed rat Cyp2d1 to 2d4 displayed similar bufuralol 1′-hydroxylase activity but are less active than human CYP2D6 (Wan et al. 1997). In this study, debrisoquine 4-hydroxylation activity is found to be specific to Cyp2d2 among rat Cyp2ds. The recombinant rat Cyp2d18 enzyme (a variant of Cyp2d4) and rat brain microsomes catalyze the *N*-demethylation, but not hydroxylation, of imipramine and desipramine, which are known substrates of human CYP2D6 (Kawashima et al. 1996; Thompson et al. 1998). Other studies also indicate that rat Cyp2d2, but not Cyp2d1, contributes to the metabolism of debrisoquine, bunitrolol, alprenolol, and bufuralol, which are known as typical human CYP2D6 substrates, in rat liver microsomes (Hiroi et al. 2002; Narimatsu et al. 1995; Suzuki et al. 1992; Yamamoto et al. 1998b). Recombinant rat Cyp2d enzymes and inhibitory chemicals and antibodies are commonly used in these studies to confirm the role of Cyp2d2 in drug metabolism.

The fluorometric compound, 3-[2-(*N*,*N*-diethyl-*N*-methyl-amino)ethyl]-7-methoxy-4-methylcoumarin (AMMC), is selectively metabolized by rat Cyp2d2 and human CYP2D6 (Stresser et al. 2002). AMMC is a selective probe for CYP2D6 (Chauret et al. 2001; Ghosal et al. 2003). Recombinant Cyp2d2, but not Cyp2d1, catalyzes AMMC-demethylation to a fluorescent metabolite. For AMMC, the apparent K_m values are 8 and 4.6 μM for rat liver microsomes and cDNA-expressed CYP2D2, respectively. Thus, AMMC can also be used as a selective probe for rat Cyp2d2.

Rat Cyp2ds are involved in the metabolism of various neurotransmitters and neurosteroids (Wang et al. 2014). A recent study has found that Cyp2ds mediate synthesis of serotonin from 5-methoxytryptamine in rat brain (Haduch et al. 2015). In rats, nicotine induces Cyp2ds and thus increases codeine activation to form morphine and tramadol metabolism (McMillan and Tyndale 2015; Wang et al. 2015). Rat Cyp2d18/2d4, which shares 85% amino acid sequence identity with mouse Cyp2d22, has been shown to be expressed in the brain and catalyzes dopamine oxidation to aminochrome through a peroxide-shunt mechanism (Thompson et al. 2000). Rat Cyp2d18/2d4 also mediates ω-hydroxylation and epoxygenation of AA, primarily leading to the formation of 8,9-, 11,12-, and 14,15-epoxyeicosatrienoic acids that have vasoactive properties in brain, kidney, and heart tissues (Thompson et al. 2000).

2.2.3 Differences in Rat Cyp2d-Dependent Metabolism

Rat Cyp2d1 has more than 95% similarity in amino acid sequence to Cyp2d5 (Kawashima and Strobel 1995). Despite the amino acid sequence similarities among rat Cyp2ds (>70%), significant differences in the ability to metabolize drugs have been observed among these Cyp2ds. Hiroi et al. (2002) have compared the catalytic activity of recombinant rat Cyp2d1, 2d2, 2d3, and 2d4 toward three probe substrates including bufuralol (1′-hydroxylation or carboxylation and 1′2′-ethenylation), debrisoquine (4-hydroxylation),

and propranolol (4-, 5- and 7-hydroxylation). In Wistar rat liver microsomes, bufuralol is metabolized to three metabolites, namely, 1′-hydroxybufuralol, 1′2′-ethenylbufuralol, and 1′-oxobufuralol. All Cyp2ds exhibited activity for bufuralol 1′-hydroxylation, with Cyp2d2 and 2d4 having the highest activity (Hiroi et al. 2002). Rat Cyp1a and Cyp2c11 also had this activity but at low levels. Since rat Cyp2d2 has an extremely low K_m value for bufuralol 1′-hydroxylation compared with other Cyp2ds (Chow et al. 1999b), Cyp2d2 is likely the predominant enzyme catalyzing the 1′-hydroxylation of bufuralol.

Rat Cyp2d4 displays the highest activity for 1′2′-ethenylbufuralol formation from bufuralol, while Cyp2d2 and 2d3 only contributed to a minor extent (Hiroi et al. 2002). Cyp2d1 did not show this activity. These results indicate that bufuralol 1′2′-ethenylation is specific to rat Cyp2d4 and human CYP2D6. Cyp2d2, 2d3, and 2d4 showed 1′-oxobufuralol formation activity. Rat Cyp1a1 had a much higher level of 1′-carboxylation activity than rat Cyp2ds. In humans, 1′-carboxylation of bufuralol is mediated by CYP1A2 (Yamazaki et al. 1994).

Debrisoquine is metabolized to 4-hydroxydebrisoquine, 3-hydroxydebrisoquine, and 1-hydroxydebrisoquine by human CYP2D6 (Eiermann et al. 1998). Recombinant rat Cyp2d2 shows a high level of debrisoquine 4-hydroxylation activity with a K_m of 6.87 μM, while Cyp2d1 also displays 4-hydroxylation activity but with a much higher (27-fold) K_m (186 μM) (Hiroi et al. 2002). Rat Cyp2d2 also exhibits 3-hydroxylation activity with a K_m of 13.1 μM (for CYP2D6: $K_m = 17.0$ μM), but it is weaker than its 4-hydroxylation activity. Only Cyp2d1 has higher 3-hydroxylation than 4-hydroxylation activity but with high K_m values for both 3- and 4-hydroxylation (152 and 186 μM, respectively) (Hiroi et al. 2002). Rat Cyp2d1 may contribute to this activity when the concentration of substrate is high. These results indicate that rat Cyp2d2 catalyzes predominantly the 4-hydroxylation of debrisoquine and both Cyp2d1 and 2d2 catalyze 3-hydroxylation of debrisoquine.

Propranolol is known to be metabolized to four main metabolites, namely, 4-, 5-, and 7-hydroxypropranolol and *N*-desisopropylpropranolol, in rat hepatic microsomes (Masubuchi et al. 1993). In humans, propranolol is metabolized to three main metabolites, including 4- and 5-hydroxypropranolol and *N*-desisopropylpropranolol (Masubuchi et al. 1994). In humans and rats, the *N*-desisopropylation of propranolol has been reported to be catalyzed by CYP1A or CYP2C enzymes, not CYP2Ds (Fujita et al. 1993; Yoshimoto et al. 1995). CYP2D6 has high 4- and 5-hydroxylation activities but very low 7-hydroxylation activity (Masubuchi et al. 1994). In rats, recombinant Cyp2d2 and 2d4 show higher levels of 4-hydroxylation activity than Cyp2d1 and 2d3 (Hiroi et al. 2002). The 5-hydroxylation of propranolol is catalyzed by Cyp2d2, 2d3, and 2d4 except Cyp2d1, but the level of activity is relatively low. Only rat Cyp2d1 has higher 3-hydroxylation than 4-hydroxylation activity with the relative activity ratio of Cyp2d1 being 7.46. The 7-hydroxylation is catalyzed only by rat Cyp2d2 (Hiroi et al. 2002).

Rat Cyp2ds 1–4 catalyze N-desisopropylation of propranolol. These findings demonstrate that rat Cyp2ds have different catalytic properties toward known CYP2D6 substrates. Cyp2d4 selectively metabolizes bufuralol via 1′,2′-ethenylation while Cyp2d2 selectively catalyzes debrisoquine 4-hydroxylation and propranolol 7-hydroxylation.

Mianserin is a tetracyclic antidepressant administered as a racemic mixture, and its pharmacological activity, metabolism, and disposition exhibit stereoselectivity (Heinig et al. 1993). In humans, its major metabolic pathways include 8-hydroxylation, N-demethylation, and N-oxidation catalyzed mainly by CYP2D6, and by CYP3A4 and 1A2 to a less extent (Dahl et al. 1994; Koyama et al. 1996). In rats, the additional metabolite, 8-hydroxy-N-desmethylmianserin, is formed from 8-hydroxymianserin and N-desmethylmianserin (Heinig et al. 1993). Chow et al. (1999a) have compared the metabolism of mianserin using rat Cyp2d1, 2d2, 2d3, and 2d4 and human CYP2D6. The five CYP2D/Cyp2ds have similar 8-hydroxylation activity toward R,S-mianserin, ranging from 1.15 to 2.53 nmol/min/nmol CYP. The K_m value of 8-hydroxylation for rat Cyp2d2 (5.9 μM) is much smaller than those for the other rat Cyp2ds (35.5–133.3 μM) and is close to that for human CYP2D6 (3.7 μM). Rat Cyp2d3 and 2d4 efficiently N-demethylate R,S-mianserin compared with the other three enzymes, with the activity of rat Cyp2d2 being negligible (Chow et al. 1999a). The K_m values for N-demethylation by rat Cyp2d1, 2d3, and 2d4 are much larger (156.5, 84.5, and 402.7 μM, respectively) than that of human CYP2D6 (5.1 μM). N-Oxidation activity for mianserin is specific to Cyp2d1, although its level is relatively low. Rat Cyp2d4 and human CYP2D6 catalyze the formation of 8-hydroxy-N-desmethylmianserin. Furthermore, rat Cyp2d1 and 2d4 selectively 8-hydroxylate the R-enantiomer, and human CYP2D6 predominately N-demethylates the R-enantiomer (Chow et al. 1999a). These results indicate that mianserin is metabolized by both rat and human CYP2D enzyme in an enzyme- and stereoselective manner.

The metabolism of carteolol, a β-adrenoceptor blocker, has been investigated in male Sprague–Dawley (SD) rat liver microsomes (Umehara et al. 1997). The formation of 8-hydroxycarteolol is the principal metabolic pathway of carteolol in vitro and followed Michaelis–Menten kinetics with a K_m of 11.0 μM. Quinine (K_i = 0.06 μM) and quinidine (K_i = 2.0 μM), selective inhibitors of rat Cyp2d1, competitively inhibit 8-hydroxycarteolol formation. Furthermore, only anti-human CYP2D6 antibody inhibits this reaction. These results suggest that carteolol is metabolized to 8-hydroxycarteolol by rat Cyp2d1. The K_m of carteolol for Cyp2d1 in male rat liver microsomes is much greater than those of propranolol or bunitrolol, indicating that carteolol has a lower affinity for rat Cyp2d1 compared with bunitrolol.

2.2.4 DARK AGOUTI RATS AS A CYP2D2-DEFICIENT MODEL

The Dark Agouti (DA) strain of rats is the inbred partner strain for a number of congenic strains. DA rats are susceptible to the induction of a number of autoimmune diseases including autoimmune thyroiditis; severe collagen-induced arthritis after immunization with bovine, chick, or rat type II collagens; and experimental allergic encephalomyelitis (Lenz et al. 1999; Stepaniak et al. 1995). As such, DA rats are commonly used in the studies of immune diseases. DA rats develop spontaneously urinary bladder tumors and are thus a feasible model of human bladder cancer (Deerberg et al. 1985). DA rats have a defective bile acid transport in females (Reichen et al. 1986) and their Cyp2d1 is deficient because of a structurally altered db1 protein (Al-Dabbagh et al. 1981; Gonzalez et al. 1987). At least one other isoform of Cyp2C or Cyp3A may also be deficient in DA rats. The female DA rats have often been used as an animal model for PM CYP2D6 phenotype, with males of other strains such as SD or Wistar serving as models for the EM phenotype (Barham et al. 1994; Schulz-Utermoehl et al. 1999; Yamamoto et al. 1998b). DA rats can be used as a preliminary screening model to explore the Cyp2d-mediated metabolism of drugs and other compounds and identify Cyp2d substrates; interspecies differences in drug metabolism suggest that DA rats may not be used to provide quantitative information regarding the contribution of CYP2D6 to a specific oxidative reaction of substrates in humans. Studies have reported that the female DA rat displays an impaired metabolism for a number of typical CYP2D6 substrates including debrisoquine, dextromethorphan, diazepam, propranolol, bufuralol, bunitrolol, alprenolol, citalopram, perhexiline, fenproporex, AMMC, 1-(3-trifluoromethylphenyl) piperazine, 1,2,3,4,-tetrahydroisoquinoline, and 1-methyl-4-phenyl-1,2,3,6-tetrahydropyridine (MPTP) (Coleman et al. 2000; Kerry et al. 1993; Kingback et al. 2011; Kraemer et al. 2004; Licari et al. 2015; Lorenc-Koci et al. 2004; Masubuchi et al. 1993; Mechan et al. 2002; Narimatsu et al. 1995, 1996; Sakai et al. 2005; Schulz-Utermoehl et al. 1999; Staack et al. 2004; Stresser et al. 2002; Suzuki et al. 1992; von Moltke et al. 2001). This may result in altered drug response and toxicities.

Debrisoquine, another probe for CYP2D6, is converted to 4-hydroxydebrisoquine and 3-hydroxydebrisoquine by this enzyme (Boobis et al. 1983; Distlerath and Guengerich 1984; Gonzalez et al. 1988; Woolhouse et al. 1979). DA rats exhibit a debrisoquine 4-hydroxylase PM phenotype, which is linked to the low level of Cyp2d2, particularly in female rats (Schulz-Utermoehl et al. 1999; Yamamoto et al. 1998b). The mRNA of both Cyp2d1 and 2d2 is low in DA rats and both Cyp2d1 and 2d2 catalyze debrisoquine 4-hydroxylase (Yamamoto et al. 1998b). It appears that DA rats show PM Cyp2d phenotype because of defective Cyp2d1 and 2d2. In Western blotting studies, the hepatic expression of Cyp2d1 is greater in DA rats than in SD or Wistar rats (Schulz-Utermoehl et al. 1999). In contrast, hepatic Cyp2d2 is 30- to 40-fold less abundant in female DA rats than female SD or Wistar rats and 6- to 8-fold less abundant in male DA rats than in male SD or Wistar rats. No hepatic Cyp2d3 is detected in either sex of any of the three strains. Hepatic Cyp2d4 expression is generally greater in male than in female rats, and higher in DA rats compared with SD or Wistar strains. Cyp2d5 is expressed in the livers of female and male DA rats but not in female SD or Wistar rats (Schulz-Utermoehl et al. 1999). Hepatic debrisoquine 4-hydroxylase activity is markedly reduced in female and male DA rats as compared to SD or Wistar rats and correlated with the hepatic Cyp2d2 content. Recombinant Cyp2d2 is 18-fold more active at catalyzing the 4-hydroxylation of debrisoquine than Cyp2d1. Furthermore, quinine markedly inhibits Cyp2d2-mediated debrisoquine and metoprolol oxidation, while quinidine, its diastereoisomer, inhibits the reactions to a lesser extent (Schulz-Utermoehl et al. 1999). These results show that impaired debrisoquine 4-hydroxylase activity in female DA rats is attributed to deficiency of Cyp2d2.

Kerry et al. (1993) have compared the enzyme kinetics of dextromethorphan O- and N-demethylation and the N- and O-demethylation of the primary metabolites dextrorphan and 3-methoxymorphinan in liver microsomes from female DA and female SD rats. The intrinsic clearance (V_{max}/K_m) of the O-demethylation of 3-methoxymorphinan to 3-hydroxymorphinan is 180-fold lower in DA rats (0.11 vs. 20.77 ml/h/mg) because of a 60-fold higher K_m (108.7 vs. 1.76 μM) and 3-fold lower V_{max} (11.5 vs. 35.95 nmol/mg/h). The kinetics for dextrorphan N-demethylation to 3-hydroxymorphinan does not differ between the two strains. The K_m for dextromethorphan N-demethylation to 3-methoxymorphinan is similar between SD and DA rats (85.04 vs. 68.99 μM). However, SD rats display a twofold higher V_{max} (83.37 vs. 35.49 nmol/mg/h) and intrinsic clearance (0.96 vs. 0.51 ml/h/mg) than DA rats (Kerry et al. 1993). The O-demethylation of dextromethorphan to dextrorphan in SD rats shows a high- and low-affinity enzyme component, with the high-affinity intrinsic clearance contributing 98% of the total intrinsic clearance. Dextromethorphan O-demethylation in DA rats is characterized by a single enzyme system. The high-affinity O-demethylating enzyme in SD rats shows a 20-fold lower K_m (2.5 vs. 55.6 μM) and a 3-fold higher V_{max} (51.04 vs. 16.84 nmol/mg/h), resulting in a 66-fold higher intrinsic clearance (20.04 vs. 0.31 ml/h/mg) compared to DA rats (Kerry et al. 1993). Quinine, dextropropoxyphene, methadone, and propafenone inhibit 3-methoxymorphinan and dextromethorphan O-demethylation but do not inhibit dextrorphan or dextromethorphan N-demethylation at similar concentrations. These results demonstrate a clear strain difference in 3-methoxymorphinan O-demethylation and dextromethorphan O-demethylation between SD and DA rats, suggesting the key role of Cyp2d2 for these two reactions. In contrast, dextrorphan N-demethylation and dextromethorphan N-demethylation do not appear to be under genetic control in SD and DA rats, suggesting a minor role of Cyp2d2 for these reactions.

Masubuchi et al. (1993) have compared the metabolism of propranolol using microsomes from DA and Wistar rats. Propranolol 4- and 5-hydroxylation followed biphasic Michaelis–Menten kinetics, and 7-hydroxylation and N-desisopropylation are monophasic in both strains. The kinetic studies show that the V_{max} for propranolol 7-hydroxylation and the V_{max} of high-affinity phases for propranolol 4- and 5-hydroxylation are significantly low in DA rats compared with Wistar rats (Masubuchi et al. 1993). The antibody against CYP2D/Cyp2d inhibits by 90% propranolol 4-, 5-, and 7-hydroxylase activities in liver microsomes from male Wistar rats at a low propranolol concentration (5 μM). However, less inhibitory effects of the antibody on propranolol 4- and 5-hydroxylase activities are observed at a high propranolol concentration (1 mM), whereas a similar inhibitory effect of the antibody on propranolol 7-hydroxylase activity is shown. The antibody inhibits propranolol N-desisopropylase activity, but less extent of the inhibition on this activity than those on ring-hydroxylase activities has been observed at the low and high propranolol concentrations. These results indicate that Cyp2ds are involved predominantly in propranolol 4-, 5-, and 7-hydroxylations at low substrate concentrations in the rat.

Narimatsu et al. (1995) have reported that alprenolol 4-hydroxylation and N-desisopropylation in liver microsomes from male Wistar rats show biphasic kinetics. For 4-hydroxylation and N-desisopropylation at a low substrate concentration (5 μM), Wistar rats show a significantly high activity than DA rats but the difference decreases at a high substrate concentration (1 mM) (Narimatsu et al. 1995). Recombinant rat Cyp2d2 exhibits a high 4-hydroxylase and N-desisopropylase activity toward 5 μM alprenolol, which is efficiently decreased by antibodies against Cyp2d2 and sparteine. Furthermore, rat Cyp2c6 and 2c11 also show 4-hydroxylase and N-desalkylase activities. These results suggest that Cyp2d2 is the primary enzyme for alprenolol 4-hydroxylation and N-desisopropylation in a lower substrate concentration range. At higher substrate concentrations, Cyp2cs play a major role in male rats.

The 1′-hydroxylation of bufuralol, a substrate of human CYP2D6, is compared in brain microsomes from male and female Wistar rats and from female DA rats (Coleman et al. 2000). The kinetics of the 1′-hydroxylation of bufuralol (1–1500 μM) by brain microsomes are biphasic. The activity of the high-affinity site of metabolism is consistent with Michaelis–Menten kinetics (apparent K_{m1} = 0.61–1.42 μM), whereas the low-affinity activity is better described by a Hill function ($K_{50\%2}$) = 253–258 μM, n = 1.2–1.3). Values for kinetic constants are similar in both rat strains. Quinine is only a weak inhibitor of both the high-affinity (apparent K_i = 90 μM) and low-affinity (210 μM) sites of metabolism. In contrast, the kinetics of 1′-hydroxylation of bufuralol by rat liver microsomes are best described by a two-site Michaelis–Menten function (Coleman et al. 2000). The V_{max} values are three to five orders of magnitude greater compared with those for brain microsomes (male and female Wistar), and liver microsomes from female DA rats are significantly less active than those from Wistar rats. These data indicate tissue- and rat strain–specific differences in the metabolism of bufuralol.

SD rats have been found to be more responsive to diazepam than the DAs (Mechan et al. 2002), probably because of the interstrain variation in diazepam metabolism. Diazepam is metabolized to three primary metabolites, 3-hydroxy-diazepam, N-desmethyl-diazepam, and p-hydroxy-diazepam, in liver microsomes of adult male Wistar rats (Neville et al.

1993). The formation of these metabolites is mainly catalyzed by Cyp3a2, 2c11, and 2d1, respectively. Saito et al. (2004a) have found the existence of EMs (17% of the rats) and PMs (83%) of diazepam metabolism in liver microsomes from Wistar rats at low concentration of substrate (3 μM). EM rats have remarkably higher activity toward diazepam p-hydroxylation than do PM rats. SD and Brown Norway rats have 300-fold higher diazepam p-hydroxylation activity than DA rats at a low concentration of diazepam (3.1 μM). DA rats show threefold higher diazepam 3-hydroxylation and twofold higher diazepam N-demethylation activities than those in the other strains. However, there is no significant difference in the expression levels of Cyp2d1 in liver microsomes between EM and PM rats (Saito et al. 2004b). DA rats show a low Cyp2d2 level and debrisoquine 4-hydroxylation activity. An antirat Cyp2d2 causes approximately 80% inhibition of the diazepam p-hydroxylation, indicating that the enzyme responsible for diazepam p-hydroxylation cross-reacts with the Cyp2d2 antibody. It does not inhibit the diazepam 3-hydroxylation and N-desmethylation. Moreover, the pattern of debrisoquine 4-hydroxylation activity, which is a Cyp2d2-mediated activity, does not coincide with that of diazepam p-hydroxylation activity in the liver of all rat strains tested (Saito et al. 2004b). Thus, Cyp2d2 is not responsible for the high-affinity enzyme activity of diazepam p-hydroxylation. These results demonstrate that DA rats eliminate diazepam more quickly from the body with a higher intrinsic clearance than SD rats. The quicker metabolic elimination of diazepam in DA rats may result in lower sensitivity to diazepam in these strains of rats than SD. The major metabolite of diazepam in SD rats is p-hydroxy-diazepam, whereas in DA rats, it is 3-hydroxy-diazepam, which may also contribute to the pharmacodynamic difference. The enzymes responsible for the activities of diazepam p-hydroxylation and diazepam 3-hydroxylation are involved in the species differences.

A further study has shown that diazepam p-hydroxylation is mediated by rat Cyp2d3 using a panel of yeast recombinant rat Cyp2ds (Sakai et al. 2005). Cyp2d3 also exhibits moderate activity toward diazepam N-desmethylation. Cyp2d4 shows a high diazepam N-desmethylation activity, while the activity of Cyp2d1 is low toward diazepam N-desmethylation and 2d2 shows no activity toward three metabolic pathways (Sakai et al. 2005). However, none of the Cyp2ds shows diazepam 3-hydroxylation activity.

The steady-state pharmacokinetics of the (+)-(S)- and (−)-(R)-enantiomers of citalopram (a known selective serotonin reuptake inhibitor and substrate of human CYP2D6) and its demethylated metabolites including demethylcitalopram and didemethylcitalopram have been compared in female SD and DA rats (Kingback et al. 2011). In humans, the N-demethylation of citalopram to demethylcitalopram is partially mediated by CYP2D6 and 2C19, whereas the formation of didemethylcitalopram from demethylcitalopram is largely by CYP2D6 (Sindrup et al. 1993). In human liver microsomes, the formation of demethylcitalopram from citalopram is mediated mainly by CYP3A4 and 2C19, with additional contributions from 2D6 (Kobayashi et al. 1997;

Olesen and Linnet 1999; Rochat et al. 1997; von Moltke et al. 1999, 2001). Furthermore, the formation of didemethylcitalopram from demethylcitalopram is exclusively mediated by CYP2D6 (Olesen and Linnet 1999). Rats are subcutaneously administered with racemic citalopram at 15 mg/kg daily for 13 days using osmotic pumps. Higher serum and brain (cortex and mesencephalon-pons) levels of citalopram and demethylcitalopram, but lower levels of didemethylcitalopram, are observed in DA rats when compared with SD rats (Kingback et al. 2011). In comparison with serum, the levels of citalopram are approximately three to six times higher in the brain in both rat strains. The enantiomeric (S/R) concentration ratios of citalopram are significantly lower in the DA rats in serum and the two brain regions when compared with the SD rats, indicating a possibly decreased capacity in the metabolism of the (−)-(R)-enantiomer in the DA rats. In addition, the ratios of parent drug over demethylcitalopram or didemethylcitalopram in serum are 2.1 and 2.3 in the SD rats and 2.1 and 9.1 in the DA rats. The ratios for brain are 8.5 and 15–19 in the SD rats and 8.0–9.5 and 28–65 in the DA rats (Kingback et al. 2011). These data indicate that Cyp2d deficiency in DA rats results in steady-state pharmacokinetic differences of the enantiomers of citalopram and its N-demethylated metabolites. Cyp2ds are present in the rat brain with catalyzing activities.

A recent study has compared the metabolism and hepatotoxicity of perhexiline, a known human CYP2D6 substrate, in DA and SD rats (Licari et al. 2015). Rats are administered with 200 mg/kg/day of racemic perhexiline maleate for 8 weeks. Plasma and liver samples are collected to determine concentrations of perhexiline and its metabolites, hepatic function, and histology. The median plasma and liver concentrations of perhexiline in SD rats are 0.09 mg/L and 5.42 ng/mg, respectively. In comparison, DA rats show significantly higher plasma (0.50 mg/L) and liver (24.5 ng/mg) perhexiline concentrations and 2.5- and 3.7-fold higher cis-OH-perhexiline concentrations, respectively, and lower plasma metabolic ratio (0.89 vs. 1.55). In both strains, the (+):(−) enantiomer ratio is 2:1. Perhexiline increases plasma LDH concentrations in DA rats but has no effect on plasma biochemistry in SD rats (Licari et al. 2015). Liver histology reveals lower glycogen content in perhexiline-treated SD rats but no effects on lipid content in either strain. These data indicate that DA rats have lower metabolism-mediated elimination of perhexiline than SD rats but a faster hydroxylation rate for perhexiline, which should be catalyzed by Cyps excluding Cyp2ds. A slower elimination of perhexiline may induce liver toxicity in DA rats.

The AMMC-demethylase activity in male and female SD rats is >15-fold higher than that in female DA rats (Stresser et al. 2002). MPTP, a parkinsonism-inducing neurotoxin, is converted to 4-phenyl-1,2,3,6-tetrahydropyridine (PTP) by DA and Wistar rat liver microsomes (Narimatsu et al. 1996). Wistar rat liver microsomes show much higher activity toward MPTP than those from DA rats. Studies using recombinant enzymes and inhibitors show that MPTP N-demethylation is mainly mediated by Cyp2d at lower substrate concentrations and by Cyp2c11 at higher substrate concentration (Narimatsu et al. 1996).

All these findings indicate that DA rats are a suitable model for human CYP2D6 PM phenotype owing to deficient Cyp2d2 activity. Like CYP2D6, Cyp2d2 can efficiently metabolize debrisoquine, dextromethorphan, bunitrolol, and propranolol.

2.2.5 REGULATION OF RAT CYP2DS

Rat Cyps have been found to be regulated by development, sex hormones, and other factors, and rat hepatic *Cyp2a2*, *2c11*, *2c12*, *3a2*, and *a2* increased with development in a sex-specific manner (DiMaio Knych and Stanley 2008; Ishizuka et al. 2002; Oka et al. 2000; Tsuneoka et al. 1992; Yamamoto et al. 1998a). Rat Cyp2ds have been found to be developmentally regulated. Rat *Cyp2d3* is shown to be a late-onset gene with its mRNA gradually being increased until rats reached puberty, while *Cyp2d5* is an early-onset gene with its peak mRNA expression being reached within 1 week after birth (Matsunaga and Gonzalez 1990). A correlation is found between mRNA expression during development and demethylation of cytosine residues located at the same position in the first exons of both the *Cyp2d3* and *2d5* genes.

Chow et al. (1999b) have studied the developmental changes in Cyp2d-mediated bufuralol 1'-hydroxylation activity in male (1-, 3-, 7-, 14-, and 34-week-old) and female (1-, 3-, 7-, 14-, and 40-week-old) Wistar rat livers. The amount of rat hepatic Cyp2d increases from 3 to 14 weeks of age in both sexes. In comparison to the amount of hepatic Cyp2d at 1 week, a significant increase is observed at 7 (1.3-fold), 14 (1.5-fold), and 34 weeks (1.3-fold) in males and at 7 (1.5-fold) and 14 weeks (1.7-fold) in females (Chow et al. 1999b). There is no gender difference in the amount of hepatic Cyp2d among the age groups. In the process of development, the amount of hepatic Cyp2d in rats increases significantly in a gender-independent manner. The bufuralol 1'-hydroxylation activity in the rat hepatic microsomes increases as the rats develop and reach a maximum level at 14 weeks in males and 7 weeks in females (Chow et al. 1999b). Compared with the activity at 1 week, a significant increase of the activity is observed at 7 (1.7-fold), 14 (1.8-fold), and 34 weeks (1.6-fold) in male rats and at 7 (1.8-fold), 14 (1.7-fold), and 40 weeks (1.7-fold) in female rats. There is no gender difference in this activity among any age groups. Eadie–Hofstee plots of bufuralol 1'-hydroxylation are obtained for monophasic kinetics ($K_m = 0.037$ μM) at 1 week and for biphasic kinetics at 7-week-old males ($K_{m1} = 0.051$ and $K_{m2} = 6.4$ μM) and females ($K_{m1} = 0.049$ and $K_{m2} = 5.3$ μM) (Chow et al. 1999b). The K_m value for bufuralol 1'-hydroxylation in 1-week-old male rats is 0.037 μM and is almost the same as that observed in the high-affinity component in 7-week-old male rats. Quinine at 100–1000 μM completely inhibits bufuralol 1'-hydroxylation activity of hepatic microsomes from 1- and 7-week-old rats (Chow et al. 1999b). These results indicate that at least two kinds of Cyp2ds, which differ markedly in their affinity for bufuralol, are present at 7 weeks of age and that the Cyp2d with a low

affinity for bufuralol is expressed with development. The K_m value of recombinant Cyp2d2 for bufuralol 1'-hydroxylation is 0.044 μM, which is nearly equal to the K_m value observed in 1-week-old rats and in the high-affinity phase in 7-week-old rats (Chow et al. 1999b). The K_m values of recombinant Cyp2d1, 2d3, and 2d4 are 9.5, 17, and 15 μM, respectively, and these values correspond well to those observed in the low-affinity phase in 7-week-old rats. Furthermore, rat hepatic *Cyp2d3* mRNA level increases until 7 weeks in males and 14 weeks in females whereas rat hepatic *Cyp2d1* and *2d2* mRNA levels remain unchanged over the range of 1 to 34 weeks in males and 1 to 40 weeks in females (Chow et al. 1999b). The *Cyp2d4* mRNA is not detected in the rat liver at any age group. These findings indicate that Cyp2d2, which has a high affinity for bufuralol, is expressed in both immature and mature rats, but Cyp2d3, which has a low affinity for bufuralol, is expressed only in mature rats. The regulation of the rat Cyp2ds during the developmental process in the rat is unaffected by gender.

Nitric oxide (NO) has been shown to be a key mediator for inflammation-induced CYP downregulation (Kingback et al. 2011; Kraemer et al. 2004; Licari et al. 2015; Staack et al. 2004). Rat Cyp2b1/2 proteins undergo an NO-dependent degradation in response to inflammatory stimuli (Licari et al. 2015). Stimulation of cultured human hepatocytes with a mixture of tumor necrosis factor-α, interleukin-1, and interferon (IFN)-γ downregulates *CYP2B6* mRNA and protein to 9% and 19% of control levels (Kraemer et al. 2004). The NO donor NOC-18 downregulates CYP2B6 protein to 30% of control, with only a small effect on *CYP2B6* mRNA. NO synthase inhibitors attenuated the downregulation of CYP2B6 protein but not mRNA by IFN-γ. No evidence is found for regulation of *CYP2E1* mRNA or protein by NO. NOC-18 treatment downregulates *CYP3A4* mRNA to 50% of control, but the NOS inhibitors fail to block the effects of interferon-γ on CYP3A4 expression (Kraemer et al. 2004). These findings indicate that the NO-dependent downregulation of human CYP2B6 is via posttranscriptional mechanism.

In isolated perfused rat livers, the catalytic activities of various rat Cyps including Cyp1a1/2, 3a2, 2b1/2, 2c11, and 2e1 except Cyp2d1 are substantially (up to 85%) decreased by treatment of sodium nitroprusside or isosorbide dinitrate, both NO donors (Vuppugalla and Mehvar 2004a). The effects of NO on rat Cyps are rapid, concentration dependent, and enzyme selective (Vuppugalla and Mehvar 2004a). Additionally, the effects of NO on rat Cyps are time dependent, consisting of both reversible and irreversible components. Further studies show that the inhibitory effects of NO on rat Cyps are mediated through selective alterations in the V_{max} or K_m of various enzymes (Vuppugalla and Mehvar 2005). In a single-pass isolated perfused rat liver model, sodium nitroprusside and isosorbide dinitrate increase the recovery of the metabolite dextrorphan 2- to 3.5-folds in the outlet perfusate when dextromethorphan is added to the inlet perfusate (Vuppugalla and Mehvar 2006). However, this is associated with a simultaneous decrease (50%–75%) in the excretion of dextrorphan into the bile, thus resulting in no change in the overall recovery of dextrorphan.

The decrease in the biliary excretion of dextrorphan is caused by NO-induced simultaneous reductions in both the conjugation of dextrorphan (75% and 50% reductions by nitroprusside and isosorbide dinitrate, respectively) and biliary clearance of the dextrorphan conjugate (Vuppugalla and Mehvar 2006). The nitroprusside treatment seems to have a more drastic effect than isosorbide dinitrate on the biliary clearance of both dextrorphan and 3-hydroxymorphinan. NO donors result in more recovery of the intact (unconjugated) species; 7%, 36%, and 26% of the administered doses of dextromethorphan is recovered as intact drug and metabolites in the control, nitroprusside-treated, and isosorbide dinitrate–treated groups, respectively. Additionally, both nitroprusside and isosorbide dinitrate significantly reduce the metabolism of dextromethorphan to 3-hydroxymorphinan, which is mostly mediated by Cyp3a2. The Cyp2e1-mediated metabolism of chlorzoxazone to 6-hydroxychlorzoxazone is also inhibited by both NO donors (Vuppugalla and Mehvar 2006). These results show that NO can inhibit rat Cyp1a1/2, 2b1/2, 2c11, 3a2, and 2e1 but upregulates Cyp2d.

Unlike rat Cyp2e1, neither the V_{max} nor the K_m value of Cyp2d1 is altered by NO donors. The differential effects of NO on the V_{max} values of these enzymes may be attributed to differences between the two enzymes in the accessibility of heme or cysteine thiolate residues to NO (Gergel et al. 1997). In addition, since NO reacts with thiol groups of amino acid residues in the apoprotein (Minamiyama et al. 1997; Vuppugalla and Mehvar 2004b; Wink et al. 1993), it may affect the binding of substrates to these enzymes and therefore affect their K_m values. The degree of involvement of the thiol-containing cysteine residues in the substrate binding may be different for various rat Cyps (Vuppugalla and Mehvar 2005). It has been shown that the critical amino acids for binding of CYP2D6 to nitrogen-containing ligands are negatively charged carboxylate-containing amino acids, such as Asp301 and Glu216 (Paine et al. 2003; Rowland et al. 2006). Therefore, a possible binding of NO with the cysteine amino acids of this enzyme is not expected to affect its substrate binding or K_m.

The rat *Cyp2d2* gene is positively regulated by the poly(C)-binding protein heterogeneous nuclear ribonucleoprotein K (hnRNP K) through a transcriptional regulatory element located in the 5′-flanking region from −94 to −113 (Sakai et al. 2009). A single substitution within the transcriptional regulatory element of the *Cyp2d2* gene is found in DA rats. The mutation is detected in the polypyrimidine sequence, which is the preferred binding site for the hnRNP K protein. The mutation within the transcriptional regulatory element attenuates the binding of the hnRNP K protein. It appears that decreased recruitment of the hnRNP K protein to the mutated sequence results in a low expression of *Cyp2d2* mRNA in DA rats.

2.2.6 Inhibitors of Rat CYP2Ds and Cyp2d-Mediated Drug–Drug Interactions

Both quinidine and quinine are known competitive human CYP2D6 inhibitors with different inhibitory potency (Newton et al. 1995). In recombinant rat Cyp2d2, quinidine

shows little inhibitory effect on bufuralol 1′-hydroxylation only at a concentration of 10 µM (Kobayashi et al. 2003). However, CYP2C6-mediated diclofenac 4-hydroxylation activity is inhibited by quinidine even at 0.1 µM. Quinine, a diastereomer of quinidine, is a more efficient inhibitor of rat Cyp2d-mediated debrisoquine 4-hydroxylase activity in rat liver microsomes (Kobayashi et al. 1989). It is approximately 80 times more potent for debrisoquine 4-hydroxylase activity than quinidine (K_i: 0.6 vs. 50 µM), while quinidine is sevenfold more potent than quinine in human liver microsomes for debrisoquine 4-hydroxylase activity (K_i: 1.7 vs. 13 µM) (Kobayashi et al. 1989). Quinine at 10 µM inhibits Cyp2d2-mediated bufuralol 1′-hydroxylation activity by more than 90%, but Cyp2c6-mediated diclofenac 4-hydroxylation and Cyp2c11-mediated testosterone 16α-hydroxylation activities are also inhibited by ~70% (Kobayashi et al. 2003). Although both quinidine and quinine are competitive inhibitors of debrisoquine 4-hydroxylase activity in rats and humans, their potencies are reversed. This suggests that the nature of the active site of CYP2D6/2ds differs between rats and humans and indicates that data on the specificity of Cyp2ds in rats should be extrapolated to humans with caution.

Several compounds are found to be mechanism-based inhibitors of rat Cyp2Ds. 4-Allyloxymethamphetamine-amine (ALLMA), an amphetamine derivative, inhibited rat Cyp2d-mediated methylenedioxymethamphetamine (MDMA) demethylation in a time- and dose-dependent manner and that the inhibition required the presence of NADPH (Cui et al. 2012). ALLMA (1–10 mg/kg body weight) is administered to adult male SD rats and liver microsomes are collected 3 h later. Cyp2d-mediated methamphetamine *p*-hydroxylation and low K_m MDMA demethylation activities are reduced by more than 80% after a dose of 10 mg/kg. Metabolic reactions mediated by Cyps other than Cyp2d, such as aniline *p*-hydroxylation, the high K_m component of MDMA demethylation, and the *N*-demethylation of methamphetamine, benzphetamine, aminopyrine, and erythromycin all appear to be minimally affected (Cui et al. 2012). The kinetics of MDMA metabolic activity in microsomes from ALLMA-pretreated rats are comparable to those from female DA rats.

Imipramine and its *N*-demethylated metabolite desipramine are inactivators of rat Cyp2ds (Lorenc-Koci et al. 2004; Topletz et al. 2013). Preincubation of microsomes from male Wistar rats with imipramine in the presence of NADPH causes a time-dependent inhibition of bunitrolol 4-hydroxylase activity, which has also been observed after in vivo administration of imipramine to rats (Topletz et al. 2013). Incubation of rat liver microsomes with [³H]imipramine in the presence of NADPH results in covalent binding of a ³H-labeled substance to liver microsomal protein. Formation rates of the reactive metabolites covalently bound to protein follow Michaelis–Menten kinetics, and the K_m value (1.1 µM) is close to that for imipramine 2-hydroxylation in rat liver microsomes (Topletz et al. 2013). The metabolism-dependent covalent binding of [³H]imipramine is lower in DA rats than in Wistar rats. The binding is inhibited by propranolol, quinidine, and an antibody against Cyp2d. Similar strain difference (DA < Wistar rats) and inhibitory effects by the compounds and the antibody are observed in imipramine 2-hydroxylase but not in *N*-demethylase activity. The proteins to which [³H]imipramine metabolites covalently bound are immunoprecipitated with the anti-Cyp2d antibody (Topletz et al. 2013). These results suggest that imipramine is converted by rat Cyp2d into a chemically reactive metabolite (probably arene-oxide) through its 2-hydroxylation, and the metabolite binds covalently to the enzyme itself and inactivates it.

The metabolite of propranolol, 4-hydroxypropranolol, is found to be an inactivator of rat Cyp2d (Ishii et al. 2014). Repetitive administration of propranolol in rats decreases the activities of hepatic Cyp2ds in hepatic microsomes. Rat microsomal bunitrolol 4-hydroxylase activity is inhibited by the addition of 4-OH-propranolol to the incubation mixture. The inhibition is greater after preincubation of microsomes with 4-OH-propranolol in the presence of NADPH than in its absence. The type of inhibition kinetics of bunitrolol 4-hydroxylase by 4-OH-propranolol is changed from a competitive type to a noncompetitive type by the preincubation. The inhibition of rat liver microsomal propranolol 5- and 7-hydroxylases by 4-OH-propranolol is blocked efficiently by co-incubation with quinine, a typical inhibitor of rat Cyp2ds, or to a lesser extent by bunitrolol (Ishii et al. 2014). However, quinidine did not significantly protect against the enzyme inactivation. Bunitrolol hydroxylase is not affected by either 1,4-naphthoquinone or 1,4-dihydroxynaphthalene, which are possible metabolites of 4-OH-propranolol. These results show that propranolol is converted by rat Cyp2ds to 4-OH-propranolol, which is further oxidized to a chemically reactive metabolite in the active site. The inactivation of rat Cyp2ds is likely the result of covalent binding of the reactive species to an amino acid residue of the active site.

The rat has been commonly used as an animal model for the investigation of Cyp2d-mediated drug–drug interactions, although in most cases the individual Cyp2d enzyme involved is unclear. For example, ondansetron increased the systemic exposure of tamoxifen via inhibition of Cyp2d and 3a1/2 in rats (Sun et al. 2015). In rats pretreated with quinine hydrochloride, the nonrenal clearance of ondansetron is significantly slower than that in control rats (Haduch et al. 2015). In Wistar rats treated with 15 mg/kg racemic metoprolol, the inhibition of Cyp2d by quinidine reduces the plasma concentrations of *O*-demethylmetoprolol and favored the formation of α-hydroxymetoprolol (McMillan and Tyndale 2015). In addition, SP-8203 (a new drug for the treatment of cerebral ischemia) significantly increases the systemic exposure of intravenously administered metoprolol in rats (Kirischian et al. 2011).

2.3 MOUSE Cyp2d SUBFAMILY: Cyp2d9–2d13, 2d22, 2d26, 2d34, AND 2d40

Mouse genome contains 102 functional *Cyp* genes and 88 pseudogenes in the genome (Nelson et al. 2004). *CYP1A* genes are conserved between mice and humans, but the *CYP2C*, *CYP2D*, and *CYP3A* gene clusters have diverged significantly between the two species, and assignment of orthologous genes between human and mouse proteins is therefore difficult. The mouse *Cyp2c*, *Cyp2d*, and *Cyp3a* gene clusters contain 15, 9, and 8 functional genes, respectively, compared with only 4, 1, and 4 genes in the corresponding gene clusters in humans (Nelson et al. 2004). In mice, there are at least nine different *Cyp2d* genes including *Cyp2d9–2d13*, *2d22*, *2d26*, *2d34*, and *2d40* and eight pseudogenes (*2d32p*, *2d33p*, *2d35p–2d39p*, and *2d41p*) (Gerhard et al. 2004; Kawai et al. 2001; Maeda et al. 2006; Nelson et al. 2004; Okazaki et al. 2002). All mouse Cyp2ds have high amino acid sequence identity (65%–75%) compared with human CYP2D6. Cyp2d22 has been suggested to be the functional ortholog of human CYP2D6. The mouse *Cyp2d22* gene contains 1515 bp and encodes a protein with 505 amino acids. Its DNA sequence shares 87% to 90% identity with rat *Cyp2d3–2d5*, with a corresponding amino acid identity of 71% to 85%. Interestingly, the amino acid sequence of the predicted mouse Cyp2d22 is less similar to mouse Cyp2d9 (69%), Cyp2d10 (69%), or Cyp2d11 (67%) than human CYP2D6 (74%) and cynomolgus monkey CYP2D17 (75%) (Nelson et al. 2004).

2.3.1 Tissue Distribution of Mouse Cyp2ds

The mice contain at least nine Cyp2d enzymes, namely, Cyp2ds 9–13, 22, 26, 34, and 40 (Nelson et al. 2004). Mouse Cyp2ds are mainly present in the liver, and other tissues including kidney, and brain at low to moderate levels (Hiroi et al. 1998; Masubuchi et al. 1996; Miksys et al. 2005). Mouse intestinal Cyp2ds follow a distinctly different expression pattern to human CYP2D6; they are highest in the colon and decrease proximally, with lowest expression in the proximal small intestine and the duodenum (Miksys et al. 2005). The pattern of expression of Cyp2ds in most brain regions in mice (Miksys et al. 2005) is similar to that in rats (Miksys et al. 2000) and humans. There is a high expression of Cyp2d in pyramidal neurons of mouse frontal cortex layers II through VI (Miksys et al. 2000, 2005). Cyp2d is expressed moderately to highly in mouse cerebellum Purkinje cells (Miksys et al. 2000).

The expression of Cyp2d9, together with Cyp2a5, 2b10, 2c29, and 3a11, is detected in the murine embryonic stem cell–derived hepatic tissue system at 16 days and 18 days after plating, and in 12-h-old and 60-h-old primary cultured adult and fetal hepatocytes (Tsutsui et al. 2006). In the murine embryonic stem cell–derived hepatic tissue system, 6β-OH-testosterone, 16β-OH-testosterone, 2α-OH-testosterone, and 2β-OH-testosterone are observed at 16 days and 18 days after plating; however, 15α-OH-testosterone, 7α-OH-testosterone, and 16α-OH-testosterone are not observed (Tsutsui et al. 2006). Specifically, the metabolic products of testosterone (15α-OH-testosterone, 6β-OH-testosterone, 7α-OH-testosterone, 16α-OH-testosterone, 16β-OH-testosterone, and 2α-OH-testosterone) are the indexes of Cyp2a4/5, Cyp3a, Cyp2a4/5 and 2d9, Cyp2d9 and 2b, Cyp2c29 and 2b, and 2d activities, respectively. This hydroxylation pattern of the murine embryonic stem cell–derived hepatic tissue system is similar to the fetal hepatocytes but is different from adult hepatocytes.

2.3.2 SUBSTRATE SPECIFICITY OF MOUSE CYP2DS

Mouse Cyp2d22 catalyzes the metabolism of dextromethorphan and shows substantially decreased O-demethylase activity as compared with human CYP2D6 (Yu and Haining 2006). However, Cyp2d22 generates more N-demethylated metabolite, 3-methoxymorphinan, than dextrorphan, an O-demethylated metabolite (Yu and Haining 2006). Cyp2d22 catalyzes dextromethorphan N-demethylation with lower K_m value than CYP2D6 (418 vs. 4750 μM) but similar to that exhibited by CYP3A4 for dextromethorphan N-demethylation (418 vs. 312 μM) or O-demethylation (171 vs. 157 μM) (Yu and Haining 2006). CYP2D6 catalyzes codeine O-demethylation, whereas Cyp2d22 and CYP3A4 mediate codeine N-demethylation. Methadone, a known CYP3A4 substrate and CYP2D6 inhibitor, is N-demethylated by Cyp2d22 with a K_m of 517 μM (Yu and Haining 2006). Quinidine and ketoconazole, potent inhibitors to CYP2D6 and CYP3A4, respectively, do not show strong inhibition toward Cyp2d22-mediated dextromethorphan O- or N-demethylation. Despite 77% amino acid identity, Cyp2d22 and CYP2D6 expressed in *Escherichia coli* are remarkably different in a range of enzymatic properties with different IC_{50} values for a panel of test compounds (McLaughlin et al. 2008).

Mouse Cyp2d22 does not metabolize any endogenous steroids (Blume et al. 2000). However, the partially purified mouse Cyp2ds, which should be the Cyp2d22 protein according to the reported amino acid sequence, catalyzes a variety of testosterone hydroxylation reactions (Masubuchi et al. 1997). However, a recent study has demonstrated that neither Cyp2d22 nor CYP2D6 catalyzes testosterone hydroxylation, whereas CYP3A4 does (Yu and Haining 2006).

These findings indicate a functional similarity between mouse Cyp2d22 and CYP3A4 while there are remarkably distinct enzyme function and substrate specificity between mouse Cyp2d22 and CYP2D6. Further studies are needed to explore the similarity and differences in the function and substrate specificity of CYP2D/Cyp2d among species.

2.3.3 CYP2D KNOCKOUT AND CYP2D6 TRANSGENIC MOUSE MODEL

In a mouse model with *Cyp2c*, *2d*, and *3a* gene clusters deleted, the mice are viable and fertile, demonstrating that these genes have evolved primarily as detoxification enzymes (Scheer et al. 2014). Although there is no overt phenotype, detailed examination has shown that *Cyp2c/2d/3a* knockout mice have a smaller body size (15%) and larger livers (20%). In addition, changes in hepatic morphology and a decreased blood glucose (30%) are also observed. A five-drug cocktail of cytochrome P450 probe substrates are used to evaluate changes in drug pharmacokinetics; marked changes are observed in either the pharmacokinetics or metabolites formed from Cyp2c, 2d, and 3a substrates, whereas the metabolism of the Cyp1a substrate

caffeine remains unchanged (Scheer et al. 2014). *Cyp2c/2d/3a* knockout mice provide a useful model to investigate the in vivo role of the P450 system in drug metabolism and efficacy, as well as in chemical toxicity.

Scheer et al. (2012) have generated several mouse lines expressing the two frequent human proteins CYP2D6.1 and CYP2D6.2 and an as yet undescribed variant of this enzyme, as well as a *Cyp2d* cluster knockout mouse. The various transgenic mouse lines cover a wide spectrum of different human CYP2D6 metabolizer phenotypes. The novel humanization strategy described here provides a robust approach for the expression of different CYP2D6 allelic variants in transgenic mice and thus can help evaluate potential CYP2D6-dependent interindividual differences in drug response in the context of personalized medicine. *CYP2D6* transgenic mice and wild-type mouse lines act as models of human EM and PM phenotypes, respectively (Corchero et al. 2001; Scheer et al. 2012; Yu et al. 2004). The transgene includes the complete human wild-type *CYP2D6* gene and its regulatory sequence. It is present in the mice at 5 ± 1 copies per haploid genome (Corchero et al. 2001). In *CYP2D6* transgenic mice, debrisoquine hydroxylase activity is significantly enhanced compared with wild-type mice (Corchero et al. 2001), indicating that CYP2D6 in humanized mice is functional. Both mouse Cyp2ds and human CYP2D6 are detected in transgenic mice. The expressions of human CYP2D6 in the liver, kidney, and intestine in humanized mice are comparable to that reported in humans. However, human CYP2D6 mRNA and protein are not expressed at detectable levels in brains of transgenic mice (Miksys et al. 2005). In the transgenic mouse, CYP2D6 cellular expression is restricted to enterocytes at the surface of the intestinal mucosa and concentrates at the tips of the villi in the kidney of transgenic mice. CYP2D6 expression is highest in the proximal tubules of the cortex in mice (Manns et al. 1989) and rats (Duclos-Vallee et al. 2000). This model has the potential for investigating human CYP2D6-mediated metabolism of xenobiotics and endogenous compounds, extrahepatic first-pass metabolism and oral bioavailability of drugs, and also the role of CYP2D6-mediated metabolism in organ-specific toxicity and pathogenesis of diseases. The humanized mouse model may represent a useful tool for studying the physiology and pharmacology of CYP2D6.

Recently, a *CYP2D6/3A4*-double transgenic mouse model has been developed by Felmlee et al. (2008). Both age and sex have considerable effects on hepatic CYP3A4 protein expression in 3- to 8-week-old transgenic mice, whereas neither factor alters CYP2D6 content. Constitutive CYP2D6 expression results in two- to threefold higher dextromethorphan O-demethylase activity in transgenic *CYP2D6/3A4* mouse liver microsomes compared with wild-type mice. In contrast, expression of CYP3A4 in transgenic mouse livers does not increase dextromethorphan N-demethylase and midazolam 1′-hydroxylase activities (Felmlee et al. 2008). Pretreatment with pregnenolone 16α-carbonitrile (PCN) and 1,4-bis-2-(3,5-dichloropyridyloxy)-benzene (TCPOBOP) increases CYP3A4 expression in double transgenic mice. Interestingly, induction of hepatic CYP3A4 is greater in females than age- and treatment-matched males. Consequently, the increase in midazolam 1′-hydroxylase activity is significantly higher in 8-week-old female mice than in corresponding males (eightfold vs. sixfold for PCN treatment and sixfold vs. fivefold for TCPOBOP). Furthermore, increases in testosterone 6β-hydroxylase activity after CYP3A induction are relatively lower compared with those in midazolam 1′-hydroxylation for age-, sex-, and treatment-matched mice. The difference in CYP3A4 expression and induction between male and female mice suggests that women may be more susceptible to CYP3A4-mediated drug–drug interactions, and the extent of drug–drug interactions could be substrate dependent.

2.4 BOVINE CYP2D14

There are a minimum of 62 *CYP* genes in the cow. Compared to the rat, cattle displays much higher coumarin 7-hydroxylase (CYP2A) and ethoxyresorufin *O*-deethylase (CYP1) activity; however, it has much lower debrisoquine 4-hydroxylase (CYP2D) and lauric acid hydroxylase activities (CYP4A) (Sivapathasundaram et al. 2001). Immunoblot analysis employing antibodies to human CYP2D6 detects a single, poorly expressed immunoreacting band in bovine microsomes. The bovine *CYP2D14* gene is located on chromosome 5 and contains 10 exons (Tsuneoka et al. 1992; Zimin et al. 2009). The *CYP2D14* gene is conserved in chimpanzee, dog, mouse, and rat. Debrisoquine 4-hydroxylase and bufuralol 1′-hydroxylase activities are detected in bovine liver as well as other tissues including kidney and brain. However, the debrisoquine 4-monooxygenase activities catalyzed by bovine and human CYP2D are one order less than those by rat and rabbit. Quinidine also acts as an inhibitor to these reactions, indicating a similar bovine CYP2D enzyme to the human CYP2D6. There are also some CYP2D transcripts and activities in the bovine adrenal cortex and medulla (Tsuneoka et al. 1992). cDNA cloning from bovine liver results in four CYP2D enzymes with only 7% difference in nucleotide sequence. Two cDNAs that are completely sequenced show 98% amino acid identity to each other and share 80% and 68% identity with human CYP2D6 and rat Cyp2d1, respectively. One of the sequence is designed CYP2D14 (Tsuneoka et al. 1992). However, the mRNA does not encode a functional protein because of a mutation that abolishes splicing at the *CYP2D14* primary transcript. It is still unclear whether CYP2D14 or a related CYP2D is involved in debrisoquine and bufuralol hydroxylase activities.

2.5 DOG CYP2D15

Dog is an important, widely used species within the pharmaceutical industry for assessing the metabolism, pharmacokinetics, safety, and efficacy of drugs and drug candidates in discovery and development. In the pharmaceutical industry, dogs are commonly used as a nonrodent species for toxicological and pharmacological studies of drug candidates. In addition, dog pharmacokinetic data along with in vitro metabolic data can be very useful for the prediction of human

in vivo pharmacokinetics and interpretation of toxicity and efficacy results in both species. However, significant interindividual difference of drug concentration in plasma is frequently observed in dogs after drug administration (Azuma et al. 2002; Paulson et al. 1999; Watanabe et al. 2004). This variability of pharmacokinetics often affects the results of toxicological and pharmacological studies. Therefore, it is important for efficient and reliable preclinical studies to clarify the mechanism of pharmacokinetic variability and to remove the factors affecting it.

In comparison to rat and human CYPs, however, relatively little is known about specific dog P450 enzymes. Dog has at least 48 *CYP* genes, and many of these genes have been isolated and cloned in the dog including CYP1A1 (Uchida et al. 1990), 1A2 (Scherr et al. 2011; Uchida et al. 1990), 1B1 (Nishida et al. 2013), 2A13 (Okamatsu et al. 2014), 2B11 (Duignan et al. 1987; Graves et al. 1990), 2C21 (Uchida et al. 1990), 2C41 (Blaisdell et al. 1998), 2D15 (Nakamura et al. 1995; Sakamoto et al. 1995), 2E1 (Lankford et al. 2000), 3A12 (Ciaccio et al. 1991), 3A26 (Fraser et al. 1997), and 26B1 (Kasimanickam and Kasimanickam 2013, 2014). The contribution of these CYPs to the interindividual variability of pharmacokinetics in dogs is unknown. A CYP4A protein has yet to be identified in dog liver (Adas et al. 1999). DUT-1, purified from liver microsomes of untreated male beagle dogs, catalyzes the 12-hydroxylation of lauric acid, but the *N*-terminal sequence of this protein is different from any other P450 characterized to date (Shiraga et al. 1994).

Dog liver microsomes contain CYP2B11 (also called PBD-2), a constitutively expressed and phenobarbital-inducible enzyme with substantial metabolic activity toward 2,4,5,2′,4′,5′-hexachlorobiphenyl, an organochlorine pesticide (Ariyoshi et al. 1992; Duignan et al. 1987, 1988), whereas the related P450 2B forms in rat (Cyp2b1) and rabbit (2B5) are unable to metabolize the compound to any significant degree (John et al. 1994). Human CYP2B6 can metabolize this compound (Ariyoshi et al. 1995). Like rat Cyp2b1, dog CYP2B11 (PDB-2) catalyzes both the 16α- and 16β-hydroxylation of testosterone (Duignan et al. 1988), but CYP2B11 preferentially catalyzes the 16α-hydroxylation of testosterone at 13 to 15 times the rate of testosterone 16β-hydroxylation (Ohmori et al. 1993). Another testosterone—16α-hydroxylase—has been purified from dog liver and identified as a member of CYP2C21 (also called DPB-3) (Uchida et al. 1990). Heterologously expressed dog CYP2B11 also shows moderate progesterone 21-hydroxylase activity (Born et al. 1995). Site-directed mutants of 2B11 expressed in *E. coli* reveal that substitution of Ile with Val at position 363 converts 2B11 into a highly active and specific progesterone, 16α-hydroxylase. Mutants Val114Ile, Asp290Ile, and Ile365Phe exhibit decreased progesterone 21- and 16α-hydroxylase activities, in accordance with decreases in androstenedione hydroxylase activities (Born et al. 1995). In contrast, replacement of Ile365 with Val or Leu results in much greater changes in progesterone than androstenedione hydroxylation. Substitution of Val114 or Asp290 with Ile decreased the product yields from 2,2′,3,3′,6,6′-hexachlorobiphenyl, and replacement of Leu363 with Val dramatically alters the profile of metabolites (Waller et al. 1999), indicating the importance of residues Val114, Asp290, and Leu363 in CYP2B11. Whereas rat CYP2B1 catalyzes the 16α- and 16β-hydroxylation of testosterone at roughly equal rates, dog CYP2B11 preferentially catalyzes the 16α-hydroxylation of testosterone at much higher rate than testosterone 16β-hydroxylation (Ohmori et al. 1993; Shou et al. 2003).

CYP2E1 cDNA has been cloned from a beagle dog liver library followed by characterization and expression of the encoded protein (Lankford et al. 2000). Interestingly, the amino acid sequence of dog CYP2E1 exhibits 77% identity to the human ortholog, which is slightly higher than the identity to the rodent or rabbit sequence (75%–76%). In rodents, CYP2E1 catalyzes the hydroxylation of 4-nitrophenol (Koop 1986), while human CYP2E1 6-hydroxylases chlorzoxazone. Characterization of the expressed CYP2E1 protein indicated that dog CYP2E1 has a lower affinity for chlorzoxazone than does human CYP2E1 (Lankford et al. 2000).

Dog liver is thought to express multiple forms of CYP3A, as has been shown in rat and human. PBD-1, a CYP3A enzyme, is purified from phenobarbital-treated dog liver (Ciaccio and Halpert 1989). Molecular and immunochemical analyses indicate the presence of at least one other CYP3A enzyme in dog liver (Ciaccio and Halpert 1989; Ciaccio et al. 1991). In contrast to rats, there are no significant sex differences in CYP3A activity in dog liver microsomes. Like the corresponding rat enzyme, the dog CYP3A12 catalyzes the 6β-hydroxylation of testosterone (Ciaccio and Halpert 1989; Shou et al. 2003). In addition, CYP3A12 catalyzes the 16β-hydroxylation of testosterone, which is also catalyzed in part by CYP2B11. A cDNA encoding a protein exhibiting 95.6% amino acid identity with CYP3A12 is isolated from phenobarbital-induced dogs (Fraser et al. 1997). This enzyme, called CYP3A26, is not as prominent as 3A12 in hydroxylating steroids.

2.5.1 CLONING AND PURIFICATION OF DOG CYP2D15

Nakamura et al. (1995) have purified a protein of approximately 49 kDa in hepatic microsomes from dogs, indicative of the presence of the CYP2D subfamily. However, the purified canine CYP2D does not catalyze propranolol 7-hydroxylation, which is almost exclusively mediated by Cyp2d in rats. Sakamoto et al. (1995) have also purified a CYP2D member from dog liver, whose amino acid sequence shares high similarity with the reported CYP2Ds (with 2D6, 74.6%; 2D14, 75.4%; 2D1, 65.4%; and 2D9, 63.6%). Moreover, the purified and cDNA-expressed CYP2D in COS-7 cells shows catalytic activity toward desipramine, metoprolol, and dextromethorphan (Sakamoto et al. 1995).

2.5.2 TISSUE DISTRIBUTION OF DOG CYP2D15

CYP2D15 is the CYP2D6 ortholog in dog (Roussel et al. 1998). Similar to human, canine CYP2D15 is expressed in the liver, with detectable levels in several other tissues. Three different CYP2D15 cDNA clones are obtained from dog

liver library. Two clones correspond to variant full-length CYP2D15 cDNAs (termed CYP2D15 WT2 and CYP2D15 V1); the third is identified as a splicing variant missing exon 3 (termed CYP2D15 V2). Recombinant baculoviruses are constructed containing full-length cDNAs and used to express CYP2D15 WT2 and CYP2D15 V1 in *Spodoptera frugiperda* (*Sf*9) cells with expression levels of up to 0.14 nmol/mg cell protein. As with human CYP2D6, the recombinant CYP2D15 enzymes exhibit bufuralol 1'-hydroxylase and dextromethorphan *O*-demethylase activities when coexpressed with rabbit NADPH:P450 oxidoreductase (Roussel et al. 1998). For bufuralol 1'-hydroxylase, apparent K_m values are 4.9, 3.7, and 2.5 µM for dog liver microsomes, CYP2D15 WT2, and the variant CYP2D15 V1, respectively. For dextromethorphan *O*-demethylase, apparent K_m values are 0.6, 0.6, and 2.0 µM for dog liver microsomes, CYP2D15 WT2, and the variant CYP2D15 V1, respectively (Roussel et al. 1998). The human CYP2D6-specific inhibitor quinidine and the rat CYP2D1-specific inhibitor quinine inhibit bufuralol 1'-hydroxylase activity for dog liver microsomes, CYP2D15 WT2, and the CYP2D15 V1 variant with nearly equal potency. Thus, dogs not only express a CYP2D ortholog that possesses enzymatic activities similar to human CYP2D6 but also are affected by the inhibitors quinine and quinidine in a manner closer to that of rat CYP2D1.

2.5.3 SUBSTRATE SPECIFICITY OF DOG CYP2D15

Infection of *Sf*9 insect cells with a recombinant dog CYP2D15 virus results in the expression of a protein that cross-reacted with a polyclonal antibody against a dog CYP2D15-specific peptide. The difference spectrum of the CO complex of reduced P450 of the infected cell microsomes has a maximal absorbance at 449 nm. The specific content of P450 is calculated to be 0.56 nmol/mg of *Sf*9 cell microsomal protein. Although the expressed dog CYP2D15 shows high catalytic activity for the hydroxylations of bunitrolol and imipramine at low substrate concentration (10 µM), the catalytic activity for that of debrisoquine (50 µM) is extremely low as compared with that of CYP2D from other species (Tasaki et al. 1998). Dog liver microsomes also show bunitrolol and imipramine hydroxylase activities, but not debrisoquine hydroxylase activity at the same substrate concentrations. In addition, the expressed CYP2D shows high catalytic activity for imipramine *N*-demethylation. Thus, the expressed dog CYP2D15 engages in high catalytic activity and has a unique substrate specificity from other CYP2D subfamilies. Western blot analysis suggests that the dog CYP2D15 contents are less than 4% of the total liver P450 content, assuming that 100% of expressed CYP2D15 incorporate heme.

To clarify the mechanism of the species difference in the metabolism of bisoprolol enantiomers, in vitro metabolic studies are performed using dog liver microsomes and human CYPs (Horikiri et al. 1998). The *O*-deisopropylation of bisoprolol enantiomers shows biphasic kinetics in dog liver microsomes. The intrinsic clearance (V_{max}/K_m) for *O*-deisopropylation of *R*-bisoprolol is higher than that of the S-isomer in both high-affinity and low-affinity components. The R/S ratios of the intrinsic clearance in high- and low-affinity components are 1.34 and 1.65, respectively. The inhibition studies in dog liver microsomes using CYP isoform-selective inhibitors indicate that the O-deisopropylation of both bisoprolol enantiomers is mediated via the CYP2D and 3A subfamilies, suggesting that high-affinity oxidation is dependent on CYP2D. The kinds of CYP subfamilies in dogs, which contribute to the metabolism of bisoprolol enantiomers, are the same as those in humans. The intrinsic clearance for O-deisopropylation of R(+)-bisoprolol by human recombinant CYP2D6 is also different from that of S-enantiomers (R/S: 1.50). However, unlike the dog microsomes, the intrinsic clearance by the human recombinant CYP3A4 does not show a stereoselective difference. Therefore, the species difference in the R/S ratio of metabolic clearance for the oxidation of bisoprolol enantiomers (dog > human) is mainly attributed to the species difference in the stereoselectivity of one of the cytochrome P450 subfamilies (CYP3A).

The pharmacokinetics of celecoxib, a selective COX-2 inhibitor, is characterized in beagle dogs (Paulson et al. 1999). Celecoxib is extensively metabolized by dogs to a hydroxymethyl metabolite with subsequent oxidization to the carboxylic acid analog. There are at least two populations of dogs, distinguished by their capacity to eliminate celecoxib from plasma at either a fast or a slow rate after intravenous administration. Within a population of 242 animals, 45.0% are of the EM phenotype, 53.5% are of the PM phenotype, and 1.65% could not be adequately characterized. The mean plasma elimination half-life and clearance of celecoxib are 1.72 h and 18.2 ml/min/kg for EM dogs and 5.18 h and 7.15 ml/min/kg for PM dogs (Paulson et al. 1999). Hepatic microsomes from EM dogs metabolize celecoxib at a higher rate than microsomes from PM dogs. The cDNA for canine CYP enzymes 2B11, 2C21, 2D15, and 3A12 are cloned and expressed in Sf9 insect cells. Three new variants of CYP2D15 as well as a novel variant of CYP3A12 are identified. Canine recombinant CYP2D15 and its variants, but not CYP2B11, 2C21, and 3A12, readily metabolize celecoxib. Quinidine, a specific CYP2D inhibitor, prevents celecoxib metabolism in dog hepatic microsomes, providing evidence of a predominant role for the CYP2D subfamily in canine celecoxib metabolism. However, the lack of a correlation between celecoxib and bufuralol metabolism in hepatic EM or PM microsomes indicates that other CYP subfamilies besides CYP2D may contribute to the polymorphism in canine celecoxib metabolism.

The metabolism of olanexidine [1-(3,4-dichlorobenzyl)-5-octylbiguanide], a new potent biguanide antiseptic, is investigated in dog liver microsomes to characterize the enzyme(s) catalyzing the biotransformation of olanexidine to C–C bond cleavage metabolites (Umehara et al. 2000). Olanexidine is initially biotransformed to monohydroxylated metabolite 2-octanol (DM-215). DM-215 is later oxidized to diol derivatives threo-2,3-octandiol (DM-221) and erythro-2,3-octandiol (DM-222). Diols are further biotransformed to a ketol derivative and C–C bond cleavage metabolite (DM-210, hexanoic acid derivative), an in vivo end product, in the incubation with

dog liver microsomes. The formation of DM-215, DM-221, DM-222, and DM-210 follows Michaelis–Menten kinetics. The K_m value for the formation of DM-210 is 2.42 μM and that for the oxidation of DM-222 is 2.48 μM (Umehara et al. 2000). The intrinsic clearance (V_{max}/K_m) of the C–C bond cleavage reactions is essentially the same with either DM-221 or DM-222 as substrate. These oxidative reactions are significantly inhibited by quinidine, indicating that the metabolic C–C bond cleavage of the octyl side chain of olanexidine is mediated via the CYP2D subfamily in dog liver microsomes. This aliphatic C–C bond cleavage by cytochrome P450s may play an important role in the metabolism of other drugs or endogenous compounds possessing aliphatic chains.

Shou et al. (2003) have expressed seven dog CYPs including CYP1A1, 2B11, 2C21, 2C41, 2D15, 3A12, and 3A26 in baculovirus-Sf21 insect cells and examined their catalytic activities for different probe substrates in comparison with dog liver microsomes. Dog CYP1A1 exhibits high 7-ethoxyresorufin O-deethylase activity to form resorufin with a K_m of 1 μM. CYP3A12 shows high activities for C3-hydroxylation of diazepam and nordiazepam, while CYP2B11 catalyzes almost exclusively the N_1-demethylation of diazepam and temazepam (Shou et al. 2003). Dog liver microsomes demonstrate both N_1-demethylase and C_3-hydroxylase activities for diazepam, with N_1-demethylation activity being threefold greater than that for C3-hydroxylation. CYP2D15 catalyzes exclusively the O-demethylation of dextromethorphan to dextrorphan, whereas 3A12 mediates dextromethorphan N-demethylation to form 3-methoxymorphinan (Shou et al. 2003). In addition, CYP1A1 and 3A26 exhibit selectivity in the formation of dextrorphan and 3-methoxymorphinan, respectively, although their activities are much lower than those of CYP2D15 and 3A12. In contrast to CYP3A12 with a high K_m (>500 μM), the CYP2D15-dependent O-demethylation of dextromethorphan has a much lower K_m (0.7 μM), similar to that seen in dog liver microsomes (K_m = 2.3 μM). The activity of dextromethorphan O-demethylase in dog liver microsomes is similar to the previous report (Chauret et al. 1997). The kinetic parameters for CYP2D15 are also reasonably comparable to those for CYP2D6 and human liver microsomes. CYP2D15 also metabolizes bufuralol via 1'-hydroxylation, with a K_m of 3.9 μM, consistent with that obtained with dog liver microsomes (K_m = 3.5 μM) (Shou et al. 2003). CYP3A12 is shown to primarily oxidize testosterone at 16α-, 2α/2β-, and 6β-positions. Selectivity of CYP3A12 is observed toward testosterone 6β-hydroxylation (K_m = 83 μM) and 2α/2β-hydroxylation (K_m = 154 μM). However, the 16α-hydroxylation of testosterone is also catalyzed by CYP2C21 with a K_m of 6.4 μM. CYP3A12 appears to be the main contributor to 2α/2β- and 6β-hydroxylation in dog liver microsomes, whereas CYP2C21 plays a major role in 16α-hydroxylation of testosterone. Substrate specificity characteristics of CYP3A isoforms in rat and human are composed of testosterone 6β-hydroxylation, diazepam 3-hydroxylation (Bertilsson et al. 1990; Ono et al. 1996; Yang et al. 1998), and dextromethorphan N-demethylation (Dayer et al. 1989; Jacqz-Aigrain et al. 1993; Kerry et al. 1994; Tenneze et al. 1999). CYP2C21 catalyzes diclofenac 4'-hydroxylation (a marker for human CYP2C9) with a K_m of 94 µM, although some activity is also observed with CYP2B11 with fivefold lower activity than 2C21 (Shou et al. 2003). The K_m value for CYP2C21 is similar to that of dog liver microsomes (K_m = 110 µM). The K_m for CYP2B11 is 5.3 µM. None of the enzymes selectively metabolize S-mephenytoin 4'-hydroxylation, a marker for human CYP2C19. Moreover, the one-carbon fragment of the octyl side chain of olanexidine could be removed by the oxidative C–C bond cleavage with the possible involvement of CYPs such as the CYP2D subfamily (Umehara et al. 2004). This oxidative C–C bond cleavage reaction by cytochrome P450s could play an important role in the removal of one-carbon fragment of other drugs or endogenous compounds containing aliphatic chains.

In dogs treated with oral dexamethasone at 2.5 and 7.5 mg for 5 days, V_{max} for bufuralol 1'-hydroxylation is decreased (Zhang et al. 2006). V_{max} for midazolam 4-hydroxylation is significantly decreased by treatment with dexamethasone at 2.5 and 7.5 mg in dogs, although K_m is not affected. In rats, V_{max} for midazolam 4-hydroxylation is significantly decreased by treatment with dexamethasone at 0.75 and 6 mg/kg but significantly increased at 48 mg/kg. Other reactions are not affected by dexamethasone treatment. These results indicate that dexamethasone downregulates CYP3A when administered at clinically relevant doses to dogs.

CYP2D-related drug metabolism in liver microsomes from animals of the Canoidea superfamily, that is, mink (*Mustela vison*), bears (*Ursus arctos*), foxes (*Vulpes vulpes*) and dogs, is compared (Ishizuka et al. 2006). Propranolol, bunitrolol, and imipramine, which are typically substrates of CYP2D subfamilies, are used in the experiment. All the animals of the Canoidea superfamily that are tested lacked the ability to catalyze 7-hydroxylation of propranolol, which is one of the major metabolic pathways in rats. Stereoselectivity of propranolol metabolism is toward S-propranolol in all the reactions of the animals tested with the exception of mink, which showed a selective tendency toward R-propranolol in N-dealkylation. As far as metabolic patterns of R- and S-propranolol are concerned, bears, foxes, and dogs are alike, but minks are somewhat different. Liver microsomes from mink showed, among the animals of the Canoidea superfamily, the lowest propranolol hydroxylase activity at 4- and 5-positions and imipramine 2-hydroxylation and N-demethylation activities.

In summary, CYP3A12, 2B11, 2D15, 2C21, and 1A1 have reaction phenotypes similar to those of corresponding rat and human P450s, although the kinetic nature of some of the enzymes has a wide range of variance. These isoform-selective reactions can be used to assess the presence of individual P450 activities in dog tissues, to evaluate drug candidates as inhibitors and inducers of P450s in the metabolism of probe substrates, and to predict drug–drug interactions from dogs to humans.

2.5.4 INDUCTION OF DOG CYP2D15

The induction of various dog CYPs has been reported. For example, phenobarbital and rifampicin induce dog CYP2B11 and 3A12, respectively, in vitro and in vivo (Ciaccio and Halpert 1989; Duignan et al. 1988; Graham et al. 2002, 2006; Klekotka

and Halpert 1995; Nishibe et al. 1998). Treatment of male beagle dog hepatocyte cultures with β-naphthoflavone or 3-methylcholanthrene results in up to a 75-fold increase in microsomal CYP1A1/2-mediated 7-ethoxyresorufin O-dealkylase activity, whereas in vivo treatment of male and female beagle dogs with β-naphthoflavone followed by ex vivo analysis causes up to a 24-fold increase (Graham et al. 2002). Phenobarbital causes a 13-fold increase in CYP2B11-mediated 7-benzyloxyresorufin O-dealkylase activity in vitro and up to a 9.9-fold increase in vivo (Graham et al. 2002). 7-Benzyloxyresorufin O-dealkylation has been shown to be a specific marker for dog CYP2B11 (Klekotka and Halpert 1995). Rifampin treatment causes a 13-fold induction of testosterone 6β-hydroxylase (CYP3A12) activity in vitro and up to a 4.5-fold increase in vivo. Another study by Graham et al. (2006) have reported a 12,700-fold and 206-fold increase in CYP1A1 and 1A2 mRNA levels, respectively, after 36 h treatment with β-naphthoflavone in primary cultures of canine hepatocytes. Phenobarbital treatment, but not rifampin treatment, induces CYP2B11 mRNA 149-fold after 48 h incubation. CYP3A12 and 3A26 mRNA levels are increased 35-fold and 18-fold, respectively, after 72 h treatment with phenobarbital, and rifampin treatment increases 35-fold and 18-fold, respectively (Graham et al. 2006). Treatment of dogs in vivo or dog hepatocytes in vitro with isoniazid has only a slight effect on chlorzoxazone 6-hydroxylase activity and immunoreactive CYP2E1 levels (Graham et al. 2002; Jayyosi et al. 1996). These studies indicate that the responses of the dog CYP1A, 2B, and 3A to these conventional inducers resemble those in rats and humans.

Grapefruit juice significantly increases the plasma concentrations of drugs that are substrates for CYP3A4 such as felodipine, triazolam, midazolam, diazepam, terfenadine, cyclosporine, and nifedipine (Bailey and Dresser 2004; Bailey et al. 2000, 2003). The principal components of grapefruit juice have been identified as the furanocoumarin bergamottin and its metabolite 6',7'-dihydroxybergamottin, as well as the flavonoids naringenin, naringin, quercetin, and kaempferol. Bergamottin is a potent inhibitor of human CYP3A4. In humans, grapefruit juice exposure decreases concentrations of intestinal CYP3A4 but does not affect CYP3A5, 1A1, 2D6 protein, or 3A4 mRNA (Lown et al. 1997). In contrast to the results described above, long-term grapefruit juice treatment in rats has been shown to increase nifedipine clearance (Mohri et al. 2000). In mice, a single dose of grapefruit juice inhibits hepatic oxidative enzyme activity, whereas multiple dosing increases activity (Dakovic-Svajcer et al. 1999). The results of these rodent studies imply that grapefruit juice is both an inhibitor and an inducer. However, this compound is both an inhibitor and an inducer of P450 enzymes in dogs. Bergamottin predosing increases the plasma levels of diazepam in beagle dogs (Sahi et al. 2002). In dog hepatic microsomes, bergamottin treatment for 10 days reduces the activity of CYP3A12 by 50% and 1A1/2 by 75%. Tolbutamide hydroxylase activity does not change, and CYP2B11 activity is moderately induced. In jejunal microsomes, CYP3A12 activity doubles with bergamottin treatment. CYP2B11 and 1A1/2 activity and tolbutamide hydroxylation are not detected.

2.5.5 Inhibitors of Dog CYP2D15

Ketoconazole, clomipramine, and loperamide are potent CYP2D15 inhibitors (Aidasani et al. 2008). Additional inhibitors belong to the antiemetic, antimitotic, and anxiolytic therapeutic classes. In addition, quinidine inhibits CYP2D15 in dog liver microsomes.

2.6 GUINEA PIG Cyp2d16

2.6.1 Tissue Distribution of Guinea Pig Cyp2d16

cDNA library from a guinea pig adrenal is cloned and sequenced a P450. The enzyme is designated Cyp2d16. Studies have shown that the primary expression site is in the inner zone of the adrenal cortex. The mRNA level is significantly higher than in guinea pig liver and kidney. It is also higher in the inner zones than the outer zones of the adrenal cortex (Jiang et al. 1996a).

2.6.2 Substrate Specificity of Guinea Pig Cyp2d16

Studies are done to characterize a guinea pig adrenal microsomal P450 that had been linked with xenobiotic metabolism in the inner zone of the gland (Jiang et al. 1995). N-terminal amino acid sequencing of the isolated protein reveals homology with members of the Cyp2d subfamily. A human CYPD2D6 cDNA probe is used to screen a guinea pig adrenal cDNA library and a full-length clone is obtained having an open-reading frame encoding a 500–amino acid protein. The sequence is found to be highly homologous with all members of the CYP2D subfamily and is designated Cyp2d16. The N-terminal sequence of 38 amino acids obtained from the protein microsequencing is identical to that deduced from the nucleotide sequence of the cloned Cyp2d16. Northern blot analysis has confirmed that Cyp2d16 is expressed at high levels in the inner zone of the guinea pig adrenal cortex. The results suggest that Cyp2d16 may account, at least in part, for the high rates of xenobiotic metabolism in the guinea pig adrenal (Jiang et al. 1995, 1996a; Yuan et al. 1997).

2.6.3 Regulation of Guinea Pig Cyp2d16

The intensity of adrenal Cyp2d16 staining is not age or gender dependent (Yuan et al. 2001). The rates of metabolism of bufuralol, a CYP2D-selective substrate, by adrenal microsomal preparations, generally correlate with the amount of the zona reticularis (and Cyp2d16) in the gland (Yuan et al. 2001). The proportion of each adrenal gland is comprised by the zona reticularis and, thus, expressing Cyp2d16 increases with aging in both sexes and is greater in males than in females. However, the rate of bufuralol metabolism declined in sexually mature females (14 weeks) from the levels found in prepubertal females (7 weeks) and then increases significantly in retired breeders (30 weeks), suggesting an inhibitory effect of estrogens on enzyme activity (Yuan et al. 2001). The results indicate that the age and gender differences in adrenal

Cyp2d16 content are largely determined by differences in the size of the zona reticularis rather than the concentrations of Cyp2d16 within cells of the zona reticularis. In addition, both adrenal enzyme activities and Cyp2d16 concentrations are similarly regulated by several physiological variables, including adrenocorticotropin, the major hormonal modulator of the adrenal cortex (Jiang et al. 1996b; Yuan et al. 1998). Metabolism of bufuralol, a Cyp2d-selective substrate, by guinea pig adrenal microsomes further implicates Cyp2d16 in adrenal xenobiotic metabolism (Jiang et al. 1996a,b).

Steroid metabolism (testosterone 6β-hydroxylation) by mouse Cyp2ds has previously been demonstrated (Wong et al. 1989). High cortisol 2α- and 6β-hydroxylase activities are found in adrenal microsomes from guinea pigs that express high levels of Cyp2d16 (Huang et al. 1996, 1998). Because the pattern of adrenal blood flow directs hormones produced in the zona fasciculata through the zona reticularis before releasing into the general circulation (Vinson et al. 1985), Cyp2d16 may have a role in the modulation of adrenal hormone secretion. Accordingly, downregulation of Cyp2d16 by adrenocorticotropin (Jiang et al. 1996b) could serve as a mechanism to increase adrenal steroid secretion. Cyp2d16 inhibition is studied using ACTH. Guinea pigs are treated with ACTH for 1, 3, or 7 days. In addition, some animals received ACTH for 7 days and are then untreated for an additional 3 or 7 days to test for reversibility of ACTH actions. ACTH treatment caused a time-dependent decrease in the rates of adrenal microsomal bufuralol metabolism, a CYP2D-catalyzed reaction. Hepatic bufuralol metabolism is unaffected by ACTH (Colby et al. 2003). Adrenal enzyme activity is significantly reduced by ACTH within 1 day and decreased by 80% after 7 days. Western blotting and in situ hybridization analyses revealed corresponding declines in adrenal Cyp2d16 protein and mRNA concentrations. Nuclear runoff assays have indicated that ACTH treatment inhibits Cyp2d16 expression at the transcriptional level. Adrenal 17 α-hydroxylase activities are increased by ACTH treatment, but CYP17 protein expression levels are not affected (Colby et al. 2003). After cessation of ACTH administration, the rates of adrenal bufuralol metabolism and Cyp2d16 protein and mRNA concentrations return to control levels within 7 days. The results demonstrate that ACTH has a relatively rapid and reversible effect to inhibit adrenal Cyp2d16 transcription, thereby decreasing adrenal xenobiotic metabolism. Thus, the actions of ACTH on Cyp2d16 expression are opposite to those on other adrenal P450 isozymes, indicating unique regulatory mechanisms.

Guinea pig adrenal microsomes convert bufuralol to 1′-hydroxybufuralol as the major metabolite and smaller amounts of a compound identified as 6-hydroxybufuralol (Colby et al. 2001). In contrast, 6-hydroxybufuralol is the major product formed by hepatic microsomal preparations. The apparent K_m values are similar for 1′-hydroxybufuralol and 6-hydroxybufuralol production in each tissue. Quinidine, a selective CYP2D inhibitor, decreases the production of both bufuralol metabolites equally in the liver and adrenal microsomes. Cortisol also causes equivalent decrease in the rates of 1′-hydroxybufuralol and 6-hydroxybufuralol formation by adrenal microsomes but has no effect on hepatic bufuralol metabolism (Colby et al. 2001). Although both bufuralol metabolites may be produced by CYP2D16, unknown factors appear to affect some differences in the catalytic characteristics of bufuralol metabolism in the adrenal and liver. The large amount of 6-hydroxybufuralol produced distinguishes bufuralol metabolism in guinea pigs from that in other species previously studied. There is an apparent bimodal distribution of adrenal xenobiotic-metabolizing activities in guinea pigs (Jiang et al. 1996a). Adrenal Cyp2d16 expression levels correlated with bufuralol 1'-hydroxylase and benzphetamine N-demethylase activities (Colby et al. 2000). The bimodal expression of adrenal CYP2D16 is at least partly responsible for the variability in adrenal xenobiotic metabolism.

2.7 MACAQUE CYP2D17, 2D29, 2D42, AND 2D44

Macaques include cynomolgus monkey (*Macaca fascicularis*), rhesus monkey (*Macaca mulatta*), and Japanese monkey (*Macaca fuscata*). The former two are used routinely as primate models in drug discovery and preclinical studies. Rhesus monkey CYP2D42 cDNA, nearly 98% homologous to cynomolgus monkey CYP2D17 cDNA, has been isolated, and its sequence can be found in GenBank. Deduced amino acid sequences of these CYP2D cDNAs share high sequence identity (93%–96%) with human CYP2D6. Cynomolgus monkey CYP2D17 and Japanese monkey CYP2D29 metabolize typical human CYP2D6 substrates; CYP2D17 metabolizes bufuralol and dextromethorphan, whereas CYP2D29 metabolizes bufuralol and debrisoquine (Hichiya et al. 2002; Mankowski et al. 1999). In addition, quinidine inhibits both cynomolgus monkey CYP2D17 and Japanese monkey 2D29.

2.7.1 CYNOMOLGUS MONKEY CYP2D17 AND 2D44

Cynomolgus monkey is commonly used in drug safety evaluation and biotransformation studies by the pharmaceutical industry. Western blot analysis reveals two CYP2D protein bands in blots of cynomolgus monkey liver microsomes. Mankowski et al. (1999) have first cloned CYP2D17 cDNA from cynomolgus monkey liver mRNA using oligonucleotide primers based on the human CYP2D6 sequence. The full-length CYP2D17 cDNA encodes a 497–amino acid protein that is 93% identical to human CYP2D6 and 90% identical to marmoset 2D19 (Mankowski et al. 1999).

In the study by Mankowski et al. (1999), the CYP2D17 cDNA is cloned into a baculovirus expression vector, and microsomes prepared from CYP2D17-infected insect cells are used to determine the catalytic properties of the recombinant enzyme. The 3'-fragment of the monkey *CYP2D17* is identical to the human *CYP2D6*. However, there are differences in codons 2 and 4 of the 5' coding region of CYP2D17 compared to human 2D6 nucleotide sequence. Monkey *CYP2D17* has A and T for nucleotides 5 and 12, respectively, whereas *CYP2D6* includes G and A in its sequence for the same nucleotides (Mankowski et al. 1999). Although the substrate

recognition sites (SRSs) in the monkey CYP2D17 contain different residues from human CYP2D6, the amino acid residues that interact with the CYP2D6 substrates and inhibitors are conserved, including Phe120, Glu216, Asp301, and Phe481. The recombinant CYP2D17 results are compared to data generated with monkey liver microsomes, human liver microsomes, and recombinant CYP2D6 and have demonstrated catalytic similarity and inhibitor selectivity using probe substrates and inhibitors. Recombinant CYP2D17 catalyzes the oxidation of bufuralol to 1'-hydroxybufuralol and dextromethorphan to dextrorphan. These reactions have been shown to be catalyzed by human CYP2D6. The apparent K_m value for bufuralol 1'-hydroxylation is 1 µM in insect cell microsomes. Similarly, using dextromethorphan as a substrate, the K_m value is 0.8 µM in insect cell microsomes. Moreover, both of these reactions are more strongly inhibited by quinidine (K_i = 0.05 µM) than by quinine (K_i = 0.67 µM) (Mankowski et al. 1999). A more complete understanding of the substrate specificities and activities of monkey CYPs will be advantageous in delineating species differences in metabolite profiles and metabolic activation of new chemical entities in the pharmaceutical industry.

Uno et al. (2014) have sequenced *CYP2D17* from 87 cynomolgus and 40 rhesus macaques and found a total of 36 nonsynonymous variants, among which 5 are located in SRSs. Twenty-two variants are unique to cynomolgus macaques, of which 11 and 9 are found only in Indochinese and Indonesian cynomolgus macaques, respectively. Eight variants are unique to rhesus macaques. The functional characterization shows that the Ser188Tyr and Val227Ile mutants heterologously expressed in *E. coli* do not show substantial differences in bufuralol 1'-hydroxylation compared with the wild type. However, the Ile297Met and Asn337Asp mutants exhibit significantly increased bufuralol 1'-hydroxylation and dextromethorphan O-demethylation activities (Uno et al. 2014).

Another 2D member, CYP2D44, has been identified in cynomolgus monkey liver (Uno et al. 2010). *CYP2D17* and *2D44* form a gene cluster in the genome, similar to human CYP2Ds. The CYP2D44 cDNA codes for a protein with 497 amino acids sharing high sequence identity (87%–93%) with other primate CYP2Ds. The encoded protein shows primary sequence structures characteristic of P450 proteins, such as six SRSs and a heme-binding region. A phylogenetic analysis using the CYP2D amino acid sequences from the human, macaque, dog, and rat indicate that cynomolgus monkey CYP2D44 and other macaque CYP2Ds are most closely clustered with human CYP2D6 (Uno et al. 2010). These results suggest the evolutionary closeness of CYP2D44 to other primate CYP2Ds, compared with CYP2Ds of dog and rat. The *CYP2D44* gene contains nine exons, and the sizes of the exons and introns range from 142 to 188 bp and from 88 to 1620 bp, respectively.

CYP2D44 mRNA is primarily expressed in the liver, similar to CYP2D17 mRNA. CYP2D44 expressed in *E. coli* metabolizes human CYP2D6 substrates bufuralol and dextromethorphan (bufuralol 1'-hydroxylation and dextromethorphan

O-demethylation) but to a lesser extent than CYP2D17 (Uno et al. 2010). Kinetic analysis of dextromethorphan metabolism indicates that the apparent K_m and V_{max} of CYP2D17- and 2D44-catalyzed *O*-demethylation are similar, and the V_{max} values of CYP2D17- and 2D44-catalyzed *N*-demethylation (which human CYP2D6 catalyzes much less effectively) are similar, but the apparent K_m for the CYP2D44-mediated reaction is higher. The CYP2D proteins are expressed in rhesus monkey liver. Similar to CYP2D6, CYP2D44 copy number varies among the eight cynomolgus monkeys and four rhesus monkeys (Uno et al. 2010). These results indicate that CYP2D17 and 2D44 have similar functional characteristics to human CYP2D6 but differs in dextromethorphan *N*-demethylation.

2.7.2 JAPANESE MONKEY CYP2D29

cDNA from a female Japanese monkey liver is cloned and expressed in the *Saccharomyces cerevisiae* AH-22 strain (Hichiya et al. 2002). The nucleotide sequence of CYP2D29 has 97%, 93%, and 89% identity to that of human 2D6, cynomolgus monkey 2D17, and marmoset monkey 2D19, respectively. CYP2D29 also shows 96% and 91% identity in the multiple amino acid sequence with human 2D6 and cynomolgus monkey 2D17, respectively. CYP2D29 is closer to human CYP2D6 than 2D17 since the amino acid sequences of the six SRSs are completely identical between CYP2D29 and 2D6 (Hichiya et al. 2002). The six SRSs are responsible for forming the active-site cavity of the CYP2D6 where the substrates are oxidized.

CYP2D29 catalyzes the debrisoquine 4-hydroxylase and bufuralol 1′-hydroxylase (Hichiya et al. 2002). The enzymatic activity of the Japanese liver microsomal fraction in catalyzing debrisoquine 4-hydroxylase is 16 times higher than that of the human liver microsomal fraction of CYP2D6. However, the recombinant CYP2D29 shows similar activity as the recombinant human CYP2D6. Similarly, both the Japanese liver microsomal fraction and recombinant CYP2D29 show higher activity than the human liver microsomes. The K_m values are similar between the Japanese monkey and human liver microsomes (Hichiya et al. 2002). However, the K_m value of the recombinant CYP2D29 is one-twentieth that of Japanese monkey liver microsomes (160.6 μM) and one-third that of human liver microsomes (25.4 μM). Quinidine (10 μM) inhibits 90% of the debrisoquine 4-hydroxylation activity in Japanese monkey liver microsomes with a K_i value of 0.53 μM. Antiserum only inhibits approximately 85% catalytic activity at a debrisoquine concentration of 100 μM (Hichiya et al. 2002). CYP2D29 is the CYP2D most homologous to the human CYP2D6 enzyme with only 20 amino acid residues different.

Site-directed mutagenesis of human CYP2D6 has shown the importance of Phe120, Glu216, Asp301, Phe481, and Phe483 in determining the interaction between ligands and the protein (de Graaf et al. 2007). These amino acid residues are well conserved between human CYP2D6, cynomolgus monkey CYP2D17 and 2D44, rhesus monkey 2D17, Japanese monkey 2D29, and marmoset 2D19 and 2D30, except that cynomolgus monkey CYP2D44 contains Val481 in SRS6. Phe481Asn and Phe481Gly mutations reduce the enzyme activities toward debrisoquine and dextromethorphan. Phe481 and Phe483 are located in the loop between the two β-strands of the fourth sheet region with Phe483 and possibly Phe482 being oriented into the active cavity (Rowland et al. 2006).

2.8 MARMOSET CYP2D8, 2D19, AND 2D30

In marmoset liver microsomes, debrisoquine is extensively hydroxylated in the 7-, 5-, 6-, and 8-positions, while 4-hydroxylation is a minor metabolic pathway (Cooke et al. 2012). In addition to the monohydroxylated metabolites, 6,7-dihydroxydebrisoquine is also formed. Bufuralol undergoes 1″-hydroxylation in marmoset liver microsomes, which is inhibited by quinidine and antibodies raised against rat CYP2D1 in a concentration-dependent manner (Hichiya et al. 2004). Propranolol is also metabolized by marmoset liver microsomes (Shimizudani et al. 2010). CYP2B, 2C, 2D, 2E, and 3A are significantly expressed in the liver of marmosets (Igarashi et al. 1997). CYP1A is also expressed but to a lesser extent and CYP3A mRNA is detectable in the small intestine of marmoset.

A cDNA encoding CYP2D8 is identified in marmosets (Uehara et al. 2015). The amino acid sequence deduced from CYP2D8 cDNA shows a high sequence identity (83%–86%) with other primate CYP2Ds. Phylogenetic analysis shows that marmoset CYP2D8 is closely clustered with human CYP2D6, unlike CYP2Ds of miniature pig, dog, rabbit, guinea pig, mouse, or rat. Marmoset CYP2D8 mRNA is predominantly expressed in the liver and small intestine (Uehara et al. 2015). Marmoset CYP2D8 expressed in *E. coli* catalyzes *O*-demethylations of metoprolol and dextromethorphan and bufuralol 1′-hydroxylation, but the catalytic efficiency is lower compared to human CYP2D6.

From liver cDNA libraries, CYP2D19 and 2D30 cDNA clones have been isolated and characterized (Hichiya et al. 2004; Igarashi et al. 1997). The deduced amino acid sequence of CYP2D19 shows 90% identities to human CYP2D6. CYP2D19 and 2D30 show homologies of 93.6% and 93.4% in their nucleotide and amino acid sequences, respectively. Marmoset CYP2D30, like human CYP2D6, exhibits high debrisoquine 4-hydroxylase activity but relatively low 5-, 6-, 7-, and 8-hydroxylase activities, while CYP2D19 lacks debrisoquine 4-hydroxylase but shows marked 5-, 6-, 7-, and 8-hydroxylase activities. The two marmoset recombinant enzymes show enantioselective bufuralol 1″-hydroxylase activities, similar to CYP2D6 (1″*R*-OH-bufuralol << 1″*S*-OH-bufuralol), whereas CYP2D19 shows a reversed selectivity. Marmoset CYP2D19 also metabolizes propranolol into 4-hydroxypropranolol (main), 5-OH-propranolol, and *N*-desisopropylpropranolol with regio- and stereoselectivity (Narimatsu et al. 2011). Like human CYP2D6, marmoset 2D19 is not induced by known CYP inducers (Schulz et al. 2001).

2.9 RABBIT CYP2D23 AND CYP2D24

CYP2D23 and 2D24 are two isomers of the CYP2D sub-family that are found in rabbits. Two full-length cDNAs from a rabbit liver cDNA library are cloned. The CYP2D23 and 2D24 contain 75.7% and 78.3% sequence identity with the human CYP2D6, respectively (Yamamoto et al. 1998a). The open-reading frame of *CYP2D23* and *2D24* encodes proteins each containing 500 amino acids, consisting of the highly conserved heme binding Cys446 and the conserved region in all CYP families around the invariant Cys residue (Yamamoto et al. 1998a). The rabbit CYP2D23 mRNA can be found in the liver, lung, small intestine, and stomach, with the strongest expression in liver and small intestine (Ishizuka et al. 2002). CYP2D24 mRNA expression level is lower in the liver and small intestine. However, CYP2D24 has higher expression in stomach compared to 2D23. The catalytic activity of CYP2D24 has been reported for the oxidation of bufuralol and bunitrolol, which are the archetypal substrates of the CYP2D subfamily. CYP2D23, however, only shows catalytic activity toward bufuralol (Ishizuka et al. 2002). Although the expression levels of CYP2D24 are lower in the liver and small intestine, the enzyme shows higher catalytic activity in bufuralol 1′-hydroxylation and bunitrolol 4-hydroxylation than CYP2D23 (Yamamoto et al. 1998a). CYP2D23 mRNA expression is also detected in the brain. However, its catalytic activity of testosterone metabolism is relatively low (Yamamoto et al. 1998a).

2.10 PIG CYP2D25

2.10.1 CLONING AND PURIFICATION OF PIG CYP2D25

Pigs have at least 58 *CYP* genes and some of them have been cloned and characterized, including CYP1A1, 1A2, 2A19, 2C33, 2D25, 2E1, 3A29, 17A1, and 19A1 (Barc et al. 2013; Brignac-Huber et al. 2013; Franczak et al. 2013; Kempisty et al. 2014; Li et al. 2014, 2015; Roelofs et al. 2013; Shang et al. 2013). CYP3A29 has been identified in porcine small intestine, and two more cDNA clones belonging to the CYP2C subfamily are also characterized from the same source (Nissen et al. 1998). A CYP2D enzyme, performing *N*-hydroxylations and *N*-reductions, is isolated and characterized (Clement et al. 1997). This enzyme has been shown to catalyze *N*-demethylation of dextromethorphan, a prototypical reaction of human CYP2D6. cDNA cloning of the entire coding sequence of CYP2D25 from domestic pigs has been reported (Postlind et al. 1997). Both DNA sequence analysis of the cDNA and protein sequence analysis show 70%–80% identity with members of the CYP2D family. Transfection of the vitamin D_3 25-hydroxylase cDNA into simian COS cells results in the synthesis of an enzyme that is recognized by a monoclonal antibody raised against purified vitamin D_3 25-hydroxylase and catalyzes 25-hydroxylation in the bioactivation of vitamin D_3 (Postlind et al. 1997). The mRNA for vitamin D_3 25-hydroxylase is found in liver and kidney.

Hosseinpour and Wikvall (2000) have purified CYP2D25 from the pig liver and found this enzyme 25-hydroxylated vitamin D_3 with an apparent K_m of 0.1 μM. The enzyme 25-hydroxylates vitamin D_3, 1α-hydroxyvitamin D_3, and vitamin D_2 and also converts tolterodine, a substrate for human CYP2D6, into its 5-hydroxymethyl metabolite (Hosseinpour and Wikvall 2000). Tolterodine inhibits the microsomal 25-hydroxylation, whereas quinidine, an inhibitor of CYP2D6, does not significantly inhibit the reaction. CYP2D25 shows 77% identity with that of human CYP2D6. CYP2D25 mRNA is expressed in higher levels in liver than in kidney and in small amounts in adrenals, brain, heart, intestine, lung, muscle, spleen, and thymus (Hosseinpour and Wikvall 2000). However, experiments with human liver microsomes and recombinantly expressed CYP2D6 indicate that the microsomal 25-hydroxylation of vitamin D3 in human liver is catalyzed by an enzyme different from CYP2D6 (Hosseinpour and Wikvall 2000).

Sakuma et al. (2004) cloned CYP2D21 (MS2D) from miniature pig livers. MS2D consists of 1613 bp including 1500 bp of an open-reading frame. Comparing the deduced amino acid sequences, CYP2D21 shows 78.3% identity to human CYP2D6. This value is 12.2%, 17.5%, 14.9%, and 16.1% lower than that of the marmoset (a New World monkey) 2D19, Japanese monkey 2D29, cynomolgus monkey (an Old World monkey) 2D17, and marmoset 2D30, respectively. The identity of miniature pig CYP2D21 to human CYP2D6 is very similar to or slightly higher than that for bovine 2D14 (77.5%), rabbit 2D24 (78.5%), dog 2D15 (75.3%), mouse 2D22 (76.1%), rat 2D4 (78.0%), hamster 2D20 (73.5%), and Guinea pig 2D16 (72.7%) (Sakuma et al. 2004). The recombinant pig CYP2D21 enzyme shows bufuralol 1′-hydroxylase activity, suggesting that miniature pig CYP2D21 is capable of metabolizing some of the same substrates associated with human CYP2D6 (Sakuma et al. 2004). The *CYP2D21* gene displays relatively lower identities between human and other experimental animals, except nonhuman primates, than do the other drug-metabolizing enzyme genes. This suggests the presence of large species differences of function as well as this gene might not be critical to life.

2.10.2 CATALYTIC ACTIVITY OF PIG CYP2D25

The amino acid sequence of domestic pig CYP2D25 shows 97.8% identity to that of miniature pig CYP2D21 (Sakuma et al. 2004). There are 10 amino acid differences between CYP2D21 and 2D25, and one located in the putative substrate recognition site for CYP2 enzyme (SRS-1 through SRS-6) described by Gotoh (1992): 204th, Q or L for CYP2D21 or 2D25, respectively. Therefore, one could not exclude the possibility that CYP2D21 and 2D25 have distinct substrate specificity.

Skaanild and Friis (1999) have reported that livers of miniature pig and domestic pig have no CYP2D-mediated debrisoquine 4-hydroxylase activity. They concluded in a later report that dextromethorphan *O*-demethylase and bufuralol 1′-hydroxylase activities, other marker activities, may be catalyzed by CYP2Bs in domestic and miniature pigs (Skaanild and Friis 2002). Low debrisoquine 4-hydroxylase activity of domestic pig CYP2D25 is confirmed by an assay using recombinant enzymes (Hosseinpour and Wikvall 2000). By contrast, Jurima-Romet et al. (2000) have demonstrated the evidence for the catalysis of dextromethorphan O-demethylation by a CYP2D6-like enzyme in domestic pig liver. These discrepancies raise the question of whether CYP2D enzymes of domestic or miniature pigs retain the functional similarity to human CYP2D6. The recombinant CYP2D21 in yeast cells show bufuralol 1′-hydroxylase activity, with an apparent K_m value of 0.98 μM (Sakuma et al. 2004). The K_m values of the human CYP2D6 expressed in the same expression system is 4.2 μM (Yokoi et al. 1996). Furthermore, a CYP2D enzyme purified from pig liver reduces sulfamethoxazole and dapsone hydroxylamines (Clement et al. 2005). These results indicate that the miniature pig possesses a CYP2D enzyme in its liver and retains the capacity to metabolize a substrate of human CYP2D6. These data indicate that the contribution of CYP2D21 to the metabolism of bufuralol is smaller than those from other CYPs, such as CYP2B.

2.11 SYRIAN HAMSTER Cyp2d27

Syrian hamster Cyp2d27 mRNA is found primarily in the liver, but not in the kidney, small intestine, and brain. The enzyme has been shown to catalyze bufuralol 1′- and debrisoquine 4-hydroxylations, which are unique to the CYP2D subfamily. The enzyme is not induced by either 3-methylcholanthrene or phenobarbital (Oka et al. 2000).

2.12 CHICKEN CYP2D49

Characterization of chicken CYPs is important for poultry pharmacology and toxicology as well as for human health because of the possible presence of toxic drug metabolites in chicken products such as meat and eggs. More than 10 functional CYPs such as CYP1A4/5, 2C23a/b, 2C45, 2H1, 3A37, 11A, 17A1 19A1, and 26B have been cloned and identified in chickens (Baader et al. 2002; Dogra et al. 1999; Goldstone and Stegeman 2006; Goriya et al. 2005; Kirischian and Wilson 2012; Lambeth et al. 2013; Martinez-Ceballos and Burdsal 2001; Mathew et al. 1990; Ourlin et al. 2000; Villalpando et al. 2000; Watanabe et al. 2013; Yang et al. 2013; Yoshida et al. 1996; Zuo et al. 2015). Among these CYPs, CYP2C45 is the most highly expressed isoform in chicken liver, while CYP2C23b is the most highly induced gene by phenobarbital in chicken xenobiotic receptors (Baader et al. 2002). Rifampicin, β-naphthoflavone, dexamethasone, and clotrimazole, all typical human CYP inducers, can induce the expression of chicken CYP3A37, 2H1, and 2C45 in LMH cells, the first continuously dividing cell line of the chicken liver (Handschin et al. 2001). Chicken liver microsomes show marked bufuralol 1′-hydroxylation activity (Khalil et al. 2001), indicating the presence of CYP2D enzymes in chickens.

Cai et al. (2012) have cloned the *CYP2D49* gene from chicken liver, which contains an open-reading frame of 502 amino acids that shares 52%–57% identities with CYP2Ds in mice, rats, rabbits, pigs, bovine, Japanese monkeys, and

humans. The gene is located on chromosome 1 and contains nine exons and eight introns. The exon–intron organization as well as the corresponding sizes of these segments and the coding region boundaries in CYP2D49 are conserved and similar to those of human CYP2D6. In addition, the neighboring genes of *CYP2D49* are conserved and similar to those of human CYP2D6. However, no CYP2D pseudogenes is found in the corresponding genomic region neighboring chicken CYP2D49. Similar to human CYP2D6, chicken 2D49 is not inducible in the liver by phenobarbital and expressed primarily in the liver, kidney, and small intestine, with lower transcription levels in the brain, lung, heart, spleen, testis, and ovary (Cai et al. 2012). CYP2D49 heterologously expressed in *E. coli* and HeLa cells metabolizes bufuralol via 1′-hydroxylation, but not debrisoquine. Moreover, quinidine, a potent inhibitor of human CYP2D6, only slightly inhibits the CYP2D49-catalyzed bufuralol 1′-hydroxylation activity (Cai et al. 2012). This is similar to the negligible inhibition of bufuralol 1′-hydroxylase activity by quinidine in mice, rats, and monkeys (Bogaards et al. 2000). Site-directed mutagenesis and circular dichroism spectroscopic studies have identified important residues including Val126, Glu222, Asp306, Phe486, and Phe488 in the enzymatic activity of CYP2D49 toward bufuralol as well as Asp306, Phe486, and Phe488 in maintaining the conformation of CYP2D49 (Cai et al. 2012). The resolved crystal structure of human CYP2D6 has confirmed the importance of Phe120, Glu216, Asp301, Phe481, and Phe483 in substrate recognition and binding (Rowland et al. 2006). Glu216 and Asp301, two negatively charged residues in the active site of human CYP2D6, facilitate the binding and orientation of the ligands in the active site through the formation of an electrostatic interaction between their carboxylate group and the basic nitrogen atom of the CYP2D6 substrates. The aromatic side chains of Phe120, Phe481, and Phe483 also bind with the aromatic moiety of the substrate via π–π stacking (Rowland et al. 2006). These findings indicate that CYP2D49 has functional characteristics similar to those of human CYP2D6 but differs in the lack of debrisoquine 4′-hydroxylation activity and quinidine inhibitory potency.

2.13 HORSE CYP2D50

CYP2D is expressed in horse liver and intestine at a low level (Tyden et al. 2014). cDNA encoding CYP2D50 is cloned from equine liver and expressed in a baculovirus expressing system in insect cells. The amino acid sequence of CYP2D50 is 77% identical to that of human 2D6 (DiMaio Knych and Stanley 2008). In a study by DiMaio Knych and Stanley (2008), the catalytic activity of CYP2D50 on O-demethylation of dextromethorphan and debrisoquine 4-hydroxylation is compared to human CYP2D6. The enzyme is not effective in metabolizing dextromethorphan with 180 times lower than CYP2D6 (DiMaio Knych and Stanley 2008). The rate of dextromethorphan O-demethylation has been shown to be 10- to 12-fold higher in comparison to CYP2D17 (Mankowski et al. 1999) and 2D15 (Shou et al. 2003). The study has also found a difference in 4-hydroxydebrisoquine activity between equine CYP2D50 and human 2D6. CYP2D50 shows very little activity, which is approximately 50 times less than CYP2D6 after 10 min of incubation at saturating substrate concentrations. Liver microsomes also have very low debrisoquine 4-hydroxylase activity, suggesting highly inefficient debrisoquine 4-hydroxylase activity (DiMaio Knych and Stanley 2008). Additional products are also found at much lower rate than the primary product. It indicates that 4-hydroxydebrisoquine is the main produce of debrisoquine in equine species although the conversion rate appears to be very slow (DiMaio Knych and Stanley 2008).

It is possible that one or more transcript variants of CYP2D50 may exist. The equine liver microsomes efficiently catalyze dextromethorphan O-demethylation. The Equine Genome Project has recently identified four additional CYP2D sequences, which have 98% identity to CYP2D50. It is possible that one of these variants have higher catalytic activities with dextromethorphan or debrisoquine (DiMaio Knych and Stanley 2008).

2.14 CONCLUSIONS AND FUTURE PERSPECTIVES

The CYP-dependent monooxygenase system has evolved as a multigene superfamily of proteins with the capacity to insert an atom of molecular oxygen into a substrate and to carry out a variety of other reactions. This property has been exploited through evolution to enable the CYP system to play a key role in xenobiotic and drug metabolism, metabolic homeostasis, and biosynthesis of endogenous compounds that play important physiological roles.

During drug discovery, the investigation of drug metabolism mediated by CYPs and the evaluation of the potential generation of toxic metabolites and harmful drug–drug interaction are essential. The experimental approach is based on animal drug-metabolizing systems and is used to predict kinetics and toxicity in human. Animals are routinely used in drug discovery and preclinical biomedical studies although a number of anatomical, biochemical, genetic, and epigenetic differences exist between humans and animals such as rodents and monkeys. Since human CYP2D6 can metabolize a large number of drugs and other compounds, which is subject to inhibition and genetic mutations, it is important to explore the use of animal models that allow us to characterize the regulation and function of CYP2Ds. Investigation of the CYP2D genomic organization is important because the position and direction of each CYP2D gene in the gene cluster provide essential information to evaluate CYP2D orthologous relationships between species. The existence of multiple functional CYP2D members in nonhuman animals reflects the divergence during evolution of CYP2D genes, but this makes the extrapolation of animal data to humans difficult. For example, animal models have been used to identify specific mechanisms by which pregnancy alters CYP2D-mediated drug disposition. There is a sixfold increase in CYP2D6 activity during pregnancy in women but Cyp2d mRNA and activity are decreased during rat pregnancy (Dickmann et al. 2008). In the rat, there is a

correlation between the Cyp2d isoform mRNA and Rarα and Hnf-3β mRNA, suggesting a potential mechanism of Cyp2d regulation by RARs. In contrast, Cyp2d11, 2d22, 2d26, and 2d40 mRNA levels are increased two- to sixfold during pregnancy in mice (Topletz et al. 2013). Upregulated Cyp2d40 is also observed in pregnant mice in another study (Shuster et al. 2013). A putative RA response element is identified within the Cyp2d40 promoter, and the mRNA of Cyp2d40 correlates with 26a1 and Rarβ. These findings suggest that mouse may provide a better and more suitable model to explore the mechanisms underlying the increased clearance of CYP2D6 probes observed during human pregnancy.

On the other hand, caution is needed when we choose animal models to explore inhibition of CYP2D-mediated reactions. Quinidine shows a different inhibition profile among species: quinidine inhibits CYP2D in man, dog, and monkey, but not in rat and mouse (Martignoni et al. 2006). In contrast, quinine inhibits CYP2D in rat, dog, and monkey, but not in man. On the basis of these findings, dog and monkey may represent proper models for the study of CYP2D inhibition. Overall, none of the animal species are completely similar to man with respect to tissue distribution, substrate specificity, inhibitor selectivity, and regulation of CYP2Ds. Further studies are certainly warranted to examine the mechanisms underlying the species differences in the structure, function, and regulation of CYP2D members.

REFERENCES

Adas, F., Berthou, F., Salaun, J. P., Dreano, Y., Amet, Y. 1999. Interspecies variations in fatty acid hydroxylations involving cytochromes P450 2E1 and 4A. Toxicol Lett 110:43–55.

Aidasani, D., Zaya, M. J., Malpas, P. B., Locuson, C. W. 2008. In vitro drug–drug interaction screens for canine veterinary medicines: Evaluation of cytochrome P450 reversible inhibition. Drug Metab Dispos 36:1512–1518.

Al-Dabbagh, S. G., Idle, J. R., Smith, R. L. 1981. Animal modelling of human polymorphic drug oxidation—The metabolism of debrisoquine and phenacetin in rat inbred strains. J Pharm Pharmacol 33:161–164.

Ariyoshi, N., Koga, N., Oguri, K., Yoshimura, H. 1992. Metabolism of 2,4,5,2′,4′,5′-hexachlorobiphenyl with liver microsomes of phenobarbital-treated dog; the possible formation of PCB 2,3-arene oxide intermediate. Xenobiotica 22:1275–1290.

Ariyoshi, N., Oguri, K., Koga, N., Yoshimura, H., Funae, Y. 1995. Metabolism of highly persistent PCB congener, 2,4,5,2′,4′,5′-hexachlorobiphenyl, by human CYP2B6. Biochem Biophys Res Commun 212:455–460.

Azuma, R., Komuro, M., Kawaguchi, Y., Okudaira, K., Hayashi, M., Kiwada, H. 2002. Comparative analysis of in vitro and in vivo pharmacokinetic parameters related to individual variability of GTS-21 in canine. Drug Metab Pharmacokinet 17:75–82.

Baader, M., Gnerre, C., Stegeman, J. J., Meyer, U. A. 2002. Transcriptional activation of cytochrome P450 CYP2C45 by drugs is mediated by the chicken xenobiotic receptor (CXR) interacting with a phenobarbital response enhancer unit. J Biol Chem 277:15647–15653.

Bailey, D. G., Dresser, G. K. 2004. Interactions between grapefruit juice and cardiovascular drugs. Am J Cardiovasc Drugs 4:281–297.

Bailey, D. G., Dresser, G. K., Kreeft, J. H., Munoz, C., Freeman, D. J., Bend, J. R. 2000. Grapefruit-felodipine interaction: Effect of unprocessed fruit and probable active ingredients. *Clin Pharmacol Ther* 68:468–477.

Bailey, D. G., Dresser, G. K., Bend, J. R. 2003. Bergamottin, lime juice, and red wine as inhibitors of cytochrome P450 3A4 activity: Comparison with grapefruit juice. *Clin Pharmacol Ther* 73:529–537.

Barc, J., Karpeta, A., Gregoraszczuk, E. L. 2013. Action of Halowax 1051 on enzymes of phase I (CYP1A1) and phase II (SULT1A and COMT) metabolism in the pig ovary. *Int J Endocrinol* 2013:590261.

Barham, H. M., Lennard, M. S., Tucker, G. T. 1994. An evaluation of cytochrome P450 isoform activities in the female dark agouti (DA) rat: Relevance to its use as a model of the CYP2D6 poor metaboliser phenotype. *Biochem Pharmacol* 47:1295–1307.

Bertilsson, L., Baillie, T. A., Reviriego, J. 1990. Factors influencing the metabolism of diazepam. *Pharmacol Ther* 45:85–91.

Blaisdell, J., Goldstein, J. A., Bai, S. A. 1998. Isolation of a new canine cytochrome P450 CDNA from the cytochrome P450 2C subfamily (CYP2C41) and evidence for polymorphic differences in its expression. *Drug Metab Dispos* 26:278–283.

Blume, N., Leonard, J., Xu, Z. J., Watanabe, O., Remotti, H., Fishman, J. 2000. Characterization of Cyp2d22, a novel cytochrome P450 expressed in mouse mammary cells. *Arch Biochem Biophys* 381:191–204.

Bogaards, J. J., Bertrand, M., Jackson, P., Oudshoorn, M. J., Weaver, R. J., van Bladeren, P. J., Walther, B. 2000. Determining the best animal model for human cytochrome P450 activities: A comparison of mouse, rat, rabbit, dog, micropig, monkey and man. *Xenobiotica* 30:1131–1152.

Bolser, D. C. 2006. Current and future centrally acting antitussives. *Respir Physiol Neurobiol* 152:349–355.

Boobis, A. R., Murray, S., Kahn, G. C., Robertz, G. M., Davies, D. S. 1983. Substrate specificity of the form of cytochrome P-450 catalyzing the 4-hydroxylation of debrisoquine in man. *Mol Pharmacol* 23:474–481.

Born, S. L., John, G. H., Harlow, G. R., Halpert, J. R. 1995. Characterization of the progesterone 21-hydroxylase activity of canine cytochrome P450 PBD-2/P450 2B11 through reconstitution, heterologous expression, and site-directed mutagenesis. *Drug Metab Dispos* 23:702–707.

Brignac-Huber, L. M., Reed, J. R., Eyer, M. K., Backes, W. L. 2013. Relationship between CYP1A2 localization and lipid microdomain formation as a function of lipid composition. *Drug Metab Dispos* 41:1896–1905.

Cai, H., Jiang, J., Yang, Q., Chen, Q., Deng, Y. 2012. Functional characterization of a first avian cytochrome P450 of the CYP2D subfamily (CYP2D49). *PLoS One* 7:e38395.

Cairns, W., Smith, C. A., McLaren, A. W., Wolf, C. R. 1996. Characterization of the human cytochrome P4502D6 promoter. A potential role for antagonistic interactions between members of the nuclear receptor family. *J Biol Chem* 271:25269–25276.

Chauret, N., Gauthier, A., Martin, J., Nicoll-Griffith, D. A. 1997. In vitro comparison of cytochrome P450-mediated metabolic activities in human, dog, cat, and horse. *Drug Metab Dispos* 25:1130–1136.

Chauret, N., Dobbs, B., Lackman, R. L., Bateman, K., Nicoll-Griffith, D. A., Stresser, D. M., Ackermann, J. M., Turner, S. D., Miller, V. P., Crespi, C. L. 2001. The use of 3-[2-(N,N-diethyl-N-methylammonium)ethyl]-7-methoxy-4-methylcoumarin (AMMC) as a specific CYP2D6 probe in human liver microsomes. *Drug Metab Dispos* 29:1196–1200.

Chen, Z. R., Irvine, R. J., Bochner, F., Somogyi, A. A. 1990. Morphine formation from codeine in rat brain: A possible mechanism of codeine analgesia. *Life Sci* 46:1067–1074.

Chow, T., Hiroi, T., Imaoka, S., Chiba, K., Funae, Y. 1999a. Isoform-selective metabolism of mianserin by cytochrome P-450 2D. *Drug Metab Dispos* 27:1200–1204.

Chow, T., Imaoka, S., Hiroi, T., Funae, Y. 1999b. Developmental changes in the catalytic activity and expression of CYP2D isoforms in the rat liver. *Drug Metab Dispos* 27:188–192.

Ciaccio, P. J., Halpert, J. R. 1989. Characterization of a phenobarbital-inducible dog liver cytochrome P450 structurally related to rat and human enzymes of the P450IIIA (steroid-inducible) gene subfamily. *Arch Biochem Biophys* 271:284–299.

Ciaccio, P. J., Graves, P. E., Bourque, D. P., Glinsmann-Gibson, B., Halpert, J. R. 1991. cDNA and deduced amino acid sequences of a dog liver cytochrome P-450 of the IIIA gene subfamily. *Biochim Biophys Acta* 1088:319–322.

Clement, B., Lomb, R., Moller, W. 1997. Isolation and characterization of the protein components of the liver microsomal O2-insensitive NADH-benzamidoxime reductase. *J Biol Chem* 272:19615–19620.

Clement, B., Behrens, D., Amschler, J., Matschke, K., Wolf, S., Havemeyer, A. 2005. Reduction of sulfamethoxazole and dapsone hydroxylamines by a microsomal enzyme system purified from pig liver and pig and human liver microsomes. *Life Sci* 77:205–219.

Colby, H. D., Huang, Y., Jiang, Q., Voigt, J. M. 2000. Bimodal expression of CYP2D16 in the guinea pig adrenal cortex. *Pharmacology* 61:78–82.

Colby, H. D., Nowak, D. M., Longhurst, P. A., Zhang, X., Hayes, J. R., Voigt, J. M. 2001. Bufuralol metabolism by guinea pig adrenal and hepatic microsomes. *Pharmacology* 62:229–233.

Colby, H. D., Longhurst, P. A., Burczynski, J. M., Hayes, J. R., Yuan, B. B., Voigt, J. V. 2003. Downregulation of CYP2D16 by ACTH in the guinea pig adrenal cortex: Time course, reversibility, and mechanism of action. *Pharmacology* 67:121–127.

Coleman, T., Spellman, E. F., Rostami-Hodjegan, A., Lennard, M. S., Tucker, G. T. 2000. The 1′-hydroxylation of Rac-bufuralol by rat brain microsomes. *Drug Metab Dispos* 28:1094–1099.

Cooke, B. R., Bligh, S. W., Cybulski, Z. R., Ioannides, C., Hall, M. 2012. Debrisoquine metabolism and CYP2D expression in marmoset liver microsomes. *Drug Metab Dispos* 40:70–75.

Corchero, J., Granvil, C. P., Akiyama, T. E., Hayhurst, G. P., Pimprale, S., Feigenbaum, L., Idle, J. R., Gonzalez, F. J. 2001. The CYP2D6 humanized mouse: Effect of the human CYP2D6 transgene and HNF4α on the disposition of debrisoquine in the mouse. *Mol Pharmacol* 60:1260–1267.

Cui, J. Y., Renaud, H. J., Klaassen, C. D. 2012. Ontogeny of novel cytochrome P450 gene isoforms during postnatal liver maturation in mice. *Drug Metab Dispos* 40:1226–1237.

Dahl, M. L., Tybring, G., Elwin, C. E., Alm, C., Andreasson, K., Gyllenpalm, M., Bertilsson, L. 1994. Stereoselective disposition of mianserin is related to debrisoquin hydroxylation polymorphism. *Clin Pharmacol Ther* 56:176–183.

Dakovic-Svajcer, K., Samojlik, I., Raskovic, A., Popovic, M., Jakovljevic, V. 1999. The activity of liver oxidative enzymes after single and multiple grapefruit juice ingestion. *Exp Toxicol Pathol* 51:304–308.

Dayer, P., Leemann, T., Striberni, R. 1989. Dextromethorphan O-demethylation in liver microsomes as a prototype reaction to monitor cytochrome P-450 db1 activity. *Clin Pharmacol Ther* 45:34–40.

de Graaf, C., Oostenbrink, C., Keizers, P. H., van Vugt-Lussenburg, B. M., van Waterschoot, R. A., Tschirret-Guth, R. A., Commandeur, J. N., Vermeulen, N. P. 2007. Molecular modeling-guided site-directed mutagenesis of cytochrome P450 2D6. *Curr Drug Metab* 8:59–77.

Deerberg, F., Rehm, S., Jostmeyer, H. H. 1985. Spontaneous urinary bladder tumors in DA/Han rats: A feasible model of human bladder cancer. *J Natl Cancer Inst* 75:1113–1121.

Dickmann, L. J., Tay, S., Senn, T. D., Zhang, H., Visone, A., Unadkat, J. D., Hebert, M. F., Isoherranen, N. 2008. Changes in maternal liver Cyp2c and Cyp2d expression and activity during rat pregnancy. *Biochem Pharmacol* 75:1677–1687.

DiMaio Knych, H. K., Stanley, S. D. 2008. Complementary DNA cloning, functional expression and characterization of a novel cytochrome P450, CYP2D50, from equine liver. *Biochem Pharmacol* 76:904–911.

Distlerath, L. M., Guengerich, F. P. 1984. Characterization of a human liver cytochrome P-450 involved in the oxidation of debrisoquine and other drugs by using antibodies raised to the analogous rat enzyme. *Proc Natl Acad Sci U S A* 81:7348–7352.

Dogra, S. C., Davidson, B. P., May, B. K. 1999. Analysis of a phenobarbital-responsive enhancer sequence located in the 5′ flanking region of the chicken CYP2H1 gene: Identification and characterization of functional protein-binding sites. *Mol Pharmacol* 55:14–22.

Duclos-Vallee, J. C., Johanet, C., Bach, J. F., Yamamoto, A. M. 2000. Autoantibodies associated with acute rejection after liver transplantation for type-2 autoimmune hepatitis. *J Hepatol* 33:163–166.

Duignan, D. B., Sipes, I. G., Leonard, T. B., Halpert, J. R. 1987. Purification and characterization of the dog hepatic cytochrome P-450 isozyme responsible for the metabolism of 2,2′,4,4′,5,5′-hexachlorobiphenyl. *Arch Biochem Biophys* 255:290–303.

Duignan, D. B., Sipes, I. G., Ciaccio, P. J., Halpert, J. R. 1988. The metabolism of xenobiotics and endogenous compounds by the constitutive dog liver cytochrome P450 PBD-2. *Arch Biochem Biophys* 267:294–304.

Eichelbaum, M., Baur, M. P., Dengler, H. J., Osikowska-Evers, B. O., Tieves, G., Zekorn, C., Rittner, C. 1987. Chromosomal assignment of human cytochrome P-450 (debrisoquine/sparteine type) to chromosome 22. *Br J Clin Pharmacol* 23:455–458.

Eiermann, B., Edlund, P. O., Tjernberg, A., Dalen, P., Dahl, M. L., Bertilsson, L. 1998. 1- and 3-hydroxylations, in addition to 4-hydroxylation, of debrisoquine are catalyzed by cytochrome P450 2D6 in humans. *Drug Metab Dispos* 26:1096–1101.

Emoto, C., Yoda, N., Uno, Y., Iwasaki, K., Umehara, K., Kashiyama, E., Yamazaki, H. 2013. Comparison of p450 enzymes between cynomolgus monkeys and humans: p450 identities, protein contents, kinetic parameters, and potential for inhibitory profiles. *Curr Drug Metab* 14:239–252.

Felmlee, M. A., Lon, H. K., Gonzalez, F. J., Yu, A. M. 2008. Cytochrome P450 expression and regulation in CYP3A4/CYP2D6 double transgenic humanized mice. *Drug Metab Dispos* 36:435–441.

Franczak, A., Wojciechowicz, B., Zmijewska, A., Kolakowska, J., Kotwica, G. 2013. The effect of interleukin 1β and interleukin 6 on estradiol-17β secretion in the endometrium of pig during early pregnancy and the estrous cycle. *Theriogenology* 80:90–98.

Fraser, D. J., Feyereisen, R., Harlow, G. R., Halpert, J. R. 1997. Isolation, heterologous expression and functional characterization of a novel cytochrome P450 3A enzyme from a canine liver cDNA library. *J Pharmacol Exp Ther* 283:1425–1432.

Fujita, S., Umeda, S., Funae, Y., Imaoka, S., Abe, H., Ishida, R., Adachi, T., Masuda, M., Kazusaka, A., Suzuki, T. 1993. Regio- and stereoselective propranolol metabolism by 15 forms of purified cytochromes P-450 from rat liver. *J Pharmacol Exp Ther* 264:226–233.

Gaedigk, A. 2013. Complexities of CYP2D6 gene analysis and interpretation. *Int Rev Psychiatry* 25:534–553.

Gergel, D., Misik, V., Riesz, P., Cederbaum, A. I. 1997. Inhibition of rat and human cytochrome P4502E1 catalytic activity and reactive oxygen radical formation by nitric oxide. *Arch Biochem Biophys* 337:239–250.

Gerhard, D. S., Wagner, L., Feingold, E. A., Shenmen, C. M., Grouse, L. H., Schuler, G., Klein, S. L. et al. 2004. The status, quality, and expansion of the NIH full-length cDNA project: The Mammalian Gene Collection (MGC). *Genome Res* 14:2121–2127.

Ghosal, A., Hapangama, N., Yuan, Y., Lu, X., Horne, D., Patrick, J. E., Zbaida, S. 2003. Rapid determination of enzyme activities of recombinant human cytochromes P450, human liver microsomes and hepatocytes. *Biopharm Drug Dispos* 24:375–384.

Goldstone, H. M., Stegeman, J. J. 2006. A revised evolutionary history of the CYP1A subfamily: Gene duplication, gene conversion, and positive selection. *J Mol Evol* 62:708–717.

Gonzalez, F. J., Matsunaga, T., Nagata, K., Meyer, U. A., Nebert, D. W., Pastewka, J., Kozak, C. A., Gillette, J., Gelboin, H. V., Hardwick, J. P. 1987. Debrisoquine 4-hydroxylase: Characterization of a new P450 gene subfamily, regulation, chromosomal mapping, and molecular analysis of the DA rat polymorphism. *DNA* 6:149–161.

Gonzalez, F. J., Skoda, R. C., Kimura, S., Umeno, M., Zanger, U. M., Nebert, D. W., Gelboin, H. V., Hardwick, J. P., Meyer, U. A. 1988. Characterization of the common genetic defect in humans deficient in debrisoquine metabolism. *Nature* 331:442–446.

Goriya, H. V., Kalia, A., Bhavsar, S. K., Joshi, C. G., Rank, D. N., Thaker, A. M. 2005. Comparative evaluation of phenobarbital-induced CYP3A and CYP2H1 gene expression by quantitative RT-PCR in Bantam, Bantamized White Leghorn and White Leghorn chicks. *J Vet Sci* 6:279–285.

Gotoh, O. 1992. Substrate recognition sites in cytochrome P450 family 2 (CYP2) proteins inferred from comparative analyses of amino acid and coding nucleotide sequences. *J Biol Chem* 267:83–90.

Graham, M. J., Lake, B. G. 2008. Induction of drug metabolism: Species differences and toxicological relevance. *Toxicology* 254:184–191.

Graham, R. A., Downey, A., Mudra, D., Krueger, L., Carroll, K., Chengelis, C., Madan, A., Parkinson, A. 2002. In vivo and in vitro induction of cytochrome P450 enzymes in beagle dogs. *Drug Metab Dispos* 30:1206–1213.

Graham, R. A., Tyler, L. O., Krol, W. L., Silver, I. S., Webster, L. O., Clark, P., Chen, L., Banks, T., LeCluyse, E. L. 2006. Temporal kinetics and concentration-response relationships for induction of CYP1A, CYP2B, and CYP3A in primary cultures of beagle dog hepatocytes. *J Biochem Mol Toxicol* 20:69–78.

Graves, P. E., Elhag, G. A., Ciaccio, P. J., Bourque, D. P., Halpert, J. R. 1990. cDNA and deduced amino acid sequences of a dog hepatic cytochrome P450IIB responsible for the metabolism of 2,2′,4,4′,5,5′-hexachlorobiphenyl. *Arch Biochem Biophys* 281:106–115.

Haduch, A., Bromek, E., Kot, M., Kaminska, K., Golembiowska, K., Daniel, W. A. 2015. The cytochrome P450 2D-mediated formation of serotonin from 5-methoxytryptamine in the brain *in vivo*: A microdialysis study. *J Neurochem* 133:83–92.

Haertter, S. 2013. Recent examples on the clinical relevance of the CYP2D6 polymorphism and endogenous functionality of CYP2D6. *Drug Metabol Drug Interact* 28:209–216.

Handschin, C., Podvinec, M., Stockli, J., Hoffmann, K., Meyer, U. A. 2001. Conservation of signaling pathways of xenobiotic-sensing orphan nuclear receptors, chicken xenobiotic receptor, constitutive androstane receptor, and pregnane X receptor, from birds to humans. *Mol Endocrinol* 15:1571–1585.

Heim, M., Meyer, U. A. 1990. Genotyping of poor metabolisers of debrisoquine by allele-specific PCR amplification. *Lancet* 336:529–532.

Heim, M. H., Meyer, U. A. 1992. Evolution of a highly polymorphic human cytochrome P450 gene cluster: *CYP2D6*. *Genomics* 14:49–58.

Heinig, R., Delbressine, L. P., Kaspersen, F. M., Blaschke, G. 1993. Enantiomeric aspects of the metabolism of mianserin in rats. *Arzneimittelforschung* 43:709–715.

Hellmold, H., Lamb, J. G., Wyss, A., Gustafsson, J. A., Warner, M. 1995. Developmental and endocrine regulation of P450 isoforms in rat breast. *Mol Pharmacol* 48:630–638.

Hichiya, H., Takemi, C., Tsuzuki, D., Yamamoto, S., Asaoka, K., Suzuki, S., Satoh, T., Shinoda, S., Kataoka, H., Narimatsu, S. 2002. Complementary DNA cloning and characterization of cytochrome P450 2D29 from Japanese monkey liver. *Biochem Pharmacol* 64:1101–1110.

Hichiya, H., Kuramoto, S., Yamamoto, S., Shinoda, S., Hanioka, N., Narimatsu, S., Asaoka, K. et al. 2004. Cloning and functional expression of a novel marmoset cytochrome P450 2D enzyme, CYP2D30: Comparison with the known marmoset CYP2D19. *Biochem Pharmacol* 68:165–175.

Hiroi, T., Imaoka, S., Chow, T., Funae, Y. 1998. Tissue distributions of CYP2D1, 2D2, 2D3 and 2D4 mRNA in rats detected by RT-PCR. *Biochim Biophys Acta* 1380:305–312.

Hiroi, T., Chow, T., Imaoka, S., Funae, Y. 2002. Catalytic specificity of CYP2D isoforms in rat and human. *Drug Metab Dispos* 30:970–976.

Horikiri, Y., Suzuki, T., Mizobe, M. 1998. Stereoselective metabolism of bisoprolol enantiomers in dogs and humans. *Life Sci* 63:1097–1108.

Hosseinpour, F., Wikvall, K. 2000. Porcine microsomal vitamin D(3) 25-hydroxylase (CYP2D25). Catalytic properties, tissue distribution, and comparison with human CYP2D6. *J Biol Chem* 275:34650–34655.

Huang, Y., Jiang, Q., Voigt, J. M., Debolt, K. M., Colby, H. D. 1996. Strain differences in adrenal CYP2D16 expression in guinea pigs. Relationship to xenobiotic metabolism. *Biochem Pharmacol* 52:1925–1929.

Huang, Y., Jiang, Q., Debolt, K. M., Voigt, J. M., Colby, H. D. 1998. Strain differences in adrenal microsomal steroid metabolism in guinea pigs. *J Steroid Biochem Mol Biol* 64:305–311.

Igarashi, T., Sakuma, T., Isogai, M., Nagata, R., Kamataki, T. 1997. Marmoset liver cytochrome P450s: Study for expression and molecular cloning of their cDNAs. *Arch Biochem Biophys* 339:85–91.

Ishii, M., Toda, T., Ikarashi, N., Kusunoki, Y., Kon, R., Ochiai, W., Machida, Y., Sugiyama, K. 2014. Total gastrectomy may result in reduced drug effectiveness due to an increase in the expression of the drug-metabolizing enzyme Cytochrome P450, in the liver. *Eur J Pharm Sci* 51:180–188.

Ishizuka, M., Yamamoto, Y., Takada, A., Kazusaka, A., Fujita, S. 2002. The loss of enzyme activities by a single amino acid substitution of a newly cloned rabbit CYP2D isozyme, CYP2D24. *Int Congr Ser* 1233:121–126.

Ishizuka, M., Lee, J. J., Masuda, M., Akahori, F., Kazusaka, A., Fujita, S. 2006. CYP2D-related metabolism in animals of the Canoidea superfamily—Species differences. *Vet Res Commun* 30:505–512.

Jacqz-Aigrain, E., Funck-Brentano, C., Cresteil, T. 1993. CYP2D6- and CYP3A-dependent metabolism of dextromethorphan in humans. *Pharmacogenetics* 3:197–204.

Jayyosi, Z., Muc, M., Erick, J., Thomas, P. E., Kelley, M. 1996. Catalytic and immunochemical characterization of cytochrome P450 isozyme induction in dog liver. *Fundam Appl Toxicol* 31:95–102.

Jiang, Q., Voigt, J. M., Colby, H. D. 1995. Molecular cloning and sequencing of a guinea pig cytochrome P4502D (CYP2D16): High level expression in adrenal microsomes. *Biochem Biophys Res Commun* 209:1149–1156.

Jiang, Q., Huang, Y., Voigt, J. M., DeBolt, B. K., Kominami, S., Takemori, S., Funae, Y., Colby, H. D. 1996a. Expression and zonal distribution of CYP2D16 in the guinea pig adrenal cortex: Relationship to xenobiotic metabolism. *Mol Pharmacol* 49:458–464.

Jiang, Q., Huang, Y., Voigt, J. M., DeBolt, K. M., Kominami, S., Takemori, S., Funae, Y., Colby, H. D. 1996b. Differential effects of adrenocorticotropin in vivo on cytochromes P4502D16 and P450c17 in the guinea pig adrenal cortex. *Endocrinology* 137:4811–4816.

John, G. H., Hasler, J. A., He, Y. A., Halpert, J. R. 1994. *Escherichia coli* expression and characterization of cytochromes P450 2B11, 2B1, and 2B5. *Arch Biochem Biophys* 314:367–375.

Jolivalt, C., Minn, A., Vincent-Viry, M., Galteau, M. M., Siest, G. 1995. Dextromethorphan O-demethylase activity in rat brain microsomes. *Neurosci Lett* 187:65–68.

Jurima-Romet, M., Casley, W. L., Leblanc, C. A., Nowakowska, M. 2000. Evidence for the catalysis of dextromethorphan O-demethylation by a CYP2D6-like enzyme in pig liver. *Toxicol In Vitro* 14:253–263.

Kasimanickam, V. R., Kasimanickam, R. K. 2013. Expression of CYP26b1 and related retinoic acid signalling molecules in young, peripubertal and adult dog testis. *Reprod Domest Anim* 48:171–176.

Kasimanickam, V., Kasimanickam, R. 2014. Exogenous retinoic acid and cytochrome P450 26B1 inhibitor modulate meiosis-associated genes expression in canine testis, an *in vitro* model. *Reprod Domest Anim* 49:315–323.

Kawai, J., Shinagawa, A., Shibata, K., Yoshino, M., Itoh, M., Ishii, Y., Arakawa, T. et al. 2001. Functional annotation of a full-length mouse cDNA collection. *Nature* 409:685–690.

Kawashima, H., Strobel, H. W. 1995. cDNA cloning of three new forms of rat brain cytochrome P450 belonging to the CYP4F subfamily. *Biochem Biophys Res Commun* 217:1137–1144.

Kawashima, H., Sequeira, D. J., Nelson, D. R., Strobel, H. W. 1996. Genomic cloning and protein expression of a novel rat brain cytochrome P-450 CYP2D18* catalyzing imipramine N-demethylation. *J Biol Chem* 271:28176–28180.

Kempisty, B., Ziolkowska, A., Ciesiolka, S., Piotrowska, H., Antosik, P., Bukowska, D., Nowicki, M., Brussow, K. P., Zabel, M. 2014. Association between the expression of *LHR*, *FSHR* and *CYP19* genes, cellular distribution of encoded proteins and proliferation of porcine granulosa cells in real-time. *J Biol Regul Homeost Agents* 28:419–431.

Kerry, N. L., Somogyi, A. A., Mikus, G., Bochner, F. 1993. Primary and secondary oxidative metabolism of dextromethorphan. In vitro studies with female Sprague–Dawley and Dark Agouti rat liver microsomes. *Biochem Pharmacol* 45:833–839.

Kerry, N. L., Somogyi, A. A., Bochner, F., Mikus, G. 1994. The role of CYP2D6 in primary and secondary oxidative metabolism of dextromethorphan: In vitro studies using human liver microsomes. *Br J Clin Pharmacol* 38:243–248.

Khalil, W. F., Saitoh, T., Shimoda, M., Kokue, E. 2001. *In vitro* cytochrome P450-mediated hepatic activities for five substrates in specific pathogen free chickens. *J Vet Pharmacol Ther* 24:343–348.

Kimura, S., Umeno, M., Skoda, R. C., Meyer, U. A., Gonzalez, F. J. 1989. The human debrisoquine 4-hydroxylase (*CYP2D*) locus: Sequence and identification of the polymorphic *CYP2D6* gene, a related gene, and a pseudogene. *Am J Hum Genet* 45:889–904.

Kingback, M., Carlsson, B., Ahlner, J., Bengtsson, F., Kugelberg, F. C. 2011. Cytochrome P450-dependent disposition of the enantiomers of citalopram and its metabolites: *In vivo* studies in Sprague–Dawley and Dark Agouti rats. *Chirality* 23:172–177.

Kirischian, N. L., Wilson, J. Y. 2012. Phylogenetic and functional analyses of the cytochrome P450 family 4. *Mol Phylogenet Evol* 62:458–471.

Kirischian, N., McArthur, A. G., Jesuthasan, C., Krattenmacher, B., Wilson, J. Y. 2011. Phylogenetic and functional analysis of the vertebrate cytochrome p450 2 family. *J Mol Evol* 72:56–71.

Klekotka, P. A., Halpert, J. R. 1995. Benzyloxyresorufin as a specific substrate for the major phenobarbital-inducible dog liver cytochrome P450 (P4502B11). *Drug Metab Dispos* 23:1434–1435.

Kobayashi, S., Murray, S., Watson, D., Sesardic, D., Davies, D. S., Boobis, A. R. 1989. The specificity of inhibition of debrisoquine 4-hydroxylase activity by quinidine and quinine in the rat is the inverse of that in man. *Biochem Pharmacol* 38:2795–2799.

Kobayashi, K., Chiba, K., Yagi, T., Shimada, N., Taniguchi, T., Horie, T., Tani, M., Yamamoto, T., Ishizaki, T., Kuroiwa, Y. 1997. Identification of cytochrome P450 isoforms involved in citalopram N-demethylation by human liver microsomes. *J Pharmacol Exp Ther* 280:927–933.

Kobayashi, K., Urashima, K., Shimada, N., Chiba, K. 2003. Selectivities of human cytochrome P450 inhibitors toward rat P450 isoforms: Study with cDNA-expressed systems of the rat. *Drug Metab Dispos* 31:833–836.

Koop, D. R. 1986. Hydroxylation of p-nitrophenol by rabbit ethanol-inducible cytochrome P-450 isozyme 3a. *Mol Pharmacol* 29:399–404.

Koyama, E., Chiba, K., Tani, M., Ishizaki, T. 1996. Identification of human cytochrome P450 isoforms involved in the stereoselective metabolism of mianserin enantiomers. *J Pharmacol Exp Ther* 278:21–30.

Kraemer, T., Pflugmann, T., Bossmann, M., Kneller, N. M., Peters, F. T., Paul, L. D., Springer, D., Staack, R. F., Maurer, H. H. 2004. Fenproporex N-dealkylation to amphetamine—Enantioselective *in vitro* studies in human liver microsomes as well as enantioselective *in vivo* studies in Wistar and Dark Agouti rats. *Biochem Pharmacol* 68:947–957.

Lambeth, L. S., Cummins, D. M., Doran, T. J., Sinclair, A. H., Smith, C. A. 2013. Overexpression of aromatase alone is sufficient for ovarian development in genetically male chicken embryos. *PLoS One* 8:e68362.

Lankford, S. M., Bai, S. A., Goldstein, J. A. 2000. Cloning of canine cytochrome P450 2E1 cDNA: Identification and characterization of two variant alleles. *Drug Metab Dispos* 28:981–986.

Lenz, D. C., Wolf, N. A., Swanborg, R. H. 1999. Strain variation in autoimmunity: Attempted tolerization of DA rats results in the induction of experimental autoimmune encephalomyelitis. *J Immunol* 163:1763–1768.

Li, X., Jin, X., Zhou, X., Wang, X., Shi, D., Xiao, Y., Bi, D. 2014. Pregnane X receptor is required for IFN-α-mediated CYP3A29 expression in pigs. *Biochem Biophys Res Commun* 445:469–474.

Li, X., Hu, X., Jin, X., Zhou, X., Wang, X., Shi, D., Bi, D. 2015. IFN-γ regulates cytochrome 3A29 through pregnane X receptor in pigs. *Xenobiotica* 45:373–379.

Licari, G., Somogyi, A. A., Milne, R. W., Sallustio, B. C. 2015. Comparison of CYP2D metabolism and hepatotoxicity of the myocardial metabolic agent perhexiline in Sprague–Dawley and Dark Agouti rats. *Xenobiotica* 45:3–9.

Lin, L. Y., Kumagai, Y., Cho, A. K. 1992. Enzymatic and chemical demethylenation of (methylenedioxy)amphetamine and (methylenedioxy)methamphetamine by rat brain microsomes. *Chem Res Toxicol* 5:401–406.

Lindell, M., Lang, M., Lennernas, H. 2003. Expression of genes encoding for drug metabolising cytochrome P450 enzymes and P-glycoprotein in the rat small intestine; comparison to the liver. *Eur J Drug Metab Pharmacokinet* 28:41–48.

Lorenc-Koci, E., Wojcikowski, J., Kot, M., Haduch, A., Boksa, J., Daniel, W. A. 2004. Disposition of 1,2,3,4,-tetrahydroisoquinoline in the brain of male Wistar and Dark Agouti rats. *Brain Res* 996:168–179.

Lown, K. S., Bailey, D. G., Fontana, R. J., Janardan, S. K., Adair, C. H., Fortlage, L. A., Brown, M. B., Guo, W., Watkins, P. B. 1997. Grapefruit juice increases felodipine oral availability in humans by decreasing intestinal CYP3A protein expression. *J Clin Invest* 99:2545–2553.

Maeda, N., Kasukawa, T., Oyama, R., Gough, J., Frith, M., Engstrom, P. G., Lenhard, B. et al. 2006. Transcript annotation in FANTOM3: Mouse gene catalog based on physical cDNAs. *PLoS Genet* 2:e62.

Mankowski, D. C., Laddison, K. J., Christopherson, P. A., Ekins, S., Tweedie, D. J., Lawton, M. P. 1999. Molecular cloning, expression, and characterization of CYP2D17 from cynomolgus monkey liver. *Arch Biochem Biophys* 372:189–196.

Manns, M. P., Johnson, E. F., Griffin, K. J., Tan, E. M., Sullivan, K. F. 1989. Major antigen of liver kidney microsomal autoantibodies in idiopathic autoimmune hepatitis is cytochrome P450db1. *J Clin Invest* 83:1066–1072.

Marathe, P. H., Rodrigues, A. D. 2006. *In vivo* animal models for investigating potential CYP3A- and Pgp-mediated drug–drug interactions. *Curr Drug Metab* 7:687–704.

Martignoni, M., Groothuis, G. M., de Kanter, R. 2006. Species differences between mouse, rat, dog, monkey and human CYP-mediated drug metabolism, inhibition and induction. *Expert Opin Drug Metab Toxicol* 2:875–894.

Martinez-Ceballos, E., Burdsal, C. A. 2001. Differential expression of chicken CYP26 in anterior versus posterior limb bud in response to retinoic acid. *J Exp Zool* 290:136–147.

Masubuchi, Y., Kagimoto, N., Narimatsu, S., Fujita, S., Suzuki, T. 1993. Regioselective contribution of the cytochrome P-450 2D subfamily to propranolol metabolism in rat liver microsomes. *Drug Metab Dispos* 21:1012–1016.

Masubuchi, Y., Hosokawa, S., Horie, T., Suzuki, T., Ohmori, S., Kitada, M., Narimatsu, S. 1994. Cytochrome P450 isozymes involved in propranolol metabolism in human liver microsomes. The role of CYP2D6 as ring-hydroxylase and CYP1A2 as N-desisopropylase. *Drug Metab Dispos* 22:909–915.

Masubuchi, Y., Yamamoto, K., Suzuki, T., Horie, T., Narimatsu, S. 1996. Characterization of the oxidation reactions catalyzed by CYP2D enzyme in rat renal microsomes. *Life Sci* 58:2431–2437.

Masubuchi, Y., Iwasa, T., Hosokawa, S., Suzuki, T., Horie, T., Imaoka, S., Funae, Y., Narimatsu, S. 1997. Selective deficiency of debrisoquine 4-hydroxylase activity in mouse liver microsomes. *J Pharmacol Exp Ther* 282:1435–1441.

Mathew, P. A., Kagawa, N., Bhasker, C. R., Waterman, M. R. 1990. Deduced amino acid sequence of heme binding region of chicken cholesterol side chain cleavage cytochrome P450. *Protein Seq Data Anal* 3:323–325.

Matsunaga, E., Gonzalez, F. J. 1990. Specific cytosine demethylations within the first exons of the rat *CYP2D3* and *CYP2D5* genes are associated with activation of hepatic gene expression during development. *DNA Cell Biol* 9:443–452.

Matsunaga, E., Zanger, U. M., Hardwick, J. P., Gelboin, H. V., Meyer, U. A., Gonzalez, F. J. 1989. The CYP2D gene subfamily: Analysis of the molecular basis of the debrisoquine 4-hydroxylase deficiency in DA rats. *Biochemistry* 28:7349–7355.

McLaughlin, L. A., Dickmann, L. J., Wolf, C. R., Henderson, C. J. 2008. Functional expression and comparative characterization of nine murine cytochromes P450 by fluorescent inhibition screening. *Drug Metab Dispos* 36:1322–1331.

McMillan, D. M., Tyndale, R. F. 2015. Nicotine increases codeine analgesia through the induction of brain CYP2D and central activation of codeine to morphine. *Neuropsychopharmacology* 40:1804–1812.

Mechan, A. O., Moran, P. M., Elliott, M., Young, A. J., Joseph, M. H., Green, R. 2002. A comparison between Dark Agouti and Sprague–Dawley rats in their behaviour on the elevated plus-maze, open-field apparatus and activity meters, and their response to diazepam. *Psychopharmacology (Berl)* 159:188–195.

Miksys, S., Rao, Y., Sellers, E. M., Kwan, M., Mendis, D., Tyndale, R. F. 2000. Regional and cellular distribution of CYP2D subfamily members in rat brain. *Xenobiotica* 30:547–564.

Miksys, S., Rao, Y., Hoffmann, E., Mash, D. C., Tyndale, R. F. 2002. Regional and cellular expression of CYP2D6 in human brain: Higher levels in alcoholics. *J Neurochem* 82:1376–1387.

Miksys, S. L., Cheung, C., Gonzalez, F. J., Tyndale, R. F. 2005. Human CYP2D6 and mouse CYP2Ds: Organ distribution in a humanized mouse model. *Drug Metab Dispos* 33:1495–1502.

Minamiyama, Y., Takemura, S., Imaoka, S., Funae, Y., Tanimoto, Y., Inoue, M. 1997. Irreversible inhibition of cytochrome P450 by nitric oxide. *J Pharmacol Exp Ther* 283:1479–1485.

Mitschke, D., Reichel, A., Fricker, G., Moenning, U. 2008. Characterization of cytochrome P450 protein expression along the entire length of the intestine of male and female rats. *Drug Metab Dispos* 36:1039–1045.

Mohri, K., Uesawa, Y., Sagawa, K. 2000. Effects of long-term grapefruit juice ingestion on nifedipine pharmacokinetics: Induction of rat hepatic P-450 by grapefruit juice. *Drug Metab Dispos* 28:482–486.

Nakamura, A., Yamamoto, Y., Tasaki, T., Sugimoto, C., Masuda, M., Kazusaka, A., Fujita, S. 1995. Purification and characterization of a dog cytochrome P450 isozyme belonging to the CYP2D subfamily and development of its antipeptide antibody. *Drug Metab Dispos* 23:1268–1273.

Narimatsu, S., Tachibana, M., Masubuchi, Y., Imaoka, S., Funae, Y., Suzuki, T. 1995. Cytochrome P450 isozymes involved in aromatic hydroxylation and side-chain *N*-desisopropylation of alprenolol in rat liver microsomes. *Biol Pharm Bull* 18:1060–1065.

Narimatsu, S., Tachibana, M., Masubuchi, Y., Suzuki, T. 1996. Cytochrome P4502D and -2C enzymes catalyze the oxidative *N*-demethylation of the parkinsonism-inducing substance 1-methyl-4-phenyl-1,2,3,6-tetrahydropyridine in rat liver microsomes. *Chem Res Toxicol* 9:93–98.

Narimatsu, S., Nakata, T., Shimizudani, T., Nagaoka, K., Nakura, H., Masuda, K., Katsu, T. et al. 2011. Regio- and stereoselective oxidation of propranolol enantiomers by human CYP2D6, cynomolgus monkey CYP2D17 and marmoset CYP2D19. *Chem Biol Interact* 189:146–152.

Nebert, D. W., Russell, D. W. 2002. Clinical importance of the cytochromes P450. *Lancet* 360:1155–1162.

Nebert, D. W., Dalton, T. P. 2006. The role of cytochrome P450 enzymes in endogenous signalling pathways and environmental carcinogenesis. *Nat Rev Cancer* 6:947–960.

Nebert, D. W., Wikvall, K., Miller, W. L. 2013. Human cytochromes P450 in health and disease. *Philos Trans R Soc Lond B Biol Sci* 368:20120431.

Nelson, D. R., Zeldin, D. C., Hoffman, S. M., Maltais, L. J., Wain, H. M., Nebert, D. W. 2004. Comparison of cytochrome P450 (CYP) genes from the mouse and human genomes, including nomenclature recommendations for genes, pseudogenes and alternative-splice variants. *Pharmacogenetics* 14:1–18.

Nelson, D. R., Goldstone, J. V., Stegeman, J. J. 2013. The cytochrome P450 genesis locus: The origin and evolution of animal cytochrome P450s. *Philos Trans R Soc Lond B Biol Sci* 368:20120474.

Neville, C. F., Ninomiya, S., Shimada, N., Kamataki, T., Imaoka, S., Funae, Y. 1993. Characterization of specific cytochrome P450 enzymes responsible for the metabolism of diazepam in hepatic microsomes of adult male rats. *Biochem Pharmacol* 45:59–65.

Newton, D. J., Wang, R. W., Lu, A. Y. 1995. Cytochrome P450 inhibitors. Evaluation of specificities in the in vitrometabolism of therapeutic agents by human liver microsomes. *Drug Metab Dispos* 23:154–158.

Nguyen, A. D., Conley, A. J. 2008. Adrenal androgens in humans and nonhuman primates: Production, zonation and regulation. *Endocr Dev* 13:33–54.

Nishibe, Y., Wakabayashi, M., Harauchi, T., Ohno, K. 1998. Characterization of cytochrome P450 (CYP3A12) induction by rifampicin in dog liver. *Xenobiotica* 28:549–557.

Nishida, C. R., Everett, S., Ortiz de Montellano, P. R. 2013. Specificity determinants of CYP1B1 estradiol hydroxylation. *Mol Pharmacol* 84:451–458.

Nissen, P. H., Wintero, A. K., Fredholm, M. 1998. Mapping of porcine genes belonging to two different cytochrome P450 subfamilies. *Anim Genet* 29:7–11.

Ohmori, S., Taniguchi, T., Rikihisa, T., Kanakubo, Y., Kitada, M. 1993. Species differences of testosterone 16-hydroxylases in liver microsomes of guinea pig, rat and dog. *Xenobiotica* 23:419–426.

Oka, T., Fukuhara, M., Ushio, F., Kurose, K. 2000. Molecular cloning and characterization of three novel cytochrome P450 2D isoforms, CYP2D20, CYP2D27, and CYP2D28 in the Syrian hamster (Mesocricetus auratus). *Comp Biochem Physiol C Toxicol Pharmacol* 127:143–152.

Okamatsu, G., Komatsu, T., Kubota, A., Onaga, T., Uchide, T., Endo, D., Kirisawa, R. et al. 2014. Identification and functional characterization of novel feline cytochrome P450 2A. *Xenobiotica* 1–8.

Okazaki, Y., Furuno, M., Kasukawa, T., Adachi, J., Bono, H., Kondo, S., Nikaido, I. et al. 2002. Analysis of the mouse transcriptome based on functional annotation of 60,770 full-length cDNAs. *Nature* 420:563–573.

Olesen, O. V., Linnet, K. 1999. Studies on the stereoselective metabolism of citalopram by human liver microsomes and cDNA-expressed cytochrome P450 enzymes. *Pharmacology* 59:298–309.

Ono, S., Hatanaka, T., Miyazawa, S., Tsutsui, M., Aoyama, T., Gonzalez, F. J., Satoh, T. 1996. Human liver microsomal diazepam metabolism using cDNA-expressed cytochrome P450s: Role of CYP2B6, 2C19 and the 3A subfamily. *Xenobiotica* 26:1155–1166.

Ourlin, J. C., Baader, M., Fraser, D., Halpert, J. R., Meyer, U. A. 2000. Cloning and functional expression of a first inducible avian cytochrome P450 of the CYP3A subfamily (CYP3A37). *Arch Biochem Biophys* 373:375–384.

Paine, M. J., McLaughlin, L. A., Flanagan, J. U., Kemp, C. A., Sutcliffe, M. J., Roberts, G. C., Wolf, C. R. 2003. Residues glutamate 216 and aspartate 301 are key determinants of substrate specificity and product regioselectivity in cytochrome P450 2D6. *J Biol Chem* 278:4021–4027.

Paulson, S. K., Engel, L., Reitz, B., Bolten, S., Burton, E. G., Maziasz, T. J., Yan, B., Schoenhard, G. L. 1999. Evidence for polymorphism in the canine metabolism of the cyclooxygenase 2 inhibitor, celecoxib. *Drug Metab Dispos* 27:1133–1142.

Postlind, H., Axen, E., Bergman, T., Wikvall, K. 1997. Cloning, structure, and expression of a cDNA encoding vitamin D3 25-hydroxylase. *Biochem Biophys Res Commun* 241:491–497.

Reichen, J., Krahenbuhl, S., Kupfer, A., Sagesser, H., Karlaganis, G. 1986. Defective bile acid transport in an animal model of defective debrisoquine hydroxylation. *Biochem Pharmacol* 35:753–759.

Riedl, A. G., Watts, P. M., Edwards, R. J., Schulz-Utermoehl, T., Boobis, A. R., Jenner, P., Marsden, C. D. 1999. Expression and localisation of CYP2D enzymes in rat basal ganglia. *Brain Res* 822:175–191.

Rochat, B., Amey, M., Gillet, M., Meyer, U. A., Baumann, P. 1997. Identification of three cytochrome P450 isozymes involved in N-demethylation of citalopram enantiomers in human liver microsomes. *Pharmacogenetics* 7:1–10.

Roelofs, M. J., Piersma, A. H., van den Berg, M. van Duursen, M. B. 2013. The relevance of chemical interactions with CYP17 enzyme activity: Assessment using a novel in vitro assay. *Toxicol Appl Pharmacol* 268:309–317.

Roussel, F., Duignan, D. B., Lawton, M. P., Obach, R. S., Strick, C. A., Tweedie, D. J. 1998. Expression and characterization of canine cytochrome P450 2D15. *Arch Biochem Biophys* 357:27–36.

Rowland, P., Blaney, F. E., Smyth, M. G., Jones, J. J., Leydon, V. R., Oxbrow, A. K., Lewis, C. J. et al. 2006. Crystal structure of human cytochrome P450 2D6. *J Biol Chem* 281:7614–7622.

Sahi, J., Reyner, E. L., Bauman, J. N., Gueneva-Boucheva, K., Burleigh, J. E., Thomas, V. H. 2002. The effect of bergamottin on diazepam plasma levels and P450 enzymes in beagle dogs. *Drug Metab Dispos* 30:135–140.

Saito, K., Kim, H. S., Sakai, N., Ishizuka, M., Kazusaka, A., Fujita, S. 2004a. Polymorphism in diazepam metabolism in Wistar rats. *J Pharm Sci* 93:1271–1278.

Saito, K., Sakai, N., Kim, H. S., Ishizuka, M., Kazusaka, A., Fujita, S. 2004b. Strain differences in diazepam metabolism at its three metabolic sites in Sprague–Dawley, Brown Norway, Dark Agouti, and Wistar strain rats. *Drug Metab Dispos* 32:959–965.

Sakai, N., Saito, K., Kim, H. S., Kazusaka, A., Ishizuka, M., Funae, Y., Fujita, S. 2005. Importance of CYP2D3 in polymorphism of diazepam p-hydroxylation in rats. *Drug Metab Dispos* 33:1657–1660.

Sakai, N., Sakamoto, K. Q., Fujita, S., Ishizuka, M. 2009. The importance of heterogeneous nuclear ribonucleoprotein K on cytochrome P450 2D2 gene regulation: Its binding is reduced in Dark Agouti rats. *Drug Metab Dispos* 37:1703–1710.

Sakamoto, K., Kirita, S., Baba, T., Nakamura, Y., Yamazoe, Y., Kato, R., Takanaka, A., Matsubara, T. 1995. A new cytochrome P450 form belonging to the CYP2D in dog liver microsomes: Purification, cDNA cloning, and enzyme characterization. *Arch Biochem Biophys* 319:372–382.

Sakuma, T., Shimojima, T., Miwa, K., Kamataki, T. 2004. Cloning CYP2D21 and CYP3A22 cDNAs from liver of miniature pigs. *Drug Metab Dispos* 32:376–378.

Scheer, N., Kapelyukh, Y., McEwan, J., Beuger, V., Stanley, L. A., Rode, A., Wolf, C. R. 2012. Modeling human cytochrome P450 2D6 metabolism and drug–drug interaction by a novel panel of knockout and humanized mouse lines. *Mol Pharmacol* 81:63–72.

Scheer, N., McLaughlin, L. A., Rode, A., Macleod, A. K., Henderson, C. J., Wolf, C. R. 2014. Deletion of 30 murine cytochrome p450 genes results in viable mice with compromised drug metabolism. *Drug Metab Dispos* 42:1022–1030.

Scherr, M. C., Lourenco, G. J., Albuquerque, D. M., Lima, C. S. 2011. Polymorphism of cytochrome P450 A2 (CYP1A2) in pure and mixed breed dogs. *J Vet Pharmacol Ther* 34:184–186.

Schulz, T. G., Thiel, R., Neubert, D., Brassil, P. J., Schulz-Utermoehl, T., Boobis, A. R., Edwards, R. J. 2001. Assessment of P450 induction in the marmoset monkey using targeted anti-peptide antibodies. *Biochim Biophys Acta* 1546:143–155.

Schulz-Utermoehl, T., Bennett, A. J., Ellis, S. W., Tucker, G. T., Boobis, A. R., Edwards, R. J. 1999. Polymorphic debrisoquine 4-hydroxylase activity in the rat is due to differences in CYP2D2 expression. *Pharmacogenetics* 9:357–366.

Shang, H., Guo, K., Liu, Y., Yang, J., Wei, H. 2013. Constitutive expression of CYP3A mRNA in Bama miniature pig tissues. *Gene* 524:261–267.

Shimizudani, T., Nagaoka, K., Hanioka, N., Yamano, S., Narimatsu, S. 2010. Comparative study of the oxidation of propranolol enantiomers in hepatic and small intestinal microsomes from cynomolgus and marmoset monkeys. *Chem Biol Interact* 183:67–78.

Shiraga, T., Iwasaki, K., Nozaki, K., Tamura, T., Yamazoe, Y., Kato, R., Takanaka, A. 1994. Isolation and characterization of four cytochrome P450 isozymes from untreated and phenobarbital-treated beagle dogs. *Biol Pharm Bull* 17:22–28.

Shou, M., Norcross, R., Sandig, G., Lu, P., Li, Y., Lin, Y., Mei, Q., Rodrigues, A. D., Rushmore, T. H. 2003. Substrate specificity and kinetic properties of seven heterologously expressed dog cytochromes p450. *Drug Metab Dispos* 31:1161–1169.

Shuster, D. L., Bammler, T. K., Beyer, R. P., Macdonald, J. W., Tsai, J. M., Farin, F. M., Hebert, M. F., Thummel, K. E., Mao, Q. 2013. Gestational age-dependent changes in gene expression of metabolic enzymes and transporters in pregnant mice. *Drug Metab Dispos* 41:332–342.

Sindrup, S. H., Brosen, K., Hansen, M. G., Aaes-Jorgensen, T., Overo, K. F., Gram, L. F. 1993. Pharmacokinetics of citalopram in relation to the sparteine and the mephenytoin oxidation polymorphisms. *Ther Drug Monit* 15:11–17.

Sivapathasundaram, S., Magnisali, P., Coldham, N. G., Howells, L. C., Sauer, M. J., Ioannides, C. 2001. A study of the expression of the xenobiotic-metabolising cytochrome P450 proteins and of testosterone metabolism in bovine liver. *Biochem Pharmacol* 62:635–645.

Skaanild, M. T., Friis, C. 1999. Cytochrome P450 sex differences in minipigs and conventional pigs. *Pharmacol Toxicol* 85:174–180.

Skaanild, M. T., Friis, C. 2002. Is cytochrome P450 CYP2D activity present in pig liver? *Pharmacol Toxicol* 91:198–203.

Staack, R. F., Paul, L. D., Springer, D., Kraemer, T., Maurer, H. H. 2004. Cytochrome P450 dependent metabolism of the new designer drug 1-(3-trifluoromethylphenyl)piperazine (TFMPP). *In vivo* studies in Wistar and Dark Agouti rats as well as *in vitro* studies in human liver microsomes. *Biochem Pharmacol* 67:235–244.

Steen, V. M., Andreassen, O. A., Daly, A. K., Tefre, T., Borresen, A. L., Idle, J. R., Gulbrandsen, A. K. 1995. Detection of the poor metabolizer-associated *CYP2D6(D)* gene deletion allele by long-PCR technology. *Pharmacogenetics* 5:215–223.

Stepaniak, J. A., Gould, K. E., Sun, D., Swanborg, R. H. 1995. A comparative study of experimental autoimmune encephalomyelitis in Lewis and DA rats. *J Immunol* 155:2762–2769.

Stresser, D. M., Turner, S. D., Blanchard, A. P., Miller, V. P., Crespi, C. L. 2002. Cytochrome P450 fluorometric substrates: Identification of isoform-selective probes for rat CYP2D2 and human CYP3A4. *Drug Metab Dispos* 30:845–852.

Sun, J., Peng, Y., Wu, H., Zhang, X., Zhong, Y., Xiao, Y., Zhang, F. et al. 2015. Guanfu base A, an antiarrhythmic alkaloid of *Aconitum coreanum*, is a CYP2D6 inhibitor of human, monkey, and dog isoforms. *Drug Metab Dispos* 43:713–724.

Suzuki, T., Narimatsu, S., Fujita, S., Masubuchi, Y., Umeda, S., Imaoka, S., Funae, Y. 1992. Purification and characterization of a cytochrome P-450 isozyme catalyzing bunitrolol 4-hydroxylation in liver microsomes of male rats. *Drug Metab Dispos* 20:367–373.

Tasaki, T., Nakamura, A., Itoh, S., Ohashi, K., Yamamoto, Y., Masuda, M., Iwata, H., Kazusaka, A., Kamataki, T., Fujita, S. 1998. Expression and characterization of dog CYP2D15 using baculovirus expression system. *J Biochem* 123:162–168.

Teh, L. K., Bertilsson, L. 2012. Pharmacogenomics of CYP2D6: Molecular genetics, interethnic differences and clinical importance. *Drug Metab Pharmacokinet* 27:55–67.

Tenneze, L., Verstuyft, C., Becquemont, L., Poirier, J. M., Wilkinson, G. R., Funck-Brentano, C. 1999. Assessment of CYP2D6 and CYP2C19 activity in vivo in humans: A cocktail study with dextromethorphan and chloroguanide alone and in combination. *Clin Pharmacol Ther* 66:582–588.

Thompson, C. M., Kawashima, H., Strobel, H. W. 1998. Isolation of partially purified P450 2D18 and characterization of activity toward the tricyclic antidepressants imipramine and desipramine. *Arch Biochem Biophys* 359:115–121.

Thompson, C. M., Capdevila, J. H., Strobel, H. W. 2000. Recombinant cytochrome P450 2D18 metabolism of dopamine and arachidonic acid. *J Pharmacol Exp Ther* 294:1120–1130.

Topletz, A. R., Le, H. N., Lee, N., Chapman, J. D., Kelly, E. J., Wang, J., Isoherranen, N. 2013. Hepatic Cyp2d and Cyp26a1 mRNAs and activities are increased during mouse pregnancy. *Drug Metab Dispos* 41:312–319.

Tsuneoka, Y., Matsuo, Y., Higuchi, R., Ichikawa, Y. 1992. Characterization of the cytochrome P-450IID subfamily in bovine liver. Nucleotide sequences and microheterogeneity. *Eur J Biochem* 208:739–746.

Tsutsui, M., Ogawa, S., Inada, Y., Tomioka, E., Kamiyoshi, A., Tanaka, S., Kishida, T. et al. 2006. Characterization of cytochrome P450 expression in murine embryonic stem cell-derived hepatic tissue system. *Drug Metab Dispos* 34:696–701.

Tyden, E., Tjalve, H., Larsson, P. 2014. Gene and protein expression and cellular localisation of cytochrome P450 enzymes of the 1A, 2A, 2C, 2D and 2E subfamilies in equine intestine and liver. *Acta Vet Scand* 56:69.

Tyndale, R. F., Li, Y., Li, N. Y., Messina, E., Miksys, S., Sellers, E. M. 1999. Characterization of cytochrome P-450 2D1 activity in rat brain: High-affinity kinetics for dextromethorphan. *Drug Metab Dispos* 27:924–930.

Uchida, T., Komori, M., Kitada, M., Kamataki, T. 1990. Isolation of cDNAs coding for three different forms of liver microsomal cytochrome P-450 from polychlorinated biphenyl-treated beagle dogs. *Mol Pharmacol* 38:644–651.

Uehara, S., Uno, Y., Hagihira, Y., Murayama, N., Shimizu, M., Inoue, T., Sasaki, E., Yamazaki, H. 2015. Marmoset cytochrome P450 2D8 in livers and small intestines metabolizes typical human P450 2D6 substrates, metoprolol, bufuralol and dextromethorphan. *Xenobiotica* 45:766–772.

Umehara, K., Kudo, S., Odomi, M. 1997. Involvement of CYP2D1 in the metabolism of carteolol by male rat liver microsomes. *Xenobiotica* 27:1121–1129.

Umehara, K., Kudo, S., Hirao, Y., Morita, S., Ohtani, T., Uchida, M., Miyamoto, G. 2000. In vitro characterization of the oxidative cleavage of the octyl side chain of olanexidine, a novel antimicrobial agent, in dog liver microsomes. *Drug Metab Dispos* 28:1417–1424.

Umehara, K., Shimokawa, Y., Koga, T., Ohtani, T., Miyamoto, G. 2004. Oxidative one-carbon cleavage of the octyl side chain of olanexidine, a novel antimicrobial agent, in dog liver microsomes. *Xenobiotica* 34:61–71.

Uno, Y., Uehara, S., Kohara, S., Murayama, N., Yamazaki, H. 2010. Cynomolgus monkey CYP2D44 newly identified in liver, metabolizes bufuralol, and dextromethorphan. *Drug Metab Dispos* 38:1486–1492.

Uno, Y., Uehara, S., Kohara, S., Murayama, N., Yamazaki, H. 2014. Polymorphisms of CYP2D17 in cynomolgus and rhesus macaques: An evidence of the genetic basis for the variability of CYP2D-dependent drug metabolism. *Drug Metab Dispos* 42:1407–1410.

Villalpando, I., Sanchez-Bringas, G., Sanchez-Vargas, I., Pedernera, E., Villafan-Monroy, H. 2000. The P450 aromatase (P450 arom) gene is asymmetrically expressed in a critical period for gonadal sexual differentiation in the chick. *Gen Comp Endocrinol* 117:325–334.

Vinson, G. P., Pudney, J. A., Whitehouse, B. J. 1985. The mammalian adrenal circulation and the relationship between adrenal blood flow and steroidogenesis. *J Endocrinol* 105:285–294.

von Moltke, L. L., Greenblatt, D. J., Grassi, J. M., Granda, B. W., Venkatakrishnan, K., Duan, S. X., Fogelman, S. M., Harmatz, J. S., Shader, R. I. 1999. Citalopram and desmethylcitalopram *in vitro*: Human cytochromes mediating transformation, and cytochrome inhibitory effects. *Biol Psychiatry* 46:839–849.

von Moltke, L. L., Greenblatt, D. J., Giancarlo, G. M., Granda, B. W., Harmatz, J. S., Shader, R. I. 2001. Escitalopram (*S*-citalopram) and its metabolites *in vitro*: Cytochromes mediating biotransformation, inhibitory effects, and comparison to *R*-citalopram. *Drug Metab Dispos* 29:1102–1109.

Vuppugalla, R., Mehvar, R. 2004a. Hepatic disposition and effects of nitric oxide donors: Rapid and concentration-dependent reduction in the cytochrome P450-mediated drug metabolism in isolated perfused rat livers. *J Pharmacol Exp Ther* 310:718–727.

Vuppugalla, R., Mehvar, R. 2004b. Short-term inhibitory effects of nitric oxide on cytochrome P450-mediated drug metabolism: Time dependency and reversibility profiles in isolated perfused rat livers. *Drug Metab Dispos* 32:1446–1454.

Vuppugalla, R., Mehvar, R. 2005. Enzyme-selective effects of nitric oxide on affinity and maximum velocity of various rat cytochromes P450. *Drug Metab Dispos* 33:829–836.

Vuppugalla, R., Mehvar, R. 2006. Selective effects of nitric oxide on the disposition of chlorzoxazone and dextromethorphan in isolated perfused rat livers. *Drug Metab Dispos* 34:1160–1166.

Waller, S. C., He, Y. A., Harlow, G. R., He, Y. Q., Mash, E. A., Halpert, J. R. 1999. 2,2′,3,3′,6,6′-hexachlorobiphenyl hydroxylation by active site mutants of cytochrome P450 2B1 and 2B11. *Chem Res Toxicol* 12:690–699.

Wan, J., Imaoka, S., Chow, T., Hiroi, T., Yabusaki, Y., Funae, Y. 1997. Expression of four rat CYP2D isoforms in *Saccharomyces cerevisiae* and their catalytic specificity. *Arch Biochem Biophys* 348:383–390.

Wang, A., Savas, U., Hsu, M. H., Stout, C. D., Johnson, E. F. 2012. Crystal structure of human cytochrome P450 2D6 with prinomastat bound. *J Biol Chem* 287:10834–10843.

Wang, X., Li, J., Dong, G., Yue, J. 2014. The endogenous substrates of brain CYP2D. *Eur J Pharmacol* 724:211–218.

Wang, Q., Han, X., Li, J., Gao, X., Wang, Y., Liu, M., Dong, G., Yue, J. 2015. Regulation of cerebral CYP2D alters tramadol metabolism in the brain: Interactions of tramadol with propranolol and nicotine. *Xenobiotica* 45:335–344.

Watanabe, T., Sugiura, T., Manabe, S., Takasaki, W., Ohashi, Y. 2004. Low glutathione S-transferase dogs. *Arch Toxicol* 78:218–225.

Watanabe, K. P., Kawai, Y. K., Ikenaka, Y., Kawata, M., Ikushiro, S., Sakaki, T., Ishizuka, M. 2013. Avian cytochrome P450 (CYP) 1–3 family genes: Isoforms, evolutionary relationships, and mRNA expression in chicken liver. *PLoS One* 8:e75689.

Wink, D. A., Osawa, Y., Darbyshire, J. F., Jones, C. R., Eshenaur, S. C., Nims, R. W. 1993. Inhibition of cytochromes P450 by nitric oxide and a nitric oxide-releasing agent. *Arch Biochem Biophys* 300:115–123.

Wong, G., Itakura, T., Kawajiri, K., Skow, L., Negishi, M. 1989. Gene family of male-specific testosterone 16 alpha-hydroxylase (C-P-450(16 alpha)) in mice. Organization, differential regulation, and chromosome localization. *J Biol Chem* 264:2920–2927.

Woolhouse, N. M., Andoh, B., Mahgoub, A., Sloan, T. P., Idle, J. R., Smith, R. L. 1979. Debrisoquin hydroxylation polymorphism among Ghanaians and Caucasians. *Clin Pharmacol Ther* 26:584–591.

Yamamoto, Y., Ishizuka, M., Takada, A., Fujita, S. 1998a. Cloning, tissue distribution, and functional expression of two novel rabbit cytochrome P450 isozymes, CYP2D23 and CYP2D24. *J Biochem* 124:503–508.

Yamamoto, Y., Tasaki, T., Nakamura, A., Iwata, H., Kazusaka, A., Gonzalez, F. J., Fujita, S. 1998b. Molecular basis of the Dark Agouti rat drug oxidation polymorphism: Importance of CYP2D1 and CYP2D2. *Pharmacogenetics* 8:73–82.

Yamazaki, H., Guo, Z., Persmark, M., Mimura, M., Inoue, K., Guengerich, F. P., Shimada, T. 1994. Bufuralol hydroxylation by cytochrome P450 2D6 and 1A2 enzymes in human liver microsomes. *Mol Pharmacol* 46:568–577.

Yang, T. J., Shou, M., Korzekwa, K. R., Gonzalez, F. J., Gelboin, H. V., Yang, S. K. 1998. Role of cDNA-expressed human cytochromes P450 in the metabolism of diazepam. *Biochem Pharmacol* 55:889–896.

Yang, J., An, J., Li, M., Hou, X., Qiu, X. 2013. Characterization of chicken cytochrome P450 1A4 and 1A5: Inter-paralog comparisons of substrate preference and inhibitor selectivity. *Comp Biochem Physiol C Toxicol Pharmacol* 157:337–343.

Yasukochi, Y., Satta, Y. 2011. Evolution of the CYP2D gene cluster in humans and four non-human primates. *Genes Genet Syst* 86:109–116.

Yasukochi, Y., Satta, Y. 2015. Molecular evolution of the CYP2D subfamily in primates: Purifying selection on substrate recognition sites without the frequent or long-tract gene conversion. *Genome Biol Evol* 7:1053–1067.

Yokoi, T., Kosaka, Y., Chida, M., Chiba, K., Nakamura, H., Ishizaki, T., Kinoshita, M., Sato, K., Gonzalez, F. J., Kamataki, T. 1996. A new *CYP2D6* allele with a nine base insertion in exon 9 in a Japanese population associated with poor metabolizer phenotype. *Pharmacogenetics* 6:395–401.

Yoshida, K., Shimada, K., Saito, N. 1996. Expression of P450(17α) hydroxylase and P450 aromatase genes in the chicken gonad before and after sexual differentiation. *Gen Comp Endocrinol* 102:233–240.

Yoshimoto, K., Echizen, H., Chiba, K., Tani, M., Ishizaki, T. 1995. Identification of human CYP isoforms involved in the metabolism of propranolol enantiomers—N-desisopropylation is mediated mainly by CYP1A2. *Br J Clin Pharmacol* 39:421–431.

Yu, A. M., Haining, R. L. 2006. Expression, purification, and characterization of mouse CYP2d22. *Drug Metab Dispos* 34:1167–1174.

Yu, A. M., Idle, J. R., Gonzalez, F. J. 2004. Polymorphic cytochrome P450 2D6: Humanized mouse model and endogenous substrates. *Drug Metab Rev* 36:243–277.

Yuan, B. B., Tchao, R., Voigt, J. M., Colby, H. D. 1997. Localization of CYP2D16 in the guinea pig adrenal cortex by immunohistochemistry and in situ hybridization. *Mol Cell Endocrinol* 134:139–146.

Yuan, B. B., Tchao, R., Funae, Y., Voigt, J. M., Colby, H. D. 1998. Effects of ACTH administration on zonation of the guinea pig adrenal cortex. *Mol Cell Endocrinol* 146:129–136.

Yuan, B. B., Tchao, R., Voigt, J. M., Colby, H. D. 2001. Maturational changes in CYP2D16 expression and xenobiotic metabolism in adrenal glands from male and female guinea pigs. *Drug Metab Dispos* 29:194–199.

Zhang, K., Kuroha, M., Shibata, Y., Kokue, E., Shimoda, M. 2006. Effect of oral administration of clinically relevant doses of dexamethasone on regulation of cytochrome P450 subfamilies in hepatic microsomes from dogs and rats. *Am J Vet Res* 67:329–334.

Zhou, S. F. 2009a. Polymorphism of human cytochrome P450 2D6 and its clinical significance: Part I. *Clin Pharmacokinet* 48:689–723.

Zhou, S. F. 2009b. Polymorphism of human cytochrome P450 2D6 and its clinical significance: Part II. *Clin Pharmacokinet* 48:761–804.

Zhou, S. F., Liu, J. P., Lai, X. S. 2009. Substrate specificity, inhibitors and regulation of human cytochrome P450 2D6 and implications in drug development. *Curr Med Chem* 16: 2661–2805.

Zimin, A. V., Delcher, A. L., Florea, L., Kelley, D. R., Schatz, M. C., Puiu, D., Hanrahan, F. et al. 2009. A whole-genome assembly of the domestic cow, *Bos taurus*. *Genome Biol* 10:R42.

Zuo, Q., Li, D., Zhang, L., Elsayed, A. K., Lian, C., Shi, Q., Zhang, Z. et al. 2015. Study on the regulatory mechanism of the lipid metabolism pathways during chicken male germ cell differentiation based on RNA-seq. *PLoS One* 10:e0109469.

3 Substrates of Human CYP2D6

3.1 INTRODUCTION

CYP2D6 metabolizes ~25% of clinical drugs in the human liver (Cascorbi 2003; Gardiner and Begg 2006; Ingelman-Sundberg 2005; Ingelman-Sundberg et al. 2007; Zhou et al. 2009). Phenotypically, the CYP2D6 ultrarapid metabolizers (UMs), extensive metabolizers (EMs), intermediate metabolizers (IMs), and poor metabolizers (PMs) compose approximately 3%–5%, 70%–80%, 10%–17%, and 3%–7% of the Caucasian population, respectively (Sachse et al. 1997). There is a large interindividual variation in the enzyme activity of CYP2D6. The *CYP2D6* gene codes for a protein with 497 amino acids (Eichelbaum et al. 1987; Gough et al. 1993; Heim and Meyer 1990; Kimura et al. 1989). *CYP2D6* belongs to a gene cluster of highly homologous inactive pseudogenes *CYP2D7P* and *2D8P* (Heim and Meyer 1992; Kimura et al. 1989; Steen et al. 1995). CYP2D6 is also expressed in human kidney (Nishimura et al. 2003), intestine (Madani et al. 1999; Nishimura et al. 2003; Prueksaritanont et al. 1995), breast (Huang et al. 1997), lung (Bernauer et al. 2006; Guidice et al. 1997), placenta (Hakkola et al. 1996), and brain (Chinta et al. 2002; Miksys et al. 2002; Siegle et al. 2001) at low levels. In the resolved crystal structure of human CYP2D6 (Protein Data Bank code: 2F9Q) (Rowland et al. 2006), the active-site cavity is bordered by the heme and lined by residues Ile106, Leu110, Phe112, Phe120, Leu121, Gln117, Gly118, Val119, Ala122, Leu213, Glu216, Ser217, Leu220, Gln244, Phe247, Leu248, Ile297, Ala300, Asp301, Ser304, Ala305, Val308, Thr309, Val370, Met374, Gly373, Phe483, and Leu484. The 2D6 structure has a well-defined active-site cavity above the heme group with a volume of ~540 Å^3, which can readily accommodate a number of substrates with distinct structures. This chapter will highlight our current knowledge on the substrate specificity of human CYP2D6.

3.2 PROBES OF CYP2D6

3.2.1 SPARTEINE AND DEBRISOQUINE

Two prototypical substrates of CYP2D6 are sparteine (Ebner et al. 1995; Eichelbaum et al. 1979, 1986b; Tyndale et al. 1990) and debrisoquine (Boobis et al. 1983; Distlerath and Guengerich 1984; Gonzalez et al. 1988; Woolhouse et al. 1979), which are widely used to determine the phenotype of CYP2D6-mediated metabolism. Sparteine, a class 1a antiarrhythmic and oxytocic agent (not approved by the Food and Drug Administration [FDA] for human use), is initially identified as a natural quinolizidine alkaloid, which is extracted from Scotch broom (*Cytisus scoparius*) (Haferkamp 1950).

Sparteine is primarily metabolized by a polymorphic cytochrome P450 to an *N*-oxide via N^1-oxidation (Figure 3.1) (Eichelbaum et al. 1975). The *N*-oxide rearranges with loss of water to 2-dehydrosparteine (i.e., 2,3-didehydrosparteine) and 5-dehydrosparteine (i.e., 5,6-didehydrosparteine).

The following equation is used to determine the metabolic ratio (MR) of sparteine:

$$MR = \frac{\text{Urinary sparteine excretion}}{\text{Urinary 2-dehydrosparteine} + \text{urinary 5-dehydrosparteine excretion}}.$$

The formation of 2-dehydrosparteine in human liver microsomes from EM (MR < 2) and PM (MR > 20) subjects showed a more than 32-fold difference in K_m between EMs and PMs (58.3 vs. 1880 μM) (Osikowska-Evers et al. 1987). Neither EMs nor PMs of sparteine are sensitive to pentobarbital treatment (Schellens et al. 1989), which is a prototypical inducer of CYP2B6 (Madan et al. 2003). A 30% increase in metabolic clearance of sparteine is found in EMs after rifampicin administration at 600 or 1200 mg/day for 7 days, whereas in PMs, no effect on the overall elimination of sparteine is observed (Eichelbaum et al. 1986a). A marginal effect of prototypical enzyme inducers on polymorphic sparteine metabolism indicates that the enzyme for sparteine metabolism is predominantly under genetic control. Further studies indicate that defective sparteine metabolism is attributed to *CYP2D6* polymorphisms (Zanger et al. 2007).

Debrisoquine is used as an antihypertensive agent, and its 4-hydroxylation is primarily mediated by CYP2D6 (Eiermann et al. 1998; Gonzalez et al. 1988; Woolhouse et al. 1979). Debrisoquine is metabolized to 4-hydroxydebrisoquine, 3-hydroxydebrisoquine, and 1-hydroxydebrisoquine by CYP2D6 (Figure 3.2) (Eiermann et al. 1998), but CYP1A1 may also play a role (Granvil et al. 2002). Mahgoub et al. (1977) have reported one subject who excreted 60.8% of the dose as debrisoquine and only 5.4% as 4-hydroxydebrisoquine, corresponding to a MR of 11.3. Subsequent studies showed this volunteer to be a phenotypic PM, while other three subjects in this study had MRs of 0.5 to 1.4, typical of what is regarded as EMs (Mahgoub et al. 1977). This is the first report showing evidence of the existence of two phenotypes of drug oxidation, EM and PM. The MR of debrisoquine (= amount of debrisoquine/amount of 4-hydroxydebrisoquine in urine collected for 8 h) is used to determine the CYP2D6 phenotype. The bimodal distribution, with an antimode at 12.6 in Caucasian populations (Alvan et al. 1990; Bertilsson et al. 1992; Evans et al. 1980), makes it possible to distinguish between PMs (MR > 12.6) and EMs (MR < 12.6). Phenotype testing with debrisoquine or sparteine provides a cheap,

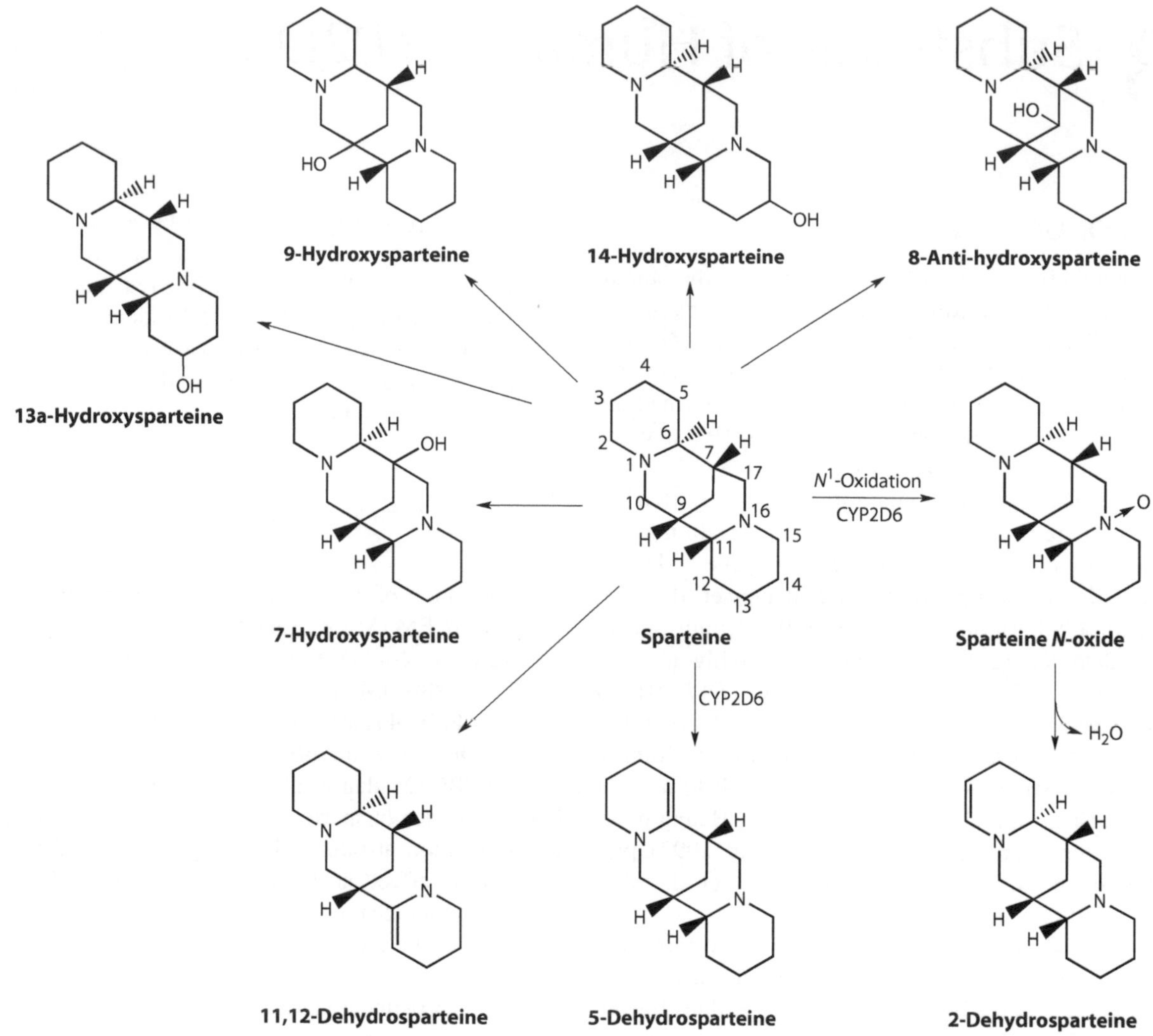

FIGURE 3.1 Metabolism of sparteine by CYP2D6. Sparteine is primarily metabolized by CYP2D6 to an *N*-oxide via *N¹*-oxidation. The *N*-oxide rearranges with loss of water to 2-dehydrosparteine (i.e., 2,3-didehydrosparteine) and 5-dehydrosparteine (i.e., 5,6-didehydrosparteine). In addition, a variety of hydroxylated (7-OH, 8-anti-OH, 9-OH, 13α-OH, and 14-OH) sparteine have been isolated from the genus *Lupinus*, but none of them have been identified in human or animal excreta.

reliable, and direct evaluation of the CYP2D6 metabolic phenotype.

3.2.2 DEXTROMETHORPHAN

Another probe substrate of CYP2D6 is dextromethorphan, which is a synthetic analog of narcotic analgesics. It is used as an over-the-counter antitussive agent that acts as an *N*-methyl-D-aspartate (NMDA) receptor antagonist (Bolser 2006; Cole et al. 1989). In humans, it is primarily excreted as unchanged parent drug and dextrorphan (Barnhart 1980), which is a pharmacologically active metabolite (Braga et al. 1994). The therapeutic activity of dextromethorphan is believed to be caused by both the drug and this metabolite (Braga et al. 1994). The formation of dextrorphan from dextromethorphan is primarily mediated by CYP2D6 via *O*-demethylation (Figure 3.3) (Dayer et al. 1989; Jacqz-Aigrain et al. 1993) and cosegregates with the debrisoquine/sparteine-type

(CYP2D6-mediated) oxidative polymorphism (Kerry et al. 1994; Sachse et al. 1997). Dextromethorphan is now established and widely used as a probe for determining CYP2D6 activity both in vivo (Kerry et al. 1994; Tenneze et al. 1999) and in vitro (Kerry et al. 1994). Dextromethorphan is also converted to 3-methoxymorphinan via *N*-demethylation and to 3-hydroxymorphinan via *N,O*-didemethylation in humans (Barnhart 1980). CYP3A4 is the principal enzyme involved in the *N*-demethylation pathway, with certain contribution from CYP2D6 (Jacqz-Aigrain et al. 1993; Yu and Haining 2001). Interestingly, dextromethorphan is a substrate of P-gp (Uhr et al. 2004).

3.2.3 BUFURALOL AND TRAMADOL

In addition, bufuralol, a β-adrenoceptor blocker, has been extensively used as a probe substrate for the in vitro study of CYP2D6 (Zanger et al. 2004). Tramadol, a synthetic opioid

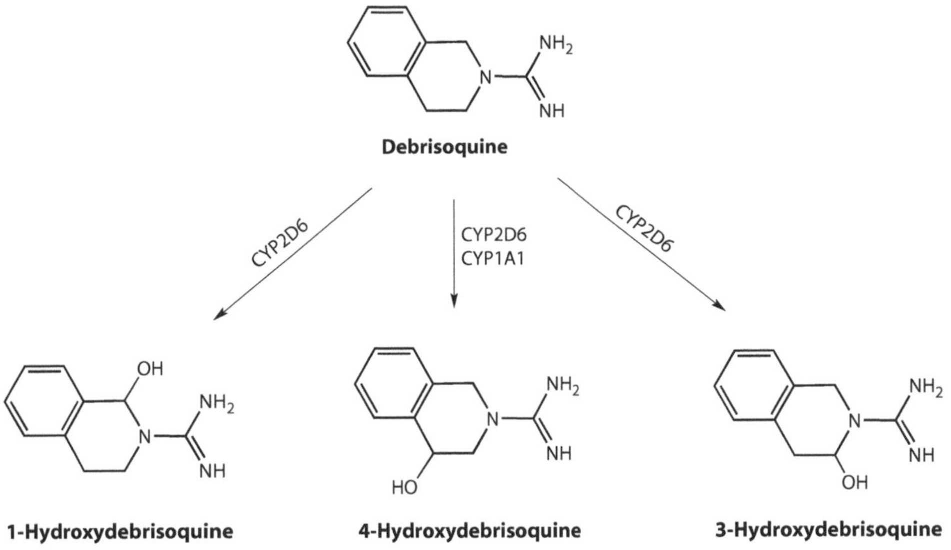

FIGURE 3.2 Metabolism of debrisoquine via 4-hydroxylation by CYP2D6. The 4-hydroxylation of debrisoquine is mainly catalyzed by CYP2D6 but with some contribution from CYP1A1. Debrisoquine is also metabolized by CYP2D6 to 1- and 3-hydroxylated metabolites to a minor extent.

analgesic of the aminocyclohexanol type that has two chiral centers, is also used as a probe substrate for CYP2D6. (−)-Tramadol is metabolized to its O-desmethyltramadol (M1) largely by CYP2D6, as is (+)-tramadol but to a lesser extent (Paar et al. 1997; Subrahmanyam et al. 2001). Among the above five probe substrates of CYP2D6, dextromethorphan (O-demethylation) and bufuralol (1′-hydroxylation) are the two commonly used CYP2D6 probe substrates in vitro preferred by pharmaceutical industry investigators, where 60% used bufuralol as the probe substrate and 30% used dextromethorphan (Yuan et al. 2002).

3.3 THERAPEUTIC DRUGS AS SUBSTRATES OF CYP2D6

3.3.1 Drugs Acting on the Central Nervous System

CYP2D6 can metabolize a number of drugs that target the central nervous system (CNS) (Gardiner and Begg 2006; Ingelman-Sundberg 2005; Zanger et al. 2004; Zhou et al. 2008, 2009), and among these are many drugs with a narrow therapeutic index. CNS drugs that are extensively metabolized by CYP2D6 include tricyclic antidepressants (e.g., clomipramine, imipramine, doxepin, desipramine, and nortriptyline), selective serotonin reuptake inhibitors (SSRIs; e.g., fluoxetine, fluvoxamine, and paroxetine), other nontricyclic antidepressants (atomoxetine, maprotiline, mianserin, and venlafaxine), neuroleptics (e.g., chlorpromazine, perphenazine, thioridazine, zotepine, zuclopenthixol, mianserin, olanzapine, risperidone, sertindole, and haloperidol), opioids (e.g., codeine, dihydrocodeine, and tramadol), and antiemetics (tropisetron, ondansetron, dolasetron, and metoclopramide) (Zanger et al. 2004; Zhou et al. 2008, 2009).

3.3.1.1 Tricyclic Antidepressants
All tricyclic antidepressants contain a core dibenzapine structure and the commonly used included imipramine, desipramine, clomipramine, amitriptyline, nortriptyline, maprotiline, trimipramine, protriptyline, amoxapine, and doxepin (Gillman 2007). Most tricyclic antidepressants have a low to moderate therapeutic index; they give rise to remarkable side effects at therapeutic concentrations and are dangerous when patients are overdosed (Gillman 2007). These agents are high-clearance drugs that undergo extensive Phase I and Phase II metabolism (Bertilsson 2007). The tertiary tricyclic antidepressants including amitriptyline (Mellstrom and von Bahr 1981; Venkatakrishnan et al. 2000, 2001), clomipramine (Nielsen et al. 1996), doxepin (Haritos et al. 2000; Hartter et al. 2002), imipramine (Lemoine et al. 1993; Venkatakrishnan et al. 1998), and trimipramine (Bolaji et al. 1993; Eap et al. 2000) are extensively metabolized by CYP2C19 and 2D6 to secondary amines via N-demethylation, with contribution from CYP2C9, 3A4, and 1A2. Both tertiary and secondary amines are active and the latter such as desipramine (from imipramine) and nortriptyline (from amitriptyline) are primarily metabolized by CYP2D6 and glucuronidation (Breyer-Pfaff et al. 1997; Olesen and Linnet 2000). Another metabolic pathway for tricyclic antidepressants (both tertiary and secondary amines) is hydroxylation of the tricyclic structure at one of the benzyl rings by CYP2D6, followed by glucuronidation.

CYP2D6 is responsible for the conversion of the tricyclic antidepressants amitriptyline and nortriptyline to E-10-hydroxyamitriptyline and E-10-hydroxynortriptyline via benzylic hydroxylation at position 10, respectively, whereas N-demethylation of amitriptyline to nortriptyline and of E-10-hydroxyamitriptyline to E-10-hydroxynortriptyline is mainly catalyzed by CYP2C19 and 2C9, 1A2, and 3A4

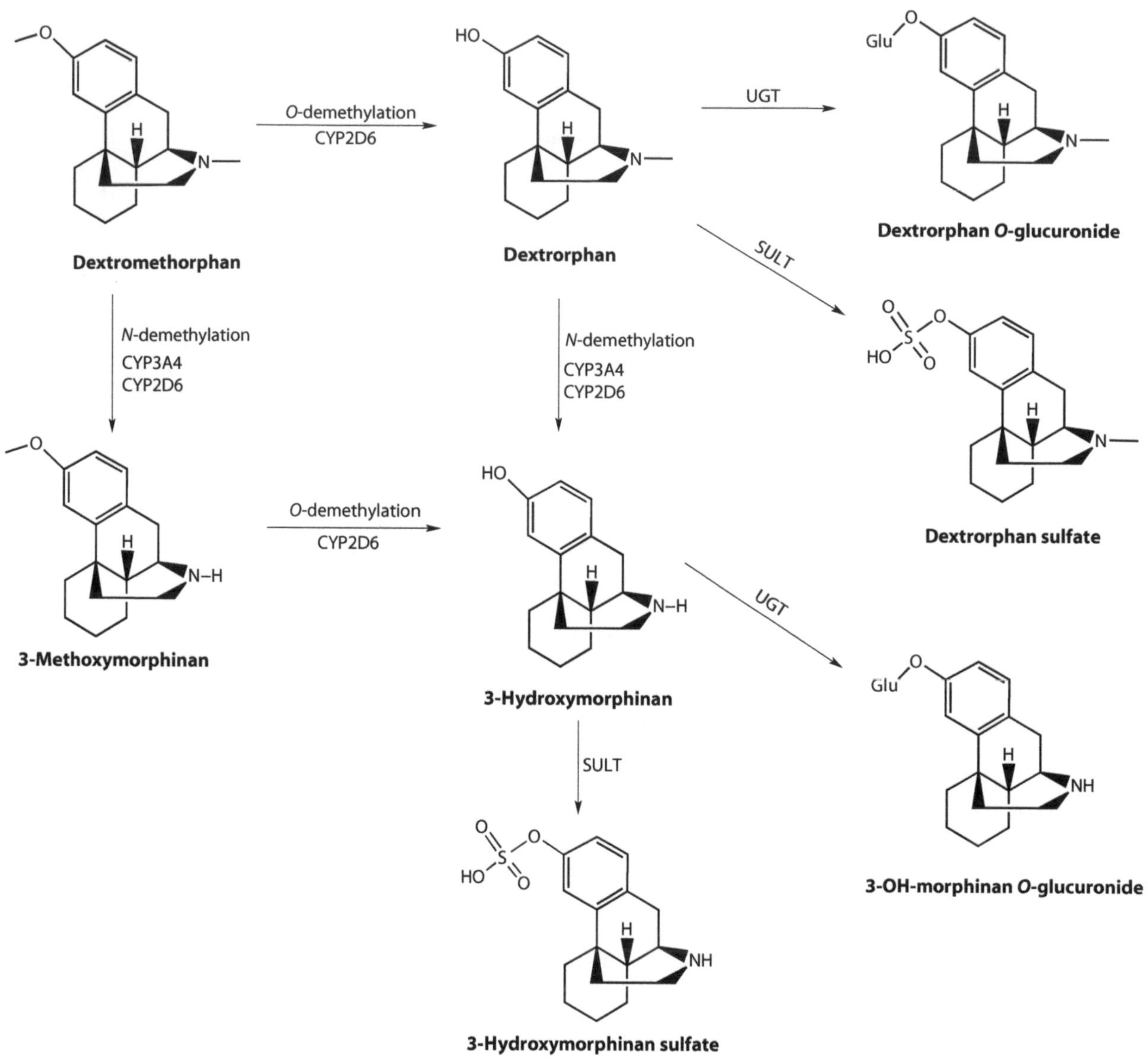

FIGURE 3.3 Metabolism of dextromethorphan by CYP2D6. In humans, dextromethorphan is primarily excreted as unchanged parent drug and dextrorphan, which is a pharmacologically active metabolite. The formation of dextrorphan from dextromethorphan is primarily mediated by CYP2D6 via *O*-demethylation and cosegregates with the debrisoquine–sparteine-type oxidative polymorphism. Dextromethorphan is also converted to 3-methoxymorphinan via *N*-demethylation and to 3-hydroxymorphinan via *N,O*-didemethylation in humans. CYP3A4 is the principal enzyme involved in the *N*-demethylation pathway, with contribution from CYP2D6.

(Figure 3.4) (Mellstrom and von Bahr 1981; Olesen and Linnet 1997; Venkatakrishnan et al. 2000, 2001). Minor pathways are *N*-demethylation and 10-hydroxylation to the *Z*-isomer of 10-OH-amitriptyline and 10-OH-nortriptyline. The hydroxylation of nortriptyline (a secondary amine) by CYP2D6 is highly stereoselective, mainly forming *R-E*-10-OH-nortriptyline (Breyer-Pfaff et al. 1992; Dahl et al. 1991b). *E*-10-hydroxynortriptyline is pharmacologically active, with approximately half the potency of the parent drug in inhibiting noradrenaline reuptake and greatly decreased anticholinergic activity (Nordin and Bertilsson 1995).

Clomipramine is converted to its active desmethylclomipramine by CYP2C19, 1A2, and 3A4, and both compounds are metabolized to respective hydroxy derivatives of the benzyl rings at positions 2, 8, and 10 largely by CYP2D6 (Figure 3.5) (Nielsen et al. 1996; Wu et al. 1998). CYP2D6 in yeast microsomes catalyzes the 8-hydroxylation of clomipramine and desmethylclomipramine (Nielsen et al. 1996). Clomipramine is also converted to 2- or 10-hydroxylated metabolite, which is further *N*-demethylated.

Doxepin is a tricyclic antidepressant that inhibits the reuptake of serotonin and noradrenaline from the synaptic cleft (Pinder et al. 1977a). Doxepin is given as a 15:85 mixture of *Z-cis-/E-trans*-isomers, with *Z-cis*-doxepin being considered to have greater antidepressive effect. Doxepin is extensively metabolized in the liver, and its Phase I and Phase II metabolites have been identified in human plasma, urine, and cerebrospinal fluid (Shu et al. 1990). The urinary metabolites

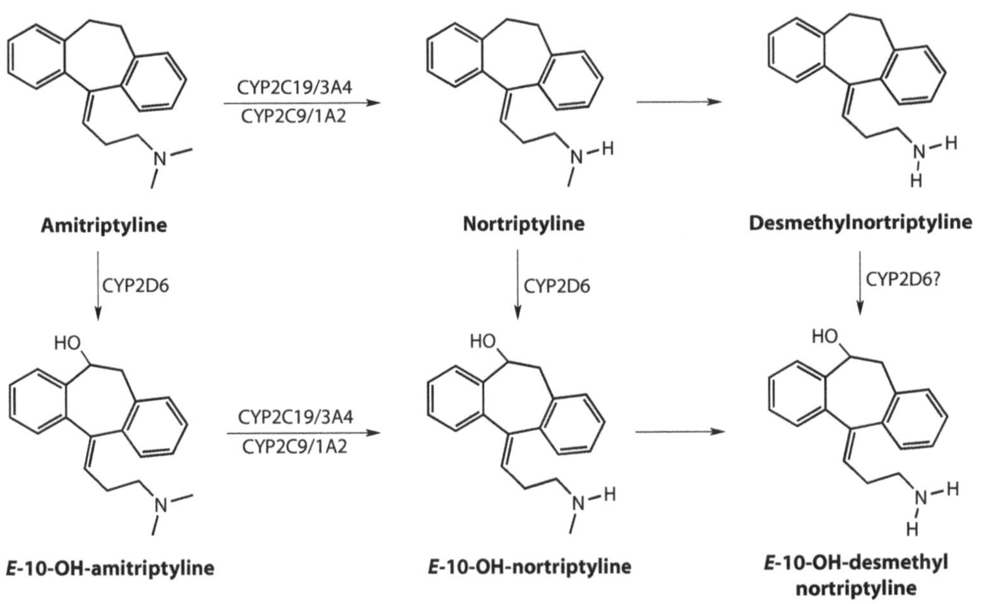

FIGURE 3.4 Metabolism of amitriptyline and nortriptyline by CYP2D6 and other enzymes. CYP2D6 is responsible for the conversion of amitriptyline and nortriptyline to *E*-10-hydroxyamitriptyline and *E*-10-hydroxynortriptyline via benzylic 10-hydroxylation, respectively, whereas demethylation of amitriptyline to nortriptyline and of *E*-10-hydroxyamitriptyline to *E*-10-hydroxynortriptyline is mainly catalyzed by CYP2C19 and 2C9, 1A2, and 3A4. The hydroxylation of nortriptyline by CYP2D6 is highly stereoselective, mainly forming *R*-*E*-10-OH-nortriptyline. Minor pathways are *N*-demethylation and 10-hydroxylation to the *Z*-isomer of 10-OH-amitriptyline and 10-OH-nortriptyline.

in humans are (*E*)-2-hydroxydoxepin, (*E*)-2-hydroxy-*N*-desmethyldoxepin, (*Z*)- and (*E*)-*N*-desmethyldoxepin, (*Z*)- and (*E*)-doxepin-*N*-oxide, (*E*)-2-*O*-glucuronyldoxepin, and a quaternary ammonium-linked glucuronide (Shu et al. 1990). The N^+-glucuronide is a major metabolite in patient urine after doxepin administration. CYP2D6 mainly converts the *E*-isomer of doxepin to its active metabolite 2-hydroxy derivative while the formation of *N*-desmethyldoxepin is mainly catalyzed by CYP2C19 and 3A4 (Figure 3.6) (Haritos et al. 2000; Hartter et al. 2002; Kirchheiner et al. 2002).

Imipramine, the first tricyclic antidepressant developed, is rapidly metabolized to desipramine via CYP2C19, 3A4, and 1A2 via *N*-demethylation, and both compounds are metabolized to their active 2-hydroxy derivatives by CYP2D6 (Figure 3.7) (Brosen et al. 1991; Koyama et al. 1997; Lemoine et al. 1993). Imipramine is also converted to 2-hydroxy metabolites, which underwent subsequent glucuronidation by UGT1A4 (Nakajima et al. 2002). Imipramine *N*-oxide and didesmethyl-imipramine are also detected in human plasma and urine as minor metabolites (Koyama et al. 1993).

Maprotiline is a tricyclic antidepressant acting as a highly selective norepinephrine reuptake inhibitor with very weak inhibitory effect on dopamine and 5-HT reuptake (Pinder et al. 1977b). It is metabolized extensively, with only 2% of the total dose being excreted unchanged in urine (Alkalay et al. 1980). Maprotiline is mainly metabolized by CYP2D6 (~83%) and 1A2 (~17%) to its major active metabolite des-methylmaprotiline (Figure 3.8) (Brachtendorf et al. 2002). Desmethylmaprotiline is further acetylated (Baumann et al. 1988). Formation of hydroxylated metabolites including 3-OH-maprotiline, 2-OH-maprotiline, and 2,3-dihydrodiol

to a lesser extent accounts for most of the remaining fraction (Breyer-Pfaff et al. 1985). Secondary metabolic steps generate a number of other metabolites. Several clinical inhibition studies indicate the important role of CYP2D6 in the metabolism of maprotiline. In depressed patients, flu-voxamine inhibits maprotiline's *N*-demethylation (Hartter et al. 1993). Risperidone also increases the plasma concentrations of maprotiline (Normann et al. 2002). In studies with therapy-resistant depressive patients treated with maprotiline, coadministration of moclobemide significantly increases maprotiline plasma levels (25%) (Konig et al. 1997). Moclobemide is a known inhibitor of CYP2D6, 1A2, and 2C19 (Gram et al. 1995).

Trimipramine, a tricyclic antidepressant used in the treatment of depression, is a moderate reuptake inhibitor of norepinephrine, and a weak reuptake inhibitor of serotonin and dopamine (Gastpar 1989). 2-Hydroxytrimipramine and 2,10- or 2,11-dihydroxytrimipramine are formed in vitro by CYP2D6 (Figure 3.9) (Bolaji et al. 1993). Acetylation of the latter metabolite results in dehydration at C10 to give 10,11-dehydro-2-acetoxytrimipramine. *N*-Glucuronide and glucuronides of the hydroxylated metabolites are also detected in human urine. A stereoselectivity in the metabolism of trimipramine has been observed, with a preferential *N*-demethylation of (D)-trimipramine and a preferential hydroxylation of (L)-trimipramine (Eap et al. 2000). CYP2D6 catalyzes the 2-hydroxylation of (L)-trimipramine and (L)- and (D)-desmethyltrimipramine, but not of (D)-trimipramine. CYP2C19, but not CYP2D6, is involved in the demethylation pathway, with a stereoselectivity toward (D)-trimipramine. CYP3A4/5 appears to be involved in the metabolism of

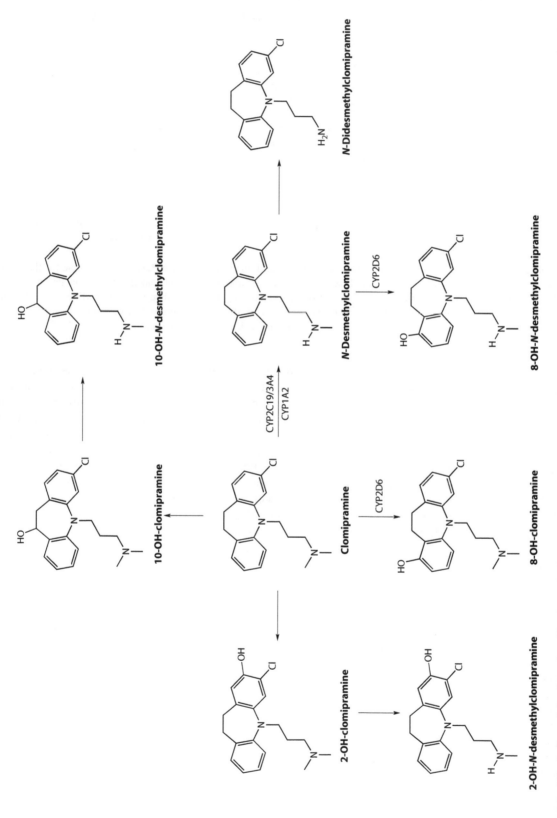

FIGURE 3.5 Metabolism of clomipramine. Clomipramine is converted to its active desmethylclomipramine by CYP2C19, 1A2, and 3A4, and both compounds are metabolized to respective hydroxy derivatives largely by CYP2D6. CYP2D6 in yeast microsomes catalyzes the 8-hydroxylation of clomipramine and desmethylclomipramine. Clomipramine is also converted to 2- or 10-hydroxylated metabolite, which is further N-demethylated.

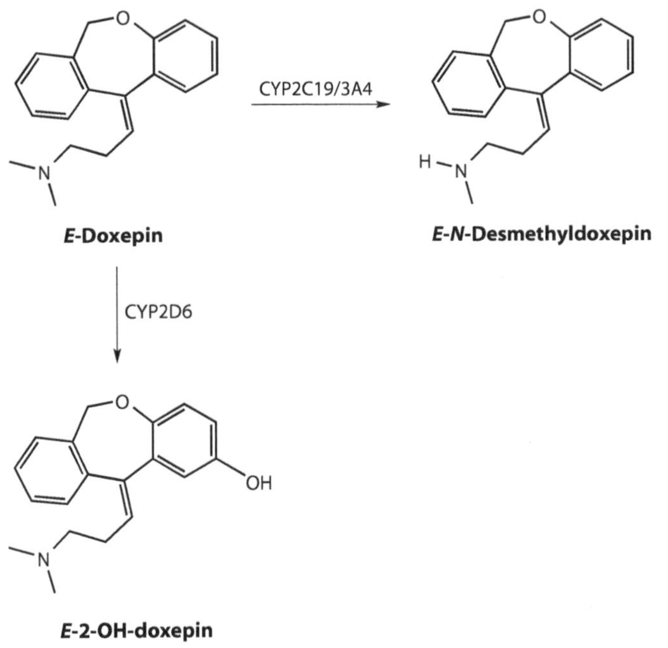

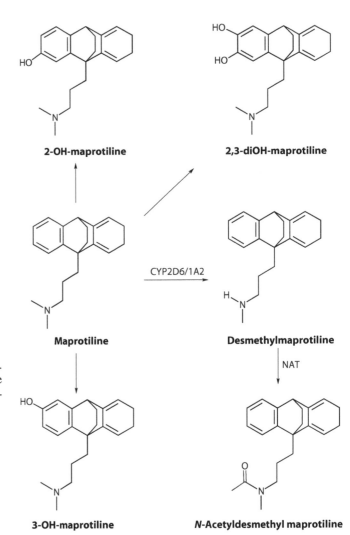

FIGURE 3.6 Metabolism of doxepin by CYP2D6, 2C19, and 3A4. CYP2D6 mainly converts the *E*-isomer of doxepin to its active metabolites 2-hydroxy derivative and *N*-desmethyldoxepin with substantial contribution from CYP2C19 and 3A4.

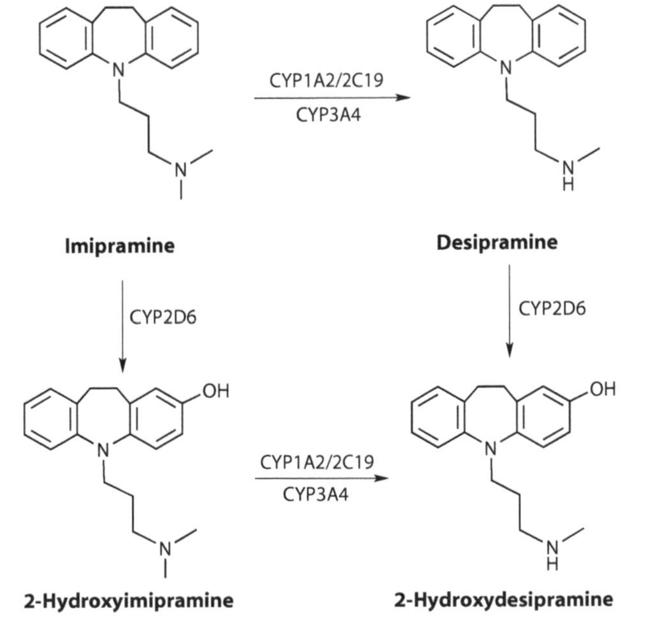

FIGURE 3.8 Metabolism of maprotiline. Maprotiline is mainly metabolized by CYP2D6 and 1A2 to its major active metabolite desmethylmaprotiline. Desmethylmaprotiline is further acetylated. Formation of hydroxylated metabolites including 3-OH-maprotiline, 2-OH-maprotiline, and 2,3-dihydrodiol to a lesser extent accounted for most of the remaining fraction.

FIGURE 3.7 Metabolism of imipramine by CYP2D6 and other CYPs. Imipramine as a tertiary amine is metabolized to desipramine via *N*-demethylation by CYP2C19, 3A4, and 1A2, and both compounds are metabolized to their active 2-hydroxy derivatives by CYP2D6, which are glucuronidated and excreted into the urine.

(L)-trimipramine to an unidentified metabolite. The CYP2D6 inhibitor quinidine increases the plasma concentration of trimipramine but the formation of 2-hydroxytrimipramine is decreased (Eap et al. 1992). Cases have been reported on increased trimipramine levels by coadministered paroxetine (Leinonen et al. 2004). A fatal case has been documented where trimipramine and citalopram are coadministered (Musshoff et al. 1999).

Overall, tricyclic antidepressants have similar metabolic routes and complex pharmacology. CYP2D6 has differing contribution to their metabolic clearance and pharmacologically active metabolites can be formed from CYP2D6-mediated pathways. There is a possible enantioselective disposition. As such, it is difficult to establish clear relationships of their pharmacokinetics and pharmacodynamic parameters with genetic variations of *CYP2D6*.

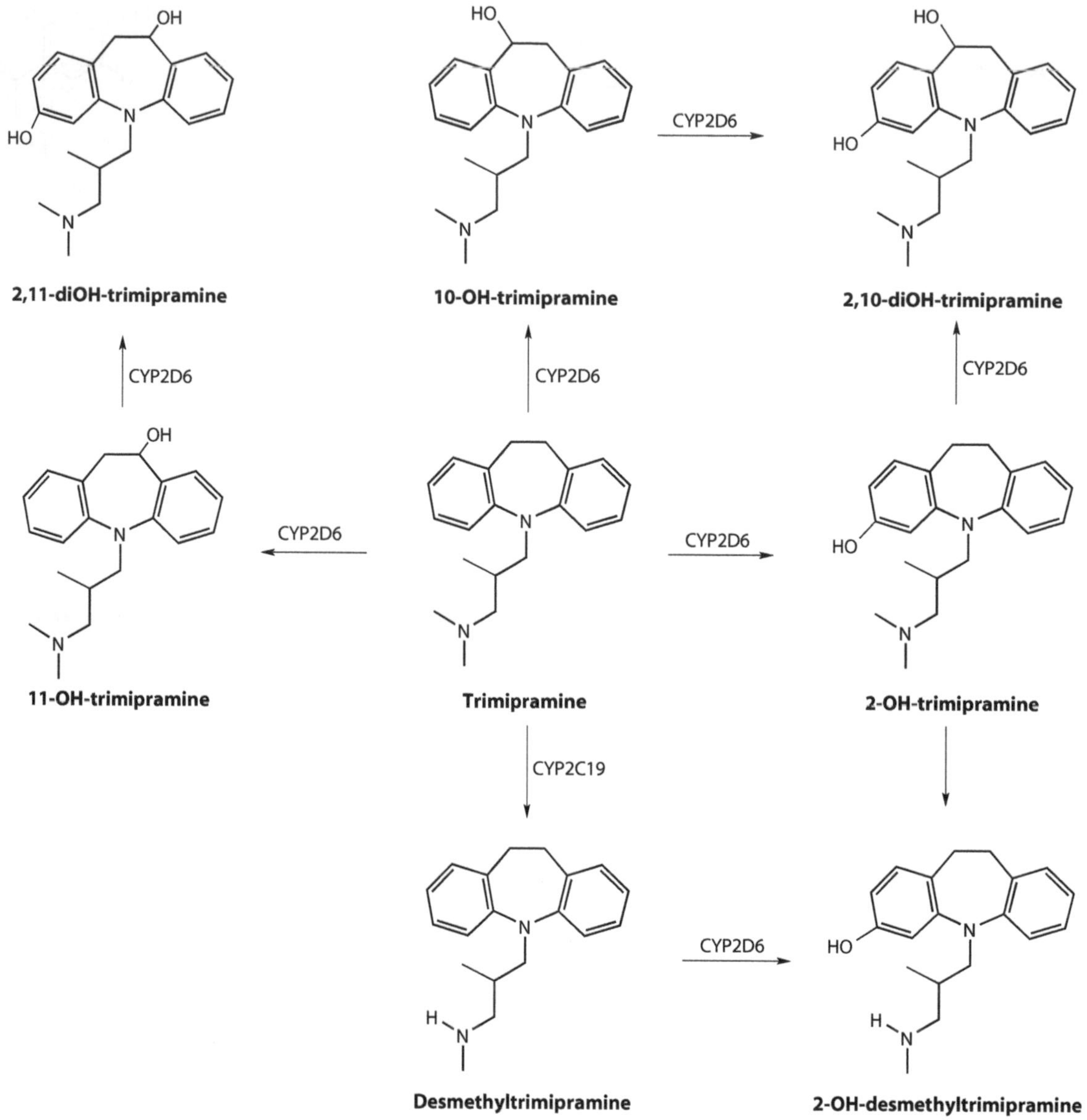

FIGURE 3.9 Metabolism of trimipramine by CYP2D6 and 2C19. CYP2D6 forms 2-hydroxytrimipramine and 2,10- or 2,11-dihydroxy-trimipramine in vitro. A stereoselectivity in the metabolism of trimipramine has been observed, with a preferential N-demethylation of (D)-trimipramine and a preferential hydroxylation of (L)-trimipramine. CYP2D6 catalyzes the 2-hydroxylation of (L)-trimipramine and (L)- and (D)-desmethyltrimipramine, but not of (D)-trimipramine. CYP2C19, but not CYP2D6, is involved in the demethylation pathway, with a stereoselectivity toward (D)-trimipramine. CYP3A4/5 appears to be involved in the metabolism of (L)-trimipramine to an unidentified metabolite.

3.3.1.2 Selective Serotonin Reuptake Inhibitors

The SSRIs including citalopram, fluoxetine, fluvoxamine, paroxetine, and sertraline possess similar mechanism of action, whereas they are chemically unrelated and show remarkable differences in their metabolism and pharmacokinetics profiles (Bertilsson 2007; Caccia 1998; Hiemke and Hartter 2000). Fluoxetine, fluvoxamine, and paroxetine, but not sertraline and citalopram, exhibit nonlinear pharmacokinetics (Hiemke and Hartter 2000). Several SSRIs including citalopram (Kobayashi et al. 1997; Olesen and Linnet 1999; Rochat et al. 1997; von Moltke et al. 2001), fluoxetine (Margolis et al. 2000; von Moltke et al. 1997), sertraline (Kobayashi et al. 1999; Obach et al. 2005; Xu et al. 1999), and

paroxetine (Bloomer et al. 1992) are metabolized by CYP2D6 and 2C19, but CYP2C9 and 3A4 also significantly contribute to their disposition in vitro. Citalopram is mainly metabolized to N-desmethylcitalopram and further demethylated to didesmethylcitalopram (Kobayashi et al. 1997; Olesen and Linnet 1999; Rochat et al. 1997; von Moltke et al. 2001), while fluoxetine is primarily metabolized via N-demethylation by CYP2D6, 2C19, 2C9, and 3A4 (Margolis et al. 2000; von Moltke et al. 1997).

Citalopram, an SSRI used to treat major depression associated with mood disorders, is converted to N-desmethylcitalopram (N-norcitalopram) and then to N,N-didesmethylcitalopram and citalopram N-oxide (Rao 2007).

Citalopram is sold as a racemic mixture, while escitalopram is the S-enantiomer of the racemic citalopram, which has been marketed as a drug as only the S-enantiomer has the desired antidepressant effect (Hyttel et al. 1992). Clinical studies have demonstrated that steady-state concentrations of R-citalopram exceed those of S-citalopram during chronic administration of racemic citalopram to humans and that the elimination half-life of R-citalopram exceeds that of S-citalopram (Bondolfi et al. 1996; Foglia et al. 1997; Rochat et al. 1995). CYP3A4, 2C19, and 2D6 are involved in the first demethylation step of citalopram, all favoring conversion of the biologically active S-enantiomer (escitalopram) (Figure 3.10) (Olesen and Linnet 1999; von Moltke et al. 2001). The approximate relative contribution at low substrate concentrations is estimated to be 34% by CYP3A4, 36% by 2C19, and 30% by 2D6 (von Moltke et al. 2001). Inhibitor studies indicated that CYP3A4 is responsible for 40%–50% of N-desmethylcitalopram formation at therapeutic citalopram concentrations, while the contribution of CYP2C19 increases and that of 2D6 tends to decrease with increasing drug concentration. CYP2D6 exclusively catalyzes the second demethylation step, and citalopram N-oxide is also exclusively formed by CYP2D6 (Olesen and Linnet 1999). However, none of the studied CYP enzymes mediates deamination to the propionic acid derivative.

Fluoxetine is a potent SSRI commonly prescribed as an antidepressant agent. Fluoxetine is metabolized extensively by the hepatic CYP enzymes, with <2.5% of the drug being found in an unchanged form in human urine (Mandrioli et al. 2006). There are two major metabolic routes for fluoxetine: N-demethylation to norfluoxetine (desmethylfluoxetine) and O-dealkylation to p-trifluoromethylphenol (van Harten 1995). CYP2D6 is largely responsible for the formation of R- and S-norfluoxetine from fluoxetine via N-demethylation at low concentrations, with increased contribution from CYP3A4 and 2C9 when substrate concentrations increase and CYP2D6 becomes saturated (Figure 3.11) (Margolis et al. 2000; Ring et al. 2001). Intrinsic clearance values for R-, S-, and racemic fluoxetine are greatest for CYP2D6, with a rank order of CYP2D6 > 2C9 > 3A4 > 2C19 for (R)-fluoxetine and CYP2D6 > 3A4 > 2C9 > 2C19 for S-fluoxetine (Margolis et al. 2000).

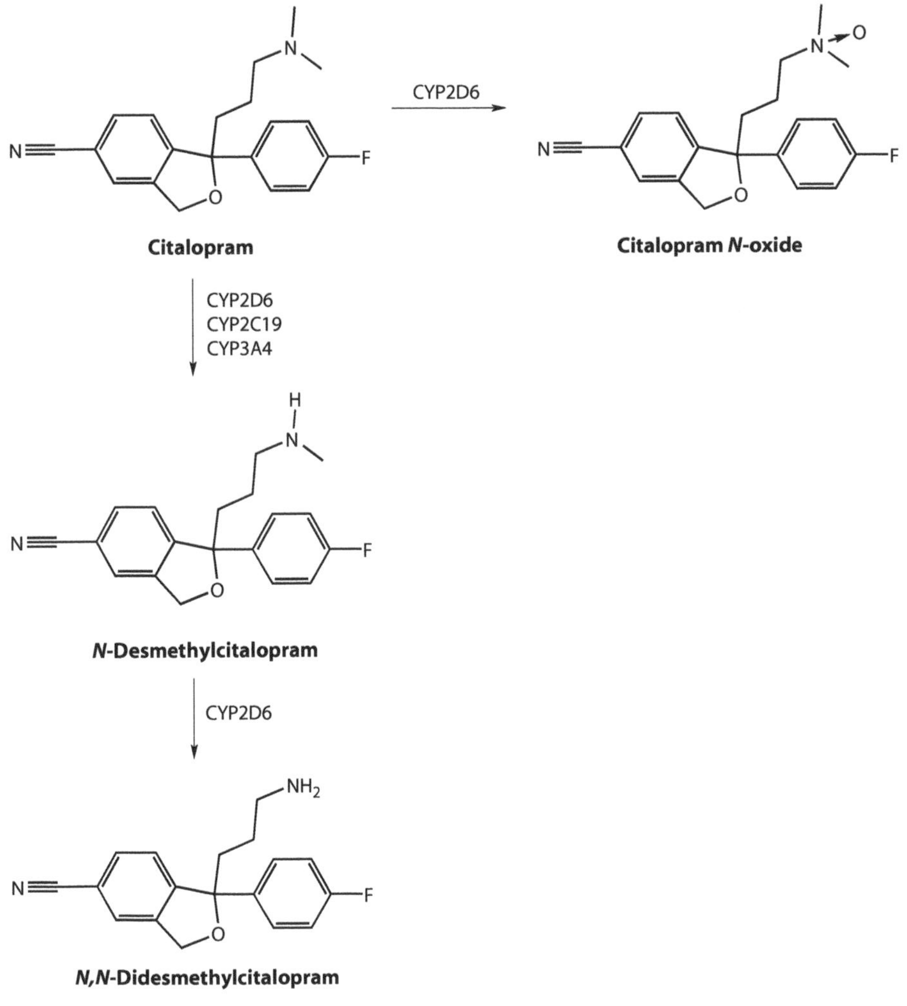

FIGURE 3.10 Metabolism of citalopram by CYP2D6, 3A4, and 2C19. Citalopram is converted to N-desmethylcitalopram (N-norcitalopram) and then to N,N-didesmethylcitalopram and citalopram N-oxide. CYP3A4, 2C19, and 2D6 are involved in the first demethylation step of citalopram, all favoring conversion of the biologically active S-enantiomer (escitalopram). CYP2D6 exclusively mediated the second demethylation step, and citalopram N-oxide is also exclusively formed by CYP2D6.

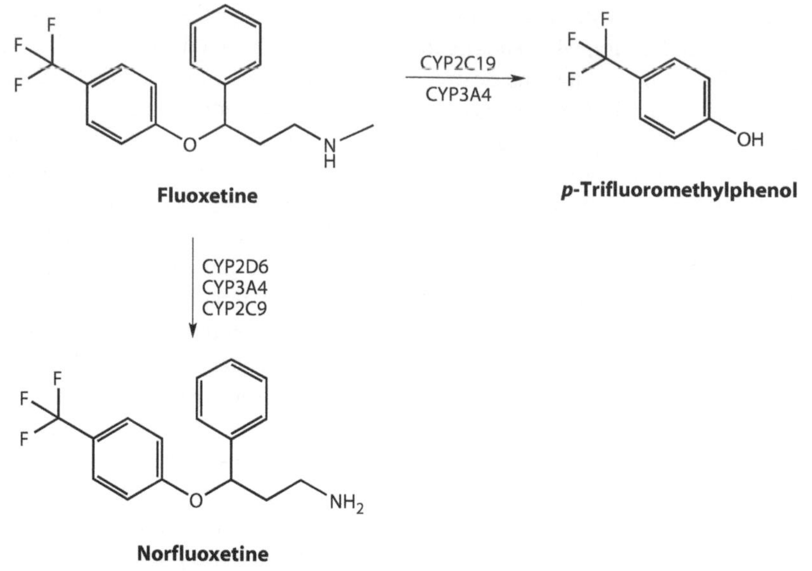

FIGURE 3.11 Metabolism of fluoxetine by CYP2D6, 3A4, 2C9, and 2C19. Fluoxetine is metabolized extensively by the hepatic CYP enzymes, and <2.5% of the drug is found unchanged in human urine. There are two major metabolic routes for fluoxetine: N-demethylation to norfluoxetine and O-dealkylation to p-trifluoromethylphenol. CYP2D6 is largely responsible for the formation of R- and S-norfluoxetine from fluoxetine via N-demethylation at low concentrations, with increased contribution from CYP3A4 and 2C9 when substrate concentrations increase and CYP2D6 becomes saturated. CYP2C19 and 3A4 are responsible for fluoxetine O-dealkylation.

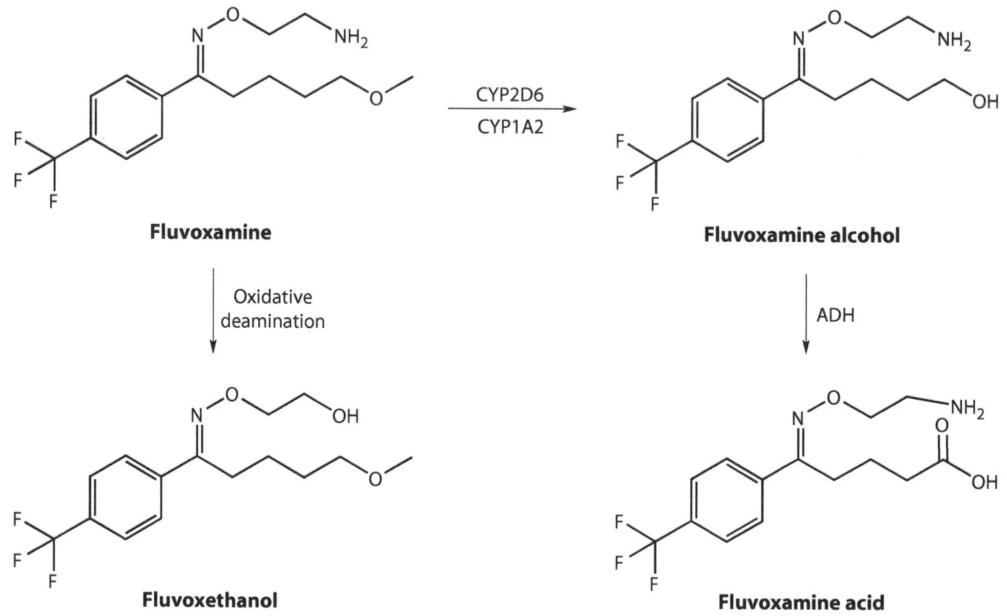

FIGURE 3.12 Metabolism of fluvoxamine by CYP2D6 and 1A2. Fluvoxamine is extensively metabolized in human liver via oxidative demethylation and oxidative deamination and N-acetylation, and <4% is excreted as unchanged form. There are at least nine metabolites detected in human urine after a 5-mg radiolabeled dose of fluvoxamine, constituting ~85% of the urinary excretion products of fluvoxamine. The main human metabolite is fluvoxamine acid, which, together with its N-acetylated analog, accounted for ~60% of the urinary excretion products. A third metabolite, fluvoxethanol, forms by oxidative deamination, which accounted for approximately 10%. Overall, ~30%–60% of the metabolites appear to be produced by oxidative demethylation of the methoxy group, whereas 20%–40% seem to be produced by degradation at the amino group or by removal of the entire ethanolamino group. Fluvoxamine appears to be converted to its major urinary metabolite, the 5-demethoxylated carboxylic acid metabolite, which is pharmacologically inactive, mainly by CYP2D6 and, to a less extent, by CYP1A2.

Fluvoxamine, one of the first SSRIs developed, is used in the treatment of major depression and anxiety disorders (Figgitt and McClellan 2000; Wilde et al. 1993). It is mainly metabolized by CYP2D6 and, to a less extent, by 1A2 (Spigset et al. 2001). Fluvoxamine is extensively metabolized in human liver via oxidative demethylation and oxidative deamination and *N*-acetylation, and <4% is excreted as unchanged form (Spigset et al. 1998). Approximately 30%–60% of the metabolites appear to be produced by oxidative demethylation of the methoxy group, whereas 20%–40% seem to be produced by oxidative deamination to fluvoxethanol or by removal of the entire ethanolamino group (Perucca et al. 1994). Fluvoxamine is converted to its major urinary metabolite, the 5-demethoxylated carboxylic acid metabolite, which is pharmacologically inactive, mainly by CYP2D6 and, to a less extent, by 1A2 (Figure 3.12) (Spigset et al. 2001). No other active metabolites are formed from fluvoxamine (DeVane and Gill 1997). Relative to other SSRIs, fluvoxamine is a potent inhibitor of CYP1A2, a moderate inhibitor of CYP2C19 and 3A4 and a weak inhibitor of CYP2D6 (van Harten 1995), and a number of drug interactions with fluvoxamine have been documented (Spina et al. 2008; Wagner and Vause 1995).

Paroxetine, an SSRI used for the treatment of depressive disorders (Gunasekara et al. 1998), is mainly inactivated by CYP2D6 via demethylenation of the methylenedioxy group, yielding a catechol metabolite and formic acid (Bloomer et al. 1992). Approximately 64% of a 30-mg oral dose of paroxetine is excreted in the urine with 62% as metabolites over a 10-day postdosing period. The principal metabolites of paroxetine are polar and conjugates of oxidative and methylated metabolites. Paroxetine is mainly (80%) metabolized by CYP2D6 via demethylenation of the methylenedioxy group, giving rise to an inactive catechol metabolite, which is then either *O*-methylated or *O*-glucuronidated, and the by-product formic acid (Figure 3.13) (Bloomer et al. 1992). Saturation and mechanism-based inhibition of CYP2D6 by paroxetine at clinical doses appears to explain for the nonlinear pharmacokinetics of paroxetine with increasing dose and increasing duration of treatment.

Sertraline is an SSRI used for the treatment of depression and mania. The predominant metabolites of sertraline in excreta include entities (e.g., hydroxyl sertraline ketone) that could arise via the further oxidative metabolism of the *N*-desmethyl metabolite or via initial deamination of the methylamino substituent in vivo (Figure 3.14). The sertraline ketone might arise from three initial possible parallel pathways: *N*-demethylation followed by *N*-deamination and hydroxylation, *N*-deamination of the methylamine substituent followed by hydroxylation, and direct *N*-deamination (Obach et al. 2005). CYP3A4 and 2C19 are involved in the

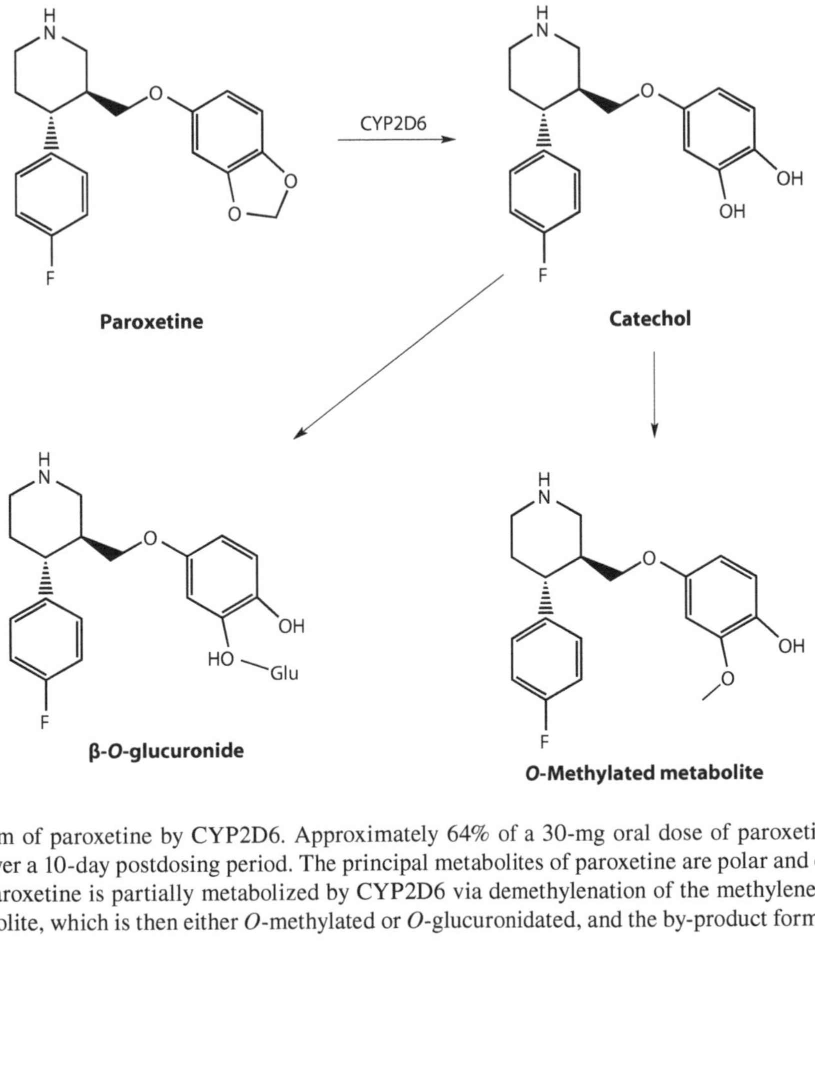

FIGURE 3.13 Metabolism of paroxetine by CYP2D6. Approximately 64% of a 30-mg oral dose of paroxetine is excreted in the urine with 62% as metabolites over a 10-day postdosing period. The principal metabolites of paroxetine are polar and conjugates of oxidative and methylated metabolites. Paroxetine is partially metabolized by CYP2D6 via demethylenation of the methylenedioxy group, giving rise to an inactive catechol metabolite, which is then either *O*-methylated or *O*-glucuronidated, and the by-product formic acid. The conjugates are excreted into the urine.

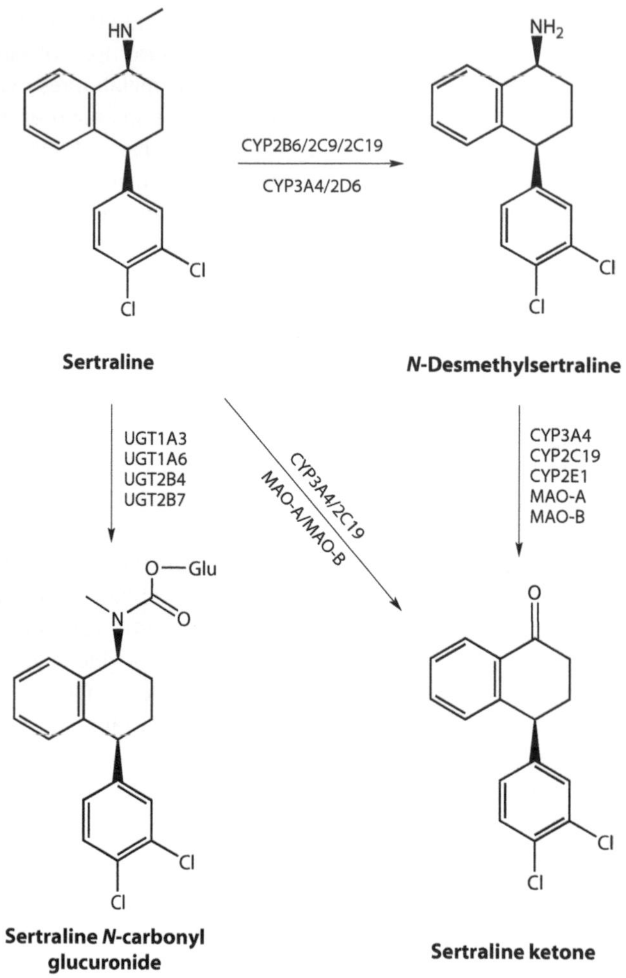

FIGURE 3.14 Metabolism of sertraline by CYPs and other enzymes. The predominant metabolites of sertraline in excreta included entities (e.g., hydroxyl sertraline ketone) that could arise via the further oxidative metabolism of the *N*-desmethyl metabolite or via initial deamination of the methylamino substituent in vivo. The sertraline ketone might arise from three initial possible parallel pathways: *N*-demethylation followed by *N*-deamination and hydroxylation, *N*-deamination of the methylamine substituent followed by hydroxylation, and direct *N*-deamination. CYP3A4 and 2C19 are involved in the *N*-deamination of sertraline, but both MAO-A and MAO-B also catalyzed sertraline deamination with comparable K_m values ranging from 230 to 270 μM. Sertraline also underwent *N*-carbamoyl glucuronidation, an unusual reaction in drug metabolism, albeit that this pathway appears to be minor in humans but major in dogs. Sertraline *N*-carbamoyl glucuronidation is catalyzed by UGT2B7 (primary), 2B4, 1A3, and 1A6. Recombinant CYP2B6, 2C9, 2C19, 2D6, and 3A4 catalyze sertraline *N*-demethylation, with a very minor role for CYP2B6.

N-deamination of sertraline, but both monoamine oxidase-A (MAO-A) and MAO-B also catalyze sertraline deamination with comparable K_m values ranging from 230 to 270 μM (Obach et al. 2005). Sertraline also undergoes *N*-carbamoyl glucuronidation, an unusual reaction in drug metabolism, albeit that this pathway appears to be minor in human but major in the dog (Tremaine et al. 1989). Sertraline *N*-carbamoyl glucuronidation is catalyzed by UGT2B7 (primary), 2B4,

1A3, and 1A6 (Obach et al. 2005). Recombinant CYP2B6, 2C9, 2C19, 2D6, and 3A4 catalyze sertraline *N*-demethylation (Kobayashi et al. 1999). Another study has identified recombinant CYP2C9, 2C19, 3A4, and 2D6 contributing to sertraline *N*-demethylation, with a very minor role for CYP2B6 (Greenblatt et al. 1999). CYP2C19 and 2C9 are the major CYP enzymes responsible for sertraline metabolism (Xu et al. 1999). Using selective chemical inhibitors and recombinant enzymes heterologously expressed in a baculovirus expression system, Obach et al. (2005) have revealed that CYP2B6 is the major enzyme for sertraline *N*-demethylation, with lesser contributions from CYP2C19, 2C9, 3A4, and 2D6. In clinical drug interaction studies, no drug has been identified to cause a large increase in sertraline exposure by inhibiting sertraline metabolism (DeVane et al. 2002).

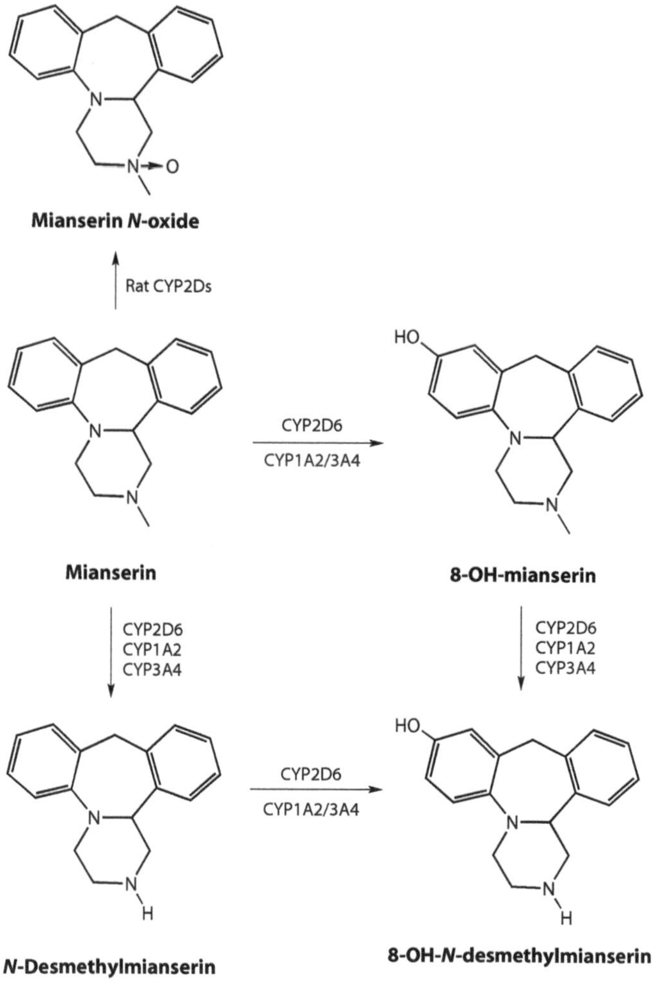

FIGURE 3.15 Metabolism of mianserin. CYP2D6 is involved in the 8-hydroxylation, *N*-demethylation, and *N*-oxidation of mianserin, with CYP3A4 and 1A2 being major contributors to the generation of desmethylmianserin (normianserin). *S*-Mianserin is more reliant on CYP2D6 and may have a greater antidepressant effect than *R*-mianserin. In rats, 8-hydroxy-*N*-desmethylmianserin is formed from 8-hydroxymianserin or *N*-desmethylmianserin by CYP2D2, 2D3, 2D4, and 2D6.

3.3.1.3 Other Antidepressants

Mianserin is a tetracyclic antidepressant administered as a racemic mixture, and its activity and disposition exhibit stereoselectivity (Pinder and Van Delft 1983). CYP2D6 is involved in the 8-hydroxylation, *N*-demethylation, and *N*-oxidation of mianserin, with 3A4 and 1A2 being major contributors to the generation of desmethylmianserin (Figure 3.15) (Koyama et al. 1996; Stormer et al. 2000). *S*-Mianserin is more reliant on CYP2D6 (Dahl et al. 1994; Yasui et al. 1997) and may have a greater antidepressant effect than *R*-mianserin. *N*-Desmethylmianserin is pharmacologically active. In rats, the additional metabolite 8-hydroxy-*N*-desmethylmianserin is formed from 8-hydroxymianserin or *N*-desmethylmianserin by Cyp2d2, 2d3, 2d4, and 2d6 (Chow et al. 1999). Thioridazine inhibits CYP2D6 and increases *S*-mianserin, *S*-desmethylmianserin, and *R*-desmethylmianserin concentrations by 1.9-, 2.1-, and 2.7-fold, respectively, whereas there is no effect on *R*-mianserin pharmacokinetics (Yasui et al. 1997).

Mirtazapine, the first noradrenergic and specific serotonergic antidepressant acting as presynaptic α_2 receptors and postsynaptic 5-HT$_2$ receptors (Anttila and Leinonen 2001), is mainly metabolized by CYP2D6, with substantial contribution from 1A2 and 3A4 (Figure 3.16) (Stormer et al. 2000). Mirtazapine is extensively metabolized in humans, and the primary oxidative metabolites are 8-hydroxymirtazapine, *N*-desmethylmirtazapine, and mirtazapine-*N*-oxide (Delbressine et al. 1998). The major metabolite in vivo is 8-hydroxymirtazapine, accounting for ~40% of the excreted dose, while *N*-desmethylmirtazapine accounts for ~25% of the excreted dose and is the only pharmacologically active metabolite. It is 5 to 10 times less potent than the parent compound and contributes 3% to 6% to the net pharmacological activity of mirtazapine. Mirtazapine-*N*-oxidation contributes ~10% to the overall clearance of mirtazapine in vivo (Delbressine et al. 1998). The primary metabolites of mirtazapine also undergo secondary metabolism and glucuronidation. An additional metabolic pathway in humans but not in animals is the formation of the quaternary N^+-glucuronide (Delbressine et al. 1998). The pharmacokinetics of mirtazapine is enantioselective, with higher plasma concentrations and longer half-life of the *R*-enantiomer compared with that of the *S*-enantiomer.

Venlafaxine is a mixed serotonin and noradrenaline reuptake inhibitor used as a first-line treatment of depressive disorders (Holliday and Benfield 1995). After oral administration, venlafaxine undergoes extensive first-pass metabolism

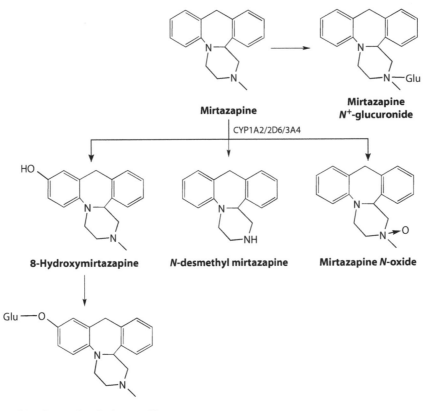

Mirtazapine

Mirtazapine *N*⁺**-glucuronide**

CYP1A2/2D6/3A4

8-Hydroxymirtazapine   ***N*-desmethyl mirtazapine**   **Mirtazapine *N*-oxide**

8-OH-mirtazapine *O*-glucuronide

FIGURE 3.16 Metabolism of mirtazapine by CYP2D6, 1A2, and 3A4. It is extensively metabolized by CYP2D6, 1A2, and 3A4 in humans, and the primary oxidative metabolites are 8-hydroxymirtazapine, *N*-desmethylmirtazapine, and mirtazapine-*N*-oxide. The major metabolite in vivo is 8-hydroxymirtazapine, accounting for ~40% of the excreted dose, while *N*-desmethylmirtazapine accounted for ~25% of excreted dose and is the only pharmacologically active metabolite. Mirtazapine-*N*-oxidation contributed ~10% to the overall clearance of mirtazapine in vivo. The primary metabolites of mirtazapine also underwent secondary metabolism and glucuronidation. An additional metabolic pathway in humans, but not in animals, is the formation of the quaternary N^+-glucuronide.

by the liver to two minor, less active metabolites, namely, N-desmethylvenlafaxine and N,O-didesmethylvenlafaxine, and a major metabolite, O-desmethylvenlafaxine, which is an active metabolite with antidepressant activity comparable with the parent drug (Figure 3.17) (Ellingrod and Perry 1994). Venlafaxine is converted by CYP2D6 to its active metabolite O-desmethylvenlafaxine, while its N-demethylation is catalyzed by CYP3A4, 2C19, and 2C9 (Fogelman et al. 1999; Otton et al. 1996).

CYP2D6 is a major enzyme that metabolizes the newly developed antidepressants including escitalopram, duloxetine, and gepirone (Caccia 2004). Duloxetine is a potent dual serotonin and norepinephrine reuptake inhibitor currently undergoing clinical investigation for treatment of depression, pain related to diabetic neuropathy, and stress urinary incontinence (Bymaster et al. 2005; Gupta et al. 2007; McCormack and Keating 2004; Muller et al. 2008). It is a substrate and an inhibitor of CYP2D6 (Skinner et al. 2003). After oral administration in humans, duloxetine is extensively metabolized to a number of metabolites primarily excreted into the urine in the conjugated form (Figure 3.18) (Lantz et al. 2003). Duloxetine accounts for only a small portion (~3% for AUC and ~9% for C_{max}) of the circulating radioactivity in plasma. Duloxetine appears to have a large apparent volume of distribution. The major metabolic routes for duloxetine include oxidation of the

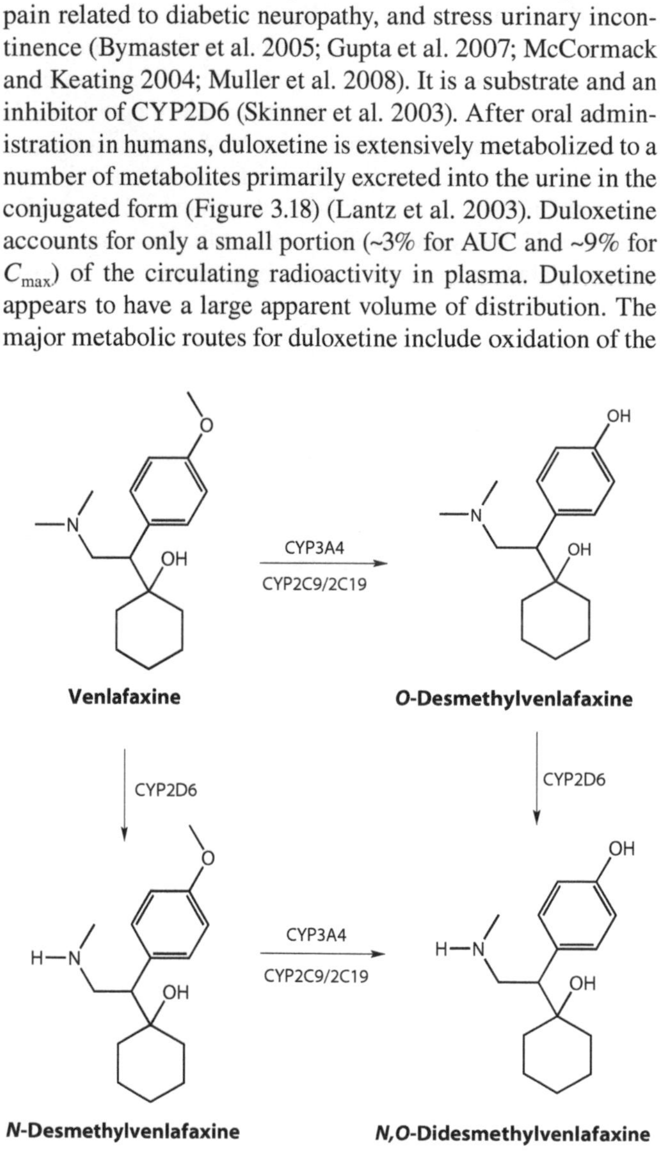

naphthyl ring at the 4-, 5-, or 6-position followed by further oxidation, methylation, and conjugation. The major metabolites found in human plasma are glucuronide conjugates of 4-hydroxy duloxetine (M6), 6-hydroxy-5-methoxy duloxetine (M10), 4,6-dihydroxy duloxetine (M9), and a sulfate conjugate of 5-hydroxy-6-methoxy duloxetine (M7) (Lantz et al. 2003). In contrast, the unconjugated forms of 4-hydroxy duloxetine (M14), 6-hydroxy-5-methoxy duloxetine (M15), and 4,6-dihydroxy duloxetine are not detected in plasma. A very minor pathway is the cleavage of duloxetine at the chiral center to form a thienyl alcohol and naphthol, with both metabolites detected in the urine (Lantz et al. 2003). In vitro studies have shown CYP2D6 and 1A2 to be the primary enzymes responsible for the oxidative metabolism at the 4-, 5-, or 6-position of the naphthyl ring of duloxetine. As an inhibitor of CYP2D6, duloxetine increases the C_{max} and AUC of desipramine 1.7- and 2.9-fold, respectively (Skinner et al. 2003), and increases the AUC of metoprolol 1.8-fold (Preskorn et al. 2007). Paroxetine increases the C_{max} and AUC of duloxetine 1.6-fold (Skinner et al. 2003).

Gepirone is a pyridinyl piperazine 5-HT$_{1A}$ agonist with antidepressant activity (McMillen and Mattiace 1983) used in the treatment of depression (Greenblatt et al. 2003). Gepirone undergoes extensive metabolism after oral administration, with the major metabolites in human plasma being 1-(2-pyrimidinyl)-piperazine and 3′-OH-gepirone (Figure 3.19) (von Moltke et al. 1998). CYP3A4 is the major enzyme for the formation of 1-(2-pyrimidinyl)-piperazine and accounts for 80%–90% of formation of hydroxylated metabolites at high substrate concentrations (≥10 μM). At low gepirone concentrations (≤1 μM), human liver microsomes form 1-(2-pyrimidinyl)-piperazine and 3′-OH-gepirone and two other hydroxylated metabolites (2-OH- and 5-OH-gepirone) (Greenblatt et al. 2003). All metabolites are formed by CYP3A4, but CYP2D6 forms 3′-OH- and 5-OH-gepirone, but not 1-(2-pyrimidinyl)-piperazine or 2-OH-gepirone (Greenblatt et al. 2003). On the basis of estimated relative abundances of the two CYPs in human liver, CYP3A4 is predicted to account for more than 95% of net clearance of gepirone in vivo at low concentrations approaching the therapeutic range, while CYP2D6 would account for less than 5% of net clearance.

Vortioxetine is an atypical antidepressant, acting as a serotonin modulator and stimulator (Alvarez et al. 2014; Citrome 2014; Dhir 2013; Garnock-Jones 2014; Pearce and Murphy 2014; Sanchez et al. 2015). It was approved in September 2013 by the FDA for the treatment of major depressive disorder in adults. Vortioxetine binds with high affinity to the human serotonin transporter (K_i = 1.6 nM), but not to the norepinephrine (K_i = 113 nM) or dopamine (K_i > 1000 nM) transporters (Bang-Andersen et al. 2011). Vortioxetine potently and selectively inhibits reuptake of serotonin (IC$_{50}$ = 5.4 nM). Vortioxetine binds to 5-HT$_3$ with a K_i of 3.7 nM, 5-HT$_{1A}$ with a K_i of 15 nM, 5-HT$_7$ with a K_i of 19 nM, 5-HT$_{1D}$ with a K_i of 54 nM, and 5-HT$_{1B}$ with a K_i of 33 nM and is a 5-HT$_3$, 5-HT$_{1D}$, and 5-HT$_7$ receptor antagonist; 5-HT$_{1B}$ receptor partial agonist; and 5-HT$_{1A}$ receptor agonist (Bang-Andersen et

FIGURE 3.17 Metabolism of venlafaxine. Venlafaxine undergoes extensive first-pass metabolism by the liver to two minor, less active metabolites, namely, N-desmethylvenlafaxine and N,O-didesmethylvenlafaxine, and a major active metabolite, O-desmethylvenlafaxine. Venlafaxine is converted by CYP2D6 to its active metabolite O-desmethylvenlafaxine, while its N-demethylation is catalyzed by CYP3A4, 2C19, and 2C9.

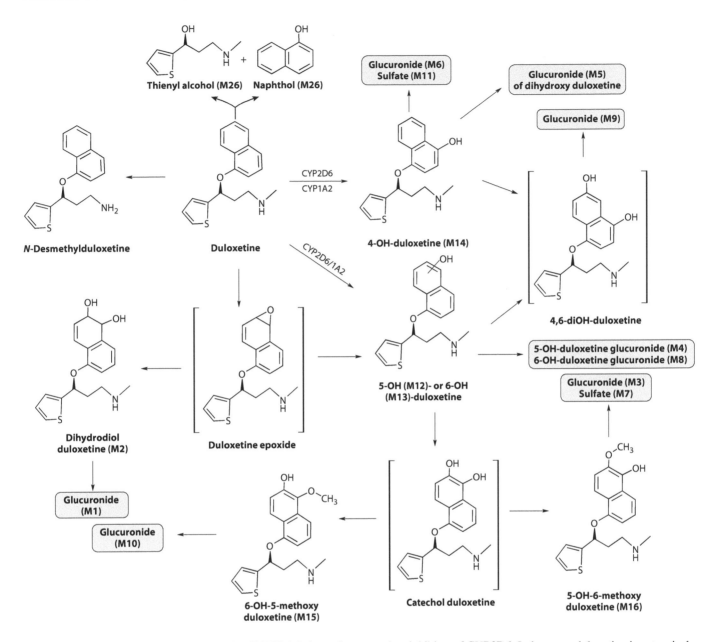

FIGURE 3.18 Metabolism of duloxetine by CYP2D6. It is a substrate and an inhibitor of CYP2D6. In humans, duloxetine is extensively metabolized to a number of metabolites primarily excreted into the urine in the conjugated form. The major metabolic routes for duloxetine included oxidation of the naphthyl ring at the 4-, 5-, or 6-positions followed by further oxidation, methylation, and conjugation. The major metabolites found in human plasma are glucuronide conjugates of 4-hydroxy duloxetine (M6), 6-hydroxy-5-methoxy duloxetine (M10), 4, 6-dihydroxy duloxetine (M9), and a sulfate conjugate of 5-hydroxy-6-methoxy duloxetine (M7). CYP2D6 and 1A2 are the primary enzymes responsible for the oxidative metabolism of duloxetine.

al. 2011). The pharmacokinetics of vortioxetine (2.5 mg to 60 mg) are linear and dose proportional when vortioxetine is administered once daily (Dubovsky 2014). Absolute oral bioavailability is 75%. The drug reaches C_{max} within 7 to 11 h postdosing, and its mean $t_{1/2\beta}$ is ~66 h (Dubovsky 2014). Steady-state plasma concentrations are typically reached within 2 weeks. Vortioxetine is extensively metabolized primarily through oxidation via hepatic CYP2D6, 3A4/5, 2C19, 2C9, 2A6, 2C8, and 2B6 and subsequent glucuronic acid conjugation (Hvenegaard et al. 2012). CYP2D6 is the primary enzyme catalyzing the metabolism of vortioxetine to its

major, pharmacologically inactive, carboxylic acid metabolite (Hvenegaard et al. 2012).

3.3.1.4 Antipsychotics

The first-generation antipsychotics are known as "typical antipsychotics," including chlorpromazine, chlorprothixene, haloperidol, flupentixol, fluphenazine, mesoridazine, perphenazine, promazine, promethazine, thioridazine, triflupromazine, and zuclopenthixol (Lieberman et al. 2008). Most of the drugs in the second generation, known as "atypical antipsychotics," have been developed. The second-generation

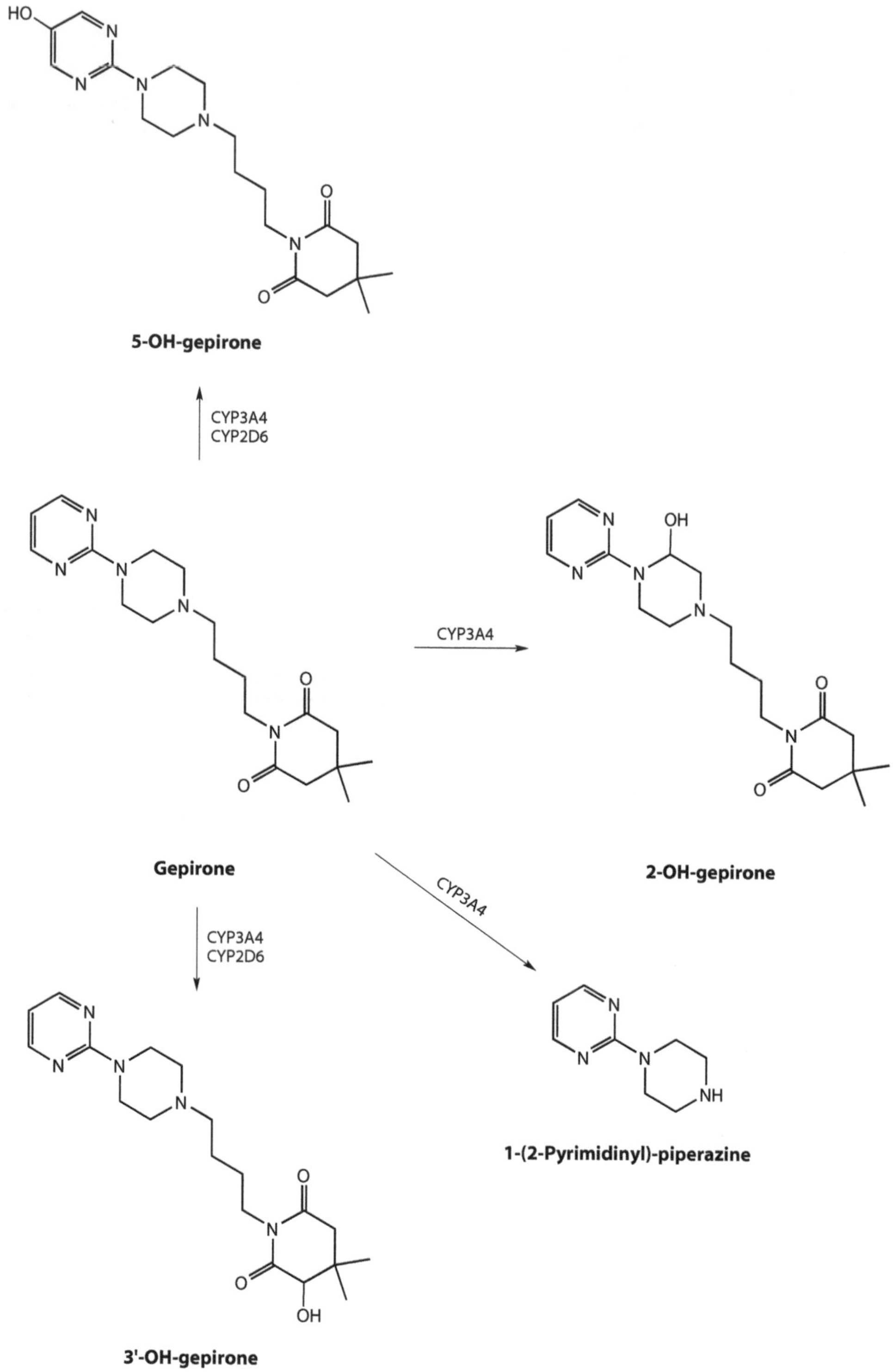

FIGURE 3.19 Metabolism of gepirone by CYP2D6 and 3A4. Gepirone undergoes extensive metabolism after oral administration, with the major metabolites in human plasma being 1-(2-pyrimidinyl)-piperazine, 3'-OH-gepirone. CYP3A4 is the major enzyme for the formation of 1-(2-pyrimidinyl)-piperazine and accounts for 80%–90% of formation of hydroxylated metabolites at high substrate concentrations (≥10 μM). At low gepirone concentrations (≤1 μM), human liver microsomes form 1-(2-pyrimidinyl)-piperazine and 3'-OH-gepirone and two other hydroxylated metabolites (2-OH- and 5-OH-gepirone). All metabolites are formed by CYP3A4, but CYP2D6 forms 3'-OH- and 5-OH-gepirone, but not 1-(2-pyrimidinyl)-piperazine or 2-OH-gepirone.

antipsychotics include amisulpride, aripiprazole, clozapine, olanzapine, paliperidone, quetiapine, risperidone, sertindole, ziprasidone, and zotepine (Lieberman et al. 2008; Vohora 2007; Worrel et al. 2000). Newer antipsychotic agents, such as asenapine, bifeprunox, norclozapine, and iloperidone, are being developed (Bishara and Taylor 2008). Antipsychotics are generally highly lipid soluble, subject to high clearance, and eliminated by metabolic rather than renal pathways. CYP2D6 is involved in the metabolism of a variety of antipsychotics, including clozapine, thioridazine, perphenazine, chlorpromazine, fluphenazine, haloperidol, zuclopenthixol, risperidone, iloperidone, and sertindole (Gardiner and Begg 2006; Ingelman-Sundberg 2005; Zhou et al. 2008). CYP1A2 and 3A are also involved in the metabolism of antipsychotic drugs including clozapine, olanzapine, pimozide, and haloperidol (Gardiner and Begg 2006; Ingelman-Sundberg 2005; Zhou et al. 2008). There is preliminary evidence that CYP2D6 phenotype status may affect the clearance of this group and thus alter the clinical response to them.

Aripiprazole, an arylpiperazine quinolinone derivative, is a novel atypical antipsychotic drug with dopamine D_2 and 5-HT_{1A} receptor agonistic and 5-HT_{2A} antagonistic property indicated for the treatment of schizophrenia in adults (Swainston Harrison and Perry 2004). Aripiprazole is also a high-affinity partial agonist at 5-HT_{2A} receptors and a low-affinity agonist at 5-HT_{2C} receptors, and has moderate affinity at α_1-adrenoceptors and H_1 receptors (Shapiro et al. 2003).

The elimination of aripiprazole is mainly mediated through the hepatic metabolism. Aripiprazole undergoes dehydrogenation and aromatic hydroxylation followed by conjugation and N-dealkylation (Figure 3.20). The main active metabolite, dehydroaripiprazole (OPC-14857 with a half-life of 90 h), has affinity for dopamine D_2 receptors and thus has some pharmacological activity similar to that of the parent drug. The formation of dehydroaripiprazole is mainly mediated by CYP2D6 and 3A4. At steady state, dehydroaripiprazole represents ~40% of aripiprazole AUC in human plasma. There is a 37-fold interindividual variability in the concentration/dose ratio for aripiprazole (Molden et al. 2006).

Chlorpromazine, a prototype phenothiazine antipsychotic drug, is used in the treatment of schizophrenia, the manic phase of bipolar disorder as well as amphetamine-induced psychoses, but its use today has been largely replaced by the newer atypical antipsychotics such as olanzapine and quetiapine. Chlorpromazine is extensively metabolized in the liver via 7-hydroxylation, N-dealkylation, N-oxidation, S-oxidation, and conjugation, resulting in more than 100 metabolites with greatly varying half-lives and pharmacological profiles (Figure 3.21). Among these metabolic pathways, 7-hydroxylation is a major metabolic pathway of chlorpromazine in human (Hartmann et al. 1983). Chlorpromazine is converted to active 7-hydroxychlorpromazine largely by CYP2D6 (Yoshii et al. 2000). This major metabolic pathway is inhibited by quinidine in EMs of debrisoquine in vivo

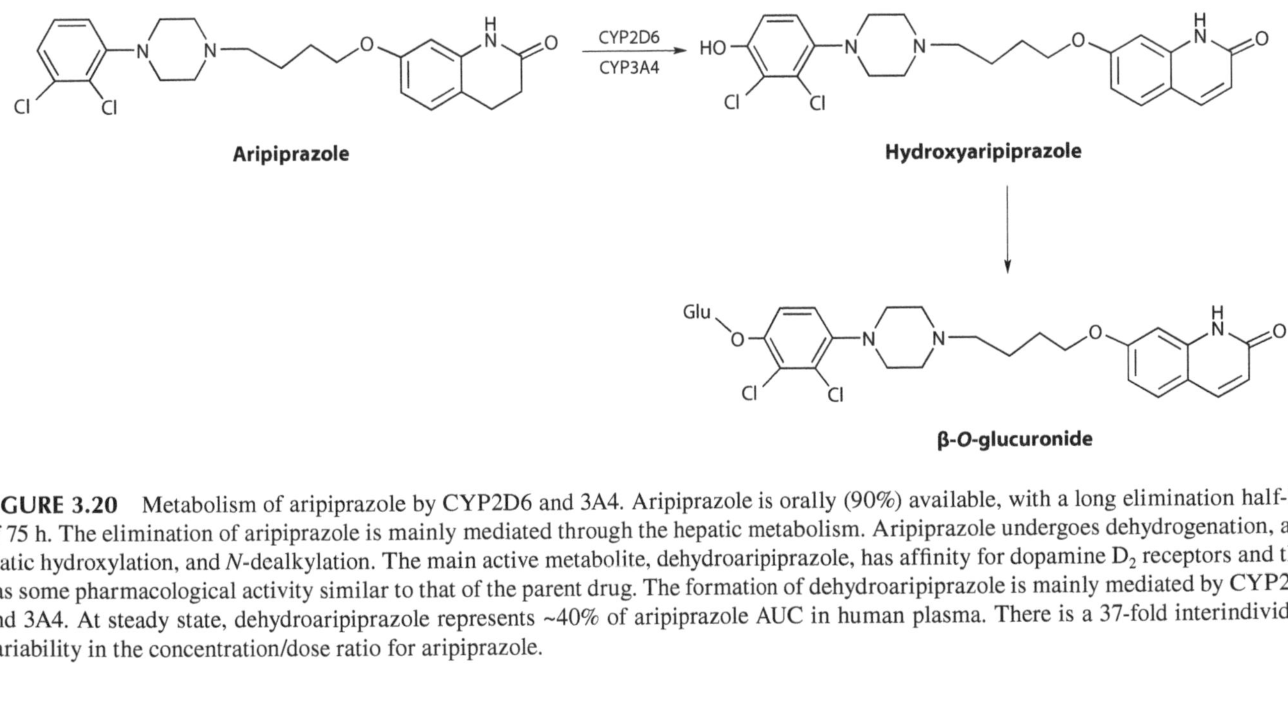

FIGURE 3.20 Metabolism of aripiprazole by CYP2D6 and 3A4. Aripiprazole is orally (90%) available, with a long elimination half-life of 75 h. The elimination of aripiprazole is mainly mediated through the hepatic metabolism. Aripiprazole undergoes dehydrogenation, aromatic hydroxylation, and N-dealkylation. The main active metabolite, dehydroaripiprazole, has affinity for dopamine D_2 receptors and thus has some pharmacological activity similar to that of the parent drug. The formation of dehydroaripiprazole is mainly mediated by CYP2D6 and 3A4. At steady state, dehydroaripiprazole represents ~40% of aripiprazole AUC in human plasma. There is a 37-fold interindividual variability in the concentration/dose ratio for aripiprazole.

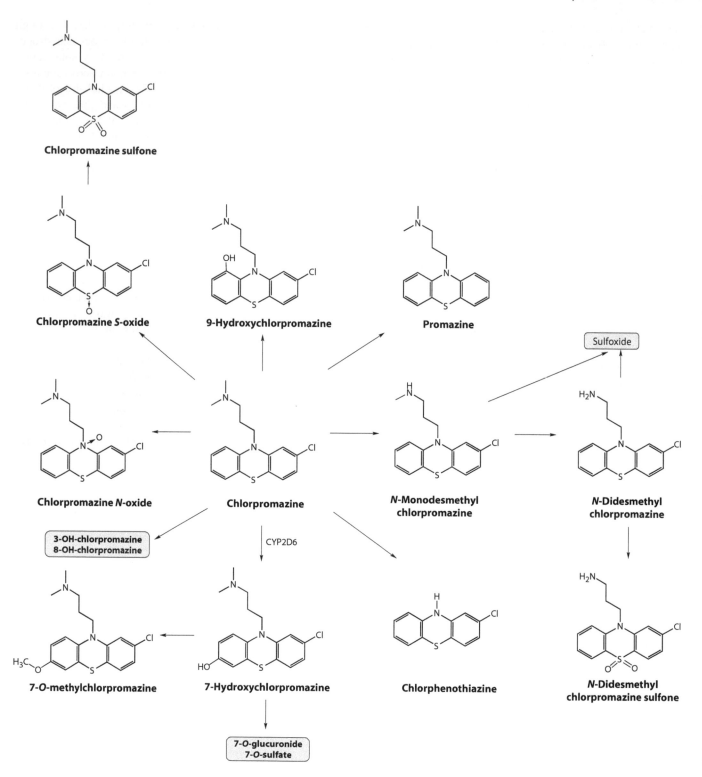

FIGURE 3.21 Metabolism of chlorpromazine by CYP2D6. Only ~32% of the administered dose is available to the systemic circulation in the active form, owing to extensive first-pass metabolism. Over time and multiple administrations, bioavailability may decrease to 20%. Less than 1% of the unchanged drug is excreted via the urine. Chlorpromazine is extensively metabolized in the liver via 7-hydroxylation, *N*-dealkylation, *N*-oxidation, *S*-oxidation, and conjugation, resulting in more than 100 metabolites with greatly varying half-lives and pharmacological profiles. Among these pathways, the 7-hydroxylation is a major metabolic pathway of chlorpromazine in human. Chlorpromazine is converted to active 7-hydroxychlorpromazine largely by CYP2D6. Two other common metabolites are *N*-didesmethylchlorpromazine and chlorphenothiazine in which the entire side chain has been removed. Only 7-hydroxychlorpromazine, promazine, 9-hydroxyrisperidone, and a few *N*-demethylated metabolites showed pharmacological activities.

(Muralidharan et al. 1996). Two other common metabolites are N-didesmethylchlorpromazine and chlorophenothiazine in which the entire side chain has been removed. Only 7-hydroxychlorpromazine, promazine, 9-hydroxyrisperidone, and a few N-demethylated metabolites show pharmacological activities.

Clozapine is the first atypical antipsychotic developed and is first introduced in Europe in 1971 but is voluntarily withdrawn by the manufacturer in 1975 after it is shown to cause agranulocytosis that leads to death in some clozapine-treated patients. In 1989, the FDA approved clozapine's use for treatment-resistant schizophrenia on the basis of the studies demonstrating that clozapine is more effective than any other antipsychotic for treating schizophrenia. The main metabolic pathways of clozapine consist of N-demethylation and N-oxide formation (Byerly and DeVane 1996). CYP1A2, 3A4, 2C9, 2C19, and 2D6 N-demethylate clozapine, while N-oxide formation is catalyzed by CYP3A4, 1A2, and flavin-containing monooxygenase 3 (FMO3) (Figure 3.22) (Eiermann et al. 1997; Fang et al. 1998; Linnet and Olesen 1997; Olesen and Linnet 2001; Tugnait et al. 1999). The estimated contribution of CYP1A2, 2C19, 3A4, 2C9, and 2D6 amounts to 30%, 24%, 22%, 12%, and 6%, respectively, with regard to the N-demethylation of clozapine in human liver microsomes

Clozapine

CYP1A2/2D6

N-desmethylclozapine

Clozapine N-oxide

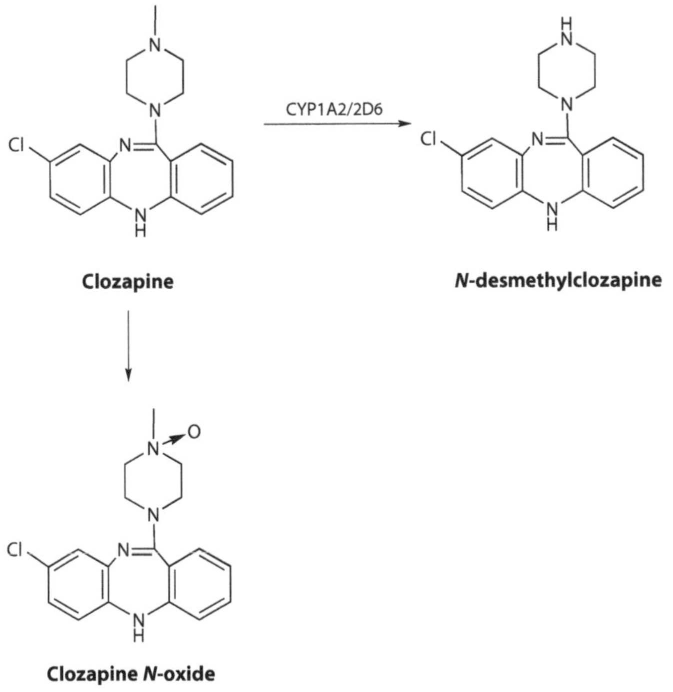

FIGURE 3.22 Metabolism of clozapine. The main metabolic pathways of clozapine consist of N-demethylation and N-oxide formation. CYP1A2, 3A4, 2C9, 2C19, and 2D6 N-demethylated clozapine, while N-oxide formation is catalyzed by CYP3A4, 1A2, and FMO3. The estimated contribution of CYP1A2, 2C19, 3A4, 2C9, and 2D6 amounted to 30%, 24%, 22%, 12%, and 6%, respectively, with regard to the N-demethylation of clozapine in human liver microsomes. CYP2D6 might play a role in the formation of metabolites other than N-desmethylclozapine and the N-oxide. In addition, clozapine is extensively glucuronidated in vitro and in vivo by UGT1A4.

(Olesen and Linnet 2001). CYP2D6 might play a role in the formation of metabolites other than N-demethylclozapine and the N-oxide (Fischer et al. 1992b). In addition, clozapine is extensively glucuronidated in vitro and in vivo by UGT1A4 (Mori et al. 2005).

Haloperidol (4-(4-chlorophenyl)-1-[4-(4-fluorophenyl)-4-oxobutyl]-4-piperidinol) is one of the most commonly used in the treatment of patients with acute and chronic schizophrenia (Kudo and Ishizaki 1999; Yatham 2002). Haloperidol undergoes extensive metabolism involving various CYPs (mainly CYP3A4), carbonyl reductase, and UGTs, with direct glucuronidation being the predominant elimination pathway followed by the reduction of haloperidol to reduced haloperidol and by CYP-mediated oxidation. Haloperidol is reduced by cytosolic carbonyl reductase to reduced form, which has 10% to 20% of the activity of the parent molecule. Reduced haloperidol is back oxidized to haloperidol primarily by CYP3A4 (Figure 3.23) (Kudo and Odomi 1998; Pan et al. 1998; Tateishi et al. 2000). N-Dealkylation of haloperidol leads to the formation of 4-(4-chlorophenyl)-4-hydroxypiperidine, while dehydration of haloperidol results in 1,2,3,6-tetrahydropyridine. Reduced haloperidol also undergoes N-dealkylation by CYP3A4 and 2D6, 1A2, 2C9, 2B6, and 2E1. Reduced haloperidol also undergoes N-dealkylation by CYP3A4, 2D6, 1A2, 2C9, 2B6, and 2E1 and glucuronidation (Kudo and Odomi 1998; Tateishi et al. 2000). Pyridinium ions can be formed from reduced haloperidol and haloperidol 1,2,3,6-tetrahydropyridine. It appears that CYP2D6 only plays a minor role for the disposition of haloperidol while CYP3A4 is the most important enzyme for its overall disposition.

Iloperidone is an atypical antipsychotic with a low propensity to produce extrapyramidal side effects (Albers et al. 2008). It is a dopamine-D_2/5-HT_2 antagonist (Szewczak et al. 1995). Iloperidone is extensively metabolized via O-dealkylation, N-dealkylation, reduction, and hydroxylation in humans (Figure 3.24) (Mutlib et al. 1995). Iloperidone is O-dealkylated to form 6-fluoro-3-[1-(3-hydroxypropyl)-4-piperidinyl]-1,2-benzisoxazole and 1-[4-[3-[4-(6-fluoro-1,2-benzisoxazol-3-yl)-1-piperidinyl]propoxy]-3-hydroxyphenyl]ethanone (M2). Oxidative N-dealkylation results in 6-fluoro-3-(4-piperidinyl)-1,2-benzisoxazol and a secondary metabolite, 3-[(4-acetyl-2-methoxy)phenoxy]propionic acid (Mutlib et al. 1995). Iloperidone is reduced to 4-[3-[4-(6-fluoro-1,2-benzisoxazol-3-yl)-1-piperidinyl]propoxy]-3-methoxy-α-methylbenzene methanol (M3), which is the major circulating metabolite in humans and rats (Mutlib et al. 1995). Hydroxylation of iloperidone yields 1-[4-[3-[4-(6-fluoro-1,2-benzisoxazol-3-yl)-1-piperidinyl]propoxy]-2-hydroxy-5-methoxyphenyl]ethanone and 1-[4-[3-[4-(6-fluoro-1,2-benzisoxazol-3-yl]-1-piperidinyl]-3-methoxyphenyl]-2-hydroxyethanone (M4) (Mutlib et al. 1995). M4 is further oxidized to 4-[3-[4-(6-fluoro-1,2-benzoisoxazol-3-yl)-piperidin1-yl]propoxy]-3-methoxyphenol (M5). In human liver microsomes, the formation of M4 is mediated by CYP2D6, while M2 is formed by CYP3A (Mutlib and Klein 1998). M3 is postulated to be produced mainly by a cytosolic enzyme, but CYP3A, 1A2, and 2E1 are involved in its formation as well (Mutlib and Klein 1998).

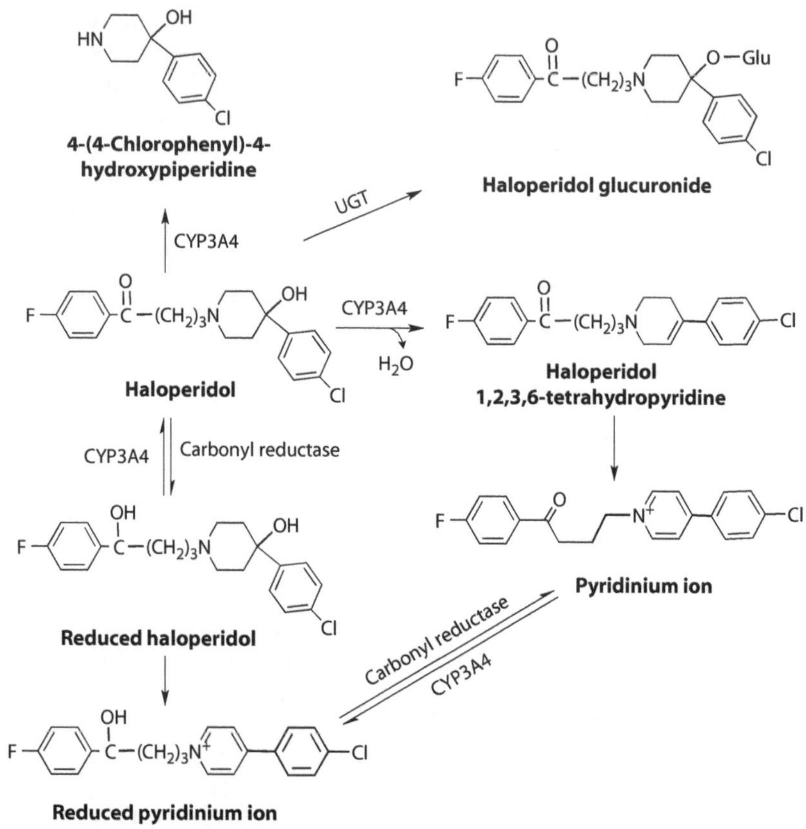

4-(4-Chlorophenyl)-4-
hydroxypiperidine

Haloperidol glucuronide

UGT

CYP3A4

Haloperidol

CYP3A4

H_2O

Haloperidol
1,2,3,6-tetrahydropyridine

CYP3A4 | Carbonyl reductase

Reduced haloperidol

Pyridinium ion

Carbonyl reductase

CYP3A4

Reduced pyridinium ion

FIGURE 3.23 Metabolism of haloperidol. Haloperidol undergoes extensive metabolism involving various CYPs (mainly CYP3A4), carbonyl reductase, and UGTs, with direct glucuronidation being the predominant elimination pathway followed by the reduction of haloperidol to reduced haloperidol and by CYP-mediated oxidation. Haloperidol is reduced by cytosolic carbonyl reductase to reduced form, which has 10% to 20% of the activity of the parent molecule. Reduced haloperidol is back oxidized to haloperidol primarily by CYP3A4. N-Dealkylation of haloperidol leads to the formation of 4-(4-chlorophenyl)-4-hydroxypiperidine, while dehydration of haloperidol results in 1,2,3,6-tetrahydropyridine. Reduced haloperidol also undergoes N-dealkylation by CYP3A4 and 2D6, 1A2, 2C9, 2B6, and 2E1 and glucuronidation. Pyridinium ions can be formed from reduced haloperidol and haloperidol 1,2,3,6-tetrahydropyridine. It appears that CYP2D6 only plays a minor role for the disposition of haloperidol while CYP3A4 is the most important enzyme for its overall disposition.

Olanzapine, an atypical antipsychotic with a thienobenzodiazepinyl structure, is used in the treatment of both positive and negative symptoms of schizophrenia, with a low incidence of extrapyramidal side effects (Beasley et al. 1996). After oral administration of olanzapine to healthy subjects, at least 65% of the dose is absorbed and approximately 87% of the dose is recovered in the urine and feces mostly within 7 days of dosing (Kassahun et al. 1997). Renal excretion is its primary route of elimination. Olanzapine is extensively metabolized in humans via hydroxylation and N-demethylation, N-oxidation, and direct glucuronidation (Figure 3.25). A major circulating metabolite of olanzapine in humans is a tertiary N-glucuronide in which the glucuronic acid moiety is attached to the nitrogen at position 10 of the molecule (Kassahun et al. 1997). The formation of this glucuronide is mainly catalyzed by UGT1A4 (Linnet 2002). Two major urinary metabolites are identified: the 4'-N- and 10-N-glucuronides of olanzapine. Oxidative metabolism on the allylic methyl group results in 2-hydroxymethyl and 2-carboxy derivatives of olanzapine (Kassahun et al. 1997). The methyl piperazine moiety is also subject to oxidation, forming the N-oxide and N-desmethyl metabolites. Other metabolites, including the N-desmethyl-2-carboxy and possibly N-oxide-2-carboxy derivatives, result from metabolic reactions at both the 4'-nitrogen and 2-methyl groups. In addition to 10-N-glucuronide, N-desmethyl, N-oxide, and 2-hydroxymethyl olanzapine are identified in human plasma. The formation of N-desmethyl and 2-hydroxymethyl olanzapine is catalyzed, by CYP1A2 and 2D6, respectively (Ring et al. 1996). The N-oxide metabolite is found to be catalyzed by FMO.

Perphenazine is a typical antipsychotic drug of the piperazinyl phenothiazines used in the treatment of schizophrenia and manic phases of bipolar disorder (Hartung et al. 2005). Perphenazine is 10 to 15 times as potent as chlorpromazine, but it causes a high incidence of early and late extrapyramidal side effects and tardive dyskinesia (Hartung et al. 2005). Perphenazine is primarily metabolized by CYP2D6 to N-dealkylperphenazine, perphenazine sulfoxide, and 7-hydroxyperphenazine (Figure 3.26), with the activity of the latter metabolite comparable with that of the parent in vitro (Olesen and Linnet 2000). Minor metabolic pathways may be N-oxidation and direct glucuronidation of the primary alcohol group. CYP2D6, 1A2, 3A4, and 2C19 are the most important

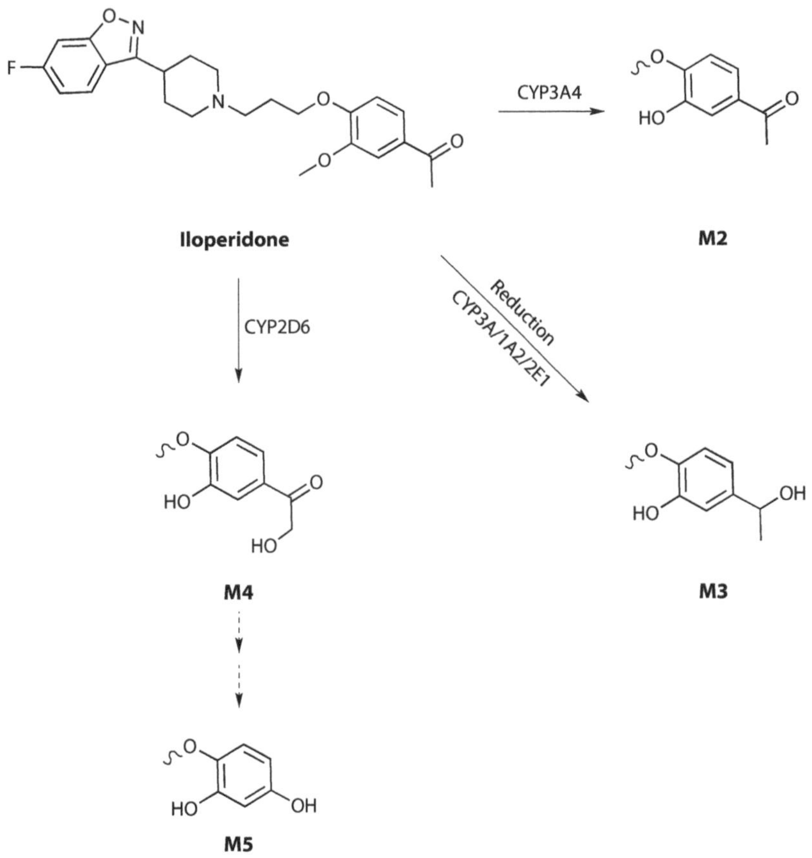

FIGURE 3.24 Metabolism of iloperidone by CYP3A4, 2D6, and other enzymes. Iloperidone is extensively metabolized via O-dealkylation, N-dealkylation, reduction, and hydroxylation in humans. Iloperidone is O-dealkylated to form 6-fluoro-3-[1-(3-hydroxypropyl)-4-piperidinyl]-1,2-benzisoxazole and 1-[4-[3-[4-(6-fluoro-1,2-benzisoxazol-3-yl)-1-piperidinyl]propoxy]-3-hydroxyphenyl]ethanone (M2). Oxidative N-dealkylation results in 6-fluoro-3-(4-piperidinyl)-1,2-benzisoxazol and a secondary metabolite, 3-[(4-acetyl-2-methoxy)phenoxy]propionic acid. Iloperidone is reduced to 4-[3-[4-(6-fluoro-1,2-benzisoxazol-3-yl)-1-piperidinyl]propoxy]-3-methoxy-α-methylbenzene methanol (M3), which is the major circulating metabolite in humans and rats. Hydroxylation of iloperidone yielded 1-[4-[3-[4-(6-fluoro-1,2-benzisoxazol-3-yl)-1-piperidinyl]propoxy]-2-hydroxy-5-methoxyphenyl]ethanone and 1-[4-[3-[4-(6-fluoro-1,2-benzisoxazol-3-yl]-1-piperidinyl]-3-methoxyphenyl]-2-hydroxyethanone (M4). M4 is further oxidized to 4-[3-[4-(6-fluoro-1,2-benzoisoxazol-3-yl)-piperidin1-yl]propoxy]-3-methoxyphenol (M5). In human liver microsomes, the formation of M4 is mediated by CYP2D6, while M2 is formed by CYP3A. M3 is postulated to be produced mainly by a cytosolic enzyme, but CYP3A, 1A2, and 2E1 are involved in its formation as well.

contributors to N-dealkylation of perphenazine (Olesen and Linnet 2000). N-Dealkylperphenazine is usually present in vivo at concentrations 1.5 to 2 times that of the parent drug. In in vitro binding studies, N-dealkylperphenazine showed a higher affinity for 5-HT$_{2A}$ receptors than for D$_2$ receptors to an extent comparable to that of some atypical neuroleptic agents.

Promazine is a member of phenothiazine class of antipsychotics. Other phenothiazine derivatives as antipsychotics include chlorpromazine, triflupromazine, levomepromazine, mesoridazine, thioridazine, fluphenazine, perphenazine, prochlorperazine, flupentixol, and trifluoperazine. Promazine is the weakest blocker of dopamine D$_2$ and HT$_2$ receptors, but it is a relatively potent antagonist of adrenergic α$_1$ and H$_1$ receptors (Daniel 2003; Di Pietro and Seamans 2007; Richelson and Nelson 1984; Wander et al. 1987). Phenothiazines such as promazine undergo mainly S-oxidation in the thiazine ring in position 5 and N-demethylation in a side chain, as well as aromatic hydroxylation and N-oxidation (Figure 3.27) (Svendsen and Bird 1986). In humans, N-demethylation and sulfoxidation are the primary metabolic routes of promazine (Goldenberg et al. 1964). CYP1A2 and 3A4 are the main enzymes responsible for 5-sulfoxidation of promazine, while CYP1A2 and 2C19 are the major enzymes catalyzing N-demethylation of promazine in human liver (Wojcikowski et al. 2003). Recombinant CYP1A1, 2B6, 1A2, 2C9, 3A4, 2E1, 2A6, 2D6, and 2C19 catalyze 5-sulfoxidation and CYP2C19, 2B6, 1A1, 1A2, 2D6, 3A4, 2C9, 2E1, and 2A6 mediate N-demethylation of promazine. The highest intrinsic clearance (V_{max}/K_m) is found for CYP1A2, 3A4, and 2B6 for 5-sulfoxidation and for CYP2C19, 1A2, and 2B6 for N-demethylation. CYP1A2 and 3A4 are the main enzymes responsible for the 5-sulfoxidation of promazine, while CYP1A2 and 2C19 are the primary enzymes that catalyze its N-demethylation in human liver (Wojcikowski et al. 2003). Overall, CYP2A6, 2B6, 2D6, and 2E1 appear to play a minor role in the metabolism of promazine.

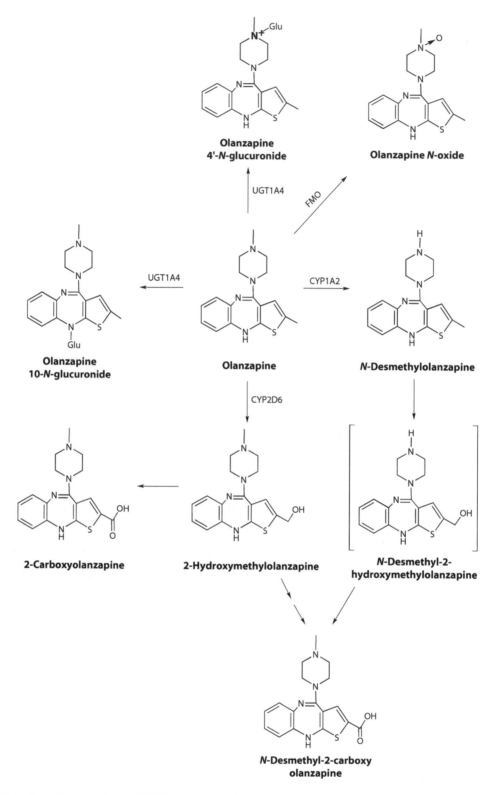

FIGURE 3.25 Metabolism of olanzapine by CYP2D6 and 1A2. Olanzapine is extensively metabolized in humans via hydroxylation and N-demethylation, N-oxidation, and direct glucuronidation. A major circulating metabolite of olanzapine in humans is a tertiary N-glucuronide in which the glucuronic acid moiety is attached to the nitrogen at position 10 of the molecule. The formation of this glucuronide is mainly catalyzed by UGT1A4. Two major urinary metabolites are identified: the 4'-N- and 10-N-glucuronides of olanzapine. Oxidative metabolism on the allylic methyl group results in 2-hydroxymethyl and 2-carboxy derivatives of olanzapine. The methyl piperazine moiety is also subject to oxidation, forming the N-oxide and N-desmethyl metabolites. Other metabolites, including the N-desmethyl-2-carboxy and possibly N-oxide-2-carboxy derivatives, result from metabolic reactions at both the 4' nitrogen and 2-methyl groups. In addition to 10-N-glucuronide, N-desmethyl, N-oxide, and 2-hydroxymethyl olanzapine are identified in human plasma. The formation of N-desmethyl and 2-hydroxymethyl olanzapine is catalyzed, by CYP1A2 and 2D6, respectively. The N-oxide metabolite is found to be catalyzed by FMO.

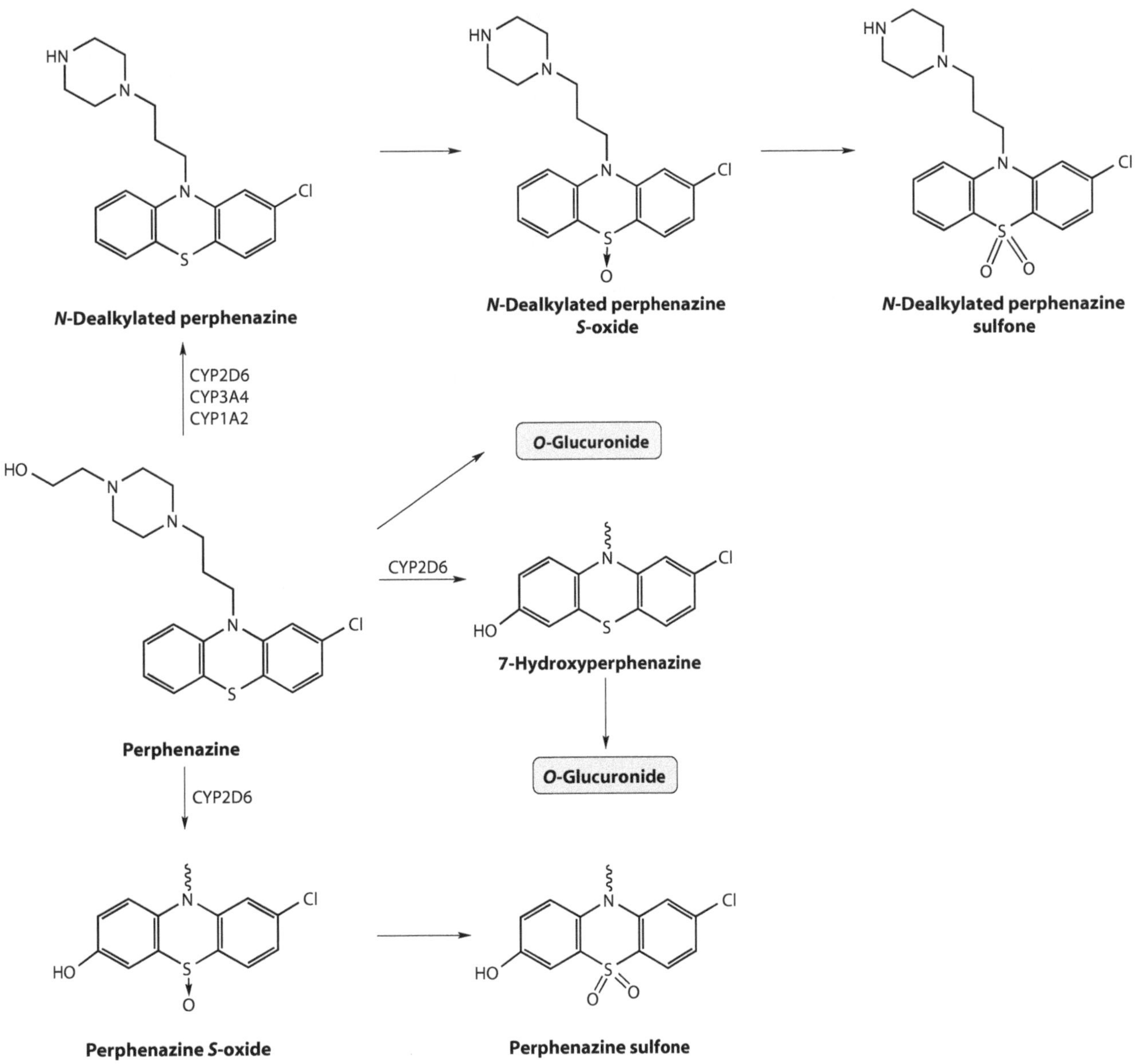

FIGURE 3.26 Metabolism of perphenazine by CYP2D6. Perphenazine has an oral bioavailability of ~40% and a half-life of 8 to 12 h (up to 20 h) and is usually given in two or three divided doses per day. Perphenazine is extensively metabolized in the liver to a number of metabolites mainly by sulfoxidation, ring hydroxylation at position 7 followed by glucuronidation, and *N*-dealkylation leading to loss of the hydroxyethyl group. Minor metabolic pathways may be *N*-oxidation and direct glucuronidation of the primary alcohol group. The plasma concentrations of *N*-dealkylated perphenazine are three times those of perphenazine, whereas perphenazine sulfoxide concentrations are in the same range as those of the parent compound. No information about the concentration of 7-OH-perphenazine in serum is available, but this compound is excreted in the urine mainly as the glucuronide. *N*-Dealkylated perphenazine is eliminated as its sulfoxide, whereas approximately 13% of a dose is excreted in the urine as perphenazine sulfoxide. Perphenazine is primarily metabolized by CYP2D6 to *N*-dealkylperphenazine, perphenazine sulfoxide, and 7-hydroxyperphenazine, with the activity of the latter metabolite comparable with that of the parent. Peak 7-hydroxyperphenazine concentrations are observed between 2 and 4 h with a terminal phase half-life ranging between 9.9 and 18.8 h.

Risperidone, an atypical antipsychotic used for the treatment of both positive and negative symptoms of schizophrenia (Fenton and Scott 2005; Moller 2005), is converted to 9-hydroxyrisperidone by CYP2D6 and 3A4 (Figure 3.28) (Yasui-Furukori et al. 2001). The latter is further metabolized by *N*-dealkylation, possibly by CYP3A4 (Jung et al. 2005; Spina et al. 2000). 9-Hydroxyrisperidone is the major metabolite in plasma and is equipotent with the parent drug

in dopamine receptor affinity and hence contributes to the overall therapeutic effect of risperidone (Schotte et al. 1996). Risperidone is also metabolized by *N*-dealkylation and 7-hydroxylation to a lesser extent (Mannens et al. 1993).

CYP2D6 and 3A4 convert thioridazine to mesoridazine (Wojcikowski et al. 2006), which correlates weakly with the debrisoquine MR (Berecz et al. 2003; Llerena et al. 2000), whereas the subsequent metabolite sulforidazine

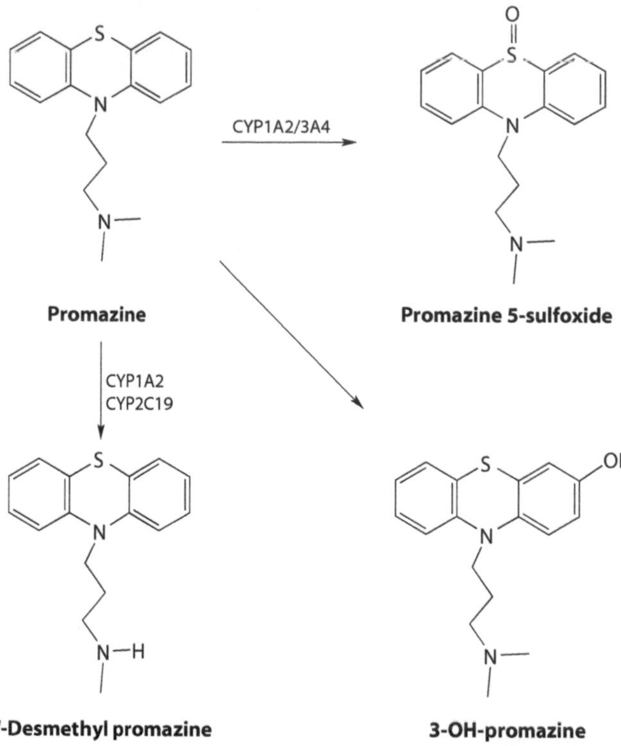

FIGURE 3.27 Metabolism of promazine by CYP1A2, 2C9, 2D6, and other CYPs. Promazine is a member of the phenothiazine class of antipsychotics and undergoes mainly S-oxidation in the thiazine ring in position 5 and N-demethylation in a side chain, as well as aromatic hydroxylation and N-oxidation. In humans, N-demethylation and sulfoxidation are the primary metabolic pathways of promazine. CYP1A2 and 3A4 are the main enzymes responsible for 5-sulfoxidation of promazine, while CYP1A2 and 2C19 are the major enzymes that catalyze N-demethylation of promazine in human liver. Recombinant CYP1A1, 2B6, 1A2, 2C9, 3A4, 2E1, 2A6, 2D6, and 2C19 catalyze 5-sulfoxidation and CYP2C19, 2B6, 1A1, 1A2, 2D6, 3A4, 2C9, 2E1, 2A6 mediate N-demethylation of promazine. The highest intrinsic clearance is found for CYP1A2, 3A4, and 2B6 for 5-sulfoxidation and for CYP2C19, 1A2, and 2B6 for N-demethylation. CYP1A2 and 3A4 are the main enzymes responsible for the 5-sulfoxidation of promazine, while CYP1A2 and 2C19 are the primary enzymes that catalyze its N-demethylation in human liver. CYP2A6, 2B6, 2D6, and 2E1 appear to play a minor role in the metabolism of promazine.

(thioridazine 2-sulfone) does not seem to be reliant on CYP2D6 (Figure 3.29) (Eap et al. 1996). There is a weak correlation between corrected QT interval and thioridazine plasma concentrations, debrisoquine MR, and thioridazine/mesoridazine concentrations (Llerena et al. 2002). The metabolites seem to have activity equal to (sulforidazine) or greater than (mesoridazine) that of the parent, whereas a further ring sulfoxide (thioridazine 5-sulfoxide) produced from thioridazine may be less active as an antipsychotic, but more arrhythmogenic.

Zuclopenthixol, a thioxanthene derivative used in the treatment of schizophrenia, has high affinity for both D_1 and D_2 receptors and α_1-adrenergic and 5-HT$_2$ receptors (Kumar and Strech 2005). It has weak histamine H$_1$ receptor blocking

activity and lower affinity for muscarinic cholinergic and α_2-adrenergic receptors. The pharmacokinetics of zuclopenthixol appears linear over the dosage range investigated. The metabolism of zuclopenthixol is mainly by sulfoxidation, side-chain N-dealkylation, and glucuronidation (Figure 3.30). The metabolites are devoid of pharmacological activity. Zuclopenthixol sulfoxidation and N-dealkylation are mainly metabolized by CYP2D6 and other enzymes (Dahl et al. 1991a). The clearance of zuclopenthixol cosegregates with debrisoquine hydroxylation in humans (Bertilsson et al. 1993; Dahl et al. 1991a), indicating the involvement of CYP2D6 in the metabolism of zuclopenthixol.

3.3.1.5 Hypnotics

Zolpidem is a short-acting nonbenzodiazepine hypnotic that potentiates γ-aminobutyric acid (GABA) by binding to benzodiazepine receptors ω1 subtype located on the GABA receptors (Holm and Goa 2000; Swainston Harrison and Keating 2005). In humans, zolpidem undergoes extensive oxidation of C-methyl groups and hydroxylation of a position on the imidazolepyridine ring system; none of the metabolites appears to be pharmacologically active (Salva and Costa 1995). It is rapidly converted to its alcohol derivatives (M3 and M4) mainly by CYP3A4, with contributions from CYP1A2 and 2D6 (Figure 3.31) (Pichard et al. 1995). The alcohol derivatives are converted to carboxylic acids (M11), representing the main method of metabolism in humans. Microsomes containing human CYP1A2, 2C9, 2C19, 2D6, and 3A4 expressed by cDNA-transfected human lymphoblastoid cells mediate M3 formation from zolpidem in vitro (von Moltke et al. 1999). The kinetic profile for zolpidem metabolite M3 formation by each individual cytochrome is combined with estimated relative abundances based on immunological quantification, yielding projected contributions to net intrinsic clearance of 61% for 3A4, 22% for 2C9, 14% for 1A2, and <3% for 2D6 and 2C19 (von Moltke et al. 1999). The formation of alcohol derivatives of zolpidem is rate limiting and principally mediated by CYP3A4 followed by CYP2C9, while CYP1A2, 2D6, and 2C19 participate in alcohol formation, but their contribution is minor. Coadministration of haloperidol, cimetidine, ranitidine, chlorpromazine, warfarin, digoxin, or flumazenil does not alter the pharmacokinetics of zolpidem (Salva and Costa 1995). However, caffeine (a CYP1A2 inhibitor) modestly increases the C_{max} and AUC of zolpidem by 30%–40% in healthy subjects (Cysneiros et al. 2007). Ketoconazole increases the AUC of zolpidem by 70% with enhanced sedative effect (Greenblatt et al. 1998). Consistently, rifampin reduces the AUC of zolpidem by 27% (Villikka et al. 1997). Itraconazole and fluconazole have a small influence on zolpidem clearance and hypnotic effect. Fluoxetine treatment also produces minimal alteration of zolpidem clearance in humans (Allard et al. 1998; Piergies et al. 1996).

3.3.1.6 Opioids and Opioid Receptor Antagonists

Codeine as a prodrug (~10% of a dose) is metabolized by CYP2D6-mediated O-demethylation into its active form, morphine, which is the key metabolite responsible for the

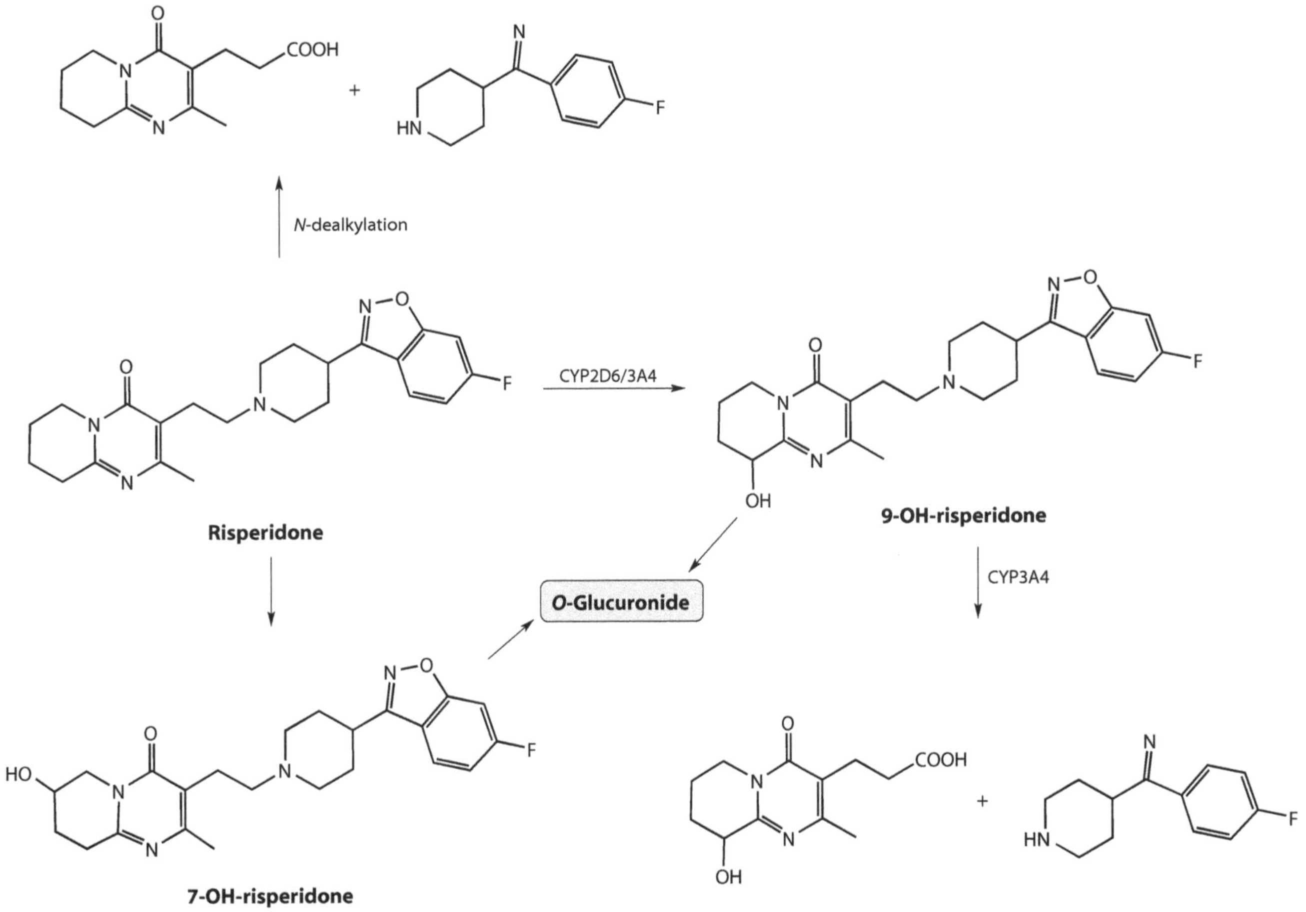

FIGURE 3.28 Metabolism of risperidone by CYP2D6 and 3A4. Approximately 70% of an oral dose of risperidone is eliminated through the renal pathway, with a half-life of approximately 3 h in EMs, but extended to 20 h in PMs. One week after a single oral dose of 1 mg ^{14}C-risperidone, 70% of the administered radioactivity is recovered in the urine and 14% in the feces. Unchanged risperidone is mainly excreted in the urine and accounted for 30%, 11%, and 4% of the administered dose in the PMs, IMs, and EMs, respectively. 9-Hydroxyrisperidone accounted for 8%, 22%, and 32% of the administered dose in the urine in PMs, IMs, and EMs, respectively. Oxidative *N*-dealkylation at the piperidine nitrogen, whether or not in combination with the 9-hydroxylation, accounted for 10%–13% of the dose. Risperidone is mainly metabolized by 9-hydroxylation of the tetrahydropyridopyrimidinone ring and to a lesser extent by *N*-dealkylation and 7-hydroxylation. Oxidative *N*-dealkylation results in two acidic metabolites, one derived from risperidone itself and the other from the 9-OH-risperidone. Risperidone is converted to 9-hydroxyrisperidone by CYP2D6 and 3A4. The latter is further metabolized by *N*-dealkylation, possibly by CYP3A4. Two enantiomers, (+)- and (−)-9-hydroxyrisperidone, might be formed. 9-Hydroxyrisperidone is the major metabolite in plasma with a half-life of 24 h and is equipotent with the parent drug in dopamine receptor affinity and hence contributes to the overall therapeutic effect of risperidone.

most antinociceptive effect of codeine, but most is glucuronidated to codeine-6-glucuronide and the remainder is metabolized by CYP3A4 to norcodeine (Figure 3.32) (Yue and Sawe 1997). Morphine is converted to its 3-*O*- and 6-*O*-glucuronide by UGT1A1, 1A8, and 2B7 (Coffman et al. 1997; King et al. 2000; Ohno et al. 2008). Morphine is *N*-demethylated to result in normorphine, which is then glucuronidated. Morphine-3-glucuronide, morphine-6-glucuronide, morphine-3,6-diglucuronide, morphine 3-ethereal sulfate, normorphine, and normorphine 6-glucuronide are found in the urine of humans (Yeh et al. 1977), whereas codeine with a methoxy group on the 3-position is converted only to the 6-glucuronide.

Dihydrocodeine is a semisynthetic opioid analog used as an analgesic and antitussive (Edwards et al. 2000). Similar to codeine, dihydrocodeine is converted by CYP2D6 to

dihydromorphine (Figure 3.33), which has activity comparable to that of morphine (Kirkwood et al. 1997). Dihydrocodeine is mostly glucuronidated by UGT2B7 to form dihydrocodeine-6-glucuronide or is *N*-demethylated by CYP3A4 to nordihydrocodeine. Dihydromorphine is converted to its 3-*O*- and 6-*O*-glucuronides by UGT1A1, 1A8, and 2B7 (Chau et al. 2014; Ohno et al. 2008) or sulfates.

Hydrocodone is a semisynthetic opioid analog used to treat moderate pain and as an antitussive. Hydrocodone differs structurally from codeine in that the C6-position is occupied by a keto group, and thus it does not undergo the extensive conjugation (>60%) that codeine undergoes. The therapeutic range of hydrocodone is 1–30 ng/ml and the toxic plasma concentration is >100 ng/ml. Approximately 26% of a single dose is excreted in a 72-h urine collection that consists of unchanged

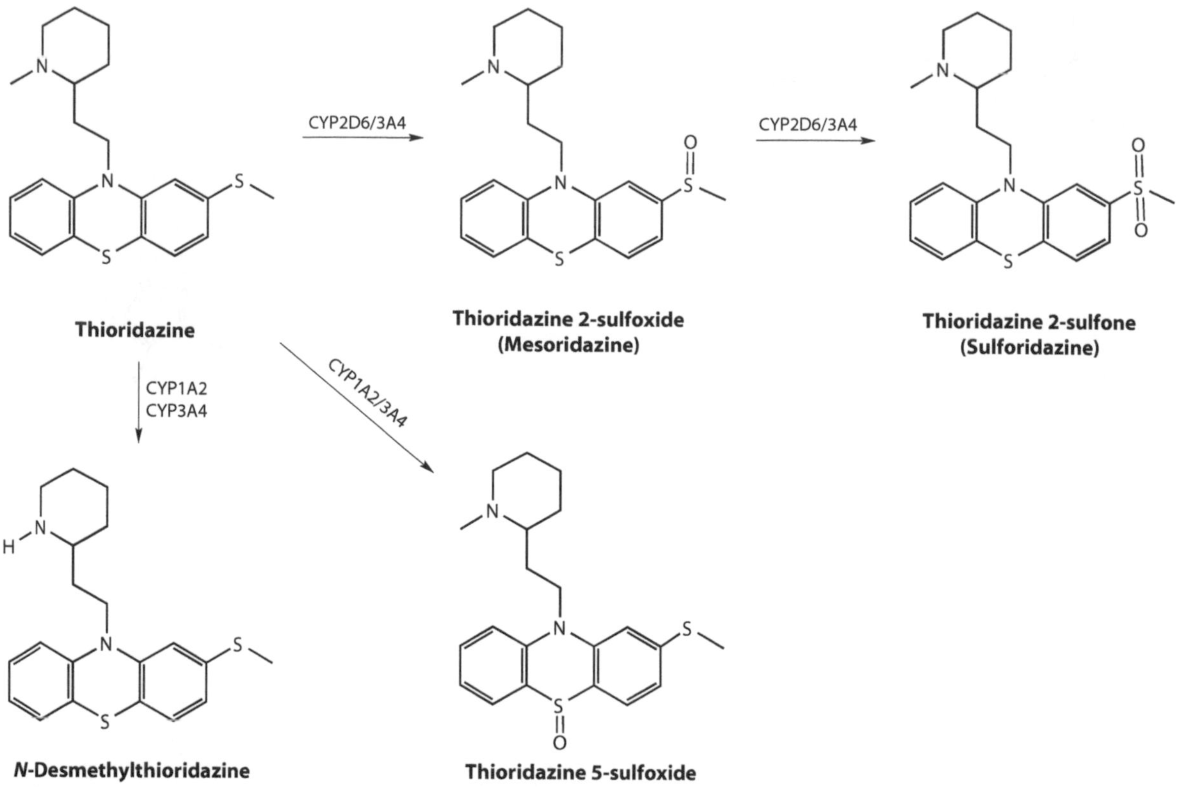

FIGURE 3.29 Metabolism of thioridazine by CYP2D6, 1A2, and 3A4. Thioridazine undergoes *S*-oxidation in the thiazine ring in position 5, as well as aromatic hydroxylation (mainly in position 7), *N*-demethylation, and *N*-oxidation. However, unlike other phenothiazines, thioridazine forms a sulfoxide in position 2 of the thiomethyl substituent (i.e., mesoridazine), which is further oxidized to a sulfone (i.e., sulforidazine). Both mesoridazine and sulforidazine are active metabolites. Thioridazine 5-sulfoxide, a ring sulfoxide, is pharmacologically inactive at dopaminergic or noradrenergic receptors, but may be associated with the cardiotoxicity of thioridazine. CYP1A2 and 3A4 are the major enzymes responsible for 5-sulfoxidation and *N*-demethylation of thioridazine, whereas CYP2D6 is the primary enzyme catalyzing mono-2- and di-2-sulfoxidation in human liver microsomes resulting in mesoridazine and sulforidazine, respectively. CYP3A4 also contributes to thioridazine mono-2-sulfoxidation.

drug (~12%), norhydrocodone (5%), conjugated hydrocodone (4%), 6-hydrocodol (3%), and conjugated 6-hydromorphol (0.1%) (Barakat et al. 2014). Approximately 40% of the clearance of hydrocodone is via non-CYP pathways. The metabolism of hydrocodone is similar to codeine, resulting in the major active metabolite hydromorphone (Figure 3.34). The μ opioid receptor binding affinity of hydromorphone is 10- to 33-fold greater than that of hydrocodone (Chen et al. 1991). Hydrocodone is metabolized by *O*- and *N*-demethylation, resulting in hydromorphone and norhydrocodone, respectively, and reduction of the 6-keto group. C6-keto reduction results in approximately equal amounts of 6-α-hydrocol and 6-β-hydrocol. The *O*-demethylation of hydrocodone is predominantly catalyzed by CYP2D6 and, to a lesser extent, by an unknown low-affinity CYP (Hutchinson et al. 2004). Norhydrocodone formation is mediated by CYP3A4.

Methadone, an opioid receptor inhibitor, is widely used for the treatment of opioid dependence and chronic pain (Garrido and Troconiz 1999). It is extensively metabolized by *N*-demethylation by CYP3A4, 2B6, 2C8, 2C19, and 2D6 to 2-ethylidene-1,5-dimethyl-3,3-diphenylpyrrolidine (EDDP) (Figure 3.35) (Foster et al. 1999; Gerber et al. 2004; Kharasch et al. 2004; Oda and Kharasch 2001; Totah et al. 2008; Wang

and DeVane 2003). In healthy volunteers, paroxetine (a potent CYP2D6 inhibitor [Crewe et al. 1992]) significantly increases the plasma concentrations of *R*- and *S*-methadone (Begre et al. 2002; Lam et al. 2002). In addition, in two PMs of CYP2D6, only *S*-methadone but not *R*-methadone is increased by paroxetine (Begre et al. 2002). Fluoxetine, another potent inhibitor of CYP2D6, stereoselectively increases the concentration of *R*- but not *S*-methadone (Eap et al. 1997). However, fluvoxamine nonselectively increases the plasma levels of both *R*- and *S*-methadone (Eap et al. 1997). Fluconazole increases plasma methadone peak and trough concentrations in healthy volunteers by 27% and 48%, respectively (Cobb et al. 1998). All these findings suggest that CYP2D6 plays an important role in the metabolism of methadone, which will be affected by the genotype of *CYP2D6*, and that they may have different stereoselectivity.

Tramadol, a synthetic opioid analgesic, is also used as a probe substrate for CYP2D6. *S*-Tramadol is metabolized to its *O*-desmethyltramadol (M1) by CYP2D6, as is *R*-tramadol but to a less extent (Figure 3.36) (Paar et al. 1997; Subrahmanyam et al. 2001). *S*-*O*-Desmethyltramadol predominantly mediates the opioid effects with a higher affinity for opioid receptors than the parent drug, whereas *S*-tramadol is more potent

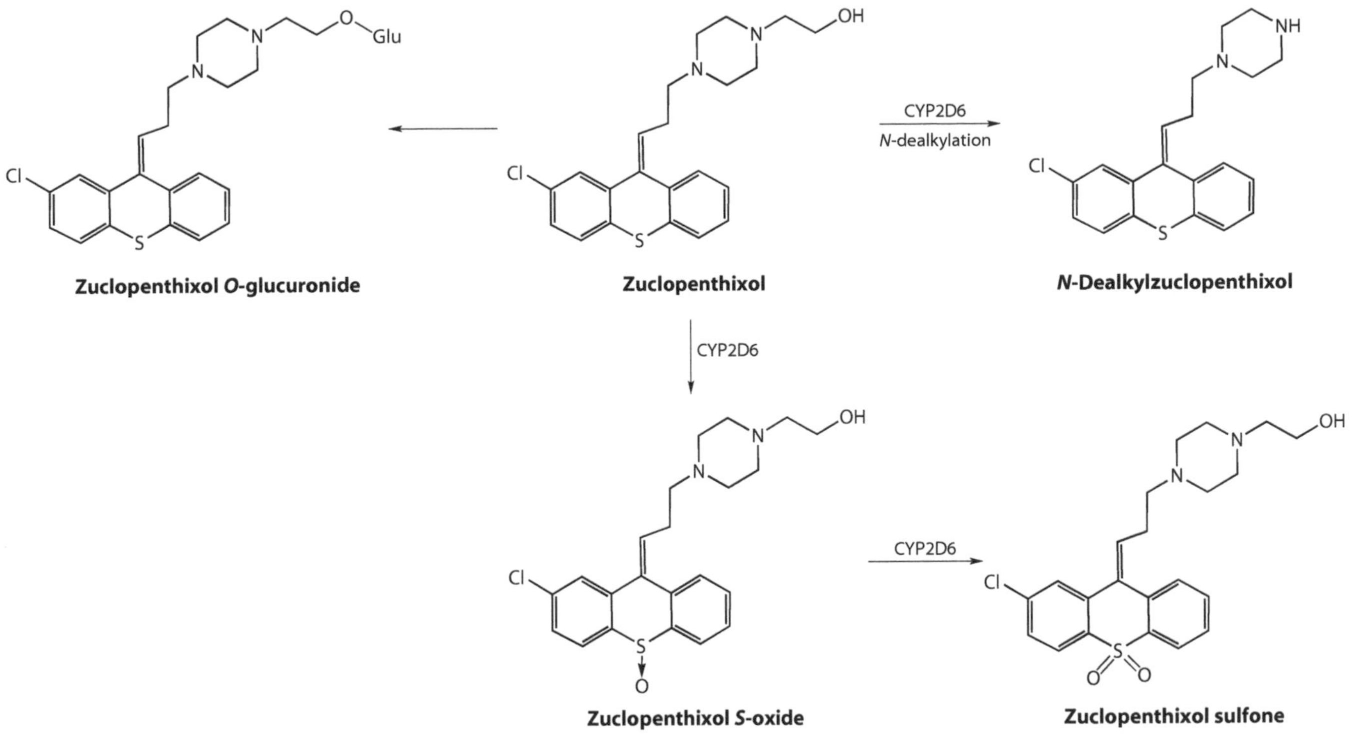

FIGURE 3.30 Metabolism of zuclopenthixol by CYP2D6. Zuclopenthixol is excreted mainly in feces with approximately 10% excreted in the urine. Approximately 0.1% of a dose is excreted unchanged in the urine. The metabolism of zuclopenthixol is mainly by sulfoxidation, side-chain *N*-dealkylation, and glucuronidation. The metabolites are devoid of pharmacological activity. Zuclopenthixol sulfoxidation and *N*-dealkylation are mainly metabolized by CYP2D6 and other enzymes.

against serotonin reuptake inhibition and *R*-tramadol against noradrenaline reuptake inhibition. After oral administration of ^{14}C-labeled tramadol to human volunteers, approximately 90% of the administered radioactivity is excreted in the urine, and some unchanged tramadol (25% or 32% of total urinary radioactivity) and a number of tramadol metabolites have been identified in the urine (Lintz et al. 1981). The primary metabolites are *O*-desmethyltramadol (M1) and *N*-desmethyltramadol (M2), which may be further metabolized to three additional secondary metabolites, namely, *N,N*-didesmethyltramadol (M3), *N,N,O*-tridesmethyltramadol (M4), and *N,O*-didesmethyltramadol (M5) (Lintz et al. 1981). These tramadol metabolites together with the conjugates (glucuronides and sulfates) of M1, M4, and M5 have been detected in human urine. In vitro studies indicate that tramadol is metabolized to M1 by cDNA-expressed CYP2D6 and to M2 by CYP2B6 and 3A4 (Paar et al. 1997; Subrahmanyam et al. 2001).

3.3.1.7 Antiemetics

CYP2D6 is involved in the metabolism of several antiemetics including tropisetron (Firkusny et al. 1995; Fischer et al. 1994; Sanwald et al. 1996a,b), ondansetron (Fischer et al. 1994; Sanwald et al. 1996a), dolasetron (Sanwald et al. 1996a), ezlopitant (Obach 2001), and metoclopramide (Desta et al. 2002). Tropisetron, ondansetron, and dolasetron are 5-HT$_3$ antagonists used in the control of chemotherapy-induced nausea and vomiting (Aapro 2005; Aapro and Blower 2005; Hesketh

2008; Schwartzberg 2007); ezlopitant and aprepitant are both nonpeptidic antagonists of the neurokinin-1 (NK$_1$) receptor (Rojas et al. 2014); and metoclopramide is mainly used as a gastroprokinetic and antiemetic agent. Tropisetron hydroxylation is primarily by CYP2D6, whereas that of ondansetron is CYP2D6 and 2E1 dependent (Sanwald et al. 1996a). Ezlopitant is metabolized by both CYP2D6 and 3A4 (Obach 2000), whereas aprepitant is mainly metabolized by CYP3A4 (Sanchez et al. 2004). The three-drug combination of a 5-HT$_3$ receptor antagonist, dexamethasone, and aprepitant is highly recommended before chemotherapy of high emetic risk (Kris et al. 2006).

Dolasetron is a 5-HT$_3$ receptor antagonist, which is used for the treatment of chemotherapy-induced emesis (Balfour and Goa 1997). It is rapidly reduced by carbonyl reductase to form its major pharmacologically active metabolite, reduced dolasetron (Shah et al. 1995). Reduced dolasetron appears in plasma within 10 min after intravenous or oral administration (Dempsey et al. 1996). Reduced dolasetron underwent oxidation of the indole aromatic ring at positions 5, 6, and 7 and also *N*-oxidation. Reduced dolasetron accounts for 17%–54% of the dose in urine, and hydroxylated metabolites of reduced dolasetron represent up to 9% of the dose in urine (Figure 3.37) (Reith et al. 1995). Most of the remaining urinary radioactivity consists of conjugated metabolites of reduced dolasetron and hydroxylated reduced dolasetron. Hydroxylation of reduced dolasetron is mediated by CYP2D6, and its *N*-oxidation is catalyzed by CYP3A4 (Sanwald et al. 1996a).

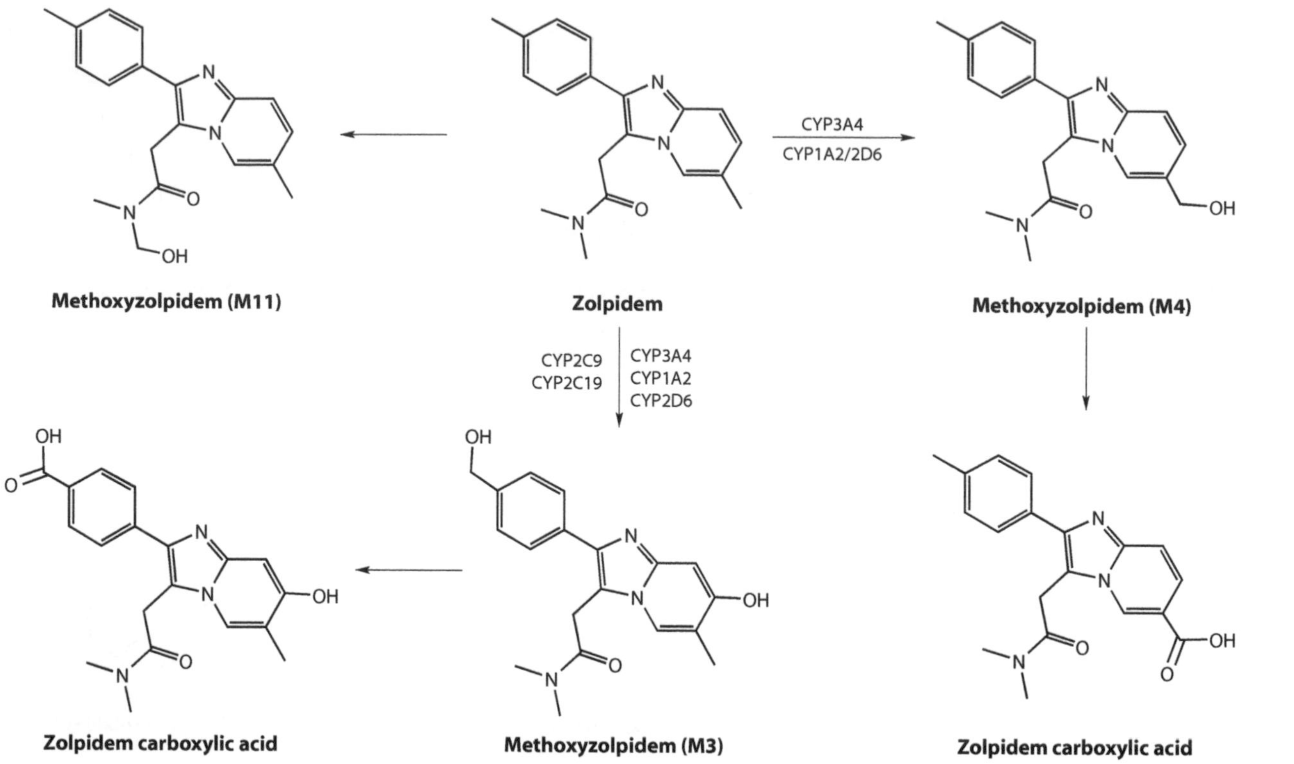

FIGURE 3.31 Metabolism of zolpidem by CYP3A4, 1A2, 2C9, and 2D6. In humans, zolpidem undergoes extensive oxidation of methyl groups and hydroxylation of a position on the imidazolepyridine ring system; none of the metabolites appears to be pharmacologically active. It is rapidly converted to its alcohol derivatives (M3 and M4) mainly by CYP3A4, with contributions from CYP1A2 and 2D6. The alcohol derivatives are converted to carboxylic acids (M11), representing the main method of metabolism in humans. Recombinant CYP1A2, 2C9, 2C19, 2D6, and 3A4 mediated M3 formation from zolpidem. The formation of alcohol derivatives of zolpidem is rate limiting and principally mediated by CYP3A4 followed by CYP2C9, while CYP1A2, 2D6, and 2C19 participate in alcohol formation, but their contribution is minor.

Ezlopitant is mainly metabolized by CYP2D6 and 3A4 (Obach 2000). Substance P is an 11–amino acid neuropeptide of the tachykinin family of peptides located in both the gut and CNS, which is thought to play a key role in emetic response. Substance P binds to NK_1 receptors, and a number of nonpeptide compounds that selectively block the NK_1 receptor have been developed and evaluated as antiemetic agents (Rojas et al. 2014). Antagonists of the human substance P receptor may be of therapeutic benefits for diseases in which substance P plays a role, such as inflammatory diseases (e.g., irritable bowel syndrome), emesis, and pain. Ezlopitant had undergone Phase II trials in the United States and Europe and Phase I in Japan for treatment of chemotherapy-induced emesis, but it appears to have been discontinued. In the rat, guinea pig, dog, and monkey, ezlopitant has a high systemic clearance and low oral bioavailability, presumably owing to high first-pass metabolism, but its clearance in humans is substantially slower than that in animals. Ezlopitant is converted primarily to two pharmacologically active metabolites, namely, CJ-12,458, an alkene metabolite, and CJ-12,764, a benzyl alcohol metabolite, both of which have been detected in the systemic circulation of preclinical species (Reed-Hagen et al. 1999) as well as in humans (Figure 3.38). CJ-12,764 is the major metabolite in preclinical species, with less CJ-12,458 observed; the

two metabolites are observed in nearly equal abundance in human blood. CJ-12,458 is further converted to two metabolites: a diol (CP-611,781) and a 1° alcohol (CP-616,762) by CYP3A4 and 2D6 (Obach 2001). Both CJ-12,458 and CJ-12,764 show substance P receptor inhibitory activities with equal potency to the parent compound, potentially contributing to the pharmacological activity of ezlopitant.

Prakash et al. (2007) have systematically investigated the disposition of ezlopitant in humans. After oral administration of ^{14}C-ezlopitant to healthy male volunteers, the total recovery of administered radioactive dose is 82.8%, with 32.0% in the urine and 50.8% in the feces. The major disposition pathway of ezlopitant in humans is the oxidation of the isopropyl side chain to form the ω-hydroxy and ω-1-hydroxy (M16) metabolites. M16 and ω,ω-1-dihydroxy (1,2-dihydroxy, M12) are identified as the major circulating metabolites, accounting for 64.6% and 15.4% of total circulating radioactivity, respectively (Prakash et al. 2007). In feces, the major metabolite M14 is the propionic acid metabolite formed by further oxidation of the ω-hydroxy metabolite. The urinary metabolites are the result of cleaved metabolites caused by oxidative dealkylation of the 2-benzhydryl-1-aza-bicyclo[2.2.2]oct-3-yl moiety. The metabolites (M1A, M1B, and M4) are as benzyl amine derivatives, accounting for ~34% of the total radioactivity in urine. The other metabolites are formed by O-demethylation,

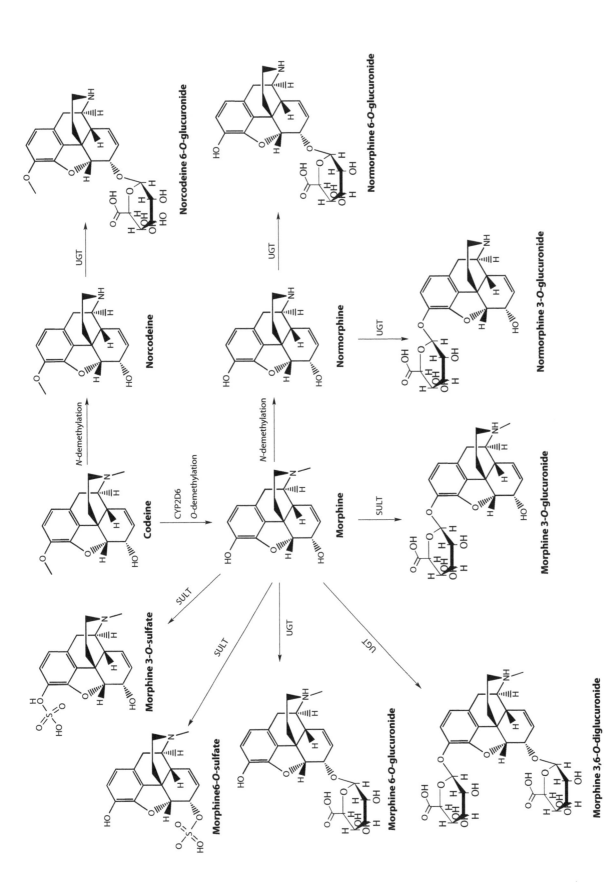

FIGURE 3.32 Metabolism of codeine. Codeine as a prodrug (~10%) is metabolized by CYP2D6-mediated *O*-demethylation into its active form, morphine, which is the key metabolite responsible for the most antinociceptive effect of codeine, but most is glucuronidated to codeine-6-glucuronide and the remainder is metabolized by CYP3A4 to norcodeine. Morphine is converted to its 3-*O*- and 6-*O*-glucuronide by UGT1A1, 1A8, and 2B7. Morphine is *N*-demethylated to result in normorphine, which is then glucuronidated. Morphine-3-glucuronide, morphine-6-glucuronide, morphine-3,6-diglucuronide, morphine 3-ethereal sulfate, normorphine, and normorphine 6-glucuronide are found in the urine of humans, whereas codeine with a methoxy group on the 3-position is converted only to 6-glucuronide.

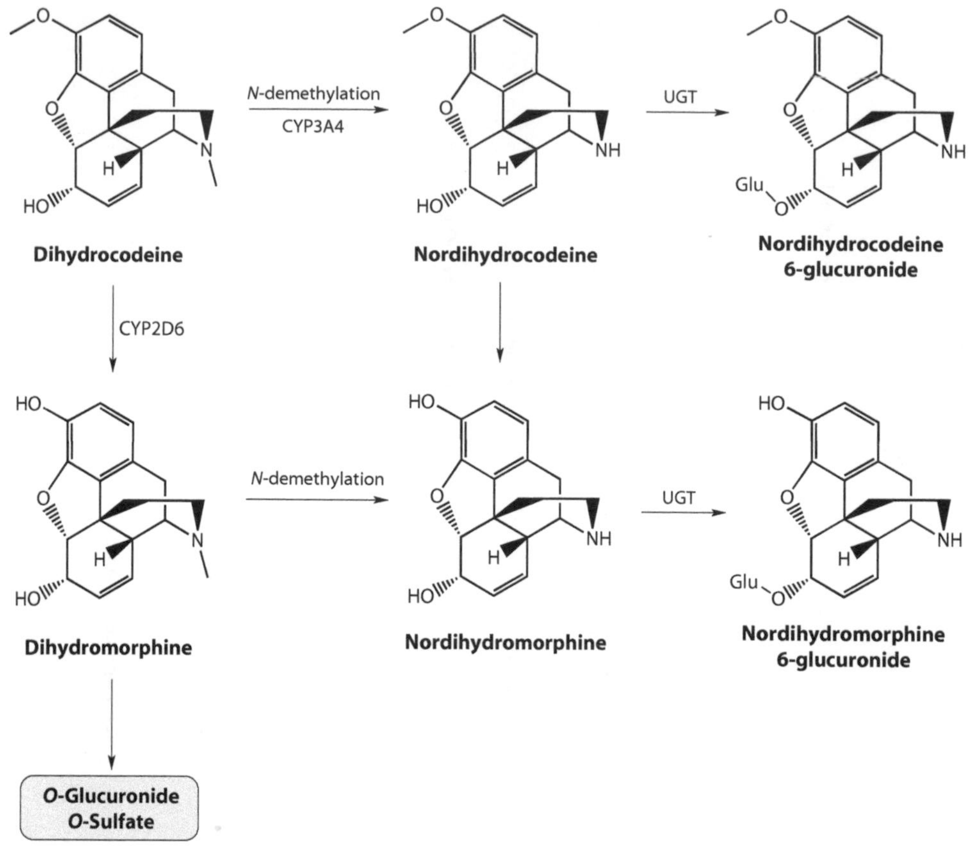

FIGURE 3.33 Metabolism of dihydrocodeine by CYP2D6. Dihydrocodeine is converted by CYP2D6 to dihydromorphine, which has activity comparable to that of morphine. Dihydrocodeine is mostly glucuronidated to form dihydrocodeine-6-glucuronide or is N-demethylated by CYP3A4 to nordihydrocodeine. Dihydromorphine is converted to its 3-O- and 6-O-glucuronides or sulfates.

dehydrogenation of the isopropyl group, and oxidation on the quinuclidine moiety (Prakash et al. 2007).

Ondansetron is a potent and selective 5-HT$_3$ receptor antagonist used mainly as an antiemetic to treat and prevent nausea and vomiting induced by chemotherapy and radiotherapy and postoperative nausea and vomiting (Christofaki and Papaioannou 2014). Ondansetron is extensively metabolized in humans, with ~5% of a radiolabeled dose recovered as the parent drug from the urine. The primary metabolic pathway is hydroxylation on the indole ring in the 6-, 7-, and 8-positions by CYP2D6, 3A4, and 1A2 followed by glucuronide or sulfate conjugation (Figure 3.39) (Fischer et al. 1994; Sanwald et al. 1996a). In terms of overall ondansetron elimination, CYP3A4 plays the predominant role. Because of the multiplicity of metabolic enzymes capable of metabolizing ondansetron, it is likely that inhibition or loss of one enzyme (e.g., CYP2D6 deficiency) will be compensated by others and may result in little change in overall rates of ondansetron elimination.

Tropisetron is a highly potent and selective 5-HT$_3$ antagonist used as an antiemetic in cancer chemotherapy (Lee et al. 1993; Simpson et al. 2000). It is extensively metabolized by hydroxylation of the indole ring in the 5-, 6-, and 7-positions followed by conjugation in humans (Figure 3.40) (Fischer et al. 1992a; Kutz 1993). The major metabolic route is 6-hydroxylation, accounting for 45% of the urinary excretion, while

12% of the dose is excreted as 5-OH-tropisetron in urine. Oxidative N-demethylation and N-oxygenation at the tropinyl nitrogen also occur but to a minor extent, with trace levels of the corresponding metabolites detected (Fischer et al. 1992a).

Metoclopramide is a common antiemetic used for preventing vomiting induced by antineoplastic drugs, particularly cisplatin (Albibi and McCallum 1983). It has also been widely used as a gastroprokinetic agent to control symptoms of upper gastrointestinal motor disorders (e.g., gastroesophageal reflux disease, dyspepsia, and diabetic gastroparesis) (Albibi and McCallum 1983). Approximately 25% of the metoclopramide dose is recovered unchanged in the urine, whereas a substantial amount of the dose appears to undergo enzymatic metabolism via oxidation and conjugation reactions (Teng et al. 1977). Metoclopramide is largely metabolized by CYP2D6 to monodeethylmetoclopramide via N-dealkylation (Figure 3.41) (Desta et al. 2002). Yu et al. (2006) have identified a novel aromatic N-hydroxy metabolite of metoclopramide based on in silico modeling studies. This metabolite is found to represent a major metabolite in human liver microsomes, and CYP2D6 is responsible for its formation (Yu et al. 2006). In their docking studies, the aromatic amino group of the drug lies in proximity to the heme, suggesting possible oxidative reactions at C$_3$, C$_4$, or the aromatic amino group. Since toxic aromatic amines are known to be produced from related compounds such as

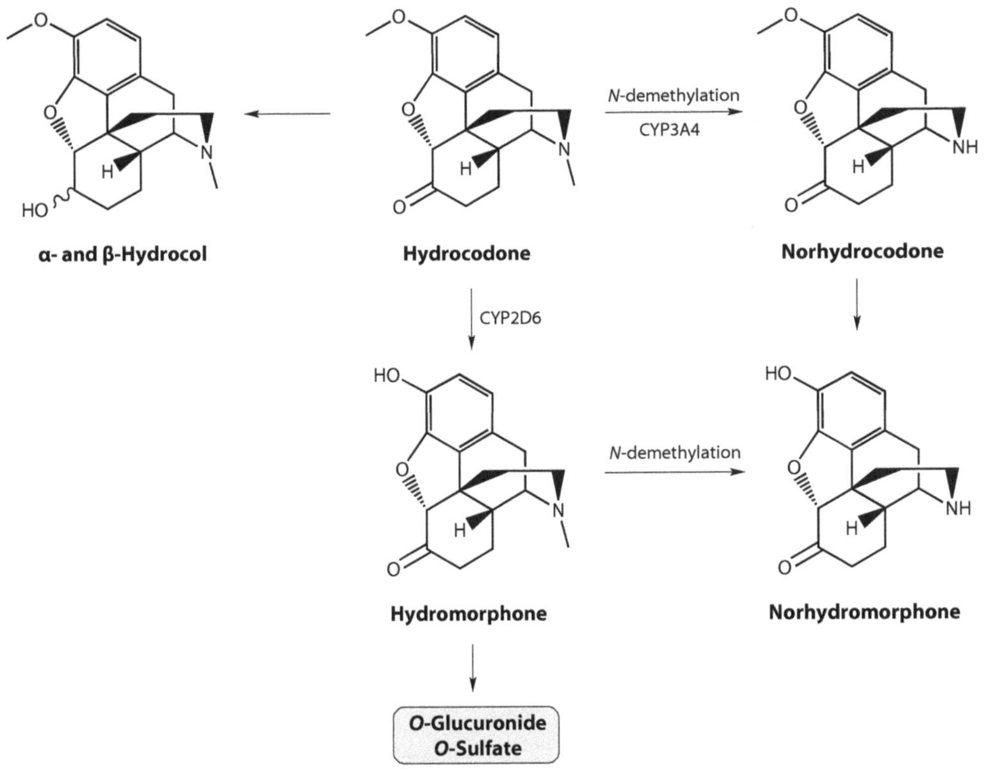

FIGURE 3.34 Metabolism of hydrocodone by CYP2D6 and 3A4. Approximately 26% of a single dose is excreted in a 72-h urine collection that consists of unchanged drug (12%), norhydrocodone (5%), conjugated hydrocodone (4%), 6-hydrocodol (3%), and conjugated 6-hydromorphol (0.1%). Approximately 40% of the clearance of hydrocodone is via non-CYP pathways. The metabolism of hydrocodone is similar to codeine, resulting in the major active metabolite hydromorphone. The μ opioid receptor binding affinity of hydromorphone is 30 times greater than that of hydrocodone. Hydrocodone is metabolized by O- and N-demethylation, resulting in hydromorphone and norhydrocodone, respectively, and reduction of the 6-keto group. C6-keto reduction results in approximately equal amounts of 6-α-hydrocol and 6-β-hydrocol. Trace amounts of reduced hydromorphone are also excreted. Hydromorphone, like morphine, has a C3-phenolic site suitable for conjugation. The O-demethylation of hydrocodone is predominantly catalyzed by CYP2D6 and, to a lesser extent, by an unknown low-affinity CYP. Norhydrocodone formation is mediated by CYP3A4.

procainamide, further investigation into the potential toxicity of the metoclopramide metabolite is needed. Therefore, examination of the docked orientations of individual new and existing drugs can provide useful insights into the metabolism of CYP2D6 substrates, as shown with metoclopramide, where new metabolites can be predicted, which may be associated with adverse reactions of metoclopramide.

3.3.1.8 Antimigraine Drugs

Almotriptan is a highly selective 5-HT$_{1B/1D}$ receptor agonist developed for the symptomatic treatment of acute migraine attacks (Gras et al. 2002; Keam et al. 2002; Pascual 2003). Almotriptan is well absorbed after oral administration and the mean oral bioavailability is 69.1% (McEnroe and Fleishaker 2005). Approximately 50% of an almotriptan dose is excreted unchanged in the urine. The predominant metabolic route of almotriptan is MAO-A–mediated oxidation, while CYP3A4/2D6-catalyzed oxidation occurs to a minor extent (Figure 3.42) (McEnroe and Fleishaker 2005). Human liver microsomes and S9 fractions metabolize almotriptan to at least six metabolites (i.e., M2–M7, with M2–M5 formation being NADPH dependent and M6–M7 formation being

non-NADPH dependent). MAO-A catalyzes oxidative deamination of almotriptan to form the indole acetic acid (M6) and indole ethyl alcohol derivatives of almotriptan (M7) (Salva et al. 2003). Almotriptan undergoes 2-hydroxylation of the pyrrolidine group to form a carbinolamine metabolite intermediate (M3), and this reaction is catalyzed by CYP2D6 and 3A4 with a minor role of 1A1 (Salva et al. 2003). M3 is further oxidized by aldehyde dehydrogenase (ALDH) to the open ring GABA metabolite (M2). The ALDH inhibitor disulfiram almost completely inhibits the formation of metabolites M2 and M6 in human liver microsomes but stimulates the formation of M3 and M7. Another ALDH inhibitor, diethyldithiocarbamate, the reduced metabolite of disulfiram, also inhibits the formation of M2 and M6 in human liver microsomes. Almotriptan is also metabolized at the dimethylaminoethyl group via N-demethylation to form N-desmethyl almotriptan (M4); this reaction is catalyzed by CYP1A1, 1A2, 2C8, 2C19, 2D6, 2E1, and 3A4. None of the cDNA-expressed CYPs is able to generate metabolites M2 or M6, whereas M5 and M7 are only formed by CYP1A1. FMO3 mediates N-oxidation of almotriptan, resulting in its N-oxide. Dopamine, a reversible inhibitor of MAO-A and MAO-B, inhibits the formation of

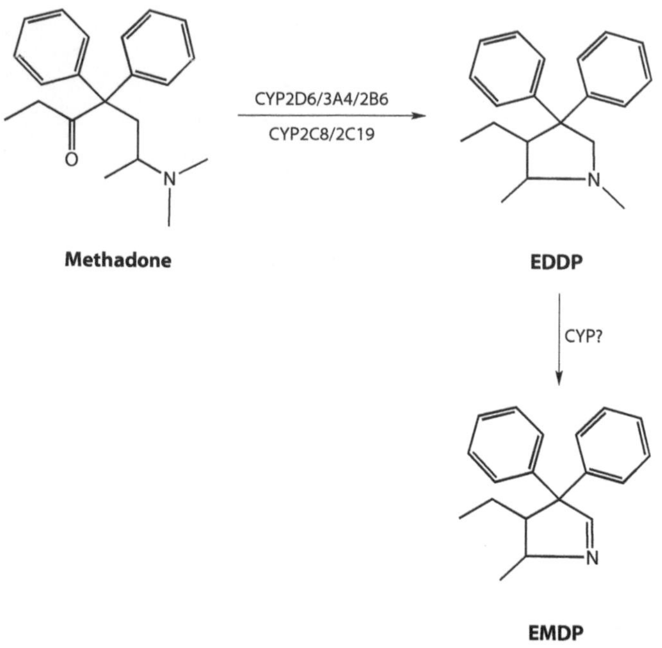

FIGURE 3.35 Metabolism of methadone by CYP2D6, 2C8, and 3A4. In humans, methadone is extensively metabolized, and one major metabolic pathway is *N*-demethylation by CYP3A4 (major), 2B6, 2C8, 2C19, and 2D6 to EDDP. The initial *N*-demethylation of methadone is accompanied by a spontaneous cyclization between the secondary amine and the ketone to form EDDP. Both EDDP and EMDP are pharmacologically inactive without opioid activity. EDDP can be converted to EMDP by further *N*-demethylation. In addition to EDDP, six other metabolites of methadone have been identified, accounting for 43% to 83% of a given dose of methadone in human urine. Approximately 75% of the urinary and fecal radioactive metabolites are unconjugated.

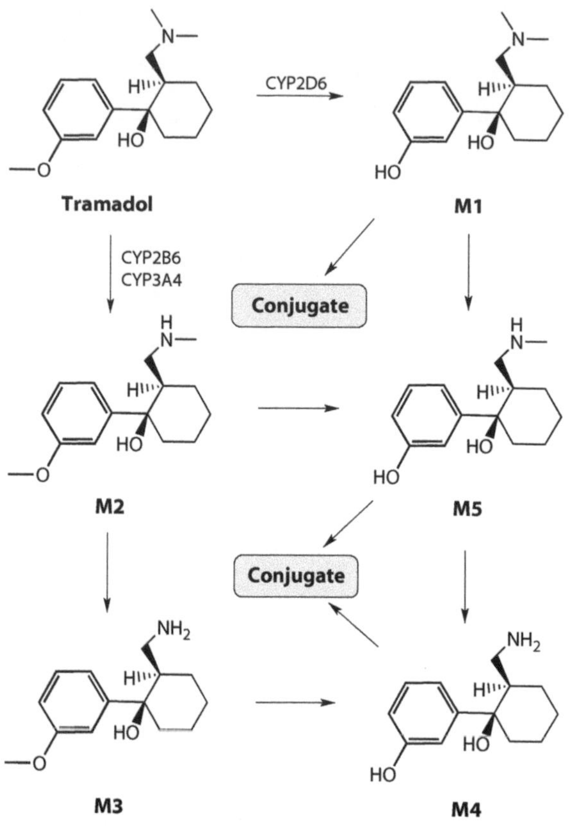

FIGURE 3.36 Metabolism of *S*-tramadol by CYP2D6, 2B6, and 3A4. The synthetic opioid analgesic *S*-tramadol is metabolized to its *O*-desmethyltramadol (M1) by CYP2D6, as is *R*-tramadol but to a lesser extent. *S*-*O*-Desmethyltramadol predominantly mediates the opioid effects with a higher affinity for opioid receptors than the parent drug. The primary metabolites are *O*-desmethyltramadol (M1) and *N*-desmethyltramadol (M2), which may be further metabolized to three additional secondary metabolites, namely, *N,N*-didesmethyltramadol (M3), *N,N,O*-tridesmethyltramadol (M4), and *N,O*-didesmethyltramadol (M5). These tramadol metabolites together with the conjugates (glucuronides and sulfates) of M1, M4, and M5 have been detected in human urine. In vitro studies indicate that tramadol is metabolized to M1 by cDNA-expressed CYP2D6 and to M2 by CYP2B6 and 3A4.

M6 and M7 by liver mitochondria with K_i values of 0.8 and 3.2 μM, respectively. Clorgyline, a metabolism-dependent inhibitor of MAO-A, completely inhibits the formation of mitochondrial metabolites within substrate concentrations from 0.1 to 5 μM (Salva et al. 2003). Deprenyl (a mechanism-based inhibitor of MAO-B) is a relatively weak inhibitor, with complete inhibition of almotriptan metabolism only at a high concentration (5 μM). These findings have confirmed the contribution of MAO-A to the metabolism of almotriptan. Both the GABA and indole acetic acid metabolites are the major in vivo metabolites of almotriptan in humans. In addition, clinical studies have shown that almotriptan clearance is decreased by 27.3% by the MAO-A inhibitor moclobemide (Fleishaker et al. 2001b) and the CYP3A inhibitor verapamil by ~20% (Fleishaker et al. 2000), which is consistent with the contribution of MAO-A and CYP3A4 to the metabolic clearance of almotriptan. Although ketoconazole has a greater effect on almotriptan clearance (reduced by 57.1%) than verapamil (Fleishaker et al. 2003), no dosage adjustment is required when almotriptan is administered concomitantly with these drugs. In contrast, the potent CYP2D6 inhibitor fluoxetine (Fleishaker et al. 2001a) and CYP1A inhibitor propranolol (Fleishaker et al. 2001c) do not alter the pharmacokinetics of almotriptan in humans.

3.3.1.9 Antiparkinsonism Agents

Selegiline is a drug used for the treatment of early-stage Parkinson's disease, depression, and senile dementia. Selegiline exhibits little therapeutic benefit when used alone but enhances and prolongs the anti-Parkinson effects of levodopa. At normal clinical doses, it is a selective irreversible MAO-B inhibitor; it also inhibits MAO-A at high doses. Selegiline possesses the classic phenethylamine skeleton with a propargyl group attached to the nitrogen. Both stereoisomers of selegiline are converted to nordeprenyl and L-methamphetamine via *N*-demethylation and *N*-depropargylation of the sole basic nitrogen in the molecule by recombinant CYP2D6 (Figure 3.43) (Grace et al. 1994). However, CYP2B6 and 2C19 are the major enzymes for the metabolism of selegiline (Hidestrand et al. 2001). L-Methamphetamine is not considered psychoactive and has little abuse potential, but the

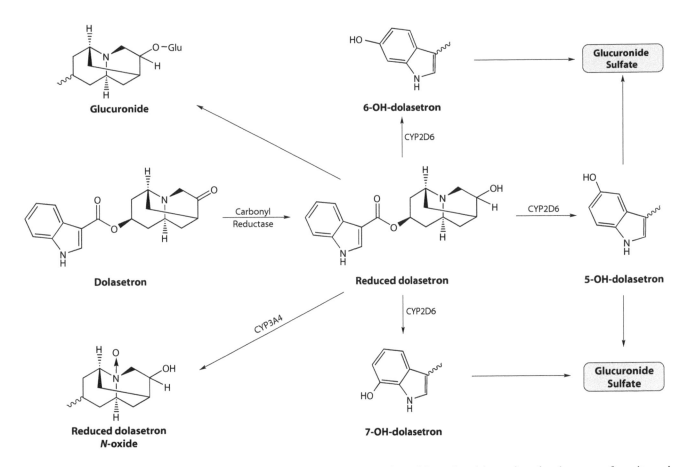

FIGURE 3.37 Metabolism of dolasetron by CYP2D6 and 3A4. Dolasetron is rapidly reduced by carbonyl reductase to form its major pharmacologically active metabolite, reduced dolasetron. Reduced dolasetron underwent oxidation of the indole aromatic ring at positions 5, 6, and 7, and also *N*-oxidation. Reduced dolasetron accounted for 17%–54% of the dose in urine, and hydroxylated metabolites of reduced dolasetron represented up to no more than 9% of the dose in urine. Hydroxylation of reduced dolasetron is mediated by CYP2D6, and *N*-oxidation is catalyzed by CYP3A4.

stimulatory effect on locomotor activity and dopamine synthesis may be attributed to L-methamphetamine. Selegiline has been classified as a controlled substance in Japan and thus can only be obtained with a prescription or special government license. Selegiline is not a controlled substance in the United States but a prescription is required to obtain it. In February 2006, the US FDA approved Emsam (selegiline), the first transdermal patch for use in treating major depression. Zelapar is a transmucosal preparation for human administration of selegiline.

3.3.1.10 Centrally Acting Cholinesterase Inhibitors

Several centrally acting cholinesterase inhibitors including tacrine (Spaldin et al. 1994), donepezil (Barner and Gray 1998), and galantamine (Bachus et al. 1999) are metabolized by CYP2D6. However, these drugs are mainly metabolized by CYP1A2 and 3A4 (Jann et al. 2002). Tacrine is the first centrally acting cholinesterase inhibitor approved for the treatment of mild to moderate Alzheimer's disease and is marketed under the trade name Cognex. However, tacrine caused a relatively high incidence of elevations in serum alanine aminotransferase, which is reversible upon cessation of drug administration (Davis et al. 1992; Farlow

et al. 1992). Other newer centrally acting cholinesterase inhibitors, such as donepezil, are now preferred over tacrine in the treatment of Alzheimer's disease. The bioavailability and plasma concentrations of tacrine are low and variable in Alzheimer's patients, with peak plasma concentrations reached in 1 to 2 h postdosing (Forsyth et al. 1989). In human liver microsomes, tacrine is mainly metabolized to 1-OH-tacrine by CYP1A2, with minor contribution from CYP2D6 (Madden et al. 1993), while 2-, 4-, and 7-OH-tacrine is also formed (Figure 3.44) (Madden et al. 1995; Spaldin et al. 1994). 7-OH-tacrine may be converted into reactive metabolites that are capable of binding proteins, and this has been suggested to contribute to tacrine-induced hepatotoxicity. In human subjects, urinary 1-OH-tacrine accounted for approximately 5% of the administered dose, while tacrine and 2-OH-tacrine comprised <1% and ~2% of the administered dose, respectively (Hooper et al. 1994). 4-OH-tacrine is not detectable. Phenol and 5,6- or 7,8-diOH-tacrine are also found at trace amounts. Another study in humans by Pool et al. (1997) found that polar metabolites accounted for 25%–26% of the total dose, whereas 1-, 2-, and 4-OH-tacrine, and tacrine accounted for only 4% and 6% of dose, respectively.

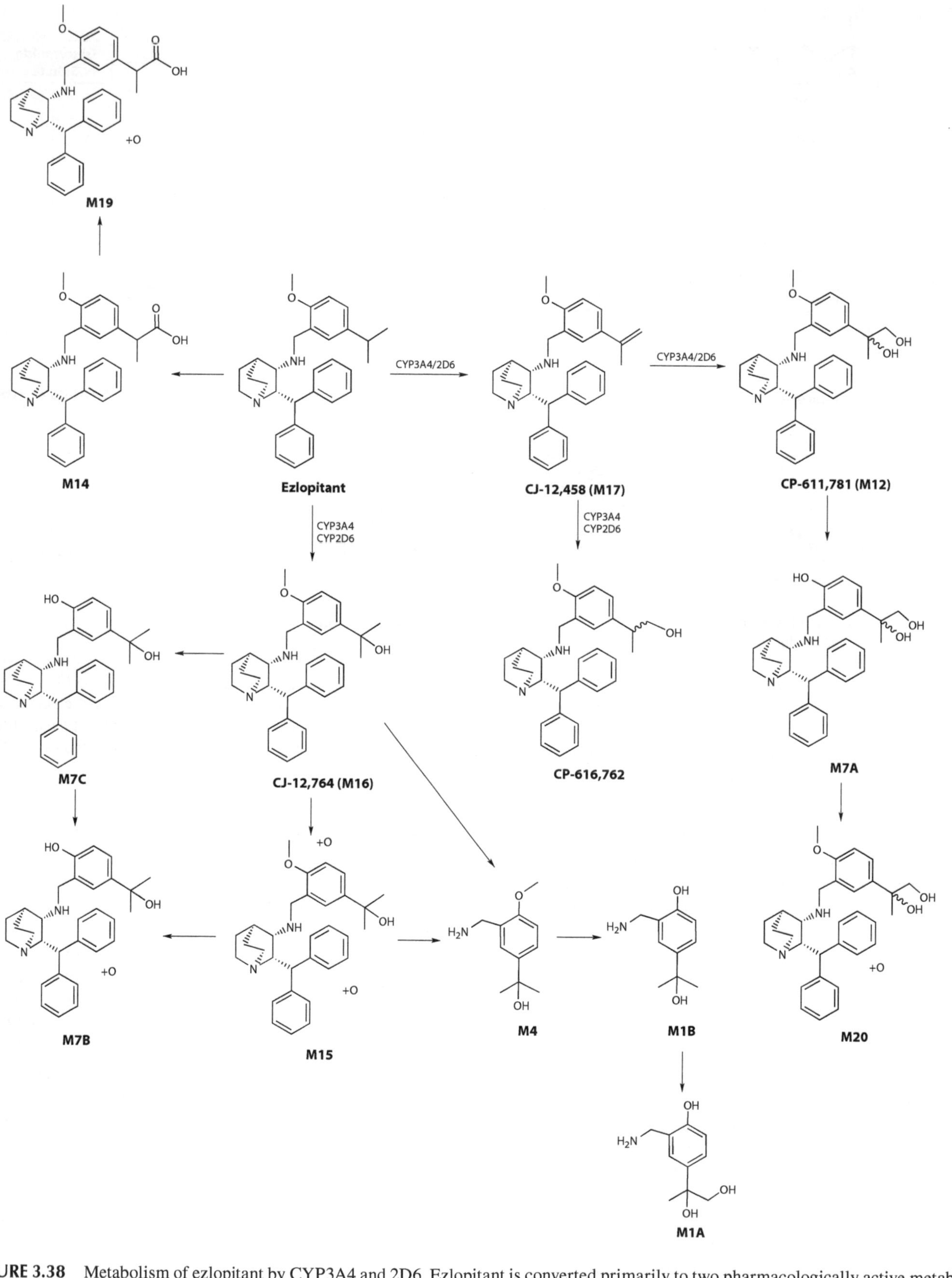

FIGURE 3.38 Metabolism of ezlopitant by CYP3A4 and 2D6. Ezlopitant is converted primarily to two pharmacologically active metabolites, namely, CJ-12,458, an alkene metabolite, and CJ-12,764, a benzyl alcohol metabolite, both of which have been detected in the systemic circulation of preclinical species as well as in humans. CJ-12,764 is the major metabolite in preclinical species, with less CJ-12,458 observed; the two metabolites are observed in nearly equal abundance in human blood. CJ-12,458 is further converted to two metabolites: a diol (CP-611,781) and a 1° alcohol (CP-616,762) by CYP3A4 and 2D6.

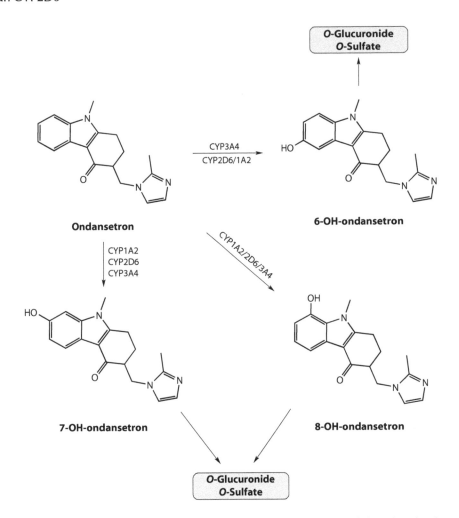

FIGURE 3.39 Metabolism of ondansetron by CYP2D6 and other enzymes. After oral administration, it takes approximately 1.5–2 h to reach C_{max}. Ondansetron is extensively metabolized in humans, with ~5% of a radiolabeled dose recovered as the parent drug from the urine. The primary metabolic pathway is hydroxylation on the indole ring in the 6-, 7-, and 8-positions by CYP2D6, 3A4, and 1A2 followed by glucuronide or sulfate conjugation. In terms of overall ondansetron elimination, CYP3A4 played the predominant role. Because of the multiplicity of metabolic enzymes capable of metabolizing ondansetron, it is likely that inhibition or loss of one enzyme (e.g., CYP2D6 deficiency) will be compensated by others and may result in little change in overall rates of ondansetron elimination.

Donepezil is a reversible cholinesterase inhibitor that exhibits high specificity for centrally active cholinesterase and is widely used in the treatment of Alzheimer's disease (Barner and Gray 1998; Benjamin and Burns 2007; Bryson and Benfield 1997; Musial et al. 2007; Seltzer 2005, 2007; Shigeta and Homma 2001; Sugimoto et al. 2000). The metabolism of donepezil is extensive in rats and involved O-demethylation at each methoxy group of the dimethoxyindan moiety followed by O-glucuronide conjugation, aromatic hydroxylation followed by sulfation, N-dealkylation at the piperidine ring, and N-oxidation at the piperidine ring (Figure 3.45) (Matsui et al. 1999a). Since the demethylate at the 6″-position of the dimethoxyindan (M1) is present in larger amounts than that at the 5″-position of the dimethoxyindan (M2) in rat urine and feces, donepezil appeared to be demethylated with position selectivity. The potency of M1 (and M3) in acetylcholinesterase inhibition activity is comparable with donepezil and is approximately 140 times higher than that of M2. M4 and M8 are formed by oxidative N-dealkylation of the piperidine ring in donepezil. Similar to rats, three metabolic pathways are identified in humans: O-dealkylation and hydroxylation to metabolites M1 and M2, with subsequent glucuronidation to metabolites M11 and M12; hydrolysis to metabolite M4; and N-oxidation to metabolite M6 (Figure 3.45) (Tiseo et al. 1998). In human urine, the parent compound accounts for 17% of the dose and the major metabolite is M4, followed by the glucuronidated conjugates M11 and M12 (Tiseo et al. 1998). In an experiment with ^{14}C-donepezil and human microsomes, the routes of metabolism are identified as N-dealkylation and O-demethylation (Matsui et al. 1999b). In human liver microsomes, M4 is formed mainly by CYP3A4 and, to a lesser extent, by CYP2C9 via N-dealkylation, while M1 and M2 are formed by CYP3A4 and 2D6 (Barner and Gray 1998). A clinical study has demonstrated that CYP2D6 UM patients had lower plasma concentrations than EM patients and showed no clinical improvement (Varsaldi et al. 2006).

Galantamine is a tertiary alkaloid obtained synthetically or extracted from the bulbs and flowers of several *Amaryllidaceae*

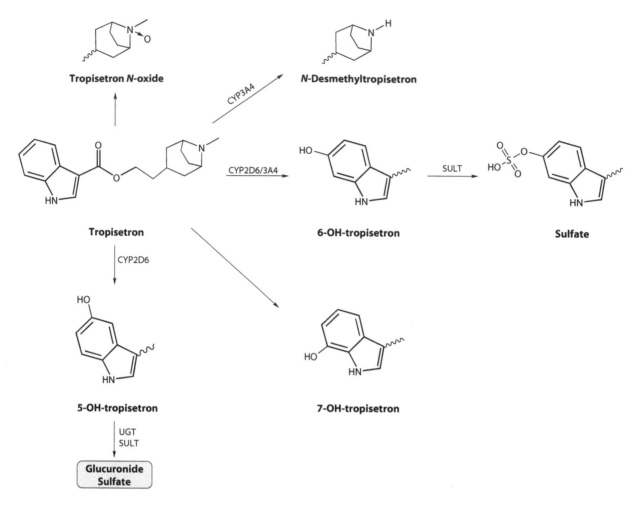

FIGURE 3.40 Metabolism of tropisetron by CYP2D6 and 3A4. Tropisetron is extensively metabolized by hydroxylation of the indole ring in the 5-, 6-, and 7-positions followed by conjugation in humans. The major metabolic route is 6-hydroxylation, accounting for 45% of the urinary excretion, while 12% of the dose is excreted as 5-OH-tropisetron in urine. Oxidative *N*-demethylation and *N*-oxygenation at the tropinyl nitrogen also occur but to a minor extent, with trace levels of the corresponding metabolites detected. CYP2D6 catalyzes 5- and 6-hydroxylation of tropisetron, while CYP3A4 forms *N*-desmethyltropisetron.

species. Galantamine has a significant therapeutic effect in the management of patients with Alzheimer's disease (Erkinjuntti et al. 2002, 2008; Kaduszkiewicz et al. 2005; Kavirajan and Schneider 2007; Loy and Schneider 2006; Raskind et al. 2000; Tariot et al. 2000; Wilcock et al. 2000). After a single oral dose of 10 mg of galantamine in healthy volunteers, Bachus et al. (1999) have identified three metabolites in urine, namely, the glucuronide of *O*-desmethylgalantamine, *N*-desmethylgalantamine, and epigalantamine (Figure 3.46). Approximately 25.1% of the dose is excreted as galantamine, 19.8% as the glucuronide of *O*-desmethylgalantamine, 5% as *N*-desmethylgalantamine, and 0.8% as epigalantamine. No glucuronides of galantamine, epigalantamine, galantaminone, and *N*-desmethylgalantamine are detected. The major route of metabolism for galantamine is through the liver, accounting for approximately 75% of the total metabolism of galantamine (Westra et al. 1986). Approximately 73% of total radioactivity is excreted within the first 24 h after dose administration in humans. Multiple metabolic pathways and renal excretion are involved in the elimination of galantamine.

Galantamine is metabolized by CYPs, glucuronidated, and excreted unchanged in the urine. Major metabolic pathways of galantamine included glucuronidation, *O*-demethylation, *N*-demethylation, *N*-oxidation, and epimerization. Epimerization may occur first, but it is also likely to occur after glucuronidation, *O*-demethylation, *N*-demethylation, and *N*-oxidation. The stereoisomeric conversion of the alcohol group of galantamine has been reported before. It is suggested to result from dehydrogenation of the alcohol group to an intermediate ketone, galantaminone, followed by rehydration (Bachus et al. 1999). Mannens et al. (2002) have also found the glucuronide of *O*-desmethylgalantamine (metabolite 2) and the glucuronide of galantamine (metabolite 5). The glucuronide of *O*-desmethylgalantamine has been shown to be inactive, whereas *O*-desmethylgalantamine is more potent than galantamine (Bachus et al. 1999). In vitro studies have shown that CYP2D6 plays a major role in galantamine *O*-demethylation (Bachus et al. 1999), whereas CYP3A4 catalyzes the *N*-oxidation of galantamine. *O*-Demethylgalantamine is 10-fold more selective for the

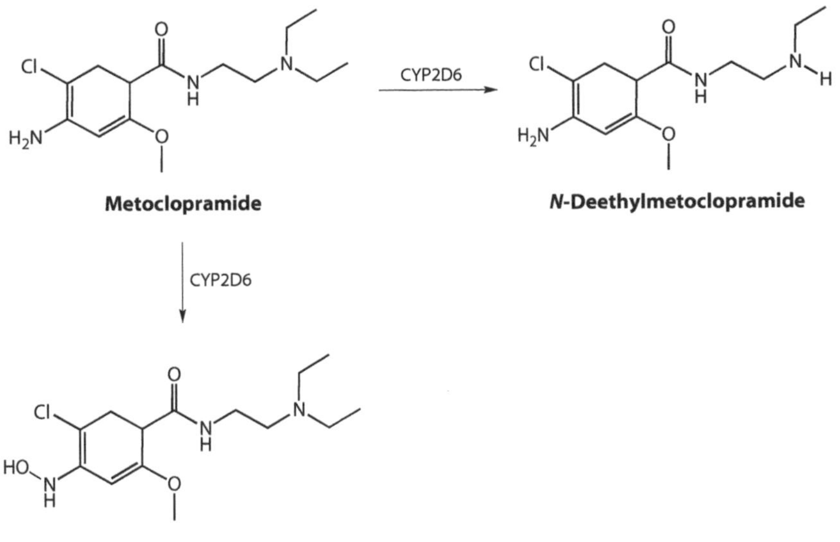

FIGURE 3.41 Metabolism of metoclopramide by CYP2D6. Approximately 25% of the metoclopramide dose is recovered unchanged in the urine, whereas a substantial amount of the dose appears to undergo enzymatic metabolism via oxidation and conjugation reactions. Metoclopramide is largely metabolized by CYP2D6 to monodeethylmetoclopramide via *N*-dealkylation. An aromatic *N*-hydroxy metabolite of metoclopramide is found to represent a major metabolite in human liver microsomes, and CYP2D6 is responsible for its formation.

inhibition of acetylcholinesterase versus butyrylcholinesterase than galantamine.

The centrally acting cholinesterase inhibitors are the first-line agents in the treatment of Alzheimer's disease, including tacrine, donepezil, galantamine, and rivastigmine. Rivastigmine is almost entirely metabolized by sulfate conjugation. Tacrine (Spaldin et al. 1994), donepezil (Barner and Gray 1998), and galantamine (Bachus et al. 1999) are metabolized by CYP2D6 to some extent. However, CYP1A2 and 3A4 play a more important role in their metabolism than 2D6 (Jann et al. 2002) and thus the impact of *CYP2D6* polymorphisms on the clearance of centrally acting cholinesterase inhibitors would be minor.

3.3.1.11 Drugs for Senile Dementia

Nicergoline is a substrate of CYP2D6 (Bottiger et al. 1996). Nicergoline, an ergot derivative previously used as a vasodilator because of its α receptor blocking activity (Moretti et al. 1979), has been used in the treatment of senile dementia (Albus et al. 1986; Baskys and Hou 2007; Fioravanti and Flicker 2001). It decreases vascular resistance and increases arterial blood flow in the brain, improving the utilization of oxygen and glucose metabolism by brain cells. Nicergoline is rapidly hydrolyzed to an alcohol derivative, 1-methyl-10α-methoxy-9,10-dihydrolysergol (MMDL), which is further *N*-demethylated to form 10α-methoxy-9,10-dihydrolysergol (MDL) or hydroxylated to 1-OH-MMDL (Figure 3.47) (Arcamone et al. 1972). In healthy subjects, the plasma levels of MDL are much higher than those of MMDL. In PMs of CYP2D6, but not CYP2C19, the AUC of MDL is significantly lower than that in EMs but the AUC of MMDL is significantly higher than that in EMs (Bottiger et al. 1996).

3.3.1.12 Drugs for the Treatment of Attention-Deficit/Hyperactivity Disorder

Atomoxetine is a nonstimulant, highly selective noradrenaline reuptake inhibitor approved by the FDA for the treatment of attention-deficit/hyperactivity disorder in children, adolescents, and adults (Corman et al. 2004; Simpson and Plosker 2004). In humans, atomoxetine is extensively metabolized via aromatic ring hydroxylation, benzylic oxidation, and *N*-demethylation (Figure 3.48). Subsequent *O*-glucuronidation of the ring-hydroxylated metabolites is the only Phase II metabolic pathway to participate in the conjugation of the hydroxylated metabolites. Atomoxetine is predominantly metabolized by CYP2D6 to 4-hydroxyatomoxetine (>80% in EMs), but multiple other CYPs including CYP2C19, 3A, 1A2, 2A6, and 2E1 also form 4-hydroxyatomoxetine at a 475-fold slower rate (Ring et al. 2002). 4-Hydroxyatomoxetine has activity similar to the parent drug.

3.3.1.13 Nonnarcotic Analgesics

Bicifadine, a nonnarcotic analgesic that inhibits norepinephrine and serotonin uptake (Epstein et al. 1982), is used in the treatment of acute dental pain and postoperative bunionectomy pain (Wang et al. 1982). It is hydroxylated at its methyl group to form M2, by CYP2D6 with lesser contribution from CYP1A2 (Figure 3.49) (Erickson et al. 2007). In recombinant enzymes, CYP2D6 has a sixfold greater activity than that of 1A2. M2 is oxidized further to the carboxylic acid metabolite (M3). Bicifadine also underwent NADPH-independent oxidation at the C2-position of the pyrrolidine ring, generating a lactam metabolite (M12), which is catalyzed by MAO-B (major) and MAO-A (Erickson et al. 2007). M12 is further oxidized to a hydroxymethyl metabolite (M8) by CYP1A2,

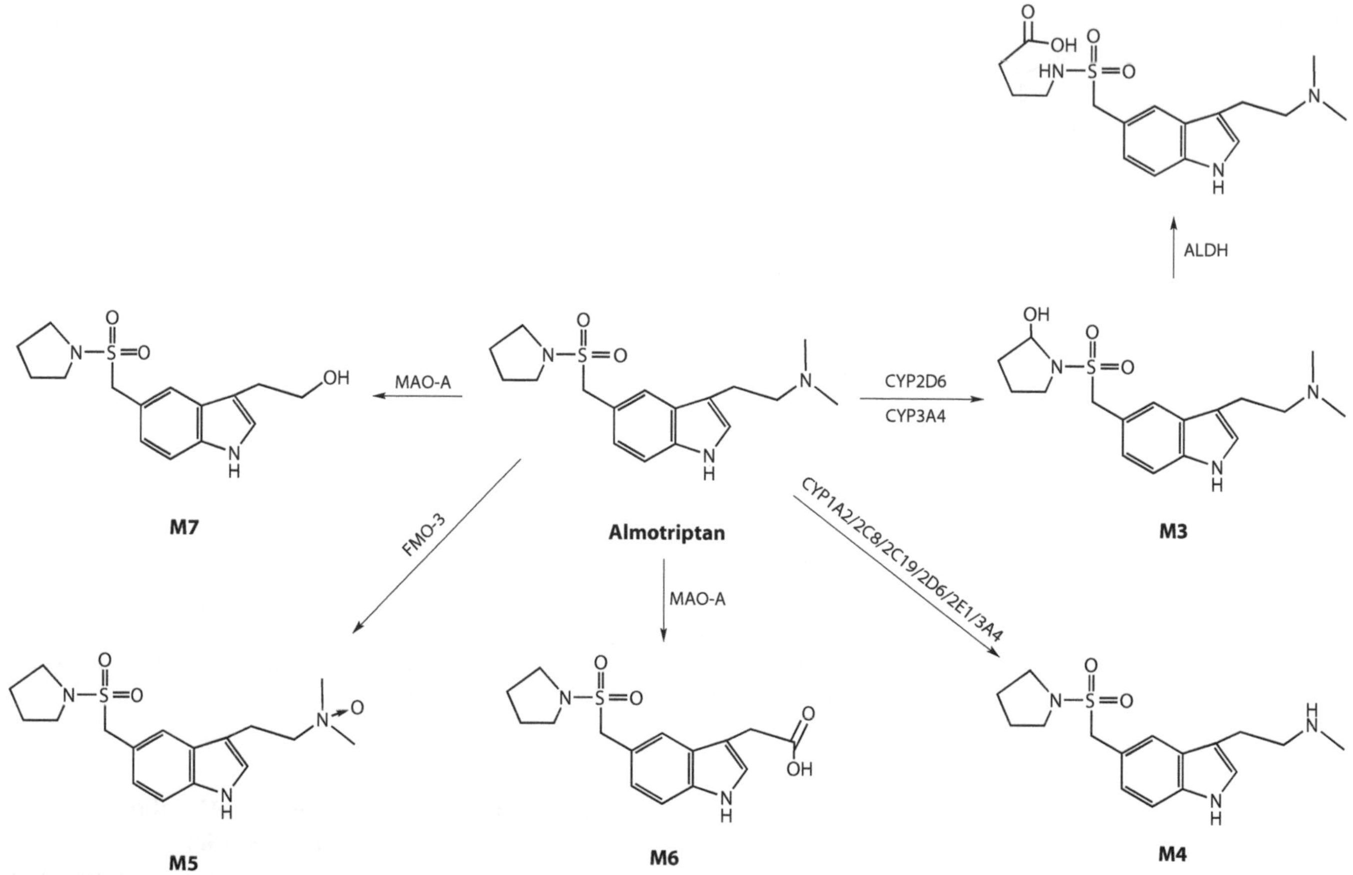

FIGURE 3.42 Metabolism of almotriptan by CYP2D6 and other enzymes. Approximately 50% of an almotriptan dose is excreted unchanged in the urine. The predominant route of metabolism is via MAO-A, and CYP3A4/2D6-mediated oxidation occurs to a minor extent. Human liver microsomes and S9 fractions metabolized almotriptan to at least six metabolites. Almotriptan undergoes 2-hydroxyl-ation of the pyrrolidine group to form a carbinolamine metabolite intermediate (M3), a reaction catalyzed by CYP2D6 and 3A4 with 1A1 playing a minor role. This metabolite is further oxidized by ALDH to the open ring GABA metabolite (M2). Almotriptan is also metabo-lized at the dimethylaminoethyl group via N-demethylation to form N-desmethyl almotriptan (M4), a reaction that is catalyzed by CYP1A1, 1A2, 2C8, 2C19, 2D6, 2E1, and 3A4. None of the cDNA-expressed CYPs is able to generate metabolites M2 and M6, whereas M5 and M7 are only produced by CYP1A1. FMO3 mediated N-oxidation of almotriptan, resulting in its N-oxide, and MAO-A catalyzed oxidative deamination to form the indole acetic acid (M6) and the indole ethyl alcohol derivatives of almotriptan (M7).

2C19, 2D6, and 2E1. M8 is also further oxidized to a carboxyl metabolite M9, and both M8 and M9 could be glucuronidated. Alternatively, bicifadine is metabolized to the carbamyl-O-glucuronide (M11). In healthy subjects receiving a single dose (200 mg), approximately 64% of the dose is metabolized to the lactam acid (M9) and its acyl glucuronide; another 23% is recovered as M3 and its acyl glucuronide (Krieter et al. 2008). Neither bicifadine nor M12 is detected in human urine or feces.

3.3.1.14 Drugs for Huntington's Disease Chorea

Tetrabenazine is a monoamine storage inhibitor that was first introduced in the 1970s for the treatment of hyperkinetic movement disorders. Despite acceptance and usage world-wide, it was only approved as an orphan drug in August 2008 by the US FDA for the treatment of Huntington cho-rea (Chen et al. 2012; de Tommaso et al. 2011; Fasano and Bentivoglio 2009; Frank 2010, 2014; Guay 2010; Jankovic 2009; Jankovic and Clarence-Smith 2011; Leung and Breden

2011; Paleacu 2007; Poon et al. 2010). It acts by depletion of the monoamines serotonin, norepinephrine, and dopamine in the CNS via reversible inhibition of vesicle monoamine transporter type 2 (K_i = 100 nM) (Guay 2010), resulting in decreased uptake of monoamines into synaptic vesicles and depletion of monoamine stores. After oral administration in humans, at least 19 metabolites of tetrabenazine have been identified, with α-hydroxy-tetrabenazine, β-hydroxy-tetrabenazine, and 9-desmethyl-dihydroxy-tetrabenazine being the major circulating metabolites, and they are later metabolized to sulfate or glucuronide conjugates. α-Hydroxy-tetrabenazine and β-hydroxy-tetrabenazine are formed by carbonyl reductase that occurs mainly in the liver. α-Hydroxy-tetrabenazine is O-dealkylated principally by CYP2D6, with some contribution from CYP1A2 to form 9-desmethyl-α-dihydroxy-tetrabenazine, a minor metabolite. β-Hydroxy-tetrabenazine is also O-dealkylated primarily by CYP2D6 to form 9-desmethyl-β-dihydroxy-tetrabenazine (Guay 2010). In a mass balance study in six healthy volunteers, approximately

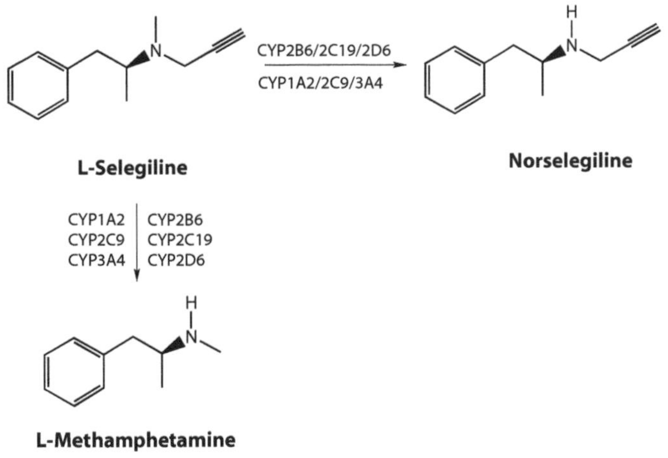

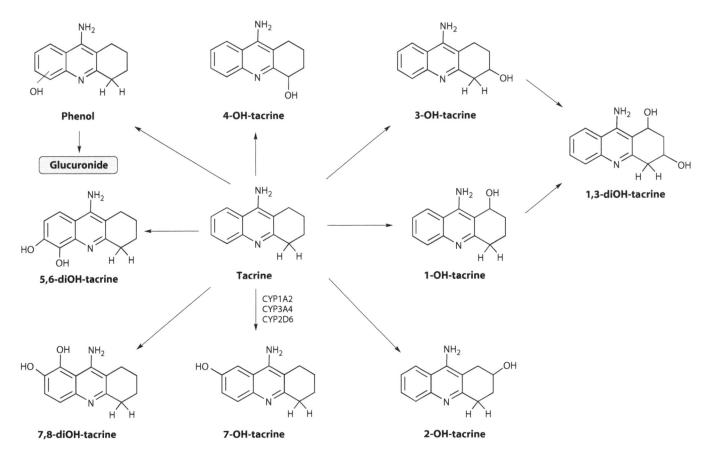

FIGURE 3.43 Metabolism of selegiline CYP2B6, 2C19, and 2D6. Selegiline is mainly metabolized by CYP2B6 and 2C19, with contribution from CYP1A2, 2D6, 2C8, and 3A4. In recombinant CYP2D6 and human liver microsomes, *N*-demethylation of L-deprenyl resulting in nordeprenyl is favored by approximately 13:1 over *N*-depropargylation, which produced methamphetamine. The stereoselectivity of CYP2D6 for deprenyl mimics MAO-B activity with the L-isomer favored. However, dopamine and pargyline, both substrates of MAO-B, are not metabolized by CYP2D6.

75% of the dose is excreted in the urine and fecal recovery accounts for approximately 7%–16% of the dose. Tetrabenazine and its active metabolites are not inhibitors for CYP2D6, 1A2, 2B6, 2C8, 2C9, 2C19, 2E1, and 3A4.

3.3.2 CARDIOVASCULAR DRUGS

CYP2D6 plays an important role in the metabolism of drugs for the treatment of cardiovascular diseases. CYP2D6 substantially metabolizes the antiarrhythmics (e.g., sparteine, propafenone, encainide, flecainide, cibenzoline, aprindine, lidocaine, procainamide, and mexiletine) and β-blockers (e.g., atenolol, bufuralol, carvedilol, metoprolol, bisoprolol, propranolol, bunitrolol, bupranolol, timolol, and alprenolol). Perhexiline, a prophylactic antianginal agent, is mainly metabolized by CYP2D6, 2B6, and 3A4 (Davies et al. 2007).

3.3.2.1 Antianginal Drugs

Perhexiline is a prophylactic antianginal drug used primarily in New Zealand and Australia (Horowitz et al. 1995). Perhexiline decreases fatty acid metabolism through the inhibition of carnitine palmitoyltransferase-1, an enzyme responsible for mitochondrial uptake of long-chain fatty acids

FIGURE 3.44 Metabolism of tacrine by CYP1A2 and 3A4. In human liver microsomes, tacrine is mainly metabolized to 1-OH-tacrine by CYP1A2, with minor contribution from CYP2D6, while 2-, 4-, and 7-OH-tacrine are also formed. In human subjects, urinary 1-OH-tacrine accounted for approximately 5% of the administered dose, while tacrine and 2-OH-tacrine comprised <1% and ~2% of the administered dose, respectively. Phenol and 5,6- or 7,8-diOH-tacrine are also found at trace amounts.

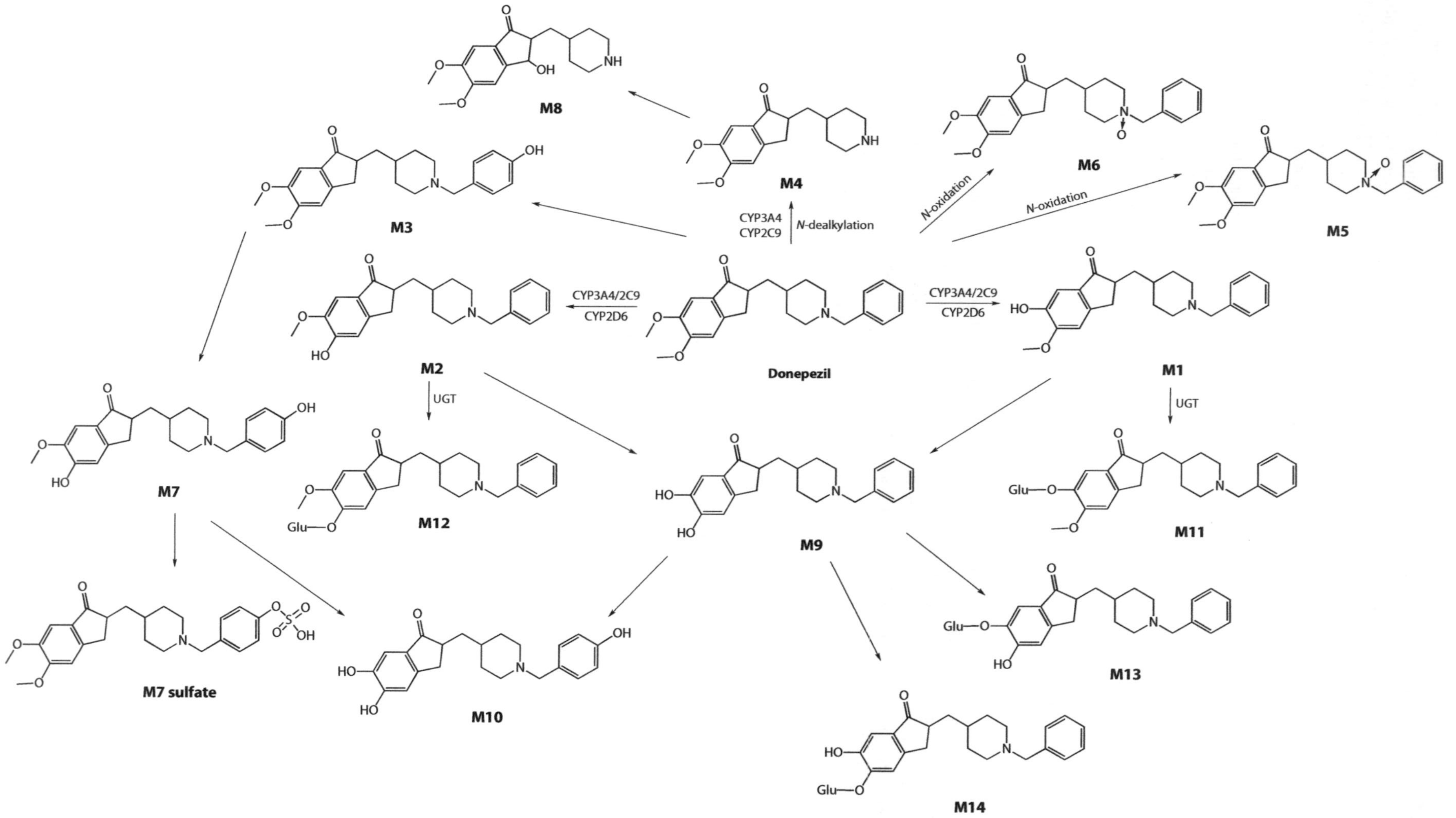

FIGURE 3.45 Metabolism of donepezil by CYP3A4, 2D6, and 2C9. Three metabolic pathways are identified in humans: *O*-dealkylation and hydroxylation to metabolites M1 and M2, with subsequent glucuronidation to metabolites M11 and M12; hydrolysis to metabolite M4; and *N*-oxidation to metabolite M6. In human liver microsomes, M4 is formed mainly by CYP3A4 and, to a lesser extent, by CYP2C9 via *N*-dealkylation, while M1 and M2 are formed by CYP3A4 and 2D6.

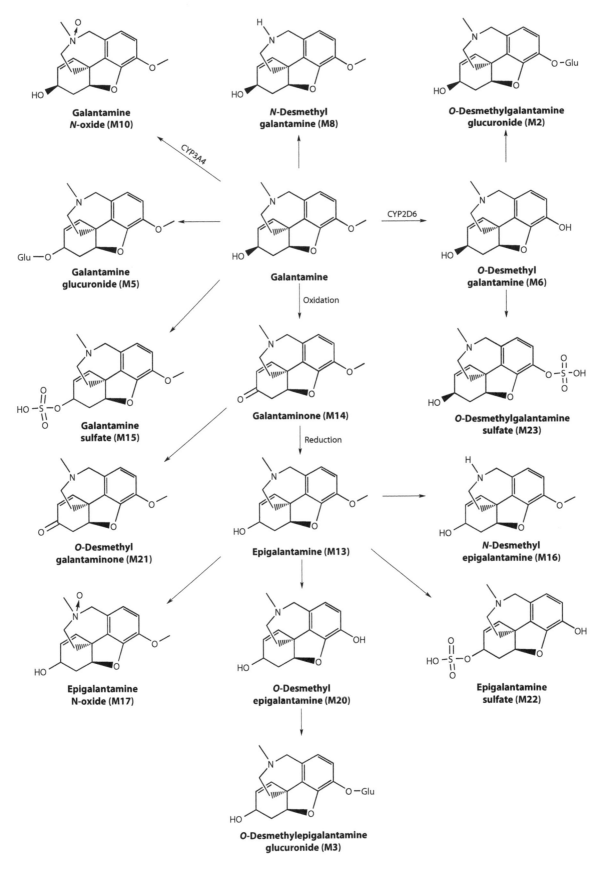

FIGURE 3.46 Metabolism of galantamine by CYP2D6 and 3A4. Major metabolic pathways of galantamine included glucuronidation, *O*-demethylation, *N*-demethylation, *N*-oxidation, and epimerization. Epimerization may occur first, but it is also likely to occur after glucuronidation, *O*-demethylation, *N*-demethylation, and *N*-oxidation. CYP2D6 plays a major role in galantamine *O*-demethylation, whereas CYP3A4 catalyzes the *N*-oxidation of galantamine.

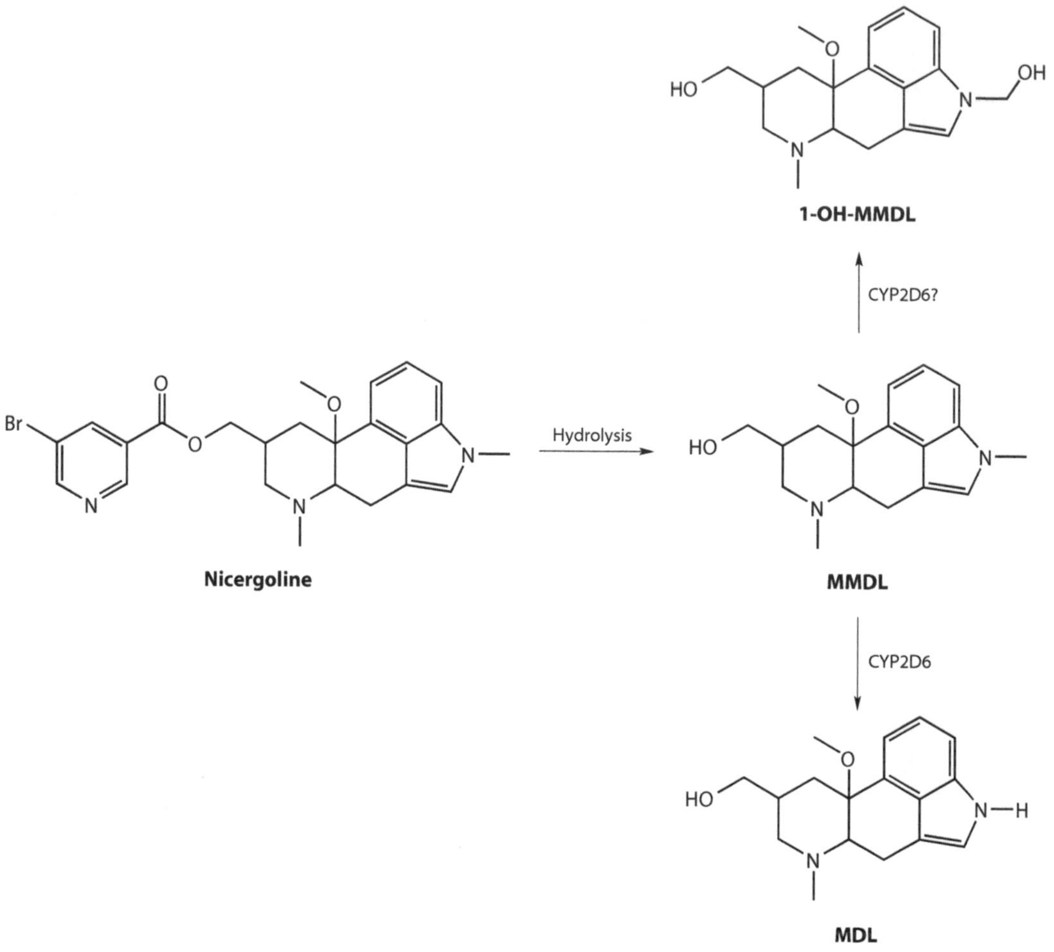

FIGURE 3.47 Metabolism of nicergoline by CYP2D6. Nicergoline is rapidly hydrolyzed to an alcohol derivative, MMDL, which is further *N*-demethylated to form MDL or hydroxylated to 1-OH-MMDL. In PMs of CYP2D6, but not CYP2C19, the AUC of MDL is significantly lower than that in EMs.

(Ashrafian et al. 2007). The corresponding shift to greater carbohydrate utilization increases myocardial efficiency, and this oxygen-sparing effect may explain its antianginal efficacy. Perhexiline is a highly lipophilic drug (Log$P$ = 5.87) with a large interindividual variability in its clearance and plasma concentrations (>100-fold). A therapeutic range of 0.15 to 0.6 mg/l has been established for plasma perhexiline at steady state (Cole et al. 1990). The saturable clearance of perhexiline shows a bimodal distribution, and the pharmacokinetics of perhexiline is nonlinear (Cole et al. 1990). Perhexiline is predominantly metabolized by CYP2D6 to two inactive metabolites *cis*-hydroxyperhexiline (M1) and *trans*-hydroxyperhexiline (M3), both existing as diastereomeric pairs (Figure 3.50) (Sorensen et al. 2003). The two main metabolites of perhexiline in the plasma and urine postdosing are M1 and M3. Dihydroxyperhexiline has also been identified in the urine. M1 is the primary determinant of perhexiline clearance and there is a large interindividual variability in metabolic clearance to M1 (Sallustio et al. 2002). The M1/ perhexiline MR has been incorporated into therapeutic drug monitoring of perhexiline and in phenotype studies as it can readily separate PMs and EMs.

3.3.2.2 Antiarrhythmics

Most antiarrhythmic drugs are extensively metabolized, and many of them are converted into pharmacologically active metabolites. Many antiarrhythmic drugs introduced into the market have a chiral center in their structure and are marketed as racemates (Mehvar et al. 2002). The hepatically eliminated antiarrhythmics are metabolized mainly by different CYP enzymes such as CYP3A4, 1A2, 2C8, 2C9, and 2D6 and partly also by Phase II conjugating enzymes (Klotz 2007). Several of them are extensively metabolized by CYP3A4, including amiodarone (Fabre et al. 1993; Trivier et al. 1993), verapamil (with contribution from CYP2C9 for *O*-demethylation), and quinidine (Nielsen et al. 1999). Amiodarone is also extensively metabolized by CYP2C8 and to a minor extent by CYP2D6 (Jaruratanasirikul and Hortiwakul 1994). Lidocaine is primarily metabolized by CYP1A2 and to a lesser extent by CYP3A4 and 2D6 (Bargetzi et al. 1989; Imaoka et al. 1990; Masubuchi et al. 1991; Wang et al. 2000). Furthermore, many antiarrhythmic agents are extensively metabolized by polymorphic CYP2D6. These include class Ia antiarrhythmic drugs such as sparteine (Ebner et al. 1995; Eichelbaum et al. 1979, 1986b; Tyndale et al. 1990) and cibenzoline (Niwa et

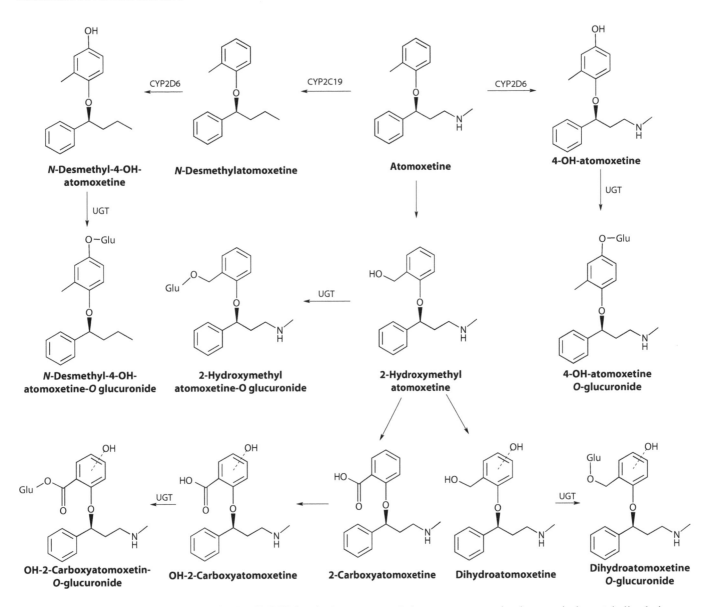

FIGURE 3.48 Metabolism of atomoxetine by CYP2D6 and other enzymes. In humans, atomoxetine is extensively metabolized via aromatic ring hydroxylation, benzylic oxidation, and *N*-demethylation. Aromatic ring hydroxylation appears to be the penultimate step for the further biotransformation of the metabolites. Subsequent *O*-glucuronidation of the ring-hydroxylated metabolites is the only Phase II metabolic pathway to participate in the conjugation of the hydroxylated metabolites. Atomoxetine is predominantly metabolized by CYP2D6 to 4-hydroxyatomoxetine (>80% in EMs), but multiple other CYPs including CYP2C19, 3A, 1A2, 2A6, and 2E1 also form 4-hydroxyatomoxetine at a 475-fold slower rate.

al. 2000), class Ib antiarrhythmic drugs such as aprindine (Ebner and Eichelbaum 1993) and mexiletine (Abolfathi et al. 1993; Broly et al. 1990), and class Ic antiarrhythmic drugs such as propafenone (Botsch et al. 1993; Kroemer et al. 1989b, 1991), encainide (Funck-Brentano et al. 1992), and flecainide (Haefeli et al. 1990). Procainamide is metabolized by CYP2D6 to a minor extent (Lessard et al. 1997). For sotalol, disopyramide, and procainamide, renal clearance contributes considerably to overall elimination (Klotz 2007).

Amiodarone is a class III antiarrhythmic acting on potassium channel, which is the most commonly used antiarrhythmic agent that can control a wide spectrum of atrial and ventricular antiarrhythmic disorders (Singh 2006). The oral bioavailability of amiodarone ranges from 22% to 86%.

Since it is a highly lipophilic drug, amiodarone is extensively distributed into tissues such as adipose tissues and skeletal muscles (Latini et al. 1984). Amiodarone is eliminated primarily by hepatic metabolism and biliary excretion and there is negligible excretion of amiodarone or desethylamiodarone in urine (<1%) (Latini et al. 1984). Amiodarone is extensively metabolized to desethylamiodarone by CYP3A4 and 2C8 and to a minor extent by CYP2D6 (Figure 3.51) (Jaruratanasirikul and Hortiwakul 1994), which can be easily detected in the patient plasma (Flanagan et al. 1982). In animals, desethylamiodarone has significant electrophysiologic and antiarrhythmic effects similar to amiodarone (Talajic et al. 1987). The development of maximal ventricular class III effects after oral amiodarone administration correlated more closely with

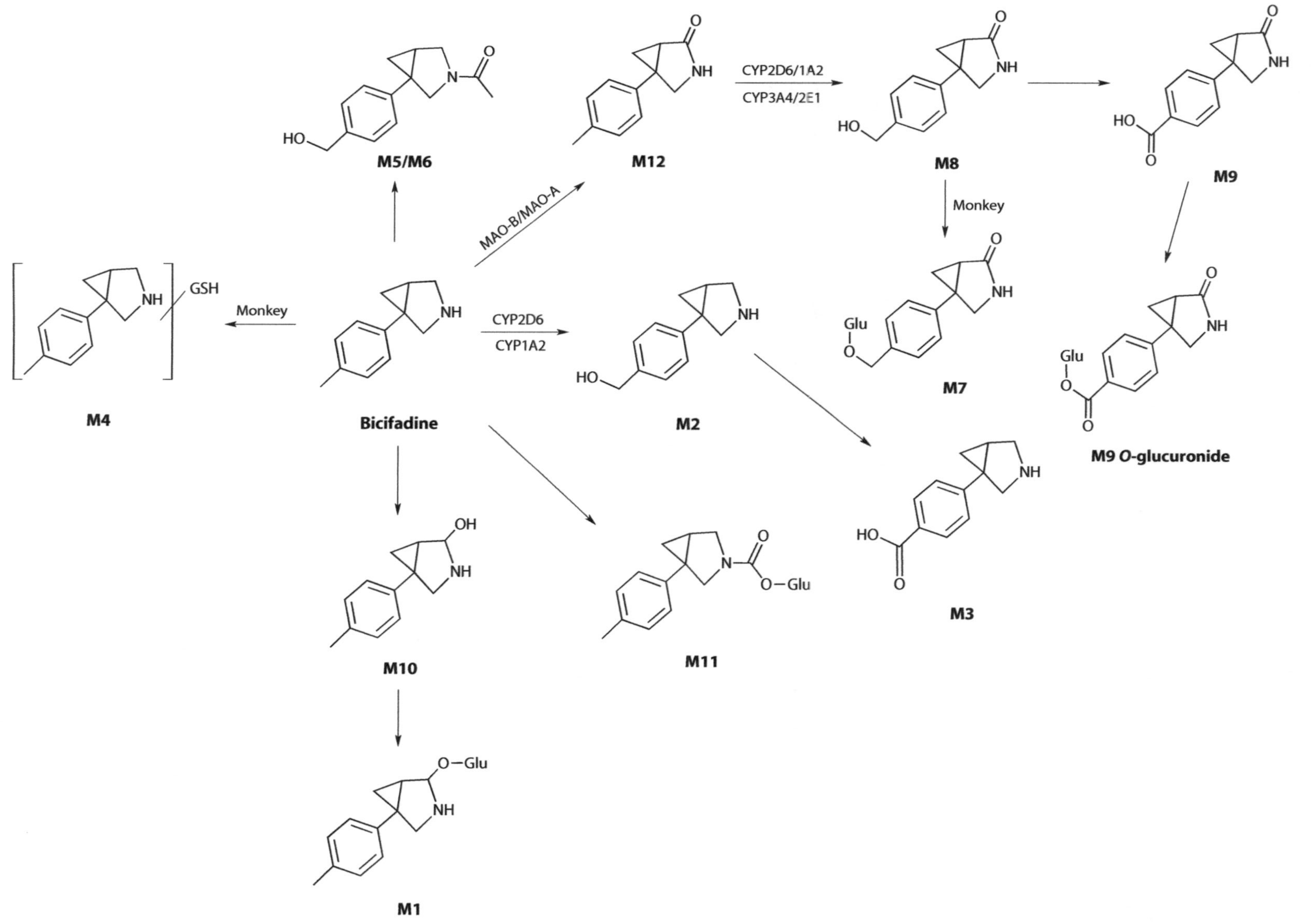

FIGURE 3.49 Metabolism of bicifadine by CYP2D6 and 1A2, and MAO-B.

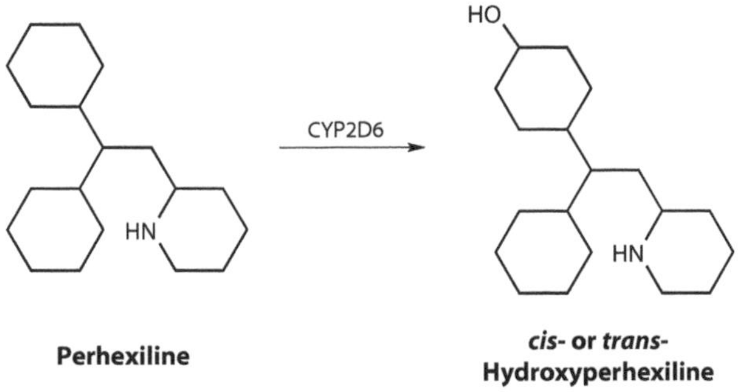

FIGURE 3.50 Metabolism of perhexiline by CYP2D6. Perhexiline is metabolized predominantly by CYP2D6 to two inactive metabolites *cis*-hydroxyperhexiline (M1) and *trans*-hydroxyperhexiline (M3), both existing as diastereomeric pairs. The rate of M1 formation in human liver microsomes from 20 livers varied 50-fold and it decreased to 5-fold in EMs. The intrinsic clearance is 112-fold lower in microsomes from PMs than those from EMs (0.026 vs. 2.9 μl/min/mg P450). The two main metabolites of perhexiline in the plasma and urine are M1 and M3. Dihydroxyperhexiline has also been identified in the urine.

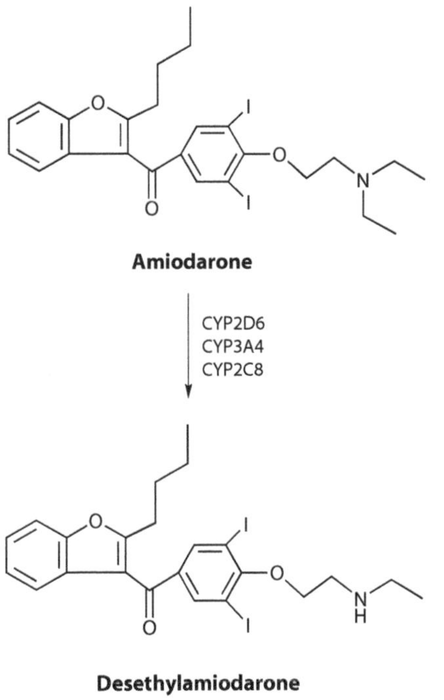

FIGURE 3.51 Metabolism of amiodarone by CYP3A4, 2C8, and 2D6. Amiodarone is extensively metabolized to desethylamiodarone by CYP3A4 and 2C8 and to a minor extent by CYP2D6.

desethylamiodarone accumulation over time than with amiodarone accumulation in humans (Lesko 1989). Amiodarone is an inhibitor of CYP3A4 and P-glycoprotein. A number of clinical drug interactions with amiodarone have been documented (Heimark et al. 1992; Lesko 1989).

Aprindine is a long-acting Ib antiarrhythmic drug used in the treatment of ventricular arrhythmias of varying etiologies, especially in the treatment of the Wolff–Parkinson–White syndrome (Danilo 1979; Fasola and Carmichael 1974).

It is extensively metabolized in humans with dose-dependent pharmacokinetics within the therapeutic dosage range after oral administration, probably attributed to saturation of metabolism (Kobari et al. 1984; Wirth et al. 1983; Yokota et al. 1987). Three main metabolites of aprindine have been identified in human urine after aprindine administration, with recovery of negligible amounts of the parent drug (Maurer 1990; Murphy 1974). More than 72% of the hydroxylated metabolites on the aromatic ring resulting in *p*-hydroxyaprindine or indan ring resulting in 5-hydroxyaprindine in urine are recovered as glucuronide or sulfate conjugates (Figure 3.52) (Murphy 1974; Shimizu et al. 1998). Another major metabolite is *N*-dealkylaprindine as a result of *N*-dealkylation, which can be detected in human plasma and urine (Ebner and Eichelbaum 1993; Murphy 1974). In vitro studies on human liver microsomes have demonstrated that there is a higher capacity for the formation of *N*-dealkylaprindine with a K_m of 41–81 μM and a lower capacity to form the hydroxylated metabolites with a K_m of 0.3–2.1 μM (Ebner and Eichelbaum 1993). Addition of quinidine or liver–kidney microsome (LKM-1) autoantibody positive sera (CYP2D6 antibody) to incubations with aprindine decreases the formation of both hydroxylated metabolites by 32%–90% whereas *N*-dealkylation is unaffected (Ebner and Eichelbaum 1993), indicating that CYP2D6 mediates hydroxylation of aprindine. Aprindine is a potent competitive inhibitor for the formation of (2*S*)-hydroxysparteine with a K_i of 17 nM and 5,6-didehydrosparteine with a K_i of 18 nM and 5-hydroxypropafenone with a K_i of 2.8 μM for the *R*-enantiomer and 2.1 μM for the *S*-enantiomer (Ebner and Eichelbaum 1993). Thus, aprindine may reduce the metabolism of drugs that are mainly metabolized by CYP2D6. Aprindine seems to reduce the oral clearance of bepridil in patients with arrhythmias carrying the *CYP2D6*10* allele (Taguchi et al. 2006).

Cibenzoline, an imidazoline derivative, is a class Ia antiarrhythmic agent with additional class III (potassium channel blocking) and IV (calcium channel blocking) activity (Harron et al. 1992; Holck and Osterrieder 1986; Kostis et al. 1984; Miura et al. 1985; Mukai et al. 1998). It is mainly used for the treatment of ventricular and supraventricular arrhythmia including drug-refractory ventricular tachycardia or ventricular arrhythmias after recent acute myocardial infarction (Harron et al. 1992; Seals et al. 1987). This drug is used as a racemic mixture of *R*- and *S*-enantiomers, and the *S*-enantiomer is approximately twice more potent than the *R*-enantiomer. Cibenzoline exhibited a linear pharmacokinetics after oral dosing ranging from 65 to 260 mg with a >90% bioavailability (Khoo et al. 1984). After oral dosing of [^{14}C]-cibenzoline to healthy volunteers, the administered radioactivity recovered in the urine is 85.7% after 6 days (Massarella et al. 1986). Approximately 55.7% of the radioactivity is found to be unchanged drug in the 0- to 72-h urine, and three metabolites representing 9% to 14% of the total radioactivity are identified in the urine. These included *p*-hydroxycibenzoline and 4,5-dehydrocibenzoline in both free and conjugated forms, and *p*-hydroxybenzophenone (Massarella et al. 1986). 4,5-Dehydrocibenzoline is pharmacologically

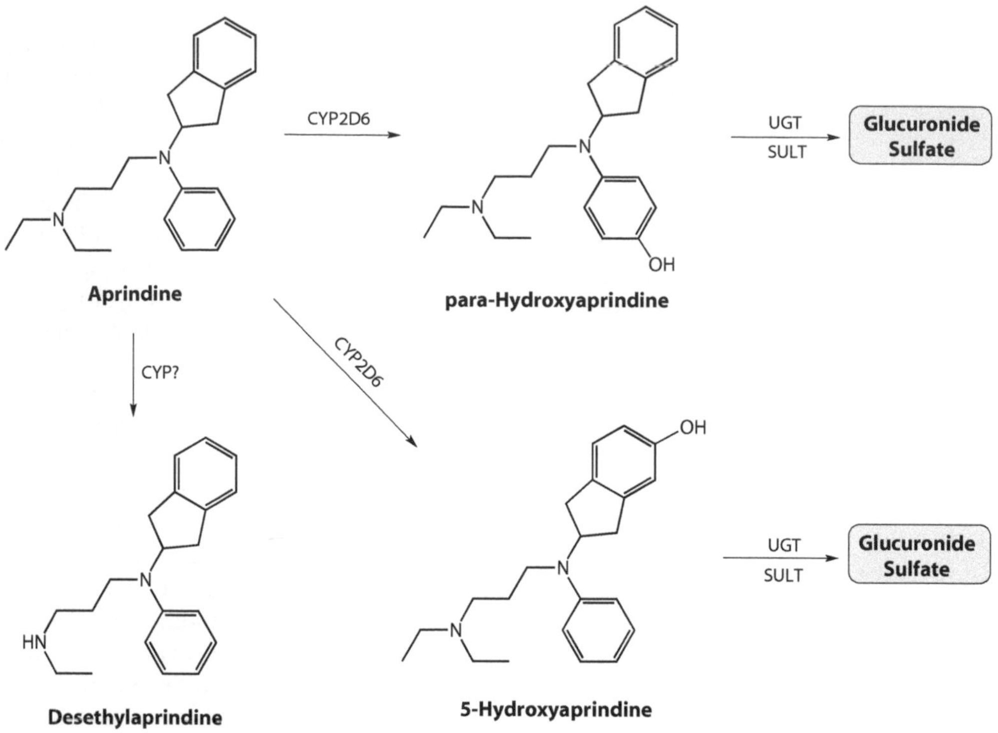

FIGURE 3.52 Metabolism of aprindine. Three main metabolites of aprindine have been identified in human urine after aprindine administration. More than 72% of the hydroxylated metabolites on the aromatic ring resulting in *p*-hydroxyaprindine, or indan ring resulting in 5-hydroxyaprindine in urine, are recovered as glucuronide or sulfate conjugates. Another major metabolite is *N*-dealkylaprindine as a result of *N*-dealkylation, which can be detected in human plasma and urine. In vitro studies in human liver microsomes have demonstrated that there is a higher capacity for the formation of *N*-dealkylaprindine and a lower capacity to form the hydroxylated metabolites by CYP2D6.

inactive. Both *p*-hydroxycibenzoline and 4,5-dehydrociben-zoline have been detected in urine after dosing to rats and dogs (Loh et al. 1986). *m*-Methoxy *p*-hydroxycibenzoline and *p*-hydroxybenzophenone have been identified as glucuronide/sulfate conjugates in bile from rats (Loh et al. 1986). Four main metabolites, *p*-hydroxycibenzoline, 4,5-dehydrociben-zoline, and unknown metabolites M3 and M4, are formed by human and rat liver microsomes (Figure 3.53) (Niwa et al. 2000). The intrinsic clearance (CL_{int}) of *p*-hydroxyciben-zoline formation from *R*-cibenzoline is 23-fold greater than that of *S*-cibenzoline in human liver microsomes, whereas the *R/S*-enantiomer ratio of CL_{int} for 4,5-dehydrobenzo-line, M3, and M4 formation is 0.39 to 0.83. Limited stereo-selective metabolism of *S*-enantiomer in man is observed (Massarella et al. 1986). The formation of *p*-hydroxyciben-zoline, 4,5-dehydrocibenzoline, and M4 from *R*-cibenzoline is catalyzed by CYP2D6, while the formation of 4,5-dehy-drocibenzoline, M3, and M4 is also catalyzed by CYP3A4 (Figure 3.53) (Niwa et al. 2000). Both CYP2D6 and 3A4 catalyze the formation of 4,5-dehydrocibenzoline at a lower concentration, and CYP3A4 plays a major role at a higher concentration. CYP3A4 appears to be mainly involved in the formation of M3 and M4, except that both CYP3A4 and 2D6 catalyze the M4 formation from *R*-cibenzoline. CYP3A and CYP2D play a major role in cibenzoline metabolism in rats (Niwa et al. 2000). The coadministration of cimetidine significantly increases the AUC of cibenzoline by 44% and

prolongs its half-life by 30% (Massarella et al. 1991), suggest-ing that cimetidine may lower the clearance of cibenzoline by inhibition of hepatic oxidative enzymes such as CYP2D6 and 3A4.

Encainide, a benzanilide derivative, is a class Ic (sodium channel blocking) antiarrhythmic agent (Antonaccio et al. 1989) but no longer used because of its frequent proarrhyth-mic side effects. Encainide is particularly effective in patients with excessive premature ventricular complexes and less so in patients with sustained ventricular tachycardia (Antonaccio et al. 1989). After encainide labeled on the carbonyl carbon with ^{14}C and at the benzylic (2′-1-ethyl) carbon with ^{13}C is admin-istered to healthy subjects, 42.0% of the radioactive dose is excreted in the urine in the first 24 h and the total urinary excretion is 47.0% and total fecal excretion is 38.7% over 5 days (Jajoo et al. 1990). At least six metabolites of encainide are identified from the hydrolyzed urine together with unchanged drug, including *O*-demethylencainide, 3-methoxy-*O*-demeth-ylencainide, and *N,O*-di-demethylencainide, *N*-demethyl-3-methoxy-*O*-demethylencainide, 3-hydroxyencainide, and *O*-demethylencainide lactam, accounting for >90% of the radio-activity excreted in the urine (Figure 3.54) (Jajoo et al. 1990). Encainide is converted by CYP2D6 to active *O*-demethyl metabolite (Funck-Brentano et al. 1992), which is 6–10 times more potent than the parent drug in blocking sodium channels (Roden et al. 1982). In humans, encainide undergoes extensive metabolism via four major routes: (a) *O*-demethylation of the

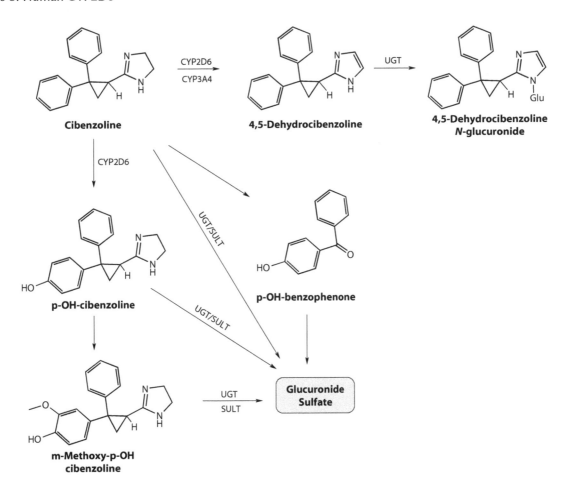

FIGURE 3.53 Metabolism of cibenzoline. Four main metabolites, *p*-hydroxycibenzoline, 4,5-dehydrocibenzoline, and unknown metabolites M3 and M4, are formed by human and rat liver microsomes. The intrinsic clearance (CL$_{int}$) of *p*-hydroxycibenzoline formation from *R*-cibenzoline is 23-fold greater than that of *S*-cibenzoline in human liver microsomes, whereas the *R*/*S*-enantiomer ratio of CL$_{int}$ for 4,5-dehydrocibenzoline, M3, and M4 formation is 0.39 to 0.83. Limited stereoselective metabolism of *S*-enantiomer in humans is observed. The formation of *p*-hydroxycibenzoline, 4,5-dehydrocibenzoline, and M4 from *R*-cibenzoline is catalyzed by CYP2D6, while the formation of 4,5-dehydrocibenzoline, M3, and M4 are catalyzed by CYP3A4. Both CYP2D6 and 3A4 catalyze the formation of 4,5-dehydrocibenzoline at a lower concentration and CYP3A4 plays a major role at a higher concentration. CYP3A4 appears to be mainly involved in the formation of M3 and M4, except that both CYP3A4 and 2D6 catalyze the M4 formation from *R*-cibenzoline.

aromatic methyl ether, (b) formation of methylated catechol derivatives, (c) *N*-demethylation of the piperidyl nitrogen, and (d) oxidation at α-carbon to the piperidyl nitrogen (Jajoo et al. 1990).

Flecainide is a class Ic antiarrhythmic agent of the local anesthetic type used to prevent and treat tachyarrhythmia, paroxysmal atrial fibrillation, and paroxysmal supraventricular tachycardia (Anderson et al. 1981; Duran et al. 1982; Falk and Fogel 1994; Holmes and Heel 1985; Mueller and Baur 1986; Somani 1980). Flecainide caused proarrhythmic effects in up to 20% of patients, which is associated with very high plasma levels of flecainide (1063 ng/ml, up to 2050 ng/ml) (Salerno et al. 1986). Flecainide elimination is impaired in patients with renal failure (Braun et al. 1987; Forland et al. 1988), chronic hepatic cirrhosis (McQuinn et al. 1988), and congestive heart failure (Cavalli et al. 1988; Nitsch et al. 1987). In humans, flecainide is rapidly and completely absorbed after oral administration with an elimination half-life of 13 h, allowing for twice-daily dosing regimens (Smith 1985). The plasma

pharmacokinetics of flecainide appear to be linear (Conard et al. 1984). In humans, 86% of a single oral dose of ^{14}C-flecainide is excreted in urine as flecainide and its metabolites; only 5% is excreted into feces (McQuinn et al. 1984). A substantial portion (27%) of a dose is excreted in urine as unchanged drug. Two major metabolites, *m*-*O*-dealkylated flecainide and the *m*-*O*-dealkylated lactam of flecainide, are found in free and conjugated forms in human urine (Figure 3.55) (Gross et al. 1989; McQuinn et al. 1984). The two major metabolites of flecainide possess little or negligible antiarrhythmic activity and are also the major metabolites present in human plasma in conjugated form. The renal elimination of flecainide is pH dependent (Funck-Brentano et al. 1994).

Lidocaine, a class Ib antiarrhythmic agent and a local anesthetic, is used for the treatment of ventricular arrhythmias occurring during cardiac surgery or myocardial infarction and as a local anesthetic (Benowitz and Meister 1978). After oral administration, lidocaine is completely absorbed with low plasma concentrations observed in man, and its elimination

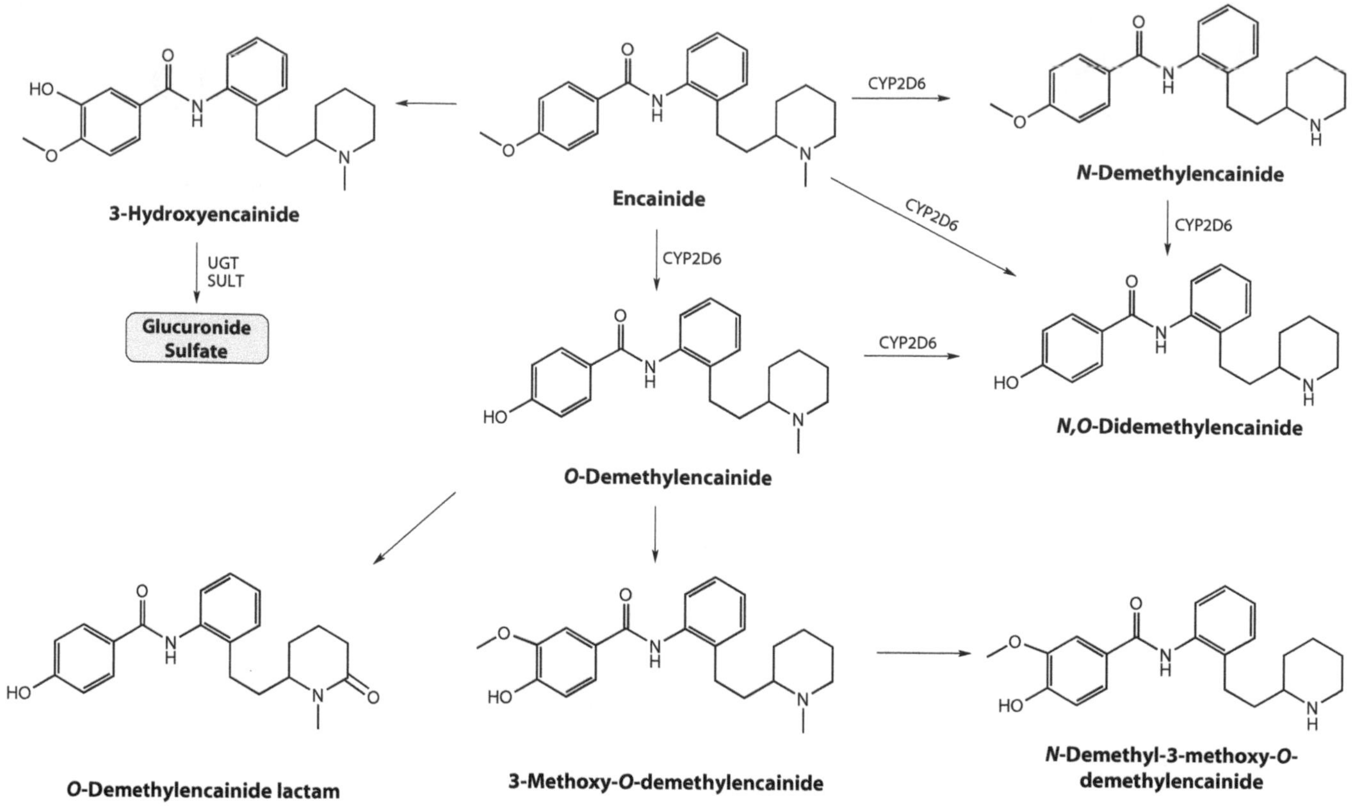

FIGURE 3.54 Metabolism of encainide by CYP2D6. At least six metabolites of encainide have been identified from the hydrolyzed human urine together with unchanged drug, including *O*-demethylencainide, 3-methoxy-*O*-demethylencainide, *N,O*-di-demethylencainide, *N*-demethyl-3-methoxy-*O*-demethylencainide, 3-hydroxyencainide, and *O*-demethylencainide lactam, accounting for >90% of the radioactivity excreted in the urine. Encainide is converted by CYP2D6 to active *O*-demethyl metabolite, which is 6–10 times more potent than the parent drug in blocking sodium channels. In humans, encainide undergoes extensive metabolism via four major routes: (a) *O*-demethylation of the aromatic methyl ether, (b) formation of methylated catechol derivatives, (c) *N*-demethylation of the piperidyl nitrogen, and (d) oxidation at α-carbon to the piperidyl nitrogen.

half-life after intravenous bolus injection is typically 1.5 to 2.0 h (Benowitz and Meister 1978). The primary metabolic pathway of lidocaine in humans is CYP1A2/3A4-mediated *N*-deethylation to the active monoethylglycinexylidide (*N*-ethylglycyl-2,6-xylidide) and further deethylation to the active glycinexylidide (glycyl-2,6-xylidide) (Figure 3.56) (Bargetzi et al. 1989; Imaoka et al. 1990). Monoethylglycinexylidide can be further 3-hydroxylated, which undergoes conjugation. Both monoethylglycinexylidide and glycinexylidide have antiarrhythmic activity. Hydroxylation of the aromatic ring to 3-hydroxylidocaine by CYP1A2, 2D6, and 3A4 is a minor pathway of lidocaine metabolism in humans (Hermansson et al. 1980; Wang et al. 1999, 2000). This is different from the metabolism in rats who undergo extensive 3-hydroxylation, and 3-hydroxylidocaine and 3-hydroxy monoethylglycinexylidide are the major urinary metabolites, whereas 4-OH-lidocaine is excreted to a lesser extent (Keenaghan and Boyes 1972). DA rats, an animal model for the CYP2D6 PM phenotype, do not 3-hydroxylate lidocaine or monoethylglycinexylidide (Masubuchi et al. 1991). 3-Hydroxylidocaine is a reactive metabolite capable of binding covalently to liver microsomal proteins (Masubuchi et al. 1992, 1993), which may be associated with lidocaine toxicity.

Lidocaine also undergoes hydrolytic conversion to result in 2,6-dimethylaniline (2,6-xylidine) (Figure 3.56) (Abdel-Rehim et al. 2000; Parker et al. 1996), a potential human carcinogen (Koujitani et al. 1999). Rat liver microsomal carboxylesterase ES-10, but not carboxylesterase ES-4, hydrolyzes lidocaine and monoethylglycinexylidide, but not glycinexylidide, to 2,6-xylidine (Alexson et al. 2002). 2,6-Dimethylaniline (2,6-Xylidine) can be further oxidized by CYP2A6 and 2E1 to 4-amino-3,5-dimethylphenol (i.e., 4-hydroxyxylidine) (Gan et al. 2001), but oxidation of the amino group to metabolites such as *N*-(2,6-dimethylphenyl) hydroxylamine is also noted in human and rats. 2,6-Xylidine can also be hydroxylated at the 3- or 4-position of the aromatic ring. In addition, 2,6-xylidine can be carboxylated to 2-amino-3-methylbenzoic acid in rabbits (Kammerer and Schmitz 1986). 4-Hydroxy-2,6-xylidine in glucuronide form is the major urinary metabolite found in man, accounting for 72.6% of an administered dose of lidocaine (Keenaghan and Boyes 1972; Tam et al. 1987). This metabolite is also the major metabolite in dogs (35.2%) but lesser in the urine of rats (12.4%) and guinea pigs (16.4%) (Keenaghan and Boyes 1972). Monoethylglycinexylidide is present in the urine in free and glucuronide- and sulfate-conjugated forms, while

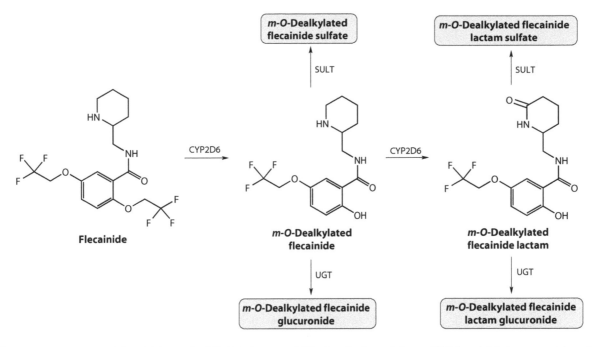

FIGURE 3.55 Metabolism of flecainide by CYP2D6. In humans, 86% of a single oral dose of ^{14}C-flecainide is excreted in urine as flecainide and its metabolites; only 5% is excreted into feces. A substantial portion (27%) of a dose is excreted in urine as unchanged drugs. Two major metabolites, *meta-O*-dealkylated flecainide and the *meta-O*-dealkylated lactam of flecainide, are found in free and conjugated forms in human urine. The two major metabolites of flecainide possess little or negligible antiarrhythmic activity and are also the major metabolites present in human plasma in conjugated form. Flecainide is a substrate of CYP2D6 and the CYP2D6-mediated biotransformation is stereoselective.

glycinexylidide is present mostly in the free form (Tam et al. 1990). Only 3% of lidocaine is excreted as unchanged drug. Phase II conjugation of *N*-(2,6-dimethylphenyl)hydroxylamine may result in reactive esters that decompose to a reactive nitrenium ion capable of reacting with protein and DNA. Reaction of the nitrenium ion with water is a second pathway for the formation of 4-amino-3,5-dimethylphenol (Gan et al. 2001). Further oxidation of 4-amino-3,5-dimethylphenol leads to the formation of the toxic iminoquinone species. Lidocaine can also be hydroxylated to 4-hydroxylidocaine to a minor extent in rats (Coutts et al. 1987) and rabbits (Kammerer and Schmitz 1986) and undergoes *N*-oxidation to its *N*-oxide seen in vitro only (Patterson et al. 1986).

Mexiletine, a class Ib antiarrhythmic drug with similar structure to lidocaine, is used for the treatment and prophylaxis of acute and long-term ventricular arrhythmias (Jarvis and Coukell 1998; Manolis et al. 1990). Unlike lidocaine, which is also a class Ib antiarrhythmic drug with a short $t_{1/2}$ (1.2–1.9 h), mexiletine is an orally effective agent with a half-life ranging from 6 to 12 h (Schrader and Bauman 1986). Mexiletine is a weakly basic drug (pK_a = 8.75) with a narrow therapeutic range. Serum concentrations must remain between 0.8 and 2.0 mg/l, and levels >2.0 mg/l can cause neurological side effects (Campbell et al. 1978). Mexiletine is administered as a racemate and undergoes extensive stereoselective metabolism with <10% of an oral dose being excreted unchanged in urine in humans (Vandamme et al. 1993). *p*-Hydroxylation (aromatic hydroxylation) is favored for *S*-mexiletine with an intrinsic clearance higher than that

for *R*-mexiletine, while the *R*-enantiomer exhibits threefold higher V_{max} value for aliphatic hydroxylation than *S*-mexiletine (Vandamme et al. 1993). After oral administration, at least 11 Phase I and Phase II metabolites of mexiletine have been identified in the urine, which are formed via oxidation, reduction, deamination, methylation, and conjugation, but none of these metabolites possesses any pharmacological activity (Beckett and Chidomere 1977; Labbe and Turgeon 1999). The major metabolites are formed via oxidation of racemic mexiletine to hydroxymethylmexiletine, *p*-hydroxymexiletine, *m*-hydroxymexiletine, and *N*-hydroxymexiletine, accounting for 20% of an administered dose (Figure 3.57) (Beckett and Chidomere 1977). Mexiletine, hydroxymethylmexiletine, and *p*-hydroxymexiletine undergo further oxidative deamination to form their corresponding alcohols. Turgeon et al. (1992) have recovered two major glucuronide metabolite forms after *N*-oxidation and deamination of racemic mexiletine, which accounts for 40% of an administered dose. Studies in human liver microsomes have indicated that the formation of hydroxymethylmexiletine, *p*-hydroxymexiletine, and *m*-hydroxymexiletine is mediated by CYP2D6 and cosegregates with polymorphic CYP2D6-mediated debrisoquine 4-hydroxylase activity (Broly et al. 1990; Vandamme et al. 1993). On the other hand, CYP1A2 is the major enzyme catalyzing *N*-oxidation of mexiletine, but CYP2E1 and 2B6 also play a role to a lesser extent (Labbe et al. 2003; Nakajima et al. 1998b). In healthy human subjects, partial metabolic clearance of *N*-hydroxymexiletine glucuronide is highly stereoselective with an *R/S* ratio of 11.3 (Abolfathi et al. 1993).

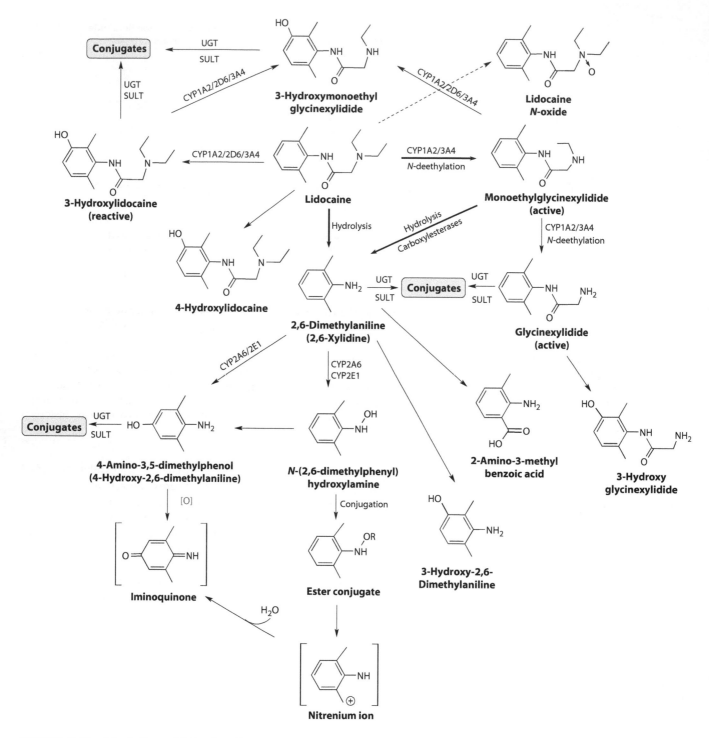

FIGURE 3.56 Metabolism of lidocaine. The primary metabolic pathway of lidocaine in humans is CYP1A2- and 3A4-mediated *N*-deethylation to the active monoethylglycinexylidide (*N*-ethylglycyl-2,6-xylidide) and further deethylation to the active glycinexylidide (glycyl-2,6-xylidide). Monoethylglycinexylidide can be further 3-hydroxylated, which undergoes conjugation. Hydroxylation of the aromatic ring to 3-hydroxylidocaine by CYP1A2, 2D6, and 3A4 is a minor pathway of lidocaine metabolism in humans. Lidocaine also undergoes hydrolytic conversion to result in 2,6-dimethylaniline (2,6-xylidine), a potential human carcinogen. 2,6-Dimethylaniline (2,6-xylidine) can be further oxidized by CYP2A6 and 2E1 to 4-amino-3,5-dimethylphenol (i.e., 4-hydroxyxylidine), but oxidation of the amino group to metabolites such as *N*-(2,6-dimethylphenyl)hydroxylamine is also noted in human and rats.

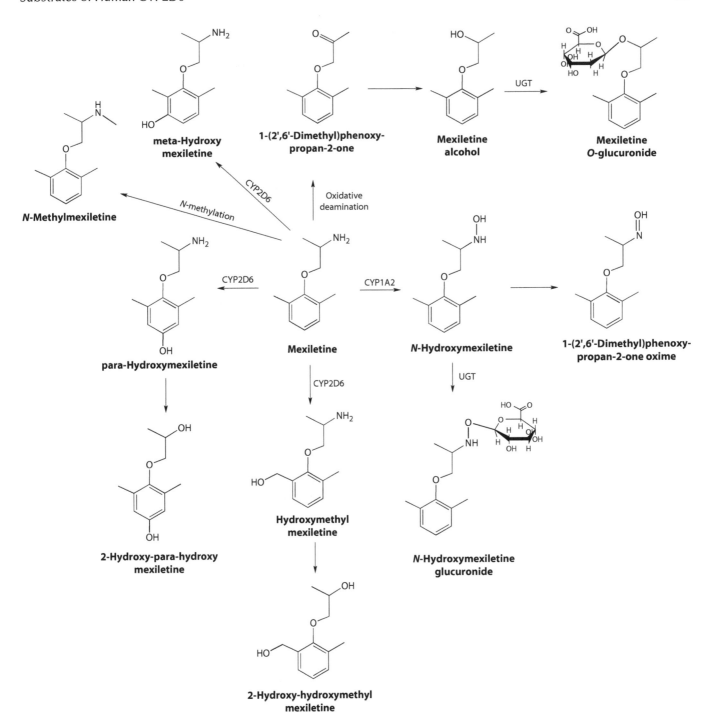

FIGURE 3.57 Metabolism of mexiletine. After oral administration, at least 11 Phase I and Phase II metabolites of mexiletine have been identified in the urine, which are formed via oxidation, reduction, deamination, methylation, and conjugation, but none of these metabolites possesses any pharmacological activity. The major metabolites are formed via oxidation of *rac*-mexiletine to hydroxymethylmexiletine, *p*-hydroxymexiletine, *m*-hydroxymexiletine, and *N*-hydroxymexiletine, accounting for 20% of an administered dose. Mexiletine, hydroxy-methylmexiletine, and *p*-hydroxymexiletine underwent further oxidative deamination to form their corresponding alcohols. Two major glucuronide metabolites formed after *N*-oxidation and deamination of *rac*-mexiletine accounted for 40% of an administered dose. Studies in human liver microsomes have indicated that the formation of hydroxymethylmexiletine, *p*-hydroxymexiletine, and *m*-hydroxymexiletine is mediated by CYP2D6. On the other hand, CYP1A2 is the major enzyme catalyzing *N*-oxidation of mexiletine, but CYP2E1 and 2B6 also play a role to a lesser extent.

Procainamide is a class Ia antiarrhythmic agent used for the treatment of both supraventricular and ventricular arrhythmias (Winkle et al. 1975). It differs from procaine, which is the *p*-aminobenzoyl ester of 2-(diethylamino)-ethanol. Procainamide has a similar effect to quinidine. A substantial but variable fraction of procainamide is metabolized by cytosolic *N*-acetyltransferase 2 (NAT2) to *N*-acetylprocainamide (Figure 3.58), ranging from 16% to 21% of an administered dose in slow acetylators and from 24% to 33% in fast acetylators (Budinsky et al. 1987; Coyle et al. 1985; du Souich and Erill 1977; Karlsson 1978). Patients treated with procainamide have plasma concentrations of *N*-acetylprocainamide generally equaling or being two to three times greater than those of the parent drug (Atkinson et al. 1988). *N*-Acetylprocainamide (acecainide) appears less potent than procainamide with regard to antiarrhythmic effect, and it is used as a class III antiarrhythmic agent (Bagwell et al. 1976; Dangman and Hoffman 1981; Harron and Brogden 1990). Procainamide also undergoes *N*-deethylation to form a desethyl derivative that can be further acetylated (Ruo et al. 1981). Trace amounts may be excreted in the urine as free and acetyl-conjugated *p*-aminobenzoic acid, which is formed by direct hydrolysis, 30% to 60% as unchanged procainamide and 6% to 52% as the *N*-acetylprocainamide derivative (Karlsson 1978). Both procainamide and *N*-acetylprocainamide are eliminated by active tubular secretion as well as by glomerular filtration. Procainamide is converted by CYP2D6 to reactive *N*-hydroxyprocainamide (Figure 3.58) (Lessard et al. 1997), which may be responsible for lupus erythematosus and skin rashes observed in patients treated with the drug (Katsutani and Shionoya 1992). Sequential oxidations at the arylamine moiety of procainamide by hepatic CYP2D6 and myeloperoxidase (MPO) in activated leukocytes result in nitrosoprocainamide, which can be conjugated with glutathione (GSH) by glutathione *S*-transferase (GST) to form a stable conjugate via an initial unstable mercaptal derivative (Freeman et al. 1981; Uetrecht 1985; Wheeler et al. 1991). The incidence of procainamide-induced lupus erythematosus has been reported to be as high as 30% of patients receiving prolonged procainamide therapy (Lawson and Jick 1977). Procainamide-induced lupus erythematosus is characterized by the production of antibodies against nuclear histones and, in particular, to the histone H2A/H2B dimer (Burlingame 1997; Burlingame and Rubin 1996; Katsutani and Shionoya 1992; Mongey and Hess 2001). Rapid acetylators required longer time to develop lupus erythematosus than slow acetylators (Woosley et al. 1978).

Propafenone, a class Ic antiarrhythmic agent, has sodium channel blocking activity and slows intracardiac conduction (Bryson et al. 1993; Grant 1996; Siddoway et al. 1984). It also has β-adrenergic, potassium channel, and slight calcium channel blocking activity (Grant 1996). It is used clinically as a racemic mixture of *S*- and *R*-enantiomers. Although both enantiomers are equally potent in their activity as sodium channel blockers, the *S*-enantiomer has ~100-fold higher β-blocking activity than the *R*-enantiomer (K_i for the affinity to the human lymphocyte β2-adrenoceptors: 571 vs. 7.2 nM) (Kroemer et al. 1989a; Stoschitzky et al.

1990). The drug is an effective agent for the management of ventricular and supraventricular arrhythmias, particularly for Wolff–Parkinson–White syndrome (Grant 1996). Both enantiomers of propafenone exhibit distinct pharmacokinetics, with the *R*-enantiomer being cleared faster than the *S*-enantiomer (X. Chen et al. 2000; Kroemer et al. 1989a; Li et al. 1998). Propafenone undergoes extensive presystemic clearance that appears to be saturable, with bioavailability increasing as dosage increases. Nonlinear pharmacokinetics is observed with propafenone in humans (Connolly et al. 1983; Komura and Iwaki 2005). In humans, conjugates of hydroxylated derivatives of propafenone are present predominantly in the plasma postdosing (Hege et al. 1984). Propafenone is extensively metabolized by Phase I and Phase II reactions, resulting in at least 11 metabolites accounting for more than 90% of the dose administered (Hege et al. 1984). Propafenone is mainly metabolized by CYP2D6 via aromatic ring hydroxylation to 5-hydroxypropafenone and also by CYP1A2 and 3A4 to *N*-desalkylpropafenone to a minor extent (Figure 3.59) (Botsch et al. 1993; Zhou et al. 2003). Subsequent metabolism is via glucuronidation. 5-Hydroxypropafenone is pharmacologically active with similar electrophysiological profile to the parent drug (Haefeli et al. 1991; Malfatto et al. 1988; von Philipsborn et al. 1984). The major metabolites in feces and urine are glucuronide and sulfate conjugates of 5-hydroxypropafenone, *N*-desalkylpropafenone, and propafenone. Furthermore, propafenone undergoes oxidative deamination, resulting in a glycol and a lactic acid derivative (Hege et al. 1984). C–C splitting yields a relatively large amount of 3-phenylpropionic acid, while cleavage of the ether group to a minor extent leads to a phenolic product.

Vernakalant (RSD1235) is a novel and relatively atrial-selective antiarrhythmic agent that is a mixed ion channel inhibitor and selectively prolongs the atrial refractory period without significantly affecting ventricular refractoriness (Naccarelli et al. 2008; Roy et al. 2008). In vitro studies in human liver microsomes and recombinant CYP2D6 have shown that vernakalant is converted by CYP2D6 into its major metabolite, RSD1385, via 4-*O*-demethylation (Figure 3.60) (see http://www.cardiome.com; Cardiome Pharma Corporation, Vancouver, British Columbia, Canada). A minor metabolite, RSD1390, is formed via 3-*O*-demethylation. RSD1385 and RSD1390 appear to be equally or less potent than vernakalant on ion channels in vitro. In contrast, the glucuronides of RSD1385 and RSD1390 are inactive at clinically relevant concentrations in vitro.

3.3.2.3 Antiplatelet Agents

Ticlopidine and prasugrel belong to the same chemical family of thienopyridine adenosine diphosphate (ADP) receptor antagonists (Flores-Runk and Raasch 1993; Kam and Nethery 2003; Savi and Herbert 2005). Ticlopidine is a potent and long-acting inhibitor of platelet aggregation acting through inhibition of the P2RY12 receptor (Quinn and Fitzgerald 1999). Although effective in preventing atherothrombotic events in cardiovascular, cerebrovascular, and peripheral vascular

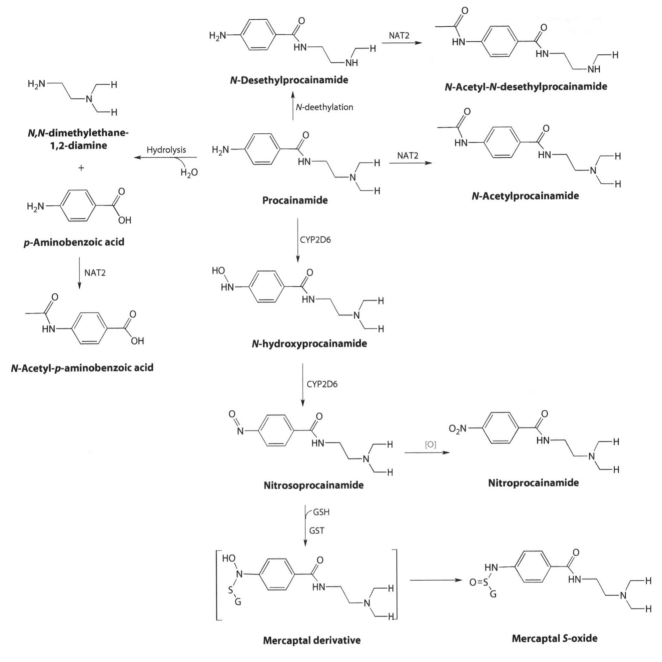

FIGURE 3.58 Metabolism of procainamide. A substantial but variable fraction of procainamide is metabolized by cytosolic NAT2 to *N*-acetylprocainamide, ranging from 16% to 21% of an administered dose in slow acetylators and from 24% to 33% in fast acetylators. Procainamide also undergoes *N*-deethylation to form a desethyl derivative that can be further acetylated. Trace amounts may be excreted in the urine as free and acetyl conjugated *p*-aminobenzoic acid, which is formed by direct hydrolysis, 30% to 60% as unchanged procainamide and 6% to 52% as the *N*-acetylprocainamide derivative. Procainamide is converted by CYP2D6 to reactive *N*-hydroxyprocainamide. Sequential oxidations at the arylamine moiety of procainamide by hepatic CYP2D6 and MPO in activated leukocytes result in nitrosoprocainamide, which can be conjugated with GSH to form a stable conjugate via an initial unstable mercaptal derivative.

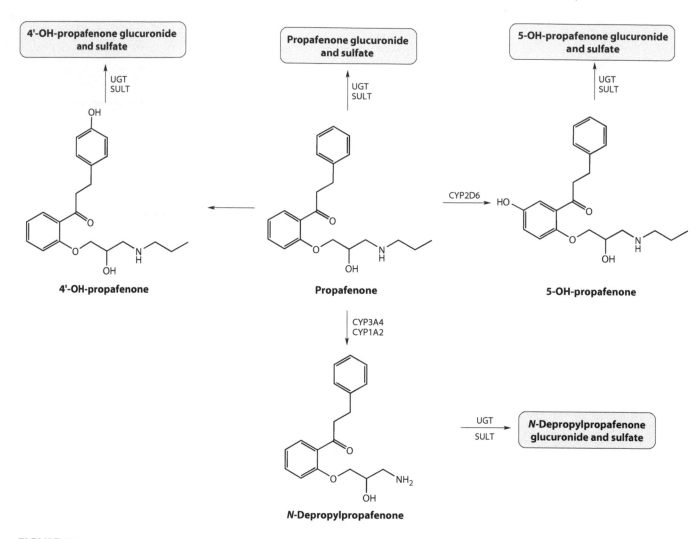

FIGURE 3.59 Metabolism of propafenone by CYP2D6 and other CYPs. In humans, conjugates of hydroxylated derivatives of propafenone are present predominantly in the plasma postdosing. Propafenone is extensively metabolized by Phase I and Phase II reactions, resulting in at least 11 metabolites accounting for more than 90% of the dose administered. Propafenone is mainly metabolized by CYP2D6 via aromatic ring hydroxylation to 5-hydroxypropafenone and also by CYP1A2 and 3A4 to *N*-desalkylpropafenone to a minor extent. Subsequent metabolism is via glucuronidation. 5-Hydroxypropafenone is pharmacologically active with similar electrophysiological profile to the parent drug. The major metabolites detected in feces and urine are glucuronide and sulfate conjugates of 5-hydroxypropafenone, *N*-desalkylpropafenone, and propafenone. Furthermore, propafenone undergoes oxidative deamination, resulting in a glycol and a lactic acid derivative. C–C splitting yields a relatively large amount of 3-phenyl-propionic acid, while cleavage of the ether group to a minor extent leads to a phenolic product.

diseases (Noble and Goa 1996), ticlopidine causes a relatively high incidence of hematological and liver toxicities (Lesesve et al. 1994; Martinez Perez-Balsa et al. 1998; Skurnik et al. 2003). The major metabolic routes of ticlopidine in vivo are *N*-dealkylation, *N*-oxidation (to form M7), and oxidation of the thiophene ring (Figure 3.61) (Desager 1994). In vitro metabolism of ticlopidine by recombinant human CYP2C19 and 2D6 has led to the identification of thiophene-*S*-oxide dimer (M2) and 2-oxoticlopidine (M4) (Ha-Duong et al. 2001). 2-Oxoticlopidine is further converted into a carboxylic acid derivative, which is the active metabolite binding to the P2RY$_{12}$ receptor. The active metabolite of ticlopidine contains a thiol group that binds to a free cysteine on the P2RY$_{12}$ receptor and irreversibly blocks ADP binding and receptor activation. Once this blocking has occurred, platelets are affected for

their entire lifespan of approximately 7–10 days. Ticlopidine undergoes MPO-catalyzed oxidation, resulting in dehydrogenated ticlopidine and 2-chloroticlopidine metabolites (Liu and Uetrecht 2000). Ticlopidine can also be oxidized to the corresponding dihydrothienopyridinium (M5) and thienopyridinium (M6) metabolites (Dalvie and O'Connell 2004). The formation of M5 and M6 is catalyzed by CYP3A4 and 2C19 as well as by peroxidases including MPO and HRP. CYP1A2, 2C9, and 2D6 also catalyzes the formation of M5 and M6 but to a lesser extent. M5 is converted to M8 (a lactam analog of ticlopidine) by CYP3A4 only. CYP2C19 is also involved in the oxidation of the tetrahydropyridine moiety of ticlopidine to form M3 and M9, in addition to the thiophene ring (Dalvie and O'Connell 2004). However, ticlopidine is a poor substrate of MAO-A and MAO-B, with low activity for M5

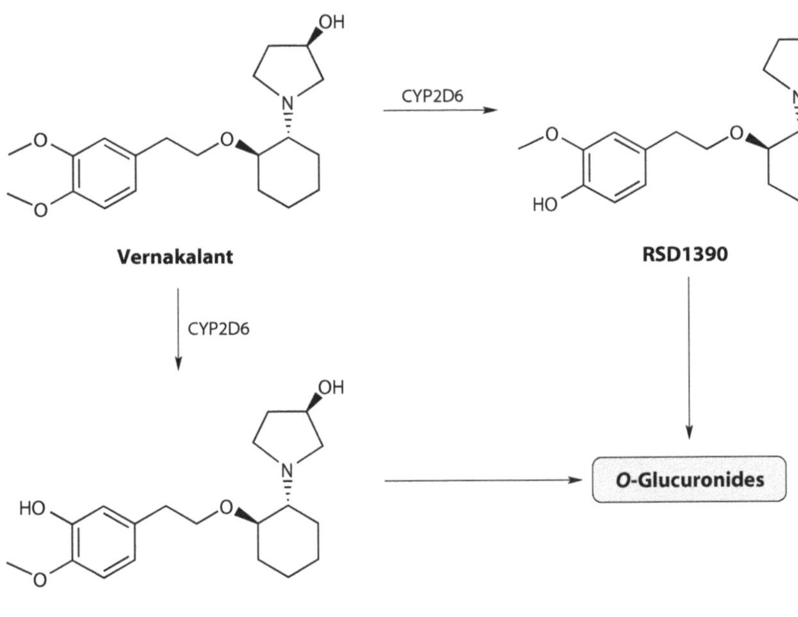

FIGURE 3.60 Metabolism of vernakalant by CYP2D6. The oral absorption of vernakalant is almost complete. In vitro studies with human liver microsomes and recombinant CYP2D6 have shown that vernakalant is converted into its major metabolite, RSD1385, via 4-*O*-demethylation by CYP2D6 (see http://www.cardiome.com). A minor metabolite, RSD1390, is produced via 3-*O*-demethylation in human liver microsomes. Both RSD1385 and RSD1390 can be further conjugated. RSD1385 and RSD1390 appear to be equally or less potent than vernakalant on ion channels in vitro. In contrast, the glucuronides of RSD1385 and RSD1390 are inactive at clinically relevant concentrations.

and M6 formation (Dalvie and O'Connell 2004). Ticlopidine is a mechanism-based (suicide) inhibitor of CYP2C19 and a known competitive inhibitor of CYP2D6 (Ha-Duong et al. 2001; Ko et al. 2000), while tienilic acid is a substrate and a suicide inhibitor of human CYP2C9 (Lopez-Garcia et al. 1994).

Prasugrel (CS-747/LY640315) is a novel third-generation thienopyridine antiplatelet prodrug that is administered orally as a racemic mixture (Angiolillo et al. 2008; Jakubowski et al. 2007; Niitsu et al. 2005; Riley et al. 2008; Tantry et al. 2006). It is rapidly metabolized in vivo to a potent ADP receptor antagonist, designated R-138727 (2-[1-[2-cyclopropyl-1-(2-fluorophenyl)-2-oxoethyl]-4-mercapto-3-piperidinylidene] acetic acid) (Wickremsinhe et al. 2007), which binds irreversibly to P2Y$_{12}$ receptors, causing inhibition of platelet aggregation that persists for the life of the platelets (Frelinger et al. 2008; Judge et al. 2008). Prasugrel is rapidly hydrolyzed by human carboxylesterases 1 and 2 (Williams et al. 2008) to yield the thiolactone, R-95913 (2-[2-oxo-6,7-dihydrothieno[3,2-c] pyridin-5(4*H*)-yl]-1-cyclopropyl-2-(2-fluorophenyl)ethanone) (Figure 3.62), which circulates in human plasma, whereas prasugrel is not detected in human plasma (Farid et al. 2007; Niitsu et al. 2005). Further metabolism of R-95913 yields a ring-opened form, R-138727 (Sugidachi et al. 2000, 2001), which has two chiral centers and is actually a mixture of four stereoisomers that vary in potency of inhibition at P2Y$_{12}$ receptors. R-138727 is further metabolized by *S*-methylation and conjugation with cysteine (Farid et al. 2007). Approximately 70% of the dose is excreted in the urine and 25% is excreted

in the feces (Farid et al. 2007). R-138727 is formed primarily by CYP3A4 and 2B6, with roles for CYP2C9, 2C19, and 2D6 (10%–20% contribution) (Rehmel et al. 2006). R-95913 inhibits CYP2C9, 2C19, 2D6, and 3A4, with K_i values ranging from 7.2 to 82 μM, but does not inhibit CYP1A2 (Rehmel et al. 2006).

3.3.2.4 β-Blockers

The β-blockers comprise a group of drugs that are mostly used to treat hypertension, cardiac arrhythmia, or ischemic heart disease. CYP2D6 and 2C19 play a major role in the metabolism of several β-blockers including bisoprolol (Horikiri et al. 1998a,b), bufuralol (Mautz et al. 1995; Narimatsu et al. 2002), bunitrolol (Narimatsu et al. 1994; Ono et al. 1995), carvedilol (Oldham and Clarke 1997), metoprolol (Belpaire et al. 1998; McGourty et al. 1985), propranolol (Johnson et al. 2000; Masubuchi et al. 1994; Tassaneeyakul et al. 1993), and timolol (Volotinen et al. 2007). Carvedilol, metoprolol, and propranolol are high-clearance drugs and timolol is a low-clearance drug. Atenolol is mainly eliminated by renal excretion, bisoprolol is in part excreted as parent compound via the renal route (50%), and the other 50% are hepatically metabolized (Brodde and Kroemer 2003).

Bisoprolol is a selective β$_1$-adrenoceptor blocker indicated for hypertension and chronic heart failure (Johns and Lopez 1995; McGavin and Keating 2002). Bisoprolol has a chiral center, with *S*-(−)-bisoprolol being 30 to 80 times more pharmacologically active than *R*-(+)-bisoprolol. The oral bioavailability of bisoprolol is high (>90%) and the drug has a long

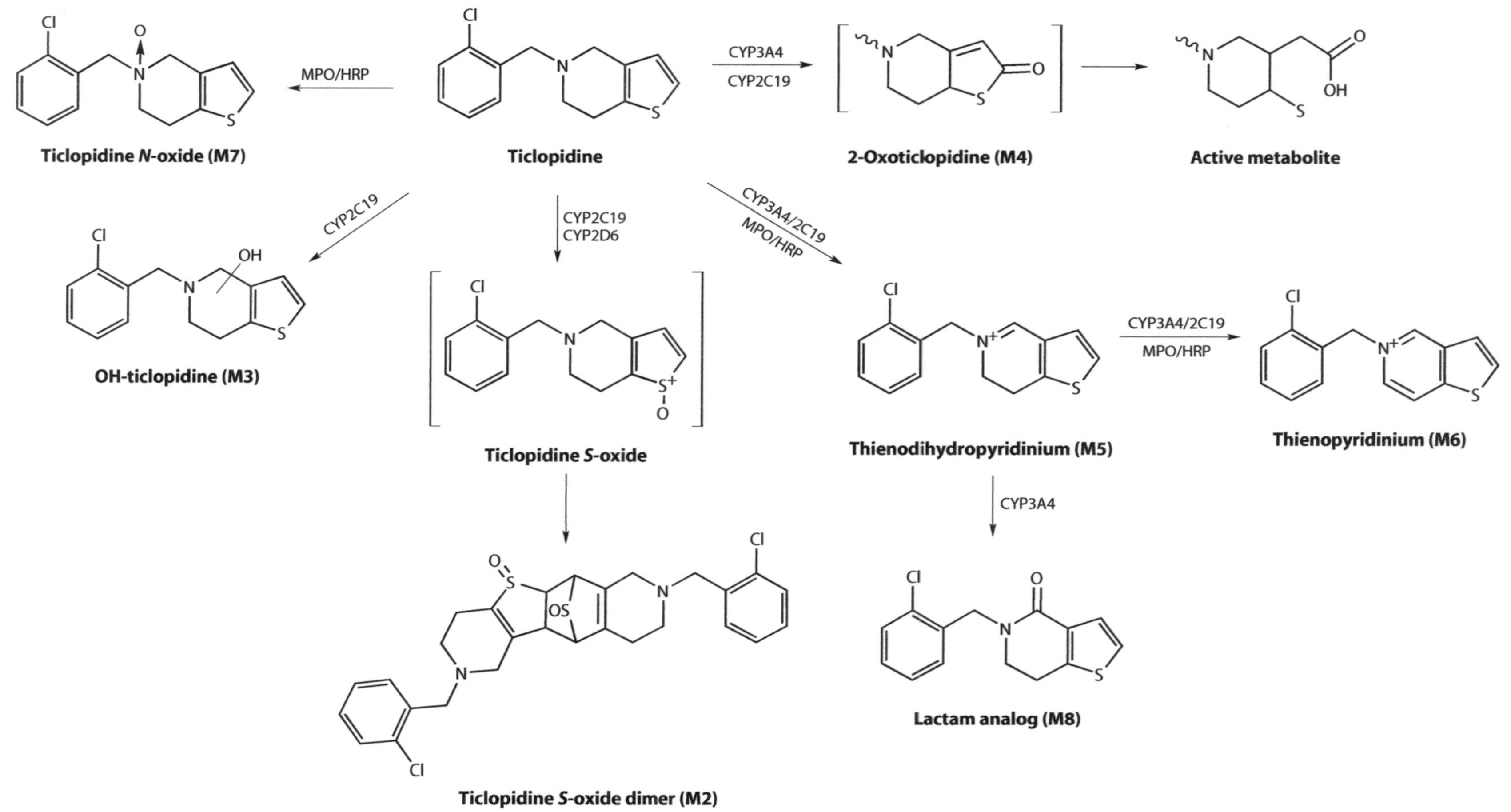

FIGURE 3.61 Metabolism of ticlopidine by CYP2C9, 2C19, 3A4, and 2D6. The major metabolic routes of ticlopidine in vivo are *N*-dealkylation, *N*-oxidation (to form M7), and oxidation of the thiophene ring. In vitro metabolism of ticlopidine by recombinant human CYP2C19 and 2D6 has led to the identification of thiophene-*S*-oxide dimer (M2) and 2-oxoticlopidine (M4). 2-Oxoticlopidine is further converted into a carboxylic acid derivative, which is the active metabolite binding to the P2RY$_{12}$ receptor. Ticlopidine undergoes MPO-catalyzed oxidation, resulting in dehydrogenated ticlopidine and 2-chloroticlopidine metabolites. Ticlopidine can also be oxidized to the corresponding dihydrothienopyridinium (M5) and thienopyridinium (M6) metabolites. The formation of M5 and M6 is catalyzed by CYP3A4 and 2C19 as well as by peroxidases including MPO and HRP. CYP1A2, 2C9, and 2D6 also catalyzed the formation of M5 and M6 but to a lesser extent. M5 is converted to M8 (a lactam analog of ticlopidine) by CYP3A4 only. CYP2C19 is also involved in the oxidation of the tetrahydropyridine moiety of ticlopidine to form M3 and M9, in addition to the thiophene ring. However, ticlopidine is a poor substrate of MAO-A and MAO-B, with low activity for M5 and M6 formation.

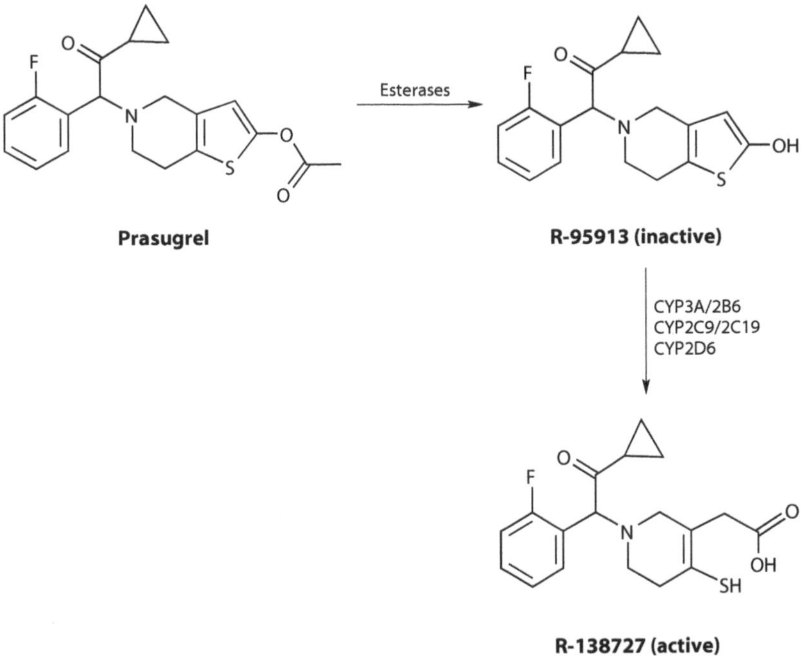

Prasugrel → Esterases → **R-95913 (inactive)**

CYP3A/2B6
CYP2C9/2C19
CYP2D6

R-138727 (active)

FIGURE 3.62 Metabolism of prasugrel by CYP3A4, 2B6, 2C9, 2C19, and 2D6. Prasugrel is rapidly hydrolyzed by human carboxylesterases 1 and 2 to yield the thiolactone, R-95913 (2-[2-oxo-6,7-dihydrothieno[3,2-c]pyridin-5(4H)-yl]-1-cyclopropyl-2-(2-fluorophenyl) ethanone), which circulates in human plasma, whereas prasugrel is not detected in human plasma. Further metabolism of R-95913 yields a ring-opened form, R-138727, which has two chiral centers and is actually a mixture of four stereoisomers that vary in potency of inhibition at P2Y$_{12}$ receptors. R-138727 is further metabolized by S-methylation and conjugation with cysteine. Approximately 70% of the dose is excreted in the urine and 25% in the feces. R-138727 is formed primarily by CYP3A4 and 2B6, with roles for CYP2C9, 2C19, and 2D6 (10%–20% contribution).

$t_{1/2\beta}$ (11 h), which allows once-daily administration (Lancaster and Sorkin 1988). In humans, 50% of the dose is eliminated renally as unchanged drug and the other 50% is eliminated metabolically, with subsequent renal excretion of the metabolites (Leopold et al. 1986). Bisoprolol is mainly metabolized to M4 via O-deisopropylation by CYP3A4 and 2D6 (Figure 3.63) (Horikiri et al. 1998a,b). CYP2D6 metabolized bisoprolol stereoselectively (R > S), whereas the metabolism of bisoprolol by CYP3A4 is not stereoselective (Horikiri et al. 1998a). In Japanese patients, the genotype of CYP2D6 and 2C9 did not affect the pharmacokinetics of bisoprolol (Taguchi et al. 2005).

Bufuralol, a β-adrenoceptor blocker, has been extensively used as a probe substrate for the in vitro study of CYP2D6 (Dayer et al. 1984; Gut et al. 1984, 1986; Minder et al. 1984). Bufuralol is metabolized to three metabolites, namely, 1′-hydroxybufuralol, 1′-oxobufuralol, and 1′2′-ethenylbufuralol (Figure 3.64) (Hiroi et al. 2002). 1′2′-Ethenylbufuralol is considered to be formed both from bufuralol by ethenylation and from 1′-hydroxybufuralol by dehydration. 1′2′-Ethenylbufuralol formation from bufuralol has been demonstrated to be mediated by CYP2D6 (Hanna et al. 2001). The level of 1′-hydroxybufuralol, a major metabolite of bufuralol, is often measured as an index of CYP2D6 activity or levels, and the amount of 1′-hydroxybufuralol formed from bufuralol is known to be small in CYP2D6-deficient metabolizers (Carcillo et al. 2003). However, bufuralol, but not

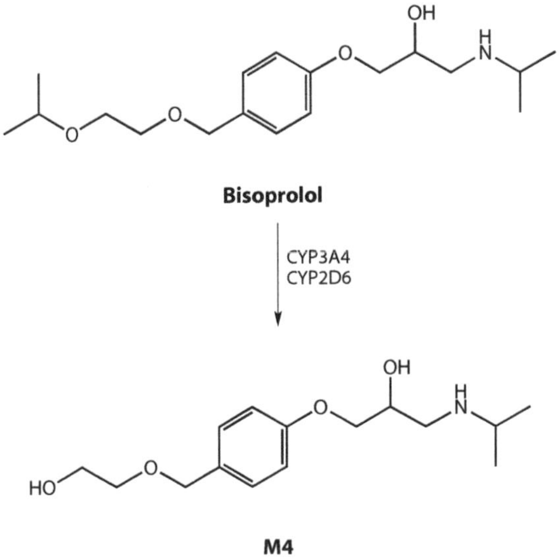

Bisoprolol

CYP3A4
CYP2D6

M4

FIGURE 3.63 Metabolism of bisoprolol by CYP2D6 and 3A4. Bisoprolol has a chiral center, with S-(−)-bisoprolol being more pharmacologically active (30 to 80 times) than R-(+)-bisoprolol. Bisoprolol is mainly metabolized to M4 via O-deisopropylation by CYP3A4 and 2D6.

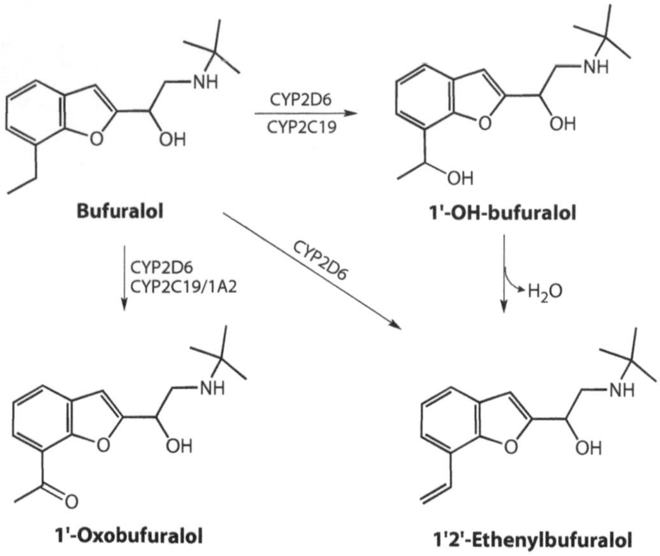

FIGURE 3.64 Metabolism of bufuralol by CYP2D6, 2C19, and 1A2. Bufuralol is metabolized to three metabolites, namely, 1′-hydroxybufuralol, 1′-oxobufuralol, and 1′2′-ethenylbufuralol. 1′2′-Ethenylbufuralol is considered to be formed both from bufuralol by ethenylation and from 1′-hydroxybufuralol by dehydration. 1′2′-Ethenylbufuralol formation from bufuralol has been demonstrated to be mediated by CYP2D6. The level of 1′-hydroxybufuralol, a major metabolite of bufuralol, is often measured as an index of CYP2D6 activity or levels, and the amount of 1′-hydroxybufuralol formed from bufuralol is known to be small in CYP2D6-deficient metabolizers. However, bufuralol, but not sparteine and debrisoquine, is also extensively metabolized by CYP2C19 and, to a lesser extent, by 1A2, and this may affect its specificity as a prototypical substrate of CYP2D6.

sparteine and debrisoquine, is also extensively metabolized by CYP2C19 (Mankowski 1999) and, to a lesser extent, by 1A2 (Yamazaki et al. 1994), and this may affect its specificity as a prototypical substrate of CYP2D6.

Bunitrol is also metabolized by CYP2D6 (Narimatsu et al. 1994; Ono et al. 1995). In human liver microsomes, bunitrol 4-hydroxylation is mainly catalyzed by CYP2D6 (Figure 3.65) (Narimatsu et al. 1994). In recombinant human CYPs, CYP2D6 and 1A2 at a substrate concentration of 5 μM, and CYP2C8 and 2C9 in addition to the two enzymes at a substrate concentration of 1 mM, show detectable 4-hydroxylase activities (Ono et al. 1995). CYP2D6 oxidizes bunitrolol to 4-hydroxybunitrolol with substrate enantioselectivity of the R-(+)-bunitrolol over the S-(−)-enantiomer (Narimatsu et al. 1999). The G42R, P34S, P34S/S486T, F120A, E216a, and E222A mutants of CYP2D6 show differential oxidation capacities toward bunitrolol as a chiral substrate compared to the wild-type CYP2D6 (Masuda et al. 2006; Tsuzuki et al. 2001, 2003). The substitution of Phe120 by Ala markedly increases the K_m and V_{max} values for enantiomeric bunitrolol 4-hydroxylation by CYP2D6, while the substitution of Glu222 as well as Glu216 by Ala remarkably decreased both the apparent K_m and V_{max} values but without changing substrate enantioselectivity or metabolite diastereoselectivity (Masuda et al. 2006).

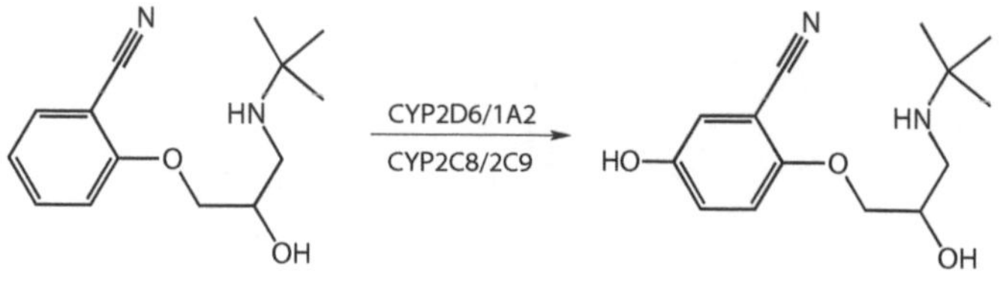

FIGURE 3.65 Metabolism of bunitrolol by CYP2D6 and other enzymes. In recombinant human CYPs, CYP2D6 and 1A2 at a substrate concentration of 5 μM, and CYP2C8 and 2C9 in addition to the two enzymes at a substrate concentration of 1 mM, showed detectable 4-hydroxylase activities (Ono et al. 1995). CYP2D6 oxidized bunitrolol to 4-hydroxybunitrolol with substrate enantioselectivity of (R)-(+)-bunitrolol over (S)-(−)-enantiomer.

Carvedilol is a β-adrenoceptor antagonist with vasodilating activity based on $α_1$-blockade, available for the treatment of hypertension and congestive heart failure. This drug is used clinically as a racemic mixture of R- and S-enantiomers. Carvedilol is extensively metabolized in man, giving products from both oxidation and conjugation pathways (Neugebauer et al. 1990). In human liver microsomes, 4′- and 5′-hydroxyphenyl carvedilol, O-desmethylcarvedilol, and 8-hydroxy carbazolyl carvedilol are detected for both the R- and S-enantiomers of carvedilol (Figure 3.66) (Oldham and Clarke 1997). 4′- and 5′-Hydroxylations are mainly catalyzed by CYP2D6, with minor contribution from CYP2E1 and 2C9 (Oldham and Clarke 1997). CYP1A2 is the most significant hepatic enzyme for 8-hydroxylation of carvedilol, with contribution from CYP3A4 and 1A1. In addition, CYP2C9 is the major enzyme for O-demethylation of carvedilol, but CYP2D6 and possibly CYP1A2 and CYP2E1 also contribute to this pathway.

Metoprolol is a cardioselective $β_1$-blocker, clinically used in the treatment of hypertension, angina pectoris, and arrhythmia. The drug is marketed as a racemic mixture, but its pharmacological effect resides in the S-enantiomer. The S-metoprolol enantiomer has approximately 500-fold more affinity for the $β_1$-adrenergic receptor than its R-antipode (Dayer et al. 1985). Metoprolol is metabolized through α-hydroxylation (~10% of dose), O-demethylation (65% of dose), and N-dealkylation by oxidative deamination (<10% of dose) (Borg et al. 1975). α-Hydroxylation of metoprolol is catalyzed almost entirely by CYP2D6 and O-demethylation via CYP2D6 and 3A4 (Figure 3.67) (Hoffmann et al. 1980; Johnson and Burlew 1996; Otton et al. 1988). Hydroxylation of the aliphatic chain of metoprolol adds a new chiral center to the corresponding α-hydroxymetoprolol, generating four optical isomers. The pharmacokinetics of metoprolol is stereoselective.

Propranolol is a nonselective β-adrenergic blocker used as a racemic mixture in the treatment of hypertension, cardiac arrhythmias, and angina pectoris. In humans, propranolol is metabolized by three main pathways: glucuronidation, aromatic ring hydroxylation, and side-chain oxidation. The aromatic ring hydroxylation may occur at position 4, 5, or 7, primarily via 4′-hydroxylation, with much less 5′-hydroxylation, smaller amounts of the 7-hydroxylation product, and very

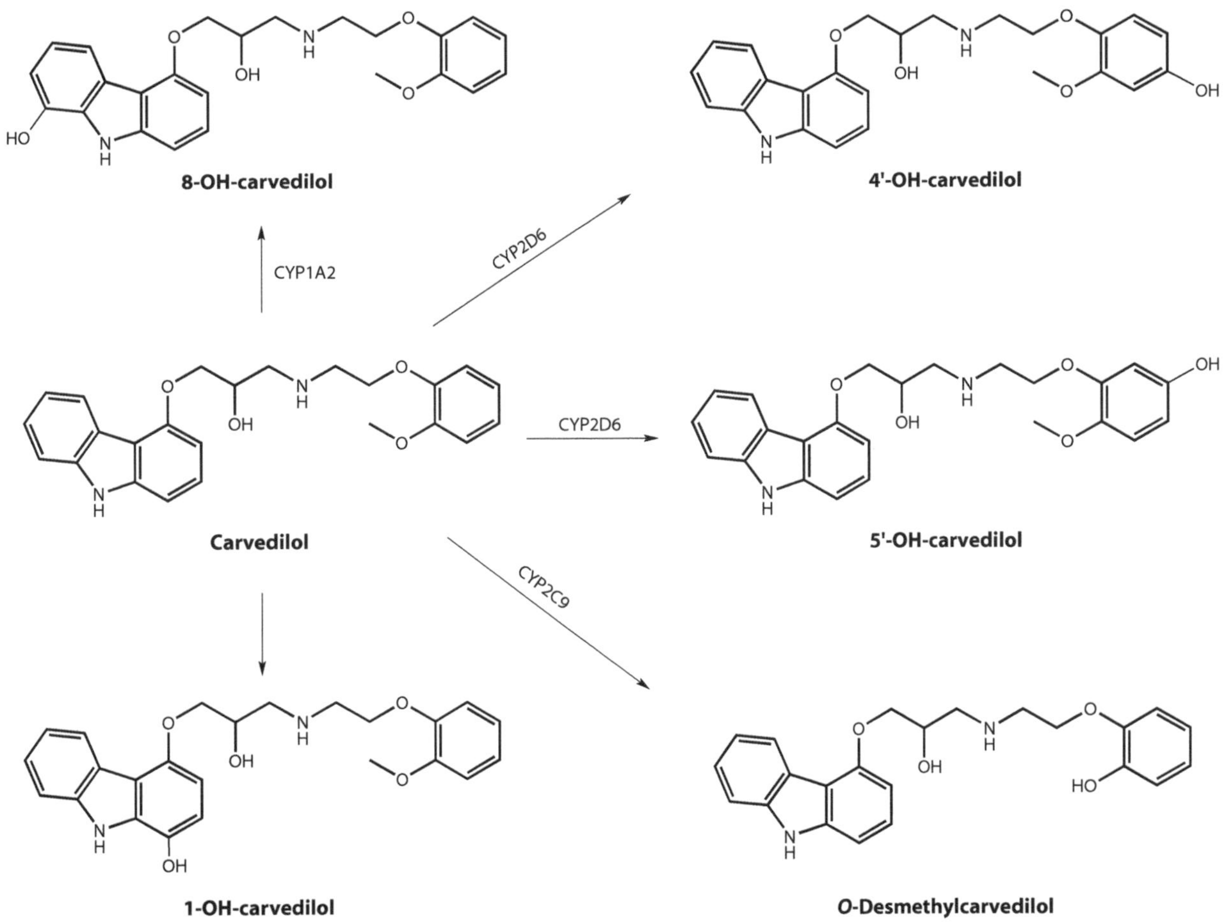

FIGURE 3.66 Metabolism of carvedilol by CYP2D6, 2C9, 1A2, and 2E1. Carvedilol is extensively metabolized in humans, giving products from both oxidation and conjugation pathways. In human liver microsomes, 4′- and 5′-hydroxyphenyl carvedilol, O-desmethylcarvedilol, and 8-hydroxy carbazolyl carvedilol are detected for both the R- and S-enantiomers of carvedilol. 4′- and 5′-Hydroxylations are mainly catalyzed by CYP2D6, with minor contribution from CYP2E1 and 2C9. CYP1A2 is the most significant hepatic enzyme for 8-hydroxylation of carvedilol, with contribution from CYP3A4 and 1A1. In addition, CYP2C9 is the major enzyme for O-demethylation of carvedilol, but CYP2D6 and possibly CYP1A2 and CYP2E1 also contribute to this pathway.

small amounts of dihydroxylated products (e.g., 4,6- and 4,8-diOH-propranolol) (Figure 3.68) (Talaat and Nelson 1988). The hydroxypropranolol is further conjugated with glucuronic acid, favoring the S-enantiomer, or sulfate, favoring the R-propranolol, before excretion into urine. Approximately 17% of the propranolol dose is metabolized in human through stereoselective glucuronidation by UGT1As and 2B7 (Silber et al. 1982; Walle et al. 1985). When racemic propranolol is administered to humans, the concentration of S-propranolol glucuronide is higher than that of R-propranolol glucuronide in both plasma and urine (Silber et al. 1982). UGT1A9 and 1A10 glucuronidate propranolol with reverse stereoselectivity (Sten et al. 2006): UGT1A9 glucuronidated S-propranolol much faster than R-propranolol, whereas UGT1A10 exhibits the opposite enantiomer preference. Propranolol is also metabolized by UGT2B7 without showing significant preference for the R- or S-enantiomer (Coffman et al. 1998). Overall, the metabolism of propranolol is stereoselective for the less active R-enantiomer, resulting in higher plasma concentrations of the S-enantiomer in humans. It appears that N-dealkylation

of propranolol is mainly governed by S-mephenytoin-4-hydroxylase (CYP1A2), whereas ring hydroxylation is predominantly related to CYP2D6 (Ward et al. 1989). Propranolol ring hydroxylation cosegregates with debrisoquine/sparteine polymorphism in vivo (Lennard et al. 1984, 1986; Raghuram et al. 1984). CYP2D6 mainly catalyzes ring hydroxylation while CYP1A2 catalyzes N-desisopropylation (Masubuchi et al. 1994). Recombinant human CYP1A2 and 2D6 exhibit comparable catalytic activity with respect to the N-desisopropylation of both propranolol enantiomers; only expressed CYP2D6 exhibits a marked catalytic activity for the 4-hydroxylation of both propranolol enantiomers (Bichara et al. 1996; Yoshimoto et al. 1995). Recombinant CYP2D6 in yeast cells catalyzes the formation of 4-OH- and 5-OH-propranolol and minor amounts of N-desisopropylpropranolol, with aromatic hydroxylation rates on the enantiomers exceeding those of N-dealkylation by 10-fold (Bichara et al. 1996). Similar results have been observed with recombinant enzymes and in human liver microsomes (Otton et al. 1990; Rowland et al. 1996; Yoshimoto et al. 1995). Propranolol has a distance between the nitrogen and

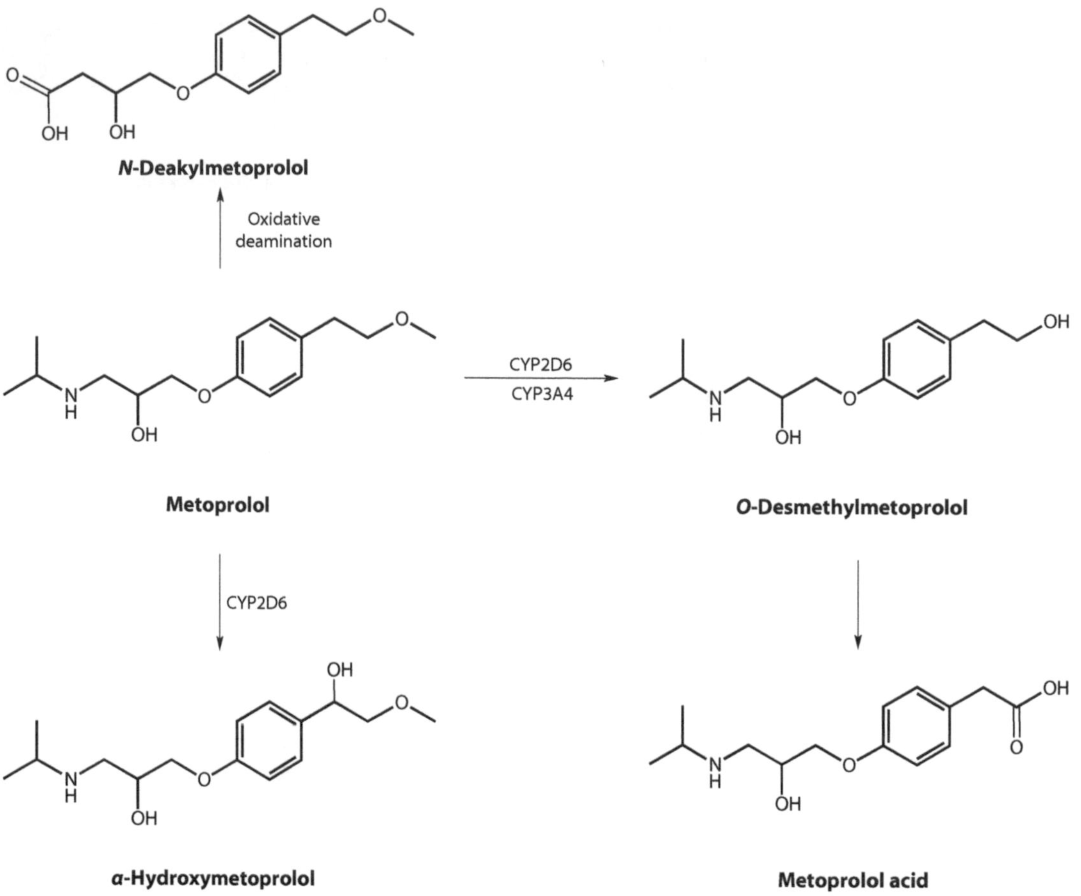

FIGURE 3.67 Metabolism of metoprolol by CYP2D6 and 3A4. Metoprolol is metabolized through α-hydroxylation (~10% of dose), O-demethylation (65% of dose), and N-dealkylation (<10% of dose). α-Hydroxylation of metoprolol is catalyzed almost entirely by CYP2D6 and O-demethylation via CYP2D6 and 3A4. In vitro studies indicate that the stereoselectivity in EMs is the result of the high-affinity component of the O-demethylation catalyzed by CYP2D6, exhibiting significant enantioselectivity for the R-enantiomer. Oxidation of R-metoprolol by CYP2D6 produced more O-demethylmetoprolol than α-hydroxymetoprolol; however, for S-metoprolol, a slight preference for α-hydroxylation is observed.

preferred site of oxidation of approximately 7.9 Å (de Groot et al. 1999a).

Timolol is a nonselective β-blocker that has mainly been used topically for the treatment of glaucoma and ocular hypertension since 1978 (Frishman et al. 1994; Heel et al. 1979). It has also been used in the treatment of hypertension and prophylaxis of migraine (Heel et al. 1979). Timolol is largely eliminated by the liver and it is primarily metabolized by CYP2D6, with minor contribution from CYP2C19 (Volotinen et al. 2007). In human liver microsomes, four metabolites have been identified, with the most abundant being a hydroxy metabolite (M1) (Figure 3.69). The contribution of CYP2D6 and 2C19 to timolol metabolism is estimated to be >90% and <10%, respectively.

3.3.2.5 Calcium Channel Blockers

Diltiazem is used in the treatment of hypertension and angina pectoris. Diltiazem undergoes complicated metabolism, including esterase-mediated deacetylation, N-demethylation, and O-demethylation. CYP2D6 and CYP3A4 are involved in O- and N-demethylation of diltiazem, respectively

(Figure 3.70) (Jones et al. 1999; Molden et al. 2000). Esterase-mediated deacetylation of diltiazem results in the first-line metabolite desacetyl diltiazem (M1), which exhibits approximately 50% of the vasodilating properties of the parent drug (Schoemaker et al. 1987). Both CYP2D6 and 3A4 are involved in the further metabolism of M1 (Molden et al. 2002a). N- and O-Demethylation of diltiazem produced M2 and M4, respectively. Coadministration of quinidine, a known potent inhibitor of CYP2D6, did not significantly increase plasma concentration of diltiazem in healthy male subjects (Laganiere et al. 1996), suggesting that CYP2D6 plays a secondary role only in the overall biotransformation of diltiazem. Consistently, the pharmacokinetics of diltiazem is not significantly different between EMs and PMs of CYP2D6 (Molden et al. 2002b). However, the systemic exposure of the pharmacologically active metabolites desacetyl diltiazem and N-demethyldesacetyl diltiazem is ≥5-fold higher in PMs than in EMs. These results indicate that the CYP2D6 phenotype/genotype does not have a major impact on the disposition of diltiazem, but desacetyl diltiazem and N-demethyldesacetyl diltiazem are markedly accumulated in PMs.

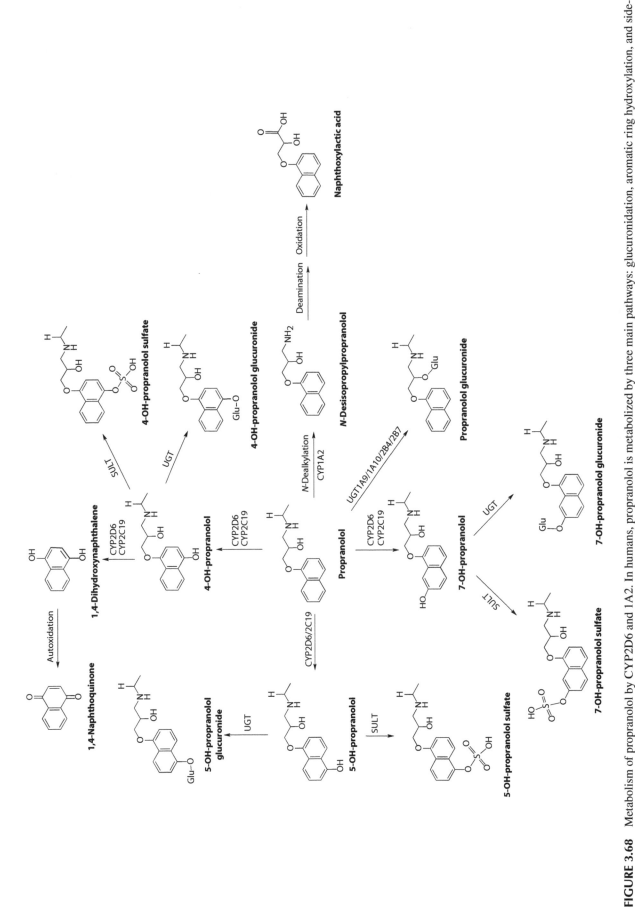

FIGURE 3.68 Metabolism of propranolol by CYP2D6 and 1A2. In humans, propranolol is metabolized by three main pathways: glucuronidation, aromatic ring hydroxylation, and side-chain oxidation. The aromatic ring hydroxylation may occur at position 4, 5, or 7, primarily via 4'-hydroxylation, with much less 5'-hydroxylation, smaller amounts of the 7-hydroxylation product, and very small amounts of dihydroxylated products (e.g., 4,6- and 4,8-diOH-propranolol). UGT1A9, 1A10, and 2B7 glucuronidated propranolol. CYP2D6 mainly catalyzes ring hydroxylation while CYP1A2 catalyzes N-desisopropylation. Recombinant human CYP1A2 and 2D6 exhibited comparable catalytic activity with respect to the N-desisopropylation while only expressed CYP2D6 exhibited a marked catalytic activity for the 4-hydroxylation of propranolol.

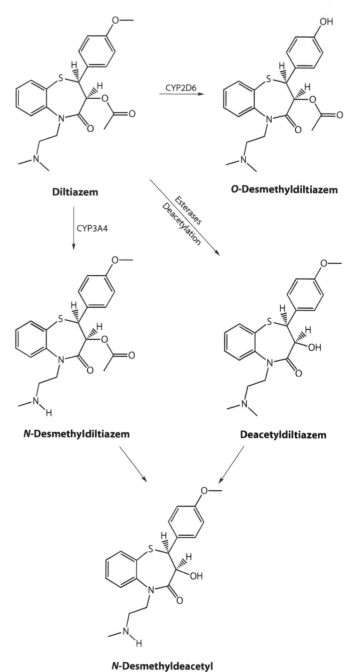

FIGURE 3.69 Metabolism of timolol by CYP2D6 and other enzymes. Detectable plasma levels of timolol occur within 1/2 h and peak plasma levels occur in approximately 1 to 2 h. Timolol is partially metabolized by the liver and timolol and its metabolites are excreted by the kidney. Earlier studies have suggested that, in humans, at least two major metabolites are formed from timolol by cleavage of the morpholine ring. In human liver microsomes, four metabolites are identified, with the most abundant being a hydroxy metabolite (M1), which is formed via hydroxylation into the morpholino ring. The M3 either is a di-hydroxylated timolol or, alternatively, is formed via hydroxylation and oxidation to ketone followed by ring opening by the addition of water molecule. M4 has a structure similar to the substrate with the exception of a double bond in the morpholine ring, which is probably formed by direct cleavage of H_2 or, alternatively, by cleavage of water from M1.

FIGURE 3.70 Metabolism of diltiazem by CYP3A4 and 2D6. Diltiazem undergoes complicated metabolism, including esterase-mediated deacetylation, N-demethylation, and O-demethylation. CYP2D6 and CYP3A4 are involved in O- and N-demethylation of diltiazem, respectively. Esterase-mediated deacetylation of diltiazem results in the first-line metabolite desacetyl diltiazem (M1), which exhibits approximately 50% of the vasodilating properties of the parent drug. Both CYP2D6 and 3A4 are involved in the further metabolism of M1. N- and O-Demethylation of diltiazem produced M2 and M4, respectively.

Pranidipine is a potent and long-acting 1,4-dihydropyridine-type calcium channel blocker used for the treatment of essential hypertension and angina pectoris (Broadhurst et al. 1991; Mori et al. 2001; Rosenthal et al. 1996). The major metabolic routes of pranidipine include dehydrogenation of the 1,4-dihydropyridine to the pyridine ring, hydrolysis of the carboxyl acid ester, hydroxylation of the methyl group accompanied by lactone formation, and glucuronidation of the Phase I metabolites (Figure 3.71) (Fujita et al. 1994). In humans, methyl 2,6-dimethyl-4-(3-nitrophenyl)-3-carboxy-5-pyridinecarboxylate (OPC-13463) is one main metabolite of pranidipine, and the plasma level of this metabolite is approximately fivefold higher than that of pranidipine (Fujita

et al. 1994). The formation of OPC-13463 from pranidipine is considered via two pathways: one is via the formation of (±)-methyl 1,4-dihydro-2,6-dimethyl-4-(3-nitrophenyl)-3-carboxy-5-pyridine carboxylate (MOP-13031) and the other is

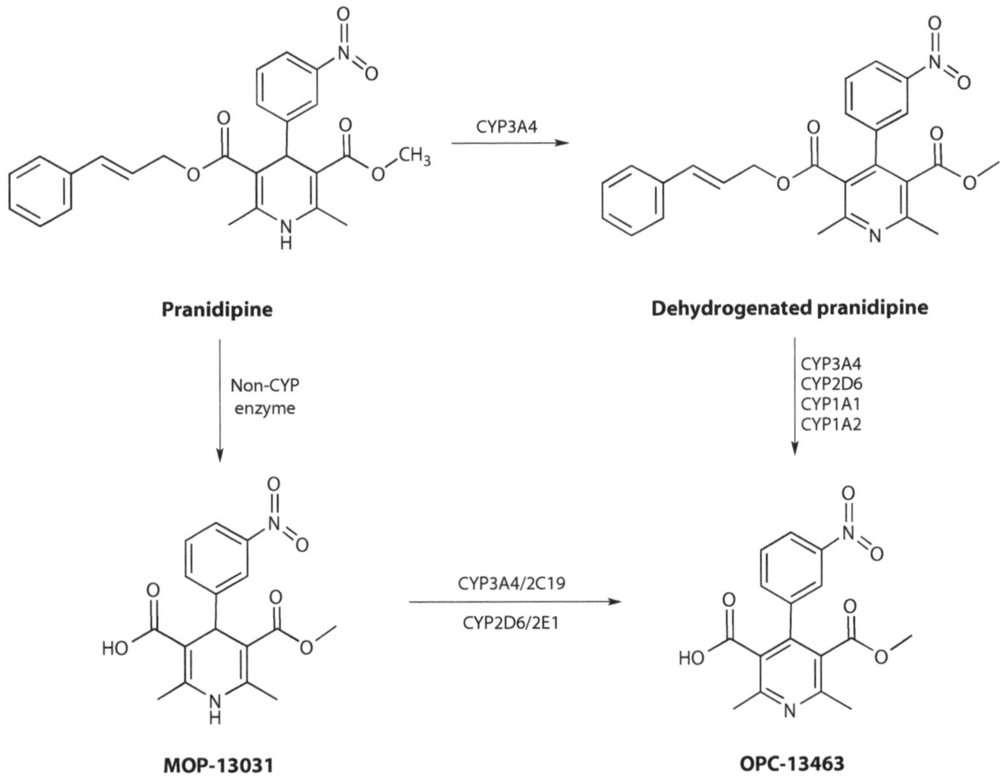

FIGURE 3.71 Metabolism of pranidipine by CYP3A4, 2C19, 2D6, and 2E1. The major metabolic routes of pranidipine include dehydrogenation of the 1,4-dihydropyridine to the pyridine ring, hydrolysis of the carboxyl acid ester, hydroxylation of the methyl group accompanied by lactone formation, and glucuronidation of the Phase I metabolites. In humans, OPC-13463 is one main metabolite of pranidipine, and the plasma level of this metabolite is ~5-fold higher than that of pranidipine. The formation of OPC-13463 from pranidipine is considered via two pathways: one is via the formation of MOP-13031 and the other is via that of dehydrogenated pranidipine. CYP3A4 is the major catalyst for the conversion of pranidipine to dehydrogenated pranidipine and of MOP-13031 to OPC-13463. Recombinant CYP1A1, 1A2, 2D6, and 3A4 cleaved the ester of dehydrogenated pranidipine to form dehydrogenated pranidipine with a catalytic activity of 5.5, 0.93, 13.1, and 22.4 nmol/30 min/nmol P-450, respectively, with CYP2D6 and 3A4 being the major enzymes for the deesterification of dehydrogenated pranidipine.

via that of dehydrogenated pranidipine. CYP3A4 is the major catalyst for the conversion of pranidipine to dehydrogenated pranidipine and of MOP-13031 to OPC-13463 (Kudo et al. 1999). Recombinant CYP1A1, 1A2, 2D6, and 3A4 cleave the ester of dehydrogenated pranidipine to form dehydrogenated pranidipine with a catalytic activity of 5.5, 0.93, 13.1, and 22.4 nmol/30 min/nmol P-450, respectively, with CYP2D6 and 3A4 being the major enzymes for the deesterification of dehydrogenated pranidipine (Kudo et al. 1999).

Verapamil, a calcium antagonist, is commonly used for the treatment of arrhythmias, angina, and hypertension. It undergoes extensive first-pass metabolism, resulting in low drug bioavailability and considerable variability of therapeutic plasma levels. The S-enantiomer of verapamil is stereoselectively cleared during first-pass metabolism at a much faster rate than its antipode (Vogelgesang et al. 1984). Verapamil is initially metabolized into two breakdown products, namely, norverapamil and D-617 [2-(3,4-dimethoxyphenyl)-5-amino-2-isopropylvaleronitrile] (Figure 3.72). These metabolites are subject to further metabolism by multiple CYPs to form additional secondary metabolites. Kroemer et al. (1993) reported

the involvement of CYP3A4 and 1A2 in the N-dealkylation (formation of D-617) and N-demethylation (formation of norverapamil) of verapamil. Further studies have indicated that CYP3A4, 3A5, 2C8, and, to a minor extent, 2E1 are involved in the metabolism of verapamil (Busse et al. 1995; Tracy et al. 1999). CYP2C8 readily metabolizes both S- and R-verapamil to D-617, norverapamil, and D-715 (also known as PR-22). Similarly, CYP3A4, 3A5, and 2C8 also mediate the metabolism of the enantiomers of norverapamil with minor contribution from CYP2D6 and 2E1. CYP3A4 and 3A5 readily form the metabolite D-620 [2-(3,4-dimethoxyphenyl)-5-methylamino-2-isopropylvaleronitrile] with generally a lower K_m and a higher V_{max} for S-norverapamil than for the R-enantiomer. CYP2C8 produces both D-620 and D-715 from the enantiomers of norverapamil, with stereoselective preference for the S-enantiomer. Verapamil is also converted to D-702 [2-(3,4-dimethoxyphenyl)-8-(4-hydroxy-3-methoxyphenyl)-6-methyl-2-isopropyl-6-azaoctanitrile] and O-demethylverapamil (D-703) via O-demethylation. CYP2C9 and 2C18 are involved in formation of the O-demethylation products D-703 and D-702, whereas

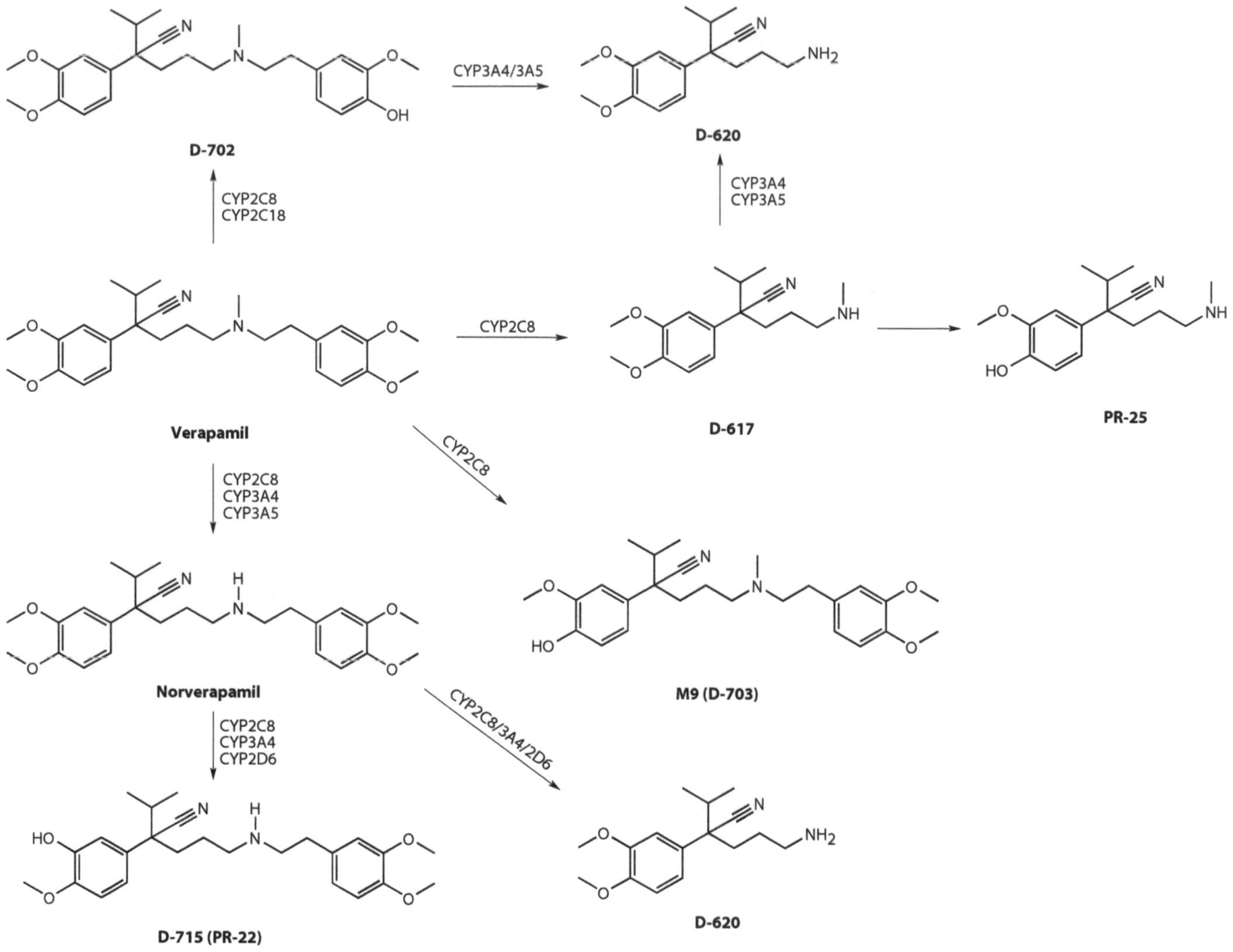

FIGURE 3.72 Metabolism of verapamil by CYP2C8, 3A4, and 2D6. The *S*-enantiomer of verapamil is stereoselectively cleared during first-pass metabolism at a much faster rate than its antipode. Verapamil is initially metabolized into two breakdown products, namely, norverapamil and D-617 [2-(3,4-dimethoxyphenyl)-5-amino-2-isopropylvaleronitrile]. These metabolites are subject to further metabolism by multiple CYPs to form additional secondary metabolites. CYP3A4 and 1A2 are involved in the *N*-dealkylation (formation of D-617) and *N*-demethylation (formation of norverapamil). Further studies indicated that CYP3A4, 3A5, 2C8, and, to a minor extent, 2E1 are involved in the metabolism of verapamil. CYP2C8 readily metabolized both *S*- and *R*-verapamil to D-617, norverapamil, and D-715 (also known as PR-22). Similarly, CYP3A4, 3A5, and 2C8 also mediated the metabolism of the enantiomers of norverapamil with minor contribution from CYP2D6 and 2E1. CYP3A4 and 3A5 readily formed the metabolite D-620 [2-(3,4-dimethoxyphenyl)-5-methylamino-2-isopropylvaleroni-trile]. CYP2C8 produced both D-620 and D-715 from the enantiomers of norverapamil, with stereoselective preference for the *S*-enantiomer. Verapamil is also converted to D-702 [2-(3,4-dimethoxyphenyl)-8-(4-hydroxy-3-methoxyphenyl)-6-methyl-2-isopropyl-6-azaoctanitrile] and *O*-demethylverapamil (D-703) via *O*-demethylation. CYP2C9 and 2C18 are involved in formation of the *O*-demethylation products D-703 and D-702, whereas CYP2C8 selectively forms D-703.

CYP2C8 selectively forms D-703 (Busse et al. 1995). Overall, CYP3A4, 2C8, and 2C9 play a major role in the metabolism of verapamil while CYP2D6 plays only a minor role in tis metabolism.

3.3.3 Antihistamines

Antihistamines are used in the treatment of allergies. First-generation histamine H$_1$ receptor antagonists, such as diphen-hydramine, chlorpheniramine (chlorphenamine), triprolidine,

and hydroxyzine, frequently cause somnolence or other CNS adverse effects. Second-generation antihistamines such as ter-fenadine, astemizole, loratadine, and cetirizine, being more lipophobic, offer the advantages of a lack of CNS and cholin-ergic effects such as sedation and dry mouth (Oppenheimer and Casale 2002). However, terfenadine and astemizole have been withdrawn from the market because both drugs have shown rare but fatal cardiotoxic side effects (e.g., QT interval prolongation). The third-generation antihistamines, metabo-lites of the earlier drugs (e.g., fexofenadine, levocetirizine, and desloratadine), have demonstrated no cardiac effects of the parent drugs and are at least as potent as their parent molecules (Devillier et al. 2008).

CYP2D6 is a major contributor to the oxidation of several antihistamines including loratadine (Yumibe et al. 1995, 1996), promethazine (Nakamura et al. 1996), astemizole (Matsumoto and Yamazoe 2001), mequitazine (Nakamura et al. 1998), terfenadine (Jones et al. 1998), azelastine (Imai et al. 1999; Nakajima et al. 1999), oxatomide (Goto et al. 2004, 2005), epinastine (Kishimoto et al. 1997), diphenhydramine (Akutsu et al. 2007), and chlorpheniramine (He et al. 2002; Yasuda et al. 2002).

3.3.3.1 Azelastine

Azelastine is a long-acting antiallergy and antiasthmatic drug, possessing histamine H_1 receptor blocking activity (Chand and Sofia 1995). It is used for the treatment of hay fever and as eye drops for allergic conjunctivitis. It also exhibits activities such as antagonism of the chemical mediators adenosine, LTC_4, LTD_4, endothelin-1, and platelet activation factor, and inhibition of the generation or release of IL-1β and LTs (Chand and Sofia 1995). Azelastine is metabolized to desmethylazelastine, 6-hydroxyazelatine, 7-oxoazelastine, 4-[(4-chlorophenyl)methyl]-2-(5-methylamino-1-carboxy-2-pentyl)-1(2H)-phthalazinone, and 4-[(4-chlorophenyl)methyl]-2-(5-methylamino-1-carboxy-3-pentyl)-1(2H)-phthalazinone in rats and guinea pigs (Adusumalli et al. 1992; Chand et al. 1993; Tatsumi et al. 1980, 1984). Desmethylazelastine has also been detected in human plasma as a metabolite of azelastine after oral administration (Pivonka et al. 1987). In vitro studies indicate that azelastine is metabolized mainly by CYP3A4 and 2D6 via N-demethylation to desmethylazelastine, with minor contribution from CYP1A2 (Figure 3.73) (Nakajima et al. 1999). The estimated contribution of these three enzymes is 76.6%, 21.8%, and 3.9%, respectively. Desmethylazelastine shows pharmacological activity equivalent to the parent drug. In microsomes from B-lymphoblast cells, CYP2D6, 1A1, 3A4, 2C19, 2C9, and 1A2 catalyze azelastine N-demethylation, while in microsomes from baculovirus-infected insect cells, CYP2D6, 1A1, 1A2, 3A4, 3A5, 2C19, 2C9, and 2C8 catalyze this reaction (Nakajima et al. 1999).

3.3.3.2 Chlorpheniramine

Chlorpheniramine, a propylamine H_1 receptor antagonist available as an over-the-counter drug, is indicated for use in the common cold and for symptomatic treatment of allergies. Chlorpheniramine has a chiral carbon and is usually given as a racemic mixture, and it demonstrates stereoselectivity in its disposition (Yasuda et al. 2002) and pharmacological response (Tran et al. 1978). The S-enantiomer is approximately 100 times more potent than the R-enantiomer with regard to the antihistamine activity (Tran et al. 1978). The metabolism of chlorpheniramine has been extensively studies in animals (Cashman et al. 1992a; Kammerer and Lampe 1987; Nomura et al. 1997; Peets et al. 1972), and it has been found that chlorpheniramine undergoes extensive N-demethylation. Rat CYP2B1 and 2C11 are involved in its N-demethylation (Nomura et al. 1997). Chlorpheniramine can also be converted to its N-oxide by FMO from hog liver (Cashman et al. 1992a). In humans, chlorpheniramine is N-demethylated by CYP2D6 (Figure 3.74). For S-chlorpheniramine, administration of quinidine, an inhibitor of CYP2D6, results in an increase in C_{max}, a reduction in oral clearance, and a prolongation of elimination half-life. Administration of quinidine decreased the oral clearance of R-chlorpheniramine. Stereoselective elimination of chlorpheniramine occurs in humans, with the most pharmacologically active S-enantiomer cleared more slowly than the R-enantiomer. In addition, a difference in receptor occupancy is observed between individuals who are CYP2D6 PMs or EMs (Yasuda et al. 1995).

3.3.3.3 Cinnarizine and Flunarizine

Both cinnarizine and flunarizine are antihistaminic drugs that are mainly used for the control of vomiting caused by motion sickness (Todd and Benfield 1989; Towse 1980). Both drugs act by interfering with the signal transmission between the vestibular apparatus of the inner ear and the vomiting center of the hypothalamus. The disparity of signal processing between inner ear motion receptors and the visual senses is abolished, such that the confusion of the brain as to whether the individual is moving or standing is reduced. Both cinnarizine and flunarizine could be regarded as nootropic drugs because of their vasorelaxing abilities resulting from calcium channel

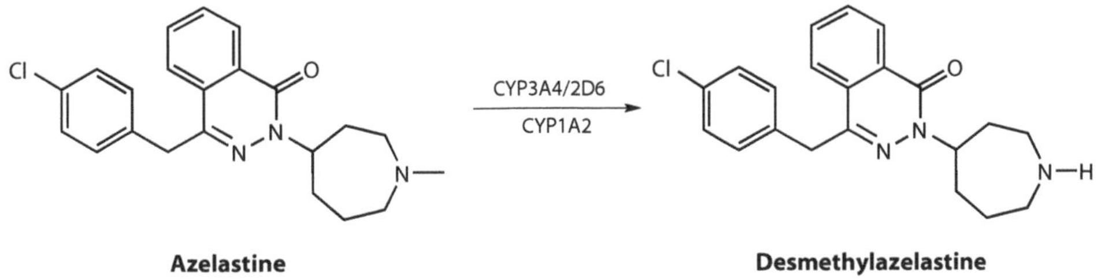

Azelastine   **Desmethylazelastine**

FIGURE 3.73 Metabolism of azelastine by CYP2D6, 3A4, and 1A2. Azelastine is metabolized to desmethylazelastine, 6-hydroxyazelatine, 7-oxoazelastine, 4-[(4-chlorophenyl)methyl]-2-(5-methylamino-1-carboxy-2-pentyl)-1(2H)-phthalazinone, and 4-[(4-chlorophenyl)methyl]-2-(5-methylamino-1-carboxy-3-pentyl)-1(2H)-phthalazinone in rats and guinea pigs. In vitro studies indicate that azelastine is metabolized mainly by CYP3A4 and 2D6 via N-demethylation to desmethylazelastine, with minor contribution from CYP1A2.

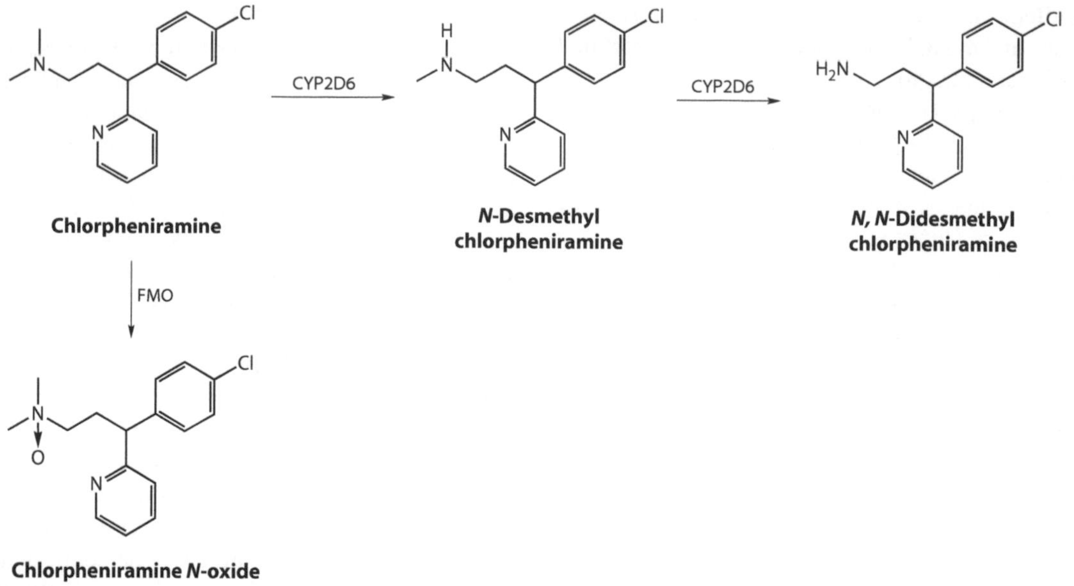

FIGURE 3.74 Metabolism of chlorpheniramine by CYP2D6. In humans, chlorpheniramine undergoes *N*-demethylation by CYP2D6. Stereoselective elimination of chlorpheniramine occurs in humans, with the most pharmacologically active *S*-enantiomer cleared more slowly than the *R*-enantiomer.

blocking activity (Cook and James 1981). The major metabolic pathway in male rats is the oxidative *N*-dealkylation to form bis(4-fluorophenyl)methanol and a number of complementary metabolites of the cinnamylpiperazine moiety, of which hippuric acid is the main one (Meuldermans et al. 1983). In female rats and male dogs, however, hydroxy-flunarizine is the main metabolite, resulting from the aromatic hydroxylation of the phenyl ring of the cinnamyl moiety. Enterohepatic circulation of bis(4-fluorophenyl)methanol and hydroxy-flunarizine is proven by "donor–acceptor" coupling in rats; in bile and urine, these two metabolites are present mainly as glucuronides. The glucuronide of hydroxy-flunarizine is also the main plasma metabolite in dogs (Meuldermans et al. 1983).

Cinnarizine and its fluorine derivative flunarizine are largely metabolized by CYP2D6, with contribution from CYP2B6 (Figures 3.75 and 3.76, respectively) (Kariya et al. 1996; Narimatsu et al. 1993). CYP2D6 mainly catalyzes *p*-hydroxylation of the cinnamyl phenyl rings of cinnarizine and flunarizine, and CYP2B6 mediates that of the diphenylmethyl group of cinnarizine. On the other hand, CYP2C9, 1A1, 1A2, and 2A6 mediate *N*-dealkylation at the 1- and 4-positions of the piperazine ring of the two drugs, resulting in M1 and M3, respectively, whereas CYP2C8, 2C19, 2E1, and 3A4 do not show detectable activity for these reactions (Kariya et al. 1996). The ring-hydroxylase activity of the cinnamyl moiety of cinnarizine and flunarizine in liver microsomes from female DA rats is significantly lower than that from female Wistar rats (Kariya et al. 1992a,b).

3.3.3.4 Diphenhydramine

Diphenhydramine is an over-the-counter antihistamine, antiemetic, sedative, and hypnotic (Banerji et al. 2007). It may also be used for the treatment of extrapyramidal side effects

of typical antipsychotics, such as the tremors caused by haloperidol (McGeer et al. 1961). Diphenhydramine is extensively metabolized by demethylation to *N*-desmethyl diphenhydramine, followed by rapid demethylation to *N*,*N*-didesmethyl diphenhydramine, which is further metabolized by oxidative deamination to diphenylmethoxyacetic acid (Figure 3.77) (Blyden et al. 1986; Chang et al. 1974). These metabolic pathways are thought to be major pathways in humans. CYP1A2, 2C9, and 2C19 are low-affinity catalyzing components for the *N*-demethylation of diphenhydramine, a member of the ethanolamine class of first-generation antihistamine agents, while CYP2D6 catalyzes this reaction as the high-affinity enzyme (Akutsu et al. 2007). In addition, CYP2C18 and 2B6 also play a role at relatively higher substrate concentration (20 μM) (Sharma and Hamelin 2003). In addition, diphenhydramine undergoes direct glucuronidation at its tertiary amino group with formation of a quaternary ammonium glucuronide (Breyer-Pfaff et al. 1997; Fischer and Breyer-Pfaff 1997; Luo et al. 1991).

3.3.3.5 Loratadine

Loratadine is an orally effective, nonsedating, long-acting tricyclic H$_1$ receptor antagonist undergoing extensive metabolism mainly by CYP3A4 and, to a lesser extent, by CYP2D6 to form its major active metabolite desloratadine (descarboethoxyloratadine, M49) (Figure 3.78) (Yumibe et al. 1996). However, the catalytic formation rate is approximately fivefold greater in cDNA-expressed CYP2D6 than in 3A4. Desloratadine is present in plasma at low levels and it is converted to several hydroxylated metabolites, including the active metabolite, 3-OH-desloratadine (M40), and 5-OH- and 6-OH-desloratadine, which are excreted as glucuronides (M13/M7/M7) (Ramanathan et al. 2007b; Yumibe

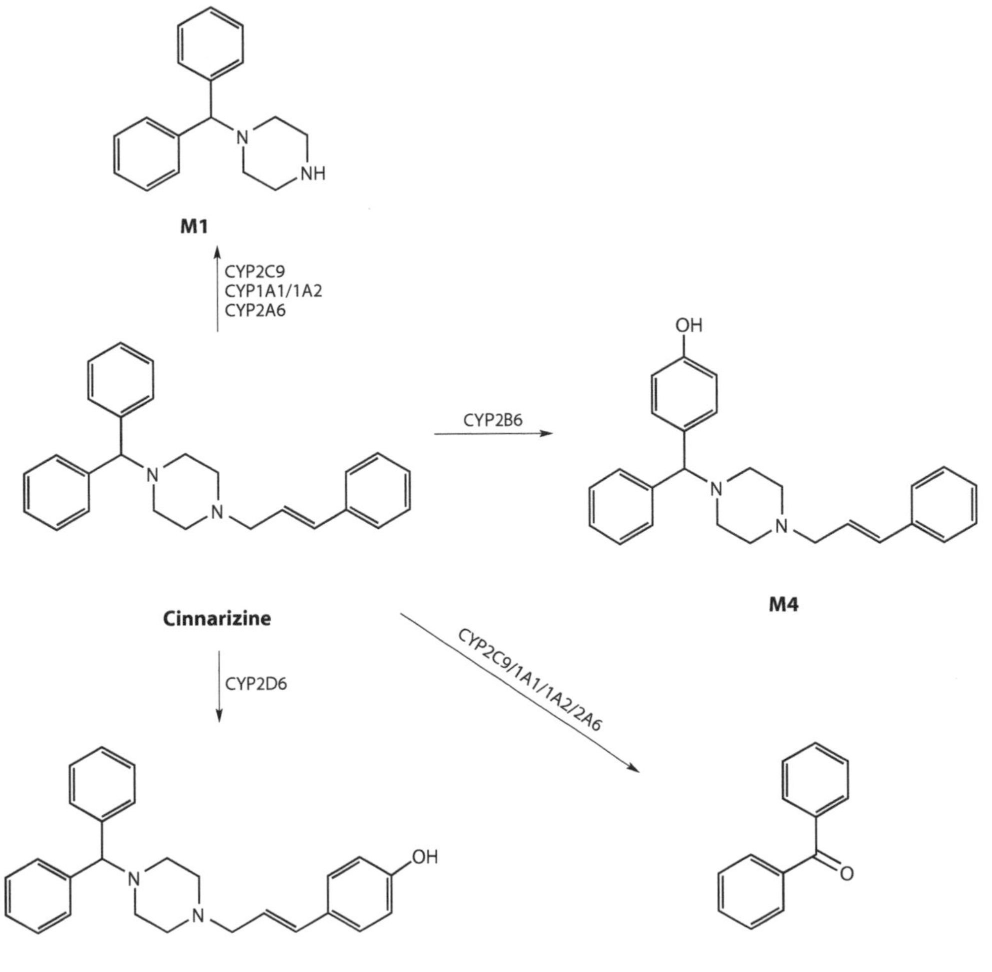

FIGURE 3.75 Metabolism of cinnarizine by CYP2D6 and other CYPs. It is largely metabolized by CYP2D6, with contribution from CYP2B6. CYP2D6 mainly catalyzes *p*-hydroxylation of the cinnamyl phenyl rings of cinnarizine, and CYP2B6 mediated that of the diphenylmethyl group of cinnarizine. On the other hand, CYP2C9 together with CYP1A1, 1A2, and 2A6 mediated *N*-dealkylation at the 1- and 4-positions of the piperazine ring of cinnarizine resulting in M1 and M3, respectively, whereas CYP2C8, 2C19, 2E1, or 3A4 did not show detectable activity for the formation of these metabolites.

et al. 1995). In humans, the major circulating metabolites of loratadine included 3-OH-desloratadine glucuronide (M13), dihydroxy-desloratadine glucuronides (M5/M8), and several metabolites resulting from descarboethoxylation and oxidation of the piperidine ring (Ramanathan et al. 2007a). Loratadine can be hydroxylated at the 3-position and other positions of the piperidine ring that are further conjugated.

3.3.3.6 Mequitazine

Mequitazine is a potent, nonsedative, and long-acting histamine H$_1$ receptor antagonist proven to be a better therapeutic drug than other conventional antihistamines (Fujimura et al. 1981). The oxidative metabolism of mequitazine is extensively studied in rats and dogs (Uzan et al. 1976a,b). The major metabolic routes of mequitazine in rats are the aromatic hydroxylation of the phenothiazine structure and *S*-oxidation and *N*-oxidation of a side chain (Hojo et al. 1981). In human liver microsomes, two metabolites, M1

(*S*-oxide) and M2 (hydroxylated metabolite), are formed by CYP2D6 (Figure 3.79) (Nakamura et al. 1998). Microsomes from human B-lymphoblastoid cells expressing CYP2D6 efficiently metabolize mequitazine to the hydroxylated and *S*-oxidized metabolites (Nakamura et al. 1998).

3.3.3.7 Oxatomide

Oxatomide is a piperazine antihistamine and antiallergic drug used in the treatment of chronic urticaria, skin itching, atopic dermatitis, allergic rhinitis, and bronchial asthma (Hayashi et al. 2003; Richards et al. 1984). Oxatomide is well absorbed and almost completely metabolized. The major metabolic pathways of oxatomide included oxidative *N*-dealkylations at the piperazine nitrogens and at the benzimidazolone nitrogen and aromatic hydroxylation at the benzimidazolone moiety (Figure 3.80) (Meuldermans et al. 1984). The main urinary metabolite in the human is 2,3-dihydro-2-oxo-1H-benzimidazole-1-propanoic acid (M5), resulting from the oxidative

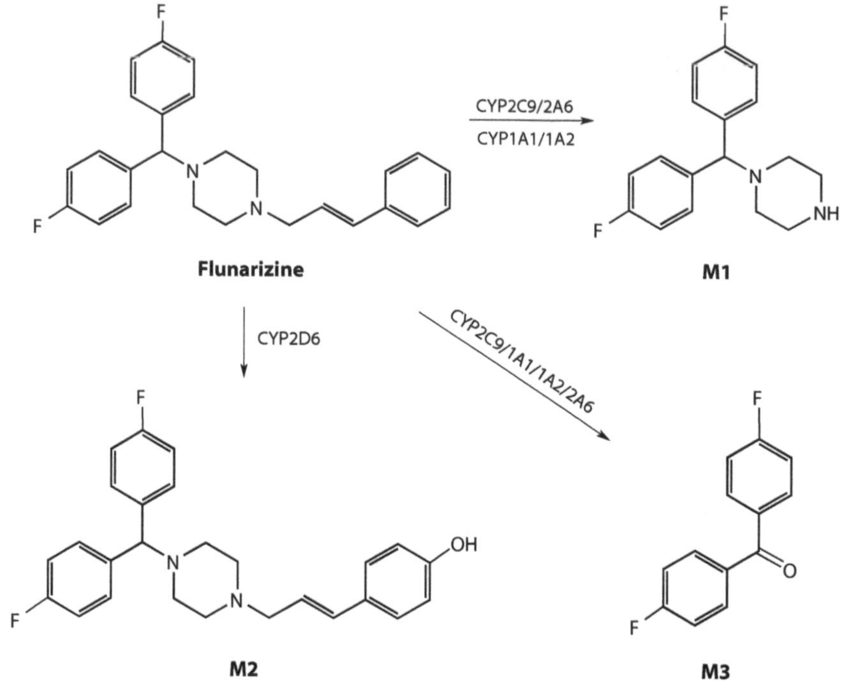

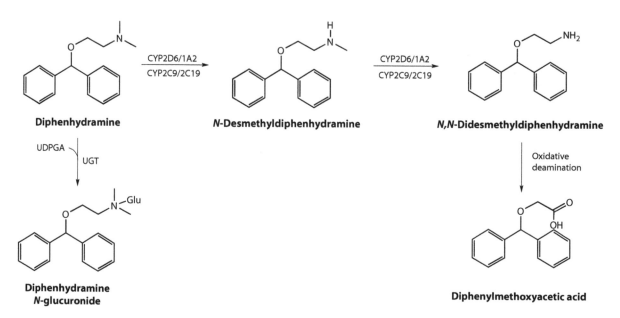

FIGURE 3.76 Metabolism of flunarizine by CYP2D6 and other CYPs. Flunarizine, a fluorine derivative of cinnarizine, is largely metabolized by CYP2D6 and 2B6. CYP2D6 mainly catalyzes *p*-hydroxylation of the cinnamyl phenyl rings of flunarizine. On the other hand, CYP2C9 together with CYP1A1, 1A2, and 2A6 mediated *N*-dealkylation at the 1- and 4-positions of the piperazine ring of flunarizine, resulting in M1 and M3, respectively, whereas CYP2C8, 2C19, 2E1, or 3A4 did not show detectable activity for these reactions.

FIGURE 3.77 Metabolism of diphenhydramine by CYP2D6 and other enzymes. Diphenhydramine is extensively metabolized by demethylation to *N*-desmethyl diphenhydramine, followed by rapid demethylation to *N,N*-didesmethyl diphenhydramine, which is further metabolized by oxidative deamination to diphenylmethoxyacetic acid. These metabolic pathways are thought to be major pathways in humans. CYP1A2, 2C9, and 2C19 are low-affinity catalyzing components for the *N*-demethylation of diphenhydramine, a member of the ethanolamine class of first-generation antihistaminergic agents, while CYP2D6 catalyzes this reaction as the high-affinity enzyme. In addition, CYP2C18 and 2B6 also play a role at relatively higher substrate concentration (20 μM). In addition, diphenhydramine undergoes direct glucuronidation at its tertiary amino group with formation of a quaternary ammonium glucuronide.

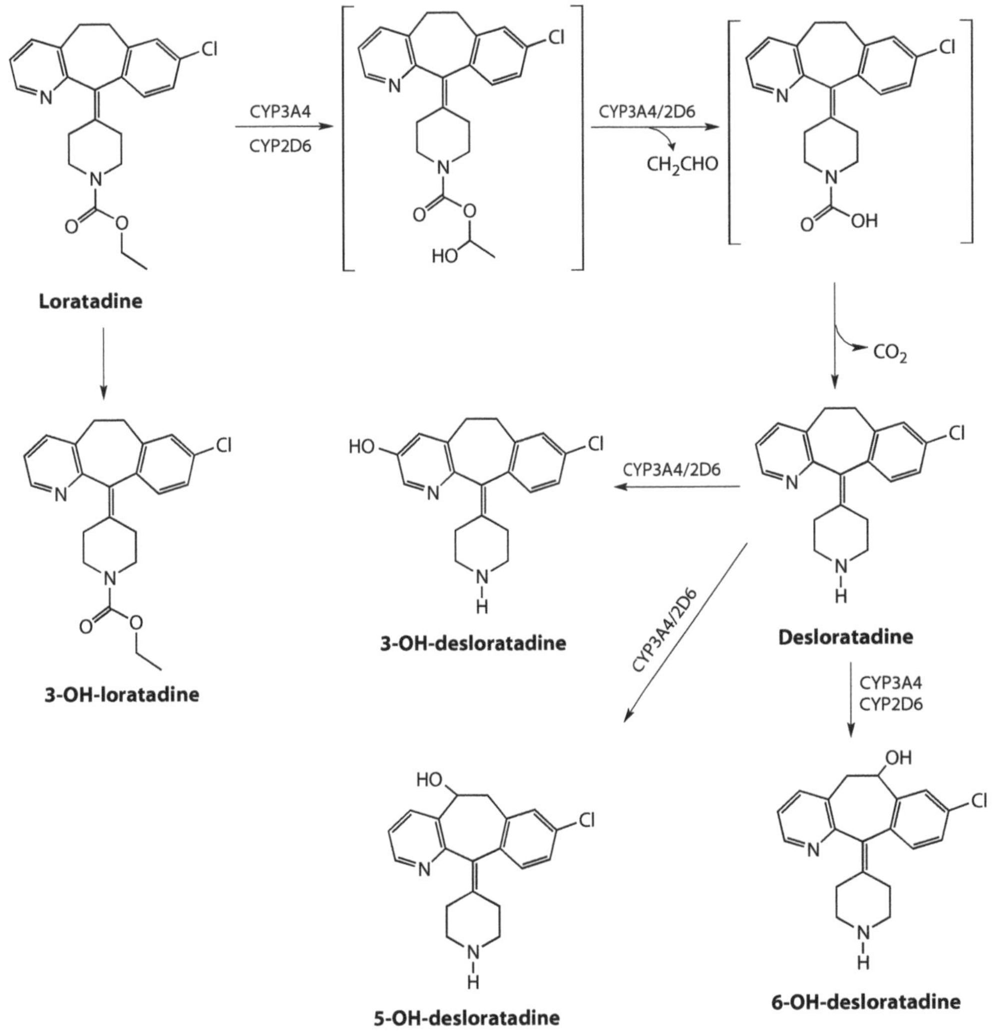

FIGURE 3.78 Metabolism of loratadine. Loratadine undergoes extensive metabolism mainly by CYP3A4 and, to a lesser extent, by CYP2D6 to form its major active metabolite desloratadine (descarboethoxyloratadine, also marketed as a drug). Desloratadine is present in the plasma at low levels and is converted to several hydroxylated metabolites, including the active metabolite, 3-OH-desloratadine (M40), and 5-OH- and 6-OH-desloratadine, which are excreted as glucuronides (M13/M7/M7). In humans, the major circulating metabolites of loratadine included 3-OH-desloratadine glucuronide (M13), dihydroxy-desloratadine glucuronides (M5/M8), and several metabolites resulting from descarboethoxylation and oxidation of the piperidine ring. Loratadine can be hydroxylated at the 3-position and other positions of the piperidine ring, which are further conjugated.

N-dealkylation at the 1-piperazine nitrogen (Meuldermans et al. 1984). Oxatomide is mainly metabolized by CYP2D6 and 3A4 (Goto et al. 2004). Glucuronidation of the hydroxy substituents on the benzimidazolone moiety is the main conjugating pathway. The aromatic hydroxylation is of greater importance in humans than in rats, since the total contribution of the hydroxylated metabolites including M1, M3, M6, M7, M9, M11, M12, and probably M4 and M10 is more abundant in humans (Meuldermans et al. 1984). Aromatic hydroxylation of oxatomide occurs mainly at the 6- but also at the 5-position of the benzimidazolone moiety. Oxidative N-dealkylation of oxatomide at the benzimidazolone nitrogen is more abundant in humans than in rats. This pathway together with aromatic hydroxylation results in the hydroxybenzimidazolone metabolite M3 and its glucuronide M1. Oxidative N-dealkylation at the 4-piperazine nitrogen is much more abundant in rats

than in humans, resulting in M8 (Meuldermans et al. 1984). In addition, oxidative N-dealkylation at the 1-piperazine nitrogen is a major metabolic pathway in rats, dogs, and humans, resulting in the acid metabolite M5, which is the main urinary metabolite (Meuldermans et al. 1984). Notably, this metabolite is also the major urinary metabolite of the structurally related drug domperidone (Meuldermans et al. 1981).

3.3.3.8 Terfenadine

Terfenadine, a nonsedating H_1 receptor antagonist, is used for the treatment of allergic conditions such as allergic rhinitis but withdrawn from the market because of fatal cardiotoxicity (e.g., *torsade de pointes*, brought about by QT prolongation and ventricular arrhythmias) (Honig et al. 1993). After oral administration, terfenadine is well absorbed and undergoes extensive first-pass metabolism in humans. It is mainly

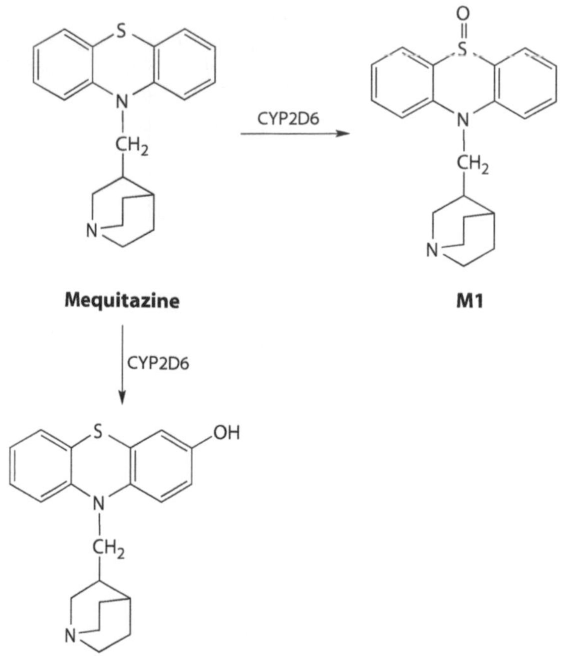

FIGURE 3.79 Metabolism of mequitazine by CYP2D6. In human liver microsomes, two metabolites, M1 (*S*-oxide) and M2 (hydroxylated metabolite), are formed by CYP2D6.

metabolized by *N*-dealkylation to azacyclonol and hydroxylation of the *t*-butyl group to hydroxyterfenadine (Figure 3.81) (Garteiz et al. 1982). Hydroxyterfenadine is further oxidized to the corresponding carboxylic acid (carboxyterfenadine; marketed as fexofenadine), which is the biologically active antihistamine (von Moltke et al. 1994). CYP3A4 is the principal enzyme responsible for the *N*-dealkylation of terfenadine to form azacyclonol and hydroxyterfenadine (Yun et al. 1993). As a lipophilic arylalkylamine, terfenadine is considered to interact with CYP2D6, either as a substrate or as an inhibitor (Smith and Jones 1992). Substrate overlays indicate that terfenadine contains a basic nitrogen atom and the site of metabolism in a spatial orientation would facilitate binding to CYP2D6, by comparison with dextromethorphan. When terfenadine is docked into a homology model of CYP2D6 with the *t*-butyl group oriented close to the heme to allow metabolism, the basic nitrogen is able to interact with Asp301, an amino acid residue critical for the ion-pair interaction (Jones et al. 1998). Amino acids that appeared to be in direct contact with the diphenyl-4-piperidinemethanol group in this model included Ala300, Leu248, Phe247, Leu208, Gly113, and Pro114 (Jones et al. 1998). Terfenadine is metabolized to hydroxyterfenadine and azacyclonol mainly by CYP3A4 and 2D6 in human liver microsomes (Jones et al. 1998). In recombinant enzymes, only CYP2D6 and 3A4 result in hydroxyterfenadine. In addition to hydroxyterfenadine, the recombinant CYP3A4 also forms significant amounts of azacyclonol. Only recombinant CYP3A4, but not 2D6, metabolizes hydroxyterfenadine to azacyclonol and carboxyterfenadine (Jones et al. 1998).

3.3.4 Anti-HIV Agents

CYP2D6 is also involved in the metabolism of several anti-HIV agents, including ritonavir (Kumar et al. 1996), nevirapine (Erickson et al. 1999), and delavirdine (Voorman et al. 1998a). Other protease inhibitors (indinavir and saquinavir) are metabolized primarily by CYP3A4/5 and, to a lesser extent, by CYP2D6 (Chiba et al. 1996; Fitzsimmons and Collins 1997). Nelfinavir is mainly metabolized by CYP3A4 and 2C19 (Hirani et al. 2004). CYP3A4 and 2B6 are the major enzymes responsible for the metabolism of efavirenz, a nonnucleoside reverse transcriptase inhibitor that is used as part of highly active antiretroviral therapy for the treatment of HIV infection, with lesser contribution from CYP1A2 (Ward et al. 2003). Saquinavir is mainly metabolized by CYP3A4. CYP3A4 and, to a lesser extent, 3A5 are major contributors for the metabolism of amprenavir (Treluyer et al. 2003).

3.3.4.1 HIV Protease Inhibitors

Indinavir and ritonavir are anti-HIV agents that selectively inhibit the HIV-1 protease. In healthy subjects, at least seven metabolites of indinavir are identified in the urine and feces (Balani et al. 1996). These included low levels of quaternary pyridine *N*-glucuronide (M1), 2′,3′-*trans*-dihydroxyindanylpyridine *N*-oxide (M2), 2′,3′-*trans*-dihydroxyindan (M3), and pyridine *N*-oxide (M4a) analogs, and despyridylmethyl analogs of M3 (M5) and indinavir (M6) (Figure 3.82). M5 and M6 are the major metabolites in urine. The metabolic profile in plasma is similar to that in urine. The metabolites in feces accounted for >47% of the dose, while the urinary excretion is only ~19%. In the feces, radioactivity is predominantly attributed to M3, M5, M6, and the parent drug. Thus, in urine and feces, the prominent metabolic pathways are oxidation and oxidative *N*-dealkylation (Balani et al. 1996). In vitro studies have identified seven metabolites of indinavir in incubations with human liver microsomes and six metabolites in incubations with recombinant CYP3A4, whereas only three metabolites appear in incubations with CYP3A5 (Koudriakova et al. 1998). Metabolites produced in incubations of indinavir with human liver microsomes result from oxidation of the indanyl, 1,1-dimethylethylaminocarbonylpiperazinyl, piperidinyl and 1,1-dimethylethylaminocarbonylpiperazinyl, phenylmethyl and dimethylethylaminocarbonylpiperazinyl, and 1,1-dimethylethylaminocarbonyl and indanyl moieties (Koudriakova et al. 1998). CYP2D6 catalyzes the oxidation of indanyl group only (M3), while CYP3A4 catalyzes the formation of M2–M5. CYP3A5 fails to catalyze the oxidation of piperidinyl and 1,1-dimethylethylaminocarbonylpiperazinyl groups, phenylmethyl and 1,1-dimethylethylaminocarbonylpiperazinyl groups, and 1,1-dimethylethylaminocarbonyl and indanyl moieties (Koudriakova et al. 1998).

Kumar et al. (1996) have identified three metabolites when ritonavir is incubated with human liver microsomes: M1, M2, and M11, resulting from decarbamoylation, hydroxylation of the isopropyl side chain, and *N*-dealkylation, respectively (Figure 3.83). Similarly, Koudriakova et al. (1998) have also identified these major metabolites of ritonavir

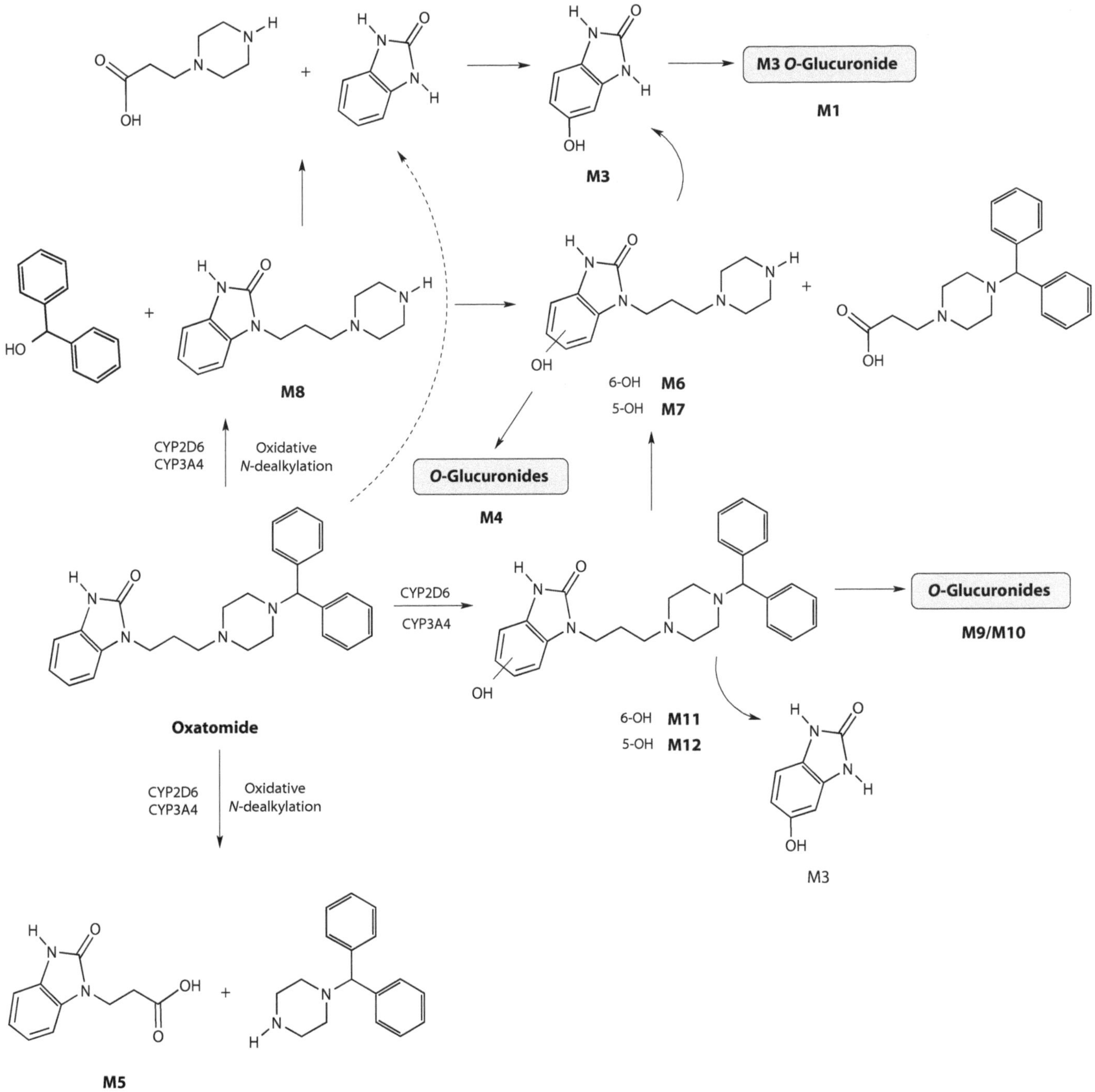

FIGURE 3.80 Metabolism of oxatomide by CYP2D6 and 3A4. The major oxatomide metabolic pathways included oxidative *N*-dealkylations and aromatic hydroxylation as well. The main urinary metabolite in the human is 2,3-dihydro-2-oxo-1H-benzimidazole-1-propanoic acid (M5), resulting from the oxidative *N*-dealkylation at the 1-piperazine nitrogen. Oxatomide is mainly metabolized by CYP2D6 and 3A4. Glucuronidation of the hydroxy substituents on the benzimidazolone moiety is the main conjugating pathway. The aromatic hydroxylation is of greater importance in humans than in rats, since the total contribution of the hydroxylated metabolites including M1, M3, M6, M7, M9, M11, M12, and probably M4 and M10 as well is more abundant in humans. Aromatic hydroxylation of oxatomide occurred mainly at the 6- but also at the 5-position of the benzimidazolone moiety. Oxidative *N*-dealkylation of oxatomide at the benzimidazolone nitrogen is more abundant in humans than in rats. This pathway together with aromatic hydroxylation results in the hydroxybenzimidazolone metabolite M3 and its glucuronide M1. Oxidative *N*-dealkylation at the 4-piperazine nitrogen is much more abundant in rats than in humans, resulting in M8. In addition, oxidative *N*-dealkylation at the 1-piperazine nitrogen is a major metabolic pathway in rats, dogs, and humans, resulting in the acid metabolite M5, which is the main urinary metabolite.

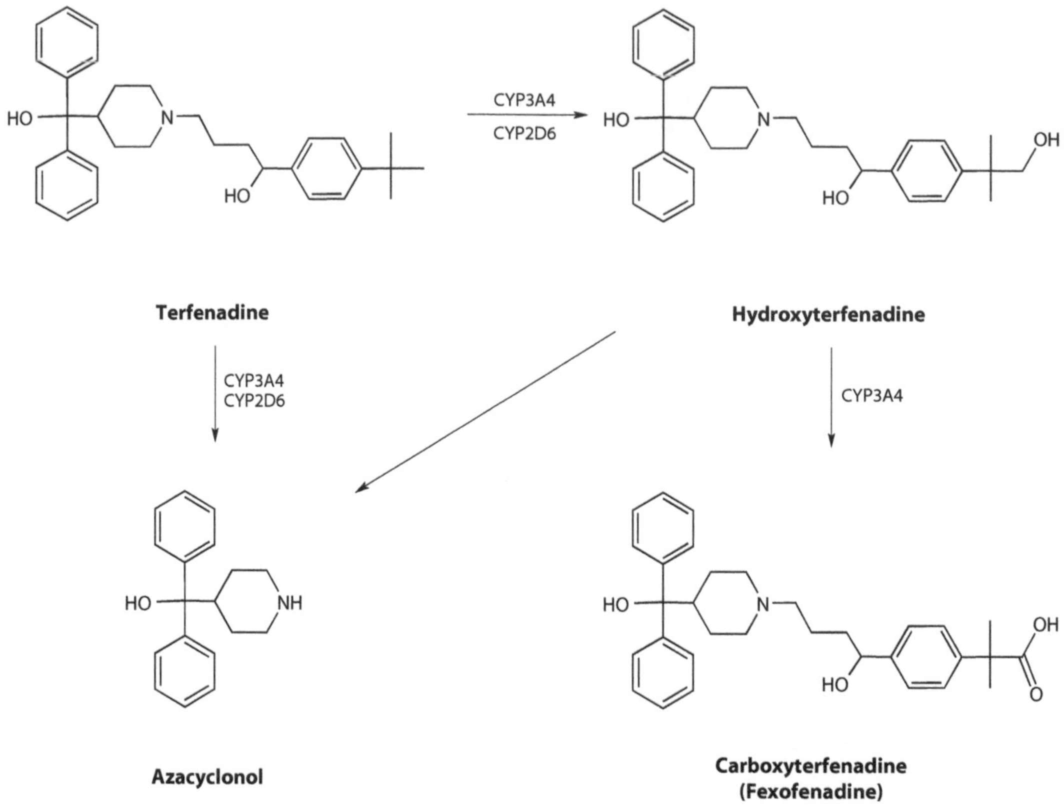

FIGURE 3.81 Metabolism of terfenadine by CYP3A4 and 2D6. After oral administration, terfenadine is well absorbed and undergoes extensive first-pass metabolism in humans. It is mainly metabolized by *N*-dealkylation to azacyclonol and hydroxylation of the *t*-butyl group to hydroxyterfenadine. Hydroxyterfenadine is further oxidized to the corresponding carboxylic acid (carboxyterfenadine; marketed as fexofenadine), which is the biologically active antihistamine. CYP3A4 is the principal enzyme responsible for the *N*-dealkylation of terfenadine to form azacyclonol and hydroxyterfenadine. As a lipophilic arylalkylamine, terfenadine is considered to interact with CYP2D6, either as a substrate or as an inhibitor. Terfenadine is metabolized to hydroxyterfenadine and azacyclonol mainly by CYP3A4 and 2D6 in human liver microsomes. In recombinant enzymes, only CYP2D6 and 3A4 result in hydroxyterfenadine. In addition to hydroxyterfenadine, the recombinant CYP3A4 also forms significant amounts of azacyclonol. Only recombinant CYP3A4, but not 2D6, metabolized hydroxyterfenadine to azacyclonol and carboxyterfenadine.

from incubations with human liver microsomes except a new product: *N*-demethylated ritonavir (M3). Ritonavir is mainly metabolized by CYP3A4/3A5 and, to a lesser extent, by 2D6 (Kumar et al. 1996). CYP3A4/3A5 forms M1, M2, M3, and M11, while CYP2D6 only forms M3. Ritonavir does not contain a basic nitrogen atom that is always present in classical substrates of CYP2D6. Ritonavir has two amide linkages and a carbamate, and these might be capable of hydrogen bonding with Asp301 or another entity of CYP2D6. Ritonavir is also a potent inhibitor of CYP3A4 and also inhibits 2D6 to a lesser extent. Ritonavir inhibits CYP2D6-mediated metabolism, causing a 145% increase in desipramine AUC (von Moltke et al. 1998). Ritonavir also appears to induce CYP3A as well as other enzymes, including UGTs, CYP1A2, and possibly 2C9 (Yeh et al. 2006).

Saquinavir is a potent HIV-1 and HIV-2 protease inhibitor with an IC_{90} of 20 nM used in the treatment of HIV-1 infection. Fitzsimmons and Collins (1997) have shown that saquinavir is oxidized by both human hepatic and small-intestinal microsomes to two major metabolites (M2 and M7) and five minor metabolites (M1, M3, M4, M5, and M6), with CYP3A4 being the predominant enzyme for its metabolism (Figure 3.84).

M2, M3, and M7 are identified as monohydroxylated products resulting from hydroxylation on the octahydro-2-(1*H*)-iso-quinolinyl (M2 and M3) and (1,1-dimethylethyl)amino (M7) groups. M4, M5, and M6 are identified as monohydroxylated products, but the site of oxidation is yet to be determined (Fitzsimmons and Collins 1997). M7 is further oxidized to the dihydroxylated metabolite M1. Recombinant CYP3A4 forms all metabolites, while incubation with recombinant CYP2D6 showed the formation of an unidentified metabolite (Fitzsimmons and Collins 1997). Eagling et al. (2002) have identified 11 metabolites from incubation of saquinavir with human liver microsomes, most of which are mono- or dihydroxylated metabolites. CYP3A4 is the predominant enzyme for the formation of all these metabolites. They also identified one of the metabolites as 6-equatorial-hydroxysaquinavir.

3.3.4.2 Nonnucleoside Reverse Transcriptase Inhibitors

Delavirdine is a potent and selective nonnucleoside reverse transcriptase inhibitor (Dueweke et al. 1993) used in the treatment of HIV-1 infection. The major metabolic pathway of delavirdine in rats involves *N*-dealkylation, pyridine

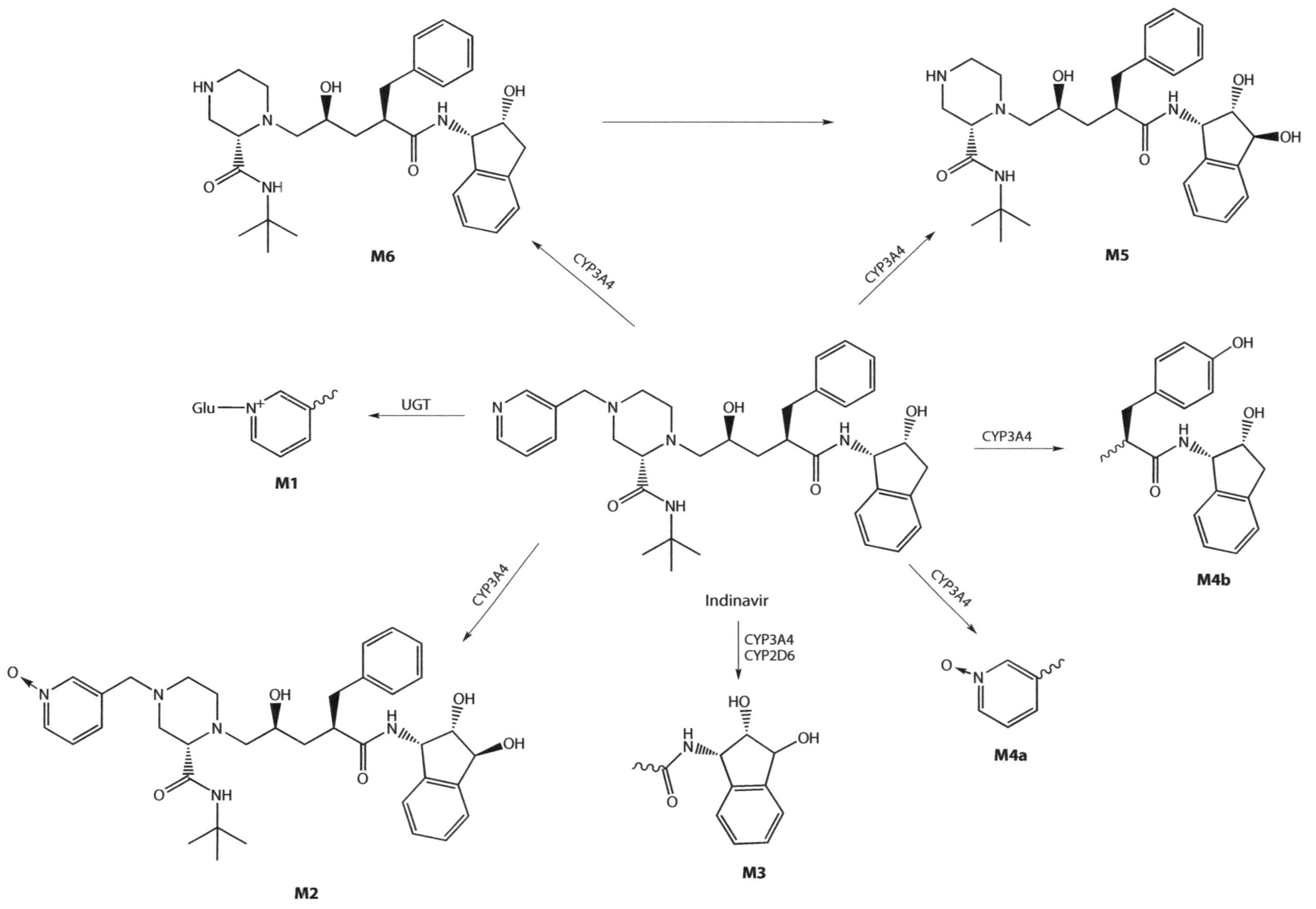

FIGURE 3.82 Metabolism of indinavir by CYP3A4, 3A5, and 2D6. In healthy subjects, at least seven metabolites of indinavir are identified in the urine and feces using HPLC radioactivity and LC-MS/MS analyses. These included low levels of quaternary pyridine *N*-glucuronide (M1), 2′,3′-*trans*-dihydroxyindanylpyridine *N*-oxide (M2), 2′,3′-*trans*-dihydroxyindan (M3) and pyridine *N*-oxide (M4a) analogs, and despyridylmethyl analogs of M3 (M5) and indinavir (M6). M5 and M6 are the major metabolites in human urine. CYP2D6 catalyzes the oxidation of indanyl group only (M3), while CYP3A4 almost catalyzes the formation of M2–M6.

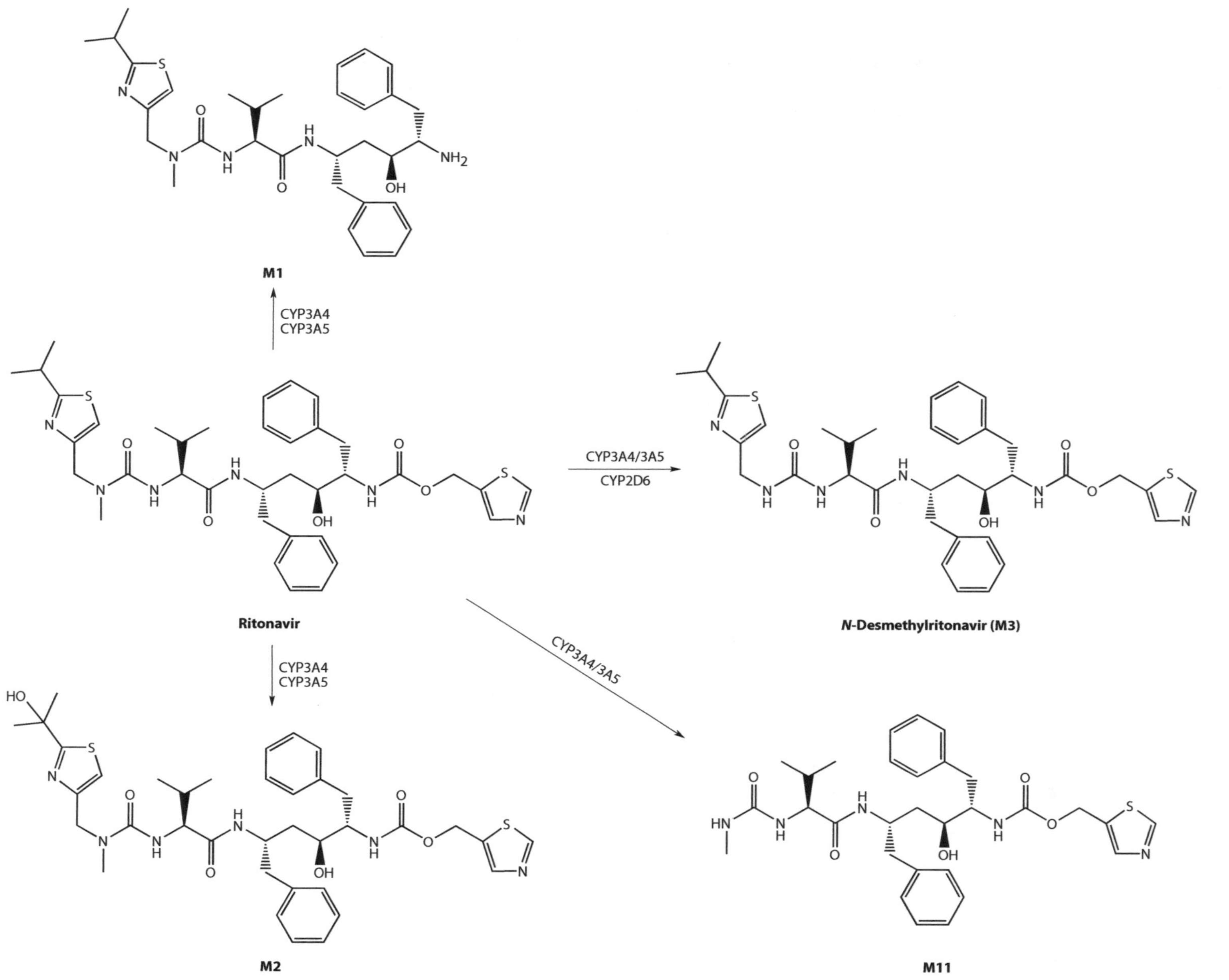

FIGURE 3.83 Metabolism of ritonavir by CYP3A4/3A5 and 2D6. When ritonavir is incubated with human liver microsomes, M1, M2, M3, and M11 are identified, resulting from the decarbamoylation, hydroxylation of the isopropyl side chain, N-demethylation, and N-dealkylation, respectively. CYP3A4/3A5 forms M1, M2, M3, and M11, while CYP2D6 only forms M3. Ritonavir is also a potent inhibitor of CYP3A4 and also inhibits CYP2D6 to a lesser extent.

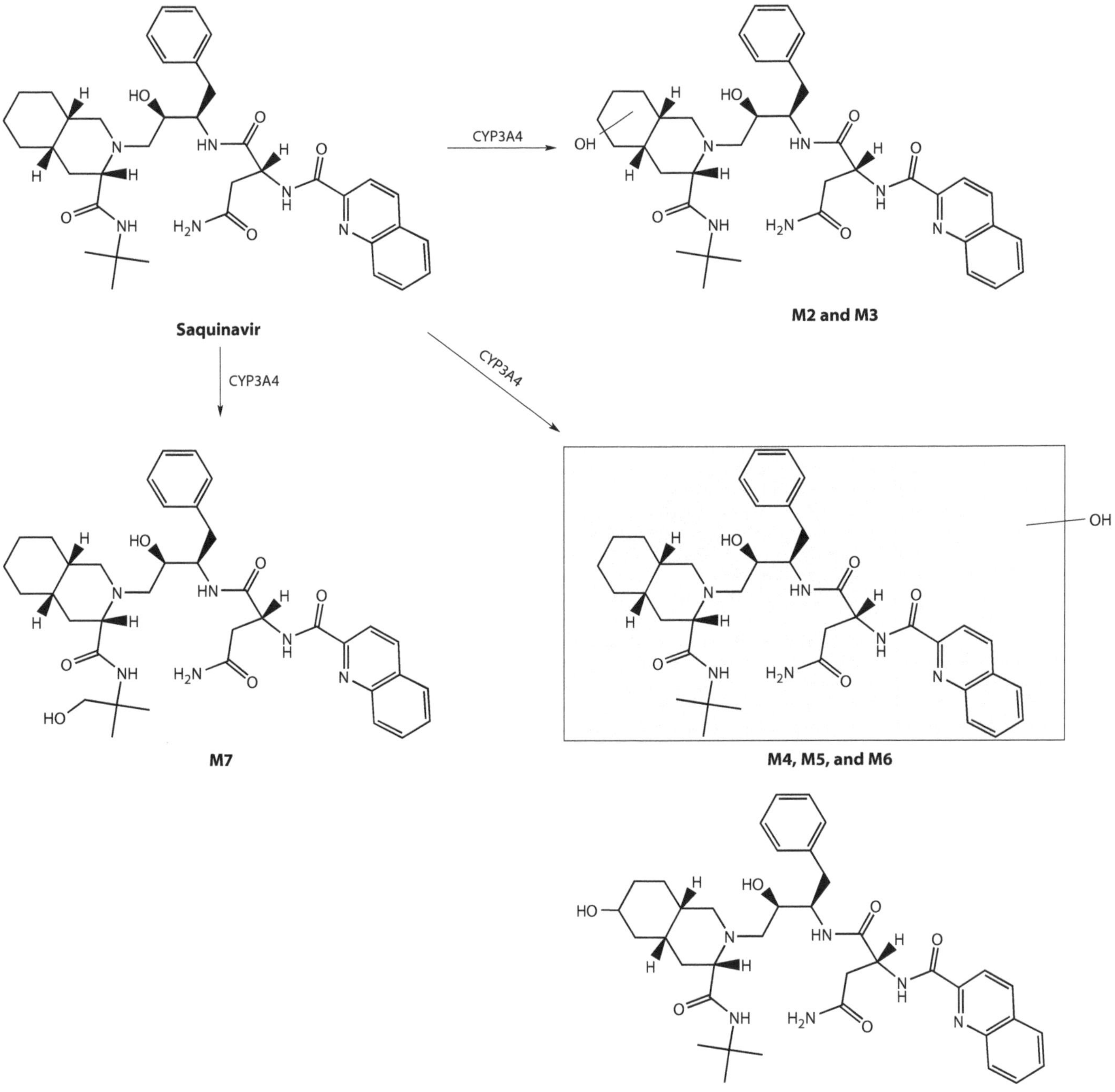

FIGURE 3.84 Metabolism of saquinavir by CYP3A4/5 and 2D6. Saquinavir is oxidized by both human hepatic and small-intestinal microsomes to two major metabolites (M2 and M7) and five minor metabolites (M1, M3, M4, M5, and M6), with CYP3A4 being the predominant enzyme involved in its metabolism. M2, M3, and M7 are identified as monohydroxylated products resulting from hydroxylation on the octahydro-2-(1*H*)-isoquinolinyl (M2 and M3) and (1,1-dimethylethyl)amino (M7) groups. M4, M5, and M6 are identified as monohydroxylated products, but the site of oxidation is yet to be determined. M7 is further oxidized to the dihydroxylated metabolite M1. Recombinant CYP3A4 forms all metabolites, while incubation with recombinant CYP2D6 showed the formation of an unidentified metabolite. One of the metabolites is identified as 6-equatorial-hydroxysaquinavir.

ring 6′-hydroxylation, pyridine ring cleavage, and amide bond cleavage (Figure 3.85) (Chang et al. 1997). Desalkyl delavirdine is the major metabolite in rat urine and plasma, which could be further hydroxylated or sulfated to form M15d and M4d, respectively. Cleavage of the amide bond in delavirdine to give *N*-isopropylpyridinepiperazine (M12)

and indole carboxylic acid (M10) constituted a minor pathway. Degradation of 6′-hydroxy delavirdine (M7) generated despyridinyl delavirdine (M2) and the pyridine ring–opened product (M14). M7 also underwent sulfation or glucuronidation to form M8 and M6, respectively. The major metabolite in rat bile is M6, whereas M7 and M8 are observed as

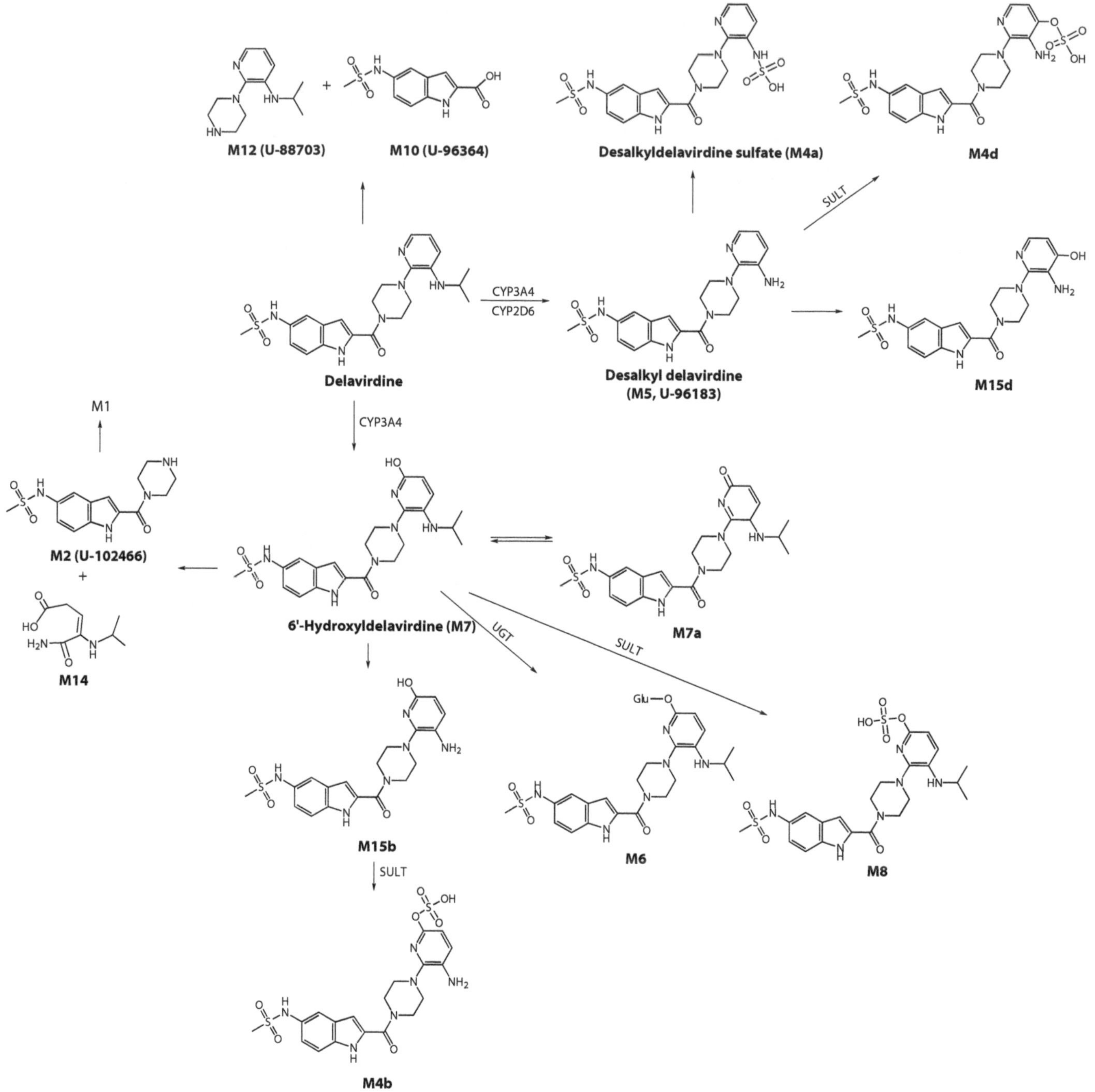

FIGURE 3.85 Proposed metabolic scheme of delavirdine. The metabolic pathway of delavirdine in rats involved *N*-desalkylation, pyridine ring 6′-hydroxylation, pyridine ring cleavage, and amide bond cleavage. Desalkyl delavirdine (M5) is the major metabolite in rat urine and plasma, which could be hydroxylated or sulfated to form M15d and M4d, respectively. Cleavage of the amide bond in delavirdine to give *N*-isopropylpyridinepiperazine (M12) and indole carboxylic acid (M10) constituted a minor pathway. Degradation of 6′-hydroxy delavirdine (M7) generated despyridinyl delavirdine (M2) and the pyridine ring–opened product (M14). M7 also underwent sulfation or glucuronidation to form M8 and M6, respectively. The major metabolite in rat bile is M6, whereas M7 and M8 are observed as minor metabolites. In human liver microsomes, at least four metabolites are formed from delavirdine: desalkyl derivative (M5), M2 (despyridinyl delavirdine), M7, and M7a. M7a and M7 may be tautomers, with M7 representing the enol form, which slowly converts to M7a, the keto form. Delavirdine is mainly metabolized by CYP3A4 and, to a lesser extent, by CYP2D6. CYP2D6 and 3A4 exhibited significant desalkylation of delavirdine, while desalkyl delavirdine is also catalyzed by CYP2C8 and 3A5 at a low activity.

minor metabolites. In human liver microsomes, at least four metabolites are formed from delavirdine: desalkyl derivative (M5), despyridinyl delavirdine (M2), M7, and M7a (Figure 3.85) (Voorman et al. 1998b). Desalkyl delavirdine is not active against HIV-1 reverse transcriptase (Genin et al. 1996). Since 6'-O-glucuronyl delavirdine (M6) is a major metabolite of delavirdine in rats (Chang et al. 1997), M7 could be the 6'-hydroxylated metabolite of delavirdine and M7a and M7 are tautomers, with M7 representing the enol form, which slowly converts to M7a, the keto form. Delavirdine is mainly metabolized by CYP3A4 and, to a lesser extent, by 2D6 (Voorman et al. 1998b). Only CYP2D6 and 3A4 exhibit significant desalkylation of delavirdine; incubation of delavirdine with increasing concentrations of CYP3A4 shows increased formation of dealkylated delavirdine as well as the formation of M7 and M7a. Desalkyl delavirdine is also catalyzed by CYP2C8 and 3A5 at a low activity.

Nevirapine is a nonnucleoside reverse transcriptase inhibitor used to treat HIV-1 infection (Freimuth 1996). It has been approved by the US FDA for use in combination with nucleoside reverse transcriptase inhibitors such as zidovudine and didanosine or protease inhibitors (e.g., saquinavir, ritonavir, and indinavir) (Milinkovic and Martinez 2004). Riska et al. (1999) have found that nevirapine is eliminated primarily (>80%) in the urine as glucuronides of 2-, 3-, 8-, and 12-hydroxynevirapine (Figure 3.86). Only a small fraction (<5%) of the radioactivity in urine (representing <3% of the total dose) is made up of parent compound; therefore, renal excretion plays a minor role in the elimination of the parent compound. In human liver microsomes, nevirapine is oxidized into at least four metabolites: 2-, 3-, 8-, and 12-hydroxynevirapine (Erickson et al. 1999; Ward et al. 2003). In studies with cDNA-expressed human CYPs, 2- and 3-hydroxynevirapine are exclusively formed by CYP3A4/3A5 and 2B6, respectively (Erickson et al. 1999). Multiple cDNA-expressed CYPs form 8- and 12-hydroxynevirapine, but CYP2D6 and 3A4 show the highest activity, respectively (Erickson et al. 1999). For the formation of 12-OH-nevirapine, CYP3A5, 2D6, and 2C9 (minor) also play a role, while CYP2A6, 2B6, and 3A4 catalyze 8-OH-nevirapine formation with a low activity.

3.3.5 ANTIMALARIAL DRUGS

3.3.5.1 Amodiaquine
Amodiaquine is a 4-aminoquinoline derivative that has been widely used for treatment of malaria and is more active than the other 4-aminoquinoline, chloroquine, against *Plasmodium falciparum* parasites, which are moderately chloroquine resistant. Upon oral administration, amodiaquine is rapidly absorbed and extensively metabolized such that very little of the parent drug is detected in the blood. The main metabolite of amodiaquine is N-desethylamodiaquine with other minor metabolites being 2-hydroxyl-N-desethylamodiaquine and N-bis-desethylamodiaquine (Figure 3.87) (Churchill et al. 1985, 1986; Mount et al. 1986). The formation of N-desethylamodiaquine is rapid, while its elimination

is very slow with a terminal half-life of more than 100 h (Laurent et al. 1993; Winstanley et al. 1987). Both amodiaquine and N-desethylamodiaquine have antimalarial activity, but amodiaquine is three times more active. Amodiaquine N-deethylation has been used as an alternative marker reaction for CYP2C8 because of its high affinity and high turnover rate ($K_m = 1.0$ μM; $V_{max} = 2.6$ pmol/min/pmol CYP2C8) (Li et al. 2002). N-Desethylamodiaquine is found to be CYP2C8 selective in the liver with a minor amount being formed by CYP2D6 and extrahepatic enzymes CYP1A1 and 1B1 (Li et al. 2002).

3.3.5.2 Chloroquine
Chloroquine is an antimalarial agent used in the treatment and prophylaxis of malaria since the 1950s. It is also used in the management of rheumatoid arthritis with maintenance doses being 7- to 20-fold greater than those as an antimalarial drug (Rintelen et al. 2006). In humans, the liver and kidneys contribute approximately equally to its body disposition (Ette et al. 1989; Frisk-Holmberg et al. 1983; Gustafsson et al. 1983). After oral or intravenous administration, chloroquine is dealkylated into two main pharmacologically active metabolites: N-desethylchloroquine and N-bis-desethylchloroquine (Figure 3.88). Blood concentrations of the S-isomer of N-desethylchloroquine always exceeded those of the R-isomer, pointing to a preferential metabolism of S-chloroquine (Ducharme and Farinotti 1996). N-Desethylchloroquine is rapidly detected in blood, and its concentrations amount to 20% to 50% of those of the parent drug, while the blood concentrations of N-bis-desethylchloroquine are usually <10% to 15% of chloroquine levels (Ducharme and Farinotti 1996; Gustafsson et al. 1983). Both chloroquine and desethylchloroquine concentrations decline slowly, with elimination half-lives of 20 to 60 days. Both parent drug and metabolite can be detected in urine months after a single dose. Recombinant CYP1A1, 2D6, 3A4, and 2C8 catalyze chloroquine N-deethylation (Figure 3.88) (Projean et al. 2003). CYP2C8 and 3A4 constitute low-affinity and high-capacity systems, whereas CYP2D6 is of a high affinity with a significantly low capacity (Projean et al. 2003). At therapeutically relevant concentrations (~100 μM in the liver), CYP2C8, 3A4, and, to a much lesser extent, 2D6 are expected to account for most of the N-desethylation of chloroquine.

3.3.5.3 Halofantrine
Halofantrine is an antimalarial drug used in the treatment of chloroquine-resistant strains of *P. falciparum*. Halofantrine is chiral and is administered as a racemate. The pharmacokinetics of halofantrine are stereoselective in humans, with the (+)-enantiomer attaining higher plasma concentrations than antipode (Karbwang and Na Bangchang 1994; Warhurst 1987). The $t_{1/2\beta}$ of halofantrine is 2–5 days in patients with malaria. Halofantrine biotransforms in the liver to its major metabolite, N-desbutylhalofantrine (Figure 3.89), which is the major circulating metabolite and is equipotent to halofantrine in antimalarial activity (Karbwang and Na Bangchang 1994).

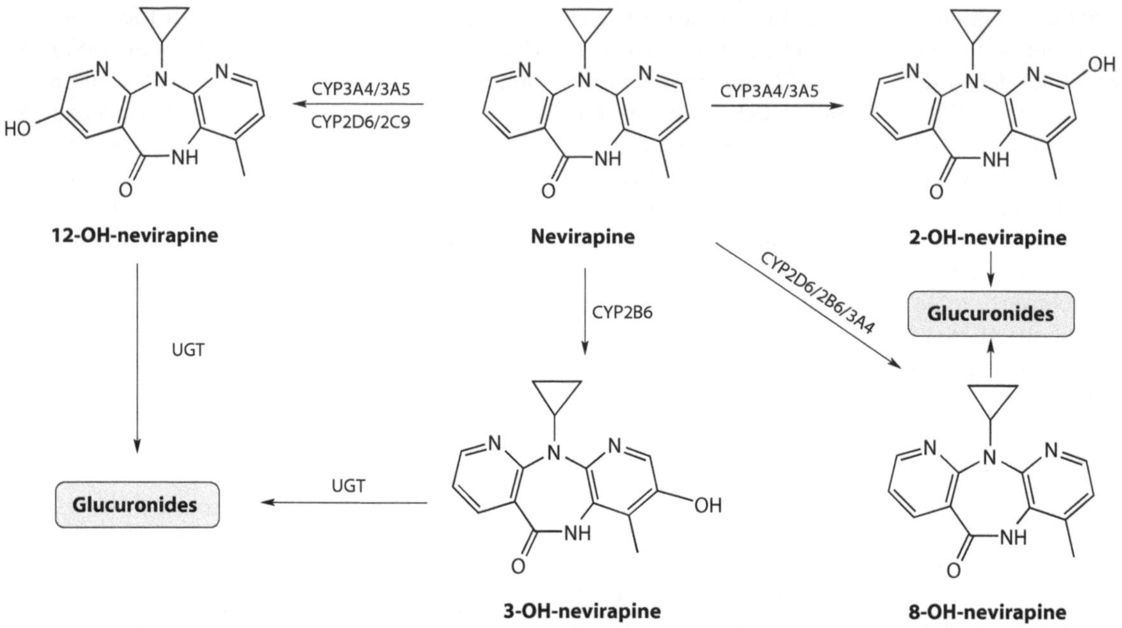

FIGURE 3.86 Metabolism of nevirapine by CYP2B6, 3A4, and 2D6. Nevirapine is eliminated primarily in the urine as glucuronide conjugates of 2-, 3-, 8-, and 12-hydroxynevirapine. In human liver microsomes, nevirapine is metabolized into at least four metabolites: 2-, 3-, 8-, and 12-hydroxynevirapine. In studies with cDNA-expressed human hepatic CYPs, 2- and 3-hydroxynevirapine are exclusively formed by CYP3A4/3A5 and CYP2B6, respectively. Multiple cDNA-expressed CYPs produced 8- and 12-hydroxynevirapine, although CYP2D6 and 3A4 primarily catalyzed their formation, respectively. For the formation of 12-OH-nevirapine, CYP3A5, 2D6, and 2C9 (minor) also play a role, while CYP2A6, 2B6, and 3A4 catalyze 8-OH-nevirapine formation with a low activity.

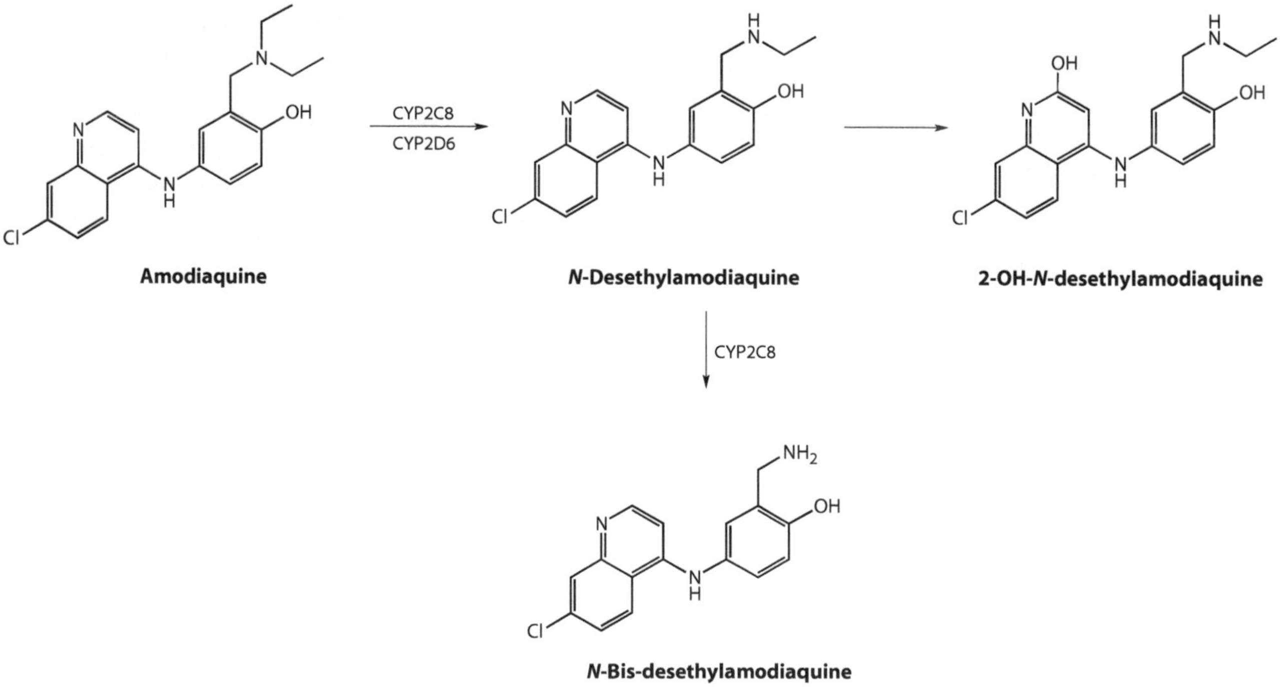

FIGURE 3.87 Metabolism of amodiaquine by CYP2C8 and 2D6. Amodiaquine is a 4-aminoquinoline derivative that has been widely used for treatment of malaria and is more active than chloroquine against *P. falciparum*. The main metabolite of amodiaquine is *N*-desethylamodiaquine with other minor metabolites being 2-hydroxyl-*N*-desethylamodiaquine and *N*-bis-desethylamodiaquine. Amodiaquine *N*-deethylation has been used as an alternative marker reaction for CYP2C8 because of its high affinity and high turnover rate. *N*-Desethylamodiaquine is found to be CYP2C8 selective in the liver with a minor amount being formed by CYP2D6 and extrahepatic enzymes CYP1A1 and 1B1.

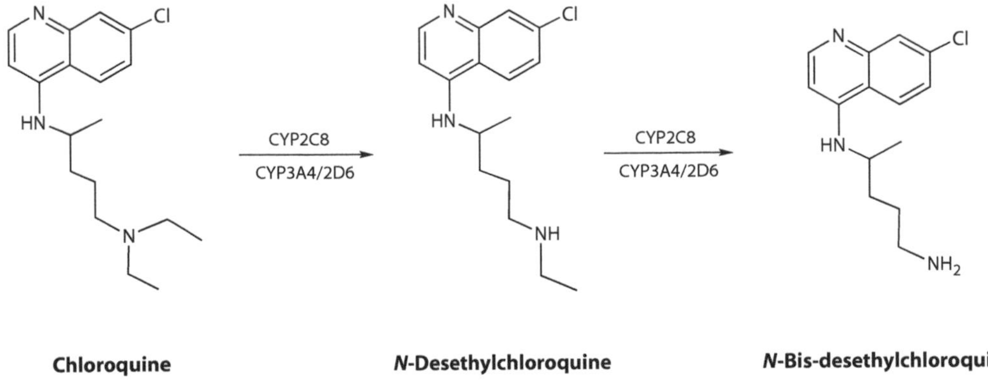

FIGURE 3.88 Metabolism of chloroquine by CYP2C8, 3A4, and 2D6. Chloroquine is an antimalarial agent used in the treatment and prophylaxis of malaria and is also used in the management of rheumatoid arthritis. After oral or intravenous administration, chloroquine is dealkylated into two main pharmacologically active metabolites: *N*-desethylchloroquine and *N*-bis-desethylchloroquine. Blood concentrations of the *S*-isomer of *N*-desethylchloroquine always exceeded those of the *R*-isomer, pointing to a preferential metabolism of *S*-chloroquine. Recombinant CYP1A1, 2D6, 3A4, and 2C8 catalyze chloroquine *N*-deethylation. CYP2C8 and 3A4 constituted low-affinity and high-capacity systems, whereas CYP2D6 is of a high affinity with a significantly low capacity.

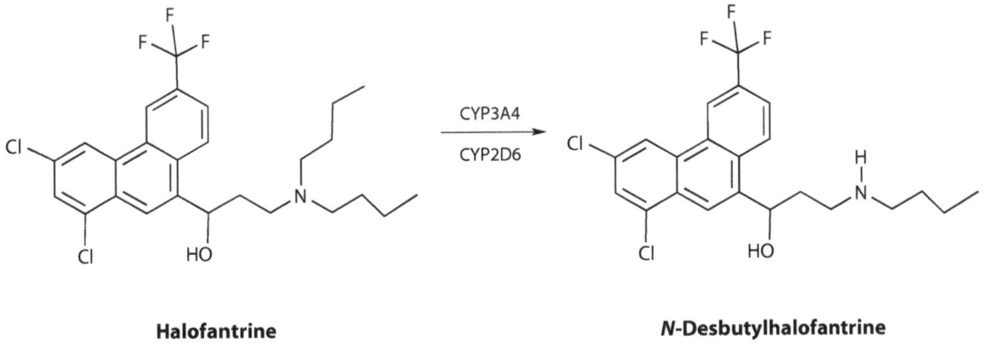

FIGURE 3.89 Metabolism of halofantrine by CYP2D6 and 3A4. Halofantrine biotransforms in the liver to its major metabolite, *N*-desbutylhalofantrine, which is the major circulating metabolite and is equipotent to halofantrine in antimalarial activity. CYP3A4 and 2D6 converted halofantrine to *N*-desbutylhalofantrine.

Recombinant CYP3A4 and 2D6 convert halofantrine to its *N*-debutylated metabolite, but the metabolism of halofantrine to its *N*-desbutyl metabolite in human liver microsomes shows no correlation with CYP2D6 genotypic or phenotypic status and there is no consistent inhibition by quinidine (Halliday et al. 1995). However, halofantrine causes a dose-dependent increase in prolongation of the electrocardiac QT interval and, in some susceptible patients, *torsade de pointes*, a potentially fatal cardiac arrhythmia (Nosten et al. 1993; Traebert and Dumotier 2005; Wesche et al. 2000; White 2007).

3.3.5.4 Phenoxypropoxybiguanides

Unlike proguanil, which is metabolized primarily by CYP2 C19, CYP3A4 plays a more important role in the metabolism of both PS-15 and JPC-2056, two phenoxypropoxybiguanides that are antimalarial prodrugs analogous to the structure of proguanil and its active metabolite cycloguanil (Diaz et al. 2008). Whereas CYP2D6 appears to play a major role in the metabolism of PS-15 to WR99210, it seems less important in the conversion of JPC-2056 to JPC-2067 (Diaz et al. 2008).

3.3.6 Hypolipidemic Agents

3.3.6.1 Fluvastatin

Fluvastatin, a commonly used HMG-CoA reductase inhibitor, is almost exclusively eliminated via metabolism, mainly hydroxylation, at the 5- and 6-position of the indole moiety and *N*-deisopropylation (Figure 3.90) (Scripture and Pieper 2001). Fluvastatin is 50%–80% metabolized by CYP2C9 to 5-hydroxy-, 6-hydroxy-, and *N*-deisopropyl-fluvastatin (Fischer et al. 1999). Recombinant CYP1A1, 2C8, CYP2C9, 2D6, and 3A4 form only 5-hydroxy-fluvastatin. Only the hydroxylated metabolites retain some HMG-CoA reductase inhibitory activity, yet they are not detected in the systemic circulation (Dain et al. 1993). However, CYP3A4 is the major enzyme metabolizing most HMG-CoA reductase inhibitors, including lovastatin (Wang et al. 1991), simvastatin (Prueksaritanont et al. 1997, 2003; Vickers et al. 1990), atorvastatin (Jacobsen et al. 2000), and cerivastatin (Boberg et al. 1997). These statins are subject to potential drug interactions with CYP3A inhibitors.

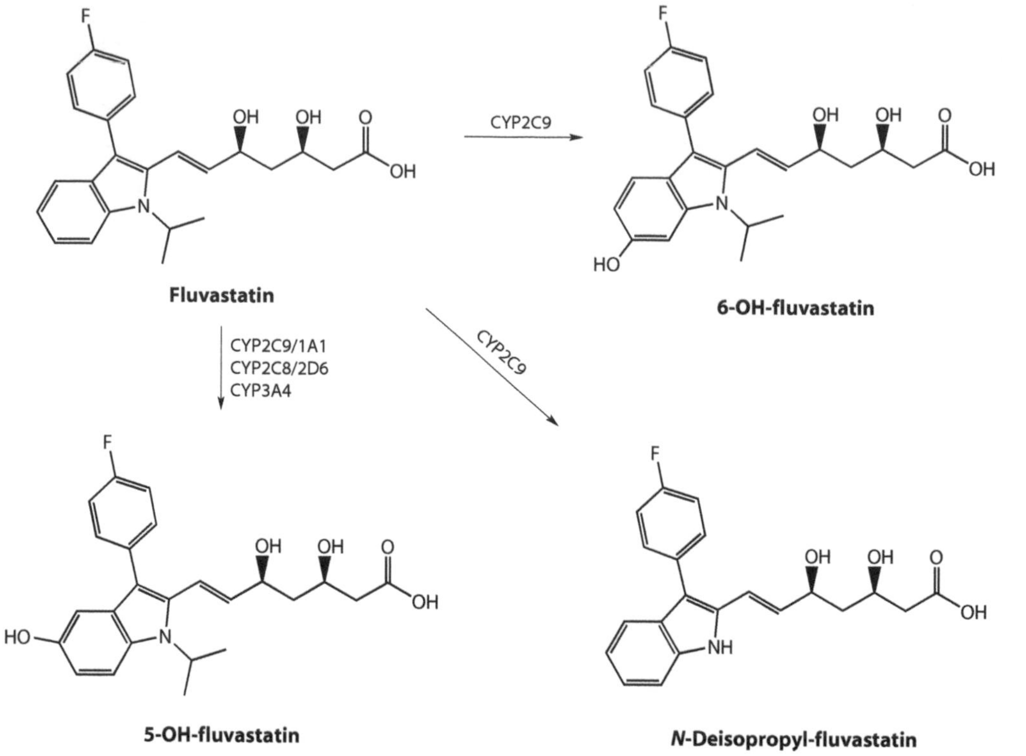

FIGURE 3.90 Metabolism of fluvastatin by CYP2C9, 3A4, 2C8, and 2D6. Fluvastatin is almost exclusively eliminated via metabolism, mainly hydroxylation, at the 5- and 6-position of the indole moiety and N-deisopropylation. Fluvastatin is 50%–80% metabolized by CYP2C9 to 5-hydroxy-, 6-hydroxy-, and N-deisopropyl-fluvastatin. Recombinant CYP1A1, 2C8, CYP2C9, 2D6, and 3A4 form only 5-hydroxy-fluvastatin. Only the hydroxylated metabolites retain some HMG-CoA reductase inhibitory activity, yet they are not detected in the systemic circulation. Fluconazole increased the AUC of fluvastatin by 84% and its elimination half-life of fluvastatin by 80%.

3.3.6.2 Pactimibe

Pactimibe, a novel acyl coenzyme A:cholesterol acyltransferase (ACAT) inhibitor developed for the treatment of hypercholesterolemia and atherosclerotic diseases, and its metabolite R-125528 are metabolized by CYP2D6 via ω-1 oxidation (Figure 3.91) (Kotsuma et al. 2008c). The K_m value of R-125528 in CYP2D6-expressing microsomes is 1.74 μM, which is comparable to those of typical basic CYP2D6 substrates (1–10 μM). Pactimibe has a lower affinity than R-125528 to CYP2D6; however, the K_m value is comparable to that of metoprolol.

3.3.7 MUSCARINIC RECEPTOR ANTAGONISTS

3.3.7.1 Tolterodine

Tolterodine is a relatively new antimuscarinic drug that is used to treat urinary incontinence and other symptoms associated with overactive bladder (Wefer et al. 2001). Tolterodine acts on M1, M2, M3, M4, and M5 subtypes of muscarinic receptors. After oral administration, tolterodine is rapidly absorbed from the gastrointestinal tract and exhibits extensive first-pass metabolism. In humans, approximately 80% of an administered oral dose of tolterodine is excreted in the urine, the major metabolites being the 5-carboxylic acids of tolterodine, N-dealkylated tolterodine, and

their glucuronides (Figure 3.92). Less than 1% of the parent compound is excreted unchanged. Tolterodine is mainly metabolized by CYP2D6 with significant contribution from CYP3A4 (Postlind et al. 1998). 5-Hydroxymethyltolterodine is the major metabolite in EMs but undetectable in the plasma of PMs (Brynne et al. 1998, 1999a,b; Olsson and Szamosi 2001a). Although tolterodine concentrations are increased 5- to 10-fold in PMs, the summed active moieties did not differ between EMs and PMs (Brynne et al. 1999a,b; Olsson and Szamosi 2001a,b). This suggests that therapeutic effects would not differ significantly between the two groups, and there is no convincing evidence of an important gene-effect correlation (Brynne et al. 1998).

3.3.8 NONSTEROIDAL ANTI-INFLAMMATORY DRUGS

3.3.8.1 Acetaminophen

Acetaminophen (APAP) is the principal p-aminophenol derivative in clinical use as a common over-the-counter analgesic and antipyretic agent (Prescott 2000). APAP is always believed to be relatively safe at therapeutic dose. However, APAP overdose may result in acute, often fatal, centrilobular liver necrosis in humans and animals (Hinson et al. 1981, 1995). When used at therapeutic doses, APAP is mainly detoxified by glucuronidation (52%–57% total urinary metabolites)

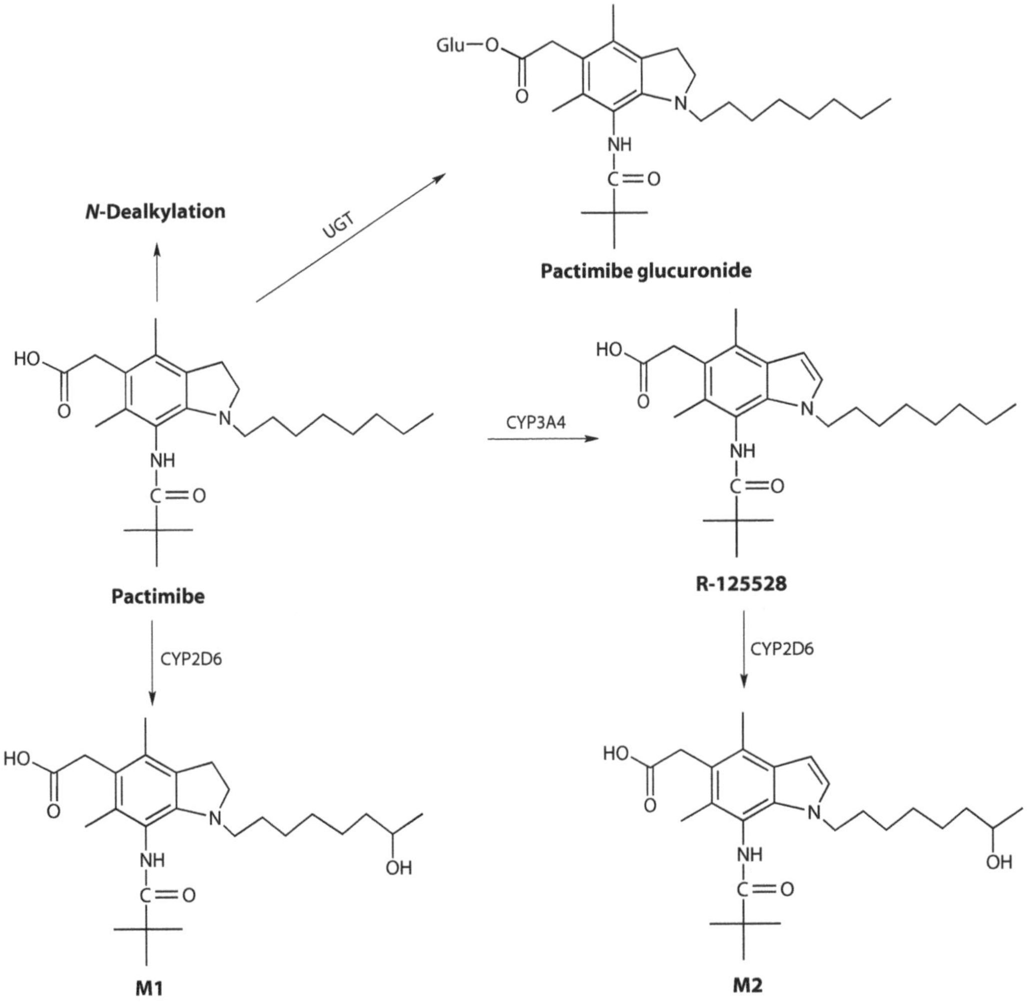

FIGURE 3.91 Metabolism of pactimibe by CYP2D6 and 3A4. Both pactimibe and its metabolite, R-125528, are metabolized by CYP2D6 via ω-1 oxidation. The indolin oxidation of pactimibe is mediated by CYP3A4. Both pactimibe and R-125528 are atypical substrates for CYP2D6 because of its acidity. Both compounds are not protonated but are negatively charged at physiological pH. Notably, their sites of metabolism, the ω-1 position of the *N*-octyl indoline/indole group, are relatively distant from the aromatic moiety. An induced-fit docking of the ligands with an x-ray crystal structure of substrate-free CYP2D6 indicated the involvement of an electrostatic interaction between the carboxyl group and Arg221 and hydrophobic interaction between the aromatic moiety and Phe483.

and sulfation (30%–44%), giving rise to glucuronides and sulfates excreted into the urine (Patel et al. 1990, 1992). In vitro studies indicated that most UGTs could catalyze APAP glucuronidation, but UGT1A1, 1A6, and 1A9 are most active (Bock et al. 1993; Court et al. 2001; Kessler et al. 2002). To a lesser extent, APAP also undergoes oxidation (<5%) at therapeutic dose (Patel et al. 1992). However, at APAP overdose, sulfation becomes saturated such that glucuronidation is the predominant pathway (66%–75% total urinary metabolites), whereas an increased amount of APAP (7%–15%) is oxidized to form reactive *N*-acetyl-*p*-benzoquinoneimine (NAPQI) (Bessems and Vermeulen 2001; Prescott 2000). There is evidence indicating that the formation of NAPQI might be via the intermediate semibenzoquinone imine by a second hydrogen atom abstraction (de Vries 1981), followed by radical recombination generating 3-hydroxy-APAP or conjugation

with GSH to form an ipso-adduct intermediate (Chen et al. 1999; Forte et al. 1984; Hoffmann et al. 1990; Koymans et al. 1989; Miner and Kissinger 1979; Myers et al. 1994). For the formation of NAPQI, recombinant CYP3A4 shows the highest activity followed by CYP2E1, 1A2, 2A6, and 2D6 (Figure 3.93) (Hinson et al. 1995; Laine et al. 2009; Manyike et al. 2000; Potter et al. 1973; Zhou et al. 1997, 2005). The resultant toxic NAPQI is generally conjugated with GSH by GSTs (mainly GSTPi) to form GSH conjugates that are readily excreted into urine without initiation of hepatotoxicity, while available GSH pools will be depleted and thus allow excessive reactive NAPQI access to liver proteins after toxic doses of APAP (Coles et al. 1988; Potter et al. 1974). NAPQI is capable of reacting with many cellular proteins and nonprotein thiols by arylation and oxidation, leading to subsequent alteration of protein structure and function (Bartolone et al.

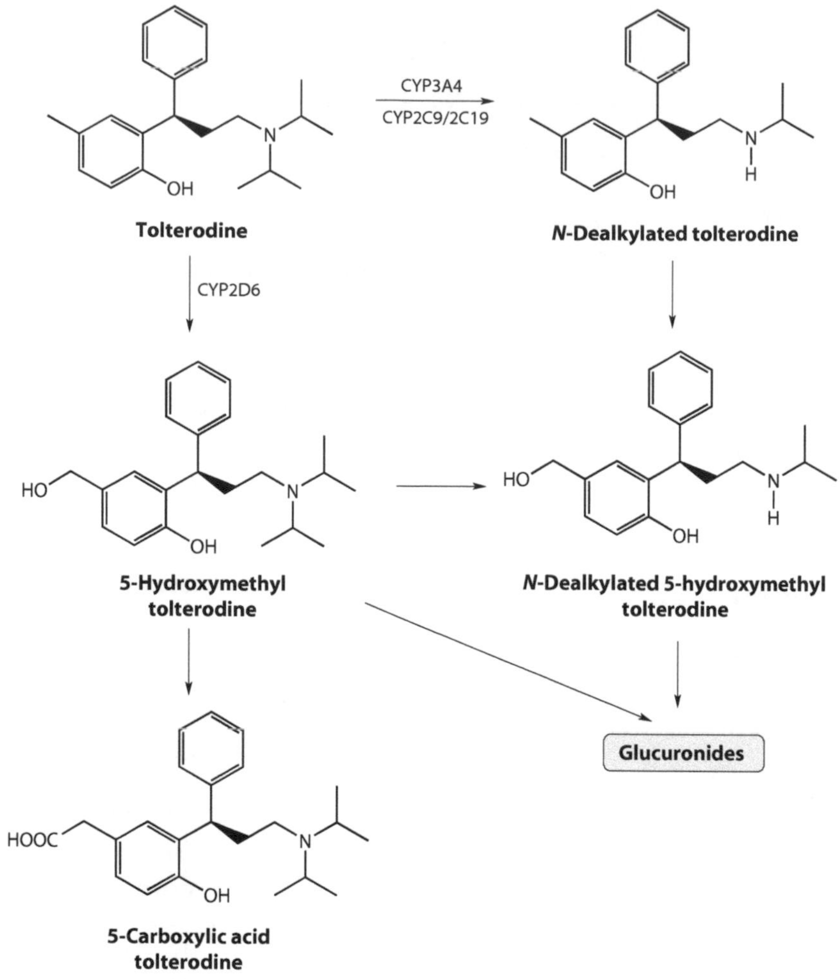

FIGURE 3.92 Metabolism of tolterodine by CYP2D6 and other enzymes. After oral administration, tolterodine is rapidly absorbed from the gastrointestinal tract and exhibits extensive first-pass metabolism. In humans, approximately 80% of an administered oral dose of tolterodine is excreted in the urine, the major metabolites being the 5-carboxylic acids of tolterodine, *N*-dealkylated tolterodine, and their glucuronides. Less than 1% of the parent compound is excreted unchanged. Tolterodine is mainly oxidized to the active 5-hydroxymethyl tolterodine by CYP2D6 and *N*-dealkylated by CYP3A4, 2C9, and 2C19.

1987; Bessems and Vermeulen 2001; Cohen and Khairallah 1997; Pumford et al. 1990).

3.3.8.2 Indomethacin

Indomethacin is a commonly used nonsteroidal anti-inflammatory drug (NSAID) of the arylalkanoic acid class used to reduce fever, pain, stiffness, and swelling (Hart and Boardman 1963). It is extensively metabolized via *O*-demethylation and *N*-deacylation into *O*-desmethylindomethacin, *N*-deschlorobenzoylindomethacin, and *O*-desmethyl-*N*-deschloro benzoylindomethacin (Figure 3.94) (Harman et al. 1964). In humans, *O*-desmethylindomethacin formation is critical to the elimination of indomethacin, accounting for 40%–55% of total drug eliminated in the urine (Duggan et al. 1972). In vitro studies have demonstrated that CYP2C9 is the major catalyst for the *O*-demethylation of indomethacin, with minor contribution from CYP1A2, 2C19, and 2D6 (Nakajima et al. 1998a). In addition, indomethacin is glucuronidated by UGT1A1 (Kuehl et al. 2005).

3.3.9 Oral Hypoglycemic Drugs

Oral hypoglycemic drugs include the sulfonylureas (e.g., glipizide, glimepiride, and glyburide) and biguanides (e.g., metformin, buformin, and phenformin), which are currently used in the treatment of type II diabetes (Marchetti et al. 1991).

3.3.9.1 Phenformin

Phenformin is a biguanide antidiabetic agent and has been withdrawn from the market because of severe lactic acidosis, which is fatal in ~50% of cases (Marchetti and Navalesi 1989). This drug has been replaced by metformin in clinical practice. Phenformin is excreted largely unchanged through the kidneys (~50%–86%), with partial metabolism mediated by polymorphic CYP2D6 via *p*-hydroxylation (Figure 3.95) (Oates et al. 1982; Shah et al. 1985). 4-Hydroxylated phenformin can be further conjugated. Phenformin and its 4-hydroxylated metabolite in a free or conjugated form are excreted into the urine. Phenformin 4-hydroxylation is highly variable

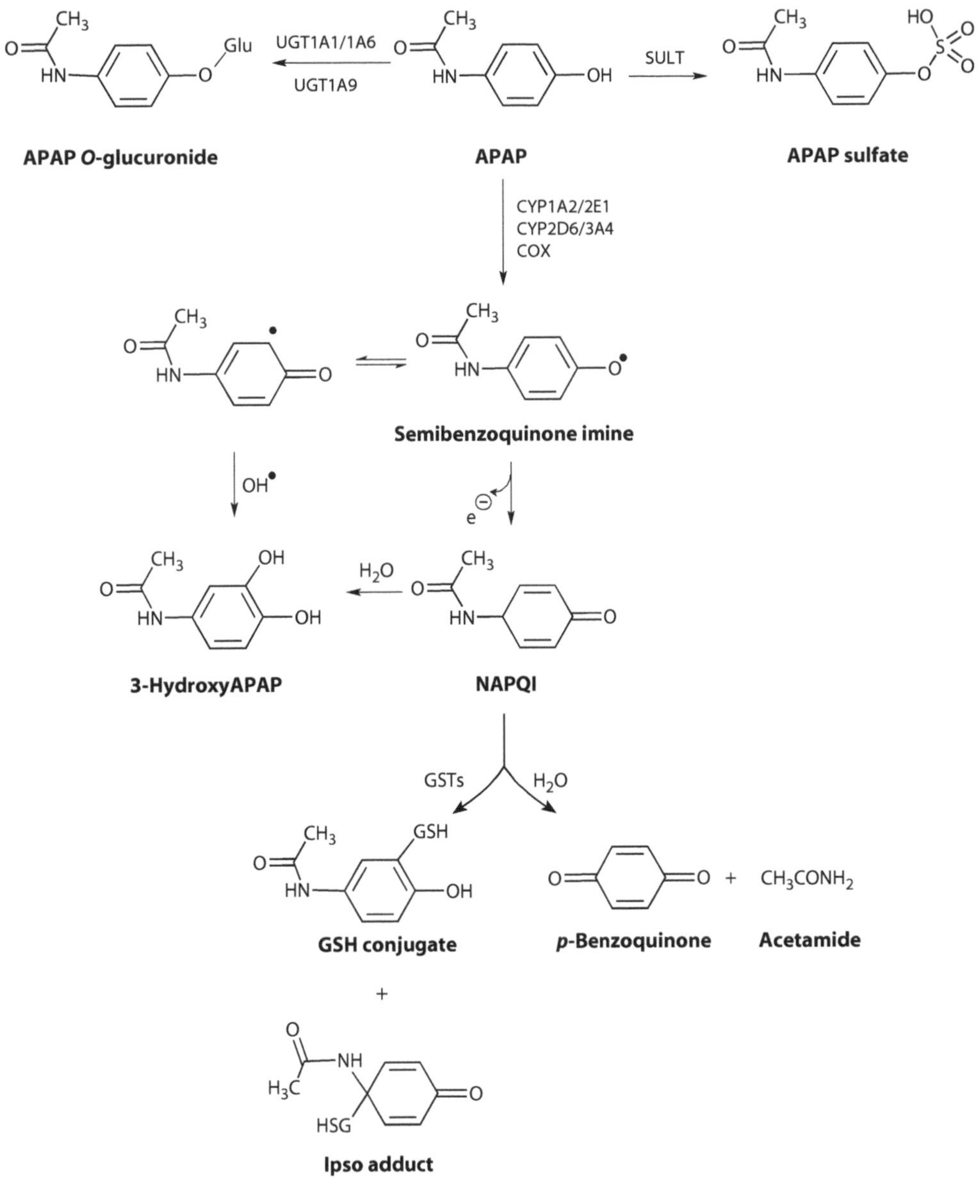

FIGURE 3.93 Formation of reactive NAPQI from APAP by multiple CYP enzymes. When used at therapeutic doses, APAP is mainly detoxified by glucuronidation (52%–57% total urinary metabolites) and sulfation (30%–44%), giving rise to glucuronides and sulfates excreted into the urine. In vitro studies indicated that most UGTs catalyze APAP glucuronidation, but UGT1A1, 1A6, and 1A9 are the most active. However, at APAP overdose, sulfation becomes saturated such that glucuronidation is the predominant pathway (66%–75% total urinary metabolites), whereas an increased amount of APAP (7%–15%) is oxidized to form reactive NAPQI. For the formation of NAPQI, recombinant CYP3A4 showed the highest activity followed by CYP2E1, 1A2, 2A6, and 2D6.

(184-fold) and cosegregated with debrisoquine MRs (Oates et al. 1982; Shah et al. 1985). 4-Hydroxyphenformin is not detected in PMs (Shah et al. 1980).

3.3.10 Selective Estrogen Receptor Modulators

Selective estrogen receptor modulators (SERMs) are structurally different compounds that interact with intracellular estrogen receptors in target organs as estrogen agonists and antagonists. The molecular basis of SERM activity involves binding of the ligand SERM to the estrogen receptor, causing conformational changes that facilitate interactions with coactivator or corepressor proteins and later initiate or suppress transcription of target genes (Shelly et al. 2008). SERMs currently approved for use in patients include tamoxifen, raloxifene, and toremifene. Tamoxifen, a triphenylethyleneamine derivative, is one of the most widely used SERMs for the treatment of postmenopausal, hormone-sensitive, estrogen receptor–positive breast cancer (Jordan 2007; Jordan and O'Malley 2007). Tamoxifen is first approved in 1977 by the US FDA for the treatment of metastatic breast cancer and in ensuing years for adjuvant treatment of breast cancer. Other triphenylethylene SERMs, analogs of tamoxifen, have been studied for breast cancer prevention and treatment, including droloxifene (3-hydroxytamoxifen)

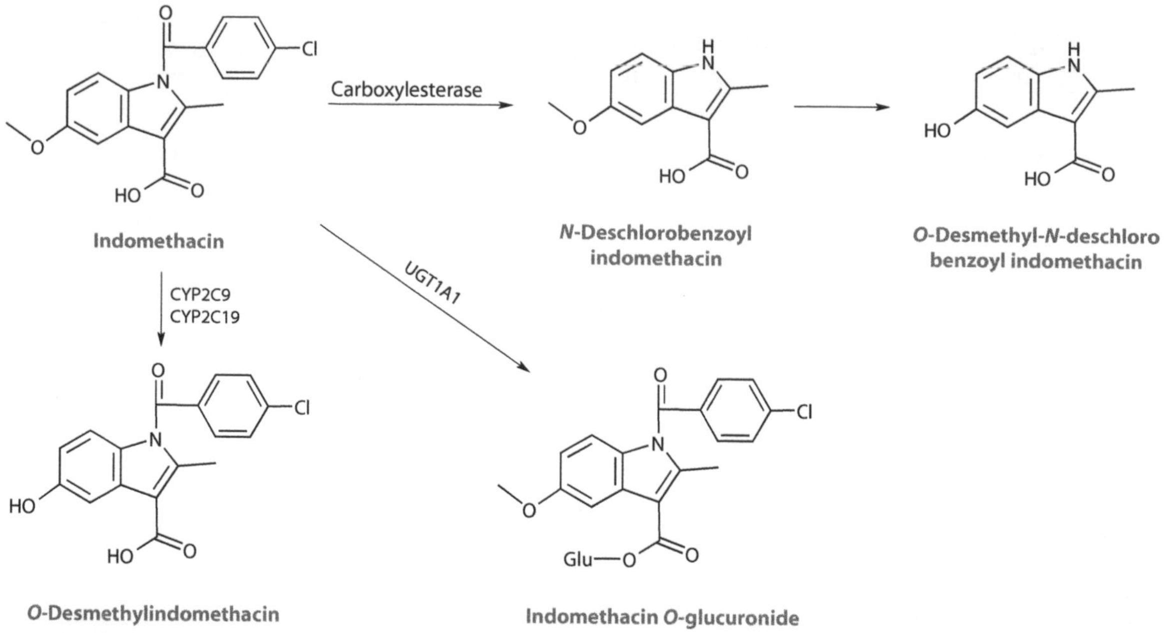

FIGURE 3.94 Metabolism of indomethacin by CYP2C9, 1A2, 2C19, and 2D6. It is extensively metabolized via O-demethylation and N-deacylation into O-desmethylindomethacin, N-deschlorobenzoylindomethacin, and O-desmethyl-N-deschlorobenzoylindomethacin. The carboxylesterase catalyzes the N-deacylation of indomethacin in pigs. In humans, O-desmethylindomethacin formation is critical to the elimination of indomethacin, accounting for 40%–55% of total drug eliminated in the urine. In vitro studies have demonstrated that CYP2C9 is the major catalyst for the O-demethylation of indomethacin, with minor contribution from CYP1A2, 2C19, and 2D6. In addition, indomethacin is glucuronidated by UGT1A1.

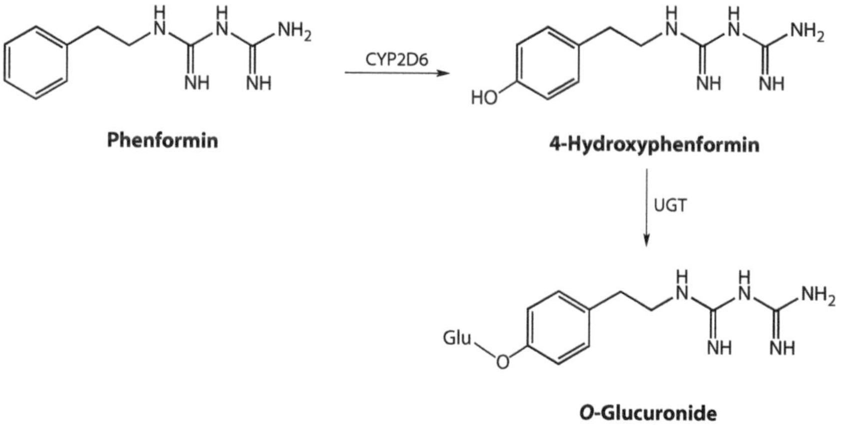

FIGURE 3.95 Metabolism of phenformin by CYP2D6 and other enzymes. Phenformin is excreted largely unchanged through the kidneys (~50%–86%), with partial metabolism mediated by CYP2D6 via p-hydroxylation (i.e., 4-hydroxylation). 4-Hydroxylated phenformin can be further conjugated. Phenformin and its 4-hydroxylated metabolite in free or conjugated forms are excreted into the urine. Phenformin 4-hydroxylation is highly variable (184-fold) and cosegregated with debrisoquine MRs.

(Buzdar et al. 2002; Roos et al. 1983; Ruenitz et al. 1982), idoxifene (pyrrolidino-4-iodotamoxifen) (Arpino et al. 2003; Chander et al. 1991; Coombes et al. 1995; Johnston et al. 2004), and ospemifene (Komi et al. 2005).

3.3.10.1 Tamoxifen

Tamoxifen undergoes extensive Phase I and Phase II metabolism (Cronin-Fenton et al. 2014; Poon et al. 1993), and the concentrations of tamoxifen and its metabolites in humans vary widely in patients. Tamoxifen undergoes extensive oxidation predominantly by various CYPs to several primary and secondary metabolites (Figure 3.96). Studies in women receiving 20 mg of ^{14}C-tamoxifen have shown that ~65% of the administered dose in the conjugate form is excreted into the feces over 2 weeks, with unchanged drug and unconjugated metabolites accounting for <30% of the total fecal

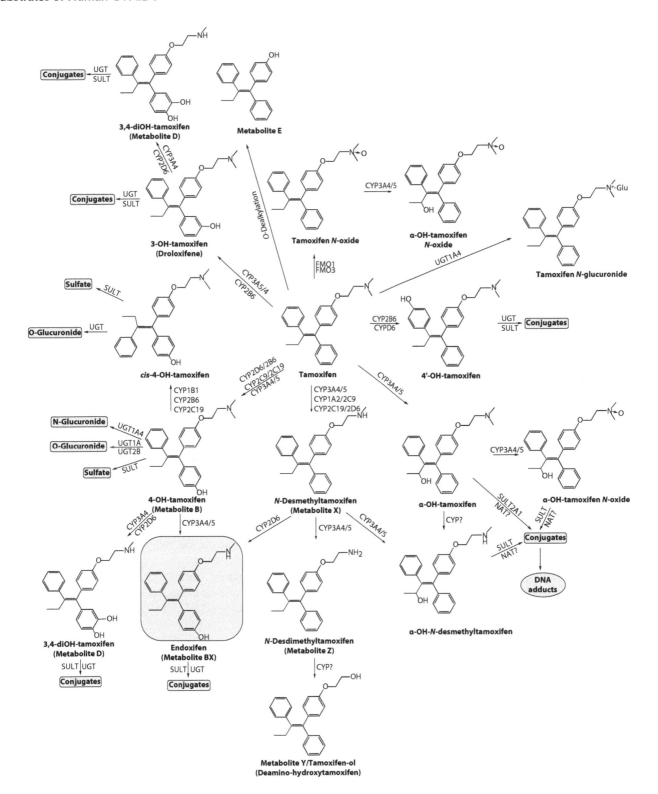

FIGURE 3.96 Metabolism of tamoxifen. Tamoxifen undergoes extensive Phase I and Phase II metabolism and the concentrations of tamoxifen and its metabolites in humans vary widely in patients. Tamoxifen undergoes extensive oxidation predominantly by various CYPs to several primary and secondary metabolites. The formation of 4-OH-tamoxifen from tamoxifen is mainly by CYP2D6, with contributions from CYP2C9 and 3A4. The main route of tamoxifen metabolism is through CYP3A4/5-mediated demethylation of the aminoethoxy side chain to N-desmethyltamoxifen. N-Desmethyltamoxifen is further demethylated by CYP3A4/5 to N-desdimethyltamoxifen and then deaminated to metabolite Y (deamino-hydroxytamoxifen). N-Desmethyltamoxifen is p-hydroxylated mainly by CYP2D6 to its most abundant and therapeutically active metabolite endoxifen. Endoxifen can also be generated from 4-hydroxytamoxifen via N-demethylation, but the 4-hydroxylation of N-desmethyltamoxifen by CYP2D6 appears to be the major source of endoxifen production in vivo. 4-OH-tamoxifen also undergoes *trans–cis* isomerization mainly by CYP1B1, with contribution from CYP2B6 and 2C19. Moreover, hydroxylation followed by conjugating reactions is a major metabolic route of tamoxifen.

radioactivity (Jordan 1982; Kisanga et al. 2005). Tamoxifen is considered as a prodrug, which must be converted into active metabolites. Early studies by Jordan (1982) and Jordan et al. (1977) have demonstrated that high first-pass metabolism of tamoxifen results in a significant increase in its antiestrogenic activity and characterized the first active primary metabolite, 4-hydroxytamoxifen (metabolite B). The formation of 4-OH-tamoxifen from tamoxifen is mainly by CYP2D6 (Beverage et al. 2007; Crewe et al. 2002; Dehal and Kupfer 1997; Jacolot et al. 1991; Stearns et al. 2003), with contributions from CYP2C9 and 3A4 (Dehal and Kupfer 1997). CYP2B6 also shows significant activity toward 4-hydroxylation at high tamoxifen concentrations (Crewe et al. 2002). Although 4-hydroxylation of tamoxifen represents less than 10% of overall tamoxifen oxidation and 4-OH-tamoxifen is a relatively minor metabolite, 4-OH-tamoxifen possesses 100-fold greater affinity for the estrogen receptor and 30- to 100-fold greater potency in inhibiting estrogen-dependent cell proliferation than tamoxifen (Katzenellenbogen et al. 1984).

The main route of tamoxifen metabolism is through CYP3A4/5-mediated demethylation of the aminoethoxy side chain to N-desmethyltamoxifen (metabolite X) (Figure 3.96) (Lim et al. 1994; Mani et al. 1993a; Ruenitz et al. 1984), which has twice the half-life (approximately 14 days) as tamoxifen (approximately 5–7 days), and plasma levels are usually two- to threefold higher than those of tamoxifen (Jordan 1982). Chronic administration of 10 mg tamoxifen twice daily for 3 months to women results in average steady-state plasma concentrations of 120 ng/ml (range, 67–183 ng/ml) for tamoxifen and 336 ng/ml (range, 148–654 ng/ml) for N-desmethyltamoxifen (Jordan 1982). The average steady-state plasma concentrations of tamoxifen and N-desmethyltamoxifen after administration of 20 mg tamoxifen once daily for 3 months are 122 ng/ml (range, 71–183 ng/ml) and 353 ng/ml (range, 152–706 ng/ml), respectively (Soininen et al. 1986). N-Desmethyltamoxifen is further demethylated by CYP3A4/5 to N-desdimethyltamoxifen (metabolite Z) and then deaminated to metabolite Y (deamino-hydroxytamoxifen; tamoxifen-ol), a glycol derivative with antiestrogenic activity (Etienne et al. 1989; Fried and Wainer 1994; Langan-Fahey et al. 1990; Milano et al. 1987; Murphy et al. 1987). N-Desdimethyltamoxifen appears to be highly concentrated in the liver and is a potent inhibitor of certain CYPs including CYP3A (Comoglio et al. 1996), and this metabolite may mediate drug interactions involving tamoxifen, or it may modify tamoxifen-induced toxicity by inhibiting formation of CYP-catalyzed reactive metabolites from tamoxifen.

In vitro studies have pointed to CYP3A4 being the major catalyst of N-demethylation of tamoxifen (Figure 3.96) (Crewe et al. 1997, 2002; Jacolot et al. 1991) with contributions from other CYPs. In vitro studies showed that erythromycin, cyclosporine, nifedipine, and diltiazem competitively inhibit formation of N-desmethyltamoxifen with apparent K_i of 20, 1, 45, and 30 μM, respectively (Jacolot et al. 1991). Studies in recombinant enzymes indicate that tamoxifen N-demethylation is mediated by CYP2D6, 1A1, 1A2, and

3A4, at low substrate concentrations, with contributions from CYP1B1, 2C9, 2C19, and 3A5 at high concentrations (Crewe et al. 2002).

N-Desmethyltamoxifen is p-hydroxylated mainly by CYP2D6 to its most abundant and therapeutically active metabolite endoxifen (4-hydroxy-N-desmethyltamoxifen, metabolite BX) (Figure 3.96) (Beverage et al. 2007; Crewe et al. 2002; Dehal and Kupfer 1997; Jacolot et al. 1991; Stearns et al. 2003). Endoxifen can also be generated from 4-hydroxytamoxifen via N-demethylation (Desta et al. 2004), but the 4-hydroxylation of N-desmethyltamoxifen by CYP2D6 appears to be the major source of endoxifen production in vivo. Plasma levels of endoxifen are 5- to 10-fold higher than those of 4-hydroxytamoxifen in most women taking tamoxifen (Johnson et al. 2004). Endoxifen exhibits potency similar to 4-hydroxytamoxifen with respect to estrogen receptor binding affinity, suppression of estrogen-dependent cell growth, and gene expression (Johnson et al. 2004). Although both 4-OH-tamoxifen and endoxifen are considered the major active metabolites of tamoxifen that have potent antiestrogenic activity, endoxifen is considered to play a more significant role in tamoxifen pharmacological activity in vivo. CYP2D6 plays a critical role in the activation of tamoxifen to endoxifen by sequential N-demethylation and 4-hydroxylation (Beverage et al. 2007; Crewe et al. 2002; Dehal and Kupfer 1997; Desta et al. 2004; Stearns et al. 2003).

4-OH-tamoxifen undergoes ortho-hydroxylated by CYP3A4 and 2D6 to 3,4-dihydroxytamoxifen catechol (metabolite D) (Figure 3.96) (Dehal and Kupfer 1999), which can be detected in the plasma and urine of women taking tamoxifen (Carter et al. 2001; Li et al. 2001). This catechol appears proximate to the reactive intermediate that binds covalently to microsomal proteins (Dehal and Kupfer 1996). 3,4-Dihydroxytamoxifen can undergo further oxidation to a highly reactive o-quinone, which is cytotoxic and has the potential to alkylate DNA (F. Zhang et al. 2000). There is evidence that the catechol metabolites of estradiol and estrone forming semiquinones and quinones (Bolton and Thatcher 2008; Shang 2006), undergoing redox cycling mediated by P450 reductase, could lead to carcinogenicity via oxidative damage to DNA and adduct formation (Yager and Liehr 1996). It has been shown that the formation of quinones by oxidation of certain semiquinones reduces O_2 to superoxide anion radical $\left(O_2^-\right)$ and metabolism of tamoxifen by hepatocytes is found to generate O_2^- (Turner et al. 1991), suggesting that tamoxifen semiquinone is generated presumably via the initial formation of tamoxifen catechol. 4-OH-tamoxifen undergoes trans–cis isomerization mainly by CYP1B1, with contribution from CYP2B6 and 2C19 (Crewe et al. 2002). trans-4-OH-tamoxifen, a potent antiestrogen, can undergo CYP-mediated isomerization to cis-4-OH-tamoxifen, a weak estrogen agonist (Williams et al. 1994). There is clinical evidence that tumor resistance to tamoxifen therapy is associated with decreased intratumoral concentrations of tamoxifen and an increase in the cis/trans ratio of 4-OH-tamoxifen (Osborne et al. 1992).

Tamoxifen is also hydroxylated at the allylic position by CYP3A4/5 to α-OH-tamoxifen in humans, and this metabolic pathway has been considered the initiation step in tamoxifen bioactivation to reactive species (Figure 3.96) (White 2003). It has been proposed that the α-hydroxylated tamoxifen metabolites, which became O-sulfated forming the unstable sulfuric acid esters, are the reactive tamoxifen intermediates that form carcinogenic DNA adducts (Dasaradhi and Shibutani 1997; Kitagawa et al. 2000; Phillips et al. 1994). An interconversion between E- and Z-α-OH-tamoxifen is possible as a result of sulfation and the subsequent reaction with water (Dasaradhi and Shibutani 1997). α-OH-tamoxifen can be O-sulfonated by rat sulfotransferase (SULT) 2A3 and human SULT2A1 (also known as human dehydroepiandrosterone sulfotransferase) and react with DNA (Apak and Duffel 2004; Glatt et al. 1998; Shibutani et al. 1998a), resulting in dG-N^2-tamoxifen adducts (Shibutani et al. 1998b). α-Acetoxy-N-desmethyltamoxifen resulting from α-hydroxytamoxifen by acetylation is highly reactive to 2′-deoxyguanosine, as is similarly observed for tamoxifen α-sulfate (Kitagawa et al. 2000). Synthetic tamoxifen α-sulfate and α-acetoxy-tamoxifen react rapidly with DNA, resulting in the formation of four diastereoisomers (two trans forms and two cis forms) of α-(N^2-deoxyguanosinyl) tamoxifen adducts (Dasaradhi and Shibutani 1997; Osborne et al. 1996). Similar adduct formation has been found using α-sulfate or α-acetyl forms of α-OH-N-desmethyltamoxifen (Gamboa da Costa et al. 2000; Kitagawa et al. 2000) and α-acetoxytamoxifen N-oxide (Umemoto et al. 1999). The study by Kim et al. (2005) has demonstrated that O-sulfonation, not O-acetylation, of α-hydroxylated tamoxifen and its metabolites contributes primarily to the formation of tamoxifen-DNA adducts in rat and human livers. In addition, a quinone methide metabolite from 4-OH-tamoxifen has been suggested as the reactive intermediate in DNA adduct formation (Fan et al. 2000). Synthetic 4-OH-tamoxifen quinone methide is stable with a half-life of approximately 3 h under physiological conditions, and its half-life is approximately 4 min in the presence of GSH (Fan et al. 2000).

α-Hydroxy N-desmethyltamoxifen can result from N-desmethyltamoxifen and tamoxifen, suggesting that tamoxifen is N-demethylated and then α-hydroxylated to α-hydroxy N-desmethyltamoxifen (Desta et al. 2004). Another possible pathway for the formation of this metabolite could be α-hydroxylation of tamoxifen followed by N-demethylation. Little is known regarding α-hydroxy N-desmethyltamoxifen, but like α-hydroxytamoxifen, it might participate in tamoxifen-induced genotoxicity. Tamoxifen undergoes sequential metabolism that includes stepwise N-demethylation (e.g., to N-didesmethyltamoxifen), hydroxylation (e.g., to 3,4-hydroxytamoxifen), or N-demethylation followed by hydroxylation or vice versa (e.g., to endoxifen and α-hydroxy N-desmethyltamoxifen). N-Desmethyltamoxifen and 4-hydroxytamoxifen formation accounts for approximately 92% and 7% of primary tamoxifen oxidation in vitro, respectively (Desta et al. 2004). 3-Hydroxytamoxifen is a minor metabolite of tamoxifen in humans, and CYP3A5 is the main enzyme responsible for tamoxifen 3-hydroxylation

(Desta et al. 2004). CYP2B6 and 2C19 are important catalysts of tamoxifen 4′-hydroxylation (Crewe et al. 2002; Desta et al. 2004). 4′-Hydroxytamoxifen has been detected in the rat (Ruenitz et al. 1984). The relevance of 4′- and 3-hydroxytamoxifen to the multiple effects of tamoxifen and their abundance in plasma of patients remains to be determined.

Alternatively, tamoxifen undergoes O-dealkylation to give cis/trans-1,2-diphenyl-1-(4-hydroxyphenyl)-but-1-ene (also known as metabolite E), which is estrogenic and readily detected in plasma (Figure 3.96) (Langan-Fahey et al. 1990). Like 4-OH-tamoxifen, cis/trans-metabolite E can be activated to form DNA adducts by rat uterine peroxidases (Pathak et al. 1996). Metabolite E has been found in tamoxifen-resistant MCF-7 human breast tumors implanted in athymic nude mice, as well as in tumors from patients with clinical resistance (Wiebe et al. 1992), suggesting a potential role of metabolite E forms by local tissues in the development of tumor resistance to tamoxifen therapy. In addition, an estrogenic metabolite of tamoxifen, bisphenol, has been detected in these resistant tumors (Wiebe et al. 1992). Because of its structural similarity to 4-OH-tamoxifen, metabolite E could also be converted to a quinone methide, which has the potential to alkylate DNA and may contribute to the genotoxic effects of tamoxifen. Synthetic metabolite E quinone methide is stable with a half-life of 4 h under physiological conditions, and its half-life is approximately 4 min in the presence of reduced GSH (Fan and Bolton 2001). However, unlike the unstable GSH adducts of 4-OH-tamoxifen quinone methide, metabolite E GSH adducts are stable enough to be isolated and characterized by NMR and liquid chromatography mass spectrometry. Reaction of metabolite E quinone methide with DNA generated exclusively deoxyguanosine adducts (Fan and Bolton 2001).

FMOs catalyze the formation of tamoxifen N-oxide in mice (Mani et al. 1993b), rats (Foster et al. 1980; Mani et al. 1993b; McCague and Seago 1986), and humans (Mani et al. 1993b; Parte and Kupfer 2005). This is considered as a detoxification pathway of tamoxifen. Other tamoxifen analogs undergoing clinical trials for anticancer treatment (e.g., idoxifene, toremifene, and droloxifene) have also been found to form the corresponding N-oxides (Jones and Lim 2002; McCague et al. 1990). Tamoxifen N-oxide has similar antiestrogenic potency to tamoxifen in vitro (Bates et al. 1982). FMO1 and FMO3 catalyze tamoxifen N-oxidation to its N-oxide (Parte and Kupfer 2005). CYP1A1, 1A2, 2A6, 2C8, 2C9, 2C19, 2D6, 2E1, and 3A4 all rapidly reduce tamoxifen N-oxide back to tamoxifen (Parte and Kupfer 2005). Tamoxifen N-oxide is converted into tamoxifen by reduced hemoglobin and NADPH-P450 oxidoreductase (Parte and Kupfer 2005), suggesting the involvement of the same heme–Fe^{2+} complex for both hemoglobin and P450s and the reductive activity may be nonenzymatic.

Hydroxylation followed by Phase II conjugating reactions (e.g., glucuronidation and sulfation) is a major metabolic route of tamoxifen (Figure 3.96) (Poon et al. 1993). After oral administration of tamoxifen, at least four glucuronides of tamoxifen metabolites in human urine are identified, including the glucuronides of 4-OH-tamoxifen, 4-OH-N-desmethyltamoxifen, 3,4-dihydroxytamoxifen, and a monohydroxy-*N*-desmethyltamoxifen (Poon et al. 1993). Tamoxifen can be converted by UGT1A4 to its *N*-glucuronide (Kaku et al. 2004). Tamoxifen *N*-glucuronide has binding affinity similar to tamoxifen (Kaku et al. 2004). *N*- and *O*-Glucuronidating activities of human liver microsomes toward *trans*-4-OH-tamoxifen are comparable, but *O*-glucuronidation is predominant for *cis*-4-OH-tamoxifen conjugation (Ogura et al. 2006). Only UGT1A4 catalyzes the *N*-linked glucuronidation of 4-OH-tamoxifen (Sun et al. 2006), but multiple UGTs including UGT1A1, 1A6, 1A7, 1A8, 1A9, 1A10, 2B4, 2B7, 2B15, and 2B17 catalyze *O*-glucuronidation of 4-OH-tamoxifen (Ogura et al. 2006). *O*-Glucuronidation of 4-OH-tamoxifen significantly decreases binding affinity for the human estrogen receptor, while 4-OH-tamoxifen *N*-glucuronide possesses binding affinity similar to 4-OH-tamoxifen (Ogura et al. 2006). 4-OH-tamoxifen is also conjugated by SULT1A1, 1E1, and 2A1 and excreted into the urine and bile (Falany et al. 2006). In patients receiving chronic tamoxifen treatment, bile and urine are rich in the conjugated metabolite Y, 4-hydroxytamoxifen, and endoxifen, whereas in feces, free 4-hydroxytamoxifen and tamoxifen are the predominating species (Lien et al. 1989).

3.3.10.2 Droloxifene

Several triphenylethylene derivatives, possibly devoid of uterine carcinogenic activity, are being developed as therapeutic agents against breast cancer, including droloxifene (3-hydroxytamoxifen, Figure 3.96) (Buzdar et al. 2002; Roos et al. 1983; Ruenitz et al. 1982), idoxifene (pyrrolidino-4-iodotamoxifen)

(Arpino et al. 2003; Chander et al. 1991; Coombes et al. 1995; Johnston et al. 2004), and ospemifene (Komi et al. 2005). Like 4-OH-tamoxifen, droloxifene undergoes *ortho*-hydroxylation by CYP3A4 and 2D6 to 3,4-dihydroxytamoxifen catechol, which is proximate to the reactive intermediate (Dehal and Kupfer 1999).

3.3.10.3 Enclomifene

Enclomifene, the more active isomer of clomifene, is mainly metabolized by CYP2D6 (Ghobadi et al. 2008). No metabolism is detected in microsomes from the PM liver. Clomifene is a diastereomeric mixture of two geometric isomers, enclomifene (*E*-clomifene) and zuclomifene (*Z*-clomifene) in the ratio of 62:38, and acts by inhibiting the action of estrogen on the gonadotrope cells in the anterior pituitary gland. Clomifene is the best initial treatment for anovulation, a common cause of infertility. Incubation of clomifene with rat liver microsomes results in 4-hydroxy-, *N*-desethyl-, and *N*-oxide metabolites (Figure 3.97) (Ruenitz et al. 1983).

3.3.10.4 Lasofoxifene

In humans, ~72% of the administered dose of lasofoxifene is recovered from the urine and feces, with majority of the dose excreted in the feces, probably via bile (Prakash et al. 2008). The absorption of lasofoxifene is slow, with T_{max} values typically exceeding 6 h. The primary metabolic routes of lasofoxifene in humans are direct conjugation (glucuronide and sulfate conjugates) and Phase I oxidation, each accounting for approximately half of the metabolism (Prakash et al. 2008). The primary Phase I metabolites result from hydroxylations

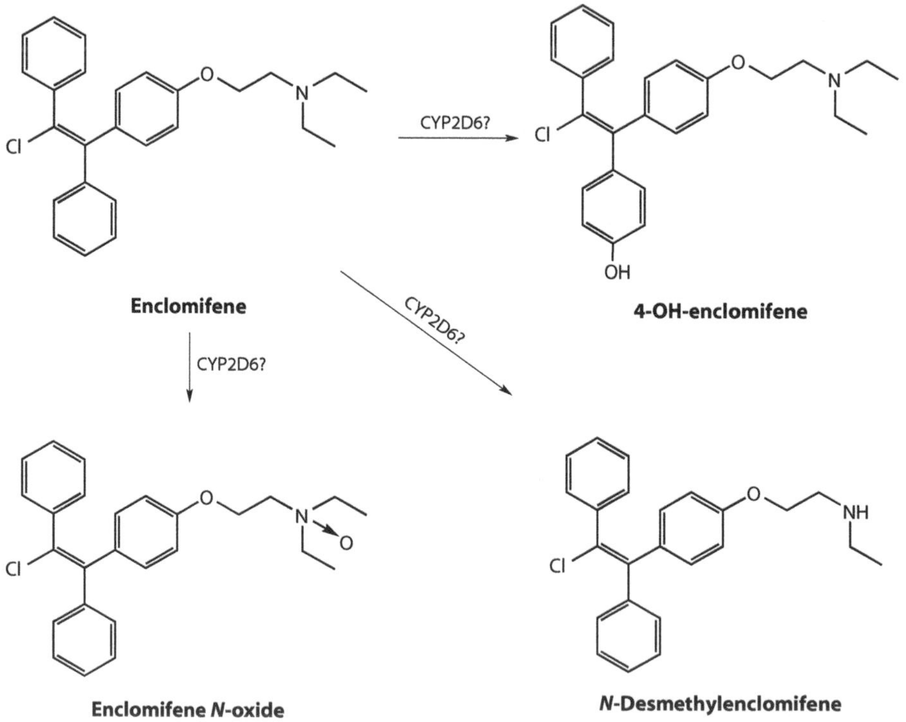

FIGURE 3.97 Metabolism of enclomifene (*E*-clomifene) by CYP2D6. Enclomifene is the more active isomer of clomifene, which is mainly metabolized by CYP2D6. No metabolism is detected in microsomes from the liver of PM of CYP2D6.

on the tetraline moiety and the phenyl rings attached to the tetraline and oxidation on the pyrrolidine moiety. The turnover of lasofoxifene is very slow in human liver microsomes, and only two metabolites are identified as two regioisomers of the catechol metabolite. Further in vitro experiments with recombinant CYPs and selective inhibitors suggested that the oxidative metabolism of lasofoxifene is catalyzed primarily by CYP3A and 2D6 (Prakash et al. 2008). In addition, its glucuronidation is catalyzed by multiple UGTs that are expressed in both the liver (UGT1A1, 1A3, A6, and 1A9) and the intestine (UGT1A8 and 1A10) (Prakash et al. 2008).

3.3.11 SELECTIVE PHOSPHODIESTERASE TYPE 5 INHIBITORS

There are currently three available oral phosphodiesterase type 5 inhibitors: sildenafil, vardenafil, and tadalafil (Gupta et al. 2005). Vardenafil is mainly metabolized by CYP3A4 and 3A5 (Ku et al. 2008), while tadalafil is a mechanism-based inhibitor of CYP3A4 (Ring et al. 2005). Sildenafil is used to treat erectile dysfunction and pulmonary arterial hypertension (Briganti et al. 2005; Cartledge and Eardley 1999; Francis and Corbin 2005; Leibovitch et al. 2007; Moreland et al. 1999; Supuran et al. 2006). It is eliminated predominantly by hepatic metabolism and is converted to an active metabolite, N-desmethylsildenafil (UK-103,320, Figure 3.98), with properties similar to the parent, sildenafil. Plasma concentrations of this metabolite are approximately 40% of those seen for sildenafil. The formation of UK-103,320 is mainly by CYP2C9, 3A4, and 3A5, with lesser contribution from CYP2C19 and 2D6 (Ku et al. 2008; Warrington et al. 2000). Both sildenafil and its metabolite have an elimination half-life of approximately 2.5 h. The concomitant use of potent CYP3A4 inhibitors such as erythromycin, ketoconazole, itraconazole, and the nonspecific CYP inhibitor, cimetidine, increases the plasma level of sildenafil (Gupta et al. 2005; Zusman et al. 1999).

3.3.12 OTHER DRUGS AND COMPOUNDS

Spirosulfonamide, a selective inhibitor of COX-2, is metabolized by CYP2D6, with contribution from CYP3A4 (Guengerich et al. 2002). When incubated with human liver microsomes, spirosulfonamide is converted into monohydroxylated syn and anti OH-spirosulfonamides (M1 and M2, respectively) and a keto metabolite, M3 (Figure 3.99) (Guengerich et al. 2002). Spirosulfonamide is an atypical substrate for CYP2D6 as it lacks the basic nitrogen atom despite a high affinity with a K_m of 7 μM. The phenylsulfonamide of spirosulfonamide contains the only nitrogen in the molecule, and at physiologically relevant pH, this moiety is not positively charged. The sulfonamide moiety is not basic owing to the strong electron-withdrawing properties of the sulfone group. Mutation of Asp301 of CYP2D6 to a neutral amino acid (Asn, Ser, or Gly) does not substantially affect the binding affinity to spirosulfonamide but reduced the metabolic turnover to the same extent as for the classical substrate such as bufuralol (90%) (Guengerich et al. 2002).

Guengerich et al. (2002) have evaluated the binding and metabolism of a series of analogs of spirosulfonamide with CYP2D6 (Figure 3.100). Analogs of spirosulfonamide in which the sulfonamide moiety is modified to an amide, thioamide, methyl sulfone, or hydrogen are ligands for CYP2D6 with K_s values of 1.7–32 μM (Guengerich et al. 2002). All the analogs are substrates of CYP2D6, and the methyl sulfone analog is oxidized to the syn spiromethylene carbinol analog of the major spirosulfonamide product (Guengerich et al. 2002). The methyl sulfone is totally devoid of nitrogen atoms, showing an order of magnitude less affinity with CYP2D6 than spirosulfonamide. Complete removal of the moiety (and the fluorine on the adjacent phenyl ring) yields a compound (analog 6) that is both a reasonably tightly bound ligand (K_d = 4 μM) and a substrate for CYP2D6 (Guengerich et al. 2002). In addition, the presence of a carboxylate (analog 5) leads to a loss of apparent binding and oxidation. This suggests that the role of Asp301 is complex rather than electrostatic interaction with a positively charged atom in the ligand. Hydrogen bonding or electrostatic interaction probably enhances binding of some substrates, but it is not required for all substrates and this explains why some predictive models fail to recognize the proclivity for many substrates, especially those without containing a basic nitrogen atom.

Imatinib is a selective and potent protein tyrosine kinase inhibitor that inhibits the BCR-ABL tyrosine kinase,

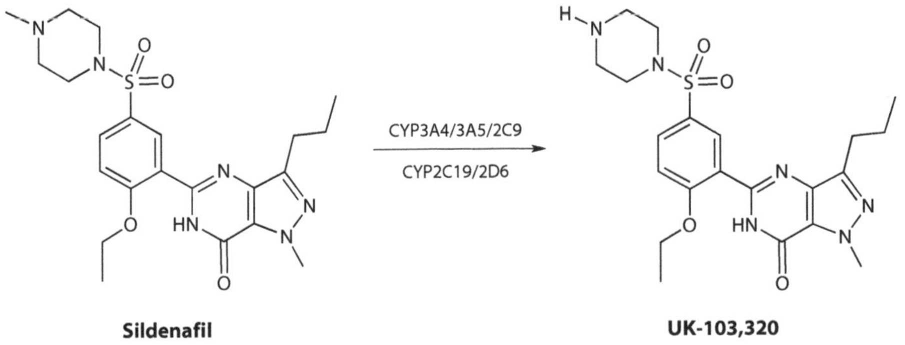

Sildenafil   CYP3A4/3A5/2C9 → CYP2C19/2D6   **UK-103,320**

FIGURE 3.98 Metabolism of sildenafil by CYP3A4, 3A5, 2C9, and 2D6. It is eliminated predominantly by hepatic metabolism and is converted to an active metabolite, N-desmethylsildenafil (UK-103,320), with properties similar to the parent, sildenafil. Plasma concentrations of this metabolite are approximately 40% of those seen for sildenafil. The formation of UK-103,320 is mainly by CYP2C9, 3A4, and 3A5, with lesser contribution from CYP2C19 and 2D6.

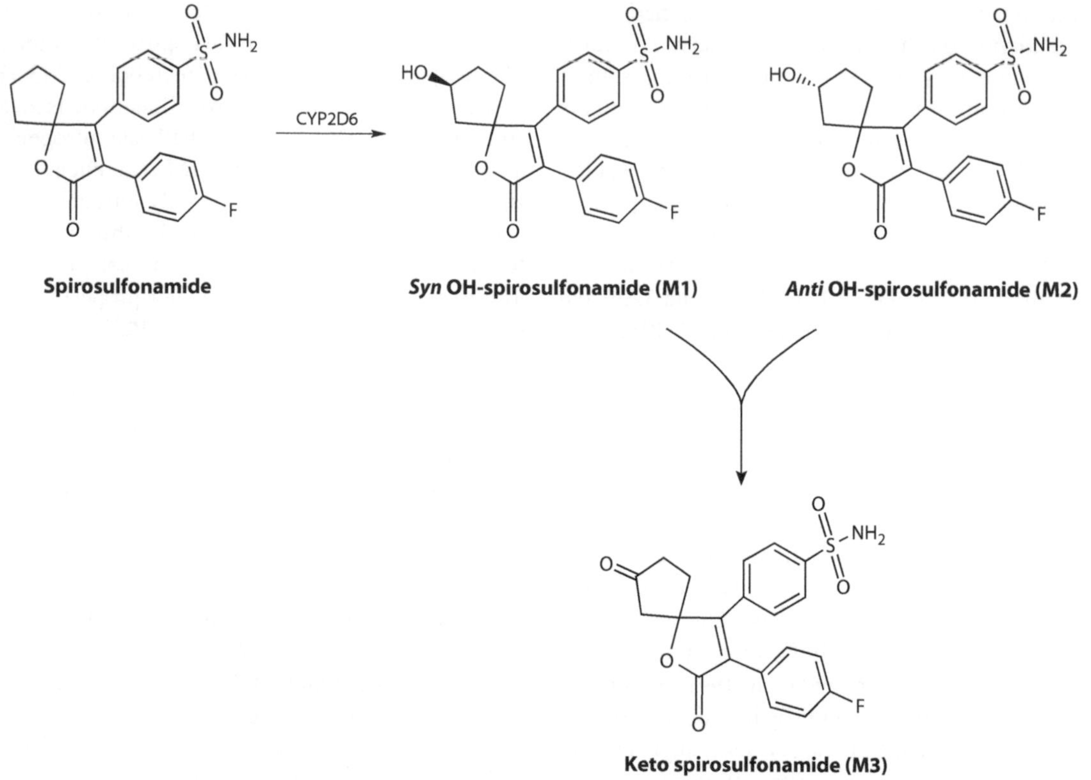

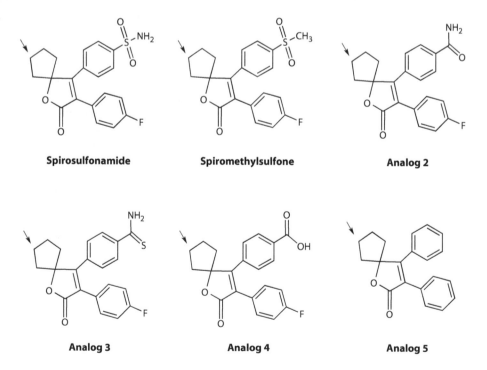

FIGURE 3.99 Metabolism of spirosulfonamide by CYP2D6. When incubated with human liver microsomes, spirosulfonamide is converted into monohydroxylated *syn* and *anti* OH-spirosulfonamides (M1 and M2, respectively) and a keto metabolite, M3. Spirosulfonamide is an atypical substrate for CYP2D6 as it lacks the basic nitrogen atom despite a high affinity with a K_m of 7 μM.

FIGURE 3.100 Structures of spirosulfonamide analogs. The analogs result from the modification of the sulfonamide moiety to an amide, thioamide, methyl sulfone, or hydrogen. All analogs are substrates of CYP2D6, and the methyl sulfone analog is oxidized to the *syn* spiromethylene carbinol analog of the major spirosulfonamide product. The methyl sulfone is totally devoid of nitrogen atoms, showing an order of magnitude less affinity with CYP2D6 than spirosulfonamide. Complete removal of the moiety (and the fluorine on the adjacent phenyl ring) yields a compound (analog 6) that is both a reasonably tightly bound ligand (K_d = 4 μM) and a substrate for CYP2D6. In addition, the presence of a carboxylate (analog 5) led to a loss of apparent binding and oxidation. The arrow indicates site of oxidation.

the constitutive abnormal tyrosine kinase created by the Philadelphia chromosome abnormality in chronic myeloid leukemia (Capdeville et al. 2002; Krause and Van Etten 2005; Muller 2009; Savage and Antman 2002). It has been approved for the treatment of chronic myelogenous leukemia and metastatic and unresectable malignant gastrointestinal stromal tumors (Croom and Perry 2003; Moen et al. 2007; O'Dwyer and Druker 2000; Savage and Antman 2002; Schiffer 2007). The metabolism of imatinib is complicated. Its Phase I metabolism pathways include *N*-demethylation (e.g., main metabolite CGP74588), piperazine ring oxidation with lactam formation (APG049, APG050, M29.6, and M28.8), piperazine-*N*-4 oxidation (CGP71422), pyridine *N*-oxidation (CGP72383), and benzylic hydroxylation (AFN911) (Figure 3.101) (Gschwind et al. 2005). Furthermore, the loss of the piperazine moiety by oxidative deamination and rapid further oxidation of the intermediate aldehyde to a carboxylic acid led to the formation of the metabolite M42.2. Phase II metabolic routes include direct conjugation of imatinib and the *N*-desmethyl metabolite (CGP74588), resulting in M21.0 and M20.0a, respectively, most probably at nitrogen, and glucuronidation

of oxidative metabolites (Gschwind et al. 2005). After oral administration in healthy volunteers, the $t_{1/2\beta}$ of imatinib and that of its major active metabolite, the *N*-demethyl derivative (CGP74588), are approximately 18 and 40 h, respectively (Gschwind et al. 2005). In vitro, imatinib is metabolized to the active CGP74588 by CYP3A4 and 3A5 and, to a lesser extent, by CYP2D6, 1A2, and 2C9 (van Erp et al. 2007). CGP74588 shows in vitro potency similar to the parent imatinib. The plasma AUC for this metabolite is approximately 15% of the AUC for imatinib (Gschwind et al. 2005). In addition, imatinib forms the major oxidative metabolite (M9) via *N*-oxidation on the piperazine ring (Ma et al. 2008, 2009). Imatinib is a potent competitive inhibitor of CYP2C9, 2D6, and 3A4/5 with K_i values of 27, 7.5, and 8 μM, respectively (van Erp et al. 2007).

Gefitinib is a selective inhibitor of the tyrosine kinases associated with epidermal growth factor receptor (EGFR) that plays a pivotal role in the control of cell growth, apoptosis, and angiogenesis; it is clinically used for the treatment of chemoresistant locally advanced or metastatic non–small cell lung cancer patients after failure of both platinum-based

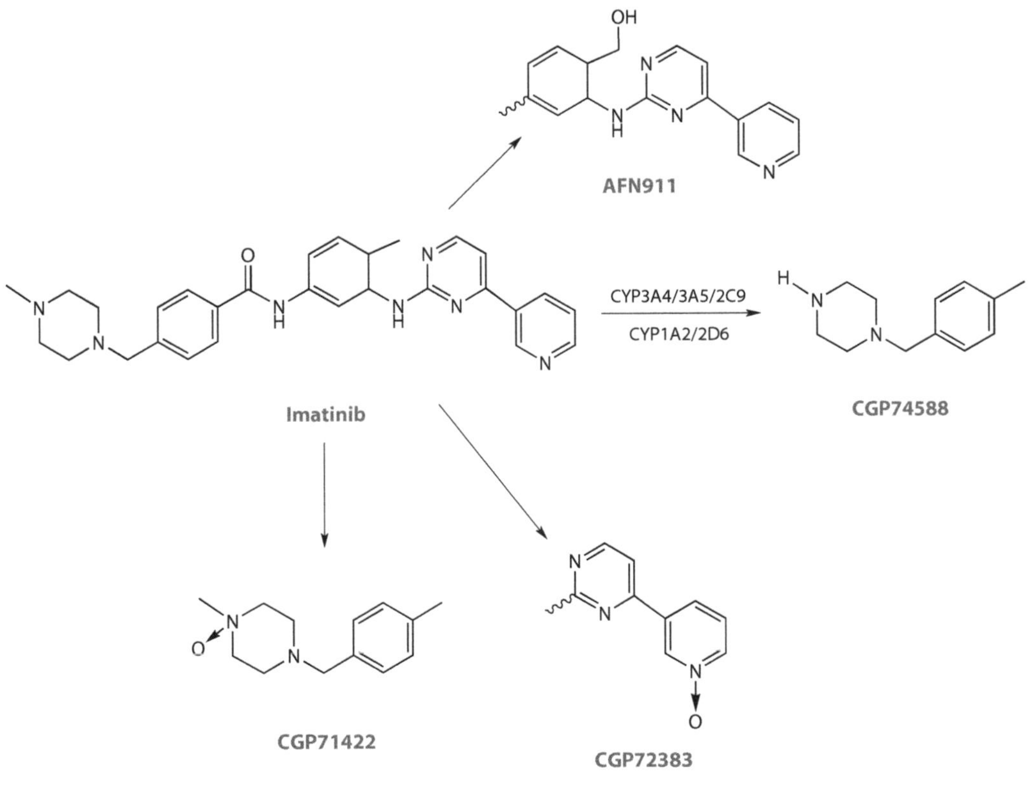

FIGURE 3.101 Metabolism of imatinib. The metabolism of imatinib is complicated. Its Phase I metabolism pathways included *N*-demethylation (e.g., main metabolite CGP74588), piperazine ring oxidation with lactam formation (APG049, APG050, M29.6, and M28.8), piperazine-*N*-4 oxidation (CGP71422), pyridine *N*-oxidation (CGP72383), and benzylic hydroxylation (AFN911). Furthermore, the loss of the piperazine moiety by oxidative deamination and rapid further oxidation of the intermediate aldehyde to a carboxylic acid led to the formation of the metabolite M42.2. Phase II metabolic routes included direct conjugation of imatinib and the *N*-desmethyl metabolite (CGP74588), resulting in M21.0 and M20.0a, respectively, most probably at nitrogen, and glucuronidation of oxidative metabolites. After oral administration in healthy volunteers, the $t_{1/2\beta}$ of imatinib and that of its major active metabolite, the *N*-demethyl derivative (CGP74588), are approximately 18 and 40 h, respectively. In vitro, imatinib is metabolized to the active CGP74588 by CYP3A4 and 3A5 and, to a lesser extent, by CYP2D6, 1A2, and 2C9. CGP74588 showed in vitro potency similar to the parent imatinib. The plasma AUC for this metabolite is approximately 15% of the AUC for imatinib. In addition, imatinib forms the major oxidative metabolite (M9) via *N*-oxidation on the piperazine ring.

and docetaxel chemotherapies (Rahman et al. 2014; Sanford and Scott 2009). After oral administration, gefitinib is slowly absorbed with bioavailability of ~60% in human. Gefitinib is metabolized extensively in the liver into five metabolites primarily by CYP3A4 and, to a lesser extent, by 3A5 and 2D6 (Li et al. 2007; McKillop et al. 2004, 2005). Three sites of biotransformation have been identified: metabolism of the N-propoxymorpholino-group, demethylation of the methoxy-substituent on the quinazoline, and oxidative defluorination of the halogenated phenyl group (McKillop et al. 2004, 2005). Five metabolites have been identified in human plasma, but only O-desmethyl gefitinib has exposure levels comparable to gefitinib. Although this metabolite has similar EGFR-inhibiting activity to gefitinib in the isolated enzyme assay, it has only 1/14 of the potency of gefitinib in one of the cell-based assays. The elimination half-life is approximately 48 h. Daily oral administration of gefitinib to lung cancer patients results in a twofold accumulation compared to single dose administration (Bergman et al. 2007; Leveque 2011; Swaisland et al. 2001). Gefitinib is eliminated mainly hepatically with total plasma clearance of 595 ml/min after intravenous administration. Excretion is predominantly via the feces (86%), with renal elimination of drug and metabolites accounting for less than 4% of the administered dose (Bergman et al. 2007; Leveque 2011; Swaisland et al. 2001, 2005b). Patients with moderately and severely elevated biochemical liver abnormalities have gefitinib pharmacokinetics similar to those of individuals without liver abnormalities. In human liver microsome, gefitinib shows no inhibitory effect on CYP1A2, 2C9, and 3A4 at concentrations ranging from 2 to 5000 ng/ml (Filppula et al. 2014; Surve et al. 2014). At the highest concentration studied (5000 ng/ml), gefitinib inhibits CYP2C19 by 24% and 2D6 by 43%. Rifampicin, an inducer of CYP3A4, reduces mean AUC of gefitinib by 85% in healthy male volunteers (Swaisland et al. 2005a). Caution should be used when administering CYP3A4 inhibitors with gefitinib. Concomitant administration of itraconazole (200 mg once daily for 12 days), an inhibitor of CYP3A4, with gefitinib (250 mg single dose) to healthy male volunteers increases mean gefitinib AUC by 88% (Swaisland et al. 2005a). In contrast, coadministration of high doses of ranitidine with sodium bicarbonate (to maintain the gastric pH above pH 5.0) reduces mean gefitinib AUC by 44%. Substances that are inducers of CYP3A4 activity increase the metabolism of gefitinib and decrease its plasma concentrations. In patients receiving a potent CYP3A4 inducer such as rifampicin or phenytoin, a dose increase to 500 mg daily should be considered in the absence of severe adverse drug reaction, and clinical response and adverse events should be carefully monitored.

Apatinib, a novel vascular endothelial growth factor receptor-2 inhibitor, is an investigational anticancer drug currently undergoing clinical III trials as a potential targeted treatment for metastatic gastric carcinoma, metastatic breast cancer, and advanced hepatocellular carcinoma (Geng and Li 2015; Hu et al. 2014a,b; Li et al. 2013; Tian et al. 2011). The primary metabolic routes of apatinib in humans include E- and Z-cyclopentyl-3-hydroxylation,

N-dealkylation, pyridyl-25-N-oxidation, 16-hydroxylation, dioxygenation, and O-glucuronidation after 3-hydroxylation (Li et al. 2010). Nine major metabolites formed in vivo are confirmed by comparison with reference standards. The total recovery of the administered dose is 76.8% within 96 h postdose, with 69.8% and 7.02% of the administered dose excreted in feces and urine, respectively (Li et al. 2010). Approximately 59.0% of the administered dose is excreted unchanged into feces. Unchanged apatinib is detected in negligible quantities in urine, indicating that systemically available apatinib is extensively metabolized. The major circulating metabolite is the pharmacologically inactive E-3-hydroxy-apatinib-O-glucuronide (M9-2), the steady-state exposure of which is 125% that of apatinib. The steady-state exposures of E-3-hydroxy-apatinib (M1-1), Z-3-hydroxy-apatinib (M1-2), and apatinib-25-N-oxide (M1-6) are 56%, 22%, and 32% of parent drug exposure, respectively. Apatinib is metabolized primarily by CYP3A4/5 and, to a lesser extent, by 2D6, 2C9, and 2E1 (Ding et al. 2013). UGT2B7 is the main enzyme responsible for M9-2 formation. Both UGT1A4 and 2B7 are responsible for the formation of Z-3-hydroxy-apatinib-O-glucuronide (M9-1) (Ding et al. 2013).

Licofelone (ML-3000), a dual inhibitor of COX-1 and COX-2 and 5-lipoxygenase under development for treatment of osteoarthritis, is hydroxylated by CYP2C8, 2J2, and 2D6 to a minor extent (Albrecht et al. 2008). In human liver microsomes, two hydroxylated metabolites, M2 and M4, are formed from licofelone. CYP2J2 is the major enzyme for the formation of M2, with minor contribution from CYP2C8 and 2C9. M4 is generated by CYP2J2, 3A4, 2C8, 2C19, and 2D6 (Albrecht et al. 2008). In human plasma, M2 achieves values of ~20% compared with that of the parent drug. Alternatively, licofelone is rapidly converted into the corresponding acyl glucuronide (M1). After glucuronidation by UGT2B7, 1A9, and 1A3, M1 is converted into the hydroxy-glucuronide M3 by CYP2C8 (Albrecht et al. 2008).

The N-demethylation of amiflamine (FLA 336), a selective and reversible inhibitor of MAO-A selective for serotonergic neurons used for the treatment of depression (Tipton et al. 1985), is correlated with debrisoquine MR in humans (Alvan et al. 1984, 1986). Amiflamine is extensively metabolized by two consecutive N-demethylations. In one clinical study, one depressed patient with extremely high levels of the parent drug (slow demethylator of amiflamine) is a slow hydroxylator of debrisoquine, and two other patients are rapid metabolizers of both drugs (Aringberg-Wistedt et al. 1987). These findings suggest that amiflamine is a substrate of CYP2D6.

CYP2D6, 1A2, 3A4, and 3A5 are the predominant enzymes in the hydroxylation of KR-62980 (a novel and selective peroxisome proliferator-activated receptor-γ agonist) and KR-63198 (Kim et al. 2008). CP-122,721, a novel NK$_1$ antagonist, undergoes O-demethylation and N-dealkylation. In human liver microsomes, O-demethylation is shown to be catalyzed by CYP2D6 with a low K_m value, while N-dealkylation is catalyzed primarily by CYP3A4 (Obach et al. 2007). ML3403, a potent inhibitor of p38 mitogen-activated protein kinase, is metabolized by CYP1A2, 2C19, 2D6, and 3A4 (Kammerer et al. 2007). DRF-4367 forms a hydroxy metabolite, DRF-6574 (a novel COX-2 inhibitor), mediated by CYP2D6 and 2C19 enzymes (Muzeeb et al. 2006). UGT1A1, 1A3, and 1A8 show the catalytic activity toward DRF-4367 (Muzeeb et al. 2006). CYP1A2, 3A4, and 2D6 are responsible for the formation of oxidized metabolites, M1, M2, M4, M5, and M9, from RWJ-53050, a novel anxiolytic agent (Wu et al. 2006). In addition, oxybutynin is metabolized by CYP2C9, 2C19, 2D6, 3A4, and 3A5 (Mizushima et al. 2007).

Tandospirone is also partially metabolized by CYP2D6 while CYP3A4 is the major catalyst for this novel antipsychotic (Natsui et al. 2007). M4 (hydroxylation of the pyrimidine ring) is the major metabolite formed with CYP2D6 while M2 (hydroxylation of the norbornan ring) and 1-PP (oxidative cleavage of the butyl chain) predominated with CYP3A4 (Natsui et al. 2007). TZB-30878, a novel 5-HT$_{1A}$ agonist/5-HT$_3$ antagonist, is currently under development for the treatment of irritable bowel syndrome. It is mainly metabolized by CYP3A, with minor contribution from CYP2D6 (Minato et al. 2008). The use of recombinant human CYP enzymes suggested that CYP3A4 is the major enzyme involved in the oxidative metabolism of CJ-036878 (a novel and potent antagonist of the N-methyl-D-aspartate receptor), with minor contributions from CYP1A2, 2C19, and 2D6 (Nishida et al. 2007).

Tiotropium is a long-acting, antimuscarinic agent, which is often referred to as an anticholinergic (Befekadu et al. 2014; Keam and Keating 2004; McKeage 2015; Rashid and Klein 2014; Rodrigo and Castro-Rodriguez 2015). It has similar affinity to the subtypes of muscarinic receptors, M$_1$ to M$_5$. In the airways, it exhibits pharmacological effects through inhibition of M$_3$ receptors at the smooth muscle leading to bronchodilation (Keam and Keating 2004). Tiotropium is used as an inhalation powder daily for the treatment of bronchospasms associated with chronic obstructive pulmonary disease. This drug is partially (25%) metabolized by CYP2D6 and 3A4, and the Phase I metabolites are subject to subsequent GSH conjugation to a variety of Phase II metabolites (Keam and Keating 2004; McKeage 2015). This enzymatic pathway is inhibited by CYP2D6 and 3A4 inhibitors, such as quinidine, ketoconazole, and gestodene. Tiotropium, an ester, is nonenzymatically cleaved to the alcohol N-methylscopine and dithienylglycolic acid, neither of which binds to muscarinic receptors (Keam and Keating 2004). Tiotropium at supratherapeutic concentrations does not inhibit CYP1A1, 1A2, 2B6, 2C9, 2C19, 2D6, 2E1, and 3A4.

Cevimeline is a muscarinic agonist that binds to muscarinic receptors (Weber and Keating 2008). Cevimeline is indicated for the treatment of symptoms of dry mouth in patients with Sjögren's syndrome (Weber and Keating 2008). After administration of a single 30-mg capsule, cevimeline is rapidly absorbed with a mean time to peak concentration of 1.5 to 2 h. After 24 h, 86.7% of the dose was recovered (16.0% unchanged, 44.5% as cis- and trans-sulfoxide, 22.3% of the dose as glucuronic acid conjugate, and 4% of the dose as N-oxide of cevimeline) (Washio et al. 2003). The mean half-life of cevimeline is ~5 h. CYP2D6 and 3A3/4 are responsible for the metabolism of cevimeline (Washio et al. 2003).

Cevimeline does not inhibit CYP1A2, 2A6, 2C9, 2C19, 2D6, 2E1, and 3A4.

Darifenacin is a competitive muscarinic antagonist indicated for treatment of overactive bladder (Abrams and Andersson 2007; Steers 2006). Darifenacin has a greater affinity for the M$_3$ receptor than for the other known muscarinic receptors (9- and 12-fold greater affinity for M$_3$ compared to M$_1$ and M$_5$, respectively, and 59-fold greater affinity for M$_3$ compared to both M$_2$ and M$_4$ receptors). Darifenacin is mainly metabolized by hepatic CYP2D6 and 3A4 (Leone Roberti Maggiore et al. 2012). The initial products of the hydroxylation and N-dealkylation pathways are the major circulating metabolites (Leone Roberti Maggiore et al. 2012), but they are unlikely to contribute significantly to the overall clinical effect of darifenacin.

Veratramine, a steroidal alkaloid originating from *Veratrum nigrum*, has shown distinct antitumor and antihypertension effects. Unlike most alkaloids, the major reactive sites of veratramine are on rings A and B instead of on the amine moiety. CYP2D6 is the major enzyme mediating hydroxylation of veratramine (Lyu et al. 2015). SULT2A1 conjugates veratramine.

Cariprazine (RGH-188) is an atypical antipsychotic in clinical development for the treatment of schizophrenia and bipolar mania/mixed episodes (Agai-Csongor et al. 2012; Altinbas et al. 2013; Calabrese et al. 2015; Choi et al. 2014; Citrome 2013a,b; Durgam et al. 2014; Gao et al. 2015; Grunder 2010; Gyertyan et al. 2011; Kiss et al. 2010; Sachs et al. 2015; Veselinovic et al. 2013). It is a D$_2$ and D$_3$ receptor partial agonist, with high selectivity toward the D$_3$ receptor (Gyertyan et al. 2011; Kiss et al. 2010). The D$_2$ and D$_3$ receptors are important targets for the treatment of schizophrenia, because the overstimulation of dopamine receptors has been implicated as a possible cause of schizophrenia. Cariprazine acts to inhibit overstimulated dopamine receptors (acting as an antagonist) and stimulate the same receptors when the endogenous dopamine levels are low (Choi et al. 2014; Kiss et al. 2010; Veselinovic et al. 2013). Cariprazine's high selectivity toward D$_3$ receptors could prove to reduce side effects associated with the other antipsychotic drugs, since D$_3$ receptors are mainly located in the ventral striatum and would not incur the same motor side effects (extrapyramidal symptoms) as drugs that act on dorsal striatum dopamine receptors (Choi et al. 2014; Tohen 2015; Veselinovic et al. 2013). Cariprazine has partial agonist as well as antagonist properties depending on the endogenous dopamine levels. When endogenous dopamine levels are high (as is hypothesized in schizophrenic patients), cariprazine acts as an antagonist by blocking dopamine receptors; when endogenous dopamine levels are low, cariprazine acts more as an agonist, increasing dopamine receptor activity. Cariprazine also acts on 5-HT$_{1A}$ receptors, though the affinity is considerably lower than the affinity to dopamine receptors (Gyertyan et al. 2011; Seneca et al. 2011). The drug was approved by the FDA for bipolar I and schizophrenia in adults on September 17, 2015. Cariprazine is converted by hepatic CYP3A4 and 2D6 to two clinically relevant metabolites: desmethyl-cariprazine and didesmethyl-cariprazine, the latter having a longer half-life than cariprazine (Citrome 2013b). It is expected that cariprazine is subject to polymorphic metabolism and inhibition by a number of other drugs. Exposure to didesmethyl-cariprazine exceeds that of the parent drug. Common adverse events of cariprazine include insomnia, extrapyramidal symptoms, akathisia, sedation, nausea, dizziness, and constipation (Citrome 2013a; Tohen 2015; Veselinovic et al. 2013).

Edivoxetine hydrochloride (LY2216684) is a highly selective and potent norepinephrine reuptake inhibitor under clinical development for the treatment of attention-deficit/hyperactivity disorder and depression (Jin et al. 2013; Kielbasa et al. 2015; Lin et al. 2014; Markowitz and Brinda 2014). In December 2013, Eli Lilly announced that the clinical development of edivoxetine will be stopped because of lack of efficacy compared to SSRI alone in three separate clinical trials. Edivoxetine is readily absorbed and metabolized by CYP2D6 and 3A4 (Kielbasa et al. 2012). The T_{max} is approximately 2 h postdose, and the plasma $t_{1/2}$ is approximately 4–6 h irrespective of the dose. Pharmacokinetic parameters are not substantially different between children and adults. Edivoxetine may be a promising nonstimulant therapeutic agent.

3.4 DRUGS OF ABUSE AS SUBSTRATES OF CYP2D6

3.4.1 AMPHETAMINE DERIVATIVES

CYP2D6 also metabolizes drugs of abuse of amphetamine type such as methamphetamine, methylenedioxymethamphetamine (MDMA), N-ethyl-3,4-methylenedioxyamphetamine, and 3,4-methylenedioxyamphetamine (tenamfetamine) (Kreth et al. 2000; Lin et al. 1997; Meyer et al. 2009; Segura et al. 2005; Wu et al. 1997). In humans, the urinary metabolites of methamphetamine included the 4-hydroxy derivative and the N-demethylation metabolite, amphetamine, which is the dominant metabolite, representing almost 50% of all metabolites excreted (Caldwell et al. 1972). CYP2D6 is the primary enzyme for the CYP2D6 in the aromatic 4-hydroxylation and N-demethylation (Figure 3.102) (Lin et al. 1997). Similarly, MDMA is metabolized to methylenedioxyamphetamine via demethylenation by CYP2D6 as a high-affinity enzyme, with low-affinity contributions from CYP1A2, 2B6, and 3A4 (Figure 3.103) (Kreth et al. 2000; Lin et al. 1997; Tucker et al. 1994). However, CYP2D6 does not N-demethylate MDMA (Lin et al. 1997). The polymorphic oxidation of amphetamine analog by CYP2D6 may be a source of interindividual variation in their abuse liability and toxicity. Genetically deficient metabolism of MDMA may help explain why some users of "Ecstasy" appear to be more sensitive to its acute effects (Lin et al. 1997). However, several studies have found no clear association between inherited *CYP2D6* deficiency and MDMA intoxication (de la Torre et al. 2000, 2004; Gilhooly and Daly 2002; Schwab et al. 1999). Further studies are needed to examine the impact of *CYP2D6* polymorphisms on amphetamine analog abuse and toxicity.

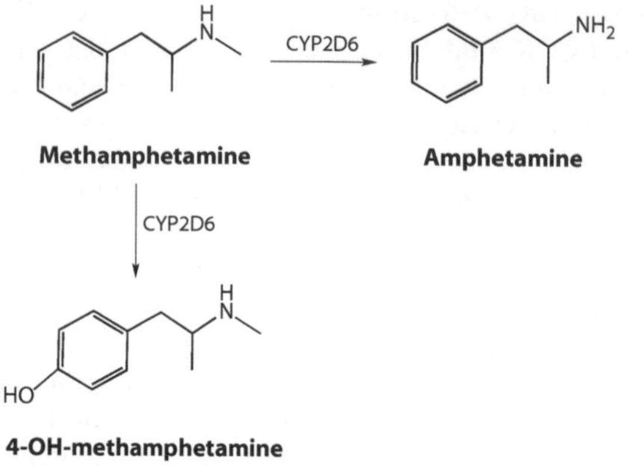

FIGURE 3.102 Metabolism of methamphetamine by CYP2D6. In humans, the urinary metabolites of methamphetamine included the 4-hydroxy derivative and the N-demethylation metabolite, amphetamine, which is the dominant metabolite, representing almost 50% of all metabolites excreted. CYP2D6 is the primary enzyme for the CYP2D6 in the aromatic 4-hydroxylation and N-demethylation. The CYP2D6-catalyzed N-demethylation in the liver microsomes appears to be stereoselective, as the metabolism of the S-isomer correlates much more strongly with CYP2D6 activity than does the R-isomer.

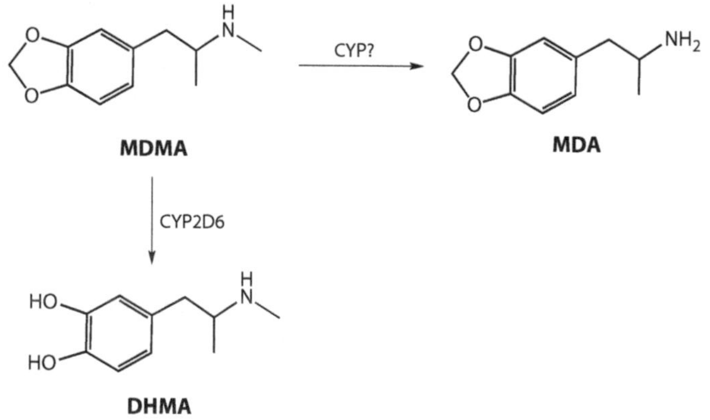

FIGURE 3.103 Metabolism of MDMA by CYP2D6. MDMA is metabolized to methylenedioxyamphetamine via demethylation by CYP2D6 as a high-affinity enzyme, with low-affinity contributions from CYP1A2, 2B6, and 3A4. However, CYP2D6 did not N-demethylate MDMA.

3.4.2 β-CARBOLINES

The structure of β-carboline is similar to that of tryptamine, with the ethylamine chain reconnected to the indole ring via an extra carbon atom, to form a three-ringed structure. β-Carbolines are good substrates for N-methyltransferase and the corresponding cation products are highly neurotoxic (Gearhart et al. 2002). In contrast, hydroxylation and O-demethylation of β-carboline alkaloids, followed by glucuronidation and sulfation (Figure 3.104), are considered major detoxification pathways that can protect the brain. Harmine has two major oxidative pathways: O-demethylation and 6-hydroxylation. Hydroxylation could also occur at the 3- or 4-position of harmine (Tweedie and

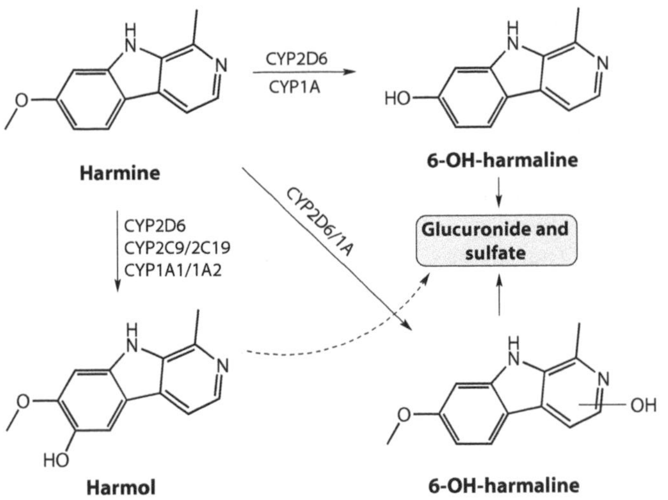

FIGURE 3.104 Metabolism of harmine by CYP2D6 and other CYPs. 6-Hydroxylation and O-demethylation of harmine, followed by glucuronidation and sulfation, are considered major detoxification processes for harmine. Hydroxylation could also occur at the 3- or 4-position of harmine. Hydroxyl β-carboline metabolite such as harmol is readily conjugated through glucuronidation and sulfation. Harmine is mainly metabolized via O-demethylation by CYP2D6, with contribution from CYP1A1, 1A2, 2C9, and 2C19.

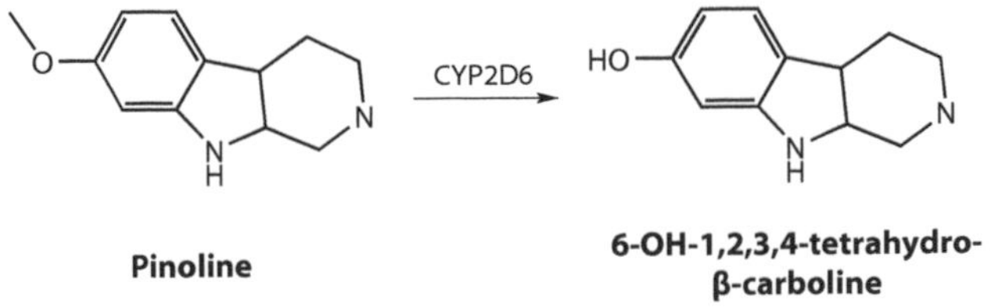

FIGURE 3.105 Metabolism of pinoline by CYP2D6. Pinoline is largely O-demethylated by CYP2D6 to 6-hydroxy-1,2,3,4-tetrahydro-β-carboline. The recombinant CYP2D6.2 exhibited five-fold lower enzyme efficiency (V_{max}/K_m) toward pinoline compared with CYP2D6.1 (wild type), and CYP2D6.10 did not exhibit any catalytic activity. The pinoline urinary MR is much higher in wild-type mice (0.29) than in CYP2D6-humanized transgenic mice (0.007).

Burke 1987). Hydroxyl β-carboline metabolites such as harmol are readily conjugated through glucuronidation and sulfation. The psychotropic β-carboline alkaloids harmaline and harmine are mainly metabolized via O-demethylation by CYP2D6, with contribution from CYP1A1, 1A2, 2C9, and 2C19 (Figure 3.104) (Yu et al. 2003d). CYP2D6 is also a major enzyme mediating the O-demethylation of pinoline, another β-carboline (Yu et al. 2003c,d). These findings indicate that CYP2D6 plays an important role in O-demethylation of the β-carbolines, although its role in their total clearance is yet to be determined. Little is known for many other β-carbolines regarding the specific enzymes that contribute to their biotransformation, although some of them have been shown to undergo aromatic hydroxylation.

Certain β-carbolines, such as pinoline, tryptoline, 6-hydroxy-tetrahydro-β-carboline, harman, and norharman, have been detected as normal constituents in human tissues and body fluids. Their levels in humans are usually increased after drinking alcohol. Harmaline and harmine exhibit potent inhibitory effect on MAO-A activity (Herraiz and Chaparro 2006), and both compounds are the principal active agents in *Peganum harmala*, a traditional Chinese medicine. The neurotoxic properties of β-carboline alkaloids may account for their associations with Parkinson's disease. Tryptophan-derived β-carbolines are similar to 1-methyl-4-phenyl-1,2,3,6-tetrahydropyridine (MPTP) in structure, which is known to induce immediate and irreversible parkinsonism through its neurotoxic metabolite, a quaternary ion (MPP+). Therefore, CYP2D6 status may be associated with Parkinson's disease.

Jiang et al. (2009) have revealed that pinoline is largely O-demethylated by CYP2D6 to 6-hydroxy-1,2,3,4-tetrahydro-β-carboline (Figure 3.105). The recombinant CYP2D6.2

exhibits fivefold lower enzyme efficiency (V_{max}/K_m) toward pinoline compared with CYP2D6.1 (wild type), and CYP2D6.10 does not exhibit any catalytic activity. The pinoline urinary MR is much higher in wild-type mice (0.29, $n = 4$) than in CYP2D6-humanized transgenic mice (0.007, $n = 4$) (Jiang et al. 2009). Pinoline O-demethylation may be regarded as a marker reaction for CYP2D6.

3.4.3 DESIGNER DRUGS

Several designer drugs of the pyrrolidonophenone type are found to be mainly metabolized by CYP2D6, with contributions from other CYPs (Maurer et al. 2004). The pyrrolidonophenones have entered the illicit drug market in Germany and other countries and are distributed among drug abusers. The major metabolic routes of all 4′-methyl pyrrolidonophenones in animals included 4′-methyl hydroxylation followed by oxidation to the respective carboxylic acids, oxidation of the pyrrolidine moiety to the respective lactam, and reduction of the keto group (Springer et al. 2002, 2003b). The new designer drug 4′-methyl-α-pyrrolidinobutyrophenone (MPBP) is mainly hydroxylated by CYP2D6, with contribution from CYP2C19 and 1A2 (Figure 3.106) (Peters et al. 2008). According to the relative activity factor approach, these enzymes account for 54%, 30%, and 16% of net clearance of MPBP. In experiments with liver microsomes from CYP2D6 and 2C19 PMs, MPBP 4′-hydroxylation is 78% and 79% lower in comparison with pooled human liver microsomes, respectively (Peters et al. 2008).

In addition, both CYP2D6 and 2C19 catalyze the 4′-hydroxylation of another designer drug 4′-methyl-α-pyrrolidinopropiophenone (MPPP) (Figure 3.107) (Springer et al. 2003a). The apparent K_m for MPPP by CYP2D6 is 9.8 μM. CYP2C19 is not saturable over the tested substrate concentration range (2–1000 μM) and showed biphasic kinetics (Springer et al. 2003a). On the basis of kinetic data corrected for the relative activity factors and inhibition studies, CYP2D6 is the enzyme mainly responsible for MPPP 4′-hydroxylation. CYP1A2, 2B6, and 2C9 exhibit only very minor activity toward MPPP. Previous rat studies have shown that MPPP is mainly metabolized by 4′-hydroxylation followed by dehydrogenation to the corresponding carboxylic acid (Springer et al. 2002).

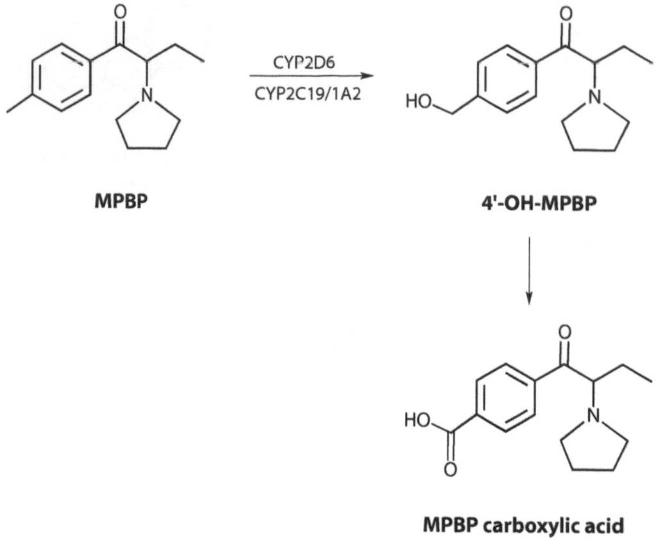

FIGURE 3.106 Metabolism of MPBP by CYP2D6, 2C19, and 1A2. The major metabolic routes of 4′-methyl pyrrolidonophenones in animals included 4′-methyl hydroxylation followed by oxidation to the respective carboxylic acids, oxidation of the pyrrolidine moiety to the respective lactam, and reduction of the keto group. MPBP is mainly hydroxylated by CYP2D6, with contribution from CYP2C19 and 1A2. According to the relative activity factor approach, these enzymes accounted for 54%, 30%, and 16% of net clearance of MPBP.

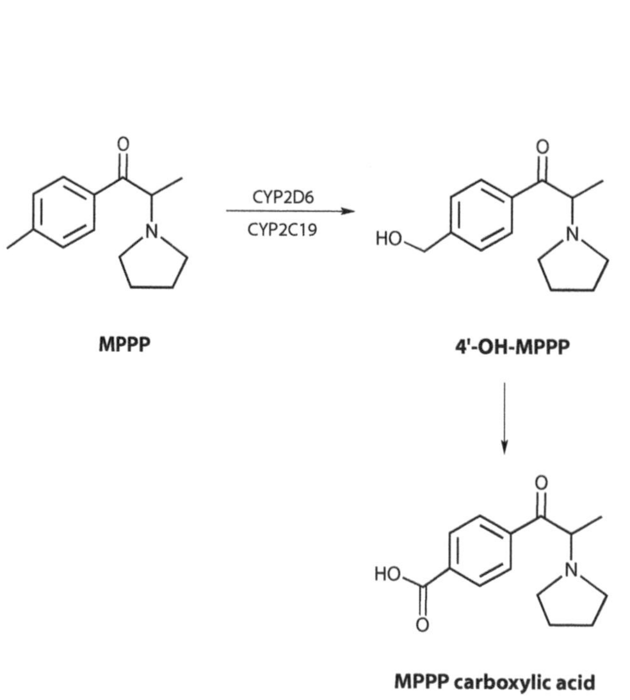

FIGURE 3.107 Metabolism of MPPP by CYP2D6 and 2C19. Both CYP2D6 and 2C19 catalyze the 4′-hydroxylation of MPPP. On the basis of kinetic data corrected for the relative activity factors, CYP2D6 is the enzyme mainly responsible for MPPP 4′-hydroxylation. CYP1A2, 2B6, and 2C9 exhibited only very minor activity toward MPPP.

3.4.4 Indolealkylamines

CYP2D6 plays an important role in the metabolism of many indolealkylamines agents, which are 5-hydroxytryptamine (5-HT/serotonin) analogs that mainly act on the serotonin system (Yu 2008). Structurally, this group of compounds contains an indole moiety and a basic nitrogen atom, which are connected by an alkyl chain usually of two carbons in length. They mainly act on the 5-HT receptors, and some indolealkylamines such as ergotamine and a series of triptans including sumatriptan, zolmitriptan, naratriptan, almotriptan, frovatriptan, and rizatriptan have been developed for the clinical treatment of migraine (Saper and Silberstein 2006). However, many other indolealkylamine agents are widely abused compounds in developed countries although some have demonstrated therapeutic potential in psychopharmacotherapy. These include the notorious lysergic acid amides such as D-lysergic acid diethylamide and ergine; tryptamine derivatives such as psilocybin (O-phosphoryl-4-hydroxy-N,N-dimethyltryptamine), N,N-dimethyltryptamine (DMT), bufotenine (5-hydroxy-dimethyltryptamine), 5-methoxy-N,N-dimethyltryptamine (5-MeO-DMT), and 5-methoxy-N,N-diisopropyltryptamine (5-MeO-DIPT, "Foxy," or "Foxy Methoxy"); and β-carbolines such as harman, harmaline, harmine, and ibogaine (Yu 2008). All these compounds produce hallucinogenic and stimulant activities and high doses of administration will cause mydriasis, nausea, jaw clenching, and overt hallucinations with auditory and visual distortions.

Structurally, DMT, psilocybin (a prodrug), 5-MeO-DMT, and bufotenine are analogous to 5-HT and tryptamine. DMT is created in small amounts by the human body during normal metabolism by the enzyme tryptamine-N-methyltransferase. Like tryptamine (Airaksinen and Kari 1981; Yu et al. 2003a), tryptamine derivatives including DMT, bufotenine, and 5-MeO-DMT are excreted primarily via oxidative deamination, which is predominantly catalyzed by MAO-A (Figure 3.108) (Sanders-Bush et al. 1976; Suzuki et al. 1981; Szara and Axelrod 1959). However, there are some species differences in the metabolism of these tryptamine derivatives. DMT and 5-MeO-DMT are metabolized through deamination, N-demethylation, O-demethylation, and N-oxygenation, and N-oxide is the major metabolite in rat tissues (Sitaram et al. 1987a,b). In contrast, oxidative deamination and O-demethylation followed by glucuronidation or deamination are the major metabolic pathways for 5-MeO-DMT in rats (Agurell et al. 1969). In human liver microsomes and cultured hepatocytes, 5-MeO-DMT undergoes extensive deamination and O-demethylation, with the latter reaction being primarily catalyzed by CYP2D6 and resulting in toxic bufotenine (Yu et al. 2003c). Bufotenine is conjugated or undergoes oxidative deamination to result in 5-OH-indole acetic acid. Alternatively, bufotenine may be converted back to 5-MeO-DMT in vivo by methyltransferase. Another pathway of 5-MeO-DMT is N-oxidation, resulting in its N-oxide, which is water soluble.

5-MeO-DMT O-demethylation is potentially a bioactivating step because the metabolite bufotenine is a biologically

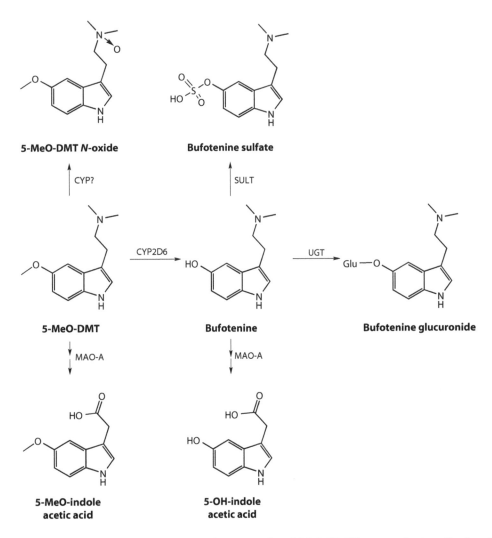

FIGURE 3.108 Metabolism of 5-MeO-DMT by CYP2D6. Like tryptamine, 5-MeO-DMT is excreted primarily via oxidative deamination resulting in 5-MeO-indole acetic acid, which is predominantly catalyzed by MAO-A. In human liver microsomes and cultured hepatocytes, 5-MeO-DMT undergoes extensive deamination and O-demethylation, with the latter reaction being primarily catalyzed by CYP2D6 and resulting in toxic bufotenine. Bufotenine is conjugated or undergoes oxidative deamination to result in 5-OH-indole acetic acid. Bufotenine may be converted back to 5-MeO-DMT in vivo by methyltransferase. Alternatively, 5-MeO-DMT can be converted to its N-oxide, which is water soluble.

active and toxic agent. Although the psychedelic activity of bufotenine has been documented (McBride 2000; Ott 2001), there are arguments about this because of its poor lipid solubility. Binding studies (McBride 2000; Roth et al. 1997) have shown that bufotenine is a potent ligand for the 5-HT_{2A} receptor with a 10-fold higher affinity compared to 5-MeO-DMT and that bufotenine is approximately three times more potent than 5-MeO-DMT in the brain (Vogel and Evans 1977). It is interesting to note that bufotenine (the metabolite of 5-MeO-DMT) is presently listed as a Schedule I controlled substance in the United States, while 5-MeO-DMT is not.

5-MeO-DIPT is another abused tryptamine derivative, which contains N,N-diisopropyl groups, rather than the N,N-dimethyl groups within 5-MeO-DMT. The abuse problem of 5-MeO-DIPT first emerged in the United States in 2001 and then in Japan, and, since then, in many other countries. 5-MeO-DIPT is amended into Schedule I of the Controlled Substances Act in 2003 in the United States and is banned in

Japan in April 2005. Analyses of urine samples collected from 5-MeO-DIPT consumers (Kamata et al. 2006; Meatherall and Sharma 2003) have indicated that this drug of abuse is mainly excreted via O- and N-dealkylation and deamination and 6-hydroxylation with the former pathway resulting in 5-OH-indole acetic acid (Figure 3.109). Because isotope-labeled 5-MeO-DIPT has not been available and 5-OH-indole acetic acid is intrinsically present in human urine, it is difficult to determine the relative contribution of these metabolic pathways in vivo. 6-OH-5-MeO-DIPT formed by 6-hydroxylation is mainly excreted as glucuronide and sulfate conjugates. Side-chain degradation by N-deisopropylation results in the corresponding secondary amine, 5-methoxy-N-isopropyltryptamine (5-MeO-NIPT). Nonetheless, extensive studies using human liver microsomes and recombinant enzymes (Narimatsu et al. 2006, 2008) have revealed that CYP2D6 is the major enzyme responsible for 5-MeO-DIPT O-dealkylation, while CYP1A2, 3A4, and 2C19 mediate its N-dealkylation.

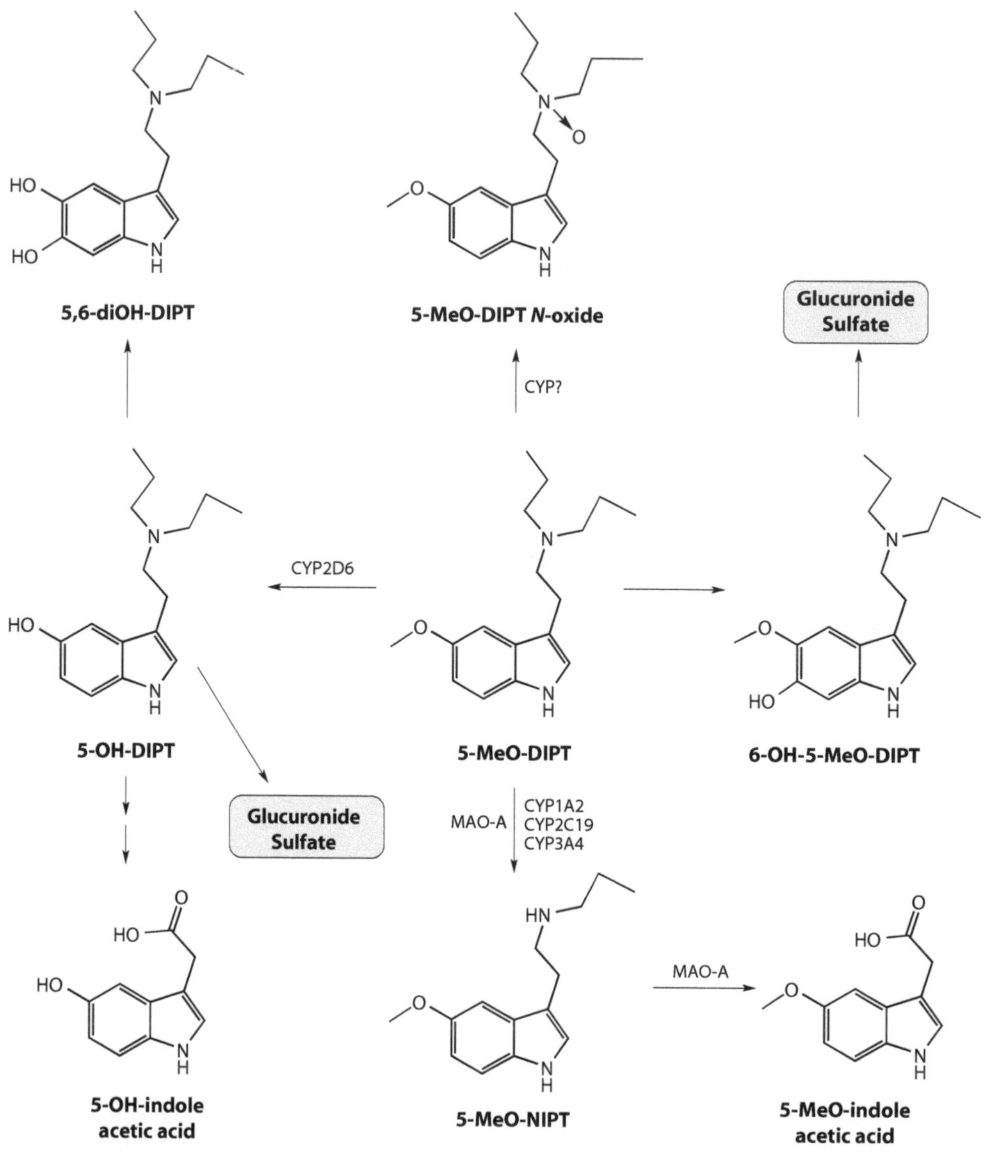

FIGURE 3.109 Metabolism of 5-MeO-DIPT. Analyses of urine samples collected from 5-MeO-DIPT consumers have indicated that this drug of abuse is mainly excreted via O- and N-dealkylation and deamination and 6-hydroxylation with the former pathway resulting in 5-OH-indole acetic acid. 6-OH-5-MeO-DIPT formed by 6-hydroxylation is mainly excreted as conjugates (both glucuronides and sulfates). Side-chain degradation by N-deisopropylation results in the corresponding secondary amine, 5-MeO-NIPT. CYP2D6 is the major enzyme responsible for 5-MeO-DIPT O-dealkylation, while CYP1A2, 3A4, and 2C19 mediate its N-dealkylation. Furthermore, 5-MeO-DIPT is mainly O-dealkylated in vitro, whereas it is primarily N-dealkylated in rat liver microsomes, reflecting the species-dependent metabolism of 5-MeO-DIPT.

The critical role for CYP2D6 in 5-MeO-DIPT O-dealkylation might provide an explanation for the considerable difference in urinary 5-hydroxy-N,N-diisopropyltryptamine levels observed between Caucasian and Japanese drug users (Kamata et al. 2006; Meatherall and Sharma 2003), when ~10% Caucasian 5-MeO-DIPT users have deficient CYP2D6 activity and most Japanese users have normal or intermediate CYP2D6 activity. Furthermore, 5-MeO-DIPT is mainly O-dealkylated in human liver microsomes, whereas it is primarily N-dealkylated in rat liver microsomes (Narimatsu et al. 2006, 2008), reflecting the species-dependent metabolism of 5-MeO-DIPT. The N-oxide of 5-MeO-DIPT is detected in vitro and not in vivo.

3.5 FLUORESCENT PROBES AS SUBSTRATES OF CYP2D6

A novel fluorogenic probe, 3-[2-(N,N-diethyl-N-methyl-ammonium)ethyl]-7-methoxy-4-methylcoumarin, is reported to be a useful substrate to probe CYP2D6 activity in vitro for high-throughput screening (Chauret et al. 2001; Stresser et al. 2002; Yu et al. 2006). CYP2D6 metabolizes 7-methoxy-4-(aminomethyl)-coumarin (MAMC), suitable for high-throughput screening (Onderwater et al. 1999), to investigate the hydroxylation of debrisoquine (Lightfoot et al. 2000). CYP2D6 is found to preferentially O-dealkylate MAMC to 7-hydroxy-4-(aminomethyl)-coumarin (HAMC)

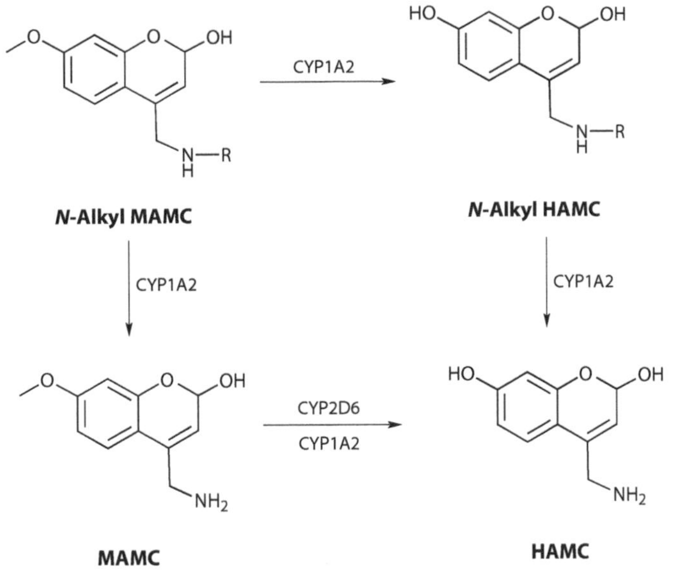

FIGURE 3.110 Metabolism of MAMC by CYP2D6 and 1A2. CYP2D6 is shown to solely *O*-dealkylate MAMC to HAMC, with a high affinity and turnover, with only CYP1A2 contributing to some extent to HAMC formation in human liver microsomes. Because of the distinctly different fluorescent properties of the parent compound and its metabolite, MAMC *O*-dealkylation can be sensitively measured using a developed microtiter assay.

(Figure 3.110), with a high affinity and turnover, with only minor contribution from CYP1A2 to HAMC formation in human liver microsomes (Onderwater et al. 1999). Because of the distinctly different fluorescent properties of the parent molecule and its metabolite, MAMC *O*-dealkylation can be sensitively measured using a developed microtiter assay with high-throughput capacity (Venhorst et al. 2000).

3.6 PLANT ALKALOIDS, TOXICANTS, AND ENVIRONMENTAL COMPOUNDS AS SUBSTRATES OF CYP2D6

3.6.1 PLANT ALKALOIDS

CYP2D6 is largely responsible for the metabolism of ibogaine to its *O*-desmethyl active metabolite 12-hydroxyibogamine (noribogaine) (Figure 3.111) (Obach et al. 1998), a psychoactive alkaloid isolated from the root of *Tabernanthe*

iboga, a rainforest shrub native to Africa. Ibogaine is used by indigenous peoples in low doses to combat fatigue, hunger, and thirst, and in higher doses as a sacrament in religious rituals. Ibogaine represents a potentially useful therapeutic agent in the treatment of opiate and psychostimulant addiction and opiate withdrawal (Alper et al. 1999, 2000; Frances et al. 1992; Glick et al. 1992; Popik et al. 1995). Ibogaine has shown preliminary efficacy for opiate detoxification and for short-term stabilization of drug-dependent persons as they prepare to enter substance abuse treatment (Mash et al. 2000). Ibogaine and noribogaine interacted with 5-HT transporters (SERT/SLC6A4) to inhibit 5-HT uptake (Baumann et al. 2001; Mash et al. 1995).

A study using human liver microsomes indicated that CYP2D6 and 3A4 are able to metabolize emetine to cephaeline (both are alkaloids from ipecac) and 9-*O*-demethylemetine, and CYP3A4 also participated in metabolizing emetine to 10-*O*-demethylemetine (Asano et al. 2001). The CYP2D6 enzyme also has high affinity for toxic plant alkaloids such as lasiocarpine and monocrotaline, both pyrrolizidine alkaloids (Wolff et al. 1985, 1987). Pyrrolizidine alkaloids are found in plants growing in most environments and all parts of the world. They have long been known to be a health hazard for livestock, wildlife, and humans (Stegelmeier et al. 1999). Plants known or suspected to contain toxic alkaloids are widely used for medicinal purposes as home remedies all over the world, without systematic testing for safety (Roeder 2000). Most pyrrolizidine alkaloids are hepatotoxic (Liddell 2001). The following are the major metabolic pathways of unsaturated pyrrolizidine alkaloids such as lasiocarpine in animals (Figures 3.112 and 3.113) (Cheeke 1988; Fu et al. 2004; Liddell 2001): (a) hydrolysis of the ester groups, (b) *N*-oxidation, and (c) dehydrogenation of the pyrrolizidine nucleus to dehydroalkaloids (pyrrolic derivatives). Routes a and b are believed to be detoxification mechanisms, while route c leads to toxic metabolites capable of binding DNA and proteins and appears to be the major activation mechanism (Fu et al. 2004; Zhou et al. 2007). Both lasiocarpine and monocrotaline are procarcinogens that are bioactivated by P450s to carcinogens (Xia et al. 2006).

3.6.2 NEUROTOXIN

CYP2D6 has also been shown to metabolize procarcinogens and neurotoxins such as MPTP (Coleman et al. 1996;

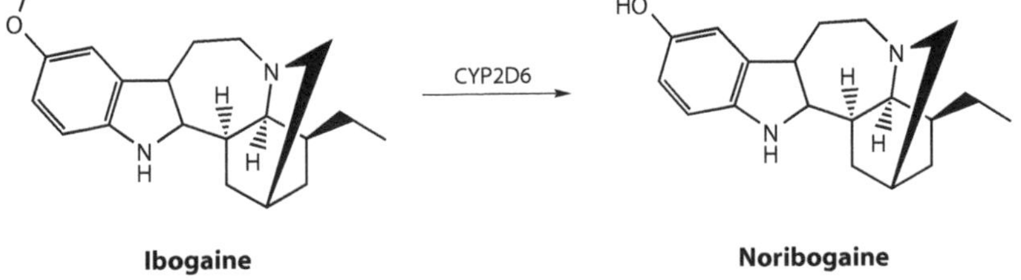

FIGURE 3.111 Metabolism of ibogaine by CYP2D6. CYP2D6 is largely responsible for the metabolism of ibogaine to its *O*-desmethyl active metabolite 12-hydroxyibogamine (noribogaine).

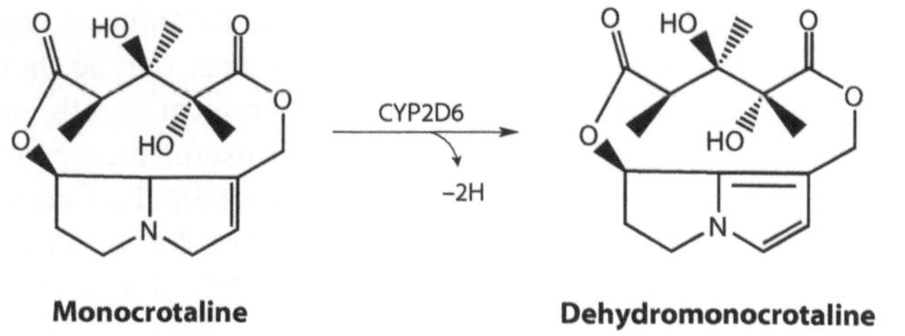

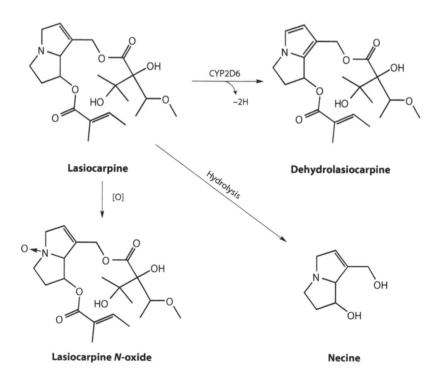

FIGURE 3.113 Metabolism of monocrotaline by CYP2D6. Monocrotaline is metabolized by CYP2D6 via dehydrogenation of the pyrrolizidine nucleus to toxic dehydromonocrotaline.

Gilham et al. 1997; Kalgutkar et al. 2003; Modi et al. 1997), 1,2,3,4-tetrahydroquinoline (Ohta et al. 1990), and indole-alkylamines (Yu et al. 2003c). MPTP is a neurotoxin and potent inducer of experimental Parkinson's disease in non-human primates (Barsoum et al. 1986; Emborg 2007; Jenner 2003). Besides the MAO-B–mediated bioactivation of MPTP to the positively charged mitochondrial neurotoxin N-methyl-4-phenylpyridinium (MPP$^+$), CYP2D6, 1A2, and 3A4 metabolize MPTP to the corresponding nonneurotoxic N-4-(4'-hydroxyphenyl)-N-methyl-1,2,3,6-tetrahydropyridine and 4-phenyl-1,2,3,6-tetrahydropyridine (PTP) metabolites via N-demethylation (Figure 3.114) (Coleman et al. 1996; Modi et al. 1997). FMO can efficiently catalyze the formation of N-oxide from MPTP. The high-affinity activity toward MPTP is absent in liver microsomes from a PM subject (Coleman et al. 1996). Rat CYP2D and 2C can N-demethylate

MPTP (Narimatsu et al. 1996) and female DA rats lacking CYP2D activity are more sensitive to MPTP neurotoxicity than other strains (Jimenez-Jimenez et al. 1991). CYP2D6 efficiently hydroxylates various β-carbolines (Herraiz et al. 2006). N^2-methyl-1,2,3,4-tetrahydro-β-carboline, a close MPTP analog, is extensively hydroxylated to 6-hydroxy-N^2-methyl-1,2,3,4-tetrahydro-β-carboline and a corresponding 7-hydroxy-derivative (Herraiz et al. 2006). CYP2D6 is also involved in the metabolism of diuron, a widely used herbicide and antifouling biocide (Abass et al. 2007).

3.6.3 Herbicides and Pesticides

CYP2D6 is involved in the metabolism of diuron (3-(3,4-dichlorophenyl)-1,1-dimethylurea), a widely used herbicide and antifouling biocide (Figure 3.115) (Abass et al. 2007). Diuron has been classified as a slightly hazardous (toxicity class III) pesticide by World Health Organization. Diuron has been characterized as a "known/likely" human carcinogen by the US Environmental Protection Agency based on the data that diuron induces urinary bladder carcinoma in both sexes of the Wistar rat, kidney carcinoma in the male rat, and breast carcinoma in female NMRI mice. In human liver microsomes, only N-demethylated diuron is formed. Recombinant CYP1A1, 1A2, 2C19, and 2D6 catalyze the N-demethylation of diuron with high activities. Relative contributions of human CYP1A2, 2C19, and 3A4 to hepatic diuron N-demethylation are estimated to be ~60%, 14%, and 13%, respectively. In studies of rats and dogs, N-(3,4-dichlorophenyl)urea is the

FIGURE 3.112 Metabolism of lasiocarpine by CYP2D6. The major metabolic pathways of unsaturated pyrrolizidine alkaloids such as lasiocarpine in animals include hydrolysis of the ester groups, N-oxidation, and dehydrogenation of the pyrrolizidine nucleus by CYP2D6 to dehydro-alkaloids (pyrrolic derivatives). Hydrolysis and N-oxidation are believed to be detoxification mechanisms, while the formation of pyrrolic metabolite leads to toxic metabolites capable of binding DNA and proteins.

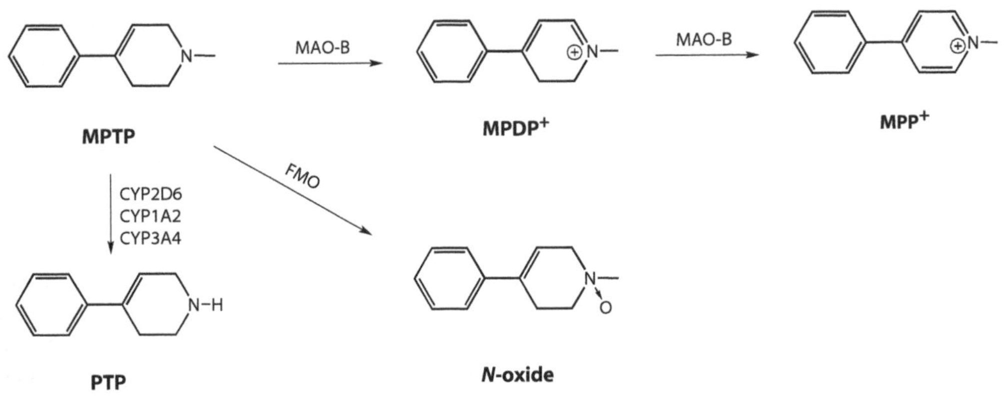

FIGURE 3.114 Metabolism of MPTP by CYP2D6 and other enzymes. Besides MAO-B–mediated bioactivation of MPTP to MPP+, CYP2D6, 1A2, and 3A4 metabolize MPTP to the corresponding nonneurotoxic N-4-(4'-hydroxyphenyl)-N-methyl-1,2,3,6-tetrahydropyridine and PTP metabolites via N-demethylation. FMO can efficiently catalyze the formation of N-oxide from MPTP.

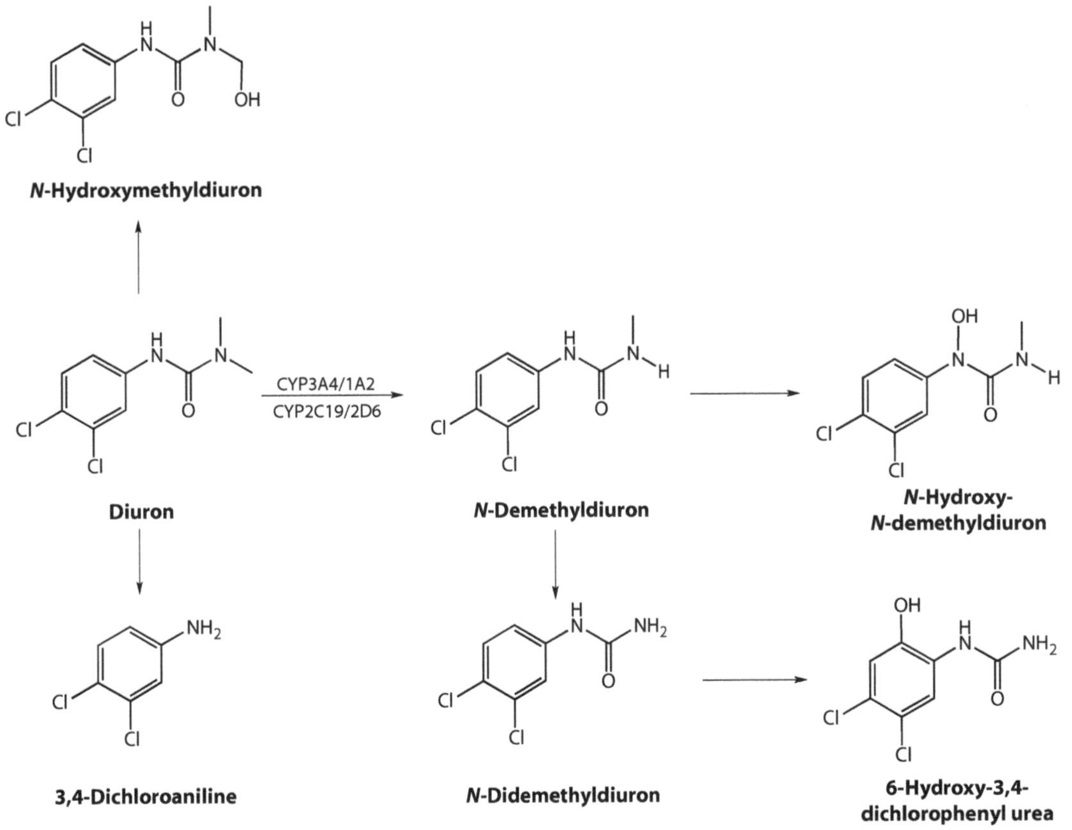

FIGURE 3.115 Metabolism of diuron in humans. Recombinant CYP1A1, 1A2, 2C19, and 2D6 catalyze the N-demethylation of diuron. In a hospitalized patient, diuron is completely metabolized, mainly via demethylation and didemethylation. In addition, urine extracts contained hydroxyphenyldiuron and 3,4-dichloroaniline. In a human postmortem case, diuron and its demethylated, didemethylated, and hydroxylated metabolites are identified in plasma and urine. In the postmortem case, N-demethyldiuron is the major metabolite, whereas in the hospitalized case, N-didemethyldiuron is the major metabolite in blood and N-demethyldiuron is dominant in urine.

predominant metabolite in the urine. Small amounts of N-(3,4-dichlorophenyl)-N-methylurea, 3,4-dichloroaniline, 3,4-dichlorophenol, and unchanged diuron are also detected (Hodge et al. 1967). In a hospitalized patient, diuron is completely metabolized, mainly via demethylation and didemethylation with the corresponding metabolites detected in the

blood and urine. In addition, high levels of hydroxyphenyldiuron and moderate levels of 3,4-dichloroaniline are detected in the urine (Van Boven et al. 1990). In a human postmortem case, diuron and its demethylated, didemethylated, and hydroxylated metabolites are all identified in plasma and urine. Diuron levels are as high as 5 mg/l in plasma and 3 mg/l in urine and the total concentration of diuron plus metabolites in plasma is approximately 100 mg/l, with an estimated ingestion of at least several grams (Verheij et al. 1989). In the postmortem case, *N*-demethyldiuron is the major metabolite, whereas in the hospitalized case, *N*-didemethyldiuron is the primary metabolite in blood and *N*-demethyldiuron is dominant in urine.

3.7 ENDOGENOUS COMPOUNDS AS SUBSTRATES OF CYP2D6

3.7.1 5-METHOXYINDOLETHYLAMINE

CYP2D6 is expressed at low levels but constitutively in human brain (Chinta et al. 2002; Miksys et al. 2002), raising the possibility of an endogenous function of CYP2D6 in the metabolism of neurochemicals. CYP2D-dependent dextromethorphan *O*-demethylase activity has been well described in rat brain tissues (Tyndale et al. 1999). A study using the CYP2D6-humanized mouse line has established that CYP2D6 is a 5-methoxyindolethylamine *O*-demethylase (Yu et al. 2003c) and 5-methoxytryptamine, a metabolite and precursor of melatonin (*N*-acetyl-5-methoxytryptamine),

is metabolized by CYP2D6 to 5-hydroxytryptamine (5-HT/serotonin) with a high turnover of 51.7 min^{-1} and a relatively low K_m of 19.5 µM (Figure 3.116) (Yu et al. 2003b). 5-HT is usually synthesized from tryptophan by hydroxylation and decarboxylation. 5-HT can be converted to melatonin via NAT and 5-hydroxyindole-*O*-methyltransferase. The production of 5-HT from 5-methoxytryptamine by CYP2D6 is significantly inhibited by SSRIs such as fluoxetine, norfluoxetine, fluvoxamine, and citalopram (Yu et al. 2003b). Liver microsomes prepared from CYP2D6-transgenic mice exhibit ~16-fold higher 5-methoxytryptamine *O*-demethylase activity than that from the wild type. Coadministration of 5-methoxytryptamine and pargyline increases serum 5-HT level approximately threefolds in human CYP2D6-transgenic mice than the wild type (Yu et al. 2003b). It appears that the CYP2D6-mediated 5-methoxytryptamine *O*-demethylation affects serotonin production and thus influences a range of neurophysiologic functions.

3.7.2 TYRAMINES

Dopamine is generated from L-tyrosine by tyrosine hydroxylase and aromatic L-amino acid decarboxylase (Udenfriend

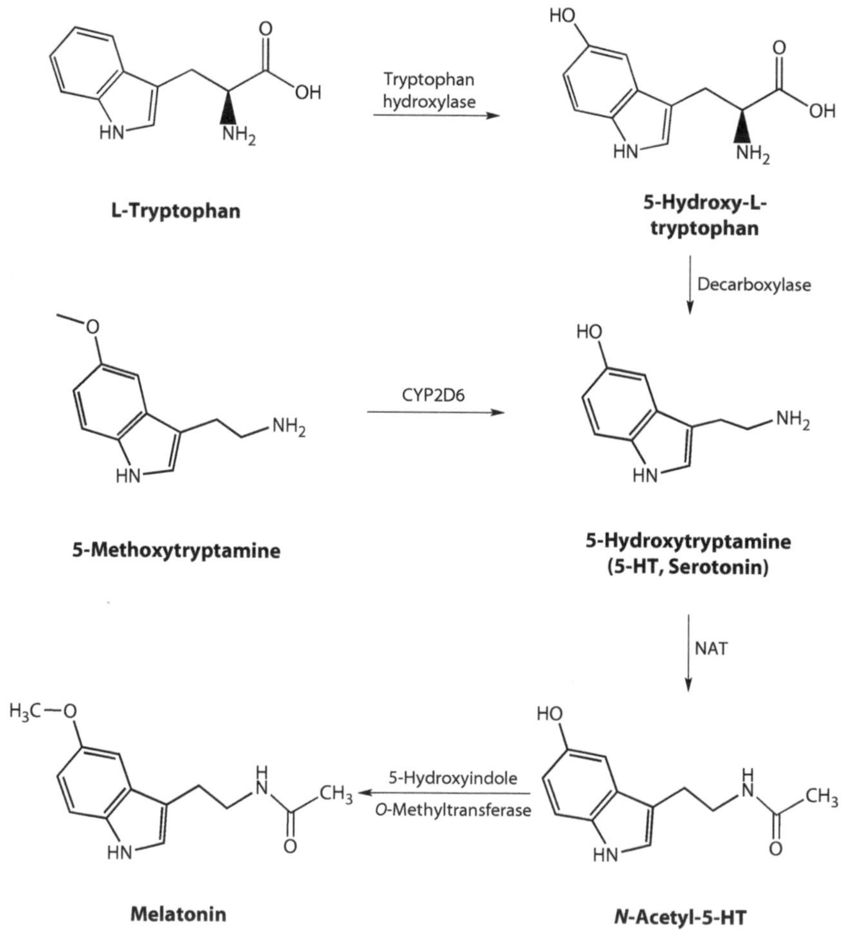

FIGURE 3.116 Metabolism of 5-methoxytryptamine. 5-Methoxytryptamine is metabolized by CYP2D6 to 5-hydroxytryptamine (5-HT/serotonin).

and Wyngaarden 1956). In addition to this pathway, the formation of dopamine from trace amines such as tyramine can be mediated by hepatic CYP2D6 (Hiroi et al. 1998). Recombinant CYP2D6 exhibits remarkable ability to convert *p*-tyramine and *m*-tyramine to dopamine. Tyramine is an endogenous compound existing in the brain not only as a trace amine but also as an exogenous compound that is found in foods such as cheese and wine. In human hepatic microsomes, the hydroxylation of tyramine to dopamine is inhibited by bufuralol and anti-CYP2D1 antiserum (Hiroi et al. 1998).

3.7.3 STEROIDS AND NEUROSTEROIDS

CYP2D6 can catalyze 2-hydroxylation of estrogens, although CYP1A1 and 1A2 have been considered as the major CYP enzymes responsible for the 2-hydroxylation of 17β-estradiol (Figure 3.117) and estrone (Figure 3.118) in extrahepatic tissues including breast (Lee et al. 2003). Subsequent metabolism of catechol estrogens involves catechol *O*-methyltransferase (COMT) and their conjugation by other Phase II enzymes. Under conditions of poor protection of catechol estrogens by Phase II enzymes, they can undergo oxidation to their reactive

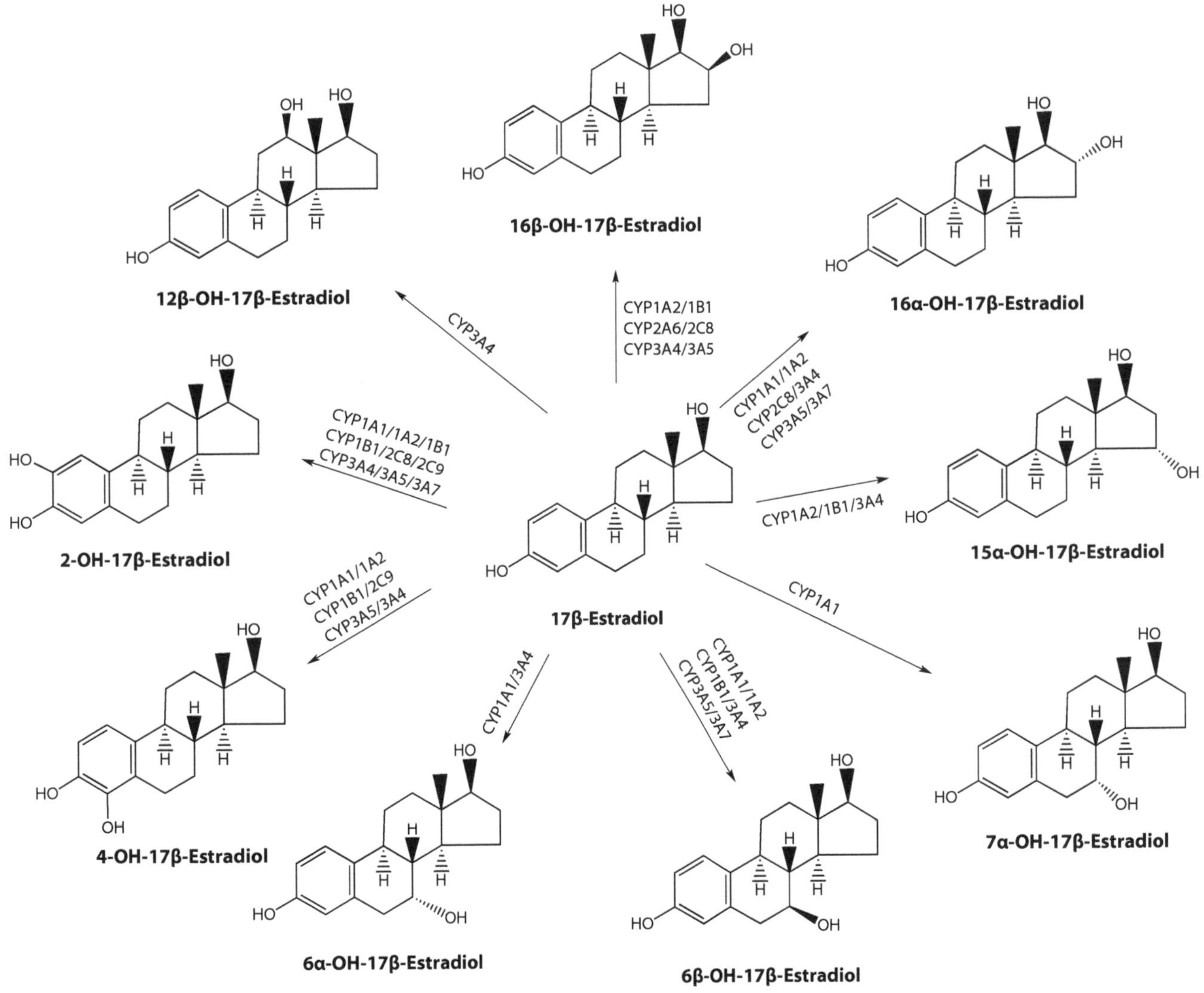

FIGURE 3.117 Metabolism of 17β-estradiol. Human CYP1A1 is primarily a 17β-estradiol 2-hydroxylase, whereas CYP1B1 is primarily a 17β-estradiol 4-hydroxylase with a lesser activity at C-2. The resultant catechol estrogens, 2- and 4-hydroxyestradiol, are converted to 2- and 4-methoxyestradiol by COMT. CYP1A1 is also efficient in steroid 16α-hydroxylation, a reaction mediated by mouse Cyp2d. CYP2A6, 2B6, 2C8, 2C9, 2C19, and 2D6 each showed a varying degree of low catalytic activity for estrogen 2-hydroxylation, whereas CYP2C18 and 2E1 did not show any detectable estrogen-hydroxylating activity. CYP3A4 had strong activity for the formation of 2-hydroxyestradiol, followed by 4-hydroxyestradiol and an unknown polar metabolite, and small amounts of 16α- and 16β-hydroxyestrogens are also formed. CYP3A5 had similar catalytic activity for the formation of 2- and 4-hydroxyestrogens; CYP3A5 had an unusually high ratio of 4- to 2-hydroxylation of 17β-estradiol. The oxidative metabolism of estrogens by CYP1A1 and 1B1 has been suggested to have a critical role in the etiology of certain cancers such as breast, ovarian, and prostate cancer.

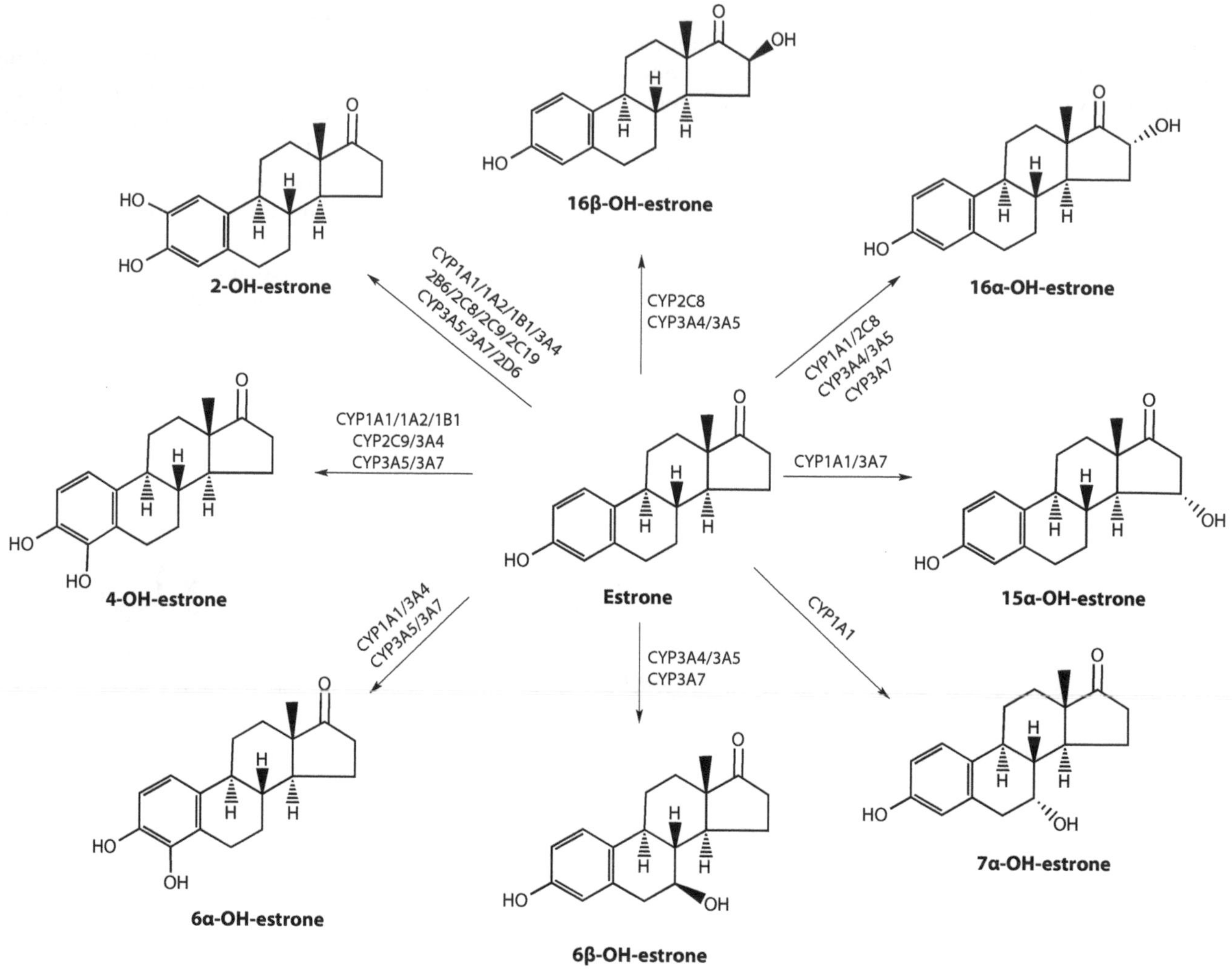

FIGURE 3.118 Metabolism of estrone. CYP1A2 had the highest activity for the 2-hydroxylation of estrone, although it also had considerable activity for its 4-hydroxylation. CYP1B1 mainly catalyzes the formation of catechol estrone, with 4-hydroxyestrone being the primary metabolite. CYP2A6, 2B6, 2C8, 2C9, 2C19, and 2D6 showed a varying degree of low catalytic activity for estrone 2-hydroxylation, whereas CYP2C18 and 2E1 did not exhibit any detectable estrone-hydroxylating activity. Notably, CYP3A5 had an unusually high ratio of 4- to 2-hydroxylation of 1 estrone. CYP3A7 had a distinct catalytic activity for the 16α-hydroxylation of estrone, but not 17β-estradiol, while CYP4A11 had little catalytic activity for the metabolism of 17β-estradiol and estrone.

semiquinone and quinone derivatives, which has been postulated to be an initiating/promoting factor in estrogen-induced carcinogenesis (Yager and Liehr 1996). CYP1A1 forms more 4-hydroxyestrone than 15α- or 6α-hydroxyestrone. CYP1A2 has the highest activity for the 2-hydroxylation of both 17β-estradiol and estrone, although it also had considerable activity for their 4-hydroxylation (9%–13% of 2-hydroxylation) (Lee et al. 2003). CYP1B1 mainly catalyzes the formation of catechol estrogens, with 4-hydroxyestrogen primary metabolites. CYP2A6, 2B6, 2C8, 2C9, 2C19, and 2D6 show a varying degree of low catalytic activity for estrogen 2-hydroxylation, whereas CYP2C18 and 2E1 do not show any detectable estrogen-hydroxylating activity. CYP3A4 has a strong activity for the formation of 2-hydroxyestradiol, followed by 4-hydroxyestradiol and an unknown polar metabolite, and small amounts of 16α- and 16β-hydroxyestrogens are also

formed. CYP3A5 has similar catalytic activity for the formation of 2- and 4- hydroxyestrogens (Lee et al. 2003). Notably, CYP3A5 has an unusually high ratio of 4- to 2-hydroxylation of 17β-estradiol or estrone. CYP3A7 has a distinct catalytic activity for the 16α-hydroxylation of estrone, but not 17β-estradiol, while CYP4A11 shows little catalytic activity for the metabolism of 17β-estradiol and estrone (Lee et al. 2003).

Hiroi et al. (2001) have observed the 2β-, 6β-, and 17-hydroxylation activity of recombinant CYP2D6 for testosterone. Testosterone plays a key role in developing and maintaining the masculine sexual organ and promotes secondary sexual characteristics, including the appearance of facial hair, sexual desire, and sexual behavior. Testosterone is synthesized from pregnenolone, which is a steroid precursor to cortisol, dehydroepiandrosterone (DHEA), progesterone, and testosterone (Freeman et al. 2001). β-Hydroxylation at either the C6- or C16-position is the major route of testosterone oxidative metabolism, whereas 1β-, 2α/β-, 11β-, and 15β-hydroxytestosterone are formed as minor metabolites (Figure 3.119). Testosterone 6β-hydroxylation is catalyzed by CYP3A4 and CYP3A5, and this accounts for 75% to 80% of all metabolites formed (Draper et al. 1998). Testosterone 6β-hydroxylation has been used as a marker reaction. Small amounts of 15β- and 2β-hydroxytestosterone are also detected in vitro (Patki et al. 2003). 11β-Hydroxylation is catalyzed by CYP3A4, 2C9, and 2C19 (Choi et al. 2005). Both CYP2C9 and 2C19 oxidize testosterone at the 17-position

to form androstenedione (Yamazaki and Shimada 1997b). In some target tissues, testosterone is 5α-hydroxylated by the cytoplasmic enzyme 5α-reductase to the more potent 5α-dihydrotestosterone or estradiol by aromatization (e.g., in the bones and the brain). 5α-Dihydrotestosterone binds with greater affinity to the androgen receptor, while estradiol binds to the estrogen receptor. 5α-Dihydrotestosterone is reversibly inactivated by 3α-reduction and a smaller amount is converted to 5α-androstanedione. Androstenedione and DHEA are metabolized to testosterone by 17β-hydroxysteroid dehydrogenase. Androstenedione may be metabolized to androsterone or etiocholanolone, which can be conjugated and

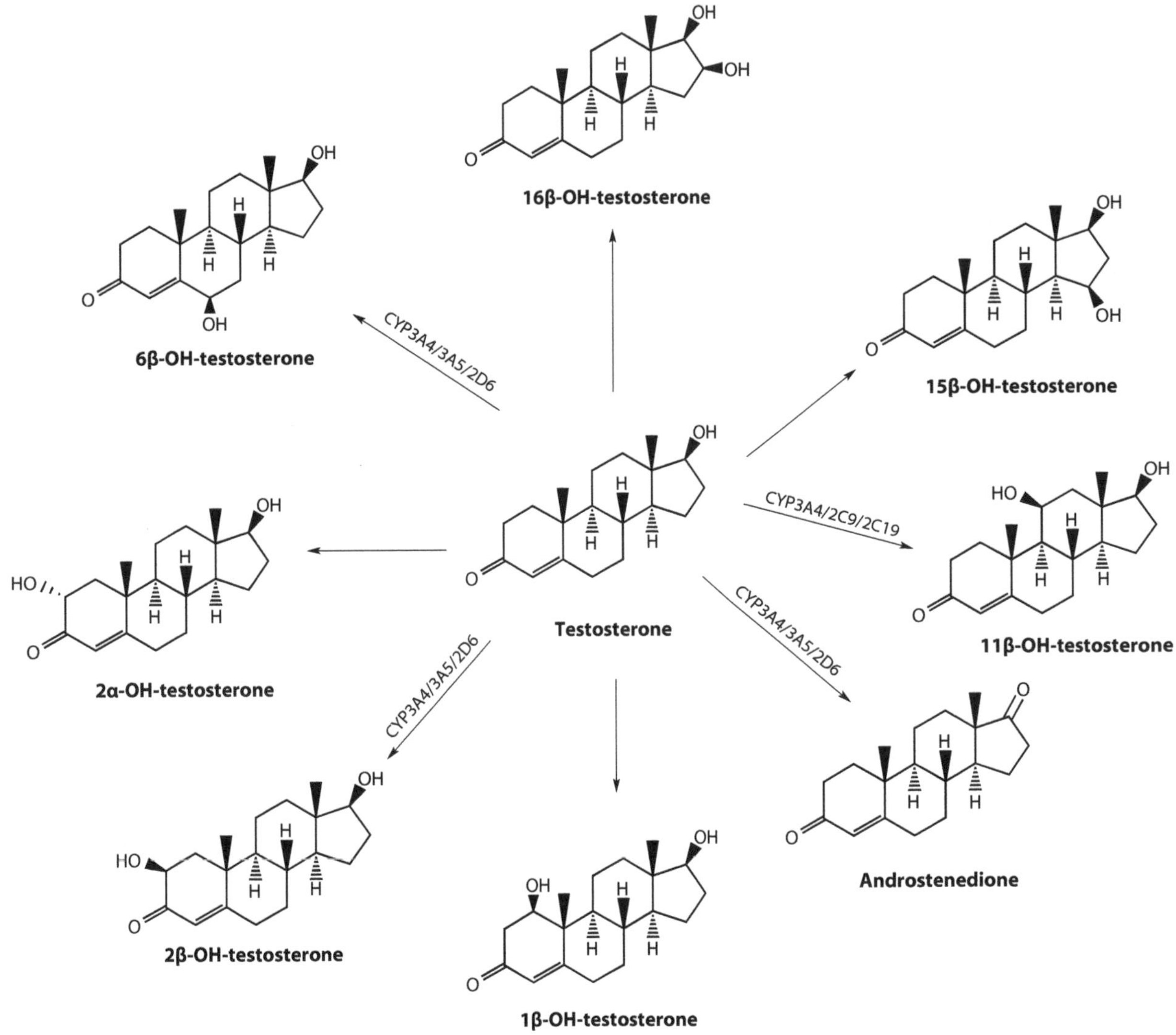

FIGURE 3.119 Metabolism of testosterone. CYP2D6 catalyzes 2β-, 6β-, and 17-hydroxylation of testosterone. Testosterone 6β-hydroxylation is mainly catalyzed by CYP3A4 and CYP3A5, and this has been used as a marker reaction. Small amounts of 15β- and 2β-hydroxytestosterone are also detected in vitro. Both CYP2C9 and 2C19 oxidized testosterone at the 17-position to form androstenedione. In some target tissues, testosterone is 5α-hydroxylated by the cytoplasmic enzyme 5α-reductase to the more potent 5α-dihydrotestosterone or estradiol by aromatization (e.g., in the bones and the brain). 5α-Dihydrotestosterone is reversibly inactivated by 3α-reduction and a smaller amount is converted to 5α-androstanedione. Androstenedione and DHEA are metabolized to testosterone by 17β-hydroxysteroid dehydrogenase. Androstenedione may be metabolized to androsterone or etiocholanolone, which can be conjugated and excreted. Most androgens are conjugated before excretion in the urine.

excreted. Most androgens are conjugated before excretion in the urine. Testosterone and androstenedione are cleared three times more rapidly in men than in women.

Human CYP2D6 and rat Cyp2d4 are the predominant CYP2D isoforms in the brain and exhibit 21-hydroxylation activity toward progesterone and its metabolite 17α-hydroxyprogesterone (Figure 3.120) (Kishimoto et al. 2004). Multiple human CYPs are involved in the hydroxylation of progesterone. CYP3A4 shows the highest progesterone 16α-hydroxylation activity, followed by CYP1A1 and 2D6 (Niwa et al. 1998). Human CYP3A4, 3A5, and 4B1 (expressed in lung but not liver) catalyze the 6β-hydroxylation of progesterone (Waxman et al. 1991; Yamakoshi et al. 1999).

Allopregnanolone (3α,5α-tetrahydroprogesterone), an important neurosteroid in the human brain and representative GABA receptor modulator (Herd et al. 2007; Mitchell et al. 2008), is also hydroxylated at the C21-position by recombinant Cyp2d4 and CYP2D6 (Figure 3.120) (Kishimoto et al. 2004). It is a metabolite of progesterone of ovarian or adrenal origin but can also be synthesized de novo in the brain from cholesterol via a side-chain cleavage enzyme that forms pregnenolone (Compagnone and Mellon 2000; Mensah-Nyagan et al. 1999), the precursor of all steroid molecules (Tsutsui 2008). Allopregnanolone has been found to be synthesized in the cerebellum and acts on Purkinje cell survival in the neonate (Tsutsui 2008). 3β-Hydroxysteroid dehydrogenase converts pregnenolone to progesterone, which is then converted via neuronal enzymes into allopregnanolone (Mensah-Nyagan et al. 1999). Allopregnanolone is a barbiturate-like modulator of central GABA receptors that can

modify a range of behaviors, including the stress response (Herd et al. 2007; Mitchell et al. 2008). Rat brain microsomal allopregnanolone 21-hydroxylation is inhibited by fluoxetine with an IC$_{50}$ value of 2 μM. It appears that the brain CYP2D isoforms can regulate levels of neurosteroids such as allopregnanolone, which is modified by CNS-acting drugs such as fluoxetine. The 3β-enantiomer of allopregnanolone is known as pregnanolone and has very similar properties to allopregnanolone. Both compounds are found endogenously and have similar hypnotic and anxiolytic effects (Herd et al. 2007; Mitchell et al. 2008).

It should be noted that steroids are atypical substrates of CYP2D6 as they often lack a basic nitrogen atom necessary for binding to Asp301. However, testosterone has only a single (conjugated) carbonyl and progesterone has two carbonyls, which may form hydrogen bonding with Asp301.

3.7.4 Endogenous Morphine

Endogenous morphine is likely to be synthesized in humans via similar routes as in the poppy plant; two of the steps involved are mediated by CYP2D6, namely, thebaine O-demethylation to oripavine and codeine O-demethylation to morphine (Sindrup et al. 1993). PMs should therefore have a much lower endogenous morphine formation than EMs. However, there are no phenotype-related differences in endogenous codeine and morphine excretion in PMs and EMs (Mikus et al. 1994). Administration of quinidine had no significant effect on endogenous codeine and morphine excretion in EMs. These findings did not support the role of CYP2D6 in the biosynthesis of morphine in humans. Interestingly, human leukocyte exposure to morphine downregulated COMT and CYP2D6 by approximately 50% compared with the control cells (Mantione et al. 2008).

3.7.5 Endocannabinoid Arachidonoylethanolamide (Anandamide)

Anandamide is the endogenous ligand to the cannabinoid receptor CB1, which is also activated by the main psychoactive component in marijuana. Snider et al. (2008) have revealed that recombinant CYP2D6 converts anandamide to 20-hydroxyeicosatetraenoic acid ethanolamide and 5,6-, 8,9-, 11,12-, and 14,15-epoxyeicosatrienoic acid ethanolamides with low micromolar K_m values. CYP2D6 together with 3A4 and 4F2 further metabolize the epoxides of anandamide to form novel dioxygenated derivatives. The 5,6-epoxide of anandamide, 5,6-epoxyeicosatrienoic acid ethanolamide, is a potent and selective CB2 agonist (Snider et al. 2009). Human brain microsomal and mitochondrial preparations metabolize anandamide to hydroxylated and epoxygenated metabolites, respectively (Snider et al. 2008). These results suggest that anandamide is a physiological substrate for brain mitochondrial CYP2D6, implicating this highly polymorphic enzyme as a potential component of the endocannabinoid system in the brain.

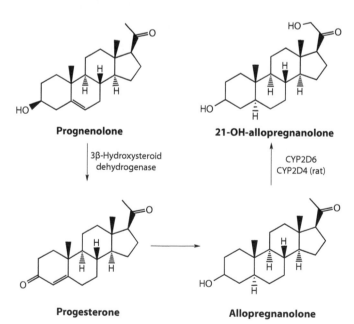

FIGURE 3.120 Metabolism of allopregnanolone (3α,5α-tetrahydroprogesterone) by CYP2D6. Allopregnanolone is hydroxylated at the C21-position by recombinant rat CYP2D4 and human CYP2D6.

3.8 STRUCTURE–ACTIVITY RELATIONSHIPS OF CYP2D6 SUBSTRATES

Pharmacophore studies have demonstrated that typical CYP2D6 substrates are usually lipophilic bases with a planar hydrophobic aromatic ring and a nitrogen atom that can be protonated at physiological pH (de Groot et al. 1999b). These compounds usually have a negative molecular electrostatic potential above the planar part of the molecule. These features are typical of a large number of drugs acting on the central nervous and cardiovascular systems. The nitrogen atom is considered to be essential for electrostatic interactions with the carboxylate group of Asp301, a candidate residue in the active site of CYP2D6 (de Groot et al. 1999b). Pharmacophore modeling studies suggest that substrate binding is generally followed by oxidation 5 to 7 Å from this proposed nitrogen–Asp301 interaction (Koymans et al. 1992; Miller et al. 2001; Wolff et al. 1985). For example, propranolol is a typical "7 Å" substrate for CYP2D6, having a distance between the nitrogen and preferred site of oxidation of approximately 7.9 Å. Besides being a basic amine, propranolol is relatively lipophilic with an aromatic ring, and it has potential hydrogen-bonding groups at the amine nitrogen and the side-chain hydroxyl group. Lipophilicity and amine basicity are thus considered to be critical determinants of substrate binding to CYP2D6. The pharmacophore models reported to date have been used to rationalize and predict many CYP2D6-catalyzed reactions.

However, both pactimibe (a novel ACAT inhibitor) and its metabolite, R-125528, are atypical substrates for CYP2D6 (Figure 3.91) because of its acidity. Both compounds are not protonated but are negatively charged at physiological pH. Interestingly, their sites of metabolism, the ω-1 position of the N-octyl indoline/indole group, are relatively distant from the aromatic moiety. An induced-fit docking of the ligands with an x-ray crystal structure of substrate-free CYP2D6 indicated the involvement of an electrostatic interaction between the carboxyl group and Arg221 and hydrophobic interaction between the aromatic moiety and Phe483 (Kotsuma et al. 2008a). With the cotreatment of ketoconazole, the AUC of pactimibe in healthy subjects increased 1.7-fold whereas the AUC of R-125528 decreased by 55%. With the cotreatment of quinidine, the AUC for pactimibe increased 1.7-fold but the AUC of R-125528 is markedly elevated 5.0-fold (Kotsuma et al. 2008b). As the estimated metabolic fraction of pactimibe by CYP3A4 and 2D6 from in vitro studies is 0.40 and 0.33, respectively, AUC increase ratios of pactimibe are estimated to be 1.7 with ketoconazole and 1.5 with quinidine. It can be expected that PMs would have significantly increased concentrations of R-125528 and moderately increased pactimibe levels.

Bonn et al. (2008) have revealed that the CYP2D6 pharmacophore and characteristic features in the active-site cavity suggest a favored substrate orientation that prevents N-dealkylation. N-Dealkylation of substrates appears to occur by CYP2D6 only when the preferred site of metabolism is blocked toward other oxidative metabolic routes. Addition of stable groups at preferred sites of metabolism generally improved the metabolic stability but also results in retained or increased inhibition of the enzyme (Bonn et al. 2008). In addition, the effect of pH on N- and O-dealkylation of dextromethorphan is shown to be in agreement with the hypothesis that an ionized amino function favored substrate dockings resulting in O-dealkylation.

The first substrate models for CYP2D6 are constructed by manual alignments based on a set of substrates containing a basic nitrogen atom at either 5 Å (Wolff et al. 1985) or 7 Å (Meyer et al. 1986) from the site of oxidation and on aromatic rings near the site of oxidation, which are fitted to be coplanar (Meyer et al. 1986; Wolff et al. 1985). In the space-filling 5-Å model, no substrates are fitted onto each other (Wolff et al. 1985). The main limitation with both of these initial models is that neither could rationalize the binding of other types of substrates. In the homology model of CYP2D6, codeine is bound with the methoxy group of the molecule closest to the heme iron (iron–methyl proton distance of 3.1 Å), consistent with the observed O-demethylation to morphine (Modi et al. 1996).

Islam et al. (1991) have derived an extended model with incorporation of the heme moiety from the crystal structure of P450cam (CYP101), which also indicated a distance between the basic nitrogen atom and the site of oxidation of between 5 and 7 Å. In their study, debrisoquine is positioned arbitrarily, in a manner similar to the orientation of camphor in the bacterial P450cam (CYP101) crystal structure (Poulos et al. 1985). Camphor, a substrate molecule, is buried in an internal pocket just above the heme distal surface adjacent to the oxygen binding site of CYP101 (Poulos et al. 1985). The model also includes the iron–oxygen complex involved in the hydroxylation, and a set of 15 compounds are fitted onto debrisoquine. However, this model fails to accommodate tamoxifen and thus Islam et al. (1991) have proposed that tamoxifen is unlikely to be metabolized by CYP2D6. In contrast, however, later studies indicate that tamoxifen is a good substrate of CYP2D6 (Dehal and Kupfer 1997).

The positioning of Asp301 in the various models studied showed that it could readily explain the "5- to 7-Å" pharmacophore model (Islam et al. 1991; Koymans et al. 1992). In the small-molecule model for CYP2D6 by Koymans et al. (1992), debrisoquine and dextromethorphan are used as templates for the 5- and 7-Å compounds, respectively. This model suggested that a hypothetical carboxylate group within the protein is responsible for a well-defined distance of either 5 or 7 Å between the basic nitrogen atom and the site of oxidation within the substrate. The oxidation sites of the debrisoquine and dextromethorphan templates are superimposed and the areas adjacent to the sites of oxidation are fitted to be coplanar, while the basic nitrogen atoms are placed at a distance of 2.5 Å so as to interact with different oxygen atoms of the postulated carboxylate group in the protein (Koymans et al. 1992). The model is constructed based on 16 substrates, accounting for 23 metabolic reactions with their sites of oxidation and basic nitrogen atoms fitted onto the site of oxidation of the templates and one of the basic nitrogen atoms of the template molecules, respectively. The model is further applied to predict the metabolism of four compounds yielding 14 possible CYP-dependent metabolites. In vivo and in vitro metabolism studies of these substrates indicated that 13 of 14 predictions are correct (Koymans et al. 1992), suggesting that the model has a high predictive value.

de Groot et al. (1996) have derived a homology model of CYP2D6 and proposed that the site of oxidation above the heme moiety occupies one of two possible sites above pyrrole ring B of the heme moiety. In a further refined model, the authors pinpointed more accurate positions of the heme moiety and helix I containing Asp301, thereby incorporating some steric restrictions and orientational preferences into their model (de Groot et al. 1997a). In this refined small-molecule model, an Asp residue is coupled to the basic nitrogen atoms, thus enhancing the model with the direction of the hydrogen bond between Asp in the protein and the protonated basic nitrogen atom. The site of oxidation for substrates is fitted onto the defined oxidation site above pyrrole ring B of the heme moiety, while the C_α and C_β atoms of the attached Asp moiety are fitted onto the C_α and C_β atoms of Asp301, respectively (de Groot et al. 1997a). A variety of substrates fitted in the original substrate model for CYP2D6 (de Groot et al. 1995, 1997b; Koymans et al. 1992) are properly fitted into the refined substrate model, indicating that the refined substrate model for CYP2D6 accommodates the same variety in molecular structures as the original substrate model. This refined model has been used to design a novel and selective CYP2D6 substrate, MAMC, suitable for high-throughput screening (Onderwater et al. 1999).

de Groot et al. (1999a,b) have further developed a combined pharmacophore and homology model for CYP2D6. This model consists of a set of two pharmacophores (one for O-dealkylation and oxidation reactions and a second one for N-dealkylation reactions catalyzed by CYP2D6) embedded in a homology model based on bacterial CYP101 (P450cam), CYP102 (P450BM3), and CYP108 (P450terp) crystal structures. This model combines for the first time the strengths of pharmacophore models (atom–atom overlap and reproducible starting points and thus the most reactive sites in the substrates could be identified) and homology models (steric interactions, conformational and stereochemical constraints imposed by the active site, and the possibility of identifying amino acids involved in substrate binding). The independent generation of the pharmacophore and protein homology models has provided the opportunity to cross-validate the approaches used. The combined model contains 51 substrates involving 72 metabolic pathways, mostly N-dealkylation. This model is also used to predict the metabolism of seven test compounds including betaxolol, fluoxetine, loratadine, MPTP, procainamide, ritonavir, and sumatriptan. The combined model correctly predicts six of the eight observed metabolites, but not the highly unusual metabolism of procainamide (N-hydroxylation) and ritonavir (marked as a nonsubstrate as it contains no basic nitrogen atom but can be metabolized by CYP2D6) (de Groot et al. 1999b; Kumar et al. 1996).

In addition, several modeling studies have pointed to a possible role for a second carboxylate group, identified as Glu216 (de Groot et al. 1999b; Lewis et al. 1997; Venhorst et al. 2003). This residue may provide an explanation for the metabolism of larger substrates with a basic nitrogen atom ≥10 Å from the site of oxidation. Basic substrates are metabolized by all four known P450s that contain a residue equivalent to both Glu216 and Asp301: CYP2D6, CYP2D14 (bovine), Cyp2d4 (rat), and CYP2J1 (rabbit). This indicates the important role for both Asp301 and Glu216 in the metabolism of these substrates.

3.9 CONCLUSIONS AND FUTURE DIRECTIONS

Although CYP2D6 accounts for only a small percentage of the total hepatic CYP content (<2%), this enzyme metabolizes a number of clinically used drugs with significant polymorphisms. In particular, a number of drugs acting on the CNS and cardiovascular system are substantially metabolized by CYP2D6. These drugs are mostly associated with G-coupled proteins when they elicit effect at the site of action. The substrates of CYP2D6 belong to several therapeutic classes that include centrally acting compounds such as tricyclic antidepressants, SSRIs, β-blockers, MAO inhibitors, neuroleptics, opioids, drugs of abuse such as MDMA, and neurotoxins (for a full list of known substrates of CYP2D6, see Table 3.1). CYP2D6 is the only active enzyme in the 2D subfamily, and it has high affinity to plant alkaloids, suggesting its important role in detoxifying these xenobiotics when human beings are dependent on plants as major food sources during evolution. The enzyme also utilizes hydroxytryptamines, steroids, and neurosteroids as endogenous substrates, but the potential physiological role of CYP2D6 is unclear. Since drug–drug interactions may occur when a CYP2D6 substrate drug is combined with a CYP2D6 inhibitor drug, it is important to know the contribution of CYP2D6 to the metabolic elimination of the drug. In addition to in silico approaches, in vitro models, in particular those with a high-throughput capacity, are useful to determine whether a new drug is a substrate of CYP2D6 and the contribution of this enzyme in their metabolic clearance or metabolic activation in the early stage of drug development. If the new drug candidate is substantially metabolized by CYP2D6 (e.g., >60%), alternative compounds that have a low affinity to CYP2D6 should be considered.

It is clear that drugs that are substrates for CYP2D6 can exhibit a large interindividual variation in their metabolism as a result of polymorphisms in CYP2D6. Variation in CYP2D6 activity has important therapeutic consequences and can play a significant role in the development of therapeutic failure or adverse events in susceptible individuals. CYP2D6 polymorphisms are likely to become increasingly important in the coming years as an increasing number of patients are prescribed multiple drugs, a proportion of which are likely to be metabolized by this isozyme.

TABLE 3.1
Reported Substrates of CYP2D6

Substrate	Drug Class	Metabolic Pathway(s)	K_m (µM)	Estimated Relative Contribution to Overall Clearance by CYP2D6 (%)[a]	M_r (Da)	LogP (cLogP)	Reference
AG7088	Antiviral	Reduction					Zhang et al. 2001
CJ-036878	Novel NMDA receptor antagonist	Oxidation					Nishida et al. 2007
CP-122,721	Novel NK₁ antagonist	O-Demethylation					Obach et al. 2007
DRF-6574	Novel COX-2 inhibitor	Hydroxylation					Muzeeb et al. 2006
KR-62980	PPARγ agonist	Hydroxylation					Kim et al. 2008
KR-63198	PPARγ agonist	Hydroxylation					Kim et al. 2008
L-775,606	Antimigraine	C-hydroxylation; N-demethylation					Prueksaritanont et al. 2000
ML3403	p38 mitogen-activated protein kinase						Kammerer et al. 2007
PCEPA	Designer drug	O-Deethylation		5–10			Sauer et al. 2008
PCMPA	Designer drug	O-Deethylation		5–10			Sauer et al. 2008
RWJ-53050	Novel anxiolytic agent	Oxidation					Wu et al. 2006
TZB-30878	Novel 5-HT$_{1A}$ agonist/5-HT$_3$ antagonist						Minato et al. 2008
1-Methyl-4-phenyl-1,2,3,6-tetrahydropyridine	Neurotoxin	N-Demethylation					Coleman et al. 1996; Gilham et al. 1997; Kalgutkar et al. 2003; Modi et al. 1997
1,2,3,4-Tetrahydroquinoline	Neurotoxin	5-Hydroxylation					Ohta et al. 1990
2-Aroylthiophene (O(CH$_2$)$_3$OH)	Chemical compound	O-Deethylation					Ekins et al. 1998
4-Chloromethyl-7-ethoxycoumarin	Chemical compound	Hydroxylation					Peters et al. 2008
4'-Methyl-α-pyrrolidinobutyrophenone	Designer drug						
4'-Methyl-α-pyrrolidinopropiophenone	Designer drug	4'-Hydroxylation					Springer et al. 2003a
5-Methoxyindolethylamine	Endogenous compound	O-Demethylation					Yu et al. 2003c
5-Methoxy-N,N-diisopropyltryptamine	Indolealkylamine	O-Dealkylation					Narimatsu et al. 2006, 2008
5-Methoxy-N,N-dimethyltryptamine	Indolealkylamine	O-Demethylation					Yu et al. 2003c
5-Methoxytryptamine							Yu et al. 2003b
7-Methoxy-4-(aminomethyl)-coumarin	Fluorescent probe	O-Dealkylation					Onderwater et al. 1999
7,8-Dihydroxybenzo[a]pyrene	Chemical compound	Tetrols and triols formation					Bjelogrlic et al. 1993

(Continued)

TABLE 3.1 (CONTINUED)
Reported Substrates of CYP2D6

Substrate	Drug Class	Metabolic Pathway(s)	K_m (μM)	Estimated Relative Contribution to Overall Clearance by CYP2D6 (%)[a]	M_r (Da)	LogP (cLogP)	Reference
7,12-Dimethylbenzo[a]anthracene	Chemical compound	Methyl hydroxylation; diol formation					Buters et al. 1999
17α-Hydroxyprogesterone	Hormone	21-Hydroxylation					Kishimoto et al. 2004
17β-Estradiol	Estrogen	2-Hydroxylation; (C4-hydroxylation)					Lee et al. 2003
Acetaminophen	NSAID	Oxidation	440	10–15	151.16	0.4 (0.51)	Bessems and Vermeulen 2001; Prescott 2000
Allopregnanolone	Neurosteroid	21-Hydroxylation		60–70	318.49	(4.89)	Kishimoto et al. 2004
Almotriptan	5-HT$_{1B/1D}$ receptor agonist	N-Demethylation		15–20	335.46	1.6 (2.04)	Salva et al. 2003
Amiflamine	MAO-A inhibitor	N-Demethylation		15–25	192.30	(2.38)	Aringberg-Wistedt et al. 1987
Aminopyrine	NSAID	N-Demethylation		5–10	231.29	(0.76)	Agundez et al. 1995
Amiodarone	Antiarrhythmic	Deethylation	8.4	5–10	645.31	7.9 (7.24)	Jaruratanasirikul and Hortiwakul 1994
Amitriptyline	Tricyclic antidepressant	E-10-Hydroxylation	5–13	40–60	277.40	4.9 (5.10)	Mellstrom and von Bahr 1981; Venkatakrishnan et al. 2000, 2001
Amodiaquine	Antimalarial	Deethylation		20–30	355.15	3.7 (4.83)	Li et al. 2002
Anandamide	Endocannabinoid	Hydroxylation		20–30	347.53	(5.67)	Snider et al. 2008
Antipyrine	NSAID	N-Demethylation	14.4	<5	188.23	0.38 (1.18)	Sharer and Wrighton 1996
Aprindine	Antiarrhythmic	Hydroxylation		20–30	322.49	4.86 (5.58)	Murphy 1974; Shimizu et al. 1998
Aripiprazole	Antipsychotic	Dehydrogenation		30–35	448.39	4.5 (5.21)	Molden et al. 2006
Artelinic acid	Antimalarial	3-Hydroxylation		15–30	418.48	(5.27)	Grace et al. 1999
Atomoxetine	Noradrenaline reuptake inhibitor	4-Hydroxylation	2.73 (149)	20–30	255.35	3.9 (3.95)	Ring et al. 2002
Azelastine	H$_1$ receptor blocker	N-Demethylation		30–40	381.90	4.9 (3.81)	Nakajima et al. 1999
Benzo[a]pyrene	Chemical compound	3- and 9-Hydroxylation		5–15	252.31	(6.4)	Gautier et al. 1996
Bicifadine	Nonnarcotic analgesic	Methyl hydroxylation		40–60	173.25	(1.97)	Erickson et al. 2007
Bisoprolol	β$_1$-Blocker	O-Deisopropylation		20–30	325.44	2.2 (2.3)	Horikiri et al. 1998a,b
Bortezomib	Anticancer agent	Deboronation; dehydrogenation		10–15	384.20	(2.03)	Uttamsingh et al. 2005
Bufuralol	β-Blocker	1,2-Ethenylation; 1-hydroxylation	3.4	60–70	216.36	(3.38)	Hanna et al. 2001; Crespi et al. 2006
Bunitrolol	Antiarrhythmic	4-Hydroxylation	0.1	60–70	248.32	(1.6)	Masuda et al. 2006
Carbamazepine	Psychotolytic	10,11-Epoxidation		20–30	236.27	2.3 (2.1)	Pirmohamed et al. 1992

(Continued)

TABLE 3.1 (CONTINUED)
Reported Substrates of CYP2D6

Substrate	Drug Class	Metabolic Pathway(s)	K_m (μM)	Estimated Relative Contribution to Overall Clearance by CYP2D6 (%)[a]	M_r (Da)	LogP (cLogP)	Reference
Carvedilol	β-Adrenoceptor antagonist	4'- and 5'-Hydroxylation		30–40	406.47	3.8 (3.05)	Oldham and Clarke 1997
Cerivastatin	HMG-CoA reductase inhibitor	O-Demethylation; 6'-methyl hydroxylation		15–20	459.55	3.4 (4.15)	Boberg et al. 1997
Chloroquine	Antimalarial	N-Deethylation		20–30	319.87	4.3 (5.29)	Projean et al. 2003
Chlorpheniramine	H₁ receptor antagonist	N-Demethylation		60–80	274.79	3.2 (3.74)	Yasuda et al. 2002
Chlorpromazine	Antipsychotic	7-Hydroxylation		30–40	318.86	4.9 (5.18)	Yoshii et al. 2000
Cibenzoline	Antiarrhythmic	p-Hydroxylation; 4,5-dehydrogenation		20–30	262.35	3.81	Niwa et al. 2000
Cinnarizine	Antihistamine	p-Hydroxylation		50–60	368.51	5.3 (5.19)	Kariya et al. 1996; Narimatsu et al. 1993
Cisapride	5-HT₄ receptor agonist	N-Dealkylation; fluorophenyl ring hydroxylation		15–20	465.95	3.3 (2.95)	Bohets et al. 2000
Citalopram	SSRI	Demethylation; N-oxidation		30	324.39	3.5 (3.58)	Olesen and Linnet 1999
Clomipramine	Tricyclic antidepressant	8-Hydroxylation		15–20	314.85	4.5 (5.04)	Nielsen et al. 1996
Clozapine	Antipsychotic	N-Demethylation; N-oxidation		6	326.82	2.7 (3.42)	Olesen and Linnet 2001
Codeine	Antimigraine (opioid)	O-Demethylation	1079	10–20	299.36	1.19 (1.2)	Yue and Sawe 1997
Cyclophosphamide	Antineoplastic	C4-hydroxylation		5–15	261.09	0.8 (0.76)	Ren et al. 1997
Dapsone	Antileprotic	N-Hydroxylation		<5	248.30	0.4 (1.19)	Gill et al. 1995
Debrisoquine	Cardiovascular agent	4-Hydroxylation	73.7	60–70	175.23	(0.14)	Distlerath et al. 1985
Delavirdine	Nonnucleoside reverse transcriptase inhibitor	Dealkylation		20–25	456.56	2.8 (2.77)	Voorman et al. 1998b
Desipramine	Tricyclic antidepressant	2-Hydroxylation		10–15	266.38	3.7 (4.02)	Brosen et al. 1991; Koyama et al. 1997; Lemoine et al. 1993
Desmethylclomipramine	Tricyclic antidepressant	8-Hydroxylation		15–20	300.85	(4.73)	Nielsen et al. 1996
Dextromethorphan	Antitussive	N-Demethylation; O-demethylation	2.97	80–90	271.40	(4.11)	Kerry et al. 1994
Diazepam	Benzodiazepine sedative	N-Demethylation; C3-hydroxylation		15–25	284.74	2.9 (2.63)	Ono et al. 1996
Diclofenac	NSAID	4'- and 5'-Hydroxylation		5–10	296.15	3.9 (4.98)	Bort et al. 1999
Dihydrocodeine	Analgesic and antitussive (opioid)	O-Demethylation		20–30	301.38	(1.58)	Kirkwood et al. 1997
Diltiazem	Calcium channel blocker	O-Demethylation	5	20–30	414.52	2.8 (3.09)	Jones et al. 1999; Molden et al. 2000

(Continued)

TABLE 3.1 (CONTINUED)
Reported Substrates of CYP2D6

Substrate	Drug Class	Metabolic Pathway(s)	K_m (μM)	Estimated Relative Contribution to Overall Clearance by CYP2D6 (%)[a]	M_r (Da)	LogP (cLogP)	Reference
Diphenhydramine	Antihistamine	N-Demethylation	1.12	30–40	255.35	3.27 (3.44)	Akutsu et al. 2007
Disulfoton	Thioether pesticides	Sulfoxidation		2–10	274.40	(3.31)	Usmani et al. 2004
Diuron	Herbicide	N-Demethylation		20–30	233.09	(2.78)	Abass et al. 2007
Dolasetron	5-HT$_3$ receptor antagonist	Hydroxylation		25–35	324.37	2.1 (2.41)	Sanwald et al. 1996a
Donepezil	Cholinesterase inhibitor	5″- and 6″-Demethylation; O-dealkylation		20–30	379.49	3.6 (4.14)	Tiseo et al. 1998
Doxepin	Tricyclic antidepressant	2-Hydroxylation		20–25	279.38	4 (4.08)	Haritos et al. 2000; Hartter et al. 2002
Droloxifene	Selective estrogen receptor modulator	Ortho-hydroxylation		15–25	387.51	(7.3)	Dehal and Kupfer 1999
Duloxetine	Antidepressant	4-, 5-, and 6-Oxidation of naphthyl ring		15–20	297.42	4 (4.72)	Skinner et al. 2003
Eletriptan	5-HT$_{1B/1D}$ receptor agonist	N-Demethylation		30–40	382.52	3.9 (3.84)	Evans et al. 2003
Encainide	Antiarrhythmic	O-Demethylation		60–80	352.47	4 (4.63)	Funck-Brentano et al. 1992
Enclomifene	Selective estrogen receptor modulator			15–20	405.96	(8.01)	Ruenitz et al. 1983
Etoricoxib	COX-2 inhibitor	6′-Methyl hydroxylation; 1′-N-oxidation		5–10	358.84	(3.7)	Kassahun et al. 2001
Estrone	Estrogen	2-Hydroxylation		5–10	270.37	3.6 (4.03)	Lee et al. 2003
Ezlopitant	Antiemetic	Dehydrogenation		20–35	486.00		Obach 2001
Flecainide	Antiarrhythmic	m-O-Dealkylation		60–80	414.34	4.6 (2.98)	Gross et al. 1989; McQuinn et al. 1984
Flunarizine	Antihistamine	p-Hydroxylation		50–60	404.49	(5.3)	Kariya et al. 1996; Narimatsu et al. 1993
Fluoxetine	SSRI	N-Demethylation	0.834	20–30	309.33	4.6 (4.1)	Margolis et al. 2000; Ring et al. 2001
Fluvastatin	HMG-CoA reductase inhibitor	5-Hydroxylation		5–15	411.18	4.5 (3.69)	Fischer et al. 1999
Fluvoxamine	SSRI	Demethylation; N-acetylation		50–60	318.33	3.2 (2.89)	Spigset et al. 1998
Galantamine	Tertiary alkaloid	O-Demethylation		30–35	287.35	1.8 (1.39)	Bachus et al. 1999
Gepirone	5-HT$_{1A}$ agonist	3′- and 5-Hydroxylation		30–35	359.47	(2.96)	Greenblatt et al. 2003
Halofantrine	Antimalarial	N-Debutylation		20–30	500.42	8.9 (7.34)	Karbwang and Na Bangchang 1994

(Continued)

TABLE 3.1 (CONTINUED)
Reported Substrates of CYP2D6

Substrate	Drug Class	Metabolic Pathway(s)	K_m (μM)	Estimated Relative Contribution to Overall Clearance by CYP2D6 (%)[a]	M_r (Da)	LogP (cLogP)	Reference
Haloperidol	Antipsychotic	N-Dealkylation		5–10	375.86	4 (3.7)	Kudo and Odomi 1998; Tateishi et al. 2000
Harmaline	β-Carboline alkaloid	O-Demethylation		10–15	214.26	(0.66)	Yu et al. 2003d
Harmine	β-Carboline alkaloid	O-Demethylation		10–15	212.25	(3.17)	Yu et al. 2003d
Hydrocodone	Opioid	O-Demethylation		25–35	299.36	1.2 (2.13)	Hutchinson et al. 2004
Ibogaine	Plant alkaloid	O-Demethylation		60–70	310.43	(4.56)	Obach et al. 1998
Ibuprofen	NSAID	2- and 3-Hydroxylation		5–10	206.28	3.6 (3.5)	Hamman et al. 1997
Ifosfamide	Antineoplastic	4-Hydroxylation		15–20	261.09	0.8 (0.57)	Preiss et al. 2002
Iloperidone	Antipsychotic	Hydroxylation		15–20	426.48	(3.81)	Mutlib et al. 1995
Imatinib	Protein tyrosine kinase inhibitor	N-Demethylation		10–15	493.60	3 (3.47)	van Erp et al. 2007
Imipramine	TCA	2-Hydroxylation		15–25	280.41	3.9 (4.53)	Brosen et al. 1991; Koyama et al. 1997; Lemoine et al. 1993
Indinavir	HIV protease inhibitor	Indanyl group oxidation		20–25	613.79	2.9 (3.26)	Koudriakova et al. 1998
Indomethacin	NSAID	O-Demethylation		20–30	357.79	3.4 (4.25)	Nakajima et al. 1998a
Lansoprazole	Proton pump inhibitor	5-Hydroxylation		15–20	369.36	1.9 (2.84)	Kim et al. 2003; Pearce et al. 1996
Lasiocarpine	Pyrrolizidine alkaloid			60–70	411.49	(2.85)	Wolff et al. 1985, 1987
Lasofoxifene	Selective estrogen receptor modulator	Oxidation		15–20	413.55	(6.51)	Prakash et al. 2008
Lauric acid	Fatty acid	ω-1-Hydroxylation		10–15	200.32	4.6 (5.13)	Imaoka et al. 1996
Licofelone	COX-1, COX-2, and 5-lipoxygenase inhibitor	Hydroxylation		10–15	379.88	(6.73)	Albrecht et al. 2008
Lidocaine	Antiarrhythmic	3-Hydroxylation		5–10	234.34	2.1 (1.81)	Hermansson et al. 1980; Wang et al. 1999, 2000
Loratadine	H1 receptor antagonist	Decarboethoxylation		15–25	382.88	3.8 (4.8)	Yumibe et al. 1996
Maprotiline	Tricyclic antidepressant	Demethylation		~83	277.40	5.1 (4.89)	Baumann et al. 1988
Mequitazine	H1 receptor antagonist	Hydroxylation; S-oxidation		60–80	322.47	4.7 (5.38)	Nakamura et al. 1998
Methadone	Opioid receptor inhibitor	N-Demethylation	0.72	15–20	309.45	3.93 (4.14)	Foster et al. 1999; Gerber et al. 2004; Kharasch et al. 2004; Oda and Kharasch 2001; Totah et al. 2008; Wang and DeVane 2003
Methamphetamine	Drug of abuse	N-Demethylation; aromatic 4-hydroxylation		60–80	149.23	2.07 (2.23)	Lin et al. 1997
Methiocarb	Thioether pesticide	Sulfoxidation		5–10	225.00		Usmani et al. 2004

(Continued)

TABLE 3.1 (CONTINUED)
Reported Substrates of CYP2D6

Substrate	Drug Class	Metabolic Pathway(s)	K_m (μM)	Estimated Relative Contribution to Overall Clearance by CYP2D6 (%)[a]	M_r (Da)	LogP (cLogP)	Reference
Methoxychlor	Insecticide	O-Demethylation		5–10	345.65	(4.56)	Stresser and Kupfer 1998
Methylenedioxymethamphetamine	Drug of abuse	Demethylation		20–30	193.24	(1.65)	Kreth et al. 2000; Lin et al. 1997; Tucker et al. 1994
Metoclopramide	Antiemetic	N-Dealkylation		60–70	299.80	1.8 (2.18)	Desta et al. 2002
Metoprolol	β₁-Blocker	α-Hydroxylation; O-demethylation	17–22	30–40	267.36	1.6 (1.8)	Hoffmann et al. 1980; Johnson and Burlew 1996; Otton et al. 1988
Mexiletine	Antiarrhythmic	p- and m-Hydroxylation		30–40	179.26	2.1 (2.17)	Broly et al. 1990; Vandamme et al. 1993
Mianserin	Tetracyclic antidepressant	8-Hydroxylation; N-demethylation; N-oxidation		10–15	264.36	(3.52)	Koyama et al. 1996; Stormer et al. 2000
Mirtazapine	Antidepressant	8-Hydroxylation; N-demethylation; N-oxidation		20–30	265.35	2.9 (2.9)	Delbressine et al. 1998
Monocrotaline	Pyrrolizidine alkaloid	Ring hydroxylation		60–70	325.15	(−0.37)	Wolff et al. 1985, 1987
Naphthalene	Chemical compound			10–20	128.17	(3.45)	Masubuchi et al. 1994
Naproxen	NSAID	O-Demethylation		15–25	230.26	2.8 (3.29)	Tracy et al. 1997
Nevirapine	Nonnucleoside reverse transcriptase inhibitor	8- and 12-Hydroxylation		20–25	266.30	2.5 (1.74)	Erickson et al. 1999
Nicardipine	Calcium channel blocker	Oxidation		5–15	479.53	3.6 (4.34)	Nakamura et al. 2005
Nicergoline	Vasodilator and anti-senile dementia	Hydroxylation		60–70	484.39	3.3 (3.99)	Bottiger et al. 1996
Nicotine	Natural compound	5′-Oxidation	0.74; 0.48	5–15	162.23	1.1 (0.87)	Cashman et al. 1992b
Nortriptyline	Tricyclic antidepressant	E-10-Hydroxylation; demethylation		40–60	263.38	4.7 (4.65)	Mellstrom and von Bahr 1981; Venkatakrishnan et al. 2000, 2001
Norverapamil	Calcium channel blocker	O-Demethylation; N-demethylation		5–10	440.58	(4.02)	Busse et al. 1995
Olanzapine	Antipsychotic	N-Demethylation; 2-hydroxymethylation		30–40	312.43	2 (2.65)	Ring et al. 1996
Omeprazole	Proton pump inhibitor	5-Pyridinyl methyl hydroxylation		5–10	345.42	0.6 (1.91)	Andersson et al. 1993
Ondansetron	5-HT₃ receptor antagonist	6-, 7-, and 8-Hydroxylation		15–20	293.36	2.4 (2.56)	Fischer et al. 1994; Sanwald et al. 1996a
Oxatomide	Antihistamine	N-Dealkylation		30–40	426.55	3.57	Goto et al. 2004
Oxybutynin	Anesthetic and adjuvant drug			10–15	357.49	4.3 (4.37)	Mizushima et al. 2007
Oxymorphone	Analgesic	O-Demethylation	130	20–30	301.34	(1.26)	Lalovic et al. 2004

(Continued)

TABLE 3.1 (CONTINUED)
Reported Substrates of CYP2D6

Substrate	Drug Class	Metabolic Pathway(s)	K_m (μM)	Estimated Relative Contribution to Overall Clearance by CYP2D6 (%)[a]	M_r (Da)	LogP (cLogP)	Reference
Paclitaxel	Antineoplastic	6-α-Hydroxylation		15–20	853.91	3 (3.2)	Sonnichsen et al. 1995
Pactimibe	Novel acyl coenzyme A:cholesterol acyltransferase inhibitor	ω-1-Oxidation	1.74 (25.1)	15–20	416.60	(6.19)	Kotsuma et al. 2008c
Paroxetine	SSRI	Demethylation		80	329.37	3.6 (3.1)	Bloomer et al. 1992
Perphenazine	Antipsychotic	N-Dealkylation; sulfoxidation; 7-hydroxylation	1.9	40–60	403.97	4.34	Olesen and Linnet 2000
Phenformin (withdrawn)	Antidiabetic	p-Hydroxylation		20–30	205.26	0.1 (0.3)	Oates et al. 1982; Shah et al. 1985
Phenoxypropoxybiguanide	Antimalarial prodrug			15–30			Diaz et al. 2008
Phenytoin	Antiepileptic	4'-Hydroxylation		5–10	252.27	2.2 (2.26)	Riley et al. 1990
Phorate	Thioether pesticide	Sulfoxidation		5–10	260.38	3.67	Usmani et al. 2004
Pinoline	β-Carboline alkaloid	O-Demethylation		60–70	202.00		Yu et al. 2003c,d
Perhexiline	Antianginal agent	Hydroxylation		5–15	277.49	6.2 (5.87)	Sorensen et al. 2003
Perphenazine	Antipsychotic	N-Dealkylation; sulfoxidation; 7-hydroxylation		15–25	409.97	3.9 (4.15)	Olesen and Linnet 2000
Pranidipine	Calcium channel blocker	Deesterification		10–20	448.47	(5.35)	Kudo et al. 1999
Prasugrel	ADP receptor antagonist	Ring-open reaction		10–15	373.44	(2.87)	Rehmel et al. 2006
Procainamide	Antiarrhythmic	N-Hydroxylation		5–10	235.33	1.3 (1.42)	Lessard et al. 1997
Progesterone	Hormone	16α- and 21-Hydroxylation	31	5–15	314.46	3.5 (3.58)	Kishimoto et al. 2004; Niwa et al. 1998
Promazine	Antipsychotic	5-Sulfoxidation; N-demethylation		10–15	284.42	4.3 (4.63)	Wojcikowski et al. 2003
Propafenone	Antiarrhythmic	Aromatic ring hydroxylation		10–20	341.44	3.2 (3.1)	Botsch et al. 1993; Zhou et al. 2003
Propofol	General anesthetic	4-Hydroxylation		15–20	178.27	3.81	Guitton et al. 1998
Propranolol	β-Adrenergic blocker	4-, 5-, and 7-Hydroxylation		20–30	259.34	3 (3.03)	Ward et al. 1989
Retinoic acid	Dermatological agent	4'-, 4-, and 7-Hydroxylation; 18- and 4-oxo-formation; 5,6-epoxidation			300.44	(6.83)	H. Chen et al. 2000; Marill et al. 2000; McSorley and Daly 2000; Q.Y. Zhang et al. 2000
Retinol	Nutritional agent	4-Hydroxylation		5–10	286.23	(6.48)	H. Chen et al. 2000
Risperidone	Antipsychotic	9-Hydroxylation		30–40	410.48	2.5 (3.28)	Yasui-Furukori et al. 2001
Ritonavir	HIV protease inhibitor	N-Demethylation		10–20	720.94	3.9 (4.24)	Kumar et al. 1996

(Continued)

TABLE 3.1 (CONTINUED)

Reported Substrates of CYP2D6

Substrate	Drug Class	Metabolic Pathway(s)	K_m (μM)	Estimated Relative Contribution to Overall Clearance by CYP2D6 (%)[a]	M_r (Da)	LogP (cLogP)	Reference
Rosiglitazone	Antidiabetic	p-Hydroxylation; N-demethylation		5–10	357.43	2.4 (2.95)	Baldwin et al. 1999
Saquinavir	HIV protease inhibitor	Oxidation		10–20	670.84	3.8 (4.04)	Fitzsimmons and Collins 1997
Selegiline	Antiparkinsonism agent	N-Demethylation; N-depropargylation	56	30–40	187.28	2.7 (3.08)	Grace et al. 1994
Seratrodast	Thromboxane A_2 receptor antagonist	5-Methyl hydroxylation; 4′-hydroxylation		5–10	354.44	(5.25)	Kumar et al. 1997
Sertraline	SSRI	N-Demethylation		10–15	306.07	5.1 (5.06)	Kobayashi et al. 1999
Sparteine	Tricyclic antidepressant	N1-Oxidation		50–70	234.21	(3.21)	Dahl et al. 1996; Gram et al. 1989; Mellstrom et al. 1981
Sildenafil	Selective phosphodiesterase type 5 inhibitor	N-Demethylation		10–15	474.58	1.3 (2.36)	Ku et al. 2008; Warrington et al. 2000
Styrene	Chemical compound	Oxidation		5–15	104.15	(2.7)	Kim et al. 1997
Sulfadiazine	Antibacterial	N-Hydroxylation		5–10	250.28	−0.2 (0.25)	Winter and Unadkat 2005
Sulprofos	Thioether pesticide	Sulfoxidation		5–10	322.45	(4.64)	Usmani et al. 2004
Tacrine	Centrally acting cholinesterase inhibitor	1-Hydroxylation		5–10	198.26	(3.32)	Madden et al. 1993
Tamoxifen	Selective estrogen receptor modulator	4- and 4′-Hydroxylation; N-demethylation		5–10	371.51	7.1 (5.93)	Beverage et al. 2007; Crewe et al. 2002; Dehal and Kupfer 1997; Jacolot et al. 1991; Parte and Kupfer 2005; Stearns et al. 2003
Tandospirone	Antipsychotic	Pyrimidine ring hydroxylation		15–20	383.49	2.02	Natsui et al. 2007
Tegafur	Antineoplastic	5′-Hydroxylation		5–15	200.17	(−0.77)	Komatsu et al. 2001
Temazepam	Benzodiazepine sedative	N-Demethylation		5–10	300.74	3 (2.16)	Yang et al. 1998
Terbinafine	Antifungal	N-Demethylation; side-chain oxidation		5–15	291.43	5.9 (5.51)	Vickers et al. 1999
Terfenadine	H_1 receptor antagonist	t-Butyl hydroxylation		25–35	471.67	7.1 (5.98)	Garteiz et al. 1982
Testosterone	Hormone	2β-, 6β-, and 17-Hydroxylation		5–15	288.42	3.6 (2.99)	Hiroi et al. 2001
Tetrahydrocannabinol	Natural compound	Oxidation			314.46	(7.68)	Matsunaga et al. 2000
Thebaine	CNS drug	O-Demethylation		15–25	311.17	(2.4)	Sindrup et al. 1993

(Continued)

TABLE 3.1 (CONTINUED)
Reported Substrates of CYP2D6

Substrate	Drug Class	Metabolic Pathway(s)	K_m (μM)	Estimated Relative Contribution to Overall Clearance by CYP2D6 (%)[a]	M_r (Da)	LogP (cLogP)	Reference
Thioridazine	Antipsychotic			15–20	370.58	5.9 (5.93)	Wojcikowski et al. 2006
Tiaramide	NSAID	N-Dealkylation; N-oxidation		15–20	355.84	(1.85)	Iwasaki et al. 1982
Ticlopidine	P2RY12 receptor inhibitor	Dehydrogenation; S-oxidation		5–10	263.79	2.9 (4.25)	Dalvie and O'Connell 2004
Tienilic acid	Diuretic	5-Hydroxylation		5–15	331.17	(3.14)	Jean et al. 1996
Timolol	β-Blocker	Hydroxylation	6.95	>90	316.42	1.2 (1.44)	Volotinen et al. 2007
Tolbutamide	Antidiabetic	p-Methyl hydroxylation		5–10	270.35	2.2 (2.04)	Miners et al. 1988
Tolterodine	Antimuscarinic	5-Hydroxymethylation		20–30	325.49	5.6 (5.39)	Postlind et al. 1998
Toluene	Chemical compound	p-Hydroxylation; methyl hydroxylation		15–20	92.14	(2.56)	Kim et al. 1997
Torasemide	Loop diuretic	Methyl tolyl hydroxylation		10–20	348.42	2.3 (1.76)	Miners et al. 1995
Tramadol	Opioid analgesic	O-Demethylation	286	15–20	263.38	2.4 (2.71)	Paar et al. 1997; Subrahmanyam et al. 2001
Trimethadione	Antiepileptic	N-Demethylation			143.14	(0.07)	Tanaka et al. 2003
Trimethoprim	Antibacterial	4'-O-Demethylation			290.32	0.6 (1.26)	Lai et al. 1999
Trimipramine	Tricyclic antidepressant	2-Hydroxylation; 2,10- and 2,11-dihydroxylation		30–40	294.43	4.2 (4.67)	Bolaji et al. 1993; Eap et al. 2000
Troglitazone	Antidiabetic	Quinone formation		5–10	441.54	3.6 (4.16)	Yamazaki et al. 1999
Tropisetron	5-HT₃ antagonist	5- and 6-Hydroxylation	3.9 and 4.66	25–35	284.15	(3.55)	Firkusny et al. 1995
Venlafaxine	Antidepressant	O- and N-Demethylation		20–30	277.20	2.8 (2.69)	Fogelman et al. 1999; Otton et al. 1996
Vernakalant	Novel antiarrhythmic	4-O-Demethylation		60–80	369.93		See http://www.cardiome.com
Verapamil	Calcium channel blocker	N- and O-Demethylation		10–15	454.60	4.7 (5.23)	Busse et al. 1995
Warfarin	Anticoagulant agent	4'- and 7-Hydroxylation		<5	308.33	3 (2.55)	Rettie et al. 1989; Yamazaki and Shimada 1997a
Zopiclone	Cyclopyrrolone sedative	N-Demethylation; N-oxidation		15–25	388.81	0.8 (0.97)	Becquemont et al. 1999
Zolpidem	Hypnotic	Oxidation of C-methyl groups and hydroxylation		10–15	307.39	1.2 (3.15)	Salva and Costa 1995
Zuclopenthixol	Antischizophrenia agent	Sulfoxidation; N-dealkylation		10–15	400.97	(4.46)	Dahl et al. 1991a

Note: COX, cyclooxygenase; HT, 5-hydroxytryptamine; MAO, monoamine oxidase; NK₁, neurokinin-1; NMDA, N-methyl-D-aspartate receptor; NSAID, nonsteroidal anti-inflammatory drug; P2RY12, purinergic receptor P2Y, G-protein coupled, 12; PCEPA, N-(1-phenylcyclohexyl)-3-ethoxypropanamine; PCMPA, N-(1-phenylcyclohexyl)-3-methoxypropanamine; PPAR, peroxisome proliferator-activated receptor; SSRI, selective serotonin reuptake inhibitor; TCA, tricyclic antidepressant.

[a] The percentage is estimated according to the drug metabolism.

REFERENCES

Aapro, M. 2005. 5-HT(3)-receptor antagonists in the management of nausea and vomiting in cancer and cancer treatment. *Oncology* 69:97–109.

Aapro, M., Blower, P. 2005. 5-hydroxytryptamine type-3 receptor antagonists for chemotherapy-induced and radiotherapy-induced nausea and emesis: Can we safely reduce the dose of administered agents? *Cancer* 104:1–18.

Abass, K., Reponen, P., Turpeinen, M., Jalonen, J., Pelkonen, O. 2007. Characterization of diuron N-demethylation by mammalian hepatic microsomes and cDNA-expressed human cytochrome P450 enzymes. *Drug Metab Dispos* 35:1634–1641.

Abdel-Rehim, M., Bielenstein, M., Askemark, Y., Tyrefors, N., Arvidsson, T. 2000. High-performance liquid chromatography-tandem electrospray mass spectrometry for the determination of lidocaine and its metabolites in human plasma and urine. *J Chromatogr B Biomed Sci Appl* 741:175–188.

Abolfathi, Z., Fiset, C., Gilbert, M., Moerike, K., Belanger, P. M., Turgeon, J. 1993. Role of polymorphic debrisoquin 4-hydroxylase activity in the stereoselective disposition of mexiletine in humans. *J Pharmacol Exp Ther* 266:1196–1201.

Abrams, P., Andersson, K. E. 2007. Muscarinic receptor antagonists for overactive bladder. *BJU Int* 100:987–1006.

Adusumalli, V. E., Wong, K. K., Kucharczyk, N., Sofia, R. D. 1992. Pharmacokinetics of azelastine and its active metabolite, desmethylazelastine, in guinea pigs. *Drug Metab Dispos* 20:530–535.

Agai-Csongor, E., Domany, G., Nogradi, K., Galambos, J., Vago, I., Keseru, G. M., Greiner, I. et al. 2012. Discovery of cariprazine (RGH-188): A novel antipsychotic acting on dopamine D_3/D_2 receptors. *Bioorg Med Chem Lett* 22:3437–3440.

Agundez, J. A., Martinez, C., Benitez, J. 1995. Metabolism of aminopyrine and derivatives in man: In vivo study of monomorphic and polymorphic metabolic pathways. *Xenobiotica* 25:417–427.

Agurell, S., Holmstedt, B., Lindgren, J. E. 1969. Metabolism of 5-methoxy-N,-N dimethyltryptamine- 14 C in the rat. *Biochem Pharmacol* 18:2771–2781.

Airaksinen, M. M., Kari, I. 1981. Beta-carbolines, psychoactive compounds in the mammalian body. Part I: Occurrence, origin and metabolism. *Med Biol* 59:21–34.

Akutsu, T., Kobayashi, K., Sakurada, K., Ikegaya, H., Furihata, T., Chiba, K. 2007. Identification of human cytochrome p450 isozymes involved in diphenhydramine *N*-demethylation. *Drug Metab Dispos* 35:72–78.

Albers, L. J., Musenga, A., Raggi, M. A. 2008. Iloperidone: A new benzisoxazole atypical antipsychotic drug. Is it novel enough to impact the crowded atypical antipsychotic market? *Expert Opin Investig Drugs* 17:61–75.

Albibi, R., McCallum, R. W. 1983. Metoclopramide: Pharmacology and clinical application. *Ann Intern Med* 98:86–95.

Albrecht, W., Unger, A., Nussler, A. K., Laufer, S. 2008. In vitro metabolism of 2-[6-(4-chlorophenyl)-2,2-dimethyl-7-phenyl-2,3-dihydro-1H-pyrrolizin-5-yl] acetic acid (licofelone, ML3000), an inhibitor of cyclooxygenase-1 and -2 and 5-lipoxygenase. *Drug Metab Dispos* 36:894–903.

Albus, M., Botschev, C., Muller-Spahn, F., Naber, D., Munch, U., Ackenheil, M. 1986. Clinical and biochemical effects of nicergoline in chronic schizophrenic patients. *Pharmacopsychiatry* 19:101–105.

Alexson, S. E., Diczfalusy, M., Halldin, M., Swedmark, S. 2002. Involvement of liver carboxylesterases in the in vitro metabolism of lidocaine. *Drug Metab Dispos* 30:643–647.

Alkalay, D., Wagner, W. E., Jr., Carlsen, S., Khemani, L., Volk, J., Bartlett, M. F., LeSher, A. 1980. Bioavailability and kinetics of maprotiline. *Clin Pharmacol Ther* 27:697–703.

Allard, S., Sainati, S., Roth-Schechter, B., MacIntyre, J. 1998. Minimal interaction between fluoxetine and multiple-dose zolpidem in healthy women. *Drug Metab Dispos* 26:617–622.

Alper, K. R., Lotsof, H. S., Frenken, G. M., Luciano, D. J., Bastiaans, J. 1999. Treatment of acute opioid withdrawal with ibogaine. *Am J Addict* 8:234–242.

Alper, K. R., Lotsof, H. S., Frenken, G. M., Luciano, D. J., Bastiaans, J. 2000. Ibogaine in acute opioid withdrawal. An open label case series. *Ann N Y Acad Sci* 909:257–259.

Altinbas, K., Guloksuz, S., Oral, E. T. 2013. Clinical potential of cariprazine in the treatment of acute mania. *Psychiatr Danub* 25:207–213.

Alvan, G., Grind, M., Graffner, C., Sjoqvist, F. 1984. Relationship of *N*-demethylation of amiflamine and its metabolite to debrisoquine hydroxylation polymorphism. *Clin Pharmacol Ther* 36:515–519.

Alvan, G., Graffner, C., Grind, M., Gustafsson, L. L., Lindgren, J. E., Nordin, C., Ross, S., Selander, H., Siwers, B. 1986. Tolerance and pilot pharmacokinetics of amiflamine after increasing single oral doses in healthy subjects. *Clin Pharmacol Ther* 40:81–85.

Alvan, G., Bechtel, P., Iselius, L., Gundert-Remy, U. 1990. Hydroxylation polymorphisms of debrisoquine and mephenytoin in European populations. *Eur J Clin Pharmacol* 39:533–537.

Alvarez, E., Perez, V., Artigas, F. 2014. Pharmacology and clinical potential of vortioxetine in the treatment of major depressive disorder. *Neuropsychiatr Dis Treat* 10:1297–1307.

Anderson, J. L., Stewart, J. R., Perry, B. A., Van Hamersveld, D. D., Johnson, T. A., Conard, G. J., Chang, S. F., Kvam, D. C., Pitt, B. 1981. Oral flecainide acetate for the treatment of ventricular arrhythmias. *N Engl J Med* 305:473–477.

Andersson, T., Miners, J. O., Veronese, M. E., Tassaneeyakul, W., Tassaneeyakul, W., Meyer, U. A., Birkett, D. J. 1993. Identification of human liver cytochrome P450 isoforms mediating omeprazole metabolism. *Br J Clin Pharmacol* 36:521–530.

Angiolillo, D. J., Bates, E. R., Bass, T. A. 2008. Clinical profile of prasugrel, a novel thienopyridine. *Am Heart J* 156:S16–S22.

Antonaccio, M. J., Gomoll, A. W., Byrne, J. E. 1989. Encainide. *Cardiovasc Drugs Ther* 3:691–710.

Anttila, S. A., Leinonen, E. V. 2001. A review of the pharmacological and clinical profile of mirtazapine. *CNS Drug Rev* 7:249–264.

Apak, T. I., Duffel, M. W. 2004. Interactions of the stereoisomers of alpha-hydroxytamoxifen with human hydroxysteroid sulfotransferase SULT2A1 and rat hydroxysteroid sulfotransferase STa. *Drug Metab Dispos* 32:1501–1508.

Arcamone, F., Glasser, A. G., Grafnetterova, J., Minghetti, A., Nicolella, V. 1972. Studies on the metabolism of ergoline derivatives. Metabolism of nicergoline in man and in animals. *Biochem Pharmacol* 21:2205–2213.

Aringberg-Wistedt, A., Alvariza, M., Bertilsson, L., Nerdrum, T., Nordin, C., Siwers, B., Uppfeldt, G. 1987. Biochemical and clinical effects of amiflamine—Determined by the debrisoquine hydroxylation phenotype? *Nordic J Psychiatr* 41:141–148.

Arpino, G., Nair Krishnan, M., Doval Dinesh, C., Bardou, V. J., Clark, G. M., Elledge, R. M. 2003. Idoxifene versus tamoxifen: A randomized comparison in postmenopausal patients with metastatic breast cancer. *Ann Oncol* 14:233–241.

Asano, T., Kushida, H., Sadakane, C., Ishihara, K., Wakui, Y., Yanagisawa, T., Kimura, M., Kamei, H., Yoshida, T. 2001. Metabolism of ipecac alkaloids cephaeline and emetine by human hepatic microsomal cytochrome P450s, and their inhibitory effects on P450 enzyme activities. *Biol Pharm Bull* 24:678–682.

Ashrafian, H., Horowitz, J. D., Frenneaux, M. P. 2007. Perhexiline. *Cardiovasc Drug Rev* 25:76–97.

Atkinson, A. J., Jr., Ruo, T. I., Piergies, A. A. 1988. Comparison of the pharmacokinetic and pharmacodynamic properties of procainamide and N-acetylprocainamide. *Angiology* 39:655–667.

Bachus, R., Bickel, U., Thomsen, T., Roots, I., Kewitz, H. 1999. The O-demethylation of the antidementia drug galanthamine is catalysed by cytochrome P450 2D6. *Pharmacogenetics* 9:661–668.

Bagwell, E. E., Walle, T., Drayer, D. E., Reidenbert, M. M., Pruett, J. K. 1976. Correlation of the electrophysiological and anti-arrhythmic properties of the N-acetyl metabolite of procainamide with plasma and tissue drug concentrations in the dog. *J Pharmacol Exp Ther* 197:38–48.

Balani, S. K., Woolf, E. J., Hoagland, V. L., Sturgill, M. G., Deutsch, P. J., Yeh, K. C., Lin, J. H. 1996. Disposition of indinavir, a potent HIV-1 protease inhibitor, after an oral dose in humans. *Drug Metab Dispos* 24:1389–1394.

Baldwin, S. J., Clarke, S. E., Chenery, R. J. 1999. Characterization of the cytochrome P450 enzymes involved in the *in vitro* metabolism of rosiglitazone. *Br J Clin Pharmacol* 48:424–432.

Balfour, J. A., Goa, K. L. 1997. Dolasetron. A review of its pharmacology and therapeutic potential in the management of nausea and vomiting induced by chemotherapy, radiotherapy or surgery. *Drugs* 54:273–298.

Banerji, A., Long, A. A., Camargo, C. A., Jr. 2007. Diphenhydramine versus nonsedating antihistamines for acute allergic reactions: A literature review. *Allergy Asthma Proc* 28:418–426.

Bang-Andersen, B., Ruhland, T., Jorgensen, M., Smith, G., Frederiksen, K., Jensen, K. G., Zhong, H. et al. 2011. Discovery of 1-[2-(2,4-dimethylphenylsulfanyl)phenyl]piperazine (Lu AA21004): A novel multimodal compound for the treatment of major depressive disorder. *J Med Chem* 54:3206–3221.

Barakat, N. H., Atayee, R. S., Best, B. M., Ma, J. D. 2014. Urinary hydrocodone and metabolite distributions in pain patients. *J Anal Toxicol* 38:404–409.

Bargetzi, M. J., Aoyama, T., Gonzalez, F. J., Meyer, U. A. 1989. Lidocaine metabolism in human liver microsomes by cytochrome P450IIIA4. *Clin Pharmacol Ther* 46:521–527.

Barner, E. L., Gray, S. L. 1998. Donepezil use in Alzheimer disease. *Ann Pharmacother* 32:70–77.

Barnhart, J. W. 1980. The urinary excretion of dextromethorphan and three metabolites in dogs and humans. *Toxicol Appl Pharmacol* 55:43–48.

Barsoum, N. J., Gough, A. W., Sturgess, J. M., de la Iglesia, F. A. 1986. Parkinson-like syndrome in nonhuman primates receiving a tetrahydropyridine derivative. *Neurotoxicology* 7:119–126.

Bartolone, J. B., Sparks, K., Cohen, S. D., Khairallah, E. A. 1987. Immunochemical detection of acetaminophen-bound liver proteins. *Biochem Pharmacol* 36:1193–1196.

Baskys, A., Hou, A. C. 2007. Vascular dementia: Pharmacological treatment approaches and perspectives. *Clin Interv Aging* 2:327–335.

Bates, D. J., Foster, A. B., Griggs, L. J., Jarman, M., Leclercq, G., Devleeshouwer, N. 1982. Metabolism of tamoxifen by isolated rat hepatocytes: Anti-estrogenic activity of tamoxifen N-oxide. *Biochem Pharmacol* 31:2823–2827.

Baumann, P., Bosshart, P., Gabris, G., Gastpar, M., Koeb, L., Woggon, B. 1988. Acetylation of maprotiline and desmethylmaprotiline in depressive patients phenotyped with sulfamidine, debrisoquine, and mephenytoin. *Arzneimittelforschung* 38:292–296.

Baumann, M. H., Rothman, R. B., Pablo, J. P., Mash, D. C. 2001. In vivo neurobiological effects of ibogaine and its O-desmethyl metabolite, 12-hydroxyibogamine (noribogaine), in rats. *J Pharmacol Exp Ther* 297:531–539.

Beasley, C. M., Jr., Tollefson, G., Tran, P., Satterlee, W., Sanger, T., Hamilton, S. 1996. Olanzapine versus placebo and haloperidol: Acute phase results of the North American double-blind olanzapine trial. *Neuropsychopharmacology* 14:111–123.

Beckett, A. H., Chidomere, E. C. 1977. The identification and analysis of mexiletine and its metabolic products in man. *J Pharm Pharmacol* 29:281–285.

Becquemont, L., Mouajjah, S., Escaffre, O., Beaune, P., Funck-Brentano, C., Jaillon, P. 1999. Cytochrome P-450 3A4 and 2C8 are involved in zopiclone metabolism. *Drug Metab Dispos* 27:1068–1073.

Befekadu, E., Onofrei, C., Colice, G. L. 2014. Tiotropium in asthma: A systematic review. *J Asthma Allergy* 7:11–21.

Begre, S., von Bardeleben, U., Ladewig, D., Jaquet-Rochat, S., Cosendai-Savary, L., Golay, K. P., Kosel, M., Baumann, P., Eap, C. B. 2002. Paroxetine increases steady-state concentrations of (R)-methadone in CYP2D6 extensive but not poor metabolizers. *J Clin Psychopharmacol* 22:211–215.

Belpaire, F. M., Wijnant, P., Temmerman, A., Rasmussen, B. B., Brosen, K. 1998. The oxidative metabolism of metoprolol in human liver microsomes: Inhibition by the selective serotonin reuptake inhibitors. *Eur J Clin Pharmacol* 54:261–264.

Benjamin, B., Burns, A. 2007. Donepezil for Alzheimer's disease. *Expert Rev Neurother* 7:1243–1249.

Benowitz, N. L., Meister, W. 1978. Clinical pharmacokinetics of lignocaine. *Clin Pharmacokinet* 3:177–201.

Berecz, R., de la Rubia, A., Dorado, P., Fernandez-Salguero, P., Dahl, M. L., LLerena, A. 2003. Thioridazine steady-state plasma concentrations are influenced by tobacco smoking and CYP2D6, but not by the CYP2C9 genotype. *Eur J Clin Pharmacol* 59:45–50.

Bergman, E., Forsell, P., Persson, E. M., Knutson, L., Dickinson, P., Smith, R., Swaisland, H., Farmer, M. R., Cantarini, M. V., Lennernas, H. 2007. Pharmacokinetics of gefitinib in humans: The influence of gastrointestinal factors. *Int J Pharm* 341:134–142.

Bernauer, U., Heinrich-Hirsch, B., Tonnies, M., Peter-Matthias, W., Gundert-Remy, U. 2006. Characterisation of the xenobiotic-metabolizing Cytochrome P450 expression pattern in human lung tissue by immunochemical and activity determination. *Toxicol Lett* 164:278–288.

Bertilsson, L. 2007. Metabolism of antidepressant and neuroleptic drugs by cytochrome p450s: Clinical and interethnic aspects. *Clin Pharmacol Ther* 82:606–609.

Bertilsson, L., Lou, Y. Q., Du, Y. L., Liu, Y., Kuang, T. Y., Liao, X. M., Wang, K. Y., Reviriego, J., Iselius, L., Sjoqvist, F. 1992. Pronounced differences between native Chinese and Swedish populations in the polymorphic hydroxylations of debrisoquin and S-mephenytoin. *Clin Pharmacol Ther* 51:388–397.

Bertilsson, L., Dahl, M. L., Ekqvist, B., Llerena, A. 1993. Disposition of the neuroleptics perphenazine, zuclopenthixol, and haloperidol cosegregates with polymorphic debrisoquine hydroxylation. *Psychopharmacol Ser* 10:230–237.

Bessems, J. G., Vermeulen, N. P. 2001. Paracetamol (acetaminophen)-induced toxicity: Molecular and biochemical mechanisms, analogues and protective approaches. *Crit Rev Toxicol* 31: 55–138.

Beverage, J. N., Sissung, T. M., Sion, A. M., Danesi, R., Figg, W. D. 2007. CYP2D6 polymorphisms and the impact on tamoxifen therapy. *J Pharm Sci* 96:2224–2231.

Bichara, N., Ching, M. S., Blake, C. L., Ghabrial, H., Smallwood, R. A. 1996. Propranolol hydroxylation and N-desisopropylation by cytochrome P4502D6: Studies using the yeast-expressed enzyme and NADPH/O2 and cumene hydroperoxide-supported reactions. *Drug Metab Dispos* 24:112–118.

Bishara, D., Taylor, D. 2008. Upcoming agents for the treatment of schizophrenia: Mechanism of action, efficacy and tolerability. *Drugs* 68:2269–2292.

Bjelogrlic, N., Peng, R., Park, S. S., Gelboin, H. V., Honkakoski, P., Pelkonen, O., Vähäkangas, K. 1993. Involvement of P450 1A1 in benzo(a)pyrene but not in benzo(a)pyrene-7,8-dihydrodiol activation by 3-methylcholanthrene-induced mouse liver microsomes. *Pharmacol Toxicol* 73:319–324.

Bloomer, J. C., Woods, F. R., Haddock, R. E., Lennard, M. S., Tucker, G. T. 1992. The role of cytochrome P4502D6 in the metabolism of paroxetine by human liver microsomes. *Br J Clin Pharmacol* 33:521–523.

Blyden, G. T., Greenblatt, D. J., Scavone, J. M., Shader, R. I. 1986. Pharmacokinetics of diphenhydramine and a demethylated metabolite following intravenous and oral administration. *J Clin Pharmacol* 26:529–533.

Boberg, M., Angerbauer, R., Fey, P., Kanhai, W. K., Karl, W., Kern, A., Ploschke, J., Radtke, M. 1997. Metabolism of cerivastatin by human liver microsomes in vitro. Characterization of primary metabolic pathways and of cytochrome P450 isozymes involved. *Drug Metab Dispos* 25:321–331.

Bock, K. W., Forster, A., Gschaidmeier, H., Bruck, M., Munzel, P., Schareck, W., Fournel-Gigleux, S., Burchell, B. 1993. Paracetamol glucuronidation by recombinant rat and human phenol UDP-glucuronosyltransferases. *Biochem Pharmacol* 45:1809–1814.

Bohets, H., Lavrijsen, K., Hendrickx, J., van Houdt, J., van Genechten, V., Verboven, P., Meuldermans, W., Heykants, J. 2000. Identification of the cytochrome P450 enzymes involved in the metabolism of cisapride: *In vitro* studies of potential co-medication interactions. *Br J Pharmacol* 129:1655–1667.

Bolaji, O. O., Coutts, R. T., Baker, G. B. 1993. Metabolism of trimipramine in vitro by human CYP2D6 isozyme. *Res Commun Chem Pathol Pharmacol* 82:111–120.

Bolser, D. C. 2006. Current and future centrally acting antitussives. *Respir Physiol Neurobiol* 152:349–355.

Bolton, J. L., Thatcher, G. R. 2008. Potential mechanisms of estrogen quinone carcinogenesis. *Chem Res Toxicol* 21:93–101.

Bondolfi, G., Chautems, C., Rochat, B., Bertschy, G., Baumann, P. 1996. Non-response to citalopram in depressive patients: Pharmacokinetic and clinical consequences of a fluvoxamine augmentation. *Psychopharmacology (Berl)* 128:421–425.

Bonn, B., Masimirembwa, C. M., Aristei, Y., Zamora, I. 2008. The molecular basis of CYP2D6-mediated N-dealkylation: Balance between metabolic clearance routes and enzyme inhibition. *Drug Metab Dispos* 36:2199–2210.

Boobis, A. R., Murray, S., Kahn, G. C., Robertz, G. M., Davies, D. S. 1983. Substrate specificity of the form of cytochrome P-450 catalyzing the 4-hydroxylation of debrisoquine in man. *Mol Pharmacol* 23:474–481.

Borg, K. O., Carlsson, E., Hoffmann, K. J., Jonsson, T. E., Thorin, H., Wallin, B. 1975. Metabolism of metoprolol-(3-h) in man, the dog and the rat. *Acta Pharmacol Toxicol (Copenh)* 36:125–135.

Bort, R., Mace, K., Boobis, A., Gomez-Lechon, M. J., Pfeifer, A., Castell, J. 1999. Hepatic metabolism of diclofenac: Role of human CYP in the minor oxidative pathways. *Biochem Pharmacol* 58:787–796.

Botsch, S., Gautier, J. C., Beaune, P., Eichelbaum, M., Kroemer, H. K. 1993. Identification and characterization of the cytochrome P450 enzymes involved in N-dealkylation of propafenone: Molecular base for interaction potential and variable disposition of active metabolites. *Mol Pharmacol* 43:120–126.

Bottiger, Y., Dostert, P., Benedetti, M. S., Bani, M., Fiorentini, F., Casati, M., Poggesti, I., Alm, C., Alvan, G., Bertilsson, L. 1996. Involvement of CYP2D6 but not CYP2C19 in nicergoline metabolism in humans. *Br J Clin Pharmacol* 42:707–711.

Brachtendorf, L., Jetter, A., Beckurts, K. T., Holscher, A. H., Fuhr, U. 2002. Cytochrome P450 enzymes contributing to demethylation of maprotiline in man. *Pharmacol Toxicol* 90:144–149.

Braga, P. C., Fossati, A., Vimercati, M. G., Caputo, R., Guffanti, E. E. 1994. Dextrorphan and dextromethorphan: Comparative antitussive effects on guinea pigs. *Drugs Exp Clin Res* 20:199–203.

Braun, J., Kollert, J. R., Becker, J. U. 1987. Pharmacokinetics of flecainide in patients with mild and moderate renal failure compared with patients with normal renal function. *Eur J Clin Pharmacol* 31:711–714.

Breyer-Pfaff, U., Kroeker, M., Winkler, T., Kriemler, P. 1985. Isolation and identification of hydroxylated maprotiline metabolites. *Xenobiotica* 15:57–66.

Breyer-Pfaff, U., Pfandl, B., Nill, K., Nusser, E., Monney, C., Jonzier-Perey, M., Baettig, D., Baumann, P. 1992. Enantioselective amitriptyline metabolism in patients phenotyped for two cytochrome P450 isozymes. *Clin Pharmacol Ther* 52:350–358.

Breyer-Pfaff, U., Fischer, D., Winne, D. 1997. Biphasic kinetics of quaternary ammonium glucuronide formation from amitriptyline and diphenhydramine in human liver microsomes. *Drug Metab Dispos* 25:340–345.

Briganti, A., Salonia, A., Gallina, A., Sacca, A., Montorsi, P., Rigatti, P., Montorsi, F. 2005. Drug Insight: Oral phosphodiesterase type 5 inhibitors for erectile dysfunction. *Nat Clin Pract Urol* 2:239–247.

Broadhurst, P., Brigden, G., Hittel, N., Lahiri, A., Raftery, E. B. 1991. Prolonged hypotensive effect of OPC-13340: A new, once-daily calcium channel blocking drug. *Eur Heart J* 12:434–438.

Brodde, O. E., Kroemer, H. K. 2003. Drug–drug interactions of beta-adrenoceptor blockers. *Arzneimittelforschung* 53:814–822.

Broly, F., Libersa, C., Lhermitte, M., Dupuis, B. 1990. Inhibitory studies of mexiletine and dextromethorphan oxidation in human liver microsomes. *Biochem Pharmacol* 39:1045–1053.

Brosen, K., Zeugin, T., Meyer, U. A. 1991. Role of P450IID6, the target of the sparteine-debrisoquin oxidation polymorphism, in the metabolism of imipramine. *Clin Pharmacol Ther* 49:609–617.

Brynne, N., Dalen, P., Alvan, G., Bertilsson, L., Gabrielsson, J. 1998. Influence of CYP2D6 polymorphism on the pharmacokinetics and pharmacodynamic of tolterodine. *Clin Pharmacol Ther* 63:529–539.

Brynne, N., Bottiger, Y., Hallen, B., Bertilsson, L. 1999a. Tolterodine does not affect the human in vivo metabolism of the probe drugs caffeine, debrisoquine and omeprazole. *Br J Clin Pharmacol* 47:145–150.

Brynne, N., Svanstrom, C., Aberg-Wistedt, A., Hallen, B., Bertilsson, L. 1999b. Fluoxetine inhibits the metabolism of tolterodine-pharmacokinetic implications and proposed clinical relevance. *Br J Clin Pharmacol* 48:553–563.

Bryson, H. M., Benfield, P. 1997. Donepezil. *Drugs Aging* 10:234–239; discussion 240–231.

Bryson, H. M., Palmer, K. J., Langtry, H. D., Fitton, A. 1993. Propafenone. A reappraisal of its pharmacology, pharmacokinetics and therapeutic use in cardiac arrhythmias. *Drugs* 45:85–130.

Budinsky, R. A., Roberts, S. M., Coats, E. A., Adams, L., Hess, E. V. 1987. The formation of procainamide hydroxylamine by rat and human liver microsomes. *Drug Metab Dispos* 15:37–43.

Burlingame, R. W. 1997. The clinical utility of antihistone antibodies. Autoantibodies reactive with chromatin in systemic lupus erythematosus and drug-induced lupus. *Clin Lab Med* 17:367–378.

Burlingame, R. W., Rubin, R. L. 1996. Autoantibody to the nucleosome subunit (H2A-H2B)-DNA is an early and ubiquitous feature of lupus-like conditions. *Mol Biol Rep* 23:159–166.

Busse, D., Cosme, J., Beaune, P., Kroemer, H. K., Eichelbaum, M. 1995. Cytochromes of the P450 2C subfamily are the major enzymes involved in the O-demethylation of verapamil in humans. *Naunyn Schmiedebergs Arch Pharmacol* 353:116–121.

Buters, J. T., Sakai, S., Richter, T., Pineau, T., Alexander, D. L., Savas, U., Doehmer, J., Ward, J. M., Jefcoate, C. R., Gonzalez, F. J. 1999. Cytochrome P450 CYP1B1 determines susceptibility to 7, 12-dimethylbenz[a]anthracene-induced lymphomas. *Proc Natl Acad Sci U S A* 96:1977–1982.

Buzdar, A., Hayes, D., El-Khoudary, A., Yan, S., Lonning, P., Lichinitser, M., Gopal, R. et al. 2002. Phase III randomized trial of droloxifene and tamoxifen as first-line endocrine treatment of ER/PgR-positive advanced breast cancer. *Breast Cancer Res Treat* 73:161–175.

Byerly, M. J., DeVane, C. L. 1996. Pharmacokinetics of clozapine and risperidone: A review of recent literature. *J Clin Psychopharmacol* 16:177–187.

Bymaster, F. P., Lee, T. C., Knadler, M. P., Detke, M. J., Iyengar, S. 2005. The dual transporter inhibitor duloxetine: A review of its preclinical pharmacology, pharmacokinetic profile, and clinical results in depression. *Curr Pharm Des* 11:1475–1493.

Caccia, S. 1998. Metabolism of the newer antidepressants. An overview of the pharmacological and pharmacokinetic implications. *Clin Pharmacokinet* 34:281–302.

Caccia, S. 2004. Metabolism of the newest antidepressants: Comparisons with related predecessors. *IDrugs* 7:143–150.

Calabrese, J. R., Keck, P. E., Jr., Starace, A., Lu, K., Ruth, A., Laszlovszky, I., Nemeth, G., Durgam, S. 2015. Efficacy and safety of low- and high-dose cariprazine in acute and mixed mania associated with bipolar I disorder: A double-blind, placebo-controlled study. *J Clin Psychiatry* 76:284–292.

Caldwell, J., Dring, L. G., Williams, R. T. 1972. Metabolism of (14 C)methamphetamine in man, the guinea pig and the rat. *Biochem J* 129:11–22.

Campbell, N. P., Kelly, J. G., Adgey, A. A., Shanks, R. G. 1978. The clinical pharmacology of mexiletine. *Br J Clin Pharmacol* 6:103–108.

Capdeville, R., Buchdunger, E., Zimmermann, J., Matter, A. 2002. Glivec (STI571, imatinib), a rationally developed, targeted anticancer drug. *Nat Rev Drug Discov* 1:493–502.

Carcillo, J. A., Adedoyin, A., Burckart, G. J., Frye, R. F., Venkataramanan, R., Knoll, C., Thummel, K. et al. 2003. Coordinated intrahepatic and extrahepatic regulation of cytochrome p4502D6 in healthy subjects and in patients after liver transplantation. *Clin Pharmacol Ther* 73:456–467.

Carter, S. J., Li, X. F., Mackey, J. R., Modi, S., Hanson, J., Dovichi, N. J. 2001. Biomonitoring of urinary tamoxifen and its metabolites from breast cancer patients using nonaqueous capillary electrophoresis with electrospray mass spectrometry. *Electrophoresis* 22:2730–2736.

Cartledge, J., Eardley, I. 1999. Sildenafil. *Expert Opin Pharmacother* 1:137–147.

Cascorbi, I. 2003. Pharmacogenetics of cytochrome P4502D6: Genetic background and clinical implication. *Eur J Clin Invest* 33:17–22.

Cashman, J. R., Celestial, J. R., Leach, A. R. 1992a. Enantioselective N-oxygenation of chlorpheniramine by the flavin-containing monooxygenase from hog liver. *Xenobiotica* 22:459–469.

Cashman, J. R., Park, S. B., Yang, Z. C., Wrighton, S. A., Jacob, P., 3rd, Benowitz, N. L. 1992b. Metabolism of nicotine by human liver microsomes: Stereoselective formation of trans-nicotine *N'*-oxide. *Chem Res Toxicol* 5:639–646.

Cavalli, A., Maggioni, A. P., Marchi, S., Volpi, A., Latini, R. 1988. Flecainide half-life prolongation in 2 patients with congestive heart failure and complex ventricular arrhythmias. *Clin Pharmacokinet* 14:187–188.

Chand, N., Sofia, R. D. 1995. Azelastine—A novel in vivo inhibitor of leukotriene biosynthesis: A possible mechanism of action: A mini review. *J Asthma* 32:227–234.

Chand, N., Adusumalli, V. E., Nolan, K., Diamantis, W., Wichmann, J. K., Pivonka, J., Langevin, C. N., Wong, K. K., Kucharczyk, N., Sofia, R. D. 1993. Pharmacodynamic and pharmacokinetic studies with azelastine in the guinea pig: Evidence for preferential distribution into the lung. *Allergy* 48:19–24.

Chander, S. K., McCague, R., Luqmani, Y., Newton, C., Dowsett, M., Jarman, M., Coombes, R. C. 1991. Pyrrolidino-4-iodotamoxifen and 4-iodotamoxifen, new analogues of the antiestrogen tamoxifen for the treatment of breast cancer. *Cancer Res* 51:5851–5858.

Chang, T., Okerholm, R. A., Glazko, A. J. 1974. Identification of diphenydramine (Benadryl) metabolities in human subjects. *Res Commun Chem Pathol Pharmacol* 9:391–404.

Chang, M., Sood, V. K., Wilson, G. J., Kloosterman, D. A., Sanders, P. E., Hauer, M. J., Fagerness, P. E. 1997. Metabolism of the human immunodeficiency virus type 1 reverse transcriptase inhibitor delavirdine in rats. *Drug Metab Dispos* 25:228–242.

Chau, N., Elliot, D. J., Lewis, B. C., Burns, K., Johnston, M. R., Mackenzie, P. I., Miners, J. O. 2014. Morphine glucuronidation and glucosidation represent complementary metabolic pathways that are both catalyzed by UDP-glucuronosyltransferase 2B7: Kinetic, inhibition, and molecular modeling studies. *J Pharmacol Exp Ther* 349:126–137.

Chauret, N., Dobbs, B., Lackman, R. L., Bateman, K., Nicoll-Griffith, D. A., Stresser, D. M., Ackermann, J. M., Turner, S. D., Miller, V. P., Crespi, C. L. 2001. The use of 3-[2-(N,N-diethyl-N-methylammonium)ethyl]-7-methoxy-4-methylcoumarin (AMMC) as a specific CYP2D6 probe in human liver microsomes. *Drug Metab Dispos* 29:1196–1200.

Cheeke, P. R. 1988. Toxicity and metabolism of pyrrolizidine alkaloids. *J Anim Sci* 66:2343–2350.

Chen, Z. R., Irvine, R. J., Somogyi, A. A., Bochner, F. 1991. μ-Receptor binding of some commonly used opioids and their metabolites. *Life Sci* 48:2165–2171.

Chen, W., Shockcor, J. P., Tonge, R., Hunter, A., Gartner, C., Nelson, S. D. 1999. Protein and nonprotein cysteinyl thiol modification by N-acetyl-p-benzoquinone imine via a novel ipso adduct. *Biochemistry* 38:8159–8166.

Chen, H., Howald, W. N., Juchau, M. R. 2000. Biosynthesis of all-*trans*-retinoic acid from all-*trans*-retinol: Catalysis of all-*trans*-retinol oxidation by human P450 cytochromes. *Drug Metab Dispos* 28:315–322.

Chen, X., Zhong, D., Blume, H. 2000. Stereoselective pharmacokinetics of propafenone and its major metabolites in healthy Chinese volunteers. *Eur J Pharm Sci* 10:11–16.

Chen, J. J., Ondo, W. G., Dashtipour, K., Swope, D. M. 2012. Tetrabenazine for the treatment of hyperkinetic movement disorders: A review of the literature. *Clin Ther* 34:1487–1504.

Chiba, M., Hensleigh, M., Nishime, J. A., Balani, S. K., Lin, J. H. 1996. Role of cytochrome P450 3A4 in human metabolism of MK-639, a potent human immunodeficiency virus protease inhibitor. *Drug Metab Dispos* 24:307–314.

Chinta, S. J., Pai, H. V., Upadhya, S. C., Boyd, M. R., Ravindranath, V. 2002. Constitutive expression and localization of the major drug metabolizing enzyme, cytochrome P4502D in human brain. *Brain Res Mol Brain Res* 103:49–61.

Choi, M. H., Skipper, P. L., Wishnok, J. S., Tannenbaum, S. R. 2005. Characterization of testosterone 11 beta-hydroxylation catalyzed by human liver microsomal cytochromes P450. *Drug Metab Dispos* 33:714–718.

Choi, Y. K., Adham, N., Kiss, B., Gyertyan, I., Tarazi, F. I. 2014. Long-term effects of cariprazine exposure on dopamine receptor subtypes. *CNS Spectr* 19:268–277.

Chow, T., Hiroi, T., Imaoka, S., Chiba, K., Funae, Y. 1999. Isoform-selective metabolism of mianserin by cytochrome P-450 2D. *Drug Metab Dispos* 27:1200–1204.

Christofaki, M., Papaioannou, A. 2014. Ondansetron: A review of pharmacokinetics and clinical experience in postoperative nausea and vomiting. *Expert Opin Drug Metab Toxicol* 10:437–444.

Churchill, F. C., Patchen, L. C., Campbell, C. C., Schwartz, I. K., Nguyen-Dinh, P., Dickinson, C. M. 1985. Amodiaquine as a prodrug: Importance of metabolite(s) in the antimalarial effect of amodiaquine in humans. *Life Sci* 36:53–62.

Churchill, F. C., Mount, D. L., Patchen, L. C., Bjorkman, A. 1986. Isolation, characterization and standardization of a major metabolite of amodiaquine by chromatographic and spectroscopic methods. *J Chromatogr* 377:307–318.

Citrome, L. 2013a. Cariprazine in bipolar disorder: Clinical efficacy, tolerability, and place in therapy. *Adv Ther* 30:102–113.

Citrome, L. 2013b. Cariprazine: Chemistry, pharmacodynamics, pharmacokinetics, and metabolism, clinical efficacy, safety, and tolerability. *Expert Opin Drug Metab Toxicol* 9:193–206.

Citrome, L. 2014. Vortioxetine for major depressive disorder: A systematic review of the efficacy and safety profile for this newly approved antidepressant—What is the number needed to treat, number needed to harm and likelihood to be helped or harmed? *Int J Clin Pract* 68:60–82.

Cobb, M. N., Desai, J., Brown, L. S., Jr., Zannikos, P. N., Rainey, P. M. 1998. The effect of fluconazole on the clinical pharmacokinetics of methadone. *Clin Pharmacol Ther* 63:655–662.

Coffman, B. L., Rios, G. R., King, C. D., Tephly, T. R. 1997. Human UGT2B7 catalyzes morphine glucuronidation. *Drug Metab Dispos* 25:1–4.

Coffman, B. L., King, C. D., Rios, G. R., Tephly, T. R. 1998. The glucuronidation of opioids, other xenobiotics, and androgens by human UGT2B7Y(268) and UGT2B7H(268). *Drug Metab Dispos* 26:73–77.

Cohen, S. D., Khairallah, E. A. 1997. Selective protein arylation and acetaminophen-induced hepatotoxicity. *Drug Metab Rev* 29:59–77.

Cole, A. E., Eccles, C. U., Aryanpur, J. J., Fisher, R. S. 1989. Selective depression of N-methyl-D-aspartate-mediated responses by dextrorphan in the hippocampal slice in rat. *Neuropharmacology* 28:249–254.

Cole, P. L., Beamer, A. D., McGowan, N., Cantillon, C. O., Benfell, K., Kelly, R. A., Hartley, L. H., Smith, T. W., Antman, E. M. 1990. Efficacy and safety of perhexiline maleate in refractory angina. A double-blind placebo-controlled clinical trial of a novel antianginal agent. *Circulation* 81:1260–1270.

Coleman, T., Ellis, S. W., Martin, I. J., Lennard, M. S., Tucker, G. T. 1996. 1-Methyl-4-phenyl-1,2,3,6-tetrahydropyridine (MPTP) is N-demethylated by cytochromes P450 2D6, 1A2 and 3A4—Implications for susceptibility to Parkinson's disease. *J Pharmacol Exp Ther* 277:685–690.

Coles, B., Wilson, I., Wardman, P., Hinson, J. A., Nelson, S. D., Ketterer, B. 1988. The spontaneous and enzymatic reaction of N-acetyl-p-benzoquinonimine with glutathione: A stopped-flow kinetic study. *Arch Biochem Biophys* 264:253–260.

Comoglio, A., Gibbs, A. H., White, I. N., Gant, T., Martin, E. A., Smith, L. L., Gamalero, S. R., DeMatteis, F. 1996. Effect of tamoxifen feeding on metabolic activation of tamoxifen by the liver of the rhesus monkey: Does liver accumulation of inhibitory metabolites protect from tamoxifen-dependent genotoxicity and cancer? *Carcinogenesis* 17:1687–1693.

Compagnone, N. A., Mellon, S. H. 2000. Neurosteroids: Biosynthesis and function of these novel neuromodulators. *Front Neuroendocrinol* 21:1–56.

Conard, G. J., Carlson, G. L., Frost, J. W., Ober, R. E., Leon, A. S., Hunninghake, D. B. 1984. Plasma concentrations of flecainide acetate, a new antiarrhythmic agent, in humans. *Clin Ther* 6: 643–652.

Connolly, S. J., Kates, R. E., Lebsack, C. S., Harrison, D. C., Winkle, R. A. 1983. Clinical pharmacology of propafenone. *Circulation* 68:589–596.

Cook, P., James, I. 1981. Cerebral vasodilators (second of two parts). *N Engl J Med* 305:1560–1564.

Coombes, R. C., Haynes, B. P., Dowsett, M., Quigley, M., English, J., Judson, I. R., Griggs, L. J., Potter, G. A., McCague, R., Jarman, M. 1995. Idoxifene: Report of a phase I study in patients with metastatic breast cancer. *Cancer Res* 55:1070–1074.

Corman, S. L., Fedutes, B. A., Culley, C. M. 2004. Atomoxetine: The first nonstimulant for the management of attention-deficit/hyperactivity disorder. *Am J Health Syst Pharm* 61: 2391–2399.

Court, M. H., Duan, S. X., von Moltke, L. L., Greenblatt, D. J., Patten, C. J., Miners, J. O., Mackenzie, P. I. 2001. Interindividual variability in acetaminophen glucuronidation by human liver microsomes: Identification of relevant acetaminophen UDP-glucuronosyltransferase isoforms. *J Pharmacol Exp Ther* 299:998–1006.

Coutts, R. T., Torok-Both, G. A., Chu, L. V., Tam, Y. K., Pasutto, F. M. 1987. In vivo metabolism of lidocaine in the rat. Isolation of urinary metabolites as pentafluorobenzoyl derivatives and their identification by combined gas chromatography-mass spectrometry. *J Chromatogr* 421:267–280.

Coyle, J. D., Boudoulas, H., Mackichan, J. J., Lima, J. J. 1985. Concentration-dependent clearance of procainamide in normal subjects. *Biopharm Drug Dispos* 6:159–165.

Crespi, C. L., Chang, T. K., Waxman, D. J. 2006. CYP2D6-dependent bufuralol 1'-hydroxylation assayed by reverse-phase ion-pair high-performance liquid chromatography with fluorescence detection. *Methods Mol Biol* 320:121–125.

Crewe, H. K., Lennard, M. S., Tucker, G. T., Woods, F. R., Haddock, R. E. 1992. The effect of selective serotonin re-uptake inhibitors on cytochrome P4502D6 (CYP2D6) activity in human liver microsomes. *Br J Clin Pharmacol* 34:262–265.

Crewe, H. K., Ellis, S. W., Lennard, M. S., Tucker, G. T. 1997. Variable contribution of cytochromes P450 2D6, 2C9 and 3A4 to the 4-hydroxylation of tamoxifen by human liver microsomes. *Biochem Pharmacol* 53:171–178.

Crewe, H. K., Notley, L. M., Wunsch, R. M., Lennard, M. S., Gillam, E. M. 2002. Metabolism of tamoxifen by recombinant human cytochrome P450 enzymes: Formation of the

4-hydroxy, 4'-hydroxy and N-desmethyl metabolites and isomerization of trans-4-hydroxytamoxifen. *Drug Metab Dispos* 30:869–874.

Cronin-Fenton, D. P., Damkier, P., Lash, T. L. 2014. Metabolism and transport of tamoxifen in relation to its effectiveness: New perspectives on an ongoing controversy. *Future Oncol* 10:107–122.

Croom, K. F., Perry, C. M. 2003. Imatinib mesylate: In the treatment of gastrointestinal stromal tumours. *Drugs* 63:513–522; discussion 523–514.

Cysneiros, R. M., Farkas, D., Harmatz, J. S., von Moltke, L. L., Greenblatt, D. J. 2007. Pharmacokinetic and pharmacodynamic interactions between zolpidem and caffeine. *Clin Pharmacol Ther* 82:54–62.

Dahl, M. L., Ekqvist, B., Widen, J., Bertilsson, L. 1991a. Disposition of the neuroleptic zuclopenthixol cosegregates with the polymorphic hydroxylation of debrisoquine in humans. *Acta Psychiatr Scand* 84:99–102.

Dahl, M. L., Nordin, C., Bertilsson, L. 1991b. Enantioselective hydroxylation of nortriptyline in human liver microsomes, intestinal homogenate, and patients treated with nortriptyline. *Ther Drug Monit* 13:189–194.

Dahl, M. L., Tybring, G., Elwin, C. E., Alm, C., Andreasson, K., Gyllenpalm, M., Bertilsson, L. 1994. Stereoselective disposition of mianserin is related to debrisoquin hydroxylation polymorphism. *Clin Pharmacol Ther* 56:176–183.

Dahl, M. L., Bertilsson, L., Nordin, C. 1996. Steady-state plasma levels of nortriptyline and its 10-hydroxy metabolite: Relationship to the CYP2D6 genotype. *Psychopharmacology (Berl)* 123:315–319.

Dain, J. G., Fu, E., Gorski, J., Nicoletti, J., Scallen, T. J. 1993. Biotransformation of fluvastatin sodium in humans. *Drug Metab Dispos* 21:567–572.

Dalvie, D. K. O'Connell, T. N. 2004. Characterization of novel dihydrothienopyridinium and thienopyridinium metabolites of ticlopidine in vitro: Role of peroxidases, cytochromes p450, and monoamine oxidases. *Drug Metab Dispos* 32:49–57.

Dangman, K. H., Hoffman, B. F. 1981. In vivo and in vitro antiarrhythmic and arrhythmogenic effects of N-acetyl procainamide. *J Pharmacol Exp Ther* 217:851–862.

Daniel, W. A. 2003. Mechanisms of cellular distribution of psychotropic drugs. Significance for drug action and interactions. *Prog Neuropsychopharmacol Biol Psychiatry* 27:65–73.

Danilo, P., Jr. 1979. Aprindine. *Am Heart J* 97:119–124.

Dasaradhi, L., Shibutani, S. 1997. Identification of tamoxifen-DNA adducts formed by alpha-sulfate tamoxifen and alpha-acetoxytamoxifen. *Chem Res Toxicol* 10:189–196.

Davies, B. J., Coller, J. K., Somogyi, A. A., Milne, R. W., Sallustio, B. C. 2007. CYP2B6, CYP2D6, and CYP3A4 catalyze the primary oxidative metabolism of perhexiline enantiomers by human liver microsomes. *Drug Metab Dispos* 35:128–138.

Davis, K. L., Thal, L. J., Gamzu, E. R., Davis, C. S., Woolson, R. F., Gracon, S. I., Drachman, D. A. et al. 1992. A double-blind, placebo-controlled multicenter study of tacrine for Alzheimer's disease. The Tacrine Collaborative Study Group. *N Engl J Med* 327:1253–1259.

Dayer, P., Gasser, R., Gut, J., Kronbach, T., Robertz, G. M., Eichelbaum, M., Meyer, U. A. 1984. Characterization of a common genetic defect of cytochrome P-450 function (debrisoquine-sparteine type polymorphism)—Increased Michaelis is Constant (Km) and loss of stereoselectivity of bufuralol 1'-hydroxylation in poor metabolizers. *Biochem Biophys Res Commun* 125:374–380.

Dayer, P., Leemann, T., Marmy, A., Rosenthaler, J. 1985. Interindividual variation of beta-adrenoceptor blocking drugs, plasma concentration and effect: Influence of genetic status on behaviour of atenolol, bopindolol and metoprolol. *Eur J Clin Pharmacol* 28:149–153.

Dayer, P., Leemann, T., Striberni, R. 1989. Dextromethorphan O-demethylation in liver microsomes as a prototype reaction to monitor cytochrome P-450 db1 activity. *Clin Pharmacol Ther* 45:34–40.

de Groot, M. J., Bijloo, G. J., Hansen, K. T., Vermeulen, N. P. 1995. Computer prediction and experimental validation of cytochrome P4502D6-dependent oxidation of GBR 12909. *Drug Metab Dispos* 23:667–669.

de Groot, M. J., Vermeulen, N. P., Kramer, J. D., van Acker, F. A., Donne-Op den Kelder, G. M. 1996. A three-dimensional protein model for human cytochrome P450 2D6 based on the crystal structures of P450 101, P450 102, and P450 108. *Chem Res Toxicol* 9:1079–1091.

de Groot, M. J., Bijloo, G. J., Martens, B. J., van Acker, F. A., Vermeulen, N. P. 1997a. A refined substrate model for human cytochrome P450 2D6. *Chem Res Toxicol* 10:41–48.

de Groot, M. J., Bijloo, G. J., van Acker, F. A., Fonseca Guerra, C., Snijders, J. G., Vermeulen, N. P. 1997b. Extension of a predictive substrate model for human cytochrome P4502D6. *Xenobiotica* 27:357–368.

de Groot, M. J., Ackland, M. J., Horne, V. A., Alex, A. A., Jones, B. C. 1999a. Novel approach to predicting P450-mediated drug metabolism: Development of a combined protein and pharmacophore model for CYP2D6. *J Med Chem* 42:1515–1524.

de Groot, M. J., Ackland, M. J., Horne, V. A., Alex, A. A., Jones, B. C. 1999b. A novel approach to predicting P450 mediated drug metabolism. CYP2D6 catalyzed N-dealkylation reactions and qualitative metabolite predictions using a combined protein and pharmacophore model for CYP2D6. *J Med Chem* 42:4062–4070.

de la Torre, R., Farre, M., Ortuno, J., Mas, M., Brenneisen, R., Roset, P. N., Segura, J., Cami, J. 2000. Non-linear pharmacokinetics of MDMA ('ecstasy') in humans. *Br J Clin Pharmacol* 49:104–109.

de la Torre, R., Farre, M., Roset, P. N., Pizarro, N., Abanades, S., Segura, M., Segura, J., Cami, J. 2004. Human pharmacology of MDMA: Pharmacokinetics, metabolism, and disposition. *Ther Drug Monit* 26:137–144.

de Tommaso, M., Serpino, C., Sciruicchio, V. 2011. Management of Huntington's disease: Role of tetrabenazine. *Ther Clin Risk Manag* 7:123–129.

de Vries, J. 1981. Hepatotoxic metabolic activation of paracetamol and its derivatives phenacetin and benorilate: Oxygenation or electron transfer? *Biochem Pharmacol* 30:399–402.

Dehal, S. S., Kupfer, D. 1996. Evidence that the catechol 3,4-dihydroxytamoxifen is a proximate intermediate to the reactive species binding covalently to proteins. *Cancer Res* 56:1283–1290.

Dehal, S. S., Kupfer, D. 1997. CYP2D6 catalyzes tamoxifen 4-hydroxylation in human liver. *Cancer Res* 57:3402–3406.

Dehal, S. S., Kupfer, D. 1999. Cytochrome P450 3A and 2D6 catalyze ortho hydroxylation of 4-hydroxytamoxifen and 3-hydroxytamoxifen (droloxifene) yielding tamoxifen catechol: Involvement of catechols in covalent binding to hepatic proteins. *Drug Metab Dispos* 27:681–688.

Delbressine, L. P., Moonen, M. E., Kaspersen, F. M., Wagenaars, G. N., Jacobs, P. L., Timmer, C. J., Paanakker, J. E., van Hal, H. J., Voortman, G. 1998. Pharmacokinetics and biotransformation of mirtazapine in human volunteers. *Clin Drug Investig* 15:45–55.

Dempsey, E., Bourque, S., Spenard, J., Landriault, H. 1996. Pharmacokinetics of single intravenous and oral doses of dolasetron mesylate in healthy elderly volunteers. *J Clin Pharmacol* 36:903–910.

Desager, J. P. 1994. Clinical pharmacokinetics of ticlopidine. *Clin Pharmacokinet* 26:347–355.

Desta, Z., Wu, G. M., Morocho, A. M., Flockhart, D. A. 2002. The gastroprokinetic and antiemetic drug metoclopramide is a substrate and inhibitor of cytochrome P450 2D6. *Drug Metab Dispos* 30:336–343.

Desta, Z., Ward, B. A., Soukhova, N. V., Flockhart, D. A. 2004. Comprehensive evaluation of tamoxifen sequential biotransformation by the human cytochrome P450 system in vitro: Prominent roles for CYP3A and CYP2D6. *J Pharmacol Exp Ther* 310:1062–1075.

DeVane, C. L., Gill, H. S. 1997. Clinical pharmacokinetics of fluvoxamine: Applications to dosage regimen design. *J Clin Psychiatry* 58 Suppl 5:7–14.

DeVane, C. L., Liston, H. L., Markowitz, J. S. 2002. Clinical pharmacokinetics of sertraline. *Clin Pharmacokinet* 41:1247–1266.

Devillier, P., Roche, N., Faisy, C. 2008. Clinical pharmacokinetics and pharmacodynamics of desloratadine, fexofenadine and levocetirizine: A comparative review. *Clin Pharmacokinet* 47: 217–230.

Dhir, A. 2013. Vortioxetine for the treatment of major depression. *Drugs Today (Barc)* 49:781–790.

Di Pietro, N. C., Seamans, J. K. 2007. Dopamine and serotonin interactions in the prefrontal cortex: Insights on antipsychotic drugs and their mechanism of action. *Pharmacopsychiatry* 40 Suppl 1:S27–S33.

Diaz, D. S., Kozar, M. P., Smith, K. S., Asher, C. O., Sousa, J. C., Schiehser, G. A., Jacobus, D. P., Milhous, W. K., Skillman, D. R., Shearer, T. W. 2008. Role of specific cytochrome P450 isoforms in the conversion of phenoxypropoxybiguanide analogs in human liver microsomes to potent antimalarial dihydrotriazines. *Drug Metab Dispos* 36:380–385.

Ding, J., Chen, X., Gao, Z., Dai, X., Li, L., Xie, C., Jiang, H., Zhang, L., Zhong, D. 2013. Metabolism and pharmacokinetics of novel selective vascular endothelial growth factor receptor-2 inhibitor apatinib in humans. *Drug Metab Dispos* 41:1195–1210.

Distlerath, L. M., Guengerich, F. P. 1984. Characterization of a human liver cytochrome P-450 involved in the oxidation of debrisoquine and other drugs by using antibodies raised to the analogous rat enzyme. *Proc Natl Acad Sci U S A* 81: 7348–7352.

Distlerath, L. M., Reilly, P. E., Martin, M. V., Davis, G. G., Wilkinson, G. R., Guengerich, F. P. 1985. Purification and characterization of the human liver cytochromes P450 involved in debrisoquine 4-hydroxylation and phenacetin *O*-deethylation, two prototypes for genetic polymorphism in oxidative drug metabolism. *J Biol Chem* 260:9057–9067.

Draper, A. J., Madan, A., Smith, K., Parkinson, A. 1998. Development of a non-high pressure liquid chromatography assay to determine testosterone hydroxylase (CYP3A) activity in human liver microsomes. *Drug Metab Dispos* 26:299–304.

du Souich, P., Erill, S. 1977. Metabolism of procainamide and p-aminobenzoic acid in patients with chronic liver disease. *Clin Pharmacol Ther* 22:588–595.

Dubovsky, S. L. 2014. Pharmacokinetic evaluation of vortioxetine for the treatment of major depressive disorder. *Expert Opin Drug Metab Toxicol* 10:759–766.

Ducharme, J., Farinotti, R. 1996. Clinical pharmacokinetics and metabolism of chloroquine. Focus on recent advancements. *Clin Pharmacokinet* 31:257–274.

Dueweke, T. J., Poppe, S. M., Romero, D. L., Swaney, S. M., So, A. G., Downey, K. M., Althaus, I. W. et al. 1993. U-90152, a potent inhibitor of human immunodeficiency virus type 1 replication. *Antimicrob Agents Chemother* 37:1127–1131.

Duggan, D. E., Hogans, A. F., Kwan, K. C., McMahon, F. G. 1972. The metabolism of indomethacin in man. *J Pharmacol Exp Ther* 181:563–575.

Duran, D., Platia, E. V., Griffith, L. S., Adhar, G., Reid, P. R. 1982. Suppression of complex ventricular arrhymias by oral flecainide. *Clin Pharmacol Ther* 32:554–561.

Durgam, S., Starace, A., Li, D., Migliore, R., Ruth, A., Nemeth, G., Laszlovszky, I. 2014. An evaluation of the safety and efficacy of cariprazine in patients with acute exacerbation of schizophrenia: A phase II, randomized clinical trial. *Schizophr Res* 152:450–457.

Eagling, V. A., Wiltshire, H., Whitcombe, I. W., Back, D. J. 2002. CYP3A4-mediated hepatic metabolism of the HIV-1 protease inhibitor saquinavir in vitro. *Xenobiotica* 32:1–17.

Eap, C. B., Laurian, S., Souche, A., Koeb, L., Reymond, P., Buclin, T., Baumann, P. 1992. Influence of quinidine on the pharmacokinetics of trimipramine and on its effect on the waking EEG of healthy volunteers. A pilot study on two subjects. *Neuropsychobiology* 25:214–220.

Eap, C. B., Guentert, T. W., Schaublin-Loidl, M., Stabl, M., Koeb, L., Powell, K., Baumann, P. 1996. Plasma levels of the enantiomers of thioridazine, thioridazine 2-sulfoxide, thioridazine 2-sulfone, and thioridazine 5-sulfoxide in poor and extensive metabolizers of dextromethorphan and mephenytoin. *Clin Pharmacol Ther* 59:322–331.

Eap, C. B., Bertschy, G., Powell, K., Baumann, P. 1997. Fluvoxamine and fluoxetine do not interact in the same way with the metabolism of the enantiomers of methadone. *J Clin Psychopharmacol* 17:113–117.

Eap, C. B., Bender, S., Gastpar, M., Fischer, W., Haarmann, C., Powell, K., Jonzier-Perey, M., Cochard, N., Baumann, P. 2000. Steady state plasma levels of the enantiomers of trimipramine and of its metabolites in CYP2D6-, CYP2C19- and CYP3A4/5-phenotyped patients. *Ther Drug Monit* 22:209–214.

Ebner, T., Eichelbaum, M. 1993. The metabolism of aprindine in relation to the sparteine/debrisoquine polymorphism. *Br J Clin Pharmacol* 35:426–430.

Ebner, T., Meese, C. O., Eichelbaum, M. 1995. Mechanism of cytochrome P450 2D6-catalyzed sparteine metabolism in humans. *Mol Pharmacol* 48:1078–1086.

Edwards, J. E., McQuay, H. J., Moore, R. A. 2000. Single dose dihydrocodeine for acute postoperative pain. *Cochrane Database Syst Rev* CD002760.

Eichelbaum, M., Spannbrucker, N., Dengler, H. J. 1975. Proceedings: *N*-oxidation of sparteine in man and its interindividual differences. *Naunyn Schmiedebergs Arch Pharmacol* 287 Suppl:R94.

Eichelbaum, M., Spannbrucker, N., Steincke, B., Dengler, H. J. 1979. Defective *N*-oxidation of sparteine in man: A new pharmacogenetic defect. *Eur J Clin Pharmacol* 16:183–187.

Eichelbaum, M., Mineshita, S., Ohnhaus, E. E., Zekorn, C. 1986a. The influence of enzyme induction on polymorphic sparteine oxidation. *Br J Clin Pharmacol* 22:49–53.

Eichelbaum, M., Reetz, K. P., Schmidt, E. K., Zekorn, C. 1986b. The genetic polymorphism of sparteine metabolism. *Xenobiotica* 16:465–481.

Eichelbaum, M., Baur, M. P., Dengler, H. J., Osikowska-Evers, B. O., Tieves, G., Zekorn, C., Rittner, C. 1987. Chromosomal assignment of human cytochrome P-450 (debrisoquine/sparteine type) to chromosome 22. *Br J Clin Pharmacol* 23:455–458.

Eiermann, B., Engel, G., Johansson, I., Zanger, U. M., Bertilsson, L. 1997. The involvement of CYP1A2 and CYP3A4 in the metabolism of clozapine. *Br J Clin Pharmacol* 44:439–446.

Eiermann, B., Edlund, P. O., Tjernberg, A., Dalen, P., Dahl, M. L., Bertilsson, L. 1998. 1- and 3-hydroxylations, in addition to 4-hydroxylation, of debrisoquine are catalyzed by cytochrome P450 2D6 in humans. *Drug Metab Dispos* 26:1096–1101.

Ekins, S., Vandenbranden, M., Ring, B. J., Gillespie, J. S., Yang, T. J., Gelboin, H. V., Wrighton, S. A. 1998. Further characterization of the expression in liver and catalytic activity of CYP2B6. *J Pharmacol Exp Ther* 286:1253–1259.

Ellingrod, V. L., Perry, P. J. 1994. Venlafaxine: A heterocyclic antidepressant. *Am J Hosp Pharm* 51:3033–3046.

Emborg, M. E. 2007. Nonhuman primate models of Parkinson's disease. *ILAR J* 48:339–355.

Epstein, J. W., Osterberg, A. C., Regan, B. A. 1982. Bicifadine: Nonnarcotic analgesic activity of 1-aryl-3-azabicyclo[3.1.0] hexanes. *NIDA Res Monogr* 41:93–98.

Erickson, D. A., Mather, G., Trager, W. F., Levy, R. H., Keirns, J. J. 1999. Characterization of the *in vitro* biotransformation of the HIV-1 reverse transcriptase inhibitor nevirapine by human hepatic cytochromes P450. *Drug Metab Dispos* 27:1488–1495.

Erickson, D. A., Hollfelder, S., Tenge, J., Gohdes, M., Burkhardt, J. J., Krieter, P. A. 2007. *In vitro* metabolism of the analgesic bicifadine in the mouse, rat, monkey, and human. *Drug Metab Dispos* 35:2232–2241.

Erkinjuntti, T., Kurz, A., Gauthier, S., Bullock, R., Lilienfeld, S., Damaraju, C. V. 2002. Efficacy of galantamine in probable vascular dementia and Alzheimer's disease combined with cerebrovascular disease: A randomised trial. *Lancet* 359:1283–1290.

Erkinjuntti, T., Gauthier, S., Bullock, R., Kurz, A., Hammond, G., Schwalen, S., Zhu, Y., Brashear, R. 2008. Galantamine treatment in Alzheimer's disease with cerebrovascular disease: Responder analyses from a randomized, controlled trial (GALINT-6). *J Psychopharmacol* 22:761–768.

Etienne, M. C., Milano, G., Fischel, J. L., Frenay, M., Francois, E., Formento, J. L., Gioanni, J., Namer, M. 1989. Tamoxifen metabolism: Pharmacokinetic and in vitro study. *Br J Cancer* 60:30–35.

Ette, E. I., Essien, E. E., Thomas, W. O., Brown-Awala, E. A. 1989. Pharmacokinetics of chloroquine and some of its metabolites in healthy volunteers: A single dose study. *J Clin Pharmacol* 29:457–462.

Evans, D. A., Mahgoub, A., Sloan, T. P., Idle, J. R., Smith, R. L. 1980. A family and population study of the genetic polymorphism of debrisoquine oxidation in a white British population. *J Med Genet* 17:102–105.

Evans, D. C., O'Connor, D., Lake, B. G., Evers, R., Allen, C., Hargreaves, R. 2003. Eletriptan metabolism by human hepatic CYP450 enzymes and transport by human P-glycoprotein. *Drug Metab Dispos* 31:861–869.

Fabre, G., Julian, B., Saint-Aubert, B., Joyeux, H., Berger, Y. 1993. Evidence for CYP3A-mediated N-deethylation of amiodarone in human liver microsomal fractions. *Drug Metab Dispos* 21:978–985.

Falany, J. L., Pilloff, D. E., Leyh, T. S., Falany, C. N. 2006. Sulfation of raloxifene and 4-hydroxytamoxifen by human cytosolic sulfotransferases. *Drug Metab Dispos* 34:361–368.

Falk, R. H., Fogel, R. I. 1994. Flecainide. *J Cardiovasc Electrophysiol* 5:964–981.

Fan, P. W., Bolton, J. L. 2001. Bioactivation of tamoxifen to metabolite E quinone methide: Reaction with glutathione and DNA. *Drug Metab Dispos* 29:891–896.

Fan, P. W., Zhang, F., Bolton, J. L. 2000. 4-Hydroxylated metabolites of the antiestrogens tamoxifen and toremifene are metabolized to unusually stable quinone methides. *Chem Res Toxicol* 13:45–52.

Fang, J., Coutts, R. T., McKenna, K. F., Baker, G. B. 1998. Elucidation of individual cytochrome P450 enzymes involved in the metabolism of clozapine. *Naunyn Schmiedebergs Arch Pharmacol* 358:592–599.

Farid, N. A., Smith, R. L., Gillespie, T. A., Rash, T. J., Blair, P. E., Kurihara, A., Goldberg, M. J. 2007. The disposition of prasugrel, a novel thienopyridine, in humans. *Drug Metab Dispos* 35:1096–1104.

Farlow, M., Gracon, S. I., Hershey, L. A., Lewis, K. W., Sadowsky, C. H., Dolan-Ureno, J. 1992. A controlled trial of tacrine in Alzheimer's disease. The Tacrine Study Group. *Jama* 268:2523–2529.

Fasano, A., Bentivoglio, A. R. 2009. Tetrabenazine. *Expert Opin Pharmacother* 10:2883–2896.

Fasola, A. F., Carmichael, R. 1974. The pharmacology and clinical evaluation of aprindine a new antiarrhythmic agent. *Acta Cardiol* Suppl 18:317–333.

Fenton, C., Scott, L. J. 2005. Risperidone: A review of its use in the treatment of bipolar mania. *CNS Drugs* 19:429–444.

Figgitt, D. P., McClellan, K. J. 2000. Fluvoxamine. An updated review of its use in the management of adults with anxiety disorders. *Drugs* 60:925–954.

Filppula, A. M., Neuvonen, P. J., Backman, J. T. 2014. *In vitro* assessment of time-dependent inhibitory effects on CYP2C8 and CYP3A activity by fourteen protein kinase inhibitors. *Drug Metab Dispos* 42:1202–1209.

Fioravanti, M., Flicker, L. 2001. Efficacy of nicergoline in dementia and other age associated forms of cognitive impairment. *Cochrane Database Syst Rev* CD003159.

Firkusny, L., Kroemer, H. K., Eichelbaum, M. 1995. In vitro characterization of cytochrome P450 catalysed metabolism of the antiemetic tropisetron. *Biochem Pharmacol* 49:1777–1784.

Fischer, D., Breyer-Pfaff, U. 1997. Variability of diphenhydramine N-glucuronidation in healthy subjects. *Eur J Drug Metab Pharmacokinet* 22:151–154.

Fischer, V., Baldeck, J. P., Tse, F. L. 1992a. Pharmacokinetics and metabolism of the 5-hydroxytryptamine antagonist tropisetron after single oral doses in humans. *Drug Metab Dispos* 20:603–607.

Fischer, V., Vogels, B., Maurer, G., Tynes, R. E. 1992b. The antipsychotic clozapine is metabolized by the polymorphic human microsomal and recombinant cytochrome P450 2D6. *J Pharmacol Exp Ther* 260:1355–1360.

Fischer, V., Vickers, A. E., Heitz, F., Mahadevan, S., Baldeck, J. P., Minery, P., Tynes, R. 1994. The polymorphic cytochrome P4502D6 is involved in the metabolism of both 5-hydroxytryptamine antagonists, tropisetron and ondansetron. *Drug Metab Dispos* 22:269–274.

Fischer, V., Johanson, L., Heitz, F., Tullman, R., Graham, E., Baldeck, J. P., Robinson, W. T. 1999. The 3-hydroxy-3-methylglutaryl coenzyme A reductase inhibitor fluvastatin: Effect on human cytochrome P450 and implications for metabolic drug interactions. *Drug Metab Dispos* 27:410–416.

Fitzsimmons, M. E., Collins, J. M. 1997. Selective biotransformation of the human immunodeficiency virus protease inhibitor saquinavir by human small-intestinal cytochrome P4503A4: Potential contribution to high first-pass metabolism. *Drug Metab Dispos* 25:256–266.

Flanagan, R. J., Storey, G. C., Holt, D. W., Farmer, P. B. 1982. Identification and measurement of desethylamiodarone in blood plasma specimens from amiodarone-treated patients. *J Pharm Pharmacol* 34:638–643.

Fleishaker, J. C., Sisson, T. A., Carel, B. J., Azie, N. E. 2000. Pharmacokinetic interaction between verapamil and almotriptan in healthy volunteers. *Clin Pharmacol Ther* 67:498–503.

Fleishaker, J. C., Ryan, K. K., Carel, B. J., Azie, N. E. 2001a. Evaluation of the potential pharmacokinetic interaction between almotriptan and fluoxetine in healthy volunteers. *J Clin Pharmacol* 41:217–223.

Fleishaker, J. C., Ryan, K. K., Jansat, J. M., Carel, B. J., Bell, D. J., Burke, M. T., Azie, N. E. 2001b. Effect of MAO-A inhibition on the pharmacokinetics of almotriptan, an antimigraine agent in humans. *Br J Clin Pharmacol* 51:437–441.

Fleishaker, J. C., Sisson, T. A., Carel, B. J., Azie, N. E. 2001c. Lack of pharmacokinetic interaction between the antimigraine compound, almotriptan, and propranolol in healthy volunteers. *Cephalalgia* 21:61–65.

Fleishaker, J. C., Herman, B. D., Carel, B. J., Azie, N. E. 2003. Interaction between ketoconazole and almotriptan in healthy volunteers. *J Clin Pharmacol* 43:423–427.

Flores-Runk, P., Raasch, R. H. 1993. Ticlopidine and antiplatelet therapy. *Ann Pharmacother* 27:1090–1098.

Fogelman, S. M., Schmider, J., Venkatakrishnan, K., von Moltke, L. L., Harmatz, J. S., Shader, R. I., Greenblatt, D. J. 1999. *O*- and *N*-demethylation of venlafaxine in vitro by human liver microsomes and by microsomes from cDNA-transfected cells: Effect of metabolic inhibitors and SSRI antidepressants. *Neuropsychopharmacology* 20:480–490.

Foglia, J. P., Pollock, B. G., Kirshner, M. A., Rosen, J., Sweet, R., Mulsant, B. 1997. Plasma levels of citalopram enantiomers and metabolites in elderly patients. *Psychopharmacol Bull* 33:109–112.

Forland, S. C., Cutler, R. E., McQuinn, R. L., Kvam, D. C., Miller, A. M., Conard, G. J., Parish, S. 1988. Flecainide pharmacokinetics after multiple dosing in patients with impaired renal function. *J Clin Pharmacol* 28:727–735.

Forsyth, D. R., Wilcock, G. K., Morgan, R. A., Truman, C. A., Ford, J. M., Roberts, C. J. 1989. Pharmacokinetics of tacrine hydrochloride in Alzheimer's disease. *Clin Pharmacol Ther* 46:634–641.

Forte, A. J., Wilson, J. M., Slattery, J. T., Nelson, S. D. 1984. The formation and toxicity of catechol metabolites of acetaminophen in mice. *Drug Metab Dispos* 12:484–491.

Foster, A. B., Griggs, L. J., Jarman, M., van Maanen, J. M., Schulten, H. R. 1980. Metabolism of tamoxifen by rat liver microsomes: Formation of the N-oxide, a new metabolite. *Biochem Pharmacol* 29:1977–1979.

Foster, D. J., Somogyi, A. A., Bochner, F. 1999. Methadone N-demethylation in human liver microsomes: Lack of stereoselectivity and involvement of CYP3A4. *Br J Clin Pharmacol* 47:403–412.

Frances, B., Gout, R., Cros, J., Zajac, J. M. 1992. Effects of ibogaine on naloxone-precipitated withdrawal in morphine-dependent mice. *Fundam Clin Pharmacol* 6:327–332.

Francis, S. H., Corbin, J. D. 2005. Sildenafil: Efficacy, safety, tolerability and mechanism of action in treating erectile dysfunction. *Expert Opin Drug Metab Toxicol* 1:283–293.

Frank, S. 2010. Tetrabenazine: The first approved drug for the treatment of chorea in US patients with Huntington disease. *Neuropsychiatr Dis Treat* 6:657–665.

Frank, S. 2014. Treatment of Huntington's disease. *Neurotherapeutics* 11:153–160.

Freeman, R. W., Uetrecht, J. P., Woosley, R. L., Oates, J. A., Harbison, R. D. 1981. Covalent binding of procainamide in vitro and in vivo to hepatic protein in mice. *Drug Metab Dispos* 9:188–192.

Freeman, E. R., Bloom, D. A., McGuire, E. J. 2001. A brief history of testosterone. *J Urol* 165:371–373.

Freimuth, W. W. 1996. Delavirdine mesylate, a potent non-nucleoside HIV-1 reverse transcriptase inhibitor. *Adv Exp Med Biol* 394:279–289.

Frelinger, A. L., 3rd, Jakubowski, J. A., Li, Y., Barnard, M. R., Linden, M. D., Tarnow, I., Fox, M. L., Sugidachi, A., Winters, K. J., Furman, M. I., Michelson, A. D. 2008. The active metabolite of prasugrel inhibits adenosine diphosphate- and collagen-stimulated platelet procoagulant activities. *J Thromb Haemost* 6:359–365.

Fried, K. M., Wainer, I. W. 1994. Direct determination of tamoxifen and its four major metabolites in plasma using coupled column high-performance liquid chromatography. *J Chromatogr B Biomed Appl* 655:261–268.

Frishman, W. H., Fuksbrumer, M. S., Tannenbaum, M. 1994. Topical ophthalmic β-adrenergic blockade for the treatment of glaucoma and ocular hypertension. *J Clin Pharmacol* 34:795–803.

Frisk-Holmberg, M., Bergqvist, Y., Domeij-Nyberg, B. 1983. Steady state disposition of chloroquine in patients with rheumatoid disease. *Eur J Clin Pharmacol* 24:837–839.

Fu, P. P., Xia, Q., Lin, G., Chou, M. W. 2004. Pyrrolizidine alkaloids—Genotoxicity, metabolism enzymes, metabolic activation, and mechanisms. *Drug Metab Rev* 36:1–55.

Fujimura, H., Tsurumi, K., Yanagihara, M., Hiramatsu, Y., Tamura, Y., Shimizu, Y., Hojo, M., Yoshida, Y., Akimoto, Y. 1981. [Pharmacological study of mequitazine (LM-209) (II). Antiallergic action (author's transl)]. *Nippon Yakurigaku Zasshi* 78:291–303.

Fujita, T., Takagi, N., Takeshige, K., Ogawa, K., Toki, T., Futaki, N., Sasabe, H., Odomi, M., Ohnishi, A. 1994. Pharmacokinetics of pranidipine in healthy volunteers (1st report): Results of single and repeated administration. *YAKURI TO RINSHO* 4:889–901.

Funck-Brentano, C., Becquemont, L., Kroemer, K. H., Kerstin, B., Knebel, N. G., Eichelbaum, M., Jaillon, P. 1994. Variable disposition kinetics and electrocardiographic effects of flecainide during repeated dosing in humans: Contribution of genetic factors, dose-dependent clearance, and interaction with amiodarone. *Clin Pharmacol Ther* 55:256–269.

Funck-Brentano, C., Thomas, G., Jacqz-Aigrain, E., Poirier, J. M., Simon, T., Bereziat, G., Jaillon, P. 1992. Polymorphism of dextromethorphan metabolism: Relationships between phenotype, genotype and response to the administration of encainide in humans. *J Pharmacol Exp Ther* 263:780–786.

Gamboa da Costa, G., Hamilton, L. P., Beland, F. A., Marques, M. M. 2000. Characterization of the major DNA adduct formed by alpha-hydroxy-N-desmethyltamoxifen in vitro and in vivo. *Chem Res Toxicol* 13:200–207.

Gan, J., Skipper, P. L. Tannenbaum, S. R. 2001. Oxidation of 2,6-dimethylaniline by recombinant human cytochrome P450s and human liver microsomes. *Chem Res Toxicol* 14:672–677.

Gao, Y., Peterson, S., Masri, B., Hougland, M. T., Adham, N., Gyertyan, I., Kiss, B., Caron, M. G., El-Mallakh, R. S. 2015. Cariprazine exerts antimanic properties and interferes with dopamine D$_2$ receptor β-arrestin interactions. *Pharmacol Res Perspect* 3:e00073.

Gardiner, S. J., Begg, E. J. 2006. Pharmacogenetics, drug-metabolizing enzymes, and clinical practice. *Pharmacol Rev* 58:521–590.

Garnock-Jones, K. P. 2014. Vortioxetine: A review of its use in major depressive disorder. *CNS Drugs* 28:855–874.

Garrido, M. J., Troconiz, I. F. 1999. Methadone: A review of its pharmacokinetic/pharmacodynamic properties. *J Pharmacol Toxicol Methods* 42:61–66.

Garteiz, D. A., Hook, R. H., Walker, B. J., Okerholm, R. A. 1982. Pharmacokinetics and biotransformation studies of terfenadine in man. *Arzneimittelforschung* 32:1185–1190.

Gastpar, M. 1989. Clinical originality and new biology of trimipramine. *Drugs* 38 Suppl 1:43–48; discussion 49–50.

Gautier, J. C., Lecoeur, S., Cosme, J., Perret, A., Urban, P., Beaune, P., Pompon, D. 1996. Contribution of human cytochrome P450 to benzo[*a*]pyrene and benzo[*a*]pyrene-7,8-dihydrodiol metabolism, as predicted from heterologous expression in yeast. *Pharmacogenetics* 6:489–499.

Gearhart, D. A., Neafsey, E. J., Collins, M. A. 2002. Phenylethanolamine N-methyltransferase has beta-carboline 2N-methyltransferase activity: Hypothetical relevance to Parkinson's disease. *Neurochem Int* 40:611–620.

Geng, R., Li, J. 2015. Apatinib for the treatment of gastric cancer. *Expert Opin Pharmacother* 16:117–122.

Genin, M. J., Poel, T. J., Yagi, Y., Biles, C., Althaus, I., Keiser, B. J., Kopta, L. A. et al. 1996. Synthesis and bioactivity of novel bis(heteroaryl)piperazine (BHAP) reverse transcriptase inhibitors: Structure-activity relationships and increased metabolic stability of novel substituted pyridine analogs. *J Med Chem* 39:5267–5275.

Gerber, J. G., Rhodes, R. J., Gal, J. 2004. Stereoselective metabolism of methadone N-demethylation by cytochrome P4502B6 and 2C19. *Chirality* 16:36–44.

Ghobadi, C., Gregory, A., Crewe, H. K., Rostami-Hodjegan, A., Lennard, M. S. 2008. CYP2D6 is primarily responsible for the metabolism of clomiphene. *Drug Metab Pharmacokinet* 23:101–105.

Gilham, D. E., Cairns, W., Paine, M. J., Modi, S., Poulsom, R., Roberts, G. C., Wolf, C. R. 1997. Metabolism of MPTP by cytochrome P450 2D6 and the demonstration of 2D6 mRNA in human foetal and adult brain by in situ hybridization. *Xenobiotica* 27:111–125.

Gilhooly, T. C., Daly, A. K. 2002. CYP2D6 deficiency, a factor in ecstasy related deaths? *Br J Clin Pharmacol* 54:69–70.

Gill, H. J., Tingle, M. D., Park, B. K. 1995. *N*-Hydroxylation of dapsone by multiple enzymes of cytochrome P450: Implications for inhibition of haemotoxicity. *Br J Clin Pharmacol* 40:531–538.

Gillman, P. K. 2007. Tricyclic antidepressant pharmacology and therapeutic drug interactions updated. *Br J Pharmacol* 151:737–748.

Glatt, H., Davis, W., Meinl, W., Hermersdorfer, H., Venitt, S., Phillips, D. H. 1998. Rat, but not human, sulfotransferase activates a tamoxifen metabolite to produce DNA adducts and gene mutations in bacteria and mammalian cells in culture. *Carcinogenesis* 19:1709–1713.

Glick, S. D., Rossman, K., Rao, N. C., Maisonneuve, I. M., Carlson, J. N. 1992. Effects of ibogaine on acute signs of morphine withdrawal in rats: Independence from tremor. *Neuropharmacology* 31:497–500.

Goldenberg, H., Fishman, V., Heaton, A., Burnett, R. 1964. A Detailed Evaluation of Promazine Metabolism. *Proc Soc Exp Biol Med* 115:1044–1051.

Gonzalez, F. J., Skoda, R. C., Kimura, S., Umeno, M., Zanger, U. M., Nebert, D. W., Gelboin, H. V., Hardwick, J. P., Meyer, U. A. 1988. Characterization of the common genetic defect in humans deficient in debrisoquine metabolism. *Nature* 331:442–446.

Goto, A., Adachi, Y., Inaba, A., Nakajima, H., Kobayashi, H., Sakai, K. 2004. Identification of human P450 isoforms involved in the metabolism of the antiallergic drug, oxatomide, and its inhibitory effect on enzyme activity. *Biol Pharm Bull* 27:684–690.

Goto, A., Ueda, K., Inaba, A., Nakajima, H., Kobayashi, H., Sakai, K. 2005. Identification of human P450 isoforms involved in the metabolism of the antiallergic drug, oxatomide, and its kinetic parameters and inhibition constants. *Biol Pharm Bull* 28:328–334.

Gough, A. C., Smith, C. A., Howell, S. M., Wolf, C. R., Bryant, S. P., Spurr, N. K. 1993. Localization of the CYP2D gene locus to human chromosome 22q13.1 by polymerase chain reaction, in situ hybridization, and linkage analysis. *Genomics* 15:430–432.

Grace, J. M., Kinter, M. T., Macdonald, T. L. 1994. Atypical metabolism of deprenyl and its enantiomer, (*S*)-(+)-*N*,α-dimethyl-*N*-propynylphenethylamine, by cytochrome P450 2D6. *Chem Res Toxicol* 7:286–290.

Grace, J. M., Skanchy, D. J., Aguilar, A. J. 1999. Metabolism of artelinic acid to dihydroqinqhaosu by human liver cytochrome P4503A. *Xenobiotica* 29:703–717.

Gram, L. F., Brosen, K., Kragh-Sorensen, P., Christensen, P. 1989. Steady-state plasma levels of E- and Z-10-OH-nortriptyline in nortriptyline-treated patients: Significance of concurrent medication and the sparteine oxidation phenotype. *Ther Drug Monit* 11:508–514.

Gram, L. F., Guentert, T. W., Grange, S., Vistisen, K., Brosen, K. 1995. Moclobemide, a substrate of CYP2C19 and an inhibitor of CYP2C19, CYP2D6, and CYP1A2: A panel study. *Clin Pharmacol Ther* 57:670–677.

Grant, A. O. 1996. Propafenone: An effective agent for the management of supraventricular arrhythmias. *J Cardiovasc Electrophysiol* 7:353–364.

Granvil, C. P., Krausz, K. W., Gelboin, H. V., Idle, J. R., Gonzalez, F. J. 2002. 4-Hydroxylation of debrisoquine by human CYP1A1 and its inhibition by quinidine and quinine. *J Pharmacol Exp Ther* 301:1025–1032.

Gras, J., Llenas, J., Jansat, J. M., Jauregui, J., Cabarrocas, X., Palacios, J. M. 2002. Almotriptan, a new anti-migraine agent: A review. *CNS Drug Rev* 8:217–234.

Greenblatt, D. J., von Moltke, L. L., Harmatz, J. S., Mertzanis, P., Graf, J. A., Durol, A. L., Counihan, M., Roth-Schechter, B., Shader, R. I. 1998. Kinetic and dynamic interaction study of zolpidem with ketoconazole, itraconazole, and fluconazole. *Clin Pharmacol Ther* 64:661–671.

Greenblatt, D. J., von Moltke, L. L., Harmatz, J. S., Shader, R. I. 1999. Human cytochromes mediating sertraline biotransformation: Seeking attribution. *J Clin Psychopharmacol* 19:489–493.

Greenblatt, D. J., Von Moltke, L. L., Giancarlo, G. M., Garteiz, D. A. 2003. Human cytochromes mediating gepirone biotransformation at low substrate concentrations. *Biopharm Drug Dispos* 24:87–94.

Gross, A. S., Mikus, G., Fischer, C., Hertrampf, R., Gundert-Remy, U., Eichelbaum, M. 1989. Stereoselective disposition of flecainide in relation to the sparteine/debrisoquine metaboliser phenotype. *Br J Clin Pharmacol* 28:555–566.

Grunder, G. 2010. Cariprazine, an orally active D_2/D_3 receptor antagonist, for the potential treatment of schizophrenia, bipolar mania and depression. *Curr Opin Investig Drugs* 11:823–832.

Gschwind, H. P., Pfaar, U., Waldmeier, F., Zollinger, M., Sayer, C., Zbinden, P., Hayes, M. et al. 2005. Metabolism and disposition of imatinib mesylate in healthy volunteers. *Drug Metab Dispos* 33:1503–1512.

Guay, D. R. 2010. Tetrabenazine, a monoamine-depleting drug used in the treatment of hyperkinetic movement disorders. *Am J Geriatr Pharmacother* 8:331–373.

Guengerich, F. P., Miller, G. P., Hanna, I. H., Martin, M. V., Leger, S., Black, C., Chauret, N. et al. 2002. Diversity in the oxidation of substrates by cytochrome P450 2D6: Lack of an obligatory role of aspartate 301-substrate electrostatic bonding. *Biochemistry* 41:11025–11034.

Guidice, J. M., Marez, D., Sabbagh, N., Legrand-Andreoletti, M., Spire, C., Alcaide, E., Lafitte, J. J., Broly, F. 1997. Evidence for CYP2D6 expression in human lung. *Biochem Biophys Res Commun* 241:79–85.

Guitton, J., Buronfosse, T., Desage, M., Flinois, J. P., Perdrix, J. P., Brazier, J. L., Beaune, P. 1998. Possible involvement of multiple human cytochrome P450 isoforms in the liver metabolism of propofol. *Br J Anaesth* 80:788–795.

Gunasekara, N. S., Noble, S., Benfield, P. 1998. Paroxetine. An update of its pharmacology and therapeutic use in depression and a review of its use in other disorders. *Drugs* 55:85–120.

Gupta, M., Kovar, A., Meibohm, B. 2005. The clinical pharmacokinetics of phosphodiesterase-5 inhibitors for erectile dysfunction. *J Clin Pharmacol* 45:987–1003.

Gupta, S., Nihalani, N., Masand, P. 2007. Duloxetine: Review of its pharmacology, and therapeutic use in depression and other psychiatric disorders. *Ann Clin Psychiatry* 19:125–132.

Gustafsson, L. L., Walker, O., Alvan, G., Beermann, B., Estevez, F., Gleisner, L., Lindstrom, B., Sjoqvist, F. 1983. Disposition of chloroquine in man after single intravenous and oral doses. *Br J Clin Pharmacol* 15:471–479.

Gut, J., Gasser, R., Dayer, P., Kronbach, T., Catin, T., Meyer, U. A. 1984. Debrisoquine-type polymorphism of drug oxidation: Purification from human liver of a cytochrome P450 isozyme with high activity for bufuralol hydroxylation. *FEBS Lett* 173:287–290.

Gut, J., Catin, T., Dayer, P., Kronbach, T., Zanger, U., Meyer, U. A. 1986. Debrisoquine/sparteine-type polymorphism of drug oxidation. Purification and characterization of two functionally different human liver cytochrome P-450 isozymes involved in impaired hydroxylation of the prototype substrate bufuralol. *J Biol Chem* 261:11734–11743.

Gyertyan, I., Kiss, B., Saghy, K., Laszy, J., Szabo, G., Szabados, T., Gemesi, L. I. et al. 2011. Cariprazine (RGH-188), a potent D$_3$/D$_2$ dopamine receptor partial agonist, binds to dopamine D$_3$ receptors in vivo and shows antipsychotic-like and procognitive effects in rodents. *Neurochem Int* 59:925–935.

Ha-Duong, N. T., Dijols, S., Macherey, A. C., Goldstein, J. A., Dansette, P. M., Mansuy, D. 2001. Ticlopidine as a selective mechanism-based inhibitor of human cytochrome P450 2C19. *Biochemistry* 40:12112–12122.

Haefeli, W. E., Bargetzi, M. J., Follath, F., Meyer, U. A. 1990. Potent inhibition of cytochrome P450IID6 (debrisoquin 4-hydroxylase) by flecainide in vitro and in vivo. *J Cardiovasc Pharmacol* 15:776–779.

Haefeli, W. E., Vozeh, S., Ha, H. R., Taeschner, W., Follath, F. 1991. Concentration-effect relations of 5-hydroxypropafenone in normal subjects. *Am J Cardiol* 67:1022–1026.

Haferkamp, H. 1950. [The common broom (*Sarothamus scoparius*) and its therapeutic application possibilities.] *Hippokrates* 21:668–671.

Hakkola, J., Raunio, H., Purkunen, R., Pelkonen, O., Saarikoski, S., Cresteil, T., Pasanen, M. 1996. Detection of cytochrome P450 gene expression in human placenta in first trimester of pregnancy. *Biochem Pharmacol* 52:379–383.

Halliday, R. C., Jones, B. C., Smith, D. A., Kitteringham, N. R., Park, B. K. 1995. An investigation of the interaction between halofantrine, CYP2D6 and CYP3A4: Studies with human liver microsomes and heterologous enzyme expression systems. *Br J Clin Pharmacol* 40:369–378.

Hamman, M. A., Thompson, G. A., Hall, S. D. 1997. Regioselective and stereoselective metabolism of ibuprofen by human cytochrome P450 2C. *Biochem Pharmacol* 54:33–41.

Hanna, I. H., Krauser, J. A., Cai, H., Kim, M. S., Guengerich, F. P. 2001. Diversity in mechanisms of substrate oxidation by cytochrome P450 2D6. Lack of an allosteric role of NADPH-cytochrome P450 reductase in catalytic regioselectivity. *J Biol Chem* 276:39553–39561.

Haritos, V. S., Ghabrial, H., Ahokas, J. T., Ching, M. S. 2000. Role of cytochrome P450 2D6 (CYP2D6) in the stereospecific metabolism of E- and Z-doxepin. *Pharmacogenetics* 10:591–603.

Harman, R. E., Meisinger, M. A., Davis, G. E., Kuehl, F. A., Jr. 1964. The metabolites of indomethacin, a new anti-inflammatory drug. *J Pharmacol Exp Ther* 143:215–220.

Harron, D. W., Brogden, R. N. 1990. Acecainide (N-acetylprocainamide). A review of its pharmacodynamic and pharmacokinetic properties, and therapeutic potential in cardiac arrhythmias. *Drugs* 39:720–740.

Harron, D. W., Brogden, R. N., Faulds, D., Fitton, A. 1992. Cibenzoline. A review of its pharmacological properties and therapeutic potential in arrhythmias. *Drugs* 43:734–759.

Hart, F. D., Boardman, P. L. 1963. Indomethacin: A new non-steroid anti-inflammatory agent. *Br Med J* 2:965–970.

Hartmann, F., Gruenke, L. D., Craig, J. C., Bissell, D. M. 1983. Chlorpromazine metabolism in extracts of liver and small intestine from guinea pig and from man. *Drug Metab Dispos* 11:244–248.

Hartter, S., Wetzel, H., Hammes, E., Hiemke, C. 1993. Inhibition of antidepressant demethylation and hydroxylation by fluvoxamine in depressed patients. *Psychopharmacology (Berl)* 110:302–308.

Hartter, S., Tybring, G., Friedberg, T., Weigmann, H., Hiemke, C. 2002. The N-demethylation of the doxepin isomers is mainly catalyzed by the polymorphic CYP2C19. *Pharm Res* 19:1034–1037.

Hartung, B., Wada, M., Laux, G., Leucht, S. 2005. Perphenazine for schizophrenia. *Cochrane Database Syst Rev* CD003443.

Hayashi, K., Yanagi, M., Wood-Baker, R., Takamatsu, I., Anami, K. 2003. Oxatomide for stable asthma in adults and children. *Cochrane Database Syst Rev* CD002179.

He, N., Zhang, W. Q., Shockley, D., Edeki, T. 2002. Inhibitory effects of H1-antihistamines on CYP2D6- and CYP2C9-mediated drug metabolic reactions in human liver microsomes. *Eur J Clin Pharmacol* 57:847–851.

Heel, R. C., Brogden, R. N., Speight, T. M., Avery, G. S. 1979. Timolol: A review of its therapeutic efficacy in the topical treatment of glaucoma. *Drugs* 17:38–55.

Hege, H. G., Hollmann, M., Kaumeier, S., Lietz, H. 1984. The metabolic fate of 2H-labelled propafenone in man. *Eur J Drug Metab Pharmacokinet* 9:41–55.

Heim, M., Meyer, U. A. 1990. Genotyping of poor metabolisers of debrisoquine by allele-specific PCR amplification. *Lancet* 336:529–532.

Heim, M. H., Meyer, U. A. 1992. Evolution of a highly polymorphic human cytochrome P450 gene cluster: CYP2D6. *Genomics* 14:49–58.

Heimark, L. D., Wienkers, L., Kunze, K., Gibaldi, M., Eddy, A. C., Trager, W. F., O'Reilly, R. A., Goulart, D. A. 1992. The mechanism of the interaction between amiodarone and warfarin in humans. *Clin Pharmacol Ther* 51:398–407.

Herd, M. B., Belelli, D., Lambert, J. J. 2007. Neurosteroid modulation of synaptic and extrasynaptic GABA(A) receptors. *Pharmacol Ther* 116:20–34.

Hermansson, J., Glaumann, H., Karlen, B., von Bahr, C. 1980. Metabolism of lidocaine in human liver in vitro. *Acta Pharmacol Toxicol (Copenh)* 47:49–52.

Herraiz, T., Chaparro, C. 2006. Human monoamine oxidase enzyme inhibition by coffee and beta-carbolines norharman and harman isolated from coffee. *Life Sci* 78:795–802.

Herraiz, T., Guillen, H., Aran, V. J., Idle, J. R., Gonzalez, F. J. 2006. Comparative aromatic hydroxylation and N-demethylation of MPTP neurotoxin and its analogs, N-methylated beta-carboline and isoquinoline alkaloids, by human cytochrome P450 2D6. *Toxicol Appl Pharmacol* 216:387–398.

Hesketh, P. J. 2008. Chemotherapy-induced nausea and vomiting. *N Engl J Med* 358:2482–2494.

Hidestrand, M., Oscarson, M., Salonen, J. S., Nyman, L., Pelkonen, O., Turpeinen, M., Ingelman-Sundberg, M. 2001. CYP2B6 and CYP2C19 as the major enzymes responsible for the metabolism of selegiline, a drug used in the treatment of Parkinson's disease, as revealed from experiments with recombinant enzymes. *Drug Metab Dispos* 29:1480–1484.

Hiemke, C., Hartter, S. 2000. Pharmacokinetics of selective serotonin reuptake inhibitors. *Pharmacol Ther* 85:11–28.

Hinson, J. A., Pohl, L. R., Monks, T. J., Gillette, J. R. 1981. Acetaminophen-induced hepatotoxicity. *Life Sci* 29:107–116.

Hinson, J. A., Pumford, N. R., Roberts, D. W. 1995. Mechanisms of acetaminophen toxicity: Immunochemical detection of drug-protein adducts. *Drug Metab Rev* 27:72–92.

Hirani, V. N., Raucy, J. L., Lasker, J. M. 2004. Conversion of the HIV protease inhibitor nelfinavir to a bioactive metabolite by human liver CYP2C19. *Drug Metab Dispos* 32:1462–1467.

Hiroi, T., Imaoka, S., Funae, Y. 1998. Dopamine formation from tyramine by CYP2D6. *Biochem Biophys Res Commun* 249:838–843.

Hiroi, T., Kishimoto, W., Chow, T., Imaoka, S., Igarashi, T., Funae, Y. 2001. Progesterone oxidation by cytochrome P450 2D isoforms in the brain. *Endocrinology* 142:3901–3908.

Hiroi, T., Chow, T., Imaoka, S., Funae, Y. 2002. Catalytic specificity of CYP2D isoforms in rat and human. *Drug Metab Dispos* 30:970–976.

Hodge, H. C., Downs, W. L., Panner, B. S., Smith, D. W., Maynard, E. A. 1967. Oral toxicity and metabolism of diuron (N-(3,4-dichlorophenyl)-N',N'-dimethylurea) in rats and dogs. *Food Cosmet Toxicol* 5:513–531.

Hoffmann, K. J., Regardh, C. G., Aurell, M., Ervik, M., Jordo, L. 1980. The effect of impaired renal function on the plasma concentration and urinary excretion of metoprolol metabolites. *Clin Pharmacokinet* 5:181–191.

Hoffmann, K. J., Axworthy, D. B., Baillie, T. A. 1990. Mechanistic studies on the metabolic activation of acetaminophen in vivo. *Chem Res Toxicol* 3:204–211.

Hojo, M., Nagasaka, Y., Katayama, O., Serizawa, I. 1981. [Pharmacological study of Mequitazine (LM-209). (V). Pharmacological actions of a main metabolite of LM-209, mequitazine sulfoxide (LM-209 SO) (author's transl)]. *Nippon Yakurigaku Zasshi* 78:431–438.

Holck, M., Osterrieder, W. 1986. Inhibition of the myocardial Ca2+ inward current by the class 1 antiarrhythmic agent, cibenzoline. *Br J Pharmacol* 87:705–711.

Holliday, S. M., Benfield, P. 1995. Venlafaxine. A review of its pharmacology and therapeutic potential in depression. *Drugs* 49:280–294.

Holm, K. J., Goa, K. L. 2000. Zolpidem: An update of its pharmacology, therapeutic efficacy and tolerability in the treatment of insomnia. *Drugs* 59:865–889.

Holmes, B., Heel, R. C. 1985. Flecainide. A preliminary review of its pharmacodynamic properties and therapeutic efficacy. *Drugs* 29:1–33.

Honig, P. K., Wortham, D. C., Zamani, K., Conner, D. P., Mullin, J. C., Cantilena, L. R. 1993. Terfenadine-ketoconazole interaction: Pharmacokinetic and electrocardiographic consequences. *JAMA* 269:1513–1518.

Hooper, W. D., Pool, W. F., Woolf, T. F., Gal, J. 1994. Stereoselective hydroxylation of tacrine in rats and humans. *Drug Metab Dispos* 22:719–724.

Horikiri, Y., Suzuki, T., Mizobe, M. 1998a. Pharmacokinetics and metabolism of bisoprolol enantiomers in humans. *J Pharm Sci* 87:289–294.

Horikiri, Y., Suzuki, T., Mizobe, M. 1998b. Stereoselective metabolism of bisoprolol enantiomers in dogs and humans. *Life Sci* 63:1097–1108.

Horowitz, J. D., Button, I. K., Wing, L. 1995. Is perhexiline essential for the optimal management of angina pectoris? *Aust N Z J Med* 25:111–113.

Hu, X., Cao, J., Hu, W., Wu, C., Pan, Y., Cai, L., Tong, Z. et al. 2014a. Multicenter phase II study of apatinib in non-triple-negative metastatic breast cancer. *BMC Cancer* 14:820.

Hu, X., Zhang, J., Xu, B., Jiang, Z., Ragaz, J., Tong, Z., Zhang, Q. et al. 2014b. Multicenter phase II study of apatinib, a novel VEGFR inhibitor in heavily pretreated patients with metastatic triple-negative breast cancer. *Int J Cancer* 135:1961–1969.

Huang, Z., Fasco, M. J., Kaminsky, L. S. 1997. Alternative splicing of CYP2D mRNA in human breast tissue. *Arch Biochem Biophys* 343:101–108.

Hutchinson, M. R., Menelaou, A., Foster, D. J., Coller, J. K., Somogyi, A. A. 2004. CYP2D6 and CYP3A4 involvement in the primary oxidative metabolism of hydrocodone by human liver microsomes. *Br J Clin Pharmacol* 57:287–297.

Hvenegaard, M. G., Bang-Andersen, B., Pedersen, H., Jorgensen, M., Puschl, A., Dalgaard, L. 2012. Identification of the cytochrome P450 and other enzymes involved in the *in vitro* oxidative metabolism of a novel antidepressant, Lu AA21004. *Drug Metab Dispos* 40:1357–1365.

Hyttel, J., Bogeso, K. P., Perregaard, J., Sanchez, C. 1992. The pharmacological effect of citalopram residues in the (S)-(+)-enantiomer. *J Neural Transm Gen Sect* 88:157–160.

Imai, T., Taketani, M., Suzu, T., Kusube, K., Otagiri, M. 1999. In vitro identification of the human cytochrome P-450 enzymes involved in the N-demethylation of azelastine. *Drug Metab Dispos* 27:942–946.

Imaoka, S., Enomoto, K., Oda, Y., Asada, A., Fujimori, M., Shimada, T., Fujita, S., Guengerich, F. P., Funae, Y. 1990. Lidocaine metabolism by human cytochrome P-450s purified from hepatic microsomes: Comparison of those with rat hepatic cytochrome P-450s. *J Pharmacol Exp Ther* 255:1385–1391.

Imaoka, S., Yamada, T., Hiroi, T., Hayashi, K., Sakaki, T., Yabusaki, Y., Funae, Y. 1996. Multiple forms of human P450 expressed in *Saccharomyces cerevisiae*. Systematic characterization and comparison with those of the rat. *Biochem Pharmacol* 51:1041–1050.

Ingelman-Sundberg, M. 2005. Genetic polymorphisms of cytochrome P450 2D6 (CYP2D6): Clinical consequences, evolutionary aspects and functional diversity. *Pharmacogenomics J* 5:6–13.

Ingelman-Sundberg, M., Sim, S. C., Gomez, A., Rodriguez-Antona, C. 2007. Influence of cytochrome P450 polymorphisms on drug therapies: Pharmacogenetic, pharmacoepigenetic and clinical aspects. *Pharmacol Ther* 116:496–526.

Islam, S. A., Wolf, C. R., Lennard, M. S., Sternberg, M. J. 1991. A three-dimensional molecular template for substrates of human cytochrome P450 involved in debrisoquine 4-hydroxylation. *Carcinogenesis* 12:2211–2219.

Iwasaki, K., Noguchi, H., Kamataki, T., Kato, R. 1982. Metabolism of tiaramide *in vitro*. I. Oxidative metabolism of tiaramide by human and rat liver microsomes. *Xenobiotica* 12:221–226.

Jacobsen, W., Kuhn, B., Soldner, A., Kirchner, G., Sewing, K. F., Kollman, P. A., Benet, L. Z., Christians, U. 2000. Lactonization is the critical first step in the disposition of the 3-hydroxy-3-methylglutaryl-CoA reductase inhibitor atorvastatin. *Drug Metab Dispos* 28:1369–1378.

Jacolot, F., Simon, I., Dreano, Y., Beaune, P., Riche, C., Berthou, F. 1991. Identification of the cytochrome P450 3A family as the enzymes involved in the *N*-demethylation of tamoxifen in human liver microsomes. *Biochem Pharmacol* 41:1911–1919.

Jacqz-Aigrain, E., Funck-Brentano, C., Cresteil, T. 1993. CYP2D6- and CYP3A-dependent metabolism of dextromethorphan in humans. *Pharmacogenetics* 3:197–204.

Jajoo, H. K., Mayol, R. F., LaBudde, J. A., Blair, I. A. 1990. Structural characterization of urinary metabolites of the anti-arrhythmic drug encainide in human subjects. *Drug Metab Dispos* 18:28–35.

Jakubowski, J. A., Winters, K. J., Naganuma, H., Wallentin, L. 2007. Prasugrel: A novel thienopyridine antiplatelet agent. A review of preclinical and clinical studies and the mechanistic basis for its distinct antiplatelet profile. *Cardiovasc Drug Rev* 25:357–374.

Jankovic, J. 2009. Treatment of hyperkinetic movement disorders. *Lancet Neurol* 8:844–856.

Jankovic, J., Clarence-Smith, K. 2011. Tetrabenazine for the treatment of chorea and other hyperkinetic movement disorders. *Expert Rev Neurother* 11:1509–1523.

Jann, M. W., Shirley, K. L., Small, G. W. 2002. Clinical pharmacokinetics and pharmacodynamics of cholinesterase inhibitors. *Clin Pharmacokinet* 41:719–739.

Jaruratanasirikul, S., Hortiwakul, R. 1994. The inhibitory effect of amiodarone and desethylamiodarone on dextromethorphan O-demethylation in human and rat liver microsomes. *J Pharm Pharmacol* 46:933–935.

Jarvis, B., Coukell, A. J. 1998. Mexiletine. A review of its therapeutic use in painful diabetic neuropathy. *Drugs* 56:691–707.

Jean, P., Lopez-Garcia, P., Dansette, P., Mansuy, D., Goldstein, J. L. 1996. Oxidation of tienilic acid by human yeast-expressed cytochromes P-450 2C8, 2C9, 2C18 and 2C19. Evidence that this drug is a mechanism-based inhibitor specific for cytochrome P-450 2C9. *Eur J Biochem* 241:797–804.

Jenner, P. 2003. The contribution of the MPTP-treated primate model to the development of new treatment strategies for Parkinson's disease. *Parkinsonism Relat Disord* 9:131–137.

Jiang, X. L., Shen, H. W., Yu, A. M. 2009. Pinoline may be used as a probe for CYP2D6 activity. *Drug Metab Dispos* 37:443–446.

Jimenez-Jimenez, F. J., Tabernero, C., Mena, M. A., Garcia de Yebenes, J., Garcia de Yebenes, M. J., Casarejos, M. J., Pardo, B., Garcia-Agundez, J. A., Benitez, J., Martinez, A., Garcia-Asenjo, J. A. L. 1991. Acute effects of 1-methyl-4-phenyl-1,2,3,6-tetrahydropyridine in a model of rat designated a poor metabolizer of debrisoquine. *J Neurochem* 57:81–87.

Jin, L., Xu, W., Krefetz, D., Gruener, D., Kielbasa, W., Tauscher-Wisniewski, S., Allen, A. J. 2013. Clinical outcomes from an open-label study of edivoxetine use in pediatric patients with attention-deficit/hyperactivity disorder. *J Child Adolesc Psychopharmacol* 23:200–207.

Johns, T. E., Lopez, L. M. 1995. Bisoprolol: Is this just another β-blocker for hypertension or angina? *Ann Pharmacother* 29:403–414.

Johnson, J. A., Burlew, B. S. 1996. Metoprolol metabolism via cytochrome P4502D6 in ethnic populations. *Drug Metab Dispos* 24:350–355.

Johnson, J. A., Herring, V. L., Wolfe, M. S., Relling, M. V. 2000. CYP1A2 and CYP2D6 4-hydroxylate propranolol and both reactions exhibit racial differences. *J Pharmacol Exp Ther* 294:1099–1105.

Johnson, M. D., Zuo, H., Lee, K. H., Trebley, J. P., Rae, J. M., Weatherman, R. V., Desta, Z., Flockhart, D. A., Skaar, T. C. 2004. Pharmacological characterization of 4-hydroxy-N-desmethyl tamoxifen, a novel active metabolite of tamoxifen. *Breast Cancer Res Treat* 85:151–159.

Johnston, S. R., Gumbrell, L. A., Evans, T. R., Coleman, R. E., Smith, I. E., Twelves, C. J., Soukop, M. et al. 2004. A cancer research (UK) randomized phase II study of idoxifene in patients with locally advanced/metastatic breast cancer resistant to tamoxifen. *Cancer Chemother Pharmacol* 53:341–348.

Jones, R. M., Lim, C. K. 2002. Toremifene metabolism in rat, mouse and human liver microsomes: Identification of alpha-hydroxytoremifene by LC-MS. *Biomed Chromatogr* 16:361–363.

Jones, B. C., Hyland, R., Ackland, M., Tyman, C. A., Smith, D. A. 1998. Interaction of terfenadine and its primary metabolites with cytochrome P450 2D6. *Drug Metab Dispos* 26:875–882.

Jones, D. R., Gorski, J. C., Hamman, M. A., Mayhew, B. S., Rider, S., Hall, S. D. 1999. Diltiazem inhibition of cytochrome P450 3A activity is due to metabolite intermediate complex formation. *J Pharmacol Exp Ther* 290:1116–1125.

Jordan, V. C. 1982. Metabolites of tamoxifen in animals and man: Identification, pharmacology, and significance. *Breast Cancer Res Treat* 2:123–138.

Jordan, V. C. 2007. Chemoprevention of breast cancer with selective oestrogen-receptor modulators. *Nat Rev Cancer* 7:46–53.

Jordan, V. C. O'Malley, B. W. 2007. Selective estrogen-receptor modulators and antihormonal resistance in breast cancer. *J Clin Oncol* 25:5815–5824.

Jordan, V. C., Collins, M. M., Rowsby, L., Prestwich, G. 1977. A monohydroxylated metabolite of tamoxifen with potent anti-oestrogenic activity. *J Endocrinol* 75:305–316.

Judge, H. M., Buckland, R. J., Sugidachi, A., Jakubowski, J. A., Storey, R. F. 2008. The active metabolite of prasugrel effectively blocks the platelet P2Y12 receptor and inhibits procoagulant and proinflammatory platelet responses. *Platelets* 19:125–133.

Jung, S. M., Kim, K. A., Cho, H. K., Jung, I. G., Park, P. W., Byun, W. T., Park, J. Y. 2005. Cytochrome P450 3A inhibitor itraconazole affects plasma concentrations of risperidone and 9-hydroxyrisperidone in schizophrenic patients. *Clin Pharmacol Ther* 78:520–528.

Kaduszkiewicz, H., Zimmermann, T., Beck-Bornholdt, H. P., van den Bussche, H. 2005. Cholinesterase inhibitors for patients with Alzheimer's disease: Systematic review of randomised clinical trials. *BMJ* 331:321–327.

Kaku, T., Ogura, K., Nishiyama, T., Ohnuma, T., Muro, K., Hiratsuka, A. 2004. Quaternary ammonium-linked glucuronidation of tamoxifen by human liver microsomes and UDP-glucuronosyltransferase 1A4. *Biochem Pharmacol* 67:2093–2102.

Kalgutkar, A. S., Zhou, S., Fahmi, O. A., Taylor, T. J. 2003. Influence of lipophilicity on the interactions of N-alkyl-4-phenyl-1,2,3,6-tetrahydropyridines and their positively charged N-alkyl-4-phenylpyridinium metabolites with cytochrome P450 2D6. *Drug Metab Dispos* 31:596–605.

Kam, P. C., Nethery, C. M. 2003. The thienopyridine derivatives (platelet adenosine diphosphate receptor antagonists), pharmacology and clinical developments. *Anaesthesia* 58:28–35.

Kamata, T., Katagi, M., Kamata, H. T., Miki, A., Shima, N., Zaitsu, K., Nishikawa, M., Tanaka, E., Honda, K., Tsuchihashi, H. 2006. Metabolism of the psychotomimetic tryptamine derivative 5-methoxy-N,N-diisopropyltryptamine in humans: Identification and quantification of its urinary metabolites. *Drug Metab Dispos* 34:281–287.

Kammerer, R. C., Schmitz, D. A. 1986. Lidocaine metabolism by rabbit-liver homogenate and detection of a new metabolite. *Xenobiotica* 16:681–690.

Kammerer, R. C., Lampe, M. A. 1987. In vitro metabolism of chlorpheniramine in the rabbit. *Biochem Pharmacol* 36:3445–3452.

Kammerer, B., Scheible, H., Albrecht, W., Gleiter, C. H., Laufer, S. 2007. Pharmacokinetics of ML3403 ({4-[5-(4-fluorophenyl)-2-methylsulfanyl-3H-imidazol-4-yl]-pyridin-2-yl}-(1-phenylethyl)-amine), a 4-Pyridinylimidazole-type p38 mitogen-activated protein kinase inhibitor. *Drug Metab Dispos* 35:875–883.

Karbwang, J., Na Bangchang, K. 1994. Clinical pharmacokinetics of halofantrine. *Clin Pharmacokinet* 27:104–119.

Kariya, S., Isozaki, S., Narimatsu, S., Suzuki, T. 1992a. Oxidative metabolism of cinnarizine in rat liver microsomes. *Biochem Pharmacol* 44:1471–1474.

Kariya, S., Isozaki, S., Narimatsu, S., Suzuki, T. 1992b. Oxidative metabolism of flunarizine in rat liver microsomes. *Res Commun Chem Pathol Pharmacol* 78:85–95.

Kariya, S., Isozaki, S., Uchino, K., Suzuki, T., Narimatsu, S. 1996. Oxidative metabolism of flunarizine and cinnarizine by microsomes from B-lymphoblastoid cell lines expressing human cytochrome P450 enzymes. *Biol Pharm Bull* 19:1511–1514.

Karlsson, E. 1978. Clinical pharmacokinetics of procainamide. *Clin Pharmacokinet* 3:97–107.

Kassahun, K., Mattiuz, E., Nyhart, E., Jr., Obermeyer, B., Gillespie, T., Murphy, A., Goodwin, R. M., Tupper, D., Callaghan, J. T., Lemberger, L. 1997. Disposition and biotransformation of the antipsychotic agent olanzapine in humans. *Drug Metab Dispos* 25:81–93.

Kassahun, K., McIntosh, I. S., Shou, M., Walsh, D. J., Rodeheffer, C., Slaughter, D. E., Geer, L. A., Halpin, R. A., Agrawal, N., Rodrigues, A. D. 2001. Role of human liver cytochrome P4503A in the metabolism of etoricoxib, a novel cyclooxygenase-2 selective inhibitor. *Drug Metab Dispos* 29:813–820.

Katsutani, N., Shionoya, H. 1992. Drug-specific immune responses induced by procainamide, hydralazine and isoniazid in guinea-pigs. *Int J Immunopharmacol* 14:673–679.

Katzenellenbogen, B. S., Norman, M. J., Eckert, R. L., Peltz, S. W., Mangel, W. F. 1984. Bioactivities, estrogen receptor interactions, and plasminogen activator-inducing activities of tamoxifen and hydroxy-tamoxifen isomers in MCF-7 human breast cancer cells. *Cancer Res* 44:112–119.

Kavirajan, H., Schneider, L. S. 2007. Efficacy and adverse effects of cholinesterase inhibitors and memantine in vascular dementia: A meta-analysis of randomised controlled trials. *Lancet Neurol* 6:782–792.

Keam, S. J., Keating, G. M. 2004. Tiotropium bromide. A review of its use as maintenance therapy in patients with COPD. *Treat Respir Med* 3:247–268.

Keam, S. J., Goa, K. L., Figgitt, D. P. 2002. Almotriptan: A review of its use in migraine. *Drugs* 62:387–414.

Keenaghan, J. B., Boyes, R. N. 1972. The tissue distribution, metabolism and excretion of lidocaine in rats, guinea pigs, dogs and man. *J Pharmacol Exp Ther* 180:454–463.

Kerry, N. L., Somogyi, A. A., Bochner, F., Mikus, G. 1994. The role of CYP2D6 in primary and secondary oxidative metabolism of dextromethorphan: *In vitro* studies using human liver microsomes. *Br J Clin Pharmacol* 38:243–248.

Kessler, F. K., Kessler, M. R., Auyeung, D. J., Ritter, J. K. 2002. Glucuronidation of acetaminophen catalyzed by multiple rat phenol UDP-glucuronosyltransferases. *Drug Metab Dispos* 30:324–330.

Kharasch, E. D., Hoffer, C., Whittington, D., Sheffels, P. 2004. Role of hepatic and intestinal cytochrome P450 3A and 2B6 in the metabolism, disposition, and miotic effects of methadone. *Clin Pharmacol Ther* 76:250–269.

Khoo, K. C., Szuna, A. J., Colburn, W. A., Aogaichi, K., Morganroth, J., Brazzell, R. K. 1984. Single-dose pharmacokinetics and dose proportionality of oral cibenzoline. *J Clin Pharmacol* 24:283–288.

Kielbasa, W., Quinlan, T., Jin, L., Xu, W., Lachno, D. R., Dean, R. A., Allen, A. J. 2012. Pharmacokinetics and pharmacodynamics of edivoxetine (LY2216684), a norepinephrine reuptake inhibitor, in pediatric patients with attention-deficit/hyperactivity disorder. *J Child Adolesc Psychopharmacol* 22:269–276.

Kielbasa, W., Pan, A., Pereira, A. 2015. A pharmacokinetic/pharmacodynamic investigation: Assessment of edivoxetine and atomoxetine on systemic and central 3,4-dihydroxyphenylglycol, a biochemical marker for norepinephrine transporter inhibition. *Eur Neuropsychopharmacol* 25:377–385.

Kim, H., Wang, R. S., Elovaara, E., Raunio, H., Pelkonen, O., Aoyama, T., Vainio, H., Nakajima, T. 1997. Cytochrome P450 isozymes responsible for the metabolism of toluene and styrene in human liver microsomes. *Xenobiotica* 27:657–665.

Kim, K. A., Kim, M. J., Park, J. Y., Shon, J. H., Yoon, Y. R., Lee, S. S., Liu, K. H., Chun, J. H., Hyun, M. H., Shin, J. G. 2003. Stereoselective metabolism of lansoprazole by human liver cytochrome P450 enzymes. *Drug Metab Dispos* 31:1227–1234.

Kim, S. Y., Laxmi, Y. R., Suzuki, N., Ogura, K., Watabe, T., Duffel, M. W., Shibutani, S. 2005. Formation of tamoxifen-DNA adducts via O-sulfonation, not O-acetylation, of alpha-hydroxytamoxifen in rat and human livers. *Drug Metab Dispos* 33:1673–1678.

Kim, K. B., Seo, K. A., Yoon, Y. J., Bae, M. A., Cheon, H. G., Shin, J. G., Liu, K. H. 2008. *In vitro* metabolism of a novel PPAR-γ agonist, KR-62980, and its stereoisomer, KR-63198, in human liver microsomes and by recombinant cytochrome P450s. *Xenobiotica* 38:1165–1176.

Kimura, S., Umeno, M., Skoda, R. C., Meyer, U. A., Gonzalez, F. J. 1989. The human debrisoquine 4-hydroxylase (CYP2D) locus: Sequence and identification of the polymorphic CYP2D6 gene, a related gene, and a pseudogene. *Am J Hum Genet* 45:889–904.

King, C., Finley, B., Franklin, R. 2000. The glucuronidation of morphine by dog liver microsomes: Identification of morphine-6-O-glucuronide. *Drug Metab Dispos* 28:661–663.

Kirchheiner, J., Meineke, I., Muller, G., Roots, I., Brockmoller, J. 2002. Contributions of CYP2D6, CYP2C9 and CYP2C19 to the biotransformation of E- and Z-doxepin in healthy volunteers. *Pharmacogenetics* 12:571–580.

Kirkwood, L. C., Nation, R. L., Somogyi, A. A. 1997. Characterization of the human cytochrome P450 enzymes involved in the metabolism of dihydrocodeine. *Br J Clin Pharmacol* 44:549–555.

Kisanga, E. R., Mellgren, G., Lien, E. A. 2005. Excretion of hydroxylated metabolites of tamoxifen in human bile and urine. *Anticancer Res* 25:4487–4492.

Kishimoto, W., Hiroi, T., Sakai, K., Funae, Y., Igarashi, T. 1997. Metabolism of epinastine, a histamine H1 receptor antagonist, in human liver microsomes in comparison with that of terfenadine. *Res Commun Mol Pathol Pharmacol* 98:273–292.

Kishimoto, W., Hiroi, T., Shiraishi, M., Osada, M., Imaoka, S., Kominami, S., Igarashi, T., Funae, Y. 2004. Cytochrome P450 2D catalyze steroid 21-hydroxylation in the brain. *Endocrinology* 145:699–705.

Kiss, B., Horvath, A., Nemethy, Z., Schmidt, E., Laszlovszky, I., Bugovics, G., Fazekas, K. et al. 2010. Cariprazine (RGH-188), a dopamine D_3 receptor-preferring, D_3/D_2 dopamine receptor antagonist-partial agonist antipsychotic candidate: *In vitro* and neurochemical profile. *J Pharmacol Exp Ther* 333:328–340.

Kitagawa, M., Ravindernath, A., Suzuki, N., Rieger, R., Terashima, I., Umemoto, A., Shibutani, S. 2000. Identification of tamoxifen-DNA adducts induced by alpha-acetoxy-N-desmethyl-tamoxifen. *Chem Res Toxicol* 13:761–769.

Klotz, U. 2007. Antiarrhythmics: Elimination and dosage considerations in hepatic impairment. *Clin Pharmacokinet* 46:985–996.

Ko, J. W., Desta, Z., Soukhova, N. V., Tracy, T., Flockhart, D. A. 2000. In vitro inhibition of the cytochrome P450 (CYP450) system by the antiplatelet drug ticlopidine: Potent effect on CYP2C19 and CYP2D6. *Br J Clin Pharmacol* 49:343–351.

Kobari, T., Itoh, T., Hirakawa, T., Namekawa, H., Suzuki, T., Satoh, T., Iida, N., Ohtsu, F., Hayakawa, H. 1984. Dose-dependent pharmacokinetics of aprindine in healthy volunteers. *Eur J Clin Pharmacol* 26:129–131.

Kobayashi, K., Chiba, K., Yagi, T., Shimada, N., Taniguchi, T., Horie, T., Tani, M., Yamamoto, T., Ishizaki, T., Kuroiwa, Y. 1997. Identification of cytochrome P450 isoforms involved in citalopram N-demethylation by human liver microsomes. *J Pharmacol Exp Ther* 280:927–933.

Kobayashi, K., Ishizuka, T., Shimada, N., Yoshimura, Y., Kamijima, K., Chiba, K. 1999. Sertraline N-demethylation is catalyzed by multiple isoforms of human cytochrome P450 *in vitro*. *Drug Metab Dispos* 27:763–766.

Komatsu, T., Yamazaki, H., Shimada, N., Nagayama, S., Kawaguchi, Y., Nakajima, M., Yokoi, T. 2001. Involvement of microsomal cytochrome P450 and cytosolic thymidine phosphorylase in 5-fluorouracil formation from tegafur in human liver. *Clin Cancer Res* 7:675–681.

Komi, J., Lankinen, K. S., Harkonen, P., DeGregorio, M. W., Voipio, S., Kivinen, S., Tuimala, R. et al. 2005. Effects of ospemifene and raloxifene on hormonal status, lipids, genital tract, and tolerability in postmenopausal women. *Menopause* 12:202–209.

Komura, H., Iwaki, M. 2005. Nonlinear pharmacokinetics of propafenone in rats and humans: Application of a substrate depletion assay using hepatocytes for assessment of nonlinearity. *Drug Metab Dispos* 33:726–732.

Konig, F., Wolfersdorf, M., Loble, M., Wossner, S., Hauger, B. 1997. Trimipramine and maprotiline plasma levels during combined treatment with moclobemide in therapy-resistant depression. *Pharmacopsychiatry* 30:125–127.

Kostis, J. B., Krieger, S., Moreyra, A., Cosgrove, N. 1984. Cibenzoline for treatment of ventricular arrhythmias: A double-blind placebo-controlled study. *J Am Coll Cardiol* 4:372–377.

Kotsuma, M., Hanzawa, H., Iwata, Y., Takahashi, K., Tokui, T. 2008a. Novel binding mode of the acidic CYP2D6 substrates pactimibe and its metabolite R-125528. *Drug Metab Dispos* 36:1938–1943.

Kotsuma, M., Tokui, T., Freudenthaler, S., Nishimura, K. 2008b. Effects of ketoconazole and quinidine on pharmacokinetics of pactimibe and its plasma metabolite, R-125528, in human. *Drug Metab Dispos* 36:1505–1511.

Kotsuma, M., Tokui, T., Ishizuka-Ozeki, T., Honda, T., Iwabuchi, H., Murai, T., Ikeda, T., Saji, H. 2008c. CYP2D6-mediated metabolism of a novel acyl coenzyme A:cholesterol acyltransferase inhibitor, pactimibe, and its unique plasma metabolite, R-125528. *Drug Metab Dispos* 36:529–534.

Koudriakova, T., Iatsimirskaia, E., Utkin, I., Gangl, E., Vouros, P., Storozhuk, E., Orza, D., Marinina, J., Gerber, N. 1998. Metabolism of the human immunodeficiency virus protease inhibitors indinavir and ritonavir by human intestinal microsomes and expressed cytochrome P4503A4/3A5: Mechanism-based inactivation of cytochrome P4503A by ritonavir. *Drug Metab Dispos* 26:552–561.

Koujitani, T., Yasuhara, K., Kobayashi, H., Shimada, A., Onodera, H., Takagi, H., Hirose, M., Mitsumori, K. 1999. Tumor-promoting activity of 2,6-dimethylaniline in a two-stage nasal carcinogenesis model in N-bis(2-hydroxypropyl)nitrosamine-treated rats. *Cancer Lett* 142:161–171.

Koyama, E., Kikuchi, Y., Echizen, H., Chiba, K., Ishizaki, T. 1993. Simultaneous high-performance liquid chromatography-electrochemical detection determination of imipramine, desipramine, their 2-hydroxylated metabolites, and imipramine N-oxide in human plasma and urine: Preliminary application to oxidation pharmacogenetics. *Ther Drug Monit* 15:224–235.

Koyama, E., Chiba, K., Tani, M., Ishizaki, T. 1996. Identification of human cytochrome P450 isoforms involved in the stereoselective metabolism of mianserin enantiomers. *J Pharmacol Exp Ther* 278:21–30.

Koyama, E., Chiba, K., Tani, M., Ishizaki, T. 1997. Reappraisal of human CYP isoforms involved in imipramine N-demethylation and 2-hydroxylation: A study using microsomes obtained from putative extensive and poor metabolizers of S-mephenytoin and eleven recombinant human CYPs. *J Pharmacol Exp Ther* 281:1199–1210.

Koymans, L., van Lenthe, J. H., van de Straat, R., Donne-Op den Kelder, G. M., Vermeulen, N. P. 1989. A theoretical study on the metabolic activation of paracetamol by cytochrome P-450: Indications for a uniform oxidation mechanism. *Chem Res Toxicol* 2:60–66.

Koymans, L., Vermeulen, N. P., van Acker, S. A., te Koppele, J. M., Heykants, J. J., Lavrijsen, K., Meuldermans, W., Donne-Op den Kelder, G. M. 1992. A predictive model for substrates of cytochrome P450-debrisoquine (2D6). *Chem Res Toxicol* 5:211–219.

Krause, D. S., Van Etten, R. A. 2005. Tyrosine kinases as targets for cancer therapy. *N Engl J Med* 353:172–187.

Kreth, K., Kovar, K., Schwab, M., Zanger, U. M. 2000. Identification of the human cytochromes P450 involved in the oxidative metabolism of "Ecstasy"-related designer drugs. *Biochem Pharmacol* 59:1563–1571.

Krieter, P. A., Gohdes, M., Musick, T. J., Duncanson, F. P., Bridson, W. E. 2008. Pharmacokinetics, disposition, and metabolism of bicifadine in humans. *Drug Metab Dispos* 36:252–259.

Kris, M. G., Hesketh, P. J., Somerfield, M. R., Feyer, P., Clark-Snow, R., Koeller, J. M., Morrow, G. R. et al. 2006. American Society of Clinical Oncology guideline for antiemetics in oncology: Update 2006. *J Clin Oncol* 24:2932–2947.

Kroemer, H. K., Funck-Brentano, C., Silberstein, D. J., Wood, A. J., Eichelbaum, M., Woosley, R. L., Roden, D. M. 1989a. Stereoselective disposition and pharmacologic activity of propafenone enantiomers. *Circulation* 79:1068–1076.

Kroemer, H. K., Mikus, G., Kronbach, T., Meyer, U. A., Eichelbaum, M. 1989b. In vitro characterization of the human cytochrome P-450 involved in polymorphic oxidation of propafenone. *Clin Pharmacol Ther* 45:28–33.

Kroemer, H. K., Fischer, C., Meese, C. O., Eichelbaum, M. 1991. Enantiomer/enantiomer interaction of (S)- and (R)-propafenone for cytochrome P450IID6-catalyzed 5-hydroxylation: In vitro evaluation of the mechanism. *Mol Pharmacol* 40:135–142.

Kroemer, H. K., Gautier, J. C., Beaune, P., Henderson, C., Wolf, C. R., Eichelbaum, M. 1993. Identification of P450 enzymes involved in metabolism of verapamil in humans. *Naunyn Schmiedebergs Arch Pharmacol* 348:332–337.

Ku, H. Y., Ahn, H. J., Seo, K. A., Kim, H., Oh, M., Bae, S. K., Shin, J. G., Shon, J. H., Liu, K. H. 2008. The contributions of cytochromes P450 3A4 and 3A5 to the metabolism of the phosphodiesterase type 5 inhibitors sildenafil, udenafil, and vardenafil. *Drug Metab Dispos* 36:986–990.

Kudo, S., Odomi, M. 1998. Involvement of human cytochrome P450 3A4 in reduced haloperidol oxidation. *Eur J Clin Pharmacol* 54:253–259.

Kudo, S., Ishizaki, T. 1999. Pharmacokinetics of haloperidol: An update. *Clin Pharmacokinet* 37:435–456.

Kudo, S., Okumura, H., Miyamoto, G., Ishizaki, T. 1999. Cytochrome P450 isoforms involved in carboxylic acid ester cleavage of Hantzsch pyridine ester of pranidipine. *Drug Metab Dispos* 27:303–308.

Kuehl, G. E., Lampe, J. W., Potter, J. D., Bigler, J. 2005. Glucuronidation of nonsteroidal anti-inflammatory drugs: Identifying the enzymes responsible in human liver microsomes. *Drug Metab Dispos* 33:1027–1035.

Kumar, A., Strech, D. 2005. Zuclopenthixol dihydrochloride for schizophrenia. *Cochrane Database Syst Rev* CD005474.

Kumar, G. N., Rodrigues, A. D., Buko, A. M., Denissen, J. F. 1996. Cytochrome P450-mediated metabolism of the HIV-1 protease inhibitor ritonavir (ABT-538) in human liver microsomes. *J Pharmacol Exp Ther* 277:423–431.

Kumar, G. N., Dubberke, E., Rodrigues, A. D., Roberts, E., Dennisen, J. F. 1997. Identification of cytochromes P450 involved in the human liver microsomal metabolism of the thromboxane A2 inhibitor seratrodast (ABT-001). *Drug Metab Dispos* 25:110–115.

Kutz, K. 1993. Pharmacology, toxicology and human pharmacokinetics of tropisetron. *Ann Oncol* 4 Suppl 3:15–18.

Labbe, L., Turgeon, J. 1999. Clinical pharmacokinetics of mexiletine. *Clin Pharmacokinet* 37:361–384.

Labbe, L., Abolfathi, Z., Lessard, E., Pakdel, H., Beaune, P., Turgeon, J. 2003. Role of specific cytochrome P450 enzymes in the N-oxidation of the antiarrhythmic agent mexiletine. *Xenobiotica* 33:13–25.

Laganiere, S., Davies, R. F., Carignan, G., Foris, K., Goernert, L., Carrier, K., Pereira, C., McGilveray, I. 1996. Pharmacokinetic and pharmacodynamic interactions between diltiazem and quinidine. *Clin Pharmacol Ther* 60:255–264.

Lai, W. G., Zahid, N., Uetrecht, J. P. 1999. Metabolism of trimethoprim to a reactive iminoquinone methide by activated human neutrophils and hepatic microsomes. *J Pharmacol Exp Ther* 291:292–299.

Laine, J. E., Auriola, S., Pasanen, M., Juvonen, R. O. 2009. Acetaminophen bioactivation by human cytochrome P450 enzymes and animal microsomes. *Xenobiotica* 39:11–21.

Lalovic, B., Phillips, B., Risler, L. L., Howald, W., Shen, D. D. 2004. Quantitative contribution of CYP2D6 and CYP3A to oxycodone metabolism in human liver and intestinal microsomes. *Drug Metab Dispos* 32:447–454.

Lam, Y. W., Gaedigk, A., Ereshefsky, L., Alfaro, C. L., Simpson, J. 2002. CYP2D6 inhibition by selective serotonin reuptake inhibitors: Analysis of achievable steady-state plasma concentrations and the effect of ultrarapid metabolism at CYP2D6. *Pharmacotherapy* 22:1001–1006.

Lancaster, S. G., Sorkin, E. M. 1988. Bisoprolol. A preliminary review of its pharmacodynamic and pharmacokinetic properties, and therapeutic efficacy in hypertension and angina pectoris. *Drugs* 36:256–285.

Langan-Fahey, S. M., Tormey, D. C., Jordan, V. C. 1990. Tamoxifen metabolites in patients on long-term adjuvant therapy for breast cancer. *Eur J Cancer* 26:883–888.

Lantz, R. J., Gillespie, T. A., Rash, T. J., Kuo, F., Skinner, M., Kuan, H. Y., Knadler, M. P. 2003. Metabolism, excretion, and pharmacokinetics of duloxetine in healthy human subjects. *Drug Metab Dispos* 31:1142–1150.

Latini, R., Tognoni, G., Kates, R. E. 1984. Clinical pharmacokinetics of amiodarone. *Clin Pharmacokinet* 9:136–156.

Laurent, F., Saivin, S., Chretien, P., Magnaval, J. F., Peyron, F., Sqalli, A., Tufenkji, A. E. et al. 1993. Pharmacokinetic and pharmacodynamic study of amodiaquine and its two metabolites after a single oral dose in human volunteers. *Arzneimittelforschung* 43:612–616.

Lawson, D. H., Jick, H. 1977. Adverse reactions to procainamide. *Br J Clin Pharmacol* 4:507–511.

Lee, C. R., Plosker, G. L., McTavish, D. 1993. Tropisetron. A review of its pharmacodynamic and pharmacokinetic properties, and therapeutic potential as an antiemetic. *Drugs* 46:925–943.

Lee, A. J., Cai, M. X., Thomas, P. E., Conney, A. H., Zhu, B. T. 2003. Characterization of the oxidative metabolites of 17β-estradiol and estrone formed by 15 selectively expressed human cytochrome P450 isoforms. *Endocrinology* 144:3382–3398.

Leibovitch, L., Matok, I., Paret, G. 2007. Therapeutic applications of sildenafil citrate in the management of paediatric pulmonary hypertension. *Drugs* 67:57–73.

Leinonen, E., Koponen, H. J., Lepola, U. 2004. Paroxetine increases serum trimipramine concentration. A report of two cases. *Human Psychopharmacol Clin Exp* 10:345–347.

Lemoine, A., Gautier, J. C., Azoulay, D., Kiffel, L., Belloc, C., Guengerich, F. P., Maurel, P., Beaune, P., Leroux, J. P. 1993. Major pathway of imipramine metabolism is catalyzed by cytochromes P450 1A2 and P450 3A4 in human liver. *Mol Pharmacol* 43:827–832.

Lennard, M. S., Jackson, P. R., Freestone, S., Tucker, G. T., Ramsay, L. E., Woods, H. F. 1984. The relationship between debrisoquine oxidation phenotype and the pharmacokinetics and pharmacodynamics of propranolol. *Br J Clin Pharmacol* 17:679–685.

Lennard, M. S., Tucker, G. T., Silas, J. H., Woods, H. F. 1986. Debrisoquine polymorphism and the metabolism and action of metoprolol, timolol, propranolol and atenolol. *Xenobiotica* 16:435–447.

Leone Roberti Maggiore, U., Salvatore, S., Alessandri, F., Remorgida, V., Origoni, M., Candiani, M., Venturini, P. L., Ferrero, S. 2012. Pharmacokinetics and toxicity of antimuscarinic drugs for overactive bladder treatment in females. *Expert Opin Drug Metab Toxicol* 8:1387–1408.

Leopold, G., Pabst, J., Ungethum, W., Buhring, K. U. 1986. Basic pharmacokinetics of bisoprolol, a new highly β_1-selective adrenoceptor antagonist. *J Clin Pharmacol* 26:616–621.

Lesesve, J. F., Callat, M. P., Lenormand, B., Monconduit, M., Noblet, C., Moore, N., Caron, F., Humbert, G., Stamatoullas, A., Tilly, H. 1994. Hematological toxicity of ticlopidine. *Am J Hematol* 47:149–150.

Lesko, L. J. 1989. Pharmacokinetic drug interactions with amiodarone. *Clin Pharmacokinet* 17:130–140.

Lessard, E., Fortin, A., Belanger, P. M., Beaune, P., Hamelin, B. A., Turgeon, J. 1997. Role of CYP2D6 in the N-hydroxylation of procainamide. *Pharmacogenetics* 7:381–390.

Leung, J. G., Breden, E. L. 2011. Tetrabenazine for the treatment of tardive dyskinesia. *Ann Pharmacother* 45:525–531.

Leveque, D. 2011. Pharmacokinetics of gefitinib and erlotinib. *Lancet Oncol* 12:1093.

Lewis, D. F., Eddershaw, P. J., Goldfarb, P. S., Tarbit, M. H. 1997. Molecular modelling of cytochrome P4502D6 (CYP2D6) based on an alignment with CYP102: Structural studies on specific CYP2D6 substrate metabolism. *Xenobiotica* 27:319–339.

Li, G., Gong, P. L., Qiu, J., Zeng, F. D., Klotz, U. 1998. Stereoselective steady state disposition and action of propafenone in Chinese subjects. *Br J Clin Pharmacol* 46:441–445.

Li, X. F., Carter, S., Dovichi, N. J., Zhao, J. Y., Kovarik, P., Sakuma, T. 2001. Analysis of tamoxifen and its metabolites in synthetic gastric fluid digests and urine samples using high-performance liquid chromatography with electrospray mass spectrometry. *J Chromatogr A* 914:5–12.

Li, X. Q., Bjorkman, A., Andersson, T. B., Ridderstrom, M., Masimirembwa, C. M. 2002. Amodiaquine clearance and its metabolism to N-desethylamodiaquine is mediated by CYP2C8: A new high affinity and turnover enzyme-specific probe substrate. *J Pharmacol Exp Ther* 300:399–407.

Li, J., Zhao, M., He, P., Hidalgo, M., Baker, S. D. 2007. Differential metabolism of gefitinib and erlotinib by human cytochrome P450 enzymes. *Clin Cancer Res* 13:3731–3737.

Li, J., Zhao, X., Chen, L., Guo, H., Lv, F., Jia, K., Yv, K. et al. 2010. Safety and pharmacokinetics of novel selective vascular endothelial growth factor receptor-2 inhibitor YN968D1 in patients with advanced malignancies. *BMC Cancer* 10:529.

Li, J., Qin, S., Xu, J., Guo, W., Xiong, J., Bai, Y., Sun, G. et al. 2013. Apatinib for chemotherapy-refractory advanced metastatic gastric cancer: Results from a randomized, placebo-controlled, parallel-arm, phase II trial. *J Clin Oncol* 31:3219–3225.

Liddell, J. R. 2001. Pyrrolizidine alkaloids. *Nat Prod Rep* 18: 441–447.

Lieberman, J. A., Bymaster, F. P., Meltzer, H. Y., Deutch, A. Y., Duncan, G. E., Marx, C. E., Aprille, J. R. et al. 2008. Antipsychotic drugs: Comparison in animal models of efficacy, neurotransmitter regulation, and neuroprotection. *Pharmacol Rev* 60:358–403.

Lien, E. A., Solheim, E., Lea, O. A., Lundgren, S., Kvinnsland, S., Ueland, P. M. 1989. Distribution of 4-hydroxy-N-desmethyltamoxifen and other tamoxifen metabolites in human biological fluids during tamoxifen treatment. *Cancer Res* 49: 2175–2183.

Lightfoot, T., Ellis, S. W., Mahling, J., Ackland, M. J., Blaney, F. E., Bijloo, G. J., De Groot, M. J. et al. 2000. Regioselective hydroxylation of debrisoquine by cytochrome P4502D6: Implications for active site modelling. *Xenobiotica* 30: 219–233.

Lim, C. K., Yuan, Z. X., Lamb, J. H., White, I. N., De Matteis, F., Smith, L. L. 1994. A comparative study of tamoxifen metabolism in female rat, mouse and human liver microsomes. *Carcinogenesis* 15:589–593.

Lin, L. Y., Di Stefano, E. W., Schmitz, D. A., Hsu, L., Ellis, S. W., Lennard, M. S., Tucker, G. T., Cho, A. K. 1997. Oxidation of methamphetamine and methylenedioxymethamphetamine by CYP2D6. *Drug Metab Dispos* 25:1059–1064.

Lin, D. Y., Kratochvil, C. J., Xu, W., Jin, L., D'Souza, D. N., Kielbasa, W., Allen, A. J. 2014. A randomized trial of edivoxetine in pediatric patients with attention-deficit/hyperactivity disorder. *J Child Adolesc Psychopharmacol* 24:190–200.

Linnet, K. 2002. Glucuronidation of olanzapine by cDNA-expressed human UDP-glucuronosyltransferases and human liver microsomes. *Hum Psychopharmacol* 17:233–238.

Linnet, K., Olesen, O. V. 1997. Metabolism of clozapine by cDNA-expressed human cytochrome P450 enzymes. *Drug Metab Dispos* 25:1379–1382.

Lintz, W., Erlacin, S., Frankus, E., Uragg, H. 1981. [Biotransformation of tramadol in man and animal (author's transl)]. *Arzneimittelforschung* 31:1932–1943.

Liu, Z. C., Uetrecht, J. P. 2000. Metabolism of ticlopidine by activated neutrophils: Implications for ticlopidine-induced agranulocytosis. *Drug Metab Dispos* 28:726–730.

Llerena, A., Berecz, R., de la Rubia, A., Norberto, M. J., Benitez, J. 2000. Use of the mesoridazine/thioridazine ratio as a marker for CYP2D6 enzyme activity. *Ther Drug Monit* 22: 397–401.

Llerena, A., Berecz, R., de la Rubia, A., Dorado, P. 2002. QTc interval lengthening is related to CYP2D6 hydroxylation capacity and plasma concentration of thioridazine in patients. *J Psychopharmacol* 16:361–364.

Loh, A. C., Williams, T. H., Tilley, J. W., Sasso, G. J., Szuna, A. J., Carbone, J. J., Toome, V., Leinweber, F. J. 1986. The metabolism of 14C-cibenzoline in dogs and rats. *Drug Metab Dispos* 14:325–330.

Lopez-Garcia, M. P., Dansette, P. M., Mansuy, D. 1994. Thiophene derivatives as new mechanism-based inhibitors of cytochromes P-450: Inactivation of yeast-expressed human liver cytochrome P-450 2C9 by tienilic acid. *Biochemistry* 33:166–175.

Loy, C., Schneider, L. 2006. Galantamine for Alzheimer's disease and mild cognitive impairment. *Cochrane Database Syst Rev* CD001747.

Luo, H., Hawes, E. M., McKay, G., Korchinski, E. D., Midha, K. K. 1991. N(+)-glucuronidation of aliphatic tertiary amines, a general phenomenon in the metabolism of H1-antihistamines in humans. *Xenobiotica* 21:1281–1288.

Lyu, C., Zhou, W., Zhang, Y., Zhang, S., Kou, F., Wei, H., Zhang, N., Zuo, Z. 2015. Identification and characterization of *in vitro* and *in vivo* metabolites of steroidal alkaloid veratramine. *Biopharm Drug Dispos* 36:308–324.

Ma, S., Subramanian, R., Xu, Y., Schrag, M., Shou, M. 2008. Structural characterization of novel adenine dinucleotide phosphate conjugates of imatinib in incubations with rat and human liver microsomes. *Drug Metab Dispos* 36:2414–2418.

Ma, S., Xu, Y., Shou, M. 2009. Characterization of imatinib metabolites in rat and human liver microsomes: Differentiation of hydroxylation from *N*-oxidation by liquid chromatography/atmospheric pressure chemical ionization mass spectrometry. *Rapid Commun Mass Spectrom* 23:1446–1450.

Madan, A., Graham, R. A., Carroll, K. M., Mudra, D. R., Burton, L. A., Krueger, L. A., Downey, A. D. et al. 2003. Effects of prototypical microsomal enzyme inducers on cytochrome P450 expression in cultured human hepatocytes. *Drug Metab Dispos* 31:421–431.

Madani, S., Paine, M. F., Lewis, L., Thummel, K. E., Shen, D. D. 1999. Comparison of CYP2D6 content and metoprolol oxidation between microsomes isolated from human livers and small intestines. *Pharm Res* 16:1199–1205.

Madden, S., Woolf, T. F., Pool, W. F., Park, B. K. 1993. An investigation into the formation of stable, protein-reactive and cytotoxic metabolites from tacrine in vitro. Studies with human and rat liver microsomes. *Biochem Pharmacol* 46:13–20.

Madden, S., Spaldin, V., Hayes, R. N., Woolf, T. F., Pool, W. F., Park, B. K. 1995. Species variation in the bioactivation of tacrine by hepatic microsomes. *Xenobiotica* 25:103–116.

Mahgoub, A., Idle, J. R., Dring, L. G., Lancaster, R., Smith, R. L. 1977. Polymorphic hydroxylation of Debrisoquine in man. *Lancet* 2:584–586.

Malfatto, G., Zaza, A., Forster, M., Sodowick, B., Danilo, P., Jr. Rosen, M. R. 1988. Electrophysiologic, inotropic and antiarrhythmic effects of propafenone, 5-hydroxypropafenone and N-depropylpropafenone. *J Pharmacol Exp Ther* 246:419–426.

Mandrioli, R., Forti, G. C., Raggi, M. A. 2006. Fluoxetine metabolism and pharmacological interactions: The role of cytochrome p450. *Curr Drug Metab* 7:127–133.

Mani, C., Gelboin, H. V., Park, S. S., Pearce, R., Parkinson, A., Kupfer, D. 1993a. Metabolism of the antimammary cancer antiestrogenic agent tamoxifen. I. Cytochrome P-450-catalyzed N-demethylation and 4-hydroxylation. *Drug Metab Dispos* 21:645–656.

Mani, C., Hodgson, E., Kupfer, D. 1993b. Metabolism of the antimammary cancer antiestrogenic agent tamoxifen. II. Flavin-containing monooxygenase-mediated N-oxidation. *Drug Metab Dispos* 21:657–661.

Mankowski, D. C. 1999. The role of CYP2C19 in the metabolism of (+/-) bufuralol, the prototypic substrate of CYP2D6. *Drug Metab Dispos* 27:1024–1028.

Mannens, G., Huang, M. L., Meuldermans, W., Hendrickx, J., Woestenborghs, R., Heykants, J. 1993. Absorption, metabolism, and excretion of risperidone in humans. *Drug Metab Dispos* 21:1134–1141.

Mannens, G. S., Snel, C. A., Hendrickx, J., Verhaeghe, T., Le Jeune, L., Bode, W., van Beijsterveldt, L. et al. 2002. The metabolism and excretion of galantamine in rats, dogs, and humans. *Drug Metab Dispos* 30:553–563.

Manolis, A. S., Deering, T. F., Cameron, J., Estes, N. A., 3rd. 1990. Mexiletine: Pharmacology and therapeutic use. *Clin Cardiol* 13:349–359.

Mantione, K. J., Cadet, P., Zhu, W., Kream, R. M., Sheehan, M., Fricchione, G. L., Goumon, Y., Esch, T., Stefano, G. B. 2008. Endogenous morphine signaling via nitric oxide regulates the expression of CYP2D6 and COMT: Autocrine/paracrine feedback inhibition. *Addict Biol* 13:118–123.

Manyike, P. T., Kharasch, E. D., Kalhorn, T. F., Slattery, J. T. 2000. Contribution of CYP2E1 and CYP3A to acetaminophen reactive metabolite formation. *Clin Pharmacol Ther* 67:275–282.

Marchetti, P., Navalesi, R. 1989. Pharmacokinetic-pharmacodynamic relationships of oral hypoglycaemic agents. An update. *Clin Pharmacokinet* 16:100–128.

Marchetti, P., Giannarelli, R., di Carlo, A., Navalesi, R. 1991. Pharmacokinetic optimisation of oral hypoglycaemic therapy. *Clin Pharmacokinet* 21:308–317.

Margolis, J. M., O'Donnell, J. P., Mankowski, D. C., Ekins, S., Obach, R. S. 2000. (R)-, (S)-, and racemic fluoxetine N-demethylation by human cytochrome P450 enzymes. *Drug Metab Dispos* 28:1187–1191.

Marill, J., Cresteil, T., Lanotte, M., Chabot, G. G. 2000. Identification of human cytochrome P450s involved in the formation of all-*trans*-retinoic acid principal metabolites. *Mol Pharmacol* 58:1341–1348.

Markowitz, J. S., Brinda, B. J. 2014. A pharmacokinetic evaluation of oral edivoxetine hydrochloride for the treatment of attention deficit-hyperactivity disorder. *Expert Opin Drug Metab Toxicol* 10:1289–1299.

Martinez Perez-Balsa, A., De Arce, A., Castiella, A., Lopez, P., Ruibal, M., Ruiz-Martinez, J., Lopez De Munain, A., Marti Masso, J. F. 1998. Hepatotoxicity due to ticlopidine. *Ann Pharmacother* 32:1250–1251.

Mash, D. C., Staley, J. K., Baumann, M. H., Rothman, R. B., Hearn, W. L. 1995. Identification of a primary metabolite of ibogaine that targets serotonin transporters and elevates serotonin. *Life Sci* 57:PL45–PL50.

Mash, D. C., Kovera, C. A., Pablo, J., Tyndale, R. F., Ervin, F. D., Williams, I. C., Singleton, E. G., Mayor, M. 2000. Ibogaine: Complex pharmacokinetics, concerns for safety, and preliminary efficacy measures. *Ann N Y Acad Sci* 914:394–401.

Massarella, J. W., Loh, A. C., Williams, T. H., Szuna, A. J., Sandor, D., Bressler, R., Leinweber, F. J. 1986. The disposition and metabolic fate of 14C-cibenzoline in man. *Drug Metab Dispos* 14:59–64.

Massarella, J. W., Defeo, T. M., Liguori, J., Passe, S., Aogaichi, K. 1991. The effects of cimetidine and ranitidine on the pharmacokinetics of cifenline. *Br J Clin Pharmacol* 31:481–483.

Masubuchi, Y., Umeda, S., Chiba, M., Fujita, S., Suzuki, T. 1991. Selective 3-hydroxylation deficiency of lidocaine and its metabolite in Dark Agouti rats. *Biochem Pharmacol* 42:693–695.

Masubuchi, Y., Araki, J., Narimatsu, S., Suzuki, T. 1992. Metabolic activation of lidocaine and covalent binding to rat liver microsomal protein. *Biochem Pharmacol* 43:2551–2557.

Masubuchi, Y., Umeda, S., Igarashi, S., Fujita, S., Narimatsu, S., Suzuki, T. 1993. Participation of the CYP2D subfamily in lidocaine 3-hydroxylation and formation of a reactive metabolite covalently bound to liver microsomal protein in rats. *Biochem Pharmacol* 46:1867–1869.

Masubuchi, Y., Hosokawa, S., Horie, T., Suzuki, T., Ohmori, S., Kitada, M., Narimatsu, S. 1994. Cytochrome P450 isozymes involved in propranolol metabolism in human liver microsomes. The role of CYP2D6 as ring-hydroxylase and CYP1A2 as N-desisopropylase. *Drug Metab Dispos* 22:909–915.

Masuda, K., Tamagake, K., Katsu, T., Torigoe, F., Saito, K., Hanioka, N., Yamano, S., Yamamoto, S., Narimatsu, S. 2006. Roles of phenylalanine at position 120 and glutamic acid at position 222 in the oxidation of chiral substrates by cytochrome P450 2D6. *Chirality* 18:167–176.

Matsui, K., Mishima, M., Nagai, Y., Yuzuriha, T., Yoshimura, T. 1999a. Absorption, distribution, metabolism, and excretion of donepezil (Aricept) after a single oral administration to rat. *Drug Metab Dispos* 27:1406–1414.

Matsui, K., Taniguchi, S., Yoshimura, T. 1999b. Correlation of the intrinsic clearance of donepezil (Aricept) between in vivo and in vitro studies in rat, dog and human. *Xenobiotica* 29:1059–1072.

Matsumoto, S., Yamazoe, Y. 2001. Involvement of multiple human cytochromes P450 in the liver microsomal metabolism of astemizole and a comparison with terfenadine. *Br J Clin Pharmacol* 51:133–142.

Matsunaga, T., Kishi, N., Higuchi, S., Watanabe, K., Ohshima, T., Yamamoto, I. 2000. CYP3A4 is a major isoform responsible for oxidation of 7-hydroxy-Δ⁸-tetrahydrocannabinol to 7-oxo-Δ⁸-tetrahydrocannabinol in human liver microsomes. *Drug Metab Dispos* 28:1291–1296.

Maurer, H. H. 1990. Identification of antiarrhythmic drugs and their metabolites in urine. *Arch Toxicol* 64:218–230.

Maurer, H. H., Kraemer, T., Springer, D., Staack, R. F. 2004. Chemistry, pharmacology, toxicology, and hepatic metabolism of designer drugs of the amphetamine (ecstasy), piperazine, and pyrrolidinophenone types: A synopsis. *Ther Drug Monit* 26:127–131.

Mautz, D. S., Nelson, W. L., Shen, D. D. 1995. Regioselective and stereoselective oxidation of metoprolol and bufuralol catalyzed by microsomes containing cDNA-expressed human P4502D6. *Drug Metab Dispos* 23:513–517.

McBride, M. C. 2000. Bufotenine: Toward an understanding of possible psychoactive mechanisms. *J Psychoactive Drugs* 32:321–331.

McCague, R., Seago, A. 1986. Aspects of metabolism of tamoxifen by rat liver microsomes. Identification of a new metabolite: E-1-[4-(2-dimethylaminoethoxy)-phenyl]-1, 2-diphenyl-1-buten-3-ol N-oxide. *Biochem Pharmacol* 35:827–834.

McCague, R., Parr, I. B., Haynes, B. P. 1990. Metabolism of the 4-iodo derivative of tamoxifen by isolated rat hepatocytes. Demonstration that the iodine atom reduces metabolic conversion and identification of four metabolites. *Biochem Pharmacol* 40:2277–2283.

McCormack, P. L., Keating, G. M. 2004. Duloxetine: In stress urinary incontinence. *Drugs* 64:2567–2573; discussion 2574–2565.

McEnroe, J. D., Fleishaker, J. C. 2005. Clinical pharmacokinetics of almotriptan, a serotonin 5-HT(1B/1D) receptor agonist for the treatment of migraine. *Clin Pharmacokinet* 44:237–246.

McGavin, J. K., Keating, G. M. 2002. Bisoprolol: A review of its use in chronic heart failure. *Drugs* 62:2677–2696.

McGeer, P. L., Boulding, J. E., Gibson, W. C., Foulkes, R. G. 1961. Drug-induced extrapyramidal reactions. Treatment with diphenhydramine hydrochloride and dihydroxyphenylalanine. *JAMA* 177:665–670.

McGourty, J. C., Silas, J. H., Lennard, M. S., Tucker, G. T., Woods, H. F. 1985. Metoprolol metabolism and debrisoquine oxidation polymorphism—population and family studies. *Br J Clin Pharmacol* 20:555–566.

McKeage, K. 2015. Tiotropium respimat®: A review of its use in asthma poorly controlled with inhaled corticosteroids and long-acting β_2-adrenergic agonists. *Drugs* 75:809–816.

McKillop, D., McCormick, A. D., Miles, G. S., Phillips, P. J., Pickup, K. J., Bushby, N., Hutchison, M. 2004. *In vitro* metabolism of gefitinib in human liver microsomes. *Xenobiotica* 34:983–1000.

McKillop, D., McCormick, A. D., Millar, A., Miles, G. S., Phillips, P. J., Hutchison, M. 2005. Cytochrome P450-dependent metabolism of gefitinib. *Xenobiotica* 35:39–50.

McMillen, B. A., Mattiace, L. A. 1983. Comparative neuropharmacology of buspirone and MJ-13805, a potential anti-anxiety drug. *J Neural Transm* 57:255–265.

McQuinn, R. L., Quarfoth, G. J., Johnson, J. D., Banitt, E. H., Pathre, S. V., Chang, S. F., Ober, R. E., Conard, G. J. 1984. Biotransformation and elimination of 14C-flecainide acetate in humans. *Drug Metab Dispos* 12:414–420.

McQuinn, R. L., Pentikainen, P. J., Chang, S. F., Conard, G. J. 1988. Pharmacokinetics of flecainide in patients with cirrhosis of the liver. *Clin Pharmacol Ther* 44:566–572.

McSorley, L. C., Daly, A. K. 2000. Identification of human cytochrome P450 isoforms that contribute to all-*trans*-retinoic acid 4-hydroxylation. *Biochem Pharmacol* 60:517–526.

Meatherall, R., Sharma, P. 2003. Foxy, a designer tryptamine hallucinogen. *J Anal Toxicol* 27:313–317.

Mehvar, R., Brocks, D. R., Vakily, M. 2002. Impact of stereoselectivity on the pharmacokinetics and pharmacodynamics of antiarrhythmic drugs. *Clin Pharmacokinet* 41:533–558.

Mellstrom, B., von Bahr, C. 1981. Demethylation and hydroxylation of amitriptyline, nortriptyline, and 10-hydroxyamitriptyline in human liver microsomes. *Drug Metab Dispos* 9:565–568.

Mellstrom, B., Bertilsson, L., Sawe, J., Schulz, H. U., Sjoqvist, F. 1981. E- and Z-10-hydroxylation of nortriptyline: Relationship to polymorphic debrisoquine hydroxylation. *Clin Pharmacol Ther* 30:189–193.

Mensah-Nyagan, A. G., Do-Rego, J. L., Beaujean, D., Luu-The, V., Pelletier, G., Vaudry, H. 1999. Neurosteroids: Expression of steroidogenic enzymes and regulation of steroid biosynthesis in the central nervous system. *Pharmacol Rev* 51:63–81.

Meuldermans, W., Hurkmans, R., Swysen, E., Hendrickx, J., Michiels, M., Lauwers, W., Heykants, J. 1981. On the pharmacokinetics of domperidone in animals and man III. Comparative study on the excretion and metabolism of domperidone in rats, dogs and man. *Eur J Drug Metab Pharmacokinet* 6:49–60.

Meuldermans, W., Hendrickx, J., Hurkmans, R., Swysen, E., Woestenborghs, R., Lauwers, W., Heykants, J. 1983. Excretion and metabolism of flunarizine in rats and dogs. *Arzneimittelforschung* 33:1142–1151.

Meuldermans, W., Hendrickx, J., Knaeps, F., Lauwers, W., Heykants, J., Grindel, J. M. 1984. Plasma levels, biotransformation and excretion of oxatomide (R 35 443) in rats, dogs and man. *Xenobiotica* 14:445–462.

Meyer, U. A., Gut, J., Kronbach, T., Skoda, C., Meier, U. T., Catin, T., Dayer, P. 1986. The molecular mechanisms of two common polymorphisms of drug oxidation—evidence for functional changes in cytochrome P-450 isozymes catalysing bufuralol and mephenytoin oxidation. *Xenobiotica* 16: 449–464.

Meyer, M. R., Peters, F. T., Maurer, H. H. 2009. The role of human hepatic cytochrome P450 isozymes in the metabolism of racemic MDEA and its single enantiomers. *Drug Metab Dispos* 37:1152–1156.

Miksys, S., Rao, Y., Hoffmann, E., Mash, D. C., Tyndale, R. F. 2002. Regional and cellular expression of CYP2D6 in human brain: Higher levels in alcoholics. *J Neurochem* 82:1376–1387.

Mikus, G., Bochner, F., Eichelbaum, M., Horak, P., Somogyi, A. A., Spector, S. 1994. Endogenous codeine and morphine in poor and extensive metabolisers of the CYP2D6 (debrisoquine/sparteine) polymorphism. *J Pharmacol Exp Ther* 268: 546–551.

Milano, G., Etienne, M. C., Frenay, M., Khater, R., Formento, J. L., Renee, N., Moll, J. L., Francoual, M., Berto, M., Namer, M. 1987. Optimised analysis of tamoxifen and its main metabolites in the plasma and cytosol of mammary tumours. *Br J Cancer* 55:509–512.

Milinkovic, A., Martinez, E. 2004. Nevirapine in the treatment of HIV. *Expert Rev Anti Infect Ther* 2:367–373.

Miller, G. P., Hanna, I. H., Nishimura, Y., Guengerich, F. P. 2001. Oxidation of phenethylamine derivatives by cytochrome P450 2D6: The issue of substrate protonation in binding and catalysis. *Biochemistry* 40:14215–14223.

Minato, K., Suzuki, R., Asagarasu, A., Matsui, T., Sato, M. 2008. Biotransformation of 3-amino-5,6,7,8-tetrahydro-2-{4-[4-(quinolin-2-yl)piperazin-1-yl]butyl}qui nazolin-4(3H)-one (TZB-30878), a novel 5-hydroxytryptamine (5-HT)1A agonist/5-HT3 antagonist, in human hepatic cytochrome P450 enzymes. *Drug Metab Dispos* 36:831–840.

Minder, E. I., Meier, P. J., Muller, H. K., Minder, C., Meyer, U. A. 1984. Bufuralol metabolism in human liver: A sensitive probe for the debrisoquine-type polymorphism of drug oxidation. *Eur J Clin Invest* 14:184–189.

Miner, D. J., Kissinger, P. T. 1979. Evidence for the involvement of N-acetyl-p- quinoneimine in acetaminophen metabolism. *Biochem Pharmacol* 28:3285–3290.

Miners, J. O., Smith, K. J., Robson, R. A., McManus, M. E., Veronese, M. E., Birkett, D. J. 1988. Tolbutamide hydroxylation by human liver microsomes. Kinetic characterisation and relationship to other cytochrome P-450 dependent xenobiotic oxidations. *Biochem Pharmacol* 37:1137–1144.

Miners, J. O., Rees, D. L., Valente, L., Veronese, M. E., Birkett, D. J. 1995. Human hepatic cytochrome P450 2C9 catalyzes the rate-limiting pathway of torsemide metabolism. *J Pharmacol Exp Ther* 272:1076–1081.

Mitchell, E. A., Herd, M. B., Gunn, B. G., Lambert, J. J., Belelli, D. 2008. Neurosteroid modulation of GABAA receptors: Molecular determinants and significance in health and disease. *Neurochem Int* 52:588–595.

Miura, D. S., Keren, G., Torres, V., Butler, B., Aogaichi, K., Somberg, J. C. 1985. Antiarrhythmic effects of cibenzoline. *Am Heart J* 109:827–833.

Mizushima, H., Takanaka, K., Abe, K., Fukazawa, I., Ishizuka, H. 2007. Stereoselective pharmacokinetics of oxybutynin and *N*-desethyloxybutynin *in vitro* and *in vivo*. *Xenobiotica* 37:59–73.

Modi, S., Paine, M. J., Sutcliffe, M. J., Lian, L. Y., Primrose, W. U., Wolf, C. R., Roberts, G. C. 1996. A model for human cytochrome P450 2D6 based on homology modeling and NMR studies of substrate binding. *Biochemistry* 35:4540–4550.

Modi, S., Gilham, D. E., Sutcliffe, M. J., Lian, L. Y., Primrose, W. U., Wolf, C. R., Roberts, G. C. 1997. 1-methyl-4-phenyl-1,2,3,6-tetrahydropyridine as a substrate of cytochrome P450 2D6: Allosteric effects of NADPH-cytochrome P450 reductase. *Biochemistry* 36:4461–4470.

Moen, M. D., McKeage, K., Plosker, G. L., Siddiqui, M. A. 2007. Imatinib: A review of its use in chronic myeloid leukaemia. *Drugs* 67:299–320.

Molden, E., Asberg, A., Christensen, H. 2000. CYP2D6 is involved in *O*-demethylation of diltiazem. An *in vitro* study with transfected human liver cells. *Eur J Clin Pharmacol* 56:575–579.

Molden, E., Asberg, A., Christensen, H. 2002a. Desacetyl-diltiazem displays severalfold higher affinity to CYP2D6 compared with CYP3A4. *Drug Metab Dispos* 30:1–3.

Molden, E., Johansen, P. W., Boe, G. H., Bergan, S., Christensen, H., Rugstad, H. E., Rootwelt, H., Reubsaet, L., Lehne, G. 2002b. Pharmacokinetics of diltiazem and its metabolites in relation to CYP2D6 genotype. *Clin Pharmacol Ther* 72:333–342.

Molden, E., Lunde, H., Lunder, N., Refsum, H. 2006. Pharmacokinetic variability of aripiprazole and the active metabolite dehydroaripiprazole in psychiatric patients. *Ther Drug Monit* 28:744–749.

Moller, H. J. 2005. Risperidone: A review. *Expert Opin Pharmacother* 6:803–818.

Mongey, A. B., Hess, E. 2001. In vitro production of antibodies to histones in patients receiving chronic procainamide therapy. *J Rheumatol* 28:1992–1998.

Moreland, R. B., Goldstein, I. I., Kim, N. N., Traish, A. 1999. Sildenafil citrate, a selective phosphodiesterase type 5 inhibitor. *Trends Endocrinol Metab* 10:97–104.

Moretti, A., Arcari, G., Pegrassi, L. 1979. [A review of pharmacological studies on nicergoline]. *Arzneimittelforschung* 29:1223–1227.

Mori, T., Takase, H., Toide, K., Hirano, T., Kambe, T., Nakayama, N., Schwartz, A. 2001. Pranidipine, a 1,4-dihydropyridine calcium channel blocker that enhances nitric oxide-induced vascular relaxation. *Cardiovasc Drug Rev* 19:1–8.

Mori, A., Maruo, Y., Iwai, M., Sato, H., Takeuchi, Y. 2005. UDP-glucuronosyltransferase 1A4 polymorphisms in a Japanese population and kinetics of clozapine glucuronidation. *Drug Metab Dispos* 33:672–675.

Mount, D. L., Patchen, L. C., Nguyen-Dinh, P., Barber, A. M., Schwartz, I. K., Churchill, F. C. 1986. Sensitive analysis of blood for amodiaquine and three metabolites by high-performance liquid chromatography with electrochemical detection. *J Chromatogr* 383:375–386.

Mueller, R. A., Baur, H. R. 1986. Flecainide: A new antiarrhythmic drug. *Clin Cardiol* 9:1–5.

Mukai, E., Ishida, H., Horie, M., Noma, A., Seino, Y., Takano, M. 1998. The antiarrhythmic agent cibenzoline inhibits KATP channels by binding to Kir6.2. *Biochem Biophys Res Commun* 251:477–481.

Muller, B. A. 2009. Imatinib and its successors—How modern chemistry has changed drug development. *Curr Pharm Des* 15:120–133.

Muller, N., Schennach, R., Riedel, M., Moller, H. J. 2008. Duloxetine in the treatment of major psychiatric and neuropathic disorders. *Expert Rev Neurother* 8:527–536.

Muralidharan, G., Cooper, J. K., Hawes, E. M., Korchinski, E. D., Midha, K. K. 1996. Quinidine inhibits the 7-hydroxylation of chlorpromazine in extensive metabolisers of debrisoquine. *Eur J Clin Pharmacol* 50:121–128.

Murphy, P. J. 1974. Metabolic pathways of aprindine. *Acta Cardiol Suppl* 18:131–142.

Murphy, C., Fotsis, T., Pantzar, P., Adlercreutz, H., Martin, F. 1987. Analysis of tamoxifen and its metabolites in human plasma by gas chromatography-mass spectrometry (GC-MS) using selected ion monitoring (SIM). *J Steroid Biochem* 26:547–555.

Musial, A., Bajda, M., Malawska, B. 2007. Recent developments in cholinesterases inhibitors for Alzheimer's disease treatment. *Curr Med Chem* 14:2654–2679.

Musshoff, F., Schmidt, P., Madea, B. 1999. Fatality caused by a combined trimipramine-citalopram intoxication. *Forensic Sci Int* 106:125–131.

Mutlib, A. E., Klein, J. T. 1998. Application of liquid chromatography/mass spectrometry in accelerating the identification of human liver cytochrome P450 isoforms involved in the metabolism of iloperidone. *J Pharmacol Exp Ther* 286:1285–1293.

Mutlib, A. E., Strupczewski, J. T., Chesson, S. M. 1995. Application of hyphenated LC/NMR and LC/MS techniques in rapid identification of *in vitro* and *in vivo* metabolites of iloperidone. *Drug Metab Dispos* 23:951–964.

Muzeeb, S., Basha, S. J., Shashikumar, D., Mullangi, R., Srinivas, N. R. 2006. Glucuronidation of DRF-6574, hydroxy metabolite of DRF-4367 (a novel COX-2 inhibitor) by pooled human liver, intestinal microsomes and recombinant human UDP-glucuronosyltransferases (UGT): Role of UGT1A1, 1A3 and 1A8. *Eur J Drug Metab Pharmacokinet* 31:299–309.

Myers, T. G., Thummel, K. E., Kalhorn, T. F., Nelson, S. D. 1994. Preferred orientations in the binding of 4'-hydroxyacetanilide (acetaminophen) to cytochrome P450 1A1 and 2B1 isoforms as determined by 13C- and 15N-NMR relaxation studies. *J Med Chem* 37:860–867.

Naccarelli, G. V., Wolbrette, D. L., Samii, S., Banchs, J. E., Penny-Peterson, E., Stevenson, R., Gonzalez, M. D. 2008. Vernakalant: Pharmacology electrophysiology, safety and efficacy. *Drugs Today (Barc)* 44:325–329.

Nakajima, M., Inoue, T., Shimada, N., Tokudome, S., Yamamoto, T., Kuroiwa, Y. 1998a. Cytochrome P450 2C9 catalyzes indomethacin O-demethylation in human liver microsomes. *Drug Metab Dispos* 26:261–266.

Nakajima, M., Kobayashi, K., Shimada, N., Tokudome, S., Yamamoto, T., Kuroiwa, Y. 1998b. Involvement of CYP1A2 in mexiletine metabolism. *Br J Clin Pharmacol* 46:55–62.

Nakajima, M., Nakamura, S., Tokudome, S., Shimada, N., Yamazaki, H., Yokoi, T. 1999. Azelastine N-demethylation by cytochrome P-450 (CYP)3A4, CYP2D6, and CYP1A2 in human liver microsomes: Evaluation of approach to predict the contribution of multiple CYPs. *Drug Metab Dispos* 27:1381–1391.

Nakajima, M., Tanaka, E., Kobayashi, T., Ohashi, N., Kume, T., Yokoi, T. 2002. Imipramine N-glucuronidation in human liver microsomes: Biphasic kinetics and characterization of UDP-glucuronosyltransferase isoforms. *Drug Metab Dispos* 30:636–642.

Nakamura, K., Yokoi, T., Inoue, K., Shimada, N., Ohashi, N., Kume, T., Kamataki, T. 1996. CYP2D6 is the principal cytochrome P450 responsible for metabolism of the histamine H1 antagonist promethazine in human liver microsomes. *Pharmacogenetics* 6:449–457.

Nakamura, K., Yokoi, T., Kodama, T., Inoue, K., Nagashima, K., Shimada, N., Shimizu, T., Kamataki, T. 1998. Oxidation of histamine H1 antagonist mequitazine is catalyzed by cytochrome P450 2D6 in human liver microsomes. *J Pharmacol Exp Ther* 284:437–442.

Nakamura, K., Ariyoshi, N., Iwatsubo, T., Fukunaga, Y., Higuchi, S., Itoh, K., Shimada, N. et al. 2005. Inhibitory effects of nicardipine to cytochrome P450 (CYP) in human liver microsomes. *Biol Pharm Bull* 28:882–885.

Narimatsu, S., Kariya, S., Isozaki, S., Ohmori, S., Kitada, M., Hosokawa, S., Masubuchi, Y., Suzuki, T. 1993. Involvement of CYP2D6 in oxidative metabolism of cinnarizine and flunarizine in human liver microsomes. *Biochem Biophys Res Commun* 193:1262–1268.

Narimatsu, S., Masubuchi, Y., Hosokawa, S., Ohmori, S., Kitada, M., Suzuki, T. 1994. Involvement of a cytochrome P4502D subfamily in human liver microsomal bunitrolol 4-hydroxylation. *Biol Pharm Bull* 17:803–807.

Narimatsu, S., Tachibana, M., Masubuchi, Y., Suzuki, T. 1996. Cytochrome P4502D and -2C enzymes catalyze the oxidative N-demethylation of the parkinsonism-inducing substance 1-methyl-4-phenyl-1,2,3,6-tetrahydropyridine in rat liver microsomes. *Chem Res Toxicol* 9:93–98.

Narimatsu, S., Kato, R. at al. 1999. Enantioselectivity of bunitrolol 4-hydroxylation is reversed by the change of an amino acid residue from valine to methionine at position 374 of cytochrome P450-2D6. *Chirality* 11:1–9.

Narimatsu, S., Takemi, C., Tsuzuki, D., Kataoka, H., Yamamoto, S., Shimada, N., Suzuki, S., Satoh, T., Meyer, U. A., Gonzalez, F. J. 2002. Stereoselective metabolism of bufuralol racemate and enantiomers in human liver microsomes. *J Pharmacol Exp Ther* 303:172–178.

Narimatsu, S., Yonemoto, R., Saito, K., Takaya, K., Kumamoto, T., Ishikawa, T., Asanuma, M. et al. 2006. Oxidative metabolism of 5-methoxy-N,N-diisopropyltryptamine (Foxy) by human liver microsomes and recombinant cytochrome P450 enzymes. *Biochem Pharmacol* 71:1377–1385.

Narimatsu, S., Yonemoto, R., Masuda, K., Katsu, T., Asanuma, M., Kamata, T., Katagi, M. et al. 2008. Oxidation of 5-methoxy-N,N-diisopropyltryptamine in rat liver microsomes and recombinant cytochrome P450 enzymes. *Biochem Pharmacol* 75:752–760.

Natsui, K., Mizuno, Y., Tani, N., Yabuki, M., Komuro, S. 2007. Identification of CYP3A4 as the primary cytochrome P450 responsible for the metabolism of tandospirone by human liver microsomes. *Eur J Drug Metab Pharmacokinet* 32:233–240.

Neugebauer, G., Akpan, W., Kaufmann, B., Reiff, K. 1990. Stereoselective disposition of carvedilol in man after intravenous and oral administration of the racemic compound. *Eur J Clin Pharmacol* 38 Suppl 2:S108–S111.

Nielsen, K. K., Flinois, J. P., Beaune, P., Brosen, K. 1996. The biotransformation of clomipramine *in vitro*, identification of the cytochrome P450s responsible for the separate metabolic pathways. *J Pharmacol Exp Ther* 277:1659–1664.

Nielsen, T. L., Rasmussen, B. B., Flinois, J. P., Beaune, P., Brosen, K. 1999. *In vitro* metabolism of quinidine: The (3S)-3-hydroxylation of quinidine is a specific marker reaction for cytochrome P-4503A4 activity in human liver microsomes. *J Pharmacol Exp Ther* 289:31–37.

Niitsu, Y., Jakubowski, J. A., Sugidachi, A., Asai, F. 2005. Pharmacology of CS-747 (prasugrel, LY640315), a novel, potent antiplatelet agent with in vivo P2Y12 receptor antagonist activity. *Semin Thromb Hemost* 31:184–194.

Nishida, H., Hirai, H., Emoto, C., Iwasaki, K. 2007. Metabolism of CJ-036878, N-(3-phenethoxybenzyl)-4-hydroxybenzamide, in liver microsomes and recombinant cytochrome P450 enzymes: Metabolite identification by LC-UV/MS(n) and ¹H-NMR. *Xenobiotica* 37:1394–1407.

Nishimura, M., Yaguti, H., Yoshitsugu, H., Naito, S., Satoh, T. 2003. Tissue distribution of mRNA expression of human cytochrome P450 isoforms assessed by high-sensitivity real-time reverse transcription PCR. *Yakugaku Zasshi* 123:369–375.

Nitsch, J., Neyses, L., Kohler, U., Luderitz, B. 1987. [Elevated plasma flecainide concentrations in heart failure]. *Dtsch Med Wochenschr* 112:1698–1700.

Niwa, T., Yabusaki, Y., Honma, K., Matsuo, N., Tatsuta, K., Ishibashi, F., Katagiri, M. 1998. Contribution of human hepatic cytochrome P450 isoforms to regioselective hydroxylation of steroid hormones. *Xenobiotica* 28:539–547.

Niwa, T., Shiraga, T., Mitani, Y., Terakawa, M., Tokuma, Y., Kagayama, A. 2000. Stereoselective metabolism of cibenzoline, an antiarrhythmic drug, by human and rat liver microsomes: Possible involvement of CYP2D and CYP3A. *Drug Metab Dispos* 28:1128–1134.

Noble, S., Goa, K. L. 1996. Ticlopidine. A review of its pharmacology, clinical efficacy and tolerability in the prevention of cerebral ischaemia and stroke. *Drugs Aging* 8:214–232.

Nomura, A., Sakurai, E., Hikichi, N. 1997. Stereoselective N-demethylation of chlorpheniramine by rat-liver microsomes and the involvement of cytochrome P450 isozymes. *J Pharm Pharmacol* 49:257–262.

Nordin, C., Bertilsson, L. 1995. Active hydroxymetabolites of antidepressants. Emphasis on E-10-hydroxy-nortriptyline. *Clin Pharmacokinet* 28:26–40.

Normann, C., Lieb, K., Walden, J. 2002. Increased plasma concentration of maprotiline by coadministration of risperidone. *J Clin Psychopharmacol* 22:92–94.

Nosten, F., ter Kuile, F. O., Luxemburger, C., Woodrow, C., Kyle, D. E., Chongsuphajaisiddhi, T., White, N. J. 1993. Cardiac effects of antimalarial treatment with halofantrine. *Lancet* 341: 1054–1056.

O'Dwyer, M. E., Druker, B. J. 2000. STI571: An inhibitor of the BCR-ABL tyrosine kinase for the treatment of chronic myelogenous leukaemia. *Lancet Oncol* 1:207–211.

Oates, N. S., Shah, R. R., Idle, J. R., Smith, R. L. 1982. Genetic polymorphism of phenformin 4-hydroxylation. *Clin Pharmacol Ther* 32:81–89.

Obach, R. S. 2000. Metabolism of ezlopitant, a nonpeptidic substance P receptor antagonist, in liver microsomes: Enzyme kinetics, cytochrome P450 isoform identity, and in vitro-in vivo correlation. *Drug Metab Dispos* 28:1069–1076.

Obach, R. S. 2001. Cytochrome P450-catalyzed metabolism of ezlopitant alkene (CJ-12,458), a pharmacologically active metabolite of ezlopitant: Enzyme kinetics and mechanism of an alkene hydration reaction. *Drug Metab Dispos* 29:1057–1067.

Obach, R. S., Pablo, J., Mash, D. C. 1998. Cytochrome P450 2D6 catalyzes the O-demethylation of the psychoactive alkaloid ibogaine to 12-hydroxyibogamine. *Drug Metab Dispos* 26:764–768.

Obach, R. S., Cox, L. M., Tremaine, L. M. 2005. Sertraline is metabolized by multiple cytochrome P450 enzymes, monoamine oxidases, and glucuronyl transferases in human: An in vitro study. *Drug Metab Dispos* 33:262–270.

Obach, R. S., Margolis, J. M., Logman, M. J. 2007. *In vitro* metabolism of CP-122,721 ((2S,3S)-2-phenyl-3-[(5-trifluoromethoxy-2-methoxy)benzylamino]piperidine), a non-peptide antagonist of the substance P receptor. *Drug Metab Pharmacokinet* 22:336–349.

Oda, Y., Kharasch, E. D. 2001. Metabolism of methadone and levo-α-acetylmethadol (LAAM) by human intestinal cytochrome P450 3A4 (CYP3A4): Potential contribution of intestinal metabolism to presystemic clearance and bioactivation. *J Pharmacol Exp Ther* 298:1021–1032.

Ogura, K., Ishikawa, Y., Kaku, T., Nishiyama, T., Ohnuma, T., Muro, K., Hiratsuka, A. 2006. Quaternary ammonium-linked glucuronidation of trans-4-hydroxytamoxifen, an active metabolite of tamoxifen, by human liver microsomes and UDP-glucuronosyltransferase 1A4. *Biochem Pharmacol* 71: 1358–1369.

Ohno, S., Kawana, K., Nakajin, S. 2008. Contribution of UDP-glucuronosyltransferase 1A1 and 1A8 to morphine-6-glucuronidation and its kinetic properties. *Drug Metab Dispos* 36:688–694.

Ohta, S., Tachikawa, O., Makino, Y., Tasaki, Y., Hirobe, M. 1990. Metabolism and brain accumulation of tetrahydroisoquinoline (TIQ) a possible parkinsonism inducing substance, in an animal model of a poor debrisoquine metabolizer. *Life Sci* 46:599–605.

Oldham, H. G., Clarke, S. E. 1997. In vitro identification of the human cytochrome P450 enzymes involved in the metabolism of R(+)- and S(–)-carvedilol. *Drug Metab Dispos* 25:970–977.

Olesen, O. V., Linnet, K. 1997. Hydroxylation and demethylation of the tricyclic antidepressant nortriptyline by cDNA-expressed human cytochrome P-450 isozymes. *Drug Metab Dispos* 25:740–744.

Olesen, O. V., Linnet, K. 1999. Studies on the stereoselective metabolism of citalopram by human liver microsomes and cDNA-expressed cytochrome P450 enzymes. *Pharmacology* 59:298–309.

Olesen, O. V., Linnet, K. 2000. Identification of the human cytochrome P450 isoforms mediating in vitro N-dealkylation of perphenazine. *Br J Clin Pharmacol* 50:563–571.

Olesen, O. V., Linnet, K. 2001. Contributions of five human cytochrome P450 isoforms to the N-demethylation of clozapine in vitro at low and high concentrations. *J Clin Pharmacol* 41:823–832.

Olsson, B., Szamosi, J. 2001a. Food does not influence the pharmacokinetics of a new extended release formulation of tolterodine for once daily treatment of patients with overactive bladder. *Clin Pharmacokinet* 40:135–143.

Olsson, B., Szamosi, J. 2001b. Multiple dose pharmacokinetics of a new once daily extended release tolterodine formulation versus immediate release tolterodine. *Clin Pharmacokinet* 40:227–235.

Onderwater, R. C., Venhorst, J., Commandeur, J. N., Vermeulen, N. P. 1999. Design, synthesis, and characterization of 7-methoxy-4-(aminomethyl)coumarin as a novel and selective cytochrome P450 2D6 substrate suitable for high-throughput screening. *Chem Res Toxicol* 12:555–559.

Ono, S., Tsutsui, M., Gonzalez, F. J., Satoh, T., Masubuchi, Y., Horie, T., Suzuki, T., Narimatsu, S. 1995. Oxidative metabolism of bunitrolol by complementary DNA-expressed human cytochrome P450 isozymes in a human hepatoma cell line (Hep G2) using recombinant vaccinia virus. *Pharmacogenetics* 5:97–102.

Ono, S., Hatanaka, T., Miyazawa, S., Tsutsui, M., Aoyama, T., Gonzalez, F. J., Satoh, T. 1996. Human liver microsomal diazepam metabolism using cDNA-expressed cytochrome P450s: Role of CYP2B6, 2C19 and the 3A subfamily. *Xenobiotica* 26:1155–1166.

Oppenheimer, J. J., Casale, T. B. 2002. Next generation antihistamines: Therapeutic rationale, accomplishments and advances. *Expert Opin Investig Drugs* 11:807–817.

Osborne, C. K., Wiebe, V. J., McGuire, W. L., Ciocca, D. R., DeGregorio, M. W. 1992. Tamoxifen and the isomers of 4-hydroxytamoxifen in tamoxifen-resistant tumors from breast cancer patients. *J Clin Oncol* 10:304–310.

Osborne, M. R., Hewer, A., Hardcastle, I. R., Carmichael, P. L., Phillips, D. H. 1996. Identification of the major tamoxifen-deoxyguanosine adduct formed in the liver DNA of rats treated with tamoxifen. *Cancer Res* 56:66–71.

Osikowska-Evers, B., Dayer, P., Meyer, U. A., Robertz, G. M., Eichelbaum, M. 1987. Evidence for altered catalytic properties of the cytochrome P-450 involved in sparteine oxidation in poor metabolizers. *Clin Pharmacol Ther* 41:320–325.

Ott, J. 2001. Pharmanopo-psychonautics: Human intranasal, sublingual, intrarectal, pulmonary and oral pharmacology of bufotenine. *J Psychoactive Drugs* 33:273–281.

Otton, S. V., Crewe, H. K., Lennard, M. S., Tucker, G. T., Woods, H. F. 1988. Use of quinidine inhibition to define the role of the sparteine/debrisoquine cytochrome P450 in metoprolol oxidation by human liver microsomes. *J Pharmacol Exp Ther* 247:242–247.

Otton, S. V., Gillam, E. M., Lennard, M. S., Tucker, G. T., Woods, H. F. 1990. Propranolol oxidation by human liver microsomes—The use of cumene hydroperoxide to probe isoenzyme specificity and regio- and stereoselectivity. *Br J Clin Pharmacol* 30:751–760.

Otton, S. V., Ball, S. E., Cheung, S. W., Inaba, T., Rudolph, R. L., Sellers, E. M. 1996. Venlafaxine oxidation *in vitro* is catalysed by CYP2D6. *Br J Clin Pharmacol* 41:149–156.

Paar, W. D., Poche, S., Gerloff, J., Dengler, H. J. 1997. Polymorphic CYP2D6 mediates *O*-demethylation of the opioid analgesic tramadol. *Eur J Clin Pharmacol* 53:235–239.

Paleacu, D. 2007. Tetrabenazine in the treatment of Huntington's disease. *Neuropsychiatr Dis Treat* 3:545–551.

Pan, L. P., De Vriendt, C., Belpaire, F. M. 1998. In-vitro characterization of the cytochrome P450 isoenzymes involved in the back oxidation and N-dealkylation of reduced haloperidol. *Pharmacogenetics* 8:383–389.

Parker, R. J., Collins, J. M., Strong, J. M. 1996. Identification of 2,6-xylidine as a major lidocaine metabolite in human liver slices. *Drug Metab Dispos* 24:1167–1173.

Parte, P., Kupfer, D. 2005. Oxidation of tamoxifen by human flavin-containing monooxygenase (FMO) 1 and FMO3 to tamoxifen-N-oxide and its novel reduction back to tamoxifen by human cytochromes P450 and hemoglobin. *Drug Metab Dispos* 33:1446–1452.

Pascual, J. 2003. Almotriptan: An effective and well-tolerated treatment for migraine pain. *Drugs Today (Barc)* 39 Suppl D:31–36.

Patel, D. K., Notarianni, L. J., Bennett, P. N. 1990. Comparative metabolism of high doses of aspirin in man and rat. *Xenobiotica* 20:847–854.

Patel, M., Tang, B. K., Kalow, W. 1992. Variability of acetaminophen metabolism in Caucasians and Orientals. *Pharmacogenetics* 2:38–45.

Pathak, D. N., Pongracz, K., Bodell, W. J. 1996. Activation of 4-hydroxytamoxifen and the tamoxifen derivative metabolite E by uterine peroxidase to form DNA adducts: Comparison with DNA adducts formed in the uterus of Sprague-Dawley rats treated with tamoxifen. *Carcinogenesis* 17:1785–1790.

Patki, K. C., Von Moltke, L. L., Greenblatt, D. J. 2003. In vitro metabolism of midazolam, triazolam, nifedipine, and testosterone by human liver microsomes and recombinant cytochromes p450: Role of cyp3a4 and cyp3a5. *Drug Metab Dispos* 31:938–944.

Patterson, L. H., Hall, G., Nijjar, B. S., Khatra, P. K., Cowan, D. A. 1986. In-vitro metabolism of lignocaine to its N-oxide. *J Pharm Pharmacol* 38:326.

Pearce, E. F., Murphy, J. A. 2014. Vortioxetine for the treatment of depression. *Ann Pharmacother* 48:758–765.

Pearce, R. E., Rodrigues, A. D., Goldstein, J. A., Parkinson, A. 1996. Identification of the human P450 enzymes involved in lansoprazole metabolism. *J Pharmacol Exp Ther* 277:805–816.

Peets, E. A., Weinstein, R., Billard, W., Symchowicz, S. 1972. The metabolism of chlorpheniramine maleate in the dog and rat. *Arch Int Pharmacodyn Ther* 199:172–190.

Perucca, E., Gatti, G., Spina, E. 1994. Clinical pharmacokinetics of fluvoxamine. *Clin Pharmacokinet* 27:175–190.

Peters, F. T., Meyer, M. R., Theobald, D. S., Maurer, H. H. 2008. Identification of cytochrome P450 enzymes involved in the metabolism of the new designer drug 4'-methyl-α-pyrrolidinobutyrophenone. *Drug Metab Dispos* 36:163–168.

Phillips, D. H., Carmichael, P. L., Hewer, A., Cole, K. J., Poon, G. K. 1994. alpha-Hydroxytamoxifen, a metabolite of tamoxifen with exceptionally high DNA-binding activity in rat hepatocytes. *Cancer Res* 54:5518–5522.

Pichard, L., Gillet, G., Bonfils, C., Domergue, J., Thenot, J. P., Maurel, P. 1995. Oxidative metabolism of zolpidem by human liver cytochrome P450s. *Drug Metab Dispos* 23:1253–1262.

Piergies, A. A., Sweet, J., Johnson, M., Roth-Schechter, B. F., Allard, S. 1996. The effect of co-administration of zolpidem with fluoxetine: Pharmacokinetics and pharmacodynamics. *Int J Clin Pharmacol Ther* 34:178–183.

Pinder, R. M., Van Delft, A. M. 1983. The potential therapeutic role of the enantiomers and metabolites of mianserin. *Br J Clin Pharmacol* 15 Suppl 2:269S–276S.

Pinder, R. M., Brogden, R. N., Speight, T. M., Avery, G. S. 1977a. Doxepin up-to-date: A review of its pharmacological properties and therapeutic efficacy with particular reference to depression. *Drugs* 13:161–218.

Pinder, R. M., Brogden, R. N., Speight, T. M., Avery, G. S. 1977b. Maprotiline: A review of its pharmacological properties and therapeutic efficacy in mental depressive states. *Drugs* 13: 321–352.

Pirmohamed, M., Kitteringham, N. R., Guenthner, T. M., Breckenridge, A. M., Park, B. K. 1992. An investigation of the formation of cytotoxic, protein-reactive and stable metabolites from carbamazepine in vitro. *Biochem Pharmacol* 43:1675–1682.

Pivonka, J., Segelman, F. H., Hartman, C. A., Segl, W. E., Kucharczyk, N., Sofia, R. D. 1987. Determination of azelastine and desmethylazelastine in human plasma by high-performance liquid chromatography. *J Chromatogr* 420:89–98.

Pool, W. F., Reily, M. D., Bjorge, S. M., Woolf, T. F. 1997. Metabolic disposition of the cognition activator tacrine in rats, dogs, and humans. Species comparisons. *Drug Metab Dispos* 25:590–597.

Poon, G. K., Chui, Y. C., McCague, R., Llnning, P. E., Feng, R., Rowlands, M. G., Jarman, M. 1993. Analysis of phase I and phase II metabolites of tamoxifen in breast cancer patients. *Drug Metab Dispos* 21:1119–1124.

Poon, L. H., Kang, G. A., Lee, A. J. 2010. Role of tetrabenazine for Huntington's disease-associated chorea. *Ann Pharmacother* 44:1080–1089.

Popik, P., Layer, R. T., Skolnick, P. 1995. 100 years of ibogaine: Neurochemical and pharmacological actions of a putative antiaddictive drug. *Pharmacol Rev* 47:235–253.

Postlind, H., Danielson, A., Lindgren, A., Andersson, S. H. 1998. Tolterodine, a new muscarinic receptor antagonist, is metabolized by cytochromes P450 2D6 and 3A in human liver microsomes. *Drug Metab Dispos* 26:289–293.

Potter, W. Z., Davis, D. C., Mitchell, J. R., Jollow, D. J., Gillette, J. R., Brodie, B. B. 1973. Acetaminophen-induced hepatic necrosis. 3. Cytochrome P-450-mediated covalent binding in vitro. *J Pharmacol Exp Ther* 187:203–210.

Potter, W. Z., Thorgeirsson, S. S., Jollow, D. J., Mitchell, J. R. 1974. Acetaminophen-induced hepatic necrosis. V. Correlation of hepatic necrosis, covalent binding and glutathione depletion in hamsters. *Pharmacology* 12:129–143.

Poulos, T. L., Finzel, B. C., Gunsalus, I. C., Wagner, G. C., Kraut, J. 1985. The 2.6-A crystal structure of Pseudomonas putida cytochrome P-450. *J Biol Chem* 260:16122–16130.

Prakash, C., O'Donnell, J., Khojasteh-Bakht, S. C. 2007. Metabolism, pharmacokinetics, and excretion of a nonpeptidic substance P receptor antagonist, ezlopitant, in normal healthy male volunteers: Characterization of polar metabolites by chemical derivatization with dansyl chloride. *Drug Metab Dispos* 35:1071–1080.

Prakash, C., Johnson, K. A., Gardner, M. J. 2008. Disposition of lasofoxifene, a next-generation selective estrogen receptor modulator, in healthy male subjects. *Drug Metab Dispos* 36:1218–1226.

Preiss, R., Schmidt, R., Baumann, F., Hanschmann, H., Hauss, J., Geissler, F., Pahlig, H., Ratzewiss, B. 2002. Measurement of 4-hydroxylation of ifosfamide in human liver microsomes using the estimation of free and protein-bound acrolein and codetermination of keto- and carboxyifosfamide. *J Cancer Res Clin Oncol* 128:385–392.

Prescott, L. F. 2000. Paracetamol: Past, present, and future. *Am J Ther* 7:143–147.

Preskorn, S. H., Greenblatt, D. J., Flockhart, D., Luo, Y., Perloff, E. S., Harmatz, J. S., Baker, B., Klick-Davis, A., Desta, Z., Burt, T. 2007. Comparison of duloxetine, escitalopram, and sertraline effects on cytochrome P450 2D6 function in healthy volunteers. *J Clin Psychopharmacol* 27:28–34.

Projean, D., Baune, B., Farinotti, R., Flinois, J. P., Beaune, P., Taburet, A. M., Ducharme, J. 2003. In vitro metabolism of chloroquine: Identification of CYP2C8, CYP3A4, and CYP2D6 as the main isoforms catalyzing N-desethylchloroquine formation. *Drug Metab Dispos* 31:748–754.

Prueksaritanont, T., Dwyer, L. M., Cribb, A. E. 1995. (+)-bufuralol 1'-hydroxylation activity in human and rhesus monkey intestine and liver. *Biochem Pharmacol* 50:1521–1525.

Prueksaritanont, T., Gorham, L. M., Ma, B., Liu, L., Yu, X., Zhao, J. J., Slaughter, D. E., Arison, B. H., Vyas, K. P. 1997. In vitro metabolism of simvastatin in humans [SBT]identification of metabolizing enzymes and effect of the drug on hepatic P450s. *Drug Metab Dispos* 25:1191–1199.

Prueksaritanont, T., Ma, B., Yu, N. 2003. The human hepatic metabolism of simvastatin hydroxy acid is mediated primarily by CYP3A, and not CYP2D6. *Br J Clin Pharmacol* 56:120–124.

Prueksaritanont, T., Lu, P., Gorham, L., Sternfeld, F., Vyas, K. P. 2000. Interspecies comparison and role of human cytochrome P450 and flavin-containing monooxygenase in hepatic metabolism of L-775,606, a potent 5-HT(1D) receptor agonist. *Xenobiotica* 30:47–59.

Pumford, N. R., Hinson, J. A., Benson, R. W., Roberts, D. W. 1990. Immunoblot analysis of protein containing 3-(cystein-S-yl) acetaminophen adducts in serum and subcellular liver fractions from acetaminophen-treated mice. *Toxicol Appl Pharmacol* 104:521–532.

Quinn, M. J., Fitzgerald, D. J. 1999. Ticlopidine and clopidogrel. *Circulation* 100:1667–1672.

Raghuram, T. C., Koshakji, R. P., Wilkinson, G. R., Wood, A. J. 1984. Polymorphic ability to metabolize propranolol alters 4-hydroxypropranolol levels but not beta blockade. *Clin Pharmacol Ther* 36:51–56.

Rahman, A. F., Korashy, H. M., Kassem, M. G. 2014. Gefitinib. *Profiles Drug Subst Excip Relat Methodol* 39:239–264.

Ramanathan, R., Reyderman, L., Kulmatycki, K., Su, A. D., Alvarez, N., Chowdhury, S. K., Alton, K. B. et al. 2007a. Disposition of loratadine in healthy volunteers. *Xenobiotica* 37:753–769.

Ramanathan, R., Reyderman, L., Su, A. D., Alvarez, N., Chowdhury, S. K., Alton, K. B., Wirth, M. A., Clement, R. P., Statkevich, P., Patrick, J. E. 2007b. Disposition of desloratadine in healthy volunteers. *Xenobiotica* 37:770–787.

Rao, N. 2007. The clinical pharmacokinetics of escitalopram. *Clin Pharmacokinet* 46:281–290.

Rashid, Q., Klein, R. 2014. Tiotropium in the treatment of patients with asthma. *South Med J* 107:330–337.

Raskind, M. A., Peskind, E. R., Wessel, T., Yuan, W. 2000. Galantamine in AD: A 6-month randomized, placebo-controlled trial with a 6-month extension. The Galantamine USA-1 Study Group. *Neurology* 54:2261–2268.

Reed-Hagen, A. E., Tsuchiya, M., Shimada, K., Wentland, J. A., Obach, R. S. 1999. Pharmacokinetics of ezlopitant, a novel non-peptidic neurokinin-1 receptor antagonist in preclinical species and metabolite kinetics of the pharmacologically active metabolites. *Biopharm Drug Dispos* 20:429–439.

Rehmel, J. L., Eckstein, J. A., Farid, N. A., Heim, J. B., Kasper, S. C., Kurihara, A., Wrighton, S. A., Ring, B. J. 2006. Interactions of two major metabolites of prasugrel, a thienopyridine antiplatelet agent, with the cytochromes P450. *Drug Metab Dispos* 34:600–607.

Reith, M. K., Sproles, G. D., Cheng, L. K. 1995. Human metabolism of dolasetron mesylate, a 5-HT3 receptor antagonist. *Drug Metab Dispos* 23:806–812.

Ren, S., Yang, J. S., Kalhorn, T. F., Slattery, J. T. 1997. Oxidation of cyclophosphamide to 4-hydroxycyclophosphamide and deschloroethylcyclophosphamide in human liver microsomes. *Cancer Res* 57:4229–4235.

Rettie, A. E., Eddy, A. C., Heimark, L. D., Gibaldi, M., Trager, W. F. 1989. Characteristics of warfarin hydroxylation catalyzed by human liver microsomes. *Drug Metab Dispos* 17:265–270.

Richards, D. M., Brogden, R. N., Heel, R. C., Speight, T. M., Avery, G. S. 1984. Oxatomide. A review of its pharmacodynamic properties and therapeutic efficacy. *Drugs* 27:210–231.

Richelson, E., Nelson, A. 1984. Antagonism by antidepressants of neurotransmitter receptors of normal human brain in vitro. *J Pharmacol Exp Ther* 230:94–102.

Riley, R. J., Roberts, P., Kitteringham, N. R., Park, B. K. 1990. Formation of cytotoxic metabolites from phenytoin, imipramine, desipramine, amitriptyline and mianserin by mouse and human hepatic microsomes. *Biochem Pharmacol* 39:1951–1958.

Riley, A. B., Tafreshi, M. J., Haber, S. L. 2008. Prasugrel: A novel antiplatelet agent. *Am J Health Syst Pharm* 65:1019–1028.

Ring, B. J., Catlow, J., Lindsay, T. J., Gillespie, T., Roskos, L. K., Cerimele, B. J., Swanson, S. P., Hamman, M. A., Wrighton, S. A. 1996. Identification of the human cytochromes P450 responsible for the in vitro formation of the major oxidative metabolites of the antipsychotic agent olanzapine. *J Pharmacol Exp Ther* 276:658–666.

Ring, B. J., Eckstein, J. A., Gillespie, J. S., Binkley, S. N., VandenBranden, M., Wrighton, S. A. 2001. Identification of the human cytochromes p450 responsible for in vitro formation of R- and S-norfluoxetine. *J Pharmacol Exp Ther* 297:1044–1050.

Ring, B. J., Gillespie, J. S., Eckstein, J. A., Wrighton, S. A. 2002. Identification of the human cytochromes P450 responsible for atomoxetine metabolism. *Drug Metab Dispos* 30:319–323.

Ring, B. J., Patterson, B. E., Mitchell, M. I., Vandenbranden, M., Gillespie, J., Bedding, A. W., Jewell, H. et al. 2005. Effect of tadalafil on cytochrome P450 3A4-mediated clearance: Studies *in vitro* and *in vivo*. *Clin Pharmacol Ther* 77:63–75.

Rintelen, B., Andel, I., Sautner, J., Leeb, B. F. 2006. Leflunomide/chloroquin combination therapy in rheumatoid arthritis: A pilot study. *Clin Rheumatol* 25:557–559.

Riska, P., Lamson, M., MacGregor, T., Sabo, J., Hattox, S., Pav, J., Keirns, J. 1999. Disposition and biotransformation of the antiretroviral drug nevirapine in humans. *Drug Metab Dispos* 27:895–901.

Rochat, B., Amey, M., Baumann, P. 1995. Analysis of enantiomers of citalopram and its demethylated metabolites in plasma of depressive patients using chiral reverse-phase liquid chromatography. *Ther Drug Monit* 17:273–279.

Rochat, B., Amey, M., Gillet, M., Meyer, U. A., Baumann, P. 1997. Identification of three cytochrome P450 isozymes involved in N-demethylation of citalopram enantiomers in human liver microsomes. *Pharmacogenetics* 7:1–10.

Roden, D. M., Duff, H. J., Altenbern, D., Woosley, R. L. 1982. Antiarrhythmic activity of the O-demethyl metabolite of encainide. *J Pharmacol Exp Ther* 221:552–557.

Rodrigo, G. J., Castro-Rodriguez, J. A. 2015. What is the role of tiotropium in asthma?: A systematic review with meta-analysis. *Chest* 147:388–396.

Roeder, E. 2000. Medicinal plants in China containing pyrrolizidine alkaloids. *Pharmazie* 55:711–726.

Rojas, C., Raje, M., Tsukamoto, T., Slusher, B. S. 2014. Molecular mechanisms of 5-HT$_3$ and NK$_1$ receptor antagonists in prevention of emesis. *Eur J Pharmacol* 722:26–37.

Roos, W., Oeze, L., Loser, R., Eppenberger, U. 1983. Antiestrogenic action of 3-hydroxytamoxifen in the human breast cancer cell line MCF-7. *J Natl Cancer Inst* 71:55–59.

Rosenthal, J., Hittel, N., Stumpe, K. O. 1996. Pranidipine, a novel calcium antagonist, once daily, for the treatment of hypertension: A multicenter, double-blind, placebo-controlled dose-finding study. *Cardiovasc Drugs Ther* 10:59–66.

Roth, B. L., Choudhary, M. S., Khan, N., Uluer, A. Z. 1997. High-affinity agonist binding is not sufficient for agonist efficacy at 5-hydroxytryptamine2A receptors: Evidence in favor of a modified ternary complex model. *J Pharmacol Exp Ther* 280:576–583.

Rowland, K., Ellis, S. W., Lennard, M. S., Tucker, G. T. 1996. Variable contribution of CYP2D6 to the N-dealkylation of S-(−)-propranolol by human liver microsomes. *Br J Clin Pharmacol* 42:390–393.

Rowland, P., Blaney, F. E., Smyth, M. G., Jones, J. J., Leydon, V. R., Oxbrow, A. K., Lewis, C. J. et al. 2006. Crystal structure of human cytochrome P450 2D6. *J Biol Chem* 281:7614–7622.

Roy, D., Pratt, C. M., Torp-Pedersen, C., Wyse, D. G., Toft, E., Juul-Moller, S., Nielsen, T. et al. 2008. Vernakalant hydrochloride for rapid conversion of atrial fibrillation: A phase 3, randomized, placebo-controlled trial. *Circulation* 117:1518–1525.

Ruenitz, P. C., Bagley, J. R., Mokler, C. M. 1982. Estrogenic and antiestrogenic activity of monophenolic analogues of tamoxifen, (Z)-2-[p-(1,2-diphenyl-1-butenyl)phenoxy]-N,N-dimethylethylamine. *J Med Chem* 25:1056–1060.

Ruenitz, P. C., Bagley, J. R., Mokler, C. M. 1983. Metabolism of clomiphene in the rat. Estrogen receptor affinity and antiestrogenic activity of clomiphene metabolites. *Biochem Pharmacol* 32:2941–2947.

Ruenitz, P. C., Bagley, J. R., Pape, C. W. 1984. Some chemical and biochemical aspects of liver microsomal metabolism of tamoxifen. *Drug Metab Dispos* 12:478–483.

Ruo, T. I., Morita, Y., Atkinson, A. J., Jr., Henthorn, T., Thenot, J. P. 1981. Identification of desethyl procainamide in patients: A new metabolite of procainamide. *J Pharmacol Exp Ther* 216: 357–362.

Sachs, G. S., Greenberg, W. M., Starace, A., Lu, K., Ruth, A., Laszlovszky, I., Nemeth, G., Durgam, S. 2015. Cariprazine in the treatment of acute mania in bipolar I disorder: A double-blind, placebo-controlled, phase III trial. *J Affect Disord* 174:296–302.

Sachse, C., Brockmoller, J., Bauer, S., Roots, I. 1997. Cytochrome P450 2D6 variants in a Caucasian population: Allele frequencies and phenotypic consequences. *Am J Hum Genet* 60:284–295.

Salerno, D. M., Granrud, G., Sharkey, P., Krejci, J., Larson, T., Erlien, D., Berry, D., Hodges, M. 1986. Pharmacodynamics and side effects of flecainide acetate. *Clin Pharmacol Ther* 40:101–107.

Sallustio, B. C., Westley, I. S., Morris, R. G. 2002. Pharmacokinetics of the antianginal agent perhexiline: Relationship between metabolic ratio and steady-state dose. *Br J Clin Pharmacol* 54:107–114.

Salva, P., Costa, J. 1995. Clinical pharmacokinetics and pharmacodynamics of zolpidem. Therapeutic implications. *Clin Pharmacokinet* 29:142–153.

Salva, M., Jansat, J. M., Martinez-Tobed, A., Palacios, J. M. 2003. Identification of the human liver enzymes involved in the metabolism of the antimigraine agent almotriptan. *Drug Metab Dispos* 31:404–411.

Sanchez, R. I., Wang, R. W., Newton, D. J., Bakhtiar, R., Lu, P., Chiu, S. H., Evans, D. C., Huskey, S. E. 2004. Cytochrome P450 3A4 is the major enzyme involved in the metabolism of the substance P receptor antagonist aprepitant. *Drug Metab Dispos* 32:1287–1292.

Sanchez, C., Asin, K. E., Artigas, F. 2015. Vortioxetine, a novel antidepressant with multimodal activity: Review of preclinical and clinical data. *Pharmacol Ther* 145:43–57.

Sanders-Bush, E., Oates, J. A., Bush, M. T. 1976. Metabolism of bufotenine-2'-14C in human volunteers. *Life Sci* 19:1407–1411.

Sanford, M., Scott, L. J. 2009. Gefitinib: A review of its use in the treatment of locally advanced/metastatic non-small cell lung cancer. *Drugs* 69:2303–2328.

Sanwald, P., David, M., Dow, J. 1996a. Characterization of the cytochrome P450 enzymes involved in the in vitro metabolism of dolasetron. Comparison with other indole-containing 5-HT3 antagonists. *Drug Metab Dispos* 24:602–609.

Sanwald, P., David, M., Dow, J. 1996b. Use of electrospray ionization liquid chromatography-mass spectrometry to study the role of CYP2D6 in the in vitro metabolism of 5-hydroxytryptamine receptor antagonists. *J Chromatogr B Biomed Appl* 678:53–61.

Saper, J. R., Silberstein, S. 2006. Pharmacology of dihydroergotamine and evidence for efficacy and safety in migraine. *Headache* 46 Suppl 4:S171–S181.

Sauer, C., Peters, F. T., Schwaninger, A. E., Meyer, M. R., Maurer, H. H. 2008. Identification of cytochrome P450 enzymes involved in the metabolism of the designer drugs *N*-(1-phenylcyclohexyl)-3-ethoxypropanamine and *N*-(1-phenylcyclohexyl)-3-methoxypropanamine. *Chem Res Toxicol* 21:1949–1955.

Savage, D. G., Antman, K. H. 2002. Imatinib mesylate—A new oral targeted therapy. *N Engl J Med* 346:683–693.

Savi, P., Herbert, J. M. 2005. Clopidogrel and ticlopidine: P2Y12 adenosine diphosphate-receptor antagonists for the prevention of atherothrombosis. *Semin Thromb Hemost* 31:174–183.

Schellens, J. H., van der Wart, J. H., Brugman, M., Breimer, D. D. 1989. Influence of enzyme induction and inhibition on the oxidation of nifedipine, sparteine, mephenytoin and antipyrine in humans as assessed by a "cocktail" study design. *J Pharmacol Exp Ther* 249:638–645.

Schiffer, C. A. 2007. BCR-ABL tyrosine kinase inhibitors for chronic myelogenous leukemia. *N Engl J Med* 357:258–265.

Schoemaker, H., Hicks, P. E., Langer, S. Z. 1987. Calcium channel receptor binding studies for diltiazem and its major metabolites: Functional correlation to inhibition of portal vein myogenic activity. *J Cardiovasc Pharmacol* 9:173–180.

Schotte, A., Janssen, P. F., Gommeren, W., Luyten, W. H., Van Gompel, P., Lesage, A. S., De Loore, K., Leysen, J. E. 1996. Risperidone compared with new and reference antipsychotic drugs: In vitro and in vivo receptor binding. *Psychopharmacology (Berl)* 124:57–73.

Schrader, B. J., Bauman, J. L. 1986. Mexiletine: A new type I antiarrhythmic agent. *Drug Intell Clin Pharm* 20:255–260.

Schwab, M., Seyringer, E., Brauer, R. B., Hellinger, A., Griese, E. U. 1999. Fatal MDMA intoxication. *Lancet* 353:593–594.

Schwartzberg, L. S. 2007. Chemotherapy-induced nausea and vomiting: Which antiemetic for which therapy? *Oncology (Williston Park)* 21:946–953; discussion 954, 959, 962 passim.

Scripture, C. D., Pieper, J. A. 2001. Clinical pharmacokinetics of fluvastatin. *Clin Pharmacokinet* 40:263–281.

Seals, A. A., Haider, R., Leon, C., Francis, M., Young, J. B., Roberts, R., Pratt, C. M. 1987. Antiarrhythmic efficacy and hemodynamic effects of cibenzoline in patients with nonsustained ventricular tachycardia and left ventricular dysfunction. *Circulation* 75:800–808.

Segura, M., Farre, M., Pichini, S., Peiro, A. M., Roset, P. N., Ramirez, A., Ortuno, J. et al. 2005. Contribution of cytochrome P450 2D6 to 3,4-methylenedioxymethamphetamine disposition in humans: Use of paroxetine as a metabolic inhibitor probe. *Clin Pharmacokinet* 44:649–660.

Seltzer, B. 2005. Donepezil: A review. *Expert Opin Drug Metab Toxicol* 1:527–536.

Seltzer, B. 2007. Donepezil: An update. *Expert Opin Pharmacother* 8:1011–1023.

Seneca, N., Finnema, S. J., Laszlovszky, I., Kiss, B., Horvath, A., Pasztor, G., Kapas, M. et al. 2011. Occupancy of dopamine D_2 and D_3 and serotonin $5-HT_{1A}$ receptors by the novel antipsychotic drug candidate, cariprazine (RGH-188), in monkey brain measured using positron emission tomography. *Psychopharmacology (Berl)* 218:579–587.

Shah, R. R., Oates, N. S., Idle, J. R., Smith, R. L. 1980. Genetic impairment of phenformin metabolism. *Lancet* 1:1147.

Shah, R. R., Evans, D. A., Oates, N. S., Idle, J. R., Smith, R. L. 1985. The genetic control of phenformin 4-hydroxylation. *J Med Genet* 22:361–366.

Shah, A., Lanman, R., Bhargava, V., Weir, S., Hahne, W. 1995. Pharmacokinetics of dolasetron following single- and multiple-dose intravenous administration to normal male subjects. *Biopharm Drug Dispos* 16:177–189.

Shang, Y. 2006. Molecular mechanisms of oestrogen and SERMs in endometrial carcinogenesis. *Nat Rev Cancer* 6:360–368.

Shapiro, D. A., Renock, S., Arrington, E., Chiodo, L. A., Liu, L. X., Sibley, D. R., Roth, B. L., Mailman, R. 2003. Aripiprazole, a novel atypical antipsychotic drug with a unique and robust pharmacology. *Neuropsychopharmacology* 28:1400–1411.

Sharer, J. E., Wrighton, S. A. 1996. Identification of the human hepatic cytochromes P450 involved in the *in vitro* oxidation of antipyrine. *Drug Metab Dispos* 24:487–494.

Sharma, A., Hamelin, B. A. 2003. Classic histamine H1 receptor antagonists: A critical review of their metabolic and pharmacokinetic fate from a bird's eye view. *Curr Drug Metab* 4:105–129.

Shelly, W., Draper, M. W., Krishnan, V., Wong, M., Jaffe, R. B. 2008. Selective estrogen receptor modulators: An update on recent clinical findings. *Obstet Gynecol Surv* 63:163–181.

Shibutani, S., Dasaradhi, L., Terashima, I., Banoglu, E., Duffel, M. W. 1998a. Alpha-hydroxytamoxifen is a substrate of hydroxysteroid (alcohol) sulfotransferase, resulting in tamoxifen DNA adducts. *Cancer Res* 58:647–653.

Shibutani, S., Shaw, P. M., Suzuki, N., Dasaradhi, L., Duffel, M. W., Terashima, I. 1998b. Sulfation of alpha-hydroxytamoxifen catalyzed by human hydroxysteroid sulfotransferase results in tamoxifen-DNA adducts. *Carcinogenesis* 19:2007–2011.

Shigeta, M., Homma, A. 2001. Donepezil for Alzheimer's disease: Pharmacodynamic, pharmacokinetic, and clinical profiles. *CNS Drug Rev* 7:353–368.

Shimizu, M., Takatori, K., Kajiwara, M., Ogata, H. 1998. Isolation of a major metabolite (i-OHAP) of aprindine and its identification as N-[3-(N,N-diethylamino)propyl]-N-phenyl-2-aminoindan-5-ol. *Biol Pharm Bull* 21:530–534.

Shu, Y. Z., Hubbard, J. W., Cooper, J. K., McKay, G., Korchinski, E. D., Kumar, R., Midha, K. K. 1990. The identification of urinary metabolites of doxepin in patients. *Drug Metab Dispos* 18:735–741.

Siddoway, L. A., Roden, D. M., Woosley, R. L. 1984. Clinical pharmacology of propafenone: Pharmacokinetics, metabolism and concentration-response relations. *Am J Cardiol* 54:9D-12D.

Siegle, I., Fritz, P., Eckhardt, K., Zanger, U. M., Eichelbaum, M. 2001. Cellular localization and regional distribution of CYP2D6 mRNA and protein expression in human brain. *Pharmacogenetics* 11:237–245.

Silber, B., Holford, N. H., Riegelman, S. 1982. Stereoselective disposition and glucuronidation of propranolol in humans. *J Pharm Sci* 71:699–704.

Simpson, D., Plosker, G. L. 2004. Atomoxetine: A review of its use in adults with attention deficit hyperactivity disorder. *Drugs* 64:205–222.

Simpson, K., Spencer, C. M., McClellan, K. J. 2000. Tropisetron: An update of its use in the prevention of chemotherapy-induced nausea and vomiting. *Drugs* 59:1297–1315.

Sindrup, S. H., Poulsen, L., Brosen, K., Arendt-Nielsen, L., Gram, L. F. 1993. Are poor metabolisers of sparteine/debrisoquine less pain tolerant than extensive metabolisers? *Pain* 53:335–339.

Singh, B. N. 2006. Amiodarone: A multifaceted antiarrhythmic drug. *Curr Cardiol Rep* 8:349–355.

Sitaram, B. R., Lockett, L., Blackman, G. L., McLeod, W. R. 1987a. Urinary excretion of 5-methoxy-N,N-dimethyltryptamine, N,N-dimethyltryptamine and their N-oxides in the rat. *Biochem Pharmacol* 36:2235–2237.

Sitaram, B. R., Talomsin, R., Blackman, G. L., McLeod, W. R. 1987b. Study of metabolism of psychotomimetic indolealkylamines by rat tissue extracts using liquid chromatography. *Biochem Pharmacol* 36:1503–1508.

Skinner, M. H., Kuan, H. Y., Pan, A., Sathirakul, K., Knadler, M. P., Gonzales, C. R., Yeo, K. P. et al. 2003. Duloxetine is both an inhibitor and a substrate of cytochrome P450 2D6 in healthy volunteers. *Clin Pharmacol Ther* 73:170–177.

Skurnik, Y. D., Tcherniak, A., Edlan, K., Sthoeger, Z. 2003. Ticlopidine-induced cholestatic hepatitis. *Ann Pharmacother* 37:371–375.

Smith, G. H. 1985. Flecainide: A new class Ic antidysrhythmic. *Drug Intell Clin Pharm* 19:703–707.

Smith, D. A., Jones, B. C. 1992. Speculations on the substrate structure-activity relationship (SSAR) of cytochrome P450 enzymes. *Biochem Pharmacol* 44:2089–2098.

Snider, N. T., Sikora, M. J., Sridar, C., Feuerstein, T. J., Rae, J. M., Hollenberg, P. F. 2008. The endocannabinoid anandamide is a substrate for the human polymorphic cytochrome P450 2D6. *J Pharmacol Exp Ther* 327:538–545.

Snider, N. T., Nast, J. A., Tesmer, L. A., Hollenberg, P. F. 2009. A cytochrome P450-derived epoxygenated metabolite of anandamide is a potent cannabinoid receptor 2-selective agonist. *Mol Pharmacol* 75:965–972.

Soininen, K., Kleimola, T., Elomaa, I., Salmo, M., Rissanen, P. 1986. The steady-state pharmacokinetics of tamoxifen and its metabolites in breast cancer patients. *J Int Med Res* 14:162–165.

Somani, P. 1980. Antiarrhythmic effects of flecainide. *Clin Pharmacol Ther* 27:464–470.

Sonnichsen, D. S., Liu, Q., Schuetz, E. G., Schuetz, J. D., Pappo, A., Relling, M. V. 1995. Variability in human cytochrome P450 paclitaxel metabolism. *J Pharmacol Exp Ther* 275:566–575.

Sorensen, L. B., Sorensen, R. N., Miners, J. O., Somogyi, A. A., Grgurinovich, N., Birkett, D. J. 2003. Polymorphic hydroxylation of perhexiline *in vitro*. *Br J Clin Pharmacol* 55:635–638.

Spaldin, V., Madden, S., Pool, W. F., Woolf, T. F., Park, B. K. 1994. The effect of enzyme inhibition on the metabolism and activation of tacrine by human liver microsomes. *Br J Clin Pharmacol* 38:15–22.

Spigset, O., Granberg, K., Hagg, S., Soderstrom, E., Dahlqvist, R. 1998. Non-linear fluvoxamine disposition. *Br J Clin Pharmacol* 45:257–263.

Spigset, O., Axelsson, S., Norstrom, A., Hagg, S., Dahlqvist, R. 2001. The major fluvoxamine metabolite in urine is formed by CYP2D6. *Eur J Clin Pharmacol* 57:653–658.

Spina, E., Avenoso, A., Facciola, G., Salemi, M., Scordo, M. G., Giacobello, T., Madia, A. G., Perucca, E. 2000. Plasma concentrations of risperidone and 9-hydroxyrisperidone: Effect of comedication with carbamazepine or valproate. *Ther Drug Monit* 22:481–485.

Spina, E., Santoro, V. D'Arrigo, C. 2008. Clinically relevant pharmacokinetic drug interactions with second-generation antidepressants: An update. *Clin Ther* 30:1206–1227.

Springer, D., Peters, F. T., Fritschi, G., Maurer, H. H. 2002. Studies on the metabolism and toxicological detection of the new designer drug 4'-methyl-alpha-pyrrolidinopropiophenone in urine using gas chromatography-mass spectrometry. *J Chromatogr B Analyt Technol Biomed Life Sci* 773:25–33.

Springer, D., Paul, L. D., Staack, R. F., Kraemer, T., Maurer, H. H. 2003a. Identification of cytochrome P450 enzymes involved in the metabolism of 4'-methyl-α-pyrrolidinopropiophenone, a novel scheduled designer drug, in human liver microsomes. *Drug Metab Dispos* 31:979–982.

Springer, D., Peters, F. T., Fritschi, G., Maurer, H. H. 2003b. New designer drug 4'-methyl-alpha-pyrrolidinohexanophenone: Studies on its metabolism and toxicological detection in urine using gas chromatography-mass spectrometry. *J Chromatogr B Analyt Technol Biomed Life Sci* 789:79–91.

Stearns, V., Johnson, M. D., Rae, J. M., Morocho, A., Novielli, A., Bhargava, P., Hayes, D. F., Desta, Z., Flockhart, D. A. 2003. Active tamoxifen metabolite plasma concentrations after coadministration of tamoxifen and the selective serotonin reuptake inhibitor paroxetine. *J Natl Cancer Inst* 95:1758–1764.

Steen, V. M., Andreassen, O. A., Daly, A. K., Tefre, T., Borresen, A. L., Idle, J. R., Gulbrandsen, A. K. 1995. Detection of the poor metabolizer-associated CYP2D6(D) gene deletion allele by long-PCR technology. *Pharmacogenetics* 5:215–223.

Steers, W. D. 2006. Darifenacin: Pharmacology and clinical usage. *Urol Clin North Am* 33:475–482, viii.

Stegelmeier, B. L., Edgar, J. A., Colegate, S. M., Gardner, D. R., Schoch, T. K., Coulombe, R. A., Molyneux, R. J. 1999. Pyrrolizidine alkaloid plants, metabolism and toxicity. *J Nat Toxins* 8:95–116.

Sten, T., Qvisen, S., Uutela, P., Luukkanen, L., Kostiainen, R., Finel, M. 2006. Prominent but reverse stereoselectivity in propranolol glucuronidation by human UDP-glucuronosyltransferases 1A9 and 1A10. *Drug Metab Dispos* 34:1488–1494.

Stormer, E., von Moltke, L. L., Shader, R. I., Greenblatt, D. J. 2000. Metabolism of the antidepressant mirtazapine *in vitro*: Contribution of cytochromes P450 1A2, 2D6, and 3A4. *Drug Metab Dispos* 28:1168–1175.

Stoschitzky, K., Klein, W., Stark, G., Stark, U., Zernig, G., Graziadei, I., Lindner, W. 1990. Different stereoselective effects of (R)- and (S)-propafenone: Clinical pharmacologic, electrophysiologic, and radioligand binding studies. *Clin Pharmacol Ther* 47:740–746.

Stresser, D. M., Kupfer, D. 1998. Human cytochrome P450-catalyzed conversion of the proestrogenic pesticide methoxychlor into an estrogen. Role of CYP2C19 and CYP1A2 in *O*-demethylation. *Drug Metab Dispos* 26:868–874.

Stresser, D. M., Turner, S. D., Blanchard, A. P., Miller, V. P., Crespi, C. L. 2002. Cytochrome P450 fluorometric substrates: Identification of isoform-selective probes for rat CYP2D2 and human CYP3A4. *Drug Metab Dispos* 30:845–852.

Subrahmanyam, V., Renwick, A. B., Walters, D. G., Young, P. J., Price, R. J., Tonelli, A. P., Lake, B. G. 2001. Identification of cytochrome P450 isoforms responsible for *cis*-tramadol metabolism in human liver microsomes. *Drug Metab Dispos* 29:1146–1155.

Sugidachi, A., Asai, F., Ogawa, T., Inoue, T., Koike, H. 2000. The in vivo pharmacological profile of CS-747, a novel antiplatelet agent with platelet ADP receptor antagonist properties. *Br J Pharmacol* 129:1439–1446.

Sugidachi, A., Asai, F., Yoneda, K., Iwamura, R., Ogawa, T., Otsuguro, K., Koike, H. 2001. Antiplatelet action of R-99224, an active metabolite of a novel thienopyridine-type G(i)-linked P2T antagonist, CS-747. *Br J Pharmacol* 132:47–54.

Sugimoto, H., Yamanishi, Y., Iimura, Y., Kawakami, Y. 2000. Donepezil hydrochloride (E2020) and other acetylcholinesterase inhibitors. *Curr Med Chem* 7:303–339.

Sun, D., Chen, G., Dellinger, R. W., Duncan, K., Fang, J. L., Lazarus, P. 2006. Characterization of tamoxifen and 4-hydroxytamoxifen glucuronidation by human UGT1A4 variants. *Breast Cancer Res* 8:R50.

Supuran, C. T., Mastrolorenzo, A., Barbaro, G., Scozzafava, A. 2006. Phosphodiesterase 5 inhibitors—drug design and differentiation based on selectivity, pharmacokinetic and efficacy profiles. *Curr Pharm Des* 12:3459–3465.

Surve, P., Ravindran, S., Acharjee, A., Rastogi, H., Basu, S., Honrao, P. 2014. Metabolite characterization of anti-cancer agent gefitinib in human hepatocytes. *Drug Metab Lett* 7:126–136.

Suzuki, O., Katsumata, Y., Oya, M. 1981. Characterization of eight biogenic indoleamines as substrates for type A and type B monoamine oxidase. *Biochem Pharmacol* 30:1353–1358.

Svendsen, C. N., Bird, E. D. 1986. HPLC with electrochemical detection to measure chlorpromazine, thioridazine and metabolites in human brain. *Psychopharmacology (Berl)* 90:316–321.

Swainston Harrison, T., Perry, C. M. 2004. Aripiprazole: A review of its use in schizophrenia and schizoaffective disorder. *Drugs* 64:1715–1736.

Swainston Harrison, T., Keating, G. M. 2005. Zolpidem: A review of its use in the management of insomnia. *CNS Drugs* 19: 65–89.

Swaisland, H., Laight, A., Stafford, L., Jones, H., Morris, C., Dane, A., Yates, R. 2001. Pharmacokinetics and tolerability of the orally active selective epidermal growth factor receptor tyrosine kinase inhibitor ZD1839 in healthy volunteers. *Clin Pharmacokinet* 40:297–306.

Swaisland, H. C., Ranson, M., Smith, R. P., Leadbetter, J., Laight, A., McKillop, D., Wild, M. J. 2005a. Pharmacokinetic drug interactions of gefitinib with rifampicin, itraconazole and metoprolol. *Clin Pharmacokinet* 44:1067–1081.

Swaisland, H. C., Smith, R. P., Laight, A., Kerr, D. J., Ranson, M., Wilder-Smith, C. H., Duvauchelle, T. 2005b. Single-dose clinical pharmacokinetic studies of gefitinib. *Clin Pharmacokinet* 44:1165–1177.

Szara, S., Axelrod, J. 1959. Hydroxylation and N-demethylation of N, N-dimethyltryptamine. *Experientia* 15:216–217.

Szewczak, M. R., Corbett, R., Rush, D. K., Wilmot, C. A., Conway, P. G., Strupczewski, J. T., Cornfeldt, M. 1995. The pharmacological profile of iloperidone, a novel atypical antipsychotic agent. *J Pharmacol Exp Ther* 274:1404–1413.

Taguchi, M., Nozawa, T., Igawa, A., Inoue, H., Takesono, C., Tahara, K., Hashimoto, Y. 2005. Pharmacokinetic variability of routinely administered bisoprolol in middle-aged and elderly Japanese patients. *Biol Pharm Bull* 28:876–881.

Taguchi, M., Fujiki, A., Iwamoto, J., Inoue, H., Tahara, K., Saigusa, K., Horiuchi, I., Oshima, Y., Hashimoto, Y. 2006. Nonlinear mixed effects model analysis of the pharmacokinetics of routinely administered bepridil in Japanese patients with arrhythmias. *Biol Pharm Bull* 29:517–521.

Talaat, R. E., Nelson, W. L. 1988. Regioisomeric aromatic dihydroxylation of propranolol. Synthesis and identification of 4,6- and 4,8-dihydroxypropranolol as metabolites in the rat and in man. *Drug Metab Dispos* 16:212–216.

Talajic, M., DeRoode, M. R., Nattel, S. 1987. Comparative electrophysiologic effects of intravenous amiodarone and desethylamiodarone in dogs: Evidence for clinically relevant activity of the metabolite. *Circulation* 75:265–271.

Tam, Y. K., Tawfik, S. R., Ke, J., Coutts, R. T., Gray, M. R., Wyse, D. G. 1987. High-performance liquid chromatography of lidocaine and nine of its metabolites in human plasma and urine. *J Chromatogr* 423:199–206.

Tam, Y. K., Ke, J., Coutts, R. T., Wyse, D. G., Gray, M. R. 1990. Quantification of three lidocaine metabolites and their conjugates. *Pharm Res* 7:504–507.

Tanaka, E., Kurata, N., Yasuhara, H. 2003. Involvement of cytochrome P450 2C9, 2E1 and 3A4 in trimethadione *N*-demethylation in human microsomes. *J Clin Pharm Ther* 28:493–496.

Tantry, U. S., Bliden, K. P., Gurbel, P. A. 2006. Prasugrel. *Expert Opin Investig Drugs* 15:1627–1633.

Tariot, P. N., Solomon, P. R., Morris, J. C., Kershaw, P., Lilienfeld, S., Ding, C. 2000. A 5-month, randomized, placebo-controlled trial of galantamine in AD. The Galantamine USA-10 Study Group. *Neurology* 54:2269–2276.

Tassaneeyakul, W., Birkett, D. J., Veronese, M. E., McManus, M. E., Tukey, R. H., Quattrochi, L. C., Gelboin, H. V., Miners, J. O. 1993. Specificity of substrate and inhibitor probes for human cytochromes P450 1A1 and 1A2. *J Pharmacol Exp Ther* 265:401–407.

Tateishi, T., Watanabe, M., Kumai, T., Tanaka, M., Moriya, H., Yamaguchi, S., Satoh, T., Kobayashi, S. 2000. CYP3A is responsible for N-dealkylation of haloperidol and bromperidol and oxidation of their reduced forms by human liver microsomes. *Life Sci* 67:2913–2920.

Tatsumi, K., Ou, T., Yamada, H., Yoshimura, H. 1980. Studies of metabolic fate of a new antiallergic agent, azelastine (4-(p-chlorobenzyl)-2-[N-methylperhydroazepinyl-(4)]-1-(2H)-phthalazinone hydrochloride). *Jpn J Pharmacol* 30: 37–48.

Tatsumi, K., Yamada, H., Yoshimura, H., Nishizawa, Y., Sakai, M., Mizuo, H., Yamato, C. 1984. Metabolism of an antiallergic agent, azelastine (4-(p-chlorobenzyl)-2-[N-methylperhydroazepinyl-(4)]-1-(2H)-p hthalazin one hydrochloride) in rats and guinea pigs. *Hiroshima J Med Sci* 33:669–678.

Teng, L., Bruce, R. B., Dunning, L. K. 1977. Metoclopramide metabolism and determination by high-pressure liquid chromatography. *J Pharm Sci* 66:1615–1618.

Tenneze, L., Verstuyft, C., Becquemont, L., Poirier, J. M., Wilkinson, G. R., Funck-Brentano, C. 1999. Assessment of CYP2D6 and CYP2C19 activity *in vivo* in humans: A cocktail study with dextromethorphan and chloroguanide alone and in combination. *Clin Pharmacol Ther* 66:582–588.

Tian, S., Quan, H., Xie, C., Guo, H., Lu, F., Xu, Y., Li, J., Lou, L. 2011. YN968D1 is a novel and selective inhibitor of vascular endothelial growth factor receptor-2 tyrosine kinase with potent activity in vitro and in vivo. *Cancer Sci* 102:1374–1380.

Tipton, K. F., O'Carroll, A. M., McCrodden, J. M., Fowler, C. J. 1985. Monoamine oxidase-A selective inhibition in human hypothalamus and liver *in-vitro* by amiflamine and its metabolites. *J Pharm Pharmacol* 37:352–354.

Tiseo, P. J., Perdomo, C. A., Friedhoff, L. T. 1998. Metabolism and elimination of 14C-donepezil in healthy volunteers: A single-dose study. *Br J Clin Pharmacol* 46 Suppl 1:19–24.

Todd, P. A., Benfield, P. 1989. Flunarizine. A reappraisal of its pharmacological properties and therapeutic use in neurological disorders. *Drugs* 38:481–499.

Tohen, M. 2015. Cariprazine in bipolar disorders. *J Clin Psychiatry* 76:e368–e370.

Totah, R. A., Sheffels, P., Roberts, T., Whittington, D., Thummel, K., Kharasch, E. D. 2008. Role of CYP2B6 in stereoselective human methadone metabolism. *Anesthesiology* 108:363–374.

Towse, G. 1980. Cinnarizine—a labyrinthine sedative. *J Laryngol Otol* 94:1009–1015.

Tracy, T. S., Marra, C., Wrighton, S. A., Gonzalez, F. J., Korzekwa, K. R. 1997. Involvement of multiple cytochrome P450 isoforms in naproxen *O*-demethylation. *Eur J Clin Pharmacol* 52:293–298.

Tracy, T. S., Korzekwa, K. R., Gonzalez, F. J., Wainer, I. W. 1999. Cytochrome P450 isoforms involved in metabolism of the enantiomers of verapamil and norverapamil. *Br J Clin Pharmacol* 47:545–552.

Traebert, M., Dumotier, B. 2005. Antimalarial drugs: QT prolongation and cardiac arrhythmias. *Expert Opin Drug Saf* 4:421–431.

Tran, V. T., Chang, R. S., Snyder, S. H. 1978. Histamine H1 receptors identified in mammalian brain membranes with [3H]mepyramine. *Proc Natl Acad Sci U S A* 75:6290–6294.

Treluyer, J. M., Bowers, G., Cazali, N., Sonnier, M., Rey, E., Pons, G., Cresteil, T. 2003. Oxidative metabolism of amprenavir in the human liver. Effect of the CYP3A maturation. *Drug Metab Dispos* 31:275–281.

Tremaine, L. M., Stroh, J. G., Ronfeld, R. A. 1989. Characterization of a carbamic acid ester glucuronide of the secondary amine sertraline. *Drug Metab Dispos* 17:58–63.

Trivier, J. M., Libersa, C., Belloc, C., Lhermitte, M. 1993. Amiodarone N-deethylation in human liver microsomes: Involvement of cytochrome P450 3A enzymes (first report). *Life Sci* 52:PL91–PL96.

Tsutsui, K. 2008. Progesterone biosynthesis and action in the developing neuron. *Endocrinology* 149:2757–2761.

Tsuzuki, D., Takemi, C., Yamamoto, S., Tamagake, K., Imaoka, S., Funae, Y., Kataoka, H., Shinoda, S., Narimatsu, S. 2001. Functional evaluation of cytochrome P450 2D6 with Gly42Arg substitution expressed in Saccharomyces cerevisiae. *Pharmacogenetics* 11:709–718.

Tsuzuki, D., Hichiya, H., Okuda, Y., Yamamoto, S., Tamagake, K., Shinoda, S., Narimatsu, S. 2003. Alteration in catalytic properties of human CYP2D6 caused by substitution of glycine-42 with arginine, lysine and glutamic acid. *Drug Metab Pharmacokinet* 18:79–85.

Tucker, G. T., Lennard, M. S., Ellis, S. W., Woods, H. F., Cho, A. K., Lin, L. Y., Hiratsuka, A., Schmitz, D. A., Chu, T. Y. 1994. The demethylenation of methylenedioxymethamphetamine ("ecstasy") by debrisoquine hydroxylase (CYP2D6). *Biochem Pharmacol* 47:1151–1156.

Tugnait, M., Hawes, E. M., McKay, G., Eichelbaum, M., Midha, K. K. 1999. Characterization of the human hepatic cytochromes P450 involved in the in vitro oxidation of clozapine. *Chem Biol Interact* 118:171–189.

Turgeon, J., Pare, J. R., Lalande, M., Grech-Belanger, O., Belanger, P. M. 1992. Isolation and structural characterization by spectroscopic methods of two glucuronide metabolites of mexiletine after N-oxidation and deamination. *Drug Metab Dispos* 20:762–769.

Turner, M. J., 3rd, Fields, C. E., Everman, D. B. 1991. Evidence for superoxide formation during hepatic metabolism of tamoxifen. *Biochem Pharmacol* 41:1701–1705.

Tweedie, D. J., Burke, M. D. 1987. Metabolism of the beta-carbolines, harmine and harmol, by liver microsomes from phenobarbitone- or 3-methylcholanthrene-treated mice. Identification and quantitation of two novel harmine metabolites. *Drug Metab Dispos* 15:74–81.

Tyndale, R. F., Gonzalez, F. J., Hardwick, J. P., Kalow, W., Inaba, T. 1990. Sparteine metabolism capacity in human liver: Structural variants of human P450IID6 as assessed by immunochemistry. *Pharmacol Toxicol* 67:14–18.

Tyndale, R. F., Li, Y., Li, N. Y., Messina, E., Miksys, S., Sellers, E. M. 1999. Characterization of cytochrome P-450 2D1 activity in rat brain: High-affinity kinetics for dextromethorphan. *Drug Metab Dispos* 27:924–930.

Udenfriend, S., Wyngaarden, J. B. 1956. Precursors of adrenal epinephrine and norepinephrine in vivo. *Biochim Biophys Acta* 20:48–52.

Uetrecht, J. P. 1985. Reactivity and possible significance of hydroxylamine and nitroso metabolites of procainamide. *J Pharmacol Exp Ther* 232:420–425.

Uhr, M., Namendorf, C., Grauer, M. T., Rosenhagen, M., Ebinger, M. 2004. P-glycoprotein is a factor in the uptake of dextromethorphan, but not of melperone, into the mouse brain: Evidence for an overlap in substrate specificity between P-gp and CYP2D6. *J Psychopharmacol* 18:509–515.

Umemoto, A., Monden, Y., Komaki, K., Suwa, M., Kanno, Y., Suzuki, M., Lin, C. X. et al. 1999. Tamoxifen-DNA adducts formed by alpha-acetoxytamoxifen N-oxide. *Chem Res Toxicol* 12:1083–1089.

Usmani, K. A., Karoly, E. D., Hodgson, E., Rose, R. L. 2004. *In vitro* sulfoxidation of thioether compounds by human cytochrome P450 and flavin-containing monooxygenase isoforms with particular reference to the CYP2C subfamily. *Drug Metab Dispos* 32:333–339.

Uttamsingh, V., Lu, C., Miwa, G., Gan, L. S. 2005. Relative contributions of the five major human cytochromes P450, 1A2, 2C9, 2C19, 2D6, and 3A4, to the hepatic metabolism of the proteasome inhibitor bortezomib. *Drug Metab Dispos* 33:1723–1728.

Uzan, A., Gueremy, C., Le Fur, G. 1976a. [Absorption, distribution and excretion of (quinuclidinyl-3 methyl)-10-phenothiazine (LM 209), a new antihistamine]. *Xenobiotica* 6:633–648.

Uzan, A., Gueremy, C., Le Fur, G. 1976b. [Biotransformation of 10-(3-quinuclidinylmethyl)phenothiazine (LM 209), a new anti-allergy agent and the distribution and excretion of its metabolites]. *Xenobiotica* 6:649–665.

Van Boven, M., Laruelle, L., Daenens, P. 1990. HPLC analysis of diuron and metabolites in blood and urine. *J Anal Toxicol* 14:231–234.

Vandamme, N., Broly, F., Libersa, C., Courseau, C., Lhermitte, M. 1993. Stereoselective hydroxylation of mexiletine in human liver microsomes: Implication of P450 2D6—a preliminary report. *J Cardiovasc Pharmacol* 21:77–83.

van Erp, N. P., Gelderblom, H., Karlsson, M. O., Li, J., Zhao, M., Ouwerkerk, J., Nortier, J. W., Guchelaar, H. J., Baker, S. D., Sparreboom, A. 2007. Influence of CYP3A4 inhibition on the steady-state pharmacokinetics of imatinib. *Clin Cancer Res* 13:7394–7400.

van Harten, J. 1995. Overview of the pharmacokinetics of fluvoxamine. *Clin Pharmacokinet* 29 Suppl 1:1–9.

Varsaldi, F., Miglio, G., Scordo, M. G., Dahl, M. L., Villa, L. M., Biolcati, A., Lombardi, G. 2006. Impact of the CYP2D6 polymorphism on steady-state plasma concentrations and clinical outcome of donepezil in Alzheimer's disease patients. *Eur J Clin Pharmacol* 62:721–726.

Venhorst, J., Onderwater, R. C., Meerman, J. H., Vermeulen, N. P., Commandeur, J. N. 2000. Evaluation of a novel high-throughput assay for cytochrome P450 2D6 using 7-methoxy-4-(aminomethyl)-coumarin. *Eur J Pharm Sci* 12:151–158.

Venhorst, J., ter Laak, A. M., Commandeur, J. N., Funae, Y., Hiroi, T., Vermeulen, N. P. 2003. Homology modeling of rat and human cytochrome P450 2D (CYP2D) isoforms and computational rationalization of experimental ligand-binding specificities. *J Med Chem* 46:74–86.

Venkatakrishnan, K., Greenblatt, D. J., von Moltke, L. L., Schmider, J., Harmatz, J. S., Shader, R. I. 1998. Five distinct human cytochromes mediate amitriptyline *N*-demethylation *in vitro*: Dominance of CYP 2C19 and 3A4. *J Clin Pharmacol* 38: 112–121.

Venkatakrishnan, K., von Moltke, L. L., Obach, R. S., Greenblatt, D. J. 2000. Microsomal binding of amitriptyline: Effect on estimation of enzyme kinetic parameters *in vitro*. *J Pharmacol Exp Ther* 293:343–350.

Venkatakrishnan, K., Schmider, J., Harmatz, J. S., Ehrenberg, B. L., von Moltke, L. L., Graf, J. A. et al. 2001. Relative contribution of CYP3A to amitriptyline clearance in humans: *In vitro* and *in vivo* studies. *J Clin Pharmacol* 41:1043–1054.

Verheij, E. R., van der Greef, J., La Vos, G. F., van der Pol, W., Niessen, W. M. 1989. Identification of diuron and four of its metabolites in human postmortem plasma and urine by LC/MS with a moving-belt interface. *J Anal Toxicol* 13:8–12.

Veselinovic, T., Paulzen, M., Grunder, G. 2013. Cariprazine, a new, orally active dopamine $D_{2/3}$ receptor partial agonist for the treatment of schizophrenia, bipolar mania and depression. *Expert Rev Neurother* 13:1141–1159.

Vickers, S., Duncan, C. A., Vyas, K. P., Kari, P. H., Arison, B., Prakash, S. R., Ramjit, H. G., Pitzenberger, S. M., Stokker, G., Duggan, D. E. 1990. In vitro and in vivo biotransformation of simvastatin, an inhibitor of HMG CoA reductase. *Drug Metab Dispos* 18:476–483.

Vickers, A. E., Sinclair, J. R., Zollinger, M., Heitz, F., Glanzel, U., Johanson, L., Fischer, V. 1999. Multiple cytochrome P450s involved in the metabolism of terbinafine suggest a limited potential for drug–drug interactions. *Drug Metab Dispos* 27:1029–1038.

Villikka, K., Kivisto, K. T., Luurila, H., Neuvonen, P. J. 1997. Rifampin reduces plasma concentrations and effects of zolpidem. *Clin Pharmacol Ther* 62:629–634.

Vogel, W. H., Evans, B. D. 1977. Structure-activity-relationships of certain hallucinogenic substances based on brain levels. *Life Sci* 20:1629–1635.

Vogelgesang, B., Echizen, H., Schmidt, E., Eichelbaum, M. 1984. Stereoselective first-pass metabolism of highly cleared drugs: Studies of the bioavailability of L- and D-verapamil examined with a stable isotope technique. *Br J Clin Pharmacol* 18:733–740.

Vohora, D. 2007. Atypical antipsychotic drugs: Current issues of safety and efficacy in the management of schizophrenia. *Curr Opin Investig Drugs* 8:531–538.

Volotinen, M., Turpeinen, M., Tolonen, A., Uusitalo, J., Maenpaa, J., Pelkonen, O. 2007. Timolol metabolism in human liver microsomes is mediated principally by CYP2D6. *Drug Metab Dispos* 35:1135–1141.

von Moltke, L. L., Greenblatt, D. J., Duan, S. X., Harmatz, J. S., Shader, R. I. 1994. *In vitro* prediction of the terfenadine-ketoconazole pharmacokinetic interaction. *J Clin Pharmacol* 34:1222–1227.

von Moltke, L. L., Greenblatt, D. J., Duan, S. X., Schmider, J., Wright, C. E., Harmatz, J. S., Shader, R. I. 1997. Human cytochromes mediating N-demethylation of fluoxetine in vitro. *Psychopharmacology (Berl)* 132:402–407.

von Moltke, L. L., Greenblatt, D. J., Duan, S. X., Daily, J. P., Harmatz, J. S., Shader, R. I. 1998. Inhibition of desipramine hydroxylation (cytochrome P450-2D6) *in vitro* by quinidine and by viral protease inhibitors: Relation to drug interactions *in vivo*. *J Pharm Sci* 87:1184–1189.

von Moltke, L. L., Greenblatt, D. J., Granda, B. W., Duan, S. X., Grassi, J. M., Venkatakrishnan, K., Harmatz, J. S., Shader, R. I. 1999. Zolpidem metabolism in vitro: Responsible cytochromes, chemical inhibitors, and in vivo correlations. *Br J Clin Pharmacol* 48:89–97.

von Moltke, L. L., Greenblatt, D. J., Giancarlo, G. M., Granda, B. W., Harmatz, J. S., Shader, R. I. 2001. Escitalopram (S-citalopram) and its metabolites in vitro: Cytochromes mediating biotransformation, inhibitory effects, and comparison to R-citalopram. *Drug Metab Dispos* 29:1102–1109.

von Philipsborn, G., Gries, J., Hofmann, H. P., Kreiskott, H., Kretzschmar, R., Muller, C. D., Raschack, M., Teschendorf, H. J. 1984. Pharmacological studies on propafenone and its main metabolite 5-hydroxypropafenone. *Arzneimittelforschung* 34:1489–1497.

Voorman, R. L., Maio, S. M., Hauer, M. J., Sanders, P. E., Payne, N. A., Ackland, M. J. 1998a. Metabolism of delavirdine, a human immunodeficiency virus type-1 reverse transcriptase inhibitor, by microsomal cytochrome P450 in humans, rats, and other species: Probable involvement of CYP2D6 and CYP3A. *Drug Metab Dispos* 26:631–639.

Voorman, R. L., Maio, S. M., Payne, N. A., Zhao, Z., Koeplinger, K. A., Wang, X. 1998b. Microsomal metabolism of delavirdine: Evidence for mechanism-based inactivation of human cytochrome P450 3A. *J Pharmacol Exp Ther* 287:381–388.

Wagner, W., Vause, E. W. 1995. Fluvoxamine. A review of global drug–drug interaction data. *Clin Pharmacokinet* 29 Suppl 1:26–31; discussion 31–22.

Walle, T., Walle, U. K., Olanoff, L. S. 1985. Quantitative account of propranolol metabolism in urine of normal man. *Drug Metab Dispos* 13:204–209.

Wander, T. J., Nelson, A., Okazaki, H., Richelson, E. 1987. Antagonism by neuroleptics of serotonin 5-HT1A and 5-HT2 receptors of normal human brain in vitro. *Eur J Pharmacol* 143:279–282.

Wang, J. S., DeVane, C. L. 2003. Involvement of CYP3A4, CYP2C8, and CYP2D6 in the metabolism of (R)- and (S)-methadone in vitro. *Drug Metab Dispos* 31:742–747.

Wang, R. I., Johnson, R. P., Lee, J. C., Waite, E. M. 1982. The oral analgesic efficacy of bicifadine hydrochloride in postoperative pain. *J Clin Pharmacol* 22:160–164.

Wang, R. W., Kari, P. H., Lu, A. Y., Thomas, P. E., Guengerich, F. P., Vyas, K. P. 1991. Biotransformation of lovastatin. IV. Identification of cytochrome P450 3A proteins as the major enzymes responsible for the oxidative metabolism of lovastatin in rat and human liver microsomes. *Arch Biochem Biophys* 290:355–361.

Wang, J. S., Backman, J. T., Wen, X., Taavitsainen, P., Neuvonen, P. J., Kivistö, K. T. 1999. Fluvoxamine is a more potent inhibitor of lidocaine metabolism than ketoconazole and erythromycin in vitro. *Pharmacol Toxicol* 85:201–205.

Wang, J. S., Backman, J. T., Taavitsainen, P., Neuvonen, P. J., Kivisto, K. T. 2000. Involvement of CYP1A2 and CYP3A4 in lidocaine N-deethylation and 3-hydroxylation in humans. *Drug Metab Dispos* 28:959–965.

Ward, S. A., Walle, T., Walle, U. K., Wilkinson, G. R., Branch, R. A. 1989. Propranolol's metabolism is determined by both mephenytoin and debrisoquin hydroxylase activities. *Clin Pharmacol Ther* 45:72–79.

Ward, B. A., Gorski, J. C., Jones, D. R., Hall, S. D., Flockhart, D. A., Desta, Z. 2003. The cytochrome P450 2B6 (CYP2B6) is the main catalyst of efavirenz primary and secondary metabolism: Implication for HIV/AIDS therapy and utility of efavirenz as a substrate marker of CYP2B6 catalytic activity. *J Pharmacol Exp Ther* 306:287–300.

Warhurst, D. C. 1987. Antimalarial drugs. An update. *Drugs* 33:50–65.

Warrington, J. S., Shader, R. I., von Moltke, L. L., Greenblatt, D. J. 2000. In vitro biotransformation of sildenafil (Viagra): Identification of human cytochromes and potential drug interactions. *Drug Metab Dispos* 28:392–397.

Washio, T., Kohsaka, K., Arisawa, H., Masunaga, H., Nagatsuka, S., Satoh, Y. 2003. Pharmacokinetics and metabolism of radiolabelled SNI-2011, a novel muscarinic receptor agonist, in healthy volunteers. Comprehensive understanding of absorption, metabolism and excretion using radiolabelled SNI-2011. *Arzneimittelforschung* 53:80–86.

Waxman, D. J., Lapenson, D. P., Aoyama, T., Gelboin, H. V., Gonzalez, F. J., Korzekwa, K. 1991. Steroid hormone hydroxylase specificities of eleven cDNA-expressed human cytochrome P450s. *Arch Biochem Biophys* 290:160–166.

Weber, J., Keating, G. M. 2008. Cevimeline. *Drugs* 68:1691–1698.

Wefer, J., Truss, M. C., Jonas, U. 2001. Tolterodine: An overview. *World J Urol* 19:312–318.

Wesche, D. L., Schuster, B. G., Wang, W. X., Woosley, R. L. 2000. Mechanism of cardiotoxicity of halofantrine. *Clin Pharmacol Ther* 67:521–529.

Westra, P., van Thiel, M. J., Vermeer, G. A., Soeterbroek, A. M., Scaf, A. H., Claessens, H. A. 1986. Pharmacokinetics of galanthamine (a long-acting anticholinesterase drug) in anaesthetized patients. *Br J Anaesth* 58:1303–1307.

Wheeler, J. F., Adams, L. E., Mongey, A. B., Roberts, S. M., Heineman, W. R., Hess, E. V. 1991. Determination of metabolically derived nitroprocainamide in the urine of procainamide-dosed humans and rats by liquid chromatography with electrochemical detection. *Drug Metab Dispos* 19:691–695.

White, I. N. 2003. Tamoxifen: Is it safe? Comparison of activation and detoxication mechanisms in rodents and in humans. *Curr Drug Metab* 4:223–239.

White, N. J. 2007. Cardiotoxicity of antimalarial drugs. *Lancet Infect Dis* 7:549–558.

Wickremsinhe, E. R., Tian, Y., Ruterbories, K. J., Verburg, E. M., Weerakkody, G. J., Kurihara, A., Farid, N. A. 2007. Stereoselective metabolism of prasugrel in humans using a novel chiral liquid chromatography-tandem mass spectrometry method. *Drug Metab Dispos* 35:917–921.

Wiebe, V. J., Osborne, C. K., McGuire, W. L., DeGregorio, M. W. 1992. Identification of estrogenic tamoxifen metabolite(s) in tamoxifen-resistant human breast tumors. *J Clin Oncol* 10:990–994.

Wilcock, G. K., Lilienfeld, S., Gaens, E. 2000. Efficacy and safety of galantamine in patients with mild to moderate Alzheimer's disease: Multicentre randomised controlled trial. Galantamine International-1 Study Group. *BMJ* 321:1445–1449.

Wilde, M. I., Plosker, G. L., Benfield, P. 1993. Fluvoxamine. An updated review of its pharmacology, and therapeutic use in depressive illness. *Drugs* 46:895–924.

Williams, M. L., Lennard, M. S., Martin, I. J., Tucker, G. T. 1994. Interindividual variation in the isomerization of 4-hydroxytamoxifen by human liver microsomes: Involvement of cytochromes P450. *Carcinogenesis* 15:2733–2738.

Williams, E. T., Jones, K. O., Ponsler, G. D., Lowery, S. M., Perkins, E. J., Wrighton, S. A., Ruterbories, K. J., Kazui, M., Farid, N. A. 2008. The biotransformation of prasugrel, a new thienopyridine prodrug, by the human carboxylesterases 1 and 2. *Drug Metab Dispos* 36:1227–1232.

Winkle, R. A., Glantz, S. A., Harrison, D. C. 1975. Pharmacologic therapy of ventricular arrhythmias. *Am J Cardiol* 36:629–650.

Winstanley, P. A., Edwards, G., Orme, M. L., Breckenridge, A. M. 1987. Effect of dose size on amodiaquine pharmacokinetics after oral administration. *Eur J Clin Pharmacol* 33:331–333.

Winter, H. R., Unadkat, J. D. 2005. Identification of cytochrome P450 and arylamine N-acetyltransferase isoforms involved in sulfadiazine metabolism. *Drug Metab Dispos* 33:969–976.

Wirth, K. E., Breithardt, G., Michaelis, L. 1983. [Detection of aprindine and its metabolites in plasma and urine]. *Herz* 8:302–308.

Wojcikowski, J., Pichard-Garcia, L., Maurel, P., Daniel, W. A. 2003. Contribution of human cytochrome p-450 isoforms to the metabolism of the simplest phenothiazine neuroleptic promazine. *Br J Pharmacol* 138:1465–1474.

Wojcikowski, J., Maurel, P., Daniel, W. A. 2006. Characterization of human cytochrome p450 enzymes involved in the metabolism of the piperidine-type phenothiazine neuroleptic thioridazine. *Drug Metab Dispos* 34:471–476.

Wolff, T., Distlerath, L. M., Worthington, M. T., Groopman, J. D., Hammons, G. J., Kadlubar, F. F., Prough, R. A., Martin, M. V., Guengerich, F. P. 1985. Substrate specificity of human liver cytochrome P450 debrisoquine 4-hydroxylase probed using immunochemical inhibition and chemical modeling. *Cancer Res* 45:2116–2122.

Wolff, T., Distlerath, L. M., Worthington, M. T., Guengerich, F. P. 1987. Human liver debrisoquine 4-hydroxylase: Test for specificity toward various monooxygenase substrates and model of the active site. *Arch Toxicol* 60:89–90.

Woolhouse, N. M., Andoh, B., Mahgoub, A., Sloan, T. P., Idle, J. R., Smith, R. L. 1979. Debrisoquin hydroxylation polymorphism among Ghanaians and Caucasians. *Clin Pharmacol Ther* 26:584–591.

Woosley, R. L., Drayer, D. E., Reidenberg, M. M., Nies, A. S., Carr, K., Oates, J. A. 1978. Effect of acetylator phenotype on the rate at which procainamide induces antinuclear antibodies and the lupus syndrome. *N Engl J Med* 298:1157–1159.

Worrel, J. A., Marken, P. A., Beckman, S. E., Ruehter, V. L. 2000. Atypical antipsychotic agents: A critical review. *Am J Health Syst Pharm* 57:238–255.

Wu, D., Otton, S. V., Inaba, T., Kalow, W., Sellers, E. M. 1997. Interactions of amphetamine analogs with human liver CYP2D6. *Biochem Pharmacol* 53:1605–1612.

Wu, Z. L., Huang, S. L., Ou-Yang, D. S., Xu, Z. H., Xie, H. G., Zhou, H. H. 1998. Clomipramine N-demethylation metabolism in human liver microsomes. *Zhongguo Yao Li Xue Bao* 19:433–436.

Wu, W. N., McKown, L. A., Reitz, A. B. 2006. Metabolism of the new anxiolytic agent, a pyrido[1,2-]benzimidazole (PBI) analog (RWJ-53050), in rat and human hepatic S9 fractions, and in dog; identification of cytochrome P450 isoforms mediated in the human microsomal metabolism. *Eur J Drug Metab Pharmacokinet* 31:277–283.

Xia, Q., Chou, M. W., Edgar, J. A., Doerge, D. R., Fu, P. P. 2006. Formation of DHP-derived DNA adducts from metabolic activation of the prototype heliotridine-type pyrrolizidine alkaloid, lasiocarpine. *Cancer Lett* 231:138–145.

Xu, Z. H., Wang, W., Zhao, X. J., Huang, S. L., Zhu, B., He, N., Shu, Y., Liu, Z. Q., Zhou, H. H. 1999. Evidence for involvement of polymorphic CYP2C19 and 2C9 in the N-demethylation of sertraline in human liver microsomes. *Br J Clin Pharmacol* 48:416–423.

Yager, J. D., Liehr, J. G. 1996. Molecular mechanisms of estrogen carcinogenesis. *Annu Rev Pharmacol Toxicol* 36:203–232.

Yamakoshi, Y., Kishimoto, T., Sugimura, K., Kawashima, H. 1999. Human prostate CYP3A5: Identification of a unique 5'-untranslated sequence and characterization of purified recombinant protein. *Biochem Biophys Res Commun* 260:676–681.

Yamazaki, H., Shimada, T. 1997a. Human liver cytochrome P450 enzymes involved in the 7-hydroxylation of *R*- and *S*-warfarin enantiomers. *Biochem Pharmacol* 54:1195–1203.

Yamazaki, H., Shimada, T. 1997b. Progesterone and testosterone hydroxylation by cytochromes P450 2C19, 2C9, and 3A4 in human liver microsomes. *Arch Biochem Biophys* 346:161–169.

Yamazaki, H., Guo, Z., Persmark, M., Mimura, M., Inoue, K., Guengerich, F. P., Shimada, T. 1994. Bufuralol hydroxylation by cytochrome P450 2D6 and 1A2 enzymes in human liver microsomes. *Mol Pharmacol* 46:568–577.

Yamazaki, H., Shibata, A., Suzuki, M., Nakajima, M., Shimada, N., Guengerich, F. P., Yokoi, T. 1999. Oxidation of troglitazone to a quinone-type metabolite catalyzed by cytochrome P-450 2C8 and P-450 3A4 in human liver microsomes. *Drug Metab Dispos* 27:1260–1266.

Yang, T. J., Shou, M., Korzekwa, K. R., Gonzalez, F. J., Gelboin, H. V., Yang, S. K. 1998. Role of cDNA-expressed human cytochromes P450 in the metabolism of diazepam. *Biochem Pharmacol* 55:889–896.

Yasuda, S. U., Wellstein, A., Likhari, P., Barbey, J. T., Woosley, R. L. 1995. Chlorpheniramine plasma concentration and histamine H1-receptor occupancy. *Clin Pharmacol Ther* 58:210–220.

Yasuda, S. U., Zannikos, P., Young, A. E., Fried, K. M., Wainer, I. W., Woosley, R. L. 2002. The roles of CYP2D6 and stereoselectivity in the clinical pharmacokinetics of chlorpheniramine. *Br J Clin Pharmacol* 53:519–525.

Yasui, N., Tybring, G., Otani, K., Mihara, K., Suzuki, A., Svensson, J. O., Kaneko, S. 1997. Effects of thioridazine, an inhibitor of CYP2D6, on the steady-state plasma concentrations of the enantiomers of mianserin and its active metabolite, desmethyl-mianserin, in depressed Japanese patients. *Pharmacogenetics* 7:369–374.

Yasui-Furukori, N., Hidestrand, M., Spina, E., Facciola, G., Scordo, M. G., Tybring, G. 2001. Different enantioselective 9-hydroxylation of risperidone by the two human CYP2D6 and CYP3A4 enzymes. *Drug Metab Dispos* 29:1263–1268.

Yatham, L. N. 2002. The role of novel antipsychotics in bipolar disorders. *J Clin Psychiatry* 63:10–14.

Yeh, S. Y., Gorodetzky, C. W., Krebs, H. A. 1977. Isolation and identification of morphine 3- and 6-glucuronides, morphine 3,6-diglucuronide, morphine 3-ethereal sulfate, normorphine, and normorphine 6-glucuronide as morphine metabolites in humans. *J Pharm Sci* 66:1288–1293.

Yeh, R. F., Gaver, V. E., Patterson, K. B., Rezk, N. L., Baxter-Meheux, F., Blake, M. J., Eron, J. J., Jr., Klein, C. E., Rublein, J. C., Kashuba, A. D. 2006. Lopinavir/ritonavir induces the hepatic activity of cytochrome P450 enzymes CYP2C9, CYP2C19, and CYP1A2 but inhibits the hepatic and intestinal activity of CYP3A as measured by a phenotyping drug cocktail in healthy volunteers. *J Acquir Immune Defic Syndr* 42:52–60.

Yokota, M., Inagaki, H., Uematsu, H., Enomoto, N., Goto, J., Sotobata, I., Takahashi, A. 1987. Non-linear pharmacokinetics of aprindine hydrochloride in oral administration. *Arzneimittelforschung* 37:184–188.

Yoshii, K., Kobayashi, K., Tsumuji, M., Tani, M., Shimada, N., Chiba, K. 2000. Identification of human cytochrome P450 isoforms involved in the 7-hydroxylation of chlorpromazine by human liver microsomes. *Life Sci* 67:175–184.

Yoshimoto, K., Echizen, H., Chiba, K., Tani, M., Ishizaki, T. 1995. Identification of human CYP isoforms involved in the metabolism of propranolol enantiomers—N-desisopropylation is mediated mainly by CYP1A2. *Br J Clin Pharmacol* 39:421–431.

Yu, A. M. 2008. Indolealkylamines: Biotransformations and potential drug–drug interactions. *AAPS J* 10:242–253.

Yu, A., Haining, R. L. 2001. Comparative contribution to dextromethorphan metabolism by cytochrome P450 isoforms *in vitro*: Can dextromethorphan be used as a dual probe for both CTP2D6 and CYP3A activities? *Drug Metab Dispos* 29:1514–1520.

Yu, A. M., Granvil, C. P., Haining, R. L., Krausz, K. W., Corchero, J., Kupfer, A., Idle, J. R., Gonzalez, F. J. 2003a. The relative contribution of monoamine oxidase and cytochrome p450 isozymes to the metabolic deamination of the trace amine tryptamine. *J Pharmacol Exp Ther* 304:539–546.

Yu, A. M., Idle, J. R., Byrd, L. G., Krausz, K. W., Kupfer, A., Gonzalez, F. J. 2003b. Regeneration of serotonin from 5-methoxytryptamine by polymorphic human CYP2D6. *Pharmacogenetics* 13:173–181.

Yu, A. M., Idle, J. R., Herraiz, T., Kupfer, A., Gonzalez, F. J. 2003c. Screening for endogenous substrates reveals that CYP2D6 is a 5-methoxyindolethylamine O-demethylase. *Pharmacogenetics* 13:307–319.

Yu, A. M., Idle, J. R., Krausz, K. W., Kupfer, A., Gonzalez, F. J. 2003d. Contribution of individual cytochrome P450 isozymes to the *O*-demethylation of the psychotropic β-carboline alkaloids harmaline and harmine. *J Pharmacol Exp Ther* 305:315–322.

Yu, J., Paine, M. J., Marechal, J. D., Kemp, C. A., Ward, C. J., Brown, S., Sutcliffe, M. J., Roberts, G. C., Rankin, E. M., Wolf, C. R. 2006. *In silico* prediction of drug binding to CYP2D6: Identification of a new metabolite of metoclopramide. *Drug Metab Dispos* 34:1386–1392.

Yuan, R., Madani, S., Wei, X. X., Reynolds, K., Huang, S. M. 2002. Evaluation of cytochrome P450 probe substrates commonly used by the pharmaceutical industry to study in vitro drug interactions. *Drug Metab Dispos* 30:1311–1319.

Yue, Q. Y., Sawe, J. 1997. Different effects of inhibitors on the O- and N-demethylation of codeine in human liver microsomes. *Eur J Clin Pharmacol* 52:41–47.

Yumibe, N., Huie, K., Chen, K. J., Snow, M., Clement, R. P., Cayen, M. N. 1996. Identification of human liver cytochrome P450 enzymes that metabolize the nonsedating antihistamine loratadine. Formation of descarboethoxyloratadine by CYP3A4 and CYP2D6. *Biochem Pharmacol* 51:165–172.

Yumibe, N., Huie, K., Chen, K. J., Clement, R. P., Cayen, M. N. 1995. Identification of human liver cytochrome P450s involved in the microsomal metabolism of the antihistaminic drug loratadine. *Int Arch Allergy Immunol* 107:420.

Yun, C. H., Okerholm, R. A., Guengerich, F. P. 1993. Oxidation of the antihistaminic drug terfenadine in human liver microsomes. Role of cytochrome P-450 3A(4) in N-dealkylation and C-hydroxylation. *Drug Metab Dispos* 21:403–409.

Zanger, U. M., Raimundo, S., Eichelbaum, M. 2004. Cytochrome P450 2D6: Overview and update on pharmacology, genetics, biochemistry. *Naunyn Schmiedebergs Arch Pharmacol* 369: 23–37.

Zanger, U. M., Klein, K., Saussele, T., Blievernicht, J., Hofmann, M. H., Schwab, M. 2007. Polymorphic CYP2B6: Molecular mechanisms and emerging clinical significance. *Pharmacogenomics* 8:743–759.

Zhang, F., Fan, P. W., Liu, X., Shen, L., van Breeman, R. B., Bolton, J. L. 2000. Synthesis and reactivity of a potential carcinogenic metabolite of tamoxifen: 3,4-dihydroxytamoxifen-o-quinone. *Chem Res Toxicol* 13:53–62.

Zhang, K. L., Hee, B., Lee, C. A., Liang, B., Potts, B. C. 2001. Liquid chromatography-mass spectrometry and liquid chromatography-NMR characterization of in vitro metabolites of a potent and irreversible peptidomimetic inhibitor of rhinovirus 3C protease. *Drug Metab Dispos* 29:729–734.

Zhang, Q. Y., Dunbar, D., Kaminsky, L. 2000. Human cytochrome P450 metabolism of retinals to retinoic acids. *Drug Metab Dispos* 28:292–297.

Zhou, L., Erickson, R. R., Hardwick, J. P., Park, S. S., Wrighton, S. A., Holtzman, J. L. 1997. Catalysis of the cysteine conjugation and protein binding of acetaminophen by microsomes from a human lymphoblast line transfected with the cDNAs of various forms of human cytochrome P450. *J Pharmacol Exp Ther* 281:785–790.

Zhou, Q., Yao, T. W., Yu, Y. N., Zeng, S. 2003. Concentration dependent stereoselectivity of propafenone N-depropylation metabolism with human hepatic recombinant CYP1A2. *Pharmazie* 58:651–653.

Zhou, S., Chan, E., Duan, W., Huang, M., Chen, Y. Z. 2005. Drug bioactivation, covalent binding to target proteins and toxicity relevance. *Drug Metab Rev* 37:41–213.

Zhou, S. F., Xue, C. C., Yu, X. Q., Wang, G. 2007. Metabolic activation of herbal and dietary constituents and its clinical and toxicological implications: An update. *Curr Drug Metab* 8:526–553.

Zhou, S. F., Di, Y. M., Chan, E., Du, Y. M., Chow, V. D., Xue, C. C., Lai, X. et al. 2008. Clinical pharmacogenetics and potential application in personalized medicine. *Curr Drug Metab* 9:738–784.

Zhou, S. F., Liu, J. P., Chowbay, B. 2009. Polymorphism of human cytochrome P450 enzymes and its clinical impact. *Drug Metab Rev* 41:89–295.

Zusman, R. M., Morales, A., Glasser, D. B., Osterloh, I. H. 1999. Overall cardiovascular profile of sildenafil citrate. *Am J Cardiol* 83:35C–44C.

4 Inhibitors of Human CYP2D6

4.1 INTRODUCTION

Human CYP2D6 is largely noninducible but it is subject to inhibition by a large number of compounds including therapeutic drugs and other substances. This is one of the most possible reasons for many clinically observed drug–drug interactions when the victim drugs are mainly metabolized by CYP2D6. The drug–drug interaction involving CYP2D6 may be minor, mild, or fatal. When the elimination of the victim drugs is compromised by CYP2D6 inhibition, drug response may be altered and adverse drug reactions (ADRs) may occur. For example, CYP2D6-mediated venlafaxine–propafenone interaction may cause hallucinations and psychomotor agitation (Gareri et al. 2008). Combined use of serotonergic agents that are substrates of CYP2D6 can also cause severe ADRs. So far, a number of endogenous compounds and xenobiotics are identified as CYP2D6 inhibitors with different inhibitory profiles and mechanisms (Table 4.1). This chapter will discuss the inhibitor selectivity of CYP2D6 and the clinical implications.

4.2 SELECTIVE INHIBITORS OF CYP2D6

Quinidine has been used clinically for more than 200 years and is still important for the treatment of atrial flutter and fibrillation. The major metabolic pathways for quinidine are 3-hydroxylation, N-oxidation, and vinylic hydroxylation in humans (Mason and Hondeghem 1984; Rakhit et al. 1984). Quinidine is metabolized to the main metabolite (3S)-3-hydroxy-quinidine, quinidine-N-oxide, and a few other minor metabolites including oxo-2′-quinidine, O-desmethylquinidine, and quinidine 10,11-dihydrodiols resulting from vinylic hydroxylation. The in vivo metabolic clearance is 15-fold faster for the 3-hydroxylation pathway than for the N-oxidation pathway, and quinidine metabolism is not associated with sparteine oxidation polymorphism (Nielsen et al. 1995), suggesting that CYP2D6 is not involved in the metabolism of quinidine.

In vitro, an anti-CYP3A4 antibody has been shown to inhibit more than 95% and 85% of the formations of 3-OH-quinidine and quinidine N-oxide, respectively (Guengerich et al. 1986). Heterologously expressed CYP3A4 has been shown to actively metabolize quinidine, whereas CYP3A5 does not (Wrighton et al. 1990). It is likely that quinidine is more specific for CYP3A4 activity than drugs like nifedipine, cortisol, and others, because these drugs are shown to be substrates for both CYP3A4 and 3A5 in the same study (Wrighton et al. 1990). Studies with yeast-expressed isozymes have revealed that only CYP3A4 actively catalyzes the (3S)-3-hydroxylation; CYP3A4 is the most active enzyme

in quinidine N-oxide formation, but CYP2C9 and 2E1 also catalyze minor proportions of the N-oxidation (Nielsen et al. 1999). An in vivo pharmacokinetic interaction between quinidine and erythromycin (Spinler et al. 1995), cimetidine (Kolb et al. 1984), and amiodarone (Saal et al. 1984) in humans has been reported. Similar to quinidine, CYP3A4 catalyzes the 3-hydroxylation of quinine (the enantiomer of quinidine) (Zhao et al. 1996).

Notably, dextromethorphan in combination with quinidine (trade name: Nuedexta) is indicated for the treatment of pseudobulbar affect. This drug was approved by the Food and Drug Administration for the treatment of pseudobulbar affect in February 2011. Dextromethorphan, the active ingredient, is a low-affinity, uncompetitive N-methyl-D-aspartate (NMDA) receptor antagonist (Yang and Deeks 2015). The drug combination is generally well tolerated in these studies, with no particular safety or tolerability concerns. Pseudobulbar affect is a common manifestation of brain pathology associated with many neurological diseases, including amyotrophic lateral sclerosis, Alzheimer's disease, stroke, multiple sclerosis, Parkinson's disease, and traumatic brain injury (Ahmed and Simmons 2013; Garnock-Jones 2011; Miller et al. 2011; Patatanian and Casselman 2014; Pioro et al. 2010; Rosen 2008; Schoedel et al. 2014; Smith 2006; Stahl 2013; Yang and Deeks 2015). Pseudobulbar affect is defined by involuntary and uncontrollable expressed emotion that is exaggerated and inappropriate and also incongruent with the underlying emotional state (Miller et al. 2011). In the combination, quinidine inhibits the CYP2D6-mediated metabolic conversion of dextromethorphan to its active metabolite, dextrorphan, thereby increasing dextromethorphan systemic bioavailability and driving the pharmacology toward that of the parent drug and away from ADRs of the dextrorphan metabolite. Clinical interactions of dextromethorphan/quinidine with paroxetine have been observed in healthy subjects (Schoedel et al. 2012). The combination increases the AUC of paroxetine by 30% and paroxetine increases the AUC of dextromethorphan and quinidine by 50% and 40%, respectively (Schoedel et al. 2012). There is a decrease of 12.3% for the AUC of dextrorphan. The side effect events are increased. Therefore, paroxetine should not be used with dextromethorphan/quinidine and patients should be monitored for potential side effects and dosage adjustment considered when combining these two agents. In contrast, memantine (NMDA receptor antagonist) does not interact with Nuedexta in healthy subjects (Pope et al. 2012). Nuedexta should be used together with drugs that both prolong the QT interval and are metabolized by CYP2D6 (e.g., thioridazine, pimozide); effects on QT interval may be increased.

TABLE 4.1

Reported Inhibitors of Human CYP2D6

Inhibitor	Category	Test System	Probe Substrate	MW (Da)	LogP (cLogP)	IC₅₀ (μM)	K_i (μM)	References
BP-11		HLM	Dextromethorphan			7.7 (no solvent); 18 (3% solvent)		Zebothsen et al. 2006
DHB	Furanocoumarin monomer	HLM	Dextromethorphan			0.9		Tassaneeyakul et al. 2000
DY-9760e	Novel calmodulin antagonist	HLM	Bufuralol				0.25	Tassaneeyakul et al. 2000
GF-I-1	Furanocoumarin dimer	HLM	Dextromethorphan			0.2		Tassaneeyakul et al. 2000
GF-I-4	Furanocoumarin dimer	HLM	Dextromethorphan			0.3		Tassaneeyakul et al. 2000
LY333531	Selective protein kinase C β inhibitor	HLM	Bufuralol				0.17	Ring et al. 2002
N-demethyl-LY333531		HLM	Bufuralol				1	Ring et al. 2002
MDMA		HLM	Bufuralol				4.85	Heydari et al. 2004; Van et al. 2007
NBPB	CYP2C19 selective inhibitor	HLM	Bufuralol			>100		Cai et al. 2004
NNC55-0396	Selective T-type Ca²⁺ channel inhibitor	cDNA	AMMC			29 nM	2.8 nM	Bui et al. 2008
Ro40-5966	Mibefradil hydrolyzed metabolite	cDNA	AMMC			46 nM	4.5 nM	Bui et al. 2008
17α-Ethinyl estradiol	Hormone							Chang et al. 2009
Ajmalicine	Indole alkaloid	HLM	Bufuralol	352.18			3.3 nM	Strobl et al. 1993
Ajmaline	Antiarrhythmic	HLM	Mexiletine	326.43	1.81 (1.72)			Broly et al. 1990
Alfentanil	Opioid	HLM	Imipramine	416.25	0.2 (2.21)			Henthorn et al. 1989
Amiodarone	Antiarrhythmic	cDNA	Bufuralol	645.31	7.9 (7.24)	54	45.1	Saal et al. 1984
Amitriptyline	Tricyclic antidepressant	cDNA	Dextromethorphan	277.40	4.9 (5.1)		31.0	Shin et al. 2002
			Sparteine				4	Crewe et al. 1992
Amphetamine	Psychostimulant drug	HLM	Dextromethorphan	135.10	1.8 (1.85)		26.5	Wu et al. 1997
Antipyrine	Analgesic and antipyretic	HLM	Sparteine	188.23	0.38 (1.18)		>3000	Crewe et al. 1992
Astemizole	Antihistamine	HLM	Dextromethorphan	458.57	5.8 (5.92)	36		Nicolas et al. 1999
Atomoxetine	Norepinephrine reuptake inhibitor	HLM	Bufuralol	255.35	3.9 (3.95)		3.6	Sauer et al. 2004
Atorvastatin	Statin	HLM	Dextromethorphan	558.64	5.7 (4.24)	192		Ciccone et al. 2006
		HLM	Bufuralol					Cohen et al. 2000
Azacyclonol	Terfenadine metabolite	HLM	Bufuralol	267.37	(3.25)	~100		Jones et al. 1998

(Continued)

TABLE 4.1 (CONTINUED)
Reported Inhibitors of Human CYP2D6

Inhibitor	Category	Test System	Probe Substrate	MW (Da)	LogP (cLogP)	IC$_{50}$ (µM)	K$_i$ (µM)	References
Azelastine	H$_1$ receptor antagonist	HLM	S-Metoprolol	381.90	4.9 (3.81)		1.7–12.1	Morganroth et al. 1997
		cDNA	Bufuralol					Nakajima et al. 1999
Benzyl-isothiocyanate		HLM	Dextromethorphan	149.21	(3.02)	4.8	1.2	Turpeinen et al. 2004
Bergamottin	Furanocoumarin monomer	HLM	Dextromethorphan	338.40	(5.92)	0.19		Tassaneeyakul et al. 2000
I3,II8-Biapigenin	Natural compound	cDNA	Bufuralol				2.3	Obach 2000
Buprenorphine	Opioid analgesic	cDNA	AMMC	467.64	3.8 (4.53)	0.25	0.1	Umeda et al. 2005
		HLM	Dextromethorphan			22.7	13	Umeda et al. 2005
		HLM	Bufuralol				21	Umehara et al. 2002
		HLM	Dextromethorphan				10	Zhang et al. 2003
Bupropion	Antidepressant	HLM	Dextromethorphan	239.74	3.6 (3.28)	58	21	Shin et al. 2002
Carboxyterfenadine	Terfenadine metabolite	HLM	Bufuralol	501.66	(4.8)	>300		Jones et al. 1998
Cathenamine	Indole alkaloid	HLM	Bufuralol	350.41	(2.47)		3.2	Strobl et al. 1993
Cephaeline	Natural alkaloid	HLM	Bufuralol	466.61	(4.09)	121	54	Asano et al. 2001
Cerivastatin	Statin	HLM	Bufuralol	459.55	3.4 (4.15)	237		Cohen et al. 2000
Cetirizine	H$_1$ antagonist	HLM	Dextromethorphan	388.89	2.8 (2.98)	>100		Nicolas et al. 1999
Chinidin (Quinidine)	Antiarrhythmic	HLM	Bufuralol				0.06	Strobl et al. 1993
Chloramphenicol	Antibiotic	HLM	Dextromethorphan	323.12	0.7 (1.15)	375.9	75.8	Park et al. 2003
Chlorpromazine	Antipsychotic	HLM	Dextromethorphan	318.86	4.9 (5.18)		6.3	Shin et al. 1999
			Desmethylimipramine				6	von Bahr et al. 1985
		HLM	Bufuralol				7.0	Strobl et al. 1993
Cimetidine[a]	H$_2$ receptor antagonist	HLM	Dextromethorphan	253.34	1 (0.44)		77	Madeira et al. 2004
		HLM	Bufuralol				50–55	Furuta et al. 2001; Knodell et al. 1991
		HLM	Dextromethorphan				38	Madeira et al. 2004
		cDNA	Dextromethorphan				103	Madeira et al. 2004
		HLM	Bufuralol				3.5	Strobl et al. 1993
Citalopram	SSRI	HLM	Dextromethorphan	324.39	3.5 (3.58)		19	Otton et al. 1996
		HLM	Imipramine				19	Skjelbo and Brosen 1992
		HLM	Sparteine				5.1	Crewe et al. 1992
		HLM	Metoprolol				25–88	Belpaire et al. 1998
Clomipramine	Tricyclic antidepressant	HLM	Sparteine	314.85	4.5 (5.074)		2.2	Crewe et al. 1992
		HLM	Imipramine				16	Skjelbo and Brosen 1992
Clozapine	Antipsychotic	HLM	Dextromethorphan	326.82	2.7 (3.42)	39.0		Shin et al. 1999
		HLM	Bufuralol				19	Ring et al. 1996

(Continued)

TABLE 4.1 (CONTINUED)
Reported Inhibitors of Human CYP2D6

Inhibitor	Category	Test System	Probe Substrate	MW (Da)	LogP (cLogP)	IC_{50} (µM)	K_i (µM)	References
Corynanthine	Yohimbine derivative	HLM	Bufuralol	354.44	(2.2)		0.080	Strobl et al. 1993
Cryptotanshinone	Herbal compound	HLM	Dextromethorphan	296.36	(4.93)	75		Qiu et al. 2008
Dalcetrapib	CETP inhibitor	HLM	Dextromethorphan	389.59	(6.83)	82		Derks et al. 2009
Danshensu	Herbal compound	HLM	Dextromethorphan	198.05	(−0.07)	>200		Qiu et al. 2008
Delavirdine	HIV-1 reverse transcriptase inhibitor	cDNA	Dextromethorphan	456.56	2.8 (2.77)		12.8	von Moltke et al. 2001
Desethylamiodarone	Dopamine D4 receptor antagonist	cDNA	Bufuralol	619.31		4.5	4.5	Ohyama et al. 2000
Desipramine	Tricyclic antidepressant	HLM	Sparteine	266.38	3.7 (4.02)		2.3	Crewe et al. 1992
		HLM	Venlafaxine	266.38	3.7 (4.02)		1.7	von Moltke et al. 1994
Desmethylcitalopram	Citalopram metabolite	cDNA	Dextromethorphan				12.5	Shin et al. 2002
Desmethylclomipramine	Clomipramine metabolite	HLM	Imipramine				1.3	Skjelbo and Brosen 1992
		HLM	Metoprolol				11–40	Belpaire et al. 1998
Desmethylsertraline	Sertraline metabolite	HLM	Imipramine				7.9	Skjelbo and Brosen 1992
		HLM	Metoprolol				18–31	Belpaire et al. 1998
Desvenlafaxine	Antidepressant	HLM	Metoprolol	263.38	(2.26)		>100	Oganesian et al. 2009
N-Desmethylatomoxetine	Atomoxetine metabolite	HLM	Bufuralol					Sauer et al. 2004
Dexmedetomidine	α_2-Adrenoceptor agonist	HLM	Dextromethorphan	200.28	2.8 (2.96)	1.8	5.3	Rodrigues and Roberts 1997
Dextropropoxyphene	Narcotic	HLM	Desmethylimipramine	339.47	(5.44)			Henthorn et al. 1989
10,11-Dihydroquinidine	Quinidine metabolite	HLM	Bufuralol				0.066	Strobl et al. 1993
Duloxetine	SSRI	HLM	Dextromethorphan				4.5	Otton et al. 1996
Emetine	Natural alkaloid	HLM	Bufuralol	480.64	(4.85)	80	43	Asano et al. 2001
19-Epiajmalicine	Ajmalicine derivative	HLM	Bufuralol				17 nM	Strobl et al. 1993
Esomeprazole	PPI	HLM	Bufuralol	345.62	0.6 (1.91)	>200		Li et al. 2004
Erythrohydrobupropion	Bupropion metabolite	HLM	Bufuralol				1.7	Reese et al. 2008
17β-Estradiol	Steroid	cDNA	Bufuralol	272.38	4.2 (3.57)			Hiroi et al. 2001
Fentanyl	Opioid	HLM	Imipramine	336.47	3.9 (4.12)			Henthorn et al. 1989
Flecainide	Antiarrhythmic	HLM	Bufuralol	414.34	4.6 (2.98)		0.954	Haefeli et al. 1990
Fluoxetine	SSRI	HLM	Dextromethorphan	309.32	4.6 (4.1)		0.2	Otton et al. 1993
		HLM	Dextromethorphan					Otton et al. 1996
		HLM	Dextromethorphan			0.24 (without preincubation)		Bertelsen et al. 2003
		HLM	Dextromethorphan			0.36 (without preincubation)		Bertelsen et al. 2003

(Continued)

TABLE 4.1 (CONTINUED)
Reported Inhibitors of Human CYP2D6

Inhibitor	Category	Test System	Probe Substrate	MW (Da)	LogP (cLogP)	IC50 (μM)	Ki (μM)	References
		HLM	Imipramine				1.6	Ball et al. 1997
		HLM	Desipramine				3.0	von Moltke et al. 1994
		HLM	Sparteine				0.6	Crewe et al. 1992
		HLM	Dextromethorphan				0.075	Brown et al. 2006
		HLM	Imipramine				0.92	Skjelbo and Brosen 1992
		HLM	Metoprolol				0.9–1.9	Belpaire et al. 1998
Fluphenazine	Antipsychotic	HLM	Dextromethorphan	437.52	4.2 (4.4)		9.4	Shin et al. 1999
Fluvastatin	Statin	HLM	Bufuralol	411.67	4.5 (3.69)	210		Cohen et al. 2000
		HLM	Dextromethorphan				>50	Transon et al. 1996
Fluvoxamine	SSRI	lHLM	Sparteine	318.33	3.2 (2.89)		8.2	Crewe et al. 1992
		HLM	Dextromethorphan					Otton et al. 1996
		HLM	Imipramine				8.0	Ball et al. 1997
		HLM	Dextromethorphan				2.47	Brown et al. 2006
		HLM	Imipramine				3.9	Skjelbo and Brosen 1992
Halofantrine	Antimalarial	HLM	Metoprolol	500.42	8.9 (7.34)		10–16	Belpaire et al. 1998
		HLM	Bufuralol			1.06	4.3	Halliday et al. 1995
Haloperidol	Antipsychotic	HLM	Dextromethorphan	375.86	4 (3.7)		7.2	Shin et al. 1999
Harmalol	Harmine analog	HLM	Bufuralol	200.24	(−1.06)		65	Strobl et al. 1993
Harman	Harmine analog	HLM	Bufuralol	182.22	(3.26)		86	Strobl et al. 1993
4-Hydroxyatomoxetine	Atomoxetine metabolite	HLM	Bufuralol				17	Sauer et al. 2004
6-Hydroxyazelastine	Azelastine metabolite	HLM	Bufuralol				3.0	Sauer et al. 2004
Hydroxybupropion	Bupropion metabolite	cDNA	Dextromethorphan			74	13	Shin et al. 2002
Hydroxyterfenadine	Terfenadine metabolite	HLM	Bufuralol			18		Jones et al. 1998
Hypericin	Herbal compound	cDNA	Bufuralol	504.44	(8.39)		2.6	Obach 2000
Hyperforin	Herbal compound	cDNA	Bufuralol	536.79	(6.32)		1.5	Obach 2000
Imipramine	Tricyclic antidepressant	HLM	Venlafaxine	280.41	3.9 (4.53)		3.9	von Moltke et al. 1994
Indinavir	Anti-HIV agent	cDNA	Dextromethorphan	613.79	2.9 (3.26)		28.6	Shin et al. 2002
		HLM	Desipramine				15.6	von Moltke et al. 1998
Lansoprazole	PPI	HLM	Dextromethorphan	369.36	1.9 (2.84)		44.7	Ko et al. 1997
		HLM	Dextromethorphan			213.4		Liu et al. 2005
		HLM	Bufuralol			>200		Li et al. 2004
S-Lansoprazole	PPI	HLM	Dextromethorphan			78.8		Liu et al. 2005
R-Lansoprazole	PPI	HLM	Dextromethorphan			136.3		Liu et al. 2005

(Continued)

TABLE 4.1 (CONTINUED)
Reported Inhibitors of Human CYP2D6

Inhibitor	Category	Test System	Probe Substrate	MW (Da)	LogP (cLogP)	IC₅₀ (µM)	Kᵢ (µM)	References
Lignocaine	Local anesthetic	HLM	Sparteine				200	Crewe et al. 1992
		HLM	Imipramine				40	Skjelbo and Brosen 1992
Lobelin	Alkaloid	HLM	Bufuralol	337.46	(3.61)		0.12	Strobl et al. 1993
Loratadine	H₁ antagonist	HLM	Dextromethorphan	382.88	3.8 (4.8)	15		Nicolas et al. 1999
Lovastatin acid	Statin	HLM	Dextromethorphan	427.53		>400	>50	Transon et al. 1996
Lovastatin sodium	Statin	HLM	Bufuralol	386.52	(4.11)	>100		Cohen et al. 2000
Medroxyprogesterone acetate	Steroid	HLM	Dextromethorphan					Zhang et al. 2006
Memantine	NMDA antagonist	HLM	Dextromethorphan	179.30	3.5 (3.31)	94.9	368.7	Micuda et al. 2004
		cDNA	Dextromethorphan			242.2	84.4	Micuda et al. 2004
Metoclopramide[a]	Antiemetic and gastroprokinetic (dopamine D₂ antagonist)	HLM	Dextromethorphan	299.80	1.8 (2.18)		0.96	Desta et al. 2002
Methylenedioxy-methamphetamine[a]	Drug of abuse	Human hepatocyte	Dextromethorphan	193.24	(1.81)		0.88 (DF: 1.2); 0.01 (DF: 5)	Van et al. 2007
Methylphenidate	Psychostimulant (for attention-deficit/hyperactivity disorder)	HLM	Dextromethorphan	233.31	2.1 (1.47)		1.23 (WR)	Ciccone et al. 2006
Metoclopramide	D₂ antagonist	HLM	Dextromethorphan	299.80	1.8 (2.18)		4.7	Desta et al. 2002
Metoprolol	β₁-Blocker	HLM	Sparteine	267.36	1.6 (1.8)		37	Crewe et al. 1992
Metyrapone	Adrenal cortex hormone	HLM	Dextromethorphan	226.27	1.8 (2.09)	1.06		Turpeinen et al. 2004
Mexiletine	Antiarrhythmic	HLM	Dextromethorphan	179.26	2.1 (2.17)			Broly et al. 1990
Mianserin	CNS drug	HLM	Imipramine	264.36	(3.52)		6.7	Skjelbo and Brosen 1992
Mibefradil	Calcium channel blocker (withdrawn)	cDNA	AMMC	495.63	(5.34)	129 nM	12.7 nM	Bui et al. 2008
Mirodenafil	Novel PDE5 inhibitor	cDNA	AMMC			77		Bui et al. 2008
Mizolastine	Antihistamine	HLM	Dextromethorphan	432.49	(3.43)	118		Nicolas et al. 1999
Moclobemide	CNS drug (RIMA)	HLM	Imipramine	268.74	1.5 (1.56)		140	Skjelbo and Brosen 1992
Nelfinavir	Anti-HIV agent	HLM	Desipramine	567.78	6 (4.62)		51.9	von Moltke et al. 1998
Nicardipine	Calcium channel blocker	HLM	Bufuralol	479.53	3.6 (4.34)		4.8	Nakamura et al. 2005
		HLM	Dextromethorphan					Otton et al. 1996
Norfluoxetine	SSRI	HLM	Desipramine	295.30	(4.36)		3.5	von Moltke et al. 1994

(Continued)

TABLE 4.1 (CONTINUED)
Reported Inhibitors of Human CYP2D6

Inhibitor	Category	Test System	Probe Substrate	MW (Da)	LogP (cLogP)	IC$_{50}$ (μM)	K$_i$ (μM)	References
		HLM	Sparteine				0.43	Crewe et al. 1992
		HLM	Imipramine				0.33	Skjelbo and Brosen 1992
		HLM	Metoprolol				0.6–2.1	Belpaire et al. 1998
Nortilidine	Narcotic	cDNA	Vivid CYP2D6 blue	259.34	(2.83)	0.15		Weiss et al. 2008
Nortriptyline	Tricyclic antidepressant	cDNA	Dextromethorphan	263.38	4.7 (4.65)		7.9	Shin et al. 2002
Olanzapine	Antipsychotic	HLM	Bufuralol	312.43	2 (2.65)		89	Ring et al. 1996
Omeprazole	PPI	HLM	Dextromethorphan	345.42	0.6 (1.91)		240.7	Ko et al. 1997
		HLM	Bufuralol				>200	Li et al. 2004
		HLM	Bufuralol				302	VandenBranden et al. 1996
Oxatomide	H$_1$ receptor antagonist	HLM	Bufuralol	426.55	(3.57)		57.4	Goto et al. 2005
Oxprenolol	β-Blocker	HLM	Mexiletine	265.35	2.1 (2.44)			Broly et al. 1990
Pantoprazole	PPI	HLM	Bufuralol	383.37	0.5 (2.11)	>200		Li et al. 2004
Paroxetine	SSRI	HLM	Dextromethorphan	329.37	3.6 (3.1)	2.85 (without preincubation)		Otton et al. 1996
		HLM	Dextromethorphan			0.36 (with preincubation)		Bertelsen et al. 2003
		HLM	Dextromethorphan				4.85	Bertelsen et al. 2003
		HLM	Dextromethorphan				3.2	Bertelsen et al. 2003
		HLM	Imipramine				0.15	Ball et al. 1997
		HLM	Sparteine				0.15	Crewe et al. 1992
		HLM	Imipramine				0.36	Skjelbo and Brosen 1992
		HLM	Metoprolol				0.9–1.4	Belpaire et al. 1998
Paroxetine metabolite I glucuronide		HLM	Sparteine				>200	Crewe et al. 1992
Paroxetine metabolite I sulfate		HLM	Sparteine				120	Crewe et al. 1992
Paroxetine metabolite I		HLM	Sparteine				16	Crewe et al. 1992
Paroxetine metabolite II		HLM	Sparteine				0.5	Crewe et al. 1992
Paroxetine metabolite III		HLM	Sparteine				>20	Crewe et al. 1992
Perphenazine	Antipsychotic	HLM	Dextromethorphan	403.97	3.9 (4.15)		0.8	Shin et al. 1999
		HLM	Imipramine				0.16	Skjelbo and Brosen 1992
Pimozide[a]	Antipsychotic	HLM	Dextromethorphan	461.55	5.6 (6.36)		0.75	Desta et al. 1998
Pravastatin	statin	HLM	Dextromethorphan	424.53	2.2 (2.23)		>50	Transon et al. 1996

(Continued)

TABLE 4.1 (CONTINUED)
Reported Inhibitors of Human CYP2D6

Inhibitor	Category	Test System	Probe Substrate	MW (Da)	LogP (cLogP)	IC$_{50}$ (μM)	K$_i$ (μM)	References
Prodipin	Dopamine-releasing compound (for Parkinson's disease)	HLM	Bufuralol	279.42	(4.29)		0.0048	Strobl et al. 1993
Progesterone	Steroid	cDNA	Bufuralol	314.46	3.5 (3.58)		33	Hiroi et al. 2001
Propafenone	Antiarrhythmic	HLM	Mexiletine	341.44	3.2 (3.1)			Broly et al. 1990
Propranolol	β-Blocker	HLM	Mexiletine	259.34	3 (3.03)			Broly et al. 1990
		HLM	Imipramine				10	Skjelbo and Brosen 1992
Protocatechuic acid	Natural compound	HLM	Dextromethorphan	154.12	(1.16)	>200		Qiu et al. 2008
Protocatechuic aldehyde	Natural compound	HLM	Dextromethorphan	138.12	(1.14)	>200		Qiu et al. 2008
Quinidine	Antiarrhythmic agent	HLM	Bufuralol	324.42	2.6 (2.82)		60 nM	Strobl et al. 1993
		HLM	Dextromethorphan			0.062 (no preincubation)	0.05 (no preincubation)	Perloff et al. 2009
		HLM	Dextromethorphan			0.026 (10 min preincubation −NADPH)		Perloff et al. 2009
		HLM	Dextromethorphan			0.029 (10 min preincubation +NADPH)		Perloff et al. 2009
		HLM	Dextromethorphan			0.031 (30 min preincubation −NADPH)		Perloff et al. 2009
		HLM	Dextromethorphan			0.06 (30 min preincubation +NADPH)		Perloff et al. 2009
		HLM	Dextromethorphan				0.12	Qiu et al. 2008
		HLM	Dextromethorphan			0.076		Volotinen et al. 2007
		HLM	Bufuralol			0.03		Li et al. 2004
		HLM	Bufuralol			0.14		Cai et al. 2004
		HLM	Dextromethorphan			0.4 (without preincubation)		Bertelsen et al. 2003
		HLM	Dextromethorphan			0.48 (with preincubation)		Bertelsen et al. 2003
		HLM	Desipramine			0.16		von Moltke et al. 1998
		HLM	Sparteine				0.03	Crewe et al. 1992
		HLM	Dextromethorphan				0.033	Brown et al. 2006
		HLM	Imipramine				9–92	Skjelbo and Brosen 1992
		HLM	Metoprolol				0.04–0.18	Belpaire et al. 1998
		HLM	Bufuralol				0.03	Ring et al. 1996
Quinine	Antimalarial agent	HLM	Bufuralol	324.42	2.6 (2.82)		4.6	Strobl et al. 1993
Quininone	Antimalarial agent	HLM	Bufuralol	322.40	(3.06)		0.72	Strobl et al. 1993

(Continued)

TABLE 4.1 (CONTINUED)

Reported Inhibitors of Human CYP2D6

Inhibitor	Category	Test System	Probe Substrate	MW (Da)	LogP (cLogP)	IC$_{50}$ (µM)	K_i (µM)	References
Rabeprazole	PPI	HLM	Bufuralol	359.44	0.6 (2.05)	>200		Li et al. 2004
Rabeprazole thioether	PPI	HLM	Bufuralol	359.44	(1.83)	12.4		Li et al. 2004
(R,R)-Reboxetine	Antidepressant	cDNA		313.39	(2.82)	5.6	2.5	Wienkers et al. 1999
(S,S)-Reboxetine	Antidepressant	cDNA				4.9		Wienkers et al. 1999
Risperidone	Antipsychotic	HLM	Dextromethorphan	410.48	2.5 (3.28)	21.9		Shin et al. 1999
Ritonavir	Anti-HIV agent	HLM	Desipramine	720.94	3.9 (4.24)		4.84	von Moltke et al. 1998
		HLM	Dextromethorphan			2.5		Kumar et al. 1999
Rosiglitazone	PPARγ agonist	HLM	Dextromethorphan	357.43	2.4 (2.95)	42.1	29.5	Sahi et al. 2003
Salvianolic acid B	Herbal compound	HLM	Dextromethorphan	718.61	(2.14)	>200		Qiu et al. 2008
Saquinavir	Anti-HIV agent	HLM	Desipramine	670.84	3.8 (4.04)		24.0	von Moltke et al. 1998
Sempervirine	Indole alkaloid	cDNA	AMMC	273.35	(2.19)		9.7	Bui et al. 2008
Serpentine	Alkaloid	cDNA	AMMC	348.40	(−0.04)		2.2	Bui et al. 2008
Sertraline	SSRI	HLM	Sparteine	305.23	5.1 (5.06)		0.7	Crewe et al. 1992
		HLM	Imipramine				24.7	Ball et al. 1997
		HLM	Desipramine				22.7	von Moltke et al. 1994
		HLM	Metoprolol				11–20	Belpaire et al. 1998
Sertraline metabolite	SSRI	HLM	Desipramine				16.0	von Moltke et al. 1994
Silibinin	Herbal compound	HLM	Dextromethorphan	482.12	(2.59)	173		Beckmann-Knopp et al. 2000
Simvastatin acid	Statin	HLM	Dextromethorphan			>400	>50	Transon et al. 1996
Simvastatin sodium	Statin	HLM	Bufuralol	441.56	4.3 (4.85)			Cohen et al. 2000
Sirolimus	Immunodepressant	HLM	Desipramine	914.17			5	Boni et al. 2009a
Tanshinone I	Herbal compound	HLM	Dextromethorphan	276.29	(4.44)	120		Qiu et al. 2008
Tanshinone IIA	Herbal compound	HLM	Dextromethorphan	276.29	(4.44)	>200		Qiu et al. 2008
Temsirolimus	Anti-RCC (renal cell carcinoma) immunodepressant	HLM	Desipramine	1030.29	(4.39)		1.5	Boni et al. 2009a
Terbinafine	Synthetic allylamine antifungal agent	cDNA	Dextromethorphan	291.43	5.9 (5.51)		0.022	Abdel-Rahman et al. 1999a
		HLM	Dextromethorphan				0.35	Abdel-Rahman et al. 1999a
		HLM	Dextromethorphan				28–44 nM	Abdel-Rahman et al. 1999b

(Continued)

TABLE 4.1 (CONTINUED)
Reported Inhibitors of Human CYP2D6

Inhibitor	Category	Test System	Probe Substrate	MW (Da)	LogP (cLogP)	IC_{50} (μM)	K_i (μM)	References
Terfenadine	H₁ receptor antagonist	HLM	Bufuralol	471.67	7.1 (5.89)		4.6	Jones et al. 1998; Smith and Jones 1992
		HLM	Dextromethorphan			18		Nicolas et al. 1999
Tetrahydroalstonine	Ajmalicine derivative	HLM	Bufuralol	352.43	(2.88)		5.0	Strobl et al. 1993
Testosterone	Steroid	cDNA	Bufuralol	288.42	3.6 (2.99)		63	Hiroi et al. 2001
Thioridazine	Antipsychotic	HLM	Dextromethorphan	370.58	5.9 (5.93)		1.4	Shin et al. 1999
		HLM	Desmethylimipramine				0.75	von Bahr et al. 1985
		HLM	Sparteine				0.52	Crewe et al. 1992
cis-Thiothixene	Antipsychotic	HLM	Dextromethorphan	433.63	(4.01)	65.0		Shin et al. 1999
Ticlopidine	Antiplatelet agent (P2Y₁₂ antagonist)	HLM	Dextromethorphan	263.79	2.9 (4.25)		3.4	Ko et al. 2000
Tilidine	Narcotic	HLM	Dextromethorphan	273.17	(3)	4.4		Turpeinen et al. 2004
Threohydrobupropion	Bupropion metabolite	cDNA	Vivid CYP2D6 blue	241.76	(2.88)	0.16		Weiss et al. 2008
Triethylenethio-phosphoramide	Chemical compound	HLM	Bufuralol	189.22	(0.52)		5.4	Reese et al. 2008
		HLM	Dextromethorphan			>1000		Turpeinen et al. 2004
Trifluperidol	Antipsychotic	HLM	Bufuralol	409.42	(2.99)		0.17	Strobl et al. 1993
β-Yohimbine	Yohimbine	HLM	Bufuralol	354.44	2.73 (2.36)		0.031	Strobl et al. 1993
Venlafaxine	Antidepressant	HLM	Dextromethorphan	277.40	2.8 (2.69)		33	Otton et al. 1996
		HLM	Imipramine				41.0	von Moltke et al. 1994
R-(+)-Venlafaxine	Antidepressant	HLM	Dextromethorphan				>100	Oganesian et al. 2009
		HLM	Dextromethorphan				52	Otton et al. 1996
S-(−)-Venlafaxine	Antidepressant	HLM	Dextromethorphan				22	Otton et al. 1996
Xanthate C8		HLM	Dextromethorphan			219		Turpeinen et al. 2004
Zafirlukast	Cysteinyl leukotriene D₄ antagonist	HLM	Dextromethorphan	575.68	5.4 (4.84)	116		Shader et al. 1999
Zileuton[a]	5-Lipoxygenase inhibitor	HLM	Bufuralol	236.29	0.9 (2.01)	>100		Lu et al. 2003
Zuclopenthixol	Antipsychotic	HLM	Dextromethorphan	400.96	(5.14)			Shin et al. 1999

Note: AMMC, 3-[2-N,N-diethyl-N-methylamino)-7-methoxy-4-methylcoumarin; BP-11, 6-amino-11,12-dihydro-11-(4-hydroxy-3,5-dimethoxyphenyl)benzo[c]phenanthridine; cDNA, cDNA-expressed enzyme; DHB, 6′,7′-dihydroxybergamottin; DY-9760e, 3-[2-[4-(3-chloro-2-methylphenyl)-1-piperazinyl]ethyl]-5,6-dimethoxy-1-(4-imidazolylmethyl)-1H-indazoledihydrochloride 3.5 hydrate; GF-I-1, (4-[[6-hydroxy-7[[1-t[(1-hydroxy-1-methyl)ethyl]-4-methyl-6-(7-oxo-7H-furo[3,2-g][1]benzopyran-4-yl)-4-hexenyl]oxy]-3,7-dimethyl-2-octenyl]oxy]-7H-furo[3,2-g][1]benzopyran-7-one; GF-I-4,4-[[6-hydroxy-7[[4-methyl-1-(1-methylethenyl)-6-(7-oxo-7H-furo[3,2-g][1]benzopyran-4-yl)-4-hexenyl]oxy]-3,7-dimethyl-2-octenyl]oxy]-7H-furo[3,2-g][1]benzopyran-7-one; HLM, human liver microsome; LY333531, (S)-13[(dimethylamino)methyl]-10,11,14,15-tetrahydro-4,9:16,21-dimetheno-1H,13H-dibenzo[e,k]pyrrolo[3,4h][1,4,13]oxadiazacyclohexadecene; NBPB, (−)-N-3-benzyl-phenobarbital; PPI, proton pump inhibitor; RIMA, reversible inhibitor of monoamine oxidase type A; SSRI, selective serotonin reuptake inhibitor.

a Indicates mechanism-based inhibitor.

4.3 MECHANISM-BASED INHIBITORS OF CYP2D6

4.3.1 PAROXETINE

Paroxetine, a selective serotonin reuptake inhibitor (SSRI), inhibits CYP2D6 activity at IC_{50} concentrations ranging from 150 nM to 2.0 μM, depending on the substrate (Figure 4.1) (Sanchez et al. 2014). Paroxetine is also a mechanism-based inhibitor of CYP2D6 (Bertelsen et al. 2003; Livezey et al. 2012). However, fluoxetine is not a mechanism-based inhibitor of CYP2D6.

A number of drug interactions with paroxetine have been documented. Paroxetine has been shown to decrease the clearance of dacomitinib (Ruiz-Garcia et al. 2014), desipramine (Alderman et al. 1997b), sparteine (Jeppesen et al. 1996), amitriptyline (Leucht et al. 2000), imipramine (Albers et al. 1996), clozapine (Joos et al. 1997), perphenazine (Ozdemir et al. 1997), metoprolol (Hemeryck et al. 2000), flecainide (Lim et al. 2010), perhexiline (Alderman et al. 1997a), risperidone (Saito et al. 2005; Spina et al. 2001), fexofenadine (Saruwatari et al. 2012), ibogaine (Glue et al. 2015), methylenedioxymethamphetamine (MDMA) (Farre et al. 2007), timolol (Maenpaa et al. 2014), carvedilol (Stout et al. 2010), metoprolol (Stout et al. 2011), bencycloquidium bromide (Agbokponto et al. 2015), nebivolol (Briciu et al. 2014), and atomoxetine (Belle et al. 2002), where the clearance of the victim drugs is impaired by 1.3- to 8-fold. These victim drugs are all partially metabolized by CYP2D6 and their inhibition by paroxetine is considered one of the major mechanisms for the observed drug–paroxetine interactions (Ereshefsky et al. 1995; Spina et al. 2008).

4.3.2 CIMETIDINE

Cimetidine is a mechanism-based inhibitor of CYP2D6, with k_{inact} and K_i of 0.03 min^{-1} and 77 μM, respectively (Figure 4.1)

(Madeira et al. 2004). It is also a weak reversible inhibitor of CYP2D6 (Furuta et al. 2001; Knodell et al. 1991; Madeira et al. 2004; Martinez et al. 1999). Cimetidine reversibly inhibits CYP2D6-mediated bufuralol 1'-hydroxylation with a K_i of 50 to 55 μM in vitro (Furuta et al. 2001; Knodell et al. 1991). When dextromethorphan O-demethylation is used as a probe of CYP2D6 activity, cimetidine competitively inhibits CYP2D6 with a K_i of 38 μM in human liver microsomes and is a mixed-type inhibitor of recombinant CYP2D6 with a K_i of 103 μM (Madeira et al. 2004).

Cimetidine has been shown to reduce the clearance of mirtazapine (Sitsen et al. 2000), imipramine (Abernethy et al. 1984; Miller and Macklin 1983), timolol (Ishii et al. 2000), nebivolol (Kamali et al. 1997), sparteine (Schellens et al. 1989), loratadine (Kosoglou et al. 2000), nortriptyline (Miller et al. 1983), gabapentin (Lal et al. 2010), and desipramine (Amsterdam et al. 1984) in humans. Cimetidine reduces the formation clearance of 2-hydroxydesipramine from desipramine by 47% in extensive metabolizers (EMs) of CYP2D6 only, but no interaction is noted in poor metabolizers (PMs) (Steiner and Spina 1987). Cimetidine reduces the formation clearance of 5-dehydrosparteine from sparteine in EMs by 37%, but plasma metabolite level is below the level of detection in PMs (Schellens et al. 1989). Cimetidine also appears to inhibit CYP2D6-mediated metabolism of debrisoquine (Philip et al. 1989) and dextromethorphan (Arnold et al. 1997). These victim drugs are all substrates of CYP2D6 and thus inhibition of this enzyme by cimetidine is considered one of the major mechanisms.

4.3.3 METOCLOPRAMIDE

Metoclopramide, a gastroprokinetic and antiemetic agent, is a substrate and inhibitor of CYP2D6 (Figure 4.1) (Desta et al.

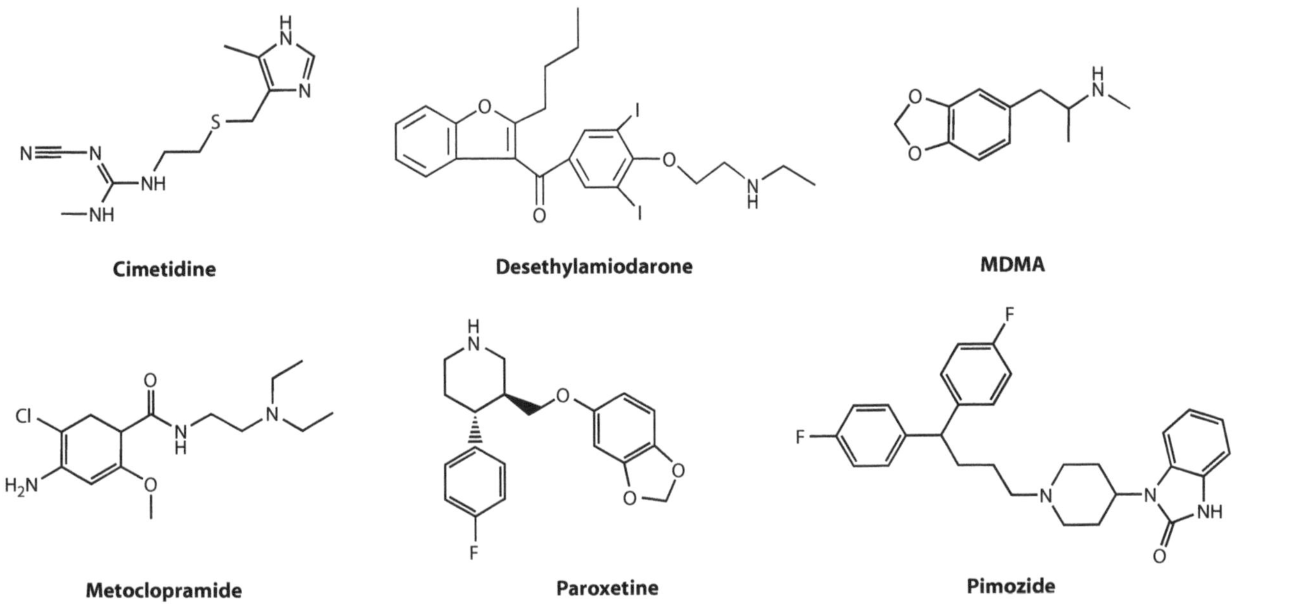

FIGURE 4.1 Chemical structures of mechanism-based inhibitors of CYP2D6. These include cimetidine, desethylamiodarone, MDMA, metoclopramide, paroxetine, and pimozide.

2002; Livezey et al. 2014). Metoclopramide is a potent inhibitor of CYP2D6 at therapeutically relevant concentrations (K_i = 4.7 µM), with negligible effect on other CYPs tested (Livezey et al. 2014). Further time-dependent inhibition of CYP2D6 is observed when metoclopramide was preincubated with human liver microsomes and NADPH-generating system before the substrate probe is added. The k_{inact} is 0.02 min^{-1} and the K_i is 0.96 µM (Desta et al. 2002).

As a CYP2D6 inhibitor, metoclopramide may reduce the clearance of itself and drugs that are substantially metabolized by CYP2D6. Metoclopramide (100 µM) decreases CYP2D6-dependent codeine O-demethylation to form morphine by 51% (Sindrup et al. 1992). Particular attention should be given when a high dose of metoclopramide is coadministered with CYP2D6 substrates in critically ill patients and in patients with renal and hepatic failure, where metoclopramide appears to exhibit excessive accumulation (Bateman 1983; Bateman et al. 1981). Metoclopramide is known to alter the pharmacokinetics of several coadministered drugs. Most of these effects can be ascribed to the ability of metoclopramide to enhance gastric emptying and increased rate of absorption of these drugs (Bateman 1983).

4.3.4 PIMOZIDE

Pimozide, a potent neuroleptic used extensively in Europe for the treatment of schizophrenia and other psychiatric diseases, is a potent inhibitor of CYP2D6 in human liver microsomes with a K_i of 0.75 µM (Figure 4.1) (Desta et al. 1998). Preincubation of pimozide with human liver microsomes and an NADPH-generating system for 15 min increases its inhibitory potency for CYP2D6 (K_i from 20.2 to 0.75 µM) (Desta et al. 1998), suggesting a metabolism-mediated inhibition by pimozide. Pimozide is known to be concentrated in the liver (~11-fold) relative to plasma (Pinder et al. 1976). The K_i (<1 µM) of pimozide for the inhibition of CYP2D6 is close to therapeutic concentrations of pimozide in the liver in vivo, which suggests that pimozide is likely to be a clinically important CYP2D6 inhibitor. The inhibitory effect of pimozide on CYP2E1, 2C9, 2C19, and 1A2 is small even at concentrations that are 100 times higher than therapeutic plasma concentrations of pimozide. Pimozide is mainly metabolized by CYP3A4 with contribution from CYP1A2, but CYP2D6 does not metabolize this drug in human liver microsomes (Pinder et al. 1976). Pimozide is superior to haloperidol in controlling symptoms of Tourette's syndrome and has less extrapyramidal symptoms (Sallee et al. 1997), and identifying potential risk factors that could modulate the efficacy and toxicity of pimozide is important to optimize the use of this otherwise effective neuroleptic drug. There have been case reports where pimozide is found to increase neurological ADRs with paroxetine (Horrigan and Barnhill 1994) and cardiac ADRs and psychomimetic effects with fluoxetine (Hansen-Grant et al. 1993). As a potent inhibitor of CYP2D6 in vitro, pimozide may increase the plasma concentrations of drugs that are mainly metabolized by CYP2D6 (e.g., tricyclic antidepressants, SSRIs, neuroleptics, and codeine) in vivo.

4.3.5 OTHER COMPOUNDS

Desethylamiodarone is a mechanism-based inhibitor of CYP2D6 (Figure 4.1) (Ohyama et al. 2000). It also inactivates CYP1A1, 1A2, and 2B6. SCH 66712 [5-fluoro-2-[4-[(2-phenyl-1H-imidazol-5-yl)methyl]-1-piperazinyl]pyrimidine], a potent and selective antagonist of dopamine D$_4$ receptor, is a potent mechanism-based inhibitor of CYP2D6 in human liver microsomes (Livezey et al. 2012; Palamanda et al. 2001). (1-[(2-Ethyl-4-methyl-1H-imidazol-5-yl)-methyl]-4-[4-(trifluoromethyl)-2-pyridinyl]piperazine is also an inactivator of CYP2D6 (Livezey et al. 2012).

MDMA is a mechanism-based inhibitor of CYP2D6 (Figure 4.1) (Heydari et al. 2004; Van et al. 2007). The values of k_{inact} (0.12–0.26 min^{-1}) and K_i (8.8–45.3 µM) for MDMA are comparable to those reported for paroxetine (k_{inact} = 0.17 min^{-1}; K_i = 4.85 µM), another mechanism-based inhibitor of CYP2D6 that also contains a methylenedioxyphenyl ring (Bertelsen et al. 2003). The presence of the methylenedioxyphenyl ring in MDMA and methylenedioxyamphetamine is most probably involved in the formation of a metabolic intermediate complex with CYP2D6, although such complexes are also generated by amine groups. CYP-mediated oxidation of the methylene carbon in both compounds would lead to formation of unstable intermediates, which are then demethylenated to catechols (3,4-dihydroxymethamphetamine and α-methyldopamine) or dehydrated to carbenes. The former compounds may be oxidized further to orthoquinones, which can react with nucleophilic groups on macromolecules or conjugate with glutathione to form neurotoxins (Easton et al. 2003). The carbene intermediates are likely to form covalent complexes with the heme iron of CYP2D6.

4.4 REVERSIBLE AND MIXED-TYPE INHIBITORS OF CYP2D6

A number of CYP2D6 substrates and other compounds have been found to inhibit CYP2D6. The inhibition of CYP2D6 by these compounds is often reversible and time independent. The inhibition of CYP2D6 by these compounds has important implications in clinical drug–drug interactions.

4.4.1 CENTRAL NERVOUS SYSTEM DRUGS

4.4.1.1 Antipsychotics

Many antipsychotic drugs including olanzapine, chlorpromazine, fluphenazine, perphenazine, haloperidol, thioridazine, risperidone, clozapine, trifluperidol, and zuclopenthixol are metabolized by CYP2D6 and also significantly inhibit this enzyme (Figure 4.2) (Shin et al. 1999). The K_i values for CYP2D6-catalyzed dextromethorphan O-demethylation by thioridazine and perphenazine are 1.4 and 0.8 µM, respectively. The K_i values of chlorpromazine, fluphenazine, and haloperidol are 6.3, 9.4, and 7.2 µM, respectively. cis-Thiothixene, clozapine, and risperidone exhibit weaker inhibition than the other drugs tested, with IC$_{50}$ values of 65.0, 39.0, and 21.9 µM, respectively (Shin et al. 1999). These antipsychotic drugs show little

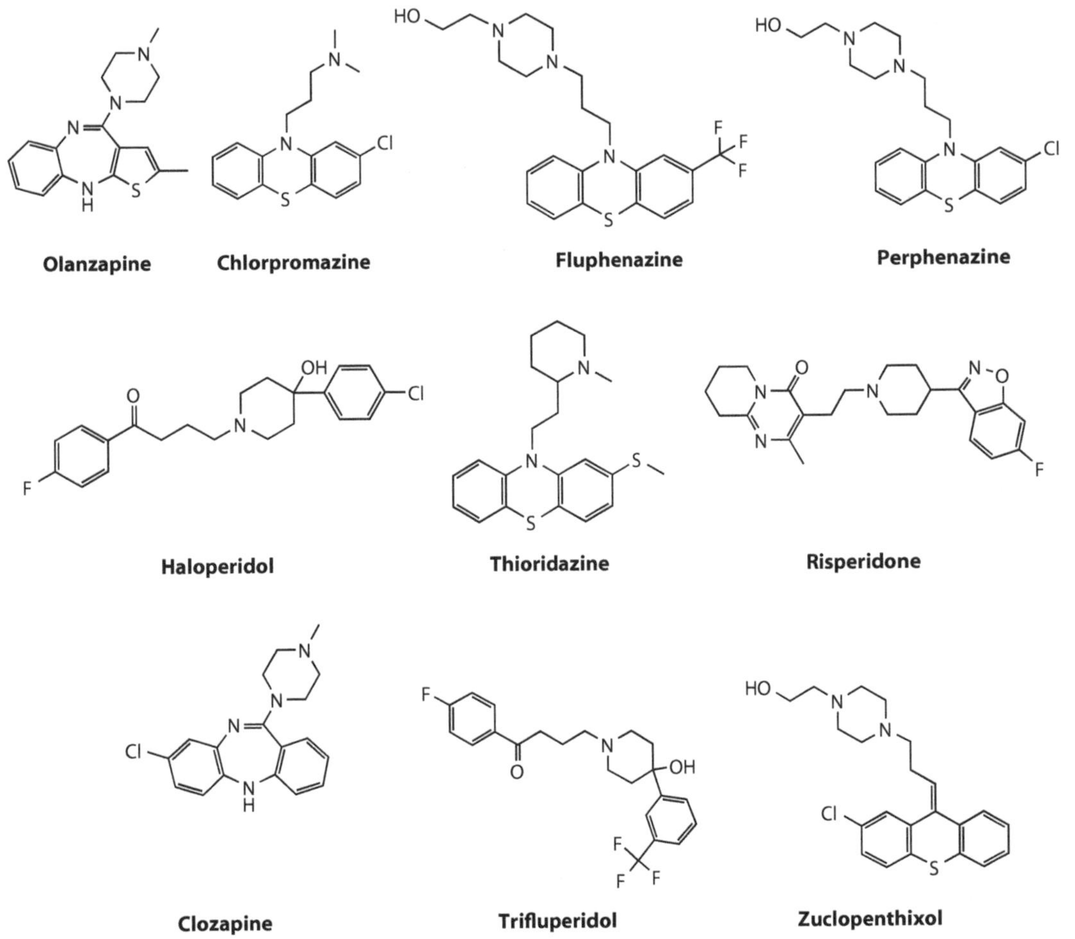

FIGURE 4.2 Chemical structures of antipsychotics that are inhibitors of CYP2D6. These include olanzapine, chlorpromazine, fluphenazine, perphenazine, haloperidol, thioridazine, risperidone, clozapine, trifluperidol, and zuclopenthixol.

inhibitory effect on CYP1A2-, 2C9-, 2C19-, or 3A4-catalyzed reactions. Most antipsychotic drugs seem to be highly selective inhibitors of CYP2D6. von Bahr et al. (1985) have reported K_i values of 0.75 and 6 µM for thioridazine and chlorpromazine for CYP2D6-catalyzed desmethylimipramine 2-hydroxylation, respectively. Clozapine inhibits CYP2D6-catalyzed bufuralol 1′-hydroxylation with a K_i of 19 µM (Ring et al. 1996).

These results suggest that antipsychotic drugs may develop pharmacokinetic drug interactions with coadministered antipsychotics and antidepressants such as amitriptyline, imipramine, nortriptyline, desipramine, clomipramine, maprotiline, trazodone, paroxetine, and fluoxetine that not only are metabolized by CYP2D6 but also inhibit this enzyme. Thioridazine, chlorpromazine, clozapine, perphenazine, haloperidol, zuclopenthixol, and fluphenazine have been reported to increase the plasma concentrations of nortriptyline, desipramine, and imipramine, which are partly metabolized by CYP2D6 (Linnet 1995; Linnoila et al. 1982; Maynard and Soni 1996; Mulsant et al. 1997; Smith and Riskin 1994).

4.4.1.2 SSRIs

Most SSRIs are reversible CYP2D6 inhibitors (Figure 4.3) (Ball et al. 1997; Crewe et al. 1992; Otton et al. 1996).

Fluoxetine inhibits CYP2D6 with a K_i of 0.2 µM (Otton et al. 1993). Otton et al. (1996) have compared the inhibitory potency of several SSRIs for CYP2D6-catalyzed dextromethorphan O-demethylation. The authors have found that paroxetine, fluoxetine, norfluoxetine, fluvoxamine, and sertraline are potent inhibitors of CYP2D6-catalyzed dextromethorphan O-demethylation with K_i values of 0.065 to 1.8 µM (Otton et al. 1996). Citalopram and duloxetine inhibit CYP2D6 with a K_i of 19 and 4.5 µM, respectively. Venlafaxine, R-(+)-venlafaxine, and S-(−)-venlafaxine cause less potent inhibition for CYP2D6, with K_i values of 33, 52, and 22 µM, respectively (Otton et al. 1996). Venlafaxine does not inhibit CYP1A2 and 2C9. Consistently, fluoxetine, paroxetine, fluvoxamine, and sertraline are found to be inhibitors of CYP2D6-mediated imipramine 2-hydroxylation, with K_i values of 1.6, 3.2, 8.0, and 24.7 µM, respectively (Ball et al. 1997). Fluoxetine and norfluoxetine also significantly inhibit desipramine 2-hydroxylation with K_i values of 3.0 and 3.5 µM, respectively, while sertraline and its metabolite desmethylsertraline inhibit the reaction with K_i of 22.7 and 16.0 µM, respectively (von Moltke et al. 1994). Venlafaxine is a less potent inhibitor of imipramine 2-hydroxylation with a K_i of 41.0 µM.

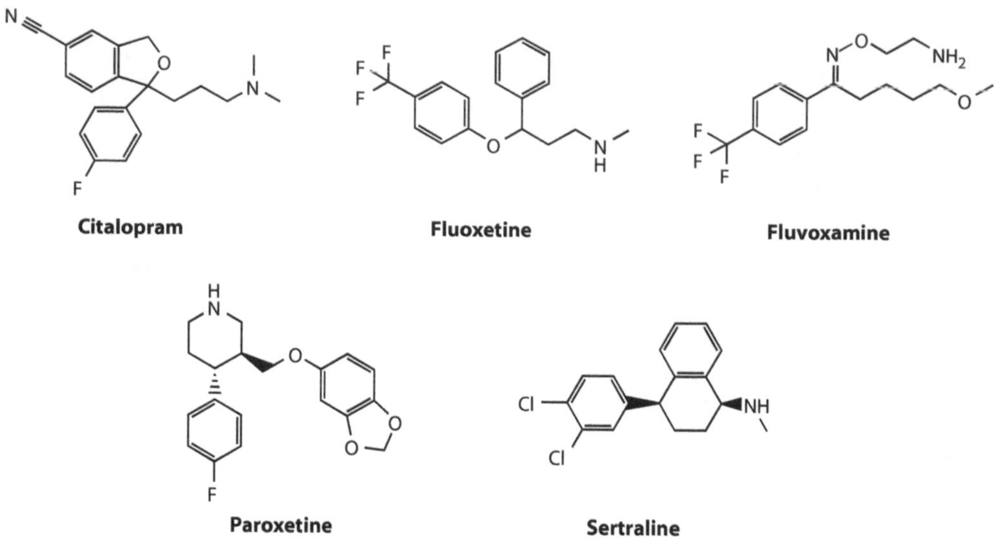

FIGURE 4.3 Chemical structures of SSRIs that are reversible inhibitors of CYP2D6. These include citalopram, fluoxetine, fluvoxamine, paroxetine, and sertraline.

A number of drug interactions with fluoxetine have been documented. Fluoxetine increased the AUC of metoclopramide 1.9-fold (Vlase et al. 2006). Fluoxetine has been shown to increase the AUC and plasma levels of tricyclic antidepressants. Coadministration of fluoxetine at 20 mg/day at steady state produces a three- to fourfold increase in C_{max} and AUC of desipramine and imipramine, while the long half-life of the active metabolite norfluoxetine is responsible for a significant and long-lasting (~3 weeks) increase in the plasma desipramine concentration after discontinuation of fluoxetine (Leroi and Walentynowicz 1996; Preskorn et al. 1994; Vandel et al. 1992).

Duloxetine coadministration increases the AUC and C_{max} of desipramine 122% and 63% in humans, respectively (Skinner et al. 2003), while desvenlafaxine has only marginal effect on the exposure of desipramine (Patroneva et al. 2008). Sertraline results in an approximately 30%–44% elevation in plasma desipramine concentrations (Alderman et al. 1997b; Barros and Asnis 1993; Preskorn et al. 1994). Citalopram causes ~50% increase in the AUC of desipramine (Gram et al. 1993).

Consistent with its minimal in vitro effect on CYP2D6, fluvoxamine shows minimal in vivo pharmacokinetic interaction with desipramine, but increases the AUC of imipramine approximately three- to fourfold, probably through inhibition of CYP3A3/4, 1A2, and 2C19. Desvenlafaxine increases the AUC of desipramine by 17%–36% (Nichols et al. 2009; Preskorn et al. 2008), and venlafaxine increases imipramine C_{max} and elevated AUC by 40% (Albers et al. 2000). Thus, the extent of the in vivo interaction between the SSRIs and tricyclic antidepressants largely reflects their in vitro inhibitory potency against CYP2D6 and other CYP enzymes (Ereshefsky et al. 1995).

4.4.1.3 Tricyclic Antidepressants

Imipramine and desipramine inhibit CYP2D6-dependent venlafaxine O-demethylation, with K_i values of 3.9 and 1.7 µM,

respectively (Ball et al. 1997). Amitriptyline, imipramine, nortriptyline, and desipramine inhibit CYP2D6-catalyzed dextromethorphan O-demethylation competitively, with estimated K_i values of 31.0, 28.6, 7.9, and 12.5 µM, respectively (Figure 4.4) (Shin et al. 2002). Imipramine reduces the plasma levels of olanzapine by 19% (Callaghan et al. 1997) but does not alter the clearance of bromperidol (Suzuki et al. 1996).

4.4.1.4 Other Antidepressants

Both bupropion and hydroxybupropion inhibit CYP2D6-mediated dextromethorphan O-demethylation, with IC_{50} values of 58 and 74 µM, respectively (Figure 4.5) (Hesse et al. 2000). When bufuralol is used as a probe in human liver microsomes, the two metabolites of bupropion (erythrohydrobupropion and threohydrobupropion) are more potent inhibitors of CYP2D6 activity (K_i = 1.7 and 5.4 µM, respectively) than hydroxybupropion (K_i = 13 µM) or bupropion (K_i = 21 µM) (Reese et al. 2008). Bupropion increases the AUC of desipramine fivefold in humans (Reese et al. 2008).

4.4.1.5 Narcotics

Alfentanil, fentanyl, and dextropropoxyphene are found to competitively inhibit CYP2D6-mediated 2-hydroxylation of desmethylimipramine (Figure 4.6) (Henthorn et al. 1989). However, other narcotics including codeine, meperidine, methadone, morphine, or nalbuphine do not inhibit this reaction. Because of the rapid elimination of atomoxetine, its average steady-state concentrations are well below the K_i. Tilidine and nortilidine (both narcotics) inhibit CYP2D6, 2C19, and 3A4 in vitro (Weiss et al. 2008).

4.4.1.6 Other Central Nervous System Drugs

Atomoxetine, N-desmethylatomoxetine, and 4-hydroxyatomoxetine competitively inhibit CYP2D6-dependent bufuralol 1′-hydroxylation, with apparent K_i values of 3.6, 5.3, and 17 µM, respectively (Figure 4.7) (Sauer et al. 2004). The

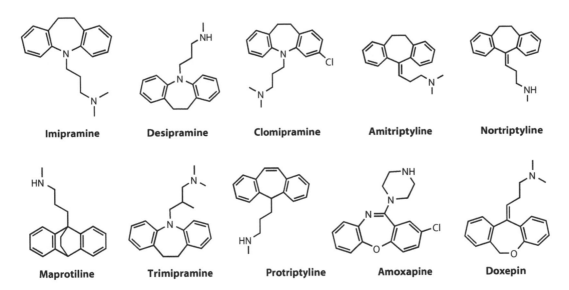

FIGURE 4.4 Chemical structures of tricyclic antidepressants that are inhibitors of CYP2D6. These include imipramine, desipramine, clomipramine, amitriptyline, nortriptyline, maprotiline, trimipramine, protriptyline, amoxapine, and doxepin.

FIGURE 4.5 Chemical structures of bupropion and hydroxybupropion, which are inhibitors of CYP2D6.

FIGURE 4.7 Chemical structures of atomoxetine, *N*-desmethylatomoxetine, and 4-hydroxyatomoxetine, which are inhibitors of CYP2D6.

additive inhibition of CYP2D6 in the EMs by atomoxetine and its two metabolites is 60%. However, coadministration of atomoxetine does not alter the C_{max} and AUC of desipramine in healthy subjects (Sauer et al. 2004). In addition, the antidepressants (*R,R*)- and (*S,S*)-reboxetine inhibit CYP2D6 with IC$_{50}$ of 5.6 and 4.9 μM, respectively (Wienkers et al. 1999).

4.4.2 H₁ Receptor Antagonists

Terfenadine, a nonsedating H₁ receptor antagonist, inhibits CYP2D6-mediated bufuralol 1′-hydroxylation with a K_i of 4.6 μM (Figure 4.8) (Jones et al. 1998; Smith and Jones 1992). Both azacyclonol and carboxyterfenadine, two metabolites of terfenadine, show weak inhibition of CYP2D6 (IC$_{50}$ values of ~100 and >300 μM, respectively), whereas

hydroxyterfenadine inhibits bufuralol 1′-hydroxylase activity with an IC$_{50}$ of 18 μM (Jones et al. 1998). Azelastine inhibits CYP2D6-dependent *S*-metoprolol hydroxylation with K_i of 1.7–12.1 μM (Figure 4.8) (Morganroth et al. 1997). Azelastine, desmethylazelastine, and 6-hydroxyazelastine competitively inhibit CYP2D6-dependent bufuralol 1′-hydroxylation, with K_i of 1.2, 1.5, and 3.0 μM, respectively (Figure 4.8) (Nakajima et al. 1999). In addition, oxatomide is a moderate inhibitor of CYP2D6-mediated bufuralol 1′-hydroxylation with a K_i of 57.4 μM (Figure 4.8) (Goto et al. 2005).

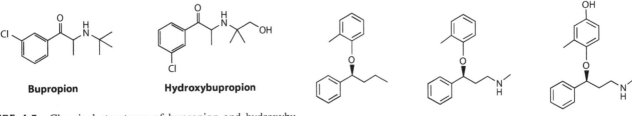

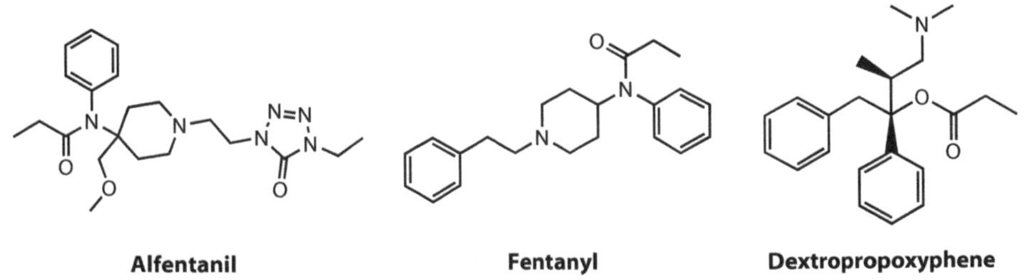

FIGURE 4.6 Chemical structures of alfentanil, fentanyl, and dextropropoxyphene, which are inhibitors of CYP2D6.

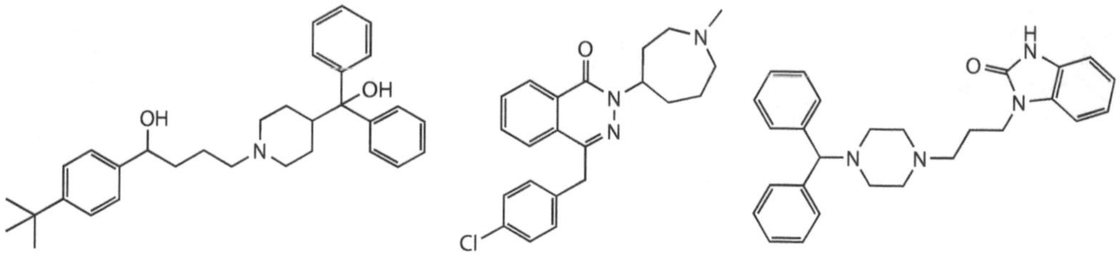

FIGURE 4.8 Chemical structures of H₁ receptor antagonists that are inhibitors of CYP2D6. These include terfenadine, azelastine, and oxatomide.

4.4.3 ANTIFUNGAL AGENTS

Commonly used antifungal agents include the azoles (e.g., ketoconazole, miconazole, itraconazole, and fluconazole), the allylamine terbinafine, and the sulfonamide sulfamethoxazole. Ketoconazole and itraconazole are potent inhibitors of CYP3A4. Coadministration of these drugs with CYP3A substrates such as cyclosporine A, tacrolimus, alprazolam, triazolam, midazolam, nifedipine, felodipine, simvastatin, lovastatin, vincristine, terfenadine, or astemizole can result in clinically significant drug interactions, some of which can be life-threatening (Niwa et al. 2014). Fluconazole, miconazole, and sulfamethoxazole are potent inhibitors of CYP2C9. Coadministration of phenytoin, warfarin, sulfamethoxazole, and losartan with fluconazole results in clinically significant drug interactions (Venkatakrishnan et al. 2000). Fluconazole is a potent inhibitor of CYP2C19 in vitro. No clinically significant drug interactions have been predicted or documented between the azoles and drugs that are primarily metabolized by CYP2D6, 1A2, or 2E1 (Venkatakrishnan et al. 2000).

Terbinafine, used for the treatment of superficial dermatophytosis, inhibits CYP2D6-mediated dextromethorphan O-demethylation with very low apparent K_i values ranging from 28 to 44 nM in human hepatic microsomes and averaging 22.4 nM for the recombinant enzyme (Figure 4.9) (Abdel-Rahman et al. 1999b). Coadministration of terbinafine results in an elevation of nortriptyline plasma concentrations to supratherapeutic levels in a patient previously stabilized on this CYP2D6 substrate (Schmutz et al. 1999; van der Kuy and Hooymans 1998). Although the in vivo significance of CYP2D6 inhibition by terbinafine remains to be characterized, sufficient caution should be warranted in the coadministration of CYP2D6 substrates with a low therapeutic index while patients are receiving therapy with terbinafine.

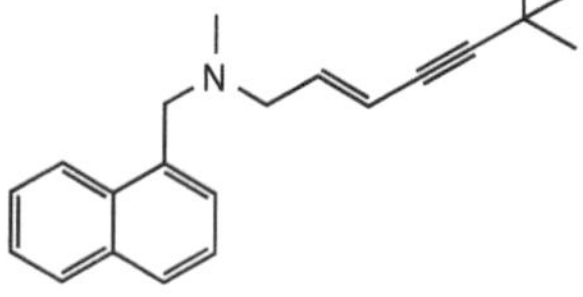

4.4.4 ANTI-HIV AGENTS

A number of anti-HIV agents are CYP2D6 inhibitors (Figure 4.10). Ritonavir inhibits CYP2D6 in vitro (von Moltke et al. 1998) and in vivo (Aarnoutse et al. 2005). Indinavir, saquinavir, nelfinavir, and delavirdine are all CYP2D6 inhibitors (von Moltke et al. 1998; Voorman et al. 2001). The K_i values for ritonavir, indinavir, saquinavir, and nelfinavir are 4.8, 15.6, 24.0, and 51.9 μM, respectively, when desipramine hydroxylation is used as a marker for CYP2D6 (von Moltke et al. 1998). In a clinical pharmacokinetic study, coadministration of ritonavir reduces the clearance of desipramine by 59% (von Moltke et al. 1998).

4.4.5 STEROIDS

Progesterone, testosterone, pregnanolone, pregnenolone, 17β-estradiol, and 17α-hydroxyprogesterone competitively inhibited CYP2D6 activity, whereas epiallopregnanolone and alfaxalone noncompetitively inhibit the activity (Figure 4.11) (Hiroi et al. 2001). Progesterone and testosterone inhibit bufuralol 1′-hydroxylation with K_i values of 33 and 63 μM, respectively. All these steroids lack the basic nitrogen atoms and are thus atypical substrates of CYP2D6.

4.4.6 OTHER DRUGS

Lansoprazole is more potent (K_i = 44.7 μM) than omeprazole (K_i = 240.7 μM) as an inhibitor of CYP2D6-mediated conversion of dextromethorphan to dextrorphan (Figure 4.12) (Ko et al. 1997). Mexiletine competitively inhibits dextromethorphan O-demethylation in human liver microsomes, and propafenone, oxprenolol, propranolol, and ajmaline inhibit the CYP2D6-mediated formation of hydroxymethylmexiletine and p-hydroxymexiletine from mexiletine (Figure 4.12) (Broly et al. 1990). Omeprazole inhibits CYP2D6-dependent 1′-hydroxylation of bufuralol with a K_i of 302 μM (VandenBranden et al. 1996). Several clinical studies have shown that omeprazole has no clinically important inhibitory effect on the clearance of drugs such as metoprolol (Andersson

FIGURE 4.9 Chemical structure of terbinafine, which is an inhibitor of CYP2D6.

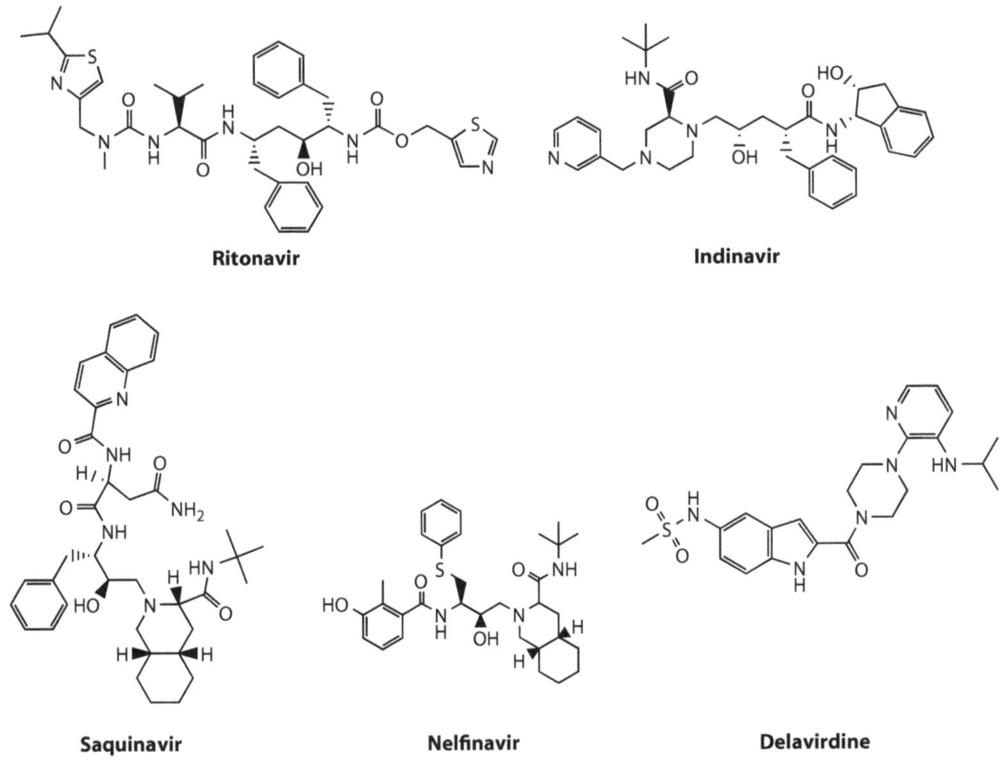

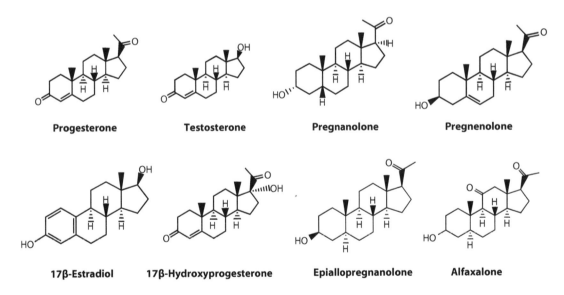

FIGURE 4.10 Chemical structures of anti-HIV agents that are inhibitors of CYP2D6. These include ritonavir, indinavir, saquinavir, nelfinavir, and delavirdine.

FIGURE 4.11 Chemical structures of steroids that are inhibitors of CYP2D6. These include progesterone, testosterone, pregnanolone, pregnenolone, 17β-estradiol, 17α-hydroxyprogesterone, epiallopregnanolone, and alfaxalone.

et al. 1991) and propranolol (Henry et al. 1987), which are partially metabolized by CYP2D6. Omeprazole seems to be a particularly selective inhibitor of CYP2C19, and it has a weak interaction with CYP2D6.

Halofantrine inhibits CYP2D6-dependent 1′-hydroxylation of bufuralol with a K_i of 4.3 μM (Figure 4.12) (Halliday et al. 1995). It appears that halofantrine competes for the substrate-binding site of CYP2D6 (Otton et al. 1988). Dexmedetomidine,

the pharmacologically active optical dextro isomer of medetomidine, which is a highly selective α_2-adrenergic receptor agonist with potent sedative, analgesic, and anesthetic effects, inhibits CYP2D6-dependent dextromethorphan O-demethylation with an IC_{50} of 1.8 μM (Rodrigues and Roberts 1997). In addition, dalcetrapib inhibits CYP2D6 with an IC_{50} of 82 μM, but it does not affect the activity of CYP2D6 in vivo (Derks et al. 2009).

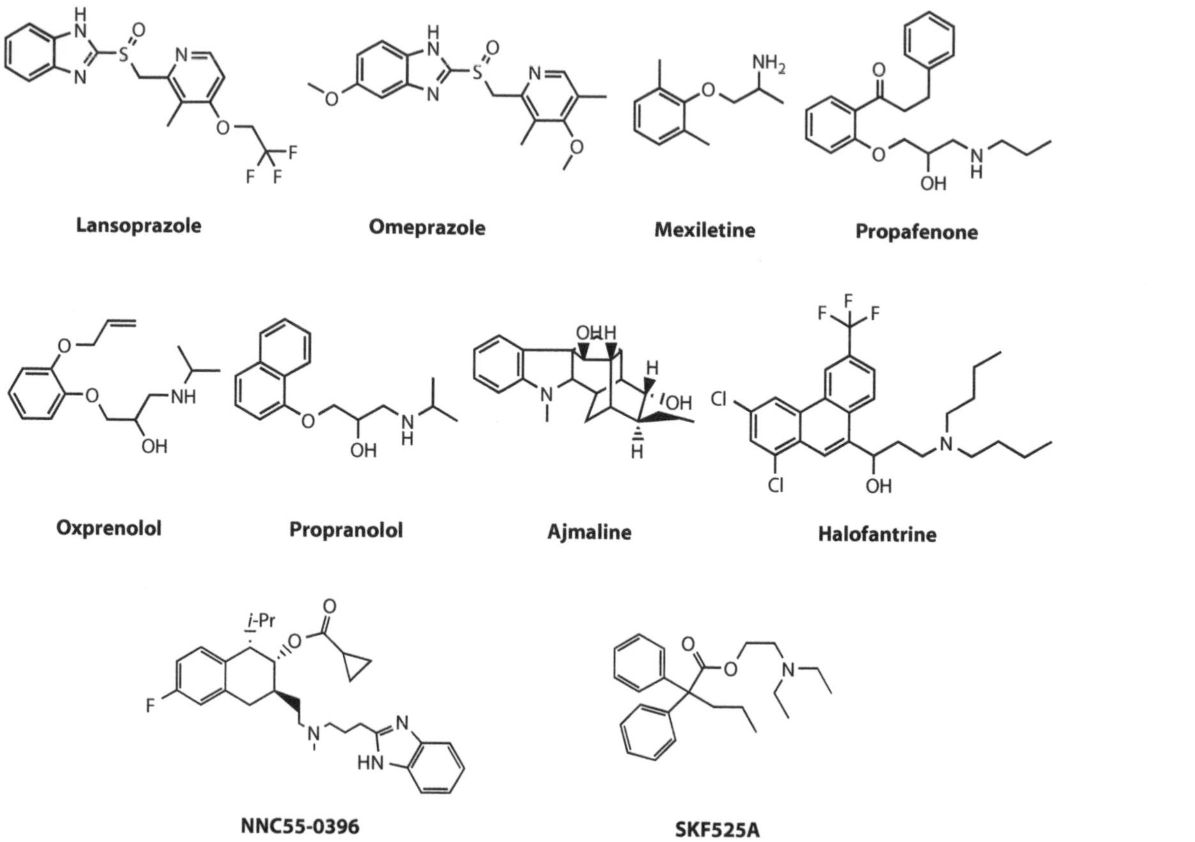

FIGURE 4.12 Chemical structures of lansoprazole, omeprazole, mexiletine, propafenone, oxprenolol, propranolol, ajmaline, halofantrine, NNC55-0396, and SKF525A, which are inhibitors of CYP2D6.

Sarpogrelate, an antagonist of 5HT$_{2A}$ and 5-HT$_{2B}$ receptors, is a potent and selective CYP2D6 inhibitor in vitro (Cho et al. 2014), and it increases the AUC of metoprolol by 53% in healthy subjects (Cho et al. 2015). Sarpogrelate potently and selectively inhibits CYP2D6-mediated dextromethorphan O-demethylation with a K_i value of 1.24 µM, in a competitive manner. Its main metabolite, M-1, also markedly inhibits CYP2D6 activity with a K_i value of 0.120 µM and is as potent as quinidine (K_i = 0.129 µM), a well-known typical CYP2D6 inhibitor. In addition, sarpogrelate and M-1 strongly inhibit both CYP2D6-catalyzed bufuralol 1′-hydroxylation and metoprolol α-hydroxylation activities (Cho et al. 2014). However, sarpogrelate and M-1 do not significantly affect CYP1A2, 2A6, 2B6, 2C8, 2C9, 2C19, 2E1, or 3A4/5 in vitro (Cho et al. 2014). Sarpogrelate may cause pharmacokinetic drug–drug interactions via inhibition of CYP2D6.

NNC55-0396 (K_i = 2.8 nM) and Ro40-5966 (K_i = 4.5 nM), two novel calcium channel blocker of mibefradil analogs, have a three- to fourfold greater inhibitory activity toward recombinant CYP2D6 than mibefradil (K_i = 12.7 nM) (Figure 4.12) (Bui et al. 2008). Temsirolimus and its principal metabolite, sirolimus, inhibit the CYP2D6 with K_i of 1.5 and 5 µM, respectively (Boni et al. 2009b). Temsirolimus does not alter the pharmacokinetics of desipramine in vivo (Boni et al. 2009b). Temsirolimus is a novel inhibitor of mammalian target of rapamycin, which has been approved for treatment

of renal cell carcinoma. Mirodenafil, a novel PDE5 inhibitor, inhibits CYP2D6 with an IC$_{50}$ of 77 µM (Lee et al. 2008).

SKF525A and its metabolite and primary amine analog all inhibit CYP2B6-, 2C9-, 2C19-, 2D6-, and 3A-selective reactions to varying degrees but have little effect on CYP1A2, 2A6, and 2E1 activities (Figure 4.12) (Franklin and Hathaway 2008). Only the inhibition of CYP3A shows major enhancement when the inhibitors are preincubated with NADPH-fortified microsomes (Franklin and Hathaway 2008). However, amobarbital, valproic acid, ethosuximide, caffeine, theophylline, disopyramide, and phenytoin do not inhibit CYP2D6 in vitro (Broly et al. 1990). Anastrozole does not inhibit CYP2D6-dependent dextromethorphan O-demethylation in human liver microsomes (Grimm and Dyroff 1997). A single dose of disulfiram does not alter the dextromethorphan metabolite ratio in healthy subjects (Kharasch et al. 1999).

4.4.7 Amphetamine Analogs

Wu et al. (1997) have investigated the inhibitory effect of 15 amphetamine analogs (Figure 4.13) on dextromethorphan O-demethylation in human liver microsomes. The authors have found that the presence of a methylenedioxy group in the 3,4-positions of both amphetamine (K_i = 26.5 µM) and methamphetamine (K_i = 25 µM) increases the affinity for CYP2D6 to 1.8 and 0.6 µM, respectively. Addition of a methoxy group

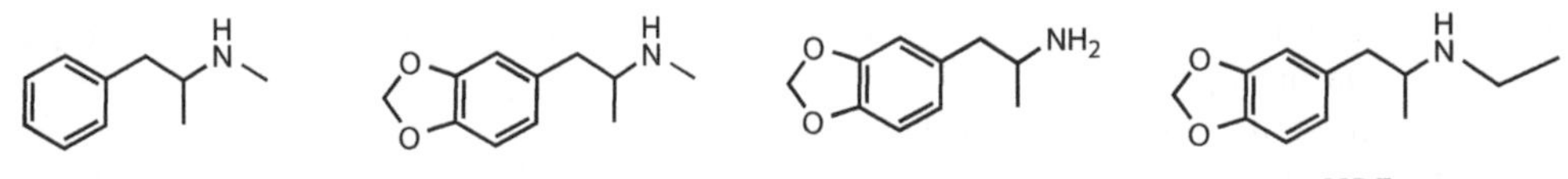

FIGURE 4.13 Chemical structures of amphetamine analogs that are inhibitors of CYP2D6. These include methamphetamine, methylene-dioxymethamphetamine (MDMA), *N*-ethyl-3,4-methylenedioxyamphetamine (MDE), and 3,4-methylenedioxyamphetamine (tenamfetamine; MDA).

to amphetamine in the 2-position also increases the affinity for CYP2D6 ($K_i = 11.5\ \mu M$) (Wu et al. 1997). The compound with the highest affinity for CYP2D6 is an amphetamine analog, MMDA-2, containing both a methoxy group in the 2-position and a methylenedioxy group ($K_i = 0.17\ \mu M$). MDMA is also a potent competitive inhibitor of CYP2D6 in human liver microsomes (Wu et al. 1997). Mescaline does not inhibit CYP2D6. Since many amphetamine analogs are metabolized by CYP2D6, they may inhibit their own metabolism and thus cause accumulation of the drugs in the body.

4.4.8 NATURAL AND HERBAL COMPOUNDS

Both cephaeline and emetine (Figure 4.14), two natural alkaloids, are potent inhibitors of CYP2D6 and CYP3A4 as indicated by the inhibition of probe substrate metabolism (Asano et al. 2001). The K_i values are 54 and 355 μM for cephaeline and 43 and 232 μM for emetine for CYP2D6 and CYP3A4, respectively (Asano et al. 2001). Hyperforin (Figure 4.14), a major active component from St. John's wort, is a potent noncompetitive inhibitor of CYP2D6-dependent bufuralol 1'-hydroxylation in recombinant enzyme with a K_i of 1.5 μM (Obach 2000). I3,II8-Biapigenin and hypericin from St. John's wort also inhibit CYP2D6 with K_i of 2.3 and 2.6 μM, respectively (Obach 2000). St. John's wort, common sage, and common valerian inhibit CYP2D6-mediated dextromethorphan metabolism, with IC_{50} of 0.07, 0.8, and 1.6 mg/ml, respectively (Hellum and Nilsen 2007). In the elderly, *Panax ginseng*, but not St. John's wort, garlic oil, and *Ginkgo biloba*, inhibits CYP2D6-mediated debrisoquine metabolism (Gurley et al. 2005). Celastrol, a main compound isolated from *Tripterygium wilfordii* Hook, is a potent CYP2D6 inhibitor with a K_i of 4.21 μM (Jin et al. 2015). Guanfu base A, a novel heterocyclic antiarrhythmic drug isolated from *Aconitum*

coreanum under Phase IV trial, is a potent noncompetitive inhibitor of CYP2D6, with K_i of 1.20 μM in human liver microsomes and K_i of 0.37 M for the recombinant enzyme (Sun et al. 2015). Guanfu base A does not inhibit human recombinant CYP1A2, 2A6, 2C8, 2C19, 3A4, or 3A5 but shows slight inhibition of 2B6 and 2E1. This natural compound is also a potent competitive inhibitor of CYP2D in monkey ($K_i = 0.38\ \mu M$) and dog ($K_i = 2.4\ \mu M$) microsomes but has no inhibitory activity on mouse or rat Cyp2ds (Sun et al. 2015).

Unlike CYP3A4, 2C9, and 1A2, the activity of CYP2D6 is not affected by other natural and herbal products. Garlic preparations (aged, odorless, oil, and freeze-dried) and three varieties of fresh garlic bulbs (common, Elephant, and Chinese) do not affect CYP2D6 activity (Foster et al. 2001). Henderson et al. (1999) did not find significant effects of seven ginsenosides and two eleutherosides (active components of the ginseng root) on the catalytic activity of recombinant CYP2D6. Soy extracts do not inhibit CYP2D6 in vitro (Anderson et al. 2003).

4.5 STRUCTURE–ACTIVITY RELATIONSHIPS OF CYP2D6 INHIBITORS

An inhibitor molecule-based model for CYP2D6 has been constructed by Strobl et al. (1993) by deriving a template from fitting six potent competitive reversible inhibitors of CYP2D6 onto each other. Ajmalicine, the most potent inhibitor for CYP2D6 with a K_i of 3.3 nM, is selected as a starting template because of its rigid structure. Other strong inhibitors used for model construction include chinidin, chlorpromazine, trifluperidol, prodipin, and lobelin, with K_i values of 0.06, 7.0, 0.17, 0.0048, and 0.12 μM, respectively. The basic nitrogen atoms of all inhibitors tested are superimposed and

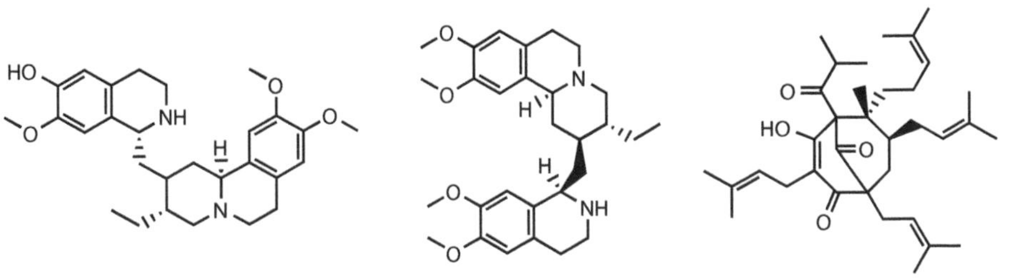

FIGURE 4.14 Chemical structures of natural and herbal compounds that are inhibitors of CYP2D6. These include cephaeline, emetine, and hyperforin.

the aromatic planes of these inhibitors are fitted coplanar. The derived preliminary pharmacophore model is characteristic of a tertiary nitrogen atom, which is protonated to a high degree at physiological pH and a flat hydrophobic region, the plane of which is almost perpendicular to the N–H axis and maximally extends up to a distance of 7.5 Å from the nitrogen atom (Strobl et al. 1993). The pharmacophore model also contains region B, in which additional functional groups with lone pairs enhanced inhibitory potency, and region C, in which hydrophobic groups are allowed but do not increase binding affinity and inhibitory effects (Strobl et al. 1993). Compounds with enhanced inhibitory potency contain additional functional groups with negative electrostatic potential and hydrogen bond acceptor properties in region B on the opposite side at distances of 4.8–5.5 Å and 6.6–7.5 Å from the nitrogen atom, respectively (Strobl et al. 1993). Compounds (e.g., reserpine) that take additional space are not inhibitors. Consequently, other inhibitors are fitted onto the derived template. These include derivatives of ajmalicine such as tetrahydroalstonine (K_i = 5.0 µM) and 19-epiajmalicine (17 nM), yohimbine derivatives such as corynanthine (0.080 µM) and β-yohimbine (0.031 µM), harmine (0 µM) and its analogs such as harmalol (65 µM) and harman (86 µM), and cinchona alkaloids including quinidine (60 nM), quinine (4.6 µM), cinchonine (3.5 µM), 10,11-dihydroquinidine (0.066 µM), and quininone (0.72 µM). Serpentine, cathenamine, and sempervirine all possess an iminium atom instead of a basic nitrogen atom but are still competitive inhibitors of CYP2D6 with K_i values of 2.2, 3.2, and 9.7 µM, respectively, suggesting a demand for a modification of the pharmacophore. The overall criteria derived for this inhibitor-based model are similar to those for the substrate-based models of CYP2D6 (Koymans et al. 1992; Strobl et al. 1993).

A set of 3D/4D quantitative structure–activity relationship (QSAR) pharmacophore models has also been derived for competitive inhibitors of CYP2D6 by Ekins et al. (1999). The first model for 20 inhibitors of CYP2D6-mediated bufuralol 1′-hydroxylation produces a positive correlation between observed and predicted K_i values, while a second model using 31 literature-derived K_i values provides a better correlation between observed and predicted K_i values with an r value of 0.91. Both pharmacophores are capable of predicting K_i values for 9 to 10 of 15 CYP2D6 inhibitors within 1 log residual (Ekins et al. 1999).

A QSAR analysis has identified a correlation between the IC_{50} value and lipophilicity in a series of close analogs of 7-methoxy-4-(aminomethyl)-coumarin (MAMC), which have been designed as specific CYP2D6 substrates (Venhorst et al. 2000). The authors have found that elongation of the alkyl chain dramatically increases the affinity of the compounds toward CYP2D6, as indicated by an up to 100-fold decrease in K_m values. The V_{max} values display a much less pronounced decrease with increasing N-alkyl chain length, resulting in as much as a 30-fold increase in the V_{max}/K_m value (Venhorst et al. 2000). In contrast to CYP2D6, N-alkylation of MAMC does not significantly affect the K_m values of O-dealkylation by CYP1A2, but it results in higher V_{max} values.

4.6 CONCLUSIONS AND FUTURE DIRECTIONS

Although CYP2D6 accounts for only a small percentage of the total hepatic CYP content (<2%), this enzyme metabolizes ~25% of clinically used drugs with significant polymorphisms. In particular, a number of drugs acting on the central nervous system and cardiovascular system are substantially metabolized by CYP2D6. These drugs are mostly associated with G-coupled proteins when they elicit effect at the site of action. CYP2D6 is subject to inhibition by many drugs, and a number of clinical drug–drug interactions attributed to CYP2D6 inhibition have been reported. Inhibition of CYP2D6-mediated drug metabolism has important clinical implications when drugs are coadministered. In drug development, if a new drug candidate is found to be a potent or mechanism-based inhibitor of CYP2D6, its further development should be halted since the compound has a high potential for causing harmful drug–drug interactions.

REFERENCES

Aarnoutse, R. E., Kleinnijenhuis, J., Koopmans, P. P., Touw, D. J., Wieling, J., Hekster, Y. A., Burger, D. M. 2005. Effect of low-dose ritonavir (100 mg twice daily) on the activity of cytochrome P450 2D6 in healthy volunteers. *Clin Pharmacol Ther* 78:664–674.

Abdel-Rahman, S. M., Gotschall, R. R., Kauffman, R. E., Leeder, J. S., Kearns, G. L. 1999a. Investigation of terbinafine as a CYP2D6 inhibitor in vivo. *Clin Pharmacol Ther* 65:465–472.

Abdel-Rahman, S. M., Marcucci, K., Boge, T., Gotschall, R. R., Kearns, G. L., Leeder, J. S. 1999b. Potent inhibition of cytochrome P450 2D6-mediated dextromethorphan O-demethylation by terbinafine. *Drug Metab Dispos* 27:770–775.

Abernethy, D. R., Greenblatt, D. J., Shader, R. I. 1984. Imipramine-cimetidine interaction: Impairment of clearance and enhanced absolute bioavailability. *J Pharmacol Exp Ther* 229:702–705.

Agbokponto, J. E., Luo, Z., Liu, R., Liu, Z., Liang, M., Ding, L. 2015. Study of pharmacokinetic interaction of paroxetine and roxithromycin on bencycloquidium bromide in healthy subjects. *Eur J Pharm Sci* 69:37–43.

Ahmed, A., Simmons, Z. 2013. Pseudobulbar affect: Prevalence and management. *Ther Clin Risk Manag* 9:483–489.

Albers, L. J., Reist, C., Helmeste, D., Vu, R., Tang, S. W. 1996. Paroxetine shifts imipramine metabolism. *Psychiatry Res* 59:189–196.

Albers, L. J., Reist, C., Vu, R. L., Fujimoto, K., Ozdemir, V., Helmeste, D., Poland, R., Tang, S. W. 2000. Effect of venlafaxine on imipramine metabolism. *Psychiatry Res* 96:235–243.

Alderman, C. P., Hundertmark, J. D., Soetratma, T. W. 1997a. Interaction of serotonin re-uptake inhibitors with perhexiline. *Aust N Z J Psychiatry* 31:601–603.

Alderman, J., Preskorn, S. H., Greenblatt, D. J., Harrison, W., Penenberg, D., Allison, J., Chung, M. 1997b. Desipramine pharmacokinetics when coadministered with paroxetine or sertraline in extensive metabolizers. *J Clin Psychopharmacol* 17:284–291.

Amsterdam, J. D., Brunswick, D. J., Potter, L., Kaplan, M. J. 1984. Cimetidine-induced alterations in desipramine plasma concentrations. *Psychopharmacology (Berl)* 83:373–375.

Anderson, G. D., Rosito, G., Mohustsy, M. A., Elmer, G. W. 2003. Drug interaction potential of soy extract and *Panax ginseng*. *J Clin Pharmacol* 43:643–648.

Andersson, T., Lundborg, P., Regardh, C. G. 1991. Lack of effect of omeprazole treatment on steady-state plasma levels of metoprolol. *Eur J Clin Pharmacol* 40:61–65.

Arnold, G. L., Griebel, M. L., Valentine, J. L., Koroma, D. M., Kearns, G. L. 1997. Dextromethorphan in nonketotic hyperglycinaemia: Metabolic variation confounds the dose-response relationship. *J Inherit Metab Dis* 20:28–38.

Asano, T., Kushida, H., Sadakane, C., Ishihara, K., Wakui, Y., Yanagisawa, T., Kimura, M., Kamei, H., Yoshida, T. 2001. Metabolism of ipecac alkaloids cephaeline and emetine by human hepatic microsomal cytochrome P450s, and their inhibitory effects on P450 enzyme activities. *Biol Pharm Bull* 24:678–682.

Ball, S. E., Ahern, D., Scatina, J., Kao, J. 1997. Venlafaxine: *In vitro* inhibition of CYP2D6 dependent imipramine and desipramine metabolism; comparative studies with selected SSRIs, and effects on human hepatic CYP3A4, CYP2C9 and CYP1A2. *Br J Clin Pharmacol* 43:619–626.

Barros, J., Asnis, G. 1993. An interaction of sertraline and desipramine. *Am J Psychiatry* 150:1751.

Bateman, D. N. 1983. Clinical pharmacokinetics of metoclopramide. *Clin Pharmacokinet* 8:523–529.

Bateman, D. N., Gokal, R., Dodd, T. R., Blain, P. G. 1981. The pharmacokinetics of single doses of metoclopramide in renal failure. *Eur J Clin Pharmacol* 19:437–441.

Beckmann-Knopp, S., Rietbrock, S., Weyhenmeyer, R., Bocker, R. H., Beckurts, K. T., Lang, W., Hunz, M., Fuhr, U. 2000. Inhibitory effects of silibinin on cytochrome P-450 enzymes in human liver microsomes. *Pharmacol Toxicol* 86:250–256.

Belle, D. J., Ernest, C. S., Sauer, J. M., Smith, B. P., Thomasson, H. R., Witcher, J. W. 2002. Effect of potent CYP2D6 inhibition by paroxetine on atomoxetine pharmacokinetics. *J Clin Pharmacol* 42:1219–1227.

Belpaire, F. M., Wijnant, P., Temmerman, A., Rasmussen, B. B., Brosen, K. 1998. The oxidative metabolism of metoprolol in human liver microsomes: Inhibition by the selective serotonin reuptake inhibitors. *Eur J Clin Pharmacol* 54:261–264.

Bertelsen, K. M., Venkatakrishnan, K., Von Moltke, L. L., Obach, R. S., Greenblatt, D. J. 2003. Apparent mechanism-based inhibition of human CYP2D6 *in vitro* by paroxetine: Comparison with fluoxetine and quinidine. *Drug Metab Dispos* 31:289–293.

Boni, J., Abbas, R., Leister, C., Burns, J., Jordan, R., Hoffmann, M., Demaio, W., Hug, B. 2009a. Disposition of desipramine, a sensitive cytochrome P450 2D6 substrate, when coadministered with intravenous temsirolimus. *Cancer Chemother Pharmacol* 64:263–270.

Boni, J., Abbas, R., Leister, C., Burns, J., Jordan, R., Hoffmann, M., DeMaio, W., Hug, B. 2009b. Disposition of desipramine, a sensitive cytochrome P450 2D6 substrate, when coadministered with intravenous temsirolimus. *Cancer Chemother Pharmacol* 64:263–270.

Briciu, C., Neag, M., Muntean, D., Vlase, L., Bocsan, C., Buzoianu, A., Gheldiu, A. M., Achim, M., Popa, A. 2014. A pharmacokinetic drug interaction study between nebivolol and paroxetine in healthy volunteers. *J Clin Pharm Ther* 39:535–540.

Broly, F., Libersa, C., Lhermitte, M., Dupuis, B. 1990. Inhibitory studies of mexiletine and dextromethorphan oxidation in human liver microsomes. *Biochem Pharmacol* 39:1045–1053.

Brown, H. S., Galetin, A., Hallifax, D., Houston, J. B. 2006. Prediction of *in vivo* drug–drug interactions from in vitro data: Factors affecting prototypic drug–drug interactions involving CYP2C9, CYP2D6 and CYP3A4. *Clin Pharmacokinet* 45:1035–1050.

Bui, P. H., Quesada, A., Handforth, A., Hankinson, O. 2008. The mibefradil derivative NNC55-0396, a specific T-type calcium channel antagonist, exhibits less CYP3A4 inhibition than mibefradil. *Drug Metab Dispos* 36:1291–1299.

Cai, X., Wang, R. W., Edom, R. W., Evans, D. C., Shou, M., Rodrigues, A. D., Liu, W., Dean, D. C., Baillie, T. A. 2004. Validation of (-)-*N*-3-benzyl-phenobarbital as a selective inhibitor of CYP2C19 in human liver microsomes. *Drug Metab Dispos* 32:584–586.

Callaghan, J. T., Cerimele, B. J., Kassahun, K. J., Nyhart, E. H., Jr., Hoyes-Beehler, P. J., Kondraske, G. V. 1997. Olanzapine: Interaction study with imipramine. *J Clin Pharmacol* 37:971–978.

Chang, S. Y., Chen, C., Yang, Z., Rodrigues, A. D. 2009. Further assessment of 17α-ethinyl estradiol as an inhibitor of different human cytochrome P450 forms *in vitro*. *Drug Metab Dispos* 37:1667–1675.

Cho, D. Y., Bae, S. H., Lee, J. K., Kim, Y. W., Kim, B. T., Bae, S. K. 2014. Selective inhibition of cytochrome P450 2D6 by sarpogrelate and its active metabolite, M-1, in human liver microsomes. *Drug Metab Dispos* 42:33–39.

Cho, D. Y., Bae, S. H., Lee, J. K., Park, J. B., Kim, Y. W., Lee, S., Oh, E., Kim, B. T., Bae, S. K. 2015. Effect of the potent CYP2D6 inhibitor sarpogrelate on the pharmacokinetics and pharmacodynamics of metoprolol in healthy male Korean volunteers. *Xenobiotica* 45:256–263.

Ciccone, P. E., Ramabadran, K., Jessen, L. M. 2006. Potential interactions of methylphenidate and atomoxetine with dextromethorphan. *J Am Pharm Assoc (2003)* 46:472–478.

Cohen, L. H., van Leeuwen, R. E., van Thiel, G. C., van Pelt, J. F., Yap, S. H. 2000. Equally potent inhibitors of cholesterol synthesis in human hepatocytes have distinguishable effects on different cytochrome P450 enzymes. *Biopharm Drug Dispos* 21:353–364.

Crewe, H. K., Lennard, M. S., Tucker, G. T., Woods, F. R., Haddock, R. E. 1992. The effect of selective serotonin re-uptake inhibitors on cytochrome P4502D6 (CYP2D6) activity in human liver microsomes. *Br J Clin Pharmacol* 34:262–265.

Derks, M., Fowler, S., Kuhlmann, O. 2009. *In vitro* and *in vivo* assessment of the effect of dalcetrapib on a panel of CYP substrates. *Curr Med Res Opin* 25:891–902.

Desta, Z., Kerbusch, T., Soukhova, N., Richard, E., Ko, J. W., Flockhart, D. A. 1998. Identification and characterization of human cytochrome P450 isoforms interacting with pimozide. *J Pharmacol Exp Ther* 285:428–437.

Desta, Z., Wu, G. M., Morocho, A. M., Flockhart, D. A. 2002. The gastroprokinetic and antiemetic drug metoclopramide is a substrate and inhibitor of cytochrome P450 2D6. *Drug Metab Dispos* 30:336–343.

Easton, N., Fry, J., O'Shea, E., Watkins, A., Kingston, S., Marsden, C. A. 2003. Synthesis, in vitro formation, and behavioural effects of glutathione regioisomers of alpha-methyldopamine with relevance to MDA and MDMA (ecstasy). *Brain Res* 987:144–154.

Ekins, S., Bravi, G., Binkley, S., Gillespie, J. S., Ring, B. J., Wikel, J. H., Wrighton, S. A. 1999. Three and four dimensional-quantitative structure activity relationship (3D/4D-QSAR) analyses of CYP2D6 inhibitors. *Pharmacogenetics* 9:477–489.

Ereshefsky, L., Riesenman, C., Lam, Y. W. 1995. Antidepressant drug interactions and the cytochrome P450 system. The role of cytochrome P450 2D6. *Clin Pharmacokinet* 29 Suppl 1:10–18; discussion 18–19.

Farre, M., Abanades, S., Roset, P. N., Peiro, A. M., Torrens, M., O'Mathuna, B., Segura, M., de la Torre, R. 2007. Pharmacological interaction between 3,4-methylenedioxymethamphetamine (ecstasy) and paroxetine: Pharmacological effects and pharmacokinetics. *J Pharmacol Exp Ther* 323:954–962.

Foster, B. C., Foster, M. S., Vandenhoek, S., Krantis, A., Budzinski, J. W., Arnason, J. T., Gallicano, K. D., Choudri, S. 2001. An *in vitro* evaluation of human cytochrome P450 3A4 and P-glycoprotein inhibition by garlic. *J Pharm Pharm Sci* 4:176–184.

Franklin, M. R., Hathaway, L. B. 2008. 2-Diethylaminoethyl-2,2-diphenylvalerate-HCl (SKF525A) revisited: Comparative cytochrome P450 inhibition in human liver microsomes by SKF525A, its metabolites, and SKF-acid and SKF-alcohol. *Drug Metab Dispos* 36:2539–2546.

Furuta, S., Kamada, E., Suzuki, T., Sugimoto, T., Kawabata, Y., Shinozaki, Y., Sano, H. 2001. Inhibition of drug metabolism in human liver microsomes by nizatidine, cimetidine and omeprazole. *Xenobiotica* 31:1–10.

Gareri, P., De Fazio, P., Gallelli, L., De Fazio, S., Davoli, A., Seminara, G., Cotroneo, A., De Sarro, G. 2008. Venlafaxine–propafenone interaction resulting in hallucinations and psychomotor agitation. *Ann Pharmacother* 42:434–438.

Garnock-Jones, K. P. 2011. Dextromethorphan/quinidine: In pseudobulbar affect. *CNS Drugs* 25:435–445.

Glue, P., Winter, H., Garbe, K., Jakobi, H., Lyudin, A., Lenagh-Glue, Z., Hung, C. T. 2015. Influence of CYP2D6 activity on the pharmacokinetics and pharmacodynamics of a single 20 mg dose of ibogaine in healthy volunteers. *J Clin Pharmacol* 55:680–687.

Goto, A., Ueda, K., Inaba, A., Nakajima, H., Kobayashi, H., Sakai, K. 2005. Identification of human P450 isoforms involved in the metabolism of the antiallergic drug, oxatomide, and its kinetic parameters and inhibition constants. *Biol Pharm Bull* 28:328–334.

Gram, L. F., Hansen, M. G., Sindrup, S. H., Brosen, K., Poulsen, J. H., Aaes-Jorgensen, T., Overo, K. F. 1993. Citalopram: Interaction studies with levomepromazine, imipramine, and lithium. *Ther Drug Monit* 15:18–24.

Grimm, S. W., Dyroff, M. C. 1997. Inhibition of human drug metabolizing cytochromes P450 by anastrozole, a potent and selective inhibitor of aromatase. *Drug Metab Dispos* 25:598–602.

Guengerich, F. P., Muller-Enoch, D., Blair, I. A. 1986. Oxidation of quinidine by human liver cytochrome P-450. *Mol Pharmacol* 30:287–295.

Gurley, B. J., Gardner, S. F., Hubbard, M. A., Williams, D. K., Gentry, W. B., Cui, Y., Ang, C. Y. 2005. Clinical assessment of effects of botanical supplementation on cytochrome P450 phenotypes in the elderly: St John's wort, garlic oil, *Panax ginseng* and *Ginkgo biloba*. *Drugs Aging* 22:525–539.

Haefeli, W. E., Bargetzi, M. J., Follath, F., Meyer, U. A. 1990. Potent inhibition of cytochrome P450 2D6 (debrisoquin 4-hydroxylase) by flecainide *in vitro* and *in vivo*. *J Cardiovasc Pharmacol* 15:776–779.

Halliday, R. C., Jones, B. C., Smith, D. A., Kitteringham, N. R., Park, B. K. 1995. An investigation of the interaction between halofantrine, CYP2D6 and CYP3A4: Studies with human liver microsomes and heterologous enzyme expression systems. *Br J Clin Pharmacol* 40:369–378.

Hansen-Grant, S., Silk, K. R., Guthrie, S. 1993. Fluoxetine-pimozide interaction. *Am J Psychiatry* 150:1751–1752.

Hellum, B. H., Nilsen, O. G. 2007. The *in vitro* inhibitory potential of trade herbal products on human CYP2D6-mediated metabolism and the influence of ethanol. *Basic Clin Pharmacol Toxicol* 101:350–358.

Hemeryck, A., Lefebvre, R. A., De Vriendt, C., Belpaire, F. M. 2000. Paroxetine affects metoprolol pharmacokinetics and pharmacodynamics in healthy volunteers. *Clin Pharmacol Ther* 67:283–291.

Henderson, G. L., Harkey, M. R., Gershwin, M. E., Hackman, R. M., Stern, J. S., Stresser, D. M. 1999. Effects of ginseng components on c-DNA-expressed cytochrome P450 enzyme catalytic activity. *Life Sci* 65:PL209-214.

Henry, D., Brent, P., Whyte, I., Mihaly, G., Devenish-Meares, S. 1987. Propranolol steady-state pharmacokinetics are unaltered by omeprazole. *Eur J Clin Pharmacol* 33:369–373.

Henthorn, T. K., Spina, E., Dumont, E., von Bahr, C. 1989. *In vitro* inhibition of a polymorphic human liver P450 isozyme by narcotic analgesics. *Anesthesiology* 70:339–342.

Hesse, L. M., Venkatakrishnan, K., Court, M. H., von Moltke, L. L., Duan, S. X., Shader, R. I., Greenblatt, D. J. 2000. CYP2B6 mediates the *in vitro* hydroxylation of bupropion: Potential drug interactions with other antidepressants. *Drug Metab Dispos* 28:1176–1183.

Heydari, A., Yeo, K. R., Lennard, M. S., Ellis, S. W., Tucker, G. T., Rostami-Hodjegan, A. 2004. Mechanism-based inactivation of CYP2D6 by methylenedioxymethamphetamine. *Drug Metab Dispos* 32:1213–1217.

Hiroi, T., Kishimoto, W., Chow, T., Imaoka, S., Igarashi, T., Funae, Y. 2001. Progesterone oxidation by cytochrome P450 2D isoforms in the brain. *Endocrinology* 142:3901–3908.

Horrigan, J. P., Barnhill, L. J. 1994. Paroxetine-pimozide drug interaction. *J Am Acad Child Adolesc Psychiatry* 33:1060–1061.

Ishii, Y., Nakamura, K., Tsutsumi, K., Kotegawa, T., Nakano, S., Nakatsuka, K. 2000. Drug interaction between cimetidine and timolol ophthalmic solution: Effect on heart rate and intraocular pressure in healthy Japanese volunteers. *J Clin Pharmacol* 40:193–199.

Jeppesen, U., Gram, L. F., Vistisen, K., Loft, S., Poulsen, H. E., Brosen, K. 1996. Dose-dependent inhibition of CYP1A2, CYP2C19 and CYP2D6 by citalopram, fluoxetine, fluvoxamine and paroxetine. *Eur J Clin Pharmacol* 51:73–78.

Jin, C., He, X., Zhang, F., He, L., Chen, J., Wang, L., An, L., Fan, Y. 2015. Inhibitory mechanisms of celastrol on human liver cytochrome P450 1A2, 2C19, 2D6, 2E1 and 3A4. *Xenobiotica* 45:571–577.

Jones, B. C., Hyland, R., Ackland, M., Tyman, C. A., Smith, D. A. 1998. Interaction of terfenadine and its primary metabolites with cytochrome P450 2D6. *Drug Metab Dispos* 26:875–882.

Joos, A. A., Konig, F., Frank, U. G., Kaschka, W. P., Morike, K. E., Ewald, R. 1997. Dose-dependent pharmacokinetic interaction of clozapine and paroxetine in an extensive metabolizer. *Pharmacopsychiatry* 30:266–270.

Kamali, F., Howes, A., Thomas, S. H., Ford, G. A., Snoeck, E. 1997. A pharmacokinetic and pharmacodynamic interaction study between nebivolol and the H_2-receptor antagonists cimetidine and ranitidine. *Br J Clin Pharmacol* 43:201–204.

Kharasch, E. D., Hankins, D. C., Jubert, C., Thummel, K. E., Taraday, J. K. 1999. Lack of single-dose disulfiram effects on cytochrome P-450 2C9, 2C19, 2D6, and 3A4 activities: Evidence for specificity toward P-450 2E1. *Drug Metab Dispos* 27:717–723.

Knodell, R. G., Browne, D. G., Gwozdz, G. P., Brian, W. R., Guengerich, F. P. 1991. Differential inhibition of individual human liver cytochromes P-450 by cimetidine. *Gastroenterology* 101:1680–1691.

Ko, J. W., Sukhova, N., Thacker, D., Chen, P., Flockhart, D. A. 1997. Evaluation of omeprazole and lansoprazole as inhibitors of cytochrome P450 isoforms. *Drug Metab Dispos* 25:853–862.

Ko, J. W., Desta, Z., Soukhova, N. V., Tracy, T., Flockhart, D. A. 2000. *In vitro* inhibition of the cytochrome P450 (CYP450) system by the antiplatelet drug ticlopidine: Potent effect on CYP2C19 and CYP2D6. *Br J Clin Pharmacol* 49:343–351.

Kolb, K. W., Garnett, W. R., Small, R. E., Vetrovec, G. W., Kline, B. J., Fox, T. 1984. Effect of cimetidine on quinidine clearance. *Ther Drug Monit* 6:306–312.

Kosoglou, T., Salfi, M., Lim, J. M., Batra, V. K., Cayen, M. N., Affrime, M. B. 2000. Evaluation of the pharmacokinetics and electrocardiographic pharmacodynamics of loratadine with concomitant administration of ketoconazole or cimetidine. *Br J Clin Pharmacol* 50:581–589.

Koymans, L., Vermeulen, N. P., van Acker, S. A., te Koppele, J. M., Heykants, J. J., Lavrijsen, K., Meuldermans, W., Donne-Op den Kelder, G. M. 1992. A predictive model for substrates of cytochrome P450-debrisoquine (2D6). *Chem Res Toxicol* 5:211–219.

Kumar, G. N., Dykstra, J., Roberts, E. M., Jayanti, V. K., Hickman, D., Uchic, J., Yao, Y., Surber, B., Thomas, S., Granneman, G. R. 1999. Potent inhibition of the cytochrome P-450 3A-mediated human liver microsomal metabolism of a novel HIV protease inhibitor by ritonavir: A positive drug–drug interaction. *Drug Metab Dispos* 27:902–908.

Lal, R., Sukbuntherng, J., Luo, W., Vicente, V., Blumenthal, R., Ho, J., Cundy, K. C. 2010. Clinical pharmacokinetic drug interaction studies of gabapentin enacarbil, a novel transported prodrug of gabapentin, with naproxen and cimetidine. *Br J Clin Pharmacol* 69:498–507.

Lee, H. S., Park, E. J., Ji, H. Y., Kim, S. Y., Im, G. J., Lee, S. M., Jang, I. J. 2008. Identification of cytochrome P450 enzymes responsible for N-dealkylation of a new oral erectogenic, mirodenafil. *Xenobiotica* 38:21–33.

Leroi, I., Walentynowicz, M. A. 1996. Fluoxetine-imipramine interaction. *Can J Psychiatry* 41:318–319.

Leucht, S., Hackl, H. J., Steimer, W., Angersbach, D., Zimmer, R. 2000. Effect of adjunctive paroxetine on serum levels and side-effects of tricyclic antidepressants in depressive inpatients. *Psychopharmacology (Berl)* 147:378–383.

Li, X. Q., Andersson, T. B., Ahlstrom, M., Weidolf, L. 2004. Comparison of inhibitory effects of the proton pump-inhibiting drugs omeprazole, esomeprazole, lansoprazole, pantoprazole, and rabeprazole on human cytochrome P450 activities. *Drug Metab Dispos* 32:821–827.

Lim, K. S., Jang, I. J., Kim, B. H., Kim, J., Jeon, J. Y., Tae, Y. M., Yi, S. et al. 2010. Changes in the QTc interval after administration of flecainide acetate, with and without coadministered paroxetine, in relation to cytochrome P450 2D6 genotype: Data from an open-label, two-period, single-sequence crossover study in healthy Korean male subjects. *Clin Ther* 32:659–666.

Linnet, K. 1995. Comparison of the kinetic interactions of the neuroleptics perphenazine and zuclopenthixol with tricyclic antidepressives. *Ther Drug Monit* 17:308–311.

Linnoila, M., George, L., Guthrie, S. 1982. Interaction between antidepressants and perphenazine in psychiatric inpatients. *Am J Psychiatry* 139:1329–1331.

Liu, K. H., Kim, M. J., Shon, J. H., Moon, Y. S., Seol, S. Y., Kang, W., Cha, I. J., Shin, J. G. 2005. Stereoselective inhibition of cytochrome P450 forms by lansoprazole and omeprazole *in vitro*. *Xenobiotica* 35:27–38.

Livezey, M., Nagy, L. D., Diffenderfer, L. E., Arthur, E. J., Hsi, D. J., Holton, J. M., Furge, L. L. 2012. Molecular analysis and modeling of inactivation of human CYP2D6 by four mechanism based inactivators. *Drug Metab Lett* 6:7–14.

Livezey, M. R., Briggs, E. D., Bolles, A. K., Nagy, L. D., Fujiwara, R., Furge, L. L. 2014. Metoclopramide is metabolized by CYP2D6 and is a reversible inhibitor, but not inactivator, of CYP2D6. *Xenobiotica* 44:309–319.

Lu, P., Schrag, M. L., Slaughter, D. E., Raab, C. E., Shou, M., Rodrigues, A. D. 2003. Mechanism-based inhibition of human liver microsomal cytochrome P450 1A2 by zileuton, a 5-lipoxygenase inhibitor. *Drug Metab Dispos* 31:1352–1360.

Madeira, M., Levine, M., Chang, T. K., Mirfazaelian, A., Bellward, G. D. 2004. The effect of cimetidine on dextromethorphan *O*-demethylase activity of human liver microsomes and recombinant CYP2D6. *Drug Metab Dispos* 32:460–467.

Maenpaa, J., Volotinen-Maja, M., Kautiainen, H., Neuvonen, M., Niemi, M., Neuvonen, P. J., Backman, J. T. 2014. Paroxetine markedly increases plasma concentrations of ophthalmic timolol; CYP2D6 inhibitors may increase the risk of cardiovascular adverse effects of 0.5% timolol eye drops. *Drug Metab Dispos* 42:2068–2076.

Martinez, C., Albet, C., Agundez, J. A., Herrero, E., Carrillo, J. A., Marquez, M., Benitez, J., Ortiz, J. A. 1999. Comparative *in vitro* and *in vivo* inhibition of cytochrome P450 CYP1A2, CYP2D6, and CYP3A by H2-receptor antagonists. *Clin Pharmacol Ther* 65:369–376.

Mason, J. W., Hondeghem, L. M. 1984. Quinidine. *Ann N Y Acad Sci* 432:162–176.

Maynard, G. L., Soni, P. 1996. Thioridazine interferences with imipramine metabolism and measurement. *Ther Drug Monit* 18:729–731.

Micuda, S., Mundlova, L., Anzenbacherova, E., Anzenbacher, P., Chladek, J., Fuksa, L., Martinkova, J. 2004. Inhibitory effects of memantine on human cytochrome P450 activities: Prediction of *in vivo* drug interactions. *Eur J Clin Pharmacol* 60:583–589.

Miller, D. D., Macklin, M. 1983. Cimetidine-imipramine interaction: A case report. *Am J Psychiatry* 140:351–352.

Miller, D. D., Sawyer, J. B., Duffy, J. P. 1983. Cimetidine's effect on steady-state serum nortriptyline concentrations. *Drug Intell Clin Pharm* 17:904–905.

Miller, A., Pratt, H., Schiffer, R. B. 2011. Pseudobulbar affect: The spectrum of clinical presentations, etiologies and treatments. *Expert Rev Neurother* 11:1077–1088.

Morganroth, J., Lyness, W. H., Perhach, J. L., Mather, G. G., Harr, J. E., Trager, W. F., Levy, R. H., Rosenberg, A. 1997. Lack of effect of azelastine and ketoconazole coadmistration on electrocardiographic parameters in healthy volunteers. *J Clin Pharmacol* 37:1065–1072.

Mulsant, B. H., Foglia, J. P., Sweet, R. A., Rosen, J., Lo, K. H., Pollock, B. G. 1997. The effects of perphenazine on the concentration of nortriptyline and its hydroxymetabolites in older patients. *J Clin Psychopharmacol* 17:318–321.

Nakajima, M., Ohyama, K., Nakamura, S., Shimada, N., Yamazaki, H., Yokoi, T. 1999. Inhibitory effects of azelastine and its metabolites on drug oxidation catalyzed by human cytochrome P450 enzymes. *Drug Metab Dispos* 27:792–797.

Nakamura, K., Ariyoshi, N., Iwatsubo, T., Fukunaga, Y., Higuchi, S., Itoh, K., Shimada, N. et al. 2005. Inhibitory effects of nicardipine to cytochrome P450 (CYP) in human liver microsomes. *Biol Pharm Bull* 28:882–885.

Nichols, A. I., Fatato, P., Shenouda, M., Paul, J., Isler, J. A., Pedersen, R. D., Jiang, Q., Ahmed, S., Patroneva, A. 2009. The effects of desvenlafaxine and paroxetine on the pharmacokinetics of the cytochrome P450 2D6 substrate desipramine in healthy adults. *J Clin Pharmacol* 49:219–228.

Nicolas, J. M., Whomsley, R., Collart, P., Roba, J. 1999. *In vitro* inhibition of human liver drug metabolizing enzymes by second generation antihistamines. *Chem Biol Interact* 123:63–79.

Nielsen, F., Rosholm, J. U., Brosen, K. 1995. Lack of relationship between quinidine pharmacokinetics and the sparteine oxidation polymorphism. *Eur J Clin Pharmacol* 48:501–504.

Nielsen, T. L., Rasmussen, B. B., Flinois, J. P., Beaune, P., Brosen, K. 1999. In vitro metabolism of quinidine: The (3S)-3-hydroxylation of quinidine is a specific marker reaction for cytochrome P-4503A4 activity in human liver microsomes. *J Pharmacol Exp Ther* 289:31–37.

Niwa, T., Imagawa, Y., Yamazaki, H. 2014. Drug interactions between nine antifungal agents and drugs metabolized by human cytochromes P450. *Curr Drug Metab* 15:651–679.

Obach, R. S. 2000. Inhibition of human cytochrome P450 enzymes by constituents of St. John's Wort, an herbal preparation used in the treatment of depression. *J Pharmacol Exp Ther* 294: 88–95.

Oganesian, A., Shilling, A. D., Young-Sciame, R., Tran, J., Watanyar, A., Azam, F., Kao, J., Leung, L. 2009. Desvenlafaxine and venlafaxine exert minimal *in vitro* inhibition of human cytochrome P450 and P-glycoprotein activities. *Psychopharmacol Bull* 42:47–63.

Ohyama, K., Nakajima, M., Suzuki, M., Shimada, N., Yamazaki, H., Yokoi, T. 2000. Inhibitory effects of amiodarone and its *N*-deethylated metabolite on human cytochrome P450 activities: Prediction of *in vivo* drug interactions. *Br J Clin Pharmacol* 49:244–253.

Otton, S. V., Crewe, H. K., Lennard, M. S., Tucker, G. T., Woods, H. F. 1988. Use of quinidine inhibition to define the role of the sparteine/debrisoquine cytochrome P450 in metoprolol oxidation by human liver microsomes. *J Pharmacol Exp Ther* 247:242–247.

Otton, S. V., Wu, D., Joffe, R. T., Cheung, S. W., Sellers, E. M. 1993. Inhibition by fluoxetine of cytochrome P450 2D6 activity. *Clin Pharmacol Ther* 53:401–409.

Otton, S. V., Ball, S. E., Cheung, S. W., Inaba, T., Rudolph, R. L., Sellers, E. M. 1996. Venlafaxine oxidation *in vitro* is catalysed by CYP2D6. *Br J Clin Pharmacol* 41:149–156.

Ozdemir, V., Naranjo, C. A., Herrmann, N., Reed, K., Sellers, E. M., Kalow, W. 1997. Paroxetine potentiates the central nervous system side effects of perphenazine: Contribution of cytochrome P4502D6 inhibition in vivo. *Clin Pharmacol Ther* 62:334–347.

Palamanda, J. R., Casciano, C. N., Norton, L. A., Clement, R. P., Favreau, L. V., Lin, C., Nomeir, A. A. 2001. Mechanism-based inactivation of CYP2D6 by 5-fluoro-2-[4-[(2-phenyl-1H-imidazol-5-yl)methyl]-1-piperazinyl]pyrimidine. *Drug Metab Dispos* 29:863–867.

Park, J. Y., Kim, K. A., Kim, S. L. 2003. Chloramphenicol is a potent inhibitor of cytochrome P450 isoforms CYP2C19 and CYP3A4 in human liver microsomes. *Antimicrob Agents Chemother* 47:3464–3469.

Patatanian, E., Casselman, J. 2014. Dextromethorphan/quinidine for the treatment of pseudobulbar affect. *Consult Pharm* 29:264–269.

Patroneva, A., Connolly, S. M., Fatato, P., Pedersen, R., Jiang, Q., Paul, J., Guico-Pabia, C., Isler, J. A., Burczynski, M. E., Nichols, A. I. 2008. An assessment of drug–drug interactions: The effect of desvenlafaxine and duloxetine on the pharmacokinetics of the CYP2D6 probe desipramine in healthy subjects. *Drug Metab Dispos* 36:2484–2491.

Perloff, E. S., Mason, A. K., Dehal, S. S., Blanchard, A. P., Morgan, L., Ho, T., Dandeneau, A. et al. 2009. Validation of cytochrome P450 time-dependent inhibition assays: A two-time point IC$_{50}$ shift approach facilitates kinact assay design. *Xenobiotica* 39:99–112.

Philip, P. A., James, C. A., Rogers, H. J. 1989. The influence of cimetidine on debrisoquine 4-hydroxylation in extensive metabolizers. *Eur J Clin Pharmacol* 36:319–321.

Pinder, R. M., Brogden, R. N., Swayer, R., Speight, T. M., Spencer, R., Avery, G. S. 1976. Pimozide: A review of its pharmacological properties and therapeutic uses in psychiatry. *Drugs* 12:1–40.

Pioro, E. P., Brooks, B. R., Cummings, J., Schiffer, R., Thisted, R. A., Wynn, D., Hepner, A., Kaye, R., Safety, Tolerability, and Efficacy Results Trial of AVP-923 in PBA Investigators. 2010. Dextromethorphan plus ultra low-dose quinidine reduces pseudobulbar affect. *Ann Neurol* 68:693–702.

Pope, L. E., Schoedel, K. A., Bartlett, C., Sellers, E. M. 2012. A study of potential pharmacokinetic and pharmacodynamic interactions between dextromethorphan/quinidine and memantine in healthy volunteers. *Clin Drug Investig* 32:e1–e15.

Preskorn, S. H., Alderman, J., Chung, M., Harrison, W., Messig, M., Harris, S. 1994. Pharmacokinetics of desipramine coadministered with sertraline or fluoxetine. *J Clin Psychopharmacol* 14:90–98.

Preskorn, S. H., Nichols, A. I., Paul, J., Patroneva, A. L., Helzner, E. C., Guico-Pabia, C. J. 2008. Effect of desvenlafaxine on the cytochrome P450 2D6 enzyme system. *J Psychiatr Pract* 14:368–378.

Qiu, F., Zhang, R., Sun, J., Jiye, A., Hao, H., Peng, Y., Ai, H., Wang, G. 2008. Inhibitory effects of seven components of danshen extract on catalytic activity of cytochrome P450 enzyme in human liver microsomes. *Drug Metab Dispos* 36:1308–1314.

Rakhit, A., Holford, N. H., Guentert, T. W., Maloney, K., Riegelman, S. 1984. Pharmacokinetics of quinidine and three of its metabolites in man. *J Pharmacokinet Biopharm* 12:1–21.

Reese, M. J., Wurm, R. M., Muir, K. T., Generaux, G. T., St John-Williams, L., McConn, D. J. 2008. An *in vitro* mechanistic study to elucidate the desipramine/bupropion clinical drug–drug interaction. *Drug Metab Dispos* 36:1198–1201.

Ring, B. J., Binkley, S. N., Vandenbranden, M., Wrighton, S. A. 1996. *In vitro* interaction of the antipsychotic agent olanzapine with human cytochromes P450 CYP2C9, CYP2C19, CYP2D6 and CYP3A. *Br J Clin Pharmacol* 41:181–186.

Ring, B. J., Gillespie, J. S., Binkley, S. N., Campanale, K. M., Wrighton, S. A. 2002. The interactions of a selective protein kinase C β inhibitor with the human cytochromes P450. *Drug Metab Dispos* 30:957–961.

Rodrigues, A. D., Roberts, E. M. 1997. The *in vitro* interaction of dexmedetomidine with human liver microsomal cytochrome P4502D6 (CYP2D6). *Drug Metab Dispos* 25:651–655.

Rosen, H. 2008. Dextromethorphan/quinidine sulfate for pseudobulbar affect. *Drugs Today (Barc)* 44:661–668.

Ruiz-Garcia, A., Giri, N., LaBadie, R. R., Ni, G., Boutros, T., Richie, N., Kocinsky, H. S., Checchio, T. M., Bello, C. L. 2014. A phase I open-label study to investigate the potential drug–drug interaction between single-dose dacomitinib and steady-state paroxetine in healthy volunteers. *J Clin Pharmacol* 54:555–562.

Saal, A. K., Werner, J. A., Greene, H. L., Sears, G. K., Graham, E. L. 1984. Effect of amiodarone on serum quinidine and procainamide levels. *Am J Cardiol* 53:1264–1267.

Sahi, J., Black, C. B., Hamilton, G. A., Zheng, X., Jolley, S., Rose, K. A., Gilbert, D., LeCluyse, E. L., Sinz, M. W. 2003. Comparative effects of thiazolidinediones on *in vitro* P450 enzyme induction and inhibition. *Drug Metab Dispos* 31:439–446.

Saito, M., Yasui-Furukori, N., Nakagami, T., Furukori, H., Kaneko, S. 2005. Dose-dependent interaction of paroxetine with risperidone in schizophrenic patients. *J Clin Psychopharmacol* 25:527–532.

Sallee, F. R., Nesbitt, L., Jackson, C., Sine, L., Sethuraman, G. 1997. Relative efficacy of haloperidol and pimozide in children and adolescents with Tourette's disorder. *Am J Psychiatry* 154:1057–1062.

Sanchez, C., Reines, E. H., Montgomery, S. A. 2014. A comparative review of escitalopram, paroxetine, and sertraline: Are they all alike? *Int Clin Psychopharmacol* 29:185–196.

Saruwatari, J., Yasui-Furukori, N., Niioka, T., Akamine, Y., Takashima, A., Kaneko, S., Uno, T. 2012. Different effects of the selective serotonin reuptake inhibitors fluvoxamine, paroxetine, and sertraline on the pharmacokinetics of fexofenadine in healthy volunteers. *J Clin Psychopharmacol* 32:195–199.

Sauer, J. M., Long, A. J., Ring, B., Gillespie, J. S., Sanburn, N. P., DeSante, K. A., Petullo, D. et al. 2004. Atomoxetine hydrochloride: Clinical drug–drug interaction prediction and outcome. *J Pharmacol Exp Ther* 308:410–418.

Schellens, J. H., van der Wart, J. H., Brugman, M., Breimer, D. D. 1989. Influence of enzyme induction and inhibition on the oxidation of nifedipine, sparteine, mephenytoin and antipyrine in humans as assessed by a "cocktail" study design. *J Pharmacol Exp Ther* 249:638–645.

Schmutz, J. L., Barbaud, A., Trechot, P. 1999. [Overdose of nortriptyline during treatment with terbinafine (1st reported case)]. *Ann Dermatol Venereol* 126:647.

Schoedel, K. A., Pope, L. E., Sellers, E. M. 2012. Randomized open-label drug–drug interaction trial of dextromethorphan/quinidine and paroxetine in healthy volunteers. *Clin Drug Investig* 32:157–169.

Schoedel, K. A., Morrow, S. A., Sellers, E. M. 2014. Evaluating the safety and efficacy of dextromethorphan/quinidine in the treatment of pseudobulbar affect. *Neuropsychiatr Dis Treat* 10:1161–1174.

Shader, R. I., Granda, B. W., von Moltke, L. L., Giancarlo, G. M., Greenblatt, D. J. 1999. Inhibition of human cytochrome P450 isoforms in vitro by zafirlukast. *Biopharm Drug Dispos* 20:385–388.

Shin, J. G., Soukhova, N., Flockhart, D. A. 1999. Effect of antipsychotic drugs on human liver cytochrome P450 (CYP) isoforms in vitro: Preferential inhibition of CYP2D6. *Drug Metab Dispos* 27:1078–1084.

Shin, J. G., Park, J. Y., Kim, M. J., Shon, J. H., Yoon, Y. R., Cha, I. J., Lee, S. S., Oh, S. W., Kim, S. W., Flockhart, D. A. 2002. Inhibitory effects of tricyclic antidepressants (TCAs) on human cytochrome P450 enzymes in vitro: Mechanism of drug interaction between TCAs and phenytoin. *Drug Metab Dispos* 30:1102–1107.

Sindrup, S. H., Arendt-Nielsen, L., Brosen, K., Bjerring, P., Angelo, H. R., Eriksen, B., Gram, L. F. 1992. The effect of quinidine on the analgesic effect of codeine. *Eur J Clin Pharmacol* 42:587–591.

Sitsen, J. M., Maris, F. A., Timmer, C. J. 2000. Concomitant use of mirtazapine and cimetidine: A drug–drug interaction study in healthy male subjects. *Eur J Clin Pharmacol* 56:389–394.

Skinner, M. H., Kuan, H. Y., Pan, A., Sathirakul, K., Knadler, M. P., Gonzales, C. R., Yeo, K. P. et al. 2003. Duloxetine is both an inhibitor and a substrate of cytochrome P4502D6 in healthy volunteers. *Clin Pharmacol Ther* 73:170–177.

Skjelbo, E., Brosen, K. 1992. Inhibitors of imipramine metabolism by human liver microsomes. *Br J Clin Pharmacol* 34:256–261.

Smith, R. A. 2006. Dextromethorphan/quinidine: A novel dextromethorphan product for the treatment of emotional lability. *Expert Opin Pharmacother* 7:2581–2598.

Smith, D. A., Jones, B. C. 1992. Speculations on the substrate structure-activity relationship (SSAR) of cytochrome P450 enzymes. *Biochem Pharmacol* 44:2089–2098.

Smith, T., Riskin, J. 1994. Effect of clozapine on plasma nortriptyline concentration. *Pharmacopsychiatry* 27:41–42.

Spina, E., Avenoso, A., Facciola, G., Scordo, M. G., Ancione, M., Madia, A. 2001. Plasma concentrations of risperidone and 9-hydroxyrisperidone during combined treatment with paroxetine. *Ther Drug Monit* 23:223–227.

Spina, E., Santoro, V. D'Arrigo, C. 2008. Clinically relevant pharmacokinetic drug interactions with second-generation antidepressants: An update. *Clin Ther* 30:1206–1227.

Spinler, S. A., Cheng, J. W., Kindwall, K. E., Charland, S. L. 1995. Possible inhibition of hepatic metabolism of quinidine by erythromycin. *Clin Pharmacol Ther* 57:89–94.

Stahl, S. M. 2013. Mechanism of action of dextromethorphan/quinidine: Comparison with ketamine. *CNS Spectr* 18:225–227.

Steiner, E., Spina, E. 1987. Differences in the inhibitory effect of cimetidine on desipramine metabolism between rapid and slow debrisoquin hydroxylators. *Clin Pharmacol Ther* 42:278–282.

Stout, S. M., Nielsen, J., Bleske, B. E., Shea, M., Brook, R., Kerber, K., Welage, L. S. 2010. The impact of paroxetine coadministration on stereospecific carvedilol pharmacokinetics. *J Cardiovasc Pharmacol Ther* 15:373–379.

Stout, S. M., Nielsen, J., Welage, L. S., Shea, M., Brook, R., Kerber, K., Bleske, B. E. 2011. Influence of metoprolol dosage release formulation on the pharmacokinetic drug interaction with paroxetine. *J Clin Pharmacol* 51:389–396.

Strobl, G. R., von Kruedener, S., Stockigt, J., Guengerich, F. P., Wolff, T. 1993. Development of a pharmacophore for inhibition of human liver cytochrome P450 2D6: Molecular modeling and inhibition studies. *J Med Chem* 36:1136–1145.

Sun, J., Peng, Y., Wu, H., Zhang, X., Zhong, Y., Xiao, Y., Zhang, F. et al. 2015. Guanfu base A, an antiarrhythmic alkaloid of *Aconitum coreanum*, is a CYP2D6 inhibitor of human, monkey, and dog isoforms. *Drug Metab Dispos* 43:713–724.

Suzuki, A., Otani, K., Ishida, M., Yasui, N., Kondo, T., Mihara, K., Kaneko, S., Inoue, Y. 1996. No interaction between desipramine and bromperidol. *Prog Neuropsychopharmacol Biol Psychiatry* 20:1265–1271.

Tassaneeyakul, W., Guo, L. Q., Fukuda, K., Ohta, T., Yamazoe, Y. 2000. Inhibition selectivity of grapefruit juice components on human cytochromes P450. *Arch Biochem Biophys* 378:356–363.

Transon, C., Leemann, T., Dayer, P. 1996. *In vitro* comparative inhibition profiles of major human drug metabolising cytochrome P450 isozymes (CYP2C9, CYP2D6 and CYP3A4) by HMG-CoA reductase inhibitors. *Eur J Clin Pharmacol* 50:209–215.

Turpeinen, M., Nieminen, R., Juntunen, T., Taavitsainen, P., Raunio, H., Pelkonen, O. 2004. Selective inhibition of CYP2B6-catalyzed bupropion hydroxylation in human liver microsomes in vitro. *Drug Metab Dispos* 32:626–631.

Umeda, S., Harakawa, N., Yamamoto, M., Ueno, K. 2005. Effect of nonspecific binding to microsomes and metabolic elimination of buprenorphine on the inhibition of cytochrome P4502D6. *Biol Pharm Bull* 28:212–216.

Umehara, K., Shimokawa, Y., Miyamoto, G. 2002. Inhibition of human drug metabolizing cytochrome P450 by buprenorphine. *Biol Pharm Bull* 25:682–685.

Van, L. M., Hargreaves, J. A., Lennard, M. S., Tucker, G. T., Rostami-Hodjegan, A. 2007. Inactivation of CYP2D6 by methylenedioxymethamphetamine in different recombinant expression systems. *Eur J Pharm Sci* 32:8–16.

Vandel, S., Bertschy, G., Bonin, B., Nezelof, S., Francois, T. H., Vandel, B., Sechter, D., Bizouard, P. 1992. Tricyclic antidepressant plasma levels after fluoxetine addition. *Neuropsychobiology* 25:202–207.

VandenBranden, M., Ring, B. J., Binkley, S. N., Wrighton, S. A. 1996. Interaction of human liver cytochromes P450 *in vitro* with LY307640, a gastric proton pump inhibitor. *Pharmacogenetics* 6:81–91.

van der Kuy, P. H., Hooymans, P. M. 1998. Nortriptyline intoxication induced by terbinafine. *BMJ* 316:441.

Venhorst, J., Onderwater, R. C., Meerman, J. H., Commandeur, J. N., Vermeulen, N. P. 2000. Influence of *N*-substitution of 7-methoxy-4-(aminomethyl)-coumarin on cytochrome P450 metabolism and selectivity. *Drug Metab Dispos* 28:1524–1532.

Venkatakrishnan, K., von Moltke, L. L., Greenblatt, D. J. 2000. Effects of the antifungal agents on oxidative drug metabolism: Clinical relevance. *Clin Pharmacokinet* 38:111–180.

Vlase, L., Leucuta, A., Farcau, D., Nanulescu, M. 2006. Pharmacokinetic interaction between fluoxetine and metoclopramide in healthy volunteers. *Biopharm Drug Dispos* 27:285–289.

Volotinen, M., Turpeinen, M., Tolonen, A., Uusitalo, J., Maenpaa, J., Pelkonen, O. 2007. Timolol metabolism in human liver microsomes is mediated principally by CYP2D6. *Drug Metab Dispos* 35:1135–1141.

von Bahr, C., Spina, E., Birgersson, C., Ericsson, O., Goransson, M., Henthorn, T., Sjoqvist, F. 1985. Inhibition of desmethylimipramine 2-hydroxylation by drugs in human liver microsomes. *Biochem Pharmacol* 34:2501–2505.

von Moltke, L. L., Greenblatt, D. J., Cotreau-Bibbo, M. M., Duan, S. X., Harmatz, J. S., Shader, R. I. 1994. Inhibition of desipramine hydroxylation *in vitro* by serotonin-reuptake-inhibitor antidepressants, and by quinidine and ketoconazole: A model system to predict drug interactions *in vivo*. *J Pharmacol Exp Ther* 268:1278–1283.

von Moltke, L. L., Greenblatt, D. J., Duan, S. X., Daily, J. P., Harmatz, J. S., Shader, R. I. 1998. Inhibition of desipramine hydroxylation (cytochrome P450 2D6) in vitro by quinidine and by viral protease inhibitors: Relation to drug interactions in vivo. *J Pharm Sci* 87:1184–1189.

von Moltke, L. L., Greenblatt, D. J., Granda, B. W., Giancarlo, G. M., Duan, S. X., Daily, J. P., Harmatz, J. S., Shader, R. I. 2001. Inhibition of human cytochrome P450 isoforms by nonnucleoside reverse transcriptase inhibitors. *J Clin Pharmacol* 41:85–91.

Voorman, R. L., Payne, N. A., Wienkers, L. C., Hauer, M. J., Sanders, P. E. 2001. Interaction of delavirdine with human liver microsomal cytochrome P450: Inhibition of CYP2C9, CYP2C19, and CYP2D6. *Drug Metab Dispos* 29:41–47.

Weiss, J., Sawa, E., Riedel, K. D., Haefeli, W. E., Mikus, G. 2008. *In vitro* metabolism of the opioid tilidine and interaction of tilidine and nortilidine with CYP3A4, CYP2C19, and CYP2D6. *Naunyn Schmiedebergs Arch Pharmacol* 378:275–282.

Wienkers, L. C., Allievi, C., Hauer, M. J., Wynalda, M. A. 1999. Cytochrome P450-mediated metabolism of the individual enantiomers of the antidepressant agent reboxetine in human liver microsomes. *Drug Metab Dispos* 27:1334–1340.

Wrighton, S. A., Brian, W. R., Sari, M. A., Iwasaki, M., Guengerich, F. P., Raucy, J. L., Molowa, D. T., Vandenbranden, M. 1990. Studies on the expression and metabolic capabilities of human liver cytochrome P450IIIA5 (HLp3). *Mol Pharmacol* 38:207–213.

Wu, D., Otton, S. V., Inaba, T., Kalow, W., Sellers, E. M. 1997. Interactions of amphetamine analogs with human liver CYP2D6. *Biochem Pharmacol* 53:1605–1612.

Yang, L. P., Deeks, E. D. 2015. Dextromethorphan/quinidine: A review of its use in adults with pseudobulbar affect. *Drugs* 75:83–90.

Zebothsen, I., Kunze, T., Clement, B. 2006. Inhibitory effects of cytostatically active 6-aminobenzo[c]phenanthridines on cytochrome P450 enzymes in human hepatic microsomes. *Basic Clin Pharmacol Toxicol* 99:37–43.

Zhang, W., Ramamoorthy, Y., Tyndale, R. F., Sellers, E. M. 2003. Interaction of buprenorphine and its metabolite norbuprenorphine with cytochromes p450 *in vitro*. *Drug Metab Dispos* 31:768–772.

Zhang, J. W., Liu, Y., Li, W., Hao, D. C., Yang, L. 2006. Inhibitory effect of medroxyprogesterone acetate on human liver cytochrome P450 enzymes. *Eur J Clin Pharmacol* 62:497–502.

Zhao, X. J., Yokoyama, H., Chiba, K., Wanwimolruk, S., Ishizaki, T. 1996. Identification of human cytochrome P450 isoforms involved in the 3-hydroxylation of quinine by human live microsomes and nine recombinant human cytochromes P450. *J Pharmacol Exp Ther* 279:1327–1334.

5 Regulation of Human CYP2D6

5.1 INTRODUCTION

Human CYP2D6 is expressed in the liver, intestine, kidney, and brain and is an important enzyme that metabolizes a number of clinically important drugs (Zhou et al. 2009). There are large interindividual variabilities in the expression and activity of CYP2D6. Changes in the expression and activity of this enzyme may cause altered drug clearance, therapeutic efficacy, risk of adverse drug reactions (ADRs), and unfavorable drug–drug interactions (Hukkanen 2012; Matoulkova et al. 2014; Zanger and Schwab 2013). In adults, differences in CYP expression and attributable metabolism are mainly determined by genetic variability and, to a lesser extent, by enzyme induction because of pharmacotherapy, xenobiotic exposure, or dietary factors (Cascorbi 2003; Fuhr 2000; Ingelman-Sundberg 2005; Ingelman-Sundberg et al. 2007; Lamba et al. 2002). Debrisoquine, sparteine, metoprolol, or dextromethorphan are well-established probe drugs for phenotyping CYP2D6 while tramadol can also be used for this purpose (Frank et al. 2007). The enzymatic activity is reflected by various pharmacokinetic metrics such as the partial clearance of a parent compound to the respective CYP2D6-mediated metabolite or metabolic ratios (MRs). A cocktail approach can be used for the in vivo and in vitro phenotyping studies of CYP2D6 together with other major CYPs (Fuhr et al. 2007; Spaggiari et al. 2014; Tanaka et al. 2003).

The mRNA expression and activity levels of CYP2D6 correlate well with each other in vitro and in vivo (Carcillo et al. 2003; Kawakami et al. 2011; Ohtsuki et al. 2012; Rodriguez-Antona et al. 2001; Sakamoto et al. 2011; Temesvari et al. 2012), and the correlation coefficients range from 0.71 to 0.91. Such a high correlation between mRNA and enzyme activity levels of CYP2D6 is also observed for CYP3A4, 1A2, 2B6, and 2C9 (Gerbal-Chaloin et al. 2002; Sy et al. 2002; Temesvari et al. 2012), whose activity levels are known to be governed by the transcriptional regulation of the genes. These results suggest that differential transcriptional regulation of CYP2D6 may contribute to the large interindividual variability in its activity. There are significant correlations observed between CYP2C9 and several regulators including pregnane X receptor (PXR), constitutive androstane receptor (CAR), hepatocyte nuclear factor-4α (HNF-4α), and aryl hydrocarbon receptor nuclear translocator at the mRNA level in a Chinese liver bank with a sample size of 96, but the correlation for CYP2D6 is weak (Wang et al. 2011), suggesting a minor to moderate role of these regulators in the regulation of CYP2D6.

Members of human CYP1A, 2B, 2C, 3A, and 4A subfamilies are highly inducible by some xenobiotics including drugs and some environmental compounds, contributing to the large interindividual activities of these enzymes and drug–drug interactions (Dickins 2004; Handschin and Meyer 2003; Masahiko and Honkakoski 2000; Nakata et al. 2006; Waxman 1999; Xu et al. 2005; Yamada et al. 2006). The xenobiotic-mediated induction of CYPs is often tissue specific, rapid, dose dependent, and reversible upon removal of the inducer. In the in vitro and in vivo studies on regulation of human CYPs, several prototypical inducers including rifampin, phenobarbital, β-naphthoflavone, and dexamethasone are commonly used (Donato and Castell 2003; Hewitt et al. 2007). Phenobarbital-type inducers activate transcription of CYP2A, 2B, 2C, and 3A members, the same CYPs activated by the dexamethasone/rifampicin-type inducers but CYP3As are more efficiently induced than CYP2Cs and 2Bs by the dexamethasone/rifampicin-type inducers. The effect of inducer drugs is not restricted to the regulation of CYPs, other drug-metabolizing enzymes, or drug transporters but involves a major pleiotropic response including the up- or downregulation of numerous genes and physiological activities. For example, a recent genome-wide study has indicated that rifampin can regulate a number of genes in human hepatocytes via binding to newly identified drug-responsive regulatory elements (Smith et al. 2014). Another genome-wide mouse study has revealed that the CAR target gene *Cyp2b10* is concomitantly hypomethylated and transcriptionally activated in a liver tissue–specific manner after phenobarbital treatment (Lempiainen et al. 2011). Furthermore, analysis of active and repressive histone modifications using chromatin immunoprecipitation shows a strong phenobarbital-mediated epigenetic switch at the *Cyp2b10* promoter. The effects of dexamethasone on CYP expression in human hepatocytes are complex, attributed in part to the physiological role of glucocorticoids in maintaining constitutive expression of PXR and CAR. Rifampin has been an effective antimicrobial agent against Gram-negative bacteria since its introduction in the 1960s and is widely used mainly in the treatment of tuberculosis and other infections such as community-acquired *Legionella pneumophila* pneumonia (Aristoff et al. 2010; Varner et al. 2011). It is known to enhance the elimination of a long list of drugs, with important clinical consequences in terms of loss or decreased drug efficacy or increased toxicity (Baciewicz et al. 2008, 2013; Niemi et al. 2003; Sousa et al. 2008). Rifampin is recognized as a pleiotropic but specific inducer that alters the activity of specific CYPs, other drug metabolizing enzymes, and drug transporters, which provides the rational basis for its broad drug–drug interactions (Baciewicz et al. 2013; Rae et al. 2001). While inhibition studies for CYPs are normally performed with either human liver microsomes or cDNA-expressed CYPs, cultured human hepatocytes have been extensively used for CYP induction studies (Dickins 2004; Gomez-Lechon et al. 2008; Lake et al. 2009; LeCluyse 2001; Luo et al. 2004; Martinez-Jimenez et al. 2007; Pavek and Dvorak 2008; Sahi et al. 2010; Sinz et al. 2008; Soars et al. 2007; Walsky and Boldt 2008).

This chapter will discuss how human CYP2D6 is regulated by internal and external factors at transcriptional, posttranscriptional, posttranslational, and epigenetic levels.

5.2 EFFECTS OF PHYSIOLOGICAL FACTORS ON CYP2D6 ACTIVITY

5.2.1 GENDER

Gender may affect CYP enzyme activity owing to endogenous hormonal fluctuations with the menstrual cycle, which are altered by oral contraceptives (Anderson 2002; Franconi and Campesi 2014; Harris et al. 1995; Scandlyn et al. 2008; Waxman and Holloway 2009). For example, the activity of CYP3A4 appears to be higher in females (Cotreau et al. 2005; Cummins et al. 2002; Waxman and Holloway 2009), whereas males seem to have a higher CYP2E1 and 1A2 activity (Gleiter and Gundert-Remy 1996; Landi et al. 1999; Scandlyn et al. 2008). Women are reported to experience more ADRs than men (Drici and Clement 2001). For example, women are more likely to experience drug-induced QT prolongation and torsade de pointes than men (Wenzel-Seifert et al. 2011).

Clinical data on the gender effects on CYP2D6 are conflicting. In a large study with 1526 healthy Caucasian men and 1539 women, age, sex, and overweight are demonstrated to have no effect on CYP2D6-mediated debrisoquine metabolism (Vincent-Viry et al. 1991). Another study in 33 healthy volunteers has shown that both CYP2D6 and 3A4 do not differ between men, women taking oral contraceptives, and regularly menstruating women not receiving oral contraceptives when dextromethorphan is used as the probe (McCune et al. 2001). In a study with 161 normal subjects (51% female subjects and 40% aged >50 years) using the cocktail approach, gender and age do not have any effects on CYP2D6 activity (Bebia et al. 2004). These results corroborate those by Mahgoub et al. (1977) and Evans et al. (1980) who do not find any influence of age and gender when debrisoquine is used as a probe for CYP2D6.

In contrast, Tamminga et al. (1999) have found a significant gender difference when 4301 healthy Dutch subjects are phenotyped with dextromethorphan as the probe substrate of CYP2D6. The investigated population mainly comprises Caucasian (98.9%) males (68%). The age ranges from 18 to 82 years. Eight percent of the subjects are poor metabolizers (PMs) of CYP2D6. Within the extensive metabolizer (EM) group, a 20% lower CYP2D6 activity is reported in females than in males (Tamminga et al. 1999). The average MR is 0.014 ± 0.033 for subjects who are EMs and 5.4 ± 7.6 for PM subjects. For PMs, there is no such difference for CYP2D6 or 2C19.

Similarly, another study in 611 Caucasian volunteers (330 males and 281 females) has also found that the median dextromethorphan MR in females without using oral contraceptives is significantly (16%) lower than that in males (Hagg et al. 2001). With respect to CYP2D6, 543 subjects are EMs and 68 are PMs. The median MR is 0.070 for EMs and 5.26 for PMs.

A study with 150 young, healthy, drug-free women and men has revealed that the dextromethorphan MR is significantly lower ($P < 0.0001$) in 56 female EMs (0.008 ± 0.021) compared to 86 male EMs (0.020 ± 0.040) (Labbe et al. 2000). pH is a significant predictor of dextromethorphan MRs in men and women ($P < 0.001$). Once-a-month phenotyping with dextromethorphan of 12 healthy young men (8 EMs and 4 PMs) over a 1-year period, as well as every-other-day phenotyping with dextromethorphan of healthy, premenopausal women (10 EMs and 2 PMs) during a complete menstrual cycle, does not follow a particular pattern and shows similar intrasubject variability ranging from 24.1% to 74.5% (mean, 50.9%) in men and from 20.5% to 96.2% (mean, 52.0%) in women, independent of the CYP2D6 phenotype ($P = 0.342$) (Labbe et al. 2000). There is a significant inverse relationship between physiological urinary pH and sequential dextromethorphan MRs as well as metoprolol MRs in men and women, with MRs varying up to 6-fold with metoprolol and up to 20-fold with dextromethorphan. These findings indicate that CYP2D6 activity is highly variable, independent of menstrual cycle phases, sex hormones, time variables, or phenotype. Up to 80% of the observed variability can be explained by variations of urinary pH within the physiological range.

It appears that gender as a single factor only has a minor to moderate effect on CYP2D6 activity in humans. Nevertheless, given the large variability in CYP2D6 activity caused by factors other than gender (e.g., polymorphism), the gender effects on CYP2D6 activity have a limited clinical significance.

5.2.2 DEVELOPMENTAL CHANGES (ONTOGENY) OF CYP2D6 EXPRESSION AND ACTIVITY

There are significant changes in hepatic CYP expression and activities during ontogeny (Rizzo et al. 2005). Such changes can have a profound effect on therapeutic efficacy in the fetus and child, as well as the risk for ADRs. Because of ethical and logistical restrictions with performing studies in pediatric populations, there have been only studies on the expression and activity of CYP2D6 in early development of humans. Treluyer et al. (1991) have found no difference between CYP2D6 activity and expression in the first week postbirth compared to the third trimester, but there is a positive correlation between postnatal age and increasing CYP2D6 activity. The authors have concluded that there is a birth-dependent mechanism regulating CYP2D6 expression. However, the small sample size representing infancy and lack of ethnic information preclude a comprehensive data analysis to understand the key factors that determine the ontogeny of CYP2D6. The mouse may represent a suitable model to investigate the mechanisms underlying the increased expression and activity of CYP2D6 during human pregnancy.

Blake et al. (2007) have reported that CYP2D6 activity change was measured using dextromethorphan as a probe in 193 healthy infants during the first year of life. CYP2D6 activity is detectable in infants aged 2 weeks and is associated with the genotype of *CYP2D6*. With postnatal age up to 1 year, there is no significant change in CYP2D6-mediated dextromethorphan *O*-demethylation in these infants. However, if the maturation of renal function is taken into account, the data from the study of Blake et al. (2007) suggest a gradual increase in hepatic CYP2D6 expression over the first year of life such that adult enzyme levels would not be achieved until approximately 6 months of age (Johnson et al. 2008).

Stevens et al. (2008) have examined the expression and activity of CYP2D6 using dextromethorphan as a probe in a large and developmentally diverse set of pediatric liver samples (*n* = 222). The age of the liver donors ranges from 8 weeks' gestation to 18 years postbirth. There are 62 fetal samples and 160 postnatal samples with 163 male and 78 female samples (unknown gender, 11 samples). For genotype, 22.1%, 15.6%, and 13.1% of the liver donors are carriers of *CYP2D6*1/*2*, **1/*4*, and **1/*1* (wild type), respectively; 8.3%, 4.8%, 4.1%, 3.4%, 3.4%, 2.8%, and 2.1% of the donors harbor *CYP2D6*2/*2*, **2/*4*, **1/*10*, **1/*17*, **1/*29*, **1×N/*2*, and **2/*10* alleles, respectively. The authors have found that there is a significant correlation between immunodetectable levels of CYP2D6 protein with the corresponding dextromethorphan *O*-demethylase activity in individual liver samples (*r* = 0.686). Of gender, ethnicity, postmortem interval, and genotype, only increasing gestational age is the only factor associated with CYP2D6 activity and protein content in prenatal samples (Stevens et al. 2008). For liver microsomes from samples of first- and second-trimester (<26 weeks of gestational age) fetuses, no or very low CYP2D6 activity is detected, while 10 of 14 samples from the third-trimester (26 to 40 weeks of gestational age) fetuses exhibit detectable CYP2D6 enzyme activity, which is approximately 3%–5% of the adult liver levels. In contrast, age, genotype, and ethnicity are associated with CYP2D6 protein expression levels for postnatal liver samples, and increasing age and genotype are associated with increasing dextromethorphan *O*-demethylation (Stevens et al. 2008). The CYP2D6 content in liver samples from neonates less than 7 days of age is higher than that observed in first- and second-trimester samples, but not significantly higher than that in third-trimester fetal samples (Stevens et al. 2008), indicating that either CYP2D6 protein or activity increases significantly and rapidly after 7 days postbirth. In contrast, CYP2D6 expression in postnatal samples greater than 7 days of age is substantially higher than that for any earlier age category. Higher CYP2D6 activity is also observed in liver samples from Caucasians (*n* = 93) when compared with those from African Americans (*n* = 83) (0.043 vs. 0.029 nmol/min/mg protein). In addition, CYP2D6 activity is higher in postnatal samples predicted to be EMs (*n* = 68) or intermediate metabolizers (IMs) (*n* = 67) versus PMs (*n* = 10) (Stevens et al. 2008). These results suggest that age and, in particular, genotype are key determinants of interindividual variability in CYP2D6 expression and activity during ontogeny. However, it is unknown why there is a remarkably lower activity in the liver samples of those aged >7 days to 18 years compared to the adult values. Sample quality issue

and prevalent loss-of-function alleles such as *CYP2D6*17* and **29* could not provide a convincing explanation for this. It appears that the *CYP2D6* gene is significantly knocked down after 7 days postbirth until the age of 18 years by unknown physiological, pathological, and environmental factors.

The data from the study by Stevens et al. (2008) are consistent with the in vivo longitudinal data reported by Blake et al. (2007) who suggest that the phenotype reflects the *CYP2D6* genotype by 2 weeks of age. The low fetal levels of CYP2D6-dependent activity and protein are also consistent with the results of Treluyer et al. (1991), as is the lack of difference between CYP2D6 activity and protein in the first week of life versus the third trimester. Because CYP2D6 protein is independent of gestational age in their sample set, Treluyer et al. (1991) have concluded that there is an important contribution from a birth-dependent regulating mechanism for CYP2D6. This is further supported by the data from Stevens et al. (2008).

5.2.3 Pregnancy Induces CYP2D6

Pregnancy is considered to be a physiological condition associated with changes in the metabolism and disposition of clinically used drugs (Anderson 2005). CYP3A4, 2D6, and 2C9, and UGT1A4 and 2B7 activities increase during the gestational period, while the activities of CYP1A2 and 2C19 are reduced (Anderson 2005). Like the liver, the placenta is also involved in drug metabolism, and CYP1A1, 2E1, 2F1, 3A4/3A5, and 4B1 are expressed at much lower levels in the term placenta than in the liver (Pasanen and Pelkonen 1989; Pavek and Smutny 2014; Ring et al. 1999; Syme et al. 2004).

Pregnancy is known to induce hepatic CYP2D6-mediated drug metabolism in women (Hogstedt et al. 1983, 1985; Hogstedt and Rane 1993; Tracy et al. 2005; Wadelius et al. 1997). For example, the plasma clearance of metoprolol increases 2- to 13-fold during pregnancy as compared to that after delivery (Hogstedt et al. 1985). In relation to the plasma drug concentration, metoprolol produces four times and twice the effect on heart rate and blood pressure during pregnancy as compared to the postpartum period (Hogstedt and Rane 1993). However, the induction of endogenous Cyp2d genes is not observed in pregnant rats (Dickmann et al. 2008), while the expression and activity of mouse Cyp2d11, 2d22, 2d26, and 2d40 are significantly increased (Topletz et al. 2013). A putative RA response element has been identified within the Cyp2d40 promoter and the mRNA of Cyp2d40 correlates with Cyp26a1 and Rarβ (Topletz et al. 2013). This likely reflects the large divergence in the regulatory region sequences of genes encoding CYP enzymes between mice, rats, and humans.

In CYP2D6-humanized transgenic mice whose genome harbors the human *CYP2D6* gene plus 2.5 kb of its upstream regulatory region, CYP2D6 expression is significantly enhanced at term pregnancy (fourfold) (Koh et al. 2014a). This is accompanied by increased recruitment of HNF-4α to the promoter of *CYP2D6*. The enhanced HNF-4α activity during pregnancy is attributed in part to the decreased expression of short heterodimer partner (SHP) (Koh et al. 2014a), a corepressor that inhibits HNF-4α activity via physical interaction. Repressed SHP expression at term pregnancy is associated with decreased hepatic levels of RA, which may be attributable to enhanced hepatic expression of CYP26A1 during pregnancy.

Among the eight Cyp2d homologs of mouse including Cyp2d9, 2d10, 2d12, 2d13, 2d22, 2d26, 2d34, and 2d40 examined in CYP2D6-humanized transgenic or wild-type mice, only Cyp2d40 expression is found induced four- to sixfold at term pregnancy as compared to prepregnancy control mice (Ning et al. 2015). In mice where hepatic HNF-4α is knocked down, the pregnancy-mediated increase in Cyp2d40 expression is abrogated. Data from transient transfection, promoter reporter assays, and electrophoretic mobility shift assays indicate that HNF-4α transactivates Cyp2d40 promoter via direct binding to −117/−105 of the gene. Chromatin immunoprecipitation assay shows a 2.3-fold increase in HNF-4α recruitment to the *Cyp2d40* promoter during pregnancy (Ning et al. 2015). Results from mice treated with an SHP inducer (i.e., GW4064) at 15 mg/kg/day for 5 days and HepG2 cells cotransfected with Krüppel-like factor 9 (KLF9) suggest that neither SHP nor KLF9 is involved in the increased HNF-4α transactivation of the *Cyp2d40* promoter during pregnancy. These data indicate that the expression of human CYP2D6 and mouse Cyp2d40 is increased during pregnancy because of HNF-4α transactivation. CYP2D40 is highly homologous (i.e., 81%) to CYP2D6, and this suggests that CYP2D40 may play similar roles as CYP2D6 by participating in hepatic drug detoxification and modulating brain functions.

Further microarray studies in CYP2D6-humanized transgenic mice have shown that seven transcription factors, namely, activating transcription factor 5 (ATF5), early growth response 1 (EGR1), forkhead box protein A3 (FOXA3), JUNB, KLF9, KLF10, and REV-ERBα, are upregulated in mouse liver during pregnancy (Koh et al. 2014b). KLF9 itself is a weak transactivator of *CYP2D6* promoter but significantly enhances *CYP2D6* promoter transactivation by HNF-4α, a known transcriptional activator of CYP2D6 expression. The results from deletion and mutation analysis of *CYP2D6* promoter activity have identified a KLF9 putative binding motif at the −22/−14 region to be critical in the potentiation of HNF-4α–induced transactivation of CYP2D6 (Koh et al. 2014b). KLF9, a member of the KLF transcription factor family of zinc finger DNA-binding proteins, can either activate or repress target gene expression in a promoter-specific context. KLF9 is involved in cell differentiation of B cells, keratinocytes, and neurons. Biologic actions of KLF9 are mediated either by its direct binding to the promoters of its target genes such as *CYP1A1* or by coactivation of other transcription factors (Kaczynski et al. 2002; Shields and Yang 1998; Zhang et al. 1998). KLF9 is also a key transcriptional regulator for uterine endometrial cell proliferation, adhesion, and differentiation, all factors that are essential during the process of pregnancy and are switched off during tumorigenesis (Pabona

et al. 2012; Shimizu et al. 2010; Simmen et al. 2008, 2015). In endometrial cells, KLF9 binds to progesterone receptors and enhances transcriptional activation of the target genes (Zhang et al. 2003).

Enhanced HNF-4α transactivation of the *CYP2D6* promoter is triggered partially by decreased SHP and increased KLF9 expression during pregnancy, but neither SHP nor KLF9 plays an important role in modulating HNF-4α transactivation of Cyp2d40 promoter in mice. This is potentially attributed to differences between CYP2D6 and mouse Cyp2d40 in the promoter sequences, resulting in altered interplay and binding affinity among SHP, KLF9, and HNF-4α. CYP2D6 promoter harbors multiple binding sites for KLF9 that enhances HNF-4α transactivation of the promoter, but mouse Cyp2d40 lacks such KLF9 binding sites in its promoter (Ning et al. 2015). Further studies are needed to further delineate the molecular mechanisms underlying pregnancy-upregulated CYP2D.

5.2.4 Fasting

A recent study has examined the effects of short-term fasting (36 h) on CYP-mediated drug metabolism in healthy subjects and rats (Lammers et al. 2015). In healthy subjects, short-term fasting increases oral caffeine clearance by 20% ($P = 0.03$) and decreases oral S-warfarin clearance by 25% ($P < 0.001$). However, short-term fasting does not alter CYP2D6-, 2C19-, and 3A4-mediated drug metabolism. In rats, short-term fasting increases mRNA expression of the orthologs of human CYP1A2, 2C19, 2D6, and 3A4 ($P < 0.05$) and decreased the mRNA expression of the ortholog of CYP2C9 ($P < 0.001$) compared with the postabsorptive state (Lammers et al. 2015). These findings demonstrate that short-term fasting alters CYP-mediated drug metabolism in a nonuniform pattern.

5.3 EFFECTS OF ENVIRONMENTAL FACTORS ON CYP2D6 ACTIVITY

5.3.1 Smoking

The frequency of PMs does not significantly vary among groups regardless of their smoking or drinking habits. These results corroborate those by Cholerton et al. (1996) who have found no statistically different genotypic frequencies between smokers and nonsmokers. There is no significant association between CYP2D6 metabolic capacity and drinking and smoking behavior in PMs, whereas slightly significant relationships are observed in EMs (Vincent-Viry et al. 2000). Smoking has been demonstrated to induce propranolol hydroxylation, which cosegregates with debrisoquine hydroxylation (Walle et al. 1987). Steiner et al. (1985) have reported that smoking does not modify CYP2D6 phenotype as determined by debrisoquine MR. In contrast, when EM subjects stop smoking, the mean MR is reduced after 1 to 3 months of abstinence (Llerena et al. 1996), which does not confirm the inducing ability of tobacco on CYP2D6.

5.3.2 Alcoholic Cirrhosis

Alcohol abuse leads to approximately 2.5 million deaths worldwide, with cirrhosis contributing to 16.6% of reported deaths (Ferraguti et al. 2015; Whiteford et al. 2013). According to Centers for Disease Control and Prevention, approximately 15,000 Americans die every year of alcoholic liver cirrhosis (Polednak 2012). A recent study has shown that alcoholic cirrhosis significantly increases ABCC4, 5, ABCG2, and solute carrier organic anion (SLCO) 2B1 mRNA expression and decreases SLCO1B3 mRNA expression in the liver (More et al. 2013). SLCO transporters are often described as uptake transporters since they are predominantly localized to the sinusoidal membrane and typically extract chemicals from blood into hepatocytes (Hagenbuch and Meier 2004). In humans, SLCO1B1, 1B3, 2B1, and 1A2 have relatively high expression in the liver. SLCO1B1, 1B3, and 2B1 transport a diverse range of drugs, including benzylpenicillin, statins, and estradiol glucuronide (Hagenbuch and Meier 2004). ABCC1, 3–5, and ABCG2 protein expression is also upregulated in alcoholic cirrhosis, and ABCC3–5 and ABCG2 protein expression is also upregulated in diabetic cirrhosis (More et al. 2013). The ABC transporter superfamily facilitates chemical efflux and includes multidrug resistance proteins (ABCB), multidrug resistance associated proteins (ABCC), bile salt export pump (ABCB11), and breast cancer resistance protein (ABCG2) (Choi and Yu 2014; ter Beek et al. 2014). In liver, ABCC2, ABCG2, and ABCBs are localized to the canalicular membrane and facilitate biliary excretion of chemicals. ABCC1 and 3–6 are localized sinusoidally or basolaterally and efflux chemicals from hepatocytes into blood. Cirrhosis increases NRF2 mRNA expression, whereas it decreases PXR and farnesoid X receptor (FXR) mRNA expression in comparison with normal livers (More et al. 2013). CYP3A4 and UGT1A3 mRNA expression is increased in livers with steatosis, while CYP2D6, UGT1A1, and 1A4 mRNA expression remains unchanged (More et al. 2013). In addition, tumor necrosis factor (TNF)-α and interleukin (IL)-1β mRNA levels are increased in alcoholic cirrhosis. These data demonstrate that alcoholic cirrhosis significantly alters the expression of multiple drug transporters but has a minor to moderate impact on the expression of CYPs (Zhou et al. 2003).

5.3.3 Herbal Medicines

Herbal medicines are becoming popular as alternative medicines in the Western world and an estimated one-third of adults in the developed countries use herbal medicines. Herbal products are often coadministered with therapeutic drugs, raising the potential of herb–drug interactions (Alissa 2014; Brazier and Levine 2003; Coxeter et al. 2004; de Lima Toccafondo Vieira and Huang 2012; Fugh-Berman and Ernst 2001; He et al. 2010; Hu et al. 2005; Izzo 2005; Liu et al. 2015; Meijerman et al. 2006; Meng and Liu 2014; Shord et al. 2009; Smolinske 1999; Ulbricht et al. 2008; Williamson 2003, 2005; A.K. Yang et al. 2010; X.X. Yang et al. 2006; Zhou

et al. 2007c). Potential interactions of herbal medicines with drugs are a major safety concern, especially for drugs with narrow therapeutic indices (e.g., warfarin and digoxin), and may lead to severe ADRs that are sometimes life-threatening. For many of the herb–drug interactions, induction or inhibition of hepatic and intestinal CYPs is one of the most possible reasons (Di et al. 2008; Gurley et al. 2012; Madabushi et al. 2006; Pal and Mitra 2006; Zhou and Lai 2008; Zhou et al. 2004a). Herbal components can act as the substrates, inducers, or inhibitors of human CYPs (Zhou et al. 2003).

St. John's wort administration for 2 weeks in healthy volunteers significantly induces intestinal and hepatic CYP3A4 activity but does not alter CYP2C9, 1A2, or 2D6 activity when a cocktail of probes is used (Roby et al. 2000). Oral administration of St. John's wort for 14 days in healthy volunteers also results in a 1.4-fold increase in P-gp expression (Durr et al. 2000). The probe substrates of P-gp, fexofenadine, and cyclosporine are found to have increased clearance in healthy subjects treated with St. John's wort (Dresser et al. 2003). Clinical studies have documented that St. John's wort reduces the plasma AUC of cyclosporine (Breidenbach et al. 2000; Moschella and Jaber 2001), midazolam (Wang et al. 2001), amitriptyline (Johne et al. 2002), digoxin (Johne et al. 1999), indinavir (Piscitelli et al. 2000), nevirapine (de Maat et al. 2001), oral contraceptives (Yue et al. 2000), warfarin (Yue et al. 2000), phenprocoumon (Maurer et al. 1999), docetaxel (Goey et al. 2014), zolpidem (Hojo et al. 2011), ambrisentan (Markert et al. 2015), theophylline (Nebel et al. 1999), S-ketamine (Peltoniemi et al. 2012), and simvastatin (Sugimoto et al. 2001). Notably, St. John's wort decreases the renal clearance of metformin but does not affect any other metformin pharmacokinetic parameter in healthy subjects (Stage et al. 2015). However, administration of St. John's wort at 900 mg/day for 14 days does not induce the metabolism of dextromethorphan in healthy subjects (Markowitz et al. 2003; Wang et al. 2001; Wenk et al. 2004).

In healthy subjects, treatment with Siberian ginseng (*Eleutherococcus senticosus*) at 485 mg twice daily for 14 days, valerian (*Valeriana officinalis*) at 1000 mg nightly for 14 days, or green tea (*Camellia sinensis*) four capsules per day for 14 days does not alter the pharmacokinetics of dextromethorphan (Donovan et al. 2003, 2004a,b). Treatment of healthy subjects with *Citrus aurantium*, *Echinacea purpurea*, milk thistle, and saw palmetto for 28 days does not significantly change CYP1A2, 2D6, 2E1, and 3A4 activity determined using a cocktail method (Gurley et al. 2004). In healthy subjects, ingestion of goldenseal (*Hydrastis canadensis*) for 28 days inhibits the activity of CYP2D6 and 3A4, kava inhibits CYP2E1, and black cohosh weakly inhibits CYP2D6 (Gurley et al. 2005b). Treatment of elderly subjects with St. John's wort for 28 days significantly induces CYP3A4 (~140%) and 2E1 activity (~28%); garlic oil inhibits CYP2E1 activity by ~22%; *Panax ginseng* inhibits CYP2D6 activity by 7% (Gurley et al. 2005a).

When debrisoquine is used as the probe for phenotyping CYP2D6 activity, treatment of healthy subjects with milk thistle, black cohosh (*Cimicifuga racemosa*), kava kava, St. John's wort, or Echinacea for 14 days does not alter CYP2D6 activity, but only goldenseal inhibits CYP2D6 activity by 50% (Gurley et al. 2008). Treatment of healthy male or female Caucasians with the oral lavender oil preparation silexan at 160 mg once daily for 11 days does not change the activities of CYP1A2, 2C9, 2D6, and 3A4 (Doroshyenko et al. 2013). Treatment of healthy men and women with EGb 761 at 120 or 240 mg twice daily for 8 days does not alter the activities of CYP1A2, 2C9, 2C19, 2D6, and 3A4 (Zadoyan et al. 2012). In healthy subjects, intake of milk thistle does not significantly change CYP1A2, 2C9, 2D6, and 3A4/5 activities (Kawaguchi-Suzuki et al. 2014).

These findings indicate that commonly used herbal medicines have minor to moderate effects on CYP2D6 activity. Therefore, the potential for clinical interactions with drugs that are CYP2D6 substrates is small.

5.4 HUMAN CYP2D6 IS LARGELY UNINDUCIBLE BY PROTOTYPICAL INDUCERS OF CYPs

5.4.1 IN VITRO STUDIES

By employing cultured human hepatocytes, the induction of CYP1A, 2A, 2B, 2C, 2E, and 3A subfamilies by prototypical inducers such as rifampin, phenobarbital, β-naphthoflavone, and dexamethasone has been reported (Donato et al. 1995; Gerbal-Chaloin et al. 2001; Meunier et al. 2000; Pascussi et al. 2001; Ramamoorthy et al. 2013; Rodriguez-Antona et al. 2000). In contrast to these CYP enzymes, none of the model inducers examined increases levels of CYP2D6, 2E1, and 4A11 in cultured human liver slices (Edwards et al. 2003; Martin et al. 2003). For CYP2D6, studies have shown that this CYP is refractory to induction by known inducers of other CYP subfamilies (Gerbal-Chaloin et al. 2001; Ramamoorthy et al. 2013; Rodriguez-Antona et al. 2000).

In cultured primary human hepatocytes, rifampin, rifapentine, and rifabutin induce CYP3A but not CYP2D6 (Li et al. 1997). Rifampicin and phenobarbital induce CYP1A and 3A4, but not CYP2C9, 2C19, P-gp/MDR-1 and MRP-1 in human hepatocytes (Runge et al. 2000). Phenobarbital and rifampin significantly induce CYP2B6, 2C8, 2C9, 2C19, 2E1, and 3A4 but fail to induce CYP2D6 activity in human hepatocytes (Madan et al. 2003). In human hepatocytes, clotrimazole, phenobarbital, rifampin, and ritonavir strongly induce CYP2B6 and activated PXR; dexamethasone and sulfinpyrazone induce CYP2B6 weakly and activate PXR moderately; paclitaxel strongly activates PXR but does not increase CYP2B6 expression; carbamazepine and phenytoin moderately or strongly increase CYP2B6 expression but weakly activate PXR; and dexamethasone, methotrexate, probenecid, sulfadimidine, and troleandomycin have weak or negligible effects on CYP2B6 and PXR (Faucette et al. 2004). Using human enterocytes collected from six healthy subjects before and after 10 days of 600 mg/day oral rifampicin administration, CYP2D6 is not induced while CYP2C8,

2C9, and 3A4 are induced (Glaeser et al. 2005). Rifampicin dose dependently induces CYP3A4 but inhibits SHP mRNA expression levels in primary human hepatocytes (Li and Chiang 2006). Rifampicin strongly stimulates PXR and HNF-4α interaction and CYP3A4 reporter activity, which is further enhanced by PPARγ coactivator 1α and steroid receptor coactivator-1 but inhibited by SHP (Li and Chiang 2006). These results indicate that PXR concomitantly inhibits its *SHP* gene transcription and maximizes PXR-mediated induction of the *CYP3A4* gene in human livers. Ritonavir and nelfinavir do not induce CYP2D6 in cultured human hepatocytes but significantly induce CYP1A2, 2B6, 2C9, 2C19, and 3A4 (Dixit et al. 2007). In cultured human hepatocytes, rifampin induces the expression of CYP3A4/5, 2B6, 2C8, 2C9, and 2A6, but it downregulates CYP2E1, 2J2, and 4A11 (Ramamoorthy et al. 2013). A very small but significant increase in expression is observed for CYP2D6, 2C19, and 1A2.

St. John's wort, a known potent inducer of CYP3A4 (Wang et al. 2001), and its major components hyperforin and hypericin do not induce CYP2D6 in cultured human hepatocytes (Komoroski et al. 2004). In human colon adenocarcinoma cell line LS-180, St. John's wort and its major constituents hyperforin and hypericin induce CYP3A4 and P-gp (Tian et al. 2005). Valerian and *Ginkgo biloba* appear to increase the activity of CYP2D6 in primary human hepatocytes (Hellum et al. 2007). Rifampin and rifabutin significantly induce CYP3A4 (80- and 20-fold, respectively) in primary human hepatocytes (Williamson et al. 2013).

In cultured human liver slices, treatment with 50 μM rifampicin induces CYP3A4 but has no effect on CYP1A2/1 (Lake et al. 1997). In cultured precision-cut human liver slices, treatment with β-naphthoflavone, lansoprazole, rifampicin, dexamethasone, methylclofenapate, or phenobarbital for 72 h does not induce CYP2D6, with little effect on CYP2C8, 2C9, 2E1, and 4A1 (Edwards et al. 2003). In this study, β-naphthoflavone significantly induces CYP1A2; phenobarbital significantly induces CYP2B6 and 3A4; and rifampicin significantly induces CYP2A6, 2B6, 2C19, and 3A4. Induction of CYP2B6 and 3A4 by phenobarbital and cyclophosphamide is observed in cultured human liver slices (Martin et al. 2003). Rifampin induces CYP3A4, 3A5, 2B6, and 2A6; UGT1A1 and UGT1A6; and BSEP, MRP2, and MDR1 and slightly downregulates OATP8 in precision-cut human liver slices (Olinga et al. 2008). Phenobarbital induces CYP3A4, 3A5, 2B6, and 2A6; UGT1A1; and all transporters examined.

5.4.2 IN VIVO STUDIES

Treatment of 28 healthy young male volunteers with rifampicin at 600 mg daily induces the activities of CYP1A2, 2C9, and 2D6 in a time-dependent manner when caffeine, mephenytoin, and debrisoquine are used as the corresponding probes (Branch et al. 2000). In elderly EMs of CYP2D6 (age, 70.5 ± 3.5 years), ingestion of 600 mg rifampin once daily for nine consecutive days induces the Phase I (mainly mediated by CYP2D6) and Phase II metabolism of propafenone (Dilger et al. 2000). In healthy subjects, treatment with nelfinavir, ritonavir, or rifampin for 2 weeks induces CYP1A2, 2B6, and 2C9, but not CYP2D6 (Kirby et al. 2011).

5.5 TRANSCRIPTIONAL AND POSTTRANSCRIPTIONAL REGULATION OF CYP2D6 BY HNF-4α AND FXR

SHP, a unique orphan nuclear receptor that contains a dimerization and ligand-binding domain but lacks the conserved DNA-binding domain, can repress other nuclear receptors through the formation of a nonproductive heterodimer that can directly compete with coactivators or acts via direct effects of its transcriptional repressor function (Bavner et al. 2005; Goodwin et al. 2000; Johansson et al. 2000; Lee et al. 1998; Seol et al. 1996, 1997, 1998). SHP is predominantly observed in the liver. Nuclear receptors that can be suppressed by SHP include RARα, RXRα, AhR, FXR, LXRα, HNF-4α, PPARAγ, androgen and estrogen receptors, growth hormone receptors, and thyroid hormone receptors (Brendel et al. 2002; Goodwin et al. 2000; Hoeke et al. 2014; Johansson et al. 2000; Kassam et al. 2001; Kim et al. 2012; Klinge et al. 2001; Lee and Moore 2002; Lee et al. 1998, 2000; Mamoon et al. 2014; Ning et al. 2015; Ourlin et al. 2003; Seol et al. 1996, 1997, 1998; Vaquero et al. 2013; Yang et al. 2015; Zhang et al. 2015; Zhi et al. 2014). As such, SHP can regulate the expression of CYP2D6 via inhibition of nuclear receptors.

5.5.1 TRANSCRIPTIONAL REGULATION OF CYP2D6 BY HNF-4α

HNF-4α is a member of the nuclear receptor superfamily and is mainly expressed in a restricted manner in the liver, intestine, kidney, and pancreas (Mendel and Crabtree 1991). It plays an important role in the regulation of many liver-specific genes, such as those encoding apolipoproteins, coagulation factors, and CYPs that are involved in lipid transport and glucose metabolism, coagulation, and drug metabolism (Chiang 2009; Crestani et al. 2004; Erdmann and Heim 1995; Gonzalez 2008; Gupta and Kaestner 2004; Hwang-Verslues and Sladek 2010; Jover et al. 2009; Mendel and Crabtree 1991; Rana et al. 2010; Walesky and Apte 2015; Wortham et al. 2007). HNF-4α is required for the PXR- and CAR-mediated transcriptional activation of CYP3A4 (Tirona et al. 2003). A DR element with a one-nucleotide spacer located in the proximal promoter region of the CYP2D6 gene plays an important role in modulating CYP2D6 expression, and HNF-4α interacts with this binding element (Cairns et al. 1996). Cotransfection of the minimal CYP2D6 promoter −CAT construct (−392 bp) with a mammalian HNF-4α expression vector results in a 30-fold induction of CAT activity in COS-7 cells. Although HNF-4α is originally identified as an orphan receptor, fatty acyl-CoA thioesters are identified to be endogenous ligands for HNF-4α (Hertz et al. 1998; Petrescu et al. 2002). The

binding of ligand may shift the oligomeric–dimeric equilibrium of HNF-4α or may modulate the affinity of HNF-4α for its cognate promoter element, resulting in either activation or inhibition of HNF-4α transcriptional activity as a function of chain length and the degree of saturation of the fatty acyl-CoA ligands (Petrescu et al. 2002). The HNF-4α binding element is conserved in the proximal promoter regions of more than 20 CYP2 genes (Chen et al. 1994; Ibeanu and Goldstein 1995). Jover et al. (2001) have demonstrated that HNF-4α plays a general role in the regulation of major P450 genes, including CYP3A4, 3A5, 2A6, 2B6, 2C9, and 2D6, in human hepatocytes using antisense technique. Using the small interfering RNA technique, Kamiyama et al. (2007) have found that suppression of HNF-4α causes a decrease in the mRNA levels of CYP2A6, 2B6, 2C8, 2C9, 2C19, 2D6, and 3A4; UGT1A1 and 1A9; SULT2A1; ABCB1; ABCB11; ABCC2; OATP1B1; and OCT1; as well as those of PXR and CAR. In addition, deletion of HNF-4α decreases debrisoquine 4-hydroxylase activity in CYP2D6-humanized mice more than 50% (Corchero et al. 2001). These findings indicate that HNF-4α may act as a common regulator of the liver-specific transcription of many P450 genes.

Similarly, the CYP2D6 gene is downregulated at the transcriptional level by nitric oxide (NO) in HepG2 cells and the NO donor (±)-N-[(E)-4-ethyl-2-[(Z)-hydroxyimino]-5-nitro-3-hexene-1-yl]-3-pyridine carboxamide decreases the expression of CYP2D6 mRNA in a concentration-dependent manner (Hara and Adachi 2002). Further mechanistic studies indicate that the DNA-binding activity of HNF-4α was directly inhibited by NO donors, S-nitrosoglutathione, and S-nitroso-N-acetyl-penicillamine in a concentration-dependent manner (Hara and Adachi 2002); mutation of the HNF-4α binding site in the CYP2D6 promoter partially restores the suppression of the promoter activity by NO donors. A guanylate cyclase inhibitor fails to prevent suppression of CYP2D6 promoter activity by S-nitrosoglutathione, indicating that the activity of the CYP2D6 promoter is suppressed via an NO-guanylate cyclase–independent pathway (Hara and Adachi 2002).

To date, a number of SNPs of the human HNF-4α gene have been described at the NCBI (National Center for Biotechnology Information) dbSNP site (http://www.ncbi .nlm.nih.gov/). Among them, four nonsynonymous SNPs have been identified: 416C>T (Thr139Ile), 449C>T (Leu289Phe), 1289C>T (Pro430Leu), and 1333C>T (Prp445Ser). Lee et al. (2008) have identified a novel mutation, Gly60Asp, in the HNF-4α gene in a Korean population. This Gly60Asp variant fails to bind to the recognition site in the CYP2D6 promoter and therefore lacks the regulatory function for this gene. Human liver specimens with the heterozygous HNF-4α Gly60Asp genotype tend to have lower levels of CYP2D6 activity than the wild-type genotype and human subjects with the Gly60Asp genotype tend to have lower CYP2D6 activity than those with the wild-type HNF-4α. The Gly60Asp variant is detected at a low frequency in Asian populations, including Koreans, Chinese, and Vietnamese, and is not detected in Africans or Caucasians (Lee et al. 2008).

5.5.2 TRANSCRIPTIONAL REGULATION OF CYP2D6 BY FXR

FXRs are nuclear hormone receptors expressed in high amounts in body tissues that participate in bilirubin metabolism including the liver, intestine, and kidney (Fiorucci et al. 2012; Gadaleta et al. 2015; Kalaany and Mangelsdorf 2006; Kemper 2011; Lee et al. 2006; Matsubara et al. 2013; Modica et al. 2010; Rader 2007; Rizzo et al. 2005; Teodoro et al. 2011; Wang et al. 2008b). Chenodeoxycholic acid and other bile acids are the natural ligands of the FXRs. FXRs downregulate the expression of the gene encoding for CYP7A1 (i.e., cholesterol 7α-hydroxylase), which is the rate-limiting enzyme in bile acid biosynthesis (Kalaany and Mangelsdorf 2006; Lee et al. 2006; Rizzo et al. 2005). FXRs induce expression of SHP, which then functions to inhibit transcription of the *CYP7A1* gene. Similar to other nuclear receptors, FXR translocates to the cell nucleus upon activation by ligand binding and forms a dimer (a heterodimer with RXR) and binds to hormone response elements on DNA, which up- or downregulates the expression of its target genes (Ding et al. 2014; Fiorucci et al. 2012; Lee et al. 2006; Matsubara et al. 2013; Rizzo et al. 2005). In addition, FXRs play a critical role in carbohydrate and lipid metabolism and regulation of insulin sensitivity (Fiorucci et al. 2007). FXRs also modulate live growth and regeneration during liver injury (Chen et al. 2011).

Recently, Koh et al. (2014a) have revealed that SHP represses HNF-4α–mediated transactivation of the *CYP2D6* promoter and thus represses hepatic CYP2D6 expression. Knockdown of SHP expression using small interfering RNA in CYP2D6-humanized transgenic mice results in enhanced hepatic CYP2D6 expression. SHP is a typical target gene of FXRs, which serve as a bile acid sensor in the liver (Goodwin et al. 2000; Hoeke et al. 2014; Li and Chiang 2014; Lu et al. 2000; Rizzo et al. 2005; Vaquero et al. 2013; Zhang et al. 2015). When hepatic bile acid levels are high under some pathological conditions such as cholestasis and cirrhosis, ligand-activated FXRs transactivate the *SHP* promoter (Goodwin et al. 2000). SHP in turn downregulates the expression of its target genes involved in bile acid homeostasis in the liver and thus protects the liver from the toxicity of excess bile acids (Li and Chiang 2014). The role of SHP and FXRs in bile acid homeostasis has been extensively characterized using selective agonists of FXRs, such as GW4064 (3-(2,6-dichlorophenyl)-4-(3′-carboxy-2-chlorostilben-4-yl) oxymethyl-5-isopropylisoxazole) (Cui et al. 2003; Hoeke et al. 2014; Kerr et al. 2002; Lee et al. 2010; Rosales et al. 2013; Vaquero et al. 2013; Wang et al. 2002; Zhang et al. 2015) and 6α-ethyl-chenodeoxycholic acid (6-ECDCA; obeticholic acid or INT-747) (Costantino et al. 2003; Li et al. 2007; Miyazaki-Anzai et al. 2010; Mudaliar et al. 2013; Pellicciari et al. 2002).

Pan et al. (2015) have further revealed that GW4064 decreases hepatic CYP2D6 expression and activity twofold while increasing SHP expression twofold and SHP recruitment to the *CYP2D6* promoter in CYP2D6-humanized transgenic mice. CYP2D6 repression by GW4064 is abrogated in Shp$^{-/-}$ CYP2D6 transgenic mice, indicating a critical role of SHP in CYP2D6 regulation by GW4064. GW4064 also decreases CYP2D6 expression and activity twofold in primary human hepatocytes using debrisoquine as a probe drug for CYP2D6 (Pan et al. 2015). These findings suggest potential drug–drug interactions between CYP2D6 substrates and FXR agonists.

In CYP2D6-humanized transgenic mice, cholestasis triggered by administration of ethinylestradiol (EE$_2$) at a high dose leads to two - to threefold decreases in CYP2D6 expression (Pan and Jeong 2015). This is accompanied by increased hepatic SHP expression and subsequent decreases in the recruitment of HNF-4α to the *CYP2D6* promoter. Estrogen-induced cholestasis also results in increased recruitment of estrogen receptor-α, but not that of FXR, to the *Shp* promoter, suggesting a predominant role of ERα in transcriptional regulation of SHP in estrogen-induced cholestasis. EE$_2$ at a low dose that does not cause cholestasis also increases SHP by ~50% and decreases CYP2D6 expression 1.5-fold in CYP2D6-humanized transgenic mice, but the magnitude of differences is much smaller than that in EE$_2$-induced cholestasis (Pan and Jeong 2015).

Presently, many FXR agonists are under development, indicated for different hepatic or metabolic diseases, including primary biliary cirrhosis, nonalcoholic steatohepatitis, and type II diabetes (Adorini et al. 2012; Ali et al. 2015; Crawley 2010; Fiorucci et al. 2007, 2010, 2012, 2014; Gioiello et al. 2014; Kemper 2011; Lindor 2011; Modica and Moschetta 2006; Mudaliar et al. 2013; Rader 2007; Silveira and Lindor 2014; Wang et al. 2008a). For example, obeticholic acid (i.e., INT-747), a potent selective FXR agonist, has completed a Phase III trial for the treatment of noncirrhotic, nonalcoholic steatohepatitis (Neuschwander-Tetri et al. 2015). The primary outcome measure is improvement in centrally scored liver histology defined as a decrease in nonalcoholic fatty liver disease activity score by at least two points without worsening of fibrosis from baseline to the end of treatment. A planned interim analysis of change in alanine aminotransferase at 24 weeks undertaken before end-of-treatment (72 weeks) biopsies supports the decision to continue the trial and shows improved efficacy of obeticholic acid. The 141 patients are randomly assigned to receive obeticholic acid and 142 to placebo. Fifty (45%) of 110 patients in the obeticholic acid group show improved liver histology compared with 23 (21%) of 109 such patients in the placebo group (relative risk, 1.9; 95% CI, 1.3 to 2.8; $P = 0.0002$). Thirty-three (23%) of 141 patients in the obeticholic acid group develop pruritus compared with 9 (6%) of 142 in the placebo group (Neuschwander-Tetri et al. 2015). Obeticholic acid is a 6α-ethyl derivative of the natural human bile acid chenodeoxycholic acid agonist that is ~100-fold more potent than chenodeoxycholic acid. In a Phase II clinical trial, administration of obeticholic acid is well tolerated, has increased insulin sensitivity, and has reduced markers of liver inflammation and fibrosis in patients with type II diabetes mellitus and nonalcoholic fatty liver disease (Mudaliar et al. 2013). In two clinical trials of obeticholic acid in patients with primary biliary cirrhosis, a progressive cholestatic liver disease, obeticholic acid significantly reduces serum alkaline phosphatase levels, an important disease marker that correlates well with clinical outcomes of patients with primary biliary cirrhosis (Hirschfield et al. 2015). FXR may represent a useful therapeutic target for the treatment of nonalcoholic fatty and cholestatic liver diseases, but more clinical studies are warranted to establish its safety and efficacy profiles.

5.6 POSTTRANSLATIONAL REGULATION OF CYP2D6

Protein posttranslational modification increases the functional diversity of the proteome by the covalent addition of functional groups or proteins, proteolytic cleavage of regulatory subunits, or degradation of entire proteins (Beltrao et al. 2012; Jensen 2004; Zhao and Jensen 2009). These modifications include phosphorylation, glycosylation, ubiquitination, nitrosylation, methylation, acetylation, lipidation, and proteolysis. Indeed, it is estimated that 5% of the proteome comprises enzymes that perform more than 200 types of posttranslational modifications (Beltrao et al. 2012). These enzymes include kinases, phosphatases, transferases, and ligases, which add or remove functional groups, proteins, lipids, or sugars to or from amino acid side chains and proteases.

Reversible protein phosphorylation, principally on serine, threonine, or tyrosine residues, is one of the most important and well-studied posttranslational modifications (Beltrao et al. 2012). Phosphorylation plays critical roles in the regulation of many cellular processes including cell cycle, growth, apoptosis, and differentiation. There are limited data on the regulation of CYP2D6 by posttranslational modifications. Using titanium dioxide resin combined with tandem mass spectrometry for phosphopeptide enrichment and sequencing, eight human CYP phosphorylation sites are identified. The data from surgical human liver samples establish that CYP1A2, 2A6, 2B6, 2E1, 2C8, 2D6, 3A4, 3A7, and 8B1 are phosphorylated in vivo (Redlich et al. 2008).

Under normal physiological conditions, the rate of de novo synthesis of CYPs should equal the rate of degradation, but such parameter values in humans in vivo are usually unavailable. The turnover half-lives of CYPs are usually derived from in vivo studies of enzyme induction or mechanism-based inhibition (Yang et al. 2008; Zhou and Zhou 2009). The estimated half-lives of CYP2D6 in humans are 46.6–51 h (Venkatakrishnan and Obach 2005; Zhou and Zhou 2009). For CYP3A4, the estimated turnover half-lives are 26–106 h (Zhou and Zhou 2009). In precision-cut human liver slices cultured for up to 72 h, the measured half-lives for CYP2C9, 2D6, 3A4, and 4A11 are 70–104 h, and those values for CYP1A2, 2A6, 2B6, 2C8, 2C19, 2E1, and 3A5 are 23–36 h (Renwick et al. 2000). Any factors that alter the stability and degradation of human CYPs will change their half-life and thus change drug metabolism. For example, paroxetine is a mechanism-based inhibitor of CYP2D6 (Bertelsen et al. 2003) and there is a delayed recovery of CYP2D6 in vivo since new CYP2D6 must be produced via de novo synthesis. Paroxetine has been found to reduce the clearance of desipramine, risperidone, perphenazine, atomoxetine, nebivolol, tramadol, methadone, carvedilol, and metoprolol three- to eightfold in humans (Begre et al. 2002; Belle et al. 2002; Briciu et al. 2014; Brosen et al. 1993; Goryachkina et al. 2008; Laugesen et al. 2005; Saito et al. 2005; Stout et al. 2010; Tang and Helmeste 2008).

5.7 GENOME-WIDE ASSOCIATION STUDIES ON THE REGULATION OF CYP2D6

A genome-wide association study (GWAS) has been used to investigate the regulating mechanisms of CYPs including CYP2D6. A recent GWAS study in 435 patients has observed genome-wide significant associations for citalopram and escitalopram concentrations with SNPs in or near the *CYP2C19* gene on chromosome 10 (rs1074145) and with S-didesmethylcitalopram concentration for SNPs near the *CYP2D6* locus on chromosome 22 (rs1065852, i.e., Pro34Ser) (Ji et al. 2014). The GWAS analyses are performed based on a total of 545,115 SNPs that are genotyped in DNA samples from each of our 435 patients, followed by imputation using "1000 Genomes Project" data as a reference, resulting in a total of 7,537,437 SNPs that are used in the GWAS analyses after imputation. Both citalopram and escitalopram are known substrates of CYP2C19, 2D6, and 3A4 (Kobayashi et al. 1997; Olesen and Linnet 1999; Rochat et al. 1997; von Moltke et al. 2001). After adjustment for the effect of *CYP2C19* functional alleles, the analyses also identified novel loci such as *CBX4* on chromosome 17 and in the *PDZD2* gene on chromosome 5 that will need further functional validation. The *CYP2C19* rs4244285 SNP also reaches genome-wide significance, with a P value of 1.6×10^{-8}. The rs1065852 SNP occurs in *CYP2D6*4*, **10*, and **14*, all of which are associated with reduced CYP2D6 activity and PM phenotypes. The rs1058172 SNP Arg345His is also associated with S-didesmethylcitalopram concentration, but it has not been associated with any *CYP2D6* haplotype.

Another study has applied a GWAS approach to explore the relationships between genetic polymorphisms and CYP expression or enzyme activities with 466 human liver samples (X. Yang et al. 2010). This interesting study has revealed sets of SNPs associated with CYP traits and suggested the existence of both *cis*-regulation of CYP expression (especially for CYP2D6) and more complex *trans*-regulation of CYP activity. CYP3A5, 2D6, 4F12, and 2E1 each have more than 30 *cis*-SNPs. *CYP2D6* is regulated by 1665 genes and *CYP2C19* is related to 3360 genes. In addition to the SNPs rs8138080, rs5751247, and rs17478227, several novel SNPs associated with CYP2D6 expression and enzyme activity are identified and validated in an independent human cohort. Of the 54 SNPs associated with CYP enzyme activities, 30 are associated with CYP2D6 activity, and they are all located within 200-kb distance of the physical location of the *CYP2D6* gene, demonstrating mainly *cis* genetic regulation of CYP2D6 enzyme activity. By constructing a weighted coexpression network and a Bayesian regulatory network, the authors have defined the human liver transcriptional network structure, uncovered subnetworks representative of the CYP regulatory system, and identified novel candidate regulatory genes, namely, *EHHADH*, *SLC10A1*, and *AKR1D1* (X. Yang et al. 2010). *EHHADH* and *ACSM3* are upstream of the pathways and their regulation is mediated through *CYP2C19*, while the downstream *CYP* targets include *CYP2C9*, *3A7*, *3A4*, and *3A43*. Another two putative *CYP* regulators, *SLC10A1* and *AKR1D1*, generate a separate branch and are upstream of eight *CYP* genes and seven known *CYP* regulators: *PGRMC1*, *CEBPD*, *FOXA2*, *NR1I3/CAR*, *NR1I2/PXR*, *PPARG*, and *HNF4A*. The authors have identified *SLC10A1*, *AKR1D1*, *GLYAT*, *ETNK2*, *LIME1*, *ZGPAT*, *BTD*, *ETNK2*, *KLKB1*, *FMO3*, *LIMK2*, *EHHADH*, and *ACSM3* to be the top global regulators, each with more than 40 downstream nodes (X. Yang et al. 2010). Therefore, *SLC10A1*, *AKR1D1*, *EHHADH*, and *ACSM3* are identified as both global regulators and P450 regulators. The CYP subnetworks are then validated using gene signatures responsive to ligands of known CYP regulators in mouse and rat. These data demonstrate networked regulating mechanisms of human CYPs in a coordinated manner in the liver. Further functional and validation studies are needed to verify these findings.

A recent genome-wide SNP study has investigated the genotypes of candidate genes associated with aspirin-induced small bowel bleeding (Shiotani et al. 2013). In this validation study involving 37 patients with small bowel bleeding and 400 controls, 4 of 27 identified SNPs, *CYP4F11* (rs1060463, 1336G>A) GG, *CYP2D6* (rs28360521, 2908G>A) GG, *CYP24A1* (rs4809957, 140T>C) T allele, and *GSTP1* (rs1695, 313A>G → Ile105Val, present in *GSTP1*2* and *GSTP1*B*) G allele, are significantly more frequent in the small bowel bleeding group compared to the controls (Shiotani et al. 2013). After adjustment for significant factors, *CYP2D6* (rs28360521) GG (odds ratio, 4.11) is associated with small bowel bleeding. *CYP4F11* and *CYP2D6* SNPs may identify patients at increased risk for aspirin-induced small bowel bleeding. It is unknown whether aspirin is metabolized by these two enzymes.

Jiang et al. (2013) have developed a novel mediation analysis approach to identify new expression quantitative trait loci (eQTLs) driving CYP2D6 activity by combining genotype, gene expression, and enzyme activity data. The authors have found 389,573 and 1,214,416 SNP–transcript–CYP2D6 activity trios that are strongly associated for two different genotype platforms, namely, Affymetrix and Illumina, respectively. In the Affymetrix data set, 295 SNPs correlate with at least 20 genes, which are used to check for overlapping with the results of mediation analysis. A total of 289 eQTL hotspots are found to correlate with 1542 gene expression profiles. The Illumina data set has found that 724 SNPs correlate with at least 20 genes, and 719 of the hotspots are significantly correlated with 2444 genes in mediation analysis. Nine hundred thirty-nine and 1420 genes are successfully mapped in the Ingenuity database for two platforms. The majority of eQTLs are *trans*-SNPs. Five (CCL16, CCL20, CMTM5, IL-6, and SPP1) and 7 (CCL16, CCL20, CKLF, CKLFSF5, EPO, FAM3C, and SPP1) cytokines, 5 (AR, NR1I2/PXR, NR1I3/CAR, NR2F6, and PPARα) and 7 (AR, ESR1, NR1I2/PXR, NR1I3/CAR, PPARα, RORα/NR1F1, and RORγ) nuclear receptors, and

80 and 113 transcription regulators are found to mediate the relationship between genetic variant and CYP2D6 activity for Affymetrix and Illumina data sets. Overlapped eQTL hotspots with the mediators lead to the identification of 64 transcription factors that can regulate CYP2D6 (Jiang et al. 2013). These transcription factors include AATF, ALYREF, ARHGAP35, ASB8, ATF4, CBX4, CEBPG, CSDA, DDIT3, E2F5, ETV7, FOXN3, FOXN3, FUBP1, GPS2, HDAC10, HMGN1, ID1, INVS, IRF9, KANK1, KAT2B, KHDRBS1, KLF12, MAF, MAML2, MEIS2, MLXIPL, MXD4, MYBBP1A, MYCL1, NCOA7, NCOR1, NFIA, NFKB2, NFYA, NOLC1, NPM1, PEX14, PYCARD, SAP18, SATB1, SIM2, SLC2A4RG, SMARCC1, SNAI3, SNW1, SOX5, TCERG1, TCF7L2, TEAD3, TEAD4, TFDP2, TFEB, TOB1, p53, YWHAB, YY1, ZGPAT, ZHX3, ZKSCAN1, ZNF132, ZNF256, and ZNF263 (Jiang et al. 2013). Among them, YY1 has been reported to putatively bind to human *CYP2D6* or rat *Cyp2d4* promoter and regulate the expression of CYP2D6 (Gong et al. 2013) and Cyp2d4 (Mizuno et al. 2003). This study has provided new insights into the complex regulatory network for hepatic CYP2D6. Addition of the p53 inhibitor cyclic PFT-α in HepG2 cells dose-dependently enhances CYP2D6 and 3A4 activity, whereas addition of the p53 activator NSC 66811 dose-dependently inhibits CYP2D6 and 3A4 activity (Xiao et al. 2015). Further functional and validation studies are certainly needed to verify the regulation of CYP2D6 by these genes.

Association between a genetic marker and a phenotype may exist as a result of linkage disequilibrium between the disease-causing allele and the marker allele. SNPs mapping within an 880-kb region flanking *CYP2D6* are identified to evaluate potential association between genetic variation and the CYP2D6 PM phenotype (Hosking et al. 2002). SNP 1 maps to base 230 and SNP 27 maps to base 879049 in the contig shown. None of the 27 SNPs maps to *CYP2D6* itself. Associations are observed across a 390-kb region between 14 SNPs and the PM phenotype. Ten of the 14 SNPs associated with the PM phenotype have *P* values of $<10^{-7}$, demonstrating a level of significance sufficient to detect a phenotype and genotype relationship within an entire genome-wide scan. Haplotype analysis reveals more significant levels of association (Hosking et al. 2002). Strong linkage disequilibrium between SNPs is observed across the same 390-kb region associated with the CYP2D6 phenotype. The observed phenotype and genotype association has reached genome-wide levels of significance and supports the strategy for potential application of linkage disequilibrium mapping and whole genome association scans to pharmacogenetic studies.

5.8 EFFECTS OF DISEASES ON CYP2D6 EXPRESSION AND ACTIVITY

5.8.1 Liver Diseases

The liver plays a significant role in the metabolism of drugs, chemicals, and endogenous substrates; thus, liver disease may have a detrimental effect on the activity of CYPs (Corsini and

Bortolini 2013; Villeneuve and Pichette 2004). CYP-mediated activation of drugs such as acetaminophen, carbamazepine, clozapine, and halothane to toxic metabolites induces hepatotoxicity (Boelsterli et al. 2006; Zhou et al. 2004b, 2005, 2007a,b). CYP2E1 plays a critical role in ROS generation, which is a key component in the pathogenesis of alcoholic and nonalcoholic fatty liver diseases (Leung and Nieto 2013). In some instances, covalent binding of the toxic metabolite to CYP leads to the formation of anti-CYP antibodies and immune-mediated hepatotoxicity (e.g., hydralazine and tienilic acid) (Corsini and Bortolini 2013; Zhou et al. 2005). Anti-CYP2D6 antibodies are present in the serum of patients with type II autoimmune hepatitis (Bogdanos and Dalekos 2008; Christen et al. 2010; Czaja and Manns 2010; Komurasaki et al. 2010; Mizutani et al. 2005), but the mechanisms for their formation and pathogenic significance remain unclear. The level and activity of CYP1A2, 2C19, and 3A4/5 appear to be particularly vulnerable to the effect of liver diseases while CYP2D6, 2C9, and 2E1 are less affected (Villeneuve and Pichette 2004). The pattern of alterations in CYP expression and activity also differs according to the etiology of liver disease. There is a marked relationship between the activity of CYPs and the severity of hepatic cirrhosis (Villeneuve and Pichette 2004).

In a study with 20 patients with different etiologies and severity of liver disease and 20 age-, sex-, and weight-matched healthy volunteers using the cocktail approach, the debrisoquine recovery ratio is 71% lower in patients than in healthy subjects (Frye et al. 2006). In the meantime, mephenytoin metabolism is significantly decreased in both patients with mild liver disease (Child-Pugh score of 5/6) (by 63%) and patients with moderate to severe liver disease (Child-Pugh score >6) (by 80%). In comparison with control subjects, the caffeine MR is 69% lower and the chlorzoxazone MR is 60% lower in patients with moderate to severe liver disease (Frye et al. 2006).

Hepatitis C virus infects at least 170 million people worldwide, and its infection is a leading cause of chronic hepatitis, liver cirrhosis, and hepatocellular carcinoma (Jeong et al. 2012; Negro 2014). Currently, the combination therapy of pegylated IFN and ribavirin is the only approved treatment for hepatitis C viral infection (Cortez and Kottilil 2015; Imran et al. 2014; Martel-Laferriere and Dieterich 2014). However, this treatment regimen is only effective in approximately 50% of all patients infected with hepatitis C. Immune-mediated liver injury, viral product-mediated cytotoxicity, and oxidative stress have been documented to play a role in the pathogenicity of hepatitis C viral infection. Liver kidney microsomal type 1 antibodies (LKM-1) targeting CYP2D6 in hepatocytes are detected in the plasma of patients with chronic hepatitis C (Bogdanos and Dalekos 2008; Bortolotti et al. 2003; Choudhuri et al. 1998; Christen et al. 2010; Dalekos et al. 2002; Hintermann et al. 2011; Kerkar et al. 2003; Komurasaki et al. 2010; Longhi et al. 2007; Ma et al. 1994, 2006; Miyakawa et al. 1999, 2000; Mizutani et al. 2005; Muratori et al. 1995, 1998; Nishioka et al. 1997; Pageaux et al. 1998; Sugimura et al. 2002; Yamamoto et al. 1993; Zachou et al. 2004; Zanger

et al. 1988). Autoimmune hepatitis type 2 is characterized to be mainly related with drug-metabolizing enzymes as autoantigens, such as anti-LKM-1 against CYP2D6, anti-LKM-2 against CYP2C9-tienilic acid, anti-LKM-3 against UGT1A, and anti-LC1 (liver cytosol antigen)-1 and anti-APS (autoimmune polyglandular syndrome type-1) against CYP1A2, CYP2A6, and others (Mizutani et al. 2005). Main antigenic epitopes on CYP2D6 are residues 193–212, 257–269, and 321–351, and Asp263 is essential. The third epitope is located on the surface of CYP2D6 and displays a hydrophobic patch that is situated between an aromatic residue Trp316 and His326 (Christen et al. 2010; Hintermann et al. 2011; Kerkar et al. 2003). Some drugs such as anticonvulsants (phenobarbital, phenytoin, and carbamazepine) and halothane can induce autoimmune hepatitis with anti-CYP3A and anti-CYP2E1, respectively (Mizutani et al. 2005). Autoantibodies against CYP11A1, CYP17, and CYP21 involved in the synthesis of steroid hormones are also detected in patients with adrenal failure, gonadal failure, or Addison disease (Mizutani et al. 2005).

Debrisoquine MR is increased in CYP2D6 EMs with acute viral hepatitis as compared with healthy EMs (median MR, 1.20 vs. 0.84; $P < 0.05$) (Joanne et al. 1994). However, there is no difference in phenotype prevalence between patients with acute viral hepatitis and healthy controls. A proportion of patients with chronic hepatitis C ranging from 1% up to 79% produce LKM-1 antibodies. Sera from LKM-1–positive patients with autoimmune hepatitis type 2 can precipitate CYP2D6 in vitro, leading to a decreased activity for debrisoquine hydroxylation (Zanger et al. 1988). Patients with chronic hepatitis C show significantly reduced CYP3A4 and 2D6 activity in comparison with healthy volunteers. The median dextromethorphan MR is sixfold higher in LKM-1–positive patients with chronic hepatitis C than in LKM-1–negative patients (0.096 vs. 0.016, $P = 0.004$), indicating that CYP2D6 activity is significantly decreased in the presence of LKM-1 antibodies (Girardin et al. 2012). In chronic hepatitis C patients with LKM-1 antibodies, the CYP2D6 metabolic activity is on average reduced by 80%. Dextromethorphan MRs do not significantly differ according to gender and display no significant association with age, body mass index, and biochemical parameters. In the LKM-1–positive group with 10 patients, only 3 patients have an EM phenotype concordant with their genotype, 1 patient with a *2/*4 genotype is phenotyped as PM, and 6 patients are phenotyped as IMs. It is unknown how LKM-1 antibodies inhibit CYP2D6 in vivo. A plausible hypothesis is that the antibodies might bring about a targeted functional inhibition via ligand-like interaction at the level of the CYP2D6 cavity, corresponding to the immunedominant epitope. Another study has reported that hepatitis C–positive subjects express approximately 2.6-fold lower CYP2D6 activity relative to hepatitis C–negative patients (Ho et al. 2011). The 2060A/A and −2053G/G variation in the CYP2D6 promoter appear to be associated with significantly lower levels of liver CYP2D6 mRNA. Two novel genetic variants, −1822A>G and −1740C>T, are also detected only in two patients with hepatitis C. In addition, the mRNA levels of CYP1A2, 2E1, and 3A4, and drug transporters including Na$^+$-taurocholate cotransporting polypeptide, organic anion transporting peptide C, and organic cation transporter 1 are decreased in chronic hepatitis C patients with progression of liver fibrosis (Nakai et al. 2008).

Several GWASs of liver biopsy specimens from patients with chronic hepatitis C have been reported (Aghemo and Colombo 2013; Asselah et al. 2005; Duggal et al. 2013; Patin et al. 2012; Tanaka et al. 2009; Urabe et al. 2013). Upregulation of IL-6 and TNF-α and downregulation of CYP2E1 are observed as compared with normal liver patients (Asselah et al. 2005; Bieche et al. 2005). It is also shown that IL-6 and TNF-α levels in patients with liver fibrosis stages 2 to 4 are significantly higher than those in patients with stage 1 fibrosis only (Asselah et al. 2005). Two SNPs near the gene *IL28B* on chromosome 19 are strongly associated with null virological response to antiviral therapy in Japanese patients with hepatitis C (Tanaka et al. 2009). One recent GWAS has been performed in a large cohort of Japanese patients with hepatitis C viral infection using 780,650 SNPs (Patin et al. 2012). In the combined cohort of patients with hepatitis C, the SNPs rs16851720 on chromosome 3 and rs4374383 on chromosome 2 in patients are associated with fibrosis progression. The SNP rs16851720 is located within ring finger protein 7 (*RNF7*), which encodes an antioxidant that protects against apoptosis. RNF7 is a probable component of the SCF (SKP1-CUL1-F-box protein) E3 ubiquitin ligase complex, which mediates the ubiquitination and subsequent proteasomal degradation of target proteins involved in cell cycle progression, signal transduction, and transcription (Mahrour et al. 2008; Sun and Li 2013; Tan et al. 2008). The SNP rs4374383, together with another replicated SNP rs9380516 on chromosome 6, is linked to the functionally related genes C-Mer proto-oncogene tyrosine kinase 1 (*MERTK*) and tubby-like protein 1 (*TULP1*), which encode factors involved in phagocytosis of apoptotic cells by macrophages (Patin et al. 2012). Mutations in *MERTK* have been associated with disruption of the retinal pigment epithelium phagocytosis pathway and onset of autosomal recessive retinitis pigmentosa (Cummings et al. 2013; Graham et al. 2014; Lemke 2013; Lemke and Rothlin 2008; van der Meer et al. 2014), where mutations in *TULP1* may be associated with juvenile retinitis pigmentosa and Leber congenital amaurosis-15 (Carroll et al. 2004; Ikeda et al. 2002; Mukhopadhyay and Jackson 2011; Weleber et al. 1993). This study has identified several susceptibility loci for hepatitis C virus–induced liver fibrosis, which are linked to genes that regulate apoptotic pathway and redox homeostasis involved in liver fibrosis.

5.8.2 Chronic Renal Failure

Chronic kidney disease is a public health problem that affects >20 million people in the United States (Coresh et al. 2007). Currently, almost 500,000 patients require chronic hemodialysis. Pharmacokinetic studies conducted in patients with chronic renal failure (CRF) demonstrate that the nonrenal clearance of many drugs that are substrates of various CYPs is decreased

(Dreisbach 2009; Dreisbach and Lertora 2003, 2008; Pichette and Leblond 2003; Sun et al. 2006; Touchette and Slaughter 1991; Yeung et al. 2014; Yuan and Venitz 2000). More than 75 commonly used drugs have been reported to exhibit altered nonrenal clearance in patients with CRF, and most of these drugs are eliminated by CYP-mediated metabolism. For example, there is a sevenfold increase in the AUC of nimodipine, which is a good substrate for CYP3A4/5. Increased oral bioavailability and decreased systemic clearance have been reported for other CYP3A4/5 substrates, such as reboxetine, nicardipine, and nitrendipine (Dreisbach 2009). Although the mechanisms for this are unclear, several studies have shown that CRF affects the metabolism of drugs by inhibiting key enzymatic systems in the liver, intestine, and kidney. Uremic toxins interfere with transcriptional activation, causing downregulation of gene expression mediated by proinflammatory cytokines, and directly inhibit the activity of CYPs and drug transporters such as P-gp, organic anion transporters, and organic cation transporters (Dreisbach 2009; Yeung et al. 2014). This is associated with major reductions in metabolism of drugs mediated by CYPs. The systemic clearance of orally administered propranolol and bufuralol is reduced in CRF patients (Tsujimoto et al. 2014), resulting in three- to fivefold increases in plasma AUC, but no such changes are observed with metoprolol and propafenone (Dreisbach 2009). A low-molecular-weight ultrafiltrate fraction (<10 kDa) of uremic plasma obtained from CRF patients inhibits the oxidative metabolism of *S*-propranolol in human liver microsomes mediated by CYP2D6 and 1A2. Liver Phase II enzymes and many drug transporters are also downregulated in CRF, and increased bioavailability of several drugs has been observed in CRF, reflecting decrease in either intestinal first-pass metabolism or extrusion of drugs mediated by P-gp and other transporters (Dreisbach and Lertora 2008; Pichette and Leblond 2003).

CRF patients have decreased sparteine partial metabolic clearance to dehydrosparteine (median of 322 ml/min and range of 62 to 670 ml/min in CRF patients vs. median of 635 ml/min and range of 77 to 1276 ml/min in normal subjects; $P < 0.02$) (Kevorkian et al. 1996). Sparteine apparent oral clearance ($P < 0.03$) and renal clearance ($P < 0.001$) are decreased in CRF patients. However, sparteine MR is not significantly altered in CRF patients and shows that all CRF patients are sparteine EMs. Although fractional urinary excretion of dextrorphan is decreased in CRF patients (median, 24.4%; range, 9.7% to 55.9%) compared with control (median, 47.5%; range, 24.1% to 72.1%) ($P = 0.02$), it also shows that all CRF subjects are EMs of dextromethorphan (Kevorkian et al. 1996). The amount of dextromethorphan excreted in urine correlates with creatinine clearance independently from CYP2D6 activity measured as sparteine partial metabolic clearance. However, it does not correlate with sparteine MR or with fractional urinary excretion of dehydrosparteine (Kevorkian et al. 1996).

5.8.3 Diabetes

Diabetes mellitus can also alter the disposition and metabolism of clinically used drugs depending on the type and time of diagnosis of the disease, as well as the substrate investigated (Dostalek et al. 2012). Diabetes mellitus can alter the pharmacokinetics by various mechanisms, including a change in the intestinal absorption, distribution, and elimination of clinically used drugs (Dostalek et al. 2012). Type 2 diabetes and gestational diabetes are found to inhibit CYP3A4/5 and 1A2, with the occurrence of a probable induction of UGT1A and 2B7 in parturients with gestational diabetes (Carvalho et al. 2011; Marques et al. 2002).

In a study with 15 patients with type 1 and 16 with type 2 diabetes and 16 healthy controls, the CYP2D6-mediated dextromethorphan metabolism is not altered (Matzke et al. 2000). Type 1 diabetes has marked effects on antipyrine metabolism whereas type 2 diabetes does not alter the metabolism of any of the probe drugs. The apparent oral clearance of antipyrine is increased 72% in patients with type 1 diabetes compared with controls; formation clearances of 4-hydroxyantipyrine and 3-hydroxymethylantipyrine are increased by 74% and 137% in those patients relative to controls. The caffeine metabolic index (CYP1A2 activity) is increased by 34% (Matzke et al. 2000). Diabetes has no marked effects on CYP2D6 activity. More children with type 1 diabetes are PMs than normal controls (Madacsy et al. 2004). In patients with gestational diabetes, the CYP2D6- and 3A4-mediated metabolism of metoprolol and transplacental distribution of metoprolol and its metabolite are not altered (Antunes Nde et al. 2015).

5.8.4 RHEUMATOID ARTHRITIS

CYP2D6 genotypes and metabolic phenotypes have been determined for 53 patients with rheumatoid arthritis and 73 healthy controls (Beyeler et al. 1994). No significant difference in the distribution of overall genotypes between the two groups is observed. When the frequency of individual alleles is investigated, a significant difference in allele frequency for the *CYP2D6D* allele ($P < 0.005$) is observed with fewer patients with rheumatoid arthritis showing this mutation. Metabolic phenotypes are broadly similar between the patients and controls (Beyeler et al. 1994). However, a number of the patients with rheumatoid arthritis show higher than expected MRs for their particular genotype because of interference by the analgesic dextropropoxyphene in the phenotyping procedure.

5.8.5 CYTOKINES AND INFLAMMATION

Infection and inflammation generally cause a decrease in hepatic capacity for drug metabolism and disposition (Shah and Smith 2015). In particular, NAT2, CYP2C19, and 2D6 are subject to changes in inflammatory conditions associated with elevated cytokines, such as human immunodeficiency virus infection, cholestasis, cancer, and liver diseases (Shah and Smith 2015). The expression of Cyp2d mRNA and protein is downregulated by the proinflammatory cytokines in the rat and mouse liver (Kurokohchi et al. 1992; Trautwein et al. 1992). Administration of IL-10 to healthy volunteers does

not alter CYP1A2, 2C9, or 2D6 activities, but it decreases CYP3A activity (Gorski et al. 2000).

5.9 CONCLUSIONS AND FUTURE DIRECTIONS

In contrast to other CYPs, CYP2D6 is generally not regulated by many known environmental agents and is not inducible by common known CYP inducers such as phenobarbital, rifampin, and dexamethasone. However, interindividual differences in response to drugs metabolized by CYP2D6 may also be influenced modestly by hormonal state, dietary factors, diseases, and xenobiotic regulation of expression of the enzyme in the liver and extrahepatic organs such as the brain, kidney, and intestine. In vitro and in vivo studies indicate that the nuclear receptors including PXR, CAR, and GR do not play a role in the regulation of CYP2D6. The large interindividual variability in CYP2D6 expression and activity is largely governed by genetic factors and coadministered drugs that may induce or inhibit CYP2D6.

While CYP2D6 is subject to inhibition by many drugs and a number of clinical drug–drug interactions attributed to CYP2D6 inhibition have been reported, this enzyme is not inducible by most known CYP inducers. This means typical CYP inducers would not cause clinically important drug interactions with drugs that are mainly metabolized by CYP2D6. CYP2D6 is insensitive to cytokines and classical CYP inducers. However, there is enough evidence to indicate that several nuclear receptors such as HNF-4α and FXR can tightly regulate CYP2D6 via SHP-mediated mechanisms. Furthermore, a few GWAS studies have revealed that there are multiple networked pathways that can tightly regulate *CYP2D6* in the liver. Further functional and validation studies are needed to clarify the role of nuclear receptors, epigenetic factors, and other factors in the regulation of CYP2D6.

REFERENCES

Adorini, L., Pruzanski, M., Shapiro, D. 2012. Farnesoid X receptor targeting to treat nonalcoholic steatohepatitis. *Drug Discov Today* 17:988–997.

Aghemo, A., Colombo, M. 2013. Hepatocellular carcinoma in chronic hepatitis C: From bench to bedside. *Semin Immunopathol* 35:111–120.

Ali, A. H., Carey, E. J., Lindor, K. D. 2015. Recent advances in the development of farnesoid X receptor agonists. *Ann Transl Med* 3:5.

Alissa, E. M. 2014. Medicinal herbs and therapeutic drugs interactions. *Ther Drug Monit* 36:413–422.

Anderson, G. D. 2002. Sex differences in drug metabolism: Cytochrome P-450 and uridine diphosphate glucuronosyltransferase. *J Gend Specif Med* 5:25–33.

Anderson, G. D. 2005. Pregnancy-induced changes in pharmacokinetics: A mechanistic-based approach. *Clin Pharmacokinet* 44:989–1008.

Antunes Nde, J., Cavalli, R. C., Marques, M. P., Moises, E. C., Lanchote, V. L. 2015. Influence of gestational diabetes on the stereoselective pharmacokinetics and placental distribution of metoprolol and its metabolites in parturients. *Br J Clin Pharmacol* 79:605–616.

Aristoff, P. A., Garcia, G. A., Kirchhoff, P. D., Showalter, H. D. 2010. Rifamycins—Obstacles and opportunities. *Tuberculosis (Edinb)* 90:94–118.

Asselah, T., Bieche, I., Laurendeau, I., Paradis, V., Vidaud, D., Degott, C., Martinot, M. et al. 2005. Liver gene expression signature of mild fibrosis in patients with chronic hepatitis C. *Gastroenterology* 129:2064–2075.

Baciewicz, A. M., Chrisman, C. R., Finch, C. K., Self, T. H. 2008. Update on rifampin and rifabutin drug interactions. *Am J Med Sci* 335:126–136.

Baciewicz, A. M., Chrisman, C. R., Finch, C. K., Self, T. H. 2013. Update on rifampin, rifabutin, and rifapentine drug interactions. *Curr Med Res Opin* 29:1–12.

Bavner, A., Sanyal, S., Gustafsson, J. A., Treuter, E. 2005. Transcriptional corepression by SHP: Molecular mechanisms and physiological consequences. *Trends Endocrinol Metab* 16:478–488.

Bebia, Z., Buch, S. C., Wilson, J. W., Frye, R. F., Romkes, M., Cecchetti, A., Chaves-Gnecco, D., Branch, R. A. 2004. Bioequivalence revisited: Influence of age and sex on CYP enzymes. *Clin Pharmacol Ther* 76:618–627.

Begre, S., von Bardeleben, U., Ladewig, D., Jaquet-Rochat, S., Cosendai-Savary, L., Golay, K. P., Kosel, M., Baumann, P., Eap, C. B. 2002. Paroxetine increases steady-state concentrations of (R)-methadone in CYP2D6 extensive but not poor metabolizers. *J Clin Psychopharmacol* 22:211–215.

Belle, D. J., Ernest, C. S., Sauer, J. M., Smith, B. P., Thomasson, H. R., Witcher, J. W. 2002. Effect of potent CYP2D6 inhibition by paroxetine on atomoxetine pharmacokinetics. *J Clin Pharmacol* 42:1219–1227.

Beltrao, P., Albanese, V., Kenner, L. R., Swaney, D. L., Burlingame, A., Villen, J., Lim, W. A., Fraser, J. S., Frydman, J., Krogan, N. J. 2012. Systematic functional prioritization of protein post-translational modifications. *Cell* 150:413–425.

Bertelsen, K. M., Venkatakrishnan, K., Von Moltke, L. L., Obach, R. S., Greenblatt, D. J. 2003. Apparent mechanism-based inhibition of human CYP2D6 in vitro by paroxetine: Comparison with fluoxetine and quinidine. *Drug Metab Dispos* 31:289–293.

Beyeler, C., Daly, A. K., Armstrong, M., Astbury, C., Bird, H. A., Idle, J. R. 1994. Phenotype/genotype relationships for the cytochrome P450 enzyme CYP2D6 in rheumatoid arthritis: Influence of drug therapy and disease activity. *J Rheumatol* 21:1034–1039.

Bieche, I., Asselah, T., Laurendeau, I., Vidaud, D., Degot, C., Paradis, V., Bedossa, P., Valla, D. C., Marcellin, P., Vidaud, M. 2005. Molecular profiling of early stage liver fibrosis in patients with chronic hepatitis C virus infection. *Virology* 332:130–144.

Blake, M. J., Gaedigk, A., Pearce, R. E., Bomgaars, L. R., Christensen, M. L., Stowe, C., James, L. P., Wilson, J. T., Kearns, G. L., Leeder, J. S. 2007. Ontogeny of dextromethorphan *O*- and *N*-demethylation in the first year of life. *Clin Pharmacol Ther* 81:510–516.

Boelsterli, U. A., Ho, H. K., Zhou, S., Leow, K. Y. 2006. Bioactivation and hepatotoxicity of nitroaromatic drugs. *Curr Drug Metab* 7:715–727.

Bogdanos, D. P., Dalekos, G. N. 2008. Enzymes as target antigens of liver-specific autoimmunity: The case of cytochromes P450s. *Curr Med Chem* 15:2285–2292.

Bortolotti, F., Muratori, L., Jara, P., Hierro, L., Verucchi, G., Giacchino, R., Barbera, C. et al. 2003. Hepatitis C virus infection associated with liver-kidney microsomal antibody type 1 (LKM1) autoantibodies in children. *J Pediatr* 142:185–190.

Branch, R. A., Adedoyin, A., Frye, R. F., Wilson, J. W., Romkes, M. 2000. *In vivo* modulation of CYP enzymes by quinidine and rifampin. *Clin Pharmacol Ther* 68:401–411.

Brazier, N. C., Levine, M. A. 2003. Drug-herb interaction among commonly used conventional medicines: A compendium for health care professionals. *Am J Ther* 10:163–169.

Breidenbach, T., Kliem, V., Burg, M., Radermacher, J., Hoffmann, M. W., Klempnauer, J. 2000. Profound drop of cyclosporin A whole blood trough levels caused by St. John's wort (*Hypericum perforatum*). *Transplantation* 69:2229–2230.

Brendel, C., Schoonjans, K., Botrugno, O. A., Treuter, E., Auwerx, J. 2002. The small heterodimer partner interacts with the liver X receptor alpha and represses its transcriptional activity. *Mol Endocrinol* 16:2065–2076.

Briciu, C., Neag, M., Muntean, D., Vlase, L., Bocsan, C., Buzoianu, A., Gheldiu, A. M., Achim, M., Popa, A. 2014. A pharmacokinetic drug interaction study between nebivolol and paroxetine in healthy volunteers. *J Clin Pharm Ther* 39:535–540.

Brosen, K., Hansen, J. G., Nielsen, K. K., Sindrup, S. H., Gram, L. F. 1993. Inhibition by paroxetine of desipramine metabolism in extensive but not in poor metabolizers of sparteine. *Eur J Clin Pharmacol* 44:349–355.

Cairns, W., Smith, C. A., McLaren, A. W., Wolf, C. R. 1996. Characterization of the human cytochrome P4502D6 promoter. A potential role for antagonistic interactions between members of the nuclear receptor family. *J Biol Chem* 271:25269–25276.

Carcillo, J. A., Adedoyin, A., Burckart, G. J., Frye, R. F., Venkataramanan, R., Knoll, C., Thummel, K. et al. 2003. Coordinated intrahepatic and extrahepatic regulation of cytochrome p4502D6 in healthy subjects and in patients after liver transplantation. *Clin Pharmacol Ther* 73:456–467.

Carroll, K., Gomez, C., Shapiro, L. 2004. Tubby proteins: The plot thickens. *Nat Rev Mol Cell Biol* 5:55–63.

Carvalho, T. M., Cavalli Rde, C., Cunha, S. P., de Baraldi, C. O., Marques, M. P., Antunes, N. J., Godoy, A. L., Lanchote, V. L. 2011. Influence of gestational diabetes mellitus on the stereoselective kinetic disposition and metabolism of labetalol in hypertensive patients. *Eur J Clin Pharmacol* 67:55–61.

Cascorbi, I. 2003. Pharmacogenetics of cytochrome p4502D6: Genetic background and clinical implication. *Eur J Clin Invest* 33 Suppl 2:17–22.

Chen, D., Lepar, G., Kemper, B. 1994. A transcriptional regulatory element common to a large family of hepatic cytochrome P450 genes is a functional binding site of the orphan receptor HNF-4. *J Biol Chem* 269:5420–5427.

Chen, W. D., Wang, Y. D., Meng, Z., Zhang, L., Huang, W. 2011. Nuclear bile acid receptor FXR in the hepatic regeneration. *Biochim Biophys Acta* 1812:888–892.

Chiang, J. Y. 2009. Hepatocyte nuclear factor 4alpha regulation of bile acid and drug metabolism. *Expert Opin Drug Metab Toxicol* 5:137–147.

Choi, Y. H., Yu, A. M. 2014. ABC transporters in multidrug resistance and pharmacokinetics, and strategies for drug development. *Curr Pharm Des* 20:793–807.

Cholerton, S., Boustead, C., Taber, H., Arpanahi, A., Idle, J. R. 1996. *CYP2D6* genotypes in cigarette smokers and non-tobacco users. *Pharmacogenetics* 6:261–263.

Choudhuri, K., Gregorio, G. V., Mieli-Vergani, G., Vergani, D. 1998. Immunological cross-reactivity to multiple autoantigens in patients with liver kidney microsomal type 1 autoimmune hepatitis. *Hepatology* 28:1177–1181.

Christen, U., Holdener, M., Hintermann, E. 2010. Cytochrome P450 2D6 as a model antigen. *Dig Dis* 28:80–85.

Corchero, J., Granvil, C. P., Akiyama, T. E., Hayhurst, G. P., Pimprale, S., Feigenbaum, L., Idle, J. R., Gonzalez, F. J. 2001. The CYP2D6 humanized mouse: Effect of the human CYP2D6 transgene and HNF4α on the disposition of debrisoquine in the mouse. *Mol Pharmacol* 60:1260–1267.

Coresh, J., Selvin, E., Stevens, L. A., Manzi, J., Kusek, J. W., Eggers, P., Van Lente, F., Levey, A. S. 2007. Prevalence of chronic kidney disease in the United States. *JAMA* 298:2038–2047.

Corsini, A., Bortolini, M. 2013. Drug-induced liver injury: The role of drug metabolism and transport. *J Clin Pharmacol* 53:463–474.

Cortez, K. J., Kottilil, S. 2015. Beyond interferon: Rationale and prospects for newer treatment paradigms for chronic hepatitis C. *Ther Adv Chronic Dis* 6:4–14.

Costantino, G., Macchiarulo, A., Entrena-Guadix, A., Camaioni, E., Pellicciari, R. 2003. Binding mode of 6ECDCA, a potent bile acid agonist of the farnesoid X receptor (FXR). *Bioorg Med Chem Lett* 13:1865–1868.

Cotreau, M. M., von Moltke, L. L., Greenblatt, D. J. 2005. The influence of age and sex on the clearance of cytochrome P450 3A substrates. *Clin Pharmacokinet* 44:33–60.

Coxeter, P. D., McLachlan, A. J., Duke, C. C., Roufogalis, B. D. 2004. Herb–drug interactions: An evidence based approach. *Curr Med Chem* 11:1513–1525.

Crawley, M. L. 2010. Farnesoid X receptor modulators: A patent review. *Expert Opin Ther Pat* 20:1047–1057.

Crestani, M., De Fabiani, E., Caruso, D., Mitro, N., Gilardi, F., Vigil Chacon, A. B., Patelli, R., Godio, C., Galli, G. 2004. LXR (liver X receptor) and HNF-4 (hepatocyte nuclear factor-4): Key regulators in reverse cholesterol transport. *Biochem Soc Trans* 32:92–96.

Cui, J., Huang, L., Zhao, A., Lew, J. L., Yu, J., Sahoo, S., Meinke, P. T., Royo, I., Pelaez, F., Wright, S. D. 2003. Guggulsterone is a farnesoid X receptor antagonist in coactivator association assays but acts to enhance transcription of bile salt export pump. *J Biol Chem* 278:10214–10220.

Cummings, C. T., Deryckere, D., Earp, H. S., Graham, D. K. 2013. Molecular pathways: MERTK signaling in cancer. *Clin Cancer Res* 19:5275–5280.

Cummins, C. L., Wu, C. Y., Benet, L. Z. 2002. Sex-related differences in the clearance of cytochrome P450 3A4 substrates may be caused by P-glycoprotein. *Clin Pharmacol Ther* 72:474–489.

Czaja, A. J., Manns, M. P. 2010. Advances in the diagnosis, pathogenesis, and management of autoimmune hepatitis. *Gastroenterology* 139:58–72.e54.

Dalekos, G. N., Zachou, K., Liaskos, C., Gatselis, N. 2002. Autoantibodies and defined target autoantigens in autoimmune hepatitis: An overview. *Eur J Intern Med* 13:293–303.

de Lima Toccafondo Vieira, M., Huang, S. M. 2012. Botanical-drug interactions: A scientific perspective. *Planta Med* 78:1400–1415.

de Maat, M. M., Hoetelmans, R. M., Mathot, R. A., van Gorp, E. C., Meenhorst, P. L., Mulder, J. W., Beijnen, J. H. 2001. Drug interaction between St John's wort and nevirapine. *AIDS* 15:420–421.

Di, Y. M., Li, C. G., Xue, C. C., Zhou, S. F. 2008. Clinical drugs that interact with St. John's wort and implication in drug development. *Curr Pharm Des* 14:1723–1742.

Dickins, M. 2004. Induction of cytochromes P450. *Curr Top Med Chem* 4:1745–1766.

Dickmann, L. J., Tay, S., Senn, T. D., Zhang, H., Visone, A., Unadkat, J. D., Hebert, M. F., Isoherranen, N. 2008. Changes in maternal liver Cyp2c and Cyp2d expression and activity during rat pregnancy. *Biochem Pharmacol* 75:1677–1687.

Dilger, K., Hofmann, U., Klotz, U. 2000. Enzyme induction in the elderly: Effect of rifampin on the pharmacokinetics and pharmacodynamics of propafenone. *Clin Pharmacol Ther* 67:512–520.

Ding, L., Pang, S., Sun, Y., Tian, Y., Yu, L., Dang, N. 2014. Coordinated actions of FXR and LXR in metabolism: From pathogenesis to pharmacological targets for type 2 diabetes. *Int J Endocrinol* 2014:751859.

Dixit, V., Hariparsad, N., Li, F., Desai, P., Thummel, K. E., Unadkat, J. D. 2007. Cytochrome P450 enzymes and transporters induced by anti-human immunodeficiency virus protease inhibitors in human hepatocytes: Implications for predicting clinical drug interactions. *Drug Metab Dispos* 35:1853–1859.

Donato, M. T., Castell, J. V. 2003. Strategies and molecular probes to investigate the role of cytochrome P450 in drug metabolism: Focus on in vitro studies. *Clin Pharmacokinet* 42:153–178.

Donato, M. T., Castell, J. V., Gomez-Lechon, M. J. 1995. Effect of model inducers on cytochrome P450 activities of human hepatocytes in primary culture. *Drug Metab Dispos* 23:553–558.

Donovan, J. L., DeVane, C. L., Chavin, K. D., Taylor, R. M., Markowitz, J. S. 2003. Siberian ginseng (*Eleutheroccus senticosus*) effects on CYP2D6 and CYP3A4 activity in normal volunteers. *Drug Metab Dispos* 31:519–522.

Donovan, J. L., Chavin, K. D., Devane, C. L., Taylor, R. M., Wang, J. S., Ruan, Y., Markowitz, J. S. 2004a. Green tea (Camellia sinensis) extract does not alter cytochrome p450 3A4 or 2D6 activity in healthy volunteers. *Drug Metab Dispos* 32:906–908.

Donovan, J. L., DeVane, C. L., Chavin, K. D., Wang, J. S., Gibson, B. B., Gefroh, H. A., Markowitz, J. S. 2004b. Multiple nighttime doses of valerian (*Valeriana officinalis*) had minimal effects on CYP3A4 activity and no effect on CYP2D6 activity in healthy volunteers. *Drug Metab Dispos* 32:1333–1336.

Doroshyenko, O., Rokitta, D., Zadoyan, G., Klement, S., Schlafke, S., Dienel, A., Gramatte, T., Luck, H., Fuhr, U. 2013. Drug cocktail interaction study on the effect of the orally administered lavender oil preparation silexan on cytochrome P450 enzymes in healthy volunteers. *Drug Metab Dispos* 41:987–993.

Dostalek, M., Akhlaghi, F., Puzanovova, M. 2012. Effect of diabetes mellitus on pharmacokinetic and pharmacodynamic properties of drugs. *Clin Pharmacokinet* 51:481–499.

Dreisbach, A. W. 2009. The influence of chronic renal failure on drug metabolism and transport. *Clin Pharmacol Ther* 86:553–556.

Dreisbach, A. W., Lertora, J. J. 2003. The effect of chronic renal failure on hepatic drug metabolism and drug disposition. *Semin Dial* 16:45–50.

Dreisbach, A. W., Lertora, J. J. 2008. The effect of chronic renal failure on drug metabolism and transport. *Expert Opin Drug Metab Toxicol* 4:1065–1074.

Dresser, G. K., Schwarz, U. I., Wilkinson, G. R., Kim, R. B. 2003. Coordinate induction of both cytochrome P4503A and MDR1 by St John's wort in healthy subjects. *Clin Pharmacol Ther* 73:41–50.

Drici, M. D., Clement, N. 2001. Is gender a risk factor for adverse drug reactions? The example of drug-induced long QT syndrome. *Drug Saf* 24:575–585.

Duggal, P., Thio, C. L., Wojcik, G. L., Goedert, J. J., Mangia, A., Latanich, R., Kim, A. Y. et al. 2013. Genome-wide association study of spontaneous resolution of hepatitis C virus infection: Data from multiple cohorts. *Ann Intern Med* 158:235–245.

Durr, D., Stieger, B., Kullak-Ublick, G. A., Rentsch, K. M., Steinert, H. C., Meier, P. J., Fattinger, K. 2000. St John's wort induces intestinal P-glycoprotein/MDR1 and intestinal and hepatic CYP3A4. *Clin Pharmacol Ther* 68:598–604.

Edwards, R. J., Price, R. J., Watts, P. S., Renwick, A. B., Tredger, J. M., Boobis, A. R., Lake, B. G. 2003. Induction of cytochrome P450 enzymes in cultured precision-cut human liver slices. *Drug Metab Dispos* 31:282–288.

Erdmann, D., Heim, J. 1995. Orphan nuclear receptor HNF-4 binds to the human coagulation factor VII promoter. *J Biol Chem* 270:22988–22996.

Evans, D. A., Mahgoub, A., Sloan, T. P., Idle, J. R., Smith, R. L. 1980. A family and population study of the genetic polymorphism of debrisoquine oxidation in a white British population. *J Med Genet* 17:102–105.

Faucette, S. R., Wang, H., Hamilton, G. A., Jolley, S. L., Gilbert, D., Lindley, C., Yan, B., Negishi, M., LeCluyse, E. L. 2004. Regulation of CYP2B6 in primary human hepatocytes by prototypical inducers. *Drug Metab Dispos* 32:348–358.

Ferraguti, G., Pascale, E., Lucarelli, M. 2015. Alcohol addiction: A molecular biology perspective. *Curr Med Chem* 22:670–684.

Fiorucci, S., Rizzo, G., Donini, A., Distrutti, E., Santucci, L. 2007. Targeting farnesoid X receptor for liver and metabolic disorders. *Trends Mol Med* 13:298–309.

Fiorucci, S., Mencarelli, A., Distrutti, E., Palladino, G., Cipriani, S. 2010. Targetting farnesoid-X-receptor: From medicinal chemistry to disease treatment. *Curr Med Chem* 17:139–159.

Fiorucci, S., Mencarelli, A., Distrutti, E., Zampella, A. 2012. Farnesoid X receptor: From medicinal chemistry to clinical applications. *Future Med Chem* 4:877–891.

Fiorucci, S., Distrutti, E., Ricci, P., Giuliano, V., Donini, A., Baldelli, F. 2014. Targeting FXR in cholestasis: Hype or hope. *Expert Opin Ther Targets* 18:1449–1459.

Franconi, F., Campesi, I. 2014. Pharmacogenomics, pharmacokinetics and pharmacodynamics: Interaction with biological differences between men and women. *Br J Pharmacol* 171:580–594.

Frank, D., Jaehde, U., Fuhr, U. 2007. Evaluation of probe drugs and pharmacokinetic metrics for CYP2D6 phenotyping. *Eur J Clin Pharmacol* 63:321–333.

Frye, R. F., Zgheib, N. K., Matzke, G. R., Chaves-Gnecco, D., Rabinovitz, M., Shaikh, O. S., Branch, R. A. 2006. Liver disease selectively modulates cytochrome P450—Mediated metabolism. *Clin Pharmacol Ther* 80:235–245.

Fugh-Berman, A., Ernst, E. 2001. Herb–drug interactions: Review and assessment of report reliability. *Br J Clin Pharmacol* 52:587–595.

Fuhr, U. 2000. Induction of drug metabolising enzymes: Pharmacokinetic and toxicological consequences in humans. *Clin Pharmacokinet* 38:493–504.

Fuhr, U., Jetter, A., Kirchheiner, J. 2007. Appropriate phenotyping procedures for drug metabolizing enzymes and transporters in humans and their simultaneous use in the "cocktail" approach. *Clin Pharmacol Ther* 81:270–283.

Gadaleta, R. M., Cariello, M., Sabba, C., Moschetta, A. 2015. Tissue-specific actions of FXR in metabolism and cancer. *Biochim Biophys Acta* 1851:30–39.

Gerbal-Chaloin, S., Pascussi, J. M., Pichard-Garcia, L., Daujat, M., Waechter, F., Fabre, J. M., Carrere, N., Maurel, P. 2001. Induction of *CYP2C* genes in human hepatocytes in primary culture. *Drug Metab Dispos* 29:242–251.

Gerbal-Chaloin, S., Daujat, M., Pascussi, J. M., Pichard-Garcia, L., Vilarem, M. J., Maurel, P. 2002. Transcriptional regulation of CYP2C9 gene. Role of glucocorticoid receptor and constitutive androstane receptor. *J Biol Chem* 277:209–217.

Gioiello, A., Cerra, B., Mostarda, S., Guercini, C., Pellicciari, R., Macchiarulo, A. 2014. Bile acid derivatives as ligands of the farnesoid x receptor: Molecular determinants for bile acid binding and receptor modulation. *Curr Top Med Chem* 14:2159–2174.

Girardin, F., Daali, Y., Gex-Fabry, M., Rebsamen, M., Roux-Lombard, P., Cerny, A., Bihl, F. et al. 2012. Liver kidney microsomal type 1 antibodies reduce the CYP2D6 activity in patients with chronic hepatitis C virus infection. *J Viral Hepat* 19:568–573.

Glaeser, H., Drescher, S., Eichelbaum, M., Fromm, M. F. 2005. Influence of rifampicin on the expression and function of human intestinal cytochrome P450 enzymes. *Br J Clin Pharmacol* 59:199–206.

Gleiter, C. H., Gundert-Remy, U. 1996. Gender differences in pharmacokinetics. *Eur J Drug Metab Pharmacokinet* 21:123–128.

Goey, A. K., Meijerman, I., Rosing, H., Marchetti, S., Mergui-Roelvink, M., Keessen, M., Burgers, J. A., Beijnen, J. H., Schellens, J. H. 2014. The effect of St John's wort on the pharmacokinetics of docetaxel. *Clin Pharmacokinet* 53:103–110.

Gomez-Lechon, M. J., Castell, J. V., Donato, M. T. 2008. An update on metabolism studies using human hepatocytes in primary culture. *Expert Opin Drug Metab Toxicol* 4:837–854.

Gong, X., Liu, Y., Zhang, X., Wei, Z., Huo, R., Shen, L., He, L., Qin, S. 2013. Systematic functional study of cytochrome P450 2D6 promoter polymorphisms in the Chinese Han population. *PLoS One* 8:e57764.

Gonzalez, F. J. 2008. Regulation of hepatocyte nuclear factor 4 alpha-mediated transcription. *Drug Metab Pharmacokinet* 23:2–7.

Goodwin, B., Jones, S. A., Price, R. R., Watson, M. A., McKee, D. D., Moore, L. B., Galardi, C. et al. 2000. A regulatory cascade of the nuclear receptors FXR, SHP-1, and LRH-1 represses bile acid biosynthesis. *Mol Cell* 6:517–526.

Gorski, J. C., Hall, S. D., Becker, P., Affrime, M. B., Cutler, D. L., Haehner-Daniels, B. 2000. In vivo effects of interleukin-10 on human cytochrome P450 activity. *Clin Pharmacol Ther* 67:32–43.

Goryachkina, K., Burbello, A., Boldueva, S., Babak, S., Bergman, U., Bertilsson, L. 2008. Inhibition of metoprolol metabolism and potentiation of its effects by paroxetine in routinely treated patients with acute myocardial infarction (AMI). *Eur J Clin Pharmacol* 64:275–282.

Graham, D. K., DeRyckere, D., Davies, K. D., Earp, H. S. 2014. The TAM family: Phosphatidylserine sensing receptor tyrosine kinases gone awry in cancer. *Nat Rev Cancer* 14:769–785.

Gupta, R. K., Kaestner, K. H. 2004. HNF-4alpha: From MODY to late-onset type 2 diabetes. *Trends Mol Med* 10:521–524.

Gurley, B. J., Gardner, S. F., Hubbard, M. A., Williams, D. K., Gentry, W. B., Carrier, J., Khan, I. A., Edwards, D. J., Shah, A. 2004. In vivo assessment of botanical supplementation on human cytochrome P450 phenotypes: *Citrus aurantium*, Echinacea purpurea, milk thistle, and saw palmetto. *Clin Pharmacol Ther* 76:428–440.

Gurley, B. J., Gardner, S. F., Hubbard, M. A., Williams, D. K., Gentry, W. B., Cui, Y., Ang, C. Y. 2005a. Clinical assessment of effects of botanical supplementation on cytochrome P450 phenotypes in the elderly: St John's wort, garlic oil, *Panax ginseng* and *Ginkgo biloba*. *Drugs Aging* 22:525–539.

Gurley, B. J., Gardner, S. F., Hubbard, M. A., Williams, D. K., Gentry, W. B., Khan, I. A., Shah, A. 2005b. In vivo effects of goldenseal, kava kava, black cohosh, and valerian on human cytochrome P450 1A2, 2D6, 2E1, and 3A4/5 phenotypes. *Clin Pharmacol Ther* 77:415–426.

Gurley, B. J., Swain, A., Hubbard, M. A., Williams, D. K., Barone, G., Hartsfield, F., Tong, Y., Carrier, D. J., Cheboyina, S., Battu, S. K. 2008. Clinical assessment of CYP2D6-mediated herb–drug interactions in humans: Effects of milk thistle, black cohosh, goldenseal, kava kava, St. John's wort, and Echinacea. *Mol Nutr Food Res* 52:755–763.

Gurley, B. J., Fifer, E. K., Gardner, Z. 2012. Pharmacokinetic herb–drug interactions (part 2): Drug interactions involving popular botanical dietary supplements and their clinical relevance. *Planta Med* 78:1490–1514.

Hagenbuch, B., Meier, P. J. 2004. Organic anion transporting polypeptides of the OATP/SLC21 family: Phylogenetic classification as OATP/SLCO superfamily, new nomenclature and molecular/functional properties. *Pflugers Arch* 447:653–665.

Hagg, S., Spigset, O., Dahlqvist, R. 2001. Influence of gender and oral contraceptives on CYP2D6 and CYP2C19 activity in healthy volunteers. *Br J Clin Pharmacol* 51:169–173.

Handschin, C., Meyer, U. A. 2003. Induction of drug metabolism: The role of nuclear receptors. *Pharmacol Rev* 55:649–673.

Hara, H., Adachi, T. 2002. Contribution of hepatocyte nuclear factor-4 to down-regulation of CYP2D6 gene expression by nitric oxide. *Mol Pharmacol* 61:194–200.

Harris, R. Z., Benet, L. Z., Schwartz, J. B. 1995. Gender effects in pharmacokinetics and pharmacodynamics. *Drugs* 50:222–239.

He, S. M., Yang, A. K., Li, X. T., Du, Y. M., Zhou, S. F. 2010. Effects of herbal products on the metabolism and transport of anticancer agents. *Expert Opin Drug Metab Toxicol* 6:1195–1213.

Hellum, B. H., Hu, Z., Nilsen, O. G. 2007. The induction of CYP1A2, CYP2D6 and CYP3A4 by six trade herbal products in cultured primary human hepatocytes. *Basic Clin Pharmacol Toxicol* 100:23–30.

Hertz, R., Magenheim, J., Berman, I., Bar-Tana, J. 1998. Fatty acyl-CoA thioesters are ligands of hepatic nuclear factor-4α. *Nature* 392:512–516.

Hewitt, N. J., Lecluyse, E. L., Ferguson, S. S. 2007. Induction of hepatic cytochrome P450 enzymes: Methods, mechanisms, recommendations, and in vitro-in vivo correlations. *Xenobiotica* 37:1196–1224.

Hintermann, E., Holdener, M., Bayer, M., Loges, S., Pfeilschifter, J. M., Granier, C., Manns, M. P., Christen, U. 2011. Epitope spreading of the anti-CYP2D6 antibody response in patients with autoimmune hepatitis and in the CYP2D6 mouse model. *J Autoimmun* 37:242–253.

Hirschfield, G. M., Mason, A., Luketic, V., Lindor, K., Gordon, S. C., Mayo, M., Kowdley, K. V. et al. 2015. Efficacy of obeticholic Acid in patients with primary biliary cirrhosis and inadequate response to ursodeoxycholic Acid. *Gastroenterology* 148:751–761.e758.

Ho, M. T., Kelly, E. J., Bodor, M., Bui, T., Kowdley, K. V., Ho, R. J. 2011. Novel cytochrome P450-2D6 promoter sequence variations in hepatitis C positive and negative subjects. *Ann Hepatol* 10:327–332.

Hoeke, M. O., Heegsma, J., Hoekstra, M., Moshage, H., Faber, K. N. 2014. Human FXR regulates SHP expression through direct binding to an LRH-1 binding site, independent of an IR-1 and LRH-1. *PLoS One* 9:e88011.

Hogstedt, S., Rane, A. 1993. Plasma concentration-effect relationship of metoprolol during and after pregnancy. *Eur J Clin Pharmacol* 44:243–246.

Hogstedt, S., Lindberg, B., Rane, A. 1983. Increased oral clearance of metoprolol in pregnancy. *Eur J Clin Pharmacol* 24:217–220.

Hogstedt, S., Lindberg, B., Peng, D. R., Regardh, C. G., Rane, A. 1985. Pregnancy-induced increase in metoprolol metabolism. *Clin Pharmacol Ther* 37:688–692.

Hojo, Y., Echizenya, M., Ohkubo, T., Shimizu, T. 2011. Drug interaction between St John's wort and zolpidem in healthy subjects. *J Clin Pharm Ther* 36:711–715.

Hosking, L. K., Boyd, P. R., Xu, C. F., Nissum, M., Cantone, K., Purvis, I. J., Khakhar, R. et al. 2002. Linkage disequilibrium mapping identifies a 390 kb region associated with CYP2D6 poor drug metabolising activity. *Pharmacogenomics J* 2:165–175.

Hu, Z., Yang, X., Ho, P. C., Chan, S. Y., Heng, P. W., Chan, E., Duan, W., Koh, H. L., Zhou, S. 2005. Herb–drug interactions: A literature review. *Drugs* 65:1239–1282.

Hukkanen, J. 2012. Induction of cytochrome P450 enzymes: A view on human in vivo findings. *Expert Rev Clin Pharmacol* 5:569–585.

Hwang-Verslues, W. W., Sladek, F. M. 2010. HNF4alpha—Role in drug metabolism and potential drug target? *Curr Opin Pharmacol* 10:698–705.

Ibeanu, G. C., Goldstein, J. A. 1995. Transcriptional regulation of human *CYP2C* genes: Functional comparison of *CYP2C9* and *CYP2C18* promoter regions. *Biochemistry* 34:8028–8036.

Ikeda, A., Nishina, P. M., Naggert, J. K. 2002. The tubby-like proteins, a family with roles in neuronal development and function. *J Cell Sci* 115:9–14.

Imran, M., Manzoor, S., Khattak, N. M., Khalid, M., Ahmed, Q. L., Parvaiz, F., Tariq, M. et al. 2014. Current and future therapies for hepatitis C virus infection: From viral proteins to host targets. *Arch Virol* 159:831–846.

Ingelman-Sundberg, M. 2005. Genetic polymorphisms of cytochrome P450 *2D6* (*CYP2D6*): Clinical consequences, evolutionary aspects and functional diversity. *Pharmacogenomics J* 5:6–13.

Ingelman-Sundberg, M., Sim, S. C., Gomez, A., Rodriguez-Antona, C. 2007. Influence of cytochrome P450 polymorphisms on drug therapies: Pharmacogenetic, pharmacoepigenetic and clinical aspects. *Pharmacol Ther* 116:496–526.

Izzo, A. A. 2005. Herb–drug interactions: An overview of the clinical evidence. *Fundam Clin Pharmacol* 19:1–16.

Jensen, O. N. 2004. Modification-specific proteomics: Characterization of post-translational modifications by mass spectrometry. *Curr Opin Chem Biol* 8:33–41.

Jeong, S. W., Jang, J. Y., Chung, R. T. 2012. Hepatitis C virus and hepatocarcinogenesis. *Clin Mol Hepatol* 18:347–356.

Ji, Y., Schaid, D. J., Desta, Z., Kubo, M., Batzler, A. J., Snyder, K., Mushiroda, T. et al. 2014. Citalopram and escitalopram plasma drug and metabolite concentrations: Genome-wide associations. *Br J Clin Pharmacol* 78:373–383.

Jiang, G., Chakraborty, A., Wang, Z., Boustani, M., Liu, Y., Skaar, T., Li, L. 2013. New aQTL SNPs for the CYP2D6 identified by a novel mediation analysis of genome-wide SNP arrays, gene expression arrays, and CYP2D6 activity. *Biomed Res Int* 2013:493019.

Joanne, C., Paintaud, G., Bresson-Hadni, S., Magnette, J., Becker, M. C., Miguet, J. P., Bechtel, P. R. 1994. Is debrisoquine hydroxylation modified during acute viral hepatitis? *Fundam Clin Pharmacol* 8:76–79.

Johansson, L., Bavner, A., Thomsen, J. S., Farnegardh, M., Gustafsson, J. A., Treuter, E. 2000. The orphan nuclear receptor SHP utilizes conserved LXXLL-related motifs for interactions with ligand-activated estrogen receptors. *Mol Cell Biol* 20:1124–1133.

Johne, A., Brockmoller, J., Bauer, S., Maurer, A., Langheinrich, M., Roots, I. 1999. Pharmacokinetic interaction of digoxin with an herbal extract from St John's wort (*Hypericum perforatum*). *Clin Pharmacol Ther* 66:338–345.

Johne, A., Schmider, J., Brockmoller, J., Stadelmann, A. M., Stormer, E., Bauer, S., Scholler, G., Langheinrich, M., Roots, I. 2002. Decreased plasma levels of amitriptyline and its metabolites on comedication with an extract from St. John's Wort (*Hypericum perforatum*). *J Clin Psychopharmacol* 22:46–54.

Johnson, T. N., Tucker, G. T., Rostami-Hodjegan, A. 2008. Development of CYP2D6 and CYP3A4 in the first year of life. *Clin Pharmacol Ther* 83:670–671.

Jover, R., Bort, R., Gomez-Lechon, M. J., Castell, J. V. 2001. Cytochrome P450 regulation by hepatocyte nuclear factor 4 in human hepatocytes: A study using adenovirus-mediated antisense targeting. *Hepatology* 33:668–675.

Jover, R., Moya, M., Gomez-Lechon, M. J. 2009. Transcriptional regulation of cytochrome p450 genes by the nuclear receptor hepatocyte nuclear factor 4-alpha. *Curr Drug Metab* 10:508–519.

Kaczynski, J. A., Conley, A. A., Fernandez Zapico, M., Delgado, S. M., Zhang, J. S., Urrutia, R. 2002. Functional analysis of basic transcription element (BTE)-binding protein (BTEB) 3 and BTEB4, a novel Sp1-like protein, reveals a subfamily of transcriptional repressors for the BTE site of the cytochrome P4501A1 gene promoter. *Biochem J* 366:873–882.

Kalaany, N. Y., Mangelsdorf, D. J. 2006. LXRS and FXR: The yin and yang of cholesterol and fat metabolism. *Annu Rev Physiol* 68:159–191.

Kamiyama, Y., Matsubara, T., Yoshinari, K., Nagata, K., Kamimura, H., Yamazoe, Y. 2007. Role of human hepatocyte nuclear factor 4α in the expression of drug-metabolizing enzymes and transporters in human hepatocytes assessed by use of small interfering RNA. *Drug Metab Pharmacokinet* 22:287–298.

Kassam, A., Capone, J. P., Rachubinski, R. A. 2001. The short heterodimer partner receptor differentially modulates peroxisome proliferator-activated receptor alpha-mediated transcription from the peroxisome proliferator-response elements of the genes encoding the peroxisomal beta-oxidation enzymes acyl-CoA oxidase and hydratase-dehydrogenase. *Mol Cell Endocrinol* 176:49–56.

Kawaguchi-Suzuki, M., Frye, R. F., Zhu, H. J., Brinda, B. J., Chavin, K. D., Bernstein, H. J., Markowitz, J. S. 2014. The effects of milk thistle (Silybum marianum) on human cytochrome P450 activity. *Drug Metab Dispos* 42:1611–1616.

Kawakami, H., Ohtsuki, S., Kamiie, J., Suzuki, T., Abe, T., Terasaki, T. 2011. Simultaneous absolute quantification of 11 cytochrome P450 isoforms in human liver microsomes by liquid chromatography tandem mass spectrometry with in silico target peptide selection. *J Pharm Sci* 100:341–352.

Kemper, J. K. 2011. Regulation of FXR transcriptional activity in health and disease: Emerging roles of FXR cofactors and post-translational modifications. *Biochim Biophys Acta* 1812:842–850.

Kerkar, N., Choudhuri, K., Ma, Y., Mahmoud, A., Bogdanos, D. P., Muratori, L., Bianchi, F., Williams, R., Mieli-Vergani, G., Vergani, D. 2003. Cytochrome P4502D6(193-212): A new immunodominant epitope and target of virus/self cross-reactivity in liver kidney microsomal autoantibody type 1-positive liver disease. *J Immunol* 170:1481–1489.

Kerr, T. A., Saeki, S., Schneider, M., Schaefer, K., Berdy, S., Redder, T., Shan, B., Russell, D. W., Schwarz, M. 2002. Loss of nuclear receptor SHP impairs but does not eliminate negative feedback regulation of bile acid synthesis. *Dev Cell* 2:713–720.

Kevorkian, J. P., Michel, C., Hofmann, U., Jacqz-Aigrain, E., Kroemer, H. K., Peraldi, M. N., Eichelbaum, M., Jaillon, P., Funck-Brentano, C. 1996. Assessment of individual CYP2D6 activity in extensive metabolizers with renal failure: Comparison of sparteine and dextromethorphan. *Clin Pharmacol Ther* 59:583–592.

Kim, Y. D., Li, T., Ahn, S. W., Kim, D. K., Lee, J. M., Hwang, S. L., Kim, Y. H. et al. 2012. Orphan nuclear receptor small heterodimer partner negatively regulates growth hormone-mediated induction of hepatic gluconeogenesis through inhibition of signal transducer and activator of transcription 5 (STAT5) transactivation. *J Biol Chem* 287:37098–37108.

Kirby, B. J., Collier, A. C., Kharasch, E. D., Dixit, V., Desai, P., Whittington, D., Thummel, K. E., Unadkat, J. D. 2011. Complex drug interactions of HIV protease inhibitors 2: In vivo induction and in vitro to in vivo correlation of induction of cytochrome P450 1A2, 2B6, and 2C9 by ritonavir or nelfinavir. *Drug Metab Dispos* 39:2329–2337.

Klinge, C. M., Jernigan, S. C., Risinger, K. E., Lee, J. E., Tyulmenkov, V. V., Falkner, K. C., Prough, R. A. 2001. Short heterodimer partner (SHP) orphan nuclear receptor inhibits the transcriptional activity of aryl hydrocarbon receptor (AHR)/AHR nuclear translocator (ARNT). *Arch Biochem Biophys* 390:64–70.

Kobayashi, K., Chiba, K., Yagi, T., Shimada, N., Taniguchi, T., Horie, T., Tani, M., Yamamoto, T., Ishizaki, T., Kuroiwa, Y. 1997. Identification of cytochrome P450 isoforms involved in citalopram N-demethylation by human liver microsomes. *J Pharmacol Exp Ther* 280:927–933.

Koh, K. H., Pan, X., Shen, H. W., Arnold, S. L., Yu, A. M., Gonzalez, F. J., Isoherranen, N., Jeong, H. 2014a. Altered expression of small heterodimer partner governs cytochrome P450 (CYP) 2D6 induction during pregnancy in CYP2D6-humanized mice. *J Biol Chem* 289:3105–3113.

Koh, K. H., Pan, X., Zhang, W., McLachlan, A., Urrutia, R., Jeong, H. 2014b. Kruppel-like factor 9 promotes hepatic cytochrome P450 2D6 expression during pregnancy in CYP2D6-humanized mice. *Mol Pharmacol* 86:727–735.

Komoroski, B. J., Zhang, S., Cai, H., Hutzler, J. M., Frye, R., Tracy, T. S., Strom, S. C., Lehmann, T., Ang, C. Y., Cui, Y. Y., Venkataramanan, R. 2004. Induction and inhibition of cytochromes P450 by the St. John's wort constituent hyperforin in human hepatocyte cultures. *Drug Metab Dispos* 32:512–518.

Komurasaki, R., Imaoka, S., Tada, N., Okada, K., Nishiguchi, S., Funae, Y. 2010. LKM-1 sera from autoimmune hepatitis patients that recognize ERp57, carboxylesterase 1 and CYP2D6. *Drug Metab Pharmacokinet* 25:84–92.

Kurokohchi, K., Yoneyama, H., Matsuo, Y., Nishioka, M., Ichikawa, Y. 1992. Effects of interleukin 1 α on the activities and gene expressions of the cytochrome P4502D subfamily. *Biochem Pharmacol* 44:1669–1674.

Labbe, L., Sirois, C., Pilote, S., Arseneault, M., Robitaille, N. M., Turgeon, J., Hamelin, B. A. 2000. Effect of gender, sex hormones, time variables and physiological urinary pH on apparent CYP2D6 activity as assessed by metabolic ratios of marker substrates. *Pharmacogenetics* 10:425–438.

Lake, B. G., Ball, S. E., Renwick, A. B., Tredger, J. M., Kao, J., Beamand, J. A., Price, R. J. 1997. Induction of CYP3A isoforms in cultured precision-cut human liver slices. *Xenobiotica* 27:1165–1173.

Lake, B. G., Price, R. J., Giddings, A. M., Walters, D. G. 2009. In vitro assays for induction of drug metabolism. *Methods Mol Biol* 481:47–58.

Lamba, J. K., Lin, Y. S., Schuetz, E. G., Thummel, K. E. 2002. Genetic contribution to variable human CYP3A-mediated metabolism. *Adv Drug Deliv Rev* 54:1271–1294.

Lammers, L. A., Achterbergh, R., de Vries, E. M., van Nierop, F. S., Klumpen, H. J., Soeters, M. R., Boelen, A., Romijn, J. A., Mathot, R. A. 2015. Short-term fasting alters cytochrome p450-mediated drug metabolism in humans. *Drug Metab Dispos* 43:819–828.

Landi, M. T., Sinha, R., Lang, N. P., Kadlubar, F. F. 1999. Human cytochrome P4501A2. *IARC Sci Publ* 173–195.

Laugesen, S., Enggaard, T. P., Pedersen, R. S., Sindrup, S. H., Brosen, K. 2005. Paroxetine, a cytochrome P450 2D6 inhibitor, diminishes the stereoselective O-demethylation and reduces the hypoalgesic effect of tramadol. *Clin Pharmacol Ther* 77:312–323.

LeCluyse, E. L. 2001. Human hepatocyte culture systems for the in vitro evaluation of cytochrome P450 expression and regulation. *Eur J Pharm Sci* 13:343–368.

Lee, Y. K., Moore, D. D. 2002. Dual mechanisms for repression of the monomeric orphan receptor liver receptor homologous protein-1 by the orphan small heterodimer partner. *J Biol Chem* 277:2463–2467.

Lee, H. K., Lee, Y. K., Park, S. H., Kim, Y. S., Park, S. H., Lee, J. W., Kwon, H. B., Soh, J., Moore, D. D., Choi, H. S. 1998. Structure and expression of the orphan nuclear receptor SHP gene. *J Biol Chem* 273:14398–14402.

Lee, Y. K., Dell, H., Dowhan, D. H., Hadzopoulou-Cladaras, M., Moore, D. D. 2000. The orphan nuclear receptor SHP inhibits hepatocyte nuclear factor 4 and retinoid X receptor transactivation: Two mechanisms for repression. *Mol Cell Biol* 20:187–195.

Lee, F. Y., Lee, H., Hubbert, M. L., Edwards, P. A., Zhang, Y. 2006. FXR, a multipurpose nuclear receptor. *Trends Biochem Sci* 31:572–580.

Lee, S. S., Cha, E. Y., Jung, H. J., Shon, J. H., Kim, E. Y., Yeo, C. W., Shin, J. G. 2008. Genetic polymorphism of hepatocyte nuclear factor-4α influences human cytochrome P450 2D6 activity. *Hepatology* 48:635–645.

Lee, J., Padhye, A., Sharma, A., Song, G., Miao, J., Mo, Y. Y., Wang, L., Kemper, J. K. 2010. A pathway involving farnesoid X receptor and small heterodimer partner positively regulates hepatic sirtuin 1 levels via microRNA-34a inhibition. *J Biol Chem* 285:12604–12611.

Lemke, G. 2013. Biology of the TAM receptors. *Cold Spring Harb Perspect Biol* 5:a009076.

Lemke, G., Rothlin, C. V. 2008. Immunobiology of the TAM receptors. *Nat Rev Immunol* 8:327–336.

Lempiainen, H., Muller, A., Brasa, S., Teo, S. S., Roloff, T. C., Morawiec, L., Zamurovic, N. et al. 2011. Phenobarbital mediates an epigenetic switch at the constitutive androstane receptor (CAR) target gene Cyp2b10 in the liver of B6C3F1 mice. *PLoS One* 6:e18216.

Leung, T. M., Nieto, N. 2013. CYP2E1 and oxidant stress in alcoholic and non-alcoholic fatty liver disease. *J Hepatol* 58:395–398.

Li, T., Chiang, J. Y. 2006. Rifampicin induction of CYP3A4 requires pregnane X receptor cross talk with hepatocyte nuclear factor 4alpha and coactivators, and suppression of small heterodimer partner gene expression. *Drug Metab Dispos* 34:756–764.

Li, T., Chiang, J. Y. 2014. Bile acid signaling in metabolic disease and drug therapy. *Pharmacol Rev* 66:948–983.

Li, A. P., Reith, M. K., Rasmussen, A., Gorski, J. C., Hall, S. D., Xu, L., Kaminski, D. L., Cheng, L. K. 1997. Primary human hepatocytes as a tool for the evaluation of structure-activity relationship in cytochrome P450 induction potential of xenobiotics: Evaluation of rifampin, rifapentine and rifabutin. *Chem Biol Interact* 107:17–30.

Li, Y. T., Swales, K. E., Thomas, G. J., Warner, T. D., Bishop-Bailey, D. 2007. Farnesoid x receptor ligands inhibit vascular smooth muscle cell inflammation and migration. *Arterioscler Thromb Vasc Biol* 27:2606–2611.

Lindor, K. D. 2011. Farnesoid X receptor agonists for primary biliary cirrhosis. *Curr Opin Gastroenterol* 27:285–288.

Liu, M. Z., Zhang, Y. L., Zeng, M. Z., He, F. Z., Luo, Z. Y., Luo, J. Q., Wen, J. G., Chen, X. P., Zhou, H. H., Zhang, W. 2015. Pharmacogenomics and herb–drug interactions: Merge of future and tradition. *Evid Based Complement Alternat Med* 2015:321091.

Llerena, A., Cobaleda, J., Martinez, C., Benitez, J. 1996. Interethnic differences in drug metabolism: Influence of genetic and environmental factors on debrisoquine hydroxylation phenotype. *Eur J Drug Metab Pharmacokinet* 21:129–138.

Longhi, M. S., Hussain, M. J., Bogdanos, D. P., Quaglia, A., Mieli-Vergani, G., Ma, Y., Vergani, D. 2007. Cytochrome P450IID6-specific CD8 T cell immune responses mirror disease activity in autoimmune hepatitis type 2. *Hepatology* 46:472–484.

Lu, T. T., Makishima, M., Repa, J. J., Schoonjans, K., Kerr, T. A., Auwerx, J., Mangelsdorf, D. J. 2000. Molecular basis for feedback regulation of bile acid synthesis by nuclear receptors. *Mol Cell* 6:507–515.

Luo, G., Guenthner, T., Gan, L. S., Humphreys, W. G. 2004. CYP3A4 induction by xenobiotics: Biochemistry, experimental methods and impact on drug discovery and development. *Curr Drug Metab* 5:483–505.

Ma, Y., Peakman, M., Lobo-Yeo, A., Wen, L., Lenzi, M., Gaken, J., Farzaneh, F., Mieli-Vergani, G., Bianchi, F. B., Vergani, D. 1994. Differences in immune recognition of cytochrome P4502D6 by liver kidney microsomal (LKM) antibody in autoimmune hepatitis and chronic hepatitis C virus infection. *Clin Exp Immunol* 97:94–99.

Ma, Y., Bogdanos, D. P., Hussain, M. J., Underhill, J., Bansal, S., Longhi, M. S., Cheeseman, P., Mieli-Vergani, G., Vergani, D. 2006. Polyclonal T-cell responses to cytochrome P450IID6 are associated with disease activity in autoimmune hepatitis type 2. *Gastroenterology* 130:868–882.

Madabushi, R., Frank, B., Drewelow, B., Derendorf, H., Butterweck, V. 2006. Hyperforin in St. John's wort drug interactions. *Eur J Clin Pharmacol* 62:225–233.

Madacsy, L., Barkai, L., Santa, A., Krikovszky, D. 2004. Altered distribution of the debrisoquine oxidative phenotypes in children with type 1 diabetes mellitus. *Horm Res* 61:176–179.

Madan, A., Graham, R. A., Carroll, K. M., Mudra, D. R., Burton, L. A., Krueger, L. A., Downey, A. D. et al. 2003. Effects of prototypical microsomal enzyme inducers on cytochrome P450 expression in cultured human hepatocytes. *Drug Metab Dispos* 31:421–431.

Mahgoub, A., Idle, J. R., Dring, L. G., Lancaster, R., Smith, R. L. 1977. Polymorphic hydroxylation of Debrisoquine in man. *Lancet* 2:584–586.

Mahrour, N., Redwine, W. B., Florens, L., Swanson, S. K., Martin-Brown, S., Bradford, W. D., Staehling-Hampton, K., Washburn, M. P., Conaway, R. C., Conaway, J. W. 2008. Characterization of Cullin-box sequences that direct recruitment of Cul2-Rbx1 and Cul5-Rbx2 modules to Elongin BC-based ubiquitin ligases. *J Biol Chem* 283:8005–8013.

Mamoon, A., Subauste, A., Subauste, M. C., Subauste, J. 2014. Retinoic acid regulates several genes in bile acid and lipid metabolism via upregulation of small heterodimer partner in hepatocytes. *Gene* 550:165–170.

Markert, C., Kastner, I. M., Hellwig, R., Kalafut, P., Schweizer, Y., Hoffmann, M. M., Burhenne, J., Weiss, J., Mikus, G., Haefeli, W. E. 2015. The effect of induction of CYP3A4 by St John's wort on ambrisentan plasma pharmacokinetics in volunteers of known CYP2C19 genotype. *Basic Clin Pharmacol Toxicol* 116:423–428.

Markowitz, J. S., Donovan, J. L., DeVane, C. L., Taylor, R. M., Ruan, Y., Wang, J. S., Chavin, K. D. 2003. Effect of St John's wort on drug metabolism by induction of cytochrome P450 3A4 enzyme. *JAMA* 290:1500–1504.

Marques, M. P., Coelho, E. B., Dos Santos, N. A., Geleilete, T. J., Lanchote, V. L. 2002. Dynamic and kinetic disposition of nisoldipine enantiomers in hypertensive patients presenting with type-2 diabetes mellitus. *Eur J Clin Pharmacol* 58:607–614.

Martel-Laferriere, V., Dieterich, D. T. 2014. Treating HCV in HIV 2013: On the cusp of change. *Liver Int* 34 Suppl 1:53–59.

Martin, H., Sarsat, J. P., de Waziers, I., Housset, C., Balladur, P., Beaune, P., Albaladejo, V., Lerche-Langrand, C. 2003. Induction of cytochrome P450 2B6 and 3A4 expression by phenobarbital and cyclophosphamide in cultured human liver slices. *Pharm Res* 20:557–568.

Martinez-Jimenez, C. P., Jover, R., Donato, M. T., Castell, J. V., Gomez-Lechon, M. J. 2007. Transcriptional regulation and expression of CYP3A4 in hepatocytes. *Curr Drug Metab* 8:185–194.

Masahiko, N., Honkakoski, P. 2000. Induction of drug metabolism by nuclear receptor CAR: Molecular mechanisms and implications for drug research. *Eur J Pharm Sci* 11:259–264.

Matoulkova, P., Pavek, P., Maly, J., Vlcek, J. 2014. Cytochrome P450 enzyme regulation by glucocorticoids and consequences in terms of drug interaction. *Expert Opin Drug Metab Toxicol* 10:425–435.

Matsubara, T., Li, F., Gonzalez, F. J. 2013. FXR signaling in the enterohepatic system. *Mol Cell Endocrinol* 368:17–29.

Matzke, G. R., Frye, R. F., Early, J. J., Straka, R. J., Carson, S. W. 2000. Evaluation of the influence of diabetes mellitus on antipyrine metabolism and CYP1A2 and CYP2D6 activity. *Pharmacotherapy* 20:182–190.

Maurer, A., Johne, A., Bauer, S. 1999. Interaction of St. John's wort extract with phenprocoumon. *Eur J Clin Pharmacol* 55:A22.

McCune, J. S., Lindley, C., Decker, J. L., Williamson, K. M., Meadowcroft, A. M., Graff, D., Sawyer, W. T., Blough, D. K., Pieper, J. A. 2001. Lack of gender differences and large intrasubject variability in cytochrome P450 activity measured by phenotyping with dextromethorphan. *J Clin Pharmacol* 41:723–731.

Meijerman, I., Beijnen, J. H., Schellens, J. H. 2006. Herb–drug interactions in oncology: Focus on mechanisms of induction. *Oncologist* 11:742–752.

Mendel, D. B., Crabtree, G. R. 1991. HNF-1, a member of a novel class of dimerizing homeodomain proteins. *J Biol Chem* 266:677–680.

Meng, Q., Liu, K. 2014. Pharmacokinetic interactions between herbal medicines and prescribed drugs: Focus on drug metabolic enzymes and transporters. *Curr Drug Metab* 15:791–807.

Meunier, V., Bourrie, M., Julian, B., Marti, E., Guillou, F., Berger, Y., Fabre, G. 2000. Expression and induction of CYP1A1/1A2, CYP2A6 and CYP3A4 in primary cultures of human hepatocytes: A 10-year follow-up. *Xenobiotica* 30:589–607.

Miyakawa, H., Kikazawa, E., Abe, K., Kikuchi, K., Fujikawa, H., Matsushita, M., Kawaguchi, N., Morizane, T., Ohya, K., Kako, M. 1999. Detection of anti-LKM-1(anti-CYP2D6) by an enzyme-linked immunosorbent assay in adult patients with chronic liver diseases. *Autoimmunity* 30:107–114.

Miyakawa, H., Kitazawa, E., Kikuchi, K., Fujikawa, H., Kawaguchi, N., Abe, K., Matsushita, M. et al. 2000. Immunoreactivity to various human cytochrome P450 proteins of sera from patients with autoimmune hepatitis, chronic hepatitis B, and chronic hepatitis C. *Autoimmunity* 33:23–32.

Miyazaki-Anzai, S., Levi, M., Kratzer, A., Ting, T. C., Lewis, L. B., Miyazaki, M. 2010. Farnesoid X receptor activation prevents the development of vascular calcification in ApoE-/- mice with chronic kidney disease. *Circ Res* 106:1807–1817.

Mizuno, D., Takahashi, Y., Hiroi, T., Imaoka, S., Kamataki, T., Funae, Y. 2003. A novel transcriptional element which regulates expression of the CYP2D4 gene by Oct-1 and YY-1 binding. *Biochim Biophys Acta* 1627:121–128.

Mizutani, T., Shinoda, M., Tanaka, Y., Kuno, T., Hattori, A., Usui, T., Kuno, N., Osaka, T. 2005. Autoantibodies against CYP2D6 and other drug-metabolizing enzymes in autoimmune hepatitis type 2. *Drug Metab Rev* 37:235–252.

Modica, S., Moschetta, A. 2006. Nuclear bile acid receptor FXR as pharmacological target: Are we there yet? *FEBS Lett* 580:5492–5499.

Modica, S., Gadaleta, R. M., Moschetta, A. 2010. Deciphering the nuclear bile acid receptor FXR paradigm. *Nucl Recept Signal* 8:e005.

More, V. R., Cheng, Q., Donepudi, A. C., Buckley, D. B., Lu, Z. J., Cherrington, N. J., Slitt, A. L. 2013. Alcohol cirrhosis alters nuclear receptor and drug transporter expression in human liver. *Drug Metab Dispos* 41:1148–1155.

Moschella, C., Jaber, B. L. 2001. Interaction between cyclosporine and *Hypericum perforatum* (St. John's wort) after organ transplantation. *Am J Kidney Dis* 38:1105–1107.

Mudaliar, S., Henry, R. R., Sanyal, A. J., Morrow, L., Marschall, H. U., Kipnes, M., Adorini, L. et al. 2013. Efficacy and safety of the farnesoid X receptor agonist obeticholic acid in patients with type 2 diabetes and nonalcoholic fatty liver disease. *Gastroenterology* 145:574–582.e571.

Mukhopadhyay, S., Jackson, P. K. 2011. The tubby family proteins. *Genome Biol* 12:225.

Muratori, L., Lenzi, M., Ma, Y., Cataleta, M., Mieli-Vergani, G., Vergani, D., Bianchi, F. B. 1995. Heterogeneity of liver/kidney microsomal antibody type 1 in autoimmune hepatitis and hepatitis C virus related liver disease. *Gut* 37:406–412.

Muratori, L., Cataleta, M., Muratori, P., Lenzi, M., Bianchi, F. B. 1998. Liver/kidney microsomal antibody type 1 and liver cytosol antibody type 1 concentrations in type 2 autoimmune hepatitis. *Gut* 42:721–726.

Nakai, K., Tanaka, H., Hanada, K., Ogata, H., Suzuki, F., Kumada, H., Miyajima, A. et al. 2008. Decreased expression of cytochromes P450 1A2, 2E1, and 3A4 and drug transporters Na+-taurocholate-cotransporting polypeptide, organic cation transporter 1, and organic anion-transporting peptide-C correlates with the progression of liver fibrosis in chronic hepatitis C patients. *Drug Metab Dispos* 36:1786–1793.

Nakata, K., Tanaka, Y., Nakano, T., Adachi, T., Tanaka, H., Kaminuma, T., Ishikawa, T. 2006. Nuclear receptor-mediated transcriptional regulation in phase I, II, and III xenobiotic metabolizing systems. *Drug Metab Pharmacokinet* 21:437–457.

Nebel, A., Schneider, B. J., Baker, R. K., Kroll, D. J. 1999. Potential metabolic interaction between St. John's wort and theophylline. *Ann Pharmacother* 33:502.

Negro, F. 2014. Hepatitis C in 2013: HCV causes systemic disorders that can be cured. *Nat Rev Gastroenterol Hepatol* 11:77–78.

Neuschwander-Tetri, B. A., Loomba, R., Sanyal, A. J., Lavine, J. E., Van Natta, M. L., Abdelmalek, M. F., Chalasani, N. et al. 2015. Farnesoid X nuclear receptor ligand obeticholic acid for non-cirrhotic, non-alcoholic steatohepatitis (FLINT): A multicentre, randomised, placebo-controlled trial. *Lancet* 385:956–965.

Niemi, M., Backman, J. T., Fromm, M. F., Neuvonen, P. J., Kivisto, K. T. 2003. Pharmacokinetic interactions with rifampicin: Clinical relevance. *Clin Pharmacokinet* 42:819–850.

Ning, M., Koh, K. H., Pan, X., Jeong, H. 2015. Hepatocyte nuclear factor (HNF) 4alpha transactivation of cytochrome P450 (Cyp) 2d40 promoter is enhanced during pregnancy in mice. *Biochem Pharmacol* 94:46–52.

Nishioka, M., Morshed, S. A., Kono, K., Himoto, T., Parveen, S., Arima, K., Watanabe, S., Manns, M. P. 1997. Frequency and significance of antibodies to P450IID6 protein in Japanese patients with chronic hepatitis C. *J Hepatol* 26:992–1000.

Ohtsuki, S., Schaefer, O., Kawakami, H., Inoue, T., Liehner, S., Saito, A., Ishiguro, N. et al. 2012. Simultaneous absolute protein quantification of transporters, cytochromes P450, and UDP-glucuronosyltransferases as a novel approach for the characterization of individual human liver: Comparison with mRNA levels and activities. *Drug Metab Dispos* 40:83–92.

Olesen, O. V., Linnet, K. 1999. Studies on the stereoselective metabolism of citalopram by human liver microsomes and cDNA-expressed cytochrome P450 enzymes. *Pharmacology* 59:298–309.

Olinga, P., Elferink, M. G., Draaisma, A. L., Merema, M. T., Castell, J. V., Perez, G., Groothuis, G. M. 2008. Coordinated induction of drug transporters and phase I and II metabolism in human liver slices. *Eur J Pharm Sci* 33:380–389.

Ourlin, J. C., Lasserre, F., Pineau, T., Fabre, J. M., Sa-Cunha, A., Maurel, P., Vilarem, M. J., Pascussi, J. M. 2003. The small heterodimer partner interacts with the pregnane X receptor and represses its transcriptional activity. *Mol Endocrinol* 17:1693–1703.

Pabona, J. M., Simmen, F. A., Nikiforov, M. A., Zhuang, D., Shankar, K., Velarde, M. C., Zelenko, Z., Giudice, L. C., Simmen, R. C. 2012. Kruppel-like factor 9 and progesterone receptor coregulation of decidualizing endometrial stromal cells: Implications for the pathogenesis of endometriosis. *J Clin Endocrinol Metab* 97:E376–E392.

Pageaux, G. P., le Bricquir, Y., Berthou, F., Bressot, N., Picot, M. C., Blanc, F., Michel, H., Larrey, D. 1998. Effects of interferon-alpha on cytochrome P-450 isoforms 1A2 and 3A activities in patients with chronic hepatitis C. *Eur J Gastroenterol Hepatol* 10:491–495.

Pal, D., Mitra, A. K. 2006. MDR- and CYP3A4-mediated drug-herbal interactions. *Life Sci* 78:2131–2145.

Pan, X., Jeong, H. 2015. Estrogen-induced cholestasis leads to repressed CYP2D6 expression in CYP2D6-humanized mice. *Mol Pharmacol* 88:106–112.

Pan, X., Lee, Y. K., Jeong, H. 2015. Farnesoid X receptor agonist represses cytochrome P450 2D6 expression by upregulating small heterodimer partner. *Drug Metab Dispos* 43:1002–1007.

Pasanen, M., Pelkonen, O. 1989. Human placental xenobiotic and steroid biotransformations catalyzed by cytochrome P450, epoxide hydrolase, and glutathione S-transferase activities and their relationships to maternal cigarette smoking. *Drug Metab Rev* 21:427–461.

Pascussi, J. M., Drocourt, L., Gerbal-Chaloin, S., Fabre, J. M., Maurel, P., Vilarem, M. J. 2001. Dual effect of dexamethasone on CYP3A4 gene expression in human hepatocytes. Sequential role of glucocorticoid receptor and pregnane X receptor. *Eur J Biochem* 268:6346–6358.

Patin, E., Kutalik, Z., Guergnon, J., Bibert, S., Nalpas, B., Jouanguy, E., Munteanu, M. et al. 2012. Genome-wide association study identifies variants associated with progression of liver fibrosis from HCV infection. *Gastroenterology* 143:1244–1252.e1–e12.

Pavek, P., Dvorak, Z. 2008. Xenobiotic-induced transcriptional regulation of xenobiotic metabolizing enzymes of the cytochrome P450 superfamily in human extrahepatic tissues. *Curr Drug Metab* 9:129–143.

Pavek, P., Smutny, T. 2014. Nuclear receptors in regulation of biotransformation enzymes and drug transporters in the placental barrier. *Drug Metab Rev* 46:19–32.

Pellicciari, R., Fiorucci, S., Camaioni, E., Clerici, C., Costantino, G., Maloney, P. R., Morelli, A., Parks, D. J., Willson, T. M. 2002. 6alpha-ethyl-chenodeoxycholic acid (6-ECDCA), a potent and selective FXR agonist endowed with anticholestatic activity. *J Med Chem* 45:3569–3572.

Peltoniemi, M. A., Saari, T. I., Hagelberg, N. M., Laine, K., Neuvonen, P. J., Olkkola, K. T. 2012. St John's wort greatly decreases the plasma concentrations of oral S-ketamine. *Fundam Clin Pharmacol* 26:743–750.

Petrescu, A. D., Hertz, R., Bar-Tana, J., Schroeder, F., Kier, A. B. 2002. Ligand specificity and conformational dependence of the hepatic nuclear factor-4α (HNF-4α). *J Biol Chem* 277:23988–23999.

Pichette, V., Leblond, F. A. 2003. Drug metabolism in chronic renal failure. *Curr Drug Metab* 4:91–103.

Piscitelli, S. C., Burstein, A. H., Chaitt, D., Alfaro, R. M., Falloon, J. 2000. Indinavir concentrations and St John's wort. *Lancet* 355:547–548.

Polednak, A. P. 2012. U.S. mortality from liver cirrhosis and alcoholic liver disease in 1999–2004: Regional and state variation in relation to per capita alcohol consumption. *Subst Use Misuse* 47:202–213.

Rader, D. J. 2007. Liver X receptor and farnesoid X receptor as therapeutic targets. *Am J Cardiol* 100:n15–n19.

Rae, J. M., Johnson, M. D., Lippman, M. E., Flockhart, D. A. 2001. Rifampin is a selective, pleiotropic inducer of drug metabolism genes in human hepatocytes: Studies with cDNA and oligonucleotide expression arrays. *J Pharmacol Exp Ther* 299:849–857.

Ramamoorthy, A., Liu, Y., Philips, S., Desta, Z., Lin, H., Goswami, C., Gaedigk, A., Li, L., Flockhart, D. A., Skaar, T. C. 2013. Regulation of microRNA expression by rifampin in human hepatocytes. *Drug Metab Dispos* 41:1763–1768.

Rana, R., Chen, Y., Ferguson, S. S., Kissling, G. E., Surapureddi, S., Goldstein, J. A. 2010. Hepatocyte nuclear factor 4{alpha} regulates rifampicin-mediated induction of CYP2C genes in primary cultures of human hepatocytes. *Drug Metab Dispos* 38:591–599.

Redlich, G., Zanger, U. M., Riedmaier, S., Bache, N., Giessing, A. B., Eisenacher, M., Stephan, C., Meyer, H. E., Jensen, O. N., Marcus, K. 2008. Distinction between human cytochrome P450 (CYP) isoforms and identification of new phosphorylation sites by mass spectrometry. *J Proteome Res* 7:4678–4688.

Renwick, A. B., Watts, P. S., Edwards, R. J., Barton, P. T., Guyonnet, I., Price, R. J., Tredger, J. M., Pelkonen, O., Boobis, A. R., Lake, B. G. 2000. Differential maintenance of cytochrome P450 enzymes in cultured precision-cut human liver slices. *Drug Metab Dispos* 28:1202–1209.

Ring, J. A., Ghabrial, H., Ching, M. S., Smallwood, R. A., Morgan, D. J. 1999. Fetal hepatic drug elimination. *Pharmacol Ther* 84:429–445.

Rizzo, G., Renga, B., Mencarelli, A., Pellicciari, R., Fiorucci, S. 2005. Role of FXR in regulating bile acid homeostasis and relevance for human diseases. *Curr Drug Targets Immune Endocr Metabol Disord* 5:289–303.

Roby, C. A., Anderson, G. D., Kantor, E., Dryer, D. A., Burstein, A. H. 2000. St John's Wort: Effect on CYP3A4 activity. *Clin Pharmacol Ther* 67:451–457.

Rochat, B., Amey, M., Gillet, M., Meyer, U. A., Baumann, P. 1997. Identification of three cytochrome P450 isozymes involved in N-demethylation of citalopram enantiomers in human liver microsomes. *Pharmacogenetics* 7:1–10.

Rodriguez-Antona, C., Jover, R., Gomez-Lechon, M. J., Castell, J. V. 2000. Quantitative RT-PCR measurement of human cytochrome P450s: Application to drug induction studies. *Arch Biochem Biophys* 376:109–116.

Rodriguez-Antona, C., Donato, M. T., Pareja, E., Gomez-Lechon, M. J., Castell, J. V. 2001. Cytochrome P-450 mRNA expression in human liver and its relationship with enzyme activity. *Arch Biochem Biophys* 393:308–315.

Rosales, R., Romero, M. R., Vaquero, J., Monte, M. J., Requena, P., Martinez-Augustin, O., Sanchez de Medina, F., Marin, J. J. 2013. FXR-dependent and -independent interaction of glucocorticoids with the regulatory pathways involved in the control of bile acid handling by the liver. *Biochem Pharmacol* 85:829–838.

Runge, D., Kohler, C., Kostrubsky, V. E., Jager, D., Lehmann, T., Runge, D. M., May, U. et al. 2000. Induction of cytochrome P450 (CYP)1A1, CYP1A2, and CYP3A4 but not of CYP2C9, CYP2C19, multidrug resistance (MDR-1) and multidrug resistance associated protein (MRP-1) by prototypical inducers in human hepatocytes. *Biochem Biophys Res Commun* 273:333–341.

Sahi, J., Grepper, S., Smith, C. 2010. Hepatocytes as a tool in drug metabolism, transport and safety evaluations in drug discovery. *Curr Drug Discov Technol* 7:188–198.

Saito, M., Yasui-Furukori, N., Nakagami, T., Furukori, H., Kaneko, S. 2005. Dose-dependent interaction of paroxetine with risperidone in schizophrenic patients. *J Clin Psychopharmacol* 25:527–532.

Sakamoto, A., Matsumaru, T., Ishiguro, N., Schaefer, O., Ohtsuki, S., Inoue, T., Kawakami, H., Terasaki, T. 2011. Reliability and robustness of simultaneous absolute quantification of drug transporters, cytochrome P450 enzymes, and Udp-glucuronosyltransferases in human liver tissue by multiplexed MRM/selected reaction monitoring mode tandem mass spectrometry with nano-liquid chromatography. *J Pharm Sci* 100:4037–4043.

Scandlyn, M. J., Stuart, E. C., Rosengren, R. J. 2008. Sex-specific differences in CYP450 isoforms in humans. *Expert Opin Drug Metab Toxicol* 4:413–424.

Seol, W., Choi, H. S., Moore, D. D. 1996. An orphan nuclear hormone receptor that lacks a DNA binding domain and heterodimerizes with other receptors. *Science* 272:1336–1339.

Seol, W., Chung, M., Moore, D. D. 1997. Novel receptor interaction and repression domains in the orphan receptor SHP. *Mol Cell Biol* 17:7126–7131.

Seol, W., Hanstein, B., Brown, M., Moore, D. D. 1998. Inhibition of estrogen receptor action by the orphan receptor SHP (short heterodimer partner). *Mol Endocrinol* 12:1551–1557.

Shah, R. R., Smith, R. L. 2015. Inflammation-induced phenoconversion of polymorphic drug metabolizing enzymes: Hypothesis with implications for personalized medicine. *Drug Metab Dispos* 43:400–410.

Shields, J. M., Yang, V. W. 1998. Identification of the DNA sequence that interacts with the gut-enriched Kruppel-like factor. *Nucleic Acids Res* 26:796–802.

Shimizu, Y., Takeuchi, T., Mita, S., Notsu, T., Mizuguchi, K., Kyo, S. 2010. Kruppel-like factor 4 mediates antiproliferative effects of progesterone with G(0)/G(1) arrest in human endometrial epithelial cells. *J Endocrinol Invest* 33:745–750.

Shiotani, A., Murao, T., Fujita, Y., Fujimura, Y., Sakakibara, T., Nishio, K., Haruma, K. 2013. Novel single nucleotide polymorphism markers for low dose aspirin-associated small bowel bleeding. *PLoS One* 8:e84244.

Shord, S. S., Shah, K., Lukose, A. 2009. Drug-botanical interactions: A review of the laboratory, animal, and human data for 8 common botanicals. *Integr Cancer Ther* 8:208–227.

Silveira, M. G., Lindor, K. D. 2014. Obeticholic acid and budesonide for the treatment of primary biliary cirrhosis. *Expert Opin Pharmacother* 15:365–372.

Simmen, F. A., Su, Y., Xiao, R., Zeng, Z., Simmen, R. C. 2008. The Kruppel-like factor 9 (KLF9) network in HEC-1-A endometrial carcinoma cells suggests the carcinogenic potential of dys-regulated KLF9 expression. *Reprod Biol Endocrinol* 6:41.

Simmen, R. C., Heard, M. E., Simmen, A. M., Montales, M. T., Marji, M., Scanlon, S., Pabona, J. M. 2015. The Kruppel-like factors in female reproductive system pathologies. *J Mol Endocrinol* 54:R89–R101.

Sinz, M., Wallace, G., Sahi, J. 2008. Current industrial practices in assessing CYP450 enzyme induction: Preclinical and clinical. *AAPS J* 10:391–400.

Smith, R. P., Eckalbar, W. L., Morrissey, K. M., Luizon, M. R., Hoffmann, T. J., Sun, X., Jones, S. L. et al. 2014. Genome-wide discovery of drug-dependent human liver regulatory elements. *PLoS Genet* 10:e1004648.

Smolinske, S. C. 1999. Dietary supplement-drug interactions. *J Am Med Womens Assoc* 54:191–192, 195.

Soars, M. G., McGinnity, D. F., Grime, K., Riley, R. J. 2007. The pivotal role of hepatocytes in drug discovery. *Chem Biol Interact* 168:2–15.

Sousa, M., Pozniak, A., Boffito, M. 2008. Pharmacokinetics and pharmacodynamics of drug interactions involving rifampicin, rifabutin and antimalarial drugs. *J Antimicrob Chemother* 62:872–878.

Spaggiari, D., Geiser, L., Daali, Y., Rudaz, S. 2014. A cocktail approach for assessing the in vitro activity of human cytochrome P450s: An overview of current methodologies. *J Pharm Biomed Anal* 101:221–237.

Stage, T. B., Pedersen, R. S., Damkier, P., Christensen, M. M., Feddersen, S., Larsen, J. T., Hojlund, K., Brosen, K. 2015. Intake of St John's wort improves the glucose tolerance in healthy subjects who ingest metformin compared with metformin alone. *Br J Clin Pharmacol* 79:298–306.

Steiner, E., Iselius, L., Alvan, G., Lindsten, J., Sjoqvist, F. 1985. A family study of genetic and environmental factors determining polymorphic hydroxylation of debrisoquin. *Clin Pharmacol Ther* 38:394–401.

Stevens, J. C., Marsh, S. A., Zaya, M. J., Regina, K. J., Divakaran, K., Le, M., Hines, R. N. 2008. Developmental changes in human liver CYP2D6 expression. *Drug Metab Dispos* 36:1587–1593.

Stout, S. M., Nielsen, J., Bleske, B. E., Shea, M., Brook, R., Kerber, K., Welage, L. S. 2010. The impact of paroxetine coadministration on stereospecific carvedilol pharmacokinetics. *J Cardiovasc Pharmacol Ther* 15:373–379.

Sugimoto, K., Ohmori, M., Tsuruoka, S., Nishiki, K., Kawaguchi, A., Harada, K., Arakawa, M. et al. 2001. Different effects of St John's Wort on the pharmacokinetics of simvastatin and pravastatin. *Clin Pharmacol Ther* 70:518–524.

Sugimura, T., Obermayer-Straub, P., Kayser, A., Braun, S., Loges, S., Alex, B., Luttig, B., Johnson, E. F., Manns, M. P., Strassburg, C. P. 2002. A major CYP2D6 autoepitope in autoimmune hepatitis type 2 and chronic hepatitis C is a three-dimensional structure homologous to other cytochrome P450 autoantigens. *Autoimmunity* 35:501–513.

Sun, Y., Li, H. 2013. Functional characterization of SAG/RBX2/ROC2/RNF7, an antioxidant protein and an E3 ubiquitin ligase. *Protein Cell* 4:103–116.

Sun, H., Frassetto, L., Benet, L. Z. 2006. Effects of renal failure on drug transport and metabolism. *Pharmacol Ther* 109:1–11.

Sy, S. K., Ciaccia, A., Li, W., Roberts, E. A., Okey, A., Kalow, W., Tang, B. K. 2002. Modeling of human hepatic CYP3A4 enzyme kinetics, protein, and mRNA indicates deviation from log-normal distribution in CYP3A4 gene expression. *Eur J Clin Pharmacol* 58:357–365.

Syme, M. R., Paxton, J. W., Keelan, J. A. 2004. Drug transfer and metabolism by the human placenta. *Clin Pharmacokinet* 43:487–514.

Tamminga, W. J., Wemer, J., Oosterhuis, B., Weiling, J., Wilffert, B., de Leij, L. F., de Zeeuw, R. A., Jonkman, J. H. 1999. CYP2D6 and CYP2C19 activity in a large population of Dutch healthy volunteers: Indications for oral contraceptive-related gender differences. *Eur J Clin Pharmacol* 55:177–184.

Tan, M., Gu, Q., He, H., Pamarthy, D., Semenza, G. L., Sun, Y. 2008. SAG/ROC2/RBX2 is a HIF-1 target gene that promotes HIF-1 alpha ubiquitination and degradation. *Oncogene* 27:1404–1411.

Tanaka, E., Kurata, N., Yasuhara, H. 2003. How useful is the "cocktail approach" for evaluating human hepatic drug metabolizing capacity using cytochrome P450 phenotyping probes in vivo? *J Clin Pharm Ther* 28:157–165.

Tanaka, Y., Nishida, N., Sugiyama, M., Kurosaki, M., Matsuura, K., Sakamoto, N., Nakagawa, M. et al. 2009. Genome-wide association of IL28B with response to pegylated interferon-alpha and ribavirin therapy for chronic hepatitis C. *Nat Genet* 41:1105–1109.

Tang, S. W., Helmeste, D. 2008. Paroxetine. *Expert Opin Pharmacother* 9:787–794.

Temesvari, M., Kobori, L., Paulik, J., Sarvary, E., Belic, A., Monostory, K. 2012. Estimation of drug-metabolizing capacity by cytochrome P450 genotyping and expression. *J Pharmacol Exp Ther* 341:294–305.

Teodoro, J. S., Rolo, A. P., Palmeira, C. M. 2011. Hepatic FXR: Key regulator of whole-body energy metabolism. *Trends Endocrinol Metab* 22:458–466.

ter Beek, J., Guskov, A., Slotboom, D. J. 2014. Structural diversity of ABC transporters. *J Gen Physiol* 143:419–435.

Tian, R., Koyabu, N., Morimoto, S., Shoyama, Y., Ohtani, H., Sawada, Y. 2005. Functional induction and de-induction of P-glycoprotein by St. John's wort and its ingredients in a human colon adenocarcinoma cell line. *Drug Metab Dispos* 33:547–554.

Tirona, R. G., Lee, W., Leake, B. F., Lan, L. B., Cline, C. B., Lamba, V., Parviz, F. et al. 2003. The orphan nuclear receptor HNF4alpha determines PXR- and CAR-mediated xenobiotic induction of CYP3A4. *Nat Med* 9:220–224.

Topletz, A. R., Le, H. N., Lee, N., Chapman, J. D., Kelly, E. J., Wang, J., Isoherranen, N. 2013. Hepatic Cyp2d and Cyp26a1 mRNAs and activities are increased during mouse pregnancy. *Drug Metab Dispos* 41:312–319.

Touchette, M. A., Slaughter, R. L. 1991. The effect of renal failure on hepatic drug clearance. *DICP* 25:1214–1224.

Tracy, T. S., Venkataramanan, R., Glover, D. D., Caritis, S. N., National Institute for Child Health and Human Development Network of Maternal-Fetal-Medicine Units. 2005. Temporal changes in drug metabolism (CYP1A2, CYP2D6 and CYP3A Activity) during pregnancy. *Am J Obstet Gynecol* 192:633–639.

Trautwein, C., Ramadori, G., Gerken, G., Meyer zum Buschenfelde, K. H., Manns, M. 1992. Regulation of cytochrome P450 2D by acute phase mediators in C3H/HeJ mice. *Biochem Biophys Res Commun* 182:617–623.

Treluyer, J. M., Jacqz-Aigrain, E., Alvarez, F., Cresteil, T. 1991. Expression of CYP2D6 in developing human liver. *Eur J Biochem* 202:583–588.

Tsujimoto, M., Sugimoto, S., Nagatomo, M., Furukubo, T., Izumi, S., Yamakawa, T., Minegaki, T., Nishiguchi, K. 2014. Possibility of decrease in CYP1A2 function in patients with end-stage renal disease. *Ther Apher Dial* 18:174–180.

Ulbricht, C., Chao, W., Costa, D., Rusie-Seamon, E., Weissner, W., Woods, J. 2008. Clinical evidence of herb–drug interactions: A systematic review by the natural standard research collaboration. *Curr Drug Metab* 9:1063–1120.

Urabe, Y., Ochi, H., Kato, N., Kumar, V., Takahashi, A., Muroyama, R., Hosono, N. et al. 2013. A genome-wide association study of HCV-induced liver cirrhosis in the Japanese population identifies novel susceptibility loci at the MHC region. *J Hepatol* 58:875–882.

van der Meer, J. H., van der Poll, T., van 't Veer, C. 2014. TAM receptors, Gas6, and protein S: Roles in inflammation and hemostasis. *Blood* 123:2460–2469.

Vaquero, J., Monte, M. J., Dominguez, M., Muntane, J., Marin, J. J. 2013. Differential activation of the human farnesoid X receptor depends on the pattern of expressed isoforms and the bile acid pool composition. *Biochem Pharmacol* 86:926–939.

Varner, T. R., Bookstaver, P. B., Rudisill, C. N., Albrecht, H. 2011. Role of rifampin-based combination therapy for severe community-acquired *Legionella pneumophila* pneumonia. *Ann Pharmacother* 45:967–976.

Venkatakrishnan, K., Obach, R. S. 2005. In vitro-in vivo extrapolation of CYP2D6 inactivation by paroxetine: Prediction of non-stationary pharmacokinetics and drug interaction magnitude. *Drug Metab Dispos* 33:845–852.

Villeneuve, J. P., Pichette, V. 2004. Cytochrome P450 and liver diseases. *Curr Drug Metab* 5:273–282.

Vincent-Viry, M., Muller, J., Fournier, B., Galteau, M. M., Siest, G. 1991. Relation between debrisoquine oxidation phenotype and morphological, biological, and pathological variables in a large population. *Clin Chem* 37:327–332.

Vincent-Viry, M., Fournier, B., Galteau, M. M. 2000. Short communication. The effects of drinking and smoking on the CYP2D6 metabolic capacity. *Drug Metab Dispos* 28:617–619.

von Moltke, L. L., Greenblatt, D. J., Giancarlo, G. M., Granda, B. W., Harmatz, J. S., Shader, R. I. 2001. Escitalopram (*S*-citalopram) and its metabolites *in vitro*: Cytochromes mediating biotransformation, inhibitory effects, and comparison to *R*-citalopram. *Drug Metab Dispos* 29:1102–1109.

Wadelius, M., Darj, E., Frenne, G., Rane, A. 1997. Induction of CYP2D6 in pregnancy. *Clin Pharmacol Ther* 62:400–407.

Walesky, C., Apte, U. 2015. Role of hepatocyte nuclear factor 4alpha (HNF4alpha) in cell proliferation and cancer. *Gene Expr* 16:101–108.

Walle, T., Walle, U. K., Cowart, T. D., Conradi, E. C., Gaffney, T. E. 1987. Selective induction of propranolol metabolism by smoking: Additional effects on renal clearance of metabolites. *J Pharmacol Exp Ther* 241:928–933.

Walsky, R. L., Boldt, S. E. 2008. In vitro cytochrome P450 inhibition and induction. *Curr Drug Metab* 9:928–939.

Wang, Z., Gorski, J. C., Hamman, M. A., Huang, S. M., Lesko, L. J., Hall, S. D. 2001. The effects of St John's wort (*Hypericum perforatum*) on human cytochrome P450 activity. *Clin Pharmacol Ther* 70:317–326.

Wang, L., Lee, Y. K., Bundman, D., Han, Y., Thevananther, S., Kim, C. S., Chua, S. S. et al. 2002. Redundant pathways for negative feedback regulation of bile acid production. *Dev Cell* 2:721–731.

Wang, Y. D., Chen, W. D., Huang, W. 2008a. FXR, a target for different diseases. *Histol Histopathol* 23:621–627.

Wang, Y. D., Chen, W. D., Moore, D. D., Huang, W. 2008b. FXR: A metabolic regulator and cell protector. *Cell Res* 18:1087–1095.

Wang, D., Jiang, Z., Shen, Z., Wang, H., Wang, B., Shou, W., Zheng, H., Chu, X., Shi, J., Huang, W. 2011. Functional evaluation of genetic and environmental regulators of p450 mRNA levels. *PLoS One* 6:e24900.

Waxman, D. J. 1999. P450 gene induction by structurally diverse xenochemicals: Central role of nuclear receptors CAR, PXR, and PPAR. *Arch Biochem Biophys* 369:11–23.

Waxman, D. J., Holloway, M. G. 2009. Sex differences in the expression of hepatic drug metabolizing enzymes. *Mol Pharmacol* 76:215–228.

Weleber, R. G., Francis, P. J., Trzupek, K. M., Beattie, C. 1993. Leber congenital amaurosis, in *GeneReviews(R)* (Pagon, R. A., Adam, M. P., Ardinger, H. H., Wallace, S. E., Amemiya, A., Bean, L. J. H., Bird, T. D. et al. eds.), Seattle, WA: University of Washington.

Wenk, M., Todesco, L., Krahenbuhl, S. 2004. Effect of *St John's wort* on the activities of CYP1A2, CYP3A4, CYP2D6, *N*-acetyltransferase 2, and xanthine oxidase in healthy males and females. *Br J Clin Pharmacol* 57:495–499.

Wenzel-Seifert, K., Wittmann, M., Haen, E. 2011. QTc prolongation by psychotropic drugs and the risk of Torsade de Pointes. *Dtsch Arztebl Int* 108:687–693.

Whiteford, H. A., Degenhardt, L., Rehm, J., Baxter, A. J., Ferrari, A. J., Erskine, H. E., Charlson, F. J. et al. 2013. Global burden of disease attributable to mental and substance use disorders: Findings from the Global Burden of Disease Study 2010. *Lancet* 382:1575–1586.

Williamson, E. M. 2003. Drug interactions between herbal and prescription medicines. *Drug Saf* 26:1075–1092.

Williamson, E. M. 2005. Interactions between herbal and conventional medicines. *Expert Opin Drug Saf* 4:355–378.

Williamson, B., Dooley, K. E., Zhang, Y., Back, D. J., Owen, A. 2013. Induction of influx and efflux transporters and cytochrome P450 3A4 in primary human hepatocytes by rifampin, rifabutin, and rifapentine. *Antimicrob Agents Chemother* 57: 6366–6369.

Wortham, M., Czerwinski, M., He, L., Parkinson, A., Wan, Y. J. 2007. Expression of constitutive androstane receptor, hepatic nuclear factor 4 alpha, and P450 oxidoreductase genes determines interindividual variability in basal expression and activity of a broad scope of xenobiotic metabolism genes in the human liver. *Drug Metab Dispos* 35:1700–1710.

Xiao, W. J., Ma, T., Ge, C., Xia, W. J., Mao, Y., Sun, R. B., Yu, X. Y., Aa, J. Y., Wang, G. J. 2015. Modulation of the pentose phosphate pathway alters phase I metabolism of testosterone and dextromethorphan in HepG2 cells. *Acta Pharmacol Sin* 36:259–267.

Xu, C., Li, C. Y., Kong, A. N. 2005. Induction of phase I, II and III drug metabolism/transport by xenobiotics. *Arch Pharm Res* 28:249–268.

Yamada, H., Ishii, Y., Yamamoto, M., Oguri, K. 2006. Induction of the hepatic cytochrome P450 2B subfamily by xenobiotics: Research history, evolutionary aspect, relation to tumorigenesis, and mechanism. *Curr Drug Metab* 7:397–409.

Yamamoto, A. M., Mura, C., De Lemos-Chiarandini, C., Krishnamoorthy, R., Alvarez, F. 1993. Cytochrome P450IID6 recognized by LKM1 antibody is not exposed on the surface of hepatocytes. *Clin Exp Immunol* 92:381–390.

Yang, X. X., Hu, Z. P., Duan, W., Zhu, Y. Z., Zhou, S. F. 2006. Drug-herb interactions: Eliminating toxicity with hard drug design. *Curr Pharm Des* 12:4649–4664.

Yang, J., Liao, M., Shou, M., Jamei, M., Yeo, K. R., Tucker, G. T., Rostami-Hodjegan, A. 2008. Cytochrome p450 turnover: Regulation of synthesis and degradation, methods for determining rates, and implications for the prediction of drug interactions. *Curr Drug Metab* 9:384–394.

Yang, A. K., He, S. M., Liu, L., Liu, J. P., Wei, M. Q., Zhou, S. F. 2010. Herbal interactions with anticancer drugs: Mechanistic and clinical considerations. *Curr Med Chem* 17:1635–1678.

Yang, X., Zhang, B., Molony, C., Chudin, E., Hao, K., Zhu, J., Gaedigk, A. et al. 2010. Systematic genetic and genomic analysis of cytochrome P450 enzyme activities in human liver. *Genome Res* 20:1020–1036.

Yang, C. S., Kim, J. J., Kim, T. S., Lee, P. Y., Kim, S. Y., Lee, H. M., Shin, D. M. et al. 2015. Small heterodimer partner interacts with NLRP3 and negatively regulates activation of the NLRP3 inflammasome. *Nat Commun* 6:6115.

Yeung, C. K., Shen, D. D., Thummel, K. E., Himmelfarb, J. 2014. Effects of chronic kidney disease and uremia on hepatic drug metabolism and transport. *Kidney Int* 85:522–528.

Yuan, R., Venitz, J. 2000. Effect of chronic renal failure on the disposition of highly hepatically metabolized drugs. *Int J Clin Pharmacol Ther* 38:245–253.

Yue, Q. Y., Bergquist, C., Gerden, B. 2000. Safety of St John's wort (*Hypericum perforatum*). *Lancet* 355:548–549.

Zachou, K., Rigopoulou, E., Dalekos, G. N. 2004. Autoantibodies and autoantigens in autoimmune hepatitis: Important tools in clinical practice and to study pathogenesis of the disease. *J Autoimmune Dis* 1:2.

Zadoyan, G., Rokitta, D., Klement, S., Dienel, A., Hoerr, R., Gramatte, T., Fuhr, U. 2012. Effect of *Ginkgo biloba* special extract EGb 761(R) on human cytochrome P450 activity: A cocktail interaction study in healthy volunteers. *Eur J Clin Pharmacol* 68:553–560.

Zanger, U. M., Schwab, M. 2013. Cytochrome P450 enzymes in drug metabolism: Regulation of gene expression, enzyme activities, and impact of genetic variation. *Pharmacol Ther* 138:103–141.

Zanger, U. M., Hauri, H. P., Loeper, J., Homberg, J. C., Meyer, U. A. 1988. Antibodies against human cytochrome P-450db1 in autoimmune hepatitis type II. *Proc Natl Acad Sci U S A* 85:8256–8260.

Zhang, W., Shields, J. M., Sogawa, K., Fujii-Kuriyama, Y., Yang, V. W. 1998. The gut-enriched Kruppel-like factor suppresses the activity of the CYP1A1 promoter in an Sp1-dependent fashion. *J Biol Chem* 273:17917–17925.

Zhang, X. L., Zhang, D., Michel, F. J., Blum, J. L., Simmen, F. A., Simmen, R. C. 2003. Selective interactions of Kruppel-like factor 9/basic transcription element-binding protein with progesterone receptor isoforms A and B determine transcriptional activity of progesterone-responsive genes in endometrial epithelial cells. *J Biol Chem* 278:21474–21482.

Zhang, S., Pan, X., Jeong, H. 2015. GW4064, an agonist of farnesoid X receptor, represses CYP3A4 expression in human hepatocytes by inducing small heterodimer partner expression. *Drug Metab Dispos* 43:743–748.

Zhao, Y., Jensen, O. N. 2009. Modification-specific proteomics: Strategies for characterization of post-translational modifications using enrichment techniques. *Proteomics* 9:4632–4641.

Zhi, X., Zhou, X. E., He, Y., Zechner, C., Suino-Powell, K. M., Kliewer, S. A., Melcher, K., Mangelsdorf, D. J., Xu, H. E. 2014. Structural insights into gene repression by the orphan nuclear receptor SHP. *Proc Natl Acad Sci U S A* 111:839–844.

Zhou, S. F., Lai, X. 2008. An update on clinical drug interactions with the herbal antidepressant St. John's wort. *Curr Drug Metab* 9:394–409.

Zhou, Z. W., Zhou, S. F. 2009. Application of mechanism-based CYP inhibition for predicting drug–drug interactions. *Expert Opin Drug Metab Toxicol* 5:579–605.

Zhou, S., Gao, Y., Jiang, W., Huang, M., Xu, A., Paxton, J. W. 2003. Interactions of herbs with cytochrome P450. *Drug Metab Rev* 35:35–98.

Zhou, S., Chan, E., Pan, S. Q., Huang, M., Lee, E. J. 2004a. Pharmacokinetic interactions of drugs with St John's wort. *J Psychopharmacol* 18:262–276.

Zhou, S., Koh, H. L., Gao, Y., Gong, Z. Y., Lee, E. J. 2004b. Herbal bioactivation: The good, the bad and the ugly. *Life Sci* 74:935–968.

Zhou, S., Chan, E., Duan, W., Huang, M., Chen, Y. Z. 2005. Drug bioactivation, covalent binding to target proteins and toxicity relevance. *Drug Metab Rev* 37:41–213.

Zhou, S. F., Xue, C. C., Yu, X. Q., Li, C., Wang, G. 2007a. Clinically important drug interactions potentially involving mechanism-based inhibition of cytochrome P450 3A4 and the role of therapeutic drug monitoring. *Ther Drug Monit* 29:687–710.

Zhou, S. F., Xue, C. C., Yu, X. Q., Wang, G. 2007b. Metabolic activation of herbal and dietary constituents and its clinical and toxicological implications: An update. *Curr Drug Metab* 8:526–553.

Zhou, S. F., Zhou, Z. W., Li, C. G., Chen, X., Yu, X., Xue, C. C., Herington, A. 2007c. Identification of drugs that interact with herbs in drug development. *Drug Discov Today* 12:664–673.

Zhou, S. F., Liu, J. P., Lai, X. S. 2009. Substrate specificity, inhibitors and regulation of human cytochrome P450 2D6 and implications in drug development. *Curr Med Chem* 16:2661–2805.

6 Structure and Function of Human CYP2D6

6.1 INTRODUCTION

Before the determination of the structure of rabbit CYP2C5, molecular and mechanistic information on how human CYPs catalyze oxidation of substrates and bind to inhibitors is mainly obtained via modeling and site-directed mutagenesis studies. This chapter will discuss the molecular determinants for the interaction between ligands and CYP2D6.

6.2 PHARMACOPHORE MODELS AND STRUCTURAL REQUIREMENTS OF CYP2D6 LIGANDS

Before the publication of the CYP2D6 crystal structure in 2006 (Rowland et al. 2006), a number of pharmacophore modeling studies had been conducted to explore the structural characteristics of CYP2D6 substrates and inhibitors and the possible ligand–enzyme interaction mode. The first substrate models for CYP2D6 are constructed by manual alignments based on a set of substrates containing a basic nitrogen atom at either 5 Å (Wolff et al. 1985) or 7 Å (Meyer et al. 1986) from the site of oxidation and planar aromatic rings close to the site of oxidation that are fitted to be coplanar (Meyer et al. 1986; Wolff et al. 1985). The distance between the basic nitrogen atom and the site of oxidation in debrisoquine is approximately 5 Å, whereas in the rigid substrate dextromethorphan, this distance is approximately 7 Å. In the 5-Å model, there are no substrates superimposed onto each other (Wolff et al. 1985). As such, these initial models failed to rationalize the binding of other types of substrates.

Islam et al. (1991) have described an extended model with incorporation of the heme moiety from the known structure of bacterial P450cam (CYP101), which indicates a distance between the basic nitrogen atom in substrate molecules and the site of oxidation of between 5 and 7 Å. The model includes the iron–oxygen complex involved in the oxidation, and a set of 15 compounds are well fitted onto debrisoquine used as a template. These include debrisoquine, sparteine, guanoxan, perhexiline, bufuralol, propranolol, desipramine, amitriptyline, nortriptyline, phenformin, methoxyamphetamine, codeine, and dextromethorphan. 2-Amino-1-methyl-6-phenylimidazo[4,5-b]pyridine (PhIP) does not fit the model and is therefore unlikely to be activated by this enzyme (Islam et al. 1991). A potent procarcinogen in tobacco smoke, NNK, fits the model but could not be modeled to form a favorable nitrogen–anion interaction and thus is not considered to be substrate of CYP2D6. Both PhIP and NNK are not metabolized by CYP2D6. However, this model fails to accommodate tamoxifen and thus Islam et al. (1991) proposed that tamoxifen

is unlikely to be metabolized by CYP2D6. In contrast, subsequent studies have indicated that tamoxifen is a good substrate of CYP2D6 (Dehal and Kupfer 1997).

In the more complete small-molecule model for CYP2D6 substrates described by Koymans et al. (1992), debrisoquine and dextromethorphan are used as templates for the 5- and 7-Å compounds, respectively. This model suggests that a hypothetical carboxylate group within CYP2D6 is responsible for a well-defined distance of either 5 or 7 Å between the basic nitrogen atom and the site of oxidation within the substrate. The model is constructed based on 16 substrates, accounting for 23 metabolic reactions with their sites of oxidation and basic nitrogen atoms fitting onto the site of oxidation of the templates. The model is further applied to predict the metabolism of four compounds (alfentanil, astemizole, risperidone, and nebivolol) involving 14 possible CYP-dependent metabolites (Koymans et al. 1992). Four out of the 14 metabolic routes appear to show the main characteristic of CYP2D6 substrates, comprising the O-demethylation of astemizole, the aromatic hydroxylation of nebivolol, and the alicyclic oxidation of risperidone at the 7- and 9-positions. The remaining 10 metabolic pathways are predicted not to be mediated by CYP2D6. In vivo and in vitro metabolism studies of these substrates indicated that 13 of 14 predictions (except for the O-demethylation of astemizole) are correct (Koymans et al. 1992).

On the basis of the substrate model developed by Koymans et al. (1992), Grace et al. (1994) could not fit deprenyl (selegiline), a drug used in the treatment of Parkinson's disease, onto debrisoquine and dextromethorphan. However, this drug contains a basic nitrogen and undergoes CYP2D6-catalyzed N-demethylation and N-depropargylation of the sole basic nitrogen in the molecule, resulting in nordeprenyl (major) and methamphetamine, respectively, with a K_m of 56–97 μM (Grace et al. 1994). Deprenyl is also an inhibitor of CYP2D6-catalyzed bufuralol 1′-hydroxylation with a K_i of 24 μM (Grace et al. 1994). Deprenyl is also mainly metabolized by CYP2B6 and 2C19 (Hidestrand et al. 2001).

An inhibitor-based model for CYP2D6 is developed by Strobl et al. (1993) by fitting six potent competitive inhibitors of CYP2D6 onto each other. Ajmalicine, the most potent inhibitor for CYP2D6 with a K_i of 3.3 nM, is selected as a starting template because of its rigid structure. Other potent inhibitors used for model construction include chinidin, chlorpromazine, trifluperidol, prodipin, and lobelin. The basic nitrogen atoms of all inhibitors examined are superimposed and the aromatic planes of these inhibitors are fitted to be coplanar. The derived preliminary pharmacophore model is characteristic of a tertiary nitrogen atom and a flat hydrophobic region, the plane of which is almost perpendicular to the N–H axis and maximally extended up to a distance of 7.5 Å from the nitrogen atom (Strobl et al. 1993). The pharmacophore model also contains region B, in which additional functional groups with lone pairs enhanced inhibitory potency, and region C, in which hydrophobic groups are allowed but do not increase binding affinity and inhibitory effects (Strobl et al. 1993). It appears that the overall criteria derived for this inhibitor-based model are similar to those for the substrate-based models of CYP2D6.

De Groot et al. (1996) have derived a homology model of CYP2D6 and proposed that the site of oxidation above the heme moiety occupies one of two possible sites above pyrrole ring B of the heme moiety. In a further refined model, the authors have pinpointed more accurate positions of the heme moiety and helix I containing Asp301, thereby incorporating some steric restrictions and orientational preferences into their model (de Groot et al. 1997a). In this refined small-molecule model, the Asp301 residue is coupled to the basic nitrogen atoms, thus enhancing the model with the direction of the hydrogen bond between Asp301 in the protein and the protonated basic nitrogen atom. The site of oxidation for substrates is fitted onto the defined oxidation site above pyrrole ring B of the heme moiety, while the C_α and C_β atoms of the attached Asp moiety are fitted onto the C_α and C_β atoms of Asp301, respectively (de Groot et al. 1997a). A variety of eight substrates (e.g., dextromethorphan, debrisoquine, methylenedioxymethamphetamine [MDMA], metoprolol, and codeine) comprising 17 metabolic pathways fitted into the original substrate model for CYP2D6 (de Groot et al. 1995, 1997b; Koymans et al. 1992) are well fitted into the refined substrate model, indicating that the refined substrate model for CYP2D6 accommodates the same variety in molecular structures as the original substrate model. This refined model has been used to design a novel and selective CYP2D6 substrate, 7-methoxy-4-(aminomethyl)-coumarin (MAMC), which has been widely used as an in vitro probe and is suitable for high-throughput screening (Onderwater et al. 1999).

A set of 3D/4D quantitative structure–activity relationship (QSAR) pharmacophore models has also been derived for competitive inhibitors of CYP2D6 by Ekins et al. (1999). The first model for 20 inhibitors of CYP2D6-mediated bufuralol 1'-hydroxylation produced a positive correlation between observed and predicted K_i values, while a second model using 31 literature-derived K_i values provided a better correlation between observed and predicted K_i values with an r value of 0.91. Both pharmacophores are capable of predicting K_i values for 9 to 10 of 15 CYP2D6 inhibitors within 1 log residual (Ekins et al. 1999).

De Groot et al. (1999a) developed a combined pharmacophore using 40 substrates of CYP2D6 undergoing O-dealkylation or hydroxylation and a homology model for CYP2D6 on the basis of bacterial CYP101 (P450cam), CYP102 (P450BM3), and CYP108 (P450terp) crystal structures. The derived model incorporated steric, electronic, and chemical stability properties. The 40 compounds could be accommodated in a pharmacophore model using one site of

oxidation and three different positions for the basic nitrogen atoms: one at 5 Å, a second at 7 Å, and a third at 10 Å from the site of oxidation (de Groot et al. 1999a). Most of the compounds are 7-Å substrates, with only three 5-Å substrates and six 10-Å substrates. The basic nitrogen atoms of the 5- and 7-Å substrates could interact with Asp301, while the basic nitrogen atoms of the 10-Å substrates interacted with Glu216. The initial pharmacophore and protein models used to construct the combined model are developed independently and showed a high level of complementarity (de Groot et al. 1999a). When the pharmacophore is inserted into the homology model, the site of oxidation is positioned 3–3.5 Å above the iron of the heme with the plane of the heme largely perpendicular to the planar region of the pharmacophore model. The combined model is in agreement with experimental results concerning the substrates used to derive the model, with site-directed mutagenesis data available for CYP2D6.

De Groot et al. (1999b) extended the combined pharmacophore and homology model by adding a pharmacophore for N-dealkylation reactions catalyzed by CYP2D6 using 14 substrates and validated the extended model using a set of seven test compounds. This model consisted of a set of two pharmacophores (one for O-dealkylation and oxidation reactions and a second one for N-dealkylation reactions catalyzed by CYP2D6) embedded in a homology model based on bacterial CYP crystal structures (de Groot et al. 1999a). The combined model contained 51 substrates involving 72 metabolic pathways, mostly N-dealkylation. This model is also employed to predict the metabolism of seven test compounds including betaxolol, fluoxetine, loratidine, 1-methyl-4-phenyl-1,2,3,6-tetrahydropyridine (MPTP), procainamide, ritonavir, and sumatriptan. The combined model correctly predicted six of the eight observed metabolites. However, the combined model failed to predict the highly unusual metabolism of procainamide for its N-hydroxylation and ritonavir, which is marked as a nonsubstrate as it contains no basic nitrogen atom but still can be metabolized by CYP2D6 (de Groot et al. 1999b; Kumar et al. 1996).

Several modeling studies have pointed to a possible role for a second carboxylate group, identified as Glu216 (de Groot et al. 1999b; Lewis et al. 1997; Venhorst et al. 2003). This residue may provide an explanation for the metabolism of relatively larger substrates (e.g., minaprine) with a basic nitrogen atom $\geq$10 Å from the site of oxidation. Basic substrates are metabolized by all four known P450s that contain a residue equivalent to both Glu216 and Asp301: CYP2D6, CYP2D14 (bovine), CYP2D4 (rat), and CYP2J1 (rabbit). This indicates the important role for both Asp301 and Glu216 in the metabolism of these substrates.

Haji-Momenian et al. (2003) described comparative molecular field analysis (CoMFA) 3D QSAR models for CYP2D6 substrates. The training set consists of 24 substrates with known K_m values and two CoMFA models are constructed, with one model with a distance constraint and another without. The model with the distance parameter is approximately equal to the CoMFA without a distance parameter. The models revealed two areas where increased positive charge on the substrate should decrease the K_m, while one area is near the site of metabolism and a second major area corresponded to 10-Å substrates. Between these two regions is an area where more negative charge on substrates should decrease K_m. The developed CoMFA model is validated by predicting the K_m values of 15 diverse CYP2D6 substrates not included in the training set resulting, giving rise to a moderate predictive ability ($R^2 = 0.62$) (Haji-Momenian et al. 2003). However, there is no correlation between $\log P$ or pK_a and K_m for the training set. This model could be employed to predict potential sites of metabolism for any amine-containing compound that has a phenyl group 5, 7, or 10 Å away.

These pharmacophore studies have demonstrated that typical CYP2D6 substrates are usually lipophilic bases with a planar hydrophobic aromatic ring and a nitrogen atom that can be protonated at physiological pH. These compounds usually have a negative molecular electrostatic potential above the planar part of the molecule. These features are typical of a large number of drugs acting on the central nervous and cardiovascular systems. The nitrogen atom is considered to be essential for electrostatic interactions with the carboxylate group of Asp301 or Glu216, two candidate residues in the active site of CYP2D6. These pharmacophore modeling studies suggest that substrate binding is generally followed by oxidation 5 to 7 Å from the proposed nitrogen–Asp301/Glu216 interaction. Lipophilicity and amine basicity are thus considered to be important determinants of substrate/inhibitor binding to CYP2D6. LogP has been correlated to the inhibition constant (K_i or log K_i) for a series of β-blockers for CYP2D6 (Ferrari et al. 1991) and there is an inverse relationship between logP and inhibitory potency of quinidine metabolites for CYP2D6 (Ching et al. 1995). Quinidine and dihydroquinidine are potent CYP2D6 inhibitors with K_i values of 0.027 and 0.013 μM, respectively, while 3-hydroxyquinidine, O-desmethylquinidine, and quinidine N-oxide inhibit this enzyme with higher K_i values (0.43 to 2.3 μM) (Ching et al. 1995). The improved pharmacophore models have been used to rationalize and predict many CYP2D6-catalyzed reactions.

Overall, a number of pharmacophore models have been developed for CYP2D6 substrates and inhibitors. These models have provided initial evidence for structural requirements of ligands that can bind to the active site of CYP2D6. However, these models have almost always been used retrospectively and have limited value in the prediction of reactions a priori.

6.3 HOMOLOGY MODELING STUDIES OF HUMAN CYP2D6

Homology modeling has been frequently used to develop structures of CYPs for which sequence information is available, but x-ray structures are lacking. Before the availability of crystal structures of mammalian CYPs, models of human CYPs are based on the structures of more distantly related bacterial CYPs including P450cam (CYP101), P450BM3 (CYP102), P450eryF (CYP107A), and P450terp (CYP108) that share less than 25% sequence identity with human CYP2D6.

Among bacterial CYPs, P450BM3 is considered to provide the most useful structural information for homology studies of eukaryotic P450s, since this well-characterized and crystallized bacterial enzyme also belongs to the so-called class II P450s (Ravichandran et al. 1993) to which many eukaryotic P450s belong. Class II P450s are bound to the endoplasmic reticulum and interact directly with a cytochrome P450 reductase, containing flavin adenine dinucleotide (FAD) and flavin mononucleotide (FMN), while class I P450s are found in the mitochondrial membranes of eukaryotes and in most bacteria and require a FAD-containing reductase and an iron–sulfur protein (putidaredoxin) (Ravichandran et al. 1993).

6.3.1 HOMOLOGY MODELS DERIVED FROM BACTERIAL CYPS

Many early models for the active site of CYP2D6 suggested the involvement of a negatively charged carboxylate group in the enzyme forming a salt bridge with the basic nitrogen atom of the substrate (de Groot et al. 1996, 1999a,b; Islam et al. 1991; Koymans et al. 1992; Lewis 1999; Lewis et al. 1997; Meyer et al. 1986; Modi et al. 1996, 1997; Venhorst et al. 2003). A number of structural models have pointed to Asp301 being the important residue in the I helix responsible for substrate recognition (Ellis et al. 1995, 1996; Islam et al. 1991; Koymans et al. 1992; Lewis 1995; Modi et al. 1996). These homology models and structure-based alignments have located Asp301 in the central region of helix I of CYP2D6, which maps to one of the SRS4 identified by Gotoh (1992) as being important in substrate binding in the CYP2 family. The central region of the I helix is one of the most conserved areas of the CYP core, which is located close to the heme moiety and runs across the distal face of the heme, completely or partially covering pyrrole ring B (Hasemann et al. 1995; Mestres 2005). In this context, it may be significant that the central region of helix I of CYP2D6, in which Asp301 is located, comprises a sequence of nonpolar residues (Leu295–Val308) interspersed at regular intervals with polar residues (Arg296, Asp301, and Ser304). Modeling indicates that these polar residues all point in the direction of the proposed active site and potentially form one of its boundaries (Wolff et al. 1985). Therefore, the introduction of a nonpolar residue (e.g., Ala) into this polar region may drastically perturb the local environment adjacent to the active site and directly or indirectly influence heme incorporation. By contrast, this may not arise when Gly is the substituent owing to the adaptable nature of this residue resulting in minimal conformational change of the active site.

Koymans et al. (1993) have developed a homology model of CYP2D6 on the basis of the structure of bacterial CYP101. Possible active-site residues are identified by docking manually debrisoquine and dextromethorphan into the proposed active site. Both substrates could be positioned into a planar pocket near the heme region formed by Val370, Pro371, Leu372, Trp316, and part of the oxygen binding site of CYP2D6. Furthermore, the carboxylate group of either Asp100 or Asp301 is identified as a possible candidate for the proposed interaction with basic nitrogen atoms of the substrates (Koymans et al. 1993). This acidic residue forms part of a region between helices B and B' of CYP2D6. In the camphor-bound CYP101 complex, the corresponding region is involved in substrate binding (Poulos et al. 1985). However, this residue is located in the peripheral region and may not be involved in substrate binding as shown in the homology model of CYP2D6. A site-directed mutagenesis study by Islam et al. (1991) has confirmed that the substitution of the residue Asp100 with neutral amino acids such as Asn or Ala did not change the catalytic activity of the enzyme.

de Groot et al. (1996) derived a homology model for CYP2D6 on the basis of the structures of bacterial CYP101, CYP102, and CYP108 and incorporating a wide variety of site-directed mutagenesis data concerning the CYP2 family members. The final model consisted of four segments: (a) the B, B', and C helices and the β1-sheet (Gly66–Lys146); (b) the F and G helices (Leu205–Asp263); (c) the I, J, J', K, and L helices, the β2-, β3-, β4-, and β5-sheets, and the heme binding domain (Pro286–Arg497); and (d) the heme moiety (de Groot et al. 1996). Three classical CYP2D6 substrates including debrisoquine, dextromethorphan, and GBR-12,909 (vanoxcrine, a dopamine reuptake inhibitor) and one potent inhibitor, ajmalicine, are consecutively docked into the active site of the protein model. Amino acids responsible for binding and orientation of the various CYP2D6 substrates and inhibitors are identified: Pro102 and Gln108 (strand leading to the B' helix, putative substrate recognition site 1 [SRS1]); Arg115, Ser116, Gln117, Leu121, and Ala122 (strand running from the B' helix, SRS1); Leu213 (F helix, SRS2); Asp301, Ser304, Ala305, and Thr309 (I helix, SRS4); Val370 (K helix, SRS5); Pro371 (β3-sheet, SRS5); and Leu484 (β5-sheet, SRS6) (de Groot et al. 1996).

The basic nitrogen atoms of the compounds are oriented within hydrogen-bonding distance of Asp301 (de Groot et al. 1996), which is demonstrated to be important for the catalytic activity of CYP2D6, and the site of oxidation of the substrates is oriented above the heme moiety, similar to the site of oxidation of camphor in the bacterial CYP101 crystal structure (Poulos et al. 1985). In particular, Asp301 is found to be a crucial amino acid responsible for forming an ionic hydrogen bond with a basic nitrogen atom from the substrate or inhibitor (de Groot et al. 1996). The energy-optimized positions of the substrates in the protein agreed well with the original relative positions of the substrates within the substrate model (de Groot et al. 1995). The substrate model incorporates only one oxidation site and two possible positions for a basic nitrogen atom of the substrates, guided by the presence of two oxygen atoms within a carboxylate (Asp301) (de Groot et al. 1995). However, the protein model for CYP2D6 suggests the presence of two possible sites of oxidation and only one position for the basic nitrogen atom (de Groot et al. 1996). Although the protein model alone is not very suitable for prediction of CYP2D6 metabolism, the derived protein model indicates new leads for experimental validation and extension of the substrate model.

Ellis et al. (1996) developed a homology model of the CYP2D6 active site on the basis of the bacterial CYP102

structure, which is similar to that derived by Lewis (1995) with a minor modification in the β1-4 region, namely, movement of the sequence by one residue, such that positions Val374 and His376 of CYP2D6 are in alignment with Phe331 and Leu333 of CYP102, respectively. In this homology model, the active site consisted of a cavity bordered by hydrophobic residues, and important active-site residues included Thr309 and Thr312 of the oxygen-binding site, with Asp301, Ser304, and Ala305 lying in the I helix and Pro371, Gly373, and Val374 of the β1-4-sheet region (Ellis et al. 1996). The active-site area is further defined by a lipophilic pocket bordered by Val480 and Phe481 of the loop and β6-2 region. This model identifies Asp301 as the critical substrate-contact residue involved in the proposed electrostatic interaction between the basic nitrogen of typical substrates of CYP2D6 and a negatively charged site in the active site. This is consistent with the data from site-directed mutagenesis studies (Ellis et al. 1995).

Venhorst et al. (2000) have evaluated the N-dealkylation and O-demethylation of six MAMC analogs as alternative substrates and inhibitors for CYP2D6 and found that by increasing the N-alkyl chain length, the affinity for the enzyme increased substantially as indicated by an up to 100-fold decrease in K_m values for CYP2D6-catalyzed O-dealkylation and IC$_{50}$ values for CYP2D6-catalyzed dextromethorphan metabolism (Venhorst et al. 2000). The K_m values are 0.20 and 0.08 μM for the O-dealkylation of N-propyl and N-butyl analog, respectively. N-Alkylation and subsequent elongation of the alkyl substituent resulted in a reduction of the maximum rate of metabolism by CYP2D6. N-Dealkylation of these MAMC analogs is also catalyzed by CYP1A2. However, the introduction of a second N-methyl group in diMMAMC yielded a decrease in the affinity for CYP2D6 compared with MAMC and MMAMC. MMpyrC showed an affinity for 2D6 similar to that of MAMC. There is a correlation between the compound lipophilicity and the log IC$_{50}$ values (Venhorst et al. 2000). This suggests that additional interaction points in the active site of CYP2D6 may play a role in binding of the longer N-alkyl chain substrates.

Venhorst et al. (2000) have also developed a homology model of the CYP2D6 active site on the basis of known structure of bacterial P450terp, P450cam, P450BM3, and P450nor and docked the MAMC analogs into the active site. The alkyl substituents of the mono-alkylated analogs mainly interacted with the hydrophobic residues located in helix I (Ile297 and Ala300) of the enzyme. Modeling studies suggested that diMMAMC interacted with Glu216, instead of Asp301; MMpyrC adopted an orientation similar to those of the mono-N-alkylated analogs.

6.3.2 HOMOLOGY MODELS DERIVED FROM RABBIT CYP2C5 STRUCTURES

The incorporation of the rabbit CYP2C5 crystal structure as a template provided more accurate information on ligand binding to CYP2D6, as rabbit CYP2C5 and human CYP2D6 share more than 40% sequence identity. Further modeling studies incorporating information from the rabbit CYP2C5 structure indicate that a second acidic residue, Glu216, is in a position where it may play an important role in binding to the basic nitrogen of CYP2D6 substrates (Kirton et al. 2002). In the model proposed by Kirton et al. (2002), it is suggested that Asp301 played a structural role through the formation of a hydrogen bond with a residue in the flexible B′–C loop.

The existence of numerous substrates of CYP2D6 such as metoprolol, which are metabolized at sites further from this nitrogen, gave rise to a different 10-Å pharmacophore (de Groot et al. 1999a) and Lewis et al. have thus suggested that Glu216 is the primary residue for recognition of basic ligands into the active site (Lewis et al. 1997), where it generates an intermediate binding site before the ligand adopting a more "reactive" position in the cavity (Marechal et al. 2008). This is similar to the intermediate binding pocket occupied by warfarin in the crystal structure of the S-warfarin/CYP2C9 complex (Williams et al. 2003) but apparently inconsistent with the Glu216Phe mutation transforming CYP2D6 into a quinidine demethylase. This issue may be resolved when cocrystal structures of substrates with CYP2D6 become available.

Homology models also suggest a role for aromatic residues in the active site to form van der Waals interactions with aromatic moieties of the substrates. Three aromatic phenylalanine residues, namely, Phe120, Phe481, and Phe483, have been proposed as important active-site residues. The aromatic moiety of Phe120 has a steric effect on the orientation of molecules in the active site of CYP2D6 and thus plays a role in controlling the regioselectivity of substrate oxidation (Flanagan et al. 2004; Kirton et al. 2002). Phe120 is positioned close to the heme iron and is a key factor in controlling substrate access to the heme, which has been confirmed by site-directed mutagenesis experiments and the recently determined CYP2D6 crystal structure (Rowland et al. 2006).

In a homology model, codeine is docked in the active site of CYP2D6 in an orientation consistent with O-demethylation (Kirton et al. 2002). Surprisingly, the docking did not position the basic nitrogen atom of the substrate close to Asp301. Instead, the basic nitrogen is observed to interact with Glu216, a second acidic residue in the active site. Early modeling studies suggested that Asp301 is not directly involved in substrate binding but plays a structural role in positioning the B–C loop, including Phe120 (Goodford 1985), and this hypothesis has subsequently been verified by the crystal structure of CYP2D6 (Rowland et al. 2006). The docking results for MPTP (Kirton et al. 2002) and dextromethorphan (Flanagan et al. 2004) also positioned the basic nitrogen atoms of the substrates close to Glu216 and away from Asp301.

Guengerich et al. (2003) derived a homology model of CYP2D6 based on rabbit CYP2C5 structure to examine the role of Glu216, Asp301, and other residues in ligand binding using the Insight II suite of programs from Accelrys Inc. (San Diego, CA) on a Silicon Graphics O2 workstation. The authors found that the residues Phe120, Leu213, Phe483, and Val374 are most likely to interact with ligands. In this model, Glu216 is positioned at the top of the proposed substrate binding cavity, with its side chain pointing down toward the heme, consistent with a role in substrate binding. In contrast, the

side chain of Asp301 is directed away from the heme and oriented in such a way as to form possible hydrogen bonds to the amide nitrogens of Val119 and Phe120. Asp301 with its oxygen atoms is 10.7 and 10.8 Å away from the ferric iron atom. The side chain of Asp301 is directed away from the active site and toward the amide protons of Val119 and Phe120, such that hydrogen bonds of 1.74 and 1.96 Å could be formed between one of the carboxylate oxygen of Asp301 and these amide protons (Guengerich et al. 2003). This is in agreement with the findings by Kirton et al. (2002), who had proposed that Asp301 may serve to stabilize the B–C loop in CYP2D6 and possibly other CYPs as well that contain an acidic residue at this position. This structural role is consistent with mutagenesis studies showing impaired holoenzyme (heme-containing) stability upon mutation of Asp301 (Hanna et al. 2001b).

Venhorst et al. (2003) generated a homology model for CYP2D6 on the basis of crystallized rabbit CYP2C5 and is validated on its ability to reproduce binding orientations corresponding to metabolic profiles of the substrates. In docking studies with a series of substrates and inhibitors, this homology model accommodated codeine well, forming a hydrogen bond with Glu216, π–π stacking with Phe120, and van der Waals interaction with Phe483. For most ligands examined except sparteine, there is an interaction with either Glu216 or Asp301 (Venhorst et al. 2003). Debrisoquine, dextromethorphan, MAMC, and phenformin appeared to interact with both Glu216 and Asp301 for their basic nitrogen atoms. Phe120 is generally involved in the π-stacking with aromatic moieties of the ligands. Other residues including Ile106, Leu213, Ala305, Val308, Val370, Val374, and Phe483 are generally involved in hydrophobic interactions. Twenty-two active-site residues in SRS1–6 are identified in the CYP2D6 binding pockets, including those from SRS1 (Pro103, Pro105, Ile106, Phe112, Phe120, and Leu121), SRS2 (Leu213, Glu216, and Ser217), SRS3 (Phe243 and Gln244), SRS4 (Asp301, Ser304, Ala305, Val308, and Thr309), SRS5 (Ile369, Val370, Val374, and Thr375), and SRS6 (Phe483 and Leu484) (Venhorst et al. 2003). Respective mutation studies have confirmed the role of Glu216, Asp301, Ser304, Thr309, Phe120, and Phe383 in ligand binding and substrate metabolism.

Vaz et al. (2005) have described a 3D QSAR comparative molecular similarity index analysis model using 36 aryloxypropanolamine compounds (e.g., propranolol) that inhibited CYP2D6-catalyzed AMMC dealkylation. These authors revealed that compounds with U-shape conformation gave optimal inhibition for CYP2D6, while substitution on either ring (phenyl or indole) decreased CYP2D6 inhibitory potency. The U-shape conformation could be stabilized by a favorable π-stacking between the two aromatic rings. Further docking studies using a homology model of CYP2D6 derived from rabbit CYP2C5 structure (Protein Data Bank [PDB] entry: 1DT6) indicated that the aryloxypropanolamines interacted with a large hydrophobic pocket defined by residues like Phe120, Val370, Met374, Phe483, and Leu484 (Vaz et al. 2005). The basic nitrogen atom of the aryloxypropanolamines could interact with Glu216 since the distance between the amine nitrogen and the carboxylate oxygen of Glu216 is approximately 4 Å. The hydroxyl group may be involved in a hydrogen bond with the backbone carbonyl group of Ser304.

De Graaf et al. (2006) have described an improved automated docking approach to predict the catalytic site of 65 substrates in a CYP2D6 homology model. This approach incorporates the water molecules at predicted positions in the active site and the rescoring of pooled docking poses from four different docking programs (AutoDock, FlexX, GOLD-Goldscore, and GOLD-Chemscore). The SCORE scoring function successful predicts experimentally reported sites of catalysis of more than 80% of the substrates (de Graaf et al. 2006). Three docking algorithms (FlexX, GOLD-Goldscore, and GOLD-Chemscore) are then employed in combination with six scoring functions (Chemscore, DOCK, FlexX, GOLD, PMF, and SCORE) to evaluate the ability of docking-based virtual screening methods to prioritize known CYP2D6 substrates seeded into a drug-like chemical database in the absence and presence of active-site water molecules. The optimized docking strategy is successfully used to identify high-affinity CYP2D6 ligands among a larger proprietary database.

De Graaf et al. (2007) have further explored and discussed the role of active-site residues in ligand binding using a refined homology model of CYP2D6 derived from the known rabbit CYP2C5 structures in complex with a dimethylsulfophenazole derivative or diclofenac (PDB codes: 1N6B and 1NR6). An interaction fingerprint analysis of docked 65 substrates in this homology model has suggested that several other active-site residues including Val104, Thr107, Leu121, Leu213, Ala305, Val308, Val370, and Leu484 probably interacted with ligands, in addition to previously identified residues (Phe120, Glu216, Ser304, Val374 and Phe383) (de Graaf et al. 2007). A number of site-directed mutagenesis studies have confirmed the role of these residues in ligand binding and substrate metabolism. Structural differences between the homology model and the 2F9Q structure are similar to those observed between substrate-free and substrate-bound structures of other CYPs, suggesting that these conformational changes are required upon substrate binding for CYP2D6.

Ito et al. (2008) have derived a homology model based on the rabbit CYP2C5 crystal structure and docked 11 substrates/inhibitors into the active site of the model. These include propranolol, metoprolol, thioridazine, R-bufuralol, MPTP, debrisoquine, dextromethorphan, nortriptyline, codeine, quinidine, and yohimbine. They have found that Glu216, Asp301, Phe120, and Phe483 are ligand-binding residues by docking and molecular dynamics simulation studies, which is in agreement with previously reported site-directed mutagenesis data and the crystal structure of CYP2D6 (Rowland et al. 2006).

A number of homology models of CYP2D6 based on bacterial CYPs and rabbit 2C5 have been developed, which have provided useful information on the active-site residues of CYP2D6. Docking studies are usually carried out on known CYP2D6 ligands (e.g., substrates or competitive inhibitors) for these homology models to explore the role of amino acids in

the active site in ligand binding. According to these models, a list of amino acid residues postulated to form the active site of CYP2D6 includes at least Asp100, Pro103, Ile106, Thr107, Leu110, Pro114, Ser116, Gln117, Phe120, Leu121, Ala122, Leu213, Glu216, Phe219, Asp301, Ser304, Ala305, Thr309, Trp316, Val370, Pro371, Gly373, Val374, Val480, Phe481, and Phe483. The negative charges at the acidic residues at both positions 301 and 216 may contribute in binding of amine substrates. However, other residues must also be involved in the binding of substrates, since both wild-type and Asp301 mutants can bind atypical substrates that are devoid of basic nitrogen (e.g., spirosulfonamide). The importance of many of these residues, in particular Glu216, Asp301, Phe120, and Phe481 has been confirmed by the crystal structure of CYP2D6.

These homology models have provided insights into the interactions of ligands with the active site of CYP2D6 and how this enzyme determines its substrate specificity. However, a major limitation of all these models is the limited information about movement of key atoms in the course of the catalytic cycle, and the rate of catalysis is probably dependent on the tightness of substrate binding in the transition state.

6.4 SITE-DIRECTED MUTAGENESIS STUDIES OF CYP2D6

6.4.1 Pro34

The naturally occurring mutation 100C>T in the *CYP2D6* gene (present in the *4, *10, *36, *37, *49, *52, *54, *56, *64, *65, and *72 alleles) caused a Pro34Ser substitution (Fukuda et al. 2000; Johansson et al. 1994) located in a region that is highly conserved in CYPs belonging to gene families 1 and 2 (Oezguen et al. 2008). This substitution caused expression of an unstable gene product, as evident from comparison of the relative levels of *CYP2D6* mRNA and protein and decreased bufuralol 1′-hydroxylase activity (Johansson et al. 1994). The Pro34Ser mutation did not drastically change K_m but decreased V_{max} for debrisoquine 4-hydroxylation, whereas K_m is increased and V_{max} is unchanged or decreased for bunitrolol 4-hydroxylation (Tsuzuki et al. 2001).

6.4.2 Gly42

The naturally occurring mutation 124G>A in the *CYP2D6*12 and *71 alleles caused a Gly42Arg or Gly42Glu substitution (Marez et al. 1996). Individuals carrying this allele showed reduced sparteine clearance. The CYP contents both in yeast microsomal and in whole-cell fractions indicated that some part of Gly42Arg protein is localized in the endoplasmic reticulum membrane fraction, whereas most of Gly42Arg protein is in some subcellular fractions other than endoplasmic reticulum (Tsuzuki et al. 2001), indicating that this mutation has interfered with subcellular fraction targeting after protein synthesis. A site-directed mutagenesis study indicated that the Gly42Arg mutant showed increased apparent K_m and decreased V_{max} value for debrisoquine 4-hydroxylation, while it increased both K_m and V_{max} for bunitrolol 4-hydroxylation (Tsuzuki et al. 2001).

Tsuzuki et al. (2003) have further investigated the effects of the substitution of Gly42 with five kinds of other amino acid residues (Lys, Val, Phe, Ser, and Glu) on the function of CYP2D6 using debrisoquine and bunitrolol as probe substrates. The replacement of Gly42 with hydrophobic amino acid residues such as Val and Phe did not change the enzymatic properties such as reduced CO-difference spectra, microsomal CYP contents, and catalytic activities toward debrisoquine and bunitrolol. The substitution of Gly42 with a polar but noncharged amino acid residue, Ser, showed a similar reduced CO-different spectrum, but the substitution with a charged basic (Lys and Arg) or acidic (Glu) amino acid residue yielded a peak at 420 nm in addition to a Soret peak at 450 nm (Tsuzuki et al. 2003). P450 and microsomal protein levels of Gly42Ser, Gly42Lys, Gly42Arg, and Gly42Glu are lower than those of the wild type. Kinetic analysis revealed that the substitution of Gly42 with charged amino acid residues such as Lys, Arg, or Glu significantly increased the apparent K_m values for debrisoquine 4-hydroxylation (47.8-, 7.8-, and 5.6-fold, respectively) and bunitrolol 1′-hydroxylation (for (+)-enantiomer: 218.0-, 63.9-, and 18.9-fold, respectively) but without remarkable changes in the V_{max} values for most mutants. The substitution of Gly42 with noncharged amino acid residues such as Ser, Val, and Phe did not substantially change the K_m values (1.1- to 2.0-fold), while Gly42Ser gave a 6- to 10-fold higher V_{max} values (Tsuzuki et al. 2003). These results indicate that the properties of amino acid residues at position 42 of CYP2D6 influence the behavior of CYP2D6 proteins such as anchoring into endoplasmic reticulum membranes, conversion of P450 to P420, and incorporation of heme into the apoprotein. Mutation of Gly42 may also interfere with the entrance of lipophilic substrates into the active site of the enzyme.

The N-terminal region of CYPs is rich in hydrophobic amino acid residues such as Gly, Phe, Leu, Ile, and Val and is proposed to be a signal-anchor sequence responsible for anchoring the CYP protein into the membrane and incorporation of heme into the apoprotein (Yamazaki et al. 1993). In addition, the Pro-rich region close to the signal-anchor sequence may be important for the proper folding of microsomal CYPs (Yamazaki et al. 1993). Cook Sangar et al. (2009) have found that CYP2D6 contains an N-terminal chimeric signal that mediates its bimodal targeting to the endoplasmic reticulum and mitochondria. In vitro mitochondrial import studies using both N-terminal deletions and point mutations suggest that the mitochondrial targeting signal is localized between residues 23 and 33 ("²³MAKKTSSKGK³³") and that the positively charged residues at positions Ala24, Lys25, Lys26, Ser28, and Lys32 are needed for mitochondrial targeting (Cook Sangar et al. 2009). The importance of the positively charged residues is confirmed by transient transfection of a CYP2D6 mitochondrial targeting signal mutant in COS-7 cells. Both the mitochondrial and microsomal CYP2D6 exhibited bufuralol 1′-hydroxylation activity, which is completely inhibited by CYP2D6 inhibitory antibody (Cook Sangar et al.

2009). It is unclear whether mitochondrial CYP2D6 contributes to drug metabolism in vivo.

6.4.3 ALA90

The naturally occurring mutation 972C>T resulted in the Ala90Val substitution, which is present in the *CYP2D6*48* allele (Soyama et al. 2004). When expressed in COS-7 cells, the mutated protein showed similar expression levels and capacity for bufuralol 1′-hydroxylation compared to the wild-type enzyme (Sakuyama et al. 2008). The residue Ala90 is located in the B helix, away from the active site (Rowland et al. 2006).

6.4.4 ASP100

Koymans et al. (1993) have previously proposed that the residue Asp100 might be an alternative candidate for the electrostatic interaction with the basic nitrogen of the substrate. However, this interaction cannot be rationalized with the homology model developed by Ellis et al. (1996) based on bacterial CYP102 structure because of its peripheral location. Consistently, replacement of Asp100 with other amino acids had little effect on ligand binding and metabolism. In the 2F9Q structure, Asp100 is located in the β1-5 region, which is outside of the active site (Rowland et al. 2006).

6.4.5 THR107

The naturally occurring mutation 1023C>T caused the Thr107Ile substitution in the *CYP2D6*17*, *40*, *58*, and *64* alleles (Gaedigk and Coetsee 2008; Gaedigk et al. 2002; Masimirembwa et al. 1996; Oscarson et al. 1997). The 1023C>T mutation (yielding the Thr107Ile substitution) is always linked to the 2850C>T (Arg296Cys) and 4180G>C (Ser486Thr) mutations on the same allele (Masimirembwa et al. 1996). The Thr107Ile substitution showed no significant effect on enzyme activity toward bufuralol but caused a significant increase in the apparent K_m for codeine (Oscarson et al. 1997). Enzymes containing both the Thr107Ile and Arg296Cys mutations exhibited a more than fivefold higher K_m for bufuralol than the wild-type enzyme.

In the homology model created by Modi et al. (1996), Ile107 (like Ile106) is located in the B′ helix of SRS1 and close to the carboxylate group of Glu216, thereby implicating an important function for this residue in substrate binding. Thr107 is involved in the binding of codeine and thus the substitution of this hydrophilic residue Thr to a hydrophobic Ile is likely to influence the substrate binding. This is compatible with the experimental data, since the Thr107Ile variant showed reduced affinity for codeine but not for bufuralol. This model also showed that Asp301 is involved in the binding of the basic nitrogen found in codeine (Koymans et al. 1992). However, unlike Ile106, Ile107 is not an active-site residue in the 2F9Q structure (Rowland et al. 2006). Ile107 may affect substrate binding through Ile106.

6.4.6 Phe120, Phe481, and Phe483

In addition to a basic nitrogen atom, most CYP2D6 substrates contain an adjacent aromatic ring that may form van der Waals interactions with Phe residues in CYP2D6. The aromatic residues Phe120, Phe481, and Phe483 in the active site of CYP2D6 have been proposed to play an important role in substrate binding using homology modeling and docking studies (de Groot et al. 1999b; Hayhurst et al. 2001; Ito et al. 2008; Kirton et al. 2002). Flanagan et al. (2004) revealed that the relative rates of *O*- versus *N*-demethylation of dextromethorphan are altered by the Phe120Ala mutation (at 1.0 mM substrate, the ratio is 1.0 for the mutant compared with 3.1 for the wild type), although the removal of the Phe120 aromatic side chain by substitution with Ala showed negligible effect on the K_m for either 1'-hydroxylation of bufuralol or the *O*-demethylation of dextromethorphan. Replacement of Phe120 with Leu, Ser, or His caused a 10- to 32-fold increase in K_m for dextromethorphan *O*-demethylation (wild type vs. Phe120Leu, Phe120Ser, or Phe120His: 1:32, 1:10, and 1:16 μM, respectively), whereas only the His substitution caused a 2.5-fold increase in K_m for bufuralol 1'-hydroxylation (wild type vs. Phe120His: 10.3 vs. 4.1 μM) (Flanagan et al. 2004). Phe120 replacement produced a greater effect on K_m values for dextromethorphan *O*-demethylation than for bufuralol 1'-hydroxylation, indicating that Phe120 is more important in dextromethorphan than in bufuralol binding.

The Phe120Ala mutant is at least as active against both substrates as the wild-type enzyme (Flanagan et al. 2004), suggesting that the phenyl ring of Phe120 is not essential for substrate binding. Furthermore, a previously unknown new metabolite, 7-hydroxydextromethorphan, is identified with this mutant (Flanagan et al. 2004). At 1 mM substrate concentration, 7-hydroxydextromethorphan became the major metabolite. 7-Hydroxydextromethorphan is also formed by all other mutants, but at much lower levels than the Phe120Ala mutant. These findings clearly suggest that the removal of the aromatic side chain of Phe120 from the active-site cavity has abolished a steric constraint on ligand binding, allowing the substrate to bind in several different orientations. Further modeling studies have demonstrated that removal of the aromatic moiety of Phe120 from the CYP2D6 binding site did not alter the ability of dextromethorphan to bind in an orientation appropriate for *O*-demethylation (Flanagan et al. 2004), consistent with the observation that the kinetic parameters for this reaction are largely unaltered by the Phe120Ala mutation. An analogous effect on substrate orientation has been observed in human CYP2C9 (corresponding to Phe114) (Haining et al. 1999) and bacterial CYP102A1/P450BM3 (corresponding to Phe87) (Chen et al. 2008) where the Phe residues equivalent to Phe120 have been mutated.

Keizers et al. (2004) have revealed that the Phe120Ala mutant abolished the *O*-demethylation activity toward MAMC, whereas bufuralol 1'-hydroxylation is not affected. Surprisingly, the mutated enzyme metabolized quinidine via *O*-demethylation and 3-hydroxylation (McLaughlin et al. 2005), unlike the wild-type CYP2D6, which does not show any activity toward quinidine. The naturally occurring mutation Phe120Ile (358T>A; rs1135822) present in *CYP2D6*10* can be found in a low percentage of the Southeast Asian population (Solus et al. 2004). All of these findings indicate that residue Phe120 in the active site is important for substrate binding and catalysis in CYP2D6. In the CYP2D6 2F9Q structure, Phe120 is located in the loop B'–C, forming part of the active-site border (Rowland et al. 2006).

Masuda et al. (2005, 2006) have also investigated the role of Phe120 using a site-directed mutagenesis approach in a yeast expression system and bufuralol and bunitrolol enantiomers as chiral probe substrates. The wild-type enzyme showed enantioselectivity of *R*- ≫ *S*-bufuralol and the Phe120Ala mutant exhibited substrate enantioselectivity of *R*- ≤ *S*-enantiomer, whereas the product diastereoselectivity of (*R*-1"-OH- ≪ *S*-1"-OH-bufuralol) is similar between the wild-type enzyme and the mutant. Kinetic analysis revealed that apparent K_m values for the formation of the four kinds of 1'-OH-bufuralol are similar between the mutant and the wild type, but V_{max} values for *S*-bufuralol 1'-hydroxylation by the mutant are ~7-fold higher than those by the wild type (Masuda et al. 2005). However, the K_m and V_{max} values for *R*-bufuralol 1'-hydroxylation are similar between the Phe120Ala mutant and wild-type enzyme. When bunitrolol enantiomers are used as substrates, the substitution of Phe120 by Ala in CYP2D6 did not change the substrate enantioselectivity but resulted in a remarkable increase in bunitrolol 4-hydroxylase activity and K_m values as compared with the wild-type enzyme (Masuda et al. 2006). A homology modeling study indicated that the hydrophobic interaction of an aromatic moiety of the substrate with Phe120 played an important role in substrate binding.

Early homology modeling studies suggested that Phe481 is an important aromatic residue associated with ligand binding (de Groot et al. 1999b). This residue appears to interact with ligands via a π–π interaction between its phenyl ring and the planar hydrophobic aromatic moiety common to many CYP2D6 substrates. Substitution of Phe481 by nonaromatic residues such as Leu or Gly reduced the affinity of several typical CYP2D6 substrates including debrisoquine, metoprolol, and dextromethorphan, with 3- to 16-fold higher K_m values compared to the wild type (Hayhurst et al. 2001). However, replacement of Phe481 with Thr did not alter the K_m and V_{max} values for *S*-metoprolol, debrisoquine, and dextromethorphan. However, homology modeling studies based on rabbit CYP2C5 have suggested that Phe481 is positioned outside the binding pocket but in close contact with the active-site residue Phe483 (Smith et al. 1998; Venhorst et al. 2003).

Smith et al. (1998) have reported that the Phe483Ile mutant, but not Phe483Trp or the wild-type enzyme, showed some spectrally detectable binding of testosterone, which does not contain any basic nitrogen atom. The Phe483Ile mutant is able to metabolize testosterone to a novel metabolite, 15α-hydroxytestosterone. However, the Phe483Ile mutation did not affect CYP2D6-catalyzed bufuralol hydroxylation. Hiroi et al. (2001) have observed 2β-, 6β-, and 17-hydroxylation activity of recombinant CYP2D6 for testosterone. Both progesterone and allopregnanolone are also substrates of CYP2D6 (Kishimoto et al. 2004). Allopregnanolone is hydroxylated at the C21-position by CYP2D6 (Kishimoto et al. 2004). In addition, CYP2D6 catalyzed 2β-, 6β-, 16α-, and 21-hydroxylation of pregnenolone (Hiroi et al. 2001). All these steroids lack the basic nitrogen atoms and are thus atypical substrates of CYP2D6.

The study by Lussenburg et al. (2005) has found that the Phe483Ile mutant catalyzed bufuralol 1′-hydroxylation with a 32-fold lower V_{max} and unchanged K_m. This mutation decreased the V_{max} 2-fold but increased K_m 14-fold for MDMA O-demethylenation. However, the mutant protein with Phe483Ile did not metabolize MAMC, a selective substrate of CYP2D6 (Lussenburg et al. 2005). The CYP2D6 Phe483Ala mutant also formed dextrorphan, with a 15-fold higher K_m and a 2-fold higher V_{max} than the wild type. For dextromethorphan, two other metabolites (i.e., the monohydroxylated and O-demethylated hydroxydextrorphan) are formed that could not be detected for the wild-type enzyme (Lussenburg et al. 2005). All these findings demonstrate a clear substrate-dependent change of enzyme activity and binding affinity when Phe483 is mutated. The location of Phe483 in the binding site has been confirmed by the crystal structure of CYP2D6 (Rowland et al. 2006).

Keizers et al. (2006) have further explored the role of Phe120 and Phe483 in the active site of CYP2D6 in the metabolism of several MAMC analogs by mutating these two residues. MAMC is a good substrate of the wild-type enzyme (Keizers et al. 2004; Lussenburg et al. 2005). Both the Phe120Ala and Phe483Ala mutants showed a significant activity toward the MAMC analogs containing an alkyl group on the 4-aminomethyl position, while the Phe120Ala and Phe483Ala mutants did not metabolize MAMC. The Phe483Ala mutant did not metabolize MAMC, and O-demethylation of MMAMC is not observed. The N-dealkylation of the substrates, however, is much more significant with the Phe120Ala mutant than with the Phe483Ala mutant, suggesting that the Phe120Ala mutant may allow more for flexible orientations of the MAMC analogs than the Phe483Ala mutant. Like in the wild-type enzyme, the catalytic efficiency of the Phe120Ala and Phe483Ala mutants toward the MAMC analogs increased with the length of the N-alkyl chain, suggesting the potential involvement of distal active-site residues of CYP2D6 in binding to these substrates. Overall, the addition and elongation of an N-alkyl chain in the MAMC analogs enhance the binding affinity of the Phe120Ala and Phe483Ala mutants of CYP2D6 to these compounds.

The Asp301Gln mutant of CYP2D6 showed the same trend in catalytic efficiency of O-demethylation of the MAMC analogs as the Phe120Ala mutant (Keizers et al. 2006). In addition, the Asp301Gln mutant had a significant increased N-dealkylation of the MAMC analogs. The correlation between the effects of mutation of Phe120 and Asp301 is consistent with the proposed role of Asp301 in positioning Phe120 (Guengerich et al. 2003; Venhorst et al. 2003). The differences in activity toward the MAMC analogs for the two mutants may be explained by the nature of the different mutations, affecting, in one case, the removal and, in the other case, redirection of the phenyl moiety of residue Phe120 combined with the removal of part of the active-site negative charge (Venhorst et al. 2003).

Keizers et al. (2006) have docked these MAMC analogs into a homology model of CYP2D6 derived from the rabbit CYP2C5 structure. A total of 50 docking poses are generated per substrate using the GOLD-Chemscore program. MAMC and the analogs had similar optimal orientations in the active site to facilitate O-demethylation, forming electrostatic interactions between their nitrogen atoms and the carboxylate of Glu216. Except for MAMC, Phe120 offered π–π stacking interactions to the aromatic moiety of all MAMC analogs. An additional difference between the binding modes of the five MAMC analogs concerns the conformation of the N-alkyl chain. N-Butyl-MAMC and, to a lesser extent, N-propyl-MAMC accommodated their relatively long N-alkyl chains in a cleft formed by the apolar side chains of residues Leu213, Val308, and Phe483 at the top of the binding pocket (Keizers et al. 2006), providing additional hydrophobic interactions with the CYP2D6 active site. Phe120 appeared to be more involved in aromatic interaction, whereas Phe483 more sterically influenced substrate binding.

6.4.7 Trp128

Trp128 in CYP2D6 is thought to be bound to the propionate of pyrrole ring D of the porphyrin via hydrogen bonds. Similar positioning of analogous Trp residues in the structures of CYP2C5, 2B4, 2C9, and 3A4 (PDB codes: 1N6B, 1PO5, 1OG2, and 1TQN, respectively) has been observed (Scott et al. 2003; Wester et al. 2003; Williams et al. 2003; Yano et al. 2004). Stortelder et al. (2006) have recently examined the changes in the structure of CYP2D6 upon binding of the substrate MAMC using steady-state and time-resolved fluorescence methods for both wild-type enzyme and the Trp128Phe mutant. The expression of Trp128Phe is equal to that of the wild-type enzyme. The Trp128Phe mutant O-demethylated MAMC with a twofold lower V_{max} compared to the wild-type enzyme, but the K_m remained unchanged (Stortelder et al. 2006). These results suggest that Trp is not a crucial residue for CYP2D6-catalyzed MAMC O-demethylation. Molecular dynamics simulations revealed that Trp128 and Arg101 form hydrogen bonds to the propionate of pyrrole ring D of the porphyrin (Stortelder et al. 2006).

The emission spectrum of the Trp128Phe is 2 nm redshifted compared with the wild-type enzyme and is ~22% lower in intensity (Stortelder et al. 2006). In the presence of MAMC, excitation and emission maxima of wild type and the Trp128Phe mutant did not shift, indicating that the overall conformational change in the enzyme structure is not large upon MAMC binding. After binding to MAMC, energy transfer from Trp128 to heme appeared to be important and its emission is associated with the shortest of the three average tryptophan fluorescence lifetimes observed for CYP2D6 (Stortelder et al. 2006). The decay times of MAMC did not change significantly when either the wild-type or the Trp128Phe mutant enzyme is present.

The accessibility of the Trp residues in CYP2D6 is further examined using acrylamide and iodide as dynamic quenchers (Stortelder et al. 2006). As expected, the Trp residues in both Trp128Phe and wild-type enzymes are less accessible to iodide than to acrylamide. The quenching constant of Trp128Phe is slightly higher than that of the wild-type enzyme. The Trp residues in CYP2D6 appeared to be less accessible for the external quenchers iodide and acrylamide in the presence of MAMC, indicating a tightening of the enzyme structure upon substrate binding (Stortelder et al. 2006). Calculation of the solvent-accessible surface areas revealed that Trp152, Trp262, and Trp75 are accessible to bulk solvent, while Trp128, Trp316, and Trp409 did not strongly interact with water.

There is a 35% reduction of the MAMC emission intensity in the presence of both the wild-type and Trp128Phe mutant enzyme. The energy transfer from MAMC to heme is very efficient since the emission of MAMC is fully abolished when bound in the active site of CYP2D6 (Stortelder et al. 2006). Steady-state anisotropy studies revealed that besides the MAMC in the active site, another 2.4% of MAMC is bound outside of the active site to the wild-type enzyme.

6.4.8 VAL136

The naturally occurring mutation 1659G>A led to the Val136Ile substitution, which is present in the *CYP2D6*29* allele (Marez et al. 1997; Wennerholm et al. 2001). When expressed in COS-1 cells, the mutated protein showed reduced capacity for bufuralol 1'-hydroxylation (Wennerholm et al. 2001). The residue Val136 is located in the C helix, away from the active site (Rowland et al. 2006).

6.4.9 GLU156

The naturally occurring mutation 1720A>C resulted in the Glu156Ala substitution, which is present in the *CYP2D6*50* allele (Soyama et al. 2004). When expressed in COS-7 cells, the mutated protein showed decreased capacity for bufuralol 1'-hydroxylation compared to the wild-type enzyme (Sakuyama et al. 2008). The residue Glu156 is located in the D helix, away from the active site (Rowland et al. 2006).

6.4.10 GLY169

The naturally occurring 1758G>A mutation of *CYP2D6* causing Gly169Arg substitution (designated as *CYP2D6*14*) is first identified in a Chinese who had reduced ability to metabolize debrisoquine (Kubota et al. 2000; Wang et al. 1999). Site-directed mutagenesis studies have demonstrated that the bufuralol 1'-hydroxylation activity of the Gly169Arg mutant is significantly lower than that of the wild-type CYP2D6 (Wang et al. 1999).

6.4.11 GLU216 AND ASP301

Ellis et al. (1995) have examined the role of Asp301 in the substrate binding of CYP2D6 by generating several

mutants substituting a variety of amino acids at this site. The Asp301Asn and Asp301Gly mutants expressed approximately equivalent amounts of recombinant apoprotein, apart from the Asp301Ala and the aspartic acid 301 deletion mutants (Asp301δ) (Ellis et al. 1995). The level of holoprotein (heme-containing) expression of the Asp301Ala mutant is only ~10% of the level of the other mutants (8 vs. 49 pmol/mg protein) and no holoprotein is detectable in microsomes from the Asp301δ mutant (Ellis et al. 1995). Variable amounts of P420 are detectable in microsomes prepared from the Asp301Ala mutant only. Ellis et al. (1995) also observed perturbed structural integrity of the active site to varying degrees when Asp301 is replaced with a neutral residue (Asn, Ala, or Gly), as there is a slight shift in the Soret absorption maximum of the carbon monoxide complex and in the different extent of heme incorporation. This may suggest that Asp301 can help maintain the integrity of the active site and that, in its absence, the topology of the site would be altered.

Ellis et al. (1995) have reported that replacement of the negatively charged residue 301 with a neutral side chain (Asn, Ala, or Gly) resulted in substantial decreases in the binding capacity of debrisoquine (loss of type I spectrum) and quinidine (1000-fold greater K_d value). Replacement of Asp301 with neutral residues in yeast-expressed CYP2D6 almost abolished (1%–2% of the wild type) catalytic activity toward debrisoquine and racemic metoprolol, two classical substrates of CYP2D6. Although the Asp301Glu mutant (which still retained the carboxylate group) retained rates of activity comparable with that of the wild type, the regioselective oxidation of metoprolol, as evaluated by the ratio of formation of *O*-desmethyl and α-OH metabolites, is significantly different in microsomes prepared from the Asp301Glu mutant compared with the wild-type enzyme (8.5:1 and 3.8:1, respectively) (Ellis et al. 1995). Since there is no gross change in the integrity of the active site in the Asp301Glu mutant as indicated by a normal Soret absorption maximum of 448 nm, proper heme incorporation, and retention of catalytic activity, the altered regioselectivity for metoprolol could be attributed to a subtle difference in the location of the substrate oxidation sites relative to the (Fe–O)³⁺ entity, as a result of the extension of the carboxylate residue by a methylene group. In contrast, enantioselective oxidation of *R*- and *S*-metoprolol is not altered by the substitution of Asp301 to Glu (Asp301Glu), indicating that a residue(s) other than Asp301 is a determinant of CYP2D6 chiral selectivity, possibly Ser304.

These findings from the study by Ellis et al. (1995) have demonstrated that substitution of Asp301 with neutral amino acids (e.g., Asn, Ala, or Gly), differing in size and polarity, results in marked reductions in enzyme catalytic activity. The attenuation of enzyme activity had been proposed to be attributed to the disruption of an electrostatic bond between Asp301 and the substrate. Substitution of the Asp301 carboxylate residue with a similar functional moiety (Glu), on the other hand, did not significantly affect the catalytic competence of the enzyme, although a subtle change in the regioselective oxidation of metoprolol and a 10-fold reduction in quinidine binding have been observed. The study has provided initial evidence that Asp301 may serve as a critical anionic charge to dock the basic nitrogen atom of ligands of CYP2D6 or to maintain the integrity of the active site.

Site-directed mutagenesis experiments have provided further evidence that Asp301 is the negatively charged amino acid responsible for binding to the nitrogen atom of substrate molecules (Mackman et al. 1996). Mackman et al. (1996) have observed a direct binding of aryldiazenes including phenyl-, 2-naphthyl-, and p-biphenyldiazene to individual pyrrole rings of the heme, suggesting a direct interaction of Asp301 with the substrate. These aryldiazenes can react with CYPs to form σ-bonded aryl–iron complexes. Substitution of Asp301 with a Glu, which preserves the carboxylate side chain, caused no detectable change in the N-aryl porphyrin regioisomer patterns and only minor changes in the catalytic activity. Replacement of Asp301 by an Asn or Gly, which removed the negatively charged side chain, suppressed migration of the aryl groups to pyrrole ring B without impairing migration to pyrrole ring A and virtually abolished catalytic activity (Mackman et al. 1996). These results indicate that the loss of activity observed when Asp301 is replaced by a neutral residue is attributed to loss of the charge-pairing interaction with the substrate positive charge or subtle structural effects in the vicinity of pyrrole ring B, but not to major structural reorganization of the active site.

Hanna et al. (2001a) have found that substitution of Asp301 with Glu, Asn, Ser, or Gly significantly decreased CYP2D6-mediated bufuralol 1′-hydroxylation with remaining enzyme activity of 61.9%, 11.4%, 9.0%, and 9.0%, respectively, compared to the wild-type enzyme. This suggests that positively charged residues are particularly disruptive in bacterial (*Escherichia coli*) and insect cell (baculovirus) expression systems. Similar reduction of 6-hydroxylation of bufuralol is observed with these mutants. With the exception of the Asp301 mutant, which had comparable expression level compared to the wild type, Asp301Gly, Asp301Ala, Asp301Leu, Asp301Ser, Asp301His, Asp301Lys, Asp301Arg, and Asp301Cys (replaced with neutral or negatively or positively charged residues) all resulted in significantly reduced levels of the holoprotein when expressed in the bacterial system (Hanna et al. 2001a), suggesting an additional role of Asp301 in protein folding and heme incorporation. Indeed, initial efforts to reverse the putative Asp301–basic substrate interaction with a Lys/Arg301–acidic substrate pair are unsuccessful because of the failure of mutants substituted with basic residues at amino acid position 301 to incorporate heme (Hanna et al. 2001a). These findings indicate that Asp301 is important for proper heme insertion and presumably protein folding; neutral and particularly basic residues are highly disruptive.

However, Guengerich et al. (2002) have found little effect of Asp301 mutations to Asn, Ser, or Gly on the binding to spirosulfonamide and its analogs, almost all high-affinity substrates of CYP2D6 but devoid of basic nitrogen atoms. The sulfonamide moiety is not basic owing to the strong electron-withdrawing properties of the sulfone group. This raises further concerns about the reliability of CYP2D6 models based on a critical electrostatic interaction with Asp301.

The Asp301Asn mutant failed to bind bufuralol and quinidine. Neutral Asp301 mutants (Asp301Asn, Asp301Ser, and Asp301Gly) showed relatively high affinity for spirosulfonamide ($K_d \approx 10^{-6}$ M) (Guengerich et al. 2002). The oxidation rate of spirosulfonamide in Asp301Asn is decreased approximately 10-fold, as with bufuralol (Hanna et al. 2001a), although the formation rate of *anti*-OH-spirosulfonamide is increased in the Asp301Gly and Asp301Ser mutants. These results argue that the loss of catalytic activity of the neutral Asp301 mutants cannot be attributed to a loss of binding affinity for a cationic substrate, because the same pattern is observed with an uncharged substrate; attenuated electrostatic interaction of substrate did not provide an explanation for the role of Asp301 in substrate binding (Guengerich et al. 2002). Asp301 may interact with spirosulfonamide through hydrogen bonding to the sulfonamide group, as opposed to electrostatic interactions. Removal of the moiety (and the fluorine on the adjacent phenyl ring) produced a compound (analog **6**) that is both a reasonably tightly bound ligand ($K_d = 4$ μM) and a substrate; the presence of a carboxylate group (analog **5**) leads to a loss of apparent binding and little oxidation (Guengerich et al. 2002). The analogous amide (**3**) and thioamide (**4**) are strong ligands. Spirosulfonamide yielded strong classic type I heme perturbation spectra with recombinant CYP2D6 with a K_s of 1.6 μM. CYP2D6 also bound and oxidized most analogs of spirosulfonamide with substitutions of the sulfonamide group. On the basis of these results with nonamine substrates, Guengerich et al. (2002) have proposed that Asp301 plays an important structural role in CYP2D6 integrity and that mutations of Asp301 cause more extensive changes in CYP2D6 than can be interpreted in the context of electrostatic interaction with ligands.

The proposed key role of Glu216 and Asp301 by homology modeling studies is further confirmed by a series of site-directed mutagenesis experiments (Paine et al. 2003). Conservative replacements of Glu216 or Asp301 with Asp and Glu, respectively, resulted in small (two- to sixfold) increases in K_m values and had negligible effects on turnover rate for both bufuralol and dextromethorphan (Paine et al. 2003). However, replacement of Asp301 by Asn or Gln led to a 130- to 145-fold increase in K_m values for bufuralol; the increase is 80-fold for the Asp301Gln mutant, but as much as 1400-fold for Asp301Asn for dextromethorphan (Paine et al. 2003). For both substrates, the effects on turnover rate are modest, ranging from a 30% decrease to a 70% increase. Substitution of Glu216 with neutral residues such as Gln, Phe, or Leu greatly decreases the K_m values by 100- to 170-fold for bufuralol 1′-hydroxylation; a smaller effect (10- to 25-fold increase in K_m) is observed with dextromethorphan O-demethylation. The wild-type enzyme and the conservative mutants Glu216Asp and Asp301Glu produced similar dextrorphan/3-methoxymorphinan ratios of ~8:1, indicating a significant preference for the O-demethylation pathway resulting in dextrorphan. However, this ratio is substantially decreased after removal of either of the negatively charged residues, consistent with a decreased preference for binding the basic nitrogen close to Glu216 and Asp301, distant from the heme iron. The Glu216Phe, Glu216Lys, Asp301Asn, Asp301Gln, and Glu216Gln/Glu301Gln mutants exhibited decreased dextrorphan/3-methoxymorphinan ratios as low as 1:1 (Paine et al. 2003). These mutations have abolished enzyme preference for *O*-demethylation.

The mutation of Glu216 altered the substrate specificity to such an extent that the mutant protein catalyzed testosterone 6β-hydroxylation (Paine et al. 2003), which is typically mediated by CYP3A4 (Yamazaki and Shimada 1997). Furthermore, the Glu216Ala/Lys and Asp301Gln mutants with removal of the negative charge from either the 216- or 301-position catalyzed the metabolism of atypical CYP2D6 substrates, including anionic compounds such as diclofenac and tolbutamide that lack a basic nitrogen atom and are used as model substrates of CYP2C9 (Paine et al. 2003). For diclofenac 4'-hydroxylation, mutants Glu216Gln, Glu216Phe, and Asp301Asn produced rates ~5-, 10-, and 22-fold higher than the wild-type enzyme, respectively, while the turnover rates of the Glu216Ala, Glu216Lys, and Asp301Gln derivatives are increased 50- to 75-fold. The catalytic activity is increased still further (>1000-fold of the wild-type enzyme) upon neutralization of both residues with double Glu216Gln/Asp301Gln mutations, but its testosterone 6β-hydroxylase activity is increased only twofold over wild type. This suggests that the binding site of CYP2D6 is thus intrinsically rather promiscuous, with Glu216 and Asp301 favoring the binding of basic substrates and discriminating against acidic substrates. The rate of formation of 4'-hydroxy diclofenac is not significantly greater with the Glu216Lys mutant than with Glu216Ala, suggesting that the carboxylate group of the substrate is not positioned near this residue.

Guengerich et al. (2003) have further investigated the role of Glu216 in ligand binding and catalysis. The authors found that all of the Glu216 mutants could be expressed with levels of holoprotein similar to the wild-type enzyme. Replacement of Glu216 with a residue (e.g., Gln, Ala, His, or Asn) other than Asp significantly decreased the binding of quinidine, bufuralol, propranolol, debrisoquine, MPTP, encainide, amitriptyline, and sparteine. The Glu216His mutant gave a K_s of 3.4 mM compared to the values of 7.3 and 7.8 μM for the wild type and the Glu216Asp mutant, respectively (Guengerich et al. 2003). Catalytic activity toward bufuralol (1'-hydroxylation) and 4-methoxyphenethylamine (*O*-demethylation) is significantly reduced or almost abolished by neutral or basic mutations at Glu216 (>95% decrease), to the same extent as the substitution of Asp301 with Asn. Unlike the Asp301 mutants, which reduced the k_{cat} by 10-fold (Guengerich et al. 2002), the Glu216Gln mutant retained 40% enzyme efficiency with the substrate spirosulfonamide (from 250 to 90 min⁻¹ mM⁻¹), an atypical substrate devoid of basic nitrogen, suggesting that the substitutions at Glu216 influence binding of amine substrates more than other catalytic steps. The wild-type enzyme and mutants with substitutions at Glu216 (to Lys) and Asp301 (to Glu) failed to show catalytic activity or preference toward potential substrates such as 4-methoxybenzylamine, 4-methoxyphenethylamine, 1-napththylacetic acid, and 4-methyoxybenzoic acid that contain a basic nitrogen or

a carboxylate (Guengerich et al. 2003). Overall, these results demonstrate an electrostatic interaction of amine substrates with Glu216 and Asp301 and at least another residue(s).

Both Glu216 and Asp301 play a critical role in the action of quinidine as an inhibitor of CYP2D6 (McLaughlin et al. 2005). Quinidine is a potent competitive inhibitor but not a substrate of CYP2D6 (Guengerich et al. 1986; Otton et al. 1988). However, a classical type I binding spectrum with CYP2D6 and quinidine is observed (Hayhurst et al. 2001), which is usually associated with substrate–enzyme binding (Schenkman et al. 1981). In addition, quinidine possesses a number of structural features seen in most typical CYP2D6 substrates, including a basic nitrogen atom, a flat hydrophobic region, and a negative electrostatic potential (Strobl et al. 1993). Studies of the relationship between structure and inhibitory activity for quinidine and its less potent stereoisomer quinine have been reported (Hutzler et al. 2003b), and substantial decreases in inhibitory potency are observed for the *N*-methyl, *N*-ethyl, and *N*-benzyl quininium salts, suggesting that the quaternary nitrogen of this antipode interacts with a distinct region of the CYP2D6 active site as compared to the corresponding nitrogen of quinidine. Notably, esterification of quinidine resulted in a substantial loss of inhibitory potency, most likely attributed to the disruption of a hydrogen-bonding interaction involving the hydroxyl group, suggesting that hydrogen bonding contributes more to the tight binding of quinidine than the charge-pair interaction of the positively charged nitrogen (Hutzler et al. 2003b). The conservative substitutions Glu216Asp and Asp301Glu showed similar inhibition by 1 μM quinidine to the wild-type enzyme, whereas enzymes with nonconservative replacements are at least 50% active with 10 μM quinidine (McLaughlin et al. 2005). The double mutant Glu216Gln/Asp301Gln, with complete removal of the charge but not the polarity, is found to be insensitive to inhibition by quinidine, retaining 80% of its bufuralol 1'-hydroxylase activity and 85% of its dextromethorphan *O*-demethylase activity in the presence of 100 μM quinidine (McLaughlin et al. 2005). However, Ala substitution of the aromatic side chain of Phe120, Phe481, or Phe483 showed only a minor effect on the enzyme inhibition by quinidine (McLaughlin et al. 2005). These findings suggest that the negative charges at Glu216 and Asp301, but not the aromatic groups of the three phenylalanine residues, are important for the binding of quinidine.

In contrast to the wild-type enzyme, the Glu216Phe mutant formed *O*-demethylated quinidine, and the mutant proteins with double Glu216Gln/Asp301Gln mutations or Phe120Ala resulted in both *O*-demethylated quinidine and 3-hydroxyquinidine (McLaughlin et al. 2005). Quinidine 3-hydroxylation turnover rates for Glu216Gln/Asp301Gln and Phe120Ala are estimated to be 0.14 and 0.07 min⁻¹, respectively, which are slower than the typical rates of 1–5 min⁻¹ observed for the wild-type enzyme for standard substrates such as bufuralol and dextromethorphan (Paine et al. 2003). 3-Hydroxyquinidine is a major metabolite of quinidine formed by CYP3A4 (Guengerich et al. 1986). The CYP2D6 mutant with double Glu216Gln/Asp301Gln substitutions is able to catalyze nifedipine *N*-oxidation (Paine et al. 2003). Substitution of Asp301 alone is not sufficient to enable CYP2D6 to metabolize quinidine, and Glu216 clearly plays an important role in determining the mode of binding; substitution of Glu216 with a bulky side chain in the Glu216Phe mutant confers on CYP2D6 the ability to catalyze the O-demethylation of quinidine and 6β-hydroxylation of testosterone (Paine et al. 2003), another marker reaction catalyzed by CYP3A4. These results indicate that both Glu216 and Asp301 have a critical role in determining the substrate specificity of CYP2D6.

Computational docking studies showed that quinidine could bind tightly to CYP2D6 but not in an orientation favorable for catalysis. The binding of quinidine to the wild-type CYP2D6 enzyme appears to be governed by interactions between the aromatic rings of quinidine and Phe120 and Phe483 and by a hydrogen bond between the hydroxyl group of quinidine and the carboxyl group of Glu216 (McLaughlin et al. 2005). Tethered docking studies demonstrated unfavorable contacts between quinidine and Phe120 and Ala305 in the orientation consistent with generation of 3-hydroxyquinidine, and with Phe120, Leu121, and Glu216 in the orientation consistent with formation of O-demethylquinidine (McLaughlin et al. 2005). This suggests that these residues are likely to be important in preventing metabolism of quinidine in wild-type CYP2D6. In contrast, the orientation of quinidine in the Glu216Phe mutant suggests an interaction between the aromatic rings of quinidine and Phe216 and Phe120, and a hydrogen bond between the basic nitrogen atom of quinidine and the side chain of Ser304, facilitating the formation of O-demethylquinidine (McLaughlin et al. 2005). However, similar docking studies on the Glu216Gln/Asp301Gln mutant produced only solutions in which the quinidine molecule is positioned away from the heme.

6.4.12 Lys281

Tyndale et al. (1991) have identified a naturally occurring mutation lacking a single codon resulting in deletion of Lys281, as a result of a 3-bp deletion of 2615_2617delAAG at the 3′ end of CYP2D6 exon 5 (CYP2D6*9). This mutation is associated with deficient microsomal metabolism of bufuralol and sparteine in vitro (Tyndale et al. 1991) and poor debrisoquine oxidation phenotype in vivo (Broly and Meyer 1993). The mutant protein expressed in HepG2 cells using vaccinia virus–mediated cDNA expression showed K_m values toward bufuralol, debrisoquine, and sparteine that are not significantly different from the wild-type enzyme (Tyndale et al. 1991). These data suggest that the poor metabolizer phenotype of CYP2D6*9 is not attributed to a catalytically defective enzyme but probably to decreased levels of the CYP2D6 in microsomal membranes resulting from deficient membrane insertion or decreased stability of the CYP. In the 2F9Q structure, Lys281 is located in helix H, away from the active site (Rowland et al. 2006).

6.4.13 Glu222

Masuda et al. (2005, 2006) have examined the role of Glu222 in the oxidation of chiral substrates bunitrolol and bufuralol.

The substitution of Glu222 by Ala remarkably decreased the V_{max} value for bufuralol but without changing the apparent K_m value, substrate enantioselectivity, or metabolite diastereoselectivity (Masuda et al. 2005). However, replacement of Glu216 by Ala increased the apparent K_m values three- to ninefold compared to the wild type, whereas V_{max} values are decreased to one-sixth to one-fourteenth those of the wild type (Masuda et al. 2005). A computer-assisted simulation study using energy minimization and molecular dynamics techniques indicated that the ionic interaction of a basic nitrogen atom of the substrate with Glu222 in combination with Glu216 played important roles in the binding of bunitrolol and bufuralol by CYP2D6 and the orientation of these substrates in the active-site cavity. In the active-site homology model, Glu222 is not within but is close to SRS2 of CYP2D6. Glu216 is located in SRS2, and the carboxylate group of Glu222 is spatially close to that of Glu216. The carboxylate group of Glu222 is the closest to the basic nitrogen of R-bufuralol with a distance of 2.8 Å followed by that of Glu216 (5.1 Å), while the distance between the carboxylate group of Asp301 and the basic nitrogen atom of R-bufuralol is 7.9 Å (Masuda et al. 2005).

6.4.14 Arg296

The natural mutation 2850C>T present in the CYP2D6*2, *11, *12, *17, *29, *30, *31, *32, *34, *35, *40, *41, *42, *45, *46, *51, *56, *58, *59, *63, *65, and *69 alleles caused an Arg296Cys substitution (Gaedigk and Coetsee 2008; Griese et al. 1998; Marez et al. 1995, 1996, 1997; Raimundo et al. 2000, 2004; Sakuyama et al. 2008). The Arg296 residue is located in helix I near Asp301. This helix is the longest one found in CYPs and is known to be involved not only in substrate binding but also in delivery of catalytic protons, with the central part of the helix coinciding with SRS4 (Lewis 1995). However, the Arg296Cys mutation did not change the catalytic activity toward debrisoquine and bunitrolol compared to the wild-type enzyme (Tsuzuki et al. 2001). In the homology model of CYP2D6 derived from the rabbit 2C5 structure, Arg296 is found to be located on the I helix close to the distal surface of the protein and remote from the active site (Allorge et al. 2005). In the 2F9Q structure (Rowland et al. 2006), Ile297 instead of Arg296 (not an active-site residue) forms part of the border of the active-site cavity. This may explain why mutation of Arg296 has little effect on debrisoquine and bunitrolol metabolism (Tsuzuki et al. 2001).

6.4.15 Ser304

Ser304, a neutral, polar residue located in helix I of CYP2D6, is found to be located in the active site in homology modeling studies (Ellis et al. 1995; Islam et al. 1991; Koymans et al. 1992; Lewis 1995; Modi et al. 1996). The position and orientation of Ser304, relative to Asp301, are compatible with the atomic distance between the hydroxyl group and the basic nitrogen of ligands such as metoprolol, propranolol, and quinidine. The modest enantioselectivity exhibited by CYP2D6 toward racemic substrates such as propranolol (Bichara et al. 1996; Ward et al. 1989; Yoshimoto et al. 1995), metoprolol (Otton et al. 1988), and bunitrolol (Narimatsu et al. 1999) may be attributed to a more favorable hydrogen bond interaction between Ser304 and the hydroxyl group of one of the enantiomers (Ellis et al. 1996). Similarly, it has been suggested that the differential inhibition of CYP2D6 activity by quinine (a weak inhibitor) and quinidine (a strong inhibitor) is caused by a more favorable hydrogen bond interaction of the hydroxyl group attached to a chiral carbon atom (C-9) of quinidine (Lewis 1995).

Ellis et al. (2000) have investigated the role of Ser304 in the regio- and enantioselective metabolism of debrisoquine, metoprolol, and propranolol using site-directed mutagenesis and heterologous yeast expression systems. The expression levels of apoprotein of Ser304Ala, Asp301Gly/Ser304Ala, and Asp301Ser/Ser304Asp are similar to those of the wild-type enzyme. However, the expression level of holoprotein in the Asp301Gly/Ser304Ala double mutant is significantly lower than that in the wild type and the Ser304Ala mutant, while no holoprotein is detected with the Asp301Ser/Ser304Asp double mutant (Ellis et al. 2000).

The kinetic parameters of formation of α-hydroxymetoprolol and 4-hydroxydebrisoquine by the wild-type enzyme and Ser 304Ala mutant are comparable (Ellis et al. 2000). However, the double mutant Asp301Gly/Ser304Ala showed much higher K_m values for debrisoquine 4-hydroxylation (2175 vs. 15 and 19 μM) and metoprolol α-hydroxylation (2346 vs. 30 and 18 μM) compared to the wild-type and Ser304Ala mutant. The V_{max} for debrisoquine 4-hydroxylation is reduced by 75% relative to Ser304Ala, whereas the V_{max} for metoprolol α-hydroxylation remained unchanged (Ellis et al. 2000).

Replacement of Ser304 with Ala (a nonpolar residue without a hydroxyl group) did not change the enantioselective oxidation toward metoprolol and propranolol compared to the wild-type enzyme (Ellis et al. 2000). With propranolol, the formation of the major 4-hydroxy metabolite is not significantly enantioselective (R/S: 0.9), while the formation of the minor 5-hydroxy metabolite exhibited an apparent shift from S-enantioselective (R/S: 0.8) with the wild-type enzyme to R-enantioselective (R/S: 1.4) with the Ser304Ala mutant. The formation of another minor metabolite of propranolol, the deisopropyl product, remained S-enantioselective with both the wild type and Ser304Ala mutant (Ellis et al. 2000).

The Ser304Ala replacement did not significantly change the regioselective oxidation of debrisoquine, metoprolol, and propranolol (Ellis et al. 2000). The alicyclic 4-hydroxylation of debrisoquine is the major metabolic route with both wild-type enzyme and the Ser304Ala mutant. However, the formation rate of the phenolic 6- and 7-hydroxy metabolites by the Ser304Ala is decreased relative to the wild-type enzyme. With metoprolol, the O-demethylation is the preferred oxidative route with its R- and S-enantiomers by both the wild-type enzyme and Ser304Ala mutant. However, there is a modest increase in the ratio of O-demethylated over α-hydroxylated metabolite with the Ser304Ala mutant relative to the wild-type enzyme (Ellis et al. 2000).

The regio- and enantioselective oxidation of an analog of metoprolol, in which the hydroxyl group attached to the chiral carbon is replaced by a methyl moiety, is similar with both the wild-type enzyme and Ser304Ala mutant (Ellis et al. 2000). However, in contrast to metoprolol, the preference for the O-demethylation is greater with the S-enantiomer than with the R-enantiomer; the O-demethylation is slightly S-enantioselective whereas α-hydroxylation is R-enantioselective in both the wild type and Ser304Ala mutant (Ellis et al. 2000). The reversal of selectivity with the methyl analog of metoprolol indicates that the hydroxyl group attached to the chiral center of ligands, such as metoprolol, is important in defining the regio- and enantioselectivity. As such, an additional hydrogen-bonding residue, other than Ser304, may be involved in this interaction.

In addition, the differential inhibition of CYP2D6 activity by quinine and quinidine is also similar with both wild-type enzyme and the Ser304Ala mutant (Ellis et al. 2000). In both wild-type enzyme and Ser304Ala mutant, quinidine is 100-fold more potent than quinine in the inhibition of α-hydroxylation and O-demethylation of metoprolol, with the IC_{50} values being 10 and 0.1 μM for quinine and quinidine, respectively.

These results demonstrate that substitution of Ser304 with Ala does not alter substrate enantioselectivity and the differential inhibitory potency of quinidine and quinine against CYP2D6, although some change of substrate regioselectivity is observed. Ser304 is not a critical ligand-binding residue, although the residue may be located in the active-site cavity of CYP2D6. However, Ser304 and Asp301 may interact in helix I to facilitate proper protein folding and heme incorporation.

6.4.16 THR309 AND THR312

The overall amino acid sequence and structure of helix I is highly conserved among CYPs in general, particularly in mammalian CYP2 family (Gotoh 1992; Hasemann et al. 1995). The crystal structures of CYP2B4, 2C5, 2C8, and 2C9 all contain a cluster of four conserved Thr residues in helix I located above the porphyrin plane (Mestres 2005). An identical cluster consisting of Thr309, Trp310, Trp312, and Trp313 has been observed in CYP2D6. According to amino acid sequence alignments and homology modeling studies (Keizers et al. 2005a; Venhorst et al. 2003), Thr309 is expected to be involved in proton delivery to the catalytic water.

Keizers et al. (2005b) have investigated the functional role of Thr309 and Thr312 of CYP2D6 by replacement with a Val residue using several classical substrates. These mutations do not significantly alter the size of the residue whereas the putative proton delivering hydroxyl moiety has been removed. Both the Thr309Val and Thr312Val mutants O-demethylated MAMC efficiently for at least 30 min, indicating that the mutated enzymes are stable for at least that time. The Thr309Val mutant showed a similar K_m for MAMC as the wild-type enzyme but a 20-fold decreased turnover rate (Keizers et al. 2005b). The Thr312Val mutant showed an activity toward MAMC similar to that of wild-type enzyme.

For 1′-hydroxylation of bufuralol, the Thr309Val mutant displayed a K_m equal to that of the wild-type enzyme, but with a 10-fold lower turnover rate (Keizers et al. 2005b). However, for 4-hydroxylation of bufuralol, both K_m and turnover rates are equal for the mutant and wild type. Bufuralol is oxidized by the Thr312Val mutant similarly as by the wild type, with only up to twofold change in the turnover rate. Interestingly, concentration dependence of 6-hydroxylation of bufuralol by the wild-type enzyme and the Thr312Val mutant showed substrate inhibition at concentrations above 20 μM (Keizers et al. 2005b).

The Thr312Val mutant showed an up to 2-fold change in K_m and V_{max} for MDMA compared to the wild-type enzyme, whereas the Thr309Val mutant O-demethylated MDMA with a 14-fold lower turnover rate (Keizers et al. 2005b). The Thr309Val mutant significantly N-dealkylated MDMA, while the wild type and the Thr312Val mutant N-dealkylated MDMA only to a minor extent. Dextromethorphan is efficiently O-demethylated by both mutants, with the Thr309Val displaying a twofold lower turnover rate and the Thr312Val mutant displaying a twofold higher turnover rate (Keizers et al. 2005b). The mutants had similar K_m values to the wild-type enzyme. However, there is a marked difference in N-demethylation of dextromethorphan. The formation of 3-methoxymorphinan by the Thr309Val mutant is eightfold higher than by the wild-type enzyme (Keizers et al. 2005b). These results demonstrate a significantly greater N-demethylation of MDMA and dextromethorphan with the Thr309Val mutant over Thr312Val and the wild-type enzyme. The affinity of the Thr309Val mutant for MDMA is decreased twofold while the affinity for dextromethorphan is increased by 30%.

To further explore the mechanism of oxidation by the Thr309Val mutant, incubation is performed with the oxene donor CuOOH, which directly forms the oxenoid-iron in CYPs (Hutzler et al. 2003a). MAMC is O-demethylated by both the wild type and Thr309Val mutant when supported by CuOOH, and the reaction rate is linear for at least 15 min and the turnover rate of the Thr309Val mutant is threefold higher than that of the wild-type enzyme (Keizers et al. 2005b). Bufuralol is oxidized by the wild-type enzyme to the same products supported by CuOOH as by O_2/NADPH. However, 1′-hydroxylation and $\Delta^{1'-2'}$ desaturation occurred with a two- to threefold higher turnover rate when compared to that of incubation supported by NADPH, while the aromatic hydroxylation rate is not affected. The Thr309Val mutant formed the same four metabolites as the wild-type enzyme, with a higher turnover rate (Keizers et al. 2005b). The rate of aromatic 4-hydroxylation is even 70-fold higher, leading to a completely different metabolite ratio compared to the wild-type enzyme.

MDMA is O-demethylenated in incubations supported by CuOOH by both the wild type and Thr309Val mutant (Keizers et al. 2005b), with the turnover rate of the Thr309Val mutant being twofold higher than that of the wild-type enzyme. However, N-dealkylation of MDMA is not detected. Dextromethorphan is O-demethylated by both enzymes, with a threefold higher rate by the Thr309Val mutant compared to

the wild-type enzyme. Furthermore, low levels of 3-methoxymorphinan and significant amounts of 3-hydroxymorphinan are formed by the Thr309Val mutant, indicating that N-dealkylation still occurred. Interestingly, the Thr309Val mutant displayed turnover rates higher than those of the wild-type enzyme in all reactions supported by CuOOH, ranging from 2-fold for MDMA O-demethylenation to 70-fold for bufuralol 4-hydroxylation (Keizers et al. 2005b).

These results indicate that the catalytic behavior of the Thr309Val mutant is clearly different from that of the wild-type CYP2D6, with significantly reduced turnover rates of MAMC, bufuralol, and MDMA. The mutation has a minor effect on substrate affinity, as bufuralol, MDMA, and dextromethorphan still bound well to the mutated enzyme. This is consistent with the data from modeling studies where Thr309 is not involved in substrate interaction (de Groot et al. 1999a; Keizers et al. 2005a). In addition, the Thr309Val mutant exhibits significantly altered product ratios of bufuralol, MDMA, and dextromethorphan metabolism compared to the wild-type enzyme. Changes in preference of certain types of reactions have been observed for other conserved Thr residue mutants of CYPs. For example, the Thr302Ala mutant of rabbit CYP2B4 showed a decreased cyclohexane hydroxylation but an increased deformylation of cyclohexane carboxaldehyde (Vaz et al. 1996). The Thr303Ala mutant of CYP2E1 had a reduced epoxidation of several model substrates but an enhanced epoxidation of olefins (Vaz et al. 1998).

CYPs can utilize three different oxygenating species in oxidation reactions, namely, a peroxo, a hydroperoxo, and an oxenoid-iron, which show different oxidative properties and thus prefer certain types of reactions (Coon et al. 1998; Vaz et al. 1998). Thr309 in CYP2D6 may be involved in dioxygen activation. The Thr309Val mutant in the presence of CuOOH is able to O-demethylate MAMC and dextromethorphan and O-demethylenated MDMA (Keizers et al. 2005b). This suggests that the altered regioselectivity of the Thr309Val mutant may be caused by the loss of its ability to generate the oxenoid-iron. Restoration of the lost activity of the Thr309Val mutant with CuOOH indicates that Thr309 is a critical determinant in maintaining the balance of reactive oxygen species. It has been proposed that the Thr252 residue in bacterial CYP101 accepts a hydrogen bond from the hydroperoxo-iron intermediate that promotes the second protonation in the distal oxygen atom, resulting in O–O cleavage and oxenoid-iron formation (Nagano and Poulos 2005).

6.4.17 His324

His324 is a highly conserved residue among P450s. The His324Pro substitution (present in the *CYP2D6*7* allele occurring in Caucasian populations with a frequency of approximately 1%), a rare *CYP2D6* missense mutation, is found to prevent normal protein folding and heme incorporation, leading to an inactive enzyme (Evert et al. 1994, 1997). This mutation is associated with the poor metabolizer phenotype of sparteine in vivo (Evert et al. 1994). When expressed in insect cells by site-directed mutagenesis, the mutant protein is expressed at amounts comparable to the wild-type enzyme. However, no spectrally detectable P450 is formed and no catalytic activity is detected. Furthermore, in contrast to the wild-type protein, the mutant protein is almost exclusively located in a detergent-insoluble insect cell fraction (Evert et al. 1997).

6.4.18 Val338

The naturally occurring mutation 3183G>A led to the Val338Met substitution, which is present in the *CYP2D6*29* allele (Marez et al. 1997; Wennerholm et al. 2001). When expressed in COS-1 cells, the mutated protein showed decreased activity for bufuralol 1′-hydroxylation (Wennerholm et al. 2001). The residue Val338 is located in the J helix, away from the active site (Rowland et al. 2006).

6.4.19 Met374

Ellis et al. (1996) have investigated the effect of Met374Val on CYP2D6 expression, ligand binding, catalysis, and stereoselective oxidation of metoprolol. The expression level of apo- and holoprotein is similar with two forms of CYP2D6 cDNA, and the binding affinities of several ligands including quinidine, debrisoquine, SKF-525A, moclobemide, metoprolol, and sparteine to the wild type and Met374Val are similar in terms of K_s. The enantioselective *O*-demethylation and α-hydroxylation of metoprolol are also similar with the two forms of CYP2D6, *O*-demethylation being *R*-enantioselective, whereas α-hydroxylation showed a preference for *S*-metoprolol (Ellis et al. 1996). However, the regioselectivity for *O*-demethylation of *R*- and *S*-metoprolol enantiomers is significantly greater for the Met374Val mutant than that for the wild type. The stereoselective properties of the Met374Val enzyme toward metoprolol are comparable to those observed with native human liver microsomes, but the regioselectivity of the wild-type enzyme is lower than that observed with human liver microsomal CYP2D6 and cDNA-expressed CYP2D6. The lower regioselectivity of the wild-type enzyme relative to human liver microsomal CYP2D6 is in agreement with data published by Mautz et al. (1995) using human lymphoblastoid-derived CYP2D6 with Met373 instead of Val374.

6.4.20 Glu410

The naturally occurring 3853G>A mutation caused Glu410Lys substitution (Marez et al. 1997). This mutation expressed in COS-7 cells did not alter the protein amount and activity toward bufuralol and dextromethorphan (Sakuyama et al. 2008). In the 2F9Q structure, Glu410 is located in the K′–K″ loop, away from the active site (Rowland et al. 2006).

6.4.21 Arg440

The naturally occurring 4042G>A mutation causing the Arg440His substitution is present in *CYP2D6*31*, which had a very low allele frequency (<0.1%) (Marez et al. 1997). Allorge et al. (2005) have examined the effect of Arg440His

on the metabolism of dextromethorphan, debrisoquine, and metoprolol using the site-directed mutagenesis approach. The holoprotein level of microsomes prepared from yeast cells expressing the wild-type and mutant forms of CYP2D6 is similar. The Arg440His mutation resulted in 2- to 3-fold higher K_m for debrisoquine 4-hydroxylation and 9- to 17-fold higher K_m for dextromethorphan *O*-demethylation, with a significant decrease in k_{cat} values for the oxidation of both substrates (Allorge et al. 2005). The rate of racemic metoprolol oxidation is also significantly decreased by ~90% in the mutant compared to the wild-type enzyme. Further docking studies suggested that the Arg440His affected binding of CYP2D6 with cytochrome P450 reductase.

6.4.22 Ser486

The natural mutation 4180G>C causing the Ser486Thr substitution is present in the *CYP2D6*2*, **4*, **10–*12*, **14*, **17*, **28–*32*, **35*, **36*, **37*, **40*, **45*, **46*, **47*, **49*, **51*, **54*, **55*, **57*, **58*, **59*, **61*, **63*, **64*, **65*, **69*, **70*, and **72* alleles (Gaedigk and Coetsee 2008; Marez et al. 1997; Matsunaga et al. 2009; Sakuyama et al. 2008). The Ser486Thr mutation did not change the expression and catalytic activity toward debrisoquine and bunitrolol compared to the wild-type enzyme (Oscarson et al. 1997; Tsuzuki et al. 2001). This is consistent with the finding from the homology model of CYP2D6 derived from the rabbit 2C5 structure where Ser486 is found to be located on a surface β-strand (β4-2), close to the C-terminal domain distant from the active site (Allorge et al. 2005).

On the basis of all the above findings, site-directed mutagenesis studies have suggested that two carboxylate-containing residues, Glu216 and Asp301, and potentially Glu222, and three phenylalanine residues, Phe120, Phe481, and Phe483, play an important role in determining the binding of ligands to the active site of CYP2D6 (Table 6.1). Thr309 and Thr312 appear to participate in proton delivery to the catalytic water. Replacement of Val374 with Met showed altered regioselectivity toward metoprolol. However, changing Asp100 and Ser304 has little effect on substrate regio- and enantioselectivity for CYP2D6.

There are a number of naturally occurring mutations in the *CYP2D6* gene with differing frequencies in different ethnic groups. Among them, the functional impact of Pro34Ser, Gly42Arg, Thr107Ile, Val136, Glu156Ala, Gly169Arg, Arg296Cys, His324Pro, Val338Met, Glu410Lys, Arg440His, and Ser486Thr has been examined using site-directed mutagenesis. The Pro34Ser substitution results in an unstable gene product, while Gly42Arg leads to mitochondrial targeting of the synthesized protein. Val136Ile, Glu156Ala, and Gly169Arg lead to reduced enzyme activity. In contrast, the Ala90Val, Arg296Cys, Glu410Lys, and Ser486Thr mutations do not change the catalytic activity toward probe substrates. Arg440 is not an active-site residue but its mutation can reduce the catalytic activity since this residue is highly likely to be involved in interaction with P450 reductase.

TABLE 6.1
Reported Site-Directed Mutagenesis Studies of Potentially Important CYP2D6 Amino Acid Residues

Residue (Wild-Type)	Mutant	Location in Secondary Structure	SRS	Residues in Close Contact	Probe Substrates Examined	k_{cat}[b]	References
Pro34	Ser (natural SNP)	N-terminus	–	–	Bunitrolol	↔	Johansson et al. 1994;
					Debrisoquine	↔	Tsuzuki et al. 2001; Wang et al. 1999
Gly42	Arg/Lys/Glu/Val/ Phe/Ser	N-terminus	–	–	Bunitrolol	↓	Tsuzuki et al. 2001
					Debrisoquine	↓	
Ala90	Val (natural SNP)	B helix	–	–	Bufuralol	↔	Sakuyama et al. 2008
Ile106	Glu	B′ helix	1	Glu216	Bufuralol	↓	van Waterschoot et al.
					Dextromethorphan	↓	2006
					MAMC	↓	
					MDMA	↓	
Thr107	Ile (natural SNP)	B′ helix	1	Glu216	Bufuralol	↔	Oscarson et al. 1997
					Codeine	↔	
Phe120	Ala (natural SNP)	B′–C loop	1	Arg68, heme	Bufuralol	↔	Flanagan et al. 2004;
					Dextromethorphan	↔	Keizers et al. 2006;
					MAMC	↓	Masuda et al. 2005,
					MDMA	↓	2006
Trp128	Phe	C helix	–	Tyr124, Arg132, Arg441, heme[a]	MAMC	↔	Stortelder et al. 2006
Val136	Ile (natural SNP)	C helix	–	–	Bufuralol	↓	Wennerholm et al. 2001
Glu156	Ala (natural SNP)	D helix	–	–	Bufuralol	↓	Sakuyama et al. 2008
Gly169	Arg (natural SNP)	D–E loop	–	–	Bufuralol	↓	Wang et al. 1999
Glu216	Asp/Gln/Ala/His/ Phe/Lys	F helix	2	Ile106, Thr107, Phe483	Bufuralol	↓	Ellis et al. 1995;
					Dextromethorphan	↓	Guengerich et al. 2003;
					Diclofenac	↑	McLaughlin et al.
					Nifedipine	↑	2005; Paine et al. 2003
					Spirosulfonamide	↔	
					Testosterone	↔	
					Quinidine	↑	
					Tolbutamide	↑	
Glu222	Ala	F–G loop	–	Arg221[a], Thr394[a]	Bufuralol	↓	Masuda et al. 2006
Lys281	δ (natural SNP)	H helix	–	–	Bufuralol	↔	Tyndale et al. 1991
					Debrisoquine	↔	
					Sparteine	↔	
Arg296	Cys (natural SNP)	I helix	–	Asp294	Bunitrolol	↔	Tsuzuki et al. 2001
					Debrisoquine	↔	
Asp301	Glu/Asn/Gly/Gln/ Ala/δ	I helix	4	Leu111, Val119[a], Phe120[a]	Bufuralol	↓	Guengerich et al. 2002;
					Dextromethorphan	↓	Hanna et al. 2001a;
					Diclofenac	↑	Keizers et al. 2006;
					Nifedipine	↑	Mackman et al. 1996;
					Spirosulfonamide	↓	McLaughlin et al.
					Testosterone	↔	2005; Paine et al. 2003
Ser304	Ala	I helix	4	Asp301[a]	Debrisoquine	↔	Ellis et al. 2000
					Metoprolol	↔	
					Propranolol	↔	
Thr309	Val	I helix	4	Val308, Ala305[a], Gly306[a], Thr310[a], heme	Bufuralol	↓	Keizers et al. 2005b
					Dextromethorphan	↔	
					MAMC	↓	
					MDMA	↓	
Thr312	Val	I helix	4	Val308	Bufuralol	↔	Keizers et al. 2005b
					Dextromethorphan	↔	
					MAMC	↔	
					MDMA	↔	

(Continued)

TABLE 6.1 (CONTINUED)

Reported Site-Directed Mutagenesis Studies of Potentially Important CYP2D6 Amino Acid Residues

Residue (Wild-Type)	Mutant	Location in Secondary Structure	SRS	Residues in Close Contact	Probe Substrates Examined	k_{cat}[b]	References
His324	Pro (natural SNP)	–	–	–	Bufuralol	↓	Evert et al. 1997
					Sparteine	↓	
Val338	Met (natural SNP)	K helix	–	–	Bufuralol	↓	Wennerholm et al. 2001
Met374	Val	β1-14 sheet	5	Arg101[a], Phe120, Phe483	Metoprolol	↔	Ellis et al. 1996
Glu410	Lys (natural SNP)	K'–K'' loop	–	–	Bufuralol	↔	Sakuyama et al. 2008
					Dextromethorphan	↔	
Arg440	His (natural SNP)	C-terminus	–	Glu99[a]	Debrisoquine	↓	Allorge et al. 2005
					Dextromethorphan	↓	
					Metoprolol	↓	
Arg441	Cys (natural SNP)	C-terminus	–	Arg440, Cys443	Propafenone	↓	Klein et al. 2007
Phe481	Tyr/Asn/Gly	β4-1 sheet	6	His477, Ser486	Debrisoquine	↓	Hayhurst et al. 2001
					Dextromethorphan	↓	
					Metoprolol	↓	
Phe483	Ala/Ile/Trp	β4-2–β4-1 sheet	6	Glu216, Leu220, Val374	Bufuralol	↓	Keizers et al. 2006;
					Dextromethorphan	↓	Lussenburg et al. 2005;
					MAMC	↑	Smith et al. 1998
					MDMA	↑	
					Testosterone	↔	
Ser486	Thr (natural SNP)	β4-2 sheet	–	Phe484	Bunitrolol	↔	Oscarson et al. 1997;
					Debrisoquine	↔	Tsuzuki et al. 2001

[a] Hydrogen bond interaction.

[b] Symbols: "↔" = unchanged; "↑" = increase; "↓" = decrease.

6.5   STUDIES USING ARYLDIAZENE PROBES

Aryldiazenes have been used to probe the nature and role of CYP2D6 active-site residues. These compounds can react with the heme iron to result in σ-bonded aryl–iron complexes. Upon oxidation of the complex with ferricyanide, the aryl group migrates from the heme iron to one of the four enamoring porphyrin nitrogens. An early study using this method has found that the active site of CYP2D6 is open above pyrrole A and to a smaller extent above pyrrole B but is closed above pyrroles C and D (Mackman et al. 1996).

van Waterschoot et al. (2006) have investigated the impact of active-site residues including Phe120, Asp301, Thr309, and Thr312 on the active-site topology of CYP2D6 using phenyl-, 2-naphthyl-, and p-biphenyldiazene. On the basis of the ratio of N-arylprotoporphyrin IX isomers obtained after treatment of the wild-type enzyme with the aryldiazenes, all three probes migrated to the nitrogen of pyrrole A and, to a lesser extent, to the nitrogen of pyrrole B, suggesting that the active-site crevice of CYP2D6 is largely located above pyrrole A and that it extends vertically at least 10 Å above the heme plane. The N-arylprotoporphyrin IX ratio in the Phe120Ala mutant indicated an increased migration of the probes to the nitrogen atoms of pyrroles C and D, which remained inaccessible in the wild-type enzyme. This is in agreement with the positioning of Phe120 indicated in homology modeling studies where

it is positioned above the heme in proximity of pyrroles C and D (Guengerich et al. 2003; Kirton et al. 2002; Venhorst et al. 2003). When the Asp301Gln mutant is treated with phenyl-, 2-naphthyl-, or p-biphenyldiazene, the overall active-site topology of this mutant remained unchanged. This supports the claim that the electrostatic interactions with the B–C loop are largely maintained and the loop retains its native orientation in this mutant (Paine et al. 2003).

Homology modeling studies have suggested that Thr309 is the closest to the heme and that it may be involved in proton transfer to the ferric hydroperoxy intermediate during the catalytic cycle (Keizers et al. 2005a; Venhorst et al. 2003). Thr309 may stabilize a distortion in helix I by providing hydrogen bonds to the carbonyl groups of Ala305 and Gly306, while Thr312 might be involved in stabilizing the distortion in helix I by providing a hydrogen bond to the carbonyl group of Val308 (Keizers et al. 2005a; Venhorst et al. 2003). Consistently, the Thr309Val mutant showed reduced turnover rate of dextromethorphan, MAMC, bufuralol, and MDMA but the activity is restored when alkylhydroperoxides are used as a source of oxidant (Keizers et al. 2005b). In the presence of the larger p-biphenyl probe, but not the naphthyl and phenyl probes, the Thr309Val mutation showed significant topological alteration (van Waterschoot et al. 2006). Since the Thr-to-Val mutation is isosteric, it is unlikely that this topological alteration is attributed to the amino acid side chain but must be caused by conformational or sterical changes at a vertical distance of ~8 Å or higher from the heme. The Thr309Val is unstable over time and a large amount of P420 is formed during expression (Keizers et al. 2005b). Collectively, these results suggest that, in addition to its potential role in dioxygen activation, Thr309 plays an important structural role within the active-site crevice. Mutation of the homologous Thr268 in bacterial CYP102 (Yeom et al. 1995) or Thr252 in CYP101 (Hishiki et al. 2000; Nagano and Poulos 2005) to Ala or Ile resulted in conformational changes, shifting helix I more above pyrrole ring A.

Homology modeling studies have suggested that Glu216 is located above pyrrole D at a vertical distance of approximately 13 Å (Keizers et al. 2005a; Venhorst et al. 2003). Within 4 Å of the γ-carboxylate group of Glu216 lies the side chain of Ile106. Replacement of Ile106 with Glu changed the topology in the active-site cavity of CYP2D6 as shown when treated with aryldiazene probes (van Waterschoot et al. 2006). When the phenyl probe is used, it migrated to the nitrogen of pyrrole A. The larger naphthyl and biphenyl probes migrated to all four heme nitrogens with migration to the nitrogens of pyrroles B, C, and D increasing with probe height. These changes are considered to be caused by electrostatic repulsion of the Glu216 by the γ-carboxylate of the introduced Glu106, causing a lateral shift of the flexible F–G loop toward the region above pyrrole A (van Waterschoot et al. 2006). Accordingly, the Ile106Glu mutant showed impaired catalytic activity toward CYP2D6 probe substrates including bufuralol and dextromethorphan. The K_m for dextromethorphan O-demethylation is increased more than 20-fold while that of bufuralol 1′-hydroxylation is increased 7-fold with a 4-fold decrease in V_{max}.

These findings indicate that the Phe120Ala, Thr309Val, Thr312Val, and Ile106Glu mutants, but not Asp301Gln, show significant altercations of the active-site topology as demonstrated by the changes in the N-arylprotoporphyrin IX isomer ratios determined when these mutants are treated with aryldiazenes. The results are largely consistent with those from homology modeling and site-directed mutagenesis studies.

6.6 ANTIBODY STUDIES OF HUMAN CYP2D6

Information on the structure of a proinhibitory site on the surface of CYP2D6 has been gained from epitope mapping studies using antisera from patients with autoimmune hepatitis or hepatitis C that inhibit CYP2D6 activity (Gueguen et al. 1991; Miyakawa et al. 1999). The sequence corresponding to residues 239–271 of CYP2D6 is recognized by all antisera tested with the core region comprising residues 261–263 (Gueguen et al. 1991). An anti–liver/kidney microsome-1 antibody present in hepatitis C virus–negative sera recognized at least two peptide regions, from residues 213–280 and residues 341–477, of human CYP2D6 (Miyakawa et al. 1998). In contrast, anti–liver/kidney microsome-1 antibody present in hepatitis C virus–positive sera recognized only the single region from residues 341–477.

Models for the structures of eukaryotic CYP enzymes predict that the region encompassing residues 254–290

of CYP2D6 lies between helices G and I (Edwards et al. 1989; Lewis 1995) and comprises a largely hydrophilic loop region on the surface of the protein. This region contains three proline residues that are also conserved in other CYP2D enzymes. Proline residues are frequently associated with loop regions and therefore such a site may be suitable for targeting by conformationally restricted antibodies. Both cyclic and linear peptides encompassing residues 254 to 290 of human CYP2D6 are used to target antibodies to CYP2D6 and the ability of the resulting anticyclic and antilinear antibodies to bind to and inhibit the activity of CYP2D6 is assessed (Schulz-Utermoehl et al. 2000). Antibodies raised against cyclic peptides bound 10 to 100 times more strongly to recombinant CYP2D6 than antibodies raised against the corresponding linear peptides (Schulz-Utermoehl et al. 2000). None of the antibodies raised against linear peptides had any effect on the debrisoquine 4-hydroxylase activity of human hepatic microsomal fraction; however, anticyclic peptide antibodies targeted against residues 254 to 273, residues 261 to 272, and residues 257 to 268 of CYP2D6 inhibited enzyme activity by a maximum of 60%, 75%, and 91%, respectively (Schulz-Utermoehl et al. 2000). In contrast, despite binding strongly to CYP2D6, an anticyclic peptide antibody directed against residues 278 to 290 did not inhibit enzyme activity. The epitope of the proinhibitory anticyclic peptide antibody directed against residues 257 to 268 of CYP2D6 included Thr261 and Trp262 and indicates a role for these residues in enzyme inhibition.

It would appear that only antibodies against cyclic peptides are able to inhibit the activity of CYP2D6. It has been suggested that cyclized peptides are more immunogenic and antigenic than linear peptides, because of the restriction of their conformational flexibility by cyclization. Cyclization of peptides greatly restricts their intramolecular movements and effectively locks them into a loop configuration. Similarly, Duclos-Vallee et al. (1995) have found that an antibody raised against a linear peptide corresponding to residues 251 to 271 of CYP2D6 failed to inhibit the O-demethylation of dextromethorphan of human hepatic microsomal fraction. Although Cribb et al. (1995) are successful in producing an antibody against a linear peptide corresponding to residues 254 to 273 of CYP2D6 that inhibited the debrisoquine 4-hydroxylase activity of human hepatic microsomal fraction up to 90%, it is necessary to use a protracted immunization schedule.

Interestingly, Gueguen et al. (1991) have determined that the essential part of the epitope of proinhibitory anti-CYP2D6 antibodies present in patients with type 1 autoimmune hepatitis antibodies comprised the tripeptide Thr–Trp–Asp (residues 261–263). This finding is supported by site-directed mutagenesis studies by Yamamoto et al. (1993), where deletion or substitution of the negatively charged aspartate residue at position Asp263 by another negatively charged amino acid, glutamate, by a neutral amino acid, asparagine, or by a basic amino acid, arginine, resulted in lack of recognition of CYP2D6 by anti–liver/kidney microsome-1 autoantibodies.

6.7 OTHER MOLECULAR MODELING STUDIES

Recently applied data visualization for multiobjective optimization techniques to CYP-mediated metabolism have revealed that CYP2D6 substrates are cationic compounds, while those of CYP2C9 are anionic (Yamashita et al. 2008). There is a clear relationship between polar surface area of ligands and binding affinity for CYP2D6, with the most potent inhibitors having a formal positive charge and a low percent polar surface area (McMasters et al. 2007). A docking study of 82 CYP2D6 substrates indicates that formal charges, number of aromatic rings, and hydrophobicity (log$P$) are the main attributes for CYP2D6 binding (Bazeley et al. 2006). Docking and molecular dynamics simulation studies indicate that Phe120, Glu216, Asp301, and Phe483, along with Phe219 and Glu222, are the binding residues for CYP2D6 inhibitors (Ito et al. 2008).

Terfloth et al. (2007) have described a series of descriptor sets for 379 compounds that can determine which CYP (2D6, 3A4, or 2C9) can metabolize them. In a four-step buildup process, 303 different descriptor components are examined for 146 compounds of a training set by various model building methods, such as multinomial logistic regression, decision tree, or support vector machine. Automatic variable selection algorithms are employed to decrease the number of descriptors. A comprehensive scheme of cross-validation experiments is used to evaluate the robustness and reliability of the four models developed. The predictive ability of the four models developed is inspected by predicting an external validation data set consisting of 233 compounds. The best model had a leave-one-out cross-validated predictivity of 89% and gives 83% correct predictions for the external validation data set (Terfloth et al. 2007). However, there are 11 compounds that are wrongly predicted by at least three of four models, including cocaine, cyclobenzaprine, dronabinol, ebastine, glibenclamide (glyburide), phenylbutazone, phenytoin, quinine, sertraline, sulfamethizole, and trimethorphan. All these compounds have been found to be metabolized by CYP2C9 or 3A4 by experimental studies, but many of them are predicted to be metabolized by CYP2D6. For those compounds that are metabolized by CYP2D6, the prediction is mostly correct.

Roy and Roy (2009) have conducted molecular shape analysis and molecular field analysis studies using a set of aryloxypropanolamine derivatives to explore the required molecular shape features as well as information on putative interactions with the active site of CYP2D6. The chemometric tools used for molecular shape analysis and molecular field analysis are genetic function approximation and genetic partial least squares (G/PLS) techniques, respectively. The G/PLS model derived in molecular field analysis using maximum common subgroup alignment is found to be the best model based on highest external predictive power with lowest RMSEP value. The molecular field analysis–derived models suggested the requirement of U-shape conformation for optimum interactions as well as the importance of different substituents on the aryloxypropanol fragment–like indolylalkyl substituent (Roy and Roy 2009). The molecular shape analysis models

indicated the importance of distribution of positive and negative charges on the surface of the molecules. The QSAR models with 2D descriptors revealed the importance of bulk, branching, and presence of different fragments. The models indicated that appropriate substituents in both the phenyl and indole ring could lead to satisfactory inhibitory potency toward the CYP2D6 enzyme.

6.8 X-RAY CRYSTALLOGRAPHIC STUDY OF HUMAN CYP2D6 AND FUNCTIONAL IMPLICATIONS

The crystal structure of human CYP2D6 has recently been determined and refined to 3.0-Å resolution by Rowland et al. (2006) (2F9Q, Figure 6.1a). The structure includes residues 52–497 (corresponding to the full-length protein sequence) and a short additional stretch of the proline-rich N-terminal region consisting of residues 34–41. CYP2D6 is truncated at residue 34, with a C-terminal His$_4$ tag and residues 2–10 of rabbit CYP2C3d (a variant that lacks the putative membrane-spanning segment of the N-terminus, residues 3–20) (von Wachenfeldt et al. 1997) inserted at the N-terminus. Modeling studies revealed a patch of hydrophobic residues in the loop region between helices F and G, which is considered to be situated on the surface of the protein and which possibly contributes to protein aggregation and membrane association. Two of these residues, Leu230 and Leu231, are selected for mutation to hydrophilic residues and 52 mutants of CYP2D6 are produced. Finally, six of the mutants showed the highest solubility and expression of holoprotein. These included Leu230Thr/Leu231Lys, Leu230Asp/Leu231Arg, Leu230Ala/Leu231Ser, Leu230Asn/Leu231Asp, Leu230Thr/Leu231Asp, and Leu230Asn/Leu231Arg. Of these, CYP2D6 Leu230Asp/Leu231Arg gave the most reliable yields and the best crystals and is subsequently used for structure determination.

6.8.1 SECONDARY STRUCTURE OF CYP2D6 IN COMPARISON WITH CYP2C9

Like other human CYP enzymes with known structures, the CYP2D6 2F9Q structure showed the typical characteristics of the mainly α-helical CYP fold as observed in other members of the CYP superfamily (Rowland et al. 2006). The lengths and orientations of the individual secondary structural elements, consisting of α-helices A to L and four β-sheets, resemble those observed in CYP2C9 and other CYP2 proteins, and helices E, I, J, K, and L constitute the core of the protein (Figure 6.1a). In CYP2D6, the F–G loop contains a small helix G', which is highly hydrophobic and probably associated with the endoplasmic reticular membrane. The long I helix, without any evident distortion, runs across the length of the entire molecule. The heme group is buried deep inside the protein and is tightly bound to CYP2D6 via hydrogen-bonding interactions.

In contrast to the structure of CYP2C9 (PDB codes: 1OG2 and 1OG5, with 40.7% amino acid sequence identity with

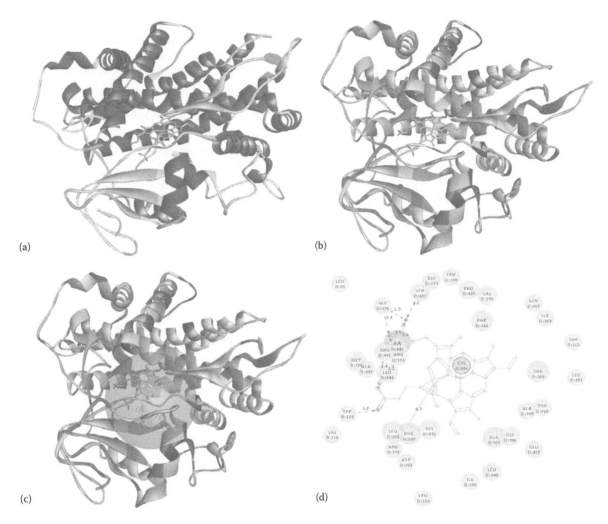

FIGURE 6.1 **(See color insert.)** The structure of human CYP2D6 (a; PDB code: 2F9Q), active-site cavity (b), and the residues around the active-site cavity (c). Color code in plot a: red, α-helix; green, β-strand; blue, loop. (With permission from the American Society for Biochemistry and Molecular Biology and Copyright Clearance Center.)

2D6) (Williams et al. 2003), there are at least six main regions in the CYP2D6 structure where remarkable differences can be observed, with the most notable distinction involving the F helix, F–G loop, B′ helix, β-sheet 4, and part of β-sheet 1 (Rowland et al. 2006). These regions are all situated on the distal face of the protein and most of them are involved in defining the shape and character of the active site. For example, two additional turns in helix F of CYP2D6 cause it to arc down much more closely over the heme pocket toward the N-terminal end of the second strand of β-sheet 1. This is considered to contribute to a marked shift in the position of strands 1 and 2 of β-sheet 1 consisting of residues 71 to 78. Helix F′ is present in CYP2C9 (Williams et al. 2003) but is absent from 2D6. In addition, helix B′ in CYP2D6 is evidently pushed away from the heme pocket, and three additional residues (residues 101–118) in the loop immediately follow it (Rowland et al. 2006). Similarly, β-sheet 4 consisting of residues 468–487 has a conformational shift in the same direction as the helix F shift.

Two of the areas located on the proximal face of CYP2D6 are different from CYP2C9. CYP2D6 contains an extra turn

at the end of helix C and thus forms a shorter loop between helices C and D, which substantially decreases the interaction between the C–D connection region and the G–H loop. Moreover, β-sheet 2 consisting of residues 380–392 in 2D6 undergoes a considerable shift in the position of the two strands relative to β-sheet 1. In CYP2D6, these strands bend up toward the underside of β-sheet 1 more evident than CYP2C9, because of the presence of more small hydrophobic side chains.

6.8.2 ACTIVE-SITE CAVITY OF CYP2D6

The CYP2D6 structure 2F9Q described by Rowland et al. (2006) has a well-defined active-site cavity over the heme group with an estimated volume of ~540 Å^3 (Figure 6.1b). This is larger than that for CYP2A6 (260 Å^3) (Yano et al. 2005), 2A13 (304 Å^3) (Smith et al. 2007), 1A2 (375 Å^3) (Sansen et al. 2007), and 2E1 (190 Å^3) (Porubsky et al. 2008), but much smaller than that for CYP2R1 (979 Å^3) (Strushkevich et al. 2008), 3A4 (1385 Å^3) (Williams et al. 2004), 2C8 (1438 Å^3) (Schoch et al. 2004), and 2C9 (1667 Å^3) (Williams et al. 2003).

Rowland et al. (2006) have found that the active-site cavity of CYP2D6 is bordered by the heme and lined by at least 28 residues from helix B′ (side chain of Ile106), loop B′–C (side chains of Leu110, Phe112, Phe120, and Leu121 and main-chain atoms of Gln117, Gly118, Val119, and Ala122), helix F (side chains of Leu213, Glu216, Ser217, and Leu220), helix G (side chains of Gln244, Phe247, and Leu248), helix I (side chains of Ile297, Ala300, Asp301, Ser304, Ala305, Val308, and Thr309), the loop between helix K and strand 4 of β-sheet 1 (side chains of Val370 and Met374 and main-chain atoms of Gly373), and residues from the loop between the strands of β-sheet 4 (side chains of Phe483 and Leu484) (Figure 6.1c).

Rowland et al. (2006) have described the active-site cavity of CYP2D6 as adopting the shape of a "right foot," with the "heel" of the cavity located above the heme and the "arch" generated by the side chain of Phe120. The "ball" of the foot is lined by residues from the B′–C loop and the N-terminal residues of helix I, which also line the entire length of the right side of the "foot." The "toe" of the foot is bordered by residues from helices B′ and G and the upper part of the foot is lined by residues in helix F. The "heel back" is formed by residues in the loop after the K helix and the "ankle" area narrows the cavity and is bordered by residues on helix F at the front and residues on helix I to the right. The cavity entrance over the ankle region is bordered by a number of long charged/hydrophilic side chains from the F helix (Gln210, Glu211, Lys214, and Arg221) and residues from the region between the two strands of β-sheet 4 (side chains of Ala482 and Ser486 and main-chain atoms of Val485). Furthermore, the upper region and right side of the foot contains the hydrophilic side chains of Glu216 in helix F, Gln244 in helix G, and Ser304 in helix I (Rowland et al. 2006).

There is a small unidentified region of residual electron density approximately 5–6 Å above the heme group, which is not close to any active-site residues but is nearest to the side chain of Phe120. This has also been observed with CYP2C9 (Williams et al. 2003). Since nearby there are no residues that provide hydrogen-bonding interactions, it is unlikely to be a water molecule. A similar situation has also been seen with CYP2C9 (Williams et al. 2004). The highly conserved Thr309 in helix I is in an ideal position to form a hydrogen bond to the water molecule generated from the cleavage of the dioxygen bond of the heme-hydroperoxy intermediate during the P450 catalytic cycle (Schlichting et al. 2000).

6.8.3 Key Active-Site Amino Acids of CYP2D6

6.8.3.1 Ile106

Ile106 is located in the B′ helix of SRS1, within 4 Å of the carboxylate group of Glu216 as shown in homology models (Keizers et al. 2005a; Venhorst et al. 2003). Substitution of Ile106 with Glu caused a profound topological effect on the higher region within the active-site cavity and decreased the catalytic activity toward MAMC, dextromethorphan, bufuralol, and MDMA (van Waterschoot et al. 2006), with significantly increased K_m values. This may be attributed to

electrostatic repulsion of the Glu216 by the introduced Glu106 leading to a lateral shift of the flexible F–G loop toward the region above pyrrole A. In the 2F9Q structure, the side chain of Ile106 from helix B′ forms part of the border of the active site (Rowland et al. 2006).

6.8.3.2 Leu213 and Val308

Docking studies of the MAMC analogs in a CYP2D6 homology model previously reported (Keizers et al. 2005a) suggested a distal hydrophobic active-site binding cleft for the substrate N-alkyl chains, consisting of the residues Leu213 and Val308 (Keizers et al. 2006). This cleft could accommodate the relatively long N-alkyl chains of MAMC analogs by the apolar side chains of residues Leu213, Val308, and Phe483 located at the top of the binding pocket (Keizers et al. 2006). In the 2F9Q structure, Phe120, Phe483, Asp301, and Leu213 are in approximately identical positions and are all active-site residues (Rowland et al. 2006) as the homology model described by Keizers et al. (2005a).

6.8.3.3 Glu216

Rowland et al. (2006) have proposed that the acidic residue Glu216 could act as an important binding residue for ligands of CYP2D6 in the 2F9Q structure, which has been shown to be critical in substrate recognition in site-directed mutagenesis studies (Guengerich et al. 2003; Paine et al. 2003). A number of studies have clearly shown that Glu216 is needed in binding of most basic substrates. The Glu216 mutants appeared to contain the same complement of heme as the wild-type enzyme, in contrast to the Asp301 mutants (Ellis et al. 1995; Guengerich et al. 2003). Glu216 sitting at the top of the active-site cavity is more likely to serve as a recognition residue that attracts basic ligands to the pocket and forms an intermediate binding site before the substrate migrating to a productive position within the cavity. Docking studies of debrisoquine into the 2F9Q structure have demonstrated that it can fit readily into the pocket at the entrance channel opening between Glu216 and Phe483 (Rowland et al. 2006). Similarly, an intermediate binding pocket has been observed for CYP2C9 in complex with warfarin (Williams et al. 2003) and CYP2R1 in complex with vitamin D_3 (Strushkevich et al. 2008).

6.8.3.4 Asp301

In the 2F9Q structure, Rowland et al. (2006) have proposed that Asp301 can act as binding residue for ligands (both substrates and competitive inhibitors) of CYP2D6. The two rotameric states, trans- and gauche-, of the aspartate can provide an explanation for various pharmacophore models of CYP2D6. Asp301 is eight residues away from the nearest Thr (>2 helix turns), and thus its carboxylic acid may not be participating in O–O bond cleavage (Gerber and Sligar 1994). However, Asp301 could be involved in a hydrogen-bonding network with water molecules, on the basis of the data for substrate-bound CYP107A1/P450eryF (Cupp-Vickery et al. 1996). Modi et al. (1997) have reported that binding of MPTP did not require Asp301 and invoked alternate residues

(Smith et al. 1998). Residues other than Asp301 (e.g., Phe481) had been proposed for binding of amine substrates that are *N*-dealkylated (de Groot et al. 1999a). Guengerich et al. (2002) have shown that Asp301 is not required to bind the spirosulfonamide, a substrate devoid of basic nitrogen, but that Asp301 is somehow needed otherwise for catalysis and effective expression of holoenzyme (heme-containing) (Hanna et al. 2001a).

Asp301 has been proposed to play a structural role in CYP2D6 and has at least part of its role in the folding and structure of CYP2D6 (Ellis et al. 1995; Guengerich et al. 2003). In this putative role, Asp301 may influence the orientation of Phe120, the side chain of which appears to be oriented into the active site in such a way that it may contribute to substrate binding. Thus, an alternative mechanism by which mutation of Asp301 may affect substrate specificity can be postulated, namely, reorientation of Phe120.

6.8.3.5 Thr309

The side chain of Thr309 together with those of Ile297, Ala300, Asp301, Ser304, Ala305, and Val308 in helix I form part of the border of the active site in the 2F9Q structure (Rowland et al. 2006). The highly conserved Thr309 in helix I of CYP2D6 is in a proper position to hydrogen bond to the water molecule formed from the cleavage of the dioxygen bond of the heme-hydroperoxy intermediate during the CYP catalytic cycle (Schlichting et al. 2000). Mutations of Thr309 have been shown to alter the metabolism of substrates and metabolite ratios (Keizers et al. 2005b).

In several other CYPs, the highly conserved Thr residue in helix I has been shown to play a role in fixation and activation of the heme-bound oxygen. A water molecule located 7 Å above the heme iron is hydrogen bonded to Thr301 of CYP2C9 (corresponding to Thr309 of CYP2D6), which is highly conserved and considered to play a role in the proton transfer process during catalytic cycle (Haines et al. 2001). In addition, Thr303 in human CYP2E1 (Blobaum et al. 2004), Thr302 in rabbit CYP2B4 (Vaz et al. 1998), Thr306 in human CYP17 (Lee-Robichaud et al. 1998), Thr252 in bacterial CYP101 (Hishiki et al. 2000; Nagano and Poulos 2005), and Thr268 in CYP102 may be important for the appropriate orientation of an oxygen molecule in the axial position over heme, which can assist in proton transfer during the catalytic cycle (Harvey et al. 2006).

In bacterial CYP101, the heme-bound oxygen is activated by proton delivery mediated by Thr252 ad Asp251 (Imai et al. 1989). The hydroxyl group of Thr252 forms a hydrogen bond to a catalytic water molecule, which is the ultimate proton donor while Asp251 forms a hydrogen bond to charged residues at the surface and thus serves as a bridge between solvent trapped in the groove of helix I and bulk solvent (Raag et al. 1991; Schlichting et al. 2000). A similar role of Thr268 and Glu267 in bacterial CYP102 in proton delivery has been proposed (Yeom et al. 1995). Substitution of these highly conserved Thr residues with Ala has been shown to alter kinetic parameters owing to disruption of proton transfer process (Blobaum et al. 2004; Hishiki et al.

2000; Imai et al. 1989; Lee-Robichaud et al. 1998; Nagano and Poulos 2005).

Besides Thr309, three other Thr residues are present in helix I of CYP2D6 at positions 310, 312, and 313. The function of the latter three Thr residues is unknown. In the homology model of CYP2D6, Thr312 is one turn of helix I further away from the heme and appears to stabilize a distortion in helix I by providing hydrogen bonds to the carbonyl group of Val308. However, replacement of Thr312 with Val did not result in remarkable changes in expression levels or in the metabolic activity toward MAMC, bufuralol, MDMA, or dextromethorphan (Keizers et al. 2005b), suggesting that Thr312 only plays a limited role in the catalytic reaction by CYP2D6. This is consistent with the 2F9Q structure where Thr312 is not an active-site residue (Rowland et al. 2006).

6.8.3.6 Met374

The enzyme encoded by the published genomic (Kimura et al. 1989) and cDNA sequence (Gonzalez et al. 1988) of *CYP2D6* differs at residue 374 and encodes Val and Met, respectively. During structure determination, Rowland et al. (2006) used Met374 as the wild type. Multiple alignment studies of the CYP2 family have shown that residue 374 lies in the conserved β1-4-sheet region of CYP2D6 (Gotoh 1992; Korzekwa and Jones 1993; Lewis 1995). Residue 374 is located in one of the SRSs inferred by Gotoh (1992) as being important in substrate binding in the CYP2 family. However, homology models did not identify 374 as a possible residue in the active site of CYP2D6 (Koymans et al. 1993; Lewis 1995).

Furthermore, the difference in the stereoselective properties of the wild type and Met374Val toward metoprolol is rationalized using a homology model of the active site of CYP2D6 derived from the crystal structure of the bacterial P450BM3 (CYP102) (Ellis et al. 1996). In this model, the Val374 residue is located in the β1-4-sheet region and formed part of the wall of the active-site cavity. This region is well defined and can be substantiated with reference to His376, as this residue is highly conserved throughout the CYP superfamily and is thought to be responsible for binding one of the propionic acid groups in the porphyrin ring. Docking studies of metoprolol binding to the active site of the enzyme with Val374 suggest that Val374 is unable to contact the substrate during the binding process and consequently is unlikely to influence the regio- or enantioselective oxidation of the substrate (Ellis et al. 1996). However, in the enzyme containing 374Met, the Met residue is larger and extended into the active site and is able to contact the substrate at the active-site center, thereby sterically affecting the regioselective oxidation of metoprolol. In the homology model of the CYP2D6 active site, residue 374 resided on the opposite wall to, and significantly remote from, the Asp301 and Ser304 residues, both of which lie in helix I. Residue Ser304 is close to, and has the potential to hydrogen bond with, the chiral alcohol of metoprolol, thus providing a possible explanation for the observed chiral selectivity of CYP2D6 toward this substrate. The chiral hydroxy group of quinidine (a potent and selective inhibitor of CYP2D6) can potentially form a hydrogen bond with Ser304, whereas quinine, its diastereoisomer, has less potential to cause this interaction (Lewis 1995).

In the structure of 2F9Q, Met374 from β-sheet 1 forms part of the border of the active-site cavity. This is in agreement with the homology model described by Ellis et al. (1996) although the authors inserted Val374 as the wild-type residue. Therefore, this residue should play a role in substrate binding.

6.8.3.7 Phe120, Phe481, and Phe483

The aromatic active-site residues Phe481 and Phe483, located in the β1-4 loop, have been proposed to be involved in ligand binding (Hayhurst et al. 2001). The location of this loop, defining the wall of the active site opposite that of the B'–C loop and Phe120, implies the potential for a concerted or independent role for the aromatic residues in ligand orientation. Surprisingly, a similar situation has been observed with CYP2C9 in complex with warfarin (Williams et al. 2003). CYP2C9 and 2C19 possess conserved Phe114 (corresponding to Phe120 of CYP2D6) and Phe476 (corresponding to Phe481 of CYP2D6) residues in SRS1 and SRS6, respectively, and mutations at these sites significantly affect warfarin and diclofenac metabolism by CYP2C9 (Haining et al. 1999; Melet et al. 2003). The Phe114Leu and Phe476Leu mutants exhibited substantially reduced S-warfarin metabolism and altered hydroxy metabolite profiles but only moderately decreased S-flurbiprofen 4'-hydroxylation while maintaining product regioselectivity (Mosher et al. 2008). The Phe114Trp and Phe476Trp mutations also had opposing effects on S-warfarin versus S-flurbiprofen 4'-hydroxylation. The Phe476Trp mutant increased the efficiency of S-warfarin metabolism 5-fold but decreased the efficiency of S-flurbiprofen 4'-hydroxylation 20-fold (Mosher et al. 2008). These results indicate that various members of the CYP2 family utilize specific aromatic interactions (π–π stacking) as a common mechanism of orientation for aromatic substrates.

In the 2F9Q structure, the substrate can contact with Phe120 and Phe483, forming a part of the border of the active site (Rowland et al. 2006), which has been found to be critical in substrate recognition in site-directed mutagenesis studies (Flanagan et al. 2004; Lussenburg et al. 2005; Smith et al. 1998).

6.8.3.8 Arg441

Klein et al. (2007) have identified the 1321C>T mutation causing Arg441Cys (designated as the *CYP2D6*62* allele) in a person with reduced sparteine oxidation phenotype when resequencing of *CYP2D6* in individuals with discrepant genotype–phenotype relationship. The frequency of the novel mutation is very low in Caucasians (<0.1%). Although the amount of protein expressed in the mutant is comparable with that of the wild type, reduced CO-difference spectroscopy of insect cell microsomes did not show absorption at 450 nm for the Arg441Cys mutant. Furthermore, the recombinantly expressed mutated protein completely abolished the activity toward propafenone as a result of missing heme incorporation (Klein et al. 2007).

In the 2F9Q structure of CYP2D6 (Rowland et al. 2006), the residue Arg441 has been mapped to the second residue N-terminal of the invariant cysteine (Cys443) in a region N-terminal to helix L. The heme iron is bound by Cys443, and the heme is anchored in the heme-binding site by hydrogen-bonding interactions between the carboxyl groups of the two heme propionate groups and the side chains of Arg101, Trp128, Arg132, His376, Ser437, and Arg441 (Rowland et al. 2006). Sequence alignment of more than 5500 P450 heme-thiolate protein sequences of various species has revealed that the positively charged Arg441 residue is part of the heme-binding signature but not strictly conserved among all the P450s.

This consensus sequence shows that the residue corresponding to Arg441 is not always occupied by a positively charged amino acid, but different residues including His, Thr, and Pro are also seen at this position. Among these, the Arg433Trp variant of *CYP2C19* (*5 allele) (Ibeanu et al. 1998) and the Arg448His variant of *CYP11B1* (Curnow et al. 1993) completely abolished S-mephenytoin and cortisol 11β-hydroxylase activities in vivo, respectively. The Arg448His mutation of *CYP11B1* is associated with congenital adrenal hyperplasia and results in a hypertensive form of the disease (Curnow et al. 1993). Another natural variant of human *CYP1A2*8* (Arg456His) expresses normal amount of protein but is functionally inactive as a result of missing heme incorporation (Saito et al. 2005). Consistent with the fact that Cys is not part of the CYP19 family or PROSITE pattern, the *CYP19* variant Arg435Cys had only residual enzyme activity (1.1% of wild type) (Ito et al. 1993). These findings suggest that Arg441 appears to be essential for heme binding and enzymatic function of CYP2D6 and that the role of this residue in other CYPs appears to be dependent on the sequence context and may also include disturbance of the interaction with cytochrome P450 reductase.

6.8.4 Heme-Binding Site

In the 2F9Q structure, the heme group is located in the binding site and coordinated by hydrogen bond interactions with the side chains of Arg101, Trp128, Arg132, His376, Ser437, and Arg441 (Rowland et al. 2006). This is similar to the structures of CYP2C9 (Williams et al. 2003) and 2R1 (Strushkevich et al. 2008). In CYP2D6, the heme iron is penta-coordinated with Cys443, without a water molecule in the sixth coordination position in the electron density maps.

6.8.5 Access and Egress Channels

In the structures of bacterial and mammalian CYPs, one or more channels are present through which substrate can enter into the active site and the metabolite can exit from the active site (Williams et al. 2000; Winn et al. 2002). CYP2D6 contains a hydrophilic access channel that is similar to that seen in CYP2C9 (Williams et al. 2004). Access to the active-site cavity passed between the second SRS2 in the F helix and the SRS6 turn region of β-sheet 4 when using the nomenclature of Gotoh (1992). This entrance is clearly located above the ankle region of the active-site cavity, which is bordered by several hydrophilic side chains from helix F including Gln210, Glu211, Lys214, and Arg221 and residues from the region between the two strands of β-sheet 4 including Ala482, Ser486, and Val485, with the side chains of Asp179 from helix E and Thr312 from helix I in the vicinity (Rowland et al. 2006). The substrate could contact with Glu216 and Phe483 when it passed the access channel.

A potential second egress channel can also be observed in the 2F9Q structure, which resembles the PW2c pathway seen in the rabbit CYP2C5 structure (Schleinkofer et al. 2005). This channel is lined by Leu110, Phe112, and Leu248 and thus is highly hydrophobic. The resultant metabolite subsequently exits through a basic opening gated by Lys245 of helix G and Arg296 of helix I. The channel in CYP2C9 exits between helices F and I and the C-terminal β4-sheet system. CYP3A4 contains three access channels (Yano et al. 2004): a channel in the same area as CYP2E1 (Porubsky et al. 2008), a second channel through the middle of the B′ loop, and a third in the same position as CYP2C9 (Williams et al. 2004). However, neither CYP2A13 nor 2A6 has a substrate access channel in the structures determined (Smith et al. 2007; Yano et al. 2005).

6.8.6 BINDING REGION FOR CYTOCHROME P450 REDUCTASE

One-third of the C-terminal residues from various P450s are highly conserved, which contains a number of basic amino acid residues, forming an interface with the redox partner, cytochrome P450 reductase protein. This region in the 2F9Q structure containing a cluster of basic residues (mainly Arg and Lys) located on the proximal surface of the protein (Rowland et al. 2006) is considered to interact with cytochrome P450 reductase.

The Arg440His mutation (a natural SNP present in *CYP2D6*31*) in the C-terminus of CYP2D6 almost abolished the enzyme activity toward debrisoquine and dextromethorphan (Allorge et al. 2005). In the homology model of CYP2D6 derived from the crystal structure of rabbit CYP2C5 (PDB code: 1DT6, with 42% sequence identity), Arg440 is located on the proximal surface of the protein. It is remote from the active site but is located on a turn along with Arg441, a highly conserved residue in most P450s that forms a salt bridge with the carboxyl group of pyrrole ring A of the heme (Lewis 1995). The side chain of Arg440 pointed outward from the proximal surface of the protein into the region that could form the binding interface with cytochrome P450 reductase (Allorge et al. 2005). This interface area contained a cluster of basic residues, including Arg62, Arg63, Arg88, Arg129, Arg132, Arg133, Arg140, Arg440, and Arg450. These basic residues could interact with the FMN binding domain of human cytochrome P450 reductase, which contains a large number of acidic residues (Zhao et al. 1999).

A series of docking studies have suggested that Arg440 is a critical member of a cluster of basic residues for binding to the FMN domain of cytochrome P450 reductase (Allorge et al. 2005). Docking Arg440 with Asp148 of the P450 reductase is shown to yield a large number of steric clashes, and the overall protein–protein interaction is largely unfavorable. The Asp148 residue is previously shown to reduce CYP2D6-mediated codeine *O*-demethylation by 70% when mutated to Asn (Zhao et al. 1999). Another region of the P450 reductase that has been found to decrease codeine *O*-demethylation when mutated is the β3–α5 loop containing residues Asp84 and Asp87 (Allorge et al. 2005). Again, docking of the Arg440 region toward these residues resulted in a number of unfavorable interactions. In contrast, docking of Arg440 onto the β2–α3 residues Asp53 and Glu55 of the reductase is highly favorable, with a 10-fold higher electrostatic contribution compared to the interactions with Asp148, Asp84, or Asp87 (Allorge et al. 2005). In this favorable orientation, Arg440 appeared to be tightly bound to Asp53 and Glu55 of the reductase and Glu96 of CYP2D6 as well (Allorge et al. 2005). Since salt bridges between Arg440 and Asp53/Glu55 are rather remote from the FMN cofactor, Arg440 is considered to be more likely to serve as a primary binding site in the interface between CYP2D6 and P450 reductase, ensuring appropriate orientation and topology for efficient electron transfer via other residue interactions. Similarly, the Lys453Glu mutation in rat Cyp1a2 reduced catalytic activity toward benzphetamine and 7-ethoxycoumarin (Shimizu et al. 1991a,b). The Lys453 of rat CYP1A2 aligns with Arg440 of CYP2D6 (Lewis 1995). This substitution apparently decreases the affinity of the reductase for CYP1A2.

6.8.7 PLASTICITY OF CYP2D6

The reported CYP2D6 structure contains a relatively small active-site volume of ~540 Å^3, which is unlikely to fit all known substrates of this enzyme with distinct structures and molecular sizes. It appears that the active site needs to be opened up to accommodate a variety of substrates. To probe the potential effect of induced fit, the conformation of Phe483, and thermal motion on the accuracy of site of metabolism prediction, Hritz et al. (2008) have docked 65 known CYP2D6 substrates into an ensemble of 2500 protein structures, which covered various conformations of the CYP2D6 active site. The protein ensemble is generated by molecular dynamics simulations of CYP2D6 in complex with five representative substrates (3,4-methylene-dioxy-*N*-ethylamphetamine, MAMC, *R*-propranolol, chlorpromazine, and tamoxifen) using the GROMOS05 simulation package in combination with the GROMOS 45A4 force field. Two different conformations of the side chain of Phe483 are considered. The docking of the known 65 substrates into the ensemble of 2500 protein structures has revealed that although thermal motion generally involves only small conformational changes, these may have a dramatic effect on the resulting docking poses (Hritz et al. 2008). The effect of the side-chain conformation of Phe483 is observed to be strongest for the cluster of five representative substrates. Finally, three most essential CYP2D6 structures (PPD_70, CHZ_170, and TMF_70) are selected, which are suitable for various kinds of ligands and a binary decision tree is developed to help decide which protein structure to dock the ligand into (Hritz et al. 2008). These data clearly demonstrate a large sensitivity of docking reliability not only owing to larger conformational changes but also owing to thermal motion.

6.8.8 A Comparison of the Homology Model and Crystal Structure of CYP2D6

Kjellander et al. (2007) have compared the homology model for CYP2D6 reported previously by De Rienzo et al. (2000) and the crystal structure for CYP2D6 (PDB code: 2F9Q) with respect to the prediction of the site of metabolism using 25 opioid analgesics including dextromethorphan, codeine, levallorphan, oxycodone, morphine, and hydrocodone as the substrates. For the prediction of opioid metabolism by MetaSite version 2.7.5 (Molecular Discovery, http://moldis covery.com), the homology model and the crystal structure gave a similar prediction rate (92% and 85%, respectively). However, there are some marked differences in the shape of active site and orientation of several key amino acids in the active site between the homology model and 2F9Q (Kjellander et al. 2007). In the homology model, the hydrophobic pocket is spread out over the heme formed by Ile106, Leu121, Leu213, Leu284, and Phe483, while the hydrophobic pocket is narrow owing to the distinct orientation of Ile106, Phe120, and Phe483, forming a channel from the heme to the surface in the 2F9Q structure. Glu216 is oriented toward the cavity in the homology model, while Asp301 did not differ much between the two structures. 2F9Q had a long F helix connected to the G helix with a loop, while the homology model has a shorter F helix and an additional F′ helix in the loop between F and G helices (Kjellander et al. 2007). In addition, Leu121, Glu216, Ser304, Ile106, Asp301, and Met374 seemed to have a greater importance in interacting with the ligand in the homology model based on interaction energy calculation, while Phe483, Leu484, Leu213, Phe120, and Val370 showed greater interaction energy with the ligands in 2F9Q.

Notably, the comparison of the two CYP2D6 structures identified differences in involvement of the two charged amino acids Glu216 and Asp301, which have been shown to be involved in substrate binding of CYP2D6 in site-directed mutagenesis studies (Guengerich et al. 2003; Paine et al. 2003). Mutated enzymes where these acidic residues have been replaced with neutral amino acids showed increased binding constants and a decreased rate of oxidation for some common basic CYP2D6 substrates. However, these two amino acids appeared to be less important for interacting with the ligand in the 2F9Q structure, and instead the hydrophobic Phe483 residue had a more apparent role. This residue has been proposed to be involved in substrate binding in a homology model described by Modi et al. (1996), but this could not be confirmed by site-directed mutagenesis using bufuralol as a probe substrate (Smith et al. 1998). On the other hand, removal of the aromatic property of Phe481 has been shown to impair the binding of CYP2D6 probe substrates (Hayhurst

et al. 2001). The importance of Phe481 seems to be ligand dependent, and the binding is not affected by this residue alone but a combination with the acidic residues. On the basis of the results of the energy calculation, hydrophobic interactions dominated the binding in 2F9Q, involving the residues Phe483, Leu213, Leu284, and Phe120. These residues are also responsible for the narrow shape of the hydrophobic pocket in the crystal structure. In the homology model, the acidic residues Glu216 and Asp301 appeared to have more impact on ligand binding.

6.8.9 Binding of Atypical Substrates to Human CYP2D6

Most CYP2D6 substrates are bases containing a basic nitrogen atom 5–10 Å from the site of metabolism. However, there are exceptions to this. For example, spirosulfonamide, a potential inhibitor of COX-2, is a high-affinity substrate of CYP2D6 with a K_m of 7 µM (Guengerich et al. 2002), which is devoid of a basic nitrogen. The phenylsulfonamide of this compound is the only nitrogen in the molecule, but this moiety is not positively charged at physiological pH because of the presence of the electron-withdrawing sulfone group. CYP3A4 is also involved in the metabolism of spirosulfonamide.

Several steroids devoid of nitrogen atom have been found to be metabolized by CYP2D6, although its relative contribution is variable. Both 17β-estradiol and estrone are metabolized by CYP2D6. CYP2D6 can catalyze 2-hydroxylation of estrogens, although CYP1A1 and 1A2 have been considered as the major CYP enzymes responsible for the 2-hydroxylation of 17β-estradiol and estrone in extrahepatic tissues such as breast (Lee et al. 2003). Hydroxylation at other positions of 17α-ethinylestradiol is catalyzed primarily by CYP3A4 and 2C9 (Wang et al. 2004).

CYP2D6 can metabolize progesterone via 6β-hydroxylation, although CYP3A4 is more efficient (Niwa et al. 1998). Rat Cyp2d1 and 2d4 catalyze progesterone 6β- and 16α-hydroxylation and 2β- and 21-hydroxylation, respectively (Hiroi et al. 2001). Human CYP2D6 and rat Cyp2d4 are the predominant CYP2Ds in the brain and exhibit 21-hydroxylation activity toward progesterone and its metabolite, 17α-hydroxyprogesterone (Kishimoto et al. 2004). Multiple human CYPs are involved in the hydroxylation of progesterone. CYP3A4 showed the highest progesterone 16α-hydroxylation activity, followed by CYP1A1 and 2D6 (Niwa et al. 1998). Human CYP3A4, 3A5, and 4B1 (expressed in the lung but not in the liver) also catalyze the 6β-hydroxylation of progesterone (Waxman et al. 1991; Yamakoshi et al. 1999). Furthermore, progesterone inhibited CYP2D6-catalyzed bufuralol 1′-hydroxylation with a K_i value of 33 µM (Hiroi et al. 2001).

CYP2D6 catalyzes the hydroxylation of pregnenolone at 2-, 6-, 16-, and 21-positions (Hiroi et al. 2001). Pregnenolone is a steroid precursor to cortisol, dehydroepiandrosterone (DHEA), progesterone, and testosterone (Freeman et al. 2001). Pregnenolone can be converted to progesterone by 3β-hydroxysteroid dehydrogenase and Δ^{4-5} isomerase. In addition, pregnenolone can be converted to 17-hydroxy-pregnenolone by 17α-hydroxylase (CYP17A1), which is then converted to DHEA by desmolase. DHEA is the precursor of androstenedione.

CYP2D6 catalyzes 2β-, 6β-, and 17-hydroxylation of testosterone, although testosterone 6β-hydroxylation is mainly catalyzed by CYP3A4 and 3A5 (Hiroi et al. 2001). Testosterone inhibited CYP2D6-catalyzed bufuralol 1'-hydroxylation with a K_i value of 63 μM (Hiroi et al. 2001). An apparent increase in active-site volume by substitution of Phe483 is sufficient to allow CYP2D6-mediated testosterone metabolism (Smith et al. 1998).

Allopregnanolone (3α,5α-tetrahydroprogesterone), an important neurosteroid in the human brain and a representative γ-aminobutyric acid receptor modulator (Herd et al. 2007; Mitchell et al. 2008), is hydroxylated at the C21-position by recombinant Cyp2d4 and CYP2D6. It is a metabolite of progesterone of ovarian or adrenal origin but can also be synthesized de novo in the brain from cholesterol via a side-chain cleavage enzyme that forms pregnenolone (Compagnone and Mellon 2000; Mensah-Nyagan et al. 1999), the precursor of all steroid molecules (Tsutsui 2008).

Ritonavir, a substrate of CYP2D6, has two amide linkages and a carbamate, and these might be capable of hydrogen bonding with Asp301 or another entity. Similarly, progesterone has two carbonyls. However, the spirosulfonamide analog and the steroid testosterone have only a single (conjugated) carbonyl. All of these compounds bind to CYP2D6 with affinities in the range of many of the amines used in modeling work and are also substrates for CYP2D6.

The ability of CYP2D6 homology models to predict the binding modes of atypical substrates devoid of a basic nitrogen (e.g., spirosulfonamide) has been questioned (Guengerich et al. 2002). However, on the basis of the homology model incorporating rabbit CYP2C5 structure (Kirton et al. 2002), Kemp et al. (2004) have successfully docked spirosulfonamide into this model. The highest ranked docked solution positioned the cyclopentyl moiety above the heme and thus correctly identified a major metabolite (Guengerich et al. 2002).

The same model is used to screen a small compound set containing 33 compounds from the National Cancer Institute database using the program GOLD version 2.0 (Kemp et al. 2004). There is a correlation between experimental IC_{50} values for CYP2D6 inhibition and the corresponding docked scores, and the docking score could discriminate between potent (IC_{50} <10 μM) and weak (IC_{50} >10 μM) inhibitors and correctly identified several new inhibitors of CYP2D6 such as NCI_17383 and NCI_249992 (Kemp et al. 2004). NCI_17383 containing a basic nitrogen atom is positioned between Glu216 and Asp301, both of which have been identified as key residues for substrate binding (Paine et al. 2003). The aromatic groups of NCI_17383 are packed between Phe120 and Phe483, two Phe residues that have been shown to be important for aromatic substrate binding in site-directed mutagenesis studies (Flanagan et al. 2004). NCI_249992 devoid of a basic nitrogen atom could also be docked into the active site close to the heme with the aromatic rings forming stacking

with Phe120 and Phe483. No hydrogen bond is formed to Glu216 and Asp301.

It has been reported that pactimibe, a novel acyl coenzyme A/cholesterol acyltransferase inhibitor developed for the treatment of hypercholesterolemia and atherosclerotic diseases, and its metabolite R-125528 are metabolized by CYP2D6 via ω-1 oxidation (Kotsuma et al. 2008c). The indolin oxidation of pactimibe is mediated by CYP3A4. Both pactimibe and R-125528 are atypical substrates for CYP2D6 because of its acidity. Both compounds are not protonated but are negatively charged at physiological pH. The K_m value of R-125528 in CYP2D6-expressing microsomes is 1.74 μM, which is comparable to those of typical basic CYP2D6 substrates (1–10 μM). Pactimibe has a lower affinity than R-125528 to CYP2D6, although the K_m value is comparable to that of metoprolol. Interestingly, their sites of metabolism, the ω-1 position of the N-octyl indoline/indole group, are relatively distant from the aromatic moiety. An induced-fit docking of the ligands to a crystal structure of substrate-free CYP2D6 indicated the involvement of an electrostatic interaction between the carboxyl group and Arg221, and hydrophobic interaction between the aromatic moiety and Phe483 (Kotsuma et al. 2008a). With the concomitant treatment of ketoconazole, AUC of pactimibe in healthy subjects increased 1.7-fold and the AUC of R-125528 decreased by 55%. With the concomitant treatment of quinidine, the AUC for pactimibe increased 1.7-fold but the AUC of R-125528 is markedly elevated 5.0-fold (Kotsuma et al. 2008b). As the estimated metabolic fraction of pactimibe by CYP3A4 and CYP2D6 from in vitro studies is 0.40 and 0.33, respectively, AUC increase ratios of pactimibe are estimated to be 1.7 with ketoconazole and 1.5 with quinidine. It can be expected that PMs would have significantly increased concentrations of R-125528 and moderately increased pactimibe levels.

6.9 BINDINGS MODES OF THE SUBSTRATES AND INHIBITORS WITH CYP2D6

The molecular determinants for ligands of CYP2D6 are critical for the ligand–enzyme interaction. The determination of the contributing factors to the molecular interactions is of great importance to distinguish the ligand of CYP2D6 and decipher the pattern of the interaction, which can eventually manipulate CYP2D6-mediated drug metabolism and generate maximal drug effect. To investigate the molecular determinants for ligands of CYP2D6, we have docked 88 substrates and 45 inhibitors of CYP2D6 into the active site of 3QM4 using the Discovery Studio 3.1 program, without crystallographic water molecules. The substrates include 17β-estradiol, 4-methoxyphenamine, 5-MeO-DIPT, 5-MeO-DMT, 5-methoxytryptamine, acetaminophen, allopregnanolone, almotriptan, amiflamine, amiodarone, antipyrine, aripiprazole, betaxolol, bicifadine, bisoprolol, bortezomib, bufuralol, bunitrolol, chlorpheniramine, cimetidine, clozapine, codeine, debrisoquine, delavirdine, dextromethorphan, dihydrocodeine, diltiazem, diuron, dolasetron, donepezil, encainide, enclomifene, estrone flecainide, fluvastatin, galantamine, gepirone, guanoxan, harmine, hydrocodone, imatinib, imipramine, indinavir, lasiocarpine, lidocaine, loratadine, MDMA, mequitazine, methamphetamine, methoxyphenamine, metoclopramide, metoprolol, mexiletine, mirtazapine, monocrotaline, morphine, MPBP, MPPP, N-desethylamiodarone, nevirapine, nicergoline, olanzapine, ondansetron, oxymorphone, pactimibe, perhexiline, phenformin, pinoline, pranidipine, procainamide, progesterone, promazine, propafenone, propranolol, saquinavir, selegiline, sildenafil, tacrine, tamoxifen, testosterone, timolol, tramadol, tropisetron, venlafaxine, verapamil, vernakalant, zolpidem, and zuclopenthixol. The molecular determinants for the substrate–CYP2D6 interaction are summarized in Table 6.2.

The inhibitors include amitriptyline, amodiaquine, aprindine, atomoxetine, azelastine, biperiden, carvedilol, chloroquine, chlorpromazine, cibenzoline, cinnarizine, citalopram, clomipramine, diphenhydramine, doxepin, duloxetine, ezlopitant, flunarizine, fluoxetine, fluvoxamine, halofantrine, haloperidol, ibogaine, iloperidone, indomethacin, maprotiline, methadone, mianserin, nebivolol, nortriptyline, oxatomide, paroxetine, perphenazine, pimozide, prasugrel, quinidine, risperidone, ritonavir, sertraline, sparteine, terfenadine, thioridazine, ticlopidine, tolterodine, and trimipramine. The molecular determinants for the inhibitor–CYP2D6 interaction are summarized in Table 6.3.

Multiple docking solutions of the CYP2D6 ligands are obtained, but only one or two docking solutions are accepted for the substrates according to their productive orientations which could lead to metabolite formation. The ionic and hydrophobic contacts between the substrates/inhibitors and the active-site residues of CYP2D6 mainly involve Phe120, Asp301, Ser304, Glu216, Gln244, Phe112, and Phe483. Phe247, Ala300, Arg221, Phe147, and Arg101 are also important in the binding of most ligands examined. Among the active-site residues, Phe120, Glu216, and Ser304 form interactions with all ligands regardless of its presence or absence of a basic nitrogen. To date, the role of Leu213 and Ala305 in the determination of substrate specificity of CYP2D6 has not been investigated using site-directed mutagenesis studies. We have also performed ADMET and hepatotoxicity prediction using Discovery Studio 3.0. The predicted ADMET profile of various substrates of CYP2D6 is presented in Table 6.4 and the predicted hepatotoxicity of various substrates of CYP2D6 is shown in Table 6.5. The predicted ADMET profile of various inhibitors of CYP2D6 is summarized in Table 6.6 and the predicted hepatotoxicity of various inhibitors of CYP2D6 is shown in Table 6.7.

6.9.1 Binding Modes of the Substrates with CYP2D6 Active Site

17β-Estradiol is able to be docked into the active site of CYP2D6 with a CDOCKER interaction energy of 37.2331 kcal/mol. 17β-Estradiol interacts with CYP2D6 active site through a hydrogen bond formation at Gln244 (Figure 6.2). 4-Methoxyphenamine binds to CYP2D6 active site and forms a hydrogen bond with Glu216 and a charge interaction at

Glu216 in the active site of CYP2D6 (Figure 6.2). 5-MeO-DIPT and 5-MeO-DMT are successfully docked into the active site of CYP2D6 with a CDOCKER interaction energy of 48.6711 and 40.0782 kcal/mol, respectively. Both 5-MeO-DIPT and 5-MeO-DMT bind to CYP2D6 mainly through π–π interaction. Phe120 is responsible for the 5-MeO-DIPT–CYP2D6 interaction and Phe120 and Phe483 account for the 5-MeO-DMT–CYP2D6 binding (Figure 6.2).

5-Methoxytryptamine binds to the active site of CYP2D6 via the formation of hydrogen bond and π–π interaction at Glu216 (Figure 6.3). The CDOCKER interaction energy for the 5-methoxytryptamine–CYP2D6 interaction is 39.7015 kcal/mol. Acetaminophen also binds to the CYP2D6 active site through the formation of a hydrogen bond at Glu216 with a CDOCKER interaction energy of 25.8107 kcal/mol (Figure 6.3). Almotriptan is docked into the active site of CYP2D6 via the hydrogen formation and π–π interaction. Asp301 is the amino acid responsible for the hydrogen formation and π–π interaction with almotriptan (Figure 6.3). The CDOCKER interaction energy is 48.1804 kcal/mol. Amiflamine is able to interact with the CYP2D6 active site via the formation of two hydrogen bonds at Asp301 and Ser304, π–π interaction at Phe120, and the charge interaction at Asp301 (Figure 6.3). The CDOCKER interaction energy is 38.3625 kcal/mol.

Amiodarone binds to CYP2D6 at the active site through the hydrogen bond formation at Cys443 (Figure 6.4). Aripiprazole is docked into the active site of CYP2D6 and interacts with CYP2D6 via the hydrogen bond formation at Arg101 (Figure 6.4). Betaxolol interacts with CYP2D6 at the active site with a CDOCKER interaction energy of 52.8065 kcal/mol. Betaxolol forms two hydrogen bonds with Glu216. Betaxolol also interacts with CYP2D6 via the π–π interaction at Phe120 and charge interaction at Glu216 (Figure 6.4). Bisoprolol also forms two hydrogen bonds at Asp301 and Ser304 at the active site of CYP2D6. The π–π interaction at Phe120 and charge interaction at Asp301 are another two molecular determinants for the bisoprolol–CYP2D6 interaction at the active site (Figure 6.4).

Bortezomib is docked into the active site of CYP2D6 and the CDOCKER interaction energy is 50.1534 kcal/mol. Bortezomib forms a hydrogen bond at Ser304 that is responsible for the interaction (Figure 6.5). Bufuralol interacts with CYP2D6 at the active site through the formation of π–π interaction (Figure 6.5). Bufuralol forms two π–π interactions at Phe120 and Phe483, and the CDOCKER interaction energy is 43.0713 kcal/mol. Bunitrolol interacts with CYP2D6 at the active site through hydrogen bond formation, π–π interaction, and charge interaction. Asp301 and Ser304 are involved in the hydrogen bond formation. Phe120 forms two π–π interactions with bunitrolol. Asp301 contributes to the charge interaction between bunitrolol and CYP2D6 (Figure 6.5). Moreover, chlorpheniramine binds to the CYP2D6 active site and forms two hydrogen bonds at Asp301, a π–π stacking interaction at Phe120, and a charge interaction at Asp301 (Figure 6.5).

Cimetidine is able to be docked into the active site of CYP2D6. Cimetidine forms two hydrogen bonds at Glu216

TABLE 6.2
Binding Modes of Known Substrates with CYP2D6 (PDB ID: 3QM4)

Substrate	CDOCKER Interaction Energy	Number of H-Bonds	Residues Involved in H-Bond Formation	Number of π–π Stacking	Residues Involved in π–π Stacking	Number of Charge Interaction	Residues Involved in Charge Interaction
17β-Estradiol	37.2331	1	O-Gln244	0	–	–	–
4-Methoxyphenamine	36.9882	1	O-Glu216	0	–	1	Glu216
5-MeO-DIPT	48.6711	0	–	1	Phe120	–	–
5-MeO-DMT	40.0782	0	–	2	Phe120 Phe483	–	–
5-Methoxytryptamine	39.7015	1	O-Glu216	0	–	1	Glu216
Acetaminophen	25.8107	1	O-Glu216	0	–	–	–
Allopregnanolone	–	–	–	–	–	–	–
Almotriptan	48.1804	1	O-Asp301	0	–	1	Asp301
Amiflamine	38.3625	2	O-Asp301 O-Ser304	1	Phe120	1	Asp301
Amiodarone	0.7993921	1	S-Cys443	0	–	–	–
Antipyrine	24.9153	0	–	0	–	–	–
Aripiprazole	62.3819	1	O-Arg101	0	–	–	–
Betaxolol	52.8065	2	O-Glu216 (2)	1	Phe120	1	Glu216
Bicifadine	26.279	0	–	0	–	–	–
Bisoprolol	57.559	2	O-Asp301 O-Ser304	1	Phe120	1	Asp301
Bortezomib	50.1534	1	O-Ser304	0	–	–	–
Bufuralol	43.0717	0	–	2	Phe120 Phe483	–	–
Bunitrolol	41.2332	2	O-Asp301 O-Ser304	2	Phe120 (2)	1	Asp301
Chlorpheniramine	46.3808	2	O-Asp301	1	Phe120	1	Asp301
Cimetidine	43.5349	2	O-Glu216 O-Asp301	2	Phe120 (2)	1	Glu216
Clozapine	64.1042	1	O-Asp301	3	Phe120 (2) Phe247	2	Glu216 Asp301
Codeine	–	–	–	–	–	–	–
Debrisoquine	36.0229	2	O-Glu216 O-Asp301	1	Phe120	1	Asp301
Delavirdine	53.9815	2	O-Arg101 O-Arg441	0	–	–	–
Dextromethorphan	–	–	–	–	–	–	–
Dihydrocodeine	–	–	–	–	–	–	–
Diltiazem	7.04228	0	–	1	Phe120	–	–
Diuron	30.6372	1	O-Ser304	0	–	–	–
Dolasetron	39.7195	0	–	2	Phe112 (2)	–	–
Donepezil	54.1521	0	–	1	Phe120	–	–
Encainide	52.3633	0	–	1	Arg221	–	–
Enclomifene	57.5954	1	O-Ser304	1	Phe120	–	–
Estrone	37.654	1	O-Gln244	0	–	–	–
Flecainide	52.5937	1	O-Asp301	0	–	1	Asp301
Fluvastatin	43.9883	0	–	1	Phe112	–	–
Galantamine	40.2228	1	O-Glu216	0	–	–	–
Gepirone	45.7063	0	–	1	Phe120	–	–
Guanoxan	29.4404	1	O-Ser304	0	–	–	–
Harmine	27.8052	1	O-Ser304	0	–	–	–
Hydrocodone	–	–	–	–	–	–	–
Imatinib	68.3682			1	Gly445		

(Continued)

TABLE 6.2 (CONTINUED)
Binding Modes of Known Substrates with CYP2D6 (PDB ID: 3QM4)

Substrate	CDOCKER Interaction Energy	Number of H-Bonds	Residues Involved in H-Bond Formation	Number of π–π Stacking	Residues Involved in π–π Stacking	Number of Charge Interaction	Residues Involved in Charge Interaction
Imipramine	41.6589	0	–	1	Phe120	–	–
Indinavir	–	–	–	–	–	–	–
Lasiocarpine	55.9236	0	–	0	–	–	–
Lidocaine	45.1015	1	O-Asp301	1	Phe120	1	Asp301
Loratadine	33.2681	0	–	1	Phe112	–	–
MDMA	36.6762	1	O-Glu216	0	–	1	Glu216
Mequitazine	49.9045	2	O-Phe120 (2)	0	–	–	–
Methamphetamine	34.6757	2	O-Asp301 O-Ser304	1	Phe120	1	Asp301
Methoxyphenamine	38.6441	2	O-Ser304 (2)	1	Phe120	1	Asp301
Metoclopramide	48.8421	1	O-Glu216	2	Phe120 Phe483	–	–
Metoprolol	46.2299	2	O-Asp301 O-Ser304	1	Phe120 (2)	1	Asp301
Mexiletine	38.2153	1	O-Asp301	1	Phe120	1	Asp301
Mirtazapine	47.4762	1	O-Glu216	0	–	2	Glu216 Arg221
Monocrotaline	–	–	–	–	–	–	–
Morphine	–	–	–	–	–	–	–
MPBP	42.8551	1	O-Glu216	1	Phe120	1	Glu216
MPPP	39.6089	0	–	2	Leu213 Phe247	–	–
N-Desethylamiodarone	37.1519	0	–	0	–	–	–
Nevirapine	36.2435	0	–	0	–	–	–
Nicergoline	57.7917	0	–	1	Phe120	1	Asp301
Olanzapine	57.2848	1	O-Glu216	1	Phe120	2	Glu216 Asp301
Ondansetron	38.1086	0	–	0	–	–	–
Oxymorphone	–	–	–	–	–	–	–
Pactimibe	15.9956	1	O-Glu216	2	Phe120 Phe483	1	Asp301
Perhexiline	52.2564	1	O-Asp301	2	Phe120 (2)	1	Asp301
Phenformin	41.2877	4	O-Glu216 O-Asp301 O-Ser304 (2)	1	Phe120	1	Asp301
Pinoline	35.0877	1	O-Glu216	0	–	2	Glu216 Arg221
Pranidipine	54.0833	1	O-Glu216	1	Phe247	–	–
Procainamide	40.7413	2	O-Glu216 O-Ser304	0	–	1	Glu216
Progesterone	–	–	–	–	–	–	–
Promazine	49.3219	0	–	1	Phe120	2	Glu216 Asp301
Propafenone	57.338	2	O-Glu216 O-Ser304	0	–	1	Glu216
Propranolol	41.1045	1	O-Ser304	1	Phe120	–	–
Saquinavir	–	–	–	–	–	–	–
Selegiline	37.4401	1	O-Asp301	0	–	1	Asp301
Sildenafil	62.179	1	O-Glu216	2	Phe120 Arg221	–	–
Tacrine	37.4595	1	O-Asp301	1	Phe120	1	Asp301

(Continued)

TABLE 6.2 (CONTINUED)
Binding Modes of Known Substrates with CYP2D6 (PDB ID: 3QM4)

Substrate	CDOCKER Interaction Energy	Number of H-Bonds	Residues Involved in H-Bond Formation	Number of π–π Stacking	Residues Involved in π–π Stacking	Number of Charge Interaction	Residues Involved in Charge Interaction
Tamoxifen	43.4693	0	–	2	Phe112 Phe120	–	–
Testosterone	–	–	–	–	–	–	–
Timolol	45.106	0	–	1	Phe120	–	–
Tramadol	49.5785	2	O-Glu216 (2)	0	–	1	Glu216
Tropisetron	41.4045	2	O-Ala300 O-Ser304	2	Phe120 Phe483	1	Glu216
Venlafaxine	51.8895	2	O-Glu216 O-Ser304	1	Phe120	1	Glu216
Verapamil	33.6044	0	–	1	Phe120	–	–
Vernakalant	59.7309	2	O-Glu216 (2)	0	–	1	Glu216
Zolpidem	41.1319	0	–	0	–	–	–
Zuclopenthixol	48.4627	1	O-Glu216	1	Phe120	–	–

and Asp301, two π–π interactions at Phe120, and a charge interaction at Glu216 (Figure 6.6). Clozapine binds to CYP2D6 with a CDOCKER interaction energy of 64.1042 kcal/mol. Clozapine forms a hydrogen bond at Asp301, three π–π interactions at Phe120 (two) and Phe247, and two charge interactions at Glu216 and Asp301 (Figure 6.6). Debrisoquine is able to interact with CYP2D6 at the active site through the hydrogen bond formation, π–π interaction, and charge interaction. Glu216 and Asp301 contribute to the hydrogen bond formation. Phe120 is involved in the π–π interaction and Asp301 accounted for the charge interaction (Figure 6.6). Delavirdine interacts with the CYP2D6 active site with a CDOCKER interaction energy of 53.9815 kcal/mol. Delavirdine forms two hydrogen bonds at Arg101 and Arg441 (Figure 6.6).

Diltiazem forms a π–π interaction at Phe120 at the active site of CYP2D6. The CDOCKER interaction energy is 7.04228 kcal/mol (Figure 6.7). Diuron forms a hydrogen bond at Ser304 at the active site of CYP2D6 (Figure 6.7). Dolasetron interacts with the CYP2D6 active site through the hydrogen bond formation. There are two π–π interactions formed at Phe112 (Figure 6.7). Donepezil forms a π–π interaction at Phe120 at the active site of CYP2D6 (Figure 6.7). Encainide is docked into the active site of CYP2D6 and forms a π–π interaction at Arg221 (Figure 6.8). Furthermore, enclomifene binds to CYP2D6 in the active site through the hydrogen bond formation at Ser304 and π–π interaction at Phe120 (Figure 6.8). Estrone binds to the CYP2D6 active site with a CDOCKER interaction energy of 37.654 kcal/mol. There is a hydrogen bond formation at Gln224 that is responsible for estrone–CYP2D6 interaction (Figure 6.8). Moreover, flecainide is oriented in the CYP2D6 active site through the hydrogen bond formation at Asp301 (Figure 6.8).

Fluvastatin is readily docked into the active site of CYP2D6 and forms a π–π interaction at Phe112 (Figure 6.9). Galantamine is positioned in the CYP2D6 active site via the hydrogen bond formation at Glu216 (Figure 6.9). Gepirone is able to be docked into the active site of CYP2D6 and forms a π–π interaction at Phe120 (Figure 6.9). Guanoxan binds to the CYP2D6 active site via the hydrogen bond formation at Ser304 (Figure 6.9). Harmine forms a hydrogen bond at Ser304 in the active site of CYP2D6 (Figure 6.10). Imatinib forms a π–π interaction at Gly445 in the active site of CYP2D6 (Figure 6.10). Imipramine binds to the CYP2D6 active site via π–π interaction. Phe120 is involved in imipramine–CYP2D6 binding (Figure 6.10). In addition, lidocaine is docked into the active site of CYP2D6 and forms a hydrogen bond at Asp301, a π–π interaction at Phe120, and a charge interaction at Asp301 (Figure 6.10).

Loratadine is positioned into the active site of CYP2D6 mainly via a π–π interaction at Phe112 (Figure 6.11). MDMA binds to the CYP2D6 active site through the hydrogen bond formation at Glu216 and charge interaction at Glu216 (Figure 6.11). Mequitazine is able to be docked in the active site of CYP2D6 and forms two π–π interactions at Phe120 (Figure 6.11). Methamphetamine binds to the CYP2D6 active site via hydrogen bond formation, π–π interaction, and charge interaction. There are two hydrogen bonds formed at Asp301 and Ser304. Phe120 and Asp301 are involved in the π–π interaction and charge interaction, respectively (Figure 6.11).

Methoxyphenamine binds to the CYP2D6 active site with a CDOCKER interaction energy of 38.6441 kcal/mol. The methoxyphenamine–CYP2D6 interaction includes two hydrogen bonds at Ser304, one π–π interaction at Phe120, and one charge interaction at Asp301 (Figure 6.12). Metoclopramide is docked into the CYP2D6 active site through hydrogen bond formation at Glu216 and two π–π interactions at Phe120 and Phe483 (Figure 6.12). In addition, metoprolol binds to the CYP2D6 active site via the hydrogen bond formation at Asp301 and Ser304. There are π–π interactions at Phe120 and one charge interaction at Asp301, which are involved in

TABLE 6.3

Binding Modes of Known Inhibitors with CYP2D6 (PDB ID: 3QM4)

Substrate	CDOCKER Interaction Energy	Number of H-Bonds	Residues Involved in H-Bond Formation	Number of π–π Stacking	Residues Involved in π–π Stacking	Number of Charge Interaction	Residues Involved in Charge Interaction
Amitriptyline	40.9778	0	–	1	Phe120	1	Glu216
Amodiaquine	48.8027	1	O-Glu216	4	Phe120 (3) Phe483	1	Glu216
Aprindine	53.336	2	O-Glu216	1	Phe120	1	Glu216
Atomoxetine	47.3775	2	O-Glu216 (2)	1	Phe120	1	Glu216
Azelastine	55.1949	2	O-Glu216	0	–	1	Glu216
Biperiden	42.1184	0	–	3	Phe112 Phe120 Phe483	–	–
Carvedilol	58.5652	0	–	2	Phe120 Phe247	–	–
Chloroquine	49.7736	1	O-Pro435	2	Phe120 Phe483	–	–
Chlorpromazine	47.9246	0	–	1	Phe120	1	Glu216
Cibenzoline	44.6437	2	O-Asp301 O-Glu216	0	–	1	Asp301
Cinnarizine	50.5242	0	–	1	Arg101	–	–
Citalopram	46.5182	1	O-Glu216	0	–	2	Glu216 Arg221
Clomipramine	51.8681	1	O-Ser304	3	Phe120 (2) Phe247	–	–
Diphenhydramine	41.7978	0	–	3	Phe112 Phe120 Phe483	–	–
Doxepin	39.7824	0	–	1	Phe120	1	Glu216
Duloxetine	49.3159	3	O-Asp301 O-Ser304	1	Phe120	1	Asp301
Ezlopitant	–	–	–	–	–	–	–
Flunarizine	56.3384	0	–	2	Phe120 Phe483	1	Glu216
Fluoxetine	49.2181	2	O-Glu216 (2)	1	Phe483	1	Glu216
Fluvoxamine	47.2923	1	O-Glu216	3	Phe120 Phe483 (2)	1	Glu216
Halofantrine	58.6801	0	–	2	Phe120 Phe483	–	–
Haloperidol	50.8447	1	O-Ser304	1	Phe120	–	–
Ibogaine	42.4151	0	–	3	Phe120 (2) Phe483	–	–
Iloperidone	57.9902	1	O-Arg101	0	–	–	–
Indomethacin	38.323	0	–	0	–	1	Glu216
Maprotiline	46.2915	1	O-Glu216	0	–	2	Glu216 Arg221
Methadone	–	–	–	–	–	–	–
Mianserin	40.2496	0	–	0	–	–	–
Nebivolol	58.9162	2	O-Glu216 O-Ser304	2	Phe120 (2)	1	Glu216
Nortriptyline	46.8642	1	O-Glu216	1	Phe120	2	Glu216 Arg221
Oxatomide	52.4162	0	–	1	Cys443	–	–
Paroxetine	50.7223	1	O-Asp301	0	–	1	Asp301
Perphenazine	58.4008	0	–	3	Phe120 (2) Phe483	1	Glu216

(Continued)

TABLE 6.3 (CONTINUED)

Binding Modes of Known Inhibitors with CYP2D6 (PDB ID: 3QM4)

Substrate	CDOCKER Interaction Energy	Number of H-Bonds	Residues Involved in H-Bond Formation	Number of π–π Stacking	Residues Involved in π–π Stacking	Number of Charge Interaction	Residues Involved in Charge Interaction
Pimozide	69.8834	1	O-Ser304	2	Arg101 Phe120	–	–
Prasugrel	47.1402	0		1	Phe112	–	–
Quinidine	49.1137	1	O-Glu216	2	Phe120 Phe483	1	Glu216
Risperidone	53.8224	0	–	1	Phe120	–	–
Ritonavir	–	–	–	–	–	–	–
Sertraline	46.7915	1	O-Glu216	1	Phe120	1	Glu216
Sparteine	–	–	–	–	–	–	–
Terfenadine	69.9005	1	O-Ser304	4	Phe112 Phe120 (2) Phe483	–	–
Thioridazine	56.2981	0	–	3	Phe120 (2) Phe483	1	Glu216
Ticlopidine	33.4758	0	–	0	–	–	–
Tolterodine	−70.4072	1	O-Glu216	3	Phe120 Phe483 (2)	1	Glu216
Trimipramine	46.3053	0	–	2	Phe120 Phe247	–	–

the metoprolol–CYP2D6 interaction as well (Figure 6.12). Moreover, mexiletine binds to the CYP2D6 active site via the hydrogen bond formation at Asp301, π–π stacking interaction at Phe120, and charge interaction at Asp301 (Figure 6.12).

Mirtazapine interacts with the CYP2D6 active site via the hydrogen bond formation at Glu216 and two charge interactions at Glu216 and Arg221 (Figure 6.13). MPBP binds to the CYP2D6 active site through the hydrogen bond formation at Glu216, π–π interaction at Phe120, and charge interaction at Glu216 (Figure 6.13). MPPP is docked into the active site of CYP2D6 and forms two π–π interactions at Leu213 and Phe247 (Figure 6.13). Nicergoline binds to the CYP2D6 active site through the π–π interaction at Phe120 and charge interaction at Asp301 (Figure 6.13). Olanzapine is docked into the CYP2D6 active site via the hydrogen bond formation at Glu216, π–π interaction at Phe120, and charge interaction at Glu216 and Asp301 (Figure 6.14). Pactimibe is also docked into the active site of CYP2D6 and forms a hydrogen bond at Glu216, a π–π interaction at Phe120 and Phe483, and a charge interaction at Asp301 (Figure 6.14). Perhexiline interacts with the CYP2D6 active site through the hydrogen bond formation at Asp301, two π–π interactions at Phe120, and a charge interaction at Asp301 (Figure 6.14). In addition, phenformin is docked into the active site of CYP2D6 via the hydrogen bond formation at Glu216, Asp301, and Ser304; π–π interaction at Phe120; and charge interaction at Asp301 (Figure 6.14).

Pinoline is docked into the active site of CYP2D6 and generates a hydrogen bond at Glu216 and two charge interactions at Glu216 and Arg221 (Figure 6.15). Pranidipine binds to the CYP2D6 active site through the hydrogen bond

formation at Glu216 and a π–π interaction at Phe247 (Figure 6.15). Procainamide is docked into the active site of CYP2D6 via two hydrogen bonds at Glu216 and Ser304 and a charge interaction at Glu216 (Figure 6.15). Promazine interacts with the CYP2D6 active site via a π–π interaction at Phe120 and two charge interactions at Glu216 and Asp301 (Figure 6.15). Propafenone is docked into the active site of CYP2D6 via the formation of two hydrogen bonds at Glu216 and Ser304 and a charge interaction at Glu216 (Figure 6.16). Propranolol binds to the CYP2D6 active site via the hydrogen bond formation at Ser304 and a π–π interaction at Phe120 (Figure 6.16). Selegiline is oriented into the CYP2D6 active site via the hydrogen bond formation at Asp301 and the charge interaction at Asp301 (Figure 6.16). In addition, sildenafil is docked into the CYP2D6 active site through a hydrogen bond formation at Glu216 and two π–π interactions at Phe120 and Arg221 (Figure 6.16).

Tacrine binds to the CYP2D6 active site via the hydrogen bond formation at Asp301, the π–π stacking at Phe120, and the charge interaction at Asp301 (Figure 6.17). Tamoxifen is docked into the active site of CYP2D6 via two π–π interactions at Phe112 and Phe120 (Figure 6.17). Timolol binds to the active site of CYP2D6 and forms a π–π interaction at Phe120 (Figure 6.17). Tramadol interacts with the CYP2D6 active site with the formation of two hydrogen bonds at Glu216 and one charge interaction at Glu216 (Figure 6.17). Tropisetron is docked into the active site of CYP2D6 via two hydrogen bonds at Ala300 and Ser304, two π–π stacking interactions at Phe120 and Phe483, and one charge interaction at Glu216 (Figure 6.18). Venlafaxine binds to the CYP2D6 active site

TABLE 6.4

Predicted ADMET Profile of Various Substrates of CYP2D6 (PDB ID: 3QM4)

Drug	Solubility ADMET Score	Solubility Level	BBB ADMET Score	BBB Level	Absorption Level	PPB ADMET Score	PPB Level
17β-Estradiol	−4.341	2	0.373	1	0	5.71108	True
4-Methoxyphenamine	−0.851	4	−0.616	3	0	−4.82701	False
5-MeO-DIPT	−3.287	3	0.156	1	0	−7.57395	False
5-MeO-DMT	−1.954	4	−0.293	2	0	−9.25839	False
5-Methoxytryptamine	−3.287	3	0.156	1	0	−7.57395	False
Acetaminophen	−0.771	4	−0.741	3	0	−11.8084	False
Allopregnanolone	−5.123	2	0.440	1	0	3.38489	True
Almotriptan	−2.274	3	−0.876	3	0	−16.4991	False
Amiflamine	−1.398	4	−0.322	2	0	−2.01874	True
Amiodarone	−6.308	1	0.937	0	0	19.6144	True
Antipyrine	−2.693	3	−0.037	2	0	−9.61909	False
Aripiprazole	−4.404	2	0.187	1	0	0.981481	True
Betaxolol	−1.222	4	−0.350	2	0	−15.6209	False
Bicifadine	−2.879	3	0.214	1	0	0.707522	True
Bisoprolol	−0.75	4	−0.660	3	0	−16.1068	False
Bortezomib	−3.166	3	0.034	1	1	−7.1193	False
Bufuralol	−2.531	3	−0.068	2	0	−1.94298	True
Bunitrolol	−0.757	4	−0.838	3	0	−4.74938	False
Chlorpheniramine	−2.957	3	0.277	1	0	−5.08852	False
Cimetidine	−0.925	4	−1.485	3	0	−35.7499	False
Clozapine	−2.072	3	−0.479	2	0	5.64331	True
Codeine	−2.015	3	−0.793	3	0	−21.9144	False
Debrisoquine	−1.025	4	−0.786	3	0	−8.9681	False
Delavirdine	−4.003	2	0.095	1	0	22.9731	True
Dextromethorphan	−3.79	3	0.307	1	0	−8.34569	False
Dihydrocodeine	−2.281	3	−0.697	3	0	−13.224	False
Diltiazem	−4.246	2	−0.134	2	0	−13.4686	False
Diuron	−3.144	3	0.083	1	0	1.92603	True
Dolasetron	−4.219	2	−0.505	2	0	−24.6568	False
Donepezil	−4.173	2	0.169	1	0	8.9683	True
Encainide	−3.611	3	0.054	1	0	−4.66108	False
Enclomifene	−5.262	2	1.175	0	0	4.56683	True
Estrone	−5.051	2	0.460	1	0	6.76303	True
Flecainide	−4.851	2	−0.191	2	0	−14.0712	False
Fluvastatin	−3.451	3	−0.589	3	0	23.388	True
Galantamine	−3.001	3	−0.373	2	0	−4.81783	False
Gepirone	−1.838	4	−1.077	3	0	−6.48596	False
Guanoxan	−1.907	4	−1.245	3	0	−7.46932	False
Harmine	−4.054	2	0.084	1	0	3.08067	True
Hydrocodone	−2.825	3	−0.644	3	0	−14.0752	False
Imatinib	−2.239	3	−1.056	3	0	−18.8144	False
Imipramine	−2.082	3	0.101	1	1	−3.9293	False
Indinavir	−3.014	3	0.035	0	1	−28.6752	False
Lidocaine	−1.471	4	−0.352	2	0	−5.84751	False
Loratadine	−6.351	1	0.745	0	0	18.1328	True
MDMA	−1.461	4	−0.250	2	1	−4.29724	False
Mequitazine	−2.898	3	0.108	1	1	−1.11299	True
Methamphetamine	−0.903	4	−3.756	0	2	−6.14962	False
Methoxyphenamine	−1.144	4	−2.968	0	1	−4.18829	False
Metoclopramide	−1.083	4	−1.176	3	0	−3.80638	False
Metoprolol	−0.507	4	−0.603	3	0	−11.8174	False
Mexiletine	−1.592	4	−0.395	2	0	−4.43085	False

(Continued)

TABLE 6.4 (CONTINUED)
Predicted ADMET Profile of Various Substrates of CYP2D6 (PDB ID: 3QM4)

Drug	Solubility ADMET Score	Solubility Level	BBB ADMET Score	BBB Level	Absorption Level	PPB ADMET Score	PPB Level
Mirtazapine	−3.287	3	0.038	1	0	−9.04282	False
Monocrotaline	−1.913	4	−1.700	3	0	−7.41446	False
Morphine	−1.354	4	−1.051	3	0	−24.959	False
MPBP	−2.739	3	0.130	1	0	−1.3592	True
MPPP	−2.323	3	−0.031	2	0	−1.84105	True
N-Desethylamiodarone	−6.153	1	0.816	0	0	19.4938	True
Nevirapine	−4.074	2	−0.331	2	0	−12.3488	False
Nicergoline	−4.276	2	−0.436	2	0	1.53591	True
Olanzapine	−2.167	3	−0.551	3	0	−0.87855	True
Ondansetron	−4.592	2	0.039	1	0	−15.1048	False
Oxymorphone	−1.193	4	−1.502	3	0	−23.1323	False
Pactimibe	−4.006	2	−0.180	2	0	−3.34332	False
Perhexiline	−4.854	2	1.234	0	1	−3.22989	False
Phenformin	−0.357	4	−1.574	3	0	−12.6896	False
Pinoline	−1.889	4	−0.358	2	0	−5.30869	False
Pranidipine	−4.936	2	−5.020	2	0	12.5472	True
Procainamide	0.283	5	−1.235	3	0	−11.7437	False
Progesterone	−5.719	2	0.492	1	0	18.2891	True
Promazine	−1.934	4	0.079	1	1	−3.14627	False
Propafenone	−2.266	3	−0.144	2	0	5.44569	True
Propranolol	−1.732	4	−0.220	2	0	−8.81116	False
Saquinavir	−4.43	2	−55.076	4	3	28.5827	True
Selegiline	−2.387	3	0.486	1	0	−3.06364	False
Sildenafil	−3.614	3	1.010	0	0	5.01499	True
Tacrine	−3.557	3	0.018	1	0	−7.27112	False
Tamoxifen	−5.211	2	1.125	0	0	14.5598	True
Testosterone	−4.75	2	0.316	1	0	8.08491	True
Timolol	−1.27	4	−1.207	3	0	−16.1044	False
Tramadol	−1.637	4	−0.323	2	0	−4.28925	False
Tropisetron	−2.857	3	−0.534	3	0	−17.3962	False
Venlafaxine	−1.856	4	−0.224	2	0	−7.59249	False
Verapamil	−4.023	2	0.095	1	0	2.64402	True
Vernakalant	−2.063	3	−0.556	3	0	−4.92668	False
Zolpidem	−4.984	2	0.378	1	0	17.235	True
Zuclopenthixol	−3.963	3	0.334	1	0	16.1336	True

via the formation of two hydrogen bonds at Glu216 and Ser304, one π–π stacking interaction at Phe120, and one charge interaction at Glu216 (Figure 6.18). Verapamil forms a π–π stacking interaction at Phe120 at the active site of CYP2D6 with a CDOCKER interaction energy of 33.6044 kcal/mol (Figure 6.18). Vernakalant forms two hydrogen bonds at Glu216 and one charge interaction at Glu216 with a CDOCKER interaction energy of 59.7309 kcal/mol (Figure 6.18). Zuclopenthixol binds to the CYP2D6 active site via the hydrogen bond formation at Glu216 and the π–π interaction at Phe120 (Figure 6.18).

However, allopregnanolone, bicifadine, codeine, antipyrine, dextromethorphan, dihydrocodeine, hydrocodone, indinavir, lasiocarpine, monocrotaline, morphine, N-desethylamiodarone, nevirapine, ondansetron, oxymorphone, progesterone, saquinavir, testosterone, and zolpidem are not able to form either a

hydrogen bond, a π–π interaction, or a charge interaction in the active site of CYP2D6 (3QM4).

6.9.2 BINDING MODES OF THE INHIBITOR WITH THE CYP2D6 ACTIVE SITE

Amitriptyline interacted with CYP2D6 at the active site via the π–π interaction at Phe120 and the charge interaction at Glu216 (Figure 6.19). Amodiaquine interacted with CYP2D6 at the active site through one hydrogen bond, four π–π interactions, and one charge interaction (Figure 6.19). Glu216 is involved in the hydrogen bond formation. Phe120 is involved in three π–π interactions and Phe483 also contributed to the π–π interaction. Glu216 accounted for the charge interaction. Aprindine interacted with CYP2D6 at the active site

TABLE 6.5

Predicted Hepatotoxicity of Various Substrates of CYP2D6 (PDB ID: 3QM4)

Drug	CYP2D6 Inhibitor Prediction	CYP2D6 ADMET Score	Hepatotoxicity Prediction	Hepatotoxicity ADMET Score
17β-Estradiol	False	−2.3631	False	−10.8076
4-Methoxyphenamine	False	−4.93777	True	−3.67278
5-MeO-DIPT	False	−2.76183	True	−2.12269
5-MeO-DMT	False	−4.85791	True	−2.91941
5-Methoxytryptamine	False	−2.76183	True	−2.12269
Acetaminophen	False	−11.3224	True	0.505529
Allopregnanolone	False	−2.25638	False	−14.6307
Almotriptan	False	−7.92392	True	−2.21809
Amiflamine	False	−4.97583	True	−2.69804
Amiodarone	False	−22.1197	True	20.5177
Antipyrine	False	−7.02733	False	−4.60525
Aripiprazole	False	−3.83394	False	−6.69453
Betaxolol	False	−5.78455	False	−8.96116
Bicifadine	False	−3.13187	False	−5.21818
Bisoprolol	False	−4.06465	False	−19.4754
Bortezomib	False	−9.53167	False	−7.9193
Bufuralol	False	−2.49007	True	−0.377323
Bunitrolol	False	−5.34392	True	−3.81777
Chlorpheniramine	False	−0.936143	False	−5.63257
Cimetidine	False	−8.12832	True	2.80307
Clozapine	False	−6.64384	False	−4.87848
Codeine	False	−4.76511	False	−16.5449
Debrisoquine	False	−6.84656	False	−5.61094
Delavirdine	False	−47.8687	True	−2.1064
Dextromethorphan	False	−3.26076	False	−12.6084
Dihydrocodeine	False	−4.1068	False	−10.629
Diltiazem	False	−4.30972	False	−6.21067
Diuron	False	−5.79378	True	1.35243
Dolasetron	False	−7.61589	False	−4.14717
Donepezil	False	−0.123058	False	−14.5574
Encainide	False	−7.50801	False	−4.11946
Enclomifene	False	−2.49654	True	0.58962
Estrone	False	−2.46931	False	−5.03846
Flecainide	False	−4.30187	False	−7.56333
Fluvastatin	False	−17.921	True	−1.30313
Galantamine	False	−3.44469	False	−9.43434
Gepirone	False	−5.71346	False	−4.1705
Guanoxan	False	−6.13294	False	−5.63057
Harmine	False	−16.6508	True	3.64259
Hydrocodone	False	−4.7127	False	−19.9619
Imatinib	False	−11.0831	True	0.107862
Imipramine	False	−0.117642	True	−3.08028
Indinavir	False	−47.4716	False	−7.20214
Lidocaine	False	−7.44735	True	−2.76758
Loratadine	False	−0.559587	True	−1.91126
MDMA	False	−5.14085	True	−1.06536
Mequitazine	False	−0.183087	True	−0.390798
Methamphetamine	False	−3.75586	False	−6.9552
Methoxyphenamine	False	−2.96789	False	−4.96726
Metoclopramide	False	−0.842753	False	−6.73149
Metoprolol	False	−5.62439	False	−8.21547
Mexiletine	False	−5.47744	True	−3.73954

(Continued)

TABLE 6.5 (CONTINUED)
Predicted Hepatotoxicity of Various Substrates of CYP2D6 (PDB ID: 3QM4)

Drug	CYP2D6 Inhibitor Prediction	CYP2D6 ADMET Score	Hepatotoxicity Prediction	Hepatotoxicity ADMET Score
Mirtazapine	False	−2.02629	True	−1.54252
Monocrotaline	False	−9.21962	True	22.6144
Morphine	False	−5.12383	False	−18.3981
MPBP	False	−2.10928	False	−4.54501
MPPP	False	−3.08028	False	−4.47162
N-Desethylamiodarone	False	−21.4817	True	20.5177
Nevirapine	False	−21.92	True	−1.31183
Nicergoline	False	−1.90821	True	−2.76697
Olanzapine	False	−16.1652	True	−2.4933
Ondansetron	False	−19.106	True	0.113426
Oxymorphone	False	−6.69733	False	−18.0098
Pactimibe	False	−4.04131	False	−6.59908
Perhexiline	False	−2.39783	True	−2.16719
Phenformin	False	−5.00219	False	−4.84304
Pinoline	False	−2.31613	True	1.45485
Pranidipine	False	−5.02024	False	−8.74828
Procainamide	False	−8.19564	True	0.0730136
Progesterone	False	−2.95189	False	−6.21105
Promazine	False	0.0792353	True	−1.2319
Propafenone	False	−0.45666	False	−8.87651
Propranolol	False	−0.241947	True	0.0820988
Saquinavir	False	−55.0758	True	−3.0622
Selegiline	False	−6.77537	False	−15.9589
Sildenafil	False	−14.696	True	1.68982
Tacrine	False	−1.98907	True	8.32944
Tamoxifen	False	−1.91669	True	8.53831
Testosterone	False	−2.66109	False	−18.7469
Timolol	False	−4.84328	True	−3.07728
Tramadol	False	−3.73337	False	−7.10128
Tropisetron	False	−8.98005	True	−1.61661
Venlafaxine	False	−6.1202	True	−2.50707
Verapamil	False	−4.21183	True	−2.3616
Vernakalant	False	−2.48668	False	−8.54195
Zolpidem	False	−1.9186	True	−0.795125
Zuclopenthixol	False	−0.5322	True	−3.77294

via a hydrogen bond formation at Glu216, a π–π interaction at Phe120, and a charge interaction at Glu216 (Figure 6.19). Atomoxetine is docked into the active site and formed two hydrogen bonds at Glu216, one π–π interaction at Phe120, and one charge interaction at Glu216 (Figure 6.19). Azelastine interacted with CYP2D6 at the active site through the formation of two hydrogen bonds at Glu216 and one charge interaction at Glu216 (Figure 6.20). Biperiden is docked into the active site of CYP2D6 and formed three π–π interactions at Phe112, Phe120, and Phe483 (Figure 6.20). Carvedilol binds to the CYP2D6 active site via two π–π interactions at Phe120 and Phe247 (Figure 6.20). Chloroquine is docked into the active site of CYP2D6 and forms one hydrogen bond at Pro435 and two π–π stacking interactions at Phe120 and Phe483 (Figure 6.20).

Chlorpromazine forms a π–π interaction at Phe120 and a charge interaction at Glu216 at the active site of CYP2D6 (Figure 6.21). Cibenzoline is oriented in the CYP2D6 active site via the formation of two hydrogen bonds at Asp301 and Glu216 and one charge interaction at Asp301 (Figure 6.21). Cinnarizine is docked into the active site of CYP2D6 and forms a π–π interaction at Arg101 (Figure 6.21). Citalopram binds to the CYP2D6 active site through the hydrogen bond formation at Glu216 and two charge interactions at Glu216 and Arg221 (Figure 6.21). Clomipramine binds to the CYP2D6 active site via the hydrogen bond formation at Ser304 and three π–π stacking interactions at Phe120 and Phe247 (Figure 6.22). In addition, diphenhydramine is docked into the CYP2D6 active site through three π–π interactions at Phe112, Phe120, and Phe483 (Figure 6.22). Doxepin binds to

TABLE 6.6

Predicted ADMET Profile of Various Inhibitors of CYP2D6 (PDB ID: 3QM4)

Drug	Solubility ADMET Score	Solubility Level	BBB ADMET Score	BBB Level	Absorption Level	PPB ADMET Score	PPB Level
Amitriptyline	−4.195	2	0.788	0	0	6.85918	True
Amodiaquine	−3.615	3	−0.18	2	0	6.71574	True
Aprindine	−3.848	3	0.774	0	0	6.98609	True
Atomoxetine	−2.773	3	0.43	1	0	1.44183	True
Azelastine	−4.217	2	0.072	1	0	−11.0549	False
Biperiden	−2.615	3	0.111	1	0	−0.607285	True
Carvedilol	−2.969	3	−0.465	2	0	10.554	True
Chloroquine	−3.045	3	0.165	1	0	−2.69874	False
Chlorpromazine	−2.765	3	0.21	1	1	0.355188	True
Cibenzoline	−3.501	3	0.211	1	0	1.71913	True
Cinnarizine	−6.209	1	1.447	0	1	18.4519	True
Citalopram	−3.446	3	−0.041	2	0	−10.5542	False
Clomipramine	−2.91	3	0.306	1	1	−0.66318	True
Diphenhydramine	−2.451	3	0.218	1	0	−3.8559	False
Doxepin	−3.526	3	0.379	1	0	−8.04329	False
Duloxetine	−3.538	3	0.443	1	0	−2.92046	False
Ezlopitant	−7.211	1	1.417	0	1	4.32841	True
Flunarizine	−5.275	2	1.094	0	0	9.76107	True
Fluoxetine	−3.877	3	0.571	0	0	4.29394	True
Fluvoxamine	−3.068	3	−0.591	3	0	−10.2569	False
Halofantrine	−7.214	1	1.518	0	1	3.56966	True
Haloperidol	−4.349	2	0.391	1	0	34.0973	True
Ibogaine	−5.254	2	0.529	1	0	−0.703092	True
Iloperidone	−3.735	3	−0.379	2	0	−2.33507	False
Indomethacin	−4.362	2	−0.348	2	0	18.5866	True
Maprotiline	−3.853	3	0.738	0	0	5.11836	True
Methadone	−3.682	3	0.417	1	0	−4.67065	False
Mianserin	−3.736	3	0.405	1	0	1.71282	True
Nebivolol	−3.5	3	−0.376	2	0	5.3737	True
Nortriptyline	−3.719	3	0.775	0	0	6.06946	True
Oxatomide	−4.292	2	0.407	1	0	26.0877	True
Paroxetine	−3.669	3	0.04	1	0	8.92422	True
Perphenazine	−2.188	3	−0.352	2	0	1.83252	True
Pimozide	−5.531	2	0.652	1	0	31.2881	True
Prasugrel	−5.205	2	0.337	1	0	4.97631	True
Quinidine	−2.579	3	−0.491	2	0	−7.19519	False
Risperidone	−3.798	3	−0.545	3	0	−14.4679	False
Ritonavir	−4.196	2	0.248	0	2	−5.96679	False
Sertraline	−4.9	2	1.01	0	0	5.01499	True
Sparteine	−2.447	3	0.458	2	1	−9.51949	False
Terfenadine	−3.933	3	0.663	1	0	2.31837	True
Thioridazine	−3.754	3	0.464	1	0	−4.74111	False
Ticlopidine	−4.677	2	0.926	0	0	17.0953	True
Tolterodine	−3.981	3	0.733	0	0	2.81166	True
Trimipramine	−2.561	3	0.24	1	1	−3.70982	False

TABLE 6.7
Predicted Hepatotoxicity of Various Inhibitors of CYP2D6 (PDB ID: 3QM4)

Drug	CYP2D6 Inhibitor Prediction	CYP2D6 ADMET Score	Hepatotoxicity Prediction	Hepatotoxicity ADMET Score
Amitriptyline	True	0.543309	True	−3.76073
Amodiaquine	True	11.4964	True	−0.939674
Aprindine	True	6.10333	False	−4.37824
Atomoxetine	True	3.55198	False	−5.59751
Azelastine	True	19.6232	True	−0.643541
Biperiden	True	1.41532	False	−12.8865
Carvedilol	True	0.597747	False	−8.03841
Chloroquine	True	3.86228	True	1.24016
Chlorpromazine	True	3.23752	True	−0.69744
Cibenzoline	True	1.03695	False	−4.36781
Cinnarizine	True	1.46046	False	−15.0659
Citalopram	True	3.32326	True	−1.21954
Clomipramine	True	2.90154	True	−3.23699
Diphenhydramine	True	2.44251	False	−8.13457
Doxepin	True	0.39685	False	−5.1379
Duloxetine	True	6.84279	True	5.32619
Ezlopitant	True	1.08951	True	−2.35843
Flunarizine	True	2.75274	False	−10.4445
Fluoxetine	True	11.0206	False	−9.51251
Fluvoxamine	True	11.2385	False	−9.06817
Halofantrine	True	16.6088	True	−2.52268
Haloperidol	True	10.3207	True	−1.08337
Ibogaine	True	0.89645	True	−4.08802
Iloperidone	True	7.8185	True	−0.563119
Indomethacin	True	0.572737	True	0.427611
Maprotiline	True	4.49406	True	−2.36985
Methadone	True	4.6415	False	−6.36538
Mianserin	True	1.14217	True	−1.98058
Nebivolol	True	1.11938	True	0.299285
Nortriptyline	True	0.543309	True	−3.76073
Oxatomide	True	1.25968	False	−14.7291
Paroxetine	True	6.91049	True	2.66663
Perphenazine	True	4.26057	True	−2.61974
Pimozide	True	7.07422	True	1.92867
Prasugrel	True	3.39143	True	−2.50901
Quinidine	True	1.2776	True	−1.80397
Risperidone	True	13.3507	False	−14.4679
Ritonavir	True	25.0031	True	28.7635
Sertraline	True	8.67827	True	−0.307459
Sparteine	True	0.457832	False	−5.69754
Terfenadine	True	3.80404	False	−11.8151
Thioridazine	True	2.37512	True	−1.67681
Ticlopidine	True	13.9868	True	−2.26449
Tolterodine	True	2.63852	False	−4.27719
Trimipramine	True	0.806646	True	−3.05617

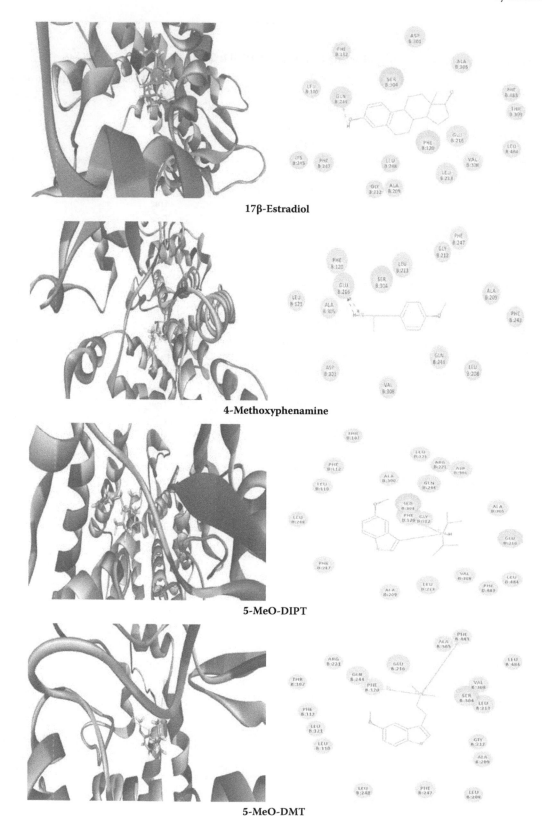

17β-Estradiol

4-Methoxyphenamine

5-MeO-DIPT

5-MeO-DMT

FIGURE 6.2 **(See color insert.)** Binding modes of 17β-estradiol–CYP2D6 (PDB code: 3QM4), estradiol–CYP2D6 (PDB code: 3QM4), 5-MeO-DIPT–CYP2D6 (PDB code: 3QM4), and 5-MeO-DMT–CYP2D6 (PDB code: 3QM4) interactions. 17β-Estradiol interacted with CYP2D6 through a hydrogen bond formation at Gln244. 4-Methoxyphenamine formed a hydrogen bond with Glu216 and a charge interaction at Glu216 in the active site of CYP2D6. 5-MeO-DIPT interacted with CYP2D6 through a π–π interaction at Phe120. 5-MeO-DMT interacted with CYP2D6 through a π–π interaction at Phe120 and Phe483.

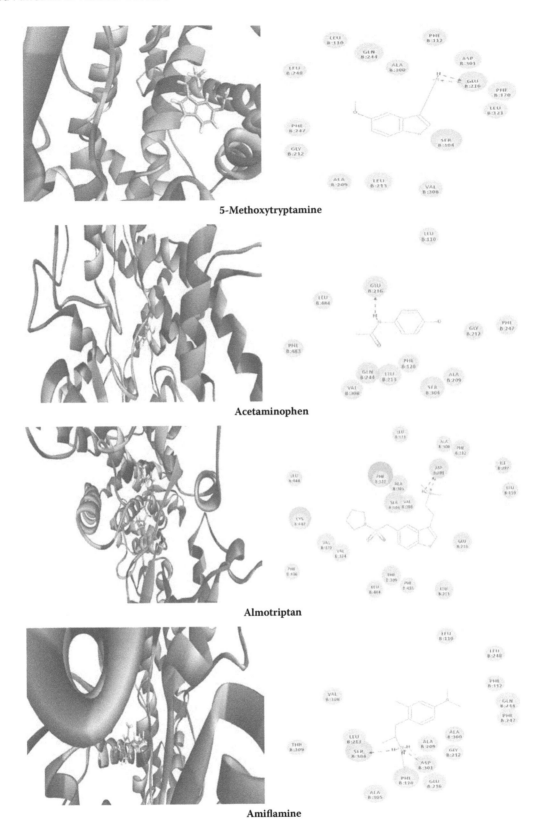

FIGURE 6.3 **(See color insert.)** Binding modes of 5-methoxytryptamine–CYP2D6 (PDB code: 3QM4), acetaminophen–CYP2D6 (PDB code: 3QM4), almotriptan–CYP2D6 (PDB code: 3QM4), and amiflamine–CYP2D6 (PDB code: 3QM4) interactions. 5-Methoxytryptamine interacted with CYP2D6 via the formation of a hydrogen bond and π–π interaction at Glu216. Acetaminophen interacted with CYP2D6 through the formation of a hydrogen bond at Glu216. Almotriptan interacted with CYP2D6 via the hydrogen formation and π–π interaction at Asp301. Amiflamine interacted with CYP2D6 via the formation of two hydrogen bonds at Asp301 and Ser304, a π–π interaction at Phe120, and a charge interaction at Asp301.

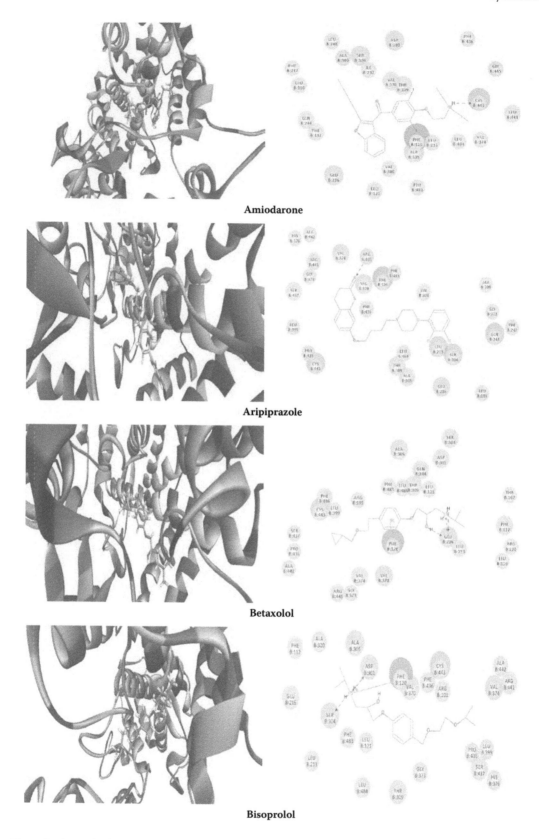

FIGURE 6.4 **(See color insert.)** Binding modes of amiodarone–CYP2D6 (PDB code: 3QM4), aripiprazole–CYP2D6 (PDB code: 3QM4), betaxolol–CYP2D6 (PDB code: 3QM4), and bisoprolol–CYP2D6 (PDB code: 3QM4) interactions. Amiodarone interacted with CYP2D6 through the hydrogen bond formation at Cys443. Aripiprazole interacted with CYP2D6 via the hydrogen bond formation at Arg101. Betaxolol formed two hydrogen bonds with Glu216. Betaxolol also interacted with CYP2D6 via the π–π interaction at Phe120 and charge interaction at Glu216. Bisoprolol formed two hydrogen bonds at Asp301 and Ser304 at the active site of CYP2D6. The π–π interaction at Phe120 and charge interaction at Asp301 are another two molecular determinants for the bisoprolol–CYP2D6 interaction at the active site.

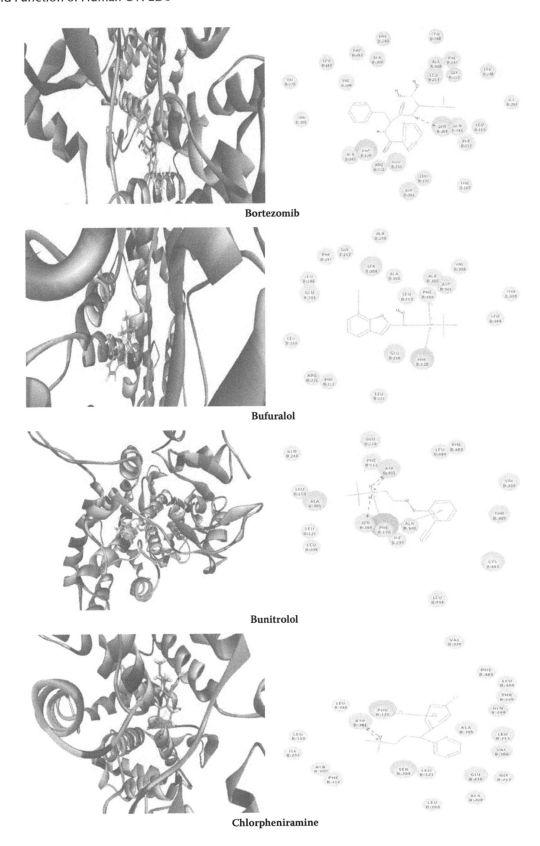

Bortezomib

Bufuralol

Bunitrolol

Chlorpheniramine

FIGURE 6.5 (See color insert.) Binding modes of bortezomib–CYP2D6 (PDB code: 3QM4), bufuralol–CYP2D6 (PDB code: 3QM4), bunitrolol–CYP2D6 (PDB code: 3QM4), and chlorpheniramine–CYP2D6 (PDB code: 3QM4) interactions. Bortezomib formed a hydrogen bond at Ser304 that is responsible for the interaction. Bufuralol formed two π–π interactions at Phe120 and Phe483. Bunitrolol interacted with CYP2D6 at the active site through hydrogen bond formation, π–π interaction, and charge interaction. Asp301 and Ser304 are involved in the hydrogen bond formation. Phe120 formed two π–π interactions with bunitrolol. Asp301 contributed to the charge interaction between bunitrolol and CYP2D6. Chlorpheniramine formed two hydrogen bonds at Asp301, a π–π interaction at Phe120, and a charge interaction at Asp301.

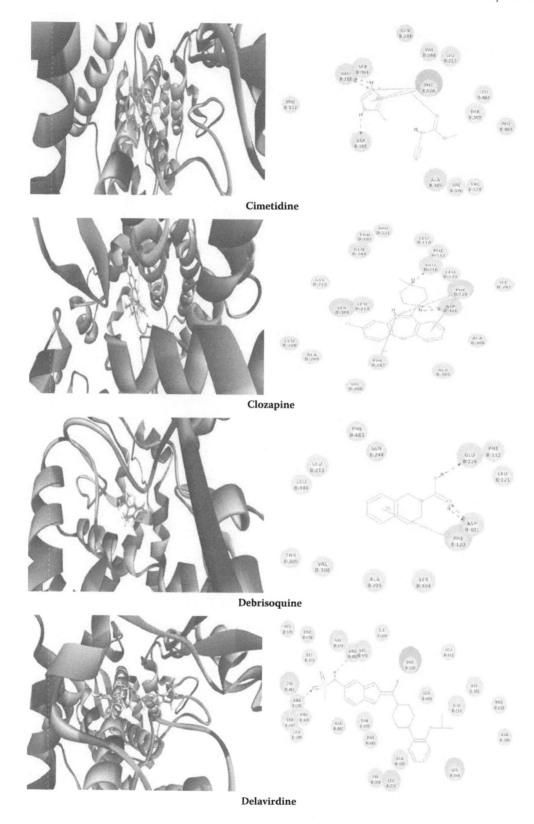

FIGURE 6.6 **(See color insert.)** Binding modes of cimetidine–CYP2D6 (PDB code: 3QM4), clozapine–CYP2D6 (PDB code: 3QM4), debrisoquine–CYP2D6 (PDB code: 3QM4), and delavirdine–CYP2D6 (PDB code: 3QM4) interactions. Cimetidine formed two hydrogen bonds at Glu216 and Asp301, two π–π interactions at Phe120, and a charge interaction at Glu216. Clozapine formed a hydrogen bond at Asp301, three π–π interactions at Phe120 (two) and Phe247, and two charge interactions at Glu216 and Asp301. Debrisoquine interacted with CYP2D6 at the active site through hydrogen bond formation, π–π interaction, and charge interaction. Glu216 and Asp301 contributed to the hydrogen bond formation. Phe120 is involved in the π–π interaction and Asp301 accounted for the charge interaction. Delavirdine interacted with CYP2D6 at the active site and formed two hydrogen bonds at Arg101 and Arg441.

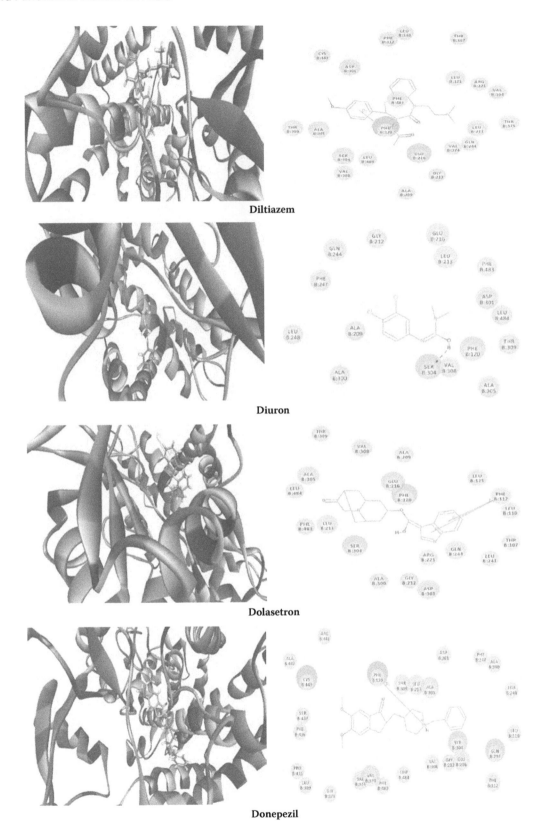

Diltiazem

Diuron

Dolasetron

Donepezil

FIGURE 6.7 **(See color insert.)** Binding modes of diltiazem–CYP2D6 (PDB code: 3QM4), diuron–CYP2D6 (PDB code: 3QM4), dolasetron–CYP2D6 (PDB code: 3QM4), and donepezil–CYP2D6 (PDB code: 3QM4) interactions. Diltiazem formed a π–π interaction at Phe120 at the active site of CYP2D6. Diuron formed a hydrogen bond at Ser304 at the active site of CYP2D6. Dolasetron interacted with CYP2D6 at the active site through the hydrogen bond formation. There are two π–π interactions formed at Phe112. Donepezil formed a π–π interaction at Phe120 at the active site of CYP2D6.

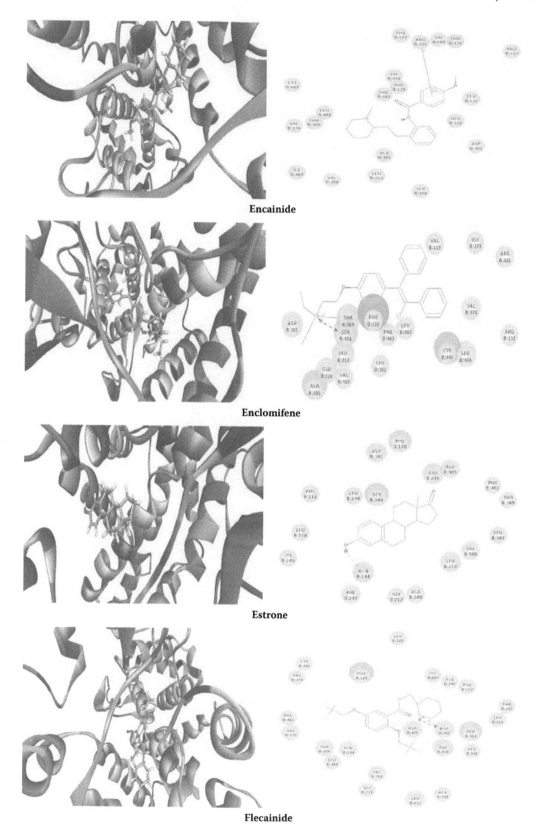

Encainide

Enclomifene

Estrone

Flecainide

FIGURE 6.8   (See color insert.) Binding modes of encainide–CYP2D6 (PDB code: 3QM4), enclomifene–CYP2D6 (PDB code: 3QM4), estrone–CYP2D6 (PDB code: 3QM4), and flecainide–CYP2D6 (PDB code: 3QM4) interactions. Encainide formed a π–π interaction at Arg221. Enclomifene interacted with CYP2D6 in the active site through the hydrogen bond formation at Ser304 and π–π interaction at Phe120. Estrone interacted with CYP2D6 with a hydrogen bond formation at Gln224. Flecainide interacted with CYP2D6 at the active site through the hydrogen bond formation at Asp301.

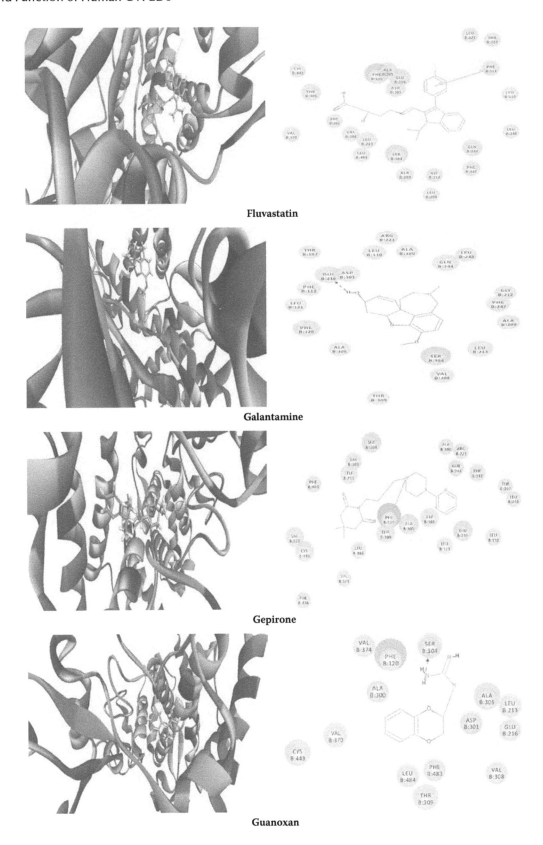

Fluvastatin

Galantamine

Gepirone

Guanoxan

FIGURE 6.9 **(See color insert.)** Binding modes of fluvastatin–CYP2D6 (PDB code: 3QM4), galantamine–CYP2D6 (PDB code: 3QM4), gepirone–CYP2D6 (PDB code: 3QM4), and guanoxan–CYP2D6 (PDB code: 3QM4) interactions. Fluvastatin formed a π–π interaction at Phe112. Galantamine interacted with CYP2D6 in the active site of CYP2D6 via the hydrogen bond formation at Glu216. Gepirone formed a π–π interaction at Phe120. Guanoxan interacted with CYP2D6 in the active site via the hydrogen bond formation at Ser304.

388

Cytochrome P450 2D6

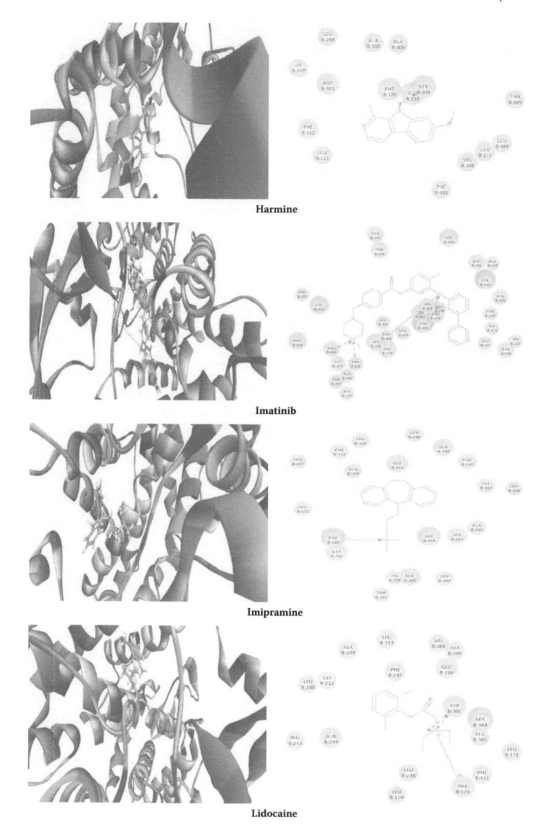

Harmine

Imatinib

Imipramine

Lidocaine

FIGURE 6.10 (See color insert.) Binding modes of harmine–CYP2D6 (PDB code: 3QM4), imatinib–CYP2D6 (PDB code: 3QM4), imipramine–CYP2D6 (PDB code: 3QM4), and lidocaine–CYP2D6 (PDB code: 3QM4) interactions. Harmine formed a hydrogen bond at Ser304 in the active site of CYP2D6. Imatinib formed a π–π interaction at Gly445 in the active site of CYP2D6. Imipramine interacted with CYP2D6 at the active site via π–π interaction. Phe120 is involved in imipramine–CYP2D6 interaction. Lidocaine formed a hydrogen bond at Asp301, a π–π interaction at Phe120, and a charge interaction at Asp301.

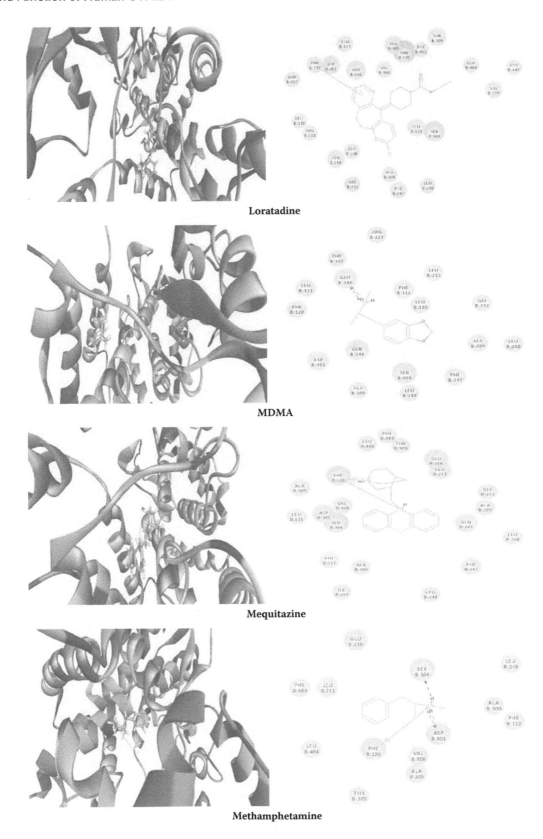

FIGURE 6.11 (See color insert.) Binding modes of loratadine–CYP2D6 (PDB code: 3QM4), MDMA–CYP2D6 (PDB code: 3QM4), mequitazine–CYP2D6 (PDB code: 3QM4), and methamphetamine–CYP2D6 (PDB code: 3QM4) interactions. Loratadine formed a π–π interaction at Phe112. MDMA interacted with CYP2D6 at the active site through the hydrogen bond formation at Glu216 and charge interaction at Glu216. Mequitazine formed two π–π interactions at Phe120. Methamphetamine interacted with CYP2D6 at the active site via hydrogen bond formation, π–π interaction, and charge interaction. There are two hydrogen bonds formed at Asp301 and Ser304. Phe120 and Asp301 are involved in π–π interaction and charge interaction, respectively.

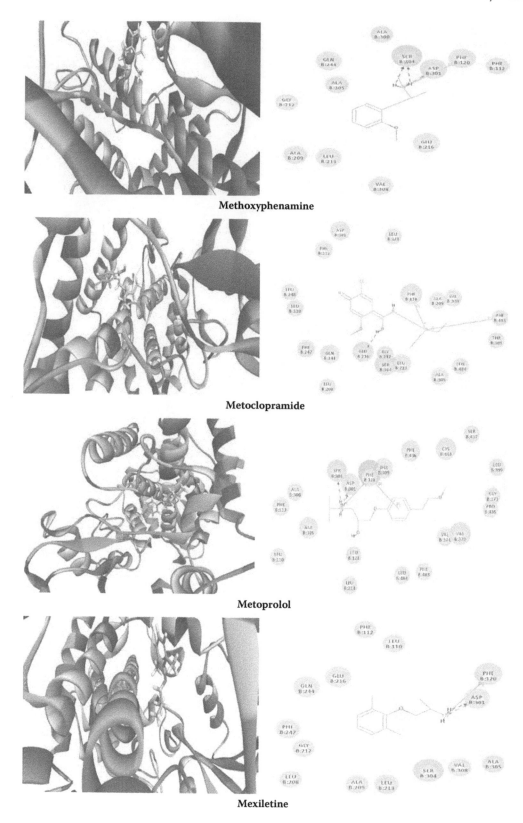

FIGURE 6.12 **(See color insert.)** Binding modes of methoxyphenamine–CYP2D6 (PDB code: 3QM4), metoclopramide–CYP2D6 (PDB code: 3QM4), metoprolol–CYP2D6 (PDB code: 3QM4), and mexiletine–CYP2D6 (PDB code: 3QM4) interactions. The methoxyphenamine–CYP2D6 interaction included two hydrogen bonds at Ser304, one π–π interaction at Phe120, and one charge interaction at Asp301. Metoclopramide interacted with CYP2D6 at the active site through the hydrogen bond formation at Glu216 and two π–π interactions at Phe120 and Phe483. Metoprolol interacted with CYP2D6 at the active site via the hydrogen bond formation at Asp301 and Ser304. There are π–π interactions at Phe120 and one charge interaction at Asp301 that are involved in the metoprolol–CYP2D6 interaction as well. Mexiletine interacted with CYP2D6 at the active site via the hydrogen bond formation at Asp301, π–π interaction at Phe120, and charge interaction at Asp301.

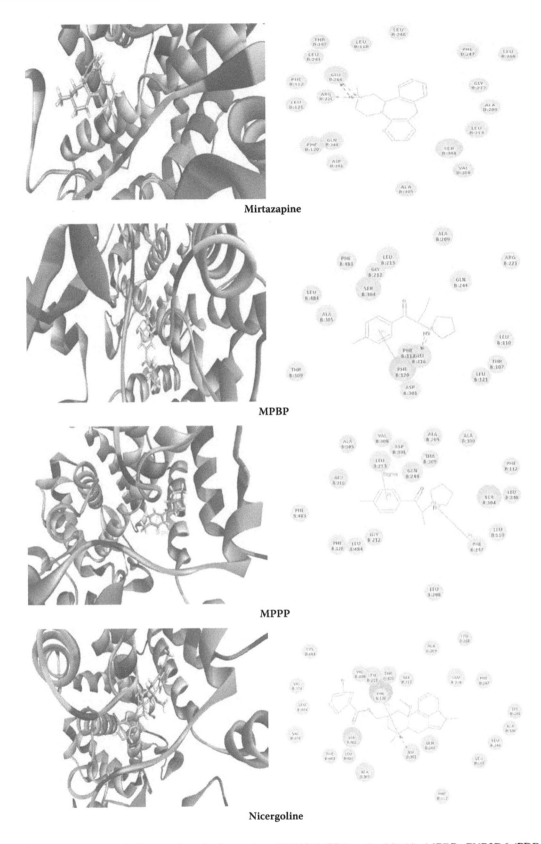

Mirtazapine

MPBP

MPPP

Nicergoline

FIGURE 6.13 **(See color insert.)** Binding modes of mirtazapine–CYP2D6 (PDB code: 3QM4), MPBP–CYP2D6 (PDB code: 3QM4), MPPP–CYP2D6 (PDB code: 3QM4), and nicergoline–CYP2D6 (PDB code: 3QM4) interactions. Mirtazapine interacted with CYP2D6 at the active site via the hydrogen bond formation at Glu216 and two charge interactions at Glu216 and Arg221. MPBP interacted with CYP2D6 at the active site through the hydrogen bond formation at Glu216, π–π interaction at Phe120, and charge interaction at Glu216. MPPP formed two π–π interactions at Leu213 and Phe247. Nicergoline interacted with CYP2D6 at the active site through the π–π interaction at Phe120 and charge interaction at Asp301.

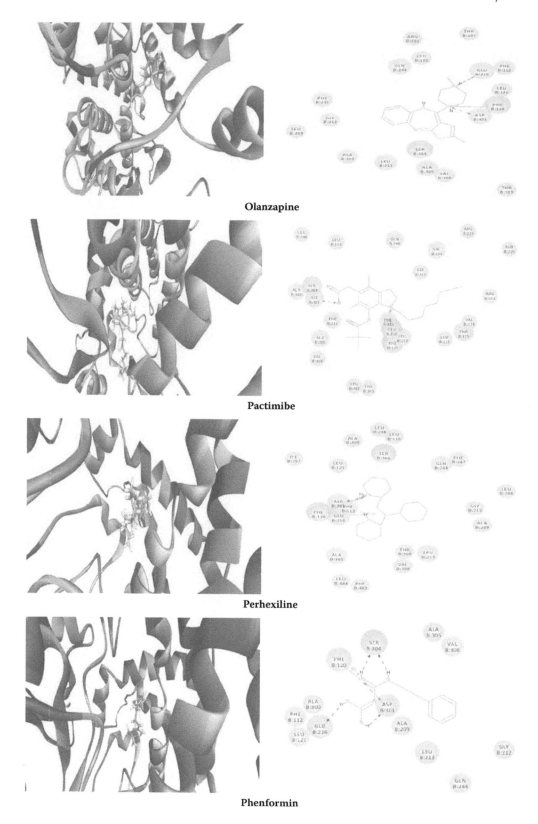

Olanzapine

Pactimibe

Perhexiline

Phenformin

FIGURE 6.14 **(See color insert.)** Binding modes of olanzapine–CYP2D6 (PDB code: 3QM4), pactimibe–CYP2D6 (PDB code: 3QM4), perhexiline–CYP2D6 (PDB code: 3QM4), and phenformin–CYP2D6 (PDB code: 3QM4) interactions. Olanzapine interacted with CYP2D6 at the active site via the hydrogen bond formation at Glu216, π–π interaction at Phe120, and charge interaction at Glu216 and Asp301. Pactimibe formed a hydrogen bond at Glu216, a π–π interaction at Phe120 and Phe483, and a charge interaction at Asp301. Perhexiline interacted with CYP2D6 at the active site through the hydrogen bond formation at Asp301, two π–π interactions at Phe120, and a charge interaction at Asp301. Phenformin is docked into the active site of CYP2D6 via the hydrogen bond formation at Glu216, Asp301, and Ser304; π–π interaction at Phe120; and charge interaction at Asp301.

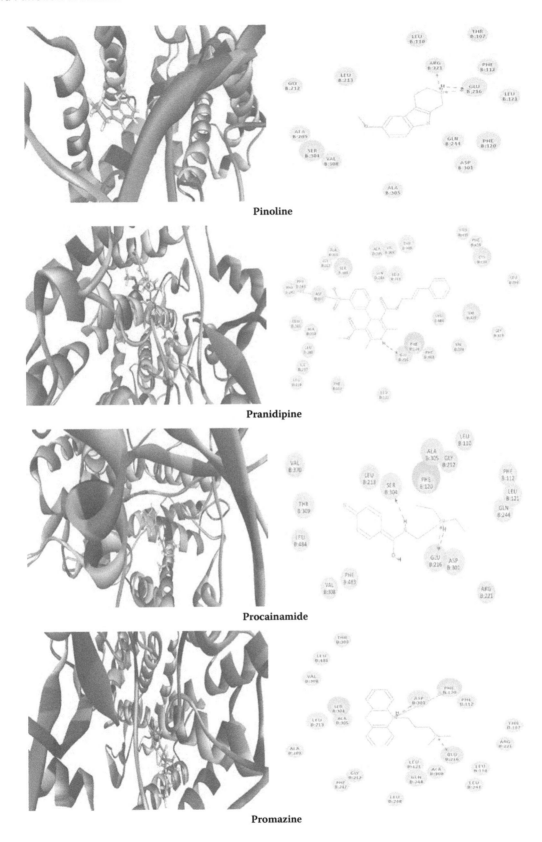

Pinoline

Pranidipine

Procainamide

Promazine

FIGURE 6.15 **(See color insert.)** Binding modes of pinoline–CYP2D6 (PDB code: 3QM4), pranidipine–CYP2D6 (PDB code: 3QM4), procainamide–CYP2D6 (PDB code: 3QM4), and promazine–CYP2D6 (PDB code: 3QM4) interactions. Pinoline formed a hydrogen bond at Glu216 and two charge interactions at Glu216 and Arg221. Pranidipine interacted with CYP2D6 at the active site through the hydrogen bond formation at Glu216 and a π–π interaction at Phe247. Procainamide formed two hydrogen bonds at Glu216 and Ser304 and a charge interaction at Glu216. Promazine interacted with CYP2D6 at the active site via a π–π interaction at Phe120 and two charge interactions at Glu216 and Asp301.

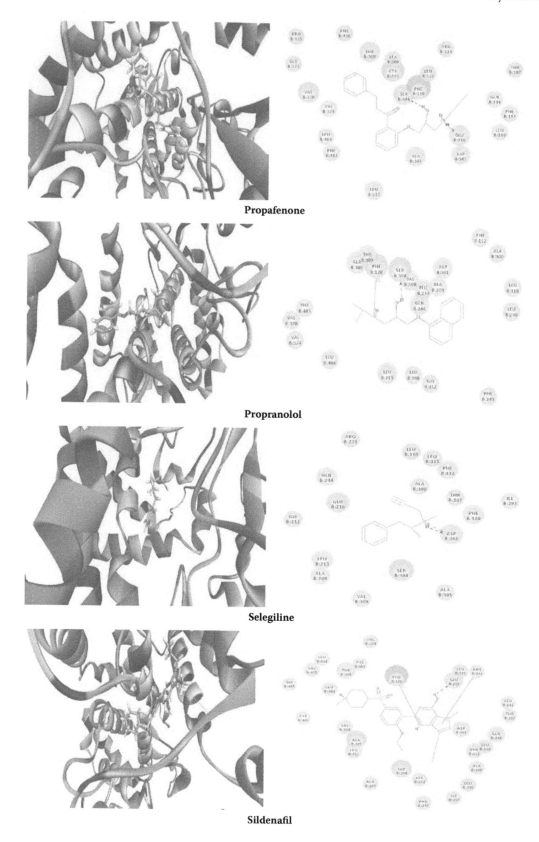

FIGURE 6.16 (See color insert.) Binding modes of propafenone–CYP2D6 (PDB code: 3QM4), propranolol–CYP2D6 (PDB code: 3QM4), selegiline–CYP2D6 (PDB code: 3QM4), and sildenafil–CYP2D6 (PDB code: 3QM4) interactions. Propafenone formed two hydrogen bonds at Glu216 and Ser304 and a charge interaction at Glu216. Propranolol interacted with CYP2D6 at the active site via the hydrogen bond formation at Ser304 and a π–π interaction at Phe120. Selegiline interacted with CYP2D6 at the active site via the hydrogen bond formation at Asp301 and the charge interaction at Asp301. Sildenafil interacted with CYP2D6 at the active site through a hydrogen bond formation at Glu216 and two π–π interactions at Phe120 and Arg221.

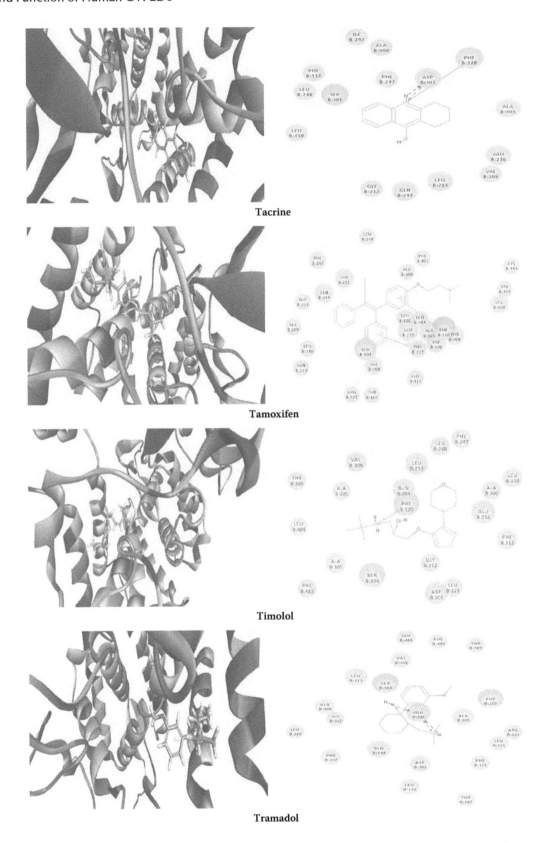

Tacrine

Tamoxifen

Timolol

Tramadol

FIGURE 6.17 **(See color insert.)** Binding modes of tacrine–CYP2D6 (PDB code: 3QM4), tamoxifen–CYP2D6 (PDB code: 3QM4), timolol–CYP2D6 (PDB code: 3QM4), and tramadol–CYP2D6 (PDB code: 3QM4) interactions. Tacrine interacted with CYP2D6 at the active site via the hydrogen bond formation at Asp301, the π–π interaction at Phe120, and the charge interaction at Asp301. Tamoxifen formed two π–π interactions at Phe112 and Phe120. Timolol is docked into the active site of CYP2D6 and formed a π–π interaction at Phe120. Tramadol interacted with CYP2D6 at the active site with the formation of two hydrogen bonds at Glu216 and one charge interaction at Glu216.

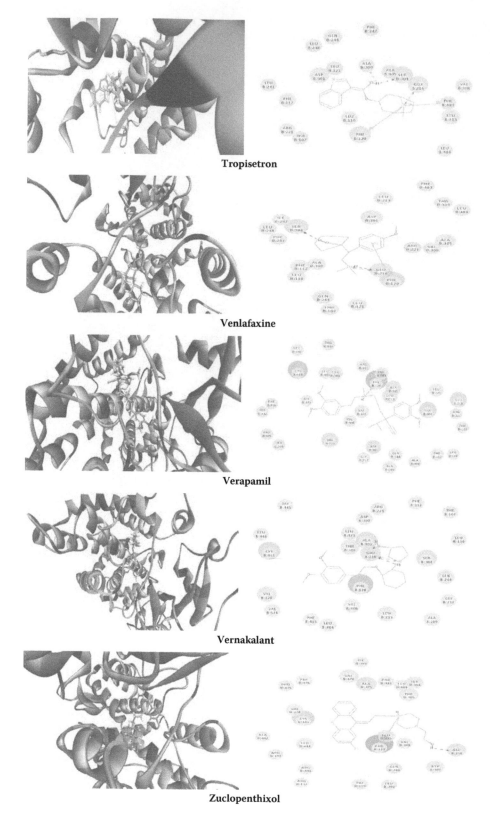

Tropisetron

Venlafaxine

Verapamil

Vernakalant

Zuclopenthixol

FIGURE 6.18 **(See color insert.)** Binding modes of tropisetron–CYP2D6 (PDB code: 3QM4), venlafaxine–CYP2D6 (PDB code: 3QM4), verapamil–CYP2D6 (PDB code: 3QM4), vernakalant–CYP2D6 (PDB code: 3QM4), and zuclopenthixol–CYP2D6 (PDB code: 3QM4) interactions. Tropisetron formed two hydrogen bonds at Ala300 and Ser304, two π–π interactions at Phe120 and Phe483, and one charge interaction at Glu216. Venlafaxine interacted with CYP2D6 at the active site via the formation of two hydrogen bonds at Glu216 and Ser304, one π–π interaction at Phe120, and one charge interaction at Glu216. Verapamil formed a π–π interaction at Phe120 at the active site of CYP2D6. Vernakalant formed two hydrogen bonds at Glu216 and one charge interaction at Glu216. Zuclopenthixol interacted with CYP2D6 at the active site via the hydrogen bond formation at Glu216 and the π–π interaction at Phe120.

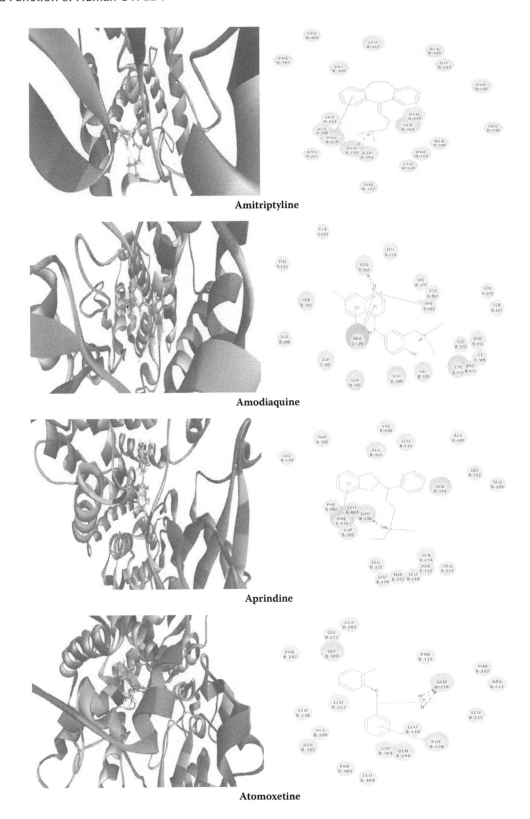

Amitriptyline

Amodiaquine

Aprindine

Atomoxetine

FIGURE 6.19 **(See color insert.)** Binding modes of amitriptyline–CYP2D6 (PDB code: 3QM4), amodiaquine–CYP2D6 (PDB code: 3QM4), aprindine–CYP2D6 (PDB code: 3QM4), and atomoxetine–CYP2D6 (PDB code: 3QM4) interactions. Amitriptyline interacted with CYP2D6 at the active site via the π–π interaction at Phe120 and the charge interaction at Glu216. Amodiaquine interacted with CYP2D6 at the active site through one hydrogen bond, four π–π interactions, and one charge interaction. Glu216 is involved in the hydrogen bond formation. Phe120 is involved in three π–π interactions and Phe483 also contributed to the π–π interaction. Glu216 accounted for the charge interaction. Aprindine interacted with CYP2D6 at the active site via a hydrogen bond formation at Glu216, a π–π interaction at Phe120, and a charge interaction at Glu216. Atomoxetine formed two hydrogen bonds at Glu216, one π–π interaction at Phe120, and one charge interaction at Glu216.

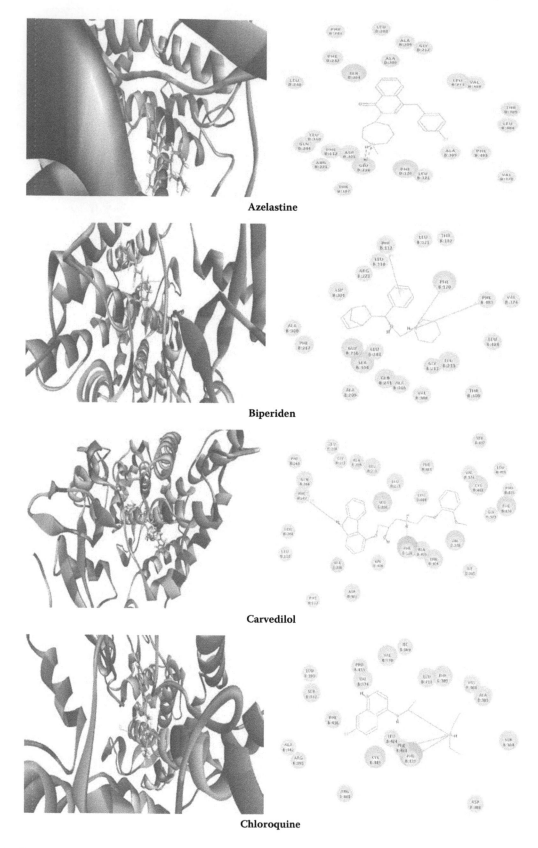

FIGURE 6.20 **(See color insert.)** Binding modes of azelastine–CYP2D6 (PDB code: 3QM4), biperiden–CYP2D6 (PDB code: 3QM4), carvedilol–CYP2D6 (PDB code: 3QM4), and chloroquine–CYP2D6 (PDB code: 3QM4) interactions. Azelastine interacted with CYP2D6 at the active site through the formation of two hydrogen bonds at Glu216 and one charge interaction at Glu216. Biperiden formed three π–π interactions at Phe112, Phe120, and Phe483. Carvedilol interacted with CYP2D6 at the active site via two π–π interactions at Phe120 and Phe247. Chloroquine is docked into the active site of CYP2D6 and formed one hydrogen bond at Pro435 and two π–π interactions at Phe120 and Phe483.

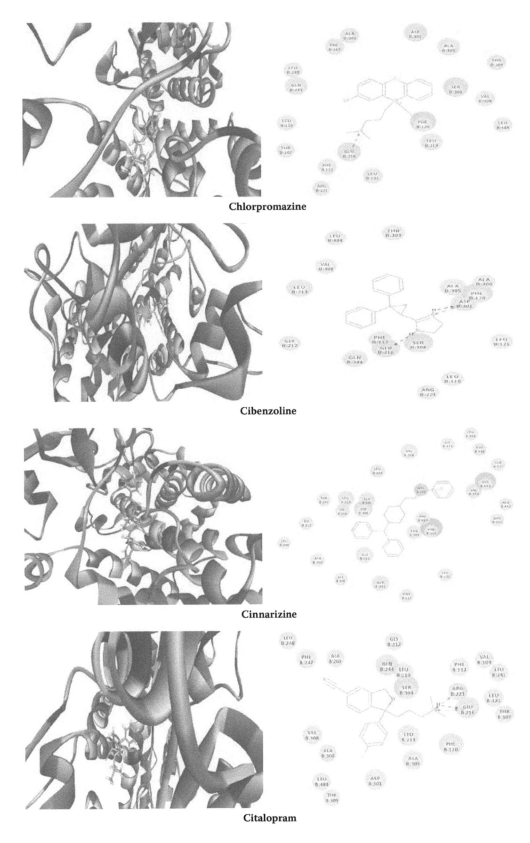

FIGURE 6.21 **(See color insert.)** Binding modes of chlorpromazine–CYP2D6 (PDB code: 3QM4), cibenzoline–CYP2D6 (PDB code: 3QM4), cinnarizine–CYP2D6 (PDB code: 3QM4), and citalopram–CYP2D6 (PDB code: 3QM4) interactions. Chlorpromazine formed a π–π interaction at Phe120 and a charge interaction at Glu216 at the active site of CYP2D6. Cibenzoline interacted with CYP2D6 at the active site via the formation of two hydrogen bonds at Asp301 and Glu216 and one charge interaction at Asp301. Cinnarizine is docked into the active site of CYP2D6 and formed a π–π interaction at Arg101. Citalopram interacted with CYP2D6 at the active site through the hydrogen bond formation at Glu216 and two charge interactions at Glu216 and Arg221.

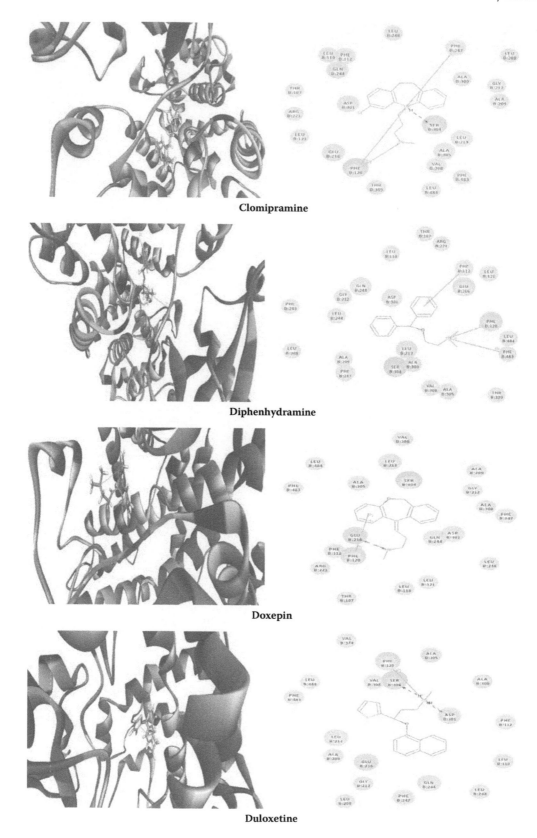

FIGURE 6.22   (See color insert.) Binding modes of clomipramine–CYP2D6 (PDB code: 3QM4), diphenhydramine–CYP2D6 (PDB code: 3QM4), doxepin–CYP2D6 (PDB code: 3QM4), and duloxetine–CYP2D6 (PDB code: 3QM4) interactions. Clomipramine interacted with CYP2D6 at the active site via the hydrogen bond formation at Ser304 and three π–π interactions at Phe120 and Phe247. Diphenhydramine interacted with CYP2D6 at the active site through three π–π interactions at Phe112, Phe120, and Phe483. Doxepin interacted with CYP2D6 at the active site via the π–π interaction at Phe120 and the charge interaction at Glu216. Duloxetine interacted with CYP2D6 at the active site through the formation of three hydrogen bonds at Asp301 and Ser304, a π–π interaction at Phe120, and a charge interaction at Asp301.

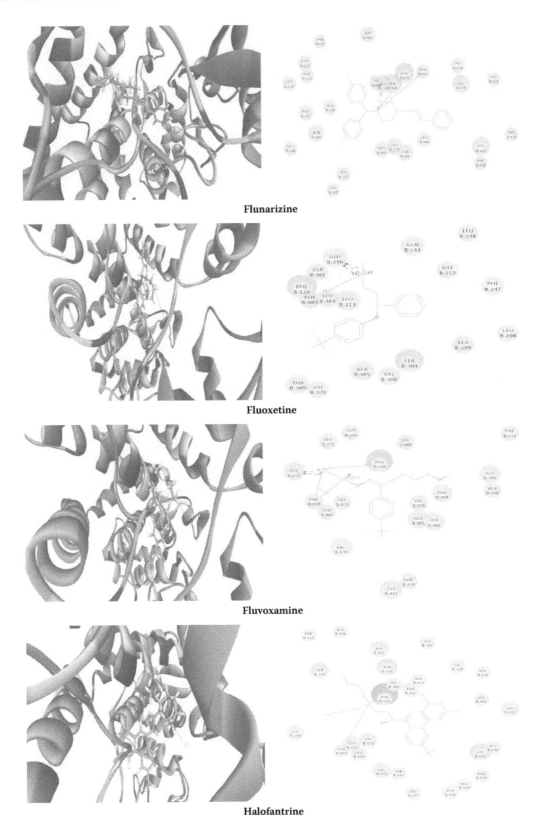

FIGURE 6.23 **(See color insert.)** Binding modes of flunarizine–CYP2D6 (PDB code: 3QM4), fluoxetine–CYP2D6 (PDB code: 3QM4), fluvoxamine–CYP2D6 (PDB code: 3QM4), and halofantrine–CYP2D6 (PDB code: 3QM4) interactions. Flunarizine formed two π–π interactions at Phe120 and Phe483 and one charge interaction at Glu216. Fluoxetine interacted with CYP2D6 at the active site through the formation of two hydrogen bonds at Glu216, one π–π interaction at Phe483, and one charge interaction at Glu216. Fluvoxamine interacted with CYP2D6 at the active site via the hydrogen bond formation at Glu216, three π–π interactions at Phe120 and Phe483, and one charge interaction at Glu216. Halofantrine is docked into the active site of CYP2D6 and formed two π–π interactions at Phe120 and Phe483.

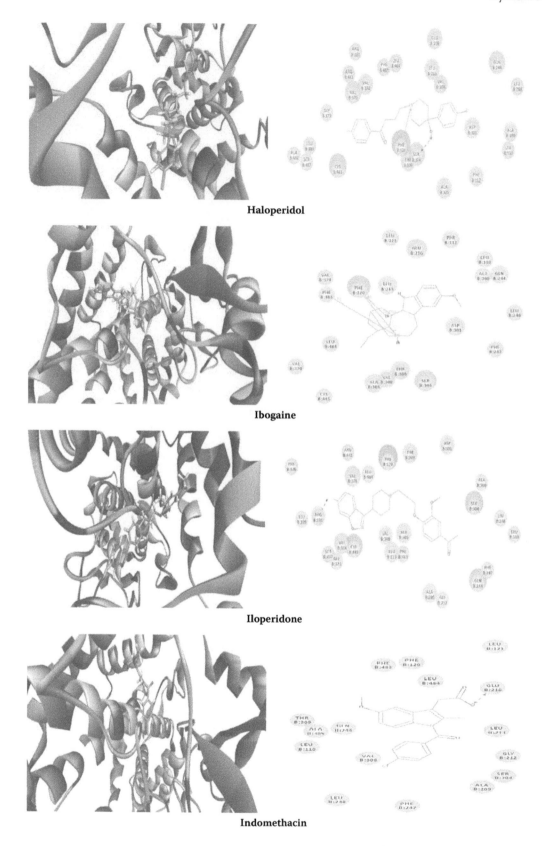

Haloperidol

Ibogaine

Iloperidone

Indomethacin

FIGURE 6.24 **(See color insert.)** Binding modes of haloperidol–CYP2D6 (PDB code: 3QM4), ibogaine–CYP2D6 (PDB code: 3QM4), iloperidone–CYP2D6 (PDB code: 3QM4), and indomethacin–CYP2D6 (PDB code: 3QM4) interactions. Haloperidol interacted with CYP2D6 at the active site via the hydrogen bond formation at Ser304 and the π–π interaction at Phe120. Ibogaine formed three π–π interactions at the active site of CYP2D6 with the involvement of Phe120 and Phe483. Iloperidone formed a hydrogen bond at Arg101. Indomethacin interacted with CYP2D6 at the active site via the charge interaction at Glu216.

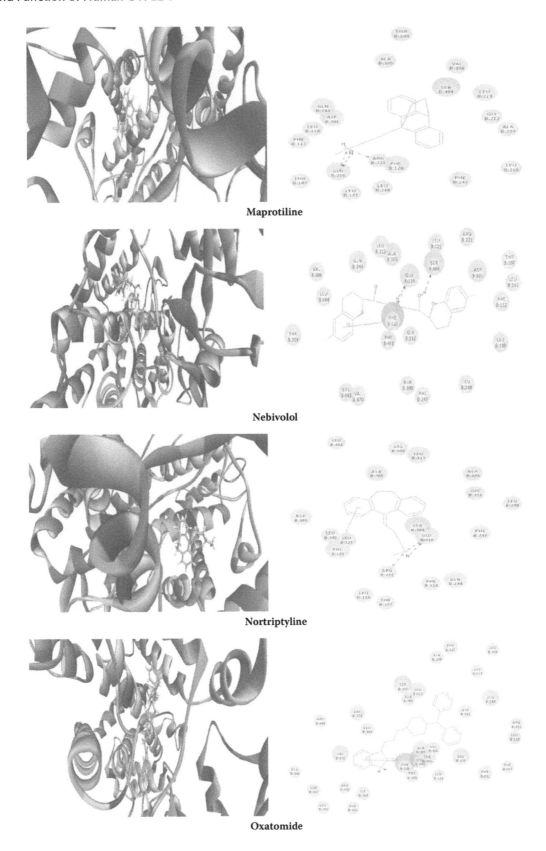

Maprotiline

Nebivolol

Nortriptyline

Oxatomide

FIGURE 6.25 **(See color insert.)** Binding modes of maprotiline–CYP2D6 (PDB code: 3QM4), nebivolol–CYP2D6 (PDB code: 3QM4), nortriptyline–CYP2D6 (PDB code: 3QM4), and oxatomide–CYP2D6 (PDB code: 3QM4) interactions. Maprotiline formed a hydrogen bond at Glu216 and two charge interactions at Glu216 and Arg221. Nebivolol interacted with CYP2D6 at the active site via the formation of two hydrogen bonds at Glu216 and Ser304, two π–π interactions at Phe120, and one charge interaction at Glu216. Nortriptyline interacted with CYP2D6 at the active site via a hydrogen bond formation at Glu216, a π–π interaction at Phe120, and two charge interactions at Glu216 and Arg221. Oxatomide formed a π–π interaction at Cys443.

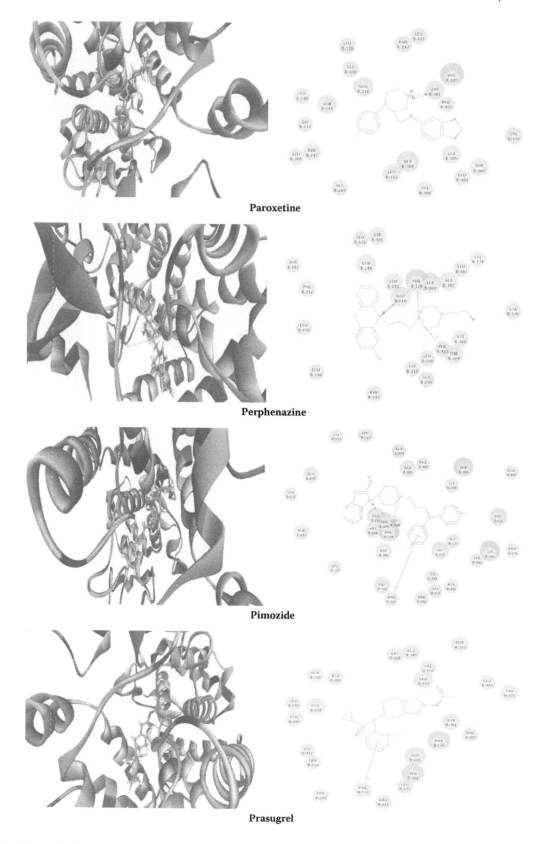

FIGURE 6.26 **(See color insert.)** Binding modes of paroxetine–CYP2D6 (PDB code: 3QM4), perphenazine–CYP2D6 (PDB code: 3QM4), pimozide–CYP2D6 (PDB code: 3QM4), and prasugrel–CYP2D6 (PDB code: 3QM4) interactions. Paroxetine interacted with CYP2D6 at the active site through the hydrogen bond formation at Asp301 and the charge interaction at Asp301. Perphenazine interacted with CYP2D6 at the active site via three π–π interactions at Phe120 and Phe483 and one charge interaction at Glu216. Pimozide interacted with CYP2D6 at the active site through the hydrogen bond formation at Ser304 and two π–π interactions at Arg101 and Phe120. Prasugrel is docked into the active site of CYP2D6 and formed a π–π interaction at Phe112.

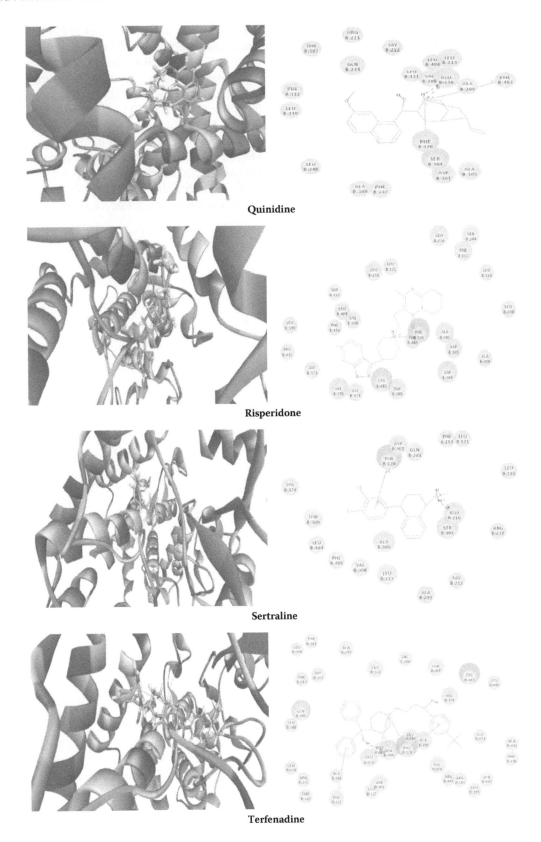

FIGURE 6.27 **(See color insert.)** Binding modes of quinidine–CYP2D6 (PDB code: 3QM4), risperidone–CYP2D6 (PDB code: 3QM4), sertraline–CYP2D6 (PDB code: 3QM4), and terfenadine–CYP2D6 (PDB code: 3QM4) interactions. Quinidine is docked into the active site of CYP2D6 and formed a hydrogen bond at Glu216, two π–π interactions at Phe120 and Phe483, and one charge interaction at Glu216. Risperidone interacted with CYP2D6 at the active site through the π–π interaction at Phe120. Sertraline interacted with CYP2D6 at the active site through a hydrogen bond formation at Glu216, a π–π interaction at Phe120, and a charge interaction at Glu216. Terfenadine formed a hydrogen bond at Ser304 and four π–π interactions at Phe112, Phe120, and Phe483.

the CYP2D6 active site via the π–π interaction at Phe120 and the charge interaction at Glu216 (Figure 6.22). Duloxetine is docked into the CYP2D6 active site through the formation of three hydrogen bonds at Asp301 and Ser304, a π–π interaction at Phe120, and a charge interaction at Asp301 (Figure 6.22).

Flunarizine is docked into the active site of CYP2D6 and forms two π–π interactions at Phe120 and Phe483 and one charge interaction at Glu216 (Figure 6.23). Fluoxetine binds to the CYP2D6 active site through the formation of two hydrogen bonds at Glu216, one π–π interaction at Phe483, and one charge interaction at Glu216 (Figure 6.23). Fluvoxamine binds to the CYP2D6 active site via the hydrogen bond formation at Glu216, three π–π interactions at Phe120 and Phe483, and one charge interaction at Glu216 (Figure 6.23). Halofantrine is docked into the active site of CYP2D6 and forms two π–π interactions at Phe120 and Phe483 (Figure 6.23). Haloperidol is oriented into the CYP2D6 active site via the hydrogen bond formation at Ser304 and the π–π interaction at Phe120 (Figure 6.24). Ibogaine forms three π–π interactions at the active site of CYP2D6 with the involvement of Phe120 and Phe483 (Figure 6.24). Iloperidone is docked into the active site of CYP2D6 and forms a hydrogen bond at Arg101 (Figure 6.24). Indomethacin binds to CYP2D6 at the active site via the charge interaction at Glu216 (Figure 6.24).

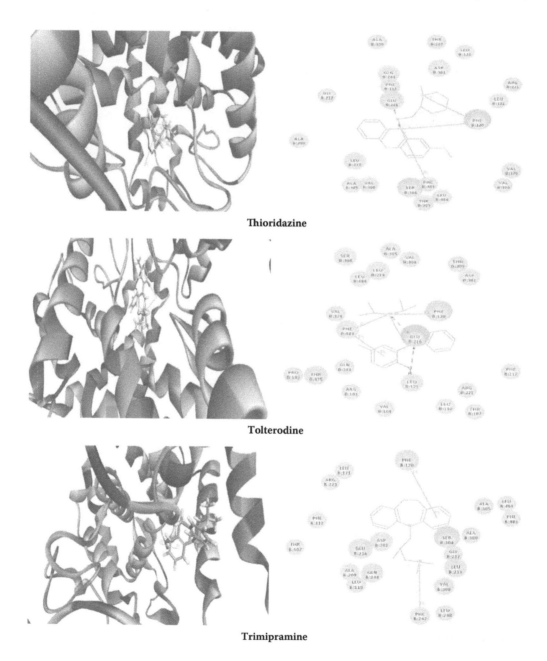

Thioridazine

Tolterodine

Trimipramine

FIGURE 6.28 **(See color insert.)** Binding modes of thioridazine–CYP2D6 (PDB code: 3QM4), tolterodine–CYP2D6 (PDB code: 3QM4), and trimipramine–CYP2D6 (PDB code: 3QM4) interactions. Thioridazine formed three π–π interactions at Phe120 and Phe483 and one charge interaction at Glu216. Tolterodine interacted with CYP2D6 at the active site through a hydrogen bond formation at Glu216, three π–π interactions at Phe120 and Phe483, and a charge interaction at Glu216. Trimipramine formed two π–π interactions at Phe120 and Phe247.

Maprotiline is docked into the active site of CYP2D6 and forms a hydrogen bond at Glu216 and two charge interactions at Glu216 and Arg221 (Figure 6.25). Nebivolol binds to the CYP2D6 active site via the formation of two hydrogen bonds at Glu216 and Ser304, two π–π interactions at Phe120, and one charge interaction at Glu216 (Figure 6.25). Nortriptyline binds to the CYP2D6 active site via a hydrogen bond formation at Glu216, a π–π interaction at Phe120, and two charge interactions at Glu216 and Arg221 (Figure 6.25). Oxatomide is docked into the active site of CYP2D6 via a π–π interaction at Cys443 (Figure 6.25). Paroxetine binds to the CYP2D6 active site through the hydrogen bond formation at Asp301 and the charge interaction at Asp301 (Figure 6.26). In addition, perphenazine is docked into the CYP2D6 active site via three π–π interactions at Phe120 and Phe483 and one charge interaction at Glu216 (Figure 6.26). Pimozide binds to CYP2D6 at the active site through the hydrogen bond formation at Ser304 and two π–π interactions at Arg101 and Phe120 (Figure 6.26). Prasugrel is docked into the active site of CYP2D6 and forms a π–π interaction at Phe112 (Figure 6.26).

Quinidine is docked into the active site of CYP2D6 via a hydrogen bond at Glu216, two π–π interactions at Phe120 and Phe483, and one charge interaction at Glu216 (Figure 6.27). Risperidone binds to the CYP2D6 active site through the π–π interaction at Phe120 (Figure 6.27). Sertraline binds to CYP2D6 at the active site through a hydrogen bond formation at Glu216, a π–π interaction at Phe120, and a charge interaction at Glu216 (Figure 6.27). In addition, terfenadine is docked into the active site of CYP2D6 via a hydrogen bond at Ser304 and four π–π interactions at Phe112, Phe120, and Phe483 (Figure 6.27). Thioridazine is docked into the active site of CYP2D6 and forms three π–π interactions at Phe120 and Phe483 and one charge interaction at Glu216 (Figure 6.28). Tolterodine binds to the CYP2D6 active site through a hydrogen bond formation at Glu216, three π–π interactions at Phe120 and Phe483, and a charge interaction at Glu216 (Figure 6.28). Trimipramine forms two π–π interactions at Phe120 and Phe247 (Figure 6.28). However, ezlopitant, methadone, mianserin, ritonavir, sparteine, and ticlopidine are not able to form either a hydrogen bond, a π–π interaction, or a charge interaction in the active site of CYP2D6 (3QM4).

6.10 CONCLUSIONS AND FUTURE DIRECTIONS

CYP2D6 is an important enzyme since it can metabolize approximately 25% of clinical drugs and is subject to polymorphism with significant clinical consequences. There are several properties that determine the metabolism of any substrate by CYP2D6. These include (a) the topography of the active site, (b) steric interactions between the ligand and the active-site pocket, and (c) key electronic interactions. The elucidation of its crystal structure has provided very useful information on how ligands (e.g., substrates or inhibitors) interact with this enzyme and how CYP2 members determine their substrate specificity. However, the resolved structure of CYP2D6 (PDB entry: 2F9Q) is ligand free, and thus how and whether ligand binding induces changes in size, shape, and hydration

of the active site are unknown. It is necessary to resolve and obtain the structures of CYP2D6 in complex with a substrate or inhibitor to better understand how ligands interact with CYP2D6 and how such interactions affect the rationalization and prediction of the catalytic activity, ligand specificity, and regiospecific metabolism of this important enzyme.

Although a number of site-directed mutagenesis studies have been conducted to examine the role of important active-site residues in ligand binding, the role of several other important residues such as Leu213, Ser217, Leu220, Ala305, Val308, Val370, and Leu484 is unclear. Our docking studies using the 2F9Q structure have revealed that these residues play an important role in interactions with ligands such as dextromethorphan, pactimibe, quinidine, and spirosulfonamide. Both pactimibe and spirosulfonamide are devoid of a basic nitrogen atom.

CYP2D6 is absent in 5%–9% of the Caucasian population, resulting in deficiencies in the oxidation of drugs that are metabolized primarily by this enzyme. Because of the potential safety concerns related to drug–drug interactions, as well as pharmacogenetic variability, small-molecule drug optimization involves optimizing compounds such that binding affinity for CYP2D6 is minimized. A better understanding of the structural information and functional relevance of CYP2D6 has important implications in drug development. It is a common practice for the pharmaceutical industry nowadays to screen drug candidates in the early stages of development as potential CYP2D6 substrates or inhibitors and drop such candidates where they have alternatives. Both in silico and in vitro models can be used for the screening for such potential ligands. The rationalization and prediction of potential CYP2D6 ligands are thus advantageous in the discovery and development of new drugs, since this would eventually lead to improved therapeutic ligand design. Further studies are needed to delineate the molecular mechanisms involved in the interaction of ligands with CYP2D6 and their clinical implications.

REFERENCES

Allorge, D., Breant, D., Harlow, J., Chowdry, J., Lo-Guidice, J. M., Chevalier, D., Cauffiez, C. et al. 2005. Functional analysis of CYP2D6.31 variant: Homology modeling suggests possible disruption of redox partner interaction by Arg440His substitution. *Proteins* 59:339–346.

Bazeley, P. S., Prithivi, S., Struble, C. A., Povinelli, R. J., Sem, D. S. 2006. Synergistic use of compound properties and docking scores in neural network modeling of CYP2D6 binding: Predicting affinity and conformational sampling. *J Chem Inf Model* 46:2698–2708.

Bichara, N., Ching, M. S., Blake, C. L., Ghabrial, H., Smallwood, R. A. 1996. Propranolol hydroxylation and N-desisopropylation by cytochrome P4502D6: Studies using the yeast-expressed enzyme and NADPH/O₂ and cumene hydroperoxide-supported reactions. *Drug Metab Dispos* 24:112–118.

Blobaum, A. L., Kent, U. M., Alworth, W. L., Hollenberg, P. F. 2004. Novel reversible inactivation of cytochrome P450 2E1 T303A by tert-butyl acetylene: The role of threonine 303 in proton delivery to the active site of cytochrome P450 2E1. *J Pharmacol Exp Ther* 310:281–290.

Broly, F., Meyer, U. A. 1993. Debrisoquine oxidation polymorphism: Phenotypic consequences of a 3-base-pair deletion in exon 5 of the *CYP2D6* gene. *Pharmacogenetics* 3:123–130.

Chen, C. K., Shokhireva, T., Berry, R. E., Zhang, H., Walker, F. A. 2008. The effect of mutation of F87 on the properties of CYP102A1-CYP4C7 chimeras: Altered regiospecificity and substrate selectivity. *J Biol Inorg Chem* 13:813–824.

Ching, M. S., Blake, C. L., Ghabrial, H., Ellis, S. W., Lennard, M. S., Tucker, G. T., Smallwood, R. A. 1995. Potent inhibition of yeast-expressed CYP2D6 by dihydroquinidine, quinidine, and its metabolites. *Biochem Pharmacol* 50:833–837.

Compagnone, N. A., Mellon, S. H. 2000. Neurosteroids: Biosynthesis and function of these novel neuromodulators. *Front Neuro-endocrinol* 21:1–56.

Cook Sangar, M., Anandatheerthavarada, H. K., Tang, W., Prabu, S. K., Martin, M. V., Dostalek, M., Guengerich, F. P., Avadhani, N. G. 2009. Human liver mitochondrial cytochrome P450 2D6—Individual variations and implications in drug metabolism. *FEBS J* 276:3440–3453.

Coon, M. J., Vaz, A. D., McGinnity, D. F., Peng, H. M. 1998. Multiple activated oxygen species in P450 catalysis: Contributions to specificity in drug metabolism. *Drug Metab Dispos* 26: 1190–1193.

Cribb, A., Nuss, C., Wang, R. 1995. Antipeptide antibodies against overlapping sequences differentially inhibit human CYP2D6. *Drug Metab Dispos* 23:671–675.

Cupp-Vickery, J. R., Han, O., Hutchinson, C. R., Poulos, T. L. 1996. Substrate-assisted catalysis in cytochrome P450eryF. *Nat Struct Biol* 3:632–637.

Curnow, K. M., Slutsker, L., Vitek, J., Cole, T., Speiser, P. W., New, M. I., White, P. C., Pascoe, L. 1993. Mutations in the *CYP11B1* gene causing congenital adrenal hyperplasia and hypertension cluster in exons 6, 7, and 8. *Proc Natl Acad Sci U S A* 90:4552–4556.

de Graaf, C., Oostenbrink, C., Keizers, P. H., van der Wijst, T., Jongejan, A., Vermeulen, N. P. 2006. Catalytic site prediction and virtual screening of cytochrome P450 2D6 substrates by consideration of water and rescoring in automated docking. *J Med Chem* 49:2417–2430.

de Graaf, C., Oostenbrink, C., Keizers, P. H., van Vugt-Lussenburg, B. M., van Waterschoot, R. A., Tschirret-Guth, R. A., Commandeur, J. N., Vermeulen, N. P. 2007. Molecular modeling-guided site-directed mutagenesis of cytochrome P450 2D6. *Curr Drug Metab* 8:59–77.

de Groot, M. J., Bijloo, G. J., Hansen, K. T., Vermeulen, N. P. 1995. Computer prediction and experimental validation of cytochrome P450 2D6-dependent oxidation of GBR 12909. *Drug Metab Dispos* 23:667–669.

de Groot, M. J., Vermeulen, N. P., Kramer, J. D., van Acker, F. A., Donne-Op den Kelder, G. M. 1996. A three-dimensional protein model for human cytochrome P450 2D6 based on the crystal structures of P450 101, P450 102, and P450 108. *Chem Res Toxicol* 9:1079–1091.

de Groot, M. J., Bijloo, G. J., Martens, B. J., van Acker, F. A., Vermeulen, N. P. 1997a. A refined substrate model for human cytochrome P450 2D6. *Chem Res Toxicol* 10:41–48.

de Groot, M. J., Bijloo, G. J., van Acker, F. A., Fonseca Guerra, C., Snijders, J. G., Vermeulen, N. P. 1997b. Extension of a predictive substrate model for human cytochrome P4502D6. *Xenobiotica* 27:357–368.

de Groot, M. J., Ackland, M. J., Horne, V. A., Alex, A. A., Jones, B. C. 1999a. Novel approach to predicting P450-mediated drug metabolism: Development of a combined protein and pharmacophore model for CYP2D6. *J Med Chem* 42:1515–1524.

de Groot, M. J., Ackland, M. J., Horne, V. A., Alex, A. A., Jones, B. C. 1999b. A novel approach to predicting P450 mediated drug metabolism. CYP2D6 catalyzed *N*-dealkylation reactions and qualitative metabolite predictions using a combined protein and pharmacophore model for CYP2D6. *J Med Chem* 42:4062–4070.

De Rienzo, F., Fanelli, F., Menziani, M. C., De Benedetti, P. G. 2000. Theoretical investigation of substrate specificity for cytochromes P450 IA2, P450 IID6 and P450 IIIA4. *J Comput Aided Mol Des* 14:93–116.

Dehal, S. S., Kupfer, D. 1997. CYP2D6 catalyzes tamoxifen 4-hydroxylation in human liver. *Cancer Res* 57:3402–3406.

Duclos-Vallee, J. C., Hajoui, O., Yamamoto, A. M., Jacz-Aigrain, E., Alvarez, F. 1995. Conformational epitopes on CYP2D6 are recognized by liver/kidney microsomal antibodies. *Gastroenterology* 108:470–476.

Edwards, R. J., Murray, B. P., Boobis, A. R., Davies, D. S. 1989. Identification and location of α-helices in mammalian cytochromes P450. *Biochemistry* 28:3762–3770.

Ekins, S., Bravi, G., Binkley, S., Gillespie, J. S., Ring, B. J., Wikel, J. H., Wrighton, S. A. 1999. Three and four dimensional-quantitative structure activity relationship (3D/4D-QSAR) analyses of CYP2D6 inhibitors. *Pharmacogenetics* 9:477–489.

Ellis, S. W., Hayhurst, G. P., Smith, G., Lightfoot, T., Wong, M. M., Simula, A. P., Ackland, M. J. et al. 1995. Evidence that aspartic acid 301 is a critical substrate-contact residue in the active site of cytochrome P450 2D6. *J Biol Chem* 270:29055–29058.

Ellis, S. W., Rowland, K., Ackland, M. J., Rekka, E., Simula, A. P., Lennard, M. S., Wolf, C. R., Tucker, G. T. 1996. Influence of amino acid residue 374 of cytochrome P450 2D6 (CYP2D6) on the regio- and enantio-selective metabolism of metoprolol. *Biochem J* 316 (Pt 2):647–654.

Ellis, S. W., Hayhurst, G. P., Lightfoot, T., Smith, G., Harlow, J., Rowland-Yeo, K., Larsson, C. et al. 2000. Evidence that serine 304 is not a key ligand-binding residue in the active site of cytochrome P450 2D6. *Biochem J* 345 Pt 3:565–571.

Evert, B., Griese, E. U., Eichelbaum, M. 1994. A missense mutation in exon 6 of the CYP2D6 gene leading to a histidine 324 to proline exchange is associated with the poor metabolizer phenotype of sparteine. *Naunyn Schmiedebergs Arch Pharmacol* 350:434–439.

Evert, B., Eichelbaum, M., Haubruck, H., Zanger, U. M. 1997. Functional properties of CYP2D6.1 (wild-type) and CYP2D6.7 (His324Pro) expressed by recombinant baculovirus in insect cells. *Naunyn Schmiedebergs Arch Pharmacol* 355:309–318.

Ferrari, S., Leemann, T., Dayer, P. 1991. The role of lipophilicity in the inhibition of polymorphic cytochrome P450IID6 oxidation by beta-blocking agents in vitro. *Life Sci* 48:2259–2265.

Flanagan, J. U., Marechal, J. D., Ward, R., Kemp, C. A., McLaughlin, L. A., Sutcliffe, M. J., Roberts, G. C., Paine, M. J., Wolf, C. R. 2004. Phe120 contributes to the regiospecificity of cytochrome P450 2D6: Mutation leads to the formation of a novel dextromethorphan metabolite. *Biochem J* 380:353–360.

Freeman, E. R., Bloom, D. A., McGuire, E. J. 2001. A brief history of testosterone. *J Urol* 165:371–373.

Fukuda, T., Nishida, Y., Imaoka, S., Hiroi, T., Naohara, M., Funae, Y., Azuma, J. 2000. The decreased *in vivo* clearance of CYP2D6 substrates by *CYP2D6*10* might be caused not only by the low-expression but also by low affinity of CYP2D6. *Arch Biochem Biophys* 380:303–308.

Gaedigk, A., Coetsee, C. 2008. The *CYP2D6* gene locus in South African Coloureds: Unique allele distributions, novel alleles and gene arrangements. *Eur J Clin Pharmacol* 64:465–475.

Gaedigk, A., Bradford, L. D., Marcucci, K. A., Leeder, J. S. 2002. Unique CYP2D6 activity distribution and genotype–phenotype discordance in black Americans. *Clin Pharmacol Ther* 72:76–89.

Gerber, N. C., Sligar, S. G. 1994. A role for Asp-251 in cytochrome P450cam oxygen activation. *J Biol Chem* 269:4260–4266.

Gonzalez, F. J., Skoda, R. C., Kimura, S., Umeno, M., Zanger, U. M., Nebert, D. W., Gelboin, H. V., Hardwick, J. P., Meyer, U. A. 1988. Characterization of the common genetic defect in humans deficient in debrisoquine metabolism. *Nature* 331:442–446.

Goodford, P. J. 1985. A computational procedure for determining energetically favorable binding sites on biologically important macromolecules. *J Med Chem* 28:849–857.

Gotoh, O. 1992. Substrate recognition sites in cytochrome P450 family 2 (CYP2) proteins inferred from comparative analyses of amino acid and coding nucleotide sequences. *J Biol Chem* 267:83–90.

Grace, J. M., Kinter, M. T., Macdonald, T. L. 1994. Atypical metabolism of deprenyl and its enantiomer, (*S*)-(+)-*N*,α-dimethyl-*N*-propynylphenethylamine, by cytochrome P450 2D6. *Chem Res Toxicol* 7:286–290.

Griese, E. U., Zanger, U. M., Brudermanns, U., Gaedigk, A., Mikus, G., Morike, K., Stuven, T., Eichelbaum, M. 1998. Assessment of the predictive power of genotypes for the *in vivo* catalytic function of CYP2D6 in a German population. *Pharmacogenetics* 8:15–26.

Gueguen, M., Boniface, O., Bernard, O., Clerc, F., Cartwright, T., Alvarez, F. 1991. Identification of the main epitope on human cytochrome P450 2D6 recognized by anti-liver kidney microsome antibody. *J Autoimmun* 4:607–615.

Guengerich, F. P., Muller-Enoch, D., Blair, I. A. 1986. Oxidation of quinidine by human liver cytochrome P450. *Mol Pharmacol* 30:287–295.

Guengerich, F. P., Miller, G. P., Hanna, I. H., Martin, M. V., Leger, S., Black, C., Chauret, N. et al. 2002. Diversity in the oxidation of substrates by cytochrome P450 2D6: Lack of an obligatory role of aspartate 301-substrate electrostatic bonding. *Biochemistry* 41:11025–11034.

Guengerich, F. P., Hanna, I. H., Martin, M. V., Gillam, E. M. 2003. Role of glutamic acid 216 in cytochrome P450 2D6 substrate binding and catalysis. *Biochemistry* 42:1245–1253.

Haines, D. C., Tomchick, D. R., Machius, M., Peterson, J. A. 2001. Pivotal role of water in the mechanism of P450BM-3. *Biochemistry* 40:13456–13465.

Haining, R. L., Jones, J. P., Henne, K. R., Fisher, M. B., Koop, D. R., Trager, W. F., Rettie, A. E. 1999. Enzymatic determinants of the substrate specificity of CYP2C9: Role of B′-C loop residues in providing the π-stacking anchor site for warfarin binding. *Biochemistry* 38:3285–3292.

Haji-Momenian, S., Rieger, J. M., Macdonald, T. L., Brown, M. L. 2003. Comparative molecular field analysis and QSAR on substrates binding to cytochrome P450 2D6. *Bioorg Med Chem* 11:5545–5554.

Hanna, I. H., Kim, M. S., Guengerich, F. P. 2001a. Heterologous expression of cytochrome P450 2D6 mutants, electron transfer, and catalysis of bufuralol hydroxylation: The role of aspartate 301 in structural integrity. *Arch Biochem Biophys* 393:255–261.

Hanna, I. H., Krauser, J. A., Cai, H., Kim, M. S., Guengerich, F. P. 2001b. Diversity in mechanisms of substrate oxidation by cytochrome P450 2D6. Lack of an allosteric role of NADPH-cytochrome P450 reductase in catalytic regioselectivity. *J Biol Chem* 276:39553–39561.

Harvey, J. N., Bathelt, C. M., Mulholland, A. J. 2006. QM/MM modeling of compound I active species in cytochrome P450, cytochrome C peroxidase, and ascorbate peroxidase. *J Comput Chem* 27:1352–1362.

Hasemann, C. A., Kurumbail, R. G., Boddupalli, S. S., Peterson, J. A., Deisenhofer, J. 1995. Structure and function of cytochromes P450: A comparative analysis of three crystal structures. *Structure* 3:41–62.

Hayhurst, G. P., Harlow, J., Chowdry, J., Gross, E., Hilton, E., Lennard, M. S., Tucker, G. T., Ellis, S. W. 2001. Influence of phenylalanine-481 substitutions on the catalytic activity of cytochrome P450 2D6. *Biochem J* 355:373–379.

Herd, M. B., Belelli, D., Lambert, J. J. 2007. Neurosteroid modulation of synaptic and extrasynaptic GABA$_A$ receptors. *Pharmacol Ther* 116:20–34.

Hidestrand, M., Oscarson, M., Salonen, J. S., Nyman, L., Pelkonen, O., Turpeinen, M., Ingelman-Sundberg, M. 2001. CYP2B6 and CYP2C19 as the major enzymes responsible for the metabolism of selegiline, a drug used in the treatment of Parkinson's disease, as revealed from experiments with recombinant enzymes. *Drug Metab Dispos* 29:1480–1484.

Hiroi, T., Kishimoto, W., Chow, T., Imaoka, S., Igarashi, T., Funae, Y. 2001. Progesterone oxidation by cytochrome P450 2D isoforms in the brain. *Endocrinology* 142:3901–3908.

Hishiki, T., Shimada, H., Nagano, S., Egawa, T., Kanamori, Y., Makino, R., Park, S. Y., Adachi, S., Shiro, Y., Ishimura, Y. 2000. X-ray crystal structure and catalytic properties of Thr252Ile mutant of cytochrome P450cam: Roles of Thr252 and water in the active center. *J Biochem* 128:965–974.

Hritz, J., de Ruiter, A., Oostenbrink, C. 2008. Impact of plasticity and flexibility on docking results for cytochrome P450 2D6: A combined approach of molecular dynamics and ligand docking. *J Med Chem* 51:7469–7477.

Hutzler, J. M., Powers, F. J., Wynalda, M. A., Wienkers, L. C. 2003a. Effect of carbonate anion on cytochrome P450 2D6-mediated metabolism in vitro: The potential role of multiple oxygenating species. *Arch Biochem Biophys* 417:165–175.

Hutzler, J. M., Walker, G. S., Wienkers, L. C. 2003b. Inhibition of cytochrome P450 2D6: Structure-activity studies using a series of quinidine and quinine analogues. *Chem Res Toxicol* 16:450–459.

Ibeanu, G. C., Blaisdell, J., Ghanayem, B. I., Beyeler, C., Benhamou, S., Bouchardy, C., Wilkinson, G. R., Dayer, P., Daly, A. K., Goldstein, J. A. 1998. An additional defective allele, *CYP2C19*5*, contributes to the *S*-mephenytoin poor metabolizer phenotype in Caucasians. *Pharmacogenetics* 8:129–135.

Imai, M., Shimada, H., Watanabe, Y., Matsushima-Hibiya, Y., Makino, R., Koga, H., Horiuchi, T., Ishimura, Y. 1989. Uncoupling of the cytochrome P-450cam monooxygenase reaction by a single mutation, threonine-252 to alanine or valine: Possible role of the hydroxy amino acid in oxygen activation. *Proc Natl Acad Sci U S A* 86:7823–7827.

Islam, S. A., Wolf, C. R., Lennard, M. S., Sternberg, M. J. 1991. A three-dimensional molecular template for substrates of human cytochrome P450 involved in debrisoquine 4-hydroxylation. *Carcinogenesis* 12:2211–2219.

Ito, Y., Fisher, C. R., Conte, F. A., Grumbach, M. M., Simpson, E. R. 1993. Molecular basis of aromatase deficiency in an adult female with sexual infantilism and polycystic ovaries. *Proc Natl Acad Sci U S A* 90:11673–11677.

Ito, Y., Kondo, H., Goldfarb, P. S., Lewis, D. F. 2008. Analysis of CYP2D6 substrate interactions by computational methods. *J Mol Graph Model* 26:947–956.

Johansson, I., Oscarson, M., Yue, Q. Y., Bertilsson, L., Sjoqvist, F., Ingelman-Sundberg, M. 1994. Genetic analysis of the Chinese cytochrome P450 *2D* locus: Characterization of variant *CYP2D6* genes present in subjects with diminished capacity for debrisoquine hydroxylation. *Mol Pharmacol* 46:452–459.

Keizers, P. H., Lussenburg, B. M., de Graaf, C., Mentink, L. M., Vermeulen, N. P., Commandeur, J. N. 2004. Influence of phenylalanine 120 on cytochrome P450 2D6 catalytic selectivity and regiospecificity: Crucial role in 7-methoxy-4-(aminomethyl)-coumarin metabolism. *Biochem Pharmacol* 68:2263–2271.

Keizers, P. H., de Graaf, C., de Kanter, F. J., Oostenbrink, C., Feenstra, K. A., Commandeur, J. N., Vermeulen, N. P. 2005a. Metabolic regio- and stereoselectivity of cytochrome P450 2D6 towards 3,4-methylenedioxy-*N*-alkylamphetamines: *In silico* predictions and experimental validation. *J Med Chem* 48:6117–6127.

Keizers, P. H., Schraven, L. H., de Graaf, C., Hidestrand, M., Ingelman-Sundberg, M., van Dijk, B. R., Vermeulen, N. P., Commandeur, J. N. 2005b. Role of the conserved threonine 309 in mechanism of oxidation by cytochrome P450 2D6. *Biochem Biophys Res Commun* 338:1065–1074.

Keizers, P. H., Van Dijk, B. R., De Graaf, C., Van Vugt-Lussenburg, B. M., Vermeulen, N. P., Commandeur, J. N. 2006. Metabolism of *N*-substituted 7-methoxy-4-(aminomethyl)-coumarins by cytochrome P450 2D6 mutants and the indication of additional substrate interaction points. *Xenobiotica* 36:763–771.

Kemp, C. A., Flanagan, J. U., van Eldik, A. J., Marechal, J. D., Wolf, C. R., Roberts, G. C., Paine, M. J., Sutcliffe, M. J. 2004. Validation of model of cytochrome P450 2D6: An *in silico* tool for predicting metabolism and inhibition. *J Med Chem* 47:5340–5346.

Kimura, S., Umeno, M., Skoda, R. C., Meyer, U. A., Gonzalez, F. J. 1989. The human debrisoquine 4-hydroxylase (CYP2D) locus: Sequence and identification of the polymorphic *CYP2D6* gene, a related gene, and a pseudogene. *Am J Hum Genet* 45:889–904.

Kirton, S. B., Kemp, C. A., Tomkinson, N. P., St-Gallay, S., Sutcliffe, M. J. 2002. Impact of incorporating the 2C5 crystal structure into comparative models of cytochrome P450 2D6. *Proteins* 49:216–231.

Kishimoto, W., Hiroi, T., Shiraishi, M., Osada, M., Imaoka, S., Kominami, S., Igarashi, T., Funae, Y. 2004. Cytochrome P450 2D catalyze steroid 21-hydroxylation in the brain. *Endocrinology* 145:699–705.

Kjellander, B., Masimirembwa, C. M., Zamora, I. 2007. Exploration of enzyme-ligand interactions in CYP2D6 & 3A4 homology models and crystal structures using a novel computational approach. *J Chem Inf Model* 47:1234–1247.

Klein, K., Tatzel, S., Raimundo, S., Saussele, T., Hustert, E., Pleiss, J., Eichelbaum, M., Zanger, U. M. 2007. A natural variant of the heme-binding signature (R441C) resulting in complete loss of function of CYP2D6. *Drug Metab Dispos* 35:1247–1250.

Korzekwa, K. R., Jones, J. P. 1993. Predicting the cytochrome P450 mediated metabolism of xenobiotics. *Pharmacogenetics* 3:1–18.

Kotsuma, M., Hanzawa, H., Iwata, Y., Takahashi, K., Tokui, T. 2008a. Novel binding mode of the acidic CYP2D6 substrates pactimibe and its metabolite R-125528. *Drug Metab Dispos* 36:1938–1943.

Kotsuma, M., Tokui, T., Freudenthaler, S., Nishimura, K. 2008b. Effects of ketoconazole and quinidine on pharmacokinetics of pactimibe and its plasma metabolite, R-125528, in human. *Drug Metab Dispos* 36:1505–1511.

Kotsuma, M., Tokui, T., Ishizuka-Ozeki, T., Honda, T., Iwabuchi, H., Murai, T., Ikeda, T., Saji, H. 2008c. CYP2D6-mediated metabolism of a novel acyl coenzyme A: Cholesterol acyltransferase inhibitor, pactimibe, and its unique plasma metabolite, R-125528. *Drug Metab Dispos* 36:529–534.

Koymans, L., Vermeulen, N. P., van Acker, S. A., te Koppele, J. M., Heykants, J. J., Lavrijsen, K., Meuldermans, W., Donne-Op den Kelder, G. M. 1992. A predictive model for substrates of cytochrome P450-debrisoquine (2D6). *Chem Res Toxicol* 5:211–219.

Koymans, L. M., Vermeulen, N. P., Baarslag, A., Donne-Op den Kelder, G. M. 1993. A preliminary 3D model for cytochrome P450 2D6 constructed by homology model building. *J Comput Aided Mol Des* 7:281–289.

Kubota, T., Yamaura, Y., Ohkawa, N., Hara, H., Chiba, K. 2000. Frequencies of CYP2D6 mutant alleles in a normal Japanese population and metabolic activity of dextromethorphan O-demethylation in different CYP2D6 genotypes. *Br J Clin Pharmacol* 50:31–34.

Kumar, G. N., Rodrigues, A. D., Buko, A. M., Denissen, J. F. 1996. Cytochrome P450-mediated metabolism of the HIV-1 protease inhibitor ritonavir (ABT-538) in human liver microsomes. *J Pharmacol Exp Ther* 277:423–431.

Lee, A. J., Cai, M. X., Thomas, P. E., Conney, A. H., Zhu, B. T. 2003. Characterization of the oxidative metabolites of 17β-estradiol and estrone formed by 15 selectively expressed human cytochrome P450 isoforms. *Endocrinology* 144:3382–3398.

Lee-Robichaud, P., Akhtar, M. E., Akhtar, M. 1998. An analysis of the role of active site protic residues of cytochrome P450s: Mechanistic and mutational studies on 17α-hydroxylase-17,20-lyase (P-45017α also CYP17). *Biochem J* 330 (Pt 2): 967–974.

Lewis, D. F. 1995. Three-dimensional models of human and other mammalian microsomal P450s constructed from an alignment with P450102 (P450bm3). *Xenobiotica* 25:333–366.

Lewis, D. F. 1999. Homology modelling of human cytochromes P450 involved in xenobiotic metabolism and rationalization of substrate selectivity. *Exp Toxicol Pathol* 51:369–374.

Lewis, D. F., Eddershaw, P. J., Goldfarb, P. S., Tarbit, M. H. 1997. Molecular modelling of cytochrome P450 2D6 (CYP2D6) based on an alignment with CYP102: Structural studies on specific CYP2D6 substrate metabolism. *Xenobiotica* 27:319–339.

Lussenburg, B. M., Keizers, P. H., de Graaf, C., Hidestrand, M., Ingelman-Sundberg, M., Vermeulen, N. P., Commandeur, J. N. 2005. The role of phenylalanine 483 in cytochrome P450 2D6 is strongly substrate dependent. *Biochem Pharmacol* 70:1253–1261.

Mackman, R., Tschirret-Guth, R. A., Smith, G., Hayhurst, G. P., Ellis, S. W., Lennard, M. S., Tucker, G. T., Wolf, C. R., Ortiz de Montellano, P. R. 1996. Active-site topologies of human CYP2D6 and its aspartate-301 → glutamate, asparagine, and glycine mutants. *Arch Biochem Biophys* 331:134–140.

Marechal, J. D., Kemp, C. A., Roberts, G. C., Paine, M. J., Wolf, C. R., Sutcliffe, M. J. 2008. Insights into drug metabolism by cytochromes P450 from modelling studies of CYP2D6-drug interactions. *Br J Pharmacol* 153 Suppl 1:S82–S89.

Marez, D., Sabbagh, N., Legrand, M., Lo-Guidice, J. M., Boone, P., Broly, F. 1995. A novel CYP2D6 allele with an abolished splice recognition site associated with the poor metabolizer phenotype. *Pharmacogenetics* 5:305–311.

Marez, D., Legrand, M., Sabbagh, N., Lo-Guidice, J. M., Boone, P., Broly, F. 1996. An additional allelic variant of the CYP2D6 gene causing impaired metabolism of sparteine. *Hum Genet* 97:668–670.

Marez, D., Legrand, M., Sabbagh, N., Guidice, J. M., Spire, C., Lafitte, J. J., Meyer, U. A., Broly, F. 1997. Polymorphism of the cytochrome P450 CYP2D6 gene in a European population: Characterization of 48 mutations and 53 alleles, their frequencies and evolution. *Pharmacogenetics* 7:193–202.

Masimirembwa, C., Persson, I., Bertilsson, L., Hasler, J., Ingelman-Sundberg, M. 1996. A novel mutant variant of the *CYP2D6* gene (*CYP2D6*17*) common in a black African population: Association with diminished debrisoquine hydroxylase activity. *Br J Clin Pharmacol* 42:713–719.

Masuda, K., Tamagake, K., Okuda, Y., Torigoe, F., Tsuzuki, D., Isobe, T., Hichiya, H., Hanioka, N., Yamamoto, S., Narimatsu, S. 2005. Change in enantioselectivity in bufuralol 1″-hydroxylation by the substitution of phenylalanine-120 by alanine in cytochrome P450 2D6. *Chirality* 17:37–43.

Masuda, K., Tamagake, K., Katsu, T., Torigoe, F., Saito, K., Hanioka, N., Yamano, S., Yamamoto, S., Narimatsu, S. 2006. Roles of phenylalanine at position 120 and glutamic acid at position 222 in the oxidation of chiral substrates by cytochrome P450 2D6. *Chirality* 18:167–176.

Matsunaga, M., Yamazaki, H., Kiyotani, K., Iwano, S., Saruwatari, J., Nakagawa, K., Soyama, A. et al. 2009. Two novel *CYP2D6*10* haplotypes as possible causes of a poor metabolic phenotype in Japanese. *Drug Metab Dispos* 37:699–701.

Mautz, D. S., Nelson, W. L., Shen, D. D. 1995. Regioselective and stereoselective oxidation of metoprolol and bufuralol catalyzed by microsomes containing cDNA-expressed human P450 2D6. *Drug Metab Dispos* 23:513–517.

McLaughlin, L. A., Paine, M. J., Kemp, C. A., Marechal, J. D., Flanagan, J. U., Ward, C. J., Sutcliffe, M. J., Roberts, G. C., Wolf, C. R. 2005. Why is quinidine an inhibitor of cytochrome P450 2D6? The role of key active-site residues in quinidine binding. *J Biol Chem* 280:38617–38624.

McMasters, D. R., Torres, R. A., Crathern, S. J., Dooney, D. L., Nachbar, R. B., Sheridan, R. P., Korzekwa, K. R. 2007. Inhibition of recombinant cytochrome P450 isoforms 2D6 and 2C9 by diverse drug-like molecules. *J Med Chem* 50:3205–3213.

Melet, A., Assrir, N., Jean, P., Pilar Lopez-Garcia, M., Marques-Soares, C., Jaouen, M., Dansette, P. M., Sari, M. A., Mansuy, D. 2003. Substrate selectivity of human cytochrome P450 2C9: Importance of residues 476, 365, and 114 in recognition of diclofenac and sulfaphenazole and in mechanism-based inactivation by tienilic acid. *Arch Biochem Biophys* 409:80–91.

Mensah-Nyagan, A. G., Do-Rego, J. L., Beaujean, D., Luu-The, V., Pelletier, G., Vaudry, H. 1999. Neurosteroids: Expression of steroidogenic enzymes and regulation of steroid biosynthesis in the central nervous system. *Pharmacol Rev* 51:63–81.

Mestres, J. 2005. Structure conservation in cytochromes P450. *Proteins* 58:596–609.

Meyer, U. A., Gut, J., Kronbach, T., Skoda, C., Meier, U. T., Catin, T., Dayer, P. 1986. The molecular mechanisms of two common polymorphisms of drug oxidation—Evidence for functional changes in cytochrome P450 isozymes catalysing bufuralol and mephenytoin oxidation. *Xenobiotica* 16:449–464.

Mitchell, E. A., Herd, M. B., Gunn, B. G., Lambert, J. J., Belelli, D. 2008. Neurosteroid modulation of GABAA receptors: Molecular determinants and significance in health and disease. *Neurochem Int* 52:588–595.

Miyakawa, H., Matsushima, H., Narita, Y., Hankins, R. W., Kitazawa, E., Fujikawa, H., Kikuchi, K. et al. 1998. Differences in antigenic sites, recognized by anti-liver-kidney microsome-1 (LKM-1) autoantibody, between HCV-positive and HCV-negative sera in Japanese patients. *J Gastroenterol* 33:529–535.

Miyakawa, H., Kikazawa, E., Abe, K., Kikuchi, K., Fujikawa, H., Matsushita, M., Kawaguchi, N., Morizane, T., Ohya, K., Kako, M. 1999. Detection of anti-LKM-1(anti-CYP2D6) by an enzyme-linked immunosorbent assay in adult patients with chronic liver diseases. *Autoimmunity* 30:107–114.

Modi, S., Paine, M. J., Sutcliffe, M. J., Lian, L. Y., Primrose, W. U., Wolf, C. R., Roberts, G. C. 1996. A model for human cytochrome P450 2D6 based on homology modeling and NMR studies of substrate binding. *Biochemistry* 35:4540–4550.

Modi, S., Gilham, D. E., Sutcliffe, M. J., Lian, L. Y., Primrose, W. U., Wolf, C. R., Roberts, G. C. 1997. 1-Methyl-4-phenyl-1,2,3,6-tetrahydropyridine as a substrate of cytochrome P450 2D6: Allosteric effects of NADPH-cytochrome P450 reductase. *Biochemistry* 36:4461–4470.

Mosher, C. M., Hummel, M. A., Tracy, T. S., Rettie, A. E. 2008. Functional analysis of phenylalanine residues in the active site of cytochrome P450 2C9. *Biochemistry* 47:11725–11734.

Nagano, S., Poulos, T. L. 2005. Crystallographic study on the dioxygen complex of wild-type and mutant cytochrome P450cam. Implications for the dioxygen activation mechanism. *J Biol Chem* 280:31659–31663.

Narimatsu, S., Kato, R., Horie, T., Ono, S., Tsutsui, M., Yabusaki, Y., Ohmori, S. et al. 1999. Enantioselectivity of bunitrolol 4-hydroxylation is reversed by the change of an amino acid residue from valine to methionine at position 374 of cytochrome P450 2D6. *Chirality* 11:1–9.

Niwa, T., Yabusaki, Y., Honma, K., Matsuo, N., Tatsuta, K., Ishibashi, F., Katagiri, M. 1998. Contribution of human hepatic cytochrome P450 isoforms to regioselective hydroxylation of steroid hormones. *Xenobiotica* 28:539–547.

Oezguen, N., Kumar, S., Hindupur, A., Braun, W., Muralidhara, B. K., Halpert, J. R. 2008. Identification and analysis of conserved sequence motifs in cytochrome P450 family 2. Functional and structural role of a motif [187]RFDYKD[192] in CYP2B enzymes. *J Biol Chem* 283:21808–21816.

Onderwater, R. C., Venhorst, J., Commandeur, J. N., Vermeulen, N. P. 1999. Design, synthesis, and characterization of 7-methoxy-4-(aminomethyl)coumarin as a novel and selective cytochrome P450 2D6 substrate suitable for high-throughput screening. *Chem Res Toxicol* 12:555–559.

Oscarson, M., Hidestrand, M., Johansson, I., Ingelman-Sundberg, M. 1997. A combination of mutations in the *CYP2D6*17* (*CYP2D6Z*) allele causes alterations in enzyme function. *Mol Pharmacol* 52:1034–1040.

Otton, S. V., Crewe, H. K., Lennard, M. S., Tucker, G. T., Woods, H. F. 1988. Use of quinidine inhibition to define the role of the sparteine/debrisoquine cytochrome P450 in metoprolol oxidation by human liver microsomes. *J Pharmacol Exp Ther* 247:242–247.

Paine, M. J., McLaughlin, L. A., Flanagan, J. U., Kemp, C. A., Sutcliffe, M. J., Roberts, G. C., Wolf, C. R. 2003. Residues glutamate 216 and aspartate 301 are key determinants of substrate specificity and product regioselectivity in cytochrome P450 2D6. *J Biol Chem* 278:4021–4027.

Porubsky, P. R., Meneely, K. M., Scott, E. E. 2008. Structures of human cytochrome P-450 2E1. Insights into the binding of inhibitors and both small molecular weight and fatty acid substrates. *J Biol Chem* 283:33698–33707.

Poulos, T. L., Finzel, B. C., Gunsalus, I. C., Wagner, G. C., Kraut, J. 1985. The 2.6-Å crystal structure of *Pseudomonas putida* cytochrome P450. *J Biol Chem* 260:16122–16130.

Raag, R., Martinis, S. A., Sligar, S. G., Poulos, T. L. 1991. Crystal structure of the cytochrome P450CAM active site mutant Thr252Ala. *Biochemistry* 30:11420–11429.

Raimundo, S., Fischer, J., Eichelbaum, M., Griese, E. U., Schwab, M., Zanger, U. M. 2000. Elucidation of the genetic basis of the common 'intermediate metabolizer' phenotype for drug oxidation by CYP2D6. *Pharmacogenetics* 10:577–581.

Raimundo, S., Toscano, C., Klein, K., Fischer, J., Griese, E. U., Eichelbaum, M., Schwab, M., Zanger, U. M. 2004. A novel intronic mutation, 2988G>A, with high predictivity for impaired function of cytochrome P450 2D6 in white subjects. *Clin Pharmacol Ther* 76:128–138.

Ravichandran, K. G., Boddupalli, S. S., Hasermann, C. A., Peterson, J. A., Deisenhofer, J. 1993. Crystal structure of hemoprotein domain of P450BM-3, a prototype for microsomal P450's. *Science* 261:731–736.

Rowland, P., Blaney, F. E., Smyth, M. G., Jones, J. J., Leydon, V. R., Oxbrow, A. K., Lewis, C. J. et al. 2006. Crystal structure of human cytochrome P450 2D6. *J Biol Chem* 281:7614–7622.

Roy, P. P., Roy, K. 2009. QSAR studies of CYP2D6 inhibitor aryloxypropanolamines using 2D and 3D descriptors. *Chem Biol Drug Des* 73:442–455.

Saito, Y., Hanioka, N., Maekawa, K., Isobe, T., Tsuneto, Y., Nakamura, R., Soyama, A. et al. 2005. Functional analysis of three CYP1A2 variants found in a Japanese population. *Drug Metab Dispos* 33:1905–1910.

Sakuyama, K., Sasaki, T., Ujiie, S., Obata, K., Mizugaki, M., Ishikawa, M., Hiratsuka, M. 2008. Functional characterization of 17 CYP2D6 allelic variants (CYP2D6.2, 10, 14A-B, 18, 27, 36, 39, 47–51, 53–55, and 57). *Drug Metab Dispos* 36:2460–2467.

Sansen, S., Yano, J. K., Reynald, R. L., Schoch, G. A., Griffin, K. J., Stout, C. D., Johnson, E. F. 2007. Adaptations for the oxidation of polycyclic aromatic hydrocarbons exhibited by the structure of human P450 1A2. *J Biol Chem* 282:14348–14355.

Schenkman, J. B., Sligar, S. G., Cinti, D. L. 1981. Substrate interaction with cytochrome P450. *Pharmacol Ther* 12:43–71.

Schleinkofer, K., Sudarko, Winn, P. J., Ludemann, S. K., Wade, R. C. 2005. Do mammalian cytochrome P450s show multiple ligand access pathways and ligand channelling? *EMBO Rep* 6:584–589.

Schlichting, I., Berendzen, J., Chu, K., Stock, A. M., Maves, S. A., Benson, D. E., Sweet, R. M., Ringe, D., Petsko, G. A., Sligar, S. G. 2000. The catalytic pathway of cytochrome P450cam at atomic resolution. *Science* 287:1615–1622.

Schoch, G. A., Yano, J. K., Wester, M. R., Griffin, K. J., Stout, C. D., Johnson, E. F. 2004. Structure of human microsomal cytochrome P450 2C8. Evidence for a peripheral fatty acid binding site. *J Biol Chem* 279:9497–9503.

Schulz-Utermoehl, T., Edwards, R. J., Boobis, A. R. 2000. Affinity and potency of proinhibitory antipeptide antibodies against CYP2D6 is enhanced using cyclic peptides as immunogens. *Drug Metab Dispos* 28:544–551.

Scott, E. E., He, Y. A., Wester, M. R., White, M. A., Chin, C. C., Halpert, J. R., Johnson, E. F., Stout, C. D. 2003. An open conformation of mammalian cytochrome P450 2B4 at 1.6-Å resolution. *Proc Natl Acad Sci U S A* 100:13196–13201.

Shimizu, T., Sadeque, A. J., Sadeque, G. N., Hatano, M., Fujii-Kuriyama, Y. 1991a. Ligand binding studies of engineered cytochrome P450d wild type, proximal mutants, and distal mutants. *Biochemistry* 30:1490–1496.

Shimizu, T., Tateishi, T., Hatano, M., Fujii-Kuriyama, Y. 1991b. Probing the role of lysines and arginines in the catalytic function of cytochrome P450d by site-directed mutagenesis. Interaction with NADPH-cytochrome P450 reductase. *J Biol Chem* 266:3372–3375.

Smith, G., Modi, S., Pillai, I., Lian, L. Y., Sutcliffe, M. J., Pritchard, M. P., Friedberg, T., Roberts, G. C., Wolf, C. R. 1998. Determinants of the substrate specificity of human cytochrome P450 CYP2D6: Design and construction of a mutant with testosterone hydroxylase activity. *Biochem J* 331 (Pt 3):783–792.

Smith, B. D., Sanders, J. L., Porubsky, P. R., Lushington, G. H., Stout, C. D., Scott, E. E. 2007. Structure of the human lung cytochrome P450 2A13. *J Biol Chem* 282:17306–17313.

Solus, J. F., Arietta, B. J., Harris, J. R., Sexton, D. P., Steward, J. Q., McMunn, C., Ihrie, P., Mehall, J. M., Edwards, T. L., Dawson, E. P. 2004. Genetic variation in eleven phase I drug metabolism genes in an ethnically diverse population. *Pharmacogenomics* 5:895–931.

Soyama, A., Kubo, T., Miyajima, A., Saito, Y., Shiseki, K., Komamura, K., Ueno, K. et al. Novel nonsynonymous single nucleotide polymorphisms in the CYP2D6 gene. *Drug Metab Pharmacokinet* 19:313–319.

Stortelder, A., Keizers, P. H., Oostenbrink, C., De Graaf, C., De Kruijf, P., Vermeulen, N. P., Gooijer, C., Commandeur, J. N., Van der Zwan, G. 2006. Binding of 7-methoxy-4-(aminomethyl)-coumarin to wild-type and W128F mutant cytochrome P450 2D6 studied by time-resolved fluorescence spectroscopy. *Biochem J* 393:635–643.

Strobl, G. R., von Kruedener, S., Stockigt, J., Guengerich, F. P., Wolff, T. 1993. Development of a pharmacophore for inhibition of human liver cytochrome P450 2D6: Molecular modeling and inhibition studies. *J Med Chem* 36:1136–1145.

Strushkevich, N., Usanov, S. A., Plotnikov, A. N., Jones, G., Park, H. W. 2008. Structural analysis of CYP2R1 in complex with vitamin D_3. *J Mol Biol* 380:95–106.

Terfloth, L., Bienfait, B., Gasteiger, J. 2007. Ligand-based models for the isoform specificity of cytochrome P450 3A4, 2D6, and 2C9 substrates. *J Chem Inf Model* 47:1688–1701.

Tsutsui, K. 2008. Progesterone biosynthesis and action in the developing neuron. *Endocrinology* 149:2757–2761.

Tsuzuki, D., Takemi, C., Yamamoto, S., Tamagake, K., Imaoka, S., Funae, Y., Kataoka, H., Shinoda, S., Narimatsu, S. 2001. Functional evaluation of cytochrome P450 2D6 with Gly42Arg substitution expressed in *Saccharomyces cerevisiae*. *Pharmacogenetics* 11:709–718.

Tsuzuki, D., Hichiya, H., Okuda, Y., Yamamoto, S., Tamagake, K., Shinoda, S., Narimatsu, S. 2003. Alteration in catalytic properties of human CYP2D6 caused by substitution of glycine-42 with arginine, lysine and glutamic acid. *Drug Metab Pharmacokinet* 18:79–85.

Tyndale, R., Aoyama, T., Broly, F., Matsunaga, T., Inaba, T., Kalow, W., Gelboin, H. V., Meyer, U. A., Gonzalez, F. J. 1991. Identification of a new variant *CYP2D6* allele lacking the codon encoding Lys-281: Possible association with the poor metabolizer phenotype. *Pharmacogenetics* 1:26–32.

van Waterschoot, R. A., Keizers, P. H., de Graaf, C., Vermeulen, N. P., Tschirret-Guth, R. A. 2006. Topological role of cytochrome P450 2D6 active site residues. *Arch Biochem Biophys* 447:53–58.

Vaz, A. D., Pernecky, S. J., Raner, G. M., Coon, M. J. 1996. Peroxoiron and oxenoid-iron species as alternative oxygenating agents in cytochrome P450-catalyzed reactions: Switching by threonine-302 to alanine mutagenesis of cytochrome P450 2B4. *Proc Natl Acad Sci U S A* 93:4644–4648.

Vaz, A. D., McGinnity, D. F., Coon, M. J. 1998. Epoxidation of olefins by cytochrome P450: Evidence from site-specific mutagenesis for hydroperoxo-iron as an electrophilic oxidant. *Proc Natl Acad Sci U S A* 95:3555–3560.

Vaz, R. J., Nayeem, A., Santone, K., Chandrasena, G., Gavai, A. V. 2005. A 3D-QSAR model for CYP2D6 inhibition in the aryloxypropanolamine series. *Bioorg Med Chem Lett* 15:3816–3820.

Venhorst, J., Onderwater, R. C., Meerman, J. H., Commandeur, J. N., Vermeulen, N. P. 2000. Influence of *N*-substitution of 7-methoxy-4-(aminomethyl)-coumarin on cytochrome P450 metabolism and selectivity. *Drug Metab Dispos* 28:1524–1532.

Venhorst, J., ter Laak, A. M., Commandeur, J. N., Funae, Y., Hiroi, T., Vermeulen, N. P. 2003. Homology modeling of rat and human cytochrome P450 2D (CYP2D) isoforms and computational rationalization of experimental ligand-binding specificities. *J Med Chem* 46:74–86.

von Wachenfeldt, C., Richardson, T. H., Cosme, J., Johnson, E. F. 1997. Microsomal P450 2C3 is expressed as a soluble dimer in *Escherichia coli* following modification of its *N*-terminus. *Arch Biochem Biophys* 339:107–114.

Wang, S. L., Lai, M. D., Huang, J. D. 1999. G169R mutation diminishes the metabolic activity of CYP2D6 in Chinese. *Drug Metab Dispos* 27:385–388.

Wang, B., Sanchez, R. I., Franklin, R. B., Evans, D. C., Huskey, S. E. 2004. The involvement of CYP3A4 and CYP2C9 in the metabolism of 17α-ethinylestradiol. *Drug Metab Dispos* 32:1209–1212.

Ward, S. A., Walle, T., Walle, U. K., Wilkinson, G. R., Branch, R. A. 1989. Propranolol's metabolism is determined by both mephenytoin and debrisoquin hydroxylase activities. *Clin Pharmacol Ther* 45:72–79.

Waxman, D. J., Lapenson, D. P., Aoyama, T., Gelboin, H. V., Gonzalez, F. J., Korzekwa, K. 1991. Steroid hormone hydroxylase specificities of eleven cDNA-expressed human cytochrome P450s. *Arch Biochem Biophys* 290:160–166.

Wennerholm, A., Johansson, I., Hidestrand, M., Bertilsson, L., Gustafsson, L. L., Ingelman-Sundberg, M. 2001. Characterization of the *CYP2D6*29* allele commonly present in a black Tanzanian population causing reduced catalytic activity. *Pharmacogenetics* 11:417–427.

Wester, M. R., Johnson, E. F., Marques-Soares, C., Dansette, P. M., Mansuy, D., Stout, C. D. 2003. Structure of a substrate complex of mammalian cytochrome P450 2C5 at 2.3 Å resolution: Evidence for multiple substrate binding modes. *Biochemistry* 42:6370–6379.

Williams, P. A., Cosme, J., Sridhar, V., Johnson, E. F., McRee, D. E. 2000. Mammalian microsomal cytochrome P450 monooxygenase: Structural adaptations for membrane binding and functional diversity. *Mol Cell* 5:121–131.

Williams, P. A., Cosme, J., Ward, A., Angove, H. C., Matak Vinkovic, D., Jhoti, H. 2003. Crystal structure of human cytochrome P450 2C9 with bound warfarin. *Nature* 424:464–468.

Williams, P. A., Cosme, J., Vinkovic, D. M., Ward, A., Angove, H. C., Day, P. J., Vonrhein, C., Tickle, I. J., Jhoti, H. 2004. Crystal structures of human cytochrome P450 3A4 bound to metyrapone and progesterone. *Science* 305:683–686.

Winn, P. J., Ludemann, S. K., Gauges, R., Lounnas, V., Wade, R. C. 2002. Comparison of the dynamics of substrate access channels in three cytochrome P450s reveals different opening mechanisms and a novel functional role for a buried arginine. *Proc Natl Acad Sci U S A* 99:5361–5366.

Wolff, T., Distlerath, L. M., Worthington, M. T., Groopman, J. D., Hammons, G. J., Kadlubar, F. F., Prough, R. A., Martin, M. V., Guengerich, F. P. 1985. Substrate specificity of human liver cytochrome P450 debrisoquine 4-hydroxylase probed using immunochemical inhibition and chemical modeling. *Cancer Res* 45:2116–2122.

Yamakoshi, Y., Kishimoto, T., Sugimura, K., Kawashima, H. 1999. Human prostate CYP3A5: Identification of a unique 5′-untranslated sequence and characterization of purified recombinant protein. *Biochem Biophys Res Commun* 260:676–681.

Yamamoto, A. M., Cresteil, D., Boniface, O., Clerc, F. F., Alvarez, F. 1993. Identification and analysis of cytochrome P450 2D6 antigenic sites recognized by anti-liver-kidney microsome Type-1 antibodies (LKM1). *Eur J Immunol* 23:1105–1111.

Yamashita, F., Hara, H., Ito, T., Hashida, M. 2008. Novel hierarchical classification and visualization method for multiobjective optimization of drug properties: Application to structure-activity relationship analysis of cytochrome P450 metabolism. *J Chem Inf Model* 48:364–369.

Yamazaki, H., Shimada, T. 1997. Progesterone and testosterone hydroxylation by cytochromes P450 2C19, 2C9, and 3A4 in human liver microsomes. *Arch Biochem Biophys* 346: 161–169.

Yamazaki, S., Sato, K., Suhara, K., Sakaguchi, M., Mihara, K., Omura, T. 1993. Importance of the proline-rich region following signal-anchor sequence in the formation of correct conformation of microsomal cytochrome P450s. *J Biochem* 114:652–657.

Yano, J. K., Wester, M. R., Schoch, G. A., Griffin, K. J., Stout, C. D., Johnson, E. F. 2004. The structure of human microsomal cytochrome P450 3A4 determined by X-ray crystallography to 2.05-Å resolution. *J Biol Chem* 279:38091–38094.

Yano, J. K., Hsu, M. H., Griffin, K. J., Stout, C. D., Johnson, E. F. 2005. Structures of human microsomal cytochrome P450 2A6 complexed with coumarin and methoxsalen. *Nat Struct Mol Biol* 12:822–823.

Yeom, H., Sligar, S. G., Li, H., Poulos, T. L., Fulco, A. J. 1995. The role of Thr268 in oxygen activation of cytochrome P450BM-3. *Biochemistry* 34:14733–14740.

Yoshimoto, K., Echizen, H., Chiba, K., Tani, M., Ishizaki, T. 1995. Identification of human CYP isoforms involved in the metabolism of propranolol enantiomers-*N*-desisopropylation is mediated mainly by CYP1A2. *Br J Clin Pharmacol* 39:421–431.

Zhao, Q., Modi, S., Smith, G., Paine, M., McDonagh, P. D., Wolf, C. R., Tew, D., Lian, L. Y., Roberts, G. C., Driessen, H. P. 1999. Crystal structure of the FMN-binding domain of human cytochrome P450 reductase at 1.93 Å resolution. *Protein Sci* 8:298–306.

7 Clinical Pharmacogenomics of Human CYP2D6

7.1 INTRODUCTION

In 1969, Alexanderson et al. (1969) provided the first direct evidence from a twin study that the metabolic clearance of nortriptyline is influenced by genetic factors. Mahgoub et al. (1977) and Eichelbaum et al. (1979) have discovered that the metabolism of debrisoquine and sparteine is polymorphic, and it is later shown that these drugs are metabolized by a common enzyme, that is, CYP2D6. To date, CYP2D6 is one of the most investigated CYPs in relation to polymorphism that accounts for only a small percentage of all hepatic CYPs (~2%–4%); however, it metabolizes ~25% of currently used drugs in the human liver (Cascorbi 2003; Gardiner and Begg 2006; Ingelman-Sundberg 2005; Ingelman-Sundberg et al. 2007; Zhou et al. 2008). Typical substrates for CYP2D6 are largely lipophilic bases and include some antidepressants, neuroleptics, antiarrhythmics, antiemetics, β-blockers, and opioids. Most CYP2D6 substrates are bases containing a basic nitrogen atom 5–10 Å from the site of metabolism (Marechal et al. 2008). CYP2D6 appears to have high affinity and low capacity for its substrates and becomes saturated at relatively low concentrations. The primarily hepatic expression of this enzyme governs first-pass metabolism after oral drug administration, whereas the low level of its intestinal expression does not appear to be important. In contrast to other drug-metabolizing CYPs (e.g., 2C9, 2B6, and 3A4), CYP2D6 is generally not regulated by many known environmental agents and is not inducible by common known CYP inducers such as steroids and rifampin. CYP2D6 is subject to inhibition by a number of drugs, resulting in clinically significant drug interactions.

Human *CYP2D6* maps to chromosome 22q13.1 and contains 12 exons (Gough et al. 1993). CYP2D6 is highly polymorphic. To date, 104 allelic variants and a series of subvariants of the *CYP2D6* gene have been reported (http://www.cypalleles.ki.se/cyp2d6.htm) and the number of alleles is still growing. Among these are fully functional alleles, alleles with reduced function, and null (nonfunctional) alleles, which convey a wide range of enzyme activity from no activity and ultrarapid metabolism and wide interindividual variability in clearance of substrates (Ingelman-Sundberg et al. 2007; Zanger et al. 2004; Zhou et al. 2008). As a consequence, drug adverse effects or lack of drug effect may occur if standard doses are applied. Phenotypically, the CYP2D6 ultrarapid metabolizers (UMs), extensive metabolizers (EMs), intermediate metabolizers (IMs), and poor metabolizers (PMs) compose approximately 3%–5%, 70%–80%, 10%–17%, and 5%–10% of Caucasians, respectively (Sachse et al. 1997). In Chinese, CYP2D6 PMs are infrequent, accounting for ~1% of the population, while 0%–19% of African Americans are PMs (Bradford 2002; Johansson et al. 1991; Wang et al. 1993). In addition, ~1%–2% of Swedish Caucasians (Dahl et al. 1995) and ~16% of black Ethiopians (Aklillu et al. 1996) have more than one extra functional gene, causing a UM phenotype. Generally, an individual's highest functioning *CYP2D6* allele predicts his or her phenotypic activity (Zanger et al. 2004). A person carrying an EM allele and a PM allele has an EM phenotype, while the presence of a UM allele and an EM allele will lead to a UM phenotype. An IM phenotype is usually found in individuals carrying one null allele and another allele with reduced function (e.g., *10)—presence of an IM allele and a PM allele leads to an IM phenotype. As such, EMs (e.g., *CYP2D6*2) possess at least one fully functional CYP2D6 allele, IMs have two reduced function or one reduced and one nonfunctional allele, and PMs (e.g., *CYP2D6*3 and *4) have two nonfunctional alleles and are not able to metabolize CYP2D6 substrates. PMs may be at increased risk for adverse drug effects owing to increased plasma levels of the parent drug or lack of efficacy caused by an inability to generate an active metabolite. There is a considerable variability in the *CYP2D6* allele distribution among different ethnic groups (Ingelman-Sundberg et al. 2007; Zanger et al. 2004; Zhou et al. 2008), resulting in variable percentage of PMs, IMs, EMs, and UMs in a given population. PMs are mainly found in Europe, and UMs are mainly present in North Africa and Oceania. Because of the high Asian prevalence of the *CYP2D6*10* allele, IMs are primarily found in Asia (Ingelman-Sundberg et al. 2007). DA rats show impaired ability to metabolize debrisoquine and other CYP2D6 substrates and therefore may be used as a model of the human CYP2D6 PM phenotype.

There is a significant difference between rodents and humans in the number of active *CYP2D* genes (see Chapter 2). Whereas the mouse has nine different active *Cyp2d* genes (Nelson et al. 2004), the rat harbors six functional *Cyp2d* genes, while the human carries only one, which indeed is absent from 7% of the Caucasian population. It is reasonable to assume that the mouse and rat have retained their *Cyp2d* genes active because of a need for a dietary detoxification potential, whereas the more restricted food taken by humans in the past including the intellectual capability to transfer information regarding suitable food between generations has resulted in the loss of a selection pressure as to keep the genes active. The evolution of the human *CYP2D* locus has involved removal of three genes, inactivation of two (*CYP2D7P* and *2D8P*), and partial inactivation of one (*CYP2D6*) (Heim and Meyer 1992; Ingelman-Sundberg 2005). On the basis of the identification and characterization of a nonfunctional *CYP2D7P* gene and a *2D8P* pseudogene, Kimura et al. (1989) have suggested that gene duplication events result in *CYP2D6* and *2D7P* and that gene conversion events occur later to generate *CYP2D8P*. There is a 2.8-kb repeated region that flanks the active *CYP2D6* gene and may have played a role in deletion and amplification of *CYP2D6* (Steen et al. 1995b). There is evidence that *CYP2D6* gene duplications occurred through unequal crossover at a breakpoint in the 3′-flanking region of the *CYP2D6*2B* allele with a specific repetitive sequence (Lundqvist et al. 1999). Alleles with 13 copies of the gene are likely formed by unequal segregation and extrachromosomal replication of acentric DNA. The high frequencies of multiduplicated genes in Saudi Arabian and Ethiopian populations indicate that a dietary selective pressure existed in those regions in the past (Lundqvist et al. 1999). These variants may have been introduced into the Spanish population during the Muslim migration.

CYP2D6 can metabolize more than 160 drugs and a number of clinical studies have indicated that the *CYP2D6* genotype has a significant impact on the clearance and response for some of these CYP2D6 substrate drugs. On the basis of available evidence for the genotype–phenotype relationships, it is necessary to develop proper clinical guidelines to maximize the therapeutic effect when minimizing the toxicity in pharmacotherapy. Several professional societies have started this work, including the Clinical Pharmacogenetics Implementation Consortium (CPIC, https://www.pharmgkb .org/page/cpic), Dutch Pharmacogenetics Working Group (DPWG, https://www.pharmgkb.org/page/dpwg), and Canadian Pharmacogenomics Network for Drug Safety (CPNDS). So far, CPIC has developed clinical guidelines for 13 CYP2D6 substrate drugs (https://www.pharmgkb.org/gene/PA128), including amitriptyline, clomipramine, citalopram, codeine, desipramine, doxepin, escitalopram, fluvoxamine, imipramine, nortriptyline, paroxetine, sertraline, and trimipramine. The DPWG has developed guidelines for 19 CYP2D6 substrate drugs (https://www.pharmgkb.org/gene/PA128), including amitriptyline, aripiprazole, atomoxetine, clomipramine, codeine, doxepin, flecainide, haloperidol, imipramine, metoprolol, nortriptyline, oxycodone, paroxetine, propafenone, risperidone, tamoxifen, tramadol, venlafaxine, and zuclopenthixol in relation to CYP2D6 phenotypes. This chapter will discuss the clinical consequences of CYP2D6 genetic mutations.

7.2 INTERINDIVIDUAL VARIABILITY IN CYP2D6 EXPRESSION AND ACTIVITY

The CYP2D6 activity ranges considerably within a population and includes individuals with UM, EM, IM, and PM status. The PM and UM differ from EMs by 5- to 15-fold if determined by rates of metabolism or by ratios of parent to metabolite concentrations. It is difficult to assess IMs as there is clear overlap between IMs and EMs. The UM phenotype has been less well characterized than the PM phenotype, though it may have significant clinical consequences, particularly in the selection of an appropriate drug dosage. There is a large interindividual variation in the enzyme activity of CYP2D6 in a given population. Even in a subpopulation who are EMs, such

a variation is large. In human liver microsomes from the liver samples of patients ($n = 20$), there is a greater than 25-fold variation in both bufuralol hydroxylation and concentration of mRNA for CYP2D6 (Carcillo et al. 2003). In healthy subjects ($n = 78$), there is a similar extent of interindividual variation in in vivo activity of CYP2D6 determined using debrisoquine as a probe drug and in mRNA for *CYP2D6* in peripheral blood mononuclear cells (Carcillo et al. 2003).

For drugs with CYP2D6 as the major contributor of clearance (>60%), large differences in AUC or oral clearance are observed between EMs and PMs (Gibbs et al. 2006). Large pharmacokinetic differences have been observed with dextromethorphan (Evans et al. 1989) and tolterodine (Brynne et al. 1998), which shows 53- and 22-fold higher oral clearances in EMs compared with PMs. Modest differences in oral clearance (3.5- to 10-fold) are observed with atomoxetine (Farid et al. 1985), propafenone (Zoble et al. 1989), desipramine (Gram and Christiansen 1975), and venlafaxine (Holliday and Benfield 1995) between EMs and PMs. For drugs that had a low estimated CYP2D6 contribution in vivo, variation in oral clearance values between EMs and PMs is <3-fold (e.g., amitriptyline [Breyer-Pfaff et al. 1992] and chlorpromazine [Sunwoo et al. 2004]).

Although *CYP2D6* polymorphisms can largely explain the wide pharmacokinetic variability of many drugs that are substantially metabolized by CYP2D6, other factors contributing to the variability should be taken into account. These include comedication, disease status, and renal and liver functions. For example, a single dose of quinidine (50–250 mg) significantly inhibits CYP2D6 activity, thus converting EMs to apparent PMs in a process known as phenocopying (Ebner and Eichelbaum 1993).

7.3 ALLELES OF THE CYP2D6 GENE

Presently, at least 104 different *CYP2D6* variant alleles (*1B* through to *105*) and a series of subvariants have been identified and designated by the human cytochrome P450 allele nomenclature committee (Table 7.1) (http:// www.cypalleles.ki.se/cyp2d6.htm). There have been at least 507 SNPs of *CYP2D6* described at NCBI dbSNP (Single Nucleotide Polymorphism Database of the National Center for Biotechnology Information), approximately 10% of which are nonsynonymous SNPs with potential functional impact on the protein (see Figure 7.1). Functionally, these alleles of *CYP2D6* can be classified into three groups: alleles resulting in increased activity, alleles resulting in decreased activity or loss of activity, and alleles with normal or increased activity. These variants result from point mutations, deletions or additions, gene rearrangements, and deletion or duplication/ multiduplication of the entire gene. The distribution of these alleles in various ethnic groups is different. The *1A* refers to the wild type or reference haplotype.

7.3.1 NULL ALLELES OF CYP2D6

Null alleles of *CYP2D6* do not encode a functional protein and there is no detectable residual enzymatic activity. It is

TABLE 7.1

Reported Alleles of Human *CYP2D6*

CYP2D6	Protein	Nucleotide Change	Amino Acid Change	Effect on Enzyme Activity	References
1A	CYP2D6.1	None (wild type)		Normal	Kimura et al. 1989
1B	CYP2D6.1	3828G>A		Normal	Marez et al. 1997
1C	CYP2D6.1	1978C>T		Normal	Marez et al. 1997
1D	CYP2D6.1	2575C>A			Marez et al. 1997
1E	CYP2D6.1	1869T>C			Sachse et al. 1997
1XN	CYP2D6.1		N active genes	↑	Dahl et al. 1995; Sachse et al. 1997
2A	CYP2D6.2	−1584C>G; −1235A>G; −740C>T; −678G>A; CYP2D7 gene conversion in intron 1; 1661G>C; 2850C>T; 4180G>C	R296C; S486T	Normal	Johansson et al. 1993; Panserat et al. 1994; Raimundo et al. 2000; Sakuyama et al. 2008
2B	CYP2D6.2	1039C>T; 1661G>C; 2850C>T; 4180G>C	R296C; S486T		Marez et al. 1997
2C	CYP2D6.2	1661G>C; 2470T>C; 2850C>T; 4180G>C	R296C; S486T		Marez et al. 1997; Sachse et al. 1997
2D	CYP2D6.2	2850C>T; 4180G>C	R296C; S486T		Marez et al. 1997
2E	CYP2D6.2	997C>G; 1661G>C; 2850C>T; 4180GC	R296C; S486T		Marez et al. 1997
2F	CYP2D6.2	1661G>C; 1724C>T; 2850C>T; 4180G>C	R296C; S486T		Marez et al. 1997
2G	CYP2D6.2	1661G>C; 2470T>C; 2575C>A; 2850C>T; 4180G>C	R296C; S486T		Marez et al. 1997
2H	CYP2D6.2	1661G>C; 2480C>T; 2850C>T; 4180G>C	R296C; S486T		Marez et al. 1997
2J		See *CYP2D6*59*			
2K	CYP2D6.2	1661G>C; 2850C>T; 4115C>T; 4180G>C	R296C; S486T		Marez et al. 1997
*2L (Formerly *41B*)	CYP2D6.2	−1584C; −1298G>A; −1235A>G; −740C>T; 310G>T; 746C>G; 843T>G; 1513C>T; 1661G>C; 1757C>T; 2850C>T; 3384A>C; 3584G>A; 3790C>T; 4180G>C	R296C; S486T		Gaedigk et al. 2005a
2M	CYP2D6.2	−1584C; −1237_−1236insAA; −1235A>G; −750_−749delGA; −740C>T; −678G>A; *CYP2D7* gene conversion in intron 1; 310G>T; 746C>G; 843T>G; 1661G>C; 2850C>T; 2988G; 3384A>C; 3584G>A; 3790C>T; 4180G>C; 4481G>A	R296C; S486T		Gaedigk et al. 2005b
2×N (N = 2, 3, 4, 5, or 13)	CYP2D6.2	1661G>C; 2850C>T; 4180G>C	R296C; S486T; **N active genes**	↑	Aklillu et al. 1996; Dahl et al. 1995; Johansson et al. 1993
3A		**2549delA**[a]	**259Frameshift**	None	Kagimoto et al. 1990
3B		1749A>G; 2549delA	N166D; 259Frameshift		Marez et al. 1997
4A		100C>T; 974C>A; 984A>G; 997C>G; 1661G>C; **1846G>A**; 4180G>C	P34S; L91M; H94R; **splicing defect**; S486T	None	Gough et al. 1990; Hanioka et al. 1990; Kagimoto et al. 1990
4B		100C>T; 974C>A; 984A>G; 997C>G; **1846G>A**; 4180G>C	P34S; L91M; H94R; **splicing defect**; S486T	None	Kagimoto et al. 1990
4C		100C>T; 1661G>C; **1846G>A**; 3887T>C; 4180G>C	P34S; **splicing defect**; L421P; S486T	None	Yokota et al. 1993
4D		100C>T; 1039C>T; 1661G>C; **1846G>A**; 4180G>C	P34S; **splicing defect**; S486T	None	Marez et al. 1997
4E		100C>T; 1661G>C; **1846G>A**; 4180G>C	P34S; **splicing defect**; S486T		Marez et al. 1997
4F		100C>T; 974C>A; 984A>G; 997C>G; 1661G>C; **1846G>A**; 1858C>T; 4180G>C	P34S; L91M; H94R; **splicing defect**; R173C; S486T		Marez et al. 1997

(Continued)

TABLE 7.1 (CONTINUED)
Reported Alleles of Human CYP2D6

CYP2D6	Protein	Nucleotide Change	Amino Acid Change	Effect on Enzyme Activity	References
*4G		100C>T; 974C>A; 984A>G; 997C>G; 1661G>C; **1846G>A**; 2938C>T; 4180G>C	P34S; L91M; H94R; **splicing defect**; P325L; S486T		Marez et al. 1997
*4H		100C>T; 974C>A; 984A>G; 997C>G; 1661G>C; **1846G>A**; 3877G>C; 4180G>C	P34S; L91M; H94R; **splicing defect**; E418Q; S486T		Marez et al. 1997
*4J		100C>T; 974C>A; 984A>G; 997C>G; 1661G>C; **1846G>A**	P34S; L91M; H94R; **splicing defect**		Marez et al. 1997
*4K		100C>T; 1661G>C; **1846G>A**; 2850C>T; 4180G>C	P34S; **splicing defect**; R296C; S486T	None	Sachse et al. 1997
*4L		100C>T; 997C>G; 1661G>C; **1846G>A**; 4180G>C	P34S; **splicing defect**; S486T		Shimada et al. 2001
*4M		−1235A>G; 746C>G; 843T>G; 974C>A; 984A>G; 997C>G; 1661G>C; **1846G>A**; 2097A>G; 3384A>C; 3582A>G; 4401C>T	L91M; H94R; **splicing defect**		Agundez et al. 1997; Fuselli et al. 2004; Gaedigk et al. 2006
*4N (Found in a gene duplication)		−1426C>T; −1235A>G; −1000G>A; 100C>T; 310G>T; 746C>G; 843T>G; 974C>A; 984A>G; 997C>G; 1661G>C; **1846G>A**; 2097A>G; 3384A>C; 3582A>G; gene conversion to CYP2D7 in exon 9; 4180G>C; 4401C>T	P34S; L91M; H94R; **splicing defect**; P469A; T470A; H478S; G479R; F481V; A482S; S486T		Gaedigk et al. 2006
*4P		−1426C>T; −1235A>G; −1000G>A; 100C>T; 310G>T; 746C>G; 843T>G; 974C>A; 984A>G; 997C>G; 1661G>C; **1846G>A**; 2097A>G; 2576C>T; 3384A>C; 3435C>A; 3582A>G; 4180G>C; 4401C>T	P34S; L91M; H94R; **splicing defect**; P268S; S486T		Dodgen et al. 2013
*4×2				None	Lovlie et al. 1997; Sachse et al. 1998
*5		*CYP2D6* deleted	*CYP2D6* deleted	None	Gaedigk et al. 1991; Steen et al. 1995b
*6A		**1707delT**	**118Frameshift**	None	Saxena et al. 1994
*6B		**1707delT**; 1976G>A	**118Frameshift**	None	Daly et al. 1995; Evert et al. 1994a
*6C		**1707delT**; 1976G>A; 4180G>C	**118Frameshift**	None	Marez et al. 1997
*6D		**1707delT**; 3288G>A	**118Frameshift**		Marez et al. 1997
*7	CYP2D6.7	**2935A>C**	**H324P**	None	Evert et al. 1994b
*8		1661G>C; 1758G>T; 2850C>T; 4180G>C	G169X	None	Broly et al. 1995
*9	CYP2D6.9	**2615_2617delAAG**	**K281del**	↓	Broly and Meyer 1993; Tyndale et al. 1991
*9×2	CYP2D6.9	**2615_2617delAAG**	**K281del**		Gaedigk et al. 2011
*10A	CYP2D6.10	**100C>T**; 1661G>C; 4180G>C	**P34S**; S486T	↓	Sakuyama et al. 2008; Yokota et al. 1993
*10B	CYP2D6.10	−1426C>T; −1237_−1236insAA; −1235A>G; −1000G>A; **100C>T**; 1039C>T; 1661G>C; 4180G>C	**P34S**; S486T	↓	Johansson et al. 1994
*10C		See *36			
*10D	CYP2D6.10	**100C>T**; 1039C>T; 1661G>C; 4180G>C, *CYP2D7*-like 3′-flanking region	**P34S**; S486T		Ishiguro et al. 2004b
*10×2	CYP2D6.10			↓	Garcia-Barcelo et al. 2000; Ishiguro et al. 2004a; Ji et al. 2002; Mitsunaga et al. 2002

(Continued)

TABLE 7.1 (CONTINUED)
Reported Alleles of Human *CYP2D6*

CYP2D6	Protein	Nucleotide Change	Amino Acid Change	Effect on Enzyme Activity	References
*11		**883G>C**; 1661G>C; 2850C>T; 4180G>C	**Splicing defect**; R296C; S486T	None	Marez et al. 1995
*12	CYP2D6.12	**124G>A**; 1661G>C; 2850C>T; 4180G>C	**G42R**; R296C; S486T	None	Marez et al. 1996
*13		All *CYP2D6*13* alleles share a *CYP2D7/2D6* hybrid gene structure, with *CYP2D7* sequence in exon 1 leading to an insertion (137_138insT) and frameshift of the open-reading frame, thereby obliterating enzyme activity.	Frameshift	None	Panserat et al. 1995

The following GenBank entries of CYP2D7/2D6 hybrid allelic variants all share a CYP2D7-derived exon 1 with the detrimental T-insertion but differ in respect to the region in which CYP2D7 switches to CYP2D6.

Some switch regions contain sequences that may be interrelated as CYP2D6 or CYP2D7; therefore, exact switch regions cannot be determined or remain open to interpretation. There is no diagnostic SNP for *CYP2D6*13*, and thus assigning it is best done by sequencing the entire gene region. See Sim, S.C., Kacevska, M., and Ingelman-Sundberg, M., 2012, for more information.

EU098008
(Originally called ***CYP2D6*13***, temporarily called *CYP2D6*13A1*)
GQ162807
(Originally called ***CYP2D6*77***, temporarily called *CYP2D6*13A2*)
HM641839
(Originally called ***CYP2D6*79***, temporarily called *CYP2D6*13B*)
HM641840
(Originally called ***CYP2D6*80***, temporarily called *CYP2D6*13C*)
GQ162808
(Originally called ***CYP2D6*78***, temporarily called *CYP2D6*13D*)
EU098009
(Originally called ***CYP2D6*67***, temporarily called *CYP2D6*13E*)
EU093102
(Originally called ***CYP2D6*16*** and ***CYP2D6*66***, temporarily called *CYP2D6*13F*)
JN618990
(Temporarily called ***CYP2D6*13G1***)
HQ670229
(Temporarily called ***CYP2D6*13G2***)
GQ162806
(Originally called ***CYP2D6*76***, temporarily called *CYP2D6*13H*)

(Continued)

TABLE 7.1 (CONTINUED)
Reported Alleles of Human *CYP2D6*

CYP2D6	Protein	Nucleotide Change	Amino Acid Change	Effect on Enzyme Activity	References
*14A	CYP2D6.14A	100C>T; **1758G>A**; 2850C>T; 4180G>C	P34S; **G169R**; R296C; S486T	None	Sakuyama et al. 2008; Wang et al. 1999
*14B	CYP2D6.14B	Intron 1 conversion with CYP2D7 (214–245); 1661G>C; **1758G>A**; 2850C>T; 4180G>C	**G169R**; R296C; S486T	↓	Ji et al. 2002; Sakuyama et al. 2008
*15		**137_138insT**	**46Frameshift**	None	Sachse et al. 1996
*16		*CYP2D7P/CYP2D6* hybrid: Exons 1–7 *CYP2D7P* related, exons 8 and 9 *CYP2D6*	Frameshift	None	Daly et al. 1996b
*17	CYP2D6.17	**1023C>T**; 1661G>C; **2850C>T**; 4180G>C	**T107I**; **R296C**; S486T	↓	Masimirembwa et al. 1996b; Oscarson et al. 1997
*17×N	CYP2D6.17			Normal (if N = 2)	Cai et al. 2006
*18	CYP2D6.18	**4125_4133dupGTGCCCACT**	**468_470dupVPT**	None	Sakuyama et al. 2008; Yokoi et al. 1996
*19		1661G>C; **2539_2542delAACT**; 2850C>T; 4180G>C	**255Frameshift**	None	Marez et al. 1997
*20		1661G>C; **1973_1974insG**; 1978C>T; 1979T>C; 2850C>T; 4180G>C	**211Frameshift**	None	Marez-Allorge et al. 1999
*21A		−1584C>G; −1426C>T; −1258_−1257insAAAAA; −1235A>G; −740C>T; −678G>A; −629A>G; 214G>C; 221C>A; 223C>G; 227T>C; 310G>T; 601delC; 1661G>C; 2573_2574insC; 2850C>T; 3584G>A; 4180G>C	267Frameshift	None	Chida et al. 1999
*21B		−1584C>G; −1235A>G; −740C>T; −678G>A; intron 1 conversion with *CYP2D7* (214–245); 1661G>C; **2573_2574insC**; 2850C>T; 4180G>C	**267Frameshift**	None	Yamazaki et al. 2003
*22	CYP2D6.22	82C>T	R28C		Marez et al. 1997
*23	CYP2D6.23	957C>T	A85V		Marez et al. 1997
*24	CYP2D6.24	2853A>C	I297L		Marez et al. 1997
*25	CYP2D6.25	3198C>G	R343G		Marez et al. 1997
*26	CYP2D6.26	3277T>C	I369T		Marez et al. 1997
*27	CYP2D6.27	3853G>A	E410K	Normal	Marez et al. 1997; Sakuyama et al. 2008
*28	CYP2D6.28	19G>A; 1661G>C; 1704C>G; 2850C>T; 4180G>C	V7M; Q151E; R296C; S486T		Marez et al. 1997
*29	CYP2D6.29	1659G>A; 1661G>C; 2850C>T; 3183G>A; 4180G>C	V136M; R296C; V338M; S486T	↓	Marez et al. 1997; Wennerholm et al. 2001, 2002
*30	CYP2D6.30	1661G>C; 1863_1864insTTTCGCCCC; 2850C>T; 4180G>C	174_175insFRP; R296C; S486T		Marez et al. 1997
*31	CYP2D6.31	1661G>C; 2850C>T; 4042G>A; 4180G>C	R296C; R440H; S486T		Marez et al. 1997
*32	CYP2D6.32	1661G>C; 2850C>T; 3853G>A; 4180G>C	R296C; E410K; S486T		Marez et al. 1997
*33	CYP2D6.33	2483G>T	A237S	Normal	Marez et al. 1997
*34	CYP2D6.34	2850C>T	R296C		Marez et al. 1997
*35	CYP2D6.35	−1584C>G; 31G>A; 1661G>C; 2850C>T; 4180G>C	V11M; R296C; S486T	Normal	Gaedigk et al. 2003b; Marez et al. 1997
*35×2	CYP2D6.35	31G>A; 1661G>C; 2850C>T; 4180G>C	V11M; R296C; S486T	↑	Griese et al. 1998

(Continued)

TABLE 7.1 (CONTINUED)
Reported Alleles of Human *CYP2D6*

CYP2D6	Protein	Nucleotide Change	Amino Acid Change	Effect on Enzyme Activity	References
*36 (Duplication or tandem)	CYP2D6.36	−1426C>T; −1237_−1236insA; −1235A>G; −1000G>A; **100C>T**; 1039C>T; 1661G>C; gene conversion to *CYP2D7* in exon 9; 4180G>C	**P34S**; P469A; T470A; H478S; G479A; F481V; A482S; S486T	Negligible	Johansson et al. 1994; Leathart et al. 1998
*36 (Single)	CYP2D6.36	−1426C>T; −1235A>G; −1000G>A; **100C>T**; 310G>T; 843T>G; 1039C>T; 1661G>C; 2097A>G; 3384A>C; 3582A>G; gene conversion to *CYP2D7* in exon 9	**P34S**; P469A; T470A; H478S; G479A; F481V; A482S; S486T	Negligible	Gaedigk et al. 2006; Sakuyama et al. 2008
*37	CYP2D6.37	100C>T; 1039C>T; 1661G>C; 1943G>A; 4180G>C	P34S; R201H; S486T		Marez et al. 1997
*38		2587_2590delGACT	**271Frameshift**	None	Leathart et al. 1998
*39	CYP2D6.39	1661G>C; 4180G>C	S486T	Normal	Sakuyama et al. 2008; Shimada et al. 2001
*40	CYP2D6.40	1023C>T; 1661G>C; **1863_1864ins(TTT CGC CCC)2**; 2850C>T; 4180G>C	T107I; **174_175ins(FRP)2**; R296C; S486T	None	Gaedigk et al. 2002
*41	CYP2D6.2	−1584C; −1235A>G; −740C>T; −678G>A; *CYP2D7* gene conversion in intron 1; 1661G>C; 2850C>T; **2988G>A**; 4180G>C	R296C; **splicing defect**; S486T	↓	Raimundo et al. 2000, 2004; Rau et al. 2006; Toscano et al. 2006a
*42		−1584C; 1661G>C; 2850C>T; **3259_3260insGT**; 4180G>C	R296C; **365Frameshift**	None	Gaedigk et al. 2003a
*43	CYP2D6.43	77G>A	R26H		Marez et al. 1997
*44		82C>T; **2950G>C**	**Splicing defect**	None	Yamazaki et al. 2003
*45A	CYP2D6.45	−1601_−1600GA>TT; −1584C; −1238_−1237delAA; −1094_−1093insA; −1011T>C; 310G>T; 746C>G; 843T>G; 1661G>C; 1716G>A; 2129A>C; 2575C>A; 2661G>A; 2850C>T; 3254T>C; 3384A>C; 3584G>A; 3790C>T; 4180G>C	E155K; R296C; S486T		Gaedigk et al. 2005a
*45B	CYP2D6.45	−1584C; −1543G>A; −1298G>A; −1235A>G; −1094_−1093insA; −740C>T; −695_−692delTGTG; 310G>T; 746C>G; 843T>G; 1661G>C; 1716G>A; 2575C>A; 2661G>A; 2850C>T; 3254T>C; 3384A>C; 3584G>A; 3790C>T; 4180G>C	E155K; R296C; S486T		Gaedigk et al. 2005a
*46	CYP2D6.46	−1584C; −1543G>A; −1298G>A; −1235A>G; −740C>T; 77G>A; 310G>T; 746C>G; 843T>G; 1661G>C; 1716G>A; 2575C>A; 2661G>A; 2850C>T; 3030G>G/A*; 3254T>C; 3384A>C; 3491G>A; 3584G>A; 3790C>T; 4180G>C	R26H; E155K; R296C; S486T		Gaedigk et al. 2005a
*47	CYP2D6.47	−1426C>T; −1235A>G; −1000G>A; 73C<T; **100C>T**; 1039C>T; 1661G>C; 4180G>C	R25W; **P34S**; S486T	Negligible	Sakuyama et al. 2008; Soyama et al. 2004
*48	CYP2D6.48	972C>T	A90V	Normal	Sakuyama et al. 2008; Soyama et al. 2004
*49	CYP2D6.49	−1426C>T; −1235A>G; −1000G>A; **100C>T**; 1039C>T; 1611T>A; 1661G>C; 4180G>C	**P34S**; F120I; S486T	↓	Sakuyama et al. 2008; Soyama et al. 2004
*50	CYP2D6.50	1720A>C	E156A	↓	Sakuyama et al. 2008; Soyama et al. 2004

(Continued)

TABLE 7.1 (CONTINUED)
Reported Alleles of Human *CYP2D6*

CYP2D6	Protein	Nucleotide Change	Amino Acid Change	Effect on Enzyme Activity	References
*51	CYP2D6.51	−1584C>G; −1235A>G; −740C>T; −678G>A; *CYP2D7* gene conversion in intron 1; 1661G>C; 2850C>T; 3172A>C; 4180G>C	R296C; E334A; S486T	Negligible	Sakuyama et al. 2008; Soyama et al. 2004
*52	CYP2D6.52	−1426C>T; −1245_−1244insGA; −1235A>G; −1028T>C; −1000G>A; −377A>G; 100C>T; 1039C>T; 1661G>C; 3877G>A; 4180G>C; 4388C>T; 4401C>T	P34S; E418K		Lee et al. 2009
*53	CYP2D6.52	1598A>G; 1611T>A; 1617G>T	F120I; A122S	↑	Ebisawa et al. 2005; Sakuyama et al. 2008
*54	CYP2D6.54	100C>T; 1039C>T; 1661G>C; 2556C>T; 4180G>C	P34S; T261I; S486T	↓	Ebisawa et al. 2005; Sakuyama et al. 2008
*55	CYP2D6.55	1661G>C; 2850C>T; 3790C>T; 3835A>C; 4180G>C	R296C; K404Q; S486T	↓	Ebisawa et al. 2005; Sakuyama et al. 2008
*56A		−1584C>G; −1235A>G; −740C>T; −678G>A; CYP2D7 gene conversion in intron 1; 1661G>C; 2850C>T; **3201C>T**; 3384A>C; 3584G>A; 3790C>T; 4180G>C	R296C; **R344X**	None	Li et al. 2006
*56B		−1426C>T; −1235A>G; −1000G>A; 100C>; 310G>T; 843T>G; 1039C>T; 1661G>C; 2097A>G; **3201C>T**; 3384A>C; 3582A>G, 4180G>C	P34S; **R344X**		Gaedigk et al. 2007a
*57 (In tandem with *10)	CYP2D6.57	100C>T; 310G>T; 843T>G; 887C>T; 1039C>T; 1661G>C; 3384A>C; 3582A>G; gene conversion to *CYP2D7* in exon 9; 4180G>C	P34S; R62W; P469A; T470A; H478S; G479A; F481V; A482S; S486T	Negligible	Sakuyama et al. 2008; Soyama et al. 2006
*58	CYP2D6.58	−1426C>T; −1235A>G; −740C>T; CYP2D7 gene conversion in intron 1; 310G>T; 843T>G; 1023C>T; 1661G>T; 1863_1864insTTTCGCCCC; 2850C>T; 3384A>C; 3584G>A; 3790C>T; 4180G>C	T107I; 174_175insFRP; R296C; S486T		
*59	CYP2D6.2	1661G>C; **2291G>A**; 2850C>T; 2939G>A; 4180G>C	R296C; S486T	↓	Marez et al. 1997; Toscano et al. 2006b
*60	CYP2D6.60				
*61	CYP2D6.61	Gene conversion to CYP2D7 in exon 9	P469A; T470A; H478S; G479A; F481V; A482S; S486T		Kramer et al. 2009
*62	CYP2D6.62	**4044C>T**	**R441C**	None	Klein et al. 2007
*63	CYP2D6.63	2850C>T; gene conversion to CYP2D7 in exon 9	R296C; P469A; T470A; H478S; G479A; F481V; A482S; S486T		Kramer et al. 2009
*64	CYP2D6.64	−1426C>T; −1235A>G; −1000G>A; 100C>T; 310G>T; 843T>G; 1023C>T; 1661G>C; 2097A>G; 3384A>C; 3582A>G; 4180G>C; 4401C>T; 4722T>G	P34S; T107I; S486T		Gaedigk and Coetsee 2008
*65	CYP2D6.65	100C>T; 310G>T; 843T>G; 1661G>C; 2850C>T; 3384A>C; 3584G>A; 3790C>T; 4180G>C; 4481G>A	P34S; R296C; S486T		Gaedigk and Coetsee 2008
*66	See *13				
*67	See *13				

(Continued)

TABLE 7.1 (CONTINUED)
Reported Alleles of Human *CYP2D6*

CYP2D6	Protein	Nucleotide Change	Amino Acid Change	Effect on Enzyme Activity	References
*68A	CYP2D6.68	−1426C>T; −1235A>G; −1000G>A; 100C>T; 310G>T; CYP2D7 sequence from intron 1 onward	P34S; CYP2D7 sequence from amino acid 61		Kramer et al. 2009
*68B	CYP2D6.68	Similar but not identical switch region compared to *CYP2D6*68A*			Gaedigk et al. 2012
*69		−1426C>T; −1235A>G; −1000G>A; 100C>T; 310G>T; 746C>G; 843T>G; 1062A>G; 1661G>C; 2850C>T; 2988G>A; 3384A>C; 3584G>A; 3790C>T; 4180G>C; 4401C>T; 4481G>A	P34S; R296C; splicing defect; S486T	↓	Gaedigk et al. 2009
*70	CYP2D6.70	−175G>A; 310G>T; 843T>G; 1608G>A; 1659G>A; 1661G>C; 3183G>A; 3384A>C; 4180G>C; 4722T>G	V119M; V136M; V338M; S486T		Matimba et al. 2009
*71	CYP2D6.71	−1584C>G; 125G>A; 1494 T>C	G42E		Zhou et al. 2009
*72	CYP2D6.72	−1426C>T; −1235A>G; −1000G>A; 100C>T; 310G>T; 843T>G; 1039C>T; 1661G>C; 2097A>G; 3318G>A; 3384A>C; 3582A>G; 4180G>C; 4401C>T	P34S; E383K; S486T		Matsunaga et al. 2009
73	CYP2D6.73	−740C>T; CYP2D7 gene conversion intron 1 (214–245); 310G>T; 746C>G; 843T>G; 1013G>A; 1661G>C; 2850C>T; 3384A>C; 3584G>A; 3790C>T; 4180G>C; 4535insT	V104M; R296C; S486T		Wright et al. 2010
*74	CYP2D6.74	974C>A; 3609G>T	L91M		Wright et al. 2010
*75	CYP2D6.75	4045G>A	R441H		Qin et al. 2008
*76		See CYP2D6*13 for details			
*77		See CYP2D6*13 for details			
*78		See CYP2D6*13 for details			
*79		See CYP2D6*13 for details			
*80		See CYP2D6*13 for details			
*81	CYP2D6.81	2579C>T; 2606G>A; 2610T>A	R269X; E278K; M279K		Chua et al. 2013
*82	CYP2D6.82	CYP2D7 gene conversion in exon 2: 974C>A; 984A>G; 997C>G; 1014T>C; 1022A>T; 1023C>A; 1028A>G; 1036T>C	L91M; H94R; V104A; T107Y; I109V		Contreras et al. 2011
*83	CYP2D6.83	843T>G; gene conversion to CYP2D7 in exon 9; 4180G>C	P469A; T470A; H478S; G479R; F481V; A482S; S486T		Gaedigk et al. 2012
*84	CYP2D6.84	−1740C>T; −1235A>G; −740C>T; −678A>G; 18G>A; intron 1 conversion with CYP2D7 (214–245); 310G>T; 746C>G; 843T>G; 1661G>C; 2574C>A; 2850C>T; 3384A>C; 3491G>A; 3584G>A; 3790C>T; 4180G>C; 4481G>A	P267H; R296C; S486T		Dodgen et al. 2013
*85	CYP2D6.85	−1740C>T; −1298G>A; −1235A>G; −740C>T; 102A>G; 310G>T; 607G>A; 746C>G; 843T>G; 1513C>T; 1661G>C; 2308G>A; 2850C>T; 3384A>C; 3584G>A; 3790C>T; 4157T>G; 4180G>C	R296C; H478Q; S486T		Dodgen et al. 2013
*86	CYP2D6.86	2606G>A; 2610T>A	E278K; M279K		Dodgen et al. 2013
*87	CYP2D6.87	14C>T; 100C>T; 310G>T; 843T>G; 1039C>T; 1661G>C; 4180G>C	A5V; P34S; S486T		Qian et al. 2013

(Continued)

TABLE 7.1 (CONTINUED)
Reported Alleles of Human *CYP2D6*

CYP2D6	Protein	Nucleotide Change	Amino Acid Change	Effect on Enzyme Activity	References
*88	CYP2D6.88	746C>G; intron 1 conversion with CYP2D7 (214–247); 843T>G; 1014T>C; 1661G>C; 3384A>C; 3584G>A; 3790C>T; 4180G>C	V104A; S486T		Qian et al. 2013
*89	CYP2D6.89	1678T>C	L142S		Qian et al. 2013
*90	CYP2D6.90	1693A>G	K147R		Qian et al. 2013
*91	CYP2D6.91	1735G>C; 2850C>T; 2988G>A	C161S; R296C; splicing defect		Qian et al. 2013
*92	CYP2D6.92	1995delC	218Frameshift		Qian et al. 2013
*93	CYP2D6.93	2519A>C	T249P		Qian et al. 2013
*94A	CYP2D6.94	100C>T; 310G>T; 843T>G; 1039C>T; 1661G>C; 3181A>G; 4180G>C	P34S; D337G; S486T		Qian et al. 2013
*94B	CYP2D6.94	100C>T; 843T>G; 1039C>T; 1661G>C; 3181A>G; 3384A>C; 4180G>C	P34S; D337G; S486T		Qian et al. 2013
*95	CYP2D6.95	100C>T; 843T>G; 1039C>T; 1661G>C; 3334G>A; 3384A>C; 4180G>C	P34S; R388H; S486T		Qian et al. 2013
*96	CYP2D6.96	3895C>T	Q424X		Qian et al. 2013
*97	CYP2D6.97	4094C>A	F457L		Qian et al. 2013
*98	CYP2D6.98	746C>G; 843T>G intron 1 conversion with CYP2D7 (214–247); 1661G>C; 2850C>T; 3384A>C; 3584G>A; 3790C>T; 4110C>G; 4180G>C	R296C; H463D; S486T		Qian et al. 2013
*99		To be released			
*100		−1426C>T; −1235A>G; −1109C>T; −1000G>A; 100C>T; 310G>T; 843T>G; 1039C>T; 1661G>C; 2097A>G; 2828delC; 3384A>C; 3582A>G; 4180G>C; 4401C>T	P34S; 288Frameshift		Montane Jaime et al. 2013
*101		−1426C>T; −1235A>G; −1000G>A; 100C>T; 310G>T; 843T>G; 1039C>T; 1661G>C; 2097A>G; 2927_2945delGATCCTACATCCGGATGTG; 3384A>C; 3582A>G; 4180G>C; 4401C>T	P34S; 321Frameshift		Montane Jaime et al. 2013
*102		Intron 1 conversion with CYP2D7 (214–245); 310G>T; 972C>T; 1661G>C; 2850C>T; 3384A>C; 3790C>T; 4180G>C; 4481G>A	A90V; R296C; S486T		Qumsieh et al. 2011
*103		Intron 1 conversion with CYP2D7 (214–245); 310G>T; 972C>T; 1661G>C; 1749A>G; 2850C>T; 3384A>C; 3790C>T; 4180G>C; 4481G>A	A90V; N166D; R296C; S486T		Qumsieh et al. 2011
*104		Intron 1 conversion with CYP2D7 (214–245); 310G>T; 843T>G; 1661G>C; 1720A>T; 2850C>T; 3384A>C; 3790C>T; 4180G>C; 4481G>A	E156V; R296C; S486T		Qumsieh et al. 2011
*105		Intron 1 conversion with CYP2D7 (214–245); 310G>T; 746C>G; 843T>G; 1661G>C; 2850C>T; 3268T>C; 3384A>C; 3790C>T; 4180G>C; 4481G>A	R296C; F366S; S486T		Qumsieh et al. 2011

Source: Data are extracted from http://www.cypalleles.ki.se/cyp2d6.htm.

[a] Nucleotide variations in bold are the major SNPs responsible for the phenotype of the corresponding allele.

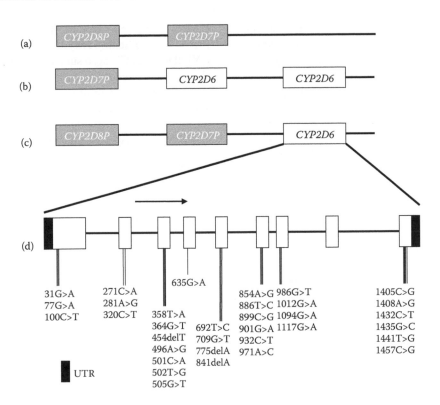

FIGURE 7.1 The common SNPs of *CYP2D6*. *CYP2D6* belongs to a gene cluster of highly homologous inactive pseudogenes *CYP2D7P* and *CYP2D8P*. The *CYP2D6* gene consists of nine exons. (a) The entire *CYP2D6* gene can be deleted (*CYP2D6*5*). (b) *CYP2D6* duplication or multiplication (tandem copy *N* = 2–12). (c) The *CYP2D* locus. (d) Common SNPs detected in the *CYP2D6* gene.

clear that alleles *3*, *4*, *5*, *6*, *7*, *8*, *11*, *12*, *13*, *14*, *15*, *16*, *18*, *19*, *20*, *21*, *38*, *40*, *42*, *44*, *56*, *62*, *68*, *92*, *100*, and *101* have no enzyme activity. They are responsible for the PM phenotype when present in homozygous or compound heterozygous constellations. These alleles are of clinical significance as they often cause altered drug clearance and drug response. Among these null alleles, *3*, *4*, *5*, and *6* account for ~97% of all alleles causing the PM phenotype in Caucasians (Sachse et al. 1997).

The mechanism for a total loss of enzyme function resulting from null alleles of *CYP2D6* includes the following: (a) single base mutations or small insertions/deletions that interrupt the reading frame or interfere with correct splicing leading to prematurely terminated protein/stop codon (e.g., *CYP2D6*3*, *4*, *6*, *7*, *8*, *11*, *15*, *19*, *20*, *38*, *40*, *42*, *44*, *56*, *62*, *92*, *100*, and *101*) (Kagimoto et al. 1990), (b) nonfunctional full-length coded alleles (e.g., *CYP2D6*12*, *14*, and *18*) (Evert et al. 1997), (c) deletion of the entire *CYP2D6* gene as a result of large sequence deletions (e.g., *5*), and (d) formation of a hybrid gene (e.g., *13* and *16*). There is a large deletion of sequence in *13* and *16*, and as a result, both contain a *CYP2D7–2D6* hybrid gene) (Gaedigk et al. 1991).

7.3.1.1 Null Alleles Attributed to Single Base Mutation or Small Insertions/Deletions

The *CYP2D6*3A* is first detected in Caucasian subjects who are PMs of debrisoquine (Kagimoto et al. 1990; Skoda et al.

1988). The *3* allele contains a deletion of A on position 2549 in exon 5, leading to a disrupted reading frame and would readily explain the absence of CYP2D6 protein and function in PMs owing to premature termination of protein synthesis when present in the homozygous state. The *CYP2D6*3B* contained an additional 1749A>G mutation resulting in an N166D substitution (Marez et al. 1997). The functional impact of this allele is unknown.

The *4* allele has a frequency of 22% and accounts for more than 75% of the mutant alleles in Swedish Caucasians (Dahl et al. 1992). However, this allele occurs at low frequency (~1%) in Chinese (Wang et al. 1993) and Africans (3.9%), and this is the reason for the low incidence of PM in Chinese (~1%) (Bertilsson et al. 1992) and Africans (0%–5%) (Masimirembwa et al. 1996a; Simooya et al. 1993) compared with 5%–10% in Caucasians. Gaedigk et al. (2006) have identified a *CYP2D6*4* allele that lacked the 100C>T and 4180G>C SNPs. The 4180G>C polymorphism is otherwise present on all defined *CYP2D6*4* alleles except *CYP2D6*4J*. The subject is a *CYP2D6*4/*4* PM who presented with a heterozygous 100C>T genotype. In addition, four other SNPs that are found on five resequenced *CYP2D6*4* alleles are also absent. This allele has been designated *CYP2D6*4M* by the P450 nomenclature committee.

The *CYP2D6*7* contained the 2935A>C transition in exon 6 of the *CYP2D6* gene, resulting in His324Pro substitution (Evert et al. 1994b). This mutation is associated with the PM phenotype of sparteine and occurs in Caucasian populations with a frequency of ~1%. When the His324Pro change is introduced into the cDNA by site-directed mutagenesis, no spectrally detectable P450 is formed and no catalytic activity toward bufuralol and sparteine is detected (Evert et al. 1994b). Furthermore, in contrast to the wild-type protein, the mutated protein is almost exclusively located in a detergent-insoluble insect cell fraction. These results demonstrate that the His324Pro substitution is responsible for the in vivo PM phenotype associated with the *CYP2D6*7* allele by preventing normal protein folding and heme incorporation.

The *CYP2D6*11* allele identified from a French population contains 883G>C, 1661G>C, 2850C>T, and 4180G>C (Marez et al. 1995). The 883G>C mutation abolishes splice recognition resulting in a deficient enzyme. The 2850C>T and 4180G>C transitions cause R296C and S486T substitutions, respectively.

*CYP2D6*19* is an allele with a four-base deletion from 2539 to 2542 (AACT) identified from a Caucasian population (Marez et al. 1997). The deletion causes frameshift at residue 255 and results in an inactive enzyme. *CYP2D6*20* is a rare allele with an insertion of G between 1973 and 1974, causing frameshift at residue 211 (Marez-Allorge et al. 1999). This mutation leads to a PM phenotype.

The *CYP2D6*56* allele has been identified from one liver sample of commercial cryopreserved human hepatocytes, containing a novel 3201C>T transition in exon 7, resulting in Arg344 (CGA) being replaced by a stop codon (TGA) and a CYP2D6 enzyme lacking the terminal 153 amino acids (Li et al. 2006). In addition to this SNP, this allele is typical of a *CYP2D6*2* allele with −1584C>G, −1235A>G, −740C>T, and −678G>A; *CYP2D7* gene conversion in intron 1: 1661G>C, 2850C>T, 3384A>C, 3584G>A, 3790C>T, and 4180G>C. No CYP2D6 activity is detected in hepatocytes from this liver sample.

*CYP2D6*62* is a recently identified new variant allele (4044C>T causing R441C) in a person with reduced sparteine oxidation phenotype, which is unexpected based on a genetic *CYP2D6*1A/*41* background (Klein et al. 2007). The recombinant CYP2D6.62 shows no activity toward propafenone as a result of missing heme incorporation. Sequence alignment reveals that the positively charged Arg441 residue is part of the heme-binding signature but not strictly conserved among all the CYPs.

7.3.1.2 Nonfunctional Full-Length Coded Alleles

The *CYP2D6*12* allele is identified from a French population, which contains 124G>A, 1661G>C, 2850C>T, and 4180G>C mutations (Marez et al. 1996). The 124G>A, 2850C>T, and 4180G>C transitions cause G42R, R296C, and S486T substitutions, respectively. The mutated protein has an almost total loss of activity for sparteine resulting in a PM phenotype.

*CYP2D6*14A* is an allele containing 100C>T, 1758G>A, 2850C>T, and 4180G>C that is first found in a Chinese subject and has since been found only in Asian populations (Kubota et al. 2000; Wang et al. 1999). There are four amino acid substitutions in CYP2D6.14: Pro34Ser, Gly169Arg, Arg296Cys, and Ser486Thr (Daly et al. 1996a; Wang et al. 1999), with

Gly169Arg being unique to this variant protein. Two of them (Pro34Ser and Ser486Thr) are overlapped with substitution in CYP2D6.10. A Chinese subject harboring *CYP2D6*5/*14* has shown a PM phenotype with a metabolic ratio (MR) of more than 12.6 for debrisoquine (Wang et al. 1999). A functional study does not find any detectable activity of CYP2D6.14A for bufura*lol and dextromethorphan* (Sakuyama et al. 2008). *CYP2D6*14B* also contains intron 1 conversion with *CYP2D7* (nucleotide bases 214–245) in addition to the above 4 SNPs in *CYP2D6* (Ji et al. 2002). This allele is also identified from a Chinese population with a frequency of 2.0%. The *14B* allele differs from the *14* allele by the absence of the 188C>T transition and by the additional 1749G>C transversion. In contrast to the null allele *14A, *14B* shows substantial enzyme activity (Sakuyama et al. 2008), probably attributed to the absence of the Pro34Ser substitution.

*CYP2D6*18* is an allele with a nine-base insertion (GTGCCCACT) from 4125 to 4133 in exon 9 first identified in a Japanese population (Yokoi et al. 1996). This allele is associated with a PM phenotype. The K_m for bufuralol 1′-hydroxylation by CYP2D6.18 is 236-fold higher than that by the wild type (990 vs. 4.2 μM) (Yokoi et al. 1996). When CYP2D6.18 is expressed in COS-7 cells, there is a decrease in protein levels and bufuralol 1′-hydroxylation activity compared with those expressing the wild type (Sakuyama et al. 2008).

7.3.1.3 Deletion of the Entire Gene

*CYP2D6*5* is a deletion of the whole gene (Gaedigk et al. 1991; Steen et al. 1995a) and its frequency is similar in different ethnic groups (4%–6%). The *CYP2D6*5* allele is the second most common inactivating allele in the UK population and the frequency of the *CYP2D6*5* allele is 4% in Caucasian populations (Nelson et al. 2004). In Chinese, its frequency is 7.2% (Ji et al. 2002). Gaedigk et al. (1991) have identified a homozygous 11.5-kb deletion associated with deletion of the entire *CYP2D6* gene and total absence of CYP2D6 protein in the liver. Steen et al. (1995b) have found a 2.8-kb repeated region that flanks the *CYP2D6* gene and contains the breakpoints associated with the generation of the *CYP2D6*5* variant. The deletion mechanism might involve unequal recombination between two homologous but nonallelic sequences, either by chromosome misalignment followed by unequal crossover or by the formation of a loop structure on a single chromosome. *CYP2D6*5* is considered to result from an unequal crossover between the two sister chromatids, while repeated unequal crossover events of chromatids will lead to gene multiplication (e.g., *CYP2D6*2×N*). Notably, Idle et al. (2000) have pointed out that the copy of chromosome 22 sequenced by the Human Genome Project has the *CYP2D6*5* deletion allele (i.e., does not contain the functional *CYP2D6* gene).

7.3.1.4 Formation of Hybrid Genes

*CYP2D6*13* is a hybrid allele consisting of exon 1 from *CYP2D7* and exons 2–9 from *CYP2D6* first identified in a French individual (Panserat et al. 1995). This allele is associated with a PM phenotype. *CYP2D6*16* is a hybrid allele with exons 1–7 from *CYP2D7P* and exons 8 and 9 from *CYP2D6*

(Daly et al. 1996b). Like *13, the hybrid is generated by large deletions in the CYP2D gene cluster owing to unequal crossover or looping-out mechanisms.

7.3.2 ALLELES WITH PARTIAL OR RESIDUAL FUNCTION

The alleles CYP2D6*10, *14, *17, *18, *36, *41, *47, *49, *50, *51, *52, *54, *55, *57, *59, *69, and *72 give rise to significantly decreased activity. The reduced enzyme activity is often attributed to decreased protein stability, disrupted substrate recognition, and reduced substrate–enzyme affinity. The enzyme activity change may be substrate dependent for some alleles such as *17. Individuals harboring either of these alleles are PMs or IMs.

CYP2D6*10 occurs in 33%–43% of Asians including Japanese, Korean, and Chinese and Pacific Islanders (Bradford 2002; Ishiguro et al. 2004b; Ji et al. 2002; Johansson et al. 1994). However, this allele occurs at low frequencies (2%–5%) in Caucasians and African Americans and other populations as well (e.g., Indians and Amerindians). (Bradford 2002; Leathart et al. 1998) The *10A allele has a 100C>T SNP causing Pro34Ser substitution in the proline-rich ("PPGP") region near the NH_2-terminal of the protein that is highly conserved among CYPs and is associated with a low in vivo sparteine clearance (i.e., with MRs > 1.5) (Yokota et al. 1993). Subjects homozygous for *10/*10 require lower dosage of metoprolol (Huang et al. 1999) and nortriptyline (Yue et al. 1998) than do subjects harboring *1/*1 to attain the same therapeutic efficacy. The residue Pro34 may act as a hinge between the hydrophobic membrane anchor and the heme-binding region of the enzyme (Yamazaki et al. 1993). The 100C>T SNP resulted in an unstable protein with significantly reduced enzyme activity toward debrisoquine, bufuralol, and dextromethorphan when expressed in COS-7 cells (Johansson et al. 1994; Sakuyama et al. 2008). This allele also contains 1661G>C and 4180G>C, resulting in Ser486Thr substitution. Shen et al. (2007) have investigated the catalytic activity of CYP2D6.10 toward a series of substrates. The authors have found that the intrinsic clearance (CL_{int}) values of bufuralol 1′-hydroxylation, dextromethorphan O-demethylation, debrisoquine 4-hydroxylation, atomoxetine 4-hydroxylation, S-fluoxetine N-demethylation, nortriptyline 10-hydroxylation, tramadol O-demethylation, and codeine O-demethylation by CYP2D6.10 are decreased to 3.65%, 5.29%, 11.84%, 8.58%, 7.54%, 1.32%, 6.90%, and 27.86%, respectively, of that seen with the wild-type protein. CYP2D6.10 demonstrates only a small change in the apparent K_m for atomoxetine, bufuralol, codeine, debrisoquine, S-fluoxetine, and tramadol compared with CYP2D6.1, and the reduced CL_{int} values are mainly attributed to a decreased V_{max}. Higher K_m values and decreased V_{max} values for dextromethorphan O-demethylation and nortriptyline 10-hydroxylation by CYP2D6.10 are observed, resulting in a significantly lower CL_{int} than CYP2D6.1. The most striking difference in CL_{int} between CYP2D6.10 and CYP2D6.1 is seen with nortriptyline, whereas the difference is the least with codeine (Shen et al. 2007).

In addition, CYP2D6*10D contained 1039C>T, 1661G>C, and CYP2D7-like 3′-flanking region in addition to 100C>T and 4180G>C, which is identified from a Japanese population (Ishiguro et al. 2004b). The functional impact of *10D is unclear. The CYP2D6*10×2 is initially found in Chinese, which contains a duplicated gene, but the enzyme activity is decreased (Garcia-Barcelo et al. 2000; Ji et al. 2002). The frequency of this allele in Japanese is low (0.6%) (Mitsunaga et al. 2002).

The *17 allele occurs at a high frequency in Africans and African Americans but is virtually absent from Caucasians (Bradford 2002). The frequency of this allele is found to be 34% in Zimbabweans (Masimirembwa et al. 1996b), 17% in Tanzanians (Wennerholm et al. 1999), 28% in Ghanaians (Griese et al. 1999), and 9% in Ethiopians (Aklillu et al. 1996). More than 10% of the Zimbabwean is homozygous for this allele. This variant appears to explain why Africans have a higher median MR than Caucasians (Leathart et al. 1998). CYP2D6*17 contains four SNPs: 1023C>T in exon 2, 1661G>C in exon 3 (a silent SNP), 2850C>T in exon 6, and 4180G>C in exon 9, resulting in three substitutions: Thr107Ile, Arg296Cys, and Ser486Thr, respectively (Masimirembwa et al. 1996b). Compared to CYP2D6*2, *17 contains one more substitution (Thr107Ile). This change occurs in a region of the β′ helix, which is conserved across species, and residue 107 may be involved in substrate recognition. The *17 allele is associated with decreased debrisoquine hydroxylation in vivo (Masimirembwa et al. 1996b). Expression of the cDNAs in COS-1 cells reveals that the CYP2D6.17 enzyme displays only 20% of the wild-type activity, whereas the Thr107Ile substitution alone has no significant effect on enzyme function (Oscarson et al. 1997). Allozymes containing both Thr107Ile and Arg296Cys show a fivefold higher K_m for bufuralol than the wild-type enzyme, whereas the Ser486Thr mutant shows no functional impact (Oscarson et al. 1997). In contrast, when codeine is used as a substrate, the Thr107Ile substitution alone is sufficient to cause a significant increase in the apparent K_m (Oscarson et al. 1997), indicating a differential effect for this substitution depending on the CYP2D6 substrate. The apparent K_m for codeine is 5- to 10-fold higher for CYP2D6.17 compared with the wild-type enzyme (Oscarson et al. 1997). These findings indicate that CYP2D6.17 displays an altered substrate affinity for substrates and that a combination of the Thr107Ile and Arg296Cys substitutions are needed for altered catalytic properties of the enzyme toward bufuralol.

Shen et al. (2007) have investigated the catalytic activity of CYP2D6.17 toward a series of substrates. The authors have reported that the CL_{int} values of bufuralol 1′-hydroxylation, dextromethorphan O-demethylation, debrisoquine 4-hydroxylation, atomoxetine 4-hydroxylation, S-fluoxetine N-demethylation, nortriptyline 10-hydroxylation, tramadol O-demethylation, and codeine O-demethylation by CYP2D6.17 are decreased to 22.69%, 16.69%, 64.25%, 21.89%, 8.17%, 7.33%, 35.70%, and 80.35%, respectively, of that seen with the wild-type enzyme. Compared with the wild-type protein (CYP2D6.1), CYP2D6.17 exhibits greater substrate-dependent changes in K_m values than CYP2D6.10.

Moderate differences in K_m values are found only with bufuralol, codeine, and tramadol; atomoxetine, codeine, debrisoquine, and dextromethorphan show comparable or slightly decreased V_{max} values. Nortriptyline has the greatest difference in CL_{int} among the substrates tested (13.6-fold), while the least difference is observed with codeine (1.24-fold). However, Bogni et al. (2005) have found that CYP2D6.17 expressed in V79 cells shows a higher activity for haloperidol than the wild-type protein CYP2D6.1. The hydroxylation of risperidone to 9-hydroxyrisperidone in carriers of *CYP2D6*17* is also more efficient than that in individuals harboring **1* (Cai et al. 2006). These results further demonstrate a clear and interesting substrate-dependent effect of the **17* allele on enzyme activity. The **17* allele may result in a protein with unusual catalytic mechanisms by which the substrate–enzyme interactions can be altered through distinct ways for different substrates.

CYP2D6*36, originally called **10C* associated with a PM phenotype, has been described in a tandem arrangement with *CYP2D6*10B* with **36* being located upstream of **10B* (Johansson et al. 1994). This allele contains multiple mutations including −1426C>T, −1235A>G, −1000G>A, 100C>T, 310G>T, 843T>G, 1039C>T, 1661G>C, 2097A>G, 3384A>C, 3582A>G, and gene conversion to *CYP2D7* in exon 9 (Gaedigk et al. 2006). The mutated protein contains Pro34Ser, Pro469Ala, Thr470Ala, His478Ser, Gly479Ala, Phe481Val, Ala482Ser, and Ser486Thr. The 100C>T transversion (Pro34Ser) in *CYP2D6*36* is the major cause for protein destabilization and that the six–amino acid difference in exon 9 (i.e., the *CYP2D7* exon 9 conversion) does not have a significant impact on enzyme activity. It appears that both CYP2D6.10 and CYP2D6.36 demonstrated decreased in vivo clearance of venlafaxine owing to reduced protein expression and increased K_m value for substrates (Fukuda et al. 2000a). The apparent K_m values of CYP2D6.1, CYP2D6.10A, and CYP2D6.10C are 1.7, 8.5, and 49.7 µM, respectively, for bufuralol 1′-hydroxylation and 9.0, 51.9, and 117.4 µM, respectively, for venlafaxine O-demethylation (Fukuda et al. 2000a). These findings indicate that the decreased CL_{int} by CYP2D6.10 and CYP2D6.36 is caused not only by low protein expression but also by increased K_m values as a result of reduced affinity to the enzyme. When expressed in COS-7 cells, CYP2D6.36 shows negligible activity toward bufuralol and dextromethorphan (Sakuyama et al. 2008). Thus, the combined activity of CYP2D6.36 and CYP2D6.10 would be expected to be comparable with that of CYP2D6.10.

The *CYP2D6*36* allele does not appear to occur by itself, and activity is conferred by the downstream **10* on this allele. However, Gaedigk et al. (2006) have found that **36* does not only exist as a gene duplication (**36×2*) or in tandem with **10* (*CYP2D6*36+*10*) but also by itself as a single copy. This "single" **36* allele is found in the most downstream position within the *CYP2D6* locus in nine African Americans and one Asian with frequencies of 0.5% and 2.6%, but is absent in the whites examined (Gaedigk et al. 2006). **36* is one of the more common PM alleles in these populations, next only to **4* and **5*. **36* may have originated in Africa and spread to Asia, but not Europe, since it has not been observed in Caucasians,

either as a single gene copy or in tandem with **10*. This allele encodes an enzyme lacking activity toward dextromethorphan (Gaedigk et al. 2006). The duplicated **36×2* allele found in a Japanese with discordant genotype and phenotype is a nonfunctional allele as determined using debrisoquine as a probe substrate (Chida et al. 2002).

CYP2D6*41 is a variant of *CYP2D6*2* with −1548C instead of G, both carrying two amino acid substitutions (Arg296Cys and Ser486Thr) (Raimundo et al. 2000, 2004; Rau et al. 2006; Toscano et al. 2006a). The former is less expressed than the latter variant. Two recent studies have found that −1584G>C is in linkage disequilibrium with another intronic SNP 2988G>A, causing splicing defect of the mRNA (Rau et al. 2006; Toscano et al. 2006a). However, individuals homozygous for *CYP2D6*41* are phenotypically similar to IMs with one deficient *CYP2D6* allele (Raimundo et al. 2000, 2004). Both *CYP2D6*2* and **41* are the most common variants of the duplicated and multiduplicated *CYP2D6* gene, and it is considered that such gene duplication and multiplication is the result of selection pressure that might have occurred in North East Africa (Ingelman-Sundberg 2005).

CYP2D6*47, **51*, and **57* encode mutated enzymes with residual activity of <5% of the wild-type protein (Sakuyama et al. 2008). The **47* allele is first identified in a Japanese population (Soyama et al. 2004). CYP2D6.47 harbors Arg25Trp, Pro34Ser, and Ser486 substitutions. The Arg25 residue is located within the transmembrane domain and is considered to serve as a halt transfer signal, which prevents the translocation of the protein into the lumen of the endoplasmic reticulum. It can be expected that individuals harboring *CYP2D6*47* would show a PM phenotype. The **51* allele is reported by Soyama et al. (2004). It entails Arg296Cys, Glu334Ala, and Ser486 substitutions. Glu334 is located in the J helix (Rowland et al. 2006) and well conserved in CYPs, suggesting its critical role in enzyme function. The **57* allele is identified in a Japanese population (Soyama et al. 2006). It harbors Pro34Ser, Arg62Trp, Pro469Ala, Thr470Ala, His478Ser, Gly479Ala, Phe481Val, Ala482Ser, and Ser486Thr. The almost loss of activity of CYP2D6.51 is probably caused by the presence of Pro34Ser, which is also present in **47* and **36* encoding proteins with extremely low enzyme activity. The **57* allele existed in a tandem arrangement with **10* and the combined activity of CYP2D6.57 and CYP2D6.10 should be comparable to that of CYP2D6.10. The in vivo functional impact of **51* and **51* is unknown but is expected to show a PM phenotype.

CYP2D6*10, **18*, **49*, **50*, **52*, **54*, and **55*, identified in Japanese populations, encode mutated enzymes with remaining activity of 7%–36% of the wild-type protein (Lee et al. 2009; Sakuyama et al. 2008). These alleles may be associated with a PM and more probably IM phenotype. CYP2D6.50 harbors Glu156Ala; CYP2D6.54 harbors Pro34Ser, Thr261Ile, and Ser486Thr; CYP2D6.55 contains Arg296Cys, Lys404Gln, and Ser486Thr. CYP2D6.2, 2D6.10, 2D6.14, 2D6.18, and 2D6.55 expressed in COS-7 cells show significantly increased apparent K_m values, while CYP2D6.49, 2D6.50, and 2D6.53 exhibit lower apparent K_m

values compared to the wild-type protein. The V_{max} values of CYP2D6.10, 2D6.49, and 2D6.54 are 20%, 17%, and 6% of that of the wild type, respectively (Sakuyama et al. 2008).

7.3.3 Alleles with Largely Normal or Increased Activity

Functional studies did not demonstrate significantly altered enzyme activity with several alleles of *CYP2D6*, including *2A, *17×2, *27, *35, *39, *41×2,* and *48.* Marez et al. (1997) have identified *2A* in Europeans and found no functional impact of this allele, which displayed metabolic activity similar to that of the wild type. They have also identified *CYP2D6*2B, *2C, *2D, *2E, *2F, *2G, *2H,* and *2K* (Marez et al. 1997). Gaedigk et al. (2005a,b) have identified two more alleles in the *2 series: *2L* and *2M,* which contain multiple mutations (e.g., 310G>T, 746C>G, and 843T>G). The −1584C>G, −1235A>G,−740C>T, and −678G>A polymorphisms are probably found in most subvariants of the *CYP2D6*2* allele (*2A* to*2H* and *2J* to *2N; *2J = *59*) (Raimundo et al. 2000). The *CYP2D7* gene conversion in intron 1 is probably also found in most subvariants of the *2 allele (Johansson et al. 1993; Raimundo et al. 2000). Their functional effect is unclear but is expected to have a minor effect on enzyme activity.

The *CYP2D6*24, *26,* and *27* alleles are rare, found in 0.1%–0.3% of Caucasians (Marez et al. 1997). CYP2D6.24 (Ile297Leu) and 2D6.26 (Ile369Thr) expressed in insect cells exhibit substantial activity for codeine and dextromethorphan O-demethylation (Zhang et al. 2009). CYP2D6.24 displays a sixfold higher intrinsic clearance for dextromethorphan O-demethylation, but it shows a 50% reduction for codeine O-demethylation compared to CYP2D6.1 owing to a threefold increase in the K_m (Zhang et al. 2009). For CYP2D6.26, the intrinsic clearance for dextromethorphan and O-demethylation is 37.7% lower than that for the wild type, but it is 42.6% higher for codeine O-demethylation. 2D7.27 also exhibits a 45.9% higher intrinsic clearance for codeine O-demethylation, but it decreases by 64.8% for dextromethorphan O-demethylation, compared to the wild-type enzyme (Zhang et al. 2009). This significant substrate-dependent effect on enzyme activity is also seen in *CYP2D6*17.*

The *CYP2D6*27, *39* and *48* alleles encode enzymes with largely normal activity compared to the wild-type protein (Sakuyama et al. 2008). CYP2D6.27 (Glu410Lys), 2D6.39 (Ser486Thr), and 2D6.48 (Ala90Val) expressed in COS-7 cells show a slightly higher intrinsic clearance than the wild-type enzyme. Ala90, Glu410, and Ser486 are located in β-sheet 3, between the K′ and K″ helices, and in B helix, respectively (Rowland et al. 2006). It appears that these residues are not important for the function of CYP2D6.

When CYP2D6.53 is expressed in COS-7 cells, it shows a 73% decrease in the apparent K_m value for bufuralol 1′-hydroxylation (2.46 vs. 9.2 μM), resulting in a fourfold increase in its intrinsic clearance compared with the wild-type enzyme (Sakuyama et al. 2008). CYP2D6.53 harbors Phe120Ile and Ala122Ser substitutions and both residues

are located in SRS1 of CYP2D6. The Phe120 residue within the B–C loop appears to play an important role in substrate binding and orientation as shown in site-directed mutagenesis studies (Flanagan et al. 2004). When Phe120 is changed to Ala, the K_m for 7-methoxy-4-(aminomethyl)-coumarin is decreased by 50% with improved enzyme affinity (Keizers et al. 2004). Surprisingly, the protein carrying the Phe120Ala replacement can metabolize quinidine via O-demethylation and 3-hydroxylation (McLaughlin et al. 2005), unlike the wild-type CYP2D6, which cannot metabolize this compound. The mutation Phe120Ile (358T>A; rs1135822) can be found in a small percentage of the Southeast Asian population (Solus et al. 2004). Individuals harboring *53 might exhibit a UM phenotype.

7.3.4 Duplication and Multiduplication of CYP2D6

The *CYP2D6* gene is subject to copy number variations. In a Caucasian family in which two siblings and their father has a UM phenotype of debrisoquine with MRs of less than 0.02, 12 extra copies of the *CYP2D6* gene are found and inherited in an autosomal dominant pattern (Johansson et al. 1993). In a second family in which two siblings had MRs of less than 0.1, the authors have found two extra copies of the *CYP2D6* gene. All affected individuals have a variant *CYP2D6* gene, termed *CYP2D6L* (later designated *CYP2D6*2A*), containing a 2850C>T transition in exon 6, leading to an Arg296Cys change, and a 4180G>C mutation in exon 9, resulting in Ser486Thr substitution (Johansson et al. 1993). This is the first description of a stably amplified active gene in humans. The MR of individuals with one copy of the *CYP2D6L* gene does not differ from those with the wild-type gene, but there is a correlation between decreased MR (i.e., increased enzyme activity) and increased copies of the *CYP2D6L* gene.

Panserat et al. (1994) are the first to identify the Arg296Cys and Ser486Thr substitutions in EMs who are previously recognized as the wild type by mistake. A functional study does not reveal any markedly altered enzyme activity in CYP2D6.2 using bufuralol and dextromethorphan as probes compared to the wild type (Sakuyama et al. 2008). However, CYP2D6.2 displays increased K_m and V_{max} values for codeine O-demethylation to form morphine compared to CYP2D6.1 (Oscarson et al. 1997). This finding is supported by the resolved crystal structure of CYP2D6 where the residue Arg296 in helix I near Asp301 and residue Ser486 in β-sheet 4 may play a role in substrate recognition of CYP2D6 (Rowland et al. 2006).

Initially, alleles with 0, 1, 2, 3, 4, 5, and 13 gene copies are reported (Aklillu et al. 1996; Johansson et al. 1993). In a Swedish family (father, daughter, and son), as many as 13 copies of a functional allele of *CYP2D6* have been identified (Johansson et al. 1993). Carriers of *CYP2D6*2×N (N = 2, 3, 4, 5, or 13) with extremely high CYP2D6 activity are identified in a Swedish population (Dahl et al. 1995) and an Ethiopian population (Aklillu et al. 1996). The gene duplication/multiduplication results from unequal crossover events and other mechanisms. Gene duplication and multiduplication of *CYP2D6* can result in enzymes that are functional, partly functional, and nonfunctional. Gene duplication events occur in*1*, *2*, *4*, *6*, *10*, *17*, *29*, *35*, *43*, and *45* (Gaedigk et al. 2007b). Duplications occur at 1.3%, 5.75%, and 2.0% in Caucasian, African American, and racially mixed populations, respectively. Most of the variant duplications except *35×N* are found in African Americans. The *4×N* is as frequent as *2×N* in African Americans (Gaedigk et al. 2007b). Extremely high CYP2D6 activity can result from gene duplication or multiduplication of functional allele *1* and *2* fused in a head-to-tail orientation (Gaedigk et al. 2007b). This is noted by a molecular characterization of the *CYP2D6* locus in patients with extremely rapid metabolisms (Bertilsson et al. 1993b).

There is a marked ethnic difference in the frequency of *CYP2D6* duplication or multiduplication. For example, the *CYP2D6*2×N* allele is more frequent in Africans (28%–56%) than in Asians and north Caucasians (Gaedigk et al. 2007b). The *4×N* variant resulting in a nonfunctional allele is found in 37% of East Asians, but rare in Africans and Caucasians. The frequency of *CYP2D6*10×2* resulting in a significantly decreased enzyme activity is low in Chinese and Japanese (0.5%–1%) (Garcia-Barcelo et al. 2000; Ji et al. 2002; Mitsunaga et al. 2002) but absent in Caucasians. The allelic incidence of high–copy number alleles varies among ethnic groups, with 1.9% in Africans (Aklillu et al. 2002), 0.5% in Japanese (Nishida et al. 2000), 0.8% in Danish (Bathum et al. 1998), 1% in Swedish (Dahl et al. 1995), and 3.3% in Tanzanians (Bathum et al. 1999).

The UM phenotype is mainly distributed in North East Africa (mainly *2×N*) and Oceania (mainly *1×N*) (Sistonen et al. 2007, 2009). In contrast to the PM phenotype where a debrisoquine MR antimode of 12.6 can be defined, there is no clear definition for UM except for the presence of more than two active gene copies. There is a substantial overlap between the EM and UM phenotypes. The UM phenotype is expected to cause a significantly increased CYP2D6 substrate clearance and result in undertherapy or lack of response. Marked decreased drug levels have been observed in UMs with tramadol (Kirchheiner et al. 2008; Stamer et al. 2007), venlafaxine (Shams et al. 2006), morphine (Kirchheiner et al. 2007), mirtazapine (Kirchheiner et al. 2004b), and metoprolol (Goryachkina et al. 2008; Kirchheiner et al. 2004a). Like the PM phenotype, the UM phenotype is also associated with adverse drug reactions, mainly attributed to the formation of high levels of active metabolites. In UMs, up to 10- to 30-fold higher amounts of metabolites can be expected (Dalen et al. 1997) and thus the potential toxicity of the metabolites has to be taken into consideration.

7.4 ETHNIC VARIATION IN THE DISTRIBUTION OF CYP2D6 POLYMORPHISMS

Between people with different ethnic backgrounds, the pattern of CYP2D6 polymorphisms and phenotypes differs dramatically. The frequency of the PM phenotype varies across different ethnic groups. In general, whites have the highest frequency of the PM phenotype, with British and Swiss whites having the highest incidences (8.9% and 10%, respectively) (Alvan et al. 1990). In contrast, the frequency of PMs in Asians is relatively low, particularly among Chinese, Korean, and Japanese populations (0%–1.2%) (Horai et al. 1989, 1990; Sohn et al. 1991). The prevalence of the PM phenotype is slightly higher in Indians than in the Asian populations of southeastern and eastern Asia, with frequencies of 1.8%–4.8% (Lamba et al. 1998). Data on the frequency of PMs in Africans differ widely, varying in the range of 0%–19% (Masimirembwa et al. 1996a; Simooya et al. 1993). There is also a wide range in the incidence of PMs in African Americans (1.9%–7.3%) (Evans et al. 1993; Gaedigk et al. 2002; Relling et al. 1991). The frequency of PMs in Hispanics is 2.2%–6.6% (Jorge et al. 1999).

The frequency of UMs also varies among ethnic groups, with low prevalence found in some European white populations (0.8% in Danish, 1%–2% in Swedish, and 3.6% in Germans) (Bathum et al. 1998; Dahl et al. 1995; Griese et al. 1998; Sachse et al. 1997). A slightly higher prevalence of the UM phenotype has been reported in American Caucasians and blacks (4.3% and 4.9%, respectively) (London et al. 1997). Higher frequencies are found in populations from countries surrounding the Mediterranean Sea, with 8.7% in Turkey, 10% on Sicily in Italy, and 7%–10% in Spain (Agundez et al. 1995; Aynacioglu et al. 1999; Bernal et al. 1999; Scordo et al. 1999). The highest frequencies of the UM phenotype are observed in black Ethiopians (16%) (Aklillu et al. 1996) and Saudi Arabians (20%) (McLellan et al. 1997).

The prevalence of the *CYP2D6* allele differs in different populations. For example, *CYP2D6*4* is by far the most frequent null allele in Caucasians. It occurs with a frequency of 20%–25% and is responsible for 70%–90% of all PMs (Sachse et al. 1997). No functional alleles are present in approximately 6% of Caucasians (Sachse et al. 1997). However, the *CYP2D6*4* allele occurs with a very low frequency of ~1% in Asians (Wang et al. 1993). Its frequency is 6%–7% in Africans and African Americans (Leathart et al. 1998).

The most common *CYP2D6* allele in the Asian population is *CYP2D6*10*, occurring with a frequency of 35%–55% in Chinese, Japanese, and Koreans (Wang et al. 1993). However, it occurs at a low frequency of ~2% in Caucasians; but it accounts for 10%–20% of individuals with the IM phenotype (Griese et al. 1998).

*CYP2D6*17* is virtually absent from European Caucasians and of low frequency in Asians, but it occurs with a high frequency in the African American/black population. This variant appears to explain why black Africans have a higher median MR (Leathart et al. 1998). Thus, there are three alleles with significantly biased distribution in different ethnic groups: *CYP2D6*4*, *10*, and *17* being prevalent in Caucasians, Asians, and Africans, respectively.

The above data would therefore explain the apparent low frequency of the PM phenotype among Asians including Chinese, Japanese, and Koreans and African populations compared to Caucasians when debrisoquine, sparteine, or metoprolol is used as a probe. These are demonstrated in studies highlighting the need for lower-dose neuroleptics that are mainly metabolized by CYP2D6 in Asian patients when compared to Caucasian subjects (Lin and Finder 1983). Also, studies have shown higher concentrations of *S*-mianserin found in Japanese patients with depression than in Caucasian patients (Mihara et al. 1997a).

7.5 ANTIANGINAL DRUGS

7.5.1 PERHEXILINE

CYP2D6 converts perhexiline to its two inactive metabolites, namely, *cis*-hydroxyperhexiline (M1) and *trans*-hydroxyperhexiline (M3) (Sorensen et al. 2003). The rate of M1 formation in human microsomes from 20 livers varied 50-fold but decreased to 5-fold in EMs (*n* = 18) (Sorensen et al. 2003). The CL_{int} is 112-fold lower in microsomes from PMs than those from EMs (0.026 vs. 2.9 µl/min/mg P450). The two main metabolites of perhexiline in the plasma and urine postdosing are M1 and M3. Dihydroxyperhexiline has also been identified in the urine. M1 is the primary determinant of perhexiline clearance and there is a large interindividual variability in metabolic clearance to M1 (Sallustio et al. 2002). The M1/perhexiline concentration ratio (MR) has been incorporated into therapeutic drug monitoring of perhexiline and in phenotype studies as it can readily separate PMs and EMs. It has been suggested that those with MR ≤ 0.4 are considered as PMs (Barclay et al. 2003). In a study with 74 patients, the PMs with an MR < 0.4 harbored nonfunctional *CYP2D6*4/*5*, **5/*6*, or **4/*6* genotypes (Barclay et al. 2003). Among the patients who are EMs, the highest MR is associated with **1* and **2* allele combinations and the MR is progressively lower in the presence of alleles with intermediate function (**9*, **10*, and **41*) followed by alleles with no functional product (**3*, **4*, **5*, and **6*) (Barclay et al. 2003). These results demonstrate a strong gene–dose effect on perhexiline.

PMs have trough concentrations up to sixfold higher than EMs after a single dose of perhexiline (300 mg) in healthy volunteers (Cooper et al. 1984, 1987). In healthy PMs (*n* = 3), the urinary excretion of M1 over 24 h is 91-fold lower than that in EMs (*n* = 47, 46 vs. 4205 µg) (Cooper et al. 1984). EMs excreted 1.4% of the dose as M1 but PMs excreted only 0.014% (100-fold lower) over 24 h. The plasma M1 level in PMs is 28-fold lower than that in EMs (0.02 vs. 0.56 µg/ml). PMs had a sixfold higher plasma level of perhexiline at 24 h postdosing than EMs (0.43 vs. 0.072 µg/ml). However, the plasma and urinary levels of M3 are comparable in PMs and EMs (Cooper et al. 1984), suggesting that the formation of this metabolite is by enzymes rather than by CYP2D6.

Perhexiline can cause frequent but reversible hepatotoxicity and peripheral neuropathy at plasma perhexiline concentrations >0.6 mg/l. The most common adverse effects include headache, dizziness, nausea, and vomiting. Both hepatotoxicity (Morgan et al. 1984) and peripheral neuropathy (Shah et al. 1982) occur more in PMs who exhibited greater plasma perhexiline levels with prolonged plasma $t_{1/2\beta}$. Fifty percent of 20 patients with perhexiline-induced neuropathy have

urinary debrisoquine MR > 12.6 compared with none out of 14 perhexiline-treated patients who do not experience neuropathy (Shah et al. 1982). Another study has found that three of four patients with perhexiline-induced liver injury are PMs, and the fourth had a severely impaired hydroxylation capacity (Morgan et al. 1984).

There is a clear gene–dose and gene–concentration effect on perhexiline. There may be a gene–toxicity effect. As such, prospective genotyping test together with phenotyping studies to determine parent-to-metabolite ratio via therapeutic drug monitoring would be useful to identify those PMs and to adjust the dosage of perhexiline in these patients. Determination of the *CYP2D6* genotype before therapy with perhexiline may help predict perhexiline dose requirements and reduce the risk of perhexiline concentration–related toxicity. The M1/perhexiline concentration ratio determined early in treatment may facilitate appropriate dose adjustment, and the daily perhexiline dose has been recommended to range from 10 to 25 mg in PMs (<50 mg/day), from 100 to 250 mg in EMs, and from 300 to 500 mg in UMs (Sallustio et al. 2002).

7.6 ANTIARRHYTHMIC DRUGS

Antiarrhythmic drugs always have a narrow therapeutic index and there is a direct relationship between their pharmacological activity and free/total plasma concentrations (Follath 1992). Hepatic metabolism plays a significant role in the elimination of many antiarrhythmics, accounting for >90% of the elimination of mexiletine, propafenone, and verapamil. Many antiarrhythmic agents are extensively metabolized by polymorphic CYP2D6. These include class Ia antiarrhythmic drugs such as sparteine (Ebner et al. 1995; Eichelbaum et al. 1979, 1986; Tyndale et al. 1990) and cibenzoline (Niwa et al. 2000), class Ib antiarrhythmic drugs such as aprindine (Ebner and Eichelbaum 1993) and mexiletine (Abolfathi et al. 1993; Broly et al. 1990), and class Ic antiarrhythmic drugs such as propafenone (Botsch et al. 1993; Kroemer et al. 1989b, 1991), encainide (Funck-Brentano et al. 1992), and flecainide (Haefeli et al. 1990). Almost all antiarrhythmics have a low therapeutic index. Most antiarrhythmic agents, including disopyramide, encainide, flecainide, mexiletine, propafenone, and tocainide, are used as racemates. Significant stereoselectivity is observed in different pathways of metabolism of these drugs (Mehvar et al. 2002). Except for encainide and flecainide, there is substantial stereoselectivity in one or more of the pharmacological effects of chiral antiarrhythmics.

Encainide is no longer used because of its frequent proarrhythmic side effects. Encainide is converted by CYP2D6 to active *O*-demethyl metabolite, which is 6–10 times more potent than the parent drug in blocking sodium channels. In EM patients receiving encainide therapy, active *O*-demethyl encainide and 3-methoxy-*O*-demethyl encainide accumulate during prolonged therapy to concentrations greater than those of the parent drug (Barbey et al. 1988; Carey et al. 1984; McAllister et al. 1986; Roden et al. 1986; Turgeon and Roden 1989; Wang et al. 1984; Woosley et al. 1986). In PMs, encainide clearance is lower, and plasma encainide concentrations are higher than those in EMs (Roden and Woosley 1988). In PMs, plasma concentrations of active metabolites are low or undetectable, and the effects of encainide therapy can be closely correlated with plasma concentrations of the parent drug.

7.6.1 FLECAINIDE

Flecainide is a substrate and inhibitor of CYP2D6. As a CYP2D6 inhibitor, it can convert an EM phenotype to a PM one (i.e., phenocopying) (Haefeli et al. 1990). In EMs, there are no significant differences in the oral clearance, $t_{1/2\beta}$, or urinary excretion of S- and R-flecainide (Gross et al. 1989). In PMs, there are significant differences in the pharmacokinetics of S- and R-flecainide. The oral clearance of R-flecainide is less than that of the S-enantiomer (467 vs. 620 ml/min) and the $t_{1/2\beta}$ of R-flecainide is longer than that of S-flecainide (12.9 vs. 9.8 h) (Gross et al. 1989). However, the renal clearance of the two enantiomers is comparable and similar to that observed in EMs. The urinary recovery of R-flecainide is greater than that of the S-enantiomer (15.6 vs. 12.0 mg). The enantioselective disposition observed in PMs is therefore attributed to a greater decrease in the metabolism of R-enantiomer than S-flecainide. The urinary recoveries of two major metabolites of flecainide, m-O-dealkylated flecainide and m-O-dealkylated lactam of flecainide, are lower in PMs, 12.0% and 8.2% of the dose administered, respectively, than in EMs, 17.7% and 16.5%, respectively (Gross et al. 1989).

PMs have ~2-fold higher plasma flecainide levels and AUCs than EMs receiving a single oral dose of flecainide (50 mg) (Gross et al. 1991; Mikus et al. 1989). The AUC is higher (1462 vs. 860 ng·h/ml), the $t_{1/2\beta}$ is prolonged (11.8 vs. 6.8 h), and the amount excreted in the urine is higher (26.7 vs. 15.4 mg) in PMs compared with EMs. Oral clearance of flecainide is significantly decreased in PMs compared to EMs (600 vs. 1041 ml/min) (Mikus et al. 1989). PMs have a lower metabolic clearance of flecainide than EMs (292 vs. 726 ml/min). Gross et al. (1991) have compared the stereoselective disposition of flecainide in healthy PMs ($n = 5$) and EMs ($n = 5$). The authors have found that the $t_{1/2\beta}$ values of R- and S-flecainide in PMs (R vs. S: 19.3 vs. 16.1 h) are approximately twice those in EMs (R vs. S: 8.8 vs. 9.1 h). The apparent oral clearances of R- and S-flecainide are lower in PMs (R vs. S: 313 vs. 379 ml/min) than in EMs (R vs. S: 783 vs. 828 ml/min). However, the renal clearance is comparable for both enantiomers in both EMs and PMs. The nonrenal clearance of both enantiomers is significantly lower in PMs (R vs. S: 123 vs. 201 ml/min) than in EMs (R vs. S: 533 vs. 586 ml/min). The partial clearance to the two major metabolites, $meta$-O-dealkylated flecainide and the $meta$-O-dealkylated lactam of flecainide, is significantly lower in PMs than in EMs (62 vs. 267 ml/min). These results indicate that the CYP2D6 phenotype has a marked impact on R-flecainide's disposition, which is mainly dependent on hepatic CYP2D6. The formation of $meta$-O-dealkylated flecainide and the $meta$-O-dealkylated lactam is at least partially mediated by CYP2D6. Flecainide is a potent inhibitor of CYP2D6 (Haefeli et al. 1990).

A study of 58 Japanese patients with supraventricular tachyarrhythmia has demonstrated that the oral clearance of flecainide is affected by the *CYP2D6* genotype (Doki et al. 2006). The ratios of oral clearance for the five *CYP2D6* phenotypes are as follows: 1.00 (EM/EM), 0.89 (EM/IM), 0.84 (EM/PM), 0.79 (IM/IM), and 0.73 (IM/PM). A population modeling study has indicated that body weight, age, sex, and serum creatinine concentration also affect the pharmacokinetics of flecainide (Doki et al. 2006). The oral clearance of flecainide in IMs is significantly lower than that in homozygous EMs among male patients under 70 years old (0.25 vs. 0.37 l/h/kg). The oral clearance in heterozygous EMs (0.32 l/h/kg) is higher than that in IMs and lower than that in homozygous EMs, though statistical significance is not achieved. These results demonstrate that the *CYP2D6* genotype affects the oral clearance of flecainide, with 11%, 16%, 21%, and 27% reduction in EM/IM, EM/PM, IM/IM, and IM/PM, respectively. This may provide an explanation for the lower clinical doses used in Japanese patients (median dose: 100 mg/day). Subtherapeutic concentrations are prevalent in younger males (67%, 22/33) and EMs (82.6%, 19/23), and thus higher doses of flecainide are recommended for these patients.

A study of 40 Spaniard patients with atrial fibrillation found that the adverse effects of flecainide are more frequent in PMs than in EMs (21.1% vs. 4.8%) (Martinez-Selles et al. 2005). Antiarrhythmic treatment is effective in 27 patients (67.5%), without significant difference between PMs and EMs. However, this study did not measure the plasma concentrations of flecainide and its metabolites.

However, there are negligible differences in the kinetics of flecainide at steady state (Funck-Brentano et al. 1994). When the subjects are treated with flecainide at 100 mg every 12 h for 5 days, mean steady-state plasma concentration of flecainide and QRS change from baseline value do not differ significantly among PMs and EMs. Except for a shortened $t_{1/2\beta}$ and nonlinear pharmacokinetics in EMs, the CYP2D6 phenotype has no significant influence on flecainide pharmacokinetics. Combination with amiodarone results in an increase in mean flecainide plasma concentration and effect in both PMs and EMs, but the extent of the increase is not significantly different between PMs and EMs (Funck-Brentano et al. 1994). In another steady-state study with a new controlled-release form and traditional form of flecainide, there is no impact of the CYP2D6 phenotype on the flecainide pharmacokinetics and QRS prolongation (Tenneze et al. 2002).

In patients with arrhythmias (symptomatic ventricular ectopic depolarizations and nonsustained ventricular tachycardia) and with EM phenotype who receive chronic flecainide treatment (50–150 mg every 12 h to EMs and 75 mg every 12 h to PMs for 15 days), quinidine at 50 mg every 6 h significantly reduces the clearance of R-flecainide by 13.8%, from 395 to 335 ml/min (Birgersdotter et al. 1992). This change is attributable to a decreased formation of the two major metabolites of flecainide, $meta$-O-dealkylated flecainide and the $meta$-O-dealkylated lactam. The renal clearance of R-flecainide increases significantly in EMs than in PMs (118 vs. 60 ml/min) with significantly increased metabolic clearance in EMs (269 vs. 70 ml/min); quinidine significantly decreases the total clearance of R-flecainide compared to monotherapy (395 vs. 335 ml/min), but quinidine does not alter the total and renal clearance of S-flecainide in EMs (total clearance: 450 vs. 419 ml/min; renal clearance: 111 vs. 129 ml/min) (Birgersdotter et al. 1992). In PMs, quinidine has only slightly increased or decreased the total clearance and renal clearance of R- and S-flecainide (total clearance for R: 131 vs. 166 ml/min; total clearance for S: 148 vs. 219 ml/min; renal clearance for R: 60 vs. 47 ml/min; renal clearance for S: 64 vs. 51 ml/min). Quinidine only slightly increases QRS in both EMs and PMs. These results indicate that quinidine inhibits the metabolism of R-flecainide and increases its renal clearance in EMs but not in PMs. The mechanism for the latter change is unclear but may be related to an interaction of flecainide and quinidine at the renal level.

A single-dose study of healthy Korean male volunteers ($n = 21$) has demonstrated that the AUC of flecainide increases significantly after paroxetine coadministration only in individuals harboring *CYP2D6*1/*1* or *1/*2* ($n = 7$, from 5717.5 to 7317.5 ng·h/ml), or *1/*10* ($n = 7$, from 6719.2 to 7828.2 ng·h/ml), but not in subjects with the nonfunctional *10/*10* or *10/*36* ($n = 7$, from 7034.9 to 6925.7 ng·h/ml) alleles, which is frequent in Asians (Lim et al. 2008). Consistently, there is a significant increase in the $t_{1/2\beta}$ values in carriers of *CYP2D6*1/*1*, *1/*2*, or *1/*10* but not in *10/*10* or *10/*36*. The apparent clearances are reduced to 77.8% and 85.7%, respectively, compared to the baseline value. When flecainide is dosed alone, a longer $t_{1/2\beta}$ is observed in carriers of *10/*10* than in carriers of *1/*1* and *1/*2* (12.5 vs. 10.0 h). There is an increasing tendency of the AUC as the number of the nonfunctional *CYP2D6*10* allele increased; however, the difference is not statistically significant. These results show that paroxetine inhibits the metabolism of flecainide and that this interaction is different among subjects with different numbers of the nonfunctional *CYP2D6*10* allele.

The above studies indicate that the *CYP2D6* polymorphism influences flecainide disposition and electrocardiographic effects after a single dose but not at steady state. Flecainide may accumulate and cause toxicity in patients with impaired renal function or with coadministered drugs that inhibit CYP2D6. The gene–concentration and gene–effect relationships between *CYP2D6* genotype and flecainide are minor. Therefore, there is limited clinical significance for CYP2D6 phenotyping and genotyping in patients on maintenance therapy of flecainide. The DPWG Guideline for doxepin recommends reducing the dose by 60% for CYP2D6 PMs and by 20% for IMs (Swen et al. 2011). An alternative drug for CYP2D6 UMs (e.g., citalopram, sertraline) should be selected.

7.6.2 MEXILETINE

Mexiletine is hydroxylated mainly by CYP2D6, which cosegregates with polymorphic CYP2D6-mediated debrisoquine 4-hydroxylase activity (Broly et al. 1990; Vandamme et al. 1993). The formation of these metabolites is lower in PMs than in EMs. Recoveries of hydroxymethylmexiletine and p-hydroxymexiletine from human urine are correlated

inversely with the urinary debrisoquine/4-hydroxydebrisoquine MRs (Broly et al. 1991b). PMs have the highest values of urinary mexiletine/p-hydroxymexiletine and mexiletine/hydroxymethylmexiletine ratios. Urinary excretion of hydroxymethylmexiletine, meta-hydroxymexiletine, and p-hydroxymexiletine in PMs is approximately one-third of that in EMs (Broly et al. 1991b; Turgeon et al. 1991). Mexiletine increases urinary debrisoquine/4-hydroxydebrisoquine MRs in EMs probably because of inhibition of CYP2D6 (Broly et al. 1991b).

In healthy subjects, PMs have significantly higher AUCs of R- and S-enantiomers (4.7 and 5.1 vs. 3.0 and 3.1 µg·h/ml) than EMs (Abolfathi et al. 1993). The total clearance and nonrenal clearance of mexiletine are twofold higher in EMs than in PMs. The $t_{1/2\beta}$ values of R- and S-mexiletine are 35% and 25% lower in EMs than in PMs. In EMs, the partial metabolic clearance to hydroxymethylmexiletine, meta-hydroxymexiletine, and para-hydroxymexiletine are increased 4.4-, 2.4-, and 18.5-fold compared with that in PMs, respectively (Abolfathi et al. 1993). However, the R/S ratio of the AUC, plasma concentrations, and total and non-renal clearances of mexiletine and the R/S ratio of the urinary recovery of R- and S-enantiomers are similar in EMs and PMs (Abolfathi et al. 1993). These ratios are not affected by quinidine coadministration (50 mg four times a day) in either PMs or EMs. However, the partial metabolic clearance of N-hydroxymexiletine glucuronide whose formation is not dependent on CYP2D6 is highly stereoselective with an R/S ratio of 11.3 (Abolfathi et al. 1993). This ratio is similar in either EMs or PMs and is not altered by quinidine coadministration. These data demonstrate a marked stereoselective disposition of mexiletine in either PMs or EMs, which is not altered by quinidine coadministration, indicating the involvement of CYPs other than CYP2D6 in the stereoselective disposition of mexiletine.

A very low dose of quinidine significantly inhibits the elimination of hydroxymethylmexiletine and p-hydroxymexiletine and decreased the total elimination of mexiletine in EMs (Broly et al. 1991a). In healthy volunteers who are EMs, quinidine decreases the total clearance of mexiletine from 621 to 471 ml/min and nonrenal clearance from 583 to 404 ml/min (Turgeon et al. 1991). Quinidine also increases mexiletine's $t_{1/2\beta}$ in EMs from 9 to 11 h. In these subjects, partial metabolic clearance to hydroxymethylmexiletine, meta-hydroxymexiletine, and p-hydroxymexiletine is decreased by quinidine coadministration five-, four-, and sevenfold, respectively, whereas partial metabolic clearance via N-hydroxylation remains unchanged (Turgeon et al. 1991). Similarly, quinidine at 200 mg/day significantly increases the plasma levels of R- and S-mexiletine to a comparable extent in EMs but not in PMs (Abolfathi et al. 1993). Quinidine decreases the total clearance and nonrenal clearance of mexiletine in EMs to the level of PMs but does not affect the clearance in PMs. In EMs, quinidine increases the amounts of R- and S-mexiletine recovered unchanged in the urine to the levels seen in PMs.

In healthy subjects who are EMs, coadministered propafenone 150 mg two times daily for 8 days significantly decreases the oral clearance of R-mexiletine from 41 to 28 l/h and of

S-mexiletine from 43 to 29 l/h to an extent seen in PMs (Labbe et al. 2000). Propafenone also decreases the partial metabolic clearance of mexiletine to hydroxymethylmexiletine, *p*-hydroxymexiletine, and *meta*-hydroxymexiletine at steady state by 71%, 67%, and 73%, respectively (Labbe et al. 2000). However, propafenone does not alter the pharmacokinetics of *R*- and *S*-mexiletine in PMs except for a slight decrease in the formation of hydroxymethylmexiletine. Quinidine coadministration impairs its oxidative carbon metabolism by 70% but has no effect on the *N*-oxidation pathway, which is not dependent on CYP2D6 in vivo (Turgeon et al. 1992). Cimetidine (Brockmeyer et al. 1989; Klein et al. 1985), amiodarone (Yonezawa et al. 2002), ranitidine (Brockmeyer et al. 1989), fluconazole (Ueno et al. 1996), and omeprazole (Kusumoto et al. 1998) do not alter the disposition and clearance of mexiletine in vivo.

The above results indicate that polymorphic CYP2D6 is involved in the metabolism of mexiletine and this is responsible for the large interindividual variability in the clearance of mexiletine in EMs. PMs lacking CYP2D6 activity show uniform values of clearance of mexiletine. Coadministered quinidine or propafenone decreases the clearance of mexiletine through inhibition of CYP2D6 and would thus produce more stable and effective plasma concentrations for mexiletine during chronic therapy in EMs. In fact, mexiletine is often used in combination with quinidine or propafenone to improve antiarrhythmic activity and to decrease incidence of side effects.

7.6.3 PROPAFENONE

Propafenone is mainly metabolized by CYP2D6, 1A2, and 3A4 (Botsch et al. 1993; Zhou et al. 2003). It is also a potent inhibitor for CYP2D6. Propafenone undergoes stereoselective disposition in patients with either the EM or PM phenotype, with 1.7 times faster clearance for the *R*-enantiomer than for the *S*-enantiomer in EMs (Kroemer et al. 1989a). However, the difference in the pharmacokinetics of *R*- and *S*-propafenone is not seen in PMs (Kroemer et al. 1989a), and quinidine does not significantly change the stereoselective steady-state disposition of propafenone (Funck-Brentano et al. 1989). The propafenone/5-hydroxypropafenone plasma concentration ratio correlates with the urinary debrisoquine MRs in patients with stable ventricular premature beats treated with a single dose (450 mg) or at steady state (Capucci et al. 1990). This ratio at 2 h postdosing also correlated well with urinary sparteine MRs in healthy volunteers after a single dose (Anzenbacherova et al. 2000). The propafenone/5-hydroxypropafenone plasma concentration ratio in patients is similar in PMs and EMs (Kroemer et al. 1989a). PMs have three- to sevenfold higher AUCs of propafenone at steady state than EMs and result in very low or undetectable plasma concentrations of 5-hydroxypropafenone at daily doses of 675 to 900 mg/day (Cai et al. 1999a; Dilger et al. 1999; Labbe et al. 2000; Lee et al. 1990; Siddoway et al. 1987). EM patients (*n* = 22) are characterized by a shorter $t_{1/2\beta}$ of propafenone (5.5 vs. 17.2 h), lower average plasma level (1.1 vs. 2.5 ng/ml/

mg dose per day), and higher oral clearance (1115 vs. 264 ml/min) than PM patients (*n* = 6) (Siddoway et al. 1987). At the lower dosages, β-blockade is present in both PMs and EMs but is significantly greater in PMs (Lee et al. 1990). The AUC of propafenone enantiomers in carriers of *CYP2D6*10/*10* (nonfunctional) is approximately 1.5–2 times of that of individuals harboring **1/*10* or **1/*1*, and the clearance of both enantiomers in **10/*10* carriers is only half of that of **1/*10* or **1/*1* carriers (Chen and Cai 2003). These results demonstrate a clear gene–concentration effect for propafenone and its 5-hydroxylation catalyzed by CYP2D6 is subject to polymorphism; the metabolic routes other than 5-hydroxylation contribute to the stereoselective disposition of propafenone during long-term oral therapy. The optimal intervals of dosage may be adjusted on the basis of the CYP2D6 phenotype. For EM patients, 24 to 30 h is required to achieve steady-state plasma concentrations of propafenone, while 72 h or more is needed for PM patients to avoid accumulation of propafenone and its active metabolites.

The results on the effect of CYP2D6 phenotype on the efficacy and adverse reactions of propafenone are conflicting and not confirmed by large studies. In patients with paroxysmal atrial fibrillation (*n* = 42), a 3-month propafenone therapy at 300–450 mg/day is 100% effective in PMs whereas 61% efficacy is observed in EMs with 0% efficacy in UMs (Jazwinska-Tarnawska et al. 2001). There is a significant correlation between propafenone oxidation phenotype determined by CYP2D6 and the ability to maintain sinus rhythm (Jazwinska-Tarnawska et al. 2001). PMs show a greater β-blocking activity compared to EMs (Lee et al. 1990; Morike and Roden 1994), but such a greater effect in PMs is not seen in another study on healthy volunteers (Labbe et al. 2000). A study of 28 patients with chronic ventricular arrhythmias does not observe a significant difference between EMs and PMs in effective propafenone dose or frequency of antiarrhythmic response (Siddoway et al. 1987). In addition, PM patients (*n* = 6) have a higher incidence of adverse effects in the central nervous system (CNS) (67% vs. 14%) than EM patients (*n* = 22) (Siddoway et al. 1987). The magnitude of QRS widening at any given propafenone concentration is greater in EMs than PMs (Siddoway et al. 1987). However, there is no significant difference between EMs and PMs in effective dose of propafenone or frequency of antiarrhythmic response (Siddoway et al. 1987). In addition, there is a case report of toxicity (e.g., dizziness and bradycardia) in a PM who received a usual dose of propafenone at 150 mg three times per day (Morike et al. 1995). In this patient, a toxic concentration of parent propafenone is observed but no 5-hydroxypropafenone is detected. Symptoms disappeared after the discontinuation of propafenone.

The CYP2D6 phenotype also influences drug interactions with propafenone. In healthy volunteers who are EMs, propafenone significantly increases the lidocaine AUC (81.7 vs. 76.3 μg·h/ml) and reduces systemic lidocaine clearance (9.53 vs. 10.27 ml/mg/kg) but does not significantly affect the volume of distribution at steady state (Ujhelyi et al. 1993). CNS adverse effects are significantly worse in severity and duration during coadministration of the two drugs. These results demonstrate a marginal effect of propafenone as a CYP2D6 inhibitor on the metabolism of lidocaine, which is mainly dependent on CYP3A4 and 1A2. Propafenone also increases the plasma levels and $t_{1/2\beta}$ of propranolol (Kowey et al. 1989) and metoprolol (Wagner et al. 1987) in humans. Patients receiving propafenone and β-blockers concurrently do not experience an increased incidence of side effects. Since the therapeutic range of β-blockers is wide, a reduction in dosage may be necessary during concomitant administration with propafenone.

Propafenone also increases the plasma levels of desipramine, a CYP2D6 substrate (Katz 1991). Vice versa, the $t_{1/2\beta}$, C_{max}, and AUC of two enantiomers of propafenone are significantly increased when fluoxetine (a CYP2D6 inhibitor) is coadministered (Cai et al. 1999b). On the other hand, rifampin induces propafenone's metabolic clearance via N-dealkylation (4.1 vs. 23.5 ml/min in EMs; 3.4 vs. 16.0 ml/min in PMs) and glucuronidation (123 vs. 457 ml/min in EMs; 43 vs. 112 ml/min in PMs) more pronounced in EMs than in PMs, but 5-hydroxylation is not induced in EMs. Overall, there is no substantial difference in the pharmacokinetics of propafenone in combination with rifampin, although the bioavailability of propafenone decreased from 30% to 10% in EMs and from 81% to 48% in PMs (Dilger et al. 1999).

Overall, there is a clear gene–concentration relationship for propafenone owing to polymorphic CYP2D6-mediated 5-hydroxylation. EMs show a nonlinear pharmacokinetics of propafenone owing to saturation of metabolism, while the formation of N-depropylpropafenone is not affected in PMs and propafenone's pharmacokinetics becomes linear. It is likely to predict the plasma concentration of propafenone and 5-hydroxypropafenone based on data of CYP2D6 phenotype/genotype. The DPWG Guideline for propafenone recommends reducing the dose of propafenone by 70% for CYP2D6 PMs and adjusting propafenone dose according to plasma concentrations or using an alternative drug for CYP2D6 IM and UM patients (e.g., sotalol, disopyramide, quinidine, and amiodarone) (Swen et al. 2011). The Food and Drug Administration (FDA)–approved drug label for this drug states that propafenone is metabolized by CYP2D6, 3A4, and 1A2; inhibitors of CYP2D6, 1A2, and 3A4 may increase propafenone levels, which may lead to cardiac arrhythmias. Simultaneous use with both a CYP3A4 and 2D6 inhibitor (or in a patient with CYP2D6 deficiency) should be avoided. The evidence for the gene–response effect of propafenone is weak. Monitoring of plasma propafenone concentrations can be used to predict CNS side effects. The establishment of clear gene–response relationships for propafenone is challenged by problems with formation of active metabolites, multiple metabolic routes, stereoselective disposition and efficacy, variable and saturable first-pass metabolism, saturable and nonlinear kinetics, and inhibition of CYP2D6 by propafenone.

7.6.4 Vernakalant

Vernakalant is a substrate of CYP2D6. The pharmacokinetics of vernakalant have recently been investigated in healthy volunteers ($n = 28$) and in patients with atrial fibrillation or atrial flutter (Mao et al. 2009). Vernakalant exhibits linear pharmacokinetics over the dose range of intravenous 0.1–5.0 mg/kg in healthy subjects with a $t_{1/2\beta}$ of 1.7 h. In a subsequent study with 35 patients with atrial fibrillation, the plasma C_{max} of RSD1385 glucuronide is high, often exceeding that of vernakalant in most patients after intravenous injection of a single dose of vernakalant (Mao et al. 2009). Unconjugated RSD1385 is found at very low levels in all patients who are EMs or PMs; however, PMs have even lower levels of unconjugated RSD1385. The plasma level of RSD1390 is also very low. In those patients with CYP2D6 genotype testing, PMs ($n = 2$) have a C_{max} of vernakalant comparable with that in EMs ($n = 10$) (1.86 vs. 1.81 µg/ml in EMs). However, the AUC of vernakalant is five times higher in PMs than in EMs (15.55 vs. 2.88 µg·h/ml in EMs), with a three times longer $t_{1/2\beta}$ in PMs (8.5 vs. 2.7 h (Mao et al. 2009). The two PM patients have very low plasma levels of RSD1385 glucuronide.

In 148 patients with atrial fibrillation or atrial flutter receiving a therapeutic regimen (3 mg/kg initially via 10-min intravenous infusion followed by a second 2-mg/kg 10-min infusion if atrial fibrillation persisted after a 15-min observation period), there is no significant difference in the C_{max} and AUC of vernakalant between PMs ($n = 7$) and EMs ($n = 141$) (C_{max}: 4.50 vs. 4.38 µg/ml in EMs; AUC: 4.48 vs. 3.95 µg·h/ml in EMs) (Mao et al. 2009). The C_{max} and AUC of vernakalant are similar in patients who received concomitant CYP2D6 inhibitor medications ($n = 24$) and in those who did not ($n = 147$) (Mao et al. 2009).

These results indicate that there is a clear gene–dose effect on vernakalant when dosed at 1.5–5.0 mg/kg, supporting a major role of CYP2D6 in the disposition of vernakalant. However, such a gene–dose effect is abolished in those patients receiving two injections of the drug (3 plus 2 mg/kg). Coadministered CYP2D6 inhibitors also do not affect the pharmacokinetics of vernakalant. A lack of differences in the pharmacokinetics of vernakalant is probably attributed to the rapid distribution of the drug and its relatively high volume of distribution (~2 l/kg), which limits the effect of CYP2D6 phenotype status on the disposition of vernakalant. Another possible explanation is vernakalant can inhibit its own metabolism and thus convert an EM phenotype into a PM phenotype. Based on this, there is no need to evaluate CYP2D6 activity when vernakalant is administered acutely and intravenously to patients with atrial fibrillation.

The clinical implications of the CYP2D6 polymorphism in antiarrhythmic therapy depend on the extent of drug elimination by CYP2D6-mediated pathways and the relative activities and potencies of the parent drug and any metabolites whose formation depends on CYP2D6. For example, encainide metabolites are more potent than the parent drug and thus QRS prolongation is more apparent in EMs than in subjects of the PMs (Funck-Brentano et al. 1989). In contrast, propafenone is a more potent β-adrenoceptor blocker than its metabolites and the β-adrenoceptor blocking activity during propafenone therapy is more prominent in PMs than in EMs (Lee et al. 1990), as the parent drug accumulates in PMs.

Since flecainide is mainly eliminated through renal excretion and both *R*- and *S*-flecainide possess equivalent potency for sodium channel inhibition, the CYP2D6 phenotype has a minor impact on the response to flecainide.

7.7 ANTIDEPRESSANTS

7.7.1 TRICYCLIC ANTIDEPRESSANTS

Tricyclic antidepressants (TCAs) are tertiary or secondary amines, and the tertiary forms are metabolized to secondary amines. Both tertiary and secondary amines are active, as are some of the resultant hydroxylated metabolites. The tertiary amines are mainly metabolized by CYP2C19 and 2C9, whereas the secondary amines are largely metabolized by CYP2D6 (Bertilsson 2007).

7.7.1.1 Amitriptyline and Nortriptyline

Both amitriptyline and nortriptyline are substrates of CYP2D6 (Mellstrom and von Bahr 1981; Olesen and Linnet 1997; Venkatakrishnan et al. 2000, 2001). The metabolism of nortriptyline is simpler than that of amitriptyline and is largely metabolized by CYP2D6 (>80% in EMs) (Breyer-Pfaff 2004). There is a significant correlation between amitriptyline clearance and the debrisoquine MR in nonsmokers (Mellstrom et al. 1986). A significant correlation between total clearance of nortriptyline via *E*-10 hydroxylation and the activity of CYP2D6 as determined by debrisoquine hydroxylation or sparteine oxidation is also observed (Dahl et al. 1996; Gram et al. 1989; Mellstrom et al. 1981).

After giving a 25-mg dose of nortriptyline to 21 healthy Swedish Caucasians, the mean plasma AUC of nortriptyline in PMs harboring *CYP2D6*4/*4* are 3.3-fold higher than that observed in EMs, with opposite changes in plasma 10-hydroxynortriptyline levels with lower concentrations in PMs (Dalen et al. 1998). The plasma concentrations of the parent drug are extremely low in one subject with 13 *CYP2D6*2* genes, but very high concentrations of 10-hydroxynortriptyline are noted (Dalen et al. 1998). This study clearly demonstrates the clinical impact of the nonfunctional *CYP2D6*4* allele as well as of the duplication/amplification of the *CYP2D6*2* gene on the disposition of nortriptyline.

There is a significant correlation between the *CYP2D6* genotype and steady-state plasma concentrations of nortriptyline in Swedish patients with depression receiving nortriptyline therapy (Dahl et al. 1996). Similar changes in AUC have also been observed in individuals carrying defective *CYP2D6* alleles leading to reduced activity (e.g., *CYP2D6*10*) in Chinese healthy subjects (Yue et al. 1998) and Japanese depressed patients (Morita et al. 2000), but the effect is less pronounced than that of the Caucasian-specific *CYP2D6*4* allele. These two studies have clearly demonstrated the influence of the *CYP2D6*10* allele on the steady-state plasma levels of nortriptyline and its 10-hydroxy metabolite. However, a study on Faroese patients has reported similar plasma levels of amitriptyline and 10-hydroxynortriptyline in PMs and EMs (Halling et al. 2008a).

Because of significantly reduced CYP2D6-mediated metabolism, PMs have higher plasma concentrations of TCAs metabolized by CYP2D6 than EMs and are therefore more likely to suffer from dose-dependent adverse drug reactions. Patients carrying one dysfunctional *CYP2D6* allele had a greater risk of adverse effects with amitriptyline at 150 mg daily than those with two functional alleles (76.5% vs. 12.1%), and this risk is associated with higher plasma nortriptyline levels (Steimer et al. 2005). However, another study on depressed patients does not find a significant effect of *CYP2D6* mutations on the incidence of adverse reactions caused by nortriptyline (Roberts et al. 2004). Since the major metabolites of nortriptyline are active, altered parent/metabolite ratios attributed to the presence of nonfunctional *CYP2D6* alleles would not significantly alter the sum of active moieties.

A number of cases have been reported where marked CNS toxicities (e.g., dizziness and sedation) with increased nortriptyline plasma concentrations in PMs and individuals receiving CYP2D6 inhibitors such as terbinafine are documented (van der Kuy and Hooymans 1998). Cases of therapeutic failures in UMs on nortriptyline have also been documented (Bertilsson et al. 1985). However, the anticholinergic effects including inhibition of salivation and accommodation disturbances, sedation, blood pressure, and pulse rate did not differ between genotypes in healthy subjects receiving a single dose of nortriptyline (25 to 50 mg) (Dalen et al. 1998).

The CPIC Dosing Guideline for amitriptyline recommends an alternative drug for CYP2D6 or CYP2C19 UMs and for CYP2D6 PMs, and one should consider a 50% dose reduction for CYP2C19 PMs and a 25% dose reduction for CYP2D6 IMs (Hicks et al. 2013). The CPIC Dosing Guideline for nortriptyline recommends a 25% dose reduction for CYP2D6 IMs. For CYP2D6 UMs or PMs, an alternative drug may be considered. The DPWG Guideline for amitriptyline recommends selecting an alternative drug or monitoring amitriptyline and nortriptyline plasma concentration for patients who are CYP2D6 PMs or UMs. One should reduce the initial dose for patients who are IMs or select an alternative drug (Swen et al. 2011). The FDA-approved drug label for amitriptyline and nortriptyline contains information regarding the metabolism of TCAs by CYP2D6. CYP2D6 PMs may have higher plasma concentrations of TCAs, and the label suggests monitoring of plasma levels if this drug is coadministrered with a CYP2D6 inhibitor.

7.7.1.2 Clomipramine

Clomipramine is a substrate of CYP2C19, 2D6, 1A2, and 3A4 (Nielsen et al. 1996). The plasma AUC of clomipramine after a single dose of 100 mg is ~1.8-fold higher in PMs than in EMs (Nielsen et al. 1994). The formation clearance of 2-hydroxyclomipramine and total clearance of clomipramine are significantly lower in PMs than in EMs. In a steady-state study with 37 depressed patients with 36 EMs and 1 PM (sparteine MR > 300) receiving clomipramine 75 mg twice daily, the sole PM has trough plasma concentrations of clomipramine, *N*-desmethylclomipramine, and the sum of clomipramine plus *N*-desmethylclomipramine that are ~3-fold higher than the median concentrations of the respective compound in EMs (Nielsen et al. 1992). The clomipramine concentration is within the range seen in EMs (570 in the PMs vs. 70–730 nM in the EMs), whereas desmethylclomipramine and summed concentrations are 20% to 40% above the upper concentrations in EMs. The only PM has the highest steady-state plasma desmethylclomipramine level and the highest desmethylclomipramine/8-hydroxydesmethylclomipramine ratio. The desmethylclomipramine and didesmethylclomipramine steady-state levels and the desmethylclomipramine/8-hydroxydesmethylclomipramine and clomipramine/8-hydroxyclomipramine ratios show a significant positive correlation with the MR. However, the steady-state plasma clomipramine levels and the clomipramine/desmethylclomipramine ratios showed no significant correlation with the MR.

In another steady-state study of 19 patients with diabetic neuropathy receiving a lower dose of clomipramine (50 mg/day), the summed concentrations of clomipramine and desmethylclomipramine are markedly higher in two PMs (590 and 750 nM) than in EMs (70–510 nM) (Sindrup et al. 1990b). Patients with a weak or absent response to clomipramine have lower plasma concentrations of clomipramine plus desmethylclomipramine less than 200 nM than patients with a better response. It appears that there is no difference in the incidence of side effect between PMs and EMs. These results indicate that proper dose adjusting based on CYP2D6 phenotype can assist in attaining target concentrations of clomipramine.

The CPIC Dosing Guideline for clomipramine recommends selecting an alternative drug for CYP2D6 or CYP2C19 UMs and for CYP2D6 PMs, and one should consider a 50% dose reduction for CYP2C19 PMs and a 25% dose reduction for CYP2D6 IMs (Hicks et al. 2013). The DPWG Guideline for clomipramine recommends reducing the dose by 50% for CYP2D6 PMs and selecting an alternative drug for UMs. One should monitor (desmethyl)clomipramine plasma concentration.

The FDA-approved drug label for clomipramine states that CYP2D6 PMs have higher than expected plasma concentrations of TCAs, such as clomipramine, when given typical doses. Additionally, certain drugs inhibit the activity of this isozyme and make normal metabolizers resemble PMs. It is therefore desirable to monitor TCA plasma levels whenever a TCA is going to be coadministered with another drug known to be a CYP2D6 inhibitor.

7.7.1.3 Doxepin

Doxepin is converted mainly by CYP2D6 and 2C19 to its active metabolites (Haritos et al. 2000; Hartter et al. 2002; Kirchheiner et al. 2002). The AUC of the active moieties after a single oral dose of 75 mg is ~3-fold higher in PMs than in EMs, whereas IMs behaved similarly to EMs (Kirchheiner et al. 2002). The median AUC of the active metabolite desmethyldoxepin is 5.28, 1.35, and 1.28 nmol·h/l in PMs, IMs, and EMs of CYP2D6, respectively. Mean E-doxepin clearance is 406, 247, and 127 l/h in EMs, IMs, and PMs of CYP2D6, respectively (Kirchheiner et al. 2002). In addition, EMs have

approximately twofold lower bioavailability compared with PMs, indicating a significant contribution of CYP2D6 to the first-pass metabolism of E-doxepin. These findings indicate that the CYP2D6 genotype has a major impact on E-doxepin pharmacokinetics and PMs might be at an increased risk of adverse drug effects when treated with recommended standard doses. Koski et al. (2007) have reported a fatal case of doxepin who carried a nonfunctional CYP2D6*3/*4 allele. The plasma concentration of doxepin and nordoxepin is 2.4 and 2.9 mg/L, respectively, with a ratio of doxepin/nordoxepin of 0.83, which is the lowest found among the 35 nordoxepin-positive postmortem cases in this study (Koski et al. 2007). It is unknown whether there is a correlation between CYP2D6 phenotype/genotype and therapeutic effect of doxepin.

The CPIC Dosing Guideline for doxepin, CYP2C19, and CYP2D6 recommends that it may be reasonable to apply the CPIC Dosing Guideline for amitriptyline, CYP2C19, and CYP2D6 to other tricyclics including doxepin (Hicks et al. 2013). In the guideline for amitriptyline, an alternative drug is recommended for CYP2D6 or CYP2C19 UMs and for CYP2D6 PMs. A 50% dose reduction for CYP2C19 PMs and a 25% dose reduction for CYP2D6 IMs should be considered. The DPWG Guideline for doxepin recommends reducing the dose by 60% for CYP2D6 PMs and by 20% for IMs. An alternative drug for CYP2D6 UMs should be selected. In addition, the FDA-approved drug label for doxepin notes that CYP2D6 and CYP2C19 PMs have higher than expected plasma concentrations of doxepin when given typical doses. Additionally, inhibitors of these CYPs may increase the exposure of doxepin. A maximum dose of doxepin in adults and elderly should be 3 mg, when doxepin is coadministered with cimetidine.

7.7.1.4 Imipramine and Desipramine

Imipramine is metabolized to desipramine by CYP1A2, 2C19, and 3A4, and both compounds are metabolized to their active 2-hydroxylated metabolites largely by CYP2D6 (Brosen et al. 1991; Koyama et al. 1997; Lemoine et al. 1993). The oral clearance of imipramine in PMs of CYP2D6 is approximately 53% of that of EMs (1.35 vs. 2.55 L/min), receiving a single oral dose of 100 mg. 2-OH-imipramine is only detectable in plasma of EMs but not in PMs, where the ratio-to-parent compound is higher in rapid EMs than in slow EMs (Brosen et al. 1986b). However, there is no significant difference in the clearance of imipramine via N-demethylation among PMs, and slow and rapid EMs (1.06 vs. 1.42 and 1.60 L/min). These findings indicate the primary role of CYP2D6 in 2-hydroxylation of imipramine and desipramine while their N-demethylation is mediated by multiple CYPs.

The plasma AUCs of desipramine after a single dose (100 mg) in Caucasians are four- to eightfold higher in PMs than in EMs (Brosen et al. 1986b; Spina et al. 1987; Steiner and Spina 1987). Mean steady-state concentrations are ~3-fold higher in PMs than in EMs (Spina et al. 1997). The oral clearance of desipramine is 0.19 l/min in PMs compared with 1.64 and 1.03 l/min in rapid EMs and slow EMs, respectively (Brosen et al. 1986b). A significantly longer $t_{1/2\beta}$ of desipramine is observed in PMs compared with EMs (81–131 vs.

13–23 h). 2-OH-desipramine is detected in EMs only. Similar effects have been seen in Japanese (Shimoda et al. 2000a). In one study, two PMs receive a greatly reduced dose compared with that given to EMs (50 mg vs. 200 mg daily) and still attain plasma desipramine concentrations at the upper end of the range observed in EMs (860 and 880 vs. 130 to 910 nM) (Sindrup et al. 1990b). The doses required by EMs (200 mg/day) would yield clearly toxic drug concentrations (>3.0 μM) in PMs.

At steady state, the plasma concentration of imipramine and desipramine is significantly higher in PMs than in EMs when treated with 100 mg imipramine per day (imipramine: 302–455 vs. 169 nmol/l; desipramine: 1148–1721 vs. 212 nmol/l) (Brosen et al. 1986a). The ratio of 2-OH-imipramine over imipramine and 2-OH-desipramine over desipramine is 5- to 10-fold higher in EMs than in PMs. Desipramine concentrations at steady-state are sevenfold higher in PMs, with the sum of imipramine plus desipramine concentrations being fivefold higher in PMs compared with EMs (Brosen et al. 1986a). The urinary MR values of sparteine and debrisoquine correlated poorly with the imipramine steady-state concentrations, but quite well with the desipramine steady-state concentrations. There are significant negative correlations between sparteine and debrisoquine MR and the 2-OH-imipramine/imipramine and 2-OH-desipramine/desipramine ratios. In addition, there is a weak negative correlation between sparteine MR and daily dose of imipramine required to achieve therapeutic range (Brosen et al. 1986a). These results further support the primary role of CYP2D6 in 2-hydroxylation of imipramine and desipramine. The sum of imipramine plus desipramine concentrations may be used to conduct therapeutic monitoring during treatment with imipramine, and PMs generally require lower doses to achieve the target concentrations. PMs needed 20 to 25 mg/day of imipramine, whereas EMs required 50 to 350 mg/day to attain summed concentrations of ~300 to 500 nM in patients with diabetic neuropathy (Sindrup et al. 1990a). It is unknown whether CYP2D6 phenotype status affects the adverse effects of imipramine.

In one steady-state study on depressed patients with 2 PMs and 29 EMs receiving 100 mg desipramine per day for 3 weeks, 3 patients including both PMs and 1 EM require a dosage decrease because of marked side effects (e.g., sedation or postural hypotension) (Spina et al. 1997). Plasma desipramine concentrations are significantly correlated with dextromethorphan MRs, with the two PMs showing the highest plasma concentrations of desipramine. At 3 weeks, there is no significant correlation between plasma levels of either desipramine or desipramine plus 2-OH-desipramine and antidepressant effect when assessed by the Hamilton Depression Rating Scale (Spina et al. 1997). These results indicate that the CYP2D6 phenotype status has a significant impact on steady-state plasma levels of desipramine in depressed patients and may identify subjects at increased risk for severe concentration-dependent CNS side effects. However, the CYP2D6 activity does not appear to predict the therapeutic effect of desipramine, which is not concentration dependent.

A retrospective study of 181 depressed patients has revealed that desipramine and the sum of imipramine plus desipramine plasma concentrations per drug dose unit, imipramine dose at steady state, and imipramine dose requirement significantly depend on CYP2D6 genotype (Schenk et al. 2008). Mean dose requirements are 131, 155, 217, 245, 326, and 509 mg imipramine per day in carriers of 0, 0.5, 1, 1.5, 2, and >2 active CYP2D6 genes, respectively. These results suggest that genotype-based drug dose recommendations might allow for the use of an adjusted starting dose and faster achievement of predefined imipramine plus desipramine plasma levels in the management of imipramine pharmacotherapy, which would improve therapeutic efficacy and minimize adverse drug reactions.

The DPWG Guideline for imipramine recommends reducing the dose of imipramine or desipramine for CYP2D6 PM (reduce dose by 70%) and IM (reduce dose by 30%) patients and monitoring imipramine and desipramine plasma concentrations (Swen et al. 2011). An alternative drug (e.g., citalopram, sertraline) or dose increase by 70% for CYP2D6 UMs should be considered.

The FDA-approved drug labels for imipramine and desipramine state that CYP2D6 PMs have higher than expected plasma concentrations of TCAs when given usual doses. Concomitant use of TCAs with drugs that can inhibit CYP2D6 may require lower doses than usually prescribed for either the TCA or the other drug. Certain drugs inhibit the activity of CYP2D6 and make normal metabolizers resemble PMs. It is therefore desirable to monitor TCA plasma levels whenever a TCA is going to be coadministered with another drug known to be a CYP2D6 inhibitor.

7.7.1.5 Maprotiline

Maprotiline is a substrate of CYP2D6 with its major active metabolite normaprotiline being formed in vivo (Brachtendorf et al. 2002). In healthy subjects, the plasma C_{max} and AUC of maprotiline is 2.7- and 3.5-fold higher in PMs ($n = 6$) than in EMs ($n = 6$) receiving 50 mg twice daily for 8 days (C_{max}: 203 vs. 73 ng/ml; $AUC_{0-48\,h}$: 8054 vs. 2289 ng·h/ml) (Firkusny and Gleiter 1994). PMs have a 2.9-fold longer $t_{1/2\beta}$ than EMs (88.3 vs. 30.4 h). The duration of the pulmonary effect (alleviation of histamine-induced bronchoconstriction) of maprotiline after cessation of 8-day maprotiline treatment in EMs is <3 weeks compared with ≥4 weeks in PMs (Firkusny and Gleiter 1994). Lower plasma concentration of maprotiline is associated with a marked decrease in bronchial sensitivity to histamine. These results demonstrate a clear gene–dose effect on maprotiline. However, no data are available on the effect of the CYP2D6 phenotype on the efficacy and adverse effects of maprotiline in depressed patients.

7.7.1.6 Trimipramine

Trimipramine is mainly metabolized (activated) by CYP2D6 and 2C19 (Bolaji et al. 1993; Eap et al. 2000). Because of the differing role of CYP2D6 and 2C19 in the metabolism of trimipramine enantiomers, the phenotype of both CYP2D6 and 2C19 would affect the disposition of racemic trimipramine.

439

In depressed patients ($n = 27$), the sole PM of CYP2D6 receiving 300 and 400 mg/day racemic trimipramine had the highest level of (L)- and (D)-desmethyltrimipramine, which are formed by CYP2C19, while the sole PM of CYP2C19 showed the highest level of (L)- and (D)-trimipramine (Eap et al. 2000). A single-dose (75 mg) study of 42 healthy volunteers indicates that the disposition of trimipramine and its demethylated metabolite is associated with the CYP2D6 phenotype (Kirchheiner et al. 2003). The oral clearance of trimipramine is 36 l/h in healthy PMs but it is 276 l/h in EMs. The bioavailability of trimipramine is threefold higher in CYP2D6 PMs than in EMs. The AUC of desmethyltrimipramine is 40-fold higher in PMs of CYP2D6 than in EMs (1.7 vs. 0.04 mg·h/l), but this metabolite is undetectable in most PMs of CYP2C19 or 2C9 (Kirchheiner et al. 2003). However, the plasma level of 2-hydroxytrimipramine is very low in PMs of CYP2D6. These findings demonstrate a major role of CYP2C9 and 2C19 in the N-demethylation of trimipramine and CYP2D6 is involved in its further metabolism (e.g., 2-hydroxylation). The genotype/phenotype of CYP2D6, 2C19, and possibly 2C9 can affect the disposition of trimipramine. CYP2D6 appears to favor trimipramine hydroxylation but not its N-demethylation. This has been confirmed by an interaction study with the CYP2D6 inhibitor quinidine, which increased the plasma concentration of trimipramine but the formation of 2-hydroxytrimipramine is decreased (Eap et al. 1992). Cases have been reported on increased trimipramine levels by coadministered paroxetine (Leinonen et al. 2004). A fatal case has been documented where trimipramine and citalopram are combined (Musshoff et al. 1999).

TCAs have similar metabolic routes and complex pharmacology. CYP2D6 has differing contribution to their metabolic clearance and pharmacologically active metabolites can be formed from CYP2D6-mediated pathways. There is a possible enantioselective disposition. As such, it is difficult to establish clear relationships of their pharmacokinetic and pharmacodynamic parameters with genetic variations of *CYP2D6*. It is unlikely to adjust the dosage on the basis of the CYP2D6 phenotype although some initial gene–dose effect has been observed for some TCAs.

The CPIC Dosing Guideline for trimipramine, CYP2C19, and the CYP2D6 phenotype recommends that it may be reasonable to apply the CPIC Dosing Guideline for amitriptyline, CYP2C19, and CYP2D6 to other TCAs including trimipramine (Hicks et al. 2013). In the guideline for amitriptyline, an alternative drug is recommended for CYP2D6 or CYP2C19 UMs and for CYP2D6 PMs. One should consider a 50% dose reduction for CYP2C19 PMs and a 25% dose reduction for CYP2D6 IMs.

7.7.2 Selective Serotonin Reuptake Inhibitors

7.7.2.1 Citalopram

Citalopram is mainly metabolized by CYP3A4, 2C19, and 2D6 (Olesen and Linnet 1999; von Moltke et al. 2001). After oral administration of racemic citalopram to healthy subjects, the AUC of S-citalopram is significantly higher in the EMs of CYP2D6 but PMs of CYP2C19 compared to the EMs of both CYP2D6 and 2C19 and PMs of CYP2D6 but EMs of CYP2C19, whereas the AUC of R-citalopram did not differ between the genotype groups (Herrlin et al. 2003). Similar differences, although they do not reach statistical significance, are observed for S- and R-desmethylcitalopram. However, a study with 1953 depressed subjects does not reveal an association between the genotype of *CYP2D6* and response and tolerance to citalopram (Peters et al. 2008). The *CYP2D6* genotype and phenotype status do not correlate with the dosage of citalopram and the length of time a subject would continue with citalopram therapy. A large ($n = 749$) Swedish study finds no difference in citalopram or desmethylcitalopram plasma concentrations between those experiencing a number of common adverse effects compared to those who did not, suggesting that adverse effects of citalopram are determined primarily by pharmacodynamic rather than by pharmacokinetic factors (Reis et al. 2003). It appears that knowledge of CYP2D6 genotype/phenotype would not assist with proper dosage adjustment of citalopram.

The CPIC Dosing Guideline for citalopram and escitalopram based on the CYP2C19 phenotype recommends considering an alternative drug not predominantly metabolized by CYP2C19 for CYP2C19 UMs (Hicks et al. 2015). For CYP2C19 PMs, one should consider a 50% reduction of recommended starting dose and titrate to response or select an alternative drug not predominantly metabolized by CYP2C19.

The FDA-approved drug label for citalopram recommends a maximum dose of 20 mg/day of citalopram in patients who are CYP2C19 PMs, owing to an increase in citalopram exposure. This increased exposure leads to a greater risk for QT prolongation, a potentially fatal abnormality in the heart's electrical activity.

7.7.2.2 Fluvoxamine

Fluvoxamine is mainly metabolized by CYP2D6 and 1A2 (Spigset et al. 2001). The clearance of the 5-demethoxylated carboxylic acid metabolite is 78% lower in CYP2D6 PMs than in the EMs after a single oral dose of fluvoxamine, but smoking and being a CYP2C19 PM did not influence its clearance (Spigset et al. 2001). There is no significant correlation between oral caffeine clearance (CYP1A2-mediated) and the clearance of this metabolite (Spigset et al. 2001). A single-dose study has observed a 1.3-fold higher AUC of fluvoxamine in PMs (Spigset et al. 2001), whereas another study does not find any difference between PMs and EMs (Christensen et al. 2002). Similar disparities are observed at steady state (Christensen et al. 2002; Spigset et al. 1998).

It appears that the *CYP2D6* genotype has an effect on the adverse effects of fluvoxamine. In a study with Japanese depressed patients ($n = 100$), the carriers of the *CYP2D6*5* and *10* alleles show increased gastrointestinal side effects induced by fluvoxamine compared to the wild type, and there is a 4.2-fold higher risk of developing gastrointestinal side effects compared with EMs when the nonfunctional *CYP2D6* allele is combined with the 5-HT$_{2A}$ receptor

A-1438G polymorphism (Suzuki et al. 2006). However, the plasma concentrations of fluvoxamine are not determined in this study.

These results suggest that the *CYP2D6* genotype has only a marginal impact on the clearance of fluvoxamine since the contribution of CYP2D6 to fluvoxamine's overall clearance is minor to moderate and other enzymes such as CYP1A2 may play a more important role in its clearance. There is weak or no evidence showing the association of the CYP2D6 phenotype with efficacy and adverse reactions of fluvoxamine. Kirchheiner et al. (2001) have suggested that homozygous EMs would need 120% of the recommended dose, heterozygous EMs would need 90% of the recommended dose, and PMs would need 60% of the recommended dose in order to achieve the same plasma concentrations of fluvoxamine. However, as fluvoxamine has a broad therapeutic index, and the interindividual metabolic variability within the EMs and PMs are high, the need for adjusting its dose on the basis of the *CYP2D6* phenotype seems to be relatively small.

The CPIC Dosing Guideline for the selective serotonin reuptake inhibitor (SSRI) fluvoxamine recommends considering a 25%–50% reduction of recommended starting dose and titrating to response or using an alternative drug not metabolized by CYP2D6 for CYP2D6 PMs who may carry *3/*4, *4/*4, *5/*5, or *5/*6 (Hicks et al. 2015). Data describing the effect of CYP2D6 ultrarapid metabolism on fluvoxamine therapy are lacking; therefore, no dosing recommendations are provided by the CPIC. Since CYP2D6 activity is fully mature by early childhood, it may be appropriate to extrapolate these recommendations to adolescents or possibly younger children with close monitoring (Hicks et al. 2015).

The FDA-approved drug label for fluvoxamine notes that this drug appears to inhibit CYP3A4 and 2D6 as it exhibits substantial interactions with drugs metabolized by the same CYP enzymes. Caution should be used in treating patients with low CYP2D6 activity and those receiving other medications known to inhibit CYP2D6. Population pharmacokinetic analysis indicates that there is a 25% decrease in median clearance in CYP2D6 PMs compared to EMs. Dosage adjustment is not necessary in patients identified as PMs as the dose of drug is individually titrated to tolerability.

7.7.2.3 Fluoxetine

Fluoxetine is mainly metabolized by CYP2D6, 3A4, and 2C9 (Margolis et al. 2000; Ring et al. 2001). The AUC of fluoxetine is 3.9-fold higher and that of norfluoxetine is 0.5-fold lower in PMs than in EMs receiving a single oral dose of 20 mg, whereas the sum of these moieties is 1.3-fold higher in PMs (Hamelin et al. 1996). When the dose is increased to 60 mg, the median AUC values of S- and R-fluoxetine are 11.5- and 2.4-fold higher in PMs, respectively, whereas S- and R-norfluoxetine are decreased by only ~20%–40% (Fjordside et al. 1999). In PMs, the oral clearance of R- and S-fluoxetine is 3.0 and 17 l/h, respectively, while the corresponding values in EMs are 36 and 40 l/h, respectively. In PMs, the $t_{1/2\beta}$ is 6.9 and 17.4 days for R- and S-norfluoxetine, respectively, and 5.5 days for both enantiomers in EMs, with a significant

phenotypical difference only for S-norfluoxetine. The $t_{1/2\beta}$ of R- and S-fluoxetine is 9.5 and 6.1 days in PMs, respectively, but decreased to 2.6 and 1.1 days, respectively, in EMs. These results indicate that CYP2D6 metabolizes extensively R- and S-fluoxetine and most likely the further metabolism of S-norfluoxetine but not of R-norfluoxetine. The CYP2D6 phenotype has a marked impact on the pharmacokinetics of fluoxetine.

The sum of racemic parent drug plus norfluoxetine trough concentrations is comparable between PMs and EMs receiving 20 mg fluoxetine per day for 23 days in a steady-state study (Fjordside et al. 1999). There is no significant difference in the plasma concentration of R-fluoxetine and R-norfluoxetine, whereas the level of S-fluoxetine is 2.2-fold higher and S-norfluoxetine is 3.4-fold lower in PMs than in EMs (Fjordside et al. 1999). In another steady-state study of 78 depressed patients, there is no significant relationship between the *CYP2D6* genotypes and the dose-normalized plasma concentrations of both enantiomers of fluoxetine or their active moieties (i.e., the sum of S-fluoxetine, R-fluoxetine, and S-norfluoxetine) (Scordo et al. 2005). However, the plasma level of S-norfluoxetine is very low in the only subject with CYP2D6 PM and the median S-norfluoxetine/S-fluoxetine ratios are higher in homozygous than in heterozygous EMs. These findings have provided further evidence that CYP2D6 mediates the demethylation of fluoxetine to norfluoxetine, with a stereoselectivity toward the S-enantiomer; *CYP2D6* polymorphisms contribute to the wide interindividual variability in fluoxetine pharmacokinetics at steady state.

It appears that the CYP2D6 phenotype status did not affect the incidence of adverse reactions to fluoxetine. In studies with small samples, a significant relationship between CYP2D6 status and adverse reactions of fluoxetine has not been observed (Roberts et al. 2004; Stedman et al. 2002). In the clinical study with 20 depressed patients, fluoxetine- or paroxetine-induced hyponatremia is not associated with the nonfunctional *CYP2D6* alleles such as *4, *5, and *6. In contrast, the trend is in the opposite, suggesting that hyponatremia induced by SSRIs is explained by factors other than genetically poor metabolism of CYP2D6, such as alterations in the renin–angiotensin system related to increased age, and other comorbidities and comedication.

The FDA-approved drug label for fluoxetine states that fluoxetine is a substrate of CYP2D6 and a potent inhibitor of this enzyme. Coadministration of fluoxetine with other drugs that are metabolized by CYP2D6, including certain antidepressants (e.g., TCAs), antipsychotics (e.g., phenothiazines and most atypicals), and antiarrhythmics (e.g., propafenone, flecainide, and others), should be approached with caution. Therapy with medications that are predominantly metabolized by CYP2D6 and that have a relatively narrow therapeutic index should be initiated at the low end of the dose range if a patient is receiving fluoxetine concurrently or has taken it in the previous 5 weeks. Because of the risk of serious ventricular arrhythmias and sudden death potentially associated with elevated plasma levels of thioridazine, thioridazine should not be administered with fluoxetine or within a minimum of 5 weeks after fluoxetine has been discontinued.

7.7.2.4 Paroxetine

Paroxetine is mainly (80%) metabolized by CYP2D6 (Bloomer et al. 1992) and it is also a mechanism-based inhibitor of CYP2D6. Several clinical studies have shown a relationship between CYP2D6 phenotype/genotype and paroxetine levels during short- or long-term dosing (Ozdemir et al. 1999; Sindrup et al. 1992a,b). The median AUC of paroxetine is sevenfold higher in PMs than in EMs treated with a single dose at 30 mg (550 vs. 3910 nmol·h/L), but the AUC difference decreases to twofold at steady state with 30 mg/day for 2 weeks (2550 vs. 4410 nmol·h/l) (Sindrup et al. 1992b). Plasma $t_{1/2\beta}$ and steady-state plasma concentration are significantly longer and higher in PMs than in EMs (41 vs. 16 h and 151 vs. 81 nmol/l). Paroxetine displays nonlinear kinetics only in EMs but not in PMs (Sindrup et al. 1992b). These results indicate that the interphenotype difference in paroxetine metabolism is less prominent at steady state than after a single dose, presumably attributed to saturable CYP2D6-mediated metabolism. Sparteine MRs increased in EMs during paroxetine treatment, and two EMs are phenotyped as PMs and the remaining EMs are converted into extremely slow EMs with MRs of 5.7 to 16.5 after 14 days of treatment.

Another steady-state study has more paroxetine dose levels (10 to 70 mg/day) in 13 EMs and three or four dose levels (10 to 40 mg/day) in 3 PMs, all treated for diabetic neuropathy symptoms (Sindrup et al. 1992a). There is 3.3-fold higher 12-h concentrations of paroxetine in PMs than in EMs and a 25-fold variation in steady-state concentrations of paroxetine (25 to 670 nmol/L). The steady-state levels of paroxetine at all dose levels showed a positive correlation with sparteine MRs in EMs. Estimates of clearance at low drug levels of the high-affinity clearing process show a significant negative correlation with the sparteine MRs, but the clearance of the low-affinity process is not related to the MRs in both EMs and PMs (Sindrup et al. 1992a). These data indicate that the metabolism of paroxetine and sparteine cosegregates and that CYP2D6 is responsible for a high-affinity saturable paroxetine elimination process.

Heterozygous EMs of CYP2D6 have nonsignificant twofold higher median steady-state trough concentrations of paroxetine than homozygous EMs, with a considerable overlap in the distribution of paroxetine concentrations between the two phenotypes (Ozdemir et al. 1999). No PMs or UMs are included in this study. Notably, UMs have very low or undetectable concentrations of paroxetine with common doses (Charlier et al. 2003; Lam et al. 2002). These subjects would be expected to undergo undertreatment when paroxetine dosage is not increased.

Evidence on the effect of the CYP2D6 phenotype on the therapeutic efficacy and adverse reactions of SSRIs are weak. One study on 30 patients has found that their sexual dysfunction is more frequent in EMs but who are converted into PMs during treatment (17/24 patients; 71%) compared with those who are phenotyped as EMs (2/6; 33%) (Zourkova and Hadasova 2002). In a prospective study of 246 elderly patients taking paroxetine together with another medication, which is a CYP2D6 substrate, the *CYP2D6* genotype is not associated with the incidence of side effect (Murphy et al. 2003). The *CYP2D6* genotype does not correlate with the incidence of paroxetine-induced nausea (Tanaka et al. 2008). The safety margin of SSRIs may be sufficient to prevent the occurrence of more side effects in IMs and PMs. However, most SSRIs are potent CYP2D6 inhibitors and can thus convert an EM into PM status, probably resulting in remarkable toxicity. Furthermore, a study has demonstrated no correlation of paroxetine concentrations with its therapeutic effect (Lam et al. 2002).

The CPIC Dosing Guideline for paroxetine recommends selecting an alternative drug not predominantly metabolized by CYP2D6 for CYP2D6 UMs and for CYP2D6 PMs (Hicks et al. 2015). For CYP2D6 PMs, if paroxetine use is warranted, one should consider a 50% reduction of recommended starting dose and titrate to response. When administered similar doses, CYP2D6 PMs have significantly greater drug exposure to paroxetine when compared to EMs. To potentially prevent an adverse effect, an alternative SSRI not extensively metabolized by CYP2D6 should be considered for CYP2D6 PMs. Because CYP2D6 activity is fully mature by early childhood, it may be appropriate to extrapolate these recommendations to adolescents or possibly younger children with close monitoring. The DPWG Guideline for paroxetine recommends selecting an alternative drug to paroxetine for CYP2D6 UM patients (e.g., citalopram, sertraline) (Swen et al. 2011).

The FDA-approved drug label for paroxetine notes that the metabolism of paroxetine is accomplished in part by CYP2D6. Coadministration of paroxetine with other drugs that are metabolized by this enzyme should be approached with caution. Because of the risk of serious ventricular arrhythmias and sudden death potentially associated with elevated plasma levels of thioridazine, paroxetine and thioridazine should not be coadministered.

7.7.2.5 Sertraline

CYP2D6 status has little impact on the clearance and effect of sertraline, which is mainly metabolized by CYP2C19, 2C9, 2B6, and 3A4 (Xu et al. 1999). Sertraline has been shown to show no differences in pharmacokinetics between CYP2D6 EMs and PMs (Hamelin et al. 1996). However, there is a 40% difference in sertraline AUC between CYP2C19 EMs and PMs (Wang et al. 2001), suggesting that this enzyme predominates in the metabolic clearance of sertraline. Side effects in CYP2C19 PMs have also been reported to be more frequent than in EMs.

The CPIC Dosing Guideline for sertraline based on the CYP2C19 phenotype recommends starting therapy with recommended starting dose for CYP2C19 UMs who carry *17/*17 or *1/*17 (Hicks et al. 2015). If the patient does not respond to recommended maintenance dosing, one should consider an alternative drug not predominantly metabolized by CYP2C19. For CYP2C19 PMs who harbor *2/*2, *2/*3, or *3/*3, one should consider a 50% reduction of recommended starting dose and titrate to response or select an alternative drug not predominantly metabolized by CYP2C19. CYP2C19 activity may be increased in children relative to adults; therefore, these recommendations should be used with caution in children and accompanied by close monitoring.

There is initial evidence for gene–dose effect on commonly used SSRIs and data on the effect of the CYP2D6 genotype/phenotype on response and adverse effects of SSRIs are scant. Therefore, recommendations for dose adjustment of prescribed SSRIs based on the CYP2D6 phenotype/genotype may not be feasible and further studies are warranted to explore the impact of the CYP2D6 genotype/phenotype on the pharmacology of SSRIs.

7.7.3 OTHER ANTIDEPRESSANTS

7.7.3.1 Mianserin

Mianserin is a substrate of CYP2D6 (Dahl et al. 1994b; Koyama et al. 1996; Stormer et al. 2000; Yasui et al. 1997). The mean AUCs of mianserin and desmethylmianserin are 1.8- and 1.5-fold higher, respectively, in PMs than in EMs (Dahl et al. 1994b). A PM for both debrisoquine and mephenytoin has the highest summed concentrations among 18 patients with diabetic neuropathy (Sindrup et al. 1992c), and three of six patients with high mianserin concentrations are phenotypic PMs in another study (Tacke et al. 1992). In contrast, only one of seven patients with slow mianserin elimination is a PM, which does not support an important role for CYP2D6 in the elimination of mianserin (Begg et al. 1989). Higher mean S-mianserin plasma concentrations and a slightly greater response are observed in Japanese depressed patients harboring CYP2D6*1/*10 compared to those harboring *1/*1 (15 vs. 8 mg/l at 12-h postdose) who receive 30 mg mianserin daily for 3 weeks (Mihara et al. 1997b). Thioridazine inhibits CYP2D6 and increases S-mianserin, S-desmethylmianserin, and R-desmethylmianserin concentrations by 1.9-, 2.1-, and 2.7-fold, respectively, whereas there is no effect on R-mianserin pharmacokinetics (Yasui et al. 1997).

7.7.3.2 Mirtazapine

Mirtazapine is mainly metabolized by CYP2D6, 1A2, and 3A4 (Stormer et al. 2000). Genetic CYP2D6 polymorphisms show different effects on the disposition of the enantiomers of mirtazapine. The plasma AUC of S-mirtazapine is 79% larger in PMs than in EMs, but there are no differences between EMs and PMs for the pharmacokinetics of the R-enantiomer (Timmer et al. 2000). A population modeling study of 49 depressed patients reports a 26% lower clearance of mirtazapine in IMs than in EMs (Grasmader et al. 2004a).

A single-dose study of 25 healthy subjects has demonstrated that the median oral clearance of racemic mirtazapine is 20.1, 39.7, and 49.8 l/h in carriers of 0, 2 (*1/*1), and 3 active genes of CYP2D6 and the median C_{max} is 129, 159, and 76 µg/l in these three groups, indicating a contribution of CYP2D6 to the first-pass metabolism of mirtazapine (Kirchheiner et al. 2004b). There is a trend with lower concentrations for the active metabolite desmethylmirtazapine in the group with three functional alleles (i.e., UMs) but does not achieve statistical significance. Mirtazapine concentrations are significantly correlated with blood pressure and the correlation is even stronger when taking the sum of mirtazapine plus desmethylmirtazapine, but the effect on blood pressure and heart rate is not correlated with the CYP2D6 genotype (Kirchheiner et al. 2004b). These results indicate that CYP2D6 contributes to ~25% of total clearance of mirtazapine in individuals carrying only one active allele of CYP2D6 and up to 55% in the UMs. The effect of the CYP2D6 gene duplication is lower than expected and high CYP2D6 activity may only explain a small percentage of the cases with therapeutic failure in mirtazapine treatment.

A study has revealed that the pharmacokinetics of R-mirtazapine is only marginally influenced by the CYP2D6 genotype, but the oral clearance of the S-enantiomer are 1.3, 2.3, and 3.4 ml/min in PMs, EMs, and UMs (Brockmoller et al. 2007), indicating a substantial first-pass metabolism of mirtazapine in EMs and UMs. Mirtazapine is enantioselectively absorbed from the small intestine, with a rate constant of 0.2 min^{-1} for S-mirtazapine but 0.08 min^{-1} for R-mirtazapine (Brockmoller et al. 2007). The effect of mirtazapine on heart rate, blood pressure, and sedation is correlated with both R- and S-mirtazapine plasma concentrations. These results demonstrate a gene–dose effect on S-mirtazapine but not the R-enantiomer since the disposition of the S-enantiomer is more dependent on CYP2D6. However, the adverse effects of mirtazapine appear to correlate with both S- and R-mirtazapine levels.

7.7.3.3 Venlafaxine

Venlafaxine is mainly metabolized by CYP2D6, 3A4, 2C19, and 2C9 (Fogelman et al. 1999; Otton et al. 1996). There is a good correlation between the debrisoquine MR and the ratio between the AUC of venlafaxine and that of O-desmethylvenlafaxine (Lindh et al. 2003). PMs of CYP2D6 have a more than fourfold lower oral clearance of venlafaxine compared to EMs, mainly attributed to a decreased capacity to form the O-demethylated metabolites (Lessard et al. 1999). In a study with 33 depressed patients receiving venlafaxine at 225 mg daily, a significant relationship between the CYP2D6 genotype and the O-desmethylvenlafaxine/venlafaxine ratio is found, with PMs having extremely low ratios and UMs having high ratios compared with homozygous and heterozygous EMs (Veefkind et al. 2000). Healthy Japanese subjects with the homozygous CYP2D6*10 allele have a 4.5-fold higher plasma AUC of venlafaxine than the wild type (EMs) (Fukuda et al. 2000b). The median oral clearance of R-venlafaxine at steady state is ninefold higher in EMs compared to PMs, while it is twofold higher for S-venlafaxine (Eap et al. 2003). The coadministration of quinidine to EMs results in an almost complete inhibition of the partial metabolic clearance of R-venlafaxine to O-demethylated metabolites, while a sevenfold decrease is observed for S-venlafaxine (Eap et al. 2003). These findings indicate that although CYP2D6 catalyzes the O-demethylation of both enantiomers of venlafaxine, it shows a marked stereoselectivity toward the R-enantiomer.

In a single-dose study with healthy subjects, the dose-corrected AUC of venlafaxine is 2.3-fold higher and that of its active metabolite O-desmethylvenlafaxine is 3.4-fold lower in PMs than in EMs (Lindh et al. 2003). In another study on 46 elderly patients with major depression receiving venlafaxine-XR, it is found that the plasma concentrations of venlafaxine is significantly higher and O-desmethylvenlafaxine concentration per unit dose is significantly lower in patients carrying one or more variant alleles compared to the wild type (Whyte et al. 2006). This study does not find an association between *CYP2D6* genotype and side effects. A study has reported that patients with O-desmethylvenlafaxine/venlafaxine ratios below 0.3 are all identified as PMs, while individuals with ratios above 5.2 are all UMs owing to gene duplications (Shams et al. 2006). Five patients with intermediate metabolic activity are heterozygous for *CYP2D6*4*. In this study, PMs experience more side effects (mainly nausea and vomiting) (Shams et al. 2006).

A steady-state study has reported that the plasma concentration of N-desmethylvenlafaxine is 5.5-fold higher in heterozygous EMs and 22-fold higher in PMs than in EMs (Hermann et al. 2008). This study suggests that there is a shift in the metabolic pathway resulting in substantially higher levels of N-desmethylvenlafaxine in heterozygous EMs and PMs. Decreased CYP2D6 activity owing to genetic mutation or inhibition by quinidine could be associated with cardiovascular toxicity as observed in four patients during treatment with the venlafaxine (Lessard et al. 1999). The side effects reported include palpitations, shortness of breath, and proarrhythmias because of the heterogeneity in cardiac repolarization. Genetic variations of the *CYP2D6* gene may contribute to interpatient variation in response to venlafaxine treatment, and an adverse reaction may indicate that a slow metabolizer and genotyping test should be considered for such patients (McAlpine et al. 2007).

The DPWG Guideline recommends selecting an alternative to venlafaxine (e.g., citalopram, sertraline) or adjusting dose to clinical response and monitoring plasma metabolite (O-desmethyl)venlafaxine level for CYP2D6 PM and IM patients (Swen et al. 2011). For CYP2D6 UMs, it recommends titrating dose to a maximum of 150% of the normal dose when monitoring venlafaxine and increased (O-desmethyl) venlafaxine plasma concentrations or selecting an alternative to venlafaxine.

7.7.3.4 Miscellaneous Antidepressants

In a retrospective study of 28 patients who experienced adverse events and 16 patients that are nonresponsive to a variety of antidepressants, there is an association with the *CYP2D6* genotype (Rau et al. 2004). Of the individuals with adverse reactions to antidepressants (most commonly TCAs), 29% are PMs, whereas 19% (3/16) of nonresponders are UMs (four- and fivefold higher than expected, respectively). A higher rate of *CYP2D6* gene duplication is observed in 108 depressed patients treated with antidepressants metabolized by CYP2D6 who failed (9.9% vs. 0.8%–1% expected) (Kawanishi et al. 2004). In another study with 136 Caucasian

depressed inpatients treated with amitriptyline, citalopram, clomipramine, doxepin, fluvoxamine, mirtazapine, paroxetine, sertraline, and venlafaxine, nonfunctional *CYP2D6* genotype and comedication with inhibitors of CYP2D6 are associated with higher plasma concentrations of antidepressants when normalized to dose; five of the six CYP2D6 PMs experienced side effects (Grasmader et al. 2004b).

A large population-based cohort study of 1198 elderly Dutch patients has examined the influence of the *CYP2D6*4* polymorphism on intolerability of antidepressants (Bijl et al. 2008). The risk of switching to another antidepressant in TCA users is higher in PMs with an adjusted odds ratio of 5.77, but not in SSRI users. Heterozygous patients do not have an increased risk of switching in both TCA and SSRI users. PMs require a lower maintenance dose of TCAs or SSRIs compared with EMs (Bijl et al. 2008).

The *CYP2D6* genotype can be used to predict the plasma concentrations of some TCAs based on a clear gene–concentration effect and SSRIs and prospective genotyping for *CYP2D6* may identify individuals at increased risk of side effects or therapeutic failure with antidepressants. However, although it is clear that toxicity and treatment failures are major issues in psychiatry, most side effects and therapeutic failures are seen in EMs rather than PMs or UMs, indicating the importance of pharmacodynamic factors and other factors affecting the response to antidepressants. Furthermore, although recommendations for dosage adjustment of antidepressant based on the *CYP2D6* genotype have been proposed (Seeringer and Kirchheiner 2008), the effectiveness of this approach has not been validated. Accurate prediction of response on the basis of variants of candidate genes such as *CYP2D6* remains elusive.

7.8 ANTIPSYCHOTICS

7.8.1 ARIPIPRAZOLE

Aripiprazole is a substrate of CYP2D6 and 3A4. Several clinical studies have found that the genotype of *CYP2D6* affects the clearance of aripiprazole (Hendset et al. 2007; Kim et al. 2008; Koue et al. 2007; Kubo et al. 2005, 2007). In psychiatric patients, the median serum concentration of aripiprazole is 1.7-fold higher in PMs than in EMs (45.5 vs. 26.3 nM/mg) and the sum of serum concentration of aripiprazole plus dehydroaripiprazole is 1.5-fold higher in PMs than in EMs (53.9 vs. 37.0 nM/mg) (Hendset et al. 2007). The oral clearance of aripiprazole in healthy Japanese subjects is estimated to be 0.0645 l/h/kg in the group with *CYP2D6*1/*1, *1/*2,* and **2/*2,* but it decreased to 0.0135 l/h/kg in the group with *CYP2D6*1/*5, *1/*10, *2/*5, *2/*10,* and **2/*41* and 0.0293 l/h/kg in the group with *CYP2D6*5/*10, *10/*10,* and **41/*41* (Koue et al. 2007). There is a case report where very high plasma levels of aripiprazole are noted in a patient who had nonfunctional CYP2D6 (Oosterhuis et al. 2007). The systemic exposure to dehydroaripiprazole is lower by 30% in IMs than in EMs (Kubo et al. 2005) as well as 30% lower in PMs than in EMs, whereas the median concentration/dose ratios of dehydroaripiprazole are similar in PMs and EMs (Hendset et al. 2007). However, only three healthy subjects with the IM phenotype completed the single-dosing study by Hendset et al. (2007), and the AUC of dehydroaripiprazole is higher in IMs than in EMs in the multiple-dosing study conducted by the same group (Kubo et al. 2007), reflecting the limited role of CYP2D6 in the formation of dehydroaripiprazole.

A study has investigated the relationship of the *CYP2D6* genotype and aripiprazole pharmacokinetics in 80 schizophrenic patients receiving multiple oral doses of aripiprazole (10–30 mg/day) (Kim et al. 2008). Covariate analysis shows that *CYP2D6* genetic polymorphisms significantly influence the oral clearance of aripiprazole and reduce the interindividual variability of oral clearance from 37.8% coefficient of variation to 30.5%. The oral clearance of aripiprazole in IMs (*n* = 27) carrying at least one partially deficient allele is ~60% of that in EMs (*n* = 53) having at least one functional allele (*1* or *2*) (Kim et al. 2008). On the basis of the *CYP2D6* genotype, the MRs are calculated at 0.20–0.34. However, the plasma concentration/dose ratio of dehydroaripiprazole is not affected by the *CYP2D6* genotype (Kim et al. 2008).

The plasma concentration of aripiprazole is increased by coadministered itraconazole, and the decrease in oral clearance is estimated to be 0.0181 l/h/kg. By coadministration of itraconazole, the oral clearance of aripiprazole in EMs is decreased by 26.6%, with an even greater decrease (47.3%) in IMs (Kubo et al. 2005). Olanzapine, alimemazine, lithium, risperidone injections, escitalopram, or lamotrigine had significant effects on aripiprazole disposition in psychiatric patients (Waade et al. 2009).

These results indicate a role of CYP2D6 in its metabolism but CYP3A4 may be more important for its disposition and the CYP2D6 phenotype marginally affects the clearance of aripiprazole. There is no need to adjust the dosage since its major metabolite dehydroaripiprazole is pharmacologically active. The FDA-approved drug label for aripiprazole contains information regarding dose adjustment in patients who are CYP2D6 PMs or in patients taking concomitant drugs that may inhibit or induce CYP2D6 or 3A4. The aripiprazole dose in PM patients should initially be reduced to one-half (50%) of the usual dose and then adjusted to achieve a favorable clinical response. The dose of aripiprazole for PM patients who are administered a strong CYP3A4 inhibitor should be reduced to one-quarter (25%) of the usual dose. Strong CYP3A4 (e.g., ketoconazole) or CYP2D6 (e.g., fluoxetine) inhibitors will increase the plasma concentrations of aripiprazole; one should reduce the dose of aripiprazole to one-half of the usual dose when used concomitantly, except when used as adjunctive treatment with antidepressants. CYP3A4 inducers (e.g., carbamazepine) will decrease drug concentrations of aripiprazole and thus one should double the dose of aripiprazole when used concomitantly.

The DPWG recommends reducing the maximum dose of aripiprazole for patients carrying PM alleles of CYP2D6 to 10 mg/day (67% of the maximum recommended daily dose). These PMs may carry PM two inactive (*3*–*8*, *11*–*16*, *19*–*21*, *38*, *40*, and *42*) alleles.

7.8.2 CHLORPROMAZINE

Chlorpromazine is converted to active 7-hydroxychlorpromazine largely by CYP2D6 (Yoshii et al. 2000) and this pathway is inhibited by quinidine in EMs of debrisoquine in vivo (Muralidharan et al. 1996). In healthy Korean volunteers, nonsignificant 1.3- and 1.7-fold higher chlorpromazine AUC values are observed in individuals heterozygous and homozygous for *CYP2D6*10*, respectively, compared to the wild type (Sunwoo et al. 2004). It appears that *CYP2D6*10* does not significantly alter the pharmacokinetics of chlorpromazine, which is metabolized not only by CYP2D6 but also by CYP1A2 and 4A11 (Yoshii et al. 2000).

7.8.3 HALOPERIDOL

Haloperidol is a substrate of CYP3A4 and 2D6 (Kudo and Odomi 1998; Pan et al. 1998; Tateishi et al. 2000). In schizophrenic patients, fluoxetine (a CYP2D6/1A2 inhibitor) significantly increases the plasma levels of haloperidol (Avenoso et al. 1997). Fluvoxamine (a CYP2D6/1A2 inhibitor) also causes a moderate increase in the plasma levels of haloperidol and its reduced metabolite in patients (Vandel et al. 1995). These results indicate that CYP2D6 plays a partial role for the disposition of haloperidol but CYP3A4 is also important for its overall disposition. There is a large interindividual variability in the pharmacokinetics and the clinical outcome of haloperidol treatment. There is a significant correlation between dextromethorphan MRs and plasma haloperidol concentrations, reduced haloperidol concentrations, and reduced haloperidol/haloperidol ratios (Lane et al. 1997). A single-dose (2–4 mg) study on healthy Caucasian volunteers has reported that PMs eliminate haloperidol twofold slower than EMs (clearance: 1.16 vs. 2.49 l/h/kg), with the mean plasma $t_{1/2\beta}$ being longer in PMs (29.4 vs. 16.3 h) (Llerena et al. 1992a). The plasma levels of reduced haloperidol from 10 to 72 h postdose are two- to fourfold higher in PMs than in EMs after a single 2- or 4-mg dose of haloperidol (Llerena et al. 1992b). A study of eight Caucasian patients shows the highest plasma concentration of haloperidol and highest dopamine D_2 receptor occupancy in the sole PM than in EMs, indicating higher risk of extrapyramidal symptoms (Nyberg et al. 1995b). This study is conducted using relatively low doses of haloperidol (30–50 mg/4 weeks). CYP2D6 appears important for the metabolism of haloperidol at low doses. The formation of reduced haloperidol from haloperidol seems to be independent of CYP2D6 activity and there is a decreased reoxidation of the reduced metabolite to haloperidol by CYP2D6.

There is a significant correlation between reduced haloperidol trough levels and haloperidol total clearance and the number of active *CYP2D6* genes in a study with 172 German psychiatric patients (Brockmoller et al. 2002). This study has also found that the rating for pseudo-parkinsonism induced by haloperidol is higher in PMs than in EMs; there is a trend toward lower therapeutic efficacy with increasing number of active *CYP2D6* genes. Another study of 26 Swedish schizophrenic patients has found a significant correlation between haloperidol plasma concentration and number of active CYP2D6 alleles (Panagiotidis et al. 2007). However, there is no correlation between plasma concentration of haloperidol or number of CYP2D6 alleles and therapeutic outcome or side effects (Panagiotidis et al. 2007).

Two studies in Japanese schizophrenic patients treated with haloperidol at a fixed daily dose of 12 mg have shown a relationship between increased steady-state haloperidol plasma concentrations and the presence of the nonfunctional CYP2D6*10 allele (Mihara et al. 1999; Suzuki et al. 1997). In one study with 67 Japanese inpatients with schizophrenia, the mean haloperidol concentrations in the patients with no, one, and two *10 alleles are 22.8, 30.1, and 31.2 nmol/l, respectively, and those values for reduced haloperidol are 6.1, 9.5, and 9.9 nmol/L, respectively (Mihara et al. 1999). The mean haloperidol concentrations is significantly higher in the patients with one *10 allele than in those with no *10 alleles, and the mean reduced haloperidol levels are significantly higher in the patients with one and two *10 alleles than in those with no *10 alleles.

In a study with 120 Korean schizophrenic patients (Roh et al. 2001a), a relationship between haloperidol concentrations normalized for dose of haloperidol and CYP2D6 genotype (CYP2D6*1/*1, *1/*10, and *10/*10) is found in patients receiving haloperidol at <20 mg daily, but not in patients receiving higher doses (>20 mg). In this study, 60% of individuals with the CYP2D6*1/*10 or *10/*10 genotypes (n = 93) require benztropine (an anticholinergic drug used to reduce the side effects of haloperidol) compared with 35% of EMs (n = 23). All four patients with a *5 allele (one together with *1 and three with *10) are found to use benztropine. However, there are no significant differences between the genotype groups with respect to the levels of reduced haloperidol. In patients with doses of haloperidol higher than 20 mg, no differences are found between the genotype groups for either haloperidol or reduced haloperidol (Roh et al. 2001a). These findings demonstrate a clear correlation between dose-corrected steady-state plasma concentrations of haloperidol, but not of reduced haloperidol, and the CYP2D6*1/*1, *1/*10, and *10/*10 genotype groups when doses are <20 mg, suggesting the involvement of CYP2D6 in the metabolism of haloperidol at low doses of haloperidol, while another enzyme, probably CYP3A4, contributes to its metabolism at higher doses.

The healthy Korean subjects with the CYP2D6*10/*10 genotype show 81% higher AUC of haloperidol compared to that of subjects with the CYP2D6*1/*1 genotype (27.6 vs. 50.2 ng·h/ml) (Park et al. 2006). When cotreated with itraconazole (a CYP3A4 inhibitor), subjects with CYP2D6*10/*10 show threefold higher AUC of haloperidol compared to that of placebo-pretreated subjects with the CYP2D6*1/*1 genotype (21.7 vs. 66.7 ng·h/ml). The CYP2D6*10 genotype and itraconazole pretreatment decrease the oral clearance of haloperidol by 24% and 25%, respectively, but without a statistical significance. However, in the subjects with both CYP2D6*10 genotype and itraconazole pretreatment, the oral clearance is significantly decreased to 42% of subjects with wild-type

genotype receiving placebo pretreatment (4.7 vs. 2.0 l/h/kg) (Park et al. 2006). Although the CYP2D6*10 genotype and itraconazole pretreatment cause higher value of haloperidol-induced side effect and scores for its therapeutic effect, it does not reach statistical significance. These results indicate that the moderate effect of the CYP2D6*10 genotype on the pharmacokinetics and pharmacodynamics of haloperidol can be augmented by the presence of itraconazole pretreatment.

A steady-state study on Chinese schizophrenic patients (n = 18) treated with oral haloperidol 10 mg/day for 2 weeks has found significant correlations of dextromethorphan MRs and plasma haloperidol concentrations, reduced haloperidol concentrations, and reduced haloperidol/haloperidol ratios (Lane et al. 1997). The authors have also reported higher reduced haloperidol concentrations and reduced haloperidol/haloperidol ratios in patients who experience more extrapyramidal side effects than other patients without these side effects (Lane et al. 1997).

However, several other studies have reported that steady-state haloperidol concentrations tend to be slightly higher in those with CYP2D6 alleles (e.g., *10 in Japanese and *4 in Caucasian) causing reduced or nonfunctional enzyme activity, or even there is no significant difference (Ohnuma et al. 2003; Pan et al. 1999; Shimoda et al. 2000b). Reduced haloperidol concentrations are more consistently increased in those with nonfunctional CYP2D6 alleles (e.g., *10 in Japanese) causing reduced or nonfunctional enzyme activity in schizophrenic patients (Mihara et al. 1999; Pan et al. 1999; Shimoda et al. 2000b; Suzuki et al. 1997). For example, a study on Japanese patients does not observe any significant difference in the plasma haloperidol levels between the subjects with no, one, and two *10 alleles (Shimoda et al. 2000b). Someya et al. (1999) also do not find an association of the CYP2D6*10A allele with plasma concentrations of haloperidol. A study of 111 Japanese patients has found no significant difference of plasma concentration of haloperidol normalized by dose between the groups classified by CYP2D6*10A and *2 genotypes, even in those patients whose daily doses are <20 mg (n = 90) (Ohnuma et al. 2003). Patients carrying duplicated CYP2D6 genes (n = 6) do not show significant difference of plasma concentrations of haloperidol compared with subjects who had no duplicated genes (Ohnuma et al. 2003). Patients with gene duplication show slightly higher daily doses of haloperidol (16 mg; range: 3–30 mg) than those without duplication (10 mg; range: 1–45), but this does not reach statistical significance.

These conflicting results suggest that CYP2D6 plays a role in the disposition of haloperidol and reduced haloperidol in Asians but other enzymes (e.g., CYP1A2) may be more important in determining the overall metabolism of haloperidol. The decrease of CYP2D6 enzyme activity caused by CYP2D6*10A might not be sufficient to affect the clinical plasma concentration of haloperidol while the nonfunctional *4 allele may cause significant change in plasma levels of haloperidol. The high-affinity, low-capacity CYP2D6 appears to play an important role at low concentrations/doses of haloperidol, while the low-affinity, high-capacity CYP3A4/1A2 is more important at higher doses of haloperidol and long-term maintenance treatment. The influence of decreased enzyme activity by *CYP2D6*10A* allele on the plasma concentration of haloperidol is yet to be confirmed by studies of larger cohorts of patients. The complexity of haloperidol pharmacokinetics compromises the clinical value of knowledge of *CYP2D6* genotype in the determination of dosage and regimen. Further studies are warranted to explore the clinical impact of *CYP2D6* polymorphisms on haloperidol therapy.

The DPWG Guideline recommends reducing haloperidol dose by 50% or selecting an alternative drug (e.g., pimozide, flupenthixol, fluphenazine, quetiapine, olanzapine, and clozapine) for CYP2D6 PM genotype patients who carry **3–*8*, **11–*16*, **19–*21*, **38*, **40*, or **42* (Swen et al. 2011). For CYP2D6 UMs, it recommends being alert to decreased haloperidol plasma concentration and adjusting maintenance dose in response to haloperidol plasma concentration or selecting an alternative drug (e.g., pimozide, flupenthixol, fluphenazine, quetiapine, olanzapine, and clozapine).

7.8.4 PERPHENAZINE

Perphenazine is metabolized by CYP2D6, 1A2, 3A4, and 2C19 (Olesen and Linnet 2000). The disposition of perphenazine cosegregates with CYP2D6-mediated debrisoquine hydroxylation (Bertilsson et al. 1993a). A fourfold higher AUC of perphenazine in PMs after a single dose is observed compared with EMs of CYP2D6 (Dahl-Puustinen et al. 1989). Similarly, steady-state studies demonstrate a twofold higher median AUC (Linnet and Wiborg 1996b) and a threefold decrease in the clearance of perphenazine (Jerling et al. 1996) in PMs. A study on nonsmoking healthy male Chinese–Canadian volunteers has reported a 2.9-fold higher AUC of perphenazine in carriers of homozygous *CYP2D6*10* than those carrying the *CYP2D6*1* allele (Ozdemir et al. 2007). In this study, individuals homozygous for *CYP2D6*10* are found to have a significantly reduced prolactin production and tissue response. The association of prolactin response to perphenazine with the *CYP2D6* genotype is also observed in another study on Swedish patients (Aklillu et al. 2007). In contrast to hyperprolactinemia, which is a common side effect of first-generation antipsychotics caused by antagonism of dopaminergic neurotransmission in the pituitary (Bushe et al. 2008; Haddad and Wieck 2004; Molitch 2008; O'Keane 2008; Peveler et al. 2008), prolactin response appears to be blunted in subjects homozygous for *CYP2D6*10*. A possible explanation is that *CYP2D6* genetic polymorphism may potentially influence pharmacodynamic tissue sensitivity in the pituitary, presumably through disposition of an endogenous substrate (e.g., 5-methoxytryptamine, which can be converted to 5-HT by CYP2D6 in the brain (Yu et al. 2003).

Paroxetine, a potent CYP2D6 inhibitor, increased the AUC of perphenazine sevenfold in EMs, which is associated with a significant increase in CNS symptoms including sedation, extrapyramidal symptoms, and psychomotor performance (Ozdemir et al. 1997). In elderly patients with dementia, perphenazine 0.05 to 0.1 mg/kg/day leads to improved psychotic symptoms overall without any difference between EMs and PMs. However, PMs of CYP2D6 have significantly more side effects (primarily extrapyramidal and sedation) early in treatment that become similar in both phenotypes by day 17 of dosing (Pollock et al. 1995).

The FDA-approved drug label for perphenazine states that CYP2D6 is involved in the metabolism of perphenazine. CTP2D6 PMs demonstrate higher plasma concentrations of antipsychotic drugs at usual doses, which may correlate with emergence of side effects. Prospective phenotyping of elderly patients before antipsychotic treatment may identify those at risk for adverse events. The concomitant administration of other drugs that inhibit the activity of CYP2D6 may acutely increase plasma concentrations of antipsychotics. Among these are TCAs and SSRIs, for example, fluoxetine, sertraline, and paroxetine. When prescribing these drugs to patients already receiving antipsychotic therapy, close monitoring is essential and dose reduction may become necessary to avoid toxicity. Lower doses than usually prescribed for either the antipsychotic or the other drug may be required.

7.8.5 RISPERIDONE

Risperidone is mainly metabolized by CYP2D6 and 3A4 (Yasui-Furukori et al. 2001). Unchanged risperidone is mainly excreted in the urine and accounted for 30%, 11%, and 4% of the administered dose in PMs, IMs, and EMs, respectively. 9-Hydroxyrisperidone accounts for 8%, 22%, and 32% of the administered dose in the urine in PMs, IMs, and EMs, respectively. After 4 weeks of paroxetine (a CYP2D6 inhibitor) treatment, the sum of the concentrations of risperidone and 9-OH-risperidone increases significantly by 45% over baseline (Spina et al. 2001). Another study of 218 schizophrenic patients has found that there are higher plasma levels of risperidone but without any change in the plasma level of 9-OH-risperidone when one or more CYP2D6 inhibitor is coadministered (Mannheimer et al. 2008). A possible explanation would be that 9-hydroxylation represents the major pathway of elimination of risperidone and similar levels of 9-OH-metabolite are found at steady state with and without reduced CYP2D6 activity. Since the pharmacological difference and distribution in the CNS between the parent compound and the 9-OH-metabolite are unclear, the clinical relevance of the *CYP2D6* polymorphism is yet to be determined.

The ratio of risperidone to 9-hydroxyrisperidone concentrations is <1 in EMs and >1 in PMs among various ethnic groups, but the sum of active moieties is comparable in EMs and PMs (Bondolfi et al. 2002; Nyberg et al. 1995a; Olesen et al. 1998; Roh et al. 2001b; Scordo et al. 1999). On the extreme, UMs have very low ratios of risperidone to 9-hydroxyrisperidone (Guzey et al. 2000; Rau et al. 2004). The study on 37 Italy schizophrenic patients has revealed that the median steady-state plasma concentrations of risperidone normalized for dose are 0.6, 1.1, 9.7, and 17.4 nmol/l/mg in UM ($n = 3$), homozygous EMs ($n = 16$), heterozygous EMs ($n = 15$), and PMs ($n = 3$), respectively, with statistically significant differences between PM and the other genotypes (Scordo et al. 1999). The concentration of 9-OH-risperidone does not correlate with the

CYP2D6 genotype. The risperidone/9-OH-risperidone ratio is associated with the CYP2D6 genotype, with the highest ratios in PMs (median 0.79). Heterozygous EMs also have significantly higher ratios than homozygous EMs (median value, 0.23 vs. 0.04) or UMs (median, 0.03) (Scordo et al. 1999). No significant differences are found in the sum of the plasma concentrations of risperidone and 9-OH-risperidone between the genotype groups. In the study of 82 Korean schizophrenic patients, the median concentrations of risperidone normalized for dose in CYP2D6*1/*1, *1/*10, and *10/*10 groups are 1.7, 2.6, and 6.7 nM/mg, respectively (Roh et al. 2001b). For 9-hydroxyriperidone, the corresponding median concentrations are 13.1, 11.9, and 13.6 nM/mg, respectively, with no significant difference between the genotypes. The medians of the ratios between risperidone and 9-hydroxyrisperidone concentrations are 0.13, 0.28, and 0.46 nM/mg in *1/*1, *1/*10, and *10/*10 genotypes, respectively ($P < 0.05$).

The CYP2D6 phenotype status also affects the extent of drug interactions when a CYP2D6 inhibitor is coadministered with risperidone. The AUC of risperidone increases from 83.1 and 398.3 ng·h/ml (monotherapy) to 345.1 and 514.0 ng·h/ml when coadministered with fluoxetine in EMs and PMs, respectively (Bondolfi et al. 2002). The AUC of the active moiety (risperidone plus 9-hydroxy-risperidone) increases from 470.0 to 663.0 ng·h/ml in EMs and from 576.3 to 788.0 ng·h/ml in PMs. In EMs, the AUC of 9-hydroxy-risperidone remains similar (monotherapy vs. combination therapy: 386.8 vs. 317.7 ng·h/ml), whereas it significantly increases in PMs (178.3 vs. 274.0 ng·h/ml).

A few studies have investigated whether CYP2D6 phenotype status is associated with the adverse reactions of risperidone. A large ($n = 500$) cross-sectional study on schizophrenic patients in southern Germany has reported a threefold increased risk of moderate to severe adverse effects in PMs than EMs (Rau et al. 2004). However, PMs comprise only a small proportion (16%) of all patients with adverse effects and only 9% of PMs discontinue risperidone because of side effects in this study. Another study on schizophrenic patients in the United States has revealed a 3.1-fold higher risk of moderate to severe adverse effects in PMs than EMs (de Leon et al. 2005). There is no correlation between the serum concentration of the active moiety and the side effects of risperidone (Olesen et al. 1998).

The above data demonstrate a clear gene–concentration effect for risperidone owing to the major role of CYP2D6 in its metabolism and activation. However, this is of minor clinical significance as there is a comparable sum of risperidone and 9-OH-risperidone in PMs and EMs and both parent and primary active metabolite formed by CYP2D6 show similar activity. The DPWG Guideline recommends selecting an alternative drug or being extra alert to adverse drug events for patients who are CYP2D6 PMs (e.g., quetiapine, olanzapine, and clozapine), IMs (e.g., quetiapine, olanzapine, and clozapine), or UMs (e.g., quetiapine, olanzapine, and clozapine) with risperidone (Swen et al. 2011). Adjusting risperidone dose to clinical response is also recommended. The FDA-approved drug label for risperidone states that risperidone is metabolized to 9-hydroxyrisperidone by CYP2D6; coadministration of drugs that are inhibitors or inducers of CYP2D6 along with risperidone may alter the plasma concentration of risperidone. Genotyping/phenotyping of CYP2D6 is helpful in identifying individuals at increased risk of toxicity but unlikely to sort out responders and nonresponders. In UMs, a dosage increase may be required to achieve therapeutically relevant concentrations.

7.8.6 THIORIDAZINE

Thioridazine is used to treat schizophrenia but has been associated with torsades de pointes and sudden death. Because of the potentially fatal side effects, it is typically reserved for patients who do not respond well to other antipsychotics. CYP2D6 and 3A4 convert thioridazine to mesoridazine (Wojcikowski et al. 2006), which correlates weakly with the debrisoquine MR (Berecz et al. 2003; Llerena et al. 2000). There is a weak correlation between corrected QT interval and thioridazine plasma concentrations, debrisoquine MR, and thioridazine/mesoridazine concentrations (Llerena et al. 2002). The metabolites seem to have activity equal to (sulforidazine) or greater (mesoridazine) than that of the parent, whereas a further ring sulfoxide (thioridazine 5-sulfoxide) produced from thioridazine may be less active as an antipsychotic, but more arrhythmogenic. After a single dose, the sum of active moieties (thioridazine, mesoridazine plus sulforidazine) is ~1.4-fold higher in PMs, largely attributed to a 4.5-fold increase in thioridazine itself (von Bahr et al. 1991). Consistent with this, the dose-corrected median steady-state plasma thioridazine concentrations are 3.8- and 1.8-fold higher in the presence of no or one active CYP2D6 allele, respectively, compared with two active alleles. Median concentrations of mesoridazine and sulforidazine are not different (Berecz et al. 2003).

The FDA-approved drug label for thioridazine states that thioridazine is contraindicated in patients with reduced CYP2D6 activity. Certain circumstances may increase the risk of torsades de pointes or sudden death in association with the use of drugs that prolong the QTc interval, including (1) bradycardia, (2) hypokalemia, (3) concomitant use of other drugs that prolong the QTc interval, (4) presence of congenital prolongation of the QT interval, and (5) for thioridazine in particular, its use in patients with reduced activity of CYP2D6 or its coadministration with drugs that may inhibit CYP2D6 or by some other mechanism that interfere with the clearance of thioridazine.

7.8.7 ZUCLOPENTHIXOL

Zuclopenthixol is mainly metabolized by CYP2D6 and other enzymes (Dahl et al. 1991). The clearance of zuclopenthixol cosegregates with debrisoquine hydroxylation in humans (Bertilsson et al. 1993a; Dahl et al. 1991), indicating the involvement of CYP2D6 in the metabolism of zuclopenthixol. A 1.9-fold higher AUC of zuclopenthixol is observed in PMs of healthy volunteers compared to the EMs ($n = 6$ for each phenotype) after a single-dose of zuclopenthixol (6 or 10 mg) (Dahl et al. 1991). The plasma $t_{1/2\beta}$ is significantly longer in PMs than in EMs (29.9 vs. 17.6 h) and the total oral plasma clearance is lower in PMs than in EMs (0.78 vs. 2.12 1/h/kg) (Dahl et al. 1991).

The median steady-state plasma concentration of zuclopenthixol normalized for dose in 12 psychiatric patients who are PMs is 60% higher than in EMs (2.0 vs. 1.25 nmol/l/mg) but is close to that of those EMs receiving potentially interacting drugs that inhibit CYP2D6 (2.0 vs. 1.80 nmol/l/mg) (Linnet and Wiborg 1996a). Another study of 52 schizophrenic patients on maintenance dosage of zuclopenthixol at 100–400 mg/4 weeks has found that the median steady-state plasma concentrations of zuclopenthixol are 1.6- and 1.4-fold higher in PMs ($n = 4$) and heterozygous EMs ($n = 13$), respectively, than in homozygous EMs ($n = 35$) (Jaanson et al. 2002). The nonfunctional CYP2D6*3 and *4 alleles tend to be more frequent in patients with neurological side effects. There is an odds ratio of 2.3 for development of parkinsonism and of 1.7 for tardive dyskinesia in an individual with at least one nonfunctional CYP2D6 allele but is not statistically significant (Jaanson et al. 2002). In addition, a steady-state study of 36 Swedish schizophrenic patients has reported that the clearance of zuclopenthixol is 2.2- and 1.5-fold higher in homozygous and heterozygous EMs, respectively, than in PMs (Jerling et al. 1996). These findings demonstrate a gene–concentration effect for zuclopenthixol but the evidence for gene–response relationship is lacking. Genotyping/phenotyping testing may be used to predict the plasma levels of zuclopenthixol.

The DPWG Guideline for zuclopenthixol recommends reducing zuclopenthixol dose or selecting an alternative drug (e.g., flupenthixol, quetiapine, olanzapine, or clozapine) for CYP2D6 PM (reduce the dose by 50%) and IM (reduce the dose by 25%) patients (Swen et al. 2011). For UMs, it recommends being alert to low zuclopenthixol plasma concentrations or selecting an alternative drug (e.g., flupenthixol, quetiapine, olanzapine, or clozapine).

7.8.8 Miscellaneous and Atypical Antipsychotics

7.8.8.1 Clozapine

Clozapine is metabolized by CYP1A2, 3A4, 2C9, 2C19, and 2D6 (Olesen and Linnet 2001). It can be expected that CYP2D6 phenotype status would not significantly change the kinetics of clozapine. In vivo studies have supported a role for CYP1A2 (Bertilsson et al. 1994), but no association has been found for metabolizer status with regard to debrisoquine (CYP2D6) or S-mephenytoin (CYP2C19) (Dahl et al. 1994a). Clozapine-induced agranulocytosis is not associated with polymorphisms of CYP2D6 and myeloperoxidase (Dettling et al. 2000). In addition, clozapine and N-desmethylclozapine concentration ratios are not related to the CYP2D6 genotype in patients (Melkersson et al. 2007).

The FDA-approved drug label for clozapine recommends using caution when administering it concomitantly with drugs that are strong CYP1A2 inhibitors (e.g., fluvoxamine, ciprofloxacin, or enoxacin), moderate or weak CYP1A2 inhibitors (e.g., oral contraceptives or caffeine), CYP2D6 or CYP3A4 inhibitors (e.g., cimetidine, escitalopram, erythromycin, paroxetine, bupropion, fluoxetine, quinidine, duloxetine, terbinafine, or sertraline), CYP3A4 inducers (e.g., phenytoin, carbamazepine, St. John's wort, and rifampin), or CYP1A2 inducers (e.g., tobacco smoking). Additionally, it may be necessary to reduce the clozapine dose in patients with significant renal or hepatic impairment or in CYP2D6 PMs. Concomitant use of clozapine with other drugs metabolized by CYP2D6 can increase levels of these CYPD26 substrates. Caution is needed when coadministering clozapine with other drugs that are metabolized by CYP2D6. It may be necessary to use lower doses of such drugs than usually prescribed. Such drugs include specific antidepressants, phenothiazines, carbamazepine, and type 1C antiarrhythmics (e.g., propafenone, flecainide, and encainide).

7.8.8.2 Olanzapine

Olanzapine is a substrate of CYP1A2 and 2D6 (Ring et al. 1996). The CYP2D6 phenotype does not correlate with the AUC of olanzapine after a single dose of 7.5 mg in healthy volunteers (Hagg et al. 2001). A similar negative result is observed at steady state in psychiatric patients (Carrillo et al. 2003). It is expected that CYP2D6 phenotype status would not significantly affect the kinetics of olanzapine.

7.8.8.3 Miscellaneous Antipsychotics

A study of 131 Slovenian schizophrenic patients on maintenance therapy with haloperidol, fluphenazine, zuclopenthixol, or risperidone has reported no significant differences in psychopathological and extrapyramidal symptoms with regard to the CYP2D6 genotype except for that PMs scored significantly higher on the negative subscale for psychopathological symptoms (Plesnicar et al. 2006). However, a pilot study of 100 psychiatric inpatients shows a trend for increasing adverse reactions with CYP2D6 substrate drugs (e.g., haloperidol, perphenazine, risperidone, and TCAs) from the UM to PM status and a higher cost of treating these two groups (Chou et al. 2000). It appears that the CYP2D6 genotype might be a factor contributing to the persistent negative symptoms of schizophrenia but not a major factor that determines the susceptibility to antipsychotic-induced extrapyramidal side effects in patients on maintenance therapy with antipsychotics that are mainly CYP2D6 substrates.

All the above results indicate significant relationships between the CYP2D6 genotype and steady-state concentrations for perphenazine, zuclopenthixol, risperidone, and haloperidol (Dahl 2002; Otani and Aoshima 2000). Other CYPs, especially CYP3A4 and 1A2, also contribute to the interindividual variability in the clearance of many antipsychotics and thus reduce such relationships. For most antipsychotics, the relative contributions of the different CYPs at therapeutic drug concentrations remain to be determined. PMs of CYP2D6 appear to be more prone to oversedation and possibly parkinsonism during treatment with classical antipsychotics, but the relationship is not conclusive.

In addition, results of the relationships between the *CYP2D6* genotype and parkinsonism or tardive dyskinesia with traditional antipsychotics are conflicting, probably attributed to small sample size, inclusion of antipsychotics with variable CYP2D6 metabolism, and comedication. A meta-analysis shows a significant 1.4-fold increased risk of tardive dyskinesia in PMs (Patsopoulos et al. 2005). On the basis of these available data, phenotyping and genotyping for CYP2D6 may be used as a complement to plasma drug concentration determination when aberrant metabolic capacity (PM or UM) of CYP2D6 substrates is suspected. CYP2D6 phenotyping and genotyping appear to be useful in predicting the steady-state concentrations of some classical antipsychotic drugs, but their usefulness in predicting clinical effects must be explored.

Therapeutic drug monitoring has been strongly recommended for many antipsychotics including haloperidol, chlorpromazine, fluphenazine, perphenazine, risperidone, and thioridazine which are all metabolized by CYP2D6 (Sjoqvist and Eliasson 2007). Clozapine and olanzapine are also included in this list, although CYP2D6 plays a minor role for their overall metabolic elimination. It is possible to merge both therapeutic drug monitoring and pharmacogenetic testing for *CYP2D6* in clinical practice.

7.9 CENTRALLY ACTING CHOLINESTERASE INHIBITORS

7.9.1 DONEPEZIL

Donepezil is a substrate of CYP3A4, 2C9, and 2D6 (Barner and Gray 1998). A clinical steady-state study of 42 patients of Caucasian ethnicity from Italy has demonstrated that UM patients ($n = 2$) have lower plasma concentrations of donepezil than EM patients ($n = 40$) and showed no clinical improvement (Varsaldi et al. 2006). Heterozygous EMs have a higher donepezil concentrations and a better clinical response than homozygous EMs. The median concentration corrected for the dose and body weight of donepezil in UMs, homozygous Ems, and heterozygous EMs is 0.13, 0.33, and 0.41 ng/ml/mg/kg, respectively, but the differences do not reach statistical significance. No PMs are found in this study. In patients treated with a multifactorial therapy, including cholinesterase inhibitors (e.g., donepezil), the best responders are the CYP2D6-related EMs and IMs, and the poorest responders are PMs and UMs (Cacabelos et al. 2007). These results suggest that the *CYP2D6* polymorphism affects donepezil metabolism and therapeutic outcome. The PM phenotype could predispose patients to higher-than-average plasma concentrations of donepezil, with an increased risk of adverse drug reactions, leading to early discontinuation of treatment. Knowledge of a patient's *CYP2D6* genotype/phenotype together with donepezil concentration measurements might be useful in the context of improving the clinical efficacy of donepezil treatment.

7.9.2 GALANTAMINE

Galantamine is a substrate of CYP2D6 and 3A4 (Bachus et al. 1999). In a study with four healthy subjects, the pharmacokinetic parameters of galantamine in PMs are similar to those seen in EMs after a single oral dose of galantamine (Mannens et al. 2002). However, there is a marked difference in the metabolism of galantamine between PMs and EMs. In EMs, six metabolites resulting from *O*-demethylation (metabolites 2, 3, 6 20, 21, and 23) represent more than 33% of the dose, whereas these metabolites amount to only 5% of the dose in PMs (Mannens et al. 2002). The lower level of excretion of metabolites formed by *O*-demethylation in PMs is compensated primarily by higher levels of unchanged galantamine and the *N*-oxide of galantamine (M10) and to a lesser extent by higher levels of the glucuronide of unchanged galantamine (M5), *N*-desmethylgalantamine (M8), *N*-desmethyl-epigalantamine (M16), the *N*-oxide of epigalantamine (M17), and epigalantamine (M13). The glucuronide of *O*-desmethylgalantamine represents up to 19% of the plasma radioactivity in EMs but could not be detected in PMs (Mannens et al. 2002). This difference is not thought to be of minor clinically relevance since it neither influences the plasma concentrations of galantamine or any pharmacologically active metabolites nor affects the rate of excretion of galantamine and its metabolites in urine and feces. These results do not demonstrate a gene–dose/concentration effect on galantamine and thus there is no need to adjust the dosage based on the CYP2D6 phenotype. The results support an important role of CYP2D6 in *O*-demethylation of galantamine.

The FDA-approved drug label for galantamine notes that population pharmacokinetic analysis indicates that there is a 25% decrease in median clearance in CYP2D6 PMs compared to EMs. Dosage adjustment is not necessary in patients identified as cyp2d6 PMs as the dose of the drug is individually titrated to tolerability.

7.10 DRUGS FOR THE TREATMENT OF ATTENTION-DEFICIT/ HYPERACTIVITY DISORDER

7.10.1 ATOMOXETINE

Atomoxetine is a substrate of CYP2C19, 3A4, 2D6, 1A2, and 2A6 (Ring et al. 2002). Liver microsomes from EMs have a CL_{int} value of 103 µl/min/mg for the formation of 4-hydroxyatomoxetine; however, microsomes from PMs exhibit a CL_{int} value of 0.2 µl/min/mg only (Ring et al. 2002). This has provided an explanation for the bimodal distribution of the clearance of atomoxetine in healthy volunteers after single and multiple dosing (Farid et al. 1985). This study has also found that 4-hydroxyatomoxetine is the major oxidative metabolite in both PMs and EMs, but its formation is greatly decreased in PMs. Individuals lacking CYP2D6 activity have a slower clearance that results in higher steady-state plasma concentrations of atomoxetine and *N*-desmethylatomoxetine compared with EMs. Paroxetine (a CYP2D6 inhibitor) increases the plasma levels and AUCs of atomoxetine 3- and 6.5-fold, respectively (Paulzen et al. 2008). This is accompanied with increased *N*-desmethylatomoxetine and decreased

4-hydroxyatomoxetine concentrations. Healthy EMs receiving paroxetine have greater increases in heart rate than those receiving atomoxetine alone.

Sauer et al. (2003) have investigated the effect of the CYP2D6 polymorphisms on the overall disposition of a 20-mg dose of atomoxetine in PMs ($n = 3$) and EMs ($n = 4$) at steady state. The biotransformation of atomoxetine is similar in either PMs or EMs undergoing aromatic ring hydroxylation, benzylic oxidation, and N-demethylation with no CYP2D6 phenotype-specific metabolites. The primary oxidative metabolite of atomoxetine is 4-hydroxyatomoxetine, which is later conjugated to 4-hydroxyatomoxetine-O-glucuronide and excreted into the urine and feces. The AUC of atomoxetine is fourfold higher in PMs than in EMs and mean $t_{1/2\beta}$ is longer (18 vs. 62 h) (Sauer et al. 2003). The C_{max} of atomoxetine is almost sixfold higher in PMs than in EMs. The oral clearance for PMs is approximately 25% of that of EMs (0.0357 vs. 0.373 l/h/kg). Only EMs have measurable 4-hydroxyatomoxetine concentrations. The plasma C_{max} for N-desmethylatomoxetine is almost 40-fold higher in PMs compared with EMs. The $t_{1/2\beta}$ of N-desmethylatomoxetine is ~9 h in EMs and 33 h in PMs. The mean $t_{1/2\beta}$ of 4-hydroxyatomoxetine-O-glucuronide is approximately 7 h in EMs and 19 h in PMs (Sauer et al. 2003). The amount from N-desmethylatomoxetine- and 2-hydroxymethylatomoxetine-derived metabolites is greater in the PMs (22% of the dose) than in EMs (3% of the dose). In EMs, atomoxetine ($t_{1/2\beta} = 5$ h) and 4-hydroxyatomoxetine-O-glucuronide ($t_{1/2\beta} = 7$ h, ~67%) are the principal circulating species, whereas atomoxetine ($t_{1/2\beta} = 20$ h) and N-desmethylatomoxetine ($t_{1/2\beta} = 33$ h) are the principal circulating species in PMs (Sauer et al. 2003). The high plasma concentrations of N-desmethylatomoxetine in PMs are not attributed to enhanced production of N-desmethylatomoxetine, but rather its slow systemic clearance. After its formation, N-desmethylatomoxetine must undergo hydroxylation and subsequent O-glucuronidation before its excretion. This hydroxylation appears to be mediated by CYP2D6 and, therefore, is slower in PMs, resulting in accumulation of N-desmethylatomoxetine in the plasma. These results demonstrate a clear effect of CYP2D6 phenotype on the formation of 4-hydroxyatomoxetine.

In healthy Chinese subjects, homozygous *CYP2D6*10* subjects have 50% lower clearance of atomoxetine compared with other EM subjects, resulting in twofold higher plasma levels (Cui et al. 2007). The CYP2D6.10 expressed in insect cells shows 11-fold lower intrinsic clearance toward atomoxetine 4-hydroxylation compared to the wild-type enzyme (Shen et al. 2007).

In children and adolescent patients with attention-deficit/hyperactivity disorder treated with atomoxetine, PMs have markedly greater reductions in mean symptom severity scores compared with EMs (Michelson et al. 2007). PMs have greater increases in heart rate and diastolic blood pressure and smaller increases in weight than EMs. Several adverse events, including decreased appetite and tremor, are more frequent in PMs (Michelson et al. 2007). These results suggest

that PMs receiving atomoxetine at doses up to 1.8 mg/kg/day are likely to have greater efficacy, greater increases in cardiovascular tone, and some differences in tolerability compared with EMs.

Trzepacz et al. (2008) have investigated whether the CYP2D6 genotype is associated with the dosage of atomoxetine in children and adolescents with attention-deficit/hyperactivity disorder. Patients are evaluated weekly up to 10 weeks and doses are titrated for efficacy and tolerability at the discretion of physicians (maximum: 1.8 mg/kg/day). The dose of atomoxetine is 0.1 mg/kg/day lower in PMs ($n = 87$) than in EMs ($n = 1239$). PMs demonstrate marginally better efficacy on the symptoms but have comparable safety profiles, except for a 4.0-beats/min greater increase in pulse rate and a 1.0-kg greater weight loss (Trzepacz et al. 2008). The discontinuation rate attributed to adverse effects in pediatric patients receiving at least 1.2 mg/kg/day is similar (~3%) in EMs and PMs. The authors suggest that genotyping is unnecessary during routine clinical management with atomoxetine, since the physicians can adjust the dosage to comparable efficacy and safety levels in all individual patients without knowledge of the CYP2D6 genotype/phenotype.

In a clinical study with depressed patients ($n = 297$), the safety and tolerability of atomoxetine in PMs are not different from those of EMs (Michelson et al. 2001), despite a greater exposure observed in PMs at comparable dose. The most common adverse events in adults are dry mouth, insomnia, nausea, decreased appetite, constipation, urinary retention or difficulties with micturition, erectile disturbance, dysmenorrhea, dizziness, and decreased libido (Wernicke and Kratochvil 2002). Another study on Latino ($n = 108$) and Caucasian ($n = 1090$) pediatric patients does not note significant difference in the incidence of common adverse effects of atomoxetine between two ethnic groups (Tamayo et al. 2008), although there is a significantly higher frequency of PMs in Caucasians compared with Latinos. However, Caucasian patients experience significantly more abdominal and throat pain, whereas Latinos reported more decreased appetite and dizziness.

These findings indicate that the bimodal distribution of atomoxetine clearance observed in vivo is attributed to the primary involvement of the polymorphic CYP2D6 in the formation of the major metabolite of atomoxetine, 4-hydroxyatomoxetine. The formation of N-desmethylatomoxetine, a minor route of atomoxetine metabolism, is mediated primarily by CYP2C19. There is a clear gene–dose effect for atomoxetine. The DPWG has recommended being alert to reduced efficacy and side effects of atomoxetine or selecting an alternative drug for CYP2D6 UMs. However, CYP2D6 phenotype does not affect the adverse reactions of atomoxetine. There are preliminary data only on the impact of the CYP2D6 phenotype/genotype on the response to atomoxetine. The FDA-approved drug label for atomoxetine states that dose adjustments of atomoxetine may be necessary in CYP2D6 PMs or if CYP2D6 inhibitors (e.g., paroxetine, fluoxetine, and quinidine) are used concomitantly. Further large-scale studies are warranted to investigate the association of the CYP2D6 genotype/phenotype and the efficacy and side effects of atomoxetine.

7.11 DRUGS FOR THE TREATMENT OF SENILE DEMENTIA

7.11.1 NICERGOLINE

Nicergoline, an ergot derivative currently used in the treatment of senile dementia and other disorders with vascular origin such as Raynaud's disease and vascular migraine, is metabolized by CYP2D6. A further study on healthy subjects has found that the AUC of MDL is 2267 nmol·h/l in EMs ($n = 10$), but this metabolite is undetectable in PMs of CYP2D6 who are not PMs of CYP2C19 ($n = 5$) (Bottiger et al. 1996). However, the AUC of MMDL is 65.7-fold higher in PMs of CYP2D6 than in EMs (9471 vs. 144 nmol·h/l). This indicates that the formation of MDL from MMDL via N-demethylation in the metabolism of nicergoline is largely by polymorphic CYP2D6 while CYP2C19 does not play a role in its metabolism. However, since it is unclear whether MMDL contributes to the clinical efficacy of nicergoline, the impact of CYP2D6 polymorphisms on the pharmacokinetics and pharmacodynamics of nicergoline remains to be determined.

7.12 ANTIMUSCARINIC DRUGS

7.12.1 TOLTERODINE

Tolterodine is a substrate of CYP2D6, 3A4, 2C9, and 2C19 (Brynne et al. 1999b; Nilvebrant et al. 1997; Postlind et al. 1998). 5-Hydroxymethyltolterodine is the major metabolite in EMs but undetectable in the plasma of PMs (Brynne et al. 1998, 1999a,c; Olsson and Szamosi 2001a). Although tolterodine concentrations are increased 5- to 10-fold in PMs, the summed active moieties do not differ between EMs and PMs (Brynne et al. 1999a,c; Olsson and Szamosi 2001a,b). This suggests that therapeutic effects would not differ significantly between the two groups, and there is no convincing evidence of an important gene–effect correlation (Brynne et al. 1998).

On the basis of the above findings, the CYP2D6 phenotype can be used to predict the hydroxylation activity of tolterodine. There is a possibility of clinical drug interaction when tolterodine is coadministered with a CYP2D6 inhibitor or to individuals associated with the CYP2D6 PM phenotype. However, the large amount of CYP3A in the liver and the fact that tolterodine is predominantly eliminated via oxidation by CYP2D6 makes it less likely that clinically significant drug–drug interactions would occur with CYP3A substrates in individuals with the CYP2D6 EM phenotype.

The FDA-approved drug label for tolterodine notes that tolterodine is used for the treatment of overactive bladder and is primarily metabolized by CYP2D6. CYP2D6 PMs may have greater plasma concentrations of the drug, which could possibly have an effect on QT interval. However, no recommendations for testing for CYP2D6 metabolizer status are provided on the label.

7.13 ANTIEMETICS

7.13.1 DOLASETRON

Dolasetron, a pseudopelletierine-derived 5-HT$_3$ receptor antagonist used for the treatment of chemotherapy-, radiotherapy-, and surgery-induced emesis, is a substrate of CYP2D6 and 3A4 (Sanwald et al. 1996). In a clinical study of 150 patients with 8 PMs, 4 IMs, 128 EMs, and 10 UMs, there are no statistically significant differences in the incidence of nausea and vomiting in PMs, IMs, and EMs treated with dolasetron or granisetron (Janicki et al. 2006). However, the UMs with the duplication of the *CYP2D6* allele ($n = 10$) experience significantly more vomiting episodes than patients in the granisetron group during the 24-h observation period (Janicki et al. 2006). However, this study does not measure the plasma levels of both dolasetron and hydrodolasetron.

7.13.2 ONDANSETRON

Ondansetron, a potent and selective 5-HT$_3$ receptor antagonist used mainly as an antiemetic to treat and prevent nausea and vomiting induced by chemotherapy and radiotherapy and postoperative nausea, is a substrate of CYP2D6, 3A4, and 1A2 (Fischer et al. 1994; Sanwald et al. 1996). In healthy volunteers, there is no difference in the AUC, C_{max}, and $t_{1/2\beta}$ of ondansetron between EMs ($n = 6$) and PMs ($n = 6$) receiving a single dose (8 mg intravenously) (Ashforth et al. 1994).

In a study of 250 patients undergoing standardized general anesthesia given 4 mg ondansetron 30 min before extubation, UMs ($n = 23$) have increased therapeutic failure and reduced response to ondansetron (Candiotti et al. 2005). Postoperative vomiting is significantly higher in UMs (46%) than in EMs (15%), IMs (17%), and PMs (8%). In patients with one, two, or three functional *CYP2D6* copies, the incidences of vomiting are 27%, 14%, and 30%, respectively. There are no differences between groups in the incidence of nausea on the basis of *CYP2D6* copy number or genotype. However, the plasma concentration of ondansetron is not determined in this study, although it could be expected that UMs metabolize ondansetron faster via CYP2D6 than other genotype groups. These results indicate that patients with three copies of the *CYP2D6* gene, a genotype consistent with the UM phenotype, show an increased incidence of ondansetron failure for the prevention of postoperative vomiting but not nausea.

7.13.3 TROPISETRON

Tropisetron, a highly potent and selective 5-HT$_3$ antagonist used as an antiemetic in cancer chemotherapy, is a substrate of CYP2D6 and 3A4 (Firkusny et al. 1995). The major route of elimination of tropisetron in EMs is via metabolism to 6-hydroxytropisetron and 5-hydroxytropisetron and their conjugates (~50%–60% of the dose excreted) whereas PMs excrete only trace amounts (Firkusny et al. 1995). CYP2D6 catalyzes 5- and 6-hydroxylation of tropisetron, while CYP3A4 forms N-desmethyltropisetron (Firkusny et al. 1995). There are five- to sevenfold higher AUCs of tropisetron in PMs than in EMs.

A Korean study ($n = 13$) has reported a 6.8-fold higher mean AUC with the *CYP2D6*10/*10* and **5/*10* genotypes compared with the wild type and a 1.9-fold higher AUC with the *CYP2D6*1/*10* genotype (Kim et al. 2003). No difference in adverse effects is seen, consistent with the high therapeutic index of this drug.

Data from a small number of UMs suggest that they have similar or slightly reduced concentrations compared with EMs (Kaiser et al. 2002; Kim et al. 2003) with a decreased antiemetic effect shown in one study (Kaiser et al. 2002). In the latter study with healthy volunteers, carriers of the duplicated *CYP2D6* allele show a decrease in the AUC, C_{max}, and plasma half-life for tropisetron compared with wild-type allele carriers (Kaiser et al. 2002).

Dolasetron, ondansetron, and tropisetron, all in part metabolized by CYP2D6, are less effective in UM patients compared with other subjects. Overall, there is a strong gene–concentration relationship for tropisetron only. *CYP2D6* genotype screening before antiemetic treatment may allow for modification of antiemetic dosing. An alternative is to use a 5-HT$_3$ agent that is metabolized independently of CYP2D6, such as granisetron, which would obviate the need for genotyping and may lead to improved drug response.

7.14 ANTIHISTAMINE

7.14.1 CHLORPHENIRAMINE

Chlorpheniramine is a substrate of CYP2D6 (He et al. 2002; Yasuda et al. 2002). In healthy EMs ($n = 6$), the mean C_{max} of chlorpheniramine is greater (12.55 vs. 5.38 ng/ml) and its oral clearance is lower (0.49 vs. 1.07 l/h/kg) for *S*-enantiomer than for *R*-chlorpheniramine after a single oral 8-mg dose (Yasuda et al. 2002). For *S*-chlorpheniramine, low-dose quinidine (a potent inhibitor of CYP2D6) causes a slight increase in C_{max} to 13.94 ng/ml, a marked decrease in oral clearance to 0.22 l/h/kg, and a prolongation of $t_{1/2\beta}$ from 18.0 to 29.3 h. Quinidine also decreases the oral clearance of *R*-chlorpheniramine to 0.60 l/h/kg. In PMs ($n = 2$), systemic exposure is greater after administration of chlorpheniramine than in EMs, and coadministration of quinidine causes only a slight increase in the oral clearance and small decrease in the AUC of both *R*- and *S*-chlorpheniramine (Yasuda et al. 2002). These findings demonstrate a stereoselective elimination of chlorpheniramine in humans, with the most pharmacologically active *S*-enantiomer being cleared more slowly than the *R*-enantiomer. Low dosages of quinidine effectively convert the EM to PM phenotype, indicating a role of CYP2D6 in the metabolic clearance of chlorpheniramine.

In addition, a marked difference in H$_1$ receptor occupancy is observed after a single 8-mg dose of chlorpheniramine between healthy PMs ($n = 5$) and EMs ($n = 6$) (Yasuda et al. 1995). In EMs, there is a >80% occupancy of H$_1$ receptors by antagonists in plasma for 12 h postdosing, but the occupancy is >60% from 12 to 30 h in PMs, when plasma concentrations have decreased to the level that produced a 50% occupancy of receptors. These findings suggest that plasma concentrations of chlorpheniramine cannot predict the extent of H$_1$ receptor occupancy and CYP2D6 appears to form a potent active metabolite from chlorpheniramine. The CYP2D6 phenotype should have minimal impact on the clinical effect of chlorpheniramine.

7.14.2 DIPHENHYDRAMINE

Diphenhydramine is a substrate of CYP1A2, 2C9, 2D6, 2C18, 2B6, and 2C19 (Akutsu et al. 2007). The clearance of diphenhydramine to its *N*-demethylated metabolite, 2-benzhydryloxy-*N*-methyl-ethanamine, is similar in EMs and PMs of CYP2D6 (Lessard et al. 2001). This suggests a limited role of CYP2D6 in diphenhydramine's metabolism. There is no need to conduct CYP2D6 genotyping/phenotyping tests for patients taking diphenhydramine.

7.14.3 LORATADINE

Loratadine is a substrate of CYP3A4 and 2D6 (Yumibe et al. 1996). After a single 20-mg dose, the oral clearance of loratadine in healthy Chinese subjects homozygous for the nonfunctional *CYP2D6*10* allele ($n = 7$), heterozygous for *CYP2D6*10* ($n = 6$), or homozygous for *CYP2D6*1* (wild type, $n = 4$) are 7.17, 11.06, and 14.59 l/h/kg, respectively (Yin et al. 2005). The corresponding MR of the AUC of desloratadine over that of loratadine is 1.55, 2.47, and 3.32, respectively. In homozygous *CYP2D6*10* carriers, the AUC of loratadine is 75.5% and 123.6% higher compared with the heterozygous *CYP2D6*10* and homozygous *CYP2D6*1* subjects, respectively. However, there is no significant difference in the plasma concentrations of desloratadine among the three genotype groups, although there is a trend of increased desloratadine concentrations in subjects with one or two *CYP2D6*10* alleles. These results indicate a clear gene–dose effect on loratadine in Chinese, but the impact of *CYP2D6* polymorphisms on the efficacy and adverse effects of loratadine is yet to be determined.

There are limited data on the association of *CYP2D6* polymorphisms and the pharmacokinetics and pharmacodynamics of antihistamines. The preliminary data indicate a gene–dose effect on diphenhydramine and loratadine, but not on diphenhydramine. It appears that the *CYP2D6* genotype is associated with the adverse effects of antihistamines. A recent study on Japanese patients has reported that patients with the nonfunctional *CYP2D6*10* allele experienced more hypersomnia than those carrying the wild-type gene (Saruwatari et al. 2006). Since antihistamines have relatively wide therapeutic ranges, there is no need to adjust their dosage based on the CYP2D6 phenotype.

7.15 β-BLOCKERS

7.15.1 CARVEDILOL

Carvedilol is a substrate of CYP2D6, 2E1, and 2C9 (Oldham and Clarke 1997). The average plasma AUC of *R*-carvedilol in PMs is 2.5-fold higher than that in EMs, with reduced clearance and the partial metabolic clearance to two metabolites 4′- and 5′-hydroxyphenyl carvedilol in PMs (Giessmann et al. 2004; Zhou and Wood 1995). A nonsignificant 1.4-fold higher AUC of S-carvedilol is seen in PMs than in EMs (Zhou and Wood 1995), whereas a significant 2.0-fold higher AUC is observed in another study (Giessmann et al. 2004). The non-functional *CYP2D6*10* allele appears to affect the disposition of both R- and S-enantiomers (Honda et al. 2005). One study has reported significantly lower systolic blood pressure in healthy PMs treated with carvedilol for 1 week (Giessmann et al. 2004), but this study does not measure the plasma levels of carvedilol.

The DPWG has evaluated therapeutic dose recommendations for carvedilol based on CYP2D6 genotypes (Swen et al. 2011). They conclude that there are no recommendations at this time.

The FDA-approved drug label states that carvedilol is affected by CYP2D6 PM resulting in two- to threefold higher plasma concentrations of R(+)-carvedilol compared to CYP2D6 EMs. Additionally, retrospective analysis of side effects in clinical trials shows that CYP2D6 PMs have a higher rate of dizziness during up-titration, presumably resulting from vasodilating effects of the higher concentrations of the R(+)-enantiomer.

7.15.2 Metoprolol

Metoprolol is metabolized by CYP2D6 and 3A4 (Hoffmann et al. 1980; Johnson and Burlew 1996; Otton et al. 1988). In hypertensive patients ($n = 143$), metoprolol α-hydroxylation and N-dealkylation are correlated with debrisoquine oxidation phenotype (McGourty et al. 1985). After oral administration of the racemate, EMs for CYP2D6 clear the R-enantiomer more rapidly than the S-enantiomer, but this stereoselectivity is not observed in PMs (Lennard et al. 1983). This leads to a higher exposure of S-metoprolol in EMs than in PMs. In vitro studies indicate that the stereoselectivity in EM subjects is the result of the high-affinity component of the O-demethylation catalyzed by CYP2D6, exhibiting significant enantioselectivity for the R-enantiomer (Kim et al. 1993). Oxidation of R-metoprolol by CYP2D6 produces more O-demethylmetoprolol than α-hydroxymetoprolol; however, for S-metoprolol, a slight preference for α-hydroxylation is observed (Mautz et al. 1995). The AUCs of metoprolol are four- to sixfold higher in PMs than in EMs after one single dose (Hamelin et al. 2000) and three- to fourfold higher after repeated dosing (Lennard et al. 1982). In healthy subjects, UMs ($n = 12$) achieve metoprolol C_{max} that is half of those observed in EMs after a single 100-mg dose ($n = 13$) (UMs vs. EMs: 67 vs. 118 μg/l), with a twofold increase in drug clearance (UMs vs. EMs: 367 vs. 168 l/h) (Kirchheiner et al. 2004a). Compared to PMs ($n = 4$), UMs have a 3.9-fold lower C_{max} (UMs vs. PMs: 67 vs. 260 μg/l) and 11.8-fold higher oral clearance (UMs vs. PMs: 367 vs. 31 l/h) (Kirchheiner et al. 2004a). In this study, the authors have found that metoprolol reduced the exercise-induced heart rate by 31, 21, and 18 beats/min in PMs, EMs, and UMs, respectively, while blood

pressure is not affected by CYP2D6 polymorphisms. These findings demonstrate a more than 10-fold difference in metoprolol clearance between PMs and UMs, but the pharmacodynamic effect differs only by less than twofold and there is only a marginal difference in metoprolol efficacy on heart rate between EMs and UMs.

There is a 1.9- and 9.5-fold lower plasma AUC of S-metoprolol in UMs after a single 100-mg dose of racemic metoprolol ($n = 12$), respectively, compared to EMs ($n = 13$) and PMs ($n = 4$) (UMs vs. EMs vs. PMs: 190 vs. 366 vs. 1804 ng·h/ml) (Seeringer et al. 2008). The AUCs of R-metoprolol are 127, 261, and 1746 ng·h/ml in UMs, EMs, and PMs, respectively. A higher proportion of R-metoprolol in the total metoprolol concentrations is observed with decreasing CYP2D6 activity, with an AUC ratio of S-metoprolol to R-metoprolol of 1.6 in UMs, 1.5 in EMs, and 1.0 in PMs. UMs and EMs have lower concentrations of R-metoprolol than S-metoprolol, whereas the concentrations of both enantiomers almost overlap in PMs (Seeringer et al. 2008). A higher ratio of S,S-α-hydroxy-metoprolol AUC to R,R-α-hydroxy-metoprolol AUC is observed in UMs (0.66) compared to EMs (0.36), while the α-hydroxyl metabolites are not detected in PMs. The concentration of S-metoprolol needed to achieve a half-maximum reduction in heart rate is estimated as 21 ng/ml in PMs, 17 ng/ml in EMs, and 11 ng/ml in UMs. The estimation for maximum percent reduction in heart rate by concentrations of R-metoprolol is 30% in PMs, 29% in EMs, and 29% in UMs; the concentration of R-metoprolol to obtain a half-maximum reduction in heart rate is estimated to be 20 ng/ml in PMs, 11 ng/ml in EMs, and 7 ng/ml in UMs (Seeringer et al. 2008). These results show that the kinetics of both enantiomers of metoprolol are influenced by CYP2D6 activity as predicted by the CYP2D6 genotype. UMs and EMs have preferential metabolism toward R-metoprolol than S-metoprolol. Because S-metoprolol is the major contributor to the β-blocking effect, EMs and UMs might benefit from the enantiopreference of CYP2D6.

PMs show a fivefold higher risk of developing adverse effects than patients who are not PMs during metoprolol treatment (Wuttke et al. 2002). However, general and dose-limiting adverse event rates do not differ by AUC quartile in PMs and EMs (Zineh et al. 2004). Adverse event rates do not differ between EMs, IMs, or PMs. There is no significant association between the CYP2D6 phenotype and the incidence of adverse effects in patients receiving chronic metoprolol treatment (Fux et al. 2005). In this study, the median dose-normalized metoprolol concentration is 0.0088, 0.047, 0.34, and 1.34 ng/ml among UMs, EMs, IMs, and PMs, respectively. There is a tendency toward a more frequent occurrence of cold extremities in the PMs plus IMs as compared with the EMs plus UMs. This study indicates that the CYP2D6 phenotype is not correlated with a propensity for adverse effects to develop during metoprolol therapy.

Bijl et al. (2009) have investigated the relationship between *CYP2D6*4* and blood pressure or heart rate changes in 1533 patients treated with metoprolol. In PMs carrying *4/*4, the adjusted heart rate is 8.5 beats/min lower compared with the

EMs harboring the wild-type*1/*1. This leads to an increased risk of bradycardia in PMs. The diastolic blood pressure in PMs is lower compared with EMs. These findings show an increased risk of bradycardia in PMs receiving metoprolol treatment. However, this study does not measure the plasma levels of metoprolol.

The DPWG Guideline for metoprolol recommends selecting another drug or reducing the dose of metoprolol for CYP2D6 PM (reduce dose by 75%) and IM (reduce dose by 50%) patients (Swen et al. 2011). It recommends using a dose titration of metoprolol to a maximum of 250% of the normal dose in response to efficacy and side effects for CYP2D6 UM patients with heart failure or selecting an alternative drug (e.g., bisoprolol, carvedilol).

The FDA-approved drug label for metoprolol notes that metoprolol is metabolized by the cytochrome P450 enzymes in the liver with a major contribution of CYP2D6. PMs and PMs who concomitantly use CYP2D6-inhibiting drugs will have increased (several-fold) metoprolol blood levels, decreasing metoprolol's cardioselectivity.

7.15.3 PROPRANOLOL

Propranolol is a substrate of CYP1A2 and 2D6 (Ward et al. 1989). Propranolol ring hydroxylation cosegregates with debrisoquine/sparteine polymorphism in vivo (Lennard et al. 1984b, 1986; Raghuram et al. 1984). The formation of the active metabolite 4-hydroxypropranolol is decreased in PMs (Lennard et al. 1984a; Raghuram et al. 1984). This pathway constitutes a minor proportion of S-propranolol elimination (~15% in EMs) (Sowinski and Burlew 1997). There are no significant differences in the AUC of S-propranolol (or racemic propranolol) or in β-blockade between EMs and PMs (Lennard et al. 1984a; Raghuram et al. 1984; Sowinski and Burlew 1997; Ward et al. 1989). However, there is an approximately 2.5-fold higher AUC of racemic propranolol observed in PMs in Caucasians (Shaheen et al. 1989) or the nonfunctional CYP2D6*10/*10 carriers in Chinese (Lai et al. 1995). Furthermore, there is a twofold increase in racemic and S-propranolol AUC with quinidine (Zhou et al. 1990). It appears that CYP2D6 polymorphisms may have only a marginal effect on propranolol overall elimination.

The FDA-approved drug label for propranolol notes that propranolol is metabolized primarily by CYP2D6. 1A2 and 2C19 may also play a role in propranolol metabolism. Propranolol should be used with caution when coadministered with drugs that have an effect on CYP2D6-, 1A2-, or 2C19-mediated metabolic pathways. Coadministration of such drugs with propranolol may lead to clinically relevant drug interactions and changes on its efficacy or toxicity.

7.15.4 TIMOLOL

Timolol is largely eliminated by CYP2D6, with a minor contribution (<10%) from CYP2C19 (Volotinen et al. 2007). After oral administration of a single 20-mg dose to healthy subjects, PMs (n = 4) have an approximately twofold higher AUC (625

vs. 294 ng·h/ml) and enhanced β-blocking activity compared with EMs (n = 6) (Lewis et al. 1985). There is a significant correlation between the debrisoquine MR and AUC of timolol or β-adrenoceptor blocking degree (Lewis et al. 1985).

Ocular administration of timolol has been associated with systemic side effects (e.g., bradycardia) (Lama 2002; Nieminen et al. 2007; Vander Zanden et al. 2001), since β-blockade is observed even at low plasma levels of timolol (~1.0 ng/ml) (Kaila et al. 1991). The change in heart rate is the most striking effect of the systematically absorbed fraction of ophthalmic timolol, with 0.5% aqueous formulations presenting larger effects than 0.1% hydrogel formulations (Nieminen et al. 2007). The variability in timolol systemic exposure is greater than differences in the debrisoquine oxidation phenotype after a single ocular timolol application (Huupponen et al. 1991). However, a crossover study of 19 glaucoma patients and 18 healthy volunteers has reported a 2.5-fold higher plasma AUC in PMs (n = 2) than in EMs (n = 8) (21.39 vs. 8.52 ng·h/ml) after application of timolol 0.5% aqueous eye drops. The plasma AUC in PMs is 3.2-fold higher than that in UMs (21.39 vs. 6.55 ng·h/ml). The PMs also have a tendency for greater β-blockade, as determined by change in heart rate with exercise. When the subjects are treated with ocular timolol 0.1% hydrogel, PMs also have a 2.5-fold higher plasma AUC than EMs, but it does not achieve statistical significance. There is also a tendency for greater β-blockade in PMs, but this effect is less pronounced compared with that for the 0.5% eye drops (Nieminen et al. 2005). In addition, application of timolol eye drops intranasally causes higher plasma concentrations and greater β-blockade in PMs than in EMs (Edeki et al. 1995).

These findings demonstrate a gene–concentration/effect relationship for timolol, but the extent is lesser compared to metoprolol. The PM phenotype may increase the systemic timolol concentration in patients. However, when plasma timolol concentrations in patients remain low (≤0.2 ng/ml), it is suggested that this type of interaction is of only minor clinical relevance. To avoid systemic effects caused by increased systemic timolol concentrations, it is important to minimize systemic absorption of timolol with topical products in the treatment of glaucoma.

The β-blockers metabolized to a significant extent by CYP2D6 and therefore potentially susceptible to genetic polymorphism include metoprolol, carvedilol, propranolol, and timolol. Since the contribution of CYP2D6 is greater for metoprolol than carvedilol, propranolol, and timolol, the strongest gene–dose effect is seen with this β-blocker, while such an effect is less or marginal in other β-blockers. For timolol, its ocular application may cause systemic side effects that are affected by the CYP2D6 phenotype, and the CYP2D6 phenotype may be used to predict the risk of developing systemic toxicity in glaucoma patients treated with timolol eye drops.

7.16 OPIOIDS

Opioids have long been used as potent analgesics. They bind to specific opioid receptors (μ, κ, and δ) in the CNS and in other tissues and produce different pharmacological response depending on which receptor they bind, the affinity for that receptor, and whether the opioid is an agonist or an antagonist. Almost all opioids are subject to *O*-dealkylation, *N*-dealkylation, ketoreduction, or deacetylation leading to oxidative metabolites (Lotsch 2005). Through glucuronidation or sulfation, Phase II metabolites are generated. Some metabolites of opioids have an activity themselves and contribute to the effects of the parent compound. CYP2D6 is involved in the oxidative metabolism of several opioids, including dextromethorphan, tramadol, codeine, dihydrocodeine, hydrocodone, and oxycodone (Leppert 2011).

7.16.1 CODEINE

Codeine must be activated by CYP2D6 to form morphine (Dayer et al. 1988). The analgesic effect of codeine is seen to be substantially less in subjects found to be PMs (Lotsch et al. 2004), or no analgesic effect is recorded in PMs (Eckhardt et al. 1998; Poulsen et al. 1996b). The AUC of codeine is similar in PMs and in EMs, whereas morphine is virtually undetectable in PMs (Caraco et al. 1996; Eckhardt et al. 1998). Clearly, a lack of CYP2D6 enzyme would be expected to result in reduced drug plasma morphine concentrations and thus reduce the effectiveness of the drug. Subjects deficient in CYP2D6 metabolism are protected against possible opioid dependence, as none of the PM subjects in the study populations show symptoms or signs of dependence (Mikus et al. 1994, 1998; Tyndale et al. 1997). However, further research is required to establish the validity of such findings. The *CYP2D6* genotype may be used to predict possible side effects associated with codeine treatments and the likelihood of their occurrence, especially in genetically PMs (Somogyi et al. 2007).

A fatal case of neonate has been reported (Koren et al. 2006), which is caused by high levels of morphine resulting from high doses of codeine taken by a breastfeeding mother who is a UM. A case–control study has found that infants fed by mothers who are EMs experience more CNS depression than those fed by PM mothers and two breastfeeding mothers whose infants exhibit severe neonatal toxicity (i.e., CNS depression) and who are CYP2D6 UMs (Madadi et al. 2009). It is clear that the UM phenotype of the mother results in excessive morphine from codeine, which is then excreted into the milk taken by the infants who would experience CNS depression or even fatality. Therefore, breastfeeding mothers should be advised not to take codeine as an analgesic whereas alternative drugs that are not CYP2D6 substrates should be chosen.

Recently, the CPIC have summarized evidence from the literature supporting the association of CYP2D6 polymorphisms and efficacy and safety of codeine and provide therapeutic recommendations for codeine based on the *CYP2D6* genotype (Crews et al. 2014). They recommend that PMs of CYP2D6 should not receive codeine for pain relief, and UMs of CYP2D6 should avoid codeine for pain relief and receive alternative analgesics that do not have potent CYP2D6 metabolites. Similarly, the CPNDS also recommend not to use PMs

and UMs of CYP2D6 (Fliegert et al. 2005). They also recommend that breastfeeding mothers and young children who are UMs of CYP2D6 should avoid codeine use. In individuals with IM or EM CYP2D6 genotypes, codeine can be used as per standard of care. Existing evidence suggests that caution is still warranted in CYP2D6 EMs receiving codeine if they are receiving maximal therapeutic doses of codeine and have additional risk factors for toxicity.

The DPWG have evaluated therapeutic dose recommendations for codeine based on *CYP2D6* genotypes (Swen et al. 2011). They recommend alternative analgesics (e.g., acetaminophen, NSAID, morphine—not tramadol or oxycodone) or being alert to symptoms of insufficient pain relief for patients carrying the PM alleles, IM alleles, or UM alleles. PMs with two inactive alleles include *3–*8, *11–*16, *19–*21, *38, *40, and *42 alleles. IMs with two decreased-activity alleles include *9, *10, *17, *29, *36, and *41 alleles or carrying one active allele (*1, *2, *33, and *35) and one inactive allele (*3–*8, *11–*16, *19–*21, *38, *40, and *42), or carrying one decreased-activity allele (*9, *10, *17, *29, *36, and *41) and one inactive allele (*3–*8, *11–*16, *19–*21, *38, *40, and *42). UMs with a gene duplication in absence of inactive alleles include *3–*8, *11–*16, *19–*21, *38, *40, and *42; those with decreased-activity alleles include *9, *10, *17, *29, *36, and *41.

7.16.2 DIHYDROCODEINE

Dihydrocodeine is converted by CYP2D6 to active dihydromorphine (Kirkwood et al. 1997). EMs have sevenfold higher dihydromorphine concentrations (Fromm et al. 1995), with quinidine producing a three- to fourfold decrease in dihydromorphine concentrations (Wilder-Smith et al. 1998). Contrary to what might be expected, dihydrocodeine 60 mg seems to produce similar analgesic effects in healthy volunteers who are EMs or PMs (Schmidt et al. 2003; Wilder-Smith et al. 1998). However, in one of the studies, no analgesic effect of dihydrocodeine is seen in the pain threshold model in either EMs or PMs, whereas pupillary diameter is reduced comparably in both EMs and PMs (Schmidt et al. 2003).

7.16.3 HYDROCODONE

Hydrocodone is a substrate of CYP2D6 and 3A4 (Hutchinson et al. 2004). Liver microsomes from the livers of PMs form substantially less hydromorphone than do the microsomes from the EM liver (0.7 vs. 3.5 µl/h/mg protein), but the *CYP2D6* genotype does not affect norhydrocodone formation (Hutchinson et al. 2004).

The urinary MR of hydrocodone/hydromorphone is correlated with the *O*-demethylation ratio for dextromethorphan in healthy subjects (Otton et al. 1993a). The production of hydromorphone is significantly decreased in PMs compared to EMs (Otton et al. 1993a). The relative contribution of norhydrocodone is increased with an increase in the amount of hydrocodone recovered unchanged in urine of PMs (18% vs. 10%). The partial clearance of hydrocodone to hydromorphone in

PMs is eightfold lower than that in EMs (3.4 vs. 28.1 ml/h/kg) (Otton et al. 1993a). Pretreatment of 100 mg quinidine decreases the partial clearance to levels similar to those seen in PMs (5.0 ± 3.6 ml/h/kg), and the C_{max} of hydromorphone is five times higher in EMs than in PMs or in EMs pretreated with quinidine (Otton et al. 1993a). Inhibition of hydromorphone formation by quinidine results in plasma hydromorphone concentrations approximately 10% of those of controls in rhesus monkeys but results in negligible effects on the antinociceptive actions of hydrocodone (Lelas et al. 1999). However, there is little evidence for a marked difference in analgesic and side effects between EMs and PMs in human studies (Kaplan et al. 1997; Otton et al. 1993a), although more positive opioid effects are seen in EMs than in PMs over the first hour postdosing (Otton et al. 1993a).

7.16.4 Oxycodone

Oxycodone (14-hydroxy-7,8-dihydrocodeinone) is a semisynthetic opioid analog used in the treatment of moderate to severe postoperative pain and pain associated with cancer (Soderberg Lofdal et al. 2013). Oral oxymorphone is 10-fold more potent than oral morphine based on dose. Oxycodone is extensively metabolized; only 10% of the dose is excreted unchanged in urine. Its metabolism is similar to codeine and hydrocodone, with O- and N-demethylation being the major pathways, resulting in oxymorphone and noroxycodone, respectively. Oxymorphone is a potent opioid that has a three to five times higher μ-opioid receptor affinity than morphine (Soderberg Lofdal et al. 2013). Noroxycodone also has binding activity. CYP3A4 and 3A5 display the highest activity for oxycodone N-demethylation (>90%), whereas CYP2D6 shows the highest activity for O-demethylation (Lalovic et al. 2004). The total intrinsic clearance for noroxycodone formation is eight times greater than that for oxymorphone formation in human liver microsomes. Experiments with human intestinal mucosal microsomes indicate a lower N-demethylation activity (20%–50%) compared with liver microsomes and negligible O-demethylation activity, indicating a minor contribution of intestinal mucosa in the first-pass oxidative metabolism of oxycodone (Lalovic et al. 2004).

CYP2D6 is the principal (70%–90%) O-demethylase of oxycodone, and oxymorphone formation in human liver microsomes is much lower in livers from PMs, as compared with EMs, and inhibited by quinidine (Otton et al. 1993b). However, inhibition of CYP2D6 by quinidine does not attenuate the opioid-induced psychomotor and subjective side effects of oxycodone in human volunteers (Heiskanen et al. 1998), although no oxymorphone is detected in the plasma of 8 of 10 subjects (Heiskanen et al. 1998). However, analgesia is not assessed because pain is not present. A significant reduction in plasma oxymorphone levels does not substantially alter the pharmacodynamic effects of oxycodone. There is little evidence of a difference in opioid effect between EMs and PMs in human studies (Heiskanen et al. 1998). It appears that oxymorphone plays a minor role in the pharmacodynamics of oxycodone.

The DPWG Guideline for oxycodone recommends using an alternative drug to oxycodone (not codeine or tramadol) for CYP2D6 PM and IM patients or being alert to insufficient pain relief (Swen et al. 2011). For CYP2D6 UM patients, it recommends using an alternative drug to oxycodone (not codeine or tramadol) or being alert to potential adverse drug events such as nausea, vomiting, constipation, respiratory depression, confusion, and urinary retention.

7.16.5 Methadone

Methadone is a substrate of CYP3A4, 2D6, 2B6, 2C8, and 2C19 (Wang and DeVane 2003). In 56 Caucasian patients on methadone maintenance treatment, the oral clearances of R-, S-, and racemic methadone varies 5.4-, 6.8-, and 6.1-fold, respectively (Coller et al. 2007). There is no significant difference in methadone oral clearance between PMs, IMs, and EMs. In another study with 245 patients receiving methadone maintenance treatment, UMs show lower trough methadone plasma levels compared with EMs plus IMs (2.4 vs. 3.3 ng/ml/mg dose), but the PM phenotype does not affect the plasma levels of methadone (Crettol et al. 2006). It appears that methadone inhibits CYP2D6 since a study of 34 Caucasian patients undergoing methadone maintenance treatment revealed a discordance between CYP2D6 genotype and phenotype (Shiran et al. 2003). While 9% of patients (3/34) are PMs carrying nonfunctional *4/*4, 57% (16/28) are PMs by phenotyping tests. Eight patients, who are genotypically EMs carrying the functional CYP2D6*1 allele, are assigned as PMs by their phenotype using dextromethorphan as a probe. The high proportion of phenotypic PMs in patients on methadone maintenance therapy is consistent with significant inhibition of CYP2D6 activity by methadone seen in vitro (Wu et al. 1993). Because CYP2D6 is involved in the metabolism of a number of clinically important drugs, the observed decrease in CYP2D6 activity during methadone treatment may affect the efficacy and toxicity of these drugs. Clinicians should be aware of the potential for CYP2D6-mediated drug interactions in EM patients on chronic methadone therapy.

In addict patients ($n = 14$), the mean increase in R-methadone concentration–dose ratio by coadministered quetiapine is 7%, 21%, and 30%, respectively, in the PMs, heterozygous EMs, and homozygous EMs (Uehlinger et al. 2007). In healthy volunteers, paroxetine (a potent CYP2D6 inhibitor) significantly increases the plasma concentrations of R- and S-methadone (Begre et al. 2002; Lam et al. 2002). In addition, in two PMs of CYP2D6, only the plasma level of S-methadone but not R-methadone is increased by coadministered paroxetine (Begre et al. 2002). Fluoxetine, another potent inhibitor of CYP2D6, stereoselectively increases the concentration of R- but not S-methadone (Eap et al. 1997). However, fluvoxamine nonselectively increases the plasma levels of both R- and S-methadone (Eap et al. 1997). In addition, fluconazole increases the plasma peak and trough concentrations of methadone by 27% and 48%, respectively, in healthy volunteers ($n = 13$) (Cobb et al. 1998).

These findings suggest that CYP2D6 only plays a minor to moderate role in the metabolism of methadone while CYP3A4 and 2B6 may play a more important role. The disposition and clearance of methadone are thus not significantly affected by the *CYP2D6* polymorphism. The *R*-enantiomer is preferentially metabolized by CYP2D6 and the *S*-enantiomer is preferentially metabolized by CYP1A2. The CYP2D6 phenotype has a moderate affect on the drug interaction with methadone.

7.16.6 TRAMADOL

Tramadol is a substrate of CYP2D6, 2B6, and 3A4 (Paar et al. 1997; Subrahmanyam et al. 2001). There is a relationship between tramadol *O*-demethylation and sparteine oxidation in human volunteers ($n = 71$) (Paar et al. 1997). The mean MR of tramadol *O*-demethylation is significantly higher in PMs than in EMs (4.4 vs. 0.8). Another study of 26 children has also reported a close correlation between tramadol/*O*-desmethyltramadol plasma concentration or AUC and the dextromethorphan MRs (Abdel-Rahman et al. 2002). The plasma levels of the pharmacologically active *O*-desmethyltramadol (M1) are markedly higher (10–100 ng/ml) in EMs than in PMs who have levels of below or around the detection limit of 3.0 ng/ml after an oral dose of tramadol at 2 mg/kg body weight (Poulsen et al. 1996a). Analgesia for experimental pains is present in both EMs and PMs but much weaker in PMs owing to less formation of M1, which acts on the opioid component. When tramadol is injected at 100 mg intravenously, tramadol decreases discomfort experienced during the cold pressor test in EMs (Enggaard et al. 2006). The pain tolerance thresholds to sural nerve stimulation are increased in PMs whose plasma levels of M1 are lower (below the limit of detection, ~5.1 ng/ml) compared to those in EMs (12–17 ng/ml) (Enggaard et al. 2006). These results indicate that M1 plays a major role in tramadol's analgesic effect but tramadol also shows some analgesic activity.

In healthy volunteers ($n = 21$) receiving a single 100-mg dose of tramadol in a slow-released formulation, the plasma C_{max} value of tramadol in PMs ($n = 7$) is 21.7% and 40.9% higher than that in heterozygous and homozygous EMs ($n = 7$ for each phenotype), respectively (Slanar et al. 2007). The AUC of tramadol in PMs is 35.8% and 56.8% higher than that in heterozygous and homozygous EMs, respectively. Consistent with this, the $t_{1/2\beta}$ in PMs is 64.3% and 75.6% longer than that in heterozygous and homozygous EMs, respectively (25.89 vs. 9.23 vs. 6.31 h). As expected, the formation of M1 and subsequently the plasma levels of this active metabolite have been significantly decreased in PMs with the C_{max} being ~3.9-fold lower than in both homozygous and heterozygous EMs (41 vs. 158 vs. 151 nmol/ml) (Slanar et al. 2007). The ratio of tramadol AUC over that of M1 is 16.8, 2.98, and 2.23, respectively, in PMs and heterozygous and homozygous EMs, respectively. In addition, PMs have an almost 2.2- and 3.2-fold lower maximal pupillary constriction at 4 h postdosing than heterozygous and homozygous EMs, respectively (0.58 vs. 1.25 vs. 1.83 Δmm) (Slanar et al. 2007). There is a significant correlation between the AUC and C_{max} values of M1 and pupillary constriction.

Fliegert et al. (2005) have investigated the effect of CYP2D6 status on the pharmacokinetics of tramadol and static and dynamic pupillometry in healthy subjects ($n = 26$) receiving a single dose of 50–150 mg tramadol. PMs ($n = 6$) exhibit 1.5- to 1.7-fold higher plasma concentrations and AUCs of both enantiomers of tramadol than in EMs ($n = 20$), and the levels of M1 are 1.5-fold lower [for (–)-M1] than those seen in EMs or below the lower limit of quantification [for (+)-M1]. PMs have 1.3- to 1.5-fold longer half-lives for tramadol and M1. In PMs, both maximum effects and the return to baseline occur much earlier (at ~3 and 8 h, respectively) than in EMs where the effects of tramadol on static and dynamic pupillometry slowly reach a maximum between 4 and 10 h postdosing and decreased until 24 h (Fliegert et al. 2005). However, the effect–time profiles, amplitude of change, velocity of constriction, and reaction duration as well as an increase of latency are comparable in PMs and EMs. The pharmacokinetic properties of tramadol in PMs and EMs may provide an explanation for most of the pupillometry findings from this study. The pupillometric response appears to be mainly mediated by the (+)-M1 in EMs, which binds to μ-receptors, whereas the non–μ-receptor component [e.g., the parent enantiomers and (–)-M1] seems to play an important role in PMs. Indeed, the analgesic effects of tramadol are attenuated but not abolished in PMs. The lack of initial miosis in PMs clearly can be attributed to the lack of formation of (+)-M1. Although the maximum plasma levels of (+)-M1 did not coincide with the maximum pupillometric effects, a delayed occurrence of the enantiomers of tramadol and M1 in the CNS owing to the action of the blood–brain barrier may explain this delay.

A study on Chinese gastric cancer patients ($n = 63$) recovering from major abdominal surgery has reported that the total consumption of tramadol over 48 h in individuals carrying the nonfunctional *CYP2D6*10/*10* allele ($n = 20$) or one **10* allele ($n = 26$) is significantly higher than that in individuals of the wild type ($n = 17$) (532.7 vs. 476.8 vs. 459.5 mg at 24 h postoperation) (Wang et al. 2006). Healthy subjects harboring at least one *CYP2D6*10* allele have significantly longer $t_{1/2\beta}$ of tramadol after a single dose compared to subjects carrying the wild-type gene (Enggaard et al. 2006; Gan et al. 2004). A recent study on 88 Faroese patients has observed an approximately 14-fold higher M1/tramadol ratio in EMs ($n = 78$) than in the PMs ($n = 10$) (Halling et al. 2008b). In Malaysian patients, the UMs and EMs have 2.6- and 1.3-fold greater clearance for tramadol, respectively, than the IMs carrying the *CYP2D6*10* allele (Gan et al. 2007). The clearance for tramadol is 16, 18, 23, and 42 l/h while mean half-lives are 7.1, 6.8, 5.6, and 3.8 h in IMs, heterozygous EMs, homozygous EMs, and UMs, respectively (Gan et al. 2007). It is found in this study that there are significant differences in the adverse effects among the various genotype groups, with the IMs experiencing more adverse effects than the EMs, and the EMs having more adverse effects than the UMs (Gan et al. 2007). Side effects may be greater in EMs, through greater opioid effects mediated by M1 (Enggaard et al. 2006; Gan et al. 2007; Poulsen et al. 1996a).

Significantly higher tramadol AUCs and lower AUCs of M1 are observed with Caucasian subjects after single oral, multiple oral, and intravenous administration of tramadol (Pedersen et al. 2005, 2006). In Spanish healthy subjects ($n = 24$), the plasma concentrations of (+)- and (−)-tramadol enantiomers are higher in PMs ($n = 5$) than in EMs ($n = 19$), with 1.98- and 1.74-fold differences in the AUC, respectively (Garcia-Quetglas et al. 2007). The oral clearance of (+)- and (−)-tramadol are 1.91- and 1.71-fold greater in PMs, which are related to difference in both O- and N-demethylation in PMs and EMs. The mean AUC for (+)- and (−)-O-desmethyltramadol (M1) is 4.33- and 0.89-fold greater in EMs than in PMs, respectively, and the values for (+)- and (−)-N-desmethyltramadol (M2) are 7.40- and 8.69-fold greater in PMs than those in EMs because of the involvement of CYP2D6 in their subsequent oxidation (Garcia-Quetglas et al. 2007). In EMs, the plasma concentrations of endogenous epinephrine are increased after tramadol administration whereas no effect is observed in PMs. These results further support the major contribution of CYP2D6 in the formation of M1 from tramadol and the N-demethylation pathway is indirectly affected by the CYP2D6 phenotype.

PMs appear to show a lower response rate to postoperative tramadol administration than EMs (Stamer et al. 2003). However, evidence is weak on the impact of CYP2D6 polymorphisms on tramadol's analgesic effect. There are case reports indicating that carriers of the *CYP2D6* duplication with the UM phenotype treated with opioids may be at a high risk for adverse events (Dalen et al. 1997; de Leon et al. 2003). The C_{max} of (+)-O-desmethyltramadol (M1) after a single dose of 100 mg tramadol is significantly higher in the UMs ($n = 11$) than in EMs ($n = 11$), with its median AUC being 448 and 416 µg·h/l, respectively (Kirchheiner et al. 2008). The median AUC of (+)-tramadol is 786 and 587 µg·h/l in EMs and UMs, respectively. There is an increased pain threshold and pain tolerance in cold pressure test and a stronger miosis after tramadol dosing in UMs compared with EMs. Almost 50% of the UMs experience nausea compared with only 9% in the EMs. These results suggest that tramadol may frequently cause adverse effects in southern Europeans and Northern Africans with a high proportion of *CYP2D6* duplication and UM phenotype.

The pharmacodynamic profiling of tramadol is very complex, and the relative contribution of tramadol and its metabolites is unknown. There is a clear gene–dose effect on tramadol and M1 formation. However, since tramadol can exert a non-µ mechanism of action that relates to a reuptake inhibition of norepinephrine and serotonin, we would expect an excess of nonopioid mechanisms of action in PMs as compared with EMs.

The polymorphic CYP2D6 catalyzes the O-demethylation pathway of many 4,5-epoxymorphinan opioids such as codeine, oxycodone, hydrocodone, and dihydrocodeine, while their N-demethylation is mediated principally by CYP3A4 (Leppert 2011). CYP2D6 inhibition studies with hydrocodone, dihydrocodeine, or oxycodone have all failed to demonstrate any significant impact on subjective opioid response in human subjects, suggesting a minor role of the active metabolites formed by CYP2D6-mediated O-demethylation. There is a clear gene–dose effect on the formation of O-demethylated metabolites, but the clinical significance may be minimal as the analgesic effect is not altered in PMs. Genetically caused inactivity of CYP2D6 renders codeine ineffective owing to lack of morphine formation, decreases the efficacy of tramadol owing to reduced formation of the active O-desmethyl-tramadol, and reduces the clearance of methadone (Zahari and Ismail 2014). Conversely, UMs may experience faster analgesic effects but may be prone to higher µ-opioid–related side effects. *CYP2D6* polymorphisms may also trigger or modify drug interactions, which in turn can alter the clinical response to opioid therapy (Lotsch et al. 2004). For example, by inhibiting CYP2D6, paroxetine increases the steady-state plasma concentrations of R-methadone in EMs but not in PMs. Genetically precipitated drug interactions might render a standard opioid dose toxic. More studies involving larger numbers of patients in different population types are needed to demonstrate genetically determined unresponsiveness and risk of developing serious adverse events for patients with pain.

The DPWG Guideline for tramadol recommends selecting an alternative to tramadol (not oxycodone or codeine) and being alert for symptoms of insufficient pain relief for CYP2D6 PMs (Swen et al. 2011). For CYP2D6 IMs, one should be alert for symptoms of insufficient pain relief and consider dose increase or select an alternative to tramadol (not oxycodone or codeine). For CYP2D6 UMs, it recommends a 30% dose decrease when being alert for adverse effects or using an alternative drug to tramadol (not oxycodone or codeine). The FDA-approved label contains information regarding metabolism of tramadol by CYP2D6: CYP2D6 PMs may have higher tramadol concentrations, and concomitant use of CYP2D6 or CYP3A4 inhibitors may reduce clearance of tramadol and increase a patient's risk for adverse events.

7.17 ORAL HYPOGLYCEMIC DRUGS

7.17.1 PHENFORMIN

Phenformin, a biguanide antidiabetic agent, is a substrate of CYP2D6 (Oates et al. 1982; Shah et al. 1985). Risk factors for phenformin-induced lactic acidosis include renal, liver or cardiac disease, alcoholism, and the CYP2D6 PM phenotype (Krentz et al. 1994; Marchetti and Navalesi 1989). Increased phenformin plasma concentrations have been observed in patients with lactic acidosis (Marchetti and Navalesi 1989), and the AUC of phenformin is ~1.4-fold higher and lactate concentrations are greater in PMs after a single dose (Oates et al. 1983). However, there is weak evidence implicating the CYP2D6 phenotype as a cause or a primary risk factor (Oates et al. 1981; Wiholm et al. 1981) and the risk of developing lactic acidosis may be related more to other factors than to CYP2D6 PM status.

7.18 SELECTIVE ESTROGEN RECEPTOR MODULATORS

7.18.1 Tamoxifen

Tamoxifen, a selective estrogen receptor modulator used for the treatment of metastatic breast cancer (Jordan 2014; Komm and Mirkin 2014), is mainly activated by CYP2D6 with contributions from CYP2C9 and 3A4 (Beverage et al. 2007; Crewe et al. 2002; Dehal and Kupfer 1997; Desta et al. 2004; Stearns et al. 2003). Women receiving tamoxifen treatment who either carry genetic variants associated with low or absent CYP2D6 activity or who receive concomitant drugs known to inhibit CYP2D6 activity have significantly lower levels of endoxifen (Jin et al. 2005; Stearns et al. 2003). Concomitant use of paroxetine is associated with lower endoxifen plasma concentrations, with the magnitude of this difference being dependent on the CYP2D6 genotype. As such, interindividual differences in the formation of these active metabolites could be an important source of variability in the response to tamoxifen.

Because of the important role of CYP2D6 in tamoxifen metabolism and activation, PMs are likely to exhibit therapeutic failure, and UMs are likely to experience adverse effects and toxicities (Binkhorst et al. 2015; Goetz and Ingle 2014; Province et al. 2014; Stingl and Viviani 2015). PMs had lower endoxifen levels than EMs (Gjerde et al. 2008). Tamoxifen-treated cancer patients carrying CYP2D6*4, *5, *10, or *41 associated with significantly decreased formation of antiestrogenic metabolites had significantly more recurrences of breast cancer and shorter relapse-free periods (Bonanni et al. 2006; Borges et al. 2006; Goetz et al. 2005, 2007; Jin et al. 2005; Schroth et al. 2007). In the study by Goetz et al. (2005) with 256 cancer patients treated with tamoxifen, the patients with the CYP2D6*4/*4 genotype (n = 13) have poorer clinical outcomes with shorter relapse-free and disease-free survival, and the 5-year disease-free survival for CYP2D6*4/*4 homozygous patients is only 46%, compared with 83% for patients who are not carriers of the CYP2D6*4 allele. None of the women with the *4/*4 genotype experiences moderate or severe hot flashes, compared with 20% of the women with either the *1/*4 or *1/*1 genotypes. However, this study does not measure the plasma levels of 4-OH-tamoxifen and endoxifen.

In a study with 206 cancer patients receiving adjuvant tamoxifen monotherapy, tamoxifen-treated patients carrying the CYP2D6*4, *5, *10, or *41 allele associated with impaired formation of antiestrogenic metabolites have significantly more recurrences of breast cancer, shorter relapse-free periods, and worse event-free survival rates compared with patients with functional CYP2D6 alleles (Schroth et al. 2007). The plasma concentrations of endoxifen and 4-OH-tamoxifen are not determined in these patients.

In Chinese breast cancer patients treated with tamoxifen (n = 152), the serum 4-OH-tamoxifen concentrations are significantly lower in women homozygous for CYP2D6*10 than those with the homozygous wild type (Zhou et al. 2008).

Patients carrying CYP2D6*10/*10 have worse disease-free survival than those with the *1/*10 and *1/*1 genotypes. In Japanese breast cancer patients (n = 67), the elevated risk of recurrence within 10 years after the operation seems to be dependent on the CYP2D6*10 allele (Kiyotani et al. 2008). Those patients homozygous for the CYP2D6*10 allele have a significantly higher incidence of recurrence compared with those homozygous for the wild-type CYP2D6*1 allele. Patients with the CYP2D6*10/*10 genotype show a significantly shorter recurrence-free survival period compared to patients with CYP2D6*1/*1 after adjustment of other prognosis factors (Kiyotani et al. 2008).

In a study with Korean breast cancer patients (n = 202), carriers of nonfunctional CYP2D6*10/*10 (n = 49) demonstrate significantly lower steady-state plasma concentrations of endoxifen and 4-OH-tamoxifen than those with other genotypes (n = 153) (endoxifen: 7.9 vs. 18.9 ng/ml; 4-OH-tamoxifen: 1.5 vs. 2.6 ng/ml) (Lim et al. 2007). In this study, there is a correlation between the CYP2D6 genotype and number of disease sites. The median time to progression for patients receiving tamoxifen is shorter in those carrying CYP2D6*10/*10 than for others (5.0 vs. 21.8 months).

PMs may also have increased risk of developing adverse effects during tamoxifen treatment, which include venous thrombosis and endometrial cancer. However, the evidence is lacking. Many studies have identified the genetic CYP2D6 status as an independent predictor for the outcome of tamoxifen treatment in women with breast cancer, but others do not observe this relationship (Nowell et al. 2005; Wegman et al. 2005, 2007). Discrepant results may be explained by differences in study designs, including size, different genetic models for the assessment of phenotypes, or stratification effects.

Although the metabolism of tamoxifen to 4-OH-tamoxifen is catalyzed by multiple enzymes, endoxifen is formed predominantly by the CYP2D6-mediated oxidation of N-desmethyltamoxifen. There is a clear gene–concentration effect for the formation of endoxifen and 4-OH-tamoxifen. Currently available clinical findings strongly suggest that CYP2D6 status is an independent predictor for the outcome with tamoxifen treatment (Goetz et al. 2008; Kirchheiner 2008; Takimoto 2007). CYP2D6*10/*10 is associated with lower steady-state plasma concentrations of active tamoxifen metabolites, which could possibly influence the clinical outcome by tamoxifen in Asian breast cancer patients. Thus, a more favorable tamoxifen treatment seems to be feasible through a priori genetic assessment of CYP2D6 and proper dose adjustment may be needed when the CYP2D6 genotype is determined in a patient. Further studies are needed in women receiving tamoxifen to fully define the effect of CYP2D6 genetic polymorphisms and medications that inhibit CYP2D6 on tamoxifen response.

The DPWG Guideline recommends considering using aromatase inhibitors for postmenopausal women because of increased risk for relapse of breast cancer with tamoxifen for CYP2D6 PMs and IMs (Swen et al. 2011). For IM patients, concomitant CYP2D6 inhibitor use should be avoided.

7.19 OTHER DRUGS

7.19.1 CEVIMELINE

Cevimeline, a drug used for the treatment of symptoms of dry mouth in Sjögren's syndrome, is mainly inactivated by CYP2D6 and 3A4 (Washio et al. 2003); therefore, the FDA-approved drug label states that drugs that inhibit CYP2D6 and 3A4 also inhibit the metabolism of cevimeline. Additionally, cevimeline should be used with caution in individuals known or suspected to be CYP2D6 PMs, since they may be at a higher risk of adverse events.

7.19.2 DARIFENACIN

Darifenacin, a drug for the treatment of overactive bladder (Abrams and Andersson 2007; Steers 2006), is primarily metabolized by CYP2D6 and 3A4 (Leone Roberti Maggiore et al. 2012). The metabolism of darifenacin in CYP2D6 PMs will be principally mediated via CYP3A4. The darifenacin ratios (PM vs. EM) for C_{max} and AUC after darifenacin 15 mg once daily at steady state are 1.9 and 1.7, respectively. The FDA-approved drug label for darifenacin notes that CYP2D6 PMs may have increased C_{max} of darifenacin, as compared to CYP2D6 EMs. However, the label does not comment on the clinical significance of these increased concentrations.

7.19.3 TETRABENAZINE

Tetrabenazine, a drug for the treatment of Huntington's disease chorea, is metabolized primarily by CYP2D6 (Fasano and Bentivoglio 2009; Frank 2014; Guay 2010). It is subject to important drug–drug interactions with inhibitors and inducers of CYP2D6, reserpine, and lithium. The effect of CYP2D6 inhibition by paroxetine on the pharmacokinetics of tetrabenazine and its metabolites is studied in 25 healthy subjects after a single 50-mg dose of tetrabenazine given after 10 days of administration of paroxetine 20 mg daily (Guay 2010). There is an approximately 30% increase in C_{max} and an approximately threefold increase in AUC for α-hydroxytetrabenazine in subjects given paroxetine before tetrabenazine compared to tetrabenazine given alone. For β-hydroxy-tetrabenazine, the C_{max} and AUC are increased 2.4- and 9-fold, respectively, in subjects given paroxetine before tetrabenazine given alone. In a recent study, it is found that CYP2D6 UM patients need a longer titration (8 vs. 3.3, 4.4, and 3 weeks, respectively; $P < 0.01$) to achieve optimal benefit and require a higher average daily dose than the other patients, but this difference does not reach statistical significance (Mehanna et al. 2013). The treatment response is less robust in the IMs when compared with the EM patients ($P = 0.013$), but there are no statistically significant differences between the various phenotype groups with regard to adverse effects (Mehanna et al. 2013).

The FDA-approved drug label for tetrabenazine states that patients requiring doses above 50 mg per day should be genotyped for CYP2D6 to determine if the patient is a PM or EM. CYP2D6 PMs should be treated with lower doses. The maximum daily dose in PMs is 50 mg with a maximum single dose of 25 mg. The maximum daily dose in EMs and IMs is 100 mg with a maximum single dose of 37.5 mg. Medications that are strong CYP2D6 inhibitors such as quinidine or antidepressants (e.g., fluoxetine, paroxetine) significantly increase the exposure to active α-hydroxy-tetrabenazine and β-hydroxy-tetrabenazine; therefore, the total dose of tetrabenazine should not exceed a maximum of 50 mg and the maximum single dose should not exceed 25 mg.

7.19.4 VORTIOXETINE

Vortioxetine, indicated as a treatment for major depressive disorder (Dhir 2013; Pearce and Murphy 2014; Sanchez et al. 2015), is mainly metabolized by CYP2D6 (Hvenegaard et al. 2012). Individuals who are CYP2D6 PMs have approximately twice the plasma concentration of vortioxetine, as compared to those who are CYP2D6 EMs. The FDA-approved drug label for vortioxetine notes that the maximum recommended dose in patients who are known CYP2D6 PMs is 10 mg/day. Additionally, patients receiving a strong inhibitor of CYP2D6 (e.g., bupropion or paroxetine) should have their dose reduced by one-half. In patients receiving concomitant CYP inducers (e.g., rifampicin, phenytoin), a dose increase can be considered, but the maximum dose is not recommended to exceed three times the original dose.

7.20 CONCLUSIONS AND FUTURE PERSPECTIVES

The clinical impact that a *CYP2D6* polymorphism has on pharmacotherapy with any of the CYP2D6 substrate drugs is determined by the resulting metabolizer status that the polymorphisms cause in the individual receiving pharmacotherapy, as well as whether the parent drug is active or if it requires CYP2D6 to metabolize it into an active metabolite. If the parent drug is active, then UMs may experience a lack of efficacy whereas IMs and PMs may suffer from complications resulting from higher-than-desired plasma concentrations of the drug. If the parent drug must be converted to an active metabolite, IMs and PMs may be deficient in the formation of the active species and thus cause therapeutic failure. A number of allelic variants of the human *CYP2D6* gene have been defined, which may result in a complete absence of enzyme activity, reduced activity, normal activity, or even increased activity phenotypically. Among the most important variants are *CYP2D6*2*, **3*, **4*, **5*, **10*, **17*, and **41*. In addition, a large number of low-frequency alleles of *CYP2D6* associated with the PM phenotype have been identified. Rearrangements within the gene *CYP2D* locus have resulted in variant alleles harboring two or multiple *CYP2D6* genes, deletion of the entire gene, or creation of fused genes. Unlike other CYPs, CYP2D6 is not inducible, and thus genetic mutations are largely responsible for the interindividual variation in enzyme expression and activity. This often presents as a barrier to reaching optimal therapeutic concentrations. The ethnicity must be considered when genotyping test is performed

TABLE 7.2

Reported Clinical Pharmacogenetic Studies on Drugs That Are Metabolized by CYP2D6

Drugs	Drug Class	Folds in Concentration/AUC Difference (PMs vs. EMs)	Response/Toxicity	References
Atomoxetine	A drug for attention-deficit/hyperactivity disorder	8-fold	More side effects in PMs	Sauer et al. 2003
Carvedilol	β-Blocker	Single dose: 2.5-fold for *R*-carvedilol; repeated dose: ~2-fold for *R*- and *S*-enantiomers	Not determined	Giessmann et al. 2004; Zhou and Wood 1995
Chlorpheniramine	Antihistamine	Single dose: 3.2- and 2.4-fold for *S*- and *R*-enantiomer, respectively	Not determined	Yasuda et al. 1995
Chlorpromazine	Antipsychotic	Single dose: nonsignificant increase in *10/*10 (1.3-fold) and *1/*10 (1.7-fold)	Not determined	Sunwoo et al. 2004
Clomipramine	Tricyclic antidepressant	Single dose: 2-fold; repeated dose: 1.6- and 3.7-fold for desmethylclomipramine + clomipramine	Little evidence for altered effect	Nielsen et al. 1992; Sindrup et al. 1990b
Codeine	Opioid	Single and repeated dosing: no difference in codeine levels; very low or undetectable morphine concentrations in PMs	Lack/reduction of analgesia; less opioid dependence	Caraco et al. 1996; Eckhardt et al. 1998; Lotsch et al. 2004
Desipramine	Tricyclic antidepressant	Single dose: 4- to 8-fold; repeated dosing: 3- to 4.5-fold (*10/*5+*10/*10), 1.5-fold (*1/*10)	Not determined	Brosen et al. 1986a, 1993; Spina et al. 1987, 1997
Dihydrocodeine	Opioid	Single dose: no difference in parent drug; 15% level of dihydromorphine levels in PMs	Similar analgesic effects in EMs and PMs	Fromm et al. 1995
Doxepin	Tricyclic antidepressant	Single dose: 2.9-fold higher in active metabolites in PMs	Little evidence for altered effect	Kirchheiner et al. 2002
Flecainide	Antiarrhythmic	Single dose: 1.7-, 2.4-, and 3.6-fold for racemic, *S*-, and *R*-enantiomer, respectively; repeated dosing: no difference	No difference in QRS prolongation; more adverse effects in PMs	Gross et al. 1991; Mikus et al. 1989
Fluoxetine	SSRI	Single dose: 3.9-, 11.5-, and 2.4-fold for racemic and *S*- and *R*-fluoxetine, respectively, 0.5-fold for norfluoxetine; repeated dosing: 2.3- and 0.3-fold for *S*-fluoxetine and *S*-norfluoxetine, respectively	Little evidence for more adverse reactions	Fjordside et al. 1999; Hamelin et al. 1996; Scordo et al. 2005
Imipramine	Tricyclic antidepressant	Single dose: 9.1-fold for desipramine; repeated dosing: 2.2-, 6.8-, and 4.8-fold for imipramine, desipramine, and the sum of both, respectively	Little evidence for altered effect/sedation or postural hypotension	Brosen et al. 1986b; Spina et al. 1987; Steiner and Spina 1987
Maprotiline	Tricyclic antidepressant	Repeated dosing: 3.5-fold	Increased pulmonary effect	Firkusny and Gleiter 1994
Metoprolol	β-Blocker	Single dose: 4- to 8-fold (PM), 0.5-fold (UM); 2.6-, 2.5-, and 3.2-fold (*10/*10); 1.4-, 1.3-, and 1.4-fold (*1/*10) for racemic and *S*- and *R*-metoprolol, respectively; repeated dosing: 3- to 4-fold	Little evidence for more adverse effect	Hamelin et al. 2000; Kirchheiner et al. 2004a; Lennard et al. 1982, 1983
Mexiletine	Antiarrhythmic	Single dose: 2.0- to 2.3-, 1.4- to 2.1-, and 1.6- to 2.2-fold for racemic and *R*- and *S*-mexiletine, respectively	Highest value of mexiletine/ *p*-hydroxymexiletine in PMs	Abolfathi et al. 1993; Broly et al. 1991b
Mianserin	Antidepressant	Single dose: 1.8- and 1.5-fold for mianserin and desmethylmianserin, respectively; repeated dosing: no difference	A slightly greater response	Dahl et al. 1994b; Mihara et al. 1997b
Nortriptyline	Tricyclic antidepressant	Single dose: 3.3-fold, no significant decrease in UM; repeated dosing: 2.1-fold (*10/*5+*10/*10)	More likely to have dose-dependent adverse reactions	Dalen et al. 1998; Morita et al. 2000; Yue et al. 1998
Paroxetine	SSRI	Single dose: 7.1-fold; 3.4-fold (*10/*10); 4-fold (*1/*10)	More sexual dysfunction in EMs	Sindrup et al. 1992b
Perhexiline	Antianginal agent	Single dose: 6-fold	More toxicity in PMs	Cooper et al. 1984

(Continued)

TABLE 7.2 (CONTINUED)

Reported Clinical Pharmacogenetic Studies on Drugs That Are Metabolized by CYP2D6

Drugs	Drug Class	Folds in Concentration/AUC Difference (PMs vs. EMs)	Response/Toxicity	References
Perphenazine	Antipsychotic	Single dose: 4.1-fold; repeated doing: 2-fold	Increase in adverse effects in PMs (primarily extrapyramidal and sedative)	Dahl-Puustinen et al. 1989; Linnet and Wiborg 1996b; Ozdemir et al. 1997, 2007; Pollock et al. 1995
Propafenone	Antiarrhythmic	Single dose: 7.9-fold; repeated dosing: 7-, 0.2-, 4.3-, and 4.8-fold for racemic parent, 5-OH-propafenone, and *R*- and *S*-enantiomers, respectively	Increased adverse effects in PMs	Cai et al. 1999a; Chen and Cai 2003; Dilger et al. 1999; Labbe et al. 2000; Lee et al. 1990; Siddoway et al. 1987
Risperidone	Antipsychotic	Repeated dosing: 4.8- and 0.5-fold for parent, 9-OH-risperidone, respectively	More adverse effects in the CNS and greater β-blockade in PMs	Bondolfi et al. 2002; Leon et al. 2007; Nyberg et al. 1995a; Olesen et al. 1998; Roh et al. 2001b; Scordo et al. 1999; van der Weide et al. 2005
Tamoxifen	Selective estrogen receptor modulator	Lower endoxifen and 4-OH-tamoxifen in PMs	More recurrences of breast cancer and shorter relapse-free periods in PMs	Bonanni et al. 2006; Borges et al. 2006; Goetz et al. 2005, 2007; Jin et al. 2005; Lim et al. 2007; Schroth et al. 2007
Thioridazine	Antipsychotic	Single dose: 4.5- and 1.4-fold for thioridazine and thioridazine + mesoridazine + sulforidazine, respectively	No evidence	von Bahr et al. 1991
Timolol	β-Blocker	Single dose: 2- to 4-fold	Increased effect	Lewis et al. 1985
Tolterodine	Muscarinic receptor antagonist	Repeated dosing: 10-fold for tolterodine; 30-fold (PM)	Antimuscarinic effect unchanged	Brynne et al. 1998
Tramadol	Opioid	Single dose: 1.2-, 1.3-, and 0.5-fold for (+)-, (−)-tramadol, and (−)-desmethyltramadol, respectively	Less analgesic effects in PMs	Fliegert et al. 2005; Poulsen et al. 1996a
Tropisetron	Antiemetic	Single dose: 6.9-; 6.8-fold (*10/*5+*10/*10); 1.9-fold (*1/*10)	Decreased antiemetic effect	Kaiser et al. 2002; Kim et al. 2003
Venlafaxine	SRI (antidepressant)	Single dose: 2.3- and 0.3-fold for venlafaxine and desmethylvenlafaxine, respectively; repeated dosing: 3.4- to 3.9-fold for venlafaxine	No increased adverse effects	Fukuda et al. 2000b; Lessard et al. 1999; Lindh et al. 2003; Whyte et al. 2006
Zuclopenthixol	Antipsychotic	Single dose: 1.9-fold; repeated dosing: 1.6-fold	Not determined	Dahl et al. 1991; Linnet and Wiborg 1996a

because there are considerable differences in the distribution of the most common alleles of *CYP2D6* among various ethnic groups. For Caucasian subjects, assessment of the most common alleles that result in loss of function would need testing for *CYP2D6*4* and should include the *3*, *5*, *6*, *10*, and *41* alleles. Determination of the *10* and *17* would be critical for the prediction of the phenotype in Asians and Africans, respectively.

To date, the functional impact of most *CYP2D6* alleles has not been systematically assessed for most clinically important drugs that are mainly metabolized by CYP2D6 (Table 7.2),

though some initial evidence has been identified for very limited drugs. The majority of reported in vivo pharmacogenetic data on CYP2D6 are from single-dose and steady-state pharmacokinetic studies for a small number of drugs. Pharmacodynamic data on *CYP2D6* polymorphisms are scant for most drug studies. Given that genotyping test for *CYP2D6* is not routinely performed in clinical practice and there is an uncertainty of genotype–phenotype, gene–concentration, and gene–dose relationships, further prospective studies on the clinical impact of CYP2D6-dependent metabolism of drugs are warranted in large cohorts of samples.

REFERENCES

Abdel-Rahman, S. M., Leeder, J. S., Wilson, J. T., Gaedigk, A., Gotschall, R. R., Medve, R., Liao, S., Spielberg, S. P., Kearns, G. L. 2002. Concordance between tramadol and dextromethorphan parent/metabolite ratios: The influence of CYP2D6 and non-CYP2D6 pathways on biotransformation. *J Clin Pharmacol* 42:24–29.

Abolfathi, Z., Fiset, C., Gilbert, M., Moerike, K., Belanger, P. M., Turgeon, J. 1993. Role of polymorphic debrisoquin 4-hydroxylase activity in the stereoselective disposition of mexiletine in humans. *J Pharmacol Exp Ther* 266:1196–1201.

Abrams, P., Andersson, K. E. 2007. Muscarinic receptor antagonists for overactive bladder. *BJU Int* 100:987–1006.

Agundez, J. A., Ledesma, M. C., Ladero, J. M., Benitez, J. 1995. Prevalence of CYP2D6 gene duplication and its repercussion on the oxidative phenotype in a white population. *Clin Pharmacol Ther* 57:265–269.

Agundez, J. A., Ramirez, R., Hernandez, M., Llerena, A., Benitez, J. 1997. Molecular heterogeneity at the CYP2D gene locus in Nicaraguans: Impact of gene-flow from Europe. *Pharmacogenetics* 7:337–340.

Aklillu, E., Persson, I., Bertilsson, L., Johansson, I., Rodrigues, F., Ingelman-Sundberg, M. 1996. Frequent distribution of ultrarapid metabolizers of debrisoquine in an ethiopian population carrying duplicated and multiduplicated functional CYP2D6 alleles. *J Pharmacol Exp Ther* 278:441–446.

Aklillu, E., Herrlin, K., Gustafsson, L. L., Bertilsson, L., Ingelman-Sundberg, M. 2002. Evidence for environmental influence on CYP2D6-catalysed debrisoquine hydroxylation as demonstrated by phenotyping and genotyping of Ethiopians living in Ethiopia or in Sweden. *Pharmacogenetics* 12:375–383.

Aklillu, E., Kalow, W., Endrenyi, L., Harper, P., Miura, J., Ozdemir, V. 2007. CYP2D6 and DRD2 genes differentially impact pharmacodynamic sensitivity and time course of prolactin response to perphenazine. *Pharmacogenet Genomics* 17:989–993.

Akutsu, T., Kobayashi, K., Sakurada, K., Ikegaya, H., Furihata, T., Chiba, K. 2007. Identification of human cytochrome p450 isozymes involved in diphenhydramine N-demethylation. *Drug Metab Dispos* 35:72–78.

Alexanderson, B., Evans, D. A., Sjoqvist, F. 1969. Steady-state plasma levels of nortriptyline in twins: Influence of genetic factors and drug therapy. *Br Med J* 4:764–768.

Alvan, G., Bechtel, P., Iselius, L., Gundert-Remy, U. 1990. Hydroxylation polymorphisms of debrisoquine and mephenytoin in European populations. *Eur J Clin Pharmacol* 39:533–537.

Anzenbacherova, E., Anzenbacher, P., Perlik, F., Kvetina, J. 2000. Use of a propafenone metabolic ratio as a measure of CYP2D6 activity. *Int J Clin Pharmacol Ther* 38:426–429.

Ashforth, E. I., Palmer, J. L., Bye, A., Bedding, A. 1994. The pharmacokinetics of ondansetron after intravenous injection in healthy volunteers phenotyped as poor or extensive metabolisers of debrisoquine. *Br J Clin Pharmacol* 37:389–391.

Avenoso, A., Spina, E., Campo, G., Facciola, G., Ferlito, M., Zuccaro, P., Perucca, E., Caputi, A. P. 1997. Interaction between fluoxetine and haloperidol: Pharmacokinetic and clinical implications. *Pharmacol Res* 35:335–339.

Aynacioglu, A. S., Sachse, C., Bozkurt, A., Kortunay, S., Nacak, M., Schroder, T., Kayaalp, S. O., Roots, I., Brockmoller, J. 1999. Low frequency of defective alleles of cytochrome P450 enzymes 2C19 and 2D6 in the Turkish population. *Clin Pharmacol Ther* 66:185–192.

Bachus, R., Bickel, U., Thomsen, T., Roots, I., Kewitz, H. 1999. The O-demethylation of the antidementia drug galanthamine is catalysed by cytochrome P450 2D6. *Pharmacogenetics* 9:661–668.

Barbey, J. T., Thompson, K. A., Echt, D. S., Woosley, R. L., Roden, D. M. 1988. Antiarrhythmic activity, electrocardiographic effects and pharmacokinetics of the encainide metabolites O-desmethyl encainide and 3-methoxy-O-desmethyl encainide in man. *Circulation* 77:380–391.

Barclay, M. L., Sawyers, S. M., Begg, E. J., Zhang, M., Roberts, R. L., Kennedy, M. A., Elliott, J. M. 2003. Correlation of CYP2D6 genotype with perhexiline phenotypic metabolizer status. *Pharmacogenetics* 13:627–632.

Barner, E. L., Gray, S. L. 1998. Donepezil use in Alzheimer disease. *Ann Pharmacother* 32:70–77.

Bathum, L., Johansson, I., Ingelman-Sundberg, M., Horder, M., Brosen, K. 1998. Ultrarapid metabolism of sparteine: Frequency of alleles with duplicated CYP2D6 genes in a Danish population as determined by restriction fragment length polymorphism and long polymerase chain reaction. *Pharmacogenetics* 8:119–123.

Bathum, L., Skjelbo, E., Mutabingwa, T. K., Madsen, H., Horder, M., Brosen, K. 1999. Phenotypes and genotypes for CYP2D6 and CYP2C19 in a black Tanzanian population. *Br J Clin Pharmacol* 48:395–401.

Begg, E. J., Sharman, J. R., Kidd, J. E., Sainsbury, R., Clark, D. W. 1989. Variability in the elimination of mianserin in elderly patients. *Br J Clin Pharmacol* 27:445–451.

Begre, S., von Bardeleben, U., Ladewig, D., Jaquet-Rochat, S., Cosendai-Savary, L., Golay, K. P., Kosel, M., Baumann, P., Eap, C. B. 2002. Paroxetine increases steady-state concentrations of (R)-methadone in CYP2D6 extensive but not poor metabolizers. *J Clin Psychopharmacol* 22:211–215.

Berecz, R., de la Rubia, A., Dorado, P., Fernandez-Salguero, P., Dahl, M. L., LLerena, A. 2003. Thioridazine steady-state plasma concentrations are influenced by tobacco smoking and CYP2D6, but not by the CYP2C9 genotype. *Eur J Clin Pharmacol* 59:45–50.

Bernal, M. L., Sinues, B., Johansson, I., McLellan, R. A., Wennerholm, A., Dahl, M. L., Ingelman-Sundberg, M., Bertilsson, L. 1999. Ten percent of North Spanish individuals carry duplicated or triplicated CYP2D6 genes associated with ultrarapid metabolism of debrisoquine. *Pharmacogenetics* 9:657–660.

Bertilsson, L. 2007. Metabolism of antidepressant and neuroleptic drugs by cytochrome p450s: Clinical and interethnic aspects. *Clin Pharmacol Ther* 82:606–609.

Bertilsson, L., Aberg-Wistedt, A., Gustafsson, L. L., Nordin, C. 1985. Extremely rapid hydroxylation of debrisoquine: A case report with implication for treatment with nortriptyline and other tricyclic antidepressants. *Ther Drug Monit* 7:478–480.

Bertilsson, L., Lou, Y. Q., Du, Y. L., Liu, Y., Kuang, T. Y., Liao, X. M., Wang, K. Y., Reviriego, J., Iselius, L., Sjoqvist, F. 1992. Pronounced differences between native Chinese and Swedish populations in the polymorphic hydroxylations of debrisoquin and S-mephenytoin. *Clin Pharmacol Ther* 51: 388–397.

Bertilsson, L., Dahl, M. L., Ekqvist, B., Llerena, A. 1993a. Disposition of the neuroleptics perphenazine, zuclopenthixol, and haloperidol cosegregates with polymorphic debrisoquine hydroxylation. *Psychopharmacol Ser* 10:230–237.

Bertilsson, L., Dahl, M. L., Sjoqvist, F., Aberg-Wistedt, A., Humble, M., Johansson, I., Lundqvist, E., Ingelman-Sundberg, M. 1993b. Molecular basis for rational megaprescribing in ultrarapid hydroxylators of debrisoquine. *Lancet* 341:63.

Bertilsson, L., Carrillo, J. A., Dahl, M. L., Llerena, A., Alm, C., Bondesson, U., Lindstrom, L., Rodriguez de la Rubia, I., Ramos, S., Benitez, J. 1994. Clozapine disposition covaries with CYP1A2 activity determined by a caffeine test. *Br J Clin Pharmacol* 38:471–473.

Beverage, J. N., Sissung, T. M., Sion, A. M., Danesi, R., Figg, W. D. 2007. CYP2D6 polymorphisms and the impact on tamoxifen therapy. *J Pharm Sci* 96:2224–2231.

Bijl, M. J., Visser, L. E., Hofman, A., Vulto, A. G., van Gelder, T., Stricker, B. H., van Schaik, R. H. 2008. Influence of the CYP2D6*4 polymorphism on dose, switching and discontinuation of antidepressants. *Br J Clin Pharmacol* 65:558–564.

Bijl, M. J., Visser, L. E., van Schaik, R. H., Kors, J. A., Witteman, J. C., Hofman, A., Vulto, A. G., van Gelder, T., Stricker, B. H. 2009. Genetic variation in the CYP2D6 gene is associated with a lower heart rate and blood pressure in beta-blocker users. *Clin Pharmacol Ther* 85:45–50.

Binkhorst, L., Mathijssen, R. H., Jager, A., van Gelder, T. 2015. Individualization of tamoxifen therapy: Much more than just CYP2D6 genotyping. *Cancer Treat Rev* 41:289–299.

Birgersdotter, U. M., Wong, W., Turgeon, J., Roden, D. M. 1992. Stereoselective genetically-determined interaction between chronic flecainide and quinidine in patients with arrhythmias. *Br J Clin Pharmacol* 33:275–280.

Bloomer, J. C., Woods, F. R., Haddock, R. E., Lennard, M. S., Tucker, G. T. 1992. The role of cytochrome P4502D6 in the metabolism of paroxetine by human liver microsomes. *Br J Clin Pharmacol* 33:521–523.

Bogni, A., Monshouwer, M., Moscone, A., Hidestrand, M., Ingelman-Sundberg, M., Hartung, T., Coecke, S. 2005. Substrate specific metabolism by polymorphic cytochrome P450 2D6 alleles. *Toxicol In Vitro* 19:621–629.

Bolaji, O. O., Coutts, R. T., Baker, G. B. 1993. Metabolism of trimipramine in vitro by human CYP2D6 isozyme. *Res Commun Chem Pathol Pharmacol* 82:111–120.

Bonanni, B., Macis, D., Maisonneuve, P., Johansson, H. A., Gucciardo, G., Oliviero, P., Travaglini, R. et al. 2006. Polymorphism in the CYP2D6 tamoxifen-metabolizing gene influences clinical effect but not hot flashes: Data from the Italian Tamoxifen Trial. *J Clin Oncol* 24:3708–3709; author reply 3709.

Bondolfi, G., Eap, C. B., Bertschy, G., Zullino, D., Vermeulen, A., Baumann, P. 2002. The effect of fluoxetine on the pharmacokinetics and safety of risperidone in psychotic patients. *Pharmacopsychiatry* 35:50–56.

Borges, S., Desta, Z., Li, L., Skaar, T. C., Ward, B. A., Nguyen, A., Jin, Y. et al. 2006. Quantitative effect of CYP2D6 genotype and inhibitors on tamoxifen metabolism: Implication for optimization of breast cancer treatment. *Clin Pharmacol Ther* 80:61–74.

Botsch, S., Gautier, J. C., Beaune, P., Eichelbaum, M., Kroemer, H. K. 1993. Identification and characterization of the cytochrome P450 enzymes involved in N-dealkylation of propafenone: Molecular base for interaction potential and variable disposition of active metabolites. *Mol Pharmacol* 43:120–126.

Bottiger, Y., Dostert, P., Benedetti, M. S., Bani, M., Fiorentini, F., Casati, M., Poggesti, I., Alm, C., Alvan, G., Bertilsson, L. 1996. Involvement of CYP2D6 but not CYP2C19 in nicergoline metabolism in humans. *Br J Clin Pharmacol* 42:707–711.

Brachtendorf, L., Jetter, A., Beckurts, K. T., Holscher, A. H., Fuhr, U. 2002. Cytochrome P450 enzymes contributing to demethylation of maprotiline in man. *Pharmacol Toxicol* 90:144–149.

Bradford, L. D. 2002. CYP2D6 allele frequency in European Caucasians, Asians, Africans and their descendants. *Pharmacogenomics* 3:229–243.

Breyer-Pfaff, U. 2004. The metabolic fate of amitriptyline, nortriptyline and amitriptylinoxide in man. *Drug Metab Rev* 36:723–746.

Breyer-Pfaff, U., Pfandl, B., Nill, K., Nusser, E., Monney, C., Jonzier-Perey, M., Baettig, D., Baumann, P. 1992. Enantioselective amitriptyline metabolism in patients phenotyped for two cytochrome P450 isozymes. *Clin Pharmacol Ther* 52:350–358.

Brockmeyer, N. H., Breithaupt, H., Ferdinand, W., von Hattingberg, M., Ohnhaus, E. E. 1989. Kinetics of oral and intravenous mexiletine: Lack of effect of cimetidine and ranitidine. *Eur J Clin Pharmacol* 36:375–378.

Brockmoller, J., Kirchheiner, J., Schmider, J., Walter, S., Sachse, C., Muller-Oerlinghausen, B., Roots, I. 2002. The impact of the CYP2D6 polymorphism on haloperidol pharmacokinetics and on the outcome of haloperidol treatment. *Clin Pharmacol Ther* 72:438–452.

Brockmoller, J., Meineke, I., Kirchheiner, J. 2007. Pharmacokinetics of mirtazapine: Enantioselective effects of the CYP2D6 ultra rapid metabolizer genotype and correlation with adverse effects. *Clin Pharmacol Ther* 81:699–707.

Broly, F., Meyer, U. A. 1993. Debrisoquine oxidation polymorphism: Phenotypic consequences of a 3-base-pair deletion in exon 5 of the CYP2D6 gene. *Pharmacogenetics* 3:123–130.

Broly, F., Libersa, C., Lhermitte, M., Dupuis, B. 1990. Inhibitory studies of mexiletine and dextromethorphan oxidation in human liver microsomes. *Biochem Pharmacol* 39:1045–1053.

Broly, F., Vandamme, N., Caron, J., Libersa, C., Lhermitte, M. 1991a. Single-dose quinidine treatment inhibits mexiletine oxidation in extensive metabolizers of debrisoquine. *Life Sci* 48:PL123–PL128.

Broly, F., Vandamme, N., Libersa, C., Lhermitte, M. 1991b. The metabolism of mexiletine in relation to the debrisoquine/sparteine-type polymorphism of drug oxidation. *Br J Clin Pharmacol* 32:459–466.

Broly, F., Marez, D., Lo Guidice, J. M., Sabbagh, N., Legrand, M., Boone, P., Meyer, U. A. 1995. A nonsense mutation in the cytochrome P450 CYP2D6 gene identified in a Caucasian with an enzyme deficiency. *Hum Genet* 96:601–603.

Brosen, K., Klysner, R., Gram, L. F., Otton, S. V., Bech, P., Bertilsson, L. 1986a. Steady-state concentrations of imipramine and its metabolites in relation to the sparteine/debrisoquine polymorphism. *Eur J Clin Pharmacol* 30:679–684.

Brosen, K., Otton, S. V., Gram, L. F. 1986b. Imipramine demethylation and hydroxylation: Impact of the sparteine oxidation phenotype. *Clin Pharmacol Ther* 40:543–549.

Brosen, K., Zeugin, T., Meyer, U. A. 1991. Role of P450IID6, the target of the sparteine-debrisoquin oxidation polymorphism, in the metabolism of imipramine. *Clin Pharmacol Ther* 49:609–617.

Brosen, K., Hansen, J. G., Nielsen, K. K., Sindrup, S. H., Gram, L. F. 1993. Inhibition by paroxetine of desipramine metabolism in extensive but not in poor metabolizers of sparteine. *Eur J Clin Pharmacol* 44:349–355.

Brynne, N., Dalen, P., Alvan, G., Bertilsson, L., Gabrielsson, J. 1998. Influence of CYP2D6 polymorphism on the pharmacokinetics and pharmacodynamic of tolterodine. *Clin Pharmacol Ther* 63:529–539.

Brynne, N., Bottiger, Y., Hallen, B., Bertilsson, L. 1999a. Tolterodine does not affect the human in vivo metabolism of the probe drugs caffeine, debrisoquine and omeprazole. *Br J Clin Pharmacol* 47:145–150.

Brynne, N., Forslund, C., Hallen, B., Gustafsson, L. L., Bertilsson, L. 1999b. Ketoconazole inhibits the metabolism of tolterodine in subjects with deficient CYP2D6 activity. *Br J Clin Pharmacol* 48:564–572.

Brynne, N., Svanstrom, C., Aberg-Wistedt, A., Hallen, B., Bertilsson, L. 1999c. Fluoxetine inhibits the metabolism of tolterodine-pharmacokinetic implications and proposed clinical relevance. *Br J Clin Pharmacol* 48:553–563.

Bushe, C., Shaw, M., Peveler, R. C. 2008. A review of the association between antipsychotic use and hyperprolactinaemia. *J Psychopharmacol* 22:46–55.

Cacabelos, R., Llovo, R., Fraile, C., Fernandez-Novoa, L. 2007. Pharmacogenetic aspects of therapy with cholinesterase inhibitors: The role of CYP2D6 in Alzheimer's disease pharmacogenetics. *Curr Alzheimer Res* 4:479–500.

Cai, W. M., Chen, B., Cai, M. H., Chen, Y., Zhang, Y. D. 1999a. The influence of CYP2D6 activity on the kinetics of propafenone enantiomers in Chinese subjects. *Br J Clin Pharmacol* 47:553–556.

Cai, W. M., Chen, B., Zhou, Y., Zhang, Y. D. 1999b. Fluoxetine impairs the CYP2D6-mediated metabolism of propafenone enantiomers in healthy Chinese volunteers. *Clin Pharmacol Ther* 66:516–521.

Cai, W. M., Nikoloff, D. M., Pan, R. M., de Leon, J., Fanti, P., Fairchild, M., Koch, W. H., Wedlund, P. J. 2006. CYP2D6 genetic variation in healthy adults and psychiatric African-American subjects: Implications for clinical practice and genetic testing. *Pharmacogenomics J* 6:343–350.

Candiotti, K. A., Birnbach, D. J., Lubarsky, D. A., Nhuch, F., Kamat, A., Koch, W. H., Nikoloff, M., Wu, L., Andrews, D. 2005. The impact of pharmacogenomics on postoperative nausea and vomiting: Do CYP2D6 allele copy number and polymorphisms affect the success or failure of ondansetron prophylaxis? *Anesthesiology* 102:543–549.

Capucci, A., Boriani, G., Marchesini, B., Strocchi, E., Tomasi, L., Balducelli, M., Frabetti, L., Ambrosioni, E., Magnani, B. 1990. Minimal effective concentration values of propafenone and 5-hydroxy-propafenone in acute and chronic therapy. *Cardiovasc Drugs Ther* 4:281–287.

Caraco, Y., Sheller, J., Wood, A. J. 1996. Pharmacogenetic determination of the effects of codeine and prediction of drug interactions. *J Pharmacol Exp Ther* 278:1165–1174.

Carcillo, J. A., Adedoyin, A., Burckart, G. J., Frye, R. F., Venkataramanan, R., Knoll, C., Thummel, K. et al. 2003. Coordinated intrahepatic and extrahepatic regulation of cytochrome p4502D6 in healthy subjects and in patients after liver transplantation. *Clin Pharmacol Ther* 73:456–467.

Carey, E. L., Jr., Duff, H. J., Roden, D. M., Primm, R. K., Wilkinson, G. R., Wang, T., Oates, J. A., Woosley, R. L. 1984. Encainide and its metabolites. Comparative effects in man on ventricular arrhythmia and electrocardiographic intervals. *J Clin Invest* 73:539–547.

Carrillo, J. A., Herraiz, A. G., Ramos, S. I., Gervasini, G., Vizcaino, S., Benitez, J. 2003. Role of the smoking-induced cytochrome P450 (CYP)1A2 and polymorphic CYP2D6 in steady-state concentration of olanzapine. *J Clin Psychopharmacol* 23:119–127.

Cascorbi, I. 2003. Pharmacogenetics of cytochrome P4502D6: Genetic background and clinical implication. *Eur J Clin Invest* 33:17–22.

Charlier, C., Broly, F., Lhermitte, M., Pinto, E., Ansseau, M., Plomteux, G. 2003. Polymorphisms in the CYP 2D6 gene: Association with plasma concentrations of fluoxetine and paroxetine. *Ther Drug Monit* 25:738–742.

Chen, B., Cai, W. M. 2003. Influence of CYP2D6*10B genotype on pharmacokinetics of propafenone enantiomers in Chinese subjects. *Acta Pharmacol Sin* 24:1277–1280.

Chida, M., Yokoi, T., Nemoto, N., Inaba, M., Kinoshita, M., Kamataki, T. 1999. A new variant CYP2D6 allele (CYP2D6*21) with a single base insertion in exon 5 in a Japanese population associated with a poor metabolizer phenotype. *Pharmacogenetics* 9:287–293.

Chida, M., Ariyoshi, N., Yokoi, T., Nemoto, N., Inaba, M., Kinoshita, M., Kamataki, T. 2002. New allelic arrangement CYP2D6*36 × 2 found in a Japanese poor metabolizer of debrisoquine. *Pharmacogenetics* 12:659–662.

Chou, W. H., Yan, F. X., de Leon, J., Barnhill, J., Rogers, T., Cronin, M., Pho, M. et al. 2000. Extension of a pilot study: Impact from the cytochrome P450 2D6 polymorphism on outcome and costs associated with severe mental illness. *J Clin Psychopharmacol* 20:246–251.

Christensen, M., Tybring, G., Mihara, K., Yasui-Furokori, N., Carrillo, J. A., Ramos, S. I., Andersson, K., Dahl, M. L., Bertilsson, L. 2002. Low daily 10-mg and 20-mg doses of fluvoxamine inhibit the metabolism of both caffeine (cytochrome P4501A2) and omeprazole (cytochrome P4502C19). *Clin Pharmacol Ther* 71:141–152.

Chua, E. W., Foulds, J., Miller, A. L., Kennedy, M. A. 2013. Novel CYP2D6 and CYP2C19 variants identified in a patient with adverse reactions towards venlafaxine monotherapy and dual therapy with nortriptyline and fluoxetine. *Pharmacogenet Genomics* 23:494–497.

Cobb, M. N., Desai, J., Brown, L. S., Jr., Zannikos, P. N., Rainey, P. M. 1998. The effect of fluconazole on the clinical pharmacokinetics of methadone. *Clin Pharmacol Ther* 63:655–662.

Coller, J. K., Joergensen, C., Foster, D. J., James, H., Gillis, D., Christrup, L., Somogyi, A. A. 2007. Lack of influence of CYP2D6 genotype on the clearance of (R)-, (S)- and racemic-methadone. *Int J Clin Pharmacol Ther* 45:410–417.

Contreras, A. V., Monge-Cazares, T., Alfaro-Ruiz, L., Hernandez-Morales, S., Miranda-Ortiz, H., Carrillo-Sanchez, K., Jimenez-Sanchez, G., Silva-Zolezzi, I. 2011. Resequencing, haplotype construction and identification of novel variants of CYP2D6 in Mexican Mestizos. *Pharmacogenomics* 12:745–756.

Cooper, R. G., Evans, D. A., Whibley, E. J. 1984. Polymorphic hydroxylation of perhexiline maleate in man. *J Med Genet* 21:27–33.

Cooper, R. G., Evans, D. A., Price, A. H. 1987. Studies on the metabolism of perhexiline in man. *Eur J Clin Pharmacol* 32:569–576.

Crettol, S., Deglon, J. J., Besson, J., Croquette-Krokar, M., Hammig, R., Gothuey, I., Monnat, M., Eap, C. B. 2006. ABCB1 and cytochrome P450 genotypes and phenotypes: Influence on methadone plasma levels and response to treatment. *Clin Pharmacol Ther* 80:668–681.

Crewe, H. K., Notley, L. M., Wunsch, R. M., Lennard, M. S., Gillam, E. M. 2002. Metabolism of tamoxifen by recombinant human cytochrome P450 enzymes: Formation of the 4-hydroxy, 4'-hydroxy and N-desmethyl metabolites and isomerization of trans-4-hydroxytamoxifen. *Drug Metab Dispos* 30:869–874.

Crews, K. R., Gaedigk, A., Dunnenberger, H. M., Leeder, J. S., Klein, T. E., Caudle, K. E., Haidar, C. E. et al. 2014. Clinical Pharmacogenetics Implementation Consortium guidelines for cytochrome P450 2D6 genotype and codeine therapy: 2014 update. *Clin Pharmacol Ther* 95:376–382.

Cui, Y. M., Teng, C. H., Pan, A. X., Yuen, E., Yeo, K. P., Zhou, Y., Zhao, X., Long, A. J., Bangs, M. E., Wise, S. D. 2007. Atomoxetine pharmacokinetics in healthy Chinese subjects and effect of the CYP2D6*10 allele. *Br J Clin Pharmacol* 64:445–449.

Dahl, M. L. 2002. Cytochrome p450 phenotyping/genotyping in patients receiving antipsychotics: Useful aid to prescribing? *Clin Pharmacokinet* 41:453–470.

Dahl, M. L., Ekqvist, B., Widen, J., Bertilsson, L. 1991. Disposition of the neuroleptic zuclopenthixol cosegregates with the polymorphic hydroxylation of debrisoquine in humans. *Acta Psychiatr Scand* 84:99–102.

Dahl, M. L., Johansson, I., Palmertz, M. P., Ingelman-Sundberg, M., Sjoqvist, F. 1992. Analysis of the CYP2D6 gene in relation to debrisoquin and desipramine hydroxylation in a Swedish population. *Clin Pharmacol Ther* 51:12–17.

Dahl, M. L., Llerena, A., Bondesson, U., Lindstrom, L., Bertilsson, L. 1994a. Disposition of clozapine in man: Lack of association with debrisoquine and *S*-mephenytoin hydroxylation polymorphisms. *Br J Clin Pharmacol* 37:71–74.

Dahl, M. L., Tybring, G., Elwin, C. E., Alm, C., Andreasson, K., Gyllenpalm, M., Bertilsson, L. 1994b. Stereoselective disposition of mianserin is related to debrisoquin hydroxylation polymorphism. *Clin Pharmacol Ther* 56:176–183.

Dahl, M. L., Johansson, I., Bertilsson, L., Ingelman-Sundberg, M., Sjoqvist, F. 1995. Ultrarapid hydroxylation of debrisoquine in a Swedish population. Analysis of the molecular genetic basis. *J Pharmacol Exp Ther* 274:516–520.

Dahl, M. L., Bertilsson, L., Nordin, C. 1996. Steady-state plasma levels of nortriptyline and its 10-hydroxy metabolite: Relationship to the CYP2D6 genotype. *Psychopharmacology (Berl)* 123:315–319.

Dahl-Puustinen, M. L., Liden, A., Alm, C., Nordin, C., Bertilsson, L. 1989. Disposition of perphenazine is related to polymorphic debrisoquin hydroxylation in human beings. *Clin Pharmacol Ther* 46:78–81.

Dalen, P., Frengell, C., Dahl, M. L., Sjoqvist, F. 1997. Quick onset of severe abdominal pain after codeine in an ultrarapid metabolizer of debrisoquine. *Ther Drug Monit* 19:543–544.

Dalen, P., Dahl, M. L., Bernal Ruiz, M. L., Nordin, J., Bertilsson, L. 1998. 10-Hydroxylation of nortriptyline in white persons with 0, 1, 2, 3, and 13 functional CYP2D6 genes. *Clin Pharmacol Ther* 63:444–452.

Daly, A. K., Leathart, J. B., London, S. J., Idle, J. R. 1995. An inactive cytochrome P450 CYP2D6 allele containing a deletion and a base substitution. *Hum Genet* 95:337–341.

Daly, A. K., Brockmoller, J., Broly, F., Eichelbaum, M., Evans, W. E., Gonzalez, F. J., Huang, J. D. et al. 1996a. Nomenclature for human CYP2D6 alleles. *Pharmacogenetics* 6:193–201.

Daly, A. K., Fairbrother, K. S., Andreassen, O. A., London, S. J., Idle, J. R., Steen, V. M. 1996b. Characterization and PCR-based detection of two different hybrid CYP2D7P/CYP2D6 alleles associated with the poor metabolizer phenotype. *Pharmacogenetics* 6:319–328.

Dayer, P., Desmeules, J., Leemann, T., Striberni, R. 1988. Bioactivation of the narcotic drug codeine in human liver is mediated by the polymorphic monooxygenase catalyzing debrisoquine 4-hydroxylation (cytochrome P-450 dbl/bufI). *Biochem Biophys Res Commun* 152:411–416.

de Leon, J., Dinsmore, L., Wedlund, P. 2003. Adverse drug reactions to oxycodone and hydrocodone in CYP2D6 ultrarapid metabolizers. *J Clin Psychopharmacol* 23:420–421.

de Leon, J., Susce, M. T., Pan, R. M., Fairchild, M., Koch, W. H., Wedlund, P. J. 2005. The CYP2D6 poor metabolizer phenotype may be associated with risperidone adverse drug reactions and discontinuation. *J Clin Psychiatry* 66:15–27.

Dehal, S. S., Kupfer, D. 1997. CYP2D6 catalyzes tamoxifen 4-hydroxylation in human liver. *Cancer Res* 57:3402–3406.

Desta, Z., Ward, B. A., Soukhova, N. V., Flockhart, D. A. 2004. Comprehensive evaluation of tamoxifen sequential biotransformation by the human cytochrome P450 system in vitro: Prominent roles for CYP3A and CYP2D6. *J Pharmacol Exp Ther* 310:1062–1075.

Dettling, M., Sachse, C., Muller-Oerlinghausen, B., Roots, I., Brockmoller, J., Rolfs, A., Cascorbi, I. 2000. Clozapine-induced agranulocytosis and hereditary polymorphisms of clozapine metabolizing enzymes: No association with myeloperoxidase and cytochrome P4502D6. *Pharmacopsychiatry* 33:218–220.

Dhir, A. 2013. Vortioxetine for the treatment of major depression. *Drugs Today (Barc)* 49:781–790.

Dilger, K., Greiner, B., Fromm, M. F., Hofmann, U., Kroemer, H. K., Eichelbaum, M. 1999. Consequences of rifampicin treatment on propafenone disposition in extensive and poor metabolizers of CYP2D6. *Pharmacogenetics* 9:551–559.

Dodgen, T. M., Hochfeld, W. E., Fickl, H., Asfaha, S. M., Durandt, C., Rheeder, P., Drogemoller, B. I. et al. 2013. Introduction of the AmpliChip CYP450 Test to a South African cohort: A platform comparative prospective cohort study. *BMC Med Genet* 14:20.

Doki, K., Homma, M., Kuga, K., Kusano, K., Watanabe, S., Yamaguchi, I., Kohda, Y. 2006. Effect of *CYP2D6* genotype on flecainide pharmacokinetics in Japanese patients with supraventricular tachyarrhythmia. *Eur J Clin Pharmacol* 62:919–926.

Eap, C. B., Laurian, S., Souche, A., Koeb, L., Reymond, P., Buclin, T., Baumann, P. 1992. Influence of quinidine on the pharmacokinetics of trimipramine and on its effect on the waking EEG of healthy volunteers. A pilot study on two subjects. *Neuropsychobiology* 25:214–220.

Eap, C. B., Bertschy, G., Powell, K., Baumann, P. 1997. Fluvoxamine and fluoxetine do not interact in the same way with the metabolism of the enantiomers of methadone. *J Clin Psychopharmacol* 17:113–117.

Eap, C. B., Bender, S., Gastpar, M., Fischer, W., Haarmann, C., Powell, K., Jonzier-Perey, M., Cochard, N., Baumann, P. 2000. Steady state plasma levels of the enantiomers of trimipramine and of its metabolites in CYP2D6-, CYP2C19- and CYP3A4/5-phenotyped patients. *Ther Drug Monit* 22:209–214.

Eap, C. B., Lessard, E., Baumann, P., Brawand-Amey, M., Yessine, M. A., O'Hara, G., Turgeon, J. 2003. Role of CYP2D6 in the stereoselective disposition of venlafaxine in humans. *Pharmacogenetics* 13:39–47.

Ebisawa, A., Hiratsuka, M., Sakuyama, K., Konno, Y., Sasaki, T., Mizugaki, M. 2005. Two novel single nucleotide polymorphisms (SNPs) of the CYP2D6 gene in Japanese individuals. *Drug Metab Pharmacokinet* 20:294–299.

Ebner, T., Eichelbaum, M. 1993. The metabolism of aprindine in relation to the sparteine/debrisoquine polymorphism. *Br J Clin Pharmacol* 35:426–430.

Ebner, T., Meese, C. O., Eichelbaum, M. 1995. Mechanism of cytochrome P450 2D6-catalyzed sparteine metabolism in humans. *Mol Pharmacol* 48:1078–1086.

Eckhardt, K., Li, S., Ammon, S., Schanzle, G., Mikus, G., Eichelbaum, M. 1998. Same incidence of adverse drug events after codeine administration irrespective of the genetically determined differences in morphine formation. *Pain* 76:27–33.

Edeki, T. I., He, H., Wood, A. J. 1995. Pharmacogenetic explanation for excessive beta-blockade following timolol eye drops. Potential for oral-ophthalmic drug interaction. *JAMA* 274:1611–1613.

Eichelbaum, M., Spannbrucker, N., Steincke, B., Dengler, H. J. 1979. Defective *N*-oxidation of sparteine in man: A new pharmacogenetic defect. *Eur J Clin Pharmacol* 16:183–187.

Eichelbaum, M., Reetz, K. P., Schmidt, E. K., Zekorn, C. 1986. The genetic polymorphism of sparteine metabolism. *Xenobiotica* 16:465–481.

Enggaard, T. P., Poulsen, L., Arendt-Nielsen, L., Brosen, K., Ossig, J., Sindrup, S. H. 2006. The analgesic effect of tramadol after intravenous injection in healthy volunteers in relation to CYP2D6. *Anesth Analg* 102:146–150.

Evans, W. E., Relling, M. V., Petros, W. P., Meyer, W. H., Mirro, J., Jr. Crom, W. R. 1989. Dextromethorphan and caffeine as probes for simultaneous determination of debrisoquin-oxidation and N-acetylation phenotypes in children. *Clin Pharmacol Ther* 45:568–573.

Evans, W. E., Relling, M. V., Rahman, A., McLeod, H. L., Scott, E. P., Lin, J. S. 1993. Genetic basis for a lower prevalence of deficient CYP2D6 oxidative drug metabolism phenotypes in black Americans. *J Clin Invest* 91:2150–2154.

Evert, B., Griese, E. U., Eichelbaum, M. 1994a. Cloning and sequencing of a new non-functional CYP2D6 allele: Deletion of T1795 in exon 3 generates a premature stop codon. *Pharmacogenetics* 4:271–274.

Evert, B., Griese, E. U., Eichelbaum, M. 1994b. A missense mutation in exon 6 of the CYP2D6 gene leading to a histidine 324 to proline exchange is associated with the poor metabolizer phenotype of sparteine. *Naunyn Schmiedebergs Arch Pharmacol* 350:434–439.

Evert, B., Eichelbaum, M., Haubruck, H., Zanger, U. M. 1997. Functional properties of CYP2D6 1 (wild-type) and CYP2D6 7 (His324Pro) expressed by recombinant baculovirus in insect cells. *Naunyn Schmiedebergs Arch Pharmacol* 355:309–318.

Farid, N. A., Bergstrom, R. F., Ziege, E. A., Parli, C. J., Lemberger, L. 1985. Single-dose and steady-state pharmacokinetics of tomoxetine in normal subjects. *J Clin Pharmacol* 25:296–301.

Fasano, A., Bentivoglio, A. R. 2009. Tetrabenazine. *Expert Opin Pharmacother* 10:2883–2896.

Firkusny, L., Gleiter, C. H. 1994. Maprotiline metabolism appears to co-segregate with the genetically-determined CYP2D6 polymorphic hydroxylation of debrisoquine. *Br J Clin Pharmacol* 37:383–388.

Firkusny, L., Kroemer, H. K., Eichelbaum, M. 1995. In vitro characterization of cytochrome P450 catalysed metabolism of the antiemetic tropisetron. *Biochem Pharmacol* 49:1777–1784.

Fischer, V., Vickers, A. E., Heitz, F., Mahadevan, S., Baldeck, J. P., Minery, P., Tynes, R. 1994. The polymorphic cytochrome P-4502D6 is involved in the metabolism of both 5-hydroxytryptamine antagonists, tropisetron and ondansetron. *Drug Metab Dispos* 22:269–274.

Fjordside, L., Jeppesen, U., Eap, C. B., Powell, K., Baumann, P., Brosen, K. 1999. The stereoselective metabolism of fluoxetine in poor and extensive metabolizers of sparteine. *Pharmacogenetics* 9:55–60.

Flanagan, J. U., Marechal, J. D., Ward, R., Kemp, C. A., McLaughlin, L. A., Sutcliffe, M. J., Roberts, G. C., Paine, M. J., Wolf, C. R. 2004. Phe120 contributes to the regiospecificity of cytochrome P450 2D6: Mutation leads to the formation of a novel dextromethorphan metabolite. *Biochem J* 380:353–360.

Fliegert, F., Kurth, B., Gohler, K. 2005. The effects of tramadol on static and dynamic pupillometry in healthy subjects—The relationship between pharmacodynamics, pharmacokinetics and *CYP2D6* metaboliser status. *Eur J Clin Pharmacol* 61:257–266.

Fogelman, S. M., Schmider, J., Venkatakrishnan, K., von Moltke, L. L., Harmatz, J. S., Shader, R. I., Greenblatt, D. J. 1999. *O*- and *N*-demethylation of venlafaxine in vitro by human liver microsomes and by microsomes from cDNA-transfected cells: Effect of metabolic inhibitors and SSRI antidepressants. *Neuropsychopharmacology* 20:480–490.

Follath, F. 1992. The utility of serum drug level monitoring during therapy with class III antiarrhythmic agents. *J Cardiovasc Pharmacol* 20 Suppl 2:S41–S43.

Frank, S. 2014. Treatment of Huntington's disease. *Neurotherapeutics* 11:153–160.

Fromm, M. F., Hofmann, U., Griese, E. U., Mikus, G. 1995. Dihydrocodeine: A new opioid substrate for the polymorphic CYP2D6 in humans. *Clin Pharmacol Ther* 58:374–382.

Fukuda, T., Nishida, Y., Imaoka, S., Hiroi, T., Naohara, M., Funae, Y., Azuma, J. 2000a. The decreased in vivo clearance of CYP2D6 substrates by CYP2D6*10 might be caused not only by the low-expression but also by low affinity of CYP2D6. *Arch Biochem Biophys* 380:303–308.

Fukuda, T., Nishida, Y., Zhou, Q., Yamamoto, I., Kondo, S., Azuma, J. 2000b. The impact of the CYP2D6 and CYP2C19 genotypes on venlafaxine pharmacokinetics in a Japanese population. *Eur J Clin Pharmacol* 56:175–180.

Funck-Brentano, C., Turgeon, J., Woosley, R. L., Roden, D. M. 1989. Effect of low dose quinidine on encainide pharmacokinetics and pharmacodynamics. Influence of genetic polymorphism. *J Pharmacol Exp Ther* 249:134–142.

Funck-Brentano, C., Thomas, G., Jacqz-Aigrain, E., Poirier, J. M., Simon, T., Bereziat, G., Jaillon, P. 1992. Polymorphism of dextromethorphan metabolism: Relationships between phenotype, genotype and response to the administration of encainide in humans. *J Pharmacol Exp Ther* 263:780–786.

Funck-Brentano, C., Becquemont, L., Kroemer, H. K., Buhl, K., Knebel, N. G., Eichelbaum, M., Jaillon, P. 1994. Variable disposition kinetics and electrocardiographic effects of flecainide during repeated dosing in humans: Contribution of genetic factors, dose-dependent clearance, and interaction with amiodarone. *Clin Pharmacol Ther* 55:256–269.

Fuselli, S., Dupanloup, I., Frigato, E., Cruciani, F., Scozzari, R., Moral, P., Sistonen, J., Sajantila, A., Barbujani, G. 2004. Molecular diversity at the CYP2D6 locus in the Mediterranean region. *Eur J Hum Genet* 12:916–924.

Fux, R., Morike, K., Prohmer, A. M., Delabar, U., Schwab, M., Schaeffeler, E., Lorenz, G., Gleiter, C. H., Eichelbaum, M., Kivisto, K. T. 2005. Impact of CYP2D6 genotype on adverse effects during treatment with metoprolol: A prospective clinical study. *Clin Pharmacol Ther* 78:378–387.

Gaedigk, A., Coetsee, C. 2008. The CYP2D6 gene locus in South African Coloureds: Unique allele distributions, novel alleles and gene arrangements. *Eur J Clin Pharmacol* 64:465–475.

Gaedigk, A., Blum, M., Gaedigk, R., Eichelbaum, M., Meyer, U. A. 1991. Deletion of the entire cytochrome P450 CYP2D6 gene as a cause of impaired drug metabolism in poor metabolizers of the debrisoquine/sparteine polymorphism. *Am J Hum Genet* 48:943–950.

Gaedigk, A., Bradford, L. D., Marcucci, K. A., Leeder, J. S. 2002. Unique CYP2D6 activity distribution and genotype–phenotype discordance in black Americans. *Clin Pharmacol Ther* 72:76–89.

Gaedigk, A., Ndjountche, L., Gaedigk, R., Leeder, J. S., Bradford, L. D. 2003a. Discovery of a novel nonfunctional cytochrome P450 2D6 allele, CYP2D642, in African American subjects. *Clin Pharmacol Ther* 73:575–576.

Gaedigk, A., Ryder, D. L., Bradford, L. D., Leeder, J. S. 2003b. CYP2D6 poor metabolizer status can be ruled out by a single genotyping assay for the −1584G promoter polymorphism. *Clin Chem* 49:1008–1011.

Gaedigk, A., Bhathena, A., Ndjountche, L., Pearce, R. E., Abdel-Rahman, S. M., Alander, S. W., Bradford, L. D., Rogan, P. K., Leeder, J. S. 2005a. Identification and characterization of novel sequence variations in the cytochrome P4502D6 (CYP2D6) gene in African Americans. *Pharmacogenomics J* 5:173–182.

Gaedigk, A., Ndjountche, L., Leeder, J. S., Bradford, L. D. 2005b. Limited association of the 2988g > a single nucleotide polymorphism with CYP2D641 in black subjects. *Clin Pharmacol Ther* 77:228–230; author reply 230–221.

Gaedigk, A., Bradford, L. D., Alander, S. W., Leeder, J. S. 2006. CYP2D6*36 gene arrangements within the cyp2d6 locus: Association of CYP2D6*36 with poor metabolizer status. *Drug Metab Dispos* 34:563–569.

Gaedigk, A., Eklund, J. D., Pearce, R. E., Leeder, J. S., Alander, S. W., Phillips, M. S., Bradford, L. D., Kennedy, M. J. 2007a. Identification and characterization of CYP2D6*56B, an allele associated with the poor metabolizer phenotype. *Clin Pharmacol Ther* 81:817–820.

Gaedigk, A., Ndjountche, L., Divakaran, K., Dianne Bradford, L., Zineh, I., Oberlander, T. F., Brousseau, D. C. et al. 2007b. Cytochrome P4502D6 (CYP2D6) gene locus heterogeneity: Characterization of gene duplication events. *Clin Pharmacol Ther* 81:242–251.

Gaedigk, A., Frank, D., Fuhr, U. 2009. Identification of a novel non-functional CYP2D6 allele, CYP2D6*69, in a Caucasian poor metabolizer individual. *Eur J Clin Pharmacol* 65:97–100.

Gaedigk, A., Hernandez, J., Garcia-Solaesa, V., Sanchez, S., Isidoro-Garcia, M. 2011. Detection and characterization of the *CYP2D6*9x2* gene duplication in two Spanish populations: Resolution of AmpliChip CYP450 test no-calls. *Pharmacogenomics* 12:1617–1622.

Gaedigk, A., Twist, G. P., Leeder, J. S. 2012. CYP2D6, SULT1A1 and UGT2B17 copy number variation: Quantitative detection by multiplex PCR. *Pharmacogenomics* 13:91–111.

Gan, S. H., Ismail, R., Wan Adnan, W. A., Zulmi, W., Jelliffe, R. W. 2004. Population pharmacokinetic modelling of tramadol with application of the NPEM algorithms. *J Clin Pharm Ther* 29:455–463.

Gan, S. H., Ismail, R., Wan Adnan, W. A., Zulmi, W. 2007. Impact of *CYP2D6* genetic polymorphism on tramadol pharmacokinetics and pharmacodynamics. *Mol Diagn Ther* 11:171–181.

Garcia-Barcelo, M., Chow, L. Y., Lam, K. L., Chiu, H. F., Wing, Y. K., Waye, M. M. 2000. Occurrence of CYP2D6 gene duplication in Hong Kong Chinese. *Clin Chem* 46:1411–1413.

Garcia-Quetglas, E., Azanza, J. R., Sadaba, B., Munoz, M. J., Gil, I., Campanero, M. A. 2007. Pharmacokinetics of tramadol enantiomers and their respective phase I metabolites in relation to CYP2D6 phenotype. *Pharmacol Res* 55:122–130.

Gardiner, S. J., Begg, E. J. 2006. Pharmacogenetics, drug-metabolizing enzymes, and clinical practice. *Pharmacol Rev* 58:521–590.

Gibbs, J. P., Hyland, R., Youdim, K. 2006. Minimizing polymorphic metabolism in drug discovery: Evaluation of the utility of in vitro methods for predicting pharmacokinetic consequences associated with CYP2D6 metabolism. *Drug Metab Dispos* 34:1516–1522.

Giessmann, T., Modess, C., Hecker, U., Zschiesche, M., Dazert, P., Kunert-Keil, C., Warzok, R. et al. 2004. CYP2D6 genotype and induction of intestinal drug transporters by rifampin predict presystemic clearance of carvedilol in healthy subjects. *Clin Pharmacol Ther* 75:213–222.

Gjerde, J., Hauglid, M., Breilid, H., Lundgren, S., Varhaug, J. E., Kisanga, E. R., Mellgren, G., Steen, V. M., Lien, E. A. 2008. Effects of CYP2D6 and SULT1A1 genotypes including SULT1A1 gene copy number on tamoxifen metabolism. *Ann Oncol* 19:56–61.

Goetz, M. P., Ingle, J. N. 2014. *CYP2D6* genotype and tamoxifen: Considerations for proper nonprospective studies. *Clin Pharmacol Ther* 96:141–144.

Goetz, M. P., Rae, J. M., Suman, V. J., Safgren, S. L., Ames, M. M., Visscher, D. W., Reynolds, C. et al. 2005. Pharmacogenetics of tamoxifen biotransformation is associated with clinical outcomes of efficacy and hot flashes. *J Clin Oncol* 23:9312–9318.

Goetz, M. P., Knox, S. K., Suman, V. J., Rae, J. M., Safgren, S. L., Ames, M. M., Visscher, D. W. et al. 2007. The impact of cytochrome P450 2D6 metabolism in women receiving adjuvant tamoxifen. *Breast Cancer Res Treat* 101:113–121.

Goetz, M. P., Kamal, A., Ames, M. M. 2008. Tamoxifen pharmacogenomics: The role of CYP2D6 as a predictor of drug response. *Clin Pharmacol Ther* 83:160–166.

Goryachkina, K., Burbello, A., Boldueva, S., Babak, S., Bergman, U., Bertilsson, L. 2008. CYP2D6 is a major determinant of metoprolol disposition and effects in hospitalized Russian patients treated for acute myocardial infarction. *Eur J Clin Pharmacol* 64:1163–1173.

Gough, A. C., Miles, J. S., Spurr, N. K., Moss, J. E., Gaedigk, A., Eichelbaum, M., Wolf, C. R. 1990. Identification of the primary gene defect at the cytochrome P450 CYP2D locus. *Nature* 347:773–776.

Gough, A. C., Smith, C. A., Howell, S. M., Wolf, C. R., Bryant, S. P., Spurr, N. K. 1993. Localization of the *CYP2D* gene locus to human chromosome 22q13.1 by polymerase chain reaction, in situ hybridization, and linkage analysis. *Genomics* 15:430–432.

Gram, L. F., Christiansen, J. 1975. First-pass metabolism of imipramine in man. *Clin Pharmacol Ther* 17:555–563.

Gram, L. F., Brosen, K., Kragh-Sorensen, P., Christensen, P. 1989. Steady-state plasma levels of E- and Z-10-OH-nortriptyline in nortriptyline-treated patients: Significance of concurrent medication and the sparteine oxidation phenotype. *Ther Drug Monit* 11:508–514.

Grasmader, K., Verwohlt, P. L., Kuhn, K. U., Dragicevic, A., von Widdern, O., Zobel, A., Hiemke, C. et al. 2004a. Population pharmacokinetic analysis of mirtazapine. *Eur J Clin Pharmacol* 60:473–480.

Grasmader, K., Verwohlt, P. L., Rietschel, M., Dragicevic, A., Muller, M., Hiemke, C., Freymann, N., Zobel, A., Maier, W., Rao, M. L. 2004b. Impact of polymorphisms of cytochrome-P450 isoenzymes 2C9, 2C19 and 2D6 on plasma concentrations and clinical effects of antidepressants in a naturalistic clinical setting. *Eur J Clin Pharmacol* 60:329–336.

Griese, E. U., Zanger, U. M., Brudermanns, U., Gaedigk, A., Mikus, G., Morike, K., Stuven, T., Eichelbaum, M. 1998. Assessment of the predictive power of genotypes for the in-vivo catalytic function of CYP2D6 in a German population. *Pharmacogenetics* 8:15–26.

Griese, E. U., Asante-Poku, S., Ofori-Adjei, D., Mikus, G., Eichelbaum, M. 1999. Analysis of the CYP2D6 gene mutations and their consequences for enzyme function in a West African population. *Pharmacogenetics* 9:715–723.

Gross, A. S., Mikus, G., Fischer, C., Hertrampf, R., Gundert-Remy, U., Eichelbaum, M. 1989. Stereoselective disposition of flecainide in relation to the sparteine/debrisoquine metaboliser phenotype. *Br J Clin Pharmacol* 28:555–566.

Gross, A. S., Mikus, G., Fischer, C., Eichelbaum, M. 1991. Polymorphic flecainide disposition under conditions of uncontrolled urine flow and pH. *Eur J Clin Pharmacol* 40:155–162.

Guay, D. R. 2010. Tetrabenazine, a monoamine-depleting drug used in the treatment of hyperkinetic movement disorders. *Am J Geriatr Pharmacother* 8:331–373.

Guzey, C., Aamo, T., Spigset, O. 2000. Risperidone metabolism and the impact of being a cytochrome P450 2D6 ultrarapid metabolizer. *J Clin Psychiatry* 61:600–601.

Haddad, P. M., Wieck, A. 2004. Antipsychotic-induced hyperprolactinaemia: Mechanisms, clinical features and management. *Drugs* 64:2291–2314.

Haefeli, W. E., Bargetzi, M. J., Follath, F., Meyer, U. A. 1990. Potent inhibition of cytochrome P450IID6 (debrisoquin 4-hydroxylase) by flecainide in vitro and in vivo. *J Cardiovasc Pharmacol* 15:776–779.

Hagg, S., Spigset, O., Lakso, H. A., Dahlqvist, R. 2001. Olanzapine disposition in humans is unrelated to CYP1A2 and CYP2D6 phenotypes. *Eur J Clin Pharmacol* 57:493–497.

Halling, J., Weihe, P., Brosen, K. 2008a. The CYP2D6 polymorphism in relation to the metabolism of amitriptyline and nortriptyline in the Faroese population. *Br J Clin Pharmacol* 65:134–138.

Halling, J., Weihe, P., Brosen, K. 2008b. CYP2D6 polymorphism in relation to tramadol metabolism: A study of faroese patients. *Ther Drug Monit* 30:271–275.

Hamelin, B. A., Turgeon, J., Vallee, F., Belanger, P. M., Paquet, F., LeBel, M. 1996. The disposition of fluoxetine but not sertraline is altered in poor metabolizers of debrisoquin. *Clin Pharmacol Ther* 60:512–521.

Hamelin, B. A., Bouayad, A., Methot, J., Jobin, J., Desgagnes, P., Poirier, P., Allaire, J., Dumesnil, J., Turgeon, J. 2000. Significant interaction between the nonprescription antihistamine diphenhydramine and the CYP2D6 substrate metoprolol in healthy men with high or low CYP2D6 activity. *Clin Pharmacol Ther* 67:466–477.

Hanioka, N., Kimura, S., Meyer, U. A., Gonzalez, F. J. 1990. The human CYP2D locus associated with a common genetic defect in drug oxidation: A G1934—A base change in intron 3 of a mutant CYP2D6 allele results in an aberrant 3′ splice recognition site. *Am J Hum Genet* 47:994–1001.

Haritos, V. S., Ghabrial, H., Ahokas, J. T., Ching, M. S. 2000. Role of cytochrome P450 2D6 (CYP2D6) in the stereospecific metabolism of E- and Z-doxepin. *Pharmacogenetics* 10:591–603.

Hartter, S., Tybring, G., Friedberg, T., Weigmann, H., Hiemke, C. 2002. The N-demethylation of the doxepin isomers is mainly catalyzed by the polymorphic CYP2C19. *Pharm Res* 19:1034–1037.

He, N., Zhang, W. Q., Shockley, D., Edeki, T. 2002. Inhibitory effects of H1-antihistamines on CYP2D6- and CYP2C9-mediated drug metabolic reactions in human liver microsomes. *Eur J Clin Pharmacol* 57:847–851.

Heim, M. H., Meyer, U. A. 1992. Evolution of a highly polymorphic human cytochrome P450 gene cluster: CYP2D6. *Genomics* 14:49–58.

Heiskanen, T., Olkkola, K. T., Kalso, E. 1998. Effects of blocking CYP2D6 on the pharmacokinetics and pharmacodynamics of oxycodone. *Clin Pharmacol Ther* 64:603–611.

Hendset, M., Hermann, M., Lunde, H., Refsum, H., Molden, E. 2007. Impact of the *CYP2D6* genotype on steady-state serum concentrations of aripiprazole and dehydroaripiprazole. *Eur J Clin Pharmacol* 63:1147–1151.

Hermann, M., Hendset, M., Fosaas, K., Hjerpset, M., Refsum, H. 2008. Serum concentrations of venlafaxine and its metabolites O-desmethylvenlafaxine and N-desmethylvenlafaxine in heterozygous carriers of the CYP2D6*3, *4 or *5 allele. *Eur J Clin Pharmacol* 64:483–487.

Herrlin, K., Yasui-Furukori, N., Tybring, G., Widen, J., Gustafsson, L. L., Bertilsson, L. 2003. Metabolism of citalopram enantiomers in CYP2C19/CYP2D6 phenotyped panels of healthy Swedes. *Br J Clin Pharmacol* 56:415–421.

Hicks, J. K., Swen, J. J., Thorn, C. F., Sangkuhl, K., Kharasch, E. D., Ellingrod, V. L., Skaar, T. C. et al. 2013. Clinical Pharmacogenetics Implementation Consortium guideline for CYP2D6 and CYP2C19 genotypes and dosing of tricyclic antidepressants. *Clin Pharmacol Ther* 93:402–408.

Hicks, J. K., Bishop, J. R., Sangkuhl, K., Muller, D. J., Ji, Y., Leckband, S. G., Leeder, J. S. et al. 2015. Clinical Pharmacogenetics Implementation Consortium (CPIC) guideline for CYP2D6 and CYP2C19 genotypes and dosing of selective serotonin reuptake inhibitors. *Clin Pharmacol Ther* 98:127–134.

Hoffmann, K. J., Regardh, C. G., Aurell, M., Ervik, M., Jordo, L. 1980. The effect of impaired renal function on the plasma concentration and urinary excretion of metoprolol metabolites. *Clin Pharmacokinet* 5:181–191.

Holliday, S. M., Benfield, P. 1995. Venlafaxine. A review of its pharmacology and therapeutic potential in depression. *Drugs* 49:280–294.

Honda, M., Nozawa, T., Igarashi, N., Inoue, H., Arakawa, R., Ogura, Y., Okabe, H., Taguchi, M., Hashimoto, Y. 2005. Effect of CYP2D6*10 on the pharmacokinetics of R- and S-carvedilol in healthy Japanese volunteers. *Biol Pharm Bull* 28:1476–1479.

Horai, Y., Nakano, M., Ishizaki, T., Ishikawa, K., Zhou, H. H., Zhou, B. I., Liao, C. L., Zhang, L. M. 1989. Metoprolol and mephenytoin oxidation polymorphisms in Far Eastern Oriental subjects: Japanese versus mainland Chinese. *Clin Pharmacol Ther* 46:198–207.

Horai, Y., Taga, J., Ishizaki, T., Ishikawa, K. 1990. Correlations among the metabolic ratios of three test probes (metoprolol, debrisoquine and sparteine) for genetically determined oxidation polymorphism in a Japanese population. *Br J Clin Pharmacol* 29:111–115.

Huang, J., Chuang, S. K., Cheng, C. L., Lai, M. L. 1999. Pharmacokinetics of metoprolol enantiomers in Chinese subjects of major CYP2D6 genotypes. *Clin Pharmacol Ther* 65:402–407.

Hutchinson, M. R., Menelaou, A., Foster, D. J., Coller, J. K., Somogyi, A. A. 2004. CYP2D6 and CYP3A4 involvement in the primary oxidative metabolism of hydrocodone by human liver microsomes. *Br J Clin Pharmacol* 57:287–297.

Huupponen, R., Kaila, T., Lahdes, K., Salminen, L., Iisalo, E. 1991. Systemic absorption of ocular timolol in poor and extensive metabolizers of debrisoquine. *J Ocul Pharmacol* 7:183–187.

Hvenegaard, M. G., Bang-Andersen, B., Pedersen, H., Jorgensen, M., Puschl, A., Dalgaard, L. 2012. Identification of the cytochrome P450 and other enzymes involved in the *in vitro* oxidative metabolism of a novel antidepressant, Lu AA21004. *Drug Metab Dispos* 40:1357–1365.

Idle, J. R., Corchero, J., Gonzalez, F. J. 2000. Medical implications of HGP's sequence of chromosome 22. *Lancet* 355:319.

Ingelman-Sundberg, M. 2005. Genetic polymorphisms of cytochrome P450 2D6 (CYP2D6): Clinical consequences, evolutionary aspects and functional diversity. *Pharmacogenomics J* 5:6–13.

Ingelman-Sundberg, M., Sim, S. C., Gomez, A., Rodriguez-Antona, C. 2007. Influence of cytochrome P450 polymorphisms on drug therapies: Pharmacogenetic, pharmacoepigenetic and clinical aspects. *Pharmacol Ther* 116:496–526.

Ishiguro, A., Kubota, T., Ishikawa, H., Iga, T. 2004a. Metabolic activity of dextromethorphan O-demethylation in healthy Japanese volunteers carrying duplicated CYP2D6 genes: Duplicated allele of CYP2D6*10 does not increase CYP2D6 metabolic activity. *Clin Chim Acta* 344:201–204.

Ishiguro, A., Kubota, T., Sasaki, H., Iga, T. 2004b. A long PCR assay to distinguish CYP2D6*5 and a novel CYP2D6 mutant allele associated with an 11-kb EcoRI haplotype. *Clin Chim Acta* 347:217–221.

Jaanson, P., Marandi, T., Kiivet, R. A., Vasar, V., Vaan, S., Svensson, J. O., Dahl, M. L. 2002. Maintenance therapy with zuclopenthixol decanoate: Associations between plasma concentrations, neurological side effects and CYP2D6 genotype. *Psychopharmacology (Berl)* 162:67–73.

Janicki, P. K., Schuler, H. G., Jarzembowski, T. M., Rossi, M., 2nd. 2006. Prevention of postoperative nausea and vomiting with granisetron and dolasetron in relation to CYP2D6 genotype. *Anesth Analg* 102:1127–1133.

Jazwinska-Tarnawska, E., Orzechowska-Juzwenko, K., Niewinski, P., Rzemislawska, Z., Loboz-Grudzien, K., Dmochowska-Perz, M., Slawin, J. 2001. The influence of CYP2D6 polymorphism on the antiarrhythmic efficacy of propafenone in patients with paroxysmal atrial fibrillation during 3 months propafenone prophylactic treatment. *Int J Clin Pharmacol Ther* 39:288–292.

Jerling, M., Dahl, M. L., Aberg-Wistedt, A., Liljenberg, B., Landell, N. E., Bertilsson, L., Sjoqvist, F. 1996. The CYP2D6 genotype predicts the oral clearance of the neuroleptic agents perphenazine and zuclopenthixol. *Clin Pharmacol Ther* 59:423–428.

Ji, L., Pan, S., Marti-Jaun, J., Hanseler, E., Rentsch, K., Hersberger, M. 2002. Single-step assays to analyze CYP2D6 gene polymorphisms in Asians: Allele frequencies and a novel *14B allele in mainland Chinese. *Clin Chem* 48:983–988.

Jin, Y., Desta, Z., Stearns, V., Ward, B., Ho, H., Lee, K. H., Skaar, T. et al. 2005. CYP2D6 genotype, antidepressant use, and tamoxifen metabolism during adjuvant breast cancer treatment. *J Natl Cancer Inst* 97:30–39.

Johansson, I., Yue, Q. Y., Dahl, M. L., Heim, M., Sawe, J., Bertilsson, L., Meyer, U. A., Sjoqvist, F., Ingelman-Sundberg, M. 1991. Genetic analysis of the interethnic difference between Chinese and Caucasians in the polymorphic metabolism of debrisoquine and codeine. *Eur J Clin Pharmacol* 40:553–556.

Johansson, I., Lundqvist, E., Bertilsson, L., Dahl, M. L., Sjoqvist, F., Ingelman-Sundberg, M. 1993. Inherited amplification of an active gene in the cytochrome P450 CYP2D locus as a cause of ultrarapid metabolism of debrisoquine. *Proc Natl Acad Sci U S A* 90:11825–11829.

Johansson, I., Oscarson, M., Yue, Q. Y., Bertilsson, L., Sjoqvist, F., Ingelman-Sundberg, M. 1994. Genetic analysis of the Chinese cytochrome P4502D locus: Characterization of variant CYP2D6 genes present in subjects with diminished capacity for debrisoquine hydroxylation. *Mol Pharmacol* 46:452–459.

Johnson, J. A., Burlew, B. S. 1996. Metoprolol metabolism via cytochrome P4502D6 in ethnic populations. *Drug Metab Dispos* 24:350–355.

Jordan, V. C. 2014. Tamoxifen as the first targeted long-term adjuvant therapy for breast cancer. *Endocr Relat Cancer* 21:R235–R246.

Jorge, L. F., Eichelbaum, M., Griese, E. U., Inaba, T., Arias, T. D. 1999. Comparative evolutionary pharmacogenetics of CYP2D6 in Ngawbe and Embera Amerindians of Panama and Colombia: Role of selection versus drift in world populations. *Pharmacogenetics* 9:217–228.

Kagimoto, M., Heim, M., Kagimoto, K., Zeugin, T., Meyer, U. A. 1990. Multiple mutations of the human cytochrome P450IID6 gene (CYP2D6) in poor metabolizers of debrisoquine. Study of the functional significance of individual mutations by expression of chimeric genes. *J Biol Chem* 265:17209–17214.

Kaila, T., Huupponen, R., Karhuvaara, S., Havula, P., Scheinin, M., Iisalo, E., Salminen, L. 1991. Beta-blocking effects of timolol at low plasma concentrations. *Clin Pharmacol Ther* 49:53–58.

Kaiser, R., Sezer, O., Papies, A., Bauer, S., Schelenz, C., Tremblay, P. B., Possinger, K., Roots, I., Brockmoller, J. 2002. Patient-tailored antiemetic treatment with 5-hydroxytryptamine type 3 receptor antagonists according to cytochrome P-450 2D6 genotypes. *J Clin Oncol* 20:2805–2811.

Kaplan, H. L., Busto, U. E., Baylon, G. J., Cheung, S. W., Otton, S. V., Somer, G., Sellers, E. M. 1997. Inhibition of cytochrome P450 2D6 metabolism of hydrocodone to hydromorphone does not importantly affect abuse liability. *J Pharmacol Exp Ther* 281:103–108.

Katz, M. R. 1991. Raised serum levels of desipramine with the antiarrhythmic propafenone. *J Clin Psychiatry* 52:432–433.

Kawanishi, C., Lundgren, S., Agren, H., Bertilsson, L. 2004. Increased incidence of CYP2D6 gene duplication in patients with persistent mood disorders: Ultrarapid metabolism of antidepressants as a cause of nonresponse. A pilot study. *Eur J Clin Pharmacol* 59:803–807.

Keizers, P. H., Lussenburg, B. M., de Graaf, C., Mentink, L. M., Vermeulen, N. P., Commandeur, J. N. 2004. Influence of phenylalanine 120 on cytochrome P450 2D6 catalytic selectivity and regiospecificity: Crucial role in 7-methoxy-4-(aminomethyl)-coumarin metabolism. *Biochem Pharmacol* 68:2263–2271.

Kim, M., Shen, D. D., Eddy, A. C., Nelson, W. L., Roskos, L. K. 1993. Inhibition of the enantioselective oxidative metabolism of metoprolol by verapamil in human liver microsomes. *Drug Metab Dispos* 21:309–317.

Kim, M. K., Cho, J. Y., Lim, H. S., Hong, K. S., Chung, J. Y., Bae, K. S., Oh, D. S. et al. 2003. Effect of the CYP2D6 genotype on the pharmacokinetics of tropisetron in healthy Korean subjects. *Eur J Clin Pharmacol* 59:111–116.

Kim, J. R., Seo, H. B., Cho, J. Y., Kang, D. H., Kim, Y. K., Bahk, W. M., Yu, K. S., Shin, S. G., Kwon, J. S., Jang, I. J. 2008. Population pharmacokinetic modelling of aripiprazole and its active metabolite, dehydroaripiprazole, in psychiatric patients. *Br J Clin Pharmacol* 66:802–810.

Kimura, S., Umeno, M., Skoda, R. C., Meyer, U. A., Gonzalez, F. J. 1989. The human debrisoquine 4-hydroxylase (CYP2D) locus: Sequence and identification of the polymorphic CYP2D6 gene, a related gene, and a pseudogene. *Am J Hum Genet* 45:889–904.

Kirchheiner, J. 2008. CYP2D6 phenotype prediction from genotype: Which system is the best? *Clin Pharmacol Ther* 83:225–227.

Kirchheiner, J., Brosen, K., Dahl, M. L., Gram, L. F., Kasper, S., Roots, I., Sjoqvist, F., Spina, E., Brockmoller, J. 2001. CYP2D6 and CYP2C19 genotype-based dose recommendations for antidepressants: A first step towards subpopulation-specific dosages. *Acta Psychiatr Scand* 104:173–192.

Kirchheiner, J., Meineke, I., Muller, G., Roots, I., Brockmoller, J. 2002. Contributions of CYP2D6, CYP2C9 and CYP2C19 to the biotransformation of E- and Z-doxepin in healthy volunteers. *Pharmacogenetics* 12:571–580.

Kirchheiner, J., Muller, G., Meineke, I., Wernecke, K. D., Roots, I., Brockmoller, J. 2003. Effects of polymorphisms in CYP2D6, CYP2C9, and CYP2C19 on trimipramine pharmacokinetics. *J Clin Psychopharmacol* 23:459–466.

Kirchheiner, J., Heesch, C., Bauer, S., Meisel, C., Seringer, A., Goldammer, M., Tzvetkov, M., Meineke, I., Roots, I., Brockmoller, J. 2004a. Impact of the ultrarapid metabolizer genotype of cytochrome P450 2D6 on metoprolol pharmacokinetics and pharmacodynamics. *Clin Pharmacol Ther* 76:302–312.

Kirchheiner, J., Henckel, H. B., Meineke, I., Roots, I., Brockmoller, J. 2004b. Impact of the CYP2D6 ultrarapid metabolizer genotype on mirtazapine pharmacokinetics and adverse events in healthy volunteers. *J Clin Psychopharmacol* 24:647–652.

Kirchheiner, J., Schmidt, H., Tzvetkov, M., Keulen, J. T., Lotsch, J., Roots, I., Brockmoller, J. 2007. Pharmacokinetics of codeine and its metabolite morphine in ultra-rapid metabolizers due to CYP2D6 duplication. *Pharmacogenomics J* 7:257–265.

Kirchheiner, J., Keulen, J. T., Bauer, S., Roots, I., Brockmoller, J. 2008. Effects of the CYP2D6 gene duplication on the pharmacokinetics and pharmacodynamics of tramadol. *J Clin Psychopharmacol* 28:78–83.

Kirkwood, L. C., Nation, R. L., Somogyi, A. A. 1997. Characterization of the human cytochrome P450 enzymes involved in the metabolism of dihydrocodeine. *Br J Clin Pharmacol* 44:549–555.

Kiyotani, K., Mushiroda, T., Sasa, M., Bando, Y., Sumitomo, I., Hosono, N., Kubo, M., Nakamura, Y., Zembutsu, H. 2008. Impact of *CYP2D6*10* on recurrence-free survival in breast cancer patients receiving adjuvant tamoxifen therapy. *Cancer Sci* 99:995–999.

Klein, A., Sami, M., Selinger, K. 1985. Mexiletine kinetics in healthy subjects taking cimetidine. *Clin Pharmacol Ther* 37:669–673.

Klein, K., Tatzel, S., Raimundo, S., Saussele, T., Hustert, E., Pleiss, J., Eichelbaum, M., Zanger, U. M. 2007. A natural variant of the heme-binding signature (R441C) resulting in complete loss of function of CYP2D6. *Drug Metab Dispos* 35:1247–1250.

Komm, B. S., Mirkin, S. 2014. An overview of current and emerging SERMs. *J Steroid Biochem Mol Biol* 143:207–222.

Koren, G., Cairns, J., Chitayat, D., Gaedigk, A., Leeder, S. J. 2006. Pharmacogenetics of morphine poisoning in a breastfed neonate of a codeine-prescribed mother. *Lancet* 368:704.

Koski, A., Ojanpera, I., Sistonen, J., Vuori, E., Sajantila, A. 2007. A fatal doxepin poisoning associated with a defective CYP2D6 genotype. *Am J Forensic Med Pathol* 28:259–261.

Koue, T., Kubo, M., Funaki, T., Fukuda, T., Azuma, J., Takaai, M., Kayano, Y., Hashimoto, Y. 2007. Nonlinear mixed effects model analysis of the pharmacokinetics of aripiprazole in healthy Japanese males. *Biol Pharm Bull* 30:2154–2158.

Kowey, P. R., Kirsten, E. B., Fu, C. H., Mason, W. D. 1989. Interaction between propranolol and propafenone in healthy volunteers. *J Clin Pharmacol* 29:512–517.

Koyama, E., Chiba, K., Tani, M., Ishizaki, T. 1996. Identification of human cytochrome P450 isoforms involved in the stereoselective metabolism of mianserin enantiomers. *J Pharmacol Exp Ther* 278:21–30.

Koyama, E., Chiba, K., Tani, M., Ishizaki, T. 1997. Reappraisal of human CYP isoforms involved in imipramine N-demethylation and 2-hydroxylation: A study using microsomes obtained from putative extensive and poor metabolizers of S-mephenytoin and eleven recombinant human CYPs. *J Pharmacol Exp Ther* 281:1199–1210.

Kramer, W. E., Walker, D. L., O'Kane, D. J., Mrazek, D. A., Fisher, P. K., Dukek, B. A., Bruflat, J. K., Black, J. L. 2009. CYP2D6: Novel genomic structures and alleles. *Pharmacogenet Genomics* 19:813–822.

Krentz, A. J., Ferner, R. E., Bailey, C. J. 1994. Comparative tolerability profiles of oral antidiabetic agents. *Drug Saf* 11: 223–241.

Kroemer, H. K., Funck-Brentano, C., Silberstein, D. J., Wood, A. J., Eichelbaum, M., Woosley, R. L., Roden, D. M. 1989a. Stereoselective disposition and pharmacologic activity of propafenone enantiomers. *Circulation* 79:1068–1076.

Kroemer, H. K., Mikus, G., Kronbach, T., Meyer, U. A., Eichelbaum, M. 1989b. In vitro characterization of the human cytochrome P-450 involved in polymorphic oxidation of propafenone. *Clin Pharmacol Ther* 45:28–33.

Kroemer, H. K., Fischer, C., Meese, C. O., Eichelbaum, M. 1991. Enantiomer/enantiomer interaction of (S)- and (R)-propafenone for cytochrome P450IID6-catalyzed 5-hydroxylation: In vitro evaluation of the mechanism. *Mol Pharmacol* 40:135–142.

Kubo, M., Koue, T., Inaba, A., Takeda, H., Maune, H., Fukuda, T., Azuma, J. 2005. Influence of itraconazole co-administration and *CYP2D6* genotype on the pharmacokinetics of the new antipsychotic aripiprazole. *Drug Metab Pharmacokinet* 20:55–64.

Kubo, M., Koue, T., Maune, H., Fukuda, T., Azuma, J. 2007. Pharmacokinetics of aripiprazole, a new antipsychotic, following oral dosing in healthy adult Japanese volunteers: Influence of CYP2D6 polymorphism. *Drug Metab Pharmacokinet* 22:358–366.

Kubota, T., Yamaura, Y., Ohkawa, N., Hara, H., Chiba, K. 2000. Frequencies of CYP2D6 mutant alleles in a normal Japanese population and metabolic activity of dextromethorphan O-demethylation in different CYP2D6 genotypes. *Br J Clin Pharmacol* 50:31–34.

Kudo, S., Odomi, M. 1998. Involvement of human cytochrome P450 3A4 in reduced haloperidol oxidation. *Eur J Clin Pharmacol* 54:253–259.

Kusumoto, M., Ueno, K., Tanaka, K., Takeda, K., Mashimo, K., Kameda, T., Fujimura, Y., Shibakawa, M. 1998. Lack of pharmacokinetic interaction between mexiletine and omeprazole. *Ann Pharmacother* 32:182–184.

Labbe, L., O'Hara, G., Lefebvre, M., Lessard, E., Gilbert, M., Adedoyin, A., Champagne, J., Hamelin, B., Turgeon, J. 2000. Pharmacokinetic and pharmacodynamic interaction between mexiletine and propafenone in human beings. *Clin Pharmacol Ther* 68:44–57.

Lai, M. L., Wang, S. L., Lai, M. D., Lin, E. T., Tse, M., Huang, J. D. 1995. Propranolol disposition in Chinese subjects of different CYP2D6 genotypes. *Clin Pharmacol Ther* 58:264–268.

Lalovic, B., Phillips, B., Risler, L. L., Howald, W., Shen, D. D. 2004. Quantitative contribution of CYP2D6 and CYP3A to oxycodone metabolism in human liver and intestinal microsomes. *Drug Metab Dispos* 32:447–454.

Lam, Y. W., Gaedigk, A., Ereshefsky, L., Alfaro, C. L., Simpson, J. 2002. CYP2D6 inhibition by selective serotonin reuptake inhibitors: Analysis of achievable steady-state plasma concentrations and the effect of ultrarapid metabolism at CYP2D6. *Pharmacotherapy* 22:1001–1006.

Lama, P. J. 2002. Systemic adverse effects of beta-adrenergic blockers: An evidence-based assessment. *Am J Ophthalmol* 134:749–760.

Lamba, V., Lamba, J. K., Dilawari, J. B., Kohli, K. K. 1998. Genetic polymorphism of CYP2D6 in North Indian subjects. *Eur J Clin Pharmacol* 54:787–791.

Lane, H. Y., Hu, O. Y., Jann, M. W., Deng, H. C., Lin, H. N., Chang, W. H. 1997. Dextromethorphan phenotyping and haloperidol disposition in schizophrenic patients. *Psychiatry Res* 69:105–111.

Leathart, J. B., London, S. J., Steward, A., Adams, J. D., Idle, J. R., Daly, A. K. 1998. CYP2D6 phenotype-genotype relationships in African-Americans and Caucasians in Los Angeles. *Pharmacogenetics* 8:529–541.

Lee, J. T., Kroemer, H. K., Silberstein, D. J., Funck-Brentano, C., Lineberry, M. D., Wood, A. J., Roden, D. M., Woosley, R. L. 1990. The role of genetically determined polymorphic drug metabolism in the beta-blockade produced by propafenone. *N Engl J Med* 322:1764–1768.

Lee, S. J., Lee, S. S., Jung, H. J., Kim, H. S., Park, S. J., Yeo, C. W., Shin, J. G. 2009. Discovery of novel functional variants and extensive evaluation of *CYP2D6* genetic polymorphisms in Koreans. *Drug Metab Dispos* 37:1464–1470.

Leinonen, E., Koponen, H. J., Lepola, U. 2004. Paroxetine increases serum trimipramine concentration. A report of two cases. *Human Psychopharmacol Clin Exp* 10:345–347.

Lelas, S., Wegert, S., Otton, S. V., Sellers, E. M., France, C. P. 1999. Inhibitors of cytochrome P450 differentially modify discriminative-stimulus and antinociceptive effects of hydrocodone and hydromorphone in rhesus monkeys. *Drug Alcohol Depend* 54:239–249.

Lemoine, A., Gautier, J. C., Azoulay, D., Kiffel, L., Belloc, C., Guengerich, F. P., Maurel, P., Beaune, P., Leroux, J. P. 1993. Major pathway of imipramine metabolism is catalyzed by cytochromes P-450 1A2 and P-450 3A4 in human liver. *Mol Pharmacol* 43:827–832.

Lennard, M. S., Silas, J. H., Freestone, S., Trevethick, J. 1982. Defective metabolism of metoprolol in poor hydroxylators of debrisoquine. *Br J Clin Pharmacol* 14:301–303.

Lennard, M. S., Tucker, G. T., Silas, J. H., Freestone, S., Ramsay, L. E., Woods, H. F. 1983. Differential stereoselective metabolism of metoprolol in extensive and poor debrisoquin metabolizers. *Clin Pharmacol Ther* 34:732–737.

Lennard, M. S., Jackson, P. R., Freestone, S., Ramsay, L. E., Tucker, G. T., Woods, H. F. 1984a. The oral clearance and beta-adrenoceptor antagonist activity of propranolol after single dose are not related to debrisoquine oxidation phenotype. *Br J Clin Pharmacol* 17 Suppl 1:106S–107S.

Lennard, M. S., Jackson, P. R., Freestone, S., Tucker, G. T., Ramsay, L. E., Woods, H. F. 1984b. The relationship between debrisoquine oxidation phenotype and the pharmacokinetics and pharmacodynamics of propranolol. *Br J Clin Pharmacol* 17:679–685.

Lennard, M. S., Tucker, G. T., Silas, J. H., Woods, H. F. 1986. Debrisoquine polymorphism and the metabolism and action of metoprolol, timolol, propranolol and atenolol. *Xenobiotica* 16:435–447.

Leon, J., Susce, M. T., Pan, R. M., Wedlund, P. J., Orrego, M. L., Diaz, F. J. 2007. A study of genetic (CYP2D6 and ABCB1) and environmental (drug inhibitors and inducers) variables that may influence plasma risperidone levels. *Pharmacopsychiatry* 40:93–102.

Leone Roberti Maggiore, U., Salvatore, S., Alessandri, F., Remorgida, V., Origoni, M., Candiani, M., Venturini, P. L., Ferrero, S. 2012. Pharmacokinetics and toxicity of antimuscarinic drugs for overactive bladder treatment in females. *Expert Opin Drug Metab Toxicol* 8:1387–1408.

Leppert, W. 2011. CYP2D6 in the metabolism of opioids for mild to moderate pain. *Pharmacology* 87:274–285.

Lessard, E., Yessine, M. A., Hamelin, B. A., O'Hara, G., LeBlanc, J., Turgeon, J. 1999. Influence of CYP2D6 activity on the disposition and cardiovascular toxicity of the antidepressant agent venlafaxine in humans. *Pharmacogenetics* 9:435–443.

Lessard, E., Yessine, M. A., Hamelin, B. A., Gauvin, C., Labbe, L., O'Hara, G., LeBlanc, J., Turgeon, J. 2001. Diphenhydramine alters the disposition of venlafaxine through inhibition of CYP2D6 activity in humans. *J Clin Psychopharmacol* 21:175–184.

Lewis, R. V., Lennard, M. S., Jackson, P. R., Tucker, G. T., Ramsay, L. E., Woods, H. F. 1985. Timolol and atenolol: Relationships between oxidation phenotype, pharmacokinetics and pharmacodynamics. *Br J Clin Pharmacol* 19:329–333.

Li, L., Pan, R. M., Porter, T. D., Jensen, N. S., Silber, P., Russo, G., Tine, J. A., Heim, J., Ring, B., Wedlund, P. J. 2006. New cytochrome P450 2D6*56 allele identified by genotype/phenotype analysis of cryopreserved human hepatocytes. *Drug Metab Dispos* 34:1411–1416.

Lim, H. S., Ju Lee, H., Seok Lee, K., Sook Lee, E., Jang, I. J., Ro, J. 2007. Clinical implications of CYP2D6 genotypes predictive of tamoxifen pharmacokinetics in metastatic breast cancer. *J Clin Oncol* 25:3837–3845.

Lim, K. S., Cho, J. Y., Jang, I. J., Kim, B. H., Kim, J., Jeon, J. Y., Tae, Y. M. et al. 2008. Pharmacokinetic interaction of flecainide and paroxetine in relation to the CYP2D6*10 allele in healthy Korean subjects. *Br J Clin Pharmacol* 66:660–666.

Lin, K. M., Finder, E. 1983. Neuroleptic dosage for Asians. *Am J Psychiatry* 140:490–491.

Lindh, J. D., Annas, A., Meurling, L., Dahl, M. L., AL-Shurbaji, A. 2003. Effect of ketoconazole on venlafaxine plasma concentrations in extensive and poor metabolisers of debrisoquine. *Eur J Clin Pharmacol* 59:401–406.

Linnet, K., Wiborg, O. 1996a. Influence of *CYP2D6* genetic polymorphism on ratios of steady-state serum concentration to dose of the neuroleptic zuclopenthixol. *Ther Drug Monit* 18:629–634.

Linnet, K., Wiborg, O. 1996b. Steady-state serum concentrations of the neuroleptic perphenazine in relation to CYP2D6 genetic polymorphism. *Clin Pharmacol Ther* 60:41–47.

Llerena, A., Alm, C., Dahl, M. L., Ekqvist, B., Bertilsson, L. 1992a. Haloperidol disposition is dependent on debrisoquine hydroxylation phenotype. *Ther Drug Monit* 14:92–97.

Llerena, A., Dahl, M. L., Ekqvist, B., Bertilsson, L. 1992b. Haloperidol disposition is dependent on the debrisoquine hydroxylation phenotype: Increased plasma levels of the reduced metabolite in poor metabolizers. *Ther Drug Monit* 14:261–264.

Llerena, A., Berecz, R., de la Rubia, A., Norberto, M. J., Benitez, J. 2000. Use of the mesoridazine/thioridazine ratio as a marker for CYP2D6 enzyme activity. *Ther Drug Monit* 22:397–401.

Llerena, A., Berecz, R., de la Rubia, A., Dorado, P. 2002. QTc interval lengthening is related to CYP2D6 hydroxylation capacity and plasma concentration of thioridazine in patients. *J Psychopharmacol* 16:361–364.

London, S. J., Daly, A. K., Leathart, J. B., Navidi, W. C., Carpenter, C. C., Idle, J. R. 1997. Genetic polymorphism of CYP2D6 and lung cancer risk in African-Americans and Caucasians in Los Angeles County. *Carcinogenesis* 18:1203–1214.

Lotsch, J. 2005. Opioid metabolites. *J Pain Symptom Manage* 29:S10–S24.

Lotsch, J., Skarke, C., Liefhold, J., Geisslinger, G. 2004. Genetic predictors of the clinical response to opioid analgesics: Clinical utility and future perspectives. *Clin Pharmacokinet* 43:983–1013.

Lovlie, R., Daly, A. K., Idle, J. R., Steen, V. M. 1997. Characterization of the 16+9 kb and 30+9 kb CYP2D6 XbaI haplotypes. *Pharmacogenetics* 7:149–152.

Lundqvist, E., Johansson, I., Ingelman-Sundberg, M. 1999. Genetic mechanisms for duplication and multiduplication of the human CYP2D6 gene and methods for detection of duplicated CYP2D6 genes. *Gene* 226:327–338.

Madadi, P., Ross, C. J., Hayden, M. R., Carleton, B. C., Gaedigk, A., Leeder, J. S., Koren, G. 2009. Pharmacogenetics of neonatal opioid toxicity following maternal use of codeine during breast-feeding: A case–control study. *Clin Pharmacol Ther* 85:31–35.

Mahgoub, A., Idle, J. R., Dring, L. G., Lancaster, R., Smith, R. L. 1977. Polymorphic hydroxylation of Debrisoquine in man. *Lancet* 2:584–586.

Mannens, G. S., Snel, C. A., Hendrickx, J., Verhaeghe, T., Le Jeune, L., Bode, W., van Beijsterveldt, L. et al. 2002. The metabolism and excretion of galantamine in rats, dogs, and humans. *Drug Metab Dispos* 30:553–563.

Mannheimer, B., Bahr, C. V., Pettersson, H., Eliasson, E. 2008. Impact of multiple inhibitors or substrates of cytochrome P450 2D6 on plasma risperidone levels in patients on polypharmacy. *Ther Drug Monit* 30:565–569.

Mao, Z. L., Wheeler, J. J., Clohs, L., Beatch, G. N., Keirns, J. 2009. Pharmacokinetics of novel atrial-selective antiarrhythmic agent vernakalant hydrochloride injection (RSD1235): Influence of CYP2D6 expression and other factors. *J Clin Pharmacol* 49:17–29.

Marchetti, P., Navalesi, R. 1989. Pharmacokinetic-pharmacodynamic relationships of oral hypoglycaemic agents. An update. *Clin Pharmacokinet* 16:100–128.

Marechal, J. D., Kemp, C. A., Roberts, G. C., Paine, M. J., Wolf, C. R., Sutcliffe, M. J. 2008. Insights into drug metabolism by cytochromes P450 from modelling studies of CYP2D6-drug interactions. *Br J Pharmacol* 153 Suppl 1:S82–S89.

Marez, D., Sabbagh, N., Legrand, M., Lo-Guidice, J. M., Boone, P., Broly, F. 1995. A novel CYP2D6 allele with an abolished splice recognition site associated with the poor metabolizer phenotype. *Pharmacogenetics* 5:305–311.

Marez, D., Legrand, M., Sabbagh, N., Lo-Guidice, J. M., Boone, P., Broly, F. 1996. An additional allelic variant of the CYP2D6 gene causing impaired metabolism of sparteine. *Hum Genet* 97:668–670.

Marez, D., Legrand, M., Sabbagh, N., Guidice, J. M., Spire, C., Lafitte, J. J., Meyer, U. A., Broly, F. 1997. Polymorphism of the cytochrome P450 CYP2D6 gene in a European population: Characterization of 48 mutations and 53 alleles, their frequencies and evolution. *Pharmacogenetics* 7:193–202.

Marez-Allorge, D., Ellis, S. W., Lo Guidice, J. M., Tucker, G. T., Broly, F. 1999. A rare G2061 insertion affecting the open reading frame of CYP2D6 and responsible for the poor metabolizer phenotype. *Pharmacogenetics* 9:393–396.

Margolis, J. M., O'Donnell, J. P., Mankowski, D. C., Ekins, S., Obach, R. S. 2000. (R)-, (S)-, and racemic fluoxetine N-demethylation by human cytochrome P450 enzymes. *Drug Metab Dispos* 28:1187–1191.

Martinez-Selles, M., Castillo, I., Montenegro, P., Martin, M. L., Almendral, J., Sanjurjo, M. 2005. [Pharmacogenetic study of the response to flecainide and propafenone in patients with atrial fibrillation]. *Rev Esp Cardiol* 58:745–748.

Masimirembwa, C., Hasler, J., Bertilssons, L., Johansson, I., Ekberg, O., Ingelman-Sundberg, M. 1996a. Phenotype and genotype analysis of debrisoquine hydroxylase (CYP2D6) in a black Zimbabwean population. Reduced enzyme activity and evaluation of metabolic correlation of CYP2D6 probe drugs. *Eur J Clin Pharmacol* 51:117–122.

Masimirembwa, C., Persson, I., Bertilsson, L., Hasler, J., Ingelman-Sundberg, M. 1996b. A novel mutant variant of the CYP2D6 gene (CYP2D6*17) common in a black African population: Association with diminished debrisoquine hydroxylase activity. *Br J Clin Pharmacol* 42:713–719.

Matimba, A., Del-Favero, J., Van Broeckhoven, C., Masimirembwa, C. 2009. Novel variants of major drug-metabolising enzyme genes in diverse African populations and their predicted functional effects. *Hum Genomics* 3:169–190.

Matsunaga, M., Yamazaki, H., Kiyotani, K., Iwano, S., Saruwatari, J., Nakagawa, K., Soyama, A. et al. 2009. Two novel *CYP2D6*10* haplotypes as possible causes of a poor metabolic phenotype in Japanese. *Drug Metab Dispos* 37:699–701.

Mautz, D. S., Nelson, W. L., Shen, D. D. 1995. Regioselective and stereoselective oxidation of metoprolol and bufuralol catalyzed by microsomes containing cDNA-expressed human P4502D6. *Drug Metab Dispos* 23:513–517.

McAllister, C. B., Wolfenden, H. T., Aslanian, W. S., Woosley, R. L., Wilkinson, G. R. 1986. Oxidative metabolism of encainide: Polymorphism, pharmacokinetics and clinical considerations. *Xenobiotica* 16:483–490.

McAlpine, D. E., O'Kane, D. J., Black, J. L., Mrazek, D. A. 2007. Cytochrome P450 2D6 genotype variation and venlafaxine dosage. *Mayo Clin Proc* 82:1065–1068.

McGourty, J. C., Silas, J. H., Lennard, M. S., Tucker, G. T., Woods, H. F. 1985. Metoprolol metabolism and debrisoquine oxidation polymorphism—Population and family studies. *Br J Clin Pharmacol* 20:555–566.

McLaughlin, L. A., Paine, M. J., Kemp, C. A., Marechal, J. D., Flanagan, J. U., Ward, C. J., Sutcliffe, M. J., Roberts, G. C., Wolf, C. R. 2005. Why is quinidine an inhibitor of cytochrome P450 2D6? The role of key active-site residues in quinidine binding. *J Biol Chem* 280:38617–38624.

McLellan, R. A., Oscarson, M., Seidegard, J., Evans, D. A., Ingelman-Sundberg, M. 1997. Frequent occurrence of CYP2D6 gene duplication in Saudi Arabians. *Pharmacogenetics* 7:187–191.

Mehanna, R., Hunter, C., Davidson, A., Jimenez-Shahed, J., Jankovic, J. 2013. Analysis of *CYP2D6* genotype and response to tetrabenazine. *Mov Disord* 28:210–215.

Mehvar, R., Brocks, D. R., Vakily, M. 2002. Impact of stereoselectivity on the pharmacokinetics and pharmacodynamics of antiarrhythmic drugs. *Clin Pharmacokinet* 41:533–558.

Melkersson, K. I., Scordo, M. G., Gunes, A., Dahl, M. L. 2007. Impact of CYP1A2 and CYP2D6 polymorphisms on drug metabolism and on insulin and lipid elevations and insulin resistance in clozapine-treated patients. *J Clin Psychiatry* 68:697–704.

Mellstrom, B., von Bahr, C. 1981. Demethylation and hydroxylation of amitriptyline, nortriptyline, and 10-hydroxyamitriptyline in human liver microsomes. *Drug Metab Dispos* 9:565–568.

Mellstrom, B., Bertilsson, L., Sawe, J., Schulz, H. U., Sjoqvist, F. 1981. E- and Z-10-hydroxylation of nortriptyline: Relationship to polymorphic debrisoquine hydroxylation. *Clin Pharmacol Ther* 30:189–193.

Mellstrom, B., Sawe, J., Bertilsson, L., Sjoqvist, F. 1986. Amitriptyline metabolism: Association with debrisoquin hydroxylation in nonsmokers. *Clin Pharmacol Ther* 39:369–371.

Michelson, D., Faries, D., Wernicke, J., Kelsey, D., Kendrick, K., Sallee, F. R., Spencer, T. 2001. Atomoxetine in the treatment of children and adolescents with attention-deficit/hyperactivity disorder: A randomized, placebo-controlled, dose-response study. *Pediatrics* 108:E83.

Michelson, D., Read, H. A., Ruff, D. D., Witcher, J., Zhang, S., McCracken, J. 2007. CYP2D6 and clinical response to atomoxetine in children and adolescents with ADHD. *J Am Acad Child Adolesc Psychiatry* 46:242–251.

Mihara, K., Otani, K., Suzuki, A., Yasui, N., Nakano, H., Meng, X., Ohkubo, T. et al. 1997a. Relationship between the CYP2D6 genotype and the steady-state plasma concentrations of trazodone and its active metabolite m-chlorophenylpiperazine. *Psychopharmacology (Berl)* 133:95–98.

Mihara, K., Otani, K., Tybring, G., Dahl, M. L., Bertilsson, L., Kaneko, S. 1997b. The CYP2D6 genotype and plasma concentrations of mianserin enantiomers in relation to therapeutic response to mianserin in depressed Japanese patients. *J Clin Psychopharmacol* 17:467–471.

Mihara, K., Suzuki, A., Kondo, T., Yasui, N., Furukori, H., Nagashima, U., Otani, K., Kaneko, S., Inoue, Y. 1999. Effects of the *CYP2D6*10* allele on the steady-state plasma concentrations of haloperidol and reduced haloperidol in Japanese patients with schizophrenia. *Clin Pharmacol Ther* 65:291–294.

Mikus, G., Gross, A. S., Beckmann, J., Hertrampf, R., Gundert-Remy, U., Eichelbaum, M. 1989. The influence of the sparteine/debrisoquin phenotype on the disposition of flecainide. *Clin Pharmacol Ther* 45:562–567.

Mikus, G., Bochner, F., Eichelbaum, M., Horak, P., Somogyi, A. A., Spector, S. 1994. Endogenous codeine and morphine in poor and extensive metabolisers of the CYP2D6 (debrisoquine/sparteine) polymorphism. *J Pharmacol Exp Ther* 268:546–551.

Mikus, G., Morike, K., Griese, E. U., Klotz, U. 1998. Relevance of deficient CYP2D6 in opiate dependence. *Pharmacogenetics* 8:565–568.

Mitsunaga, Y., Kubota, T., Ishiguro, A., Yamada, Y., Sasaki, H., Chiba, K., Iga, T. 2002. Frequent occurrence of CYP2D6*10 duplication allele in a Japanese population. *Mutat Res* 505:83–85.

Molitch, M. E. 2008. Drugs and prolactin. *Pituitary* 11:209–218.

Montane Jaime, L. K., Lalla, A., Steimer, W., Gaedigk, A. 2013. Characterization of the *CYP2D6* gene locus and metabolic activity in Indo- and Afro-Trinidadians: Discovery of novel allelic variants. *Pharmacogenomics* 14:261–276.

Morgan, M. Y., Reshef, R., Shah, R. R., Oates, N. S., Smith, R. L., Sherlock, S. 1984. Impaired oxidation of debrisoquine in patients with perhexiline liver injury. *Gut* 25:1057–1064.

Morike, K. E., Roden, D. M. 1994. Quinidine-enhanced beta-blockade during treatment with propafenone in extensive metabolizer human subjects. *Clin Pharmacol Ther* 55:28–34.

Morike, K., Magadum, S., Mettang, T., Griese, E. U., Machleidt, C., Kuhlmann, U. 1995. Propafenone in a usual dose produces severe side-effects: The impact of genetically determined metabolic status on drug therapy. *J Intern Med* 238:469–472.

Morita, S., Shimoda, K., Someya, T., Yoshimura, Y., Kamijima, K., Kato, N. 2000. Steady-state plasma levels of nortriptyline and its hydroxylated metabolites in Japanese patients: Impact of CYP2D6 genotype on the hydroxylation of nortriptyline. *J Clin Psychopharmacol* 20:141–149.

Muralidharan, G., Cooper, J. K., Hawes, E. M., Korchinski, E. D., Midha, K. K. 1996. Quinidine inhibits the 7-hydroxylation of chlorpromazine in extensive metabolisers of debrisoquine. *Eur J Clin Pharmacol* 50:121–128.

Murphy, G. M., Jr., Kremer, C., Rodrigues, H. E., Schatzberg, A. F. 2003. Pharmacogenetics of antidepressant medication intolerance. *Am J Psychiatry* 160:1830–1835.

Musshoff, F., Schmidt, P., Madea, B. 1999. Fatality caused by a combined trimipramine-citalopram intoxication. *Forensic Sci Int* 106:125–131.

Nelson, D. R., Zeldin, D. C., Hoffman, S. M., Maltais, L. J., Wain, H. M., Nebert, D. W. 2004. Comparison of cytochrome P450 (CYP) genes from the mouse and human genomes, including nomenclature recommendations for genes, pseudogenes and alternative-splice variants. *Pharmacogenetics* 14:1–18.

Nielsen, K. K., Brosen, K., Gram, L. F. 1992. Steady-state plasma levels of clomipramine and its metabolites: Impact of the sparteine/debrisoquine oxidation polymorphism. Danish University Antidepressant Group. *Eur J Clin Pharmacol* 43:405–411.

Nielsen, K. K., Brosen, K., Hansen, M. G., Gram, L. F. 1994. Single-dose kinetics of clomipramine: Relationship to the sparteine and *S*-mephenytoin oxidation polymorphisms. *Clin Pharmacol Ther* 55:518–527.

Nielsen, K. K., Flinois, J. P., Beaune, P., Brosen, K. 1996. The biotransformation of clomipramine *in vitro*, identification of the cytochrome P450s responsible for the separate metabolic pathways. *J Pharmacol Exp Ther* 277:1659–1664.

Nieminen, T., Uusitalo, H., Maenpaa, J., Turjanmaa, V., Rane, A., Lundgren, S., Ropo, A., Rontu, R., Lehtimaki, T., Kahonen, M. 2005. Polymorphisms of genes CYP2D6, ADRB1 and GNAS1 in pharmacokinetics and systemic effects of ophthalmic timolol. A pilot study. *Eur J Clin Pharmacol* 61:811–819.

Nieminen, T., Lehtimaki, T., Maenpaa, J., Ropo, A., Uusitalo, H., Kahonen, M. 2007. Ophthalmic timolol: Plasma concentration and systemic cardiopulmonary effects. *Scand J Clin Lab Invest* 67:237–245.

Nilvebrant, L., Gillberg, P. G., Sparf, B. 1997. Antimuscarinic potency and bladder selectivity of PNU-200577, a major metabolite of tolterodine. *Pharmacol Toxicol* 81:169–172.

Nishida, Y., Fukuda, T., Yamamoto, I., Azuma, J. 2000. CYP2D6 genotypes in a Japanese population: Low frequencies of CYP2D6 gene duplication but high frequency of CYP2D6*10. *Pharmacogenetics* 10:567–570.

Niwa, T., Shiraga, T., Mitani, Y., Terakawa, M., Tokuma, Y., Kagayama, A. 2000. Stereoselective metabolism of cibenzoline, an antiarrhythmic drug, by human and rat liver microsomes: Possible involvement of CYP2D and CYP3A. *Drug Metab Dispos* 28:1128–1134.

Nowell, S. A., Ahn, J., Rae, J. M., Scheys, J. O., Trovato, A., Sweeney, C., MacLeod, S. L., Kadlubar, F. F., Ambrosone, C. B. 2005. Association of genetic variation in tamoxifen-metabolizing enzymes with overall survival and recurrence of disease in breast cancer patients. *Breast Cancer Res Treat* 91:249–258.

Nyberg, S., Dahl, M. L., Halldin, C. 1995a. A PET study of D2 and 5-HT2 receptor occupancy induced by risperidone in poor metabolizers of debrisoquin and risperidone. *Psychopharmacology (Berl)* 119:345–348.

Nyberg, S., Farde, L., Halldin, C., Dahl, M. L., Bertilsson, L. 1995b. D2 dopamine receptor occupancy during low-dose treatment with haloperidol decanoate. *Am J Psychiatry* 152:173–178.

O'Keane, V. 2008. Antipsychotic-induced hyperprolactinaemia, hypogonadism and osteoporosis in the treatment of schizophrenia. *J Psychopharmacol* 22:70–75.

Oates, N. S., Shah, R. R., Idle, J. R., Smith, R. L. 1981. Phenformin-induced lacticacidosis associated with impaired debrisoquine hydroxylation. *Lancet* 1:837–838.

Oates, N. S., Shah, R. R., Idle, J. R., Smith, R. L. 1982. Genetic polymorphism of phenformin 4-hydroxylation. *Clin Pharmacol Ther* 32:81–89.

Oates, N. S., Shah, R. R., Idle, J. R., Smith, R. L. 1983. Influence of oxidation polymorphism on phenformin kinetics and dynamics. *Clin Pharmacol Ther* 34:827–834.

Ohnuma, T., Shibata, N., Matsubara, Y., Arai, H. 2003. Haloperidol plasma concentration in Japanese psychiatric subjects with gene duplication of CYP2D6. *Br J Clin Pharmacol* 56:315–320.

Oldham, H. G., Clarke, S. E. 1997. In vitro identification of the human cytochrome P450 enzymes involved in the metabolism of R(+)- and S(−)-carvedilol. *Drug Metab Dispos* 25:970–977.

Olesen, O. V., Linnet, K. 1997. Hydroxylation and demethylation of the tricyclic antidepressant nortriptyline by cDNA-expressed human cytochrome P-450 isozymes. *Drug Metab Dispos* 25:740–744.

Olesen, O. V., Linnet, K. 1999. Studies on the stereoselective metabolism of citalopram by human liver microsomes and cDNA-expressed cytochrome P450 enzymes. *Pharmacology* 59:298–309.

Olesen, O. V., Linnet, K. 2000. Identification of the human cytochrome P450 isoforms mediating in vitro N-dealkylation of perphenazine. *Br J Clin Pharmacol* 50:563–571.

Olesen, O. V., Linnet, K. 2001. Contributions of five human cytochrome P450 isoforms to the *N*-demethylation of clozapine *in vitro* at low and high concentrations. *J Clin Pharmacol* 41:823–832.

Olesen, O. V., Licht, R. W., Thomsen, E., Bruun, T., Viftrup, J. E., Linnet, K. 1998. Serum concentrations and side effects in psychiatric patients during risperidone therapy. *Ther Drug Monit* 20:380–384.

Olsson, B., Szamosi, J. 2001a. Food does not influence the pharmacokinetics of a new extended release formulation of tolterodine for once daily treatment of patients with overactive bladder. *Clin Pharmacokinet* 40:135–143.

Olsson, B., Szamosi, J. 2001b. Multiple dose pharmacokinetics of a new once daily extended release tolterodine formulation versus immediate release tolterodine. *Clin Pharmacokinet* 40:227–235.

Oosterhuis, M., Van De Kraats, G., Tenback, D. 2007. Safety of aripiprazole: High serum levels in a CYP2D6 mutated patient. *Am J Psychiatry* 164:175.

Oscarson, M., Hidestrand, M., Johansson, I., Ingelman-Sundberg, M. 1997. A combination of mutations in the CYP2D6*17 (CYP2D6Z) allele causes alterations in enzyme function. *Mol Pharmacol* 52:1034–1040.

Otani, K., Aoshima, T. 2000. Pharmacogenetics of classical and new antipsychotic drugs. *Ther Drug Monit* 22:118–121.

Otton, S. V., Crewe, H. K., Lennard, M. S., Tucker, G. T., Woods, H. F. 1988. Use of quinidine inhibition to define the role of the sparteine/debrisoquine cytochrome P450 in metoprolol oxidation by human liver microsomes. *J Pharmacol Exp Ther* 247:242–247.

Otton, S. V., Schadel, M., Cheung, S. W., Kaplan, H. L., Busto, U. E., Sellers, E. M. 1993a. CYP2D6 phenotype determines the metabolic conversion of hydrocodone to hydromorphone. *Clin Pharmacol Ther* 54:463–472.

Otton, S. V., Wu, D., Joffe, R. T., Cheung, S. W., Sellers, E. M. 1993b. Inhibition by fluoxetine of cytochrome P450 2D6 activity. *Clin Pharmacol Ther* 53:401–409.

Otton, S. V., Ball, S. E., Cheung, S. W., Inaba, T., Rudolph, R. L., Sellers, E. M. 1996. Venlafaxine oxidation *in vitro* is catalysed by CYP2D6. *Br J Clin Pharmacol* 41:149–156.

Ozdemir, V., Naranjo, C. A., Herrmann, N., Reed, K., Sellers, E. M., Kalow, W. 1997. Paroxetine potentiates the central nervous system side effects of perphenazine: Contribution of cytochrome P4502D6 inhibition in vivo. *Clin Pharmacol Ther* 62:334–347.

Ozdemir, V., Tyndale, R. F., Reed, K., Herrmann, N., Sellers, E. M., Kalow, W., Naranjo, C. A. 1999. Paroxetine steady-state plasma concentration in relation to CYP2D6 genotype in extensive metabolizers. *J Clin Psychopharmacol* 19:472–475.

Ozdemir, V., Bertilsson, L., Miura, J., Carpenter, E., Reist, C., Harper, P., Widen, J. et al. 2007. CYP2D6 genotype in relation to perphenazine concentration and pituitary pharmacodynamic tissue sensitivity in Asians: CYP2D6-serotonin-dopamine crosstalk revisited. *Pharmacogenet Genomics* 17:339–347.

Paar, W. D., Poche, S., Gerloff, J., Dengler, H. J. 1997. Polymorphic CYP2D6 mediates *O*-demethylation of the opioid analgesic tramadol. *Eur J Clin Pharmacol* 53:235–239.

Pan, L. P., De Vriendt, C., Belpaire, F. M. 1998. In-vitro characterization of the cytochrome P450 isoenzymes involved in the back oxidation and N-dealkylation of reduced haloperidol. *Pharmacogenetics* 8:383–389.

Pan, L., Vander Stichele, R., Rosseel, M. T., Berlo, J. A., De Schepper, N., Belpaire, F. M. 1999. Effects of smoking, CYP2D6 genotype, and concomitant drug intake on the steady state plasma concentrations of haloperidol and reduced haloperidol in schizophrenic inpatients. *Ther Drug Monit* 21:489–497.

Panagiotidis, G., Arthur, H. W., Lindh, J. D., Dahl, M. L., Sjoqvist, F. 2007. Depot haloperidol treatment in outpatients with schizophrenia on monotherapy: Impact of CYP2D6 polymorphism on pharmacokinetics and treatment outcome. *Ther Drug Monit* 29:417–422.

Panserat, S., Mura, C., Gerard, N., Vincent-Viry, M., Galteau, M. M., Jacqz-Aigrain, E., Krishnamoorthy, R. 1994. DNA haplotype-dependent differences in the amino acid sequence of debrisoquine 4-hydroxylase (CYP2D6): Evidence for two major allozymes in extensive metabolisers. *Hum Genet* 94:401–406.

Panserat, S., Mura, C., Gerard, N., Vincent-Viry, M., Galteau, M. M., Jacoz-Aigrain, E., Krishnamoorthy, R. 1995. An unequal crossover event within the CYP2D gene cluster generates a chimeric CYP2D7/CYP2D6 gene which is associated with the poor metabolizer phenotype. *Br J Clin Pharmacol* 40:361–367.

Park, J. Y., Shon, J. H., Kim, K. A., Jung, H. J., Shim, J. C., Yoon, Y. R., Cha, I. J., Shin, J. G. 2006. Combined effects of itraconazole and CYP2D6*10 genetic polymorphism on the pharmacokinetics and pharmacodynamics of haloperidol in healthy subjects. *J Clin Psychopharmacol* 26:135–142.

Patsopoulos, N. A., Ntzani, E. E., Zintzaras, E., Ioannidis, J. P. 2005. CYP2D6 polymorphisms and the risk of tardive dyskinesia in schizophrenia: A meta-analysis. *Pharmacogenet Genomics* 15:151–158.

Paulzen, M., Clement, H. W., Grunder, G. 2008. Enhancement of atomoxetine serum levels by co-administration of paroxetine. *Int J Neuropsychopharmacol* 11:289–291.

Pearce, E. F., Murphy, J. A. 2014. Vortioxetine for the treatment of depression. *Ann Pharmacother* 48:758–765.

Pedersen, R. S., Damkier, P., Brosen, K. 2005. Tramadol as a new probe for cytochrome P450 2D6 phenotyping: A population study. *Clin Pharmacol Ther* 77:458–467.

Pedersen, R. S., Damkier, P., Brosen, K. 2006. Enantioselective pharmacokinetics of tramadol in CYP2D6 extensive and poor metabolizers. *Eur J Clin Pharmacol* 62:513–521.

Peters, E. J., Slager, S. L., Kraft, J. B., Jenkins, G. D., Reinalda, M. S., McGrath, P. J., Hamilton, S. P. 2008. Pharmacokinetic genes do not influence response or tolerance to citalopram in the STAR*D sample. *PLoS One* 3:e1872.

Peveler, R. C., Branford, D., Citrome, L., Fitzgerald, P., Harvey, P. W., Holt, R. I., Howard, L. et al. 2008. Antipsychotics and hyperprolactinaemia: Clinical recommendations. *J Psychopharmacol* 22:98–103.

Plesnicar, B. K., Zalar, B., Breskvar, K., Dolzan, V. 2006. The influence of the CYP2D6 polymorphism on psychopathological and extrapyramidal symptoms in the patients on long-term antipsychotic treatment. *J Psychopharmacol* 20:829–833.

Pollock, B. G., Mulsant, B. H., Sweet, R. A., Rosen, J., Altieri, L. P., Perel, J. M. 1995. Prospective cytochrome P450 phenotyping for neuroleptic treatment in dementia. *Psychopharmacol Bull* 31:327–331.

Postlind, H., Danielson, A., Lindgren, A., Andersson, S. H. 1998. Tolterodine, a new muscarinic receptor antagonist, is metabolized by cytochromes P450 2D6 and 3A in human liver microsomes. *Drug Metab Dispos* 26:289–293.

Poulsen, L., Arendt-Nielsen, L., Brosen, K., Sindrup, S. H. 1996a. The hypoalgesic effect of tramadol in relation to CYP2D6. *Clin Pharmacol Ther* 60:636–644.

Poulsen, L., Brosen, K., Arendt-Nielsen, L., Gram, L. F., Elbaek, K., Sindrup, S. H. 1996b. Codeine and morphine in extensive and poor metabolizers of sparteine: Pharmacokinetics, analgesic effect and side effects. *Eur J Clin Pharmacol* 51:289–295.

Province, M. A., Altman, R. B., Klein, T. E. 2014. Interpreting the *CYP2D6* results from the International Tamoxifen Pharmacogenetics Consortium. *Clin Pharmacol Ther* 96:144–146.

Qian, J. C., Xu, X. M., Hu, G. X., Dai, D. P., Xu, R. A., Hu, L. M., Li, F. H., Zhang, X. H., Yang, J. F., Cai, J. P. 2013. Genetic variations of human *CYP2D6* in the Chinese Han population. *Pharmacogenomics* 14:1731–1743.

Qin, S., Shen, L., Zhang, A., Xie, J., Shen, W., Chen, L., Tang, J. et al. 2008. Systematic polymorphism analysis of the *CYP2D6* gene in four different geographical Han populations in mainland China. *Genomics* 92:152–158.

Qumsieh, R. Y., Ali, B. R., Abdulrazzaq, Y. M., Osman, O., Akawi, N. A., Bastaki, S. M. 2011. Identification of new alleles and the determination of alleles and genotypes frequencies at the *CYP2D6* gene in Emiratis. *PLoS One* 6:e28943.

Raghuram, T. C., Koshakji, R. P., Wilkinson, G. R., Wood, A. J. 1984. Polymorphic ability to metabolize propranolol alters 4-hydroxypropranolol levels but not beta blockade. *Clin Pharmacol Ther* 36:51–56.

Raimundo, S., Fischer, J., Eichelbaum, M., Griese, E. U., Schwab, M., Zanger, U. M. 2000. Elucidation of the genetic basis of the common 'intermediate metabolizer' phenotype for drug oxidation by CYP2D6. *Pharmacogenetics* 10:577–581.

Raimundo, S., Toscano, C., Klein, K., Fischer, J., Griese, E. U., Eichelbaum, M., Schwab, M., Zanger, U. M. 2004. A novel intronic mutation, 2988G>A, with high predictivity for impaired function of cytochrome P450 2D6 in white subjects. *Clin Pharmacol Ther* 76:128–138.

Rau, T., Wohlleben, G., Wuttke, H., Thuerauf, N., Lunkenheimer, J., Lanczik, M., Eschenhagen, T. 2004. *CYP2D6* genotype: Impact on adverse effects and nonresponse during treatment with antidepressants—A pilot study. *Clin Pharmacol Ther* 75:386–393.

Rau, T., Diepenbruck, S., Diepenbruck, I., Eschenhagen, T. 2006. The 2988G>A polymorphism affects splicing of a CYP2D6 minigene. *Clin Pharmacol Ther* 80:555–558; author reply 558–560.

Reis, M., Lundmark, J., Bengtsson, F. 2003. Therapeutic drug monitoring of racemic citalopram: A 5-year experience in Sweden, 1992–1997. *Ther Drug Monit* 25:183–191.

Relling, M. V., Cherrie, J., Schell, M. J., Petros, W. P., Meyer, W. H., Evans, W. E. 1991. Lower prevalence of the debrisoquin oxidative poor metabolizer phenotype in American black versus white subjects. *Clin Pharmacol Ther* 50:308–313.

Ring, B. J., Catlow, J., Lindsay, T. J., Gillespie, T., Roskos, L. K., Cerimele, B. J., Swanson, S. P., Hamman, M. A., Wrighton, S. A. 1996. Identification of the human cytochromes P450 responsible for the in vitro formation of the major oxidative metabolites of the antipsychotic agent olanzapine. *J Pharmacol Exp Ther* 276:658–666.

Ring, B. J., Eckstein, J. A., Gillespie, J. S., Binkley, S. N., VandenBranden, M., Wrighton, S. A. 2001. Identification of the human cytochromes p450 responsible for in vitro formation of R- and S-norfluoxetine. *J Pharmacol Exp Ther* 297:1044–1050.

Ring, B. J., Gillespie, J. S., Eckstein, J. A., Wrighton, S. A. 2002. Identification of the human cytochromes P450 responsible for atomoxetine metabolism. *Drug Metab Dispos* 30:319–323.

Roberts, R. L., Mulder, R. T., Joyce, P. R., Luty, S. E., Kennedy, M. A. 2004. No evidence of increased adverse drug reactions in cytochrome P450 CYP2D6 poor metabolizers treated with fluoxetine or nortriptyline. *Hum Psychopharmacol* 19:17–23.

Roden, D. M., Woosley, R. L. 1988. Clinical pharmacokinetics of encainide. *Clin Pharmacokinet* 14:141–147.

Roden, D. M., Wood, A. J., Wilkinson, G. R., Woosley, R. L. 1986. Disposition kinetics of encainide and metabolites. *Am J Cardiol* 58:4C–9C.

Roh, H. K., Chung, J. Y., Oh, D. Y., Park, C. S., Svensson, J. O., Dahl, M. L., Bertilsson, L. 2001a. Plasma concentrations of haloperidol are related to CYP2D6 genotype at low, but not high doses of haloperidol in Korean schizophrenic patients. *Br J Clin Pharmacol* 52:265–271.

Roh, H. K., Kim, C. E., Chung, W. G., Park, C. S., Svensson, J. O., Bertilsson, L. 2001b. Risperidone metabolism in relation to CYP2D6*10 allele in Korean schizophrenic patients. *Eur J Clin Pharmacol* 57:671–675.

Rowland, P., Blaney, F. E., Smyth, M. G., Jones, J. J., Leydon, V. R., Oxbrow, A. K., Lewis, C. J. et al. 2006. Crystal structure of human cytochrome P450 2D6. *J Biol Chem* 281:7614–7622.

Sachse, C., Brockmoller, J., Bauer, S., Reum, T., Roots, I. 1996. A rare insertion of T226 in exon 1 of CYP2D6 causes a frameshift and is associated with the poor metabolizer phenotype: CYP2D6*15. *Pharmacogenetics* 6:269–272.

Sachse, C., Brockmoller, J., Bauer, S., Roots, I. 1997. Cytochrome P450 2D6 variants in a Caucasian population: Allele frequencies and phenotypic consequences. *Am J Hum Genet* 60:284–295.

Sachse, C., Brockmoller, J., Hildebrand, M., Muller, K., Roots, I. 1998. Correctness of prediction of the CYP2D6 phenotype confirmed by genotyping 47 intermediate and poor metabolizers of debrisoquine. *Pharmacogenetics* 8:181–185.

Sakuyama, K., Sasaki, T., Ujiie, S., Obata, K., Mizugaki, M., Ishikawa, M., Hiratsuka, M. 2008. Functional characterization of 17 CYP2D6 allelic variants (CYP2D6.2, 10, 14A-B, 18, 27, 36, 39, 47–51, 53–55, and 57). *Drug Metab Dispos* 36:2460–2467.

Sallustio, B. C., Westley, I. S., Morris, R. G. 2002. Pharmacokinetics of the antianginal agent perhexiline: Relationship between metabolic ratio and steady-state dose. *Br J Clin Pharmacol* 54:107–114.

Sanchez, C., Asin, K. E., Artigas, F. 2015. Vortioxetine, a novel antidepressant with multimodal activity: Review of preclinical and clinical data. *Pharmacol Ther* 145:43–57.

Sanwald, P., David, M., Dow, J. 1996. Characterization of the cytochrome P450 enzymes involved in the in vitro metabolism of dolasetron. Comparison with other indole-containing 5-HT3 antagonists. *Drug Metab Dispos* 24:602–609.

Saruwatari, J., Matsunaga, M., Ikeda, K., Nakao, M., Oniki, K., Seo, T., Mihara, S., Marubayashi, T., Kamataki, T., Nakagawa, K. 2006. Impact of *CYP2D6*10* on H1-antihistamine-induced hypersomnia. *Eur J Clin Pharmacol* 62:995–1001.

Sauer, J. M., Ponsler, G. D., Mattiuz, E. L., Long, A. J., Witcher, J. W., Thomasson, H. R., Desante, K. A. 2003. Disposition and metabolic fate of atomoxetine hydrochloride: The role of CYP2D6 in human disposition and metabolism. *Drug Metab Dispos* 31:98–107.

Saxena, R., Shaw, G. L., Relling, M. V., Frame, J. N., Moir, D. T., Evans, W. E., Caporaso, N., Weiffenbach, B. 1994. Identification of a new variant CYP2D6 allele with a single base deletion in exon 3 and its association with the poor metabolizer phenotype. *Hum Mol Genet* 3:923–926.

Schenk, P. W., van Fessem, M. A., Verploegh-Van Rij, S., Mathot, R. A., van Gelder, T., Vulto, A. G., van Vliet, M., Lindemans, J., Bruijn, J. A., van Schaik, R. H. 2008. Association of graded allele-specific changes in CYP2D6 function with imipramine dose requirement in a large group of depressed patients. *Mol Psychiatry* 13:597–605.

Schmidt, H., Vormfelde, S. V., Walchner-Bonjean, M., Klinder, K., Freudenthaler, S., Gleiter, C. H., Gundert-Remy, U., Skopp, G., Aderjan, R., Fuhr, U. 2003. The role of active metabolites in dihydrocodeine effects. *Int J Clin Pharmacol Ther* 41:95–106.

Schroth, W., Antoniadou, L., Fritz, P., Schwab, M., Muerdter, T., Zanger, U. M., Simon, W., Eichelbaum, M., Brauch, H. 2007. Breast cancer treatment outcome with adjuvant tamoxifen relative to patient CYP2D6 and CYP2C19 genotypes. *J Clin Oncol* 25:5187–5193.

Scordo, M. G., Spina, E., Facciola, G., Avenoso, A., Johansson, I., Dahl, M. L. 1999. Cytochrome P450 2D6 genotype and steady state plasma levels of risperidone and 9-hydroxyrisperidone. *Psychopharmacology (Berl)* 147:300–305.

Scordo, M. G., Spina, E., Dahl, M. L., Gatti, G., Perucca, E. 2005. Influence of CYP2C9, 2C19 and 2D6 genetic polymorphisms on the steady-state plasma concentrations of the enantiomers of fluoxetine and norfluoxetine. *Basic Clin Pharmacol Toxicol* 97:296–301.

Seeringer, A., Kirchheiner, J. 2008. Pharmacogenetics-guided dose modifications of antidepressants. *Clin Lab Med* 28:619–626.

Seeringer, A., Brockmoller, J., Bauer, S., Kirchheiner, J. 2008. Enantiospecific pharmacokinetics of metoprolol in CYP2D6 ultra-rapid metabolizers and correlation with exercise-induced heart rate. *Eur J Clin Pharmacol* 64:883–888.

Shah, R. R., Oates, N. S., Idle, J. R., Smith, R. L., Lockhart, J. D. 1982. Impaired oxidation of debrisoquine in patients with perhexiline neuropathy. *Br Med J (Clin Res Ed)* 284:295–299.

Shah, R. R., Evans, D. A., Oates, N. S., Idle, J. R., Smith, R. L. 1985. The genetic control of phenformin 4-hydroxylation. *J Med Genet* 22:361–366.

Shaheen, O., Biollaz, J., Koshakji, R. P., Wilkinson, G. R., Wood, A. J. 1989. Influence of debrisoquin phenotype on the inducibility of propranolol metabolism. *Clin Pharmacol Ther* 45:439–443.

Shams, M. E., Arneth, B., Hiemke, C., Dragicevic, A., Muller, M. J., Kaiser, R., Lackner, K., Hartter, S. 2006. CYP2D6 polymorphism and clinical effect of the antidepressant venlafaxine. *J Clin Pharm Ther* 31:493–502.

Shen, H., He, M. M., Liu, H., Wrighton, S. A., Wang, L., Guo, B., Li, C. 2007. Comparative metabolic capabilities and inhibitory profiles of CYP2D6.1, CYP2D6.10, and CYP2D6.17. *Drug Metab Dispos* 35:1292–1300.

Shimada, T., Tsumura, F., Yamazaki, H., Guengerich, F. P., Inoue, K. 2001. Characterization of (+/-)-bufuralol hydroxylation activities in liver microsomes of Japanese and Caucasian subjects genotyped for CYP2D6. *Pharmacogenetics* 11:143–156.

Shimoda, K., Morita, S., Hirokane, G., Yokono, A., Someya, T., Takahashi, S. 2000a. Metabolism of desipramine in Japanese psychiatric patients: The impact of CYP2D6 genotype on the hydroxylation of desipramine. *Pharmacol Toxicol* 86:245–249.

Shimoda, K., Morita, S., Yokono, A., Someya, T., Hirokane, G., Sunahara, N., Takahashi, S. 2000b. CYP2D6*10 alleles are not the determinant of the plasma haloperidol concentrations in Asian patients. *Ther Drug Monit* 22:392–396.

Shiran, M. R., Chowdry, J., Rostami-Hodjegan, A., Ellis, S. W., Lennard, M. S., Iqbal, M. Z., Lagundoye, O., Seivewright, N., Tucker, G. T. 2003. A discordance between cytochrome P450 2D6 genotype and phenotype in patients undergoing methadone maintenance treatment. *Br J Clin Pharmacol* 56:220–224.

Siddoway, L. A., Thompson, K. A., McAllister, C. B., Wang, T., Wilkinson, G. R., Roden, D. M., Woosley, R. L. 1987. Polymorphism of propafenone metabolism and disposition in man: Clinical and pharmacokinetic consequences. *Circulation* 75:785–791.

Simooya, O. O., Njunju, E., Hodjegan, A. R., Lennard, M. S., Tucker, G. T. 1993. Debrisoquine and metoprolol oxidation in Zambians: A population study. *Pharmacogenetics* 3:205–208.

Sindrup, S. H., Brosen, K., Gram, L. F. 1990a. Nonlinear kinetics of imipramine in low and medium plasma level ranges. *Ther Drug Monit* 12:445–449.

Sindrup, S. H., Gram, L. F., Skjold, T., Grodum, E., Brosen, K., Beck-Nielsen, H. 1990b. Clomipramine vs desipramine vs placebo in the treatment of diabetic neuropathy symptoms. A double-blind cross-over study. *Br J Clin Pharmacol* 30:683–691.

Sindrup, S. H., Brosen, K., Gram, L. F. 1992a. Pharmacokinetics of the selective serotonin reuptake inhibitor paroxetine: Non-linearity and relation to the sparteine oxidation polymorphism. *Clin Pharmacol Ther* 51:288–295.

Sindrup, S. H., Brosen, K., Gram, L. F., Hallas, J., Skjelbo, E., Allen, A., Allen, G. D. et al. 1992b. The relationship between paroxetine and the sparteine oxidation polymorphism. *Clin Pharmacol Ther* 51:278–287.

Sindrup, S. H., Tuxen, C., Gram, L. F., Grodum, E., Skjold, T., Brosen, K., Beck-Nielsen, H. 1992c. Lack of effect of mianserin on the symptoms of diabetic neuropathy. *Eur J Clin Pharmacol* 43:251–255.

Sistonen, J., Sajantila, A., Lao, O., Corander, J., Barbujani, G., Fuselli, S. 2007. CYP2D6 worldwide genetic variation shows high frequency of altered activity variants and no continental structure. *Pharmacogenet Genomics* 17:93–101.

Sistonen, J., Fuselli, S., Palo, J. U., Chauhan, N., Padh, H., Sajantila, A. 2009. Pharmacogenetic variation at CYP2C9, CYP2C19, and CYP2D6 at global and microgeographic scales. *Pharmacogenet Genomics* 19:170–179.

Sjoqvist, F., Eliasson, E. 2007. The convergence of conventional therapeutic drug monitoring and pharmacogenetic testing in personalized medicine: Focus on antidepressants. *Clin Pharmacol Ther* 81:899–902.

Skoda, R. C., Gonzalez, F. J., Demierre, A., Meyer, U. A. 1988. Two mutant alleles of the human cytochrome P-450db1 gene (P450C2D1) associated with genetically deficient metabolism of debrisoquine and other drugs. *Proc Natl Acad Sci U S A* 85:5240–5243.

Slanar, O., Nobilis, M., Kvetina, J., Mikoviny, R., Zima, T., Idle, J. R., Perlik, F. 2007. Miotic action of tramadol is determined by CYP2D6 genotype. *Physiol Res* 56:129–136.

Soderberg Lofdal, K. C., Andersson, M. L., Gustafsson, L. L. 2013. Cytochrome P450-mediated changes in oxycodone pharmacokinetics/pharmacodynamics and their clinical implications. *Drugs* 73:533–543.

Sohn, D. R., Shin, S. G., Park, C. W., Kusaka, M., Chiba, K., Ishizaki, T. 1991. Metoprolol oxidation polymorphism in a Korean population: Comparison with native Japanese and Chinese populations. *Br J Clin Pharmacol* 32:504–507.

Solus, J. F., Arietta, B. J., Harris, J. R., Sexton, D. P., Steward, J. Q., McMunn, C., Ihrie, P., Mehall, J. M., Edwards, T. L., Dawson, E. P. 2004. Genetic variation in eleven phase I drug metabolism genes in an ethnically diverse population. *Pharmacogenomics* 5:895–931.

Someya, T., Suzuki, Y., Shimoda, K., Hirokane, G., Morita, S., Yokono, A., Inoue, Y., Takahashi, S. 1999. The effect of cytochrome P450 2D6 genotypes on haloperidol metabolism: A preliminary study in a psychiatric population. *Psychiatry Clin Neurosci* 53:593–597.

Somogyi, A. A., Barratt, D. T., Coller, J. K. 2007. Pharmacogenetics of opioids. *Clin Pharmacol Ther* 81:429–444.

Sorensen, L. B., Sorensen, R. N., Miners, J. O., Somogyi, A. A., Grgurinovich, N., Birkett, D. J. 2003. Polymorphic hydroxylation of perhexiline *in vitro*. *Br J Clin Pharmacol* 55:635–638.

Sowinski, K. M., Burlew, B. S. 1997. Impact of CYP2D6 poor metabolizer phenotype on propranolol pharmacokinetics and response. *Pharmacotherapy* 17:1305–1310.

Soyama, A., Kubo, T., Miyajima, A., Saito, Y., Shiseki, K., Komamura, K., Ueno, K. et al. 2004. Novel nonsynonymous single nucleotide polymorphisms in the CYP2D6 gene. *Drug Metab Pharmacokinet* 19:313–319.

Soyama, A., Saito, Y., Kubo, T., Miyajima, A., Ohno, Y., Komamura, K., Ueno, K. et al. 2006. Sequence-based analysis of the CYP2D6*36-CYP2D6*10 tandem-type arrangement, a major CYP2D6*10 haplotype in the Japanese population. *Drug Metab Pharmacokinet* 21:208–216.

Spigset, O., Granberg, K., Hagg, S., Soderstrom, E., Dahlqvist, R. 1998. Non-linear fluvoxamine disposition. *Br J Clin Pharmacol* 45:257–263.

Spigset, O., Axelsson, S., Norstrom, A., Hagg, S., Dahlqvist, R. 2001. The major fluvoxamine metabolite in urine is formed by CYP2D6. *Eur J Clin Pharmacol* 57:653–658.

Spina, E., Steiner, E., Ericsson, O., Sjoqvist, F. 1987. Hydroxylation of desmethylimipramine: Dependence on the debrisoquin hydroxylation phenotype. *Clin Pharmacol Ther* 41:314–319.

Spina, E., Gitto, C., Avenoso, A., Campo, G. M., Caputi, A. P., Perucca, E. 1997. Relationship between plasma desipramine levels, CYP2D6 phenotype and clinical response to desipramine: A prospective study. *Eur J Clin Pharmacol* 51:395–398.

Spina, E., Avenoso, A., Facciola, G., Scordo, M. G., Ancione, M., Madia, A. 2001. Plasma concentrations of risperidone and 9-hydroxyrisperidone during combined treatment with paroxetine. *Ther Drug Monit* 23:223–227.

Stamer, U. M., Lehnen, K., Hothker, F., Bayerer, B., Wolf, S., Hoeft, A., Stuber, F. 2003. Impact of CYP2D6 genotype on postoperative tramadol analgesia. *Pain* 105:231–238.

Stamer, U. M., Musshoff, F., Kobilay, M., Madea, B., Hoeft, A., Stuber, F. 2007. Concentrations of tramadol and O-desmethyltramadol enantiomers in different CYP2D6 genotypes. *Clin Pharmacol Ther* 82:41–47.

Stearns, V., Johnson, M. D., Rae, J. M., Morocho, A., Novielli, A., Bhargava, P., Hayes, D. F., Desta, Z., Flockhart, D. A. 2003. Active tamoxifen metabolite plasma concentrations after coadministration of tamoxifen and the selective serotonin reuptake inhibitor paroxetine. *J Natl Cancer Inst* 95:1758–1764.

Stedman, C. A., Begg, E. J., Kennedy, M. A., Roberts, R., Wilkinson, T. J. 2002. Cytochrome P450 2D6 genotype does not predict SSRI (fluoxetine or paroxetine) induced hyponatraemia. *Hum Psychopharmacol* 17:187–190.

Steen, V. M., Andreassen, O. A., Daly, A. K., Tefre, T., Borresen, A. L., Idle, J. R., Gulbrandsen, A. K. 1995a. Detection of the poor metabolizer-associated CYP2D6(D) gene deletion allele by long-PCR technology. *Pharmacogenetics* 5:215–223.

Steen, V. M., Molven, A., Aarskog, N. K., Gulbrandsen, A. K. 1995b. Homologous unequal cross-over involving a 2.8 kb direct repeat as a mechanism for the generation of allelic variants of human cytochrome P450 CYP2D6 gene. *Hum Mol Genet* 4:2251–2257.

Steers, W. D. 2006. Darifenacin: Pharmacology and clinical usage. *Urol Clin North Am* 33:475–482, viii.

Steimer, W., Zopf, K., von Amelunxen, S., Pfeiffer, H., Bachofer, J., Popp, J., Messner, B., Kissling, W., Leucht, S. 2005. Amitriptyline or not, that is the question: Pharmacogenetic testing of CYP2D6 and CYP2C19 identifies patients with low or high risk for side effects in amitriptyline therapy. *Clin Chem* 51:376–385.

Steiner, E., Spina, E. 1987. Differences in the inhibitory effect of cimetidine on desipramine metabolism between rapid and slow debrisoquin hydroxylators. *Clin Pharmacol Ther* 42:278–282.

Stingl, J., Viviani, R. 2015. Polymorphism in *CYP2D6* and *CYP2C19*, members of the cytochrome P450 mixed-function oxidase system, in the metabolism of psychotropic drugs. *J Intern Med* 277:167–177.

Stormer, E., von Moltke, L. L., Shader, R. I., Greenblatt, D. J. 2000. Metabolism of the antidepressant mirtazapine in vitro: Contribution of cytochromes P-450 1A2, 2D6, and 3A4. *Drug Metab Dispos* 28:1168–1175.

Subrahmanyam, V., Renwick, A. B., Walters, D. G., Young, P. J., Price, R. J., Tonelli, A. P., Lake, B. G. 2001. Identification of cytochrome P-450 isoforms responsible for *cis*-tramadol metabolism in human liver microsomes. *Drug Metab Dispos* 29:1146–1155.

Sunwoo, Y. E., Ryu, J., Jung, H., Kang, K., Liu, K., Yoon, Y.-R., Lee, S.-H. Shin, J. 2004. Disposition of chlorpromazine in Korean healthy subjects with CYP2D6 wild type and *10B mutation (Abstract). *Clin Pharmacol Ther* 73:PII-146.

Suzuki, A., Otani, K., Mihara, K., Yasui, N., Kaneko, S., Inoue, Y., Hayashi, K. 1997. Effects of the CYP2D6 genotype on the steady-state plasma concentrations of haloperidol and reduced haloperidol in Japanese schizophrenic patients. *Pharmacogenetics* 7:415–418.

Suzuki, Y., Sawamura, K., Someya, T. 2006. Polymorphisms in the 5-hydroxytryptamine 2A receptor and cytochromeP4502D6 genes synergistically predict fluvoxamine-induced side effects in japanese depressed patients. *Neuropsychopharmacology* 31:825–831.

Swen, J. J., Nijenhuis, M., de Boer, A., Grandia, L., Maitland-van der Zee, A. H., Mulder, H., Rongen, G. A. et al. 2011. Pharmacogenetics: From bench to byte—An update of guidelines. *Clin Pharmacol Ther* 89:662–673.

Tacke, U., Leinonen, E., Lillsunde, P., Seppala, T., Arvela, P., Pelkonen, O., Ylitalo, P. 1992. Debrisoquine hydroxylation phenotypes of patients with high versus low to normal serum antidepressant concentrations. *J Clin Psychopharmacol* 12: 262–267.

Takimoto, C. H. 2007. Can tamoxifen therapy be optimized for patients with breast cancer on the basis of CYP2D6 activity assessments? *Nat Clin Pract Oncol* 4:152–153.

Tamayo, J. M., Pumariega, A., Rothe, E. M., Kelsey, D., Allen, A. J., Velez-Borras, J., Williams, D., Anderson, S. G., Durell, T. M. 2008. Latino versus Caucasian response to atomoxetine in attention-deficit/hyperactivity disorder. *J Child Adolesc Psychopharmacol* 18:44–53.

Tanaka, M., Kobayashi, D., Murakami, Y., Ozaki, N., Suzuki, T., Iwata, N., Haraguchi, K. et al. 2008. Genetic polymorphisms in the 5-hydroxytryptamine type 3B receptor gene and paroxetine-induced nausea. *Int J Neuropsychopharmacol* 11:261–267.

Tateishi, T., Watanabe, M., Kumai, T., Tanaka, M., Moriya, H., Yamaguchi, S., Satoh, T., Kobayashi, S. 2000. CYP3A is responsible for N-dealkylation of haloperidol and bromperidol and oxidation of their reduced forms by human liver microsomes. *Life Sci* 67:2913–2920.

Tenneze, L., Tarral, E., Ducloux, N., Funck-Brentano, C. 2002. Pharmacokinetics and electrocardiographic effects of a new controlled-release form of flecainide acetate: Comparison with the standard form and influence of the CYP2D6 polymorphism. *Clin Pharmacol Ther* 72:112–122.

Timmer, C. J., Sitsen, J. M., Delbressine, L. P. 2000. Clinical pharmacokinetics of mirtazapine. *Clin Pharmacokinet* 38:461–474.

Toscano, C., Klein, K., Blievernicht, J., Schaeffeler, E., Saussele, T., Raimundo, S., Eichelbaum, M., Schwab, M., Zanger, U. M. 2006a. Impaired expression of CYP2D6 in intermediate metabolizers carrying the *41 allele caused by the intronic SNP 2988G>A: Evidence for modulation of splicing events. *Pharmacogenet Genomics* 16:755–766.

Toscano, C., Raimundo, S., Klein, K., Eichelbaum, M., Schwab, M., Zanger, U. M. 2006b. A silent mutation (2939G>A, exon 6; CYP2D6*59) leading to impaired expression and function of CYP2D6. *Pharmacogenet Genomics* 16:767–770.

Trzepacz, P. T., Williams, D. W., Feldman, P. D., Wrishko, R. E., Witcher, J. W., Buitelaar, J. K. 2008. CYP2D6 metabolizer status and atomoxetine dosing in children and adolescents with ADHD. *Eur Neuropsychopharmacol* 18:79–86.

Turgeon, J., Roden, D. M. 1989. Pharmacokinetic profile of encainide. *Clin Pharmacol Ther* 45:692–694.

Turgeon, J., Fiset, C., Giguere, R., Gilbert, M., Moerike, K., Rouleau, J. R., Kroemer, H. K., Eichelbaum, M., Grech-Belanger, O., Belanger, P. M. 1991. Influence of debrisoquine phenotype and of quinidine on mexiletine disposition in man. *J Pharmacol Exp Ther* 259:789–798.

Turgeon, J., Pare, J. R., Lalande, M., Grech-Belanger, O., Belanger, P. M. 1992. Isolation and structural characterization by spectroscopic methods of two glucuronide metabolites of mexiletine after N-oxidation and deamination. *Drug Metab Dispos* 20:762–769.

Tyndale, R. F., Gonzalez, F. J., Hardwick, J. P., Kalow, W., Inaba, T. 1990. Sparteine metabolism capacity in human liver: Structural variants of human P450IID6 as assessed by immunochemistry. *Pharmacol Toxicol* 67:14–18.

Tyndale, R., Aoyama, T., Broly, F., Matsunaga, T., Inaba, T., Kalow, W., Gelboin, H. V., Meyer, U. A., Gonzalez, F. J. 1991. Identification of a new variant CYP2D6 allele lacking the codon encoding Lys-281: Possible association with the poor metabolizer phenotype. *Pharmacogenetics* 1:26–32.

Tyndale, R. F., Droll, K. P., Sellers, E. M. 1997. Genetically deficient CYP2D6 metabolism provides protection against oral opiate dependence. *Pharmacogenetics* 7:375–379.

Uehlinger, C., Crettol, S., Chassot, P., Brocard, M., Koeb, L., Brawand-Amey, M., Eap, C. B. 2007. Increased (R)-methadone plasma concentrations by quetiapine in cytochrome P450s and ABCB1 genotyped patients. *J Clin Psychopharmacol* 27:273–278.

Ueno, K., Yamaguchi, R., Tanaka, K., Sakaguchi, M., Morishima, Y., Yamauchi, K., Iwai, A. 1996. Lack of a kinetic interaction between fluconazole and mexiletine. *Eur J Clin Pharmacol* 50:129–131.

Ujhelyi, M. R., O'Rangers, E. A., Fan, C., Kluger, J., Pharand, C., Chow, M. S. 1993. The pharmacokinetic and pharmacodynamic interaction between propafenone and lidocaine. *Clin Pharmacol Ther* 53:38–48.

van der Kuy, P. H., Hooymans, P. M. 1998. Nortriptyline intoxication induced by terbinafine. *BMJ* 316:441.

van der Weide, J., van Baalen-Benedek, E. H., Kootstra-Ros, J. E. 2005. Metabolic ratios of psychotropics as indication of cytochrome P450 2D6/2C19 genotype. *Ther Drug Monit* 27:478–483.

Vandamme, N., Broly, F., Libersa, C., Courseau, C., Lhermitte, M. 1993. Stereoselective hydroxylation of mexiletine in human liver microsomes: Implication of P450IID6—A preliminary report. *J Cardiovasc Pharmacol* 21:77–83.

Vandel, S., Bertschy, G., Baumann, P., Bouquet, S., Bonin, B., Francois, T., Sechter, D., Bizouard, P. 1995. Fluvoxamine and fluoxetine: Interaction studies with amitriptyline, clomipramine and neuroleptics in phenotyped patients. *Pharmacol Res* 31:347–353.

Vander Zanden, J. A., Valuck, R. J., Bunch, C. L., Perlman, J. I., Anderson, C., Wortman, G. I. 2001. Systemic adverse effects of ophthalmic beta-blockers. *Ann Pharmacother* 35:1633–1637.

Varsaldi, F., Miglio, G., Scordo, M. G., Dahl, M. L., Villa, L. M., Biolcati, A., Lombardi, G. 2006. Impact of the CYP2D6 polymorphism on steady-state plasma concentrations and clinical outcome of donepezil in Alzheimer's disease patients. *Eur J Clin Pharmacol* 62:721–726.

Veefkind, A. H., Haffmans, P. M., Hoencamp, E. 2000. Venlafaxine serum levels and CYP2D6 genotype. *Ther Drug Monit* 22:202–208.

Venkatakrishnan, K., von Moltke, L. L., Obach, R. S., Greenblatt, D. J. 2000. Microsomal binding of amitriptyline: Effect on estimation of enzyme kinetic parameters *in vitro*. *J Pharmacol Exp Ther* 293:343–350.

Venkatakrishnan, K., Schmider, J., Harmatz, J. S., Ehrenberg, B. L., von Moltke, L. L., Graf, J. A., Mertzanis, P. et al. 2001. Relative contribution of CYP3A to amitriptyline clearance in humans: *In vitro* and *in vivo* studies. *J Clin Pharmacol* 41:1043–1054.

Volotinen, M., Turpeinen, M., Tolonen, A., Uusitalo, J., Maenpaa, J., Pelkonen, O. 2007. Timolol metabolism in human liver microsomes is mediated principally by CYP2D6. *Drug Metab Dispos* 35:1135–1141.

von Bahr, C., Movin, G., Nordin, C., Liden, A., Hammarlund-Udenaes, M., Hedberg, A., Ring, H., Sjoqvist, F. 1991. Plasma levels of thioridazine and metabolites are influenced by the debrisoquin hydroxylation phenotype. *Clin Pharmacol Ther* 49:234–240.

von Moltke, L. L., Greenblatt, D. J., Giancarlo, G. M., Granda, B. W., Harmatz, J. S., Shader, R. I. 2001. Escitalopram (S-citalopram) and its metabolites in vitro: Cytochromes mediating biotransformation, inhibitory effects, and comparison to R-citalopram. *Drug Metab Dispos* 29:1102–1109.

Waade, R. B., Christensen, H., Rudberg, I., Refsum, H., Hermann, M. 2009. Influence of comedication on serum concentrations of aripiprazole and dehydroaripiprazole. *Ther Drug Monit* 31:233–238.

Wagner, F., Kalusche, D., Trenk, D., Jahnchen, E., Roskamm, H. 1987. Drug interaction between propafenone and metoprolol. *Br J Clin Pharmacol* 24:213–220.

Wang, J. S., DeVane, C. L. 2003. Involvement of CYP3A4, CYP2C8, and CYP2D6 in the metabolism of (R)- and (S)-methadone in vitro. *Drug Metab Dispos* 31:742–747.

Wang, T., Roden, D. M., Wolfenden, H. T., Woosley, R. L., Wood, A. J., Wilkinson, G. R. 1984. Influence of genetic polymorphism on the metabolism and disposition of encainide in man. *J Pharmacol Exp Ther* 228:605–611.

Wang, S. L., Huang, J. D., Lai, M. D., Liu, B. H., Lai, M. L. 1993. Molecular basis of genetic variation in debrisoquin hydroxylation in Chinese subjects: Polymorphism in RFLP and DNA sequence of CYP2D6. *Clin Pharmacol Ther* 53:410–418.

Wang, S. L., Lai, M. D., Huang, J. D. 1999. G169R mutation diminishes the metabolic activity of CYP2D6 in Chinese. *Drug Metab Dispos* 27:385–388.

Wang, J. H., Liu, Z. Q., Wang, W., Chen, X. P., Shu, Y., He, N., Zhou, H. H. 2001. Pharmacokinetics of sertraline in relation to genetic polymorphism of CYP2C19. *Clin Pharmacol Ther* 70:42–47.

Wang, G., Zhang, H., He, F., Fang, X. 2006. Effect of the CYP2D6*10 C188T polymorphism on postoperative tramadol analgesia in a Chinese population. *Eur J Clin Pharmacol* 62:927–931.

Ward, S. A., Walle, T., Walle, U. K., Wilkinson, G. R., Branch, R. A. 1989. Propranolol's metabolism is determined by both mephenytoin and debrisoquin hydroxylase activities. *Clin Pharmacol Ther* 45:72–79.

Washio, T., Kohsaka, K., Arisawa, H., Masunaga, H., Nagatsuka, S., Satoh, Y. 2003. Pharmacokinetics and metabolism of radiolabelled SNI-2011, a novel muscarinic receptor agonist, in healthy volunteers. Comprehensive understanding of absorption, metabolism and excretion using radiolabelled SNI-2011. *Arzneimittelforschung* 53:80–86.

Wegman, P., Vainikka, L., Stal, O., Nordenskjold, B., Skoog, L., Rutqvist, L. E., Wingren, S. 2005. Genotype of metabolic enzymes and the benefit of tamoxifen in postmenopausal breast cancer patients. *Breast Cancer Res* 7:R284–R290.

Wegman, P., Elingarami, S., Carstensen, J., Stal, O., Nordenskjold, B., Wingren, S. 2007. Genetic variants of CYP3A5, CYP2D6, SULT1A1, UGT2B15 and tamoxifen response in postmenopausal patients with breast cancer. *Breast Cancer Res* 9:R7.

Wennerholm, A., Johansson, I., Massele, A. Y., Lande, M., Alm, C., Aden-Abdi, Y., Dahl, M. L., Ingelman-Sundberg, M., Bertilsson, L., Gustafsson, L. L. 1999. Decreased capacity for debrisoquine metabolism among black Tanzanians: Analyses of the CYP2D6 genotype and phenotype. *Pharmacogenetics* 9:707–714.

Wennerholm, A., Johansson, I., Hidestrand, M., Bertilsson, L., Gustafsson, L. L., Ingelman-Sundberg, M. 2001. Characterization of the CYP2D6*29 allele commonly present in a black Tanzanian population causing reduced catalytic activity. *Pharmacogenetics* 11:417–427.

Wennerholm, A., Dandara, C., Sayi, J., Svensson, J. O., Abdi, Y. A., Ingelman-Sundberg, M., Bertilsson, L., Hasler, J., Gustafsson, L. L. 2002. The African-specific CYP2D617 allele encodes an enzyme with changed substrate specificity. *Clin Pharmacol Ther* 71:77–88.

Wernicke, J. F., Kratochvil, C. J. 2002. Safety profile of atomoxetine in the treatment of children and adolescents with ADHD. *J Clin Psychiatry* 63 Suppl 12:50–55.

Whyte, E. M., Romkes, M., Mulsant, B. H., Kirshne, M. A., Begley, A. E., Reynolds, C. F., 3rd, Pollock, B. G. 2006. CYP2D6 genotype and venlafaxine-XR concentrations in depressed elderly. *Int J Geriatr Psychiatry* 21:542–549.

Wiholm, B. E., Alvan, G., Bertilsson, L., Sawe, J., Sjoqvist, F. 1981. Hydroxylation of debrisoquine in patients with lacticacidosis after phenformin. *Lancet* 1:1098–1099.

Wilder-Smith, C. H., Hufschmid, E., Thormann, W. 1998. The visceral and somatic antinociceptive effects of dihydrocodeine and its metabolite, dihydromorphine. A cross-over study with extensive and quinidine-induced poor metabolizers. *Br J Clin Pharmacol* 45:575–581.

Wojcikowski, J., Maurel, P., Daniel, W. A. 2006. Characterization of human cytochrome p450 enzymes involved in the metabolism of the piperidine-type phenothiazine neuroleptic thioridazine. *Drug Metab Dispos* 34:471–476.

Woosley, R. L., Roden, D. M., Dai, G. H., Wang, T., Altenbern, D., Oates, J., Wilkinson, G. R. 1986. Co-inheritance of the polymorphic metabolism of encainide and debrisoquin. *Clin Pharmacol Ther* 39:282–287.

Wright, G. E., Niehaus, D. J., Drogemoller, B. I., Koen, L., Gaedigk, A., Warnich, L. 2010. Elucidation of CYP2D6 genetic diversity in a unique African population: Implications for the future application of pharmacogenetics in the Xhosa population. *Ann Hum Genet* 74:340–350.

Wu, D., Otton, S. V., Sproule, B. A., Busto, U., Inaba, T., Kalow, W., Sellers, E. M. 1993. Inhibition of human cytochrome P450 2D6 (CYP2D6) by methadone. *Br J Clin Pharmacol* 35:30–34.

Wuttke, H., Rau, T., Heide, R., Bergmann, K., Bohm, M., Weil, J., Werner, D., Eschenhagen, T. 2002. Increased frequency of cytochrome P450 2D6 poor metabolizers among patients with metoprolol-associated adverse effects. *Clin Pharmacol Ther* 72:429–437.

Xu, Z. H., Wang, W., Zhao, X. J., Huang, S. L., Zhu, B., He, N., Shu, Y., Liu, Z. Q., Zhou, H. H. 1999. Evidence for involvement of polymorphic CYP2C19 and 2C9 in the N-demethylation of sertraline in human liver microsomes. *Br J Clin Pharmacol* 48:416–423.

Yamazaki, S., Sato, K., Suhara, K., Sakaguchi, M., Mihara, K., Omura, T. 1993. Importance of the proline-rich region following signal-anchor sequence in the formation of correct conformation of microsomal cytochrome P-450s. *J Biochem* 114:652–657.

Yamazaki, H., Kiyotani, K., Tsubuko, S., Matsunaga, M., Fujieda, M., Saito, T., Miura, J., Kobayashi, S., Kamataki, T. 2003. Two novel haplotypes of CYP2D6 gene in a Japanese population. *Drug Metab Pharmacokinet* 18:269–271.

Yasuda, S. U., Wellstein, A., Likhari, P., Barbey, J. T., Woosley, R. L. 1995. Chlorpheniramine plasma concentration and histamine H1-receptor occupancy. *Clin Pharmacol Ther* 58:210–220.

Yasuda, S. U., Zannikos, P., Young, A. E., Fried, K. M., Wainer, I. W., Woosley, R. L. 2002. The roles of CYP2D6 and stereoselectivity in the clinical pharmacokinetics of chlorpheniramine. *Br J Clin Pharmacol* 53:519–525.

Yasui, N., Tybring, G., Otani, K., Mihara, K., Suzuki, A., Svensson, J. O., Kaneko, S. 1997. Effects of thioridazine, an inhibitor of CYP2D6, on the steady-state plasma concentrations of the enantiomers of mianserin and its active metabolite, desmethylmianserin, in depressed Japanese patients. *Pharmacogenetics* 7:369–374.

Yasui-Furukori, N., Hidestrand, M., Spina, E., Facciola, G., Scordo, M. G., Tybring, G. 2001. Different enantioselective 9-hydroxylation of risperidone by the two human CYP2D6 and CYP3A4 enzymes. *Drug Metab Dispos* 29:1263–1268.

Yin, O. Q., Shi, X. J., Tomlinson, B., Chow, M. S. 2005. Effect of CYP2D6*10 allele on the pharmacokinetics of loratadine in chinese subjects. *Drug Metab Dispos* 33:1283–1287.

Yokoi, T., Kosaka, Y., Chida, M., Chiba, K., Nakamura, H., Ishizaki, T., Kinoshita, M., Sato, K., Gonzalez, F. J., Kamataki, T. 1996. A new CYP2D6 allele with a nine base insertion in exon 9 in a Japanese population associated with poor metabolizer phenotype. *Pharmacogenetics* 6:395–401.

Yokota, H., Tamura, S., Furuya, H., Kimura, S., Watanabe, M., Kanazawa, I., Kondo, I., Gonzalez, F. J. 1993. Evidence for a new variant CYP2D6 allele CYP2D6J in a Japanese population associated with lower in vivo rates of sparteine metabolism. *Pharmacogenetics* 3:256–263.

Yonezawa, E., Matsumoto, K., Ueno, K., Tachibana, M., Hashimoto, H., Komamura, K., Kamakura, S., Miyatake, K., Tanaka, K. 2002. Lack of interaction between amiodarone and mexiletine in cardiac arrhythmia patients. *J Clin Pharmacol* 42:342–346.

Yoshii, K., Kobayashi, K., Tsumuji, M., Tani, M., Shimada, N., Chiba, K. 2000. Identification of human cytochrome P450 isoforms involved in the 7-hydroxylation of chlorpromazine by human liver microsomes. *Life Sci* 67:175–184.

Yu, A. M., Idle, J. R., Byrd, L. G., Krausz, K. W., Kupfer, A., Gonzalez, F. J. 2003. Regeneration of serotonin from 5-methoxytryptamine by polymorphic human CYP2D6. *Pharmacogenetics* 13:173–181.

Yue, Q. Y., Zhong, Z. H., Tybring, G., Dalen, P., Dahl, M. L., Bertilsson, L., Sjoqvist, F. 1998. Pharmacokinetics of nortriptyline and its 10-hydroxy metabolite in Chinese subjects of different CYP2D6 genotypes. *Clin Pharmacol Ther* 64:384–390.

Yumibe, N., Huie, K., Chen, K. J., Snow, M., Clement, R. P., Cayen, M. N. 1996. Identification of human liver cytochrome P450 enzymes that metabolize the nonsedating antihistamine loratadine. Formation of descarboethoxyloratadine by CYP3A4 and CYP2D6. *Biochem Pharmacol* 51:165–172.

Zahari, Z., Ismail, R. 2014. Influence of Cytochrome P450, Family 2, Subfamily D, Polypeptide 6 (CYP2D6) polymorphisms on pain sensitivity and clinical response to weak opioid analgesics. *Drug Metab Pharmacokinet* 29:29–43.

Zanger, U. M., Raimundo, S., Eichelbaum, M. 2004. Cytochrome P450 2D6: Overview and update on pharmacology, genetics, biochemistry. *Naunyn Schmiedebergs Arch Pharmacol* 369:23–37.

Zhang, W. Y., Tu, Y. B., Haining, R. L., Yu, A. M. 2009. Expression and functional analysis of CYP2D6.24, CYP2D6.26, CYP2D6.27, and CYP2D7 isozymes. *Drug Metab Dispos* 37:1–4.

Zhou, H. H., Wood, A. J. 1995. Stereoselective disposition of carvedilol is determined by CYP2D6. *Clin Pharmacol Ther* 57:518–524.

Zhou, H. H., Anthony, L. B., Roden, D. M., Wood, A. J. 1990. Quinidine reduces clearance of (+)-propranolol more than (−)-propranolol through marked reduction in 4-hydroxylation. *Clin Pharmacol Ther* 47:686–693.

Zhou, Q., Yao, T. W., Yu, Y. N., Zeng, S. 2003. Concentration dependent stereoselectivity of propafenone N-depropylation metabolism with human hepatic recombinant CYP1A2. *Pharmazie* 58:651–653.

Zhou, S. F., Di, Y. M., Chan, E., Du, Y. M., Chow, V. D., Xue, C. C., Lai, X. et al. 2008. Clinical pharmacogenetics and potential application in personalized medicine. *Curr Drug Metab* 9:738–784.

Zhou, Q., Yu, X. M., Lin, H. B., Wang, L., Yun, Q. Z., Hu, S. N., Wang, D. M. 2009. Genetic polymorphism, linkage disequilibrium, haplotype structure and novel allele analysis of CYP2C19 and CYP2D6 in Han Chinese. *Pharmacogenomics J* 9:380–394.

Zineh, I., Beitelshees, A. L., Gaedigk, A., Walker, J. R., Pauly, D. F., Eberst, K., Leeder, J. S., Phillips, M. S., Gelfand, C. A., Johnson, J. A. 2004. Pharmacokinetics and CYP2D6 genotypes do not predict metoprolol adverse events or efficacy in hypertension. *Clin Pharmacol Ther* 76:536–544.

Zoble, R. G., Kirsten, E. B., Brewington, J. 1989. Pharmacokinetic and pharmacodynamic evaluation of propafenone in patients with ventricular arrhythmia. Propafenone Research Group. *Clin Pharmacol Ther* 45:535–541.

Zourkova, A., Hadasova, E. 2002. Relationship between CYP 2D6 metabolic status and sexual dysfunction in paroxetine treatment. *J Sex Marital Ther* 28:451–461.

8 General Discussion about Human CYP2D6

CYP enzymes are involved in the metabolism of various endogenous and exogenous chemicals, including steroids, bile acids, fatty acids, eicosanoids, xenobiotics, environmental pollutants, and carcinogens (Nebert and Dalton 2006; Nebert and Russell 2002; Nebert et al. 2013). Human CYPs are important in physiology, drug metabolism, and pathogenesis of certain diseases, and thus there is a strong need to explore the genetic variations and regulatory mechanisms of these enzymes. Mutations in *CYP* genes leading to deficiency in the enzymes result in a wide spectrum of human diseases such as glaucoma (*CYP1B1*), elevated cholesterol (*CYP7A1*), congenital adrenal hyperplasia, congenital hypoaldosteronism (*CYP11B2*), essential hypertension (*CYP4A11*), Bietti's crystalline corneoretinal dystrophy (*CYP4V2*), lamellar ichthyosis (*CYP4F22*), and cancers (Nebert et al. 2013). Members of families CYP1, CYP2, and CYP3, especially CYP1A2, 2B6, 2C9, 2C19, 2D6, and 3A4 in liver, metabolize approximately 95% of clinical drugs (Hodgson and Rose 2007; Nebert et al. 2013), which are also active in the formation of carcinogens from procarcinogens (Ghoshal et al. 2014; Go et al. 2015; Panigrahy et al. 2010; Raunio and Rahnasto-Rilla 2012; Rodriguez and Potter 2013; Xiang et al. 2015; Xu et al. 2013). Many CYPs are linked to cancer growth, development, and metastasis via multiple mechanisms (Choong et al. 2015; Go et al. 2015; Ye et al. 2015). The CYP4 family members are involved in the disposition of fatty acids and eicosanoids (Hsu et al. 2007; Simpson 1997), whereas members of CYP11, 17, 19, and 21 families are steroidogenic enzymes in tissues such as adrenal cortex, ovaries, and testes (Niwa et al. 2009; Snider et al. 2010). A number of clinical studies have shown that polymorphisms in *CYP2D6* and *CYP2C19* are linked to altered response or occurrence of adverse drug reactions (ADRs) of various drugs used in the treatment of cancer, psychiatric disorders, and cardiovascular diseases (Zhou 2009a,b). In addition to genetic polymorphisms within *CYP* genes, regulatory mechanisms at transcriptional, translational, posttranslational, and epigenetic levels also contribute to the large variability in CYP expression and activities observed across ethnic groups and individuals (Polimanti et al. 2012; Sinz et al. 2008; Tralau and Luch 2013).

Pharmacogenomics is the study on how genetic factors influence individual responses to different drugs, which may affect drug efficacy, drug side effects, and adverse events related to the therapy (Ashley et al. 2010; Bartlett et al. 2013; Hertz and McLeod 2013; Jensen and McLeod 2013; Lee and Morton 2008; McCarthy et al. 2013; McLeod 2013; Meyer 2000; Meyers et al. 2014; Meyerson 2003; Nakamura 2008; Ramos et al. 2014; Roederer et al. 2011; Samani et al. 2010; Wang et al. 2009a, 2011; Weinshilboum 2003; Wilkinson 2005; Zhou 2009a,b; Zhou et al. 2010, 2015). The clinical goals of pharmacogenomics are to minimize adverse drug events and maximize drug efficacy. By incorporating genetic information, health professionals can identify the patient's polymorphisms and disease subtypes to determine the most advantageous management. This can include the immediate administration of the most efficacious and least toxic drugs at the correct doses. Understanding and identifying an individual's genetic variations has the potential to decrease both the time expended in achieving effective therapy and the number of visits required for proper dose adjustment (Haghgoo et al. 2015; Ma and Lu 2011; McCarthy et al. 2013; Wang et al. 2011). Thus, there are two general goals for the clinical application of pharmacogenomics: the ability to predict those patients at high risk of toxicity (and in whom a lower dose or a different drug would be administered) and the ability to predict those patients who are most likely to obtain the desired therapeutic effect from the drug (Ma and Lu 2011; McCarthy et al. 2013; Wang et al. 2011).

ADRs are a major concern in both clinical practice and pharmaceutical industry, because of the resultant morbidity and mortality in patients (Bienfait et al. 2013; Daly 2013). In the United Kingdom, 6.5% of patients in 18,820 admissions to two National Health Service hospitals were caused by ADRs (Pirmohamed et al. 2004). In Japan, 1.6%, 4.9%, and 33% of 3459 adult patients experienced fatal, life-threatening, and serious ADRs in 3459 adult patients, respectively (Morimoto et al. 2011). In China, 692,904 cases of ADRs were reported to the National Center for ADR Monitoring in 2010 (http://www.sfda.gov.cn). In the United States, there are approximately 6.7% of hospitalized patients experiencing serious ADRs, of which 0.32% of ADRs are fatal reactions. It renders ADRs ranking between the fourth and sixth leading causes of death (Lazarou et al. 1998). An estimation of the annual cost for ADRs is more than $136 billion in the United States. There are multiple factors that can cause ADRs, including drug coadministration, lifestyle, age, and diet. Notably, increasing evidence shows that interindividual genetic difference is an important contributing factor to ADRs. Genetic variations in drug-metabolizing enzymes, drug transporters, and drug targets substantially contribute to the alteration of pharmacokinetics and pharmacodynamics (Evans and McLeod 2003; Phillips et al. 2001). Extensive effort has been made to characterize the genetic factors, identify the interindividual genetic differences, and establish the association between genetic variations and ADRs.

CYP2D6 is an important and highly polymorphic Phase I enzyme in the liver since it can metabolize more than 160 clinical drugs, and the metabolism of these drugs is subject to polymorphism attributed to genetic mutations of the *CYP2D6* gene and inhibition by many drugs. This will have significant clinical implication in patients who take drugs that are substrates or inhibitors of CYP2D6. In Food and Drug Administration (FDA)–approved drug labels, at least 50 of them contain information on the impact of CYP2D6 polymorphism on drug kinetics and response and dose adjustment recommendations.

It is a well-recognized fact that individuals respond differently to drug therapy and that no single drug is 100% effective in all treated patients (Wilke and Dolan 2011). While some individuals obtain the desired effects, others can have little or no therapeutic response. Additionally, certain patients might experience adverse effects that vary from mild and tolerable to life-threatening events. The remarkable interindividual variability in drug response is thought to be a consequence of multiple factors such as disease determinants, genetic and environmental factors, variability in drug target response or idiosyncratic response, and other factors including age, gender, disease status, concomitant therapies, and lifestyle factors such as smoking and alcohol consumption (Daly 2012).

The CYP2D6 activity ranges considerably within a population and includes individuals with ultrarapid metabolizer (UM), extensive metabolizer (EM), intermediate metabolizer (IM), and poor metabolizer (PM) status. There is a large interindividual variation in the enzyme expression and activity of CYP2D6. Unlike other CYPs, CYP2D6 is not inducible, and thus genetic mutations are largely responsible for the interindividual variation in enzyme expression and activity. This often presents as a barrier to reaching optimal therapeutic concentrations. For drugs with CYP2D6 as the major contributor of clearance (>60%), large differences in AUC or oral clearance were observed between EMs and PMs (Zhou 2009a,b). The most extreme pharmacokinetic differences were observed with dextromethorphan (Evans et al. 1989) and tolterodine (Brynne et al. 1998), which showed 53- and 22-fold higher oral clearances in EMs compared with PMs. Modest differences in oral clearance (3.5- to 10-fold) were observed with atomoxetine (Farid et al. 1985), propafenone (Zoble et al. 1989), desipramine (Gram and Christiansen 1975), and venlafaxine (Holliday and Benfield 1995) between EMs and PMs. For drugs that had a low estimated CYP2D6 contribution in vivo, variation in oral clearance values between EMs and PMs was <2–3 (e.g., amitriptyline [Breyer-Pfaff et al. 1992]). Drugs that are most affected by *CYP2D6* polymorphisms are commonly those in which CYP2D6 represents a substantial metabolic pathway either in the activation or clearance of the agent.

Minimizing interindividual variability in drug exposure is an important goal in drug development and discovery (Gibbs et al. 2006). CYP2D6 is considered one of the most important CYPs, with substrate specificity typical of many new chemical entities (i.e., lipophilic bases). An estimated 20% to 30% of all drugs in clinical use are metabolized at least in part by CYP2D6. The primarily hepatic expression of this enzyme governs the first-pass metabolism of many orally administered drugs, whereas the low levels of intestinal expression do not appear to be important (Madani et al. 1999).

Presently, at least 104 different human *CYP2D6* variant alleles and a series of subvariants have been identified and designated by the Human Cytochrome P450 Allele Nomenclature Committee (http://www.imm.ki.se/CYPalleles). In particular, there are a number of null alleles of *CYP2D6* that do not encode a functional protein and there is no detectable residual enzymatic activity. They are responsible for the PM phenotype when present in homozygous or compound heterozygous constellations. The mechanism that may lead to a total loss of function includes the following: (a) single-base mutations or small insertions/deletions that interrupt the reading frame or interfere with correct splicing leading to prematurely terminated protein/stop codon (e.g., *CYP2D6*3*, *4*, *6*, *8*, *11*, *15*, *19*, *20*, *38*, *40*, *42*, and *44*) (Kagimoto et al. 1990), (b) nonfunctional full-length coded alleles (e.g., *CYP2D6*12*, *14*, and *18*) (Evert et al. 1997), (c) deletion of the entire *CYP2D6* gene as a result of large sequence deletions (e.g., *CYP2D6*5*), and (d) formation of hybrid genes (e.g., *CYP2D6*13* and *16*) (Gaedigk et al. 1991). On the other hand, extremely high CYP2D6 activity results from gene duplication of functional allele *1* and *2* fused in a head-to-tail orientation, as a result of unequal crossover events and other mechanisms. This was noted by a molecular characterization of the *CYP2D6* locus in patients with extremely rapid metabolisms (Bertilsson et al. 1993).

A number of clinical studies have documented the impact of *CYP2D6* polymorphisms on drug clearance (Zhou 2009a,b). The clinical consequence of the CYP2D6 polymorphism can be either occurrence of ADRs or altered drug response. For example, tamoxifen-treated patients carrying *CYP2D6*4*, *5*, *10*, or *41* associated with decreased formation of antiestrogenic metabolites have significantly more recurrences of breast cancer and shorter relapse-free periods (Bonanni et al. 2006; Borges et al. 2006; Goetz et al. 2005, 2007; Jin et al. 2005; Schroth et al. 2007). The analgesic effect of codeine was seen to be substantially less in subjects found to be PMs (Lotsch et al. 2004), or no analgesic effect was noted in PMs (Eckhardt et al. 1998; Poulsen et al. 1996). A large (*n* = 500) cross-sectional study on schizophrenic patients in southern Germany reported a threefold increased risk of moderate to severe adverse effects in PMs than in EMs receiving risperidone treatment (Rau et al. 2004).

Ethnicity should be considered in drug development when the candidates are substrates of CYP2D6. Between people with different ethnic backgrounds, the pattern of *CYP2D6* polymorphisms differs dramatically. *2* occurs with allele frequencies of 1%–2% in Swedish Caucasians (Griese et al. 1998), 3.6% in Germans (Sachse et al. 1997), 7%–10% in white Spaniards (Agundez et al. 1995; Bernal et al. 1999), and 10% on Sicily in Italy (Scordo et al. 1999). However, significantly higher frequencies have been observed in Saudi Arabians (20%) (McLellan et al. 1997) and black Ethiopians (up to 29%) (Aklillu et al. 1996). This allele is essentially absent in Asians. *4* is by far the most frequent null allele in Caucasians. It occurs with a frequency of 20%–25% and is responsible for 70%–90% of all PMs (Sachse et al. 1997). No functional alleles are present in approximately 6% of

Caucasians (Sachse et al. 1997). However, *4 occurs with a very low frequency of ~1% in the Asian population (Wang et al. 1993). Its frequency is 6%–7% in Africans and African Americans (Leathart et al. 1998). The most common *CYP2D6* allele in the Asian population is *10, occurring with a frequency of ~50% in Chinese, Japanese, and Koreans (Wang et al. 1993). However, it occurs at a low frequency of ~2% in Caucasians, but it accounts for 10%–20% of individuals with the IM phenotype (Griese et al. 1998).

CYP2D6 has an important role in drug development and it is a common practice for the pharmaceutical industry nowadays to a great extent screen drug candidates early in development as possible CYP2D6 substrates and drop such candidates where they have alternatives. This candidate selection might eventually lead to a less prominent role of this enzyme in the future for drug metabolism and minimize the potential for significant polymorphic metabolism in humans and drug–drug interactions when used in combination with CYP2D6 inhibitors.

Phenotype prediction based on genotype of *CYP2D6* is important in clinical practice. Several psychiatric hospitals have already adopted *CYP2D6* testing before treating a patient with antidepressant or antipsychotic drug therapy. However, the large number of allelic variants and even the larger number of potential combinations of these alleles certainly challenge our phenotype prediction. The clinical phenotype of the *CYP2D6* polymorphisms is variable, depending on a number of factors associated with the drug, patients, disease, and other factors. This makes the prediction of phenotype more difficult. Important factors affecting the phenotype include the extent of CYP2D6-mediated metabolism, activity of the parent drug and its metabolites, and the overall contribution of the CYP2D6-dependent pathway to the clearance of the drug from the body. Furthermore, the therapeutic index of the drug, possible saturation of the CYP2D6-dependent pathway, the contribution of other pathways of elimination, and coadministered drugs need to be considered when performing phenotype prediction in a clinical setting.

Traditionally, the phenotype of CYP2D6 is defined as PM, IM, EM, and UM, which is largely dependent on "the number of active genes." The substantial overlap among the phenotypes in particular between IM and EM and between EM and UM reduces its value in practical application. The determined phenotype data are highly variable within a given ethnic group and become more variable across ethnic groups, which often cover one to two orders of magnitude (Griese et al. 1998). As such, a few additional systems have been proposed to establish and predict more reliable genotype–phenotype relationships. Steimer et al. (2004) applied the concept of "semiquantitative gene dose" that defined three categories for functional, reduced, and nonfunctional alleles on the basis of the relationship of functional gene dose and steady-state plasma amitriptyline or nortriptyline concentrations in a cohort of 50 Caucasians. When this model is used to venlafaxine, fluoxetine, and risperidone, the phenotype prediction rate is only slightly lower than the metabolic ratio (MR) approach (Hinrichs et al. 2008). However, this concept has not been

validated in large study populations, in other ethnic groups, or for other drugs that are mainly metabolized by CYP2D6. Recently, Gaedigk et al. (2008) proposed a six-grade "activity score" (grade: 0, 0.5, 1, 1.5, 2, and >2) approach for phenotype prediction in Caucasians and African Americans. The prediction appears more accurate if ethnicity is taken into account. Ethnicity and genotype as indicator variables explained 60%, whereas ethnicity and activity score as indicators explained 59% of the variability of MRs when dextromethorphan is used as a probe. However, this "activity score" approach has not been further validated in Asians and there is limited value in clinical practice when urinary data must be obtained.

CYP2D6 genotypes can readily be determined with currently available commercial genotyping techniques. The genotype can usually be unambiguously assigned on the basis of analysis of approximately 20–25 polymorphic sites. However, some individuals show a phenotype that is discordant with their determined genotype. Most of these unusual cases are PMs but carrying at least one functional or reduced-function allele indicating an EM or IM phenotype. Subsequent studies of such cases have led to the identification of novel alleles such as CYP2D6*36, *40, and *42 in African Americans (Gaedigk et al. 2002, 2003, 2006) and CYP2D6*21 and *44 in Japanese (Yamazaki et al. 2003).

The structure of human CYP2D6 has been recently solved (Rowland et al. 2006), which may provide insights into the complex genotype–phenotype relationship and ligand–enzyme interactions of CYP2D6. The structure of human CYP2D6 shows the characteristic CYP fold as observed in other CYPs but with remarkable differences compared to other CYPs. The structure will help in determining the structural factors that govern its substrate specificity and inhibitor selectivity. The 2D6 structure has a well-defined active-site cavity above the heme group with a volume of ~540 Å^3. Many important residues in CYP2D6 have been implicated in substrate recognition and binding, including Asp301, Glu216, Phe483, and Phe120 (Rowland et al. 2006). The heme is anchored in the binding site by hydrogen-bonding interactions with the side chains of Arg101, Trp128, Arg132, His376, Ser437, and Arg441. The importance of some of these residues has been confirmed by site-directed mutagenesis studies (Mackman et al. 1996; Paine et al. 2003). However, there are no natural mutations in many of these residues and thus functional studies are needed.

Since the data on the safety of a new drug candidate are often scant, it is beneficial to limit the contribution of polymorphic enzymes such as CYP2D6 below some cutoff to avoid the requirement of phenotyping/genotyping testing before the commencement of drug treatment and different dosing regimens in EM and PM subjects. A variety of in vitro models are available and have been widely used to determine the relative contribution or percent contribution to metabolism by a polymorphic enzyme such as CYP2D6 ($f_{m(cyp)}$). These include coincubation with specific enzyme inhibitors such as selective chemical inhibitors and inhibitory antibodies; use of human liver microsomes, cultured hepatocytes, or precision-cut slices; and recombinant enzyme systems (Lin and Lu 1997). Gibbs et al. (2006) have proposed using the following equation to calculate the AUC ratio between PMs and EMs:

$$\frac{\text{AUC}_{\text{PM}}}{\text{AUC}_{\text{EM}}} = \frac{1}{1 - f_{\text{m(CYP2D6)}}}$$

On the basis of the above equation, the relationship between fraction metabolized ($f_{\text{m(CYP2D6)}}$) and EM/PM AUC or clearance differences is almost flat when CYP2D6 contribution is below 60% (Gibbs et al. 2006). For example, 20% and 50% contribution from CYP2D6 will be predicted to result in 1.25- and 2.0-fold increases in PM AUC relative to EM, respectively. Above 60% contribution by CYP2D6, supraproportional increases in the ratio of PM/EM AUC are predicted as the f_{m} by CYP2D6 goes to unity. Because of the nonlinear shape of the relationship between CYP2D6 contribution and EM/PM AUC or clearance differences, small changes in percentage contribution can result in large EM/PM pharmacokinetic differences when the fraction metabolized exceeds 60%. Only drugs with a very narrow therapeutic index (e.g., warfarin) would need dosage adjustments for a twofold increase in exposure.

Given the large interindividual variability in clearance observed for a typical CYP2D6 or CYP3A4 substrate, a 2- to 2.5-fold EM/PM AUC or clearance difference should be manageable for the majority of new drugs (Gibbs et al. 2006). Alternatively, with safety information or different assumptions based on previous experience within a therapeutic class, a more precise limit could be selected to enable a single dose level appropriate for both EMs and PMs (e.g., if the new drug has a narrow therapeutic index, this cutoff should be smaller). Until the use of pharmacogenetic tools becomes a common clinical practice, it will be necessary to exclude compounds from drug discovery that do not meet preset criteria for interindividual variability.

Concomitant use of a drug that affects the activity of the same CYP responsible for biotransformation of another drug can lead to significant increases in plasma concentrations and potentially important drug–drug interactions (Ito et al. 1998). Such interactions may be associated with poor tolerability or increased risk for toxicity. On the other hand, for drugs/prodrugs requiring biotransformation via CYP enzymes from an inactive/less active parent compound to a pharmacologically active metabolite, drug interactions may manifest as a reduction in efficacy. Cases of fatal drug interactions have been reported and several prominent drugs (e.g., cerivastatin, mibefradil, sorivudine, and terfenadine) have been withdrawn from the market because of severe adverse reactions related to drug interactions (Li 2001). Both pharmacokinetic and pharmacodynamic components may be involved in these toxic drug interactions. Because of the clinical significance of drug interactions, it is important to identify drugs and compounds in development that may interact with other drugs, and timely identification of such drugs using proper in vitro and in vivo approaches has important implications for drug development (Zhou et al. 2007).

Drugs that are most affected by CYP2D6 polymorphisms are commonly those in which CYP2D6 represents a substantial metabolic pathway either in the activation to form active metabolites or clearance of the agent. However, such a functional impact is substrate dependent. For example, CYP2D6*17 is generally considered as an allele with reduced function, but it displays remarkable variability in activity toward substrates including dextromethorphan, risperidone, codeine, and haloperidol (Wennerholm et al. 2002). On the other hand, it is expected that for drugs that are metabolized by CYP2D6 to a minor extent (e.g., chlorpromazine), CYP2D6 mutations would not alter their clearance significantly.

Identification of drugs as CYP2D6 inhibitors has important clinical implications when drugs are coadministered. Drugs that inhibit CYP2D6 would be expected to cause increases in the plasma concentration of these drugs extensively metabolized by CYP2D6. In particular, drugs that cause mechanism-based inhibition of CYP2D6 will cause enzyme inactivation and consequently result in significant drug interactions. Because of a critical role of CYP2D6 in drug metabolism, significant inactivation of this enzyme could result in marked pharmacokinetic drug–drug interactions. The in vivo inhibitory effect of a mechanistic inactivator is more prominent after multiple dosing and lasts longer than that of a reversible inhibitor (Lin and Lu 1998; Zhou et al. 2004, 2005).

Paroxetine, a mechanism-based and reversible CYP2D6 inhibitor, has been shown to decrease the clearance of a number of CYP2D6 substrates such as desipramine (Alderman et al. 1997; Nichols et al. 2009), sparteine (Jeppesen et al. 1996), atomoxetine (Belle et al. 2002), risperidone (Saito et al. 2005; Spina et al. 2001), dextromethorphan/quinidine (Schoedel et al. 2012), pimozide (Horrigan and Barnhill 1994), and perphenazine (Ozdemir et al. 1997). For example, the effect of paroxetine on the pharmacokinetics of atomoxetine has been evaluated when both drugs are at steady state (Belle et al. 2002). In healthy volunteers who are EMs of CYP2D6, paroxetine 20 mg daily is given in combination with 20 mg atomoxetine every 12 h. This results in increases in steady-state atomoxetine AUC values that are six- to eightfold greater and in atomoxetine C_{max} values that are three- to fourfold greater than when atomoxetine is given alone (Belle et al. 2002). Dosage adjustment of atomoxetine may be necessary, and it is recommended that atomoxetine be initiated at a reduced dose when it is given with paroxetine. In a controlled study of healthy volunteers, after paroxetine is titrated to 60 mg daily, coadministration of a single dose of 2 mg pimozide is associated with mean increases in pimozide AUC of 151% and C_{max} of 62%, compared to pimozide administered alone (Horrigan and Barnhill 1994). Because of the narrow therapeutic index of pimozide and its known ability to prolong the QT interval (Wenzel-Seifert et al. 2011), concomitant use of pimozide and paroxetine is contraindicated. In most patients (>90%), CYP2D6 is saturated early during dosing with paroxetine. Similarly, as a potent inhibitor of CYP2D6, fluoxetine can increase serum concentrations of many neuroleptics and antidepressants, and numerous case reports have documented concomitant adverse events (Ciraulo and Shader 1990; Fuller et al. 1991; Naranjo et al. 1999). These CYP2D6 inhibitors can readily convert an EM or IM phenotype to a PM phenotype (i.e., phenocopying or phenoconversion) and thus alter drug clearance and response. Because of the risk of serious ventricular arrhythmias and sudden death potentially associated with elevated plasma levels of thioridazine, paroxetine and thioridazine should not be coadministered. In order to avoid or minimize concerns related to CYP2D6-mediated drug–drug interactions, pharmaceutical companies routinely screen for potential CYP2D6 liability of lead candidates in the early stage of the drug discovery process (Le Bourdonnec and Leister 2009). Those drug candidates with high affinity to CYP2D6 should be discontinued at the early stage.

Early and timely identification of candidate drugs that inactivate CYP2D6 is important in drug development, as this can save large expenditures in time and money. In cases where new drugs have already been launched, and where a large population of patients has been exposed to the drug, severe interactions and even fatal toxicity may occur. Therefore, lessons should be learned from such cases (e.g., mibefradil) where a drug was withdrawn from the market because of an unacceptable risk of potentially dangerous drug–drug interactions resulting from CYP3A inactivation. With the current techniques, it is increasingly likely to predict which new drugs will be associated with the formation of reactive metabolites and CYP2D6 inactivation, leading to potentially toxic responses. Thus, by screening drug candidates for possible formation of reactive metabolites that inhibit CYP2D6 in an irreversible manner, and by establishing structure–activity relationships, it is possible to identify and eliminate such chemicals at an early stage of development. The application of genomic and proteomic approaches to the study of drug–CYP interactions, as well as the use of heterogeneous expression systems for specific CYPs including CYP2D6, has the potential to lead to a more effective screen owing to a high-throughput capacity.

Because of frequent pharmacokinetic drug–drug interactions by CYP2D6 inactivators/inhibitors, monitoring of plasma concentrations of concurrently administered drugs that are CYP2D6 substrates or have narrow therapeutic indices and observation for signs of clinical toxicity are always imperative. This will allow early identification of potential drug–drug interactions and subsequent toxicities.

Although CYP2D6 accounts for only a small percentage of the total hepatic CYP content (<2%), this enzyme metabolizes ~25% of clinically used drugs with significant polymorphisms. In particular, a number of drugs acting on the central nervous system and cardiovascular system are substantially metabolized by CYP2D6. These drugs are mostly associated with G-coupled proteins when they elicit effect at the site of action. The substrates of CYP2D6 belong to several therapeutic classes that include centrally acting compounds such as tricyclic antidepressants, selective serotonin reuptake inhibitors, β-blockers, monoamine oxidase inhibitors, neuroleptics, opioids, drugs of abuse such as MDMA (methylenedioxymethamphetamine), and neurotoxins. CYP2D6 is the only active enzyme in the 2D subfamily, and it has high affinity to plant alkaloids, suggesting its important role in detoxifying these xenobiotics when human beings were dependent on plants as major food sources during evolution. The enzyme also utilizes hydroxytryptamines, steroids, and neurosteroids as endogenous substrates, but the potential physiological role of CYP2D6 is unclear.

CYP2D6 is subject to inhibition by many drugs and a number of clinical drug–drug interactions attributed to CYP2D6 inhibition have been reported. Inhibition of CYP2D6-mediated drug metabolism has important clinical implications when drugs are coadministered. In drug development, if a new drug candidate is found to be a potent or mechanism-based inhibitor of CYP2D6, its further development should be halted since the compound has a high potential for causing adverse drug interactions.

Unlike other genetic testing, pharmacogenomics does not aim to specifically determine or predict the risk of disease but rather characterizes an individual on the basis of disease susceptibility, risk of severe adverse effects, or even efficacy of certain drugs (Ma and Lu 2011). Pharmacogenomic testing has the ability to give an estimate of the likely effectiveness, thereby removing much of the uncertainty surrounding current pharmacotherapy. The ultimate goal of pharmacogenetic testing is to aid physicians in the prescription of the appropriate medication at the correct dose before the initiation of the therapy in an attempt to minimize adverse events and toxicity and maximize efficacy by excluding those who are unlikely to benefit (nonresponders) or who may be harmed (adverse responders) (Daly 2012; Ma and Lu 2011). The promise of pharmacogenomics lies in its potential to identify the right drug at the right dose for the right individual. The application of pharmacogenomics also aims to discover better drugs and improve the efficacy and safety of both prospective as well as licensed drugs (Cong et al. 2012). Minimal pharmacogenomic testing is required for all new drug applications to the FDA, including a requirement for germline DNA to be prospectively collected from all subjects participating in preapproval clinical trials and genotyping studies for drugs that are metabolized by enzymes whose genes contain polymorphisms with significant functional impacts.

There is increasing evidence that clinical pharmacogenetics is beginning to take on a role in health care with the emergence of commercially provided services, such as AmpliChip (http://www.amplichip.us), which claims to offer personalized, fact-based pharmacogenetic information to assist physicians in optimizing an individual patient's drug therapy. AmpliChip is the first FDA-cleared test for genotype analysis of *CYP2D6* and *2C19* using a microarray hybridization method (Heller et al. 2006). The AmpliChip tests the DNA from patients' white blood cells collected in a standard anticoagulated blood sample for 29 polymorphisms and mutations from the *CYP2D6* gene and two polymorphisms from the *CYP2C19* gene. Recently, the FDA has also approved the clinical use of genotyping kits for *CYP2D6* (xTAG CYP2D6 Kit v3), *CYP2C19* (Spartan RX CYP2C19 Test System, Verigene CYP2C 19 Nucleic Acid Test, and Infiniti CYP2C19 Assay), *UGT1A1* (Invader UGT1A1 Molecular Assay), and

CYP2C9 and *VKORC1* (Gentris Rapid Genotyping Assay, INFINITI 2C9 and VKORC1 Multiplex Assay for Warfarin, eQ-PCR LC Warfarin Genotyping kit, and Verigene Warfarin Metabolism Nucleic Acid Test and Verigene System). As such, the paradigm of medicine is progressively changing from merely treating individuals on the basis of their symptoms to being able to predict disease susceptibility and patient tolerability to pharmacotherapy in order to customize diagnosis and pharmacotherapy. By incorporating genotyping and phenotyping information on CYP2D6, health professionals can identify the patient's polymorphisms and disease subtypes to determine the most advantageous management. This can include the immediate administration of the most efficacious and least toxic drugs at the correct doses. Understanding and identifying an individual's genetic variations has the potential to decrease both the time expended in achieving effective therapy and the number of visits required for proper dose adjustment.

The FDA has approved four warfarin pharmacogenetic test kits, but most third-party payers are reluctant to reimburse for such testing because it is not currently considered a standard of care (Myburgh et al. 2012). These tests typically cost a few hundred dollars, but it should become less expensive as it becomes more commonplace. The current FDA-approved product label for warfarin does not recommend routine pharmacogenomic testing for determining initial or maintenance doses, but it does acknowledge that dose requirements are significantly affected by *CYP2C9* and *VKORC1* and states that genotype information can assist in selecting the proper starting dose. A well-developed warfarin-dosing algorithm incorporating conventional clinical factors and genetic status is available at the website http://www.warfarindosing.org.

In silico screening can also possibly identify high-affinity ligands for CYP2D6. In addition to in silico approaches (Bell and Wang 2012; Carosati 2013; de Graaf et al. 2005; de Groot et al. 2004; DeLisle et al. 2011; Ekins and Wrighton 2001; Foti et al. 2010; Greer et al. 2010; Guguen-Guillouzo and Guillouzo 2010; Handschin et al. 2003; Hutter 2009; Kamel and Harriman 2013; Lake and Price 2013; Liu et al. 2013; Martiny and Miteva 2013; Mudra et al. 2011; Stoll et al. 2011; Tarcsay and Keseru 2011; Zhang et al. 2011), in vitro models (Donato et al. 2008; Gomez-Lechon et al. 2007, 2008; Iwatsubo et al. 1997; Kamel and Harriman 2013; Kramer and Tracy 2008; LeCluyse 2001; Ong et al. 2013; von Moltke et al. 1998), in particular those with a high-throughput capacity, are very useful to determine whether a new drug is a substrate or inhibitor of CYP2D6 and the contribution of this enzyme in their metabolic clearance or metabolic activation in the early stage of drug development. If the new drug candidate is substantially metabolized by CYP2D6 (e.g., >60%) or it is a very potent inhibitor of this enzyme (e.g., $K_i < 0.1$ μM), alternative compounds that have a low affinity to CYP2D6 should be considered.

Allelic variations in the genes encoding drug targets, drug transporters, and drug-metabolizing enzymes as a result of polymorphism have the potential to have a substantial effect on drug clearance and response (Ma and Lu 2011). It is expected that personalized treatments will be offered in the near future on the basis of the genotypes of individuals, therefore optimizing the dosage and decreasing the frequency of ADRs. Personalized medicine is the use of detailed information about a patient's genotype or level of gene expression and a patient's clinical data in order to select a medication, therapy, or preventative measure that is particularly suited to that patient at the time of administration (Hamburg and Collins 2010). The benefits of this approach include accuracy, efficacy, safety, and speed.

Pharmacogenomic profiling may consequently affect drug labeling to limit prescriptions only to those individuals with the appropriate genetic profile. To date, approximately 15% of labels for FDA-approved drugs (>146 drugs, see http://www.fda.gov/drugs/scienceresearch/researchareas/pharmacogenetics/ucm083378.htm) contain pharmacogenomic information—a substantial increase since the 1990s. The label information involves at least 50 important genes that encode drug-metabolizing enzymes (e.g., *CYP2D6*), important proteins, or drug targets. This means that, for drugs already approved by the regulatory agencies, succeeding discoveries that individuals with certain genetic profiles might experience adverse drug effects would require addition of this information to the label as well as a warning that genetic screening is necessary. For example, the first label update for warfarin was issued in August 2007 and was in reference to its sensitivity in CYP2C9 PMs. The second warfarin label update was issued in January 2010, and this label included the effect of a second gene, *VKORC1*, as well as a table with pharmacogenomics-guided dosing ranges for the drug. One could then take the view that pharmacogenomics is not about individualized drug therapy but rather about reclassifying risk factors. For example, instead of hypertension being a coexisting risk factor that should be considered before starting a new therapy, the risk factor will be patients with *CYP2D6/2C9* allelic variants who may have a reduced ability to eliminate the administered drugs.

Tamoxifen lacks efficacy in those female patients who are CYP2D6 PMs (i.e., ~7% in the white population) (Zhou 2009a,b), but the FDA has not made firm recommendations about *CYP2D6* testing before prescribing tamoxifen since the evidence of benefit has been considered insufficient. Importantly, clinicians should be aware that the clinical efficacy of tamoxifen is greatly decreased by concomitant drugs that are potent CYP2D6 inhibitors (Wang et al. 2009b; Zhou et al. 2009a).

Regrettably, the pharmacogenomic approach to clinical medicine is currently very limited despite its discovery dating back to the early 1960s and the presence of a large amount of pharmacogenetic information. At present, prescription genetic screening is largely confined to teaching hospitals and specialized laboratories and is not yet a part of routine practice. The strategy of prescription genotyping is seldom practiced in the clinic, even for substrates of extensively characterized SNPs such as *CYP2D6* and codeine or tamoxifen, *CYP2C19* and phenytoin, and *CYP2C9* and warfarin. Its limited use may be attributed to the fact that the *CYP2C9* genotype contributes to <10% of the total variability in an individual (Myburgh et al. 2012).

To date, there is an increasing demand in the application of pharmacogenomics to predict ADRs in clinical practice. With the prediction of how certain patients interact with a drug, it can prevent the unwanted ADRs that are associated with genetic variations. There is a growing number of genetic variations in genes encoding drug-metabolizing enzymes, drug transporters, and drug targets that have been identified. These genetic variations have shown an association with ADRs with varying strength of evidence (International Transporter Consortium et al. 2010; Roden et al. 2011; Sim et al. 2013). A significant association between genetic variations and the risk of ADRs requires strong and comprehensive clinical evidence; however, there are many challenges and hurdles along the way to fully understand and elucidate the contribution of genetic variations to ADRs and to translate it into clinical practice. Herein, we focus on the association between genetic variations and ADRs and the clinical evidence for the genetic variation–ADR association and delineate how pharmacogenomics can improve drug therapy, reduce and minimize the risk of ADRs, and eventually achieve optimal therapeutic outcomes in the future.

Genome-wide association studies (GWASs) for pharmacogenomics-related traits are increasingly being conducted to identify loci that affect either drug response or susceptibility to ADRs (Daly 2010, 2012; Jiang et al. 2013; Nebert et al. 2008). With the application of GWASs, there are increased reports on the identification of genetic mutations that are associated with ADRs. Although this approach is often limited by the hurdle of a large number of cases and controls collection, many GWASs have been conducted and a number of common and rare genetic mutations significantly associated with ADRs have been identified (Daly 2010, 2012; Low et al. 2014; Wheeler et al. 2013), such as *HLA-B*57:01*, *HLA-DRB1*15:01*, *SLCO1B1*, *KCNE1*, and phosphodiesterase 4B (PDE4B). Most of these GWASs have been replicated and confirmed with independent studies (Daly 2012). Genetic variation of *HLA* loci is one of the most significant pharmacogenomic biomarkers associated with ADRs. It strongly associates with flucloxacillin-induced hepatotoxicities (*HLA-B*57:01*), carbamazepine-induced hypersensitivity and severe cutaneous adverse reactions (*HLA-A*31:01*), and lumiracoxib-induced hepatotoxicity (*HLA-DRB1*15:01* and *HLA-DQB1*06:02*) (McCormack et al. 2011). Simvastatin-induced myopathy is also significantly associated with *SLCO1B1* polymorphism with an odds ratio of 16.9 (Voora et al. 2009). In addition, several other GWASs have been published including an association between lamotrigine-induced skin injury and an *ADAM22* polymorphism (Shen et al. 2012), bisphosphate-induced osteonecrosis of the jaw and the *CYP2C8* SNP rs1934951, as well as T-cell leukemia 1A (*TCL1A*) SNPs and musculoskeletal adverse events induced by the aromatase inhibitors such as anastrozole and exemestane in women with early breast cancer. A recent GWAS on 435 patients has identified significant associations for citalopram and escitalopram concentrations with SNPs in or near the *CYP2C19* gene (rs1074145)

and with *S*-didesmethylcitalopram concentration for SNPs near the *CYP2D6* locus (rs1065852) (Ji et al. 2014). Another GWAS has examined the relationships between genetic polymorphisms and CYP expression or enzyme activities using 466 human liver samples (Yang et al. 2010).

Although there are well-established techniques for molecular genotyping and phenotyping in major research institutes, the facilities for genetic testing and measurement of parent drug and pharmacologically active metabolite concentrations are not always accessible in the diagnostic laboratory. What's more, mutational screening using the current state-of-the-art technology is still laborious and time-consuming. Furthermore, there is a lack of adequate education provided to physicians in clinical practice as well as medical undergraduates and trainee physicians, and it is yet to be incorporated into the curriculum of medical courses worldwide. Thus, the bridging between medicine and basic science requires the collaborative efforts of both clinicians and researchers.

REFERENCES

Agundez, J. A., Ledesma, M. C., Ladero, J. M., Benitez, J. 1995. Prevalence of *CYP2D6* gene duplication and its repercussion on the oxidative phenotype in a white population. *Clin Pharmacol Ther* 57:265–269.

Aklillu, E., Persson, I., Bertilsson, L., Johansson, I., Rodrigues, F., Ingelman-Sundberg, M. 1996. Frequent distribution of ultra-rapid metabolizers of debrisoquine in an ethiopian population carrying duplicated and multiduplicated functional *CYP2D6* alleles. *J Pharmacol Exp Ther* 278:441–446.

Alderman, J., Preskorn, S. H., Greenblatt, D. J., Harrison, W., Penenberg, D., Allison, J., Chung, M. 1997. Desipramine pharmacokinetics when coadministered with paroxetine or sertraline in extensive metabolizers. *J Clin Psychopharmacol* 17:284–291.

Ashley, E. A., Butte, A. J., Wheeler, M. T., Chen, R., Klein, T. E., Dewey, F. E., Dudley, J. T. 2010. Clinical assessment incorporating a personal genome. *Lancet* 375:1525–1535.

Bartlett, M. J., Green, D. W., Shephard, E. A. 2013. Pharmacogenetic testing in the UK clinical setting. *Lancet* 381:1903.

Bell, L. C., Wang, J. 2012. Probe ADME and test hypotheses: A PATH beyond clearance *in vitro-in vivo* correlations in early drug discovery. *Expert Opin Drug Metab Toxicol* 8:1131–1155.

Belle, D. J., Ernest, C. S., Sauer, J. M., Smith, B. P., Thomasson, H. R., Witcher, J. W. 2002. Effect of potent CYP2D6 inhibition by paroxetine on atomoxetine pharmacokinetics. *J Clin Pharmacol* 42:1219–1227.

Bernal, M. L., Sinues, B., Johansson, I., McLellan, R. A., Wennerholm, A., Dahl, M. L., Ingelman-Sundberg, M., Bertilsson, L. 1999. Ten percent of North Spanish individuals carry duplicated or triplicated *CYP2D6* genes associated with ultrarapid metabolism of debrisoquine. *Pharmacogenetics* 9:657–660.

Bertilsson, L., Dahl, M. L., Sjoqvist, F., Aberg-Wistedt, A., Humble, M., Johansson, I., Lundqvist, E., Ingelman-Sundberg, M. 1993. Molecular basis for rational megaprescribing in ultrarapid hydroxylators of debrisoquine. *Lancet* 341:63.

Bienfait, K. L., Shaw, P. M., Murthy, G., Warner, A. W. 2013. Mobilizing pharmacogenomic analyses during clinical trials in drug development. *Pharmacogenomics* 14:1227–1235.

Bonanni, B., Macis, D., Maisonneuve, P., Johansson, H. A., Gucciardo, G., Oliviero, P., Travaglini, R. et al. 2006. Polymorphism in the CYP2D6 tamoxifen-metabolizing gene influences clinical effect but not hot flashes: Data from the Italian Tamoxifen Trial. *J Clin Oncol* 24:3708–3709; author reply 3709.

Borges, S., Desta, Z., Li, L., Skaar, T. C., Ward, B. A., Nguyen, A., Jin, Y. et al. 2006. Quantitative effect of *CYP2D6* genotype and inhibitors on tamoxifen metabolism: Implication for optimization of breast cancer treatment. *Clin Pharmacol Ther* 80:61–74.

Breyer-Pfaff, U., Pfandl, B., Nill, K., Nusser, E., Monney, C., Jonzier-Perey, M., Baettig, D., Baumann, P. 1992. Enantioselective amitriptyline metabolism in patients phenotyped for two cytochrome P450 isozymes. *Clin Pharmacol Ther* 52:350–358.

Brynne, N., Dalen, P., Alvan, G., Bertilsson, L., Gabrielsson, J. 1998. Influence of *CYP2D6* polymorphism on the pharmacokinetics and pharmacodynamic of tolterodine. *Clin Pharmacol Ther* 63:529–539.

Carosati, E. 2013. Modelling cytochromes P450 binding modes to predict P450 inhibition, metabolic stability and isoform selectivity. *Drug Discov Today Technol* 10:e167–e175.

Choong, E., Guo, J., Persson, A., Virding, S., Johansson, I., Mkrtchian, S., Ingelman-Sundberg, M. 2015. Developmental regulation and induction of cytochrome P450 2W1, an enzyme expressed in colon tumors. *PLoS One* 10:e0122820.

Ciraulo, D. A., Shader, R. I. 1990. Fluoxetine drug–drug interactions: I. Antidepressants and antipsychotics. *J Clin Psychopharmacol* 10:48–50.

Cong, F., Cheung, A. K., Huang, S. M. 2012. Chemical genetics-based target identification in drug discovery. *Annu Rev Pharmacol Toxicol* 52:57–78.

Daly, A. K. 2010. Genome-wide association studies in pharmacogenomics. *Nat Rev Genet* 11:241–246.

Daly, A. K. 2012. Using genome-wide association studies to identify genes important in serious adverse drug reactions. *Annu Rev Pharmacol Toxicol* 52:21–35.

Daly, A. K. 2013. Pharmacogenomics of adverse drug reactions. *Genome Med* 5:5.

de Graaf, C., Vermeulen, N. P., Feenstra, K. A. 2005. Cytochrome p450 *in silico*: An integrative modeling approach. *J Med Chem* 48:2725–2755.

de Groot, M. J., Kirton, S. B., Sutcliffe, M. J. 2004. *In silico* methods for predicting ligand binding determinants of cytochromes P450. *Curr Top Med Chem* 4:1803–1824.

DeLisle, R. K., Otten, J., Rhodes, S. 2011. *In silico* modeling of p450 substrates, inhibitors, activators, and inducers. *Comb Chem High Throughput Screen* 14:396–416.

Donato, M. T., Lahoz, A., Castell, J. V., Gomez-Lechon, M. J. 2008. Cell lines: A tool for *in vitro* drug metabolism studies. *Curr Drug Metab* 9:1–11.

Eckhardt, K., Li, S., Ammon, S., Schanzle, G., Mikus, G., Eichelbaum, M. 1998. Same incidence of adverse drug events after codeine administration irrespective of the genetically determined differences in morphine formation. *Pain* 76:27–33.

Ekins, S., Wrighton, S. A. 2001. Application of *in silico* approaches to predicting drug–drug interactions. *J Pharmacol Toxicol Methods* 45:65–69.

Evans, W. E., McLeod, H. L. 2003. Pharmacogenomics—Drug disposition, drug targets, and side effects. *N Engl J Med* 348:538–549.

Evans, W. E., Relling, M. V., Petros, W. P., Meyer, W. H., Mirro, J., Jr., Crom, W. R. 1989. Dextromethorphan and caffeine as probes for simultaneous determination of debrisoquin-oxidation and *N*-acetylation phenotypes in children. *Clin Pharmacol Ther* 45:568–573.

Evert, B., Eichelbaum, M., Haubruck, H., Zanger, U. M. 1997. Functional properties of CYP2D6 1 (wild-type) and CYP2D6 7

(His324Pro) expressed by recombinant baculovirus in insect cells. *Naunyn Schmiedebergs Arch Pharmacol* 355:309–318.

Farid, N. A., Bergstrom, R. F., Ziege, E. A., Parli, C. J., Lemberger, L. 1985. Single-dose and steady-state pharmacokinetics of tomoxetine in normal subjects. *J Clin Pharmacol* 25:296–301.

Foti, R. S., Wienkers, L. C., Wahlstrom, J. L. 2010. Application of cytochrome P450 drug interaction screening in drug discovery. *Comb Chem High Throughput Screen* 13:145–158.

Fuller, R. W., Wong, D. T., Robertson, D. W. 1991. Fluoxetine, a selective inhibitor of serotonin uptake. *Med Res Rev* 11:17–34.

Gaedigk, A., Blum, M., Gaedigk, R., Eichelbaum, M., Meyer, U. A. 1991. Deletion of the entire cytochrome P450 *CYP2D6* gene as a cause of impaired drug metabolism in poor metabolizers of the debrisoquine/sparteine polymorphism. *Am J Hum Genet* 48:943–950.

Gaedigk, A., Bradford, L. D., Marcucci, K. A., Leeder, J. S. 2002. Unique CYP2D6 activity distribution and genotype–phenotype discordance in black Americans. *Clin Pharmacol Ther* 72:76–89.

Gaedigk, A., Ndjountche, L., Gaedigk, R., Leeder, J. S., Bradford, L. D. 2003. Discovery of a novel nonfunctional cytochrome P450 2D6 allele, CYP2D642, in African American subjects. *Clin Pharmacol Ther* 73:575–576.

Gaedigk, A., Bradford, L. D., Alander, S. W., Leeder, J. S. 2006. CYP2D6*36 gene arrangements within the cyp2d6 locus: Association of CYP2D6*36 with poor metabolizer status. *Drug Metab Dispos* 34:563–569.

Gaedigk, A., Simon, S. D., Pearce, R. E., Bradford, L. D., Kennedy, M. J., Leeder, J. S. 2008. The CYP2D6 activity score: Translating genotype information into a qualitative measure of phenotype. *Clin Pharmacol Ther* 83:234–242.

Ghoshal, U., Tripathi, S., Kumar, S., Mittal, B., Chourasia, D., Kumari, N., Krishnani, N., Ghoshal, U. C. 2014. Genetic polymorphism of cytochrome P450 (CYP) 1A1, CYP1A2, and CYP2E1 genes modulate susceptibility to gastric cancer in patients with Helicobacter pylori infection. *Gastric Cancer* 17:226–234.

Gibbs, J. P., Hyland, R., Youdim, K. 2006. Minimizing polymorphic metabolism in drug discovery: Evaluation of the utility of in vitro methods for predicting pharmacokinetic consequences associated with CYP2D6 metabolism. *Drug Metab Dispos* 34:1516–1522.

Go, R. E., Hwang, K. A., Choi, K. C. 2015. Cytochrome P450 1 family and cancers. *J Steroid Biochem Mol Biol* 147:24–30.

Goetz, M. P., Rae, J. M., Suman, V. J., Safgren, S. L., Ames, M. M., Visscher, D. W., Reynolds, C. et al. 2005. Pharmacogenetics of tamoxifen biotransformation is associated with clinical outcomes of efficacy and hot flashes. *J Clin Oncol* 23:9312–9318.

Goetz, M. P., Knox, S. K., Suman, V. J., Rae, J. M., Safgren, S. L., Ames, M. M., Visscher, D. W. et al. 2007. The impact of cytochrome P450 2D6 metabolism in women receiving adjuvant tamoxifen. *Breast Cancer Res Treat* 101:113–121.

Gomez-Lechon, M. J., Castell, J. V., Donato, M. T. 2007. Hepatocytes—The choice to investigate drug metabolism and toxicity in man: *In vitro* variability as a reflection of *in vivo*. *Chem Biol Interact* 168:30–50.

Gomez-Lechon, M. J., Castell, J. V., Donato, M. T. 2008. An update on metabolism studies using human hepatocytes in primary culture. *Expert Opin Drug Metab Toxicol* 4:837–854.

Gram, L. F., Christiansen, J. 1975. First-pass metabolism of imipramine in man. *Clin Pharmacol Ther* 17:555–563.

Greer, M. L., Barber, J., Eakins, J., Kenna, J. G. 2010. Cell based approaches for evaluation of drug-induced liver injury. *Toxicology* 268:125–131.

Griese, E. U., Zanger, U. M., Brudermanns, U., Gaedigk, A., Mikus, G., Morike, K., Stuven, T., Eichelbaum, M. 1998. Assessment of the predictive power of genotypes for the in-vivo catalytic function of CYP2D6 in a German population. *Pharmacogenetics* 8:15–26.

Guguen-Guillouzo, C., Guillouzo, A. 2010. General review on *in vitro* hepatocyte models and their applications. *Methods Mol Biol* 640:1–40.

Haghgoo, S. M., Allameh, A., Mortaz, E., Garssen, J., Folkerts, G., Barnes, P. J., Adcock, I. M. 2015. Pharmacogenomics and targeted therapy of cancer: Focusing on non-small cell lung cancer. *Eur J Pharmacol* 754:82–91.

Hamburg, M. A., Collins, F. S. 2010. The path to personalized medicine. *N Engl J Med* 363:301–304.

Handschin, C., Podvinec, M., Meyer, U. A. 2003. *In silico* approaches, and *in vitro* and *in vivo* experiments to predict induction of drug metabolism. *Drug News Perspect* 16:423–434.

Heller, T., Kirchheiner, J., Armstrong, V. W., Luthe, H., Tzvetkov, M., Brockmoller, J., Oellerich, M. 2006. AmpliChip CYP450 GeneChip: A new gene chip that allows rapid and accurate CYP2D6 genotyping. *Ther Drug Monit* 28:673–677.

Hertz, D. L., McLeod, H. L. 2013. Use of pharmacogenetics for predicting cancer prognosis and treatment exposure, response and toxicity. *J Hum Genet* 58:346–352.

Hinrichs, J. W., Loovers, H. M., Scholten, B., van der Weide, J. 2008. Semi-quantitative CYP2D6 gene doses in relation to metabolic ratios of psychotropics. *Eur J Clin Pharmacol* 64:979–986.

Hodgson, E., Rose, R. L. 2007. The importance of cytochrome P450 2B6 in the human metabolism of environmental chemicals. *Pharmacol Ther* 113:420–428.

Holliday, S. M., Benfield, P. 1995. Venlafaxine. A review of its pharmacology and therapeutic potential in depression. *Drugs* 49:280–294.

Horrigan, J. P., Barnhill, L. J. 1994. Paroxetine-pimozide drug interaction. *J Am Acad Child Adolesc Psychiatry* 33:1060–1061.

Hsu, M. H., Savas, U., Griffin, K. J., Johnson, E. F. 2007. Human cytochrome p450 family 4 enzymes: Function, genetic variation and regulation. *Drug Metab Rev* 39:515–538.

Hutter, M. C. 2009. *In silico* prediction of drug properties. *Curr Med Chem* 16:189–202.

International Transporter Consortium, Giacomini, K. M., Huang, S. M., Tweedie, D. J., Benet, L. Z., Brouwer, K. L., Chu, X., Dahlin, A. et al. 2010. Membrane transporters in drug development. *Nat Rev Drug Discov* 9:215–236.

Ito, K., Iwatsubo, T., Kanamitsu, S., Ueda, K., Suzuki, H., Sugiyama, Y. 1998. Prediction of pharmacokinetic alterations caused by drug–drug interactions: Metabolic interaction in the liver. *Pharmacol Rev* 50:387–412.

Iwatsubo, T., Hirota, N., Ooie, T., Suzuki, H., Shimada, N., Chiba, K., Ishizaki, T., Green, C. E., Tyson, C. A., Sugiyama, Y. 1997. Prediction of *in vivo* drug metabolism in the human liver from *in vitro* metabolism data. *Pharmacol Ther* 73:147–171.

Jensen, B. C., McLeod, H. L. 2013. Pharmacogenomics as a risk mitigation strategy for chemotherapeutic cardiotoxicity. *Pharmacogenomics* 14:205–213.

Jeppesen, U., Gram, L. F., Vistisen, K., Loft, S., Poulsen, H. E., Brosen, K. 1996. Dose-dependent inhibition of CYP1A2, CYP2C19 and CYP2D6 by citalopram, fluoxetine, fluvoxamine and paroxetine. *Eur J Clin Pharmacol* 51:73–78.

Ji, Y., Schaid, D. J., Desta, Z., Kubo, M., Batzler, A. J., Snyder, K., Mushiroda, T., Kamatani, N. et al. 2014. Citalopram and escitalopram plasma drug and metabolite concentrations: Genome-wide associations. *Br J Clin Pharmacol* 78:373–383.

Jiang, J., Fridley, B. L., Feng, Q., Abo, R. P., Brisbin, A., Batzler, A., Jenkins, G., Long, P. A., Wang, L. 2013. Genome-wide association study for biomarker identification of Rapamycin and Everolimus using a lymphoblastoid cell line system. *Front Genet* 4:166.

Jin, Y., Desta, Z., Stearns, V., Ward, B., Ho, H., Lee, K. H., Skaar, T. et al. 2005. CYP2D6 genotype, antidepressant use, and tamoxifen metabolism during adjuvant breast cancer treatment. *J Natl Cancer Inst* 97:30–39.

Kagimoto, M., Heim, M., Kagimoto, K., Zeugin, T., Meyer, U. A. 1990. Multiple mutations of the human cytochrome P450IID6 gene (CYP2D6) in poor metabolizers of debrisoquine. Study of the functional significance of individual mutations by expression of chimeric genes. *J Biol Chem* 265:17209–17214.

Kamel, A., Harriman, S. 2013. Inhibition of cytochrome P450 enzymes and biochemical aspects of mechanism-based inactivation (MBI). *Drug Discov Today Technol* 10:e177–e189.

Kramer, M. A., Tracy, T. S. 2008. Studying cytochrome P450 kinetics in drug metabolism. *Expert Opin Drug Metab Toxicol* 4:591–603.

Lake, B. G., Price, R. J. 2013. Evaluation of the metabolism and hepatotoxicity of xenobiotics utilizing precision-cut slices. *Xenobiotica* 43:41–53.

Lazarou, J., Pomeranz, B. H., Corey, P. N. 1998. Incidence of adverse drug reactions in hospitalized patients: A meta-analysis of prospective studies. *JAMA* 279:1200–1205.

Le Bourdonnec, B., Leister, L. K. 2009. Medicinal chemistry strategies to reduce CYP2D6 inhibitory activity of lead candidates. *Curr Med Chem* 16:3093–3121.

Leathart, J. B., London, S. J., Steward, A., Adams, J. D., Idle, J. R., Daly, A. K. 1998. CYP2D6 phenotype-genotype relationships in African-Americans and Caucasians in Los Angeles. *Pharmacogenetics* 8:529–541.

LeCluyse, E. L. 2001. Human hepatocyte culture systems for the *in vitro* evaluation of cytochrome P450 expression and regulation. *Eur J Pharm Sci* 13:343–368.

Lee, C., Morton, C. C. 2008. Structural genomic variation and personalized medicine. *N Engl J Med* 358:740–741.

Li, A. P. 2001. Screening for human ADME/Tox drug properties in drug discovery. *Drug Discov Today* 6:357–366.

Lin, J. H., Lu, A. Y. 1997. Role of pharmacokinetics and metabolism in drug discovery and development. *Pharmacol Rev* 49:403–449.

Lin, J. H., Lu, A. Y. 1998. Inhibition and induction of cytochrome P450 and the clinical implications. *Clin Pharmacokinet* 35:361–390.

Liu, X., Shen, Q., Li, J., Li, S., Luo, C., Zhu, W., Luo, X., Zheng, M., Jiang, H. 2013. *In silico* prediction of cytochrome P450-mediated site of metabolism (SOM). *Protein Pept Lett* 20:279–289.

Lotsch, J., Skarke, C., Liefhold, J., Geisslinger, G. 2004. Genetic predictors of the clinical response to opioid analgesics: Clinical utility and future perspectives. *Clin Pharmacokinet* 43:983–1013.

Low, S. K., Takahashi, A., Mushiroda, T., Kubo, M. 2014. Genome-wide association study: A useful tool to identify common genetic variants associated with drug toxicity and efficacy in cancer pharmacogenomics. *Clin Cancer Res* 20:2541–2552.

Ma, Q., Lu, A. Y. 2011. Pharmacogenetics, pharmacogenomics, and individualized medicine. *Pharmacol Rev* 63:437–459.

Mackman, R., Tschirret-Guth, R. A., Smith, G., Hayhurst, G. P., Ellis, S. W., Lennard, M. S., Tucker, G. T., Wolf, C. R., Ortiz de Montellano, P. R. 1996. Active-site topologies of human CYP2D6 and its aspartate-301 —> glutamate, asparagine, and glycine mutants. *Arch Biochem Biophys* 331:134–140.

Madani, S., Paine, M. F., Lewis, L., Thummel, K. E., Shen, D. D. 1999. Comparison of CYP2D6 content and metoprolol oxidation between microsomes isolated from human livers and small intestines. *Pharm Res* 16:1199–1205.

Martiny, V. Y., Miteva, M. A. 2013. Advances in molecular modeling of human cytochrome P450 polymorphism. *J Mol Biol* 425:3978–3992.

McCarthy, J. J., McLeod, H. L., Ginsburg, G. S. 2013. Genomic medicine: A decade of successes, challenges, and opportunities. *Sci Transl Med* 5:189sr184.

McCormack, M., Alfirevic, A., Bourgeois, S., Farrell, J. J., Kasperaviciute, D., Carrington, M., Sills, G. J. et al. 2011. *HLA-A*3101* and carbamazepine-induced hypersensitivity reactions in Europeans. *N Engl J Med* 364:1134–1143.

McLellan, R. A., Oscarson, M., Seidegard, J., Evans, D. A., Ingelman-Sundberg, M. 1997. Frequent occurrence of *CYP2D6* gene duplication in Saudi Arabians. *Pharmacogenetics* 7:187–191.

McLeod, H. L. 2013. Cancer pharmacogenomics: Early promise, but concerted effort needed. *Science* 339:1563–1566.

Meyer, U. A. 2000. Pharmacogenetics and adverse drug reactions. *Lancet* 356:1667–1671.

Meyers, D. A., Bleecker, E. R., Holloway, J. W., Holgate, S. T. 2014. Asthma genetics and personalised medicine. *Lancet Respir Med* 2:405–415.

Meyerson, M. 2003. Human genetic variation and disease. *Lancet* 362:259–260.

Morimoto, T., Sakuma, M., Matsui, K., Kuramoto, N., Toshiro, J., Murakami, J., Fukui, T., Saito, M., Hiraide, A., Bates, D. W. 2011. Incidence of adverse drug events and medication errors in Japan: The JADE study. *J Gen Intern Med* 26:148–153.

Mudra, D. R., Desino, K. E., Desai, P. V. 2011. *In silico, in vitro* and *in situ* models to assess interplay between CYP3A and P-gp. *Curr Drug Metab* 12:750–773.

Myburgh, R., Hochfeld, W. E., Dodgen, T. M., Ker, J., Pepper, M. S. 2012. Cardiovascular pharmacogenetics. *Pharmacol Ther* 133:280–290.

Nakamura, Y. 2008. Pharmacogenomics and drug toxicity. *N Engl J Med* 359:856–858.

Naranjo, C. A., Sproule, B. A., Knoke, D. M. 1999. Metabolic interactions of central nervous system medications and selective serotonin reuptake inhibitors. *Int Clin Psychopharmacol* 14 Suppl 2:S35–S47.

Nebert, D. W., Russell, D. W. 2002. Clinical importance of the cytochromes P450. *Lancet* 360:1155–1162.

Nebert, D. W., Dalton, T. P. 2006. The role of cytochrome P450 enzymes in endogenous signalling pathways and environmental carcinogenesis. *Nat Rev Cancer* 6:947–960.

Nebert, D. W., Zhang, G., Vesell, E. S. 2008. From human genetics and genomics to pharmacogenetics and pharmacogenomics: Past lessons, future directions. *Drug Metab Rev* 40:187–224.

Nebert, D. W., Wikvall, K., Miller, W. L. 2013. Human cytochromes P450 in health and disease. *Philos Trans R Soc Lond B Biol Sci* 368:20120431.

Nichols, A. I., Fatato, P., Shenouda, M., Paul, J., Isler, J. A., Pedersen, R. D., Jiang, Q., Ahmed, S., Patroneva, A. 2009. The effects of desvenlafaxine and paroxetine on the pharmacokinetics of the cytochrome P450 2D6 substrate desipramine in healthy adults. *J Clin Pharmacol* 49:219–228.

Niwa, T., Murayama, N., Yamazaki, H. 2009. Oxidation of endobiotics mediated by xenobiotic-metabolizing forms of human cytochrome. *Curr Drug Metab* 10:700–712.

Ong, C. E., Pan, Y., Mak, J. W., Ismail, R. 2013. *In vitro* approaches to investigate cytochrome P450 activities: Update on current status and their applicability. *Expert Opin Drug Metab Toxicol* 9:1097–1113.

Ozdemir, V., Naranjo, C. A., Herrmann, N., Reed, K., Sellers, E. M., Kalow, W. 1997. Paroxetine potentiates the central nervous system side effects of perphenazine: Contribution of cytochrome P4502D6 inhibition in vivo. *Clin Pharmacol Ther* 62:334–347.

Paine, M. J., McLaughlin, L. A., Flanagan, J. U., Kemp, C. A., Sutcliffe, M. J., Roberts, G. C., Wolf, C. R. 2003. Residues glutamate 216 and aspartate 301 are key determinants of substrate specificity and product regioselectivity in cytochrome P450 2D6. *J Biol Chem* 278:4021–4027.

Panigrahy, D., Kaipainen, A., Greene, E. R., Huang, S. 2010. Cytochrome P450-derived eicosanoids: The neglected pathway in cancer. *Cancer Metastasis Rev* 29:723–735.

Phillips, K. A., Veenstra, D. L., Oren, E., Lee, J. K., Sadee, W. 2001. Potential role of pharmacogenomics in reducing adverse drug reactions: A systematic review. *JAMA* 286:2270–2279.

Pirmohamed, M., James, S., Meakin, S., Green, C., Scott, A. K., Walley, T. J., Farrar, K., Park, B., K. Breckenridge, A. M. 2004. Adverse drug reactions as cause of admission to hospital: Prospective analysis of 18 820 patients. *BMJ* 329:15–19.

Polimanti, R., Piacentini, S., Manfellotto, D., Fuciarelli, M. 2012. Human genetic variation of CYP450 superfamily: Analysis of functional diversity in worldwide populations. *Pharmacogenomics* 13:1951–1960.

Poulsen, L., Brosen, K., Arendt-Nielsen, L., Gram, L. F., Elbaek, K., Sindrup, S. H. 1996. Codeine and morphine in extensive and poor metabolizers of sparteine: Pharmacokinetics, analgesic effect and side effects. *Eur J Clin Pharmacol* 51:289–295.

Ramos, E., Doumatey, A., Elkahloun, A. G., Shriner, D., Huang, H., Chen, G., Zhou, J., McLeod, H., Adeyemo, A., Rotimi, C. N. 2014. Pharmacogenomics, ancestry and clinical decision making for global populations. *Pharmacogenomics J* 14:217–222.

Rau, T., Wohlleben, G., Wuttke, H., Thuerauf, N., Lunkenheimer, J., Lanczik, M., Eschenhagen, T. 2004. *CYP2D6* genotype: Impact on adverse effects and nonresponse during treatment with antidepressants—A pilot study. *Clin Pharmacol Ther* 75:386–393.

Raunio, H., Rahnasto-Rilla, M. 2012. CYP2A6: Genetics, structure, regulation, and function. *Drug Metabol Drug Interact* 27:73–88.

Roden, D. M., Wilke, R. A., Kroemer, H. K., Stein, C. M. 2011. Pharmacogenomics: The genetics of variable drug responses. *Circulation* 123:1661–1670.

Rodriguez, M., Potter, D. A. 2013. CYP1A1 regulates breast cancer proliferation and survival. *Mol Cancer Res* 11:780–792.

Roederer, M. W., Sanchez-Giron, F., Kalideen, K., Kudzi, W., McLeod, H. L., Zhang, W., Pharmacogenetics for Every Nation Initiative. 2011. Pharmacogenetics and rational drug use around the world. *Pharmacogenomics* 12:897–905.

Rowland, P., Blaney, F. E., Smyth, M. G., Jones, J. J., Leydon, V. R., Oxbrow, A. K., Lewis, C. J. et al. 2006. Crystal structure of human cytochrome P450 2D6. *J Biol Chem* 281:7614–7622.

Sachse, C., Brockmoller, J., Bauer, S., Roots, I. 1997. Cytochrome P450 2D6 variants in a Caucasian population: Allele frequencies and phenotypic consequences. *Am J Hum Genet* 60:284–295.

Saito, M., Yasui-Furukori, N., Nakagami, T., Furukori, H., Kaneko, S. 2005. Dose-dependent interaction of paroxetine with risperidone in schizophrenic patients. *J Clin Psychopharmacol* 25:527–532.

Samani, N. J., Tomaszewski, M., Schunkert, H. 2010. The personal genome—The future of personalised medicine? *Lancet* 375:1497–1498.

Schoedel, K. A., Pope, L. E., Sellers, E. M. 2012. Randomized open-label drug–drug interaction trial of dextromethorphan/quinidine and paroxetine in healthy volunteers. *Clin Drug Investig* 32:157–169.

Schroth, W., Antoniadou, L., Fritz, P., Schwab, M., Muerdter, T., Zanger, U. M., Simon, W., Eichelbaum, M., Brauch, H. 2007. Breast cancer treatment outcome with adjuvant tamoxifen relative to patient *CYP2D6* and *CYP2C19* genotypes. *J Clin Oncol* 25:5187–5193.

Scordo, M. G., Spina, E., Facciola, G., Avenoso, A., Johansson, I., Dahl, M. L. 1999. Cytochrome P450 *2D6* genotype and steady state plasma levels of risperidone and 9-hydroxyrisperidone. *Psychopharmacology (Berl)* 147:300–305.

Shen, Y., Nicoletti, P., Floratos, A., Pirmohamed, M., Molokhia, M., Geppetti, P., Benemei, S. et al. 2012. Genome-wide association study of serious blistering skin rash caused by drugs. *Pharmacogenomics J* 12:96–104.

Sim, S. C., Kacevska, M., Ingelman-Sundberg, M. 2013. Pharmacogenomics of drug-metabolizing enzymes: A recent update on clinical implications and endogenous effects. *Pharmacogenomics J* 13:1–11.

Simpson, A. E. 1997. The cytochrome P450 4 (CYP4) family. *Gen Pharmacol* 28:351–359.

Sinz, M., Wallace, G., Sahi, J. 2008. Current industrial practices in assessing CYP450 enzyme induction: Preclinical and clinical. *AAPS J* 10:391–400.

Snider, N. T., Walker, V. J., Hollenberg, P. F. 2010. Oxidation of the endogenous cannabinoid arachidonoyl ethanolamide by the cytochrome P450 monooxygenases: Physiological and pharmacological implications. *Pharmacol Rev* 62:136–154.

Spina, E., Avenoso, A., Facciola, G., Scordo, M. G., Ancione, M., Madia, A. 2001. Plasma concentrations of risperidone and 9-hydroxyrisperidone during combined treatment with paroxetine. *Ther Drug Monit* 23:223–227.

Steimer, W., Zopf, K., von Amelunxen, S., Pfeiffer, H., Bachofer, J., Popp, J., Messner, B., Kissling, W., Leucht, S. 2004. Allele-specific change of concentration and functional gene dose for the prediction of steady-state serum concentrations of amitriptyline and nortriptyline in CYP2C19 and CYP2D6 extensive and intermediate metabolizers. *Clin Chem* 50:1623–1633.

Stoll, F., Goller, A. H., Hillisch, A. 2011. Utility of protein structures in overcoming ADMET-related issues of drug-like compounds. *Drug Discov Today* 16:530–538.

Tarcsay, A., Keseru, G. M. 2011. *In silico* site of metabolism prediction of cytochrome P450-mediated biotransformations. *Expert Opin Drug Metab Toxicol* 7:299–312.

Tralau, T., Luch, A. 2013. The evolution of our understanding of endo-xenobiotic crosstalk and cytochrome P450 regulation and the therapeutic implications. *Expert Opin Drug Metab Toxicol* 9:1541–1554.

von Moltke, L. L., Greenblatt, D. J., Schmider, J., Wright, C. E., Harmatz, J. S., Shader, R. I. 1998. *In vitro* approaches to predicting drug interactions *in vivo*. *Biochem Pharmacol* 55:113–122.

Voora, D., Shah, S. H., Spasojevic, I., Ali, S., Reed, C. R., Salisbury, B. A., Ginsburg, G. S. 2009. The *SLCO1B1*5* genetic variant is associated with statin-induced side effects. *J Am Coll Cardiol* 54:1609–1616.

Wang, S. L., Huang, J. D., Lai, M. D., Liu, B. H., Lai, M. L. 1993. Molecular basis of genetic variation in debrisoquin hydroxylation in Chinese subjects: Polymorphism in RFLP and DNA sequence of CYP2D6. *Clin Pharmacol Ther* 53:410–418.

Wang, B., Wang, J., Huang, S. Q., Su, H. H., Zhou, S. F. 2009a. Genetic polymorphism of the human cytochrome P450 2C9 gene and its clinical significance. *Curr Drug Metab* 10:781–834.

Wang, B., Yang, L. P., Zhang, X. Z., Huang, S. Q., Bartlam, M., Zhou, S. F. 2009b. New insights into the structural characteristics and functional relevance of the human cytochrome P450 2D6 enzyme. *Drug Metab Rev* 41:573–643.

Wang, L., McLeod, H. L., Weinshilboum, R. M. 2011. Genomics and drug response. *N Engl J Med* 364:1144–1153.

Weinshilboum, R. 2003. Inheritance and drug response. *N Engl J Med* 348:529–537.

Wennerholm, A., Dandara, C., Sayi, J., Svensson, J. O., Abdi, Y. A., Ingelman-Sundberg, M., Bertilsson, L., Hasler, J., Gustafsson, L. L. 2002. The African-specific CYP2D617 allele encodes an enzyme with changed substrate specificity. *Clin Pharmacol Ther* 71:77–88.

Wenzel-Seifert, K., Wittmann, M., Haen, E. 2011. QTc prolongation by psychotropic drugs and the risk of Torsade de Pointes. *Dtsch Arztebl Int* 108:687–693.

Wheeler, H. E., Maitland, M. L., Dolan, M. E., Cox, N. J., Ratain, M. J. 2013. Cancer pharmacogenomics: Strategies and challenges. *Nat Rev Genet* 14:23–34.

Wilke, R. A., Dolan, M. E. 2011. Genetics and variable drug response. *JAMA* 306:306–307.

Wilkinson, G. R. 2005. Drug metabolism and variability among patients in drug response. *N Engl J Med* 352:2211–2221.

Xiang, C., Wang, J., Kou, X., Chen, X., Qin, Z., Jiang, Y., Sun, C. et al. 2015. Pulmonary expression of CYP2A13 and ABCB1 is regulated by FOXA2, and their genetic interaction is associated with lung cancer. *FASEB J* 29:1986–1998.

Xu, M., Ju, W., Hao, H., Wang, G., Li, P. 2013. Cytochrome P450 2J2: Distribution, function, regulation, genetic polymorphisms and clinical significance. *Drug Metab Rev* 45:311–352.

Yamazaki, H., Kiyotani, K., Tsubuko, S., Matsunaga, M., Fujieda, M., Saito, T., Miura, J., Kobayashi, S., Kamataki, T. 2003. Two novel haplotypes of CYP2D6 gene in a Japanese population. *Drug Metab Pharmacokinet* 18:269–271.

Yang, X., Zhang, B., Molony, C., Chudin, E., Hao, K., Zhu, J., Gaedigk, A. et al. 2010. Systematic genetic and genomic analysis of cytochrome P450 enzyme activities in human liver. *Genome Res* 20:1020–1036.

Ye, X. H., Song, L., Peng, L., Bu, Z., Yan, S. X., Feng, J., Zhu, X. L., Liao, X. B., Yu, X. L., Yan, D. 2015. Association between the CYP2E1 polymorphisms and lung cancer risk: A meta-analysis. *Mol Genet Genomics* 290:545–558.

Zhang, T., Chen, Q., Li, L., Liu, L. A., Wei, D. Q. 2011. *In silico* prediction of cytochrome P450-mediated drug metabolism. *Comb Chem High Throughput Screen* 14:388–395.

Zhou, S. F. 2009a. Polymorphism of human cytochrome P450 2D6 and its clinical significance: Part I. *Clin Pharmacokinet* 48:689–723.

Zhou, S. F. 2009b. Polymorphism of human cytochrome P450 2D6 and its clinical significance: Part II. *Clin Pharmacokinet* 48:761–804.

Zhou, S., Chan, E., Lim, L. Y., Boelsterli, U. A., Li, S. C., Wang, J., Zhang, Q., Huang, M., Xu, A. 2004. Therapeutic drugs that behave as mechanism-based inhibitors of cytochrome P450 3A4. *Curr Drug Metab* 5:415–442.

Zhou, S., Yung Chan, S., Cher Goh, B., Chan, E., Duan, W., Huang, M., McLeod, H. L. 2005. Mechanism-based inhibition of cytochrome P450 3A4 by therapeutic drugs. *Clin Pharmacokinet* 44:279–304.

Zhou, S. F., Xue, C. C., Yu, X. Q., Li, C., Wang, G. 2007. Clinically important drug interactions potentially involving mechanism-based inhibition of cytochrome P450 3A4 and the role of therapeutic drug monitoring. *Ther Drug Monit* 29:687–710.

Zhou, S. F., Liu, J. P., Chowbay, B. 2009a. Polymorphism of human cytochrome P450 enzymes and its clinical impact. *Drug Metab Rev* 41:89–295.

Zhou, S. F., Liu, J. P., Lai, X. S. 2009b. Substrate specificity, inhibitors and regulation of human cytochrome P450 2D6 and implications in drug development. *Curr Med Chem* 16:2661–2805.

Zhou, S. F., Zhou, Z. W., Huang, M. 2010. Polymorphisms of human cytochrome P450 2C9 and the functional relevance. *Toxicology* 278:165–188.

Zhou, Z. W., Chen, X. W., Sneed, K. B., Yang, Y. X., Zhang, X., He, Z. X., Chow, K., Yang, T., Duan, W., Zhou, S. F. 2015. Clinical association between pharmacogenomics and adverse drug reactions. *Drugs* 75:589–631.

Zoble, R. G., Kirsten, E. B., Brewington, J. 1989. Pharmacokinetic and pharmacodynamic evaluation of propafenone in patients with ventricular arrhythmia. Propafenone Research Group. *Clin Pharmacol Ther* 45:535–541.

Index

Page numbers followed by f and t indicate figures and tables, respectively.

GWAS on, 323–324, 326
overview, 315–316
physiological factors on, 316–318
   developmental changes (ontogeny), 316–317
   fasting, 318
   gender, 316
   pregnancy, 317–318
transcriptional and posttranscriptional
   by FXR, 322–323
   by HNF-4α, 321
   uninducible by prototypical inducers of
      CYPs, 320–321
   in vitro studies, 320
   in vivo studies, 320–321
Renal clearance, of R- and S-flecainide, 432–433
Repaglinide, 33
Residual function, alleles with, 427–429
Resveratrol, 25
Retinoic acid, 49
Retinol, intracellular processing of, 55
Reversible and mixed-type inhibitors, of
   CYP2D6, 300–307
amphetamine analogs, 306–307, 307f
anastrozole, 306
antifungal agents, 304, 304f
anti-HIV agents, 304, 305f
CNS drugs, 300–303
   antipsychotics, 300–301, 301f
   narcotics, 302, 303f
   other antidepressants, 302, 303f
   others, 302, 303, 303f
   SSRIs, 301–302, 302f
   tricyclic antidepressants, 302, 303f
halofantrine, 305, 306f
H₁ receptor antagonist, 303, 304f
lansoprazole, 304, 306f
mexiletine, 304, 306f
mibefradil, 306
natural and herbal compounds, 307, 307f
omeprazole, 304, 305, 306f
sarpogrelate, 306
steroids, 304, 305f
temsirolimus, 306
Reversible inhibition, defined, 19
R-flecainide, 432–433
Rheumatoid arthritis, CYP2D6 expression and
   activity, 327
Rifabutin, 42, 320
Rifampicin, 24–25, 35, 44, 47, 124, 139, 230, 320
Rifampin, 30, 33, 42, 124, 315
Riluzole, 23
Risperidone, 161, 163f, 405f, 407, 428, 446–447
Ritonavir
   chemical structure, 304, 305f
   HIV protease inhibitors, 41, 208, 210, 212f
   hydrogen bonding with Asp301, 367
Rivastigmine, 175
Rosiglitazone, 33, 56
Rubia cordifolia, 24
R-warfarin, 23

S

Saccharomyces cerevisiae, 127
S-acenocoumarol, 34
Salt-wasting syndrome, 54
Saquinavir, 210, 213f
Sarpogrelate, 306
Schizophrenia
   aripiprazole, 155
   cariprazine, 231

clozapine, 157
haloperidol, 444–445
olanzapine, 158
pimozide, 300
risperidone, 161, 446–447
thioridazine, 447
zuclopenthixol, 162, 448
Selective estrogen receptor modulators (SERMs),
   221–227
droloxifene, 226
enclomifene, 226, 226f, 371, 386f
lasofoxifene, 226, 227
tamoxifen, 53, 119, 221, 222–226, 223f, 341,
   373, 395f, 459
Selective inhibitors, of CYP2D6, 289
Selective phosphodiesterase type 5 inhibitors,
   227, 227f
Selective serotonin reuptake inhibitors (SSRIs),
   146–150, 439–442
chemical structures, 301–302, 302f
citalopram, 116, 145, 146–147, 147f, 377,
   399f, 439
fluoxetine, see Fluoxetine
fluvoxamine, 24, 148f, 149, 164, 302, 302f,
   401f, 406, 439–440
overview, 141, 146
paroxetine, see Paroxetine
sertraline, see Sertraline
Selegiline, 31, 170, 171, 177f, 341, 373, 394f
Semiquantitative gene dose, 485
S-enantiomer, 147
Senile dementia, drugs for
   nicergoline, 175, 180f, 451
   perphenazine, 446
   selegiline, 170, 171
Ser304, 353–354, 357t
Ser486, 356, 358t
Seratrodast, 29
SERMs, see Selective estrogen receptor
   modulators (SERMs)
Sertraline
   CYP2D6, binding modes, 405f, 407
   for depression and mania, 149
   metabolism, 150, 150f
   metabolites of, 149
   SSRIs, 35, 441–442
S-flecainide, 432–433
Short heterodimer partner (SHP), 318, 320, 321
Siberian ginseng, 319
Sildenafil, 41, 227, 227f, 373, 394f
Simvastatin acid, 32, 41
Single base mutation, null alleles attributed to,
   425–426
Site-directed mutagenesis studies, of CYP2D6,
   346–358, 357t–358t
Ala90, 347
Arg296, 353
Arg440, 356
Asp100, 347
Glu156, 350
Glu222, 353
Glu410, 356
Glu216 and Asp301, 350–353
Gly42, 346–347
Gly169, 350
His324, 355–356
Lys281, 353
Met374, 356
Phe120, Phe481, and Phe483, 348–349
Pro34, 346
Ser304, 353–354

Ser486, 356
Thr107, 347
Thr309, 354–355, 358–359
Thr312, 354–355
Trp128, 349–350
Val136, 350
Val338, 356
Sjögren's syndrome, 231, 460
Skin cancer, 55
S-mephenytoin, 35, 124, 364, 448
S-methyl-N,N-diethylthiolcarbamate sulfoxide,
   26
Smoking
   CYP1A2 activity, 24
   CYP2D6 activity, 318
Sodium nitroprusside, 117–118
Solute carrier organic anion (SLCO), 319
Sparteine, 139–140, 140f, 141f, 326
Spastic paraplegia, 39, 49
Spirosulfonamide, 227, 228f, 351–352, 367
Spodoptera frugiperda, 122
SSRIs, see Selective serotonin reuptake
   inhibitors (SSRIs)
St. John's wort, 20, 319, 320
Steroids
   biosynthesis, 17
   chemical structures of, 304, 305f
   endogenous compounds, 241–244, 241f,
      242f, 243f, 244f
   mechanism-based CYP3A4 inhibitor, 41
   metabolism, 16, 125
Streptomyces avermitilis, 3
Structural features, of CYPs, 17–19, 18t
Structural requirements, CYP2D6 ligands,
   341–343
Structure–activity relationships, of CYP2D6
   inhibitors, 307, 308
   substrates, 245–246
Structure and function, of human CYP2D6,
   341–407
antibody studies, 359
bindings modes of inhibitors with CYP2D6,
   367–407
   active site, see Active site, of CYP2D6
   known inhibitors, 368, 373t
   predicted ADMET profile, 368, 378t
   predicted hepatotoxicity, 368, 379t
bindings modes of substrates with CYP2D6,
   367–407
   active site, 368, 371, 373, 375, 380f–396f;
      see also Active site, of CYP2D6
   known substrates, 368, 369t–371t
   predicted ADMET profile, 368, 374t–375t
   predicted hepatotoxicity, 368, 376t–377t
homology modeling studies, 343–346
   derived from bacterial CYPs, 343–344
   derived from rabbit CYP2C5 structures,
      344–346
   overview, 343
ligands, pharmacophore models and
   structural requirements of, 341–343
other molecular modeling studies, 360
overview, 341
site-directed mutagenesis studies, 346–358,
   357t–358t
   Ala90, 347
   Arg296, 353
   Arg440, 356
   Asp100, 347
   Glu156, 350
   Glu222, 353